Universitext

Universitext

Universitext is a series of textbooks that presents material from a wide variety of mathematical disciplines at master's level and beyond. The books, often well class-tested by their author, may have an informal, personal, even experimental approach to their subject matter. Some of the most successful and established books in the series have evolved through several editions, always following the evolution of teaching curricula, into very polished texts.

Thus as research topics trickle down into graduate-level teaching, first textbooks written for new, cutting-edge courses may make their way into *Universitext*.

For further volumes:
www.springer.com/series/223

Boris Makarov • Anatolii Podkorytov

Real Analysis: Measures, Integrals and Applications

Springer

Boris Makarov
Mathematics and Mechanics Faculty
St Petersburg State University
St Petersburg, Russia

Anatolii Podkorytov
Mathematics and Mechanics Faculty
St Petersburg State University
St Petersburg, Russia

Translated from the Russian language edition:

Лекции по вещественному анализу
by Б.М. Макаров, А.Н. Подкорытов (B.M. Makarov, A.N. Podkorytov)

ISSN 0172-5939
ISSN 2191-6675 (electronic)
Universitext
ISBN 978-1-4471-5121-0
ISBN 978-1-4471-5122-7 (eBook)
DOI 10.1007/978-1-4471-5122-7
Springer London Heidelberg New York Dordrecht

Library of Congress Control Number: 2013940613

Mathematics Subject Classification: 28A12, 28A20, 28A25, 28A35, 28A75, 28A78, 28B05, 31B05, 42A20, 42B05, 42B10

Printed on acid-free paper

Springer is part of Springer Science+Business Media (www.springer.com)

Diagram of Chapter Dependency

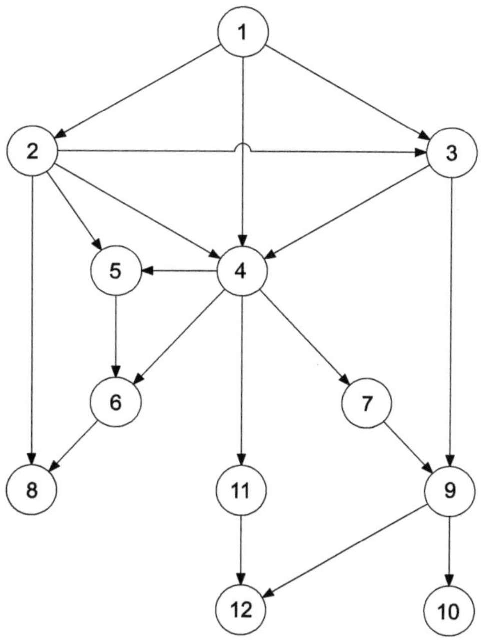

Preface to the English Translation

This book reflects our experience in teaching at the Department of Mathematics and Mechanics of St. Petersburg State University. It is aimed primarily at readers making their first acquaintance with the subject.

Lecture courses on measure theory and integration are often confined to abstract measure theory, with little attention paid to such topics as integration with respect to Lebesgue measure, its transformation under a diffeomorphism and so on—that is, topics that are more special but no less important for applications. Believing that such reticence is counterproductive, we choose an approach that avoids it and combines general notions with classical special cases.

A substantial part of the book is devoted to examples illustrating the obtained results both in and beyond the framework of mathematical analysis, in particular, in geometry. The exercises appearing at the end of almost every section serve the same purpose.

In the English translation we use three-digit numbers for sections. The first digit refers to a Chapter, the second to a Section within the Chapter and the third to Subsection. When referencing to a statement we give the number of a Subsection which contains it. E.g., Lemma 7.5.4 would mean a lemma from Sect. 7.5.4.

Comparing with Russian edition, we have extended the book by adding, in particular, the new Sects. 6.1.3 and 6.2.6.

Taking into account the difference between curricula in Russia and the West, as well as the considerable volume of our book, we think it necessary to say several words about how to use it, and we draw the reader's attention to the chapter dependency chart. A reader interested only in an introduction to the foundations of measure theory and integration may prefer to read only those sections of Chaps. 1–5 that are not marked with a $\star$. This symbol indicates sections that contain either some illustrative material (e.g., Sects. 2.8, 6.6–6.7, 7.2–7.3, 8.7, 10.2, 10.6), or some optional information that can be omitted in the first reading (e.g., Sects. 1.6, 4.11, 5.5–5.6, 6.5, 7.4, 8.8, 10.4, 12.1–12.3), or else material used outside Chaps. 1–5 (Sects. 2.6, 3.4, 4.9). The material of Sects. 1.1–1.4, 2.1–2.5, 3.1–3.2, 4.1–4.8, 5.1–5.4 can be taken as a basis for a two semester course on the foundations of measure theory and integration. Time permitting, the course can be extended by including the material of Sects. 6.1–6.2, 3.3, 4.9, 6.4.

The book can also be used for courses aimed at students familiar with the notion of integration with respect to a measure. There is a sufficiently wide choice of such courses devoted to relatively narrow topics of real analysis. For example:

- The maximal function and differentiation of measures (Sects. 2.7, 4.9, 11.2, 11.3).
- Surface integrals (Sects. 2.6, 8.1–8.6).
- Functionals in spaces of measurable and continuous functions (Sects. 11.1–11.2, Chap. 12).
- Approximate identities and their applications (Sects. 7.5–7.6, Chap. 9).
- Fourier series and the Fourier transform (Chaps. 9, 10).
- A course covering only the preliminaries of the theory of Fourier series and the Fourier transform may be based, for example, on Sects. 9.1.1–9.1.3, 10.1.1–10.1.4, 10.3.1–10.3.6, 10.5.1–10.5.4.

Acknowledgments We are deeply indebted to Springer for publishing our book and we are happy to see it reach a much wider audience via its English translation. We are grateful to V.P. Havin who attracted the publisher's attention to the Russian edition of our book soon after its publication.

In the process of preparing and publishing the volume, we have been helped by several people to whom we would like to express our thanks. In particular, the anonymous referees provided us with constructive comments and suggestions. B.M. Bekker, A.A. Lodkin, F.L. Nazarov and N.V. Tsilevich undertook the hard task of translation. We were also most fortunate to receive feedback from A.I. Nazarov, F.L. Nazarov and O.L. Vinogradov, which helped us to improve the text in many places. Our special thanks go to Joerg Sixt, the Springer Editor, for his invaluable help and encouragement in preparing this manuscript.

We also wish to acknowledge the help and support we received from our friends, families and colleagues who read and commented upon various drafts and contributed to the translation.

Our special thanks go to O.B. Makarova (who happened to be a granddaughter of the first of co-authors) who very competently and patiently conducted our correspondence related to publication of this book and helped us enormously with proofreading.

The translation was carried out with the support of the St. Petersburg State University program "Function theory, operator theory and their applications" 6.38.78.2011.

St Petersburg, Russia

Boris Makarov
Anatolii Podkorytov

Preface

Measure theory has been an integral part of undergraduate and graduate curricula in mathematics for a long time now. A number of texts in this subject area have become well-established and widely used. For example, one might recall books by B.Z. Vulih [Vu], A.N. Kolmogorov and S.V. Fomin [KF], not to mention the classical monograph by P. Halmos [H]. However, books on measure theory typically treat it as an isolated subject, which makes it difficult to include it in a general course in analysis in a natural and seamless way. For example, the invariance of the Lebesgue measure is either omitted entirely, or considered as a special case of the invariance of the Haar measure. Quite often, the question of how Lebesgue measure transforms under diffeomorphisms is left out. On the other hand, most introductory courses on integration are still based on the theory of the Riemann integral. As a result, the students are forced to absorb numerous, however similar, definitions based on Riemann sums corresponding to various situations, such as double integrals, triple integrals, line integrals, surface integrals and so on. They must also overcome the unnecessary technical complications caused by the lack of a sufficiently general approach. Typical examples of such difficulties include justifying the change of the order of integration and taking limits under the integral sign.

For this reason, one often faces a two-tier exposition of the theory of integration, where at the first stage the notion of measure is not discussed at all, and later the elementary topics are never revisited, leaving the task of reconciling the various approaches to the student. The authors aim to eliminate this divide and provide an exposition of the theory of the integral that is modern, yet easily integrated into a general course in analysis. This encapsulates in a textbook the established practice at the Department of Mathematics and Mechanics of the University of St. Petersburg. This practice is based on an idea introduced in the early 60s by G.P. Akilov and first implemented by V.P. Havin during the academic year 1963–1964.

The main emphasis of the book is on the exposition of the properties of the Lebesgue integral and its various applications. This approach determined the style of exposition as well as the choice of the material. It is our hope that the reader who masters the first third of the book will be sufficiently prepared to study any area of mathematics that relies upon the general theory of measure, such as, among others, probability theory, functional analysis and mathematical physics.

Applications of the theory of integration constitute a substantial part of this book. In addition to some elements of harmonic analysis, they also include geometric applications, among which the reader will find both classical inequalities, such as the Brunn–Minkowski and isoperimetric inequalities, and more recent results, such as the proof of Brouwer's theorem on vector fields on the sphere based on a change of variables, the K. Ball inequality and others. In order to illustrate the effectiveness and applicability of the theorems presented, and to give the reader an opportunity to absorb the material in a hands-on fashion, the book includes numerous examples and exercises of various degrees of difficulty.

Pedagogical considerations caused us to refrain from stating some of the results in their full generality. In such cases, references to the appropriate literature are provided for the interested reader. The notion of surface area is discussed in more detail than is common in analysis texts. Using a descriptive definition, we prove its uniqueness on Borel subsets of smooth and Lipschitz manifolds.

It is desirable that the reader be familiar with the notion of an integral of a continuous function of one variable on an interval prior to being exposed to the basics of measure theory. However, we do not feel that this prerequisite necessarily needs to be fulfilled in the context of the Riemann integral, which we view to be primarily of historical interest. A possible alternative approach is outlined in Appendix 13.1.

This book is based on a series of lectures delivered by the authors at the Department of Mathematics and Mechanics of St. Petersburg State University. The majority of the material in Chaps. 1–8 approximately corresponds to the fourth and fifth semester analysis program for mathematics majors in our department. The material from Chaps. 9–12 and some other parts of the book was previously included by the authors in advanced courses and lectures in functional analysis. Some additional information is presented in Appendices 13.2–13.6. Appendix 13.7, dedicated to smooth mappings, is included for the sake of completeness.

The reader is expected to have the necessary mathematical background. The students entering the fourth semester at the Department of Mathematics and Mechanics of St. Petersburg State University are familiar with multivariable calculus and basic linear algebra. This prerequisite material is used throughout the book without any additional explanations. In Chap. 8, familiarity with the basics of smooth manifold theory is assumed. In Appendices 13.2 and 13.3, the rudiments of the theory of metric spaces are taken for granted.

The authors have previously encountered texts where a definition or notation, once introduced, is never repeated and is used without any further comments or references many pages later. We believe that such manner of presentation, possibly appropriate in monographs of an encyclopedic nature, puts too much strain on the reader's memory and attention span. Taking into account the fact that this is a textbook intended for relatively inexperienced readers, many of whom will be encountering the subject matter for the first time, the authors find it useful to include some repetitions and reminders. However, they are unable to measure the degree to which they have succeeded in this direction.

In the process of writing this book, the authors have frequently sought advice from their colleagues. The comments and suggestions of D.A. Vladimirov, A.A. Lodkin, A.I. Nazarov, F.L. Nazarov, A.A. Florinsky and V.P. Havin proved especially useful. We are grateful to them as well as to A.L. Gromov, who kindly agreed to produce computer generated graphics and K.P. Kohas, who handled the type-setting of the book.

The chapters are numbered using Roman numerals. They are divided into sections consisting of subsections which are numbered using two Arabic numerals. The first of these indicates the number of the section, and the second the number of the subsection. The subsections in Appendices are numbered by two numerals, one Roman (Appendix number) and the other Arabic, with the addition of the letter A when referencing.

All the assertions contained in a given subsection are numbered in the same way as the subsection itself. In the case of references within a given chapter, only the number of the subsection is indicated. For example, the reference "by Theorem 2.1" refers to a theorem in subsection 2.1 of a given chapter. When referencing material from another chapter, the number of the chapter is also indicated. For example, the reference "Corollary II.3.4" refers to a corollary contained in subsection 3.4 of Chapter II. The enumeration of the formulas is consecutive within each section. The end of a proof is indicated by black triangle ►.

Basic Notation

Logical Symbols	
$P \Rightarrow Q,\ Q \Leftarrow P$	P implies Q
$\forall$	Universal quantifier ("for every")
$\exists$	Existential quantifier ("there exists")
Sets	
$x \in X$	An element x belongs to a set X
$x \notin X$	An element x does not belong to a set X
$A \subset B$, $B \supset A$	A is a subset of a set B
$A \cap B$	The intersection of sets A and B
$A \cup B$	The union of sets A and B
$A \vee B$	The union of disjoint sets A and B
$A \setminus B$	The difference of sets A and B
$A \times B$	The direct (Cartesian) product of sets A and B
$\mathrm{card}(A)$	The cardinality of a set A
$\{x \in X \mid P(x)\}$	The subset of a set X whose elements have the property P
$\varnothing$	The empty set
Sets of Numbers	
$\mathbb{N}$	The set of positive integers
$\mathbb{Z}$	The set of integers
$\mathbb{Q}$	The set of rational numbers
$\mathbb{R}$	The set of real numbers
$\overline{\mathbb{R}} = [-\infty, +\infty]$	The extended real line
$\mathbb{C}$	The set of complex numbers
$\mathbb{R}^m$	The arithmetic m-dimensional space
$\mathbb{R}_+$	The set of positive numbers
$\mathbb{R}^m_+$	The subset of $\mathbb{R}^m$ consisting of all points with positive coordinates
$\mathbb{Q}^m$, $\mathbb{Z}^m$	The subsets of $\mathbb{R}^m$ consisting of all points with rational and integer coordinates, respectively

(a,b), $[a,b)$, $[a,b]$	An open, half-open, and closed interval, respectively
$\langle a,b\rangle$	An arbitrary interval with endpoints a and b
$\inf A$ ($\sup A$)	The greatest lower (least upper) bound of a number set A
Sets in Topological and Metric Spaces	
$\overline{A}$	The closure of a set A
$\mathrm{Int}(A)$	The interior of a set A
$B(a,r)$	The open ball of radius $r>0$ centered at a
$\overline{B}(a,r)$	The closed ball of radius $r>0$ centered at a
$B(r)$ or $B^m(r)$	The ball $B(0,r)$ in the space $\mathbb{R}^m$
B^m	The ball $B^m(1)$
S^{m-1}	The unit sphere (the boundary of B^m) in the space $\mathbb{R}^m$
$\mathrm{diam}(A)$	The diameter of a set A
$\mathrm{dist}(x,A)$	The distance from a point x to a set A
Systems of Sets	
$\mathfrak{A}^m$	The σ-algebra of Lebesgue measurable subsets of $\mathbb{R}^m$
$\mathfrak{B}^m$	The σ-algebra of Borel subsets of $\mathbb{R}^m$
$\mathfrak{B}(\mathcal{E})$	The Borel hull of a system $\mathcal{E}$
$\mathfrak{B}_X$	The σ-algebra of Borel subsets of a space X
$\mathscr{P}^m$	The semiring of m-dimensional cells
$\mathscr{P}\odot\mathcal{Q}$	The product of semirings $\mathscr{P}$ and $\mathcal{Q}$
Maps and Functions	
$\det(A)$	The determinant of a square matrix A
f_+, f_-	The functions $\max\{f,0\}$, $\max\{-f,0\}$
$f_n \rightrightarrows f$	A sequence of functions f_n uniformly converges to a function f
$J_\Phi(x)=\det(\Phi'(x))$	The Jacobian of a map Φ at a point x
$\mathrm{supp}(f)$	The support of a function f
$T: X\to Y$	A map T acting from X to Y
$T(A)$	The image of a set A under a map T
$T^{-1}(B)$	The inverse image of a set B under a map T
$T\circ S$	The composition of maps T and S
$T\vert_A$	The restriction of a map T to a set A
$\operatorname*{esssup}_X f$	The essential supremum of a function f on a set X
$x\mapsto T(x)$	A map T sends a point x to $T(x)$
$\Gamma_f(E)$	The graph of a function $f: E\to\overline{\mathbb{R}}$
$\mathscr{P}_f(E)$	The region under the graph of a non-negative function f over a set E
χ_E	The characteristic function of a set E
$\Phi'(x)$	The Jacobi matrix of a map Φ at a point x
$\Vert\cdot\Vert$	The Euclidean norm of a vector, or the norm of a function in $\mathscr{L}^2(X,\mu)$, or the norm in a Banach space
$\Vert f\Vert_p$	The norm of a function f in $\mathscr{L}^p(X,\mu)$
$\Vert f\Vert_\infty$	$=\operatorname*{esssup}_X \vert f\vert$

$\langle \cdot, \cdot \rangle$	The inner product of vectors in a Euclidean space, or of functions in $\mathscr{L}^2(X, \mu)$

Measures

$(X, \mathfrak{A}, \mu)$	A measure space
$(X, \mathfrak{A})$	A measurable space
α_m	The volume (Lebesgue measure) of the unit ball in $\mathbb{R}^m$
λ_m	m-dimensional Lebesgue measure
$\mu \times \nu$	The product of measures μ and ν
σ_k	k-dimensional area

Sets of Functions

$C(X)$	The set of continuous functions on a topological space X
$C_0(X)$	The set of compactly supported continuous functions on a locally compact topological space X
$C^r(\mathcal{O})$ $(C^r(\mathcal{O}; \mathbb{R}^n))$	The set of r times ($r = 0, 1, \dots, +\infty$) differentiable functions ($\mathbb{R}^n$-valued maps) defined on an open subset $\mathcal{O}$ of $\mathbb{R}^m$
$C_0^\infty(\mathcal{O})$	The set of infinitely differentiable compactly supported functions defined on an open subset $\mathcal{O}$ of $\mathbb{R}^m$
$\mathscr{L}^0(X, \mu)$	The set of measurable functions defined on X and finite almost everywhere with respect to a measure μ
$\mathscr{L}^p(X, \mu)$	The set of functions from $\mathscr{L}^0(X, \mu)$ satisfying the condition $\int_X |f|^p \, d\mu < +\infty$
$\mathscr{L}^\infty(X, \mu)$	The set of functions each of which is bounded on a subset of full measure
$\mathscr{L}(X, \mu) = \mathscr{L}^1(X, \mu)$	The set of functions summable on X with respect to a measure μ

Chapter 1
Measure

1.1 Systems of Sets

In classical analysis, one usually works with functions that depend on one or several numerical variables, but here we will study functions whose argument is a set. Our main focus will be on measures, i.e., set functions that generalize the notions of length, area and volume. Dealing with such generalizations, it is natural to aim at defining a measure on a sufficiently "good" class of sets. We would like this class to have a number of natural properties, namely, to contain, with any two elements, their union, intersection and set-theoretic difference. In order for a measure to be of interest, its domain must also be sufficiently rich in sets. Aiming to satisfy these requirements, we arrive at the notions of an algebra and a σ-algebra of sets.

As a synonym for "a set of sets", we use the term "a system of sets". The sets constituting a system are called its elements. The phrase "a set A is contained in a given system of sets $\mathfrak{A}$" means that A belongs to $\mathfrak{A}$, i.e., A is an element of $\mathfrak{A}$. To avoid notational confusion, we usually denote sets by upper case Latin letters $A, B, \ldots$, and points belonging to these sets by lower case Latin letters $a, b, \ldots$. For systems of sets, we use Gothic and calligraphic letters. The symbol $\varnothing$ stands for the empty set.

1.1.1 We assume that the reader is familiar with the basics of naive set theory. In particular, we leave the proofs of set-theoretic identities as easy exercises. Some of these identities, which will be used especially often, are summarized in the following lemma for the reader's convenience.

Lemma *Let A, A_ω $(\omega \in \Omega)$ be arbitrary subsets of a set X. Then*

(1) $X \setminus \bigcup_{\omega\in\Omega} A_\omega = \bigcap_{\omega\in\Omega}(X \setminus A_\omega)$;
(2) $X \setminus \bigcap_{\omega\in\Omega} A_\omega = \bigcup_{\omega\in\Omega}(X \setminus A_\omega)$;
(3) $A \cap \bigcup_{\omega\in\Omega} A_\omega = \bigcup_{\omega\in\Omega}(A \cap A_\omega)$.

Equations (1) and (2) are called De Morgan's laws. Equation (3) is the distributive law of intersection over union. Associating union with addition and intersection with

B. Makarov, A. Podkorytov, *Real Analysis: Measures, Integrals and Applications*, Universitext, DOI 10.1007/978-1-4471-5122-7_1, © Springer-Verlag London 2013

multiplication, the reader can easily see the analogy between this property and the usual distributivity for numbers.

Considering the union and intersection of a family of sets with a countable set of indices Ω, we usually assume that the indices are positive integers. This does not affect the generality of our results, since for every "numbering" of Ω (i.e., every bijection $n \mapsto \omega_n$ from the set of positive integers onto Ω), we have the equalities

$$\bigcup_{\omega\in\Omega} A_\omega = \bigcup_{n\in\mathbb{N}} A_{\omega_n}, \qquad \bigcap_{\omega\in\Omega} A_\omega = \bigcap_{n\in\mathbb{N}} A_{\omega_n},$$

which follow directly from the definition of the union and intersection.

In what follows, we often write a set as the union of pairwise disjoint subsets. Thus it is convenient to introduce the following definition.

Definition A family of sets $\{E_\omega\}_{\omega\in\Omega}$ is called a *partition of a set* E if E_ω are pairwise disjoint and $\bigcup_{\omega\in\Omega} E_\omega = E$.

We do not exclude the case where some elements of a partition coincide with the empty set.

A union of disjoint sets will be called a *disjoint union* and denoted by $\vee$. Thus $A \vee B$ stands for the union $A \cup B$ in the case where $A \cap B = \varnothing$. Correspondingly, $\bigvee_{\omega\in\Omega} E_\omega$ stands for the union of a family of sets E_ω in the case where all these sets are pairwise disjoint.

We always assume that the system of sets under consideration consists of subsets of a fixed non-empty set, which will be called the ground set. The complement of a set A in the ground set X, i.e., the set-theoretic difference $X \setminus A$, is denoted by A^c.

Definition A system of sets $\mathfrak{A}$ is called *symmetric* if it contains the complement A^c of every element $A \in \mathfrak{A}$.

Consider the following four properties of a system of sets $\mathfrak{A}$:

(σ_0) the union of any two elements of $\mathfrak{A}$ belongs to $\mathfrak{A}$;
(δ_0) the intersection of any two elements of $\mathfrak{A}$ belongs to $\mathfrak{A}$;
(σ) the union of any sequence of elements of $\mathfrak{A}$ belongs to $\mathfrak{A}$;
(δ) the intersection of any sequence of elements of $\mathfrak{A}$ belongs to $\mathfrak{A}$.

The following result holds.

Proposition *If $\mathfrak{A}$ is a symmetric system of sets, then* (σ_0) *is equivalent to* (δ_0) *and* (σ) *is equivalent to* (δ).

Proof The proof follows immediately from De Morgan's laws. Let us prove, for example, that $(\delta) \Rightarrow (\sigma)$. Consider an arbitrary sequence $\{A_n\}_{n\geqslant 1}$ of elements of $\mathfrak{A}$. Their union can be written in the form

$$\bigcup_{n\geqslant 1} A_n = \Bigl(\bigcap_{n\geqslant 1} A_n^c\Bigr)^c.$$

Since $A_n^c \in \mathfrak{A}$ for all n (by the symmetry of $\mathfrak{A}$), it follows from (δ) that the intersection of these complements also belongs to $\mathfrak{A}$. It remains to use again the symmetry of $\mathfrak{A}$, which implies that $\mathfrak{A}$ also contains the complement of this intersection, i.e., the union of the original sets.

The reader can easily establish the remaining implications. □

1.1.2 Now we introduce systems of sets that are of great importance for us.

Definition A non-empty symmetric system of sets $\mathfrak{A}$ is called an *algebra* if it satisfies the (equivalent) conditions (σ_0) and (δ_0). An algebra is called a *σ-algebra* (sigma-algebra) if it satisfies the (equivalent) conditions (σ) and (δ).

Note the following three properties of an algebra $\mathfrak{A}$.

(1) $\varnothing, X \in \mathfrak{A}$. Indeed, let $A \in \mathfrak{A}$. Then $\varnothing = A \cap A^c \in \mathfrak{A}$ and $X = A \cup A^c \in \mathfrak{A}$ directly by the definition of an algebra.
(2) For any two sets $A, B \in \mathfrak{A}$, their set-theoretic difference $A \setminus B$ also belongs to $\mathfrak{A}$. This follows from the identity $A \setminus B = A \cap B^c$ and the definition of an algebra.
(3) If $A_1, \dots, A_n$ are elements of $\mathfrak{A}$, then their union and intersection also belong to $\mathfrak{A}$. This property can be proved by induction.

Examples

(1) The system that consists of all bounded subsets of the plane $\mathbb{R}^2$ and their complements is an algebra (but not a σ-algebra!).
(2) The system that consists of only two sets, X and $\varnothing$, is obviously an algebra and a σ-algebra. It is often called the trivial algebra on X.
(3) The other extreme case (as compared to the trivial algebra) is the system of all subsets of X. It is obviously a σ-algebra.
(4) If $\mathfrak{A}$ is an algebra (σ-algebra) of subsets of a set X and $Y \subset X$, then the system of sets $\{A \cap Y \mid A \in \mathfrak{A}\}$ is an algebra (respectively, σ-algebra) of subsets of Y. We call it the *induced algebra* (on Y) and denote it by $\mathfrak{A} \cap Y$.

More generally, if $\mathcal{E}$ is an arbitrary system of subsets of a set X and $Y \subset X$, then $\{E \cap Y \mid E \in \mathcal{E}\}$ is called the system induced on Y by $\mathcal{E}$ and is denoted by $\mathcal{E} \cap Y$. The part of $\mathcal{E} \cap Y$ that consists of the sets belonging to $\mathcal{E}$ and lying in Y is denoted by $\mathcal{E}_Y$. Note that if $\mathcal{E}$ is an algebra, then $\mathcal{E}_Y$ is an algebra if and only if $Y \in \mathcal{E}$.

Proposition *Let $\{\mathfrak{A}_\omega\}_{\omega \in \Omega}$ be an arbitrary family of algebras (σ-algebras) consisting of subsets of some set. Then the system $\bigcap_{\omega \in \Omega} \mathfrak{A}_\omega$ is again an algebra (σ-algebra).*

Proof The proof is left to the reader. □

It is sometimes convenient to consider, along with algebras, related systems of sets that do not satisfy the symmetry requirement. A system of sets $\mathfrak{A}$ is called a *ring* if for any two elements $A, B \in \mathfrak{A}$, the sets $A \cup B$, $A \cap B$ and $A \setminus B$ also belong to $\mathfrak{A}$. A ring that contains the union of any sequence of elements is called a σ-ring.

Clearly, every algebra (σ-algebra) is also a ring (σ-ring).

1.1.3 Every system of sets is contained in some σ-algebra, for example, in the σ-algebra of all subsets of the ground set X. But this σ-algebra usually contains "too many" sets, and it is often useful to embed the given system of sets into an algebra in the most economical way, so that the ambient algebra does not contain "superfluous" elements.

It turns out that every finite collection of subsets $\{A_k\}_{k=1}^n$ of a set X is a part of an algebra consisting of finitely many elements. This is obvious if the sets under consideration form a partition of X. Then all finite unions of these sets, together with the empty set (which, in set theory, is considered the union over an empty set of indices), constitute an algebra. But if the sets A_k do not form a partition, there is a standard procedure for constructing an auxiliary partition that generates an algebra containing these sets. This procedure is as follows: to each collection $\varepsilon = \{\varepsilon_1, \dots, \varepsilon_n\}$, where $\varepsilon_k = 0$ or $\varepsilon_k = 1$, we associate the intersection $B_\varepsilon = A_1^{\varepsilon_1} \cap \dots \cap A_n^{\varepsilon_n}$, where $A_k^0 = A_k$ and $A_k^1 = A_k^c$ $(= X \setminus A_k)$. Note that, by Property (3), the sets B_ε must belong to every algebra containing $A_1, \dots, A_n$. The reader can easily check that the sets B_ε form a partition of X, which we will call the *canonical partition* corresponding to the sets $A_1, \dots, A_n$. We encourage the reader to find the sets B_ε in the case where the original collection of sets is already a partition of X. It is clear that B_ε is either contained in A_k (if $\varepsilon_k = 0$), or is disjoint with it. Hence $A_k = \bigcup_{\varepsilon_k=0} B_\varepsilon$. All finite unions of the sets B_ε (together with the empty set) form an algebra containing all A_k. This algebra contains at most 2^{2^n} sets (see Exercise 6) and (like any algebra consisting of finitely many sets) is a σ-algebra. Clearly, it is the smallest σ-algebra containing all A_k.

The description of the sets that constitute the minimal σ-algebra containing a given infinite system of sets is very complicated; we will not consider this question, instead restricting ourselves to the proof that such a σ-algebra exists. This important result will often be used in what follows.

Theorem *For every system $\mathcal{E}$ of subsets of a set X there exists a minimal σ-algebra containing $\mathcal{E}$.*

This σ-algebra is called the *Borel*[1] hull of $\mathcal{E}$ and is denoted by $\mathfrak{B}(\mathcal{E})$. It consists of subsets of the same ground set as $\mathcal{E}$.

Proof Clearly, there exists a σ-algebra containing $\mathcal{E}$ (for example, the σ-algebra of all subsets of X). Consider the intersection of all such σ-algebras. This system of sets contains $\mathcal{E}$ and is a σ-algebra by Proposition 1.1.2. Its minimality follows from the construction. □

[1] Émile Borel (1871–1956)—French mathematician.

Definition An element of the minimal σ-algebra containing all open subsets of the space $\mathbb{R}^m$ is called a *Borel subset of* $\mathbb{R}^m$ or merely a Borel set. The σ-algebra of Borel subsets of $\mathbb{R}^m$ is denoted by $\mathfrak{B}^m$.

Remarks

(1) The simplest examples of Borel sets, along with open and closed sets, are countable intersections of open sets and countable unions of closed sets. They are called G_δ and F_σ sets, respectively.

(2) It is not at all obvious that the σ-algebra $\mathfrak{B}^m$ does not coincide with the σ-algebra of all subsets of $\mathbb{R}^m$, but this is indeed the case. Moreover, these σ-algebras have different cardinalities. One can prove that $\mathfrak{B}^m$ has the cardinality of the continuum, i.e., the same cardinality as $\mathbb{R}^m$ while the cardinality of the σ-algebra of all subsets of $\mathbb{R}^m$, by Cantor's theorem, is strictly greater than the cardinality of $\mathbb{R}^m$. We will not dwell on the proofs of these results; the reader can find them, for example, in the books [Bo, Bou].

1.1.4 Before proceeding to the definition of another system of sets, we establish an auxiliary result, which will be repeatedly used in what follows.

Lemma (Disjoint decomposition) *Let* $\{A_n\}_{n\geqslant 1}$ *be an arbitrary sequence of sets. Then*

$$\bigcup_{n=1}^{\infty} A_n = \bigvee_{n=1}^{\infty}\left(A_n \setminus \bigcup_{k=0}^{n-1} A_k\right) \tag{1}$$

(for uniformity, we assume that $A_0 = \varnothing$*).*

Proof Let $E_n = A_n \setminus \bigcup_{k=0}^{n-1} A_k$. It is clear that these sets are pairwise disjoint: if, say, $m < n$, then $E_m \subset A_m$, while $E_n \subset A_n \setminus A_m$.

To verify (1), take an arbitrary point x from $\bigcup_{n=1}^{\infty} A_n$. Let m be the smallest of the indices n such that $x \in A_n$, i.e., $x \in A_m$ and $x \notin A_k$ for $k < m$. Then $x \in E_m \subset \bigcup_{n\geqslant 1} E_n$. Thus $\bigcup_{n=1}^{\infty} A_n \subset \bigcup_{n=1}^{\infty}(A_n \setminus \bigcup_{k=0}^{n-1} A_k)$. The reverse inclusion is trivial. □

Note that every finite collection of sets $A_1, \dots, A_N$ satisfies a similar identity:

$$\bigcup_{n=1}^{N} A_n = \bigvee_{n=1}^{N}\left(A_n \setminus \bigcup_{k=0}^{n-1} A_k\right). \tag{1$'$}$$

The proof is almost a literal repetition of that of the lemma (one can also apply the lemma to the sequence of sets $\{A_n\}_{n=1}^{\infty}$ with $A_n = \varnothing$ for $n > N$).

Along with algebras and σ-algebras, it will also be convenient to use systems of sets that are not so "good", but are often more tractable; namely, so-called semirings.

Definition A system of subsets $\mathscr{P}$ is called a *semiring* if the following conditions are satisfied:

(I) $\varnothing \in \mathscr{P}$;
(II) if $A, B \in \mathscr{P}$, then $A \cap B \in \mathscr{P}$;
(III) if $A, B \in \mathscr{P}$, then the set-theoretic difference $A \setminus B$ can be written as a finite union of pairwise disjoint elements of $\mathscr{P}$, i.e.,

$$A \setminus B = \bigvee_{j=1}^{m} Q_j, \quad \text{where } Q_j \in \mathscr{P}.$$

Example The system $\mathscr{P}^1$ of all half-open intervals of the form $[a, b)$, where $a, b \in \mathbb{R}$, $a \leqslant b$, and the part $\mathscr{P}_r^1$ of $\mathscr{P}^1$ that consists of intervals with rational endpoints, are semirings.

We leave the reader to prove these simple but important facts.

Every algebra is a semiring, but, as one can see from the above example, the converse is not true. If $\mathscr{P}$ is a semiring, then, for arbitrary Y, the systems $\mathscr{P} \cap Y$ and $\mathscr{P}_Y$ are, obviously, semirings too. Also, every system of pairwise disjoint sets containing the empty set is a semiring.

The union and the set-theoretic difference of elements of a semiring $\mathscr{P}$ may not belong to $\mathscr{P}$. However, they have partitions consisting of elements of $\mathscr{P}$. We will prove this result in a slightly stronger form.

Theorem *Let $\mathscr{P}$ be a semiring and $P, P_1, \ldots, \ldots, P_n, \ldots \in \mathscr{P}$. Then for every N the sets $P \setminus \bigcup_{n=1}^{N} P_n$ and $\bigcup_{n=1}^{N} P_n$ have decompositions of the form*

$$P \setminus \bigcup_{n=1}^{N} P_n = \bigvee_{j=1}^{m} Q_j, \quad \text{where } Q_j \in \mathscr{P}; \tag{2}$$

$$\bigcup_{n=1}^{N} P_n = \bigvee_{n=1}^{N} \bigvee_{j=1}^{m_n} Q_{nj}, \quad \text{where } Q_{nj} \in \mathscr{P} \text{ and } Q_{nj} \subset P_n. \tag{3}$$

Furthermore,

$$\bigcup_{n=1}^{\infty} P_n = \bigvee_{n=1}^{\infty} \bigvee_{j=1}^{m_n} Q_{nj}, \quad \text{where } Q_{nj} \in \mathscr{P} \text{ and } Q_{nj} \subset P_n. \tag{4}$$

It follows from (3) and (4) that the union of an arbitrary (finite or infinite) sequence of elements of a semiring can be written as a finite or countable disjoint union of "finer" sets (i.e., subsets of the original sets) that are pairwise disjoint and still belong to the semiring.

Proof Formula (2) can be proved by induction. To prove (3) and (4), we use (2) and formulas (1) and (1′). □

Corollary *Let $\mathscr{P}$ be a semiring of subsets of a set X and $\mathcal{R}$ be the system of sets that can be written as finite unions of elements of $\mathscr{P}$. Then the union, intersection, and set-theoretic difference of two elements of $\mathcal{R}$ also belongs to $\mathcal{R}$. If $X \in \mathscr{P}$ (or at least $X \in \mathcal{R}$), then $\mathcal{R}$ is an algebra.*

Thus the system $\mathcal{R}$ of finite unions of elements of a semiring $\mathscr{P}$ is a ring. It is obviously the smallest ring containing $\mathscr{P}$.

Remark Equality (3) can be strengthened as follows: the union of P_n can be written in the form

$$\bigcup_{n=1}^{N} P_n = \bigvee_{k=1}^{K} R_k, \quad \text{where } R_1, \ldots, R_K \in \mathscr{P}$$

and for any k and n the following alternative holds: either R_k is contained in P_n, or these sets are disjoint.

To prove this for $N = 2$, use the identity

$$P_1 \cup P_2 = (P_1 \setminus P_2) \vee (P_1 \cap P_2) \vee (P_2 \setminus P_1)$$

and write each of the differences $P_1 \setminus P_2$ and $P_2 \setminus P_1$ as a disjoint union according to the definition of a semiring. The general case can be proved by induction (to prove the inductive step from N to $N + 1$, replace P_1 with $\bigcup_{n=1}^{N} P_n$ in the above argument).

1.1.5 Let $\mathscr{P}$ and $\mathcal{Q}$ be semirings of subsets of sets X and Y, respectively. Consider the Cartesian product $X \times Y$ and the system $\mathscr{P} \odot \mathcal{Q}$ of subsets of $X \times Y$ that consists of the products of elements of $\mathscr{P}$ and $\mathcal{Q}$:

$$\mathscr{P} \odot \mathcal{Q} = \{P \times Q \mid P \in \mathscr{P},\ Q \in \mathcal{Q}\}.$$

We call this system the *product of the semirings* $\mathscr{P}$ and $\mathcal{Q}$.

Theorem *The product of semirings is a semiring.*

Proof The system $\mathscr{P} \odot \mathcal{Q}$ obviously satisfies condition I from the definition of a semiring. Let $A = P \times Q$ and $B = P_0 \times Q_0$, where $P, P_0 \in \mathscr{P}$ and $Q, Q_0 \in \mathcal{Q}$. It follows from the identity $A \cap B = (P \cap P_0) \times (Q \cap Q_0)$ that the system $\mathscr{P} \odot \mathcal{Q}$ also satisfies condition II.

To verify condition III, we may assume that $B \subset A$, i.e., $P_0 \subset P$ and $Q_0 \subset Q$ (otherwise replace B with $B \cap A$). Then, by the definition of a semiring, we have

$$P = P_0 \vee P_1 \vee \cdots \vee P_m \quad \text{and} \quad Q = Q_0 \vee Q_1 \vee \cdots \vee Q_n$$

for some $P_1, \ldots, P_m \in \mathscr{P}$ and $Q_1, \ldots, Q_n \in \mathcal{Q}$. Hence all "rectangles" $P_k \times Q_j$, $0 \leqslant k \leqslant m$, $0 \leqslant j \leqslant n$, form a partition of the product $A = P \times Q$. Removing from them the set $B = P_0 \times Q_0$, we obtain a partition of the set-theoretic difference $A \setminus B$ into elements of the system $\mathscr{P} \odot \mathcal{Q}$, as required in condition III. □

1.1.6 Now consider two very important examples of semirings of subsets of $\mathbb{R}^m$. We identify the space $\mathbb{R}^m$ with the Cartesian product $\mathbb{R} \times \cdots \times \mathbb{R}$ (m factors). The coordinates of a point $x \in \mathbb{R}^m$ are denoted by the same letter with subscripts. Thus $x \equiv (x_1, x_2, \ldots, x_m)$. In some cases, we will also canonically identify $\mathbb{R}^m$ with the product of spaces of smaller dimension: $\mathbb{R}^m = \mathbb{R}^k \times \mathbb{R}^{m-k}$ for $1 \leqslant k < m$.

Recall that, by definition, the distance $\rho(x, y)$ between points $x, y \in \mathbb{R}^m$ is equal to $(\sum_{k=1}^m (x_k - y_k)^2)^{1/2}$. The function $x \mapsto (\sum_{k=1}^m x_k^2)^{1/2} \equiv \|x\|$ is called the (Euclidean) *norm*. Clearly, $\rho(x, y) = \|x - y\|$. Given a set $A \subset \mathbb{R}^m$, the value $\sup\{\|x - y\| \mid x, y \in A\}$ is called the *diameter* of A and is denoted by $\operatorname{diam}(A)$.

The systems of sets we are going to consider first consist of rectangular parallelepipeds. As is well known, an open *parallelepiped* in $\mathbb{R}^m$ spanned by linearly independent vectors $\{v_j\}_{j=1}^m$ is the set (hereafter $a \in \mathbb{R}^m$)

$$P(a; v_1, \ldots, v_m) = \left\{ a + \sum_{j=1}^m t_j v_j \,\middle|\, 0 < t_j < 1 \text{ for } j = 1, 2, \ldots, m \right\}.$$

Replacing the conditions $0 < t_j < 1$ by the conditions $0 \leqslant t_j \leqslant 1$, we obtain the closed parallelepiped $\overline{P}(a; v_1, \ldots, v_m)$, which is obviously the closure of $P(a; v_1, \ldots, v_m)$. Every set P such that

$$P(a; v_1, \ldots, v_m) \subset P \subset \overline{P}(a; v_1, \ldots, v_m)$$

is also called a parallelepiped.

The vectors v_j are called the edges of $P(a; v_1, \ldots, v_m)$. If they are pairwise orthogonal, then the parallelepiped is called *rectangular*. The vectors of the form $a + \sum_{j \in J} v_j$, where J is an arbitrary subset of $\{1, \ldots, m\}$, are called the *vertices* of $P(a; v_1, \ldots, v_m)$, and the vector $a + \frac{1}{2} \sum_{j=1}^m v_j$ is the *center* of $P(a; v_1, \ldots, v_m)$.

A key role in our considerations is played by rectangular parallelepipeds of a special form, with edges parallel to the coordinate axes. Let us describe them in more detail.

Let $a = (a_1, \ldots, a_m) \in \mathbb{R}^m$, $b = (b_1, \ldots, b_m) \in \mathbb{R}^m$. We write $a \leqslant b$ if $a_j \leqslant b_j$ for all $j = 1, \ldots, m$. The notation $a < b$ means that $a_j < b_j$ for all $j = 1, 2, \ldots, m$. Generalizing the notion of a one-dimensional interval, we set, for $a \leqslant b$,

$$(a, b) = \prod_{j=1}^m (a_j, b_j) = \{x = (x_1, \ldots, x_m) \mid a_j < x_j < b_j \text{ for all } j = 1, \ldots, m\}.$$

Thus, for $a < b$, we may say that $(a, b) = P(a; v_1, \ldots, v_m)$, where $v_j = (b_j - a_j)e_j$ for $j = 1, \ldots, m$. Obviously, the edge lengths of this parallelepiped are equal to $b_1 - a_1, \ldots, b_m - a_m$.

The corresponding closed parallelepiped, which is nothing else than $\prod_{j=1}^m [a_j, b_j]$, will be denoted by $[a, b]$, by analogy with the one-dimensional case.

Unfortunately, neither open nor closed parallelepipeds form a semiring. Hence in what follows we are mainly interested in parallelepipeds $[a,b)$ of another form, which we call *cells* (of dimension m). By definition,

$$[a,b) = \prod_{j=1}^{m} [a_j, b_j) = \{x = (x_1, \dots, x_m) \mid a_j \leqslant x_j < b_j \text{ for all } j = 1, \dots, m\}.$$

If $a_j = b_j$ for at least one j, then the sets (a,b) and $[a,b)$ are empty. Thus (a,b), $[a,b) \neq \varnothing$ if and only if $a < b$. Note also that the Cartesian product of cells of dimension m and l is again a cell (of dimension $m+l$).

Proposition *Every non-empty cell is the intersection of a decreasing sequence of open parallelepipeds and the union of an increasing sequence of closed parallelepipeds.*

Proof Let $[a,b)$ be a non-empty cell and $h > 0$ be a vector such that $b - h \in [a,b)$. Consider the parallelepipeds $I_k = (a - \frac{1}{k}h, b)$ and $S_k = [a, b - \frac{1}{k}h]$. Then $[a,b) = \bigcup_{k \geqslant 1} S_k = \bigcap_{k \geqslant 1} I_k$. The details are left to the reader. □

As follows from the proposition, every cell is simultaneously a G_δ and an F_σ set. In particular, every cell is a Borel set.

If all edge lengths of a cell are equal, then it is called a cubic cell. If all vertices of a cell have rational coordinates, we call it a cell with rational vertices. Note the following simple but important fact: every cell with rational vertices is the disjoint union of finitely many cubic cells.

Indeed, since the coordinates of the vertices of such a cell can be written as fractions with a common denominator n, it can be split into cubes with edge length $\frac{1}{n}$.

The system of all m-dimensional cells will be denoted by $\mathscr{P}^m$, and its part consisting of cells with rational vertices, by $\mathscr{P}_r^m$.

Theorem *The systems $\mathscr{P}^m$ and $\mathscr{P}_r^m$ are semirings.*

Proof The proof is by induction on the dimension. In the one-dimensional case, the assertion is obvious (see Example 1.1.4). The inductive step is based on Theorem 1.1.5 and the fact that, by the definition of cells, $\mathscr{P}^m = \mathscr{P}^{m-1} \odot \mathscr{P}^1$ and $\mathscr{P}_r^m = \mathscr{P}_r^{m-1} \odot \mathscr{P}_r^1$. □

Remark In some cases (see the proof of Theorem 10.5.5), instead of $\mathscr{P}_r^m$ we need to consider the system $\mathscr{P}_E^m$ consisting of all cells for which the coordinates of all vertices belong to a fixed set $E \subset \mathbb{R}$. As one can easily see, this system is also a semiring.

1.1.7 The next theorem will be repeatedly used in what follows.

Theorem *Every non-empty open subset G of the space $\mathbb{R}^m$ is the union of a countable family of pairwise disjoint cells whose closures are contained in G. All these cells may be assumed to have rational vertices.*

Proof For each point $x \in G$, find a cell $R_x \in \mathscr{P}_r^m$ such that $x \in R_x$ and $\overline{R_x} \subset G$. Obviously, $G = \bigcup_{x \in G} R_x$. Since the semiring $\mathscr{P}_r^m$ is countable, among R_x there are only countably many distinct cells. Numbering them, we obtain a sequence of cells P_k ($k \in \mathbb{N}$) with the following properties:

$$\bigcup_{k=1}^{\infty} P_k = G, \quad \overline{P_k} \subset G \text{ for all } k \in \mathbb{N}.$$

To obtain a decomposition of G into disjoint cells with rational vertices, it remains to use decomposition (4) from Theorem 1.1.4 on the properties of semirings. □

Corollary $\mathfrak{B}(\mathscr{P}^m) = \mathfrak{B}(\mathscr{P}_r^m) = \mathfrak{B}^m$.

Proof The inclusions $\mathfrak{B}(\mathscr{P}_r^m) \subset \mathfrak{B}(\mathscr{P}^m) \subset \mathfrak{B}^m$ are obvious. The reverse inclusion $\mathfrak{B}^m \subset \mathfrak{B}(\mathscr{P}_r^m)$ follows from the definition of $\mathfrak{B}^m$, since, by the above theorem, the σ-algebra $\mathfrak{B}(\mathscr{P}_r^m)$ contains all open sets. □

Remark The proof of the theorem remains valid for every semiring $\mathscr{P}_E^m$ provided that the set E is dense. The corollary also remains valid in this case.

EXERCISES

1. Show that the system of all (one-dimensional) open intervals and the system of all closed intervals are not semirings.

2. Verify that the circular arcs (including degenerate ones) of angle less than π form a semiring; show that without this additional restriction the assertion is false.

3. What is the Borel hull of the system of all half-lines of the form $(-\infty, a)$, where $a \in \mathbb{R}$? Does the answer change if we consider only rational a or if we consider closed rather than open half-lines?

4. For sets A, B, their symmetric difference is the set $A \triangle B = (A \setminus B) \cup (B \setminus A)$. Show that $A \triangle B = (A \cup B) \setminus (A \cap B)$. Give an example of a symmetric system of sets $\mathfrak{A}$ that contains the symmetric difference of any two elements $A, B \in \mathfrak{A}$, but is not an algebra. *Hint*. Assuming that $X = \{a, b, c, d\}$, consider the system of all subsets of X consisting of an even number of points.

5. Let $\mathfrak{A}$ be the algebra of all subsets of a two-point set. Show that the semiring $\mathfrak{A} \odot \mathfrak{A}$ does not contain the complements of one-point sets and hence is not an algebra.

6. Show that the minimal algebra containing n sets has at most 2^{2^n} elements. Show that this bound is sharp.

7. Show that all subsets of $\mathbb{R}^m$ that are simultaneously G_δ and F_σ sets form an algebra containing all open sets. Verify that it is not a σ-algebra (for instance, it does not contain $\mathbb{Q}^m$).

8. Refine Theorem 1.1.7 by proving that it suffices to use only cubic cells satisfying the additional condition that the diameter of each cell is substantially smaller than the distance to the boundary of the set:

$$\operatorname{diam}(P) \leqslant C \min\{\|x - y\| \mid x \in P,\ y \in \partial G\}$$

(here $C > 0$ is a predefined arbitrarily small coefficient).

9. Let $P_1, \dots, P_n$ be elements of a semiring $\mathscr{P}$. Show that all elements of the canonical partition corresponding to these sets, except possibly for the set $\bigcap_{k=1}^{n} P_k^c$, can be written as disjoint unions of elements of $\mathscr{P}$. Deduce the result mentioned in Remark 1.1.4.

10. A symmetric system of sets $\mathcal{E}$ is called a D-system if it contains the unions of all at most countable families of pairwise disjoint elements $A_1, A_2, \dots \in \mathcal{E}$. Let $\mathcal{E}$ be a D-system and $A, B \in \mathcal{E}$. Show that:

(a) if $A \subset B$, then $B \setminus A \in \mathcal{E}$;
(b) each of the inclusions $A \cap B \in \mathcal{E}$, $A \cup B \in \mathcal{E}$ and $A \setminus B \in \mathcal{E}$ implies the other two.

11. Let a D-system contain all finite intersections of sets $A_1, \dots, A_n$. Show that it also contains the minimal algebra generated by these sets.

12. A system $\mathfrak{F}$ of non-empty subsets of a set X is called a *filter* (in X) if it contains the intersection of any elements $A, B \in \mathfrak{F}$. For example, the system of all neighborhoods of a given point is a filter. A filter $\mathfrak{U}$ is called an *ultrafilter* if every filter containing $\mathfrak{U}$ coincides with $\mathfrak{U}$. An example of an ultrafilter is the system of all sets containing a given point (a trivial ultrafilter).
Show that a filter $\mathfrak{F}$ in X is an ultrafilter if and only if for every set $A \subset X$ the following alternative holds: either A or $X \setminus A$ belongs to $\mathfrak{F}$. Using Zorn's lemma, show that for every filter there exists an ultrafilter that contains it.

1.2 Volume

In this section, we embark on the study of the main topic of this chapter. Namely, we will investigate the properties of so-called additive set functions. The assertion that some quantity is additive means that the value corresponding to a whole object is equal to the sum of the values corresponding to the parts of this object for "every" partition of the object into disjoint parts. Numerous examples of additive quantities appearing in mathematics, as well as their prototypes in mechanics and physics, are well known. They include, in particular, length, area, probability, mass, moment of inertia about a fixed axis, quantity of electricity, etc. In this chapter, we restrict ourselves to the study of additive functions with non-negative numerical (possibly infinite) values. The properties of additive functions of an arbitrary sign will be studied in Chap. 11. Let us proceed to more precise statements.

1.2.1 Let X be an arbitrary set and $\mathcal{E}$ be a system of subsets of X.

Definition A function $\varphi : \mathcal{E} \to (-\infty, +\infty]$ defined on $\mathcal{E}$ is called *additive* if

$$\varphi(A \vee B) = \varphi(A) + \varphi(B) \quad \text{provided that } A, B \in \mathcal{E} \text{ and } A \vee B \in \mathcal{E}. \tag{1}$$

It is called *finitely additive* if for every set $A \in \mathcal{E}$ and every finite partition of A into elements $A_1, \ldots, A_n$ of $\mathcal{E}$,

$$\varphi(A) = \sum_{k=1}^{n} \varphi(A_k). \tag{1'}$$

The sums on the right-hand sides of (1) and (1′) always make sense, because the corresponding terms cannot take infinite values of opposite sign (by definition, $\varphi > -\infty$).

Remark If φ is defined on an algebra (or a ring) $\mathfrak{A}$, then the additivity of φ implies its finite additivity. This can be proved by induction using (1).

1.2.2 We define the concept to which this paragraph is devoted.

Definition A finitely additive function μ defined on a semiring of subsets of a set X is called a *volume*[2] (in X) if μ is non-negative and $\mu(\varnothing) = 0$.

According to the definition of an additive function, a volume may take infinite values. It is called *finite* if X belongs to the semiring and $\mu(X) < +\infty$. A volume is called *σ-finite* if X can be written as the union of a sequence of sets of finite volume.

Examples

(1) The length of an interval is a volume on the semiring $\mathscr{P}^1$.
We leave the reader to verify this.

(2) Another very important example of a volume is a generalization of the length, the *ordinary volume* λ_m, which is defined on the semiring $\mathscr{P}^m$ of m-dimensional cells by the following formula:

$$\text{if } P = \prod_{k=1}^{m} [a_k, b_k), \quad \text{then } \lambda_m(P) = \prod_{k=1}^{m} (b_k - a_k).$$

It is obvious that for $m = 1$, the ordinary volume coincides with the length of an interval; for $m = 2$, with the area of a rectangle; and for $m = 3$, with the volume of a parallelepiped. The additivity of the ordinary volume will be proved in Corollary 1.2.4.

[2]This term is not widely accepted, but we temporarily use it, for lack of a better one, instead of the lengthy expression "a non-negative finitely additive set function".

(3) Let g be a non-decreasing function defined on $\mathbb{R}$. We define a function ν_g on the semiring $\mathscr{P}^1$ as follows: $\nu_g([a,b)) = g(b) - g(a)$. It is a volume, as the reader can easily verify.

(4) Let $\mathfrak{A}$ be an arbitrary algebra of subsets of a set X, $x_0 \in X$ and $a \in [0,+\infty]$. Given $A \in \mathfrak{A}$, put

$$\mu(A) = \begin{cases} a & \text{if } x_0 \in A, \\ 0 & \text{if } x_0 \notin A. \end{cases}$$

One can easily check that μ is a volume. We will say that μ is the volume generated by a point mass of size a at x_0.

More generally, if the volume μ of a one-point set $\{x_0\}$ is equal to $a > 0$, we say that μ has a point mass of size a at x_0.

To obtain a generalization of the last example, we use the notion of the sum of a family of numbers. For brevity, a family of non-negative numbers is called positive. Recall that $\operatorname{card}(E)$ stands for the cardinality of a set E.

Definition The sum of a positive family $\{\omega_x\}_{x\in X}$ is the value

$$\sup\left\{\sum_{x\in E}\omega_x \;\middle|\; E \subset X, \operatorname{card}(E) < +\infty\right\},$$

which is denoted by $\sum_{x\in X}\omega_x$.

A family $\{\omega_x\}_{x\in X}$ of numbers of arbitrary sign is called *summable* if

$$\sum_{x\in X}|\omega_x| < +\infty.$$

The *sum* of such a family is the value

$$\sum_{x\in X}\omega_x = \sum_{x\in X}\omega_x^+ - \sum_{x\in X}\omega_x^-, \quad \text{where } \omega_x^\pm = \max\{\pm\omega_x, 0\}.$$

For a summable family, the set $\{x \in X \mid \omega_x \neq 0\}$ is at most countable. Indeed, it can be exhausted by the sets $X_n = \{x \in X \mid |\omega_x| \geqslant \frac{1}{n}\}$ $(n \in \mathbb{N})$, each of which is finite, because

$$\operatorname{card}(X_n) \leqslant n \sum_{x\in X_n}|\omega_x| \leqslant n \sum_{x\in X}|\omega_x| < +\infty.$$

Since, obviously, for every positive family we have

$$\sum_{x\in X}\omega_x = \sum_{\{x\in X \mid \omega_x > 0\}}\omega_x,$$

the obtained result allows one to reduce the computation of the sum of an arbitrary summable family to the computation of the sum of a family with a countable set of indices. The latter problem can be reduced to the computation of the sum of a series.

If X is a countable set, a bijection $\varphi : \mathbb{N} \to X$ will be called a numbering of X and denoted by $\{x_n\}_{n\geqslant 1}$, where $x_n = \varphi(n)$.

Lemma *Let $\{\omega_x\}_{x\in X}$ be an arbitrary positive family. If the set X is countable, then for an arbitrary numbering $\{x_n\}_{n\geqslant 1}$ of X,*

$$\sum_{x\in X} \omega_x = \sum_{n=1}^{\infty} \omega_{x_n}.$$

Proof Denote by S_1 and S_2 the left- and right-hand sides of this equality, respectively. On the one hand, for every finite set $E \subset X$, we have $\sum_{x\in E} \omega_x \leqslant \sum_{n=1}^{\infty} \omega_{x_n}$ (since for every $x \in E$, the number ω_x is an element of the series). Hence $S_1 \leqslant S_2$.

On the other hand, for every k we have $\sum_{n=1}^{k} \omega_{x_n} \leqslant S_1$, by the definition of the sum of a family, whence $S_2 \leqslant S_1$. Since $S_1 \leqslant S_2$, this completes the proof. □

We leave the reader to check that the equality we have proved is valid for the sum of every summable family with a countable set of indices.

Now consider the following example.

(5) Let $\{\omega_x\}_{x\in X}$ be an arbitrary positive family. Assuming that $\mathfrak{A}$ is an algebra of subsets of X that contains all one-point sets, define a function μ on $\mathfrak{A}$ as follows:

$$\mu(A) = \sum_{x\in A} \omega_x \quad (A \in \mathfrak{A})$$

(by definition, we assume that $\sum_{x\in\varnothing} \omega_x = 0$). Note that since $\mu(E) = \omega_{x_1} + \dots + \omega_{x_N}$ for every finite set $E = \{x_1, \dots, x_N\}$, we have

$$\mu(A) = \sup\bigl\{\mu(E) \mid E \subset A,\ \operatorname{card}(E) < +\infty\bigr\}.$$

The reader can easily verify that μ is additive.

(6) An example of a volume defined on the algebra of bounded sets and their complements (see Sect. 1.1.2, Example (1)) can be obtained as follows. Given $a > 0$, put

$$\mu(A) = \begin{cases} 0 & \text{if } A \text{ is bounded,} \\ a & \text{if } A \text{ is unbounded.} \end{cases}$$

This volume will be useful for constructing various counterexamples.

1.2.3 We establish the basic properties of volume.

Theorem *Let μ be a volume on a semiring $\mathscr{P}$, and let $P, P', P_1, \dots, P_n \in \mathscr{P}$. Then*

(1) *if $P' \subset P$, then $\mu(P') \leqslant \mu(P)$;*
(2) *if $\bigvee_{k=1}^{n} P_k \subset P$, then $\sum_{k=1}^{n} \mu(P_k) \leqslant \mu(P)$;*
(3) *if $P \subset \bigcup_{k=1}^{n} P_k$, then $\mu(P) \leqslant \sum_{k=1}^{n} \mu(P_k)$.*

Properties (1) and (2) are called the *monotonicity* and the *strong monotonicity* of μ, respectively; property (3) is called the *subadditivity* of μ.

Proof Obviously, the monotonicity of μ follows from its strong monotonicity, so we will prove the latter property.

By the theorem on the properties of semirings, the set-theoretic difference $P \setminus \bigvee_{k=1}^{n} P_k$ can be written in the form $P \setminus \bigvee_{k=1}^{n} P_k = \bigvee_{j=1}^{m} Q_j$, where $Q_j \in \mathscr{P}$. Therefore, $P = (\bigvee_{k=1}^{n} P_k) \vee (\bigvee_{j=1}^{m} Q_j)$, and, by the additivity of μ,

$$\mu(P) = \sum_{k=1}^{n} \mu(P_k) + \sum_{j=1}^{m} \mu(Q_j) \geqslant \sum_{k=1}^{n} \mu(P_k).$$

To prove the subadditivity of μ, put $P_k' = P \cap P_k$. Then $P = \bigcup_{k=1}^{n} P_k'$, $P_k' \in \mathscr{P}$. By the theorem on the properties of semirings,

$$P = \bigvee_{k=1}^{n} \bigvee_{j=1}^{m_k} Q_{kj},$$

where $Q_{kj} \in \mathscr{P}$ and $Q_{kj} \subset P_k' \subset P_k$ for $1 \leqslant k \leqslant n$ and $1 \leqslant j \leqslant m_k$. It follows from the strong monotonicity of μ that $\sum_{j=1}^{m_k} \mu(Q_{kj}) \leqslant \mu(P_k)$. Therefore,

$$\mu(P) = \sum_{k=1}^{n} \sum_{j=1}^{m_k} \mu(Q_{kj}) \leqslant \sum_{k=1}^{n} \mu(P_k).$$

□

Note that if a volume is defined on an algebra (or a ring) $\mathfrak{A}$, then $\mu(A \setminus B) = \mu(A) - \mu(B)$ provided that $A, B \in \mathfrak{A}$, $B \subset A$ and $\mu(B) < +\infty$. Indeed, since $A \setminus B \in \mathfrak{A}$, we have $\mu(A) = \mu(B) + \mu(A \setminus B)$.

Remark A volume μ defined on a semiring $\mathscr{P}$ can be uniquely extended to the ring $\mathcal{R}$ consisting of all finite unions of elements of $\mathscr{P}$. Indeed, let $E = \bigcup_{k=1}^{n} P_k$, where $P_k \in \mathscr{P}$. We may assume without loss of generality that the sets P_k are pairwise disjoint (see Theorem 1.1.4). Put $\widetilde{\mu}(E) = \sum_{k=1}^{n} \mu(P_k)$. We leave the reader to show that this function is well defined and that $\widetilde{\mu}$ is a volume that coincides with μ on $\mathscr{P}$.

1.2.4 Now let us check that the ordinary volume is indeed a volume in the sense of our definition. Since $\mathscr{P}^m = \mathscr{P}^1 \odot \mathscr{P}^{m-1}$, this is a corollary of the following general theorem, in which we use the notion of the product of arbitrary semirings (see Sect. 1.1.5).

Theorem *Let X, Y be non-empty sets, $\mathscr{P}$, $\mathcal{Q}$ be semirings of subsets of these sets, and μ, ν be volumes defined on $\mathscr{P}$ and $\mathcal{Q}$, respectively. We define a function λ on the semiring $\mathscr{P} \odot \mathcal{Q}$ by the formula*

$$\lambda(P \times Q) = \mu(P) \cdot \nu(Q) \quad \textit{for any } P \in \mathscr{P}, Q \in \mathcal{Q}$$

(the products $0 \cdot (+\infty)$ *and* $(+\infty) \cdot 0$ *are assumed to vanish).*
Then λ *is a volume on* $\mathscr{P} \odot \mathscr{Q}$.

The volume λ is called the *product of the volumes* μ and ν.

Proof We need to check only the finite additivity of λ. First consider a partition of $P \times Q$ of a special form. Let P and Q be partitioned into disjoint sets:

$$P = P_1 \vee \cdots \vee P_I, \qquad Q = Q_1 \vee \cdots \vee Q_J \quad (P_i \in \mathscr{P},\ Q_j \in \mathscr{Q}).$$

Then the sets $P_i \times Q_j$ ($1 \leqslant i \leqslant I, 1 \leqslant j \leqslant J$) belong to the semiring $\mathscr{P} \odot \mathscr{Q}$ and form a partition of $P \times Q$, which we will call a grid partition. For such a partition, the desired equality is obvious:

$$\lambda(P \times Q) = \mu(P)\nu(Q) = \sum_{i=1}^{I} \mu(P_i) \sum_{j=1}^{J} \nu(Q_j) = \sum_{\substack{1 \leqslant i \leqslant I \\ 1 \leqslant j \leqslant J}} \lambda(P_i \times Q_j).$$

Now consider an arbitrary partition of the set $P \times Q$ into elements of the semiring $\mathscr{P} \odot \mathscr{Q}$:

$$P \times Q = (P_1 \times Q_1) \vee \cdots \vee (P_N \times Q_N) \quad (P_n \in \mathscr{P}, Q_n \in \mathscr{Q}).$$

In general, it is not a grid partition, but, refining it, we can reduce the problem to such a partition. Clearly, $P = P_1 \cup \cdots \cup P_N$ and $Q = Q_1 \cup \cdots \cup Q_N$, where the sets $P_1, \ldots, P_N$ and $Q_1, \ldots, Q_n$, respectively, may not be disjoint. However, as we observed in Sect. 1.1 (see the remark in Sect. 1.1.4), there exist partitions

$$P = A_1 \vee \cdots \vee A_I \quad (A_i \in \mathscr{P}) \quad \text{and} \quad Q = B_1 \vee \cdots \vee B_J \quad (B_j \in \mathscr{Q})$$

such that

$$\text{for all } i, n, \quad \text{either } A_i \subset P_n \quad \text{or} \quad A_i \cap P_n = \varnothing;$$
$$\text{for all } j, n, \quad \text{either } B_j \subset Q_n \quad \text{or} \quad B_j \cap Q_n = \varnothing.$$

Since the sets $A_i \times B_j$ form a grid partition of the product $P \times Q$, we have

$$\lambda(P \times Q) = \sum_{\substack{1 \leqslant i \leqslant I \\ 1 \leqslant j \leqslant J}} \lambda(A_i \times B_j). \tag{2}$$

On the other hand, it is clear that for every n the families $\{A_i \mid A_i \subset P_n\}$ and $\{B_j \mid B_j \subset Q_n\}$ are partitions of the sets P_n and Q_n, respectively. Hence $\{A_i \times B_j \mid A_i \subset P_n,\ B_j \subset Q_n\}$ is a grid partition of the product $P_n \times Q_n$. Therefore,

$$\lambda(P_n \times Q_n) = \sum_{\substack{i:\, A_i \subset P_n \\ j:\, B_j \subset Q_n}} \lambda(A_i \times B_j).$$

Rearranging the terms on the right-hand side of (2), we obtain the desired equality:

$$\lambda(P\times Q)=\sum_{\substack{1\leqslant i\leqslant I\\1\leqslant j\leqslant J}}\lambda(A_i\times B_j)=\sum_{1\leqslant n\leqslant N}\sum_{\substack{i:A_i\subset P_n\\j:B_j\subset Q_n}}\lambda(A_i\times B_j)=\sum_{1\leqslant n\leqslant N}\lambda(P_n\times Q_n).$$ □

Corollary *The ordinary volume λ_m is a volume in the sense of Definition* 1.2.2.

Proof The proof proceeds by induction on the dimension. The one-dimensional case is left to the reader. Now the additivity of λ_m follows immediately from the theorem, since $\mathscr{P}^m=\mathscr{P}^1\odot\mathscr{P}^{m-1}$ and λ_m is the product of the volumes λ_1 and λ_{m-1}. □

EXERCISES In Exercises 1–3, μ is a finite volume defined on an algebra $\mathfrak{A}$ of subsets of a set X.

1. Show that for any elements of $\mathfrak{A}$,

$$\mu(A\cup B)=\mu(A)+\mu(B)-\mu(A\cap B);$$

$$\mu(A\cup B\cup C)=\mu(A)+\mu(B)+\mu(C)-\mu(A\cap B)-\mu(B\cap C)-\mu(A\cap C)$$
$$+\mu(A\cap B\cap C).$$

Generalize these equalities to the case of four and more sets.

2. Let $\mu(X)=1$, and let $A_1,\dots,A_n\in\mathfrak{A}$. Show that if $\sum_{k=1}^n\mu(A_k)>n-1$, then $\bigcap_{k=1}^n A_k\neq\varnothing$.

3. Show that every partition of X into subsets of positive volume is at most countable.

1.3 Properties of Measure

The key property in the definition of a volume is its finite additivity, i.e., the assertion that "the volume of a whole object is the sum of the volumes of its parts" provided that the number of these "parts" is finite. As we will see below, this rule may be violated if the "parts" form an infinite sequence. Of course, infinite partitions arise only as an idealization of real-life situations, so it is hard to provide a natural scientifically motivated explanation of why we need to consider volumes with such a strong additivity property, which is called countable additivity.

However, intuitively, a violation of the rule "the volume of a whole object is the sum of the volumes of its parts" for a countable set of parts seems to be quite unnatural if, for example, by the volume we mean the length or the area. It is the countable additivity that allows one to develop a deep theory that comes close to the theory of integration. This and the next sections are devoted to the theory of countably additive volumes, which is usually called measure theory. It has numerous important applications. First of all, it is worth mentioning that measure theory lies at the foundations of modern probability theory.

1.3.1 Let us proceed to precise definitions.

Definition A volume μ defined on a semiring $\mathscr{P}$ is called *countably additive* if for every set $P \in \mathscr{P}$ and every partition $\{P_k\}_{k=1}^{\infty}$ of P into elements of $\mathscr{P}$,

$$\mu(P) = \sum_{k \geqslant 1} \mu(P_k).$$

A countably additive volume is called a *measure*.

Using the notion of the sum of a family and Lemma 1.2.2, we can formulate the definition of countable additivity in an equivalent, though formally more general form: a volume μ defined on a semiring $\mathscr{P}$ is countably additive if for every set $P \in \mathscr{P}$ and every countable partition $\{P_\omega\}_{\omega \in \Omega}$ of P into elements of $\mathscr{P}$,

$$\mu(P) = \sum_{\omega \in \Omega} \mu(P_\omega).$$

Countable additivity does not follow from finite additivity, so that not every volume is a measure. In particular, the volume from Example (6) of Sect. 1.2.2 is not a measure, as the reader can easily check.

Examples

(1) The ordinary volume is a measure (see Theorem 2.1.1).
(2) Consider the volume $\nu_g([a,b)) = g(b) - g(a)$ defined in Example (3) of Sect. 1.2.2. Its countable additivity means, in particular, that if $[b_0, b) = \bigvee_{n=0}^{\infty}[b_n, b_{n+1})$, where $b_n \to b, b_n < b_{n+1}$, then $\nu_g([b_0,b)) = \sum_{n=0}^{\infty} \nu_g([b_n, b_{n+1}))$. Since $\nu([b_n, b_{n+1})) = g(b_{n+1}) - g(b_n)$, this is equivalent to the condition $g(b_n) \underset{n\to\infty}{\longrightarrow} g(b)$.

Thus for ν_g to be countably additive, it is necessary that the function g be continuous from the left.

Given an arbitrary increasing function g, one can obtain a measure by setting $\mu_g([a,b)) = g(b-0) - g(a-0)$, where $g(a-0)$ and $g(b-0)$ are the left limits of g at the points a and b, respectively. We will prove the countable additivity of μ_g in Theorem 4.10.2. It implies, in particular, that the continuity of the function g from the left is not only a necessary, but also a sufficient condition for the volume ν_g to be a measure.
(3) The volume generated by a positive point mass (see Example (4) in Sect. 1.2.2) is a measure.
(4) Let X be an arbitrary set and $\mathfrak{A}$ be a σ-algebra of subsets of X containing all one-point sets. We define a function μ on $\mathfrak{A}$ as follows:

$$\mu(A) = \begin{cases} \text{the number of points in } A & \text{if } A \text{ is finite;} \\ +\infty & \text{if } A \text{ is infinite.} \end{cases}$$

We leave the reader to verify that the function μ thus defined is indeed a measure. It is called the *counting measure*.

(5) Let us verify that the volume μ constructed in Example (5) of Sect. 1.2.2 is countably additive, i.e., that μ is a measure.

Indeed, let $A = \bigvee_{k=1}^{\infty} A_k$, where $A, A_k \in \mathfrak{A}$. It is clear that for every $n \in \mathbb{N}$,

$$\mu(A) \geqslant \mu\left(\bigvee_{k=1}^{n} A_k\right) = \sum_{k=1}^{n} \mu(A_k),$$

whence $\mu(A) \geqslant \sum_{k=1}^{\infty} \mu(A_k)$. On the other hand, if E is an arbitrary finite subset of A, then for some n we have $E \subset \bigvee_{k=1}^{n} A_k$. Therefore,

$$\mu(E) \leqslant \mu\left(\bigvee_{k=1}^{n} A_k\right) = \sum_{k=1}^{n} \mu(A_k) \leqslant \sum_{k=1}^{\infty} \mu(A_k).$$

It follows that

$$\mu(A) = \sup\bigl\{\mu(E) \mid E \subset A,\ \operatorname{card}(E) < +\infty\bigr\} \leqslant \sum_{k=1}^{\infty} \mu(A_k).$$

Together with the reverse inequality obtained above, this proves the countable additivity of μ.

We will say that μ is the *discrete measure* generated by the masses ω_x. If $\omega_x \equiv 1$, then, obviously, μ is the counting measure.

1.3.2 We establish an important characteristic property of measures.

Theorem *A volume μ defined on a semiring $\mathscr{P}$ is a measure if and only if it is countably subadditive, i.e.,*

$$\text{the conditions } P \subset \bigcup_{k\geqslant 1} P_k, \quad P, P_k \in \mathscr{P} \quad \text{imply that } \mu(P) \leqslant \sum_{k\geqslant 1} \mu(P_k). \tag{1}$$

Proof [3] Let μ be a countably additive volume. Replacing the sets P_k in condition (1) by the sets $P'_k = P \cap P_k$, we see that

$$P = \bigcup_{k\geqslant 1} P'_k, \quad P'_k \in \mathscr{P} \quad (k \in \mathbb{N}).$$

[3] It is instructive to compare this argument with the proof of Theorem 1.2.3.

By Theorem 1.1.4, P can be written in the form

$$P = \bigvee_{k \geqslant 1} \bigvee_{j=1}^{n_k} Q_{kj} \quad (Q_{kj} \in \mathscr{P}).$$

Furthermore, $\bigvee_{j=1}^{n_k} Q_{kj} \subset P'_k$. Hence, by the strong monotonicity of a volume, $\sum_{j=1}^{n_k} \mu(Q_{kj}) \leqslant \mu(P'_k) \leqslant \mu(P_k)$. Using the countable additivity, we obtain

$$\mu(P) = \sum_{k \geqslant 1} \sum_{j=1}^{n_k} \mu(Q_{kj}) \leqslant \sum_{k \geqslant 1} \mu(P_k),$$

as required.

Now let us prove that countable subadditivity implies countable additivity. Let $\{P_k\}_{k=1}^{\infty} \subset \mathscr{P}$ be a partition of a set $P \in \mathscr{P}$. By the countable subadditivity of μ,

$$\mu(P) \leqslant \sum_{k \geqslant 1} \mu(P_k). \tag{2}$$

On the other hand, the strong monotonicity of a volume implies that $\mu(P) \geqslant \sum_{k=1}^{n} \mu(P_k)$ for every $n \in \mathbb{N}$. Passing to the limit as $n \to \infty$, we see that $\mu(P) \geqslant \sum_{k \geqslant 1} \mu(P_k)$. Together with (2), this proves the countable additivity of μ. □

The last theorem implies a result that we will often use in what follows.

Corollary *Let μ be a measure defined on a σ-algebra $\mathfrak{A}$. Then a countable union of sets of zero measure is again a set of zero measure.*

Indeed, if e_n are sets from $\mathfrak{A}$ that have zero measure, then their union also belongs to $\mathfrak{A}$ and $\mu(\bigcup_{n \geqslant 1} e_n) \leqslant \sum_{n \geqslant 1} \mu(e_n) = 0$.

1.3.3 We will check that for a volume defined on the algebra, countable additivity is equivalent to a property analogous to continuity.

Theorem *A volume μ defined on an algebra $\mathfrak{A}$ is a measure if and only if it is continuous from below, i.e.,*

$$\text{the conditions } A, A_k \in \mathfrak{A}, \quad A_k \subset A_{k+1}, \quad A = \bigcup_{k \geqslant 1} A_k$$

$$\text{imply that } \mu(A_k) \underset{k \to \infty}{\longrightarrow} \mu(A). \tag{3}$$

Remark If the algebra $\mathfrak{A}$ from the statement of the theorem is a σ-algebra, then the condition $A \in \mathfrak{A}$ in the definition of continuity from below can be omitted, because it follows from the equality $A = \bigcup_{k \geqslant 1} A_k$.

Proof Let μ be a countably additive volume and A, A_k be sets satisfying conditions (3). Putting $B_1 = A_1$, $B_k = A_k \setminus A_{k-1}$ for $k > 1$, we see that $B_k \in \mathfrak{A}$, $B_k \cap B_j = \varnothing$ for $k \neq j$ $(j, k \in \mathbb{N})$, and

$$A_k = \bigvee_{j=1}^{k} B_j, \qquad A = \bigvee_{j \geqslant 1} B_j.$$

Therefore, $\mu(A_k) = \sum_{j=1}^{k} \mu(B_j)$ and

$$\mu(A) = \sum_{j \geqslant 1} \mu(B_j) = \lim_{k \to \infty} \sum_{j=1}^{k} \mu(B_j) = \lim_{k \to \infty} \mu(A_k).$$

Now let us prove that continuity from below implies countable additivity. Let $\{E_j\}_{j=1}^{\infty} \subset \mathfrak{A}$ be a partition of a set $A \in \mathfrak{A}$. Put $A_k = \bigvee_{j=1}^{k} E_j$. Then

$$A_k \in \mathfrak{A}, \quad A_k \subset A_{k+1}, \quad A = \bigcup_{k \geqslant 1} A_k,$$

and $\mu(A_k) = \sum_{j=1}^{k} \mu(E_j)$. Since μ is continuous from below, we obtain

$$\mu(A) = \lim_{k \to \infty} \mu(A_k) = \lim_{k \to \infty} \sum_{j=1}^{k} \mu(E_j) = \sum_{j \geqslant 1} \mu(E_j).$$ □

1.3.4 Recall that a volume μ defined on a semiring $\mathscr{P}$ of subsets of a set X is called finite if $X \in \mathscr{P}$ and $\mu(X) < +\infty$ (see Definition 1.2.2).

Theorem *Let μ be a finite volume defined on an algebra $\mathfrak{A}$. The following conditions are equivalent:*

(1) *μ is a measure;*
(2) *μ is continuous from above, i.e.,*

$$\text{the conditions } A, A_k \in \mathfrak{A}, \quad A_k \supset A_{k+1}, \quad \bigcap_{k \geqslant 1} A_k = A \tag{4}$$

$$\text{imply} \quad \mu(A_k) \underset{k \to \infty}{\longrightarrow} \mu(A);$$

(3) *μ is continuous from above at the empty set, i.e.,*

$$\text{the conditions } A_k \in \mathfrak{A}, \quad A_k \supset A_{k+1}, \quad \bigcap_{k \geqslant 1} A_k = \varnothing \tag{4$'$}$$

$$\text{imply} \quad \mu(A_k) \underset{k \to \infty}{\longrightarrow} 0.$$

Proof (1) ⇒ (2). Let A_k be sets satisfying conditions (4). Put $B = A_1 \setminus A$, $B_k = A_1 \setminus A_k$. Then $B_k \subset B_{k+1}$ and $B = \bigcup_{k\geqslant 1} B_k$. By the continuity of a measure from below,

$$\mu(A_1) - \mu(A_k) = \mu(B_k) \underset{k\to\infty}{\longrightarrow} \mu(B) = \mu(A_1) - \mu(A),$$

i.e., $\mu(A_k) \underset{k\to\infty}{\longrightarrow} \mu(A)$.

The implication (2) ⇒ (3) is trivial. Let us prove that (3) ⇒ (1). Let $\{E_j\}_{j=1}^{\infty} \subset \mathfrak{A}$ be a partition of a set $A \in \mathfrak{A}$. Put $A_k = \bigvee_{j=k+1}^{\infty} E_j$. Then $A_k \in \mathfrak{A}$, since $A_k = A \setminus \bigvee_{j=1}^{k} E_j$, and the sets A_k obviously satisfy all conditions (4′). Hence $\mu(A_k) \underset{k\to\infty}{\longrightarrow} 0$. Furthermore, $A = A_k \vee \bigvee_{j=1}^{k} E_j$. Thus $\mu(A) = \mu(A_k) + \sum_{j=1}^{k} \mu(E_j)$, $\mu(A_k) \underset{k\to\infty}{\longrightarrow} 0$, and, consequently, $\mu(A) = \sum_{j\geqslant 1} \mu(E_j)$, as required. □

Corollary *Every measure is conditionally continuous from above. The latter means that the conditions* $A,\ A_k \in \mathfrak{A},\ A_k \supset A_{k+1},\ \bigcap_{k\geqslant 1} A_k = A$ *and* $\mu(A_m) < +\infty$ *for some* m *imply that* $\mu(A_k) \underset{k\to\infty}{\longrightarrow} \mu(A)$.

To prove this, it suffices to consider the restriction of the measure μ to the induced algebra $\mathfrak{A} \cap A_m$ (see Example (4) in Sect. 1.1.2) and use the continuity from above of the obtained finite measure.

Remarks

(1) If a volume is infinite, then continuity from above does not imply countable additivity (see Exercise 1).

(2) If a volume is defined on a semiring, then in Theorems 1.3.3 and 1.3.4 only the "only if" parts are true (see Exercise 2).

In what follows, we usually consider measures defined on σ-algebras. The collection consisting of three objects—a set X, a σ-algebra $\mathfrak{A}$ of subsets of X, and a measure μ defined on $\mathfrak{A}$—is usually denoted by $(X, \mathfrak{A}, \mu)$ and is called a *measure space*. The sets for which the measure is defined, i.e., the elements of the σ-algebra $\mathfrak{A}$, are called *measurable*, or, more precisely, measurable with respect to $\mathfrak{A}$.

EXERCISES

1. Show that the infinite volume from Example (6) in Sect. 1.2.2 ($a = +\infty$) is conditionally continuous from above, but is not a measure.

2. Let $X = [0, 1) \cap \mathbb{Q}$, and let $\mathscr{P}$ be the system of all sets P of the form $P \equiv [a, b) \cap \mathbb{Q}$, where $0 \leqslant a \leqslant b \leqslant 1$. Put $\mu(P) = b - a$. Show that $\mathscr{P}$ is a semiring and μ is a volume that is continuous from above and from below, but is not a measure.

3. Let $(X, \mathfrak{A}, \mu)$ be a measure space, and let E_k be measurable sets such that $\sum_{k=1}^{\infty} \mu(E_k) < +\infty$. Consider the sets

$$A_n = \{x \in X \mid x \in E_k \text{ for exactly } n \text{ values of } k\},$$
$$B_n = \{x \in X \mid x \in E_k \text{ for at least } n \text{ values of } k\}.$$

Show that the sets A_n, B_n are measurable and

$$\sum_{n=1}^{\infty} n\mu(A_n) = \sum_{n=1}^{\infty} \mu(B_n) = \sum_{n=1}^{\infty} \mu(E_n).$$

4. Using the counting measure on $\mathbb{N}$, show that continuity from above at the empty set does not follow from countable additivity.

5. Show that a finite volume μ defined on an algebra $\mathfrak{A}$ is countably additive provided that it is "continuous from below at X", i.e., the conditions $A_k \subset A_{k+1}$, $\bigcup_{k\geqslant 1} A_k = X$, $A_k \in \mathfrak{A}$ imply $\mu(A_k) \underset{k\to\infty}{\longrightarrow} \mu(X)$.

6. Show that for a σ-finite measure, every partition into sets of positive measure is at most countable.

7. Assume that a measure is such that there exist arbitrarily (finitely) many pairwise disjoint subsets of positive measure. Show that there exists an infinite family of such subsets.

1.4 Extension of Measure

1.4.1 Although we have considered characteristic properties of measures, with the exception of the counting measure, we still have not produced a non-trivial example of a measure defined on a σ-algebra.

The reason is that we are presently able to define measures only on "poor" systems of sets, such as most semirings. Due to the tractability of these systems, it is comparatively easy to define volumes on them (see Examples (1)–(3) in Sect. 1.2.2). But we cannot yet define measures on wider systems of sets, e.g., on σ-algebras, except for several quite trivial cases. This situation is, of course, highly unsatisfactory.

Indeed, even if we know that the ordinary volume in m-dimensional space defined on the semiring of cells is countably additive (this will be proved in Theorem 2.1.1), we certainly cannot consider the problem of constructing a measure on $\mathbb{R}^m$ completely solved, since it is highly dubious whether a measure on Euclidean space that cannot be used to "measure" pyramids, balls, and other important bodies has any value; and this is exactly the situation we find ourselves in. The very tractability of semirings, their being poor in sets, which allowed us, in the cases considered above, to easily define volumes on them, now demonstrates its downsides. Thus we must learn to construct measures on richer systems of sets. This problem is difficult even if we restrict ourselves to the σ-algebra of Borel sets of the real line and try to assign a length to every Borel set (speaking more formally, try to extend the one-dimensional ordinary volume to the Borel σ-algebra). It was the solution of this problem suggested by Lebesgue[4] in 1902 that marked the beginning of

[4]Henri Léon Lebesgue (1875–1941)—French mathematician.

measure theory. This result, inspired by the needs of several areas of mathematics, was a major breakthrough in the theory of integration.

Lebesgue's construction of an extension of the length (the one-dimensional ordinary volume) to a measure defined on a σ-algebra of subsets of the real line was based on clear geometric considerations. It splits into several steps. First, Lebesgue assigns a measure $m(G)$ to all open sets $G \subset \mathbb{R}$, where $m(G)$ is the sum of the lengths of the intervals constituting G. Then he introduces a quantity called the outer measure; for an arbitrary set $E \subset \mathbb{R}$, it is defined by the formula

$$m_e(E) = \inf\{m(G) \mid G \supset E,\ G \text{ is an open set}\}.$$

The inner measure $m_i(E)$ of a bounded set E is equal to $m_i(E) = m(\Delta) - m_e(\Delta \setminus E)$, where Δ is an arbitrary interval containing E. A bounded set is called measurable if its inner and outer measures coincide. The common value of the inner and outer measures of a measurable set E is declared to be the measure of E. Then one checks that the system of measurable sets contained in a fixed interval is a σ-algebra and that the constructed measure is countably additive. Thus Lebesgue's method of extending a measure is not altogether direct. It contains an important intermediate step, the construction of the outer measure. So to speak, we "cross a chasm in two jumps". A detailed realization of this program (which is described in a slightly modified form, e.g., in [N]) is not at all easy.

Along with some advantages (first of all, the geometric clarity of the construction), this approach also has its disadvantages. Of course, since every open subset of a Euclidean space is the union of a sequence of cells, the analogy we should follow in order to extend a measure from the semiring of cells is clear. However, it is still not clear how one should act to extend a measure defined on a semiring of subsets of a ground set that has no topology and, consequently, no open sets. This question is all the more relevant, because in the axiomatization of probability theory in the framework of measure theory, the ground set is the space of "elementary events", which is not necessarily a topological space.

Later, due mainly to Carathéodory's[5] results, it became clear that the crucial elements of Lebesgue's construction are the following two facts. First, that the outer measure is countably subadditive, and, second, that it can be constructed without involving open sets, i.e., without using the topology. For this (bearing in mind that an open set is the union of a sequence of cells), one should only interpret the inclusion $E \subset G$ used in the one-dimensional case as the fact that E can be covered by a sequence of elements of the semiring. This observation allows one to construct the outer measure for an arbitrary measure, regardless of whether or not the ground set is a topological space.

The method suggested by Carathéodory shows that it is useful, especially from a technical point of view, not to restrict ourselves to additive functions, but instead to consider countably subadditive functions defined on all subsets of the ground set. These functions are now called outer measures. Here we must warn the reader that

[5]Constantin Carathéodory (1873–1950)—a German mathematician of Greek origin.

the terminology is slightly confusing: in general, an outer measure is not a measure in the sense of Definition 1.3.1.

The key point of Carathéodory's construction is the fact that every outer measure gives rise in a natural way to a σ-algebra (which in non-degenerate cases is quite wide) on which this outer measure is additive and hence countably additive. Thus every outer measure generates a measure. Since outer measures are much easier to construct, this approach turns out to be useful not only for extending measures, but also in other cases when we need to find a measure with given properties. We will encounter such examples when proving the existence of the surface area (which reduces to constructing the Hausdorff measure of appropriate dimension) and when describing positive functionals on the space of continuous functions (Sect. 12.2).

We preface a detailed description of Carathéodory's method with the definition of outer measures and the study of their basic properties.

1.4.2 Here we will consider subsets of a fixed non-empty set X, which we call the ground set. Recall that by A^c we denote the complement of a set $A \subset X$, i.e., the set-theoretic difference $X \setminus A$.

Definition 1 Let $\mathfrak{A}(X)$ be the σ-algebra of all subsets of the ground set X. An *outer measure on* X is a function $\tau : \mathfrak{A}(X) \to [0, +\infty]$ such that:

I. $\tau(\varnothing) = 0$ and
II. $\tau(A) \leqslant \sum_{n=1}^{\infty} \tau(A_n)$ if $A \subset \bigcup_{n=1}^{\infty} A_n$.

Property II is called *countable subadditivity.*

We mention two simple properties of outer measures.

(1) An outer measure is *finitely subadditive*, i.e., *the inclusion* $A \subset A_1 \cup \cdots \cup A_N$ *implies that* $\tau(A) \leqslant \tau(A_1) + \cdots + \tau(A_N)$.

This property follows immediately from the countable subadditivity of τ if we assume that the sets A_n are empty for all $n > N$.

(2) An outer measure is *monotone*, i.e., *the inclusion* $A \subset B$ *implies that* $\tau(A) \leqslant \tau(B)$.

This is a special case of property 1 (corresponding to $N = 1$).

As we will see below, outer measures naturally appear in various situations (see Sects. 2.1, 2.6, 12.2). Here we only mention that an example of an outer measure is any measure defined on all subsets of the ground set, in particular, a discrete measure (see Example (5) in Sect. 1.3.1).

The next definition is motivated by our desire to single out an algebra of sets on which an outer measure τ is additive. If A and E are such sets, then

$$\tau(E) = \tau(E \cap A) + \tau(E \setminus A). \tag{1}$$

To construct a desired system of sets, we let it contain those subsets A of the ground set that "split every set E additively". Thus we arrive at the following definition.

Definition 2 Let τ be an outer measure on X. A set A is called *measurable*, or, more exactly, *τ-measurable* if (1) holds for every set $E \subset X$.

The system of all τ-measurable sets will be denoted by $\mathfrak{A}_\tau$.

Let us illustrate this definition by the following informal example. Consider a commuter rail system divided into fare zones. Let X be the collection of intervals between neighboring stations. An arbitrary collection of intervals (a subset of X) will be called a path. If the price of a trip along a connected path is proportional to the number of zones through which it travels, and for an unconnected path it is the sum of the prices of the connected components, then the price of a trip is an outer measure on the set of intervals. A path is measurable if and only if it consists of entire zones.

Note that since $E = (E \cap A) \cup (E \setminus A)$ and an outer measure is countably subadditive, the inequality $\tau(E) \leqslant \tau(E \cap A) + \tau(E \setminus A)$ always holds. Hence, to verify (1), it only suffices to establish the inequality

$$\tau(E) \geqslant \tau(E \cap A) + \tau(E \setminus A), \tag{1'}$$

and usually we will do exactly this.

Remark If $\tau(A) = 0$, then $\tau(E \cap A) = 0$, and hence (1′) holds for every set E. Thus all sets of zero outer measure are measurable.

1.4.3 The main result of this subsection is the following theorem.

Theorem *Let τ be an outer measure on X. Then $\mathfrak{A}_\tau$ is a σ-algebra and the restriction of τ to this σ-algebra is a measure.*

Proof First of all, observe that the system of τ-measurable sets is symmetric, i.e., together with every set A it also contains its complement A^c. This follows from the fact that, in view of the identity $E \setminus A = E \cap A^c$, condition (1) can be written in a symmetric form: $\tau(E) = \tau(E \cap A) + \tau(E \cap A^c)$.

Now let us prove that $\mathfrak{A}_\tau$ is an algebra of sets. According to Definition 1.1.2, it suffices to check that $\mathfrak{A}_\tau$ contains the union of any two elements of $\mathfrak{A}_\tau$.

Let $A, B \in \mathfrak{A}_\tau$, and let E be an arbitrary set. Using successively the measurability of A and B, we obtain

$$\tau(E) = \tau(E \cap A) + \tau(E \setminus A) = \tau(E \cap A) + \tau\big((E \setminus A) \cap B\big) + \tau\big((E \setminus A) \setminus B\big).$$

The third term on the right-hand side of this inequality is obviously equal to $\tau(E \setminus (A \cup B))$, and the sum of the first two terms can be estimated using the subadditivity of τ:

$$\tau(E \cap A) + \tau\big((E \setminus A) \cap B\big) \geqslant \tau\big((E \cap A) \cup \big((E \setminus A) \cap B\big)\big) = \tau\big(E \cap (A \cup B)\big).$$

Thus

$$\tau(E) \geqslant \tau\big(E \cap (A \cup B)\big) + \tau\big(E \setminus (A \cup B)\big),$$

i.e., the union $A \cup B$ satisfies (1′) for every set E. Hence $A \cup B \in \mathfrak{A}_\tau$ for any A and B in $\mathfrak{A}_\tau$. So, $\mathfrak{A}_\tau$ is an algebra.

If A and B are disjoint measurable sets, then $(E \cap (A \vee B)) \cap A = E \cap A$ and $(E \cap (A \vee B)) \setminus A = E \cap B$ for an arbitrary set E. Hence $\tau(E \cap (A \vee B)) = \tau(E \cap A) + \tau(E \cap B)$. Then, by induction, for every $n \in \mathbb{N}$, for pairwise disjoint sets $A_1, \ldots, A_n$ and an arbitrary set E,

$$\tau\left(E \cap \bigvee_{j=1}^{n} A_j\right) = \sum_{j=1}^{n} \tau(E \cap A_j). \tag{2}$$

Taking $E = X$, we see that the outer measure is additive on $\mathfrak{A}_\tau$:

$$\tau\left(\bigvee_{j=1}^{n} A_j\right) = \sum_{j=1}^{n} \tau(A_j). \tag{2′}$$

Now let us check that $\mathfrak{A}_\tau$ is a σ-algebra. For this we must show that $\mathfrak{A}_\tau$ contains the union $A = \bigcup_{j=1}^{\infty} A_j$ of an arbitrary sequence of measurable sets A_j. First assume that the sets A_j are pairwise disjoint. Then for every set E and every n it follows from (2) that

$$\begin{aligned}\tau(E) &= \tau\left(E \cap \bigvee_{j=1}^{n} A_j\right) + \tau\left(E \setminus \bigvee_{j=1}^{n} A_j\right) = \sum_{j=1}^{n} \tau(E \cap A_j) + \tau\left(E \setminus \bigvee_{j=1}^{n} A_j\right) \\ &\geqslant \sum_{j=1}^{n} \tau(E \cap A_j) + \tau(E \setminus A).\end{aligned}$$

Passing to the limit as $n \to \infty$ and using the countable subadditivity of τ, we obtain

$$\begin{aligned}\tau(E) &\geqslant \sum_{j=1}^{\infty} \tau(E \cap A_j) + \tau(E \setminus A) \geqslant \tau\left(\bigvee_{j=1}^{\infty} (E \cap A_j)\right) + \tau(E \setminus A) \\ &= \tau(E \cap A) + \tau(E \setminus A).\end{aligned}$$

Thus we have confirmed that A satisfies (1′), so that $A \in \mathfrak{A}_\tau$.

The general case can be reduced to that considered above by using a disjoint decomposition (see Lemma 1.1.4): $A = \bigvee_{j=1}^{\infty} B_j$, where $B_1 = A_1$ and $B_j = A_j \setminus (A_1 \cup \cdots \cup A_{j-1})$ for $j \geqslant 2$ (the sets B_j are measurable, since $\mathfrak{A}_\tau$ is an algebra).

It remains to prove the second claim of the theorem. Let μ be the restriction of τ to $\mathfrak{A}_\tau$. It follows from (2′) that μ is a volume. It is countably subadditive, since τ is. By Theorem 1.3.2, μ is a measure. □

The remark at the end of Sect. 1.4.2 suggests to single out the measures satisfying an important additional property. In view of monotonicity, it is natural to expect that every subset of a set of zero measure also has zero measure. However, this is not always the case, because this subset may not be measurable (for instance, if the measure is defined only on Borel sets). Measures for which subsets of sets of zero measure also have zero measure are of special interest.

Definition A measure μ defined on a semiring $\mathscr{P}$ is called *complete* if the conditions $E \in \mathscr{P}$ and $\mu(E) = 0$ imply that every subset $\widetilde{E}$ of E also belongs to $\mathscr{P}$ (and, consequently, $\mu(\widetilde{E}) = 0$).

Using this definition and the remark from Sect. 1.4.2, we can refine the theorem by saying that an outer measure generates a complete measure. In other words, we have the following corollary.

Corollary *The restriction of an outer measure τ to the σ-algebra $\mathfrak{A}_\tau$ is a complete measure.*

1.4.4 We now proceed to the description of Carathéodory's method of extending a measure. Like Lebesgue's original construction, it consists of two steps. At the first step, given a measure μ_0, we construct an auxiliary function μ^* that extends μ_0 from the original semiring to the system of all subsets. It is no longer countably additive, but we can prove that it has a weaker property, countable subadditivity, so that μ^* is an outer measure. At the second step, we restrict the constructed outer measure to the system of μ^*-measurable sets; as a result, according to Theorem 1.4.3, we obtain a new measure defined on a σ-algebra. To verify that this measure is an extension of μ_0, it remains to show that the original semiring is contained in the σ-algebra of μ^*-measurable sets. Let us proceed to the realization of this program.

Let μ_0 be a measure defined on a semiring $\mathscr{P}$ of subsets of a set X. For every set $E \subset X$, put

$$\mu^*(E) = \inf\left\{ \sum_{j=1}^{\infty} \mu_0(P_j) \,\middle|\, E \subset \bigcup_{j=1}^{\infty} P_j,\ P_j \in \mathscr{P} \text{ for all } j \in \mathbb{N} \right\} \tag{3}$$

(if E cannot be covered by a sequence of elements of $\mathscr{P}$, we put $\mu^*(E) = +\infty$).

Note that instead of $\{P_j\}_{j \geqslant 1}$ in (3) we may consider an arbitrary countable family $\{P_\omega\}_{\omega \in \Omega}$, since the sum $\sum_{\omega \in \Omega} \mu_0(P_\omega)$ coincides with $\sum_{j=1}^{\infty} \mu_0(P_{\omega_j})$ for every numbering of Ω.

Theorem *The function μ^* defined by formula* (3) *is an outer measure that coincides with μ_0 on $\mathscr{P}$.*

We will say that μ^* is the *outer measure generated by* μ_0.

Proof Let $E \in \mathscr{P}$. Then the sequence $E, \varnothing, \varnothing, \ldots$ is a cover of E by elements of $\mathscr{P}$. It follows that $\mu^*(E) \leqslant \mu_0(E)$. On the other hand, if $E \subset \bigcup_{j=1}^{\infty} P_j$, where $P_j \in \mathscr{P}$ for all $j \in \mathbb{N}$, then $\mu_0(E) \leqslant \sum_{j=1}^{\infty} \mu_0(P_j)$ by the countable subadditivity of a measure (Theorem 1.3.2). Since $\{P_j\}_{j\geqslant 1}$ is an arbitrary sequence, it follows that $\mu_0(E) \leqslant \mu^*(E)$. Thus $\mu^*(E) = \mu_0(E)$; in particular, $\mu^*(\varnothing) = 0$.

It remains to show that μ^* is countably subadditive, i.e., that

$$\mu^*(E) \leqslant \sum_{n=1}^{\infty} \mu^*(E_n)$$

if $E \subset \bigcup_{n=1}^{\infty} E_n$. We may assume that the right-hand side is finite, since otherwise the inequality is trivial. Fix an arbitrary $\varepsilon > 0$, and for every n find sets $P_j^{(n)} \in \mathscr{P}$ ($j \in \mathbb{N}$) such that

$$E_n \subset \bigcup_{j=1}^{\infty} P_j^{(n)} \quad \text{and} \quad \sum_{j=1}^{\infty} \mu_0\bigl(P_j^{(n)}\bigr) < \mu^*(E_n) + \frac{\varepsilon}{2^n}.$$

In this case,

$$E \subset \bigcup_{n=1}^{\infty} E_n \subset \bigcup_{n=1}^{\infty} \bigcup_{j=1}^{\infty} P_j^{(n)}.$$

Hence, by the definition of $\mu^*(E)$,

$$\mu^*(E) \leqslant \sum_{n=1}^{\infty} \sum_{j=1}^{\infty} \mu_0\bigl(P_j^{(n)}\bigr) < \sum_{n=1}^{\infty} \left(\mu^*(E_n) + \frac{\varepsilon}{2^n}\right) = \sum_{n=1}^{\infty} \mu^*(E_n) + \varepsilon.$$

Since ε is arbitrary, it follows that μ^* is countably subadditive. □

1.4.5 Now we are in a position to prove the theorem on extension of measures, which is our main goal in this section.

Theorem *Let μ_0 be a measure defined on a semiring $\mathscr{P}$, μ^* be the outer measure generated by μ_0, and $\mathfrak{A}_{\mu^*}$ be the σ-algebra of μ^*-measurable sets. Then $\mathscr{P} \subset \mathfrak{A}_{\mu^*}$, and the restriction of μ^* to $\mathfrak{A}_{\mu^*}$ is an extension of μ_0.*

Proof By Theorem 1.4.3, the restriction of μ^* to $\mathfrak{A}_{\mu^*}$ is a measure. Since, by Theorem 1.4.4, μ^* coincides with μ_0 on $\mathscr{P}$, we need only to prove that $\mathscr{P} \subset \mathfrak{A}_{\mu^*}$, i.e., that every set $P \in \mathscr{P}$ is μ^*-measurable. For this we must check inequality (1′) from Sect. 1.4.2, which in our notation takes the following form: for every set E,

$$\mu^*(E) \geqslant \mu^*(E \cap P) + \mu^*(E \setminus P). \tag{4}$$

We verify this inequality in two steps. First assume that $E \in \mathscr{P}$. Then, by the definition of a semiring, $E \setminus P = \bigvee_{j=1}^{N} Q_j$, where $Q_j \in \mathscr{P}$. Hence E splits into disjoint elements of $\mathscr{P}$: $E = (E \cap P) \vee \bigvee_{j=1}^{N} Q_j$. Therefore, by the additivity of μ_0

and the subadditivity of μ^*

$$\mu^*(E) = \mu_0(E) = \mu_0(E \cap P) + \sum_{j=1}^{N} \mu_0(Q_j) = \mu^*(E \cap P) + \sum_{j=1}^{N} \mu^*(Q_j)$$

$$\geqslant \mu^*(E \cap P) + \mu^*\left(\bigvee_{j=1}^{N} Q_j\right) = \mu^*(E \cap P) + \mu^*(E \setminus P).$$

Thus in the case under consideration (4) is proved.

To prove (4) in the general case, we may assume that $\mu^*(E) < +\infty$. Fix an arbitrary $\varepsilon > 0$ and choose sets $P_j \in \mathscr{P}$ such that $E \subset \bigcup_{j=1}^{\infty} P_j$ and $\sum_{j=1}^{\infty} \mu_0(P_j) < \mu^*(E) + \varepsilon$. As we have already proved,

$$\mu_0(P_j) = \mu^*(P_j) \geqslant \mu^*(P_j \cap P) + \mu^*(P_j \setminus P).$$

Hence

$$\mu^*(E) + \varepsilon > \sum_{j=1}^{\infty} \mu_0(P_j) \geqslant \sum_{j=1}^{\infty} \big(\mu^*(P_j \cap P) + \mu^*(P_j \setminus P)\big).$$

Using the countable subadditivity and monotonicity of μ^*, we obtain

$$\mu^*(E)+\varepsilon > \mu^*\left(\left(\bigcup_{j=1}^{\infty} P_j\right) \cap P\right) + \mu^*\left(\left(\bigcup_{j=1}^{\infty} P_j\right) \setminus P\right) \geqslant \mu^*(E \cap P) + \mu^*(E \setminus P).$$

Since ε is arbitrary, this implies (4). Thus we have proved the μ^*-measurability of every set $P \in \mathscr{P}$, and hence the inclusion $\mathscr{P} \subset \mathfrak{A}_{\mu^*}$. □

The measure constructed in the theorem is called the *Carathéodory extension* of μ_0. Since such an extension always exists, we may always assume without loss of generality that a measure under consideration is defined on a σ-algebra.

We draw the reader's attention to the fact that the theorem not only guarantees the existence of an extension, but provides formula (3), i.e., a method for computing the extended measure μ from the original measure μ_0. Of course, since these measures coincide on $\mathscr{P}$, we can also rewrite formula (3) for measurable sets, replacing μ_0 by μ, in the form

$$\mu(A) = \inf\left\{ \sum_{j=1}^{\infty} \mu(P_j) \,\middle|\, A \subset \bigcup_{j=1}^{\infty} P_j,\ P_j \in \mathscr{P} \text{ for all } j \in \mathbb{N} \right\}. \tag{3'}$$

We will often use this equality in what follows.

In conclusion, observe that the repeated application of the Carathéodory extension procedure yields the same result as the first one. To check this, let us show that the measures μ_0 and μ generate the same outer measure. Indeed, the right-hand side of (3) does not increase if we replace the semiring $\mathscr{P}$ by the σ-algebra $\mathfrak{A}_{\mu^*}$ and the measure μ_0 by the measure μ. This means that the outer measure generated by μ is not greater than μ^*. To obtain the reverse inequality, it suffices to observe that

$$\mu^*(E) \leqslant \sum_{j=1}^{\infty} \mu^*(A_j) = \sum_{j=1}^{\infty} \mu(A_j)$$

for every cover of E by sets A_j from the σ-algebra $\mathfrak{A}_{\mu^*}$.

EXERCISES

1. We define a function τ on subsets of the set $X = \{1, 2, 3\}$ as follows:

$$\tau(\varnothing) = 0, \qquad \tau(X) = 2, \qquad \tau(E) = 1 \qquad \text{otherwise.}$$

Show that τ is an outer measure. Which sets are τ-measurable?

2. Let $\mathcal{E}$ be an arbitrary system of sets containing $\varnothing$, and let $\alpha : \mathcal{E} \to [0, +\infty]$ be a non-negative function with $\alpha(\varnothing) = 0$. Put

$$\tau(E) = \inf\left\{\sum_{j=1}^{\infty} \alpha(E_j) \,\middle|\, E \subset \bigcup_{j=1}^{\infty} E_j,\ E_j \in \mathcal{E} \text{ for all } j \in \mathbb{N}\right\}$$

(in the case where E cannot be covered by a sequence of elements of $\mathcal{E}$, we assume that $\tau(E) = +\infty$). Show that τ is an outer measure, and that it is an extension of α if and only if the function α is countably subadditive.

3. Let τ be an outer measure. Show that a set A is τ-measurable if and only if $\tau(B \cup C) = \tau(B) + \tau(C)$ for any sets B and C satisfying the conditions $B \subset A$ and $C \cap A = \varnothing$.

1.5 Properties of the Carathéodory Extension

We keep the notation of the previous section and assume that μ is the Carathéodory extension of a measure μ_0 defined on a semiring $\mathscr{P}$ and μ^* is the outer measure generated by μ_0. We will call μ^*-measurable sets just measurable and denote the σ-algebra of measurable sets by $\mathfrak{A}$.

1.5.1 We begin with the main question of this section: do there exist extensions of μ_0 other than the Carathéodory extension? This breaks down into two questions. First, does the measure μ_0 have an extension to a σ-algebra wider than $\mathfrak{A}$? Secondly, do there exist other extensions of μ_0 to the algebra $\mathfrak{A}$ or to some part of this algebra, for example, the Borel hull of the semiring $\mathscr{P}$?

We leave the first question aside. One can prove (see [Bo, Vol. 1]), that it is usually possible to further extend the measure μ, but such an extension is neither motivated by any application nor even by the needs of "pure" mathematics. The σ-algebra $\mathfrak{A}$ is usually so wide that one has no need to extend it.

The second question is of quite a different nature, and the importance of the answer to it cannot be overestimated. It is of crucial importance to know whether an extension of the original measure at least to the minimal σ-algebra generated by the semiring $\mathscr{P}$ is unique. As we will show, in a wide class of cases (in particular, for all finite measures), the answer to this question is positive. The existence of "non-standard" extensions should be considered as a pathology, which usually appears in some artificial situations; we will encounter them only in several counterexamples.

The extension is unique if we restrict ourselves to σ-finite measures (introduced in Sect. 1.2.2). Obviously, both a measure and its Carathéodory extension are σ-finite, or not σ-finite.

Theorem (Uniqueness of an extension) *Let μ be the Carathéodory extension of a measure μ_0 defined on a semiring $\mathscr{P}$, $\mathfrak{A}$ be the σ-algebra of measurable sets, and ν be a measure extending μ_0 to a σ-algebra $\mathfrak{A}'$ containing $\mathscr{P}$. Then:*
(1) *$\nu(A) \leqslant \mu(A)$ for every set $A \in \mathfrak{A} \cap \mathfrak{A}'$; if $\mu(A) < +\infty$, then $\nu(A) = \mu(A)$;*
(2) *if μ_0 is σ-finite, then μ and ν coincide on $\mathfrak{A} \cap \mathfrak{A}'$.*

In particular, a σ-finite measure has a unique extension from the semiring $\mathscr{P}$ to the σ-algebras $\mathfrak{A}$ and $\mathfrak{B}(\mathscr{P})$.

Proof Let $\{P_j\}_{j\geqslant 1}$ be a countable cover of a set A by elements of $\mathscr{P}$. Then $\nu(A) \leqslant \sum_{j=1}^{\infty} \nu(P_j) = \sum_{j=1}^{\infty} \mu_0(P_j)$. Since this inequality holds for every cover, $\nu(A) \leqslant \mu(A)$.

It follows that $\nu(P \cap A) = \mu(P \cap A)$ if $P \in \mathscr{P}$ and $\mu(P) < +\infty$. Indeed, otherwise we have $\nu(P \cap A) < \mu(P \cap A)$, which leads to a contradiction:

$$\mu(P) = \nu(P) = \nu(P \cap A) + \nu(P \setminus A) < \mu(P \cap A) + \mu(P \setminus A) = \mu(P).$$

If $\mu(A) < +\infty$ or the measure μ_0 is σ-finite, then A can be covered by elements P_j of the semiring $\mathscr{P}$ that have finite measure. By Theorem 1.1.4, we may assume that they are pairwise disjoint. Then

$$\nu(A) = \sum_{j=1}^{\infty} \nu(A \cap P_j) = \sum_{j=1}^{\infty} \mu(A \cap P_j) = \mu(A).$$

Thus we have proved both claims of the theorem. □

Simple examples show that the σ-finiteness assumption in the second claim of the theorem is indispensable. Indeed, let X be the set consisting of two points a and b, $\mathscr{P}$ be the semiring consisting of the empty set and the one-point set $\{a\}$, μ_0 be the measure identically equal to zero, and μ be the Carathéodory extension of μ_0. Then, by the definition of the Carathéodory extension, $\mu(X) = \mu(\{b\}) = +\infty$. On the other hand, it is clear that μ_0 has another extension, identically equal to zero. In the case under consideration, Theorem 1.5.1 cannot be applied, because the measure μ_0 is not σ-finite (the set X cannot be written as a countable union of elements of $\mathscr{P}$).

Another example showing that an extension of a non-σ-finite measure is not always unique can be obtained using the discrete measure generated by a summable family of masses (see Exercise 4).

1.5.2 Let us now consider the properties of measurable sets appearing in the Carathéodory extension procedure. To describe them, it is convenient to introduce several new terms.

Definition Let $\mathcal{E}$ be an arbitrary system of subsets of a ground set X. A set H is called an $\mathcal{E}_\sigma$ set (an $\mathcal{E}_\delta$ set) if $H = \bigcup_{n\geqslant 1} A_n$ (respectively, $H = \bigcap_{n\geqslant 1} A_n$), where $A_n \in \mathcal{E}$ for all $n \in \mathbb{N}$. Sets of the type $(\mathcal{E}_\sigma)_\delta$, i.e., sets that can be written in the form $\bigcap_{n\geqslant 1} H_n$, where H_n are $\mathcal{E}_\sigma$ sets for all $n \in \mathbb{N}$, will be called $\mathcal{E}_{\sigma\delta}$ sets.

It is clear that both $\mathcal{E}_\sigma$ and $\mathcal{E}_\delta$ sets, as well as $\mathcal{E}_{\sigma\delta}$ sets, belong to the σ-algebra $\mathfrak{B}(\mathcal{E})$, the Borel hull of $\mathcal{E}$.

Theorem *Let μ be the Carathéodory extension of a measure μ_0 from a semiring $\mathscr{P}$. If $\mu^*(E) < +\infty$, then there exists a $\mathscr{P}_{\sigma\delta}$ set C such that*

$$E \subset C \quad \textit{and} \quad \mu^*(E) = \mu(C).$$

Proof By the definition of μ^*, for every positive integer n there exist sets $P_j^{(n)} \in \mathscr{P}$ $(j \in \mathbb{N})$ such that

$$\bigcup_{j\geqslant 1} P_j^{(n)} \supset E, \qquad \sum_{j\geqslant 1} \mu\big(P_j^{(n)}\big) < \mu^*(E) + \frac{1}{n}.$$

Put $C_n = \bigcup_{j\geqslant 1} P_j^{(n)}$. It is clear that

$$E \subset C_n \in \mathscr{P}_\sigma, \quad \mu^*(E) \leqslant \mu(C_n) \leqslant \sum_{j\geqslant 1} \mu\big(P_j^{(n)}\big) < \mu^*(E) + \frac{1}{n}$$

for every $n \in \mathbb{N}$. Hence the set $C = \bigcap_{n\geqslant 1} C_n$ obviously has the desired properties. □

Now we can prove that every measurable set of finite measure can be approximated, up to sets of zero measure, from the inside and from the outside by elements of $\mathfrak{B}(\mathscr{P})$.

Corollary *Let A be a measurable set of finite measure. Then there exist sets B and C from $\mathfrak{B}(\mathscr{P})$ such that*

$$B \subset A \subset C \quad \textit{and} \quad \mu(C \setminus B) = 0.$$

In particular, $\mu(A) = \mu(B) = \mu(C)$.

Proof Let C be the set constructed in the theorem. Put $e = C \setminus A$. By the above, there is a set $\widetilde{e} \in \mathfrak{B}(\mathscr{P})$ containing e such that $\mu(\widetilde{e}) = 0$. The reader can easily check that the set $B = C \setminus \widetilde{e}$ has all the required properties. □

1.5.3 Now let us establish the minimality of the Carathéodory extension. It turns out that in the case of a σ-finite measure, it is the most "economic" extension (provided that we want to obtain a complete measure).

Theorem *Let μ be the Carathéodory extension of a σ-finite measure μ_0 and $\mathfrak{A}$ be the σ-algebra of measurable sets. If μ' is a complete measure that is an extension of μ_0 to a σ-algebra $\mathfrak{A}'$, then $\mathfrak{A} \subset \mathfrak{A}'$.*

Proof First of all, observe that $\mathfrak{A}' \supset \mathfrak{B}(\mathscr{P})$, since $\mathfrak{A}' \supset \mathscr{P}$. Now let us check that if $\mu(e) = 0$, then $e \in \mathfrak{A}'$. Indeed, as we have established in Theorem 1.5.2, the set e is contained in a set $\widetilde{e}$ that is also of zero measure and belongs to $\mathfrak{B}(\mathscr{P})$. By Theorem 1.5.1, $\nu(\widetilde{e}) = \mu(\widetilde{e}) = 0$. Hence $e \in \mathfrak{A}'$ by the completeness of ν. If A is a measurable set of finite measure, then, by Corollary 1.5.2, it can be written in the form $A = C \setminus e$, where $C \in \mathfrak{B}(\mathscr{P})$ and $\mu(e) = 0$. Hence $A \in \mathfrak{A}'$.

Finally, if A is an arbitrary measurable set, then, using the σ-finiteness of μ, we can write it as the union of a sequence of measurable sets of finite measure belonging to $\mathfrak{A}'$. Therefore, in this case we also see that $A \in \mathfrak{A}'$. □

In conclusion, we give a convenient measurability criterion which is valid not only for the Carathéodory extension, but also for an arbitrary complete measure.

Lemma *Let $(X, \mathfrak{A}, \mu)$ be an arbitrary space with a complete measure, and let $E \subset X$. If for every $\varepsilon > 0$ there exist measurable sets A_ε and B_ε such that $A_\varepsilon \subset E \subset B_\varepsilon$ and $\mu(B_\varepsilon \setminus A_\varepsilon) < \varepsilon$, then E is measurable.*

In particular, if for every $\varepsilon > 0$ there exists a measurable set E_ε such that $E \subset E_\varepsilon$ and $\mu(E_\varepsilon) < \varepsilon$, then E is measurable (and $\mu(E) = 0$).

Proof Taking ε equal to $1/n$ ($n = 1, 2, \ldots$), consider the sets $A_{1/n}$ and $B_{1/n}$. Then the sets $A = \bigcup_{n=1}^{\infty} A_{1/n}$ and $B = \bigcap_{n=1}^{\infty} B_{1/n}$ are measurable and $A \subset E \subset B$. Furthermore, $\mu(B \setminus A) = 0$, since $\mu(B \setminus A) \leqslant \mu(B_{1/n} \setminus A_{1/n}) < \frac{1}{n}$ for every n. Thus the set $E \setminus A$ is contained in the set $B \setminus A$ of zero measure, and, consequently, it is measurable by the completeness of μ. Then the set $E = A \cup (E \setminus A)$ is also measurable. □

EXERCISES

1. Let X and $\mathscr{P}$ be as in the example from Sect. 1.5.1, i.e., $X = \{a, b\}$ is a two-point set and $\mathscr{P} = \{\varnothing, \{a\}\}$; let μ_0 be an arbitrary finite measure on $\mathscr{P}$. Show that for every α ($0 \leqslant \alpha \leqslant +\infty$), the formulas

$$\nu_\alpha(\varnothing) = 0, \qquad \nu_\alpha(\{a\}) = \mu_0(\{a\}),$$
$$\nu_\alpha(\{b\}) = \alpha, \qquad \nu_\alpha(X) = \alpha + \mu_0(\{a\})$$

define a measure ν_α that is an extension of μ_0 to the algebra of all subsets of X. Which of the measures ν_α is the Carathéodory extension of μ_0? Why do the (finite) measures ν_1 and ν_2, which coincide on $\mathscr{P}$, fail to coincide on $\mathfrak{B}(\mathscr{P})$?

In the next series of exercises, μ is the Carathéodory extension of a measure μ_0 from a semiring $\mathscr{P}$ to the σ-algebra $\mathfrak{A}$ of measurable sets.

2. Show that if μ_0 is σ-finite, then the condition $\mu(A) < +\infty$ in Corollary 1.5.2 can be dropped. The set C can still be assumed to be a $\mathscr{P}_{\sigma\delta}$ set.

3. Show that for every set A there exists a set $B \in \mathfrak{B}(\mathscr{P})$ such that $A \subset B$ and $\mu^*(A) = \mu(B)$.

4. Consider the discrete measure ν generated by a countable family of masses on an uncountable set X. Let μ_0 be its restriction to the semiring of at most countable subsets. Show that the Carathéodory extension of μ_0 is defined, like ν, on the σ-algebra of all subsets of X, but, in contrast to ν, is infinite on all uncountable sets.

5. Let μ_0 be a measure taking only finite values. For $A \in \mathfrak{A}$, put

$$\widetilde{\mu}(A) = \sup\bigl\{\mu(B) \mid B \subset A,\ B \in \mathfrak{A},\ \mu(B) < +\infty\bigr\}.$$

Show that $\widetilde{\mu}$ is a measure extending μ_0 and that this extension is minimal in the sense that $\widetilde{\mu} \leqslant \nu$ for every extension ν of μ_0 to $\mathfrak{A}$.
Using Exercise 1, give an example of a measure that extends μ_0, but does not coincide with μ and $\widetilde{\mu}$.

6. Denote by $\mathcal{N}$ the system of all sets of zero outer measure. Show that:

(a) $\mathfrak{B}(\mathscr{P} \cup \mathcal{N}) \subset \mathfrak{A}$ and the restriction of μ to $\mathfrak{B}(\mathscr{P} \cup \mathcal{N})$ is a complete measure;

(b) if μ is σ-finite, then $\mathfrak{A} = \mathfrak{B}(\mathscr{P} \cup \mathcal{N})$.

Give an example of a measure for which $\mathfrak{A} \neq \mathfrak{B}(\mathscr{P} \cup \mathcal{N})$ (consider the counting measure defined on the semiring of finite subsets of an uncountable set).

7. Show that the Carathéodory extension of a σ-finite complete measure μ defined on a σ-algebra coincides with μ.

1.6 *Properties of the Borel Hull of a System of Sets

1.6.1 Let X, Y be arbitrary sets, $\varphi : X \to Y$ be a map from X to Y, and $\mathcal{E}$ be a system of subsets of Y. By $\varphi^{-1}(\mathcal{E})$ we denote the "inverse image of $\mathcal{E}$", i.e., the system of sets $\{\varphi^{-1}(E) \mid E \in \mathcal{E}\}$. It turns out that the inverse image of a σ-algebra is a σ-algebra. More precisely, the following result holds.

Lemma

(1) *The inverse image of a σ-algebra (algebra) is again a σ-algebra (algebra).*
(2) *If $\mathfrak{A}$ is a σ-algebra (algebra) of subsets of X, then the system $\{B \subset Y \mid \varphi^{-1}(B) \in \mathfrak{A}\}$ is also a σ-algebra (algebra).*

Proof Both claims follow immediately from the equalities

$$\varphi^{-1}(Y \setminus B) = X \setminus \varphi^{-1}(B), \qquad \varphi^{-1}\left(\bigcup_{n=1}^{\infty} B_n\right) = \bigcup_{n=1}^{\infty} \varphi^{-1}(B_n).$$

The details are left to the reader. □

The main result of this section is the following theorem.

Theorem *Let X, Y be arbitrary sets, $\mathfrak{A}$ be a σ-algebra of subsets of X, $\mathcal{E}$ be a system of subsets of Y, and φ be an arbitrary map from X to Y. Then:*
(1) *if $\varphi^{-1}(\mathcal{E}) \subset \mathfrak{A}$, then $\varphi^{-1}(\mathfrak{B}(\mathcal{E})) \subset \mathfrak{A}$;*
(2) $\mathfrak{B}(\varphi^{-1}(\mathcal{E})) = \varphi^{-1}(\mathfrak{B}(\mathcal{E}))$.

Proof (1) Consider the system of sets $\mathfrak{A}' = \{B \subset Y \mid \varphi^{-1}(B) \in \mathfrak{A}\}$. By the lemma, $\mathfrak{A}'$ is a σ-algebra. Since $\mathfrak{A}' \supset \mathcal{E}$, it follows from the definition of the Borel hull that $\mathfrak{A}' \supset \mathfrak{B}(\mathcal{E})$.

(2) Assuming that $\mathfrak{A} = \mathfrak{B}(\varphi^{-1}(\mathcal{E}))$, the first claim implies that

$$\varphi^{-1}\big(\mathfrak{B}(\mathcal{E})\big) \subset \mathfrak{B}\big(\varphi^{-1}(\mathcal{E})\big). \tag{1}$$

On the other hand, by the lemma, the system $\varphi^{-1}(\mathfrak{B}(\mathcal{E}))$ is a σ-algebra. Since $\varphi^{-1}(\mathcal{E}) \subset \varphi^{-1}(\mathfrak{B}(\mathcal{E}))$, it follows from the definition of the Borel hull that

$$\mathfrak{B}\big(\varphi^{-1}(\mathcal{E})\big) \subset \varphi^{-1}\big(\mathfrak{B}(\mathcal{E})\big).$$

Together with (1), this yields the desired equality. □

1.6.2 Let us mention several corollaries of the theorem we have just proved. The first four of them are just rephrasings or special cases of the theorem, as the reader can easily check.

In the corollaries, by $\mathcal{E}$ we denote an arbitrary system of subsets of a set Y (as in the theorem).

Let $X \subset Y$, and let $\varphi = \mathrm{id} : X \to Y$ be the identity map (that assigns to each point $x \in X$ the same point regarded as an element of Y). Obviously, $\varphi^{-1}(E) = E \cap X$ for every set $E \subset Y$. It is clear that the induced system $\mathcal{E} \cap X$ coincides with $\varphi^{-1}(\mathcal{E})$. If $\mathcal{E}$ consists of subsets of X, where $X \subset Y$, then, in order to distinguish between the Borel hulls of $\mathcal{E}$ that consist of subsets of X and of Y, we will denote them by $\mathfrak{B}^{(X)}(\mathcal{E})$ and $\mathfrak{B}^{(Y)}(\mathcal{E})$, respectively. The following result holds.

Corollary 1 $\mathfrak{B}^{(X)}(\mathcal{E} \cap X) = \mathfrak{B}^{(Y)}(\mathcal{E}) \cap X$.

Note that, by definition, the left-hand side is a system of subsets of X, since $\mathcal{E} \cap X$ is such a system.

To prove the corollary, it suffices to apply the theorem assuming that $\varphi = \mathrm{id} : X \to Y$ is the identity map. Then the induced system $\mathcal{E} \cap X$ coincides with $\varphi^{-1}(\mathcal{E})$, since $\varphi^{-1}(E) = E \cap X$ for every set $E \subset Y$.

Generalizing the notion of a Borel set in $\mathbb{R}^m$ (see Sect. 1.1.3), we say that a subset of a topological space X is a Borel set if it belongs to the minimal σ-algebra containing all open sets. This σ-algebra will be denoted by $\mathfrak{B}_X$.

Corollary 2 *Let X and Y be topological spaces and $\varphi : X \to Y$ be a continuous map. Then the inverse image of every Borel subset of Y is a Borel subset of X, i.e., $\varphi^{-1}(\mathfrak{B}_Y) \subset \mathfrak{B}_X$.*

Note that Corollary 2 is no longer true if we replace the inverse images by the images. For example, one can prove that the image of a Borel set under the orthogonal projection of the plane onto a line is not always Borel. This non-trivial result is due to M.Ya. Suslin.[6]

Corollary 3 *Let Y be a topological space and X be a subspace of Y. Then every Borel subset A of X is a trace of a Borel subset of Y, i.e., $A = X \cap B$, where B is an element of $\mathfrak{B}_Y$.*

Using Theorem 1.1.7, one can easily obtain the following result.

Corollary 4 *Let G be an open subset of $\mathbb{R}^m$ and $\mathcal{P}^m_G = \{P \in \mathcal{P}^m \mid \overline{P} \subset G\}$. Then $\mathfrak{B}(\mathcal{P}^m_G) = \mathfrak{B}_G$ (here $\mathcal{P}^m_G$ is regarded as a system of subsets of G).*

We write $X \times \mathcal{E}$ for the system $\{X \times E \mid E \in \mathcal{E}\}$. The following lemma holds.

Lemma $\mathfrak{B}(X \times \mathcal{E}) = X \times \mathfrak{B}(\mathcal{E})$.

Proof To prove the lemma, take φ to be the canonical projection of $X \times Y$ to Y and apply the theorem. □

Let $\mathcal{E}'$ and $\mathcal{E}$ be arbitrary systems of subsets of X and Y, respectively. The system $\{E' \times E \mid E' \in \mathcal{E}',\ E \in \mathcal{E}\}$ of subsets of the Cartesian product $X \times Y$ will be denoted by $\mathcal{E}' \odot \mathcal{E}$.

Corollary 5 *Let $\mathcal{E}'$ be a system of subsets of a set X. Then*

$$\mathfrak{B}(\mathcal{E}' \odot \mathcal{E}) = \mathfrak{B}\big(\mathfrak{B}(\mathcal{E}') \odot \mathfrak{B}(\mathcal{E})\big).$$

---

[6]Mikhail Yakovlevich Suslin (1894–1919)—Russian mathematician.

Proof Let us first check that

$$\mathcal{E}' \odot \mathfrak{B}(\mathcal{E}) \subset \mathfrak{B}(\mathcal{E}' \odot \mathcal{E}). \tag{2}$$

For this it suffices to observe that, by the lemma (with X replaced by $E' \in \mathcal{E}'$),

$$E' \times \mathfrak{B}(\mathcal{E}) = \mathfrak{B}(E' \times \mathcal{E}) \subset \mathfrak{B}(\mathcal{E}' \odot \mathcal{E}).$$

Now fix some sets $U \in \mathfrak{B}(\mathcal{E}')$ and $V \in \mathfrak{B}(\mathcal{E})$. Then, by the lemma and inclusion (2),

$$U \times Y \in \mathfrak{B}(\mathcal{E}' \times Y) \subset \mathfrak{B}(\mathcal{E}' \odot \mathfrak{B}(\mathcal{E})) \subset \mathfrak{B}(\mathcal{E}' \odot \mathcal{E}).$$

Analogously, $X \times V \in \mathfrak{B}(\mathcal{E}' \odot \mathcal{E})$. Hence

$$U \times V = (U \times Y) \cap (X \times V) \in \mathfrak{B}(\mathcal{E}' \odot \mathcal{E}).$$

Therefore, $\mathfrak{B}(\mathfrak{B}(\mathcal{E}') \odot \mathfrak{B}(\mathcal{E})) \subset \mathfrak{B}(\mathcal{E}' \odot \mathcal{E})$. The reverse inclusion is obvious, since $\mathcal{E}' \odot \mathcal{E} \subset \mathfrak{B}(\mathcal{E}') \odot \mathfrak{B}(\mathcal{E})$. □

Corollary 6 *If X and Y are topological spaces, then*

$$\mathfrak{B}(\mathfrak{B}_X \odot \mathfrak{B}_Y) \subset \mathfrak{B}_{X \times Y}.$$

In particular, the product of Borel subsets of X and Y is a Borel subset of $X \times Y$. If X and Y are second-countable, then $\mathfrak{B}(\mathfrak{B}_X \odot \mathfrak{B}_Y) = \mathfrak{B}_{X \times Y}$.

Proof Let $\mathfrak{G}_X$, $\mathfrak{G}_Y$ and $\mathfrak{G}$ be the systems of open sets in the spaces X, Y and $X \times Y$, respectively. By Corollary 5, $\mathfrak{B}(\mathfrak{B}_X \odot \mathfrak{B}_Y) = \mathfrak{B}(\mathfrak{G}_X \odot \mathfrak{G}_Y)$. Since $\mathfrak{G}_X \odot \mathfrak{G}_Y \subset \mathfrak{G}$, we have

$$\mathfrak{B}(\mathfrak{B}_X \odot \mathfrak{B}_Y) = \mathfrak{B}(\mathfrak{G}_X \odot \mathfrak{G}_Y) \subset \mathfrak{B}_{X \times Y}.$$

The second axiom of countability implies that every element of $\mathfrak{G}$ is an at most countable union of elements of $\mathfrak{G}_X \odot \mathfrak{G}_Y$. Hence $\mathfrak{G} \subset \mathfrak{B}(\mathfrak{G}_X \odot \mathfrak{G}_Y) \subset \mathfrak{B}(\mathfrak{B}_X \odot \mathfrak{B}_Y)$ and, consequently, $\mathfrak{B}_{X \times Y} \subset \mathfrak{B}(\mathfrak{B}_X \odot \mathfrak{B}_Y)$. The reverse inclusion, as we have already established, always holds. □

1.6.3 Another property of the Borel hull is related to the notion of a monotone class of sets.

Definition A system of sets is called a *monotone class* if it contains the unions of all increasing sequences and the intersections of all decreasing sequences of its elements.

Theorem (On a monotone class) *If a monotone class contains an algebra $\mathfrak{A}$ of subsets of a set X, then it contains the Borel hull of this algebra.*

Proof Consider a minimal monotone class $\mathcal{M}$ containing $\mathfrak{A}$. Such a class obviously exists: it suffices to consider the intersection of all monotone classes containing $\mathfrak{A}$. Let us check that $\mathcal{M} = \mathfrak{B}(\mathfrak{A})$. Clearly, $\mathcal{M} \subset \mathfrak{B}(\mathfrak{A})$, since a σ-algebra is a monotone class. Hence it remains to establish the inclusion $\mathcal{M} \supset \mathfrak{B}(\mathfrak{A})$. For this it suffices to check that the class $\mathcal{M}$ is a σ-algebra.

First let us prove that

$$\text{if } A \in \mathfrak{A}, \quad \text{then } A \cap B \in \mathcal{M} \text{ and } A \cap B^c \in \mathcal{M} \text{ for every } B \in \mathcal{M} \tag{3}$$

(by B^c we mean the complement of a set B with respect to $X: \; B^c = X \setminus B$).

Indeed, given a set $A \in \mathfrak{A}$, put

$$\mathcal{M}_A = \left\{B \in \mathcal{M} \mid A \cap B \in \mathcal{M}, \; A \cap B^c \in \mathcal{M}\right\}.$$

One can easily check that $\mathcal{M}_A$ is a monotone class containing $\mathfrak{A}$; by construction, $\mathcal{M}_A \subset \mathcal{M}$. Hence, by the minimality of $\mathcal{M}$, we have $\mathcal{M}_A = \mathcal{M}$, which proves (3).

For $A = X$, it follows from (3) that the system $\mathcal{M}$ contains the complement of each of its elements, i.e., $\mathcal{M}$ is symmetric.

Now let us check that the class $\mathcal{M}$ contains the intersection of any two of its elements. Let $B \in \mathcal{M}$. Consider the system of sets

$$\mathcal{N}_B = \{E \in \mathcal{M} \mid B \cap E \in \mathcal{M}\}.$$

As at the previous step, it is clear that $\mathcal{N}_B$ is a monotone class. It follows from (3) that it contains $\mathfrak{A}$. Hence, by the minimality of $\mathcal{M}$, the sets $\mathcal{N}_B$ and $\mathcal{M}$ coincide. Since B is arbitrary, this means that $\mathcal{M}$ contains the intersection of any two of its elements B and E. By the symmetry of $\mathcal{M}$, it follows that it also contains finite unions of its elements. Together with the monotonicity, this implies that $\mathcal{M}$ also contains countable unions of its elements, i.e., $\mathcal{M}$ is a σ-algebra. Thus $\mathcal{M}$ is a σ-algebra containing $\mathfrak{A}$. Hence $\mathcal{M} \supset \mathfrak{B}(\mathfrak{A})$ by the definition of the Borel hull. □

EXERCISES

1. Show that Corollary 4 from Sect. 1.6.2 remains valid if we replace the semiring $\mathscr{P}^m_G$ by the semiring $\{P \in \mathscr{P}^m_r \mid \overline{P} \subset G\}$.
2. Show that the map $(t_1, \dots, t_m) \mapsto (e^{it_1}, \dots, e^{it_m}) \in \mathbb{C}^m \quad ((t_1, \dots, t_m) \in \mathbb{R}^m)$ sends Borel sets to Borel sets.
3. Let X be a set that consists of at least two points and $\mathscr{P}$ be the system of subsets of X that consist of at most one point. Show that $\mathscr{P}$ is a semiring and a monotone class that does not coincide with its Borel hull.
4. Show that every D-system (see Sect. 1.1, Exercise 10) is a monotone class.
5. Let $\mathcal{E}$ be a D-system of subsets of $\mathbb{R}^m$ that contains all finite intersections of open balls. Show that it also contains all finite unions of balls. Using Exercise 4, deduce that $\mathcal{E}$ contains all Borel sets.

Chapter 2
The Lebesgue Measure

2.1 Definition and Basic Properties of the Lebesgue Measure

This chapter is devoted to the most important and historically the first example of a measure: the Carathéodory extension of the ordinary volume.

2.1.1 In order to apply the general extension procedure described in Sect. 1.4 to the ordinary volume, we should make sure that it is a measure.

Theorem *The ordinary volume λ_m on the semiring $\mathscr{P}^m$ is a σ-finite measure.*

Proof We only need to prove that the volume λ_m is countably additive, since it is clearly σ-finite. For this it suffices to check that it is countably subadditive (see Theorem 1.3.2), i.e., that if $P, P_n \in \mathscr{P}^m$ $(n \in \mathbb{N})$, $P \subset \bigcup_{n\geqslant 1} P_n$, then

$$\lambda_m(P) \leqslant \sum_{n\geqslant 1} \lambda_m(P_n). \tag{1}$$

Let us prove this inequality up to an arbitrary positive number ε. Let $P = [a, b) \neq \varnothing$ and $P_n = [a_n, b_n)$. We will use the fact that, as follows from the definition, the ordinary volume of a cell is a continuous function of its vertices. Choose vectors $a'_n < a_n$ in such a way that

$$\lambda_m\big([a'_n, b_n)\big) < \lambda_m\big([a_n, b_n)\big) + \frac{\varepsilon}{2^n} \quad \text{for all } n \in \mathbb{N}. \tag{2}$$

Let us estimate the volume $\lambda_m([a, t))$ from above for an arbitrary t, $a < t < b$.

Since $[a, t] \subset [a, b) = P$ and $P_n = [a_n, b_n) \subset (a'_n, b_n)$, it is clear that

$$[a, t] \subset P \subset \bigcup_{n\geqslant 1} P_n \subset \bigcup_{n\geqslant 1} (a'_n, b_n).$$

B. Makarov, A. Podkorytov, *Real Analysis: Measures, Integrals and Applications*, Universitext, DOI 10.1007/978-1-4471-5122-7_2, © Springer-Verlag London 2013

The parallelepiped $[a,t]$ is compact, hence the cover of $[a,t]$ by the sets (a'_n, b_n) contains a finite subcover. Thus for some $N \in \mathbb{N}$ we have

$$[a,t] \subset \bigcup_{n=1}^{N} (a'_n, b_n).$$

Even more so,

$$[a,t) \subset \bigcup_{n=1}^{N} [a'_n, b_n).$$

Using the (finite) subadditivity of the ordinary volume, we obtain

$$\lambda_m\big([a,t)\big) \leqslant \sum_{n=1}^{N} \lambda_m\big([a'_n, b_n)\big).$$

Together with (2) this yields

$$\lambda_m\big([a,t)\big) < \sum_{n=1}^{N} \left(\lambda_m(P_n) + \frac{\varepsilon}{2^n}\right) < \sum_{n \geqslant 1} \lambda_m(P_n) + \varepsilon.$$

Again using the fact that the volume of a cell depends continuously on its vertices and passing to the limit as $t \to b$, we see that

$$\lambda_m(P) = \lim_{t \to b} \lambda_m\big([a,t)\big) \leqslant \sum_{n \geqslant 1} \lambda_m(P_n) + \varepsilon.$$

Since ε is arbitrary, the last inequality implies (1). □

2.1.2 Now we are in a position to introduce, using the measure extension theorem 1.4.5, the very important notion of the Lebesgue measure.

Definition The *Lebesgue measure* on the space $\mathbb{R}^m$ (the m-dimensional Lebesgue measure) is the Carathéodory extension of the ordinary volume from the semiring $\mathscr{P}^m$.

The m-dimensional Lebesgue measure is denoted by the same symbol λ_m as the ordinary volume. If the dimension is fixed, we sometimes omit the subscript and write simply λ, especially in the one-dimensional case. Hereafter in this section, the term "measure" refers to the Lebesgue measure.

The σ-algebra of sets on which the m-dimensional Lebesgue measure is defined is denoted by $\mathfrak{A}^m$; sets from this σ-algebra are called *Lebesgue measurable*, or simply *measurable*.

As follows from the definition of the Carathéodory extension, for a measurable set A,

$$\lambda_m(A) = \inf\Big\{\sum_{k\geqslant 1} \lambda_m(P_k) \,\Big|\, P_k \in \mathscr{P}^m, \bigcup_{k\geqslant 1} P_k \supset A\Big\}.$$

Since every cell is contained in a cell of arbitrarily close measure with rational vertices, all cells in the last formula may be assumed to have rational vertices. Thus

$$\lambda_m(A) = \inf\Big\{\sum_{k\geqslant 1} \lambda_m(P_k) \,\Big|\, P_k \in \mathscr{P}^m_r, \bigcup_{k\geqslant 1} P_k \supset A\Big\}. \tag{3}$$

Hence the Lebesgue measure can also be regarded as the Carathéodory extension of the ordinary volume from the semiring $\mathscr{P}^m_r$.

2.1.3 *Basic properties of the Lebesgue measure.*

(1) *Open sets are measurable*; *the measure of a non-empty open set is strictly positive.*

The first claim follows from Theorem 1.1.7; the second one is obvious, since a non-empty open set contains a non-degenerate cell.

(2) *Closed sets are measurable; the measure of a one-point set is zero.*

The first claim follows from Property (1); the second one is obvious, since every point is contained in a cell of arbitrarily small measure.

The following important property is obvious.

(3) *The measure of a measurable bounded set is finite. Every measurable set is the union of a sequence of sets of finite measure.*

The next property shows that a set that can be well approximated by measurable sets both from the inside and from the outside, is itself measurable.

(4) *Let $E \subset \mathbb{R}^m$. If for every $\varepsilon > 0$ there exist measurable sets A_ε and B_ε such that $A_\varepsilon \subset E \subset B_\varepsilon$ and $\lambda_m(B_\varepsilon \setminus A_\varepsilon) < \varepsilon$,* then *$E$ is measurable.*

In particular, if for every $\varepsilon > 0$ there exists a measurable set E_ε such that $E \subset E_\varepsilon$ and $\lambda_m(E_\varepsilon) < \varepsilon$, then E is measurable (and $\lambda_m(E) = 0$).

This property follows from the fact that the Lebesgue measure is complete. It is a special case of Lemma 1.5.3.

(5) *A countable union of sets of zero measure is again a set of zero measure.*

This is a general property of all measures defined on a σ-algebra (see the corollary of Theorem 1.3.2).

In particular,

(5′) *Every countable set has zero measure.*

Since a non-empty open set is of positive measure, we see that

(6) *A set of zero measure has no interior points.*

(7) *If $\lambda_m(e) = 0$, then for every $\varepsilon > 0$ there exist cubic cells Q_j such that*

$$\bigcup_{j\geqslant 1} Q_j \supset e, \qquad \sum_{j\geqslant 1} \lambda_m(Q_j) < \varepsilon.$$

Indeed, it follows from (3) that e can be covered by cells P_n with rational vertices in such a way that $\sum_{n\geqslant 1} \lambda_m(P_n) < \varepsilon$. It remains to recall that every cell with rational vertices is a disjoint union of finitely many cubic cells. Hence $P_n = \bigvee_{j=1}^{k_n} Q_{nj}$ and $\lambda_m(P_n) = \sum_{j=1}^{k_n} \lambda_m(Q_{nj})$. Therefore,

$$e \subset \bigcup_{n\geqslant 1} P_n = \bigcup_{n\geqslant 1} \bigvee_{j=1}^{k_n} Q_{nj} \quad \text{and} \quad \sum_{n\geqslant 1} \sum_{j=1}^{k_n} \lambda_m(Q_{nj}) = \sum_{n\geqslant 1} \lambda_m(P_n) < \varepsilon.$$

Do there exist uncountable sets of zero measure? Such sets are easy to construct if the dimension of the space is greater than one. In particular, examples of such sets are provided by arbitrary proper affine subspaces. Such subspaces of maximal dimension will be called *planes*. We will prove this result in full generality at the end of Sect. 2.3.1, but now we establish it only for planes of a special form.

(8) *Let m and k be positive integers, $m \geqslant 2$, $1 \leqslant k \leqslant m$, and let $c \in \mathbb{R}$. Consider the plane $H_k(c)$ orthogonal to the kth coordinate axis:*

$$H_k(c) = \{x = (x_1, \ldots, x_m) \in \mathbb{R}^m \mid x_k = c\}.$$

Then $\lambda_m(H_k(c)) = 0$.

It suffices to prove that every bounded part of $H_k(c)$ has zero measure. The latter is true since such a part is contained in a cell of arbitrarily small measure (the kth edge of the cell can be made arbitrarily small).

(9) *Every set contained in a finite or countable union of planes perpendicular to the coordinate axes has zero measure.*

It follows that the measures of an open parallelepiped (a, b), the cell $[a, b)$, and the closed parallelepiped $[a, b]$ coincide, because the boundary of a parallelepiped has zero measure.

(10) *There exist Lebesgue non-measurable sets.*

We will prove a somewhat stronger assertion:

Every set of positive measure contains a Lebesgue non-measurable subset.

Indeed, let $A \in \mathfrak{A}^m$ and $\lambda_m(A) > 0$. We may assume without loss of generality that the set A is bounded: $\|x\| < R$ for $x \in A$. Let us introduce an equivalence relation on A by assuming that $x \sim y$ if the difference $x - y$ is a vector with rational coordinates, i.e., if $x - y \in \mathbb{Q}^m$. Then A is partitioned into pairwise disjoint non-empty classes consisting of equivalent points. Clearly, each such class is at most countable. Using the axiom of choice, take a subset E in A that contains exactly one point in common with each class. Let us check that E is not Lebesgue measurable. Consider the rational translations of E, i.e., the sets of the form $r + E = \{r + x \mid x \in E\}$ with $r \in \mathbb{Q}^m$ (we retain the notation r for vectors in $\mathbb{Q}^m$ up to the end of the proof). They are pairwise disjoint (otherwise E would contain two points from the same equivalence class). Furthermore, since $\|x - y\| < 2R$ for $x, y \in A$, it is clear that A is contained in the bounded set $W = \bigvee_{\|r\|<2R}(r + E)$.

Assume that E is measurable. As we will see below (see Theorem 2.4.1), a translation of a measurable set is again a measurable set of the same measure. Hence the set W is measurable. Its measure is positive, because $A \subset W$ and $\lambda_m(A) > 0$. In addition, it is finite, since W is bounded. Thus $0 < \lambda_m(W) < +\infty$. At the same time, by the countable additivity of the Lebesgue measure,

$$\lambda_m(W) = \sum_{\|r\|<2R} \lambda_m(r + E) = \sum_{\|r\|<2R} \lambda_m(E).$$

But the sum on the right-hand side is either zero (if $\lambda_m(E) = 0$) or infinite (if $\lambda_m(E) > 0$), and this is incompatible with the double inequality $0 < \lambda_m(W) < +\infty$. Thus the assumption that the set E is measurable leads to a contradiction.

Note that we have proved a more general fact than the existence of Lebesgue non-measurable sets. Indeed, our construction does not use any specific properties of the Lebesgue measure except for the fact that it is finite on bounded sets and translation-invariant. This means that non-measurable sets exist for any (non-zero) measure that enjoys these two properties. In other words, such a measure cannot be defined on all subsets of $\mathbb{R}^m$. In this connection, observe that if we drop the condition of countable additivity and content ourselves only with finite additivity, then the situation is different: there exists a translation-invariant volume defined on the system of all subsets of $\mathbb{R}^m$ that coincides with the Lebesgue measure on $\mathfrak{A}^m$.

The complicated construction and the somewhat mysterious character of the constructed Lebesgue non-measurable set should not obscure the essence of the matter: in a typical situation, when we apply the Carathéodory extension, not all sets turn out to be measurable. An everyday illustration of this phenomenon is the following ingenious example communicated to us by D.A. Vladimirov.

A number of shoes of the same color, model and size are heaped in a pile X. Each proper pair (consisting of one left and one right shoe) has a price, and there is a collection of several such pairs. Thus we have a measure (price) on a system of subsets of X. However, it cannot be extended in a natural way to the system of all subsets. Indeed, if we split the set formed by two proper pairs into two parts, one consisting of the left shoes and the other one consisting of the right shoes, then the total price of these parts (assuming that they have a price) should be the same as for the original set. But then one of the parts must cost at least as much as a proper pair, which is absurd.

2.1.4 As follows from property (9), the space $\mathbb{R}^m$ with $m \geqslant 2$ contains uncountable sets of zero measure. In the one-dimensional case, it is not as easy to give examples of such sets. Here we will discuss an interesting example of this kind: the Cantor set, which is obtained by deleting a countable set of open intervals from the segment [0, 1]. First we delete one interval, the middle third of the initial segment [0, 1] (i.e., the interval $(1/3, 2/3)$), then the middle thirds of the remaining two segments, and so on. The points in [0, 1] that do not belong to any of the deleted intervals form the Cantor set $\mathcal{C}$. Let us consider this construction in more detail.

Example (The Cantor[1] set) Let $\Delta = [0, 1]$, and let C_1 be the set obtained from Δ by deleting the open interval $\delta = (1/3, 2/3)$:

$$C_1 = \Delta \setminus \delta = \left[0, \frac{1}{3}\right] \cup \left[\frac{2}{3}, 1\right].$$

We will call $\Delta_0 = [0, 1/3]$ and $\Delta_1 = [2/3, 1]$ the segments of the first rank. The set C_2 is obtained by deleting from Δ_0 and Δ_1 their middle thirds, i.e., the intervals $\delta_0 = (1/9, 2/9)$, $\delta_1 = (7/9, 8/9)$. The set-theoretic difference $\Delta_\varepsilon \setminus \delta_\varepsilon$ $(\varepsilon = 0,\ 1)$ consists of two segments; denote the left one by $\Delta_{\varepsilon 0}$ and the right one by $\Delta_{\varepsilon 1}$. Thus C_2 is the union of four segments Δ_{00}, Δ_{01}, Δ_{10}, Δ_{11}, which will be called the segments of the second rank. For future use, we note that the segments of the second rank are indexed by the pairs $\varepsilon_1\varepsilon_2$, where ε_1 and ε_2 independently take the values 0 and 1. Note also that $\Delta_{\varepsilon_1\varepsilon_2} \subset \Delta_{\varepsilon_1}$.

Now the construction proceeds by induction. Assume that we have already constructed the set C_n consisting of 2^n pairwise disjoint segments of the nth rank. It is convenient to index these segments by the sequences $\varepsilon_1, \dots, \varepsilon_n$, where ε_j may take the value 0 or 1. For the segments of the first and the second rank, we have already described these indices. We then proceed as follows. When constructing the segments of the $(n+1)$th rank, we delete the middle third $\delta_{\varepsilon_1,\dots,\varepsilon_n}$ from each segment $\Delta_{\varepsilon_1,\dots,\varepsilon_n}$ of the nth rank. The set-theoretic difference $\Delta_{\varepsilon_1\dots\varepsilon_n} \setminus \delta_{\varepsilon_1\dots\varepsilon_n}$ consists of two segments of the $(n+1)$th rank; we denote the left one by $\Delta_{\varepsilon_1\dots\varepsilon_n 0}$ and the right one by $\Delta_{\varepsilon_1\dots\varepsilon_n 1}$. Thus when passing from n to $n+1$ the number of segments doubles and the length of these segments becomes three times less. The segments of the $(n+1)$th rank are pairwise disjoint, and $\Delta_{\varepsilon_1\dots\varepsilon_n\varepsilon_{n+1}} \subset \Delta_{\varepsilon_1\dots\varepsilon_n}$. Let C_{n+1} be the union of all segments of the $(n+1)$th rank. The intersection $\mathcal{C} = \bigcap_{n\geqslant 1} C_n$ is called the *Cantor set*. It has zero measure. Indeed, the length of each segment of the nth rank is clearly equal to $1/3^n$. Hence the measure of the set C_n is equal to $(2/3)^n$, and the measure of the set $\mathcal{C} = \bigcap_{n\geqslant 1} C_n$ vanishes.

Now let us prove that the set $\mathcal{C}$ has the same cardinality as the set $\mathcal{E}$ of all binary sequences, i.e., the cardinality of the continuum. Recall that a binary sequence is a sequence every element of which is equal to 0 or 1.

Since for every binary sequence $\varepsilon = \{\varepsilon_n\}_{n\geqslant 1}$, the segment $\Delta_{\varepsilon_1\dots\varepsilon_n\varepsilon_{n+1}}$ is contained in $\Delta_{\varepsilon_1\dots\varepsilon_n}$, the sequence $\{\Delta_{\varepsilon_1\dots\varepsilon_n}\}_{n\geqslant 1}$ has a non-empty intersection, which

[1] Georg Ferdinand Ludwig Philipp Cantor (1845–1918)—German mathematician.

obviously consists of a single point $t(\varepsilon)$. For distinct binary sequences ε and ε', the points $t(\varepsilon)$ and $t(\varepsilon')$ are distinct. Indeed, let $\varepsilon = \{\varepsilon_n\}_{n\geqslant 1}$, $\varepsilon' = \{\varepsilon'_n\}_{n\geqslant 1}$, and let k be the first index such that $\varepsilon_n \neq \varepsilon'_n$. In other words,

$$\varepsilon = \{\varepsilon_1, \dots, \varepsilon_{k-1}, \varepsilon_k, \dots\}, \qquad \varepsilon' = \{\varepsilon_1, \dots, \varepsilon_{k-1}, \varepsilon'_k, \dots\} \quad \text{and} \quad \varepsilon_k \neq \varepsilon'_k.$$

By the construction of the points $t(\varepsilon)$ and $t(\varepsilon')$,

$$t(\varepsilon) \in \Delta_{\varepsilon_1 \dots \varepsilon_{k-1}\varepsilon_k}, \qquad t(\varepsilon') \in \Delta_{\varepsilon_1 \dots \varepsilon_{k-1}\varepsilon'_k}.$$

Since distinct segments of the kth rank are disjoint, $t(\varepsilon)$ and $t(\varepsilon')$ cannot coincide, which proves that the map $\varepsilon \mapsto t(\varepsilon)$ is one-to-one. But every point in $\mathcal{C}$ belongs to the intersection of a sequence of segments $\{\Delta_{\varepsilon_1 \dots \varepsilon_n}\}_{n\geqslant 1}$, so the constructed map is onto. This completes the proof of the bijectivity of the map $\varepsilon \mapsto t(\varepsilon)$ from $\mathcal{E}$ onto $\mathcal{C}$.

EXERCISES In Exercises 1–12, by measurability we mean Lebesgue measurability and λ stands for the Lebesgue measure of appropriate dimension.

1. Let $E \subset \mathbb{R}^m$ be a measurable set, $0 < \lambda(E) < +\infty$, and $\varepsilon \in (0, 1)$. Show that there exists a cube Q such that $\lambda(E \cap Q) > (1 - \varepsilon)\lambda_m(Q)$.

2. Let $E \subset \mathbb{R}^m$ and $0 < t < \lambda_m(E)$. Show that in E there is a bounded subset A such that $\lambda_m(A) = t$.

3. If the Lebesgue measure of a set $A \subset \mathbb{R}^m$ is greater than 1, then there exist distinct points $x, y \in A$ such that $x - y \in \mathbb{Z}^m$.

4. Let $r > 2$. Show that for almost all numbers x there exists a coefficient $c_x > 0$ such that $|x - \frac{k}{n}| \geqslant \frac{c_x}{n^r}$ for all fractions $\frac{k}{n}$.

5. Give an example of (pairwise distinct) subsets $A_1, \dots, A_N$ in $[0, 1]$ of measure $\frac{1}{2}$ such that all elements of the corresponding canonical partition (see Sect. 1.1.3) are intervals of equal length.

6. Show that the union of any (even uncountable) family of non-degenerate intervals is measurable.

7. Show that a point t belongs to the Cantor set $\mathcal{C}$ if and only if it can be written in the form $t = 2\sum_{n=1}^{\infty} \varepsilon_n 3^{-n}$, where ε_n is equal to 0 or 1. Show that such a representation is unique. Verify the equalities $\mathcal{C} + \mathcal{C} = \{s + t \mid s, t \in \mathcal{C}\} = [0, 2]$ and $\mathcal{C} - \mathcal{C} = \{s - t \mid s, t \in \mathcal{C}\} = [-1, 1]$.

8. Let $a_n > 0$ ($n \geqslant 0$) be a sequence of numbers such that $\sum_{n=0}^{\infty} 2^n a_n < 1$. Let us imitate the construction of the Cantor set. First delete from the segment $[0, 1]$ its middle part of length a_0, i.e., the open interval $\delta = (\frac{1-a_0}{2}, \frac{1+a_0}{2})$. From the two remaining segments delete their middle parts of length a_1, and so on. Show that this construction yields a set of positive measure that has no interior points.

9. Using sets similar to those constructed in the previous exercise, show that there exists a measurable set $E \subset (0, 1)$ such that for every non-empty interval $\Delta \subset (0, 1)$, the sets $\Delta \cap E$ and $\Delta \setminus E$ have positive measure.

10. Show that the boundary of an open subset of the line can have positive measure.

11. Using the result of Exercise 1, show that if a set A is of positive measure, then zero is an interior point of the set $A - A = \{x - y \mid x, y \in A\}$.

12. Let $\mathfrak{U}$ be an ultrafilter in $\mathbb{N}$ that consists of infinite sets (see Sect. 1.1, Exercise 12). With each set $U \in \mathfrak{U}$ we associate the point $x_U = \sum_{n\in U} 2^{-n} \in [0,1]$ and consider the set $E = \{x_U \mid U \in \mathfrak{U}\}$. Show that it is not measurable. *Hint.* Show that for every interval $(k2^{-N}, (k+1)2^{-N}) \subset (0,1)$ and every irrational point z in this interval, the following alternative holds: either $z \in E$, $z' \notin E$, or $z \notin E$, $z' \in E$, where z' is the point symmetric to z with respect to the middle of this interval. Assume the contrary and use the result of Exercise 1 with $\varepsilon < 1/2$.

2.2 Regularity of the Lebesgue Measure

In this section, we establish an important property of the Lebesgue measure, which shows that it agrees with the topology. We will denote the Lebesgue measure on $\mathbb{R}^m$ by λ without indicating the dimension.

2.2.1 We prove that every measurable set can be approximated by open sets.

Theorem *For every measurable set $E \subset \mathbb{R}^m$ and every $\varepsilon > 0$ there exists an open set G such that*

$$G \supset E \quad \text{and} \quad \lambda(G \setminus E) < \varepsilon.$$

Proof First assume that $\lambda(E) < +\infty$. Using formula (3) from Sect. 2.1.2, find cells $P_n = [a_n, b_n)$ such that

$$\bigcup_{n\geqslant 1} P_n \supset E, \qquad \sum_{n=1}^{\infty} \lambda(P_n) < \lambda(E) + \varepsilon. \tag{1}$$

Since the measure of a cell depends continuously on its vertices, we can choose points $a'_n < a_n$ sufficiently close to a_n so that

$$\lambda\big([a'_n, b_n)\big) < \lambda(P_n) + \frac{\varepsilon}{2^n} \quad \text{for all } n \text{ in } \mathbb{N}.$$

Set $G = \bigcup_{n\geqslant 1}(a'_n, b_n)$. Obviously,

$$E \subset \bigcup_{n\geqslant 1} P_n \subset \bigcup_{n\geqslant 1}(a'_n, b_n) = G \subset \bigcup_{n\geqslant 1}[a'_n, b_n).$$

By the countable subadditivity of the Lebesgue measure,

$$\lambda(G) \leqslant \sum_{n\geqslant 1} \lambda\big([a'_n, b_n)\big) < \sum_{n\geqslant 1}\Big(\lambda(P_n) + \frac{\varepsilon}{2^n}\Big) = \sum_{n\geqslant 1} \lambda(P_n) + \varepsilon < \lambda(E) + 2\varepsilon \tag{2}$$

(in the last transition we have used inequality (1)). Therefore,

$$\lambda(G \setminus E) = \lambda(G) - \lambda(E) < 2\varepsilon.$$

Since ε is arbitrary, the theorem is proved for a set of finite measure.

In the general case, we write E as the union of a sequence of sets of finite measure: $E = \bigcup_{n\geqslant 1} E_n$. As we have already proved, for each n there is an open set G_n such that $E_n \subset G_n$ and $\lambda(G_n \setminus E_n) < \varepsilon/2^n$. Let us check that the set $G = \bigcup_{n\geqslant 1} G_n$ satisfies the desired conditions. Indeed,

$$E = \bigcup_{n\geqslant 1} E_n \subset \bigcup_{n\geqslant 1} G_n = G \quad \text{and} \quad G \setminus E = \bigcup_{n\geqslant 1} (G_n \setminus E) \subset \bigcup_{n\geqslant 1} (G_n \setminus E_n).$$

Using the countable subadditivity of λ, we obtain

$$\lambda(G \setminus E) \leqslant \sum_{n\geqslant 1} \lambda(G_n \setminus E_n) < \sum_{n\geqslant 1} \frac{\varepsilon}{2^n} = \varepsilon.$$

□

2.2.2 Let us mention several important corollaries of Theorem 2.2.1.

Corollary 1 *For every measurable set E and every $\varepsilon > 0$ there exists a closed set F such that $F \subset E$ and $\lambda(E \setminus F) < \varepsilon$.*

Proof To prove this corollary, consider an open set G such that

$$G \supset E^c = \mathbb{R}^m \setminus E, \qquad \lambda\big(G \setminus E^c\big) < \varepsilon.$$

Then the set $F = G^c$ is of the desired form, since it is closed, contained in E, and $E \setminus F = G \setminus E^c$. □

Corollary 2 *For every measurable set E the following equalities hold:*

$$\lambda(E) = \inf\big\{\lambda(G) \mid G \supset E,\ G \text{ is an open set}\big\},$$

$$\lambda(E) = \sup\big\{\lambda(F) \mid F \subset E,\ F \text{ is a closed set}\big\}.$$

The second formula can be refined:

$$\lambda(E) = \sup\big\{\lambda(K) \mid K \subset E,\ K \text{ is a compact set}\big\}.$$

Proof The proof of the first two equalities follows immediately from the theorem and Corollary 1. The fact that we may use only compact subsets follows from the formula $\lambda(F) = \lim_{n\to\infty} \lambda(F \cap [-n, n]^m)$, which ensues from the continuity of λ from below (see Sect. 1.3.3). It allows one to exhaust every closed subset $F \subset E$, and hence the whole set E, by the compact sets $F \cap [-n, n]^m$ with an arbitrary accuracy. □

The property established in Corollary 2 is called the *regularity* of the Lebesgue measure. It means that every measurable set can be approximated, with an arbitrarily small change in the measure, by closed sets from the inside and by open sets from the outside. Observe that we cannot swap the roles of closed and open sets. For instance, let E be the set that consists of all rational points of the interval (0, 1); obviously, it has zero (one-dimensional) Lebesgue measure. This set cannot be approximated by ambient closed sets, because every such set contains the segment [0, 1], so that its measure is at least one. In a similar way, the complement of E in [0, 1], which has measure 1 but an empty interior, cannot be approximated by smaller open sets.

Remark The first equality in Corollary 2 remains valid for any (not necessarily measurable) set if one replaces $\lambda(E)$ by the outer measure $\lambda^*(E)$.

Indeed, if $\lambda^*(E) = +\infty$, then it is obvious by the monotonicity of the outer measure, and if $\lambda^*(E) < +\infty$, then one can argue in exactly the same way as in the proof of inequality (2), but replacing $\lambda(E)$ with $\lambda^*(E)$.

The value $\lambda_*(E) = \sup\{\lambda(F) \mid F \subset E,\ F \text{ is a closed set}\}$ is sometimes called the *inner measure* of E. As we have seen, the equality of the outer and the inner measures is a necessary condition for a set to be measurable. One can prove (see Exercise 1) that if $\lambda_*(E) < +\infty$, then this condition is also sufficient. It was this condition that Lebesgue used to define the measurability of a bounded set.

Corollary 3 *Every measurable set E can be written in the form $E = e \cup \bigcup_{n\geqslant 1} K_n$, where $\{K_n\}_{n\in\mathbb{N}}$ is an increasing sequence of compact sets and $\lambda(e) = 0$.*

Proof It suffices to consider the case where E is bounded. By Corollary 2, there exist compact sets $K_n \subset E$ such that $\lambda(E \setminus K_n) \underset{n\to\infty}{\longrightarrow} 0$. We may assume that $K_n \subset K_{n+1}$ (otherwise replace the set K_n by the union $K_1 \cup \cdots \cup K_n$). Put

$$e = E \setminus \bigcup_{n\geqslant 1} K_n.$$

Then $E = e \cup \bigcup_{n\geqslant 1} K_n$ and $\lambda(e) = 0$, because $\lambda(e) \leqslant \lambda(E \setminus K_n) \underset{n\to\infty}{\longrightarrow} 0$. □

Corollary 4 *Every measurable set E can be written in the form $E = (\bigcap_{n\geqslant 1} G_n) \setminus e$, where G_n are open sets and $\lambda(e) = 0$.*

The proof of this corollary is left to the reader.

Corollaries 3 and 4 show that, up to sets of zero measure, every measurable set is the union of a sequence of closed sets (i.e., an F_σ set) and the intersection of a sequence of open sets (i.e., a G_δ set).

Recall that the elements of the minimal σ-algebra containing all open sets are called Borel sets. Corollaries 3 and 4 imply the following.

Corollary 5 *Every measurable set can be approximated from the inside and from the outside by Borel sets of the same measure. In other words, if E is a measurable set, then there exist Borel sets A and B such that*

$$A \subset E \subset B, \quad \lambda(B \setminus A) = 0.$$

If $\lambda(E) < +\infty$, then this corollary is a special case of Corollary 1.5.2.

2.2.3 If we want to generalize the notion of regularity to other measures on $\mathbb{R}^m$, we must assume that these measures are defined on all open and closed sets, and hence on the minimal σ-algebra containing these sets, i.e., on the σ-algebra of Borel sets. Thus we introduce the following definition.

Definition A measure defined on the σ-algebra of Borel subsets of a topological space X is called a *Borel* measure on X.

Theorem 2.2.1 remains valid for every Borel measure μ on an open set $\mathcal{O}$ ($\mathcal{O} \subset \mathbb{R}^m$) provided that this measure is finite on cells whose closures are contained in $\mathcal{O}$.

Indeed, the only specific property of the Lebesgue measure that we have used in the proof of the theorem is that the measure of a cell depends continuously on its vertices. In the general case, we can use instead the continuity of the measure from above and argue in the following way. A cell $P = [a, b)$ is the intersection of the decreasing sequence of cells $[a - \frac{1}{n}h, b)$, where $h = b - a > 0$. It is clear that $[a - \frac{1}{n}h, b] \subset \mathcal{O}$ for sufficiently large n (recall that $\overline{P} \subset \mathcal{O}$). By the continuity of μ from above, $\mu([a - \frac{1}{n}h, b)) \underset{n\to\infty}{\longrightarrow} \mu(P)$. Hence for every $\varepsilon > 0$ there is a cell $[a', b)$ such that $P \subset (a', b)$, $[a', b] \subset \mathcal{O}$, and $\mu([a', b)) < \mu(P) + \varepsilon$ (for instance, we can put $a' = a - \frac{1}{n}h$ for sufficiently large n). Using this fact, we can construct cells $[a'_n, b_n)$, $a'_n < a_n$, satisfying (2) (with μ in place of λ), and then the proof of Theorem 2.2.1 for the measure μ works without any modification.

All corollaries of Theorem 2.2.1 also remain valid in this more general situation. As in the case of the Lebesgue measure, the property from Corollary 2 is called the *regularity* of measure. Thus the following theorem holds.

Theorem *Let $\mathcal{O}$ be an arbitrary open subset of the space $\mathbb{R}^m$. If a Borel measure μ on $\mathcal{O}$ is finite on cells whose closures are contained in $\mathcal{O}$, then it is regular, i.e., for every Borel set E, $E \subset \mathcal{O}$, the following equalities hold*:

$$\mu(E) = \inf\{\mu(G) \mid G \supset E,\ G \text{ is an open set},\ G \subset \mathcal{O}\},$$
$$\mu(E) = \sup\{\mu(F) \mid F \subset E,\ F \text{ is a closed set}\}.$$

Corollary *Let μ be a Borel measure on the space $\mathbb{R}^m$. Then for every Borel set $E \subset \mathbb{R}^m$ of finite measure, the following equality holds*:

$$\mu(E) = \sup\{\mu(K) \mid K \subset E,\ K \text{ is a compact set}\}.$$

Proof Indeed, we may assume without loss of generality that μ is a finite measure (otherwise replace it with the measure $\widetilde{\mu}$ defined by the formula $\widetilde{\mu}(A) = \mu(A \cap E)$). For a finite measure, the claim can be proved by analogy with the proof of Corollary 2 from Sect. 2.2.2. □

Note that a σ-finite Borel measure on the space $\mathbb{R}^m$ is not necessarily regular (see Exercise 3). For further results on the regularity of Borel measures in metrizable spaces, see Appendix 13.3.

EXERCISES

1. Show that the set whose inner and outer measures coincide and are finite is measurable.

2. Show that the Carathéodory extension of an arbitrary regular measure is a regular measure.

3. Show that the Borel measure on $\mathbb{R}$ generated by the unit masses at the points $1, \frac{1}{2}, \dots, \frac{1}{n}, \dots$ is not regular.

2.3 Preservation of Measurability Under Smooth Maps

Let $\mathcal{O}$ be an open subset of the space $\mathbb{R}^m$. By $C^1(\mathcal{O}, \mathbb{R}^n)$ we denote the set of all smooth maps (i.e., maps that have one continuous derivative) from $\mathcal{O}$ to $\mathbb{R}^n$. The derivative of a smooth map Φ at a point x is denoted by $\Phi'(x)$. The open ball of radius r centered at a point x is denoted by $B(x, r)$.

2.3.1 Let us establish a simple sufficient condition for the measurability to be preserved. For brevity, we denote the Lebesgue measure on $\mathbb{R}^m$ by λ, without indicating the dimension.

Theorem *Let $\mathcal{O}$ be an open subset of the space $\mathbb{R}^m$, and let $\Phi \in C^1(\mathcal{O}, \mathbb{R}^m)$. Then for every measurable set A, $A \subset \mathcal{O}$, the set $\Phi(A)$ is also measurable. If $\lambda(A) = 0$, then $\lambda(\Phi(A)) = 0$.*

Proof As follows from the regularity of the Lebesgue measure (see Sect. 2.2.2, Corollary 3), a measurable set A can be written in the form $A = e \cup \bigcup_{n\geqslant 1} K_n$, where K_n are compact sets and e is a set of zero measure. Since the sets $\Phi(K_n)$ are compact and

$$\Phi(A) = \Phi(e) \cup \bigcup_{n\geqslant 1} \Phi(K_n),$$

it suffices to verify the last assertion of the theorem.

So, let $\lambda(A) = 0$. First assume that

$$A \subset P, \qquad \overline{P} \subset \mathcal{O}, \quad \text{where } P \in \mathscr{P}^m.$$

Let L be the Lipschitz constant corresponding to $\overline{P}$ (see Lagrange's inequality in Sect. 13.7.2). Fix an arbitrary $\varepsilon > 0$ and, using Property (7) from Sect. 2.1.3, find a sequence of cubic cells $\{Q_n\}_{n\geqslant 1}$ such that

$$A \subset \bigcup_{n\geqslant 1} Q_n, \qquad \sum_{n\geqslant 1} \lambda(Q_n) < \varepsilon.$$

Obviously, $A \subset \bigcup_{n\geqslant 1}(Q_n \cap P)$. Let h_n be the edge length of Q_n. Then $\|x-y\| \leqslant h_n\sqrt{m}$ for all x, y in $\overline{Q}_n$, and hence $\|\Phi(x)-\Phi(y)\| \leqslant L\,\|x-y\| \leqslant L\,h_n\sqrt{m}$ for $x, y \in \overline{Q}_n \cap \overline{P}$. Thus the set $\Phi(\overline{Q}_n \cap \overline{P})$ is contained in a ball of radius $Lh_n\sqrt{m}$ and, consequently, in a cube with edge length $2Lh_n\sqrt{m}$. Hence $\lambda(\Phi(\overline{Q}_n \cap \overline{P})) \leqslant (2Lh_n\sqrt{m})^m \equiv C\lambda(Q_n)$. The set $H = \bigcup_{n\geqslant 1}\Phi(\overline{Q}_n \cap \overline{P})$ contains $\Phi(A)$ and, being a union of compact sets, is measurable. Furthermore,

$$\lambda(H) \leqslant \sum_{n\geqslant 1}\lambda\big(\Phi(\overline{Q}_n \cap \overline{P})\big) \leqslant C\sum_{n\geqslant 1}\lambda(Q_n) < C\varepsilon.$$

Thus the set $\Phi(A)$ is contained in a set of arbitrarily small measure. Since the Lebesgue measure is complete, $\Phi(A)$ is measurable and has zero measure (see Sect. 2.1.3, Property (4)).

Now consider the general case. By Theorem 1.1.7, the open set $\mathcal{O}$ can be written as the union of a sequence of cells P_n whose closures are contained in $\mathcal{O}$: $\mathcal{O} = \bigcup_{n\geqslant 1} P_n$, $\overline{P}_n \subset \mathcal{O}$. In this case,

$$A = \bigcup_{n\geqslant 1}(A \cap P_n), \qquad \Phi(A) = \bigcup_{n\geqslant 1}\Phi(A \cap P_n).$$

As we have already proved, the sets $\Phi(A \cap P_n)$ have zero measure. Therefore the measure of the whole set $\Phi(A)$ is also zero. □

Corollary *Let G be an open subset of the space $\mathbb{R}^m$, let $f \in C^1(G)$, and let $\Gamma_f = \{(x, f(x)) \mid x \in G\} \subset \mathbb{R}^{m+1}$ be the graph of the function f (we identify the spaces $\mathbb{R}^{m+1}$ and $\mathbb{R}^m \times \mathbb{R}$ in the natural way). Then $\lambda_{m+1}(\Gamma_f) = 0$.*

Proof Let $\mathcal{O} = G \times \mathbb{R}$. Let $\Phi(x, y) = (x, f(x))$ for points (x, y) in $\mathcal{O}$. It is clear that $\mathcal{O}$ is an open subset of the space $\mathbb{R}^{m+1}$ and $\Phi \in C^1(\mathcal{O}, \mathbb{R}^{m+1})$. Obviously, $\Gamma_f = \Phi(e)$, where $e = G \times \{0\}$. Since $\lambda_{m+1}(e) = 0$, the equality $\lambda_{m+1}(\Gamma_f) = 0$ follows from the theorem. □

The corollary implies, in particular, that the m-dimensional Lebesgue measure of every proper affine subspace of $\mathbb{R}^m$ vanishes. Hence the measure of every parallelepiped coincides, as we have already observed in Sect. 2.1.3, with the measure of its closure and its interior. In a similar way, the measure of an open ball coincides with the measure of its closure.

Remark Let us introduce an important class of maps which will be repeatedly used in what follows.

Definition Given a set $E \subset \mathbb{R}^m$, one says that a map $\Phi : E \to \mathbb{R}^n$ satisfies the *Lipschitz condition*[2] on E if there exists a constant L such that

$$\|\Phi(x) - \Phi(x')\| \leqslant L\|x - x'\| \quad \text{for all } x, x' \text{ in } E.$$

The number L is called the *Lipschitz constant* for Φ.

As follows from Lagrange's inequality (see Sect. 13.7.2), a smooth map locally satisfies the Lipschitz condition.

As one can see from the proof of the theorem, we have not used the smoothness in full strength, but need only the Lipschitz condition. Hence the theorem remains valid for every map that locally satisfies this condition. In particular, such maps send sets of zero measure to sets of zero measure, i.e., have Luzin's[3] property (N).

2.3.2 Here we will show that the Lebesgue measurability of a set is not in general preserved under a continuous map. Thus the Lipschitz condition, which guarantees the measurability of the image of a measurable set (see the remark in the previous section) cannot be replaced by the weaker continuity condition.

In order to check this, it suffices to construct a continuous map φ that sends a set e of zero measure to a set $\varphi(e)$ of positive measure. Indeed, in this case, taking a non-measurable subset E of $\varphi(e)$ (see Sect. 2.1.3), we will obtain that $E = \varphi(e_0)$ with $e_0 \subset e$. The set e_0 is measurable (the Lebesgue measure is complete, so all subsets of a set of zero measure are measurable), while its image E is not.

In order to construct such an example, restricting ourselves to the one-dimensional case, we use the *Cantor function* φ, which often turns out to be useful in similar situations, because it has rather unusual properties. For example, it is continuous and its derivative vanishes almost everywhere, but $\varphi \not\equiv \text{const}$ (for other properties of the Cantor function, see Exercises 3–5).

This function, defined on $[0, 1]$ and closely related to the Cantor set $\mathcal{C}$ (see Sect. 2.1.4), is constructed as follows. By definition, $\varphi(0) = 0$, $\varphi(1) = 1$, and on the middle third of the interval $(0, 1)$, i.e., for $x \in [\frac{1}{3}, \frac{2}{3}]$, the function φ is constant and equal to the half-sum of its values at the endpoints of the interval: $\varphi(x) = \frac{1}{2}(\varphi(0) + \varphi(1)) = \frac{1}{2}$. For each of the remaining intervals $(0, \frac{1}{3})$ and $(\frac{2}{3}, 1)$, we repeat the same procedure: at the middle third of the interval, the function φ is constant and equal to the half-sum of its values at the endpoints of the interval (i.e., $\varphi(x) = \frac{1}{4}$ on $[\frac{1}{9}, \frac{2}{9}]$ and $\varphi(x) = \frac{3}{4}$ on $[\frac{7}{9}, \frac{8}{9}]$). Repeating this construction ad infinitum, we will define φ on a dense subset of $[0, 1]$. It remains to define it on the complement of this set, i.e., on the set obtained from the Cantor set $\mathcal{C}$ by deleting the endpoints of all complementary intervals. If we want to preserve the continuity or the monotonicity on the whole interval $[0, 1]$, this can be done in a unique way. To see this, it suffices to observe that at each step of our construction we obtain an increasing function whose increments on intervals of length $\frac{1}{3^n}$ do not exceed $\frac{1}{2^n}$. The graph of the function φ (see Fig. 2.1) is sometimes called the Cantor staircase.

---

[2]Rudolf Otto Sigismund Lipschitz (1832–1903)—German mathematician.

[3]Nikolai Nikolaevich Luzin (1883–1950)—Russian mathematician.

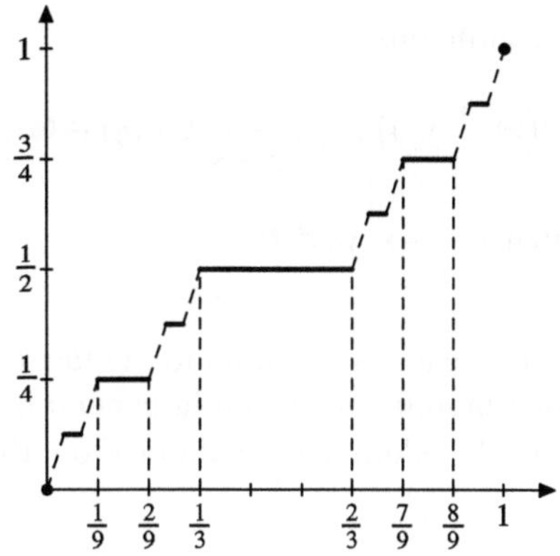

Fig. 2.1 Graph of the Cantor function

It follows from the construction that φ is constant on complementary intervals of the Cantor set. Since their endpoints belong to $\mathcal{C}$, we have $\varphi(\mathcal{C}) = \varphi([0, 1])$. Thus the φ-image of the Cantor set, which is of zero measure, coincides with $[0, 1]$. As we have observed above, this implies that the image of some part of the Cantor set is not measurable.

2.3.3 In conclusion of this section, we briefly discuss the preservation of Borel measurability. There is a general result according to which a homeomorphic image of a Borel set is again a Borel set. We confine ourselves to the proof of this assertion under an additional assumption; this suffices for our purposes. The general result can be proved using Theorem 13.2.3; we encourage the reader to do this (see Exercise 8).

Proposition *Let Θ be a homeomorphism defined on a Borel set A. If the inverse map Θ^{-1} satisfies the Lipschitz condition, then $B = \Theta(A)$ is a Borel set.*

Proof Since the inverse map satisfies the Lipschitz condition and, consequently, is uniformly continuous on B, it can be extended to a continuous map $\Psi : \overline{B} \to \mathbb{R}^m$. Let us check that

$$\Psi^{-1}(A) = B. \tag{1}$$

If this equality is true, then, by Corollary 1 from Sect. 1.6.2, B is a Borel subset of $\overline{B}$ (as the inverse image of a Borel set under a continuous map), and hence a Borel subset of $\mathbb{R}^m$ (see Corollary 2 in Sect. 1.6.2).

Obviously, $\Psi^{-1}(A) \supset B$. Hence, when proving equality (1), it suffices to check the reverse inclusion. If it is false, then there is a point $y_0 \in \overline{B} \setminus B$ such that $x_0 = \Psi(y_0) \in A$. Consider points $y_j \in B$ converging to y_0 and set $x_j = \Psi(y_j) = \Theta^{-1}(y_j)$. Then, since Ψ is continuous, we have $x_j = \Psi(y_j) \underset{j\to\infty}{\longrightarrow} \Psi(y_0) = x_0$. At the same time, since Θ is continuous,

$$\Theta(x_j) = \Theta\big(\Theta^{-1}(y_j)\big) = y_j \underset{j\to\infty}{\longrightarrow} \Theta(x_0) \in \Theta(A) = B.$$

This contradicts the fact that $y_j \underset{j\to\infty}{\longrightarrow} y_0 \notin B$. □

We would like to draw the reader's attention to the fact that a homeomorphism, while preserving Borel measurability, does not in general preserve Lebesgue measurability, even if it satisfies the additional condition from the proposition (see Exercise 5).

EXERCISES

1. Show that the graph of a function continuous in an open subset of $\mathbb{R}^m$ has zero $(m+1)$-dimensional measure.

2. Let X be a measurable subset of $\mathbb{R}^m$ and $F \in C(X, \mathbb{R}^m)$. Show that F preserves measurability if and only if it sends every set of zero (Lebesgue) measure to a set of zero measure.

3. Establish the following properties of the Cantor function φ (see Sect. 2.3.2):

(a) $\varphi(x) + \varphi(1-x) = 1$ for $0 \leqslant x \leqslant 1$;
(b) $\varphi(x/3) = \varphi(x)/2$ for $0 \leqslant x \leqslant 1$;
(c) $\varphi(x + \frac{2}{3^n}) = \varphi(x) + \frac{1}{2^n}$ for $0 \leqslant x \leqslant \frac{1}{3^n}$;
(d) $(\frac{x}{2})^\alpha \leqslant \varphi(x) \leqslant x^\alpha$ for $0 \leqslant x \leqslant 1$, where $\alpha = \log_3 2$.

4. What is the area of the region under the graph of the Cantor function?

5. Let $\mathcal{C}$ be the Cantor set, φ be the Cantor function (see Sect. 2.3.2), and $g(x) = x + \varphi(x)$ $(x \in [0,1])$. Show that:

(a) g is a homeomorphism;
(b) the measure of the set $g(\mathcal{C})$ is equal to one;
(c) among the images of sets of zero measure there are non-measurable sets.

Thus the homeomorphism g does not preserve measurability.

6. One says that a function f on an interval $[a,b]$ satisfies the Lipschitz condition of order α $(\alpha > 0)$ if there exists a positive constant L such that $|f(x) - f(y)| \leqslant L|x-y|^\alpha$ for all x, y in $[a,b]$. Show that the Cantor function satisfies the Lipschitz condition of order $\alpha = \log_3 2$.

7. Show that for every $\alpha \in (0,1)$ there exists a function that satisfies the Lipschitz condition of order α and sends some set of zero measure to a set of positive measure (and, therefore, does not preserve Lebesgue measurability). *Hint*. Generalize the construction of the Cantor function using the set described in Exercise 8 from Sect. 2.1 instead of $\mathcal{C}$.

8. Show that a homeomorphic image of a Borel set is again a Borel set. *Hint*. Use Theorem 13.2.3 and the scheme of the proof of Proposition 2.3.3.

2.4 Invariance of the Lebesgue Measure Under Rigid Motions

Recall that a *rigid motion of the space* $\mathbb{R}^m$ is the composition of a translation and an orthogonal transformation. We begin with the study of the behavior of the Lebesgue measure under translations.

Everywhere in this section, except for Sects. 2.4.5 and 2.4.6, we denote the Lebesgue measure by λ, without indicating the dimension.

2.4.1 The *translation* by a vector $v \in \mathbb{R}^m$ is the map $x \mapsto v + x$ $(x \in \mathbb{R}^m)$. The image of a set E under this map will be denoted by $v + E$.

Theorem *A translation sends a measurable set to a measurable set and preserves the measure of a set. In other words, if $v \in \mathbb{R}^m$, $E \in \mathfrak{A}^m$, then $v + E \in \mathfrak{A}^m$ and $\lambda(v + E) = \lambda(E)$.*

Proof The measurability of $v + E$ follows immediately from Theorem 2.3.1, since a translation is a smooth map. Hence, fixing an arbitrary vector v, we can define a function μ on the σ-algebra $\mathfrak{A}^m$ by the formula $\mu(E) = \lambda(v + E)$ $(E \in \mathfrak{A}^m)$. We leave the reader to check that μ is a measure. Since the translation by v sends a cell $[a, b)$ to the cell $[a + v, b + v)$ with the same edge lengths, the measures μ and λ coincide on the semiring of cells, and, by the uniqueness theorem for the extension of a measure (see Sect. 1.5.1), they coincide on the whole σ-algebra $\mathfrak{A}^m$. □

2.4.2 Now let us consider the problem of describing all translation-invariant measures in $\mathbb{R}^m$. In order to exclude pathological cases (for instance, the counting measure, which is obviously invariant under any bijection), we impose a natural restriction on the measures in question. It then turns out that every translation-invariant measure is proportional to the Lebesgue measure.

Theorem *Let μ be a measure defined on the algebra $\mathfrak{A}^m$ of Lebesgue measurable sets. Assume that:*

(a) *μ is translation-invariant, i.e., $\mu(v + E) = \mu(E)$ for every v in $\mathbb{R}^m$ and every E in $\mathfrak{A}^m$;*
(b) *the measure of every bounded measurable set is finite.*

Then there exists a constant k, $0 \leqslant k < +\infty$, such that $\mu = k\lambda$, i.e.,

$$\mu(E) = k\lambda(E) \quad \textit{for every set } E \textit{ in } \mathfrak{A}^m. \tag{1}$$

It is easy to see that condition (b) is equivalent to the assumption that the measures of all cells are finite, and, in view of condition (a), to the assumption that the measure of at least one non-empty cell is finite. Equivalently, one might also require that the measures of compact sets be finite.

Proof Set $Q = [0, 1)^m$. If equality (1) holds, then, obviously, $k = \mu(Q)$. It is this number k that we will consider.

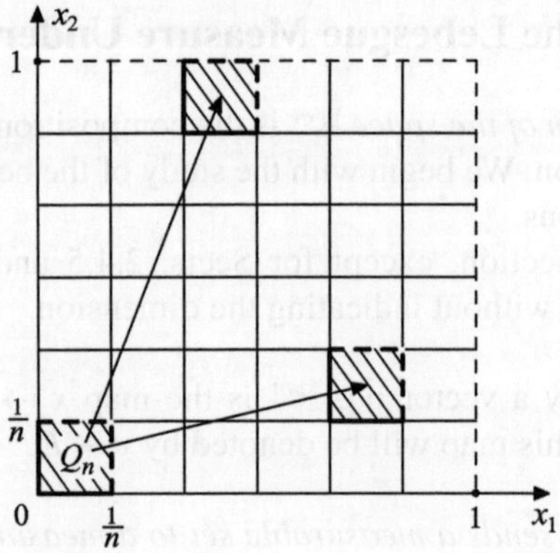

Fig. 2.2 Partition of the unit square into congruent parts

(1) First let $k = 1$, i.e., $\mu(Q) = \lambda(Q) = 1$. Let us check that $\mu = \lambda$. As we noted after formula (3) in Sect. 2.1.2, λ is the Carathéodory extension of the ordinary volume from the semiring $\mathscr{P}_r^m$ of cells with rational vertices. Hence, by the uniqueness theorem, in order to prove that the measures λ and μ coincide, it suffices to verify that they coincide on $\mathscr{P}_r^m$. Since every cell with rational vertices is a disjoint union of cubic cells with rational vertices, it suffices to check that λ and μ coincide on such cells. Since every cell is a translation of a cell having a vertex at the origin, it suffices to prove that the measures μ and λ coincide on cells of the form $Q_n = [0, \frac{1}{n})^m$ with $n \in \mathbb{N}$.

The cell Q is the union of n^m pairwise disjoint translations of the cell Q_n (see Fig. 2.2).

Hence $n^m \mu(Q_n) = \mu(Q) = 1$ and, consequently, $\mu(Q_n) = n^{-m} = \lambda(Q_n)$. Thus in the case under consideration the proof is complete.

(2) Now let $k = \mu(Q)$ be an arbitrary positive number. Consider the auxiliary measure $\widetilde{\mu} = \mu/k$. Clearly, it is also translation-invariant, and $\widetilde{\mu}(Q) = 1$. As we have already proved, such a measure coincides with λ, and hence (1) holds.

(3) If $\mu(Q) = 0$, then $\mu(\mathbb{R}^m) = 0$, since the space $\mathbb{R}^m$ can be covered by a countable family of translations of the cell Q. Thus in this case μ is the zero measure, and (1) holds with $k = 0$. □

Remark If a measure μ satisfying conditions (a) and (b) is defined not on the whole σ-algebra $\mathfrak{A}^m$, but on a subalgebra that contains all cells and the translations of all sets belonging to this subalgebra (for example, on all Borel sets), then, as one can see from the proof, Eq. (1) remains valid for all sets on which μ is defined.

2.4.3 From the above theorem and the arguments used in Sect. 2.1.3 when proving the existence of Lebesgue non-measurable sets, it follows that there does not exist a non-zero measure defined on all subsets of $\mathbb{R}^m$ that is finite on all cells and translation-invariant.

If we drop the condition of countable additivity, then the situation changes. As Banach[4] proved, in every space $\mathbb{R}^m$ there exists a (non-unique) volume defined on the ring of all bounded sets that is translation-invariant and coincides with the ordinary volume on cells. In the two-dimensional case, one can even ensure that such a volume is invariant not only under translations, but under all rotations. In spaces of higher dimension, rotation-invariant volumes defined on all bounded sets cannot exist, since there are "too many" rotations and the group of motions is "too non-commutative" (see [N, Chap. III, Sect. 7, and Appendices]; for a discussion of this question from a more general point of view, see [G]).

2.4.4 It turns out that the Lebesgue measure is invariant not only under translations, but also under all orthogonal transformations.

Theorem *An orthogonal transformation sends a measurable set to a measurable set and preserves the measure of a set. In other words, if $U : \mathbb{R}^m \to \mathbb{R}^m$ is an orthogonal transformation and $E \in \mathfrak{A}^m$, then $U(E) \in \mathfrak{A}^m$ and $\lambda(U(E)) = \lambda(E)$.*

Proof The fact that an orthogonal transformation preserves measurability follows from Theorem 2.3.1. In order to prove that it preserves the measure of a set, we will use the theorem on translation-invariant measures.

On the σ-algebra $\mathfrak{A}^m$ consider the set function μ defined by the formula

$$\mu(E) = \lambda\bigl(U(E)\bigr) \quad \bigl(E \in \mathfrak{A}^m\bigr).$$

The reader can easily verify that μ is indeed a measure and that it is finite on cells. Our aim is to prove that $\mu = \lambda$. Let us check that the measure μ is translation-invariant. Since $U(v+E) = U(v)+U(E)$, it follows from the translation invariance of λ that

$$\mu(v + E) = \lambda\bigl(U(v + E)\bigr) = \lambda\bigl(U(v) + U(E)\bigr) = \lambda\bigl(U(E)\bigr) = \mu(E).$$

By Theorem 2.4.2, the measure μ is proportional to the Lebesgue measure: $\mu = k\lambda$. Finally, let us check that $k = 1$. Let B be an arbitrary ball centered at the origin. Then $U(B) = B$, whence

$$k\lambda(B) = \mu(B) = \lambda\bigl(U(B)\bigr) = \lambda(B) > 0.$$

Therefore, $k = 1$, and the measures μ and λ coincide. □

Comparing the above theorem with Theorem 2.4.1, we obtain an important result.

Corollary *The Lebesgue measure is invariant under rigid motions.*

---

[4]Stefan Banach (1892–1945)—Polish mathematician.

This invariance property of the Lebesgue measure allows one to compute the volumes of rectangular parallelepipeds, since every such parallelepiped can be transformed by a rigid motion into a parallelepiped with edges parallel to the coordinate axes.

Example The measure of a rectangular parallelepiped is equal to the product of the lengths of its edges.

First observe that the volumes of all parallelepipeds with a fixed vertex and fixed edge lengths coincide (see the remark after Corollary 2.3.1). Since the Lebesgue measure is translation-invariant, it suffices to compute the volume of an open parallelepiped of the form

$$P = \left\{ \sum_{j=1}^{m} t_j v_j \,\middle|\, 0 < t_j < 1 \text{ for } j = 1, 2, \ldots, m \right\}$$

whose edges $v_1, \ldots, v_m$ are pairwise orthogonal. Let us normalize the vectors v_j by setting $g_j = \frac{v_j}{s_j}$, where $s_j = \|v_j\|$ $(j = 1, \ldots, m)$. Clearly, the vectors $g_1, \ldots, g_m$ form an orthonormal basis in $\mathbb{R}^m$. Consider the linear transformation U that sends the canonical basis vectors $e_1, \ldots, e_m$ to the vectors $g_1, \ldots, g_m$. This is an orthogonal transformation, and $v_j = s_j U(e_j)$. By the definition of a parallelepiped, $P = U(R)$, where R is the parallelepiped $\prod_{j=1}^{m}(0, s_j)$. Since orthogonal transformations preserve measure,

$$\lambda(P) = \lambda(R) = \prod_{j=1}^{m} s_j = \prod_{j=1}^{m} \|v_j\|.$$

In Sect. 2.5.3, we will consider the problem of computing the volume of a (not necessarily rectangular) parallelepiped with given edge lengths in full generality.

2.4.5 By Theorem 2.4.2, we can speak about the Lebesgue measure on any finite-dimensional vector space X. Indeed, since X is algebraically isomorphic to $\mathbb{R}^m$ for $m = \dim X$, we can use this isomorphism to "transfer" the Lebesgue measure from $\mathbb{R}^m$ to X and obtain a measure μ that is translation-invariant and finite on bounded subsets. As follows from Theorem 2.4.2, any other measure satisfying these properties is proportional to μ. Thus, applying this construction with another isomorphism, we will obtain a measure proportional to μ. If X is a Euclidean space, and we consider only isomorphisms that preserve the inner product, then the measure μ is determined uniquely, since, by Theorem 2.4.4, a linear isometry preserves the Lebesgue measure.

Let us mention one important fact, which will be used in Sect. 2.5 and then in Chap. 8. For $k < m$, we can naturally define the k-dimensional Lebesgue measure on all k-dimensional affine subspaces of $\mathbb{R}^m$. By definition, it is the image of the Lebesgue measure λ_k in $\mathbb{R}^k$ under some rigid motion (we identify $\mathbb{R}^k$ with the subspace consisting of all points whose last $m - k$ coordinates are equal to zero). In other words, if L is a k-dimensional affine subspace in $\mathbb{R}^m$ and $T : \mathbb{R}^m \to \mathbb{R}^m$ is a rigid motion such that $L = T(\mathbb{R}^k)$, then a set $E \subset L$ is called measurable if its inverse image $T^{-1}(E) \subset \mathbb{R}^k$ is measurable, and we set the Lebesgue measure of E equal to $\lambda_k(T^{-1}(E))$. Since the Lebesgue measure is invariant under rigid motions (see Corollary 2.4.4), the Lebesgue measure on a subspace does not depend on the motion used in its construction. It also follows immediately from the definition that the Lebesgue measures in subspaces transform into each other under rigid motions; in this sense, they form a coherent family. The Lebesgue measure on a k-dimensional affine subspace will be denoted by the same symbol λ_k as the measure on $\mathbb{R}^k$. It will always be clear from the context on which subspace the measure is considered.

2.4.6 Let us find out how the measure of a set in an affine subspace $L \subset \mathbb{R}^m$ of dimension $m-1$ is related to the measure of its orthogonal projection to $\mathbb{R}^{m-1}$ (as usual, we regard $\mathbb{R}^{m-1}$ as a subspace in $\mathbb{R}^m$, identifying a vector $(x_1, \dots, x_{m-1})$ in $\mathbb{R}^{m-1}$ with the vector $(x_1, \dots, x_{m-1}, 0)$ in $\mathbb{R}^m$). In both subspaces, the $(m-1)$-dimensional Lebesgue measures will be denoted by λ_{m-1}.

Let P be the orthogonal projection from $\mathbb{R}^m$ to $\mathbb{R}^{m-1}$. We exclude the trivial case where $P(L) \neq \mathbb{R}^{m-1}$, i.e., assume that the normal N to L is not orthogonal to the vector $e_m = (0, \dots, 0, 1)$, which is the normal to $\mathbb{R}^{m-1}$. Let θ be the angle between these normals. Then $\cos\theta = \langle N, e_m\rangle / \|N\| \neq 0$. Let us establish a relationship between the measure of a set and the measure of its projection which generalizes a well-known fact from elementary geometry.

Proposition *For every measurable set $E \subset L$,*

$$\lambda_{m-1}\big(P(E)\big) = |\cos\theta|\,\lambda_{m-1}(E).$$

Proof We will assume without loss of generality that L is a linear subspace (otherwise we can translate it). Since the restriction of the projection P to L is a linear isomorphism, the function

$$\mu : E \mapsto \lambda_{m-1}\big(P(E)\big) \quad (E \subset L)$$

is obviously a measure on the σ-algebra of Lebesgue measurable sets that is translation-invariant and finite on bounded sets. Hence (see Theorem 2.4.2) $\mu = k\,\lambda_{m-1}$, where k is a positive coefficient. In other words,

$$\lambda_{m-1}\big(P(E)\big) = k\,\lambda_{m-1}(E)$$

for every measurable set E, $E \subset L$.

In order to find k, consider an orthonormal basis $v_1, \dots, v_{m-1}$ in L with $v_1, \dots, v_{m-2} \in \mathbb{R}^{m-1}$. Then $v_1, \dots, v_{m-2}, P(v_{m-1})$ is an orthogonal basis in $\mathbb{R}^{m-1}$. Moreover, $\|P(v_{m-1})\| = |\cos\theta|$. The unit cube Q spanned by the edges $v_1, \dots, v_{m-1}$ lies in L, and its projection is the rectangular parallelepiped spanned by the edges $v_1, \dots, v_{m-2}, P(v_{m-1})$, whose measure is equal to $\lambda_{m-1}(P(Q)) = \|P(v_{m-1})\| = |\cos\theta|$. Therefore,

$$|\cos\theta| = \lambda_{m-1}\big(P(Q)\big) = k\,\lambda_{m-1}(Q) = k. \qquad \square$$

Note that this proposition obviously remains valid in the case where $\cos\theta = 0$.

EXERCISES

1. The homothety in the space $\mathbb{R}^m$ with ratio $k > 0$ is the map $x \mapsto kx$. Arguing as in the proof of Theorem 2.4.1, show that it sends a measurable set E to a measurable set and the measure of the image of E is equal to $k^m \lambda_m(E)$.

2. Show that if a set $A \subset \mathbb{R}$ is measurable, then the set $B = \{(x, y) \in \mathbb{R}^2 \mid x - y \in A\}$ is also measurable.

3. How large can the area of a measurable set contained in the square $[0, 6]^2$ be if this set is disjoint with its translation by the vector $(1, 2)$?

4. We say that a set $E \subset \mathbb{R}^m$ generates a tiling if the translations of E by all vectors with integer coordinates form a partition of $\mathbb{R}^m$, i.e.,

$$\mathbb{R}^m = \bigvee_{n \in \mathbb{Z}^m} (n + E).$$

Show that the measure of a measurable set that generates a tiling is equal to one.

5. Let $E \subset \mathbb{R}^m$, $\lambda_m(E) > 0$, and let A be a dense set in $\mathbb{R}^m$. Show that $\lambda_m(\mathbb{R}^m \setminus \bigcup_{a \in A}(a + E)) = 0$. Show that we can drop the assumption of E being measurable by replacing the condition $\lambda_m(E) > 0$ with $\lambda_m^*(E) > 0$.

Given a number a, the translation by a modulo 1 is the map $x \mapsto \{x + a\}$, where $\{x + a\}$ is the fractional part of $x + a$ (i.e., $\{x + a\} = x + a - [x + a]$). Two subsets of the interval $[0, 1)$ are said to be congruent modulo 1 if one of them can be obtained from the other by a translation modulo 1.

6. By analogy with Theorem 2.4.1, show that the Lebesgue measure on $[0, 1)$ is invariant under translations modulo 1: they send measurable sets to measurable sets and preserve the measure of a set. Extend this result to the multi-dimensional case, replacing the interval $[0, 1)$ by the cubic cell $[0, 1)^m$.

7. Using the construction of a non-measurable set (see Sect. 2.1.3), show that there exists a set $E \subset [0, 1)$ with the following properties:

(a) the outer measure of E is equal to one;
(b) there exists a sequence of pairwise disjoint subsets of $[0, 1)$ congruent to E modulo 1.

2.5 Behavior of the Lebesgue Measure Under Linear Maps

Now we turn to the question of how the Lebesgue measure changes under an arbitrary linear transformation $V : \mathbb{R}^m \to \mathbb{R}^m$. If V is not invertible, then $V(\mathbb{R}^m)$ is a proper subspace of $\mathbb{R}^m$ and $\lambda_m(V(\mathbb{R}^m)) = 0$ (see the remark after Corollary 2.3.1), so that the image of every set has zero measure. In what follows, we exclude this degenerate case and consider only invertible linear transformations. Recall that the determinant of a linear transformation acting on a finite-dimensional space is, by definition, the determinant of the matrix of this transformation (in an arbitrary basis).

2.5.1 Let us first prove one auxiliary result.

Lemma *Let $V : \mathbb{R}^m \to \mathbb{R}^m$ be an invertible linear transformation. Then there exist orthonormal bases $\{g_j\}_{j=1}^m$, $\{h_j\}_{j=1}^m$ and positive numbers $s_1, \dots, s_m$ such that*

$$V(x) = \sum_{j=1}^m s_j \langle x, g_j \rangle h_j \quad \textit{for all } x \in \mathbb{R}^m. \tag{1}$$

Moreover, $|\det V| = s_1 \cdots s_m$.

The notation $\langle x, y \rangle$ denotes the inner product of x and y.

Proof Let V^* be the adjoint of V. Consider the self-adjoint transformation $W = V^*V$. As we know from linear algebra, there exists an orthonormal basis $g_1, \dots, g_m$ consisting of the eigenvectors of W. Let $c_1, \dots, c_m$ be the corresponding eigenvalues. They are positive, because the quadratic form $\langle W(x), x \rangle = \|V(x)\|^2$ is positive definite. Set $s_j = \sqrt{c_j}$ $(1 \leqslant j \leqslant m)$. For every vector x we have

$$x = \sum_{j=1}^m \langle x, g_j \rangle g_j \quad \text{and} \quad V(x) = \sum_{j=1}^m \langle x, g_j \rangle V(g_j) = \sum_{j=1}^m s_j \langle x, g_j \rangle h_j,$$

where $h_j = \frac{1}{s_j} V(g_j)$. These vectors form an orthonormal system, because

$$\langle h_k, h_j \rangle = \frac{1}{s_k s_j} \bigl\langle V(g_k), V(g_j) \bigr\rangle = \frac{1}{s_k s_j} \bigl\langle W(g_k), g_j \bigr\rangle = \frac{1}{s_k s_j} \bigl\langle s_k^2 g_k, g_j \bigr\rangle$$
$$= \begin{cases} 0 & \text{if } k \neq j, \\ 1 & \text{if } k = j. \end{cases}$$

Since the determinant $\det W$ is equal to the product of all eigenvalues,

$$(\det V)^2 = \det V^*V = \det W = \prod_{j=1}^m c_j.$$

Therefore, $|\det V| = \prod_{j=1}^m s_j$. □

2.5.2 Now we can find out how the Lebesgue measure changes under a linear transformation. In this and the next subsection, we denote the Lebesgue measure by λ without indicating the dimension.

Theorem *If $V : \mathbb{R}^m \to \mathbb{R}^m$ is a linear transformation and a set E, $E \subset \mathbb{R}^m$, is measurable, then the set $V(E)$ is also measurable and $\lambda(V(E)) = |\det(V)|\,\lambda(E)$.*

Thus the absolute value of the determinant has a simple geometric interpretation: it is the ratio of the measure of $V(E)$ to the measure of E for any measurable E.

Proof The fact that the image of a measurable set under any linear transformation is also measurable is a special case of Theorem 2.3.1. We have already seen that the desired assertion is true for non-invertible transformations. So in what follows we assume that V is invertible.

We define a measure μ on $\mathfrak{A}^m$ by the formula

$$\mu(E) = \lambda\bigl(V(E)\bigr) \quad \bigl(E \in \mathfrak{A}^m\bigr).$$

We leave the reader to check that μ is indeed a measure. It is translation-invariant:

$$\mu(c+E) = \lambda\bigl(V(c+E)\bigr) = \lambda\bigl(V(c) + V(E)\bigr) = \lambda\bigl(V(E)\bigr) = \mu(E).$$

Hence, by Theorem 2.4.2, μ is proportional to the Lebesgue measure: $\mu = k\lambda$, where k is a non-negative coefficient. In order to find this coefficient, we use the lemma to represent V in the form (1) and observe how the unit cube Q spanned by the vectors $g_1, \dots, g_m$ is being transformed. Since $V(g_j) = s_j h_j$, the image of Q is the rectangular parallelepiped with edges $s_j h_j$. Since $|\det V| = \prod_{j=1}^m s_j$, we see that

$$k = k\lambda(Q) = \mu(Q) = \lambda\bigl(V(Q)\bigr) = \prod_{j=1}^{m} \|s_j h_j\| = \prod_{j=1}^{m} s_j = |\det V|. \qquad \square$$

Let us mention a special case of this result which is constantly used in elementary geometry for computing areas and volumes: relation between the measures of similar sets.

The *homothety* in the space $\mathbb{R}^m$ with ratio k, $k > 0$, is the map $x \mapsto kx$ $(x \in \mathbb{R}^m)$. The image of a set E under this map will be denoted by kE.

It is obvious that a homothety is a linear map which in every basis is represented by the diagonal matrix with all diagonal entries equal to k. We obtain the following special case of the above theorem.

Corollary 1 *Let k be an arbitrary positive number. Then $kE \in \mathfrak{A}^m$ and $\lambda(kE) = k^m\lambda(E)$ for any measurable set E.*

Corollary 2 *The measure of an m-dimensional ball with an arbitrary center and radius r is equal to $\alpha_m r^m$, where α_m is the measure of the unit ball.*

This assertion follows from the fact that an arbitrary ball $B(x_0, r)$ can be obtained from the unit ball by a homothety and a translation: $B(x_0, r) = x_0 + rB(0, 1)$.

Since an ellipsoid with semi-axes $a_1, \dots, a_m$ can be obtained from the unit ball by dilations (with ratio a_i along the ith axis, $i = 1, \dots, m$), its volume is equal to $\alpha_m a_1 \cdots a_m$.

Note also that the measure of an open convex set C is equal to the measure of its closure. Indeed, we may assume that $0 \in C$. Then $C \subset \overline{C} = C \cup \partial C \subset kC$ for every $k > 1$. Therefore, $\lambda(C) \leqslant \lambda(\overline{C}) \leqslant \lambda(kC) = k^m \lambda(C)$. Taking the limit as $k \to 1$, we obtain the desired equality. It easily implies that $\lambda(\partial C) = 0$ (even if $\lambda(C) = +\infty$).

2.5.3 Extending the $a = 0$ case of the definition from Sect. 1.1.6, we define the n-dimensional *parallelepiped* in $\mathbb{R}^m$ ($n \leqslant m$) spanned by linearly independent vectors $\{v_j\}_{j=1}^n$ as the set

$$P(v_1, \dots, v_n) = \left\{ \sum_{j=1}^{n} t_j v_j \,\middle|\, 0 < t_j < 1 \text{ for } j = 1, 2, \dots, n \right\}.$$

As before, the vectors v_j will be called the edges of the parallelepiped P.

To avoid unnecessary stipulations, we keep the notation $P(v_1, \dots, v_n)$ in the case where the vectors $v_1, \dots, v_n$ are linearly dependent, even though such a set cannot actually be called a parallelepiped.

Let us compute the n-dimensional volume of the parallelepiped $P(v_1, \dots, v_n)$. First consider the case where $n = m$. Let $e_1, \dots, e_m$ be the canonical basis vectors, and let V be the linear transformation that sends them to the vectors $v_1, \dots, v_m$. Obviously, $P(v_1, \dots, v_m)$ is the V-image of the open cube $Q = (0, 1)^m$. Using Theorem 2.5.2, we obtain

$$\lambda\big(P(v_1, \dots, v_m)\big) = \lambda\big(V(Q)\big) = \big|\det(V)\big|. \tag{2}$$

In order to express the volume of the parallelepiped $P(v_1, \dots, v_m)$ directly in terms of the vectors $v_1, \dots, v_m$, we need to use the notion of Gram determinant (which is perhaps familiar to the reader from algebra). Recall the corresponding definition.

Definition The *Gram*[5] *determinant* $\Gamma(v_1, \dots, v_n)$ of a set of vectors $v_1, \dots, v_n \in \mathbb{R}^m$ is the determinant of the *Gram matrix*

$$\begin{pmatrix} \langle v_1, v_1\rangle & \langle v_1, v_2\rangle & \dots & \langle v_1, v_n\rangle \\ \langle v_2, v_1\rangle & \langle v_2, v_2\rangle & \dots & \langle v_2, v_n\rangle \\ \vdots & \vdots & \ddots & \vdots \\ \langle v_n, v_1\rangle & \langle v_n, v_2\rangle & \dots & \langle v_n, v_n\rangle \end{pmatrix},$$

whose entries are the pairwise inner products of the vectors $v_1, v_2, \dots, v_n$.

[5] Jørgen Pedersen Gram (1850–1916)—Danish mathematician.

The Gram matrix is the matrix of the positive semidefinite quadratic form

$$\sum_{j,k=1}^{n} \langle v_j, v_k \rangle t_j t_k = \left\| \sum_{j=1}^{n} t_j v_j \right\|^2.$$

Hence the Gram determinant is non-negative (this result also follows from the theorem proved below). It is clear that if the vectors $v_1, \dots, v_n$ are linearly dependent, then the rows of the Gram matrix are also linearly dependent and, consequently, $\Gamma(v_1, \dots, v_n) = 0$.

For $n = m$, the Gram determinant has a simple geometric interpretation.

Theorem $\Gamma(v_1, \dots, v_m) = \lambda^2(P(v_1, \dots, v_m))$.

Note that this equality is also obviously true in the case where the vectors $v_1, \dots, v_m$ are linearly dependent.

Proof Consider the linear transformation $V : \mathbb{R}^m \to \mathbb{R}^m$ that sends the canonical basis vectors to the vectors $v_1, \dots, v_m$. In the canonical basis, V is represented by the matrix W whose columns are the vectors $v_1, \dots, v_m$. By (2),

$$\lambda\big(P(v_1, \dots, v_m)\big) = \big|\det(V)\big| = \big|\det(W)\big|.$$

On the other hand, the product $W^T W$ is precisely the Gram matrix of the system under consideration (here W^T is the transpose of W). Hence

$$\lambda^2\big(P(v_1 \dots, v_m)\big) = \det\big(W^T\big)\det(W) = \det\big(W^T W\big) = \Gamma(v_1, \dots, v_m). \qquad \square$$

The geometric interpretation of the Gram determinant also remains valid in the case where the number of vectors $v_1, \dots, v_n$ is less than the dimension of the space. Indeed, these vectors, as well as the set $P(v_1, \dots, v_n)$, lie in a subspace L, their linear hull. If they are linearly independent, then $\dim L = n$. Since L is isomorphic to $\mathbb{R}^n$ as a Euclidean space, the Lebesgue measure is defined in L, and the theorem continues to hold:

$$\Gamma(v_1, \dots, v_n) = \lambda_n^2\big(P(v_1, \dots, v_n)\big).$$

Thus the volume of a parallelepiped with edges $v_1, \dots, v_n$ is the square root of the corresponding Gram determinant.

Knowing the geometric interpretation of the Gram determinant, we can describe how the (n-dimensional) measure changes under a linear transformation from $\mathbb{R}^n$ to $\mathbb{R}^m$ for $n < m$.

Proposition *Let $V : \mathbb{R}^n \to \mathbb{R}^m$ $(n \leqslant m)$ be a linear map. If $E \in \mathfrak{A}^n$, then*

$$\lambda_n\big(V(E)\big) = \sqrt{\det\big(W^T W\big)} \cdot \lambda_n(E)$$

(here W is the matrix of V in the canonical basis).

Proof First let $\operatorname{rank}(V) = n$. The set $V(E)$ is measurable by the definition of the Lebesgue measure on the space $X = V(\mathbb{R}^n)$ (see Sect. 2.4.5). In order to compute $\lambda_n(V(E))$, we introduce an auxiliary measure μ by setting $\mu(E) = \lambda_n(V(E))$ ($E \in \mathfrak{A}^n$). As the reader can easily verify, this measure is translation-invariant and hence proportional to the Lebesgue measure.

It is clear that the proportionality coefficient is equal to $\mu(Q)$, where $Q = [0, 1)^n$. Let us find it using the geometric interpretation of the Gram determinant. Let $v_1, \dots, v_n$ be the V-images of the canonical basis vectors of V (obviously, they are the columns of W). Hence $W^T W$ is the Gram matrix of the vectors $v_1, \dots, v_n$. On the other hand, one can easily see that $V(Q)$ is simply the parallelepiped $P(v_1, \dots, v_n)$ spanned by the vectors $v_1, \dots, v_n$. By the theorem,

$$\lambda_n^2\bigl(P(v_1, \dots, v_n)\bigr) = \Gamma(v_1, \dots, v_n) = \det\bigl(W^T W\bigr).$$

Therefore, $\mu(Q) = \lambda_n(V(Q)) = \sqrt{\det(W^T W)}$.

If $\operatorname{rank}(V) < n$, then the set $V(E)$ is contained in a subspace of dimension less than n, and hence its n-dimensional measure vanishes. The value $\det(W^T W)$ also vanishes, since it is the Gram determinant of the linearly dependent vectors $v_1, \dots, v_n$. □

As is well known from linear algebra, for a matrix W with m rows and n ($n \leqslant m$) columns, the following Binet[6]–Cauchy[7] formula holds. Let $A \subset \{1, 2, \dots, m\}$, $\operatorname{card} A = n$, and let W_A be the $n \times n$ matrix obtained from W by deleting all rows with indices not in A. The Binet–Cauchy formula says that

$$\det\bigl(W^T W\bigr) = \sum_A \det^2(W_A).$$

This equality has a beautiful geometric interpretation. By the above proposition, the left-hand side is simply the squared measure of the set $C = V(Q)$, where $V : \mathbb{R}^n \to \mathbb{R}^m$ is the map corresponding to the matrix W and Q is an arbitrary set of measure one. Consider the orthogonal projection P_A from $\mathbb{R}^m$ to the n-dimensional subspace L_A spanned by the canonical basis vectors with indices in A. Clearly, $P_A \circ V$ is the map corresponding to the matrix W_A. Hence, up to sign, $\det(W_A)$ is precisely the measure of the projection $P_A(C)$. Thus the Binet–Cauchy formula can be rewritten in the form

$$\lambda_n^2(C) = \sum_A \lambda_n^2\bigl(P_A(C)\bigr). \tag{3}$$

In particular, the squared volume of an n-dimensional parallelepiped contained in the space $\mathbb{R}^m$ ($m \geqslant n$) is the sum of the squared volumes of its projections to all

---

[6] Jacques Philippe Marie Binet (1786–1856)—French mathematician.

[7] Augustin-Louis Cauchy (1789–1857)—French mathematician.

possible subspaces L_A. If $n = 1$, then such a parallelepiped is just an interval and $P_A(C)$ are its projections to the coordinate axes, so that formula (3) turns into the Pythagorean theorem. In the case $n = m - 1$, formula (3) (and hence the Binet–Cauchy formula) can be proved as follows. Let N be the unit vector orthogonal to the subspace containing the parallelepiped C. Its coordinates are $\cos\theta_1, \dots, \cos\theta_m$, where $\theta_1, \dots, \theta_m$ are the angles between N and the coordinate axes. As we proved in Proposition 2.4.6, the area of the projection of C to the subspace orthogonal to the ith coordinate axis is equal to $|\cos\theta_i|\lambda_{m-1}(C)$. Since $\sum_{i=1}^m \cos\theta_i^2 = \|N\|^2 = 1$, multiplying this equation by $\lambda_{m-1}^2(C)$ yields formula (3) in the case under consideration.

We leave the reader to check that formula (3) is valid not only for a parallelepiped, but also for any measurable set lying in an n-dimensional subspace of $\mathbb{R}^m$.

2.5.4 Using the geometric interpretation of the Gram determinant, we can generalize a well-known fact from elementary geometry: the volume of a parallelepiped is the area of its base multiplied by the height.

Consider a parallelepiped $P = P(v_1, \dots, v_m)$ and write v_m in the form $v_m = y + z$, where y is the projection of v_m to the subspace spanned by $v_1, \dots, v_{m-1}$ and z (the "height" of P) is perpendicular to $v_1, \dots, v_{m-1}$. It is natural to say that the parallelepiped $P(v_1, \dots, v_{m-1})$ (of dimension $m - 1$) is the base of P. Since $y = c_1v_1 + \dots + c_{m-1}v_{m-1}$, multiplying the rows of the Gram matrix with indices $1, \dots, m - 1$ by the numbers $c_1, \dots, c_{m-1}$ and subtracting them from the last row, we see that $\Gamma(v_1, \dots, v_m)$ is equal to the determinant of the matrix

$$\begin{pmatrix} \langle v_1, v_1\rangle & \langle v_1, v_2\rangle & \dots & \langle v_1, v_m\rangle \\ \langle v_2, v_1\rangle & \langle v_2, v_2\rangle & \dots & \langle v_2, v_m\rangle \\ \vdots & \vdots & \ddots & \vdots \\ \langle z, v_1\rangle & \langle z, v_2\rangle & \dots & \langle z, v_m\rangle \end{pmatrix}.$$

We have $\langle z, v_j\rangle = 0$ for $j = 1, \dots, m - 1$ and $\langle z, v_m\rangle = \langle z, z\rangle = \|z\|^2$, whence

$$\Gamma(v_1, \dots, v_m) = \Gamma(v_1, \dots, v_{m-1})\|z\|^2.$$

According to the geometric interpretation of the Gram determinant, this means that the volume of the m-dimensional parallelepiped $P(v_1, \dots, v_m)$ is equal, just as in the three-dimensional case familiar to the reader, to the $(m - 1)$-dimensional volume of its base multiplied by the height:

$$\lambda_m\big(P(v_1, \dots, v_m)\big) = \lambda_{m-1}\big(P(v_1, \dots, v_{m-1})\big) \cdot \|z\|. \tag{4}$$

Let us obtain a natural and important bound on the volume of a parallelepiped P, which is an easy corollary of (4). Since, by the Pythagorean theorem, $\|v_m\|^2 = \|y\|^2 + \|z\|^2 \geqslant \|z\|^2$, it follows from (4) that

$$\lambda_m\big(P(v_1, \dots, v_m)\big) \leqslant \lambda_{m-1}\big(P(v_1, \dots, v_{m-1})\big) \cdot \|v_m\|.$$

Repeating this estimate, we obtain the important *Hadamard*[8] *inequality*:

$$\lambda_m\bigl(P(v_1,\dots,v_m)\bigr) \leqslant \|v_1\|\cdots\|v_m\|. \tag{5}$$

In other words, the volume of the parallelepiped with edges $v_1,\dots,v_m$ does not exceed the product of their lengths. Clearly, this bound is sharp: a parallelepiped with edges of given lengths has the largest volume if its edges are pairwise perpendicular.

We can write the Hadamard inequality in purely analytic terms, without involving the notion of volume. Let A be an arbitrary $m\times m$ matrix, and let a_k be its kth column ($k=1,\dots,m$). Then

$$\bigl|\det(A)\bigr| \leqslant \|a_1\|\cdots\|a_m\|.$$

This inequality is also called the Hadamard inequality. It follows from inequality (2) applied to the map V corresponding to the matrix A and inequality (5):

$$\bigl|\det(A)\bigr| = \lambda_m\bigl(P(a_1,\dots,a_m)\bigr) \leqslant \|a_1\|\cdots\|a_m\|.$$

2.5.5 In conclusion of this section, we consider an interesting geometric problem related to *convex bodies*, i.e., convex compact sets with a non-empty interior. Considerable information about a convex body can be obtained if we know an ellipsoid of maximal volume contained in it (by an ellipsoid we mean an affine image of a closed ball). The problem of the existence and uniqueness of such an ellipsoid is solved by the following theorem.

Theorem *Among the ellipsoids contained in a convex body $K\subset\mathbb{R}^m$, there exists a unique ellipsoid of maximal volume.*

Proof Let us first verify that such an ellipsoid exists. If a sequence of ellipsoids $E_n\subset K$ is such that

$$\lambda(E_n) \underset{n\to\infty}{\longrightarrow} V = \sup\bigl\{\lambda(E)\mid E\subset K,\ E \text{ is an ellipsoid}\bigr\},$$

then, passing if necessary to a subsequence, we may assume that both the centers c_n of these ellipsoids and the vectors $v_i^{(n)}$ ($i=1,\dots,m$) corresponding to their semi-axes have limits: $c_n\to c$ and $v_1^{(n)}\to v_1,\dots,v_m^{(n)}\to v_m$ as $n\to\infty$. It follows that K contains the ellipsoid $\mathcal{E}$ with center c and semi-axes $v_1,\dots,v_m$. As we have noted in Sect. 2.5.2, its volume is equal to (hereafter α_m is the volume of the unit ball)

$$\lambda(\mathcal{E}) = \alpha_m\,\|v_1\|\cdots\|v_m\| = \alpha_m \lim_{n\to\infty}\bigl\|v_1^{(n)}\bigr\|\cdots\bigl\|v_m^{(n)}\bigr\| = \lim_{n\to\infty}\lambda(E_n) = V.$$

Hence $\mathcal{E}$ is an ellipsoid of maximal volume for K.

[8] Jacques Salomon Hadamard (1865–1963)—French mathematician.

Now assume that there exist two ellipsoids of maximal volume. Since an affine transformation sends an ellipsoid to an ellipsoid and preserves the ratio of volumes, we may assume without loss of generality that one of the ellipsoids coincides with the unit ball $\overline{B}$ centered at the origin and the semi-axes of the other ellipsoid (denoted by $\mathcal{E}$) are parallel to the coordinate axes. Let c be the center of $\mathcal{E}$ and $a_1, \dots, a_m$ be the lengths of its semi-axes. Then $y \in \mathcal{E}$ if and only if y can be written in the form $y = c + (a_1 x_1, \dots, a_m x_m)$, where $x = (x_1, \dots, x_m) \in \overline{B}$.

Consider the new ellipsoid E with center $\frac{c}{2}$ and semi-axes (parallel to the coordinate axes) of lengths $\frac{1+a_1}{2}, \dots, \frac{1+a_m}{2}$. Each point z of E can be written in the form $z = \frac{1}{2}c + (\frac{1+a_1}{2}x_1, \dots, \frac{1+a_m}{2}x_m)$, where $x = (x_1, \dots, x_m) \in \overline{B}$. Hence $z = \frac{x+y}{2}$, where $y = c + (a_1 x_1, \dots, a_m x_m) \in \mathcal{E}$. Thus $E \subset \frac{1}{2}\overline{B} + \frac{1}{2}\mathcal{E} \subset K$. At the same time,

$$\alpha_m = \lambda_m(\overline{B}) \geqslant \lambda(E) = \alpha_m \prod_{i=1}^{m} \frac{1+a_i}{2} \geqslant \alpha_m \prod_{i=1}^{m} \sqrt{a_i} = \alpha_m$$

(the product $a_1 \cdots a_m$ is equal to 1, because $\alpha_m = \lambda(\mathcal{E}) = \alpha_m \, a_1 \cdots a_m$). Since the outer terms of the last inequality coincide, it is an equality. Hence $\frac{1+a_i}{2} = \sqrt{a_i}$ and, consequently, $a_i = 1$ for all i. Thus $\mathcal{E}$ is a unit ball. If it does not coincide with $\overline{B}$, then, as the reader can easily verify, the convex hull of these balls contains an ellipsoid of revolution (obtained by rotating about the axis passing through their centers) whose volume is greater than α_m, a contradiction. □

This theorem makes it possible to prove that for a "sufficiently symmetric" body, the ellipsoid of maximal volume is a ball. This is the case, for example, for the cube, for the octahedron determined by the inequality $\sum_{i=1}^{m} |x_i| \leqslant 1$, and for the regular simplex.

It turns out that the ellipsoid of maximal volume occupies a sufficiently large part of a convex body. The following theorem holds.

Theorem (John[9]) *Let $\mathcal{E}$ be the ellipsoid of maximal volume for a convex body $K \subset \mathbb{R}^m$. Then:*

(1) *if the center of $\mathcal{E}$ is at the origin, then $K \subset m\mathcal{E}$;*
(2) *if the body K is centrally symmetric, then $K \subset \sqrt{m}\mathcal{E}$.*

Considering a simplex and a cube shows that the inequalities in the theorem are sharp.

Proof (1) As in the proof of the previous theorem, we may assume that $\mathcal{E}$ coincides with the unit ball. To prove the inclusion $K \subset m\mathcal{E}$, assume to the contrary that $\|x\| > m$ for some point $x \in K$. We may assume that $x = (c, 0, \dots, 0)$ with $c > m$. Let T be the convex hull of the ball $\mathcal{E}$ and the point x. Obviously, $T \subset K$. Take a

[9]Fritz John (1910–1994)—German mathematician.

small number $\varepsilon \geqslant 0$ and consider the ellipse $\frac{(x_1-\varepsilon)^2}{(1+\varepsilon)^2} + \frac{x_2^2}{b^2} \leqslant 1$ in the plane OX_1X_2 with $b^2 = \frac{c-1-2\varepsilon}{c-1}$. We leave the reader to check that this ellipse is contained in the section of T by the plane OX_1X_2. Hence the ellipsoid $E(\varepsilon)$ obtained by rotating the constructed ellipse about the axis OX_1 is contained in T and, consequently, in K. Its first semi-axis has length $1+\varepsilon$, and the other semi-axes have length b. The volume $V(\varepsilon) = \lambda(E(\varepsilon))$ can be computed by the formula

$$V(\varepsilon) = \alpha_m(1+\varepsilon)b^{m-1} = \alpha_m(1+\varepsilon)\left(\frac{c-1-2\varepsilon}{c-1}\right)^{\frac{m-1}{2}}.$$

Clearly, $V(0) = \alpha_m$ and $V'(0) = \alpha_m \frac{c-m}{c-1} > 0$. Hence for $\varepsilon > 0$ close to zero, $\lambda(E(\varepsilon)) = V(\varepsilon) > \alpha_m$, but $E(\varepsilon) \subset K$. This contradicts the fact that the ellipsoid of maximal volume for K is the unit ball.

(2) If the body K is centrally symmetric with respect to the origin, then the same is true for the ellipsoid of maximal volume $\mathcal{E}$. Indeed, since the "reflected" ellipsoid $-\mathcal{E}$ is contained in K, it follows from the uniqueness of the ellipsoid of maximal volume that $\mathcal{E} = -\mathcal{E}$, i.e., the center of $\mathcal{E}$ coincides with the origin.

The remaining part of the proof for the case of a centrally symmetric body is similar to the above arguments. Again assuming that $\mathcal{E}$ is the unit ball, we now define a body T as the convex hull of the ball $\mathcal{E}$ and the points $\pm(c, 0, \dots, 0)$ for $c > \sqrt{m}$, and consider the ellipse $\frac{x_1^2}{(1+\varepsilon)^2} + \frac{x_2^2}{b^2} \geqslant 1$ with $b^2 = \frac{c^2-(1+\varepsilon)^2}{c^2-1}$ inscribed into the two-dimensional section of T by the plane OX_1X_2. We leave the reader to complete the proof. □

EXERCISES

1. Let us regard $\mathbb{R}^2$ as the set of complex numbers. How does the Lebesgue measure change under the transformation $z \mapsto az$, where a is a fixed complex number?
2. Let $A, B : \mathbb{R}^m \to \mathbb{R}^m$ be linear maps. Show that if $\|A(x)\| \leqslant \|B(x)\|$ for every $x \in \mathbb{R}^m$, then $\lambda_m(A(E)) \leqslant \lambda_m(B(E))$ for every measurable set E.
3. Let $E \subset \mathbb{R}_+$ and $S = \{x \in \mathbb{R}^m \mid \|x\| \in E\}$. Show that these sets are either both Lebesgue measurable or both non-measurable. Show that each of the equalities $\lambda_1(E) = 0$ and $\lambda_m(S) = 0$ implies the other.

2.6 *Hausdorff Measures

Here we will construct a family of measures μ_p $(p > 0)$ generalizing the Lebesgue measure. For $p = m$, the measure μ_p in $\mathbb{R}^m$ will be proportional to λ_m, and for $p = 1, 2, \dots, m-1$, we will obtain generalizations of the Lebesgue measures defined so far only on (measurable) subsets of p-dimensional subspaces.

The construction of the measures μ_p is based on an important geometric characteristic of a set, its diameter. Recall that the diameter of a set E is the value

$$\operatorname{diam}(E) = \sup\bigl\{\|x-y\| \mid x, y \in E\bigr\}.$$

The diameter of the empty set is assumed to be zero.

2.6.1 Let $\varepsilon > 0$. A family of sets $\{e_\alpha\}_{\alpha\in A}$ is called an *ε-cover* of a set $E \subset \mathbb{R}^m$ if

$$E \subset \bigcup_{\alpha\in A} e_\alpha \quad \text{and} \quad \operatorname{diam}(e_\alpha) \leqslant \varepsilon \quad \text{for every } \alpha \in A.$$

In what follows, we will need only covers that are at most countable, so hereafter we assume that the set A is countable without stating this explicitly. We may assume without loss of generality that $A = \mathbb{N}$. We do this in most cases, but sometimes it is convenient to use other sets of indices. It is clear that for every $\varepsilon > 0$, the space $\mathbb{R}^m$ and, consequently, every subset of $\mathbb{R}^m$, has an ε-cover.

For arbitrary $p > 0$ and $E \subset \mathbb{R}^m$, set

$$\mu_p(E,\varepsilon) = \inf\left\{\sum_{j=1}^{\infty}\left(\frac{\operatorname{diam}(e_j)}{2}\right)^p \,\middle|\, \{e_j\}_{j\geqslant 1} \text{ is an } \varepsilon\text{-cover of } E\right\}.$$

Obviously, the function $\varepsilon \mapsto \mu_p(E,\varepsilon)$ (which may take infinite values) is decreasing, and hence the limit

$$\lim_{\varepsilon\to+0} \mu_p(E,\varepsilon) = \sup_{\varepsilon>0} \mu_p(A,\varepsilon)$$

exists.

Definition The function

$$E \mapsto \mu_p^*(E) = \lim_{\varepsilon\to+0} \mu_p(E,\varepsilon),$$

defined on all subsets of $\mathbb{R}^m$, is called the *p-dimensional outer Hausdorff*[10] *measure*.

We will soon see that μ_p^* is indeed an outer measure in the sense of Definition 1.4.2.

Note also that, interpreting the space $\mathbb{R}^m$ as a subspace of $\mathbb{R}^n$ $(n > m)$, we may regard every set E contained in $\mathbb{R}^m$ as a subset of $\mathbb{R}^n$. The diameters of a set computed in the spaces $\mathbb{R}^m$ and $\mathbb{R}^n$, obviously, coincide, so that the value $\mu_p^*(E)$ does not depend on the ambient space. Thus, speaking about the outer Hausdorff measure of a set E, we may, and shall, omit reference to the space in which we regard it to be embedded. When it is necessary to specify the domain of the function μ_p^*, we mention it explicitly.

In this connection, note that for subsets of the space $\mathbb{R}^m$, the outer measures μ_p^* are of interest only for $p \leqslant m$, since otherwise $\mu_p^* \equiv 0$ (see the end of Sect. 2.6.6).

[10] Felix Hausdorff (1868–1942)—German mathematician.

2.6.2 Let us establish the basic properties of the function μ_p^*.

(1) $0 \leqslant \mu_p^*(E) \leqslant +\infty$, $\mu_p^*(\varnothing) = 0$.
(2) Monotonicity: *if* $E \subset F$, *then* $\mu_p^*(E) \leqslant \mu_p^*(F)$.

These properties are obvious.

(3) μ_p^* *is an outer measure*: *if* $E \subset \bigcup_{n=1}^{\infty} E_n$, *then* $\mu_p^*(E) \leqslant \sum_{n=1}^{\infty} \mu_p^*(E_n)$.

Proof We will assume that $\sum_{n=1}^{\infty} \mu_p^*(E_n) < +\infty$, since otherwise the inequality in question is trivial. Fix a number $\varepsilon > 0$ and consider ε-covers $\{e_j^{(n)}\}_{j \geqslant 1}$ of the sets E_n such that

$$\sum_{j=1}^{\infty} \left(\frac{\operatorname{diam}(e_j^{(n)})}{2} \right)^p < \mu_p(E_n, \varepsilon) + \frac{\varepsilon}{2^n} \quad (n = 1, 2, \ldots).$$

Obviously, the family $\{e_j^{(n)}\}_{n,j \geqslant 1}$ is an ε-cover of E, and, therefore,

$$\mu_p(E, \varepsilon) \leqslant \sum_{n,j=1}^{\infty} \left(\frac{\operatorname{diam}(e_j^{(n)})}{2} \right)^p \leqslant \sum_{n=1}^{\infty} \left(\mu_p(E_n, \varepsilon) + \frac{\varepsilon}{2^n} \right) \leqslant \sum_{n=1}^{\infty} \mu_p^*(E_n) + \varepsilon.$$

Passing to the limit as $\varepsilon \to 0$, we obtain the desired result. □

On sets that are sufficiently far from each other, the function μ_p^* is additive. More precisely, sets E and F are called *separated* if

$$\inf\{ \|x - y\| \mid x \in E,\ y \in F \} > 0.$$

(4) *For separated sets*, $\mu_p^*(E \vee F) = \mu_p^*(E) + \mu_p^*(F)$.

Proof Since $\mu_p^*(E \vee F) \leqslant \mu_p^*(E) + \mu_p^*(F)$ by the subadditivity of μ_p^*, we only need to prove the reverse inequality.

Let $0 < \varepsilon < \inf\{\|x - y\| \mid x \in E,\ y \in F\}$. Consider an arbitrary ε-cover $\{e_j\}_{j \geqslant 1}$ of the set $E \vee F$. By the choice of ε, for every index j at least one of the intersections $e_j \cap E$, $e_j \cap F$ is empty, whence

$$\sum_{j=1}^{\infty} \left(\frac{\operatorname{diam}(e_j)}{2} \right)^p \geqslant \sum_{e_j \cap E \neq \varnothing} \left(\frac{\operatorname{diam}(e_j)}{2} \right)^p + \sum_{e_j \cap F \neq \varnothing} \left(\frac{\operatorname{diam}(e_j)}{2} \right)^p.$$

Since the families $\{e_j\}_{e_j \cap E \neq \varnothing}$ and $\{e_j\}_{e_j \cap F \neq \varnothing}$ are ε-covers of the sets E and F, respectively, we have

$$\sum_{j=1}^{\infty} \left(\frac{\operatorname{diam}(e_j)}{2} \right)^p \geqslant \mu_p(E, \varepsilon) + \mu_p(F, \varepsilon).$$

Taking the lower boundary of the left-hand side over all ε-covers, we see that $\mu_p(E \vee F, \varepsilon) \geqslant \mu_p(E, \varepsilon) + \mu_p(F, \varepsilon)$. To complete the proof, it suffices to let $\varepsilon \to 0$. □

(5) *Let $E \subset \mathbb{R}^m$, and let $\Phi : E \to \mathbb{R}^n$ be a map satisfying the Lipschitz condition*:

$$\|\Phi(x) - \Phi(y)\| \leqslant L\|x - y\| \quad \text{for } x, y \in E,$$

where L is a constant. Then

$$\mu_p^*(\Phi(E)) \leqslant L^p \mu_p^*(E).$$

In particular, $\mu_p^*(\Phi(E)) = 0$ if $\mu_p^*(E) = 0$.

Proof Let $\mu_p^*(E) < +\infty$, and let $\{e_j\}_{j \geqslant 1}$ be an ε-cover of E such that

$$\sum_{j=1}^{\infty} \left(\frac{\operatorname{diam}(e_j)}{2}\right)^p < \mu_p(E, \varepsilon) + \varepsilon.$$

We will assume that $e_j \subset E$ for all j (otherwise replace e_j by $e_j \cap E$). Since $\operatorname{diam}(\Phi(e_j)) \leqslant L \operatorname{diam}(e_j)$, the sets $\Phi(e_j)$ form an $L\varepsilon$-cover of the set $\Phi(E)$, whence

$$\mu_p\big(\Phi(E), L\varepsilon\big) \leqslant \sum_{j=1}^{\infty} \left(\frac{\operatorname{diam}(\Phi(e_j))}{2}\right)^p$$

$$\leqslant L^p \sum_{j=1}^{\infty} \left(\frac{\operatorname{diam}(e_j)}{2}\right)^p < L^p\big(\mu_p(E, \varepsilon) + \varepsilon\big).$$

Passing to the limit as $\varepsilon \to 0$, we obtain the desired inequality. □

Remark For $\mu_p^*(E) = 0$, the equality $\mu_p^*(\Phi(E)) = 0$ can be obtained under less restrictive assumptions on the map Φ. It suffices to require that it is only locally Lipschitz (this condition is satisfied, in particular, for maps that are smooth in a neighborhood of E). Then one should split E into countably many parts on which Φ satisfies the Lipschitz condition (with a separate constant for each part), apply the obtained result to each of them, and then use the countable subadditivity of μ_p^*.

To formulate the next property, we introduce two important classes of continuous maps.

Definition Let $E \subset \mathbb{R}^m$. We say that a map $\Theta : E \to \mathbb{R}^n$ is a *weak contraction* of E if $\|\Theta(x) - \Theta(y)\| \leqslant \|x - y\|$ for all x, y in E.

We say that a continuous map $\Theta : E \to \mathbb{R}^n$ is *expanding* on E if $\|\Theta(x) - \Theta(y)\| \geqslant \|x - y\|$ for all x, y from E.

In other words, a weak contraction is a map that satisfies the Lipschitz condition with Lipschitz constant 1. It is not necessarily invertible. However, an expanding map is invertible, and its inverse is a weak contraction. In particular, any expanding map is a homeomorphism. We emphasize that the image of a Borel set under an expanding map is again a Borel set (this is a direct corollary of the proposition from Sect. 2.3.3).

(6) If Θ *is a weak contraction of a set* E, *then* $\mu_p^*(\Theta(E)) \leqslant \mu_p^*(E)$. *For an expanding map, the reverse inequality holds.*

This follows immediately from Property (5).

(7) *If a map* Φ *preserves the distances between points of a set* E, *then* $\mu_p^*(\Phi(E)) = \mu_p^*(E)$. *In particular, the outer Hausdorff measure is invariant under translations and orthogonal transformations.*

The next result follows from Property (5).

(8) *The outer Hausdorff measures of similar sets are proportional. More precisely,*

$$\mu_p^*(a\,E) = |a|^p\,\mu_p^*(E) \quad \textit{where} \quad a\,E = \{ax \mid x \in E\} \quad (a \in \mathbb{R}).$$

2.6.3 As we know (see Sect. 1.4.3), every outer measure generates a measure on the σ-algebra of measurable sets. The measure obtained by restricting the outer measure μ_p^* to the σ-algebra of measurable (i.e., μ_p^*-measurable) sets is called the *Hausdorff measure* and is denoted by μ_p. Which sets are measurable with respect to this measure? The theorem below provides a wide class of such sets. In its proof it is convenient to use the simple and important geometric notion of the ε-neighborhood of a set.

Definition Let $\varepsilon > 0$ and $E \subset \mathbb{R}^m$. The set E_ε formed by the points that lie at distance at most ε from E is called the *ε-neighborhood* of E:

$$E_\varepsilon = \bigcup_{x \in E} B(x, \varepsilon).$$

Obviously, E_ε are open sets that grow with ε: $\overline{E_\varepsilon} \subset E_\delta$ if $0 < \varepsilon < \delta$. Note also that

$$(\overline{E}\,)_\varepsilon = E_\varepsilon, \qquad (E_\varepsilon)_\delta = E_{\varepsilon+\delta} \quad \text{for any } \varepsilon > 0, \delta > 0, \quad \text{and} \quad \bigcap_{\varepsilon > 0} E_\varepsilon = \overline{E}.$$

All these equalities are easy to verify.

Theorem *Borel sets are* μ_p^**-measurable.*

Proof Since the measurable sets form a σ-algebra, it suffices to check that any closed set F is measurable. By the definition of measurability, we must check that

$\mu_p^*(E) = \mu_p^*(E \cap F) + \mu_p^*(E \setminus F)$ for every set $E \subset \mathbb{R}^m$. By the subadditivity, $\mu_p^*(E) \leqslant \mu_p^*(E \cap F) + \mu_p^*(E \setminus F)$, so it remains to show that

$$\mu_p^*(E) \geqslant \mu_p^*(E \cap F) + \mu_p^*(E \setminus F). \tag{1}$$

When proving this inequality, we may assume that $\mu_p^*(E) < +\infty$.

Let $\varepsilon > 0$, and let F_ε be the ε-neighborhood of F. Put $A_n = E \setminus F_{1/n}$. Then the sets A_n and $E \cap F$ are obviously separated, and, by Property (4),

$$\mu_p^*(E) \geqslant \mu_p^*\big((E \cap F) \cup A_n\big) = \mu_p^*(E \cap F) + \mu_p^*(A_n).$$

To obtain (1) by passing to the limit in this inequality, we should check that

$$\mu_p^*(A_n) \to \mu_p^*(E \setminus F) \quad \text{as } n \to \infty. \tag{2}$$

Since F is closed, $\bigcap_{\varepsilon>0} F_\varepsilon = F$, whence $E \setminus F = \bigcup_{n\geqslant 1} A_n$. Set $B_j = A_{j+1} \setminus A_j$. Now $E \setminus F = A_n \vee \bigvee_{j\geqslant n} B_j$ and, since μ_p^* is monotone and countably subadditive,

$$\mu_p^*(A_n) \leqslant \mu_p^*(E \setminus F) \leqslant \mu_p^*(A_n) + \sum_{j\geqslant n} \mu_p^*(B_j) \quad \text{(for every } n \in \mathbb{N}\text{)}.$$

Hence if the series

$$\sum_{j\geqslant 1} \mu_p^*(B_j) \tag{3}$$

converges, then the difference $\mu_p^*(E \setminus F) - \mu_p^*(A_n)$ can be bounded by the remainder of a convergent series, which implies (2). To prove that the series (3) converges, we use the fact that the sets B_k and B_l are separated for $|k - l| > 1$ (which is left to the reader to check). It follows that for every N

$$\sum_{j=1}^{N} \mu_p^*(B_{2j}) = \mu_p^*\left(\bigvee_{j=1}^{N} B_{2j}\right) \leqslant \mu_p^*(E) < +\infty.$$

Hence the series $\sum_{j=1}^{\infty} \mu_p^*(B_{2j})$ converges. In a similar way we verify that the series $\sum_{j=1}^{\infty} \mu_p^*(B_{2j+1})$ converges. This ensures the convergence of the series (3), which, as we have already observed, suffices to complete the proof of the theorem. □

We complement the obtained result with an assertion showing that any set (not necessarily measurable) is contained in a Borel set of the same Hausdorff measure. For the Lebesgue measure, we have already met a similar result (for a measurable set) at the end of Sect. 2.2.2.

Proposition *For every set E, $E \subset \mathbb{R}^m$, there exists a Borel set C such that $E \subset C$ and $\mu_p^*(E) = \mu_p(C)$.*

Proof We will assume that $\mu_p^*(E) < +\infty$ (otherwise we can take $\mathbb{R}^m$ as C). For every $n \in \mathbb{N}$, find a $\frac{1}{n}$-cover $\{e_j^{(n)}\}_{j=1}^\infty$ of E such that

$$\sum_{j=1}^\infty \left(\frac{\operatorname{diam}(e_j^{(n)})}{2}\right)^p < \mu_p\left(E, \frac{1}{n}\right) + \frac{1}{n},$$

and let $C_n = \bigcup_{j=1}^\infty \overline{e_j^{(n)}}$. Since the diameter of a set coincides with the diameter of its closure,

$$\mu_p\left(C_n, \frac{1}{n}\right) \leqslant \sum_{j=1}^\infty \left(\frac{\operatorname{diam}(e_j^{(n)})}{2}\right)^p < \mu_p\left(E, \frac{1}{n}\right) + \frac{1}{n}.$$

It is clear that the Borel set $C = \bigcap_{n=1}^\infty C_n$ contains E, and for every n,

$$\mu_p\left(C, \frac{1}{n}\right) \leqslant \mu_p\left(C_n, \frac{1}{n}\right) \leqslant \mu_p\left(E, \frac{1}{n}\right) + \frac{1}{n} \leqslant \mu_p\left(C, \frac{1}{n}\right) + \frac{1}{n}.$$

Passing to the limit as $n \to \infty$, we see that $\mu_p^*(E) = \mu_p(C)$. □

2.6.4 Now let us show that in the case $p = m$ the Hausdorff measure essentially coincides with the m-dimensional Lebesgue measure. We will need the following easy estimate.

Lemma *If $Q = [0, 1]^m$ is the unit cube, then $0 < \mu_m^*(Q) < +\infty$.*

Proof To verify the left inequality, observe that every set e is contained in a closed ball of radius $\operatorname{diam}(e)$. Hence every cover $\{e_j\}_{j\geqslant 1}$ of the cube Q generates a cover of Q by closed balls B_j of radii $r_j = \operatorname{diam}(e_j)$. By the countable subadditivity of the Lebesgue measure, we have

$$1 = \lambda_m(Q) \leqslant \sum_{j=1}^\infty \lambda_m(B_j) = \sum_{j=1}^m \alpha_m r_j^m = 2^m \alpha_m \sum_{j=1}^\infty \left(\frac{\operatorname{diam}(e_j)}{2}\right)^m,$$

where $\alpha_m = \lambda_m(B(0, 1))$. Hence

$$\frac{1}{2^m \alpha_m} \leqslant \sum_{j=1}^\infty \left(\frac{\operatorname{diam}(e_j)}{2}\right)^m$$

for an arbitrary ε-cover $\{e_j\}_{j\geqslant 1}$ of the cube Q. Therefore, $\mu_m(Q, \varepsilon) \geqslant 2^{-m}/\alpha_m$, whence $\mu_m(Q) = \sup_{\varepsilon>0} \mu_m(Q, \varepsilon) \geqslant 2^{-m}/\alpha_m$.

To prove the right inequality, split the cube Q into N^m congruent cubes Q_j. The diameter of each of them is equal to $\frac{\sqrt{m}}{N}$. Hence they form a $\frac{\sqrt{m}}{N}$-cover of the cube Q. Then

$$\mu_m\left(Q,\frac{\sqrt{m}}{N}\right)\leqslant\sum_{j=1}^{N^m}\left(\frac{\operatorname{diam}(Q_j)}{2}\right)^m=N^m\left(\frac{\sqrt{m}}{2N}\right)^m=2^{-m}m^{\frac{m}{2}}.$$

Therefore,

$$\mu_m^*(Q)=\lim_{N\to\infty}\mu_m\left(Q,\frac{\sqrt{m}}{N}\right)\leqslant 2^{-m}m^{\frac{m}{2}}<+\infty. \qquad \square$$

Theorem *The Hausdorff measure μ_m is proportional to the Lebesgue measure λ_m.*

Proof Let $\mathfrak{A}^m$ be the σ-algebra of Lebesgue measurable subsets of $\mathbb{R}^m$ and $\widetilde{\mathfrak{A}}^m$ be the σ-algebra of μ_m^*-measurable sets.

Both measures λ_m and μ_m are translation-invariant, and $\mu_m([0,1]^m)<+\infty$ by the lemma. By Theorem 2.4.2 and the remark after it, these measures are proportional at least on the σ-algebra of Borel sets, and the proportionality coefficient is positive because $\mu_m([0,1]^m)>0$. It follows that on Borel sets they vanish or do not vanish simultaneously. Since both measures are complete, Proposition 2.6.3 implies that $\mathfrak{A}^m=\widetilde{\mathfrak{A}}^m$, and on this σ-algebra the measures are proportional. $\square$

It easily follows from this theorem that for $k=1,2,\ldots,m-1$, a similar result holds for the restrictions of the Hausdorff measure μ_k to k-dimensional affine subspaces.

Later, in Chap. 6, we will derive a precise formula that shows how the Lebesgue measure changes under a diffeomorphic transformation. Now we only mention a qualitative result following from the theorem and Property (6) from Sect. 2.6.2.

Corollary *The outer Lebesgue measure does not increase under weak contractions and does not decrease under expanding maps.*

One should be careful when considering the problem of whether the image of a measurable set under an expanding map is measurable. Of course, this is only a problem in the case of a non-smooth expanding map. The inverse of an expanding map, which is Lipschitz, preserves Lebesgue measurability (see Sect. 2.3.1). But the map itself does not necessarily have this property: it can expand a set of zero measure too much (see Exercise 5 in Sect. 2.3). At the same time, as we have already observed, the narrower class of Borel sets is preserved under expanding maps.

2.6.5 As we have proved in Theorem 2.6.4, the measures λ_m and μ_m are proportional. The computation of the proportionality coefficient is based on two geometric results, which are of independent interest.

Lemma (On exhaustion by balls) *Every non-empty open subset G of the space $\mathbb{R}^m$ can be written as the union of a sequence of pairwise disjoint balls B_n and a set of* *zero measure e*:

$$G = e \vee \bigvee_{n=1}^{\infty} B_n.$$

The diameters of the balls may be chosen arbitrarily small.

Proof The proof will be divided into two steps. First we show that *in every bounded open set G, $G \neq \varnothing$, one can find pairwise disjoint balls $B_1, \dots, B_N$ such that*

$$\lambda_m\big(G \setminus (B_1 \vee \dots \vee B_N)\big) < \theta \lambda_m(G)$$

(*the coefficient $\theta = \theta_m \in (0, 1)$ depends only on the dimension of the space*).

Let us split the set G into cubic cells Q_n with rational vertices (see Sect. 1.1.7). Since they can be further split into smaller parts, we may assume that the diameters of these cells are arbitrarily small. Since $\lambda_m(G) < +\infty$, for sufficiently large N we have

$$\lambda_m(G) = \sum_{n=1}^{\infty} \lambda_m(Q_n) < 2 \sum_{n=1}^{N} \lambda_m(Q_n).$$

Let B_n be the open ball inscribed into the cell Q_n (the centers of B_n and Q_n coincide, and the radius r_n of the ball is equal to half the length of the edge). The volume of the ball constitutes a fraction of the volume of the cell that depends only on the dimension:

$$\lambda_m(B_n) = \alpha_m r_n^m = \frac{\alpha_m}{2^m} \lambda_m(Q_n) \equiv \widetilde{\alpha}_m \lambda_m(Q_n),$$

where $\alpha_m = \lambda_m(B(0, 1))$. Hence

$$\sum_{n=1}^{N} \lambda_m(B_n) = \widetilde{\alpha}_m \sum_{n=1}^{N} \lambda_m(Q_n) > \frac{\widetilde{\alpha}_m}{2} \lambda_m(G).$$

Therefore,

$$\lambda_m\big(G \setminus (B_1 \vee \dots \vee B_N)\big) = \lambda_m(G) - \sum_{n=1}^{N} \lambda_m(B_n) < \lambda_m(G) - \frac{\widetilde{\alpha}_m}{2} \lambda_m(G).$$

Thus we may set $\theta = 1 - \widetilde{\alpha}_m/2$.

Let us proceed to the second step of the proof, first assuming that the set G is bounded. As we have just seen, we can remove from G a finite collection of pairwise disjoint balls $B_1, \dots, B_{N_1}$ so that the measure of the remaining set is less than $\theta \lambda_m(G)$. Removing from G the closures of these balls, we obtain an open set $G_1 \subset G$ with $\lambda_m(G_1) < \theta \lambda_m(G)$. Now we can repeat this construction with G_1, finding a finite collection of pairwise disjoint balls $B_{N_1+1}, \dots, B_{N_2}$ such that the measure of the remaining part of the set G_1 is less than $\theta \lambda_m(G_1)$. Removing from G_1 the closures of these balls, we obtain an open set $G_2 \subset G_1$ with $\lambda_m(G_2) <$

$\theta\lambda_m(G_1) < \theta^2\lambda_m(G)$. Continuing by induction, we construct a sequence of pairwise disjoint balls B_n, $B_n \subset G$, and a sequence of nested open sets G_j, $G \supset G_1 \supset G_2 \supset \dots$, such that

$$G \setminus \bigcup_{n\geqslant 1} \overline{B}_n \subset G_j \quad \text{and} \quad \lambda_m(G_j) < \theta^j \lambda_m(G) \quad \text{for every } j.$$

It remains to observe that the set $e = G \setminus \bigvee_{n\geqslant 1} B_n$ has zero measure, since it is contained in the union of the sets $\bigcap_{j\geqslant 1} G_j$ and $\bigcup_{n\geqslant 1} \partial \overline{B}_n$.

If the set G is not bounded, it can be written as the union of a set of zero measure and a sequence of pairwise disjoint bounded open sets. We will obtain the desired decomposition applying the assertion already proved to each of these parts. □

Another proof of this lemma can be obtained from the Vitali theorem (see Corollary 2 in Sect. 2.7.3).

We will also need another geometric fact. Namely, the so-called isodiametric inequality, which can be stated as follows (see Sect. 2.8.3):

Among all compact sets of a given diameter, the ball has the largest volume.

Now we can find the proportionality coefficient between the measures λ_m and μ_m.

Proposition $\lambda_m = \alpha_m \mu_m$.

Proof It suffices to establish the equality $\lambda_m(E) = \alpha_m \mu_m(E)$ for at least one set of positive finite measure.

Let $\{e_j\}_{j\geqslant 1}$ be an ε-cover of a non-empty open bounded subset G in $\mathbb{R}^m$. Note that, by the isodiametric inequality, $\lambda_m(\overline{e}_j) \leqslant \alpha_m (\frac{\operatorname{diam}(e_j)}{2})^m$. Hence

$$\lambda_m(G) \leqslant \sum_{j=1}^{\infty} \lambda_m(\overline{e}_j) \leqslant \sum_{j=1}^{\infty} \alpha_m \left(\frac{\operatorname{diam}(e_j)}{2}\right)^m = \alpha_m \sum_{j=1}^{\infty} \left(\frac{\operatorname{diam}(e_j)}{2}\right)^m.$$

Taking the lower boundary of the right-hand side over all ε-covers, and then passing to the limit in ε, we obtain

$$\lambda_m(G) \leqslant \alpha_m \mu_m(G). \tag{4}$$

On the other hand, by the lemma, the set G can be written as the union of a sequence of pairwise disjoint balls $B_j = B(x_j, r_j)$ and a set e of zero Lebesgue measure. Then

$$\lambda_m(G) = \sum_{j=1}^{\infty} \lambda_m(B_j) = \alpha_m \sum_{j=1}^{\infty} r_j^m.$$

The radii of the balls may be chosen arbitrarily small. We will assume that all of them are less than ε.

Since $\mu_m(e) = \lambda_m(e) = 0$, we have $\mu_m(e, \varepsilon) = 0$, and hence there exists an ε-cover $\{e_j\}_{j\geqslant 1}$ of the set e such that

$$\sum_{j=1}^{\infty}\left(\frac{\operatorname{diam}(e_j)}{2}\right)^m < \varepsilon.$$

Thus the sequences $\{B_j\}_{j\geqslant 1}$ and $\{e_j\}_{j\geqslant 1}$ together form an ε-cover of G, and, consequently,

$$\mu_m(G, \varepsilon) \leqslant \sum_{j=1}^{\infty} r_j^m + \sum_{j=1}^{\infty}\left(\frac{\operatorname{diam}(e_j)}{2}\right)^m < \frac{1}{\alpha_m}\lambda_m(G) + \varepsilon.$$

Passing to the limit as $\varepsilon \to 0$, we obtain an upper bound on $\mu_m(G)$: $\mu_m(G) \leqslant \frac{1}{\alpha_m}\lambda_m(G)$. Together with (4) this yields the desired result. □

2.6.6 In conclusion let us discuss the dependence of the value $\mu_p^*(E)$ on p. Obviously, $\mu_p^*(E)$ decreases as p grows. Moreover, it turns out that $\mu_q^*(E) = 0$ if $\mu_p^*(E) < +\infty$ for some $p < q$. Indeed, let $0 < \varepsilon < 1$, and let $\{e_j\}_{j\geqslant 1}$ be an ε-cover of E such that

$$\sum_{j=1}^{\infty}\left(\frac{\operatorname{diam}(e_j)}{2}\right)^p < 1 + \mu_p(E, \varepsilon) \leqslant 1 + \mu_p^*(E) < +\infty.$$

Then

$$\begin{aligned}\mu_q(E, \varepsilon) &\leqslant \sum_{j=1}^{\infty}\left(\frac{\operatorname{diam}(e_j)}{2}\right)^q \leqslant \left(\frac{\varepsilon}{2}\right)^{q-p}\sum_{j=1}^{\infty}\left(\frac{\operatorname{diam}(e_j)}{2}\right)^p \\ &< \left(\frac{\varepsilon}{2}\right)^{q-p}\left(1 + \mu_p^*(E)\right).\end{aligned}$$

Passing to the limit as $\varepsilon \to 0$, we see that

$$\mu_q^*(E) = \lim_{\varepsilon\to 0}\mu_q(E, \varepsilon) = 0.$$

The obtained result can also be interpreted as follows: if $0 < \mu_p^*(E) < +\infty$, then $\mu_q^*(E) = +\infty$ for $q < p$ and $\mu_q^*(E) = 0$ for $q > p$. It follows that for every set E we have

$$\inf\{q > 0 \mid \mu_q^*(E) = 0\} = \sup\{q > 0 \mid \mu_q^*(E) = +\infty\}.$$

This critical value characterizing the set E is of special importance. It is called the *Hausdorff dimension* of E and is denoted by $\dim_H(E)$ (if $\mu_q^*(E) = 0$ for all $q > 0$, then, by definition, $\dim_H(E) = 0$). It follows from Lemma 2.6.4 that $\mu_q^*(E) = 0$ if $E \subset \mathbb{R}^m$ and $q > m$. Thus the Hausdorff dimension of every subset of $\mathbb{R}^m$ does not exceed m. It is equal to m if the outer Lebesgue measure of the set is positive.

EXERCISES

1. Without using Proposition 2.6.5, show directly that $\mu_1([a,b]) = (b-a)/2$ and, consequently, $\lambda_1 = 2\mu_1$.

2. What is the Hausdorff dimension of a countable set? Show that

$$\dim_H\Big(\bigcup_{n\geqslant 1} E_n\Big) = \sup_n \dim_H(E_n).$$

3. Two points $x, y \in \mathbb{R}^m$ are called ε-distinguishable if $\|x-y\| \geqslant \varepsilon$. Show that

$$\dim_H(E) \leqslant \varliminf_{\varepsilon\to+0} \frac{\log(N_E(\varepsilon))}{|\log\varepsilon|},$$

where $N_E(\varepsilon)$ is the maximum number of pairwise ε-distinguishable points contained in a bounded set $E \subset \mathbb{R}^m$. Considering the set $E = \{1, 2^{-p}, 3^{-p}, \ldots\}$ with $p > 0$, show that this inequality cannot be replaced by an equality.

4. Show that the Hausdorff dimension does not increase under a map satisfying the Lipschitz condition and hence is preserved under a diffeomorphism.

5. Show that the Hausdorff dimension of the Cantor set is equal to $\log_3 2$.

6. Show that for $x \geqslant 0$, the Cantor function φ satisfies the equality $\varphi(x) = 2^p\mu_p([0,x] \cap \mathcal{C})$, where $\mathcal{C}$ is the Cantor set and $p = \dim_H(\mathcal{C})$.

7. Modifying the construction of the Cantor set, show that for every p, $0 < p < 1$, there exists a compact set E contained in $[0,1]$ whose Hausdorff dimension is equal to p. Illustrate with examples that each of the following three cases is possible: $\mu_p(E) = 0$, $\mu_p(E) = +\infty$, $0 < \mu_p(E) < +\infty$.

8. Show that there exists a set contained in $[0,1]$ for which the Lebesgue measure is equal to zero and the Hausdorff dimension is equal to one.

9. Show that in the lemma on exhaustion by balls (see Sect. 2.6.5), the ball can be replaced with a bounded measurable set whose measure is positive and coincides with the measure of its closure (e.g., a convex body).

10. Consider a sequence of balls in $\mathbb{R}^m$ whose radii tend to zero and whose total volume is infinite. Show that one can put a finite number of such balls into the cube so that they fill at least 99 % of its volume.

11. Let G and A be bounded open subsets of $\mathbb{R}^m$, A being convex. Consider a special method of exhaustion of G which successively removes from G the maximum possible sets similar to A. That is, at the first step we find the maximum coefficient $c_1 > 0$ such that some translation $x_1 + c_1 A$ of the set $c_1 A$ is contained in G (such a coefficient exists). Then we put $G_1 = G \setminus (x_1 + c_1\overline{A})$ and, repeating the procedure, construct a set G_2, and so on. Show that $\lambda_m(G_n) \underset{n\to\infty}{\longrightarrow} 0$.

12. A subset E of the space $\mathbb{R}^m$ is called negligible if $\lambda_m(E_\varepsilon) = o(\varepsilon)$ as $\varepsilon \to 0$ (here E_ε is the ε-neighborhood of E). Show that if E is negligible, then $\mu^*_{m-1}(E) = 0$.

2.7 *The Vitali Theorem

In this section, we prove two theorems on covers used in the study of the properties of measurable sets and functions (see Chap. 4). We denote the Lebesgue measure on $\mathbb{R}^m$ by λ without indicating the dimension; given a ball B, we write $r(B)$ for its radius and B^* for the ball of radius $5r(B)$ with the same center.

2.7.1 We will establish one fact of independent interest before proving the Vitali theorem which is the main result of this section.

Theorem *Let $\mathcal{B}$ be a collection of balls that form a cover of a bounded set E ($E \subset \mathbb{R}^m$). If the radii of the balls are bounded, then we can extract from this collection a sequence (finite or not) of pairwise disjoint balls B_k such that*

$$E \subset \bigcup_{k\geqslant 1} B_k^*.$$

Proof We will assume without loss of generality that $E \cap B \neq \varnothing$ for all balls B from $\mathcal{B}$. It is clear that in this case the set $\bigcup_{B\in\mathcal{B}} B$ is bounded.

We will construct the desired sequence of balls $B_k = B(x_k, r_k)$ by induction. For the sake of uniformity, let $\mathcal{B} = \mathcal{B}_1$ and $R_1 = \sup\{r(B) \mid B \in \mathcal{B}_1\}$. Choose $B_1 \in \mathcal{B}_1$ so that $r_1 = r(B_1) > R_1/2$. Assume that pairwise disjoint balls $B_1, \dots, B_n$ and subsets $\mathcal{B}_1, \dots, \mathcal{B}_n$ of the initial collection $\mathcal{B}$ have already been constructed. Let

$$\mathcal{B}_{n+1} = \left\{ B \in \mathcal{B}_n \,\middle|\, B \cap \bigcup_{k=1}^{n} B_k = \varnothing \right\}.$$

If $\mathcal{B}_{n+1} \neq \varnothing$, then we put $R_{n+1} = \sup\{r(B) \mid B \in \mathcal{B}_{n+1}\}$ and choose a ball B_{n+1} so that $r_{n+1} = r(B_{n+1}) > R_{n+1}/2$, and so on. Thus either the set $\mathcal{B}_{n+1}$ is non-empty at each step and we obtain an infinite sequence of balls, or $\mathcal{B}_{n+1} = \varnothing$ at some step and the process terminates. Let us consider both possibilities, starting with the second one.

Let $\mathcal{B}_{n+1} = \varnothing$ and x be an arbitrary point from E. It belongs to some ball $B = B(a, r) \in \mathcal{B}$, and $B \cap \bigcup_{k=1}^{n} B_k \neq \varnothing$. Let j be the smallest of the indices k such that $B \cap B_k \neq \varnothing$. Then $r \leqslant R_j$ (for $j = 1$ this inequality is trivial, and for $j > 1$ it follows from the fact that B is disjoint with the union $\bigcup_{k=1}^{j-1} B_k$). Let us check that $x \in B_j^*$. Indeed, since the balls B and B_j have a non-empty intersection,

$$\|x - x_j\| < \operatorname{diam}(B) + r(B_j) = 2r + r_j \leqslant 2R_j + r_j < 5r_j$$

(the last inequality holds, because $r_j > R_j/2$ by construction).

Now consider the main case, where the sequence of balls $\{B_k\}_{k\geqslant 1}$ is infinite. First of all, observe that the series $\sum_{k\geqslant 1} \lambda(B_k)$ converges. Indeed, the balls B_k are pairwise disjoint by construction. Hence the sum of the series is simply the measure of the bounded set $\bigcup_{k\geqslant 1} B_k$ (at the beginning of the proof, we have observed that the union of all balls from $\mathcal{B}$ is bounded). From the convergence of the series it follows immediately that $r_k \to 0$.

Let $x \in E$, and let B be a ball from $\mathcal{B}$ such that $x \in B$. Let us check that

$$B \cap \bigcup_{k=1}^{\infty} B_k \neq \varnothing. \tag{1}$$

Indeed, otherwise $B \cap \bigcup_{k=1}^{n} B_k = \varnothing$. Then $0 < r(B) \leqslant R_{n+1} < 2r_{n+1}$ for every n, which is impossible since $r_n \to 0$. It follows from (1) that the intersection $B \cap B_k$ is not empty for some indices k. Let j be the smallest of them. Repeating the above argument, we see that $x \in B_j^*$. □

Note that, as one can see from the proof, the conclusion of the theorem holds for every sequence of balls $\{B_k\}_{k\geqslant 1}$ from the cover $\mathcal{B}$ satisfying the following condition for every n:

$$B_{n+1} \cap \bigcup_{k=1}^{n} B_k = \varnothing, \qquad 2r(B_{n+1}) > \sup\left\{ r(B) \,\middle|\, B \cap \bigcup_{k=1}^{n} B_k = \varnothing \right\}. \tag{2}$$

2.7.2 The theorem can be substantially refined if the cover satisfies an additional condition.

Definition A collection $\mathcal{B}$ of open balls is called a *Vitali*[11] *cover* of a set E ($E \subset \mathbb{R}^m$) if for every point x in E, there is an arbitrarily small ball in $\mathcal{B}$ containing x.

Theorem (Vitali) *In every Vitali cover $\mathcal{B}$ of a bounded set E there exists a sequence (finite or not) of balls B_k satisfying the following conditions*:

(1) *the balls B_k are pairwise disjoint*;
(2) $E \subset \bigcup_{k\geqslant 1} B_k^*$;
(3) $\lambda(E \setminus \bigcup_{k\geqslant 1} B_k) = 0$.

Note that we do not assume that the set E is measurable.

Proof Discarding, if necessary, balls with too large radii, we assume that $r(B) < 1$ for all balls B in $\mathcal{B}$. Then we may apply Theorem 2.7.1. Let $\{B_k\}_{k\geqslant 1}$ be the sequence of balls constructed in that theorem. It satisfies conditions (1) and (2). Let us check that it also has Property (3).

If this sequence is finite and consists of n balls, then $E \subset \bigcup_{k=1}^{n} \overline{B_k}$. Indeed, the finiteness means that $B \cap \bigcup_{k=1}^{n} B_k \neq \varnothing$ for every $B \in \mathcal{B}$. Since every point in E belongs to a ball with arbitrarily small radius, this would be impossible unless there

[11] Giuseppe Vitali (1875–1932)—Italian mathematician.

are points in E not belonging to $\bigcup_{k=1}^n \overline{B_k}$. Therefore, $E \setminus \bigcup_{k=1}^n B_k \subset \bigcup_{k=1}^n \partial B_k$, and hence condition (3) is satisfied.

Now consider the case where the sequence $\{B_k\}_{k\geqslant 1}$ is infinite. Let us check that for every n

$$E \setminus \bigcup_{k=1}^{n} \overline{B_k} \subset \bigcup_{k=n+1}^{\infty} B_k^*. \tag{3}$$

Let $\mathcal{B}_{n+1}$ be the set of balls constructed in the proof of Theorem 2.7.1. It forms a Vitali cover of the set $E_n = E \setminus \bigcup_{k=1}^n \overline{B_k}$, and the sequence of balls $\{B_{n+k}\}_{k\geqslant 1}$ satisfies condition (2). Hence, by Theorem 2.7.1, $E_n \subset \bigcup_{k=1}^{\infty} B_{n+k}^*$. Therefore, for every n we have

$$E \setminus \bigcup_{k=1}^{\infty} B_k \subset \left(\bigcup_{k=1}^{n} \partial B_k\right) \cup E_n \subset \left(\bigcup_{k=1}^{n} \partial B_k\right) \cup \left(\bigcup_{k=n+1}^{\infty} B_k^*\right).$$

Moreover,

$$\lambda\left(\bigcup_{k=1}^{n} \partial B_k \cup \bigcup_{k=n+1}^{\infty} B_k^*\right) \leqslant \sum_{k=n+1}^{\infty} \lambda(B_k^*) = 5^m \sum_{k=n+1}^{\infty} \lambda(B_k) \underset{n\to\infty}{\longrightarrow} 0$$

(the last sum tends to zero as the remainder of a convergent series). Thus the set $E \setminus \bigcup_{k=1}^{\infty} B_k$ is contained in a set of arbitrarily small measure, which implies (3). □

Remark Splitting an arbitrary set into bounded parts, one can easily show that claims (1) and (3) of the theorem also remain valid for an unbounded set E.

2.7.3 One of the important corollaries of the Vitali theorem is related to density points.

Definition A point x_0 is called a *density point* of a set E if

$$\lambda^*\big(E \cap B(x_0, r)\big)/\lambda\big(B(x_0, r)\big) \to 1 \quad \text{as } r \to +0.$$

Corollary 1 *Let E' be the set of density points of an arbitrary set E. Then $\lambda(E \setminus E') = 0$. In particular, almost every point of a measurable set is a density point of this set.*

Proof Let $E_0 = E \setminus E'$. If $x \in E_0$, then there exists a sequence of radii $\{r_n(x)\}_{n\geqslant 1}$ decreasing to zero such that

$$\lim_{n\to\infty} \frac{\lambda^*(E \cap B(x_0, r_n(x)))}{\lambda(B(x_0, r_n(x)))} < 1.$$

For $\theta \in (0,1)$ put

$$E_\theta = \left\{ x \in E_0 \,\middle|\, \lim_{n\to\infty} \frac{\lambda^*(E \cap B(x, r_n(x)))}{\lambda(B(x, r_n(x)))} < \theta \right\}.$$

Since $E_\theta \subset E_{\theta'}$ for $\theta < \theta'$ and $E_0 = \bigcup_{\theta\in(0,1)} E_\theta$, it suffices to verify that $\lambda(E_\theta) = 0$ (note that we do not know anything about the measurability of the sets E_θ yet). Fixing $\theta \in (0,1)$ and an arbitrarily small positive number ε, let us find an open set G containing E_θ such that $\lambda(G) < \lambda^*(E_\theta) + \varepsilon$ (see the remark in Sect. 2.2.2). All balls $B(x, r_n(x))$, $x \in E_\theta$, that are contained in G and satisfy the condition $\lambda^*(E \cap B(x, r_n(x))) \leqslant \theta\lambda(B(x, r_n(x)))$ form a Vitali cover of the set E_θ. By the Vitali theorem, there exists a subsystem of pairwise disjoint balls $B_k = B(x_k, r_{n_k}(x_k))$ such that the set-theoretic difference $e = E_\theta \setminus \bigcup_{k\geqslant 1} B_k$ has zero measure. By the countable subadditivity of the outer measure, we obtain

$$\lambda^*(E_\theta) \leqslant \lambda^*(e) + \sum_{k\geqslant 1} \lambda^*(E_\theta \cap B_k) \leqslant \sum_{k\geqslant 1} \lambda^*(E \cap B_k)$$

$$\leqslant \sum_{k\geqslant 1} \theta\lambda(B_k) \leqslant \theta\lambda(G) < \theta\big(\lambda^*(E_\theta) + \varepsilon\big).$$

Thus $\lambda^*(E_\theta) < \frac{\varepsilon\theta}{1-\theta}$, and, since ε is arbitrary, it follows that $\lambda^*(E_\theta) = 0$. This means that the set E_θ is measurable and $\lambda(E_\theta) = 0$. □

The Vitali theorem easily implies the result on exhaustion of an open set by balls obtained in Lemma 2.6.5.

Corollary 2 *Every non-empty open subset G in the space $\mathbb{R}^m$ can be written as the union of a sequence of disjoint balls B_n and a set of zero measure e:*

$$G = e \cup \bigcup_{n=1}^{\infty} B_n.$$

Proof Consider the system of all balls contained in G. It obviously forms a Vitali cover for G. Hence, if the set G is bounded, it suffices to use Claim (2) of the Vitali theorem and put $e = G \setminus \bigcup_{k\geqslant 1} B_k$. In the case where G is not bounded, one should refer to Remark 2.7.2. □

2.7.4 The Vitali theorem has various generalizations. To describe one of them, we introduce the notion of a regular cover. A system of sets $\mathcal{B} = \{E_j(x) \,|\, x \in E,\ j \in \mathbb{N}\}$ is called a *regular cover* of a set E if the following conditions hold:

(1) $E_j(x) \subset B(x, r_j(x))$, $r_j(x) \underset{j\to\infty}{\longrightarrow} 0$;

(2) $\inf_{j\in\mathbb{N}} \frac{\lambda(E_j(x))}{\lambda(B(x, r_j(x)))} > 0$ for every $x \in E$.

For example, as $E_j(x)$ one can take cubes etc. that are "not too small" compared to $B(x, r_j(x))$. For regular covers, an analog of the Vitali theorem holds (see, for example, [S, Chap. IV, Sect. 3]).

Theorem *In every regular cover of a set E there exists a sequence of pairwise disjoint sets $E_k = E_{j_k}(x_k)$ such that $\lambda(E \setminus \bigcup_{k\geqslant 1} E_k) = 0$.*

One can prove that the Vitali theorem remains valid for every Borel measure μ in a metric space if it is finite on balls and "quasihomogeneous", i.e., there exist constants $K > 1$ and $a > 1$ such that $\mu(B(x, ar)) \leqslant K\,\mu(B(x,r))$ for all x and $r > 0$. For example, this condition is satisfied for the surface area on a compact smooth manifold (see Chap. 8).

EXERCISES

1. Show that the Vitali theorem is also valid for an unbounded set.
2. Show that every differentiable function on an interval preserves Lebesgue measurability.
3. Extend the result of Exercise 6 in Sect. 2.1 to the two-dimensional case by showing that the union of an arbitrary family of non-degenerate triangles on the plane is measurable. Is the same true if we replace triangles by their boundaries?
4. Let G be an open subset of $\mathbb{R}^m$, $\Phi \in C^1(G, \mathbb{R}^m)$ and $E \subset G$. Show that if Φ is expanding on E, then $|\det \Phi'(x)| \geqslant 1$ almost everywhere on E. Can we assert that $|\det \Phi'(x)| \geqslant 1$ everywhere on E provided that it is connected? *Hint*. Show that the desired inequality holds at every density point of E.
5. Let $\mathcal{O}$ be an open subset of $\mathbb{R}^m$, and let $\Phi \in C^1(\mathcal{O}, \mathbb{R}^m)$ with $\det \Phi' \neq 0$ (the last assumption can be dropped by Sard's theorem, see Appendix 13.5). Show that there exists an open set $G \subset \mathcal{O}$ such that the restriction of Φ to G is one-to-one and $\Phi(\mathcal{O}) = \Phi(G) \cup e$, where $\lambda(e) = 0$. *Hint*. Splitting the set $\mathcal{O}$ into parts, reduce the assertion to the case where the closure of $\mathcal{O}$ is compact, $\lambda(\partial\mathcal{O}) = 0$, and Φ is smooth in a neighborhood of $\overline{\mathcal{O}}$. Show that the set of inverse images of every point from $\Phi(\mathcal{O})$ is finite. Show that if a point x does not belong to $\Phi(\partial\mathcal{O})$, then it has a neighborhood whose full inverse image breaks into n connected components, where n is the number of inverse images of x. Using the Vitali theorem, find a sequence of such neighborhoods that "almost cover" $\Phi(\mathcal{O})$ and form G from the components of their inverse images.

2.8 *The Brunn–Minkowski Inequality

In this section, by λ we denote the Lebesgue measure on $\mathbb{R}^m$, which we also call the volume.

2.8.1 The main result of this section is the following statement.

Theorem *For compact sets $A, B \subset \mathbb{R}^m$, the following inequality holds:*

$$\lambda^{\frac{1}{m}}(A+B) \geqslant \lambda^{\frac{1}{m}}(A) + \lambda^{\frac{1}{m}}(B).$$

Here $A+B$ is the algebraic sum of A and B, i.e., $A+B = \{x+y \mid x \in A,\ y \in B\}$, $A, B \neq \varnothing$.

This is the Brunn[12]–Minkowski[13] inequality.

If A and B are sets of positive measure, then the Brunn–Minkowski inequality becomes an equality only in the case where A and B are similar. The proof of this fact is not easy even for convex bodies (cf. Exercise 3). For a discussion of this and related results, see, for example, [BZ].

Proof The proof splits into several steps, with the sets A and B becoming more and more complicated.

(1) Let A and B be parallelepipeds with edge lengths $\alpha_1, \dots, \alpha_m$ and $\beta_1, \dots, \beta_m$, respectively. Then $A+B$ is a parallelepiped with edge lengths $\alpha_1+\beta_1, \dots, \alpha_1+\beta_m$. We will assume that $\alpha_j + \beta_j \equiv 1$ (the general case can then be obtained by scaling along the coordinate axes). Thus

$$\lambda(A) = \alpha_1 \cdots \alpha_m, \qquad \lambda(B) = \beta_1 \cdots \beta_m, \qquad \lambda(A+B) = 1.$$

It remains to apply the inequality of arithmetic and geometric means:

$$\lambda^{\frac{1}{m}}(A) + \lambda^{\frac{1}{m}}(B) = (\alpha_1 \cdots \alpha_m)^{\frac{1}{m}} + (\beta_1 \cdots \beta_m)^{\frac{1}{m}} \leqslant \frac{1}{m}\sum_{j=1}^{m} \alpha_j + \frac{1}{m}\sum_{j=1}^{m} \beta_j = 1$$

$$= \lambda^{\frac{1}{m}}(A+B).$$

(2) Now let each of the sets A and B be a finite union of cells. By the theorem on properties of semirings (see Sect. 1.1.4), such unions can be assumed disjoint:

$$A = \bigvee_{k=1}^{r} P_k, \qquad B = \bigvee_{j=1}^{s} Q_j.$$

We will argue by induction on the sum $n = r + s$, assuming that $P_k, Q_j \neq \varnothing$. The inductive base (for $n = 2$) was proved in the previous step.

Assume that the desired inequality is true for $r+s < n$. Let us prove the inductive step for $n \geqslant 3$. Since r and s are interchangeable, we may assume that $r \geqslant 2$. The cells P_1 and P_r have no common points, hence their projections to at least one coordinate axis, say x_1, have no common points either. This means that P_1 and P_r lie

[12] Hermann Karl Brunn (1862–1939)—German mathematician.

[13] Hermann Minkowski (1864–1909)—German mathematician.

on opposite sides of some plane $x_1 = a$. We may assume without loss of generality that $P_1 \subset H^+ = \{(x_1, \dots, x_m) \mid x_1 \geqslant a\}$ and $P_r \subset H^- = \{(x_1, \dots, x_m) \mid x_1 < a\}$. Put

$$A^{\pm} = A \cap H^{\pm}, \qquad P_k^{\pm} = P_k \cap H^{\pm}.$$

Each of the sets $A^{\pm}$ can be written as the union of at most $(r-1)$ cells:

$$A^+ = \bigcup_{k=1}^{r-1} P_k^+, \qquad A^- = \bigcup_{k=2}^{r} P_k^-.$$

Now consider a plane $x_1 = b$ that divides the set B in the same ratio as the plane $x_1 = a$ divides the set A. More precisely, we mean that the measures of the sets $B^+ = B \cap \{(x_1, \dots, x_m) \mid x_1 \geqslant b\}$ and $B^- = B \cap \{(x_1, \dots, x_m) \mid x_1 < b\}$ are in the same ratio as the measures of the sets $A^{\pm}$. The latter condition is equivalent to the following one: for some $\theta \in (0, 1)$,

$$\frac{\lambda(B^+)}{\lambda(B)} = \frac{\lambda(A^+)}{\lambda(A)} = \theta \quad \text{and} \quad \frac{\lambda(B^-)}{\lambda(B)} = \frac{\lambda(A^-)}{\lambda(A)} = 1 - \theta.$$

Note that each of the sets $B^{\pm}$ (as well as B) is the union of at most s pairwise disjoint cells. Obviously, $A + B \supset (A^+ + B^+) \cup (A^- + B^-)$, and the sets $A^+ + B^+$ and $A^- + B^-$ are disjoint (since they lie on opposite sides of the plane $x_1 = a + b$). Hence

$$\lambda(A+B) \geqslant \lambda\big((A^+ + B^+) \cup (A^- + B^-)\big) = \lambda(A^+ + B^+) + \lambda(A^- + B^-).$$

The measures on the right-hand side can be estimated from below by the induction hypothesis:

$$\lambda(A^{\pm} + B^{\pm}) \geqslant \big(\lambda^{\frac{1}{m}}(A^{\pm}) + \lambda^{\frac{1}{m}}(B^{\pm})\big)^m.$$

Together with the previous inequality, this yields

$$\begin{aligned}
\lambda(A+B) &\geqslant \big(\lambda^{\frac{1}{m}}(A^+) + \lambda^{\frac{1}{m}}(B^+)\big)^m + \big(\lambda^{\frac{1}{m}}(A^-) + \lambda^{\frac{1}{m}}(B^-)\big)^m \\
&= \theta\big(\lambda^{\frac{1}{m}}(A) + \lambda^{\frac{1}{m}}(B)\big)^m + (1-\theta)\big(\lambda^{\frac{1}{m}}(A) + \lambda^{\frac{1}{m}}(B)\big)^m \\
&= \big(\lambda^{\frac{1}{m}}(A) + \lambda^{\frac{1}{m}}(B)\big)^m,
\end{aligned}$$

which completes the inductive step.

(3) Now let A and B be arbitrary compact sets. Obviously, the set $A + B$ is also compact. We will obtain the desired result by approximation.

The sets A and B have finite covers by open parallelepipeds (and hence cells) lying in the δ-neighborhoods of these sets. Let A' and B' be the unions of the cells covering A and B, respectively. Clearly, $A' + B' \subset (A + B)_{2\delta}$. As we have already observed (see Sect. 2.6.3), the intersection of all δ-neighborhoods of a set coincides with its closure. Hence, by the continuity of λ from above, we have $\lambda((A+B)_{2\delta}) \to \lambda(A+B)$ as $\delta \to 0$. By the result proved at the previous step,

$$\lambda^{\frac{1}{m}}((A+B)_{2\delta}) \geqslant \lambda^{\frac{1}{m}}(A'+B') \geqslant \lambda^{\frac{1}{m}}(A') + \lambda^{\frac{1}{m}}(B') \geqslant \lambda^{\frac{1}{m}}(A) + \lambda^{\frac{1}{m}}(B).$$

Now the desired inequality can be obtained by passing to the limit. □

Remark We have considered the Brunn–Minkowski inequality in the main special case, namely, for compact sets. Using similar arguments, one can easily prove it, for example, for open sets. However, one should bear in mind that for arbitrary measurable sets A and B, the set $A+B$ is not necessarily measurable (see Exercise 6). Accordingly, the Brunn–Minkowski inequality for non-empty measurable sets takes the form

$$\big(\lambda^*(A+B)\big)^{\frac{1}{m}} \geqslant \lambda^{\frac{1}{m}}(A) + \lambda^{\frac{1}{m}}(B),$$

where λ^* is the outer Lebesgue measure. To prove it, recall (see Corollary 3 in Sect. 2.2.2) that, by the regularity of the Lebesgue measure, the sets A and B can be written in the form

$$A = e \cup \bigcup_{n=1}^{\infty} A_n, \qquad B = e' \cup \bigcup_{n=1}^{\infty} B_n,$$

where $\lambda(e) = \lambda(e') = 0$ and $\{A_n\}_{n\geqslant 1}$ and $\{B_n\}_{n\geqslant 1}$ are increasing sequences of compact sets. Since $A+B \supset A_n + B_n$ for every n, we have

$$\big(\lambda^*(A+B)\big)^{\frac{1}{m}} \geqslant \lambda^{\frac{1}{m}}(A_n+B_n) \geqslant \lambda^{\frac{1}{m}}(A_n) + \lambda^{\frac{1}{m}}(B_n).$$

Now $\lambda(A_n) \to \lambda(A)$ and $\lambda(B_n) \to \lambda(B)$ as $n \to \infty$, so that passing to the limit yields the desired result.

One can also prove (see [F]) that

$$\big(\lambda^*(A+B)\big)^{\frac{1}{m}} \geqslant \big(\lambda^*(A)\big)^{\frac{1}{m}} + \big(\lambda^*(B)\big)^{\frac{1}{m}}$$

for arbitrary sets, but we will not dwell on this here.

2.8.2 The Brunn–Minkowski inequality easily implies an inequality relating the volume of a body and its surface area (by a body we mean a compact set with a non-empty interior). The notion of surface area is discussed in detail in Chap. 8; here we restrict ourselves to defining the Minkowski surface area needed for stating this inequality. The definition is based on the following apparent observation: when we pass from a body K to its ε-neighborhood (see Sect. 2.6.3), the increment of the volume for small $\varepsilon > 0$ must be almost proportional to the area of the boundary of K.

Definition Let $K \subset \mathbb{R}^m$ be an arbitrary body and K_ε be its ε-neighborhood. The *lower Minkowski area* of ∂K is the value

$$\Sigma^-_{m-1}(\partial K) = \varliminf_{\varepsilon\to+0} \frac{\lambda(K_\varepsilon \setminus K)}{\varepsilon}.$$

The limit $\lim_{\varepsilon\to 0} \frac{\lambda(K_\varepsilon \setminus K)}{\varepsilon}$ (if it exists) is called the *Minkowski area of* ∂K. We will denote it by $\Sigma_{m-1}(\partial K)$.

Simple calculations show that for a sphere $S(r) \subset \mathbb{R}^m$ of radius r, we have $\Sigma_{m-1}(S(r)) = m\alpha_m r^{m-1}$, where α_m is the volume of the unit ball in $\mathbb{R}^m$. As we will see (cf. Sects. 8.4.4 and 13.4.7), for bodies with sufficiently smooth boundary and for convex bodies, the Minkowski area of the boundary is proportional to the Hausdorff measure μ_{m-1}.

Theorem (Isoperimetric inequality) *For every body* $K \subset \mathbb{R}^m$,

$$\Sigma^-_{m-1}(\partial K) \geqslant m\alpha_m^{\frac{1}{m}} \lambda^{\frac{m-1}{m}}(K).$$

If K is a ball, this inequality becomes an equality, which implies the isoperimetric inequality in its classical form, where by the surface area we mean the lower Minkowski area:

Among all bodies of a given volume, the ball has the smallest surface area.
Among all bodies of a given surface area, the ball has the greatest volume.

As the reader can easily check, the isoperimetric inequality can also be written in the following form (hereafter B is a ball in $\mathbb{R}^m$):

$$\left(\frac{\Sigma^-_{m-1}(\partial K)}{\Sigma^-_{m-1}(\partial B)}\right)^{\frac{1}{m-1}} \geqslant \left(\frac{\lambda(K)}{\lambda(B)}\right)^{\frac{1}{m}}.$$

Proof Let $B = B(0, 1)$, $\alpha_m = \lambda(B)$. By the Brunn–Minkowski inequality,

$$\lambda^{\frac{1}{m}}(K_\varepsilon) = \lambda^{\frac{1}{m}}(K + \varepsilon B) \geqslant \lambda^{\frac{1}{m}}(K) + \varepsilon\, \alpha_m^{\frac{1}{m}}.$$

Raising to the power m yields

$$\lambda(K_\varepsilon \setminus K) = \lambda(K_\varepsilon) - \lambda(K) \geqslant m\,\varepsilon\, \alpha_m^{\frac{1}{m}} \lambda^{\frac{m-1}{m}}(K) + O\big(\varepsilon^2\big).$$

The desired inequality can be obtained by dividing by ε and passing to the limit:

$$\Sigma^-_{m-1}(\partial K) = \varliminf_{\varepsilon\to+0} \frac{1}{\varepsilon}\lambda(K_\varepsilon \setminus K) \geqslant m\, \alpha_m^{\frac{1}{m}} \lambda^{\frac{m-1}{m}}(K).$$

□

2.8.3 As another application of the Brunn–Minkowski inequality, we will also prove the *isodiametric*, or *Bieberbach*[14] *inequality.*

Theorem *Among all measurable sets of given diameter, the ball has the greatest volume.*

Proof Let $A \subset \mathbb{R}^m$ with $\operatorname{diam}(A) = d$. Since the diameter of a set coincides with the diameter of its closure, we may assume that A is closed and hence compact. Consider the sets $A' = -A$ and $E = \frac{1}{2}(A + A')$. By the Brunn–Minkowski inequality, the volume of E is not less than the volume of A:

$$\lambda^{\frac{1}{m}}(E) = \frac{1}{2}\lambda^{\frac{1}{m}}(A + A') \geqslant \frac{1}{2}\big(\lambda^{\frac{1}{m}}(A) + \lambda^{\frac{1}{m}}(A')\big) = \lambda^{\frac{1}{m}}(A).$$

Let us check that the set E is contained in a closed ball of radius $d/2$. Indeed, if $x \in E$, then $x = (s-t)/2$, where $s, t \in A$. Hence $\|x\| = \frac{1}{2}\|s - t\| \leqslant \frac{d}{2}$. Thus $\lambda(A) \leqslant \lambda(E)$, and E is contained in a ball B of radius $d/2$. Therefore, $\lambda(A) \leqslant \lambda(B)$. □

EXERCISES In what follows, A and B are subsets of $\mathbb{R}^m$.

1. Let A and B be compact sets. Show that the function $t \mapsto \lambda^{\frac{1}{m}}(tA + (1-t)B)$ is concave on [0, 1], i.e., that

$$\lambda^{\frac{1}{m}}\big(tA + (1-t)B\big) \geqslant t\lambda^{\frac{1}{m}}(A) + (1-t)\lambda^{\frac{1}{m}}(B)$$

for every t, $0 \leqslant t \leqslant 1$. Using the fact that the logarithm is concave, deduce that

$$\lambda\big(tA + (1-t)B\big) \geqslant \lambda^t(A)\lambda^{1-t}(B).$$

2. Arguing as in Remark 2.8.1, show that for arbitrary (possibly, non-measurable) sets A and B,

$$\lambda_*^{\frac{1}{m}}(A + B) \geqslant \lambda_*^{\frac{1}{m}}(A) + \lambda_*^{\frac{1}{m}}(B),$$

where λ_* is the inner measure (for the definition, see Sect. 2.2.2).

3. Show that for ellipsoids, the Brunn–Minkowski inequality becomes an equality only in the case where they are similar. *Hint.* Apply the method used in the proof of Theorem 2.5.5 on the uniqueness of an ellipsoid of maximal volume.

4. Let $[a, b]$ be the projection to the first coordinate axis of a convex body lying in $\mathbb{R}^m$ ($m \geqslant 2$). Let $S(t)$ be the area of the section of this body by the plane $x_1 = t$ and $V(t)$ be the volume of its part lying in the half-space $x_1 \leqslant t$. Show that the function $S^{\frac{1}{m-1}}$ is concave and the ratio $V^{\frac{1}{m}}/S^{\frac{1}{m-1}}$ does not decrease on $(a, b]$.

5. Show that the arguments of Sect. 2.8.3 remain valid if we replace the Euclidean norm $\|\cdot\|$ by an arbitrary norm $\|\cdot\|_*$: if a measurable set $A \subset \mathbb{R}^m$ is such that $\|x - y\|_* \leqslant 2r$ for all $x, y \in A$, then $\lambda(A) \leqslant \lambda(B_*(r))$, where $B_*(r) = \{x \in \mathbb{R}^m \mid \|x\|_* < r\}$.

[14] Ludwig Georg Elias Moses Bieberbach (1886–1982)—German mathematician.

6. Show that the algebraic sum of sets of zero Lebesgue measure can be non-measurable. *Hint*. Consider the set $C + 2E$, where $C = \{\sum_{n=1}^{\infty} \varepsilon_n 4^{-n} \,|\, \varepsilon_n = 0 \text{ or } 1\}$ and the set $E \subset C$ is constructed from an ultrafilter $\mathfrak{U}$ in $\mathbb{N}$ consisting of infinite sets: $E = \{\sum_{n \in U} 4^{-n} \,|\, U \in \mathfrak{U}\}$. Use the same trick as in the solution of Exercise 12 in Sect. 2.1.

Chapter 3
Measurable Functions

The introduction of the notion of a measure is a necessary step towards the solution of the main problem, that of defining the integral. However, even now, having become familiar with measures, we cannot proceed directly to this task. The problem is that without specifying in advance for which functions the integral is being constructed we will inevitably run into difficulties. To illustrate this, consider the following very simple situation.

It is natural to try to define the integral of a bounded function defined on the interval $[a, b]$ as the limit of the (Riemann) integral sums, i.e., sums of the form

$$\sum_{k=1}^{n} f(\xi_k)(x_k - x_{k-1}), \qquad \text{where } x_0 = a < x_1 < \cdots < x_n = b, \ x_{k-1} \leqslant \xi_k \leqslant x_k.$$

The limit is taken as the maximum of the differences $x_k - x_{k-1}$ tends to zero, and it should not depend on the choice of the points ξ_k.

If the function f is continuous, then this limit exists (see Theorem 4.7.3). But an attempt to apply this procedure to functions with "many" discontinuities fails. For example, if f is the *Dirichlet function*, which is equal to one at rational points and zero at irrational points, we see that the integral sum vanishes if all ξ_k are irrational and equals $b - a$ if all ξ_k are rational. This is true for an arbitrarily fine partition of the interval, so that the integral sums have no limit.

To understand the reasons why this approach to the definition of the integral fails, we should notice that for a discontinuous function, the procedure of partitioning the interval into "small subintervals" and constructing the corresponding integral sum is not at all as natural as for a continuous function. Indeed, in the latter case, the limit of the integral sums does not depend on the choice of the points ξ_k because a continuous function changes very little on the subintervals $[x_{k-1}, x_k]$. Of course, we cannot expect this to hold for a discontinuous function. Hence, if we want to construct the integral of such a function, a natural idea, first conceived by Lebesgue, is to partition the interval $[a, b]$ not into subintervals (on which, in spite of their "smallness", the function may still vary considerably), but into some other sets. And the "smallness" of a set should be determined not in terms of its size, but in terms

B. Makarov, A. Podkorytov, *Real Analysis: Measures, Integrals and Applications*, Universitext, DOI 10.1007/978-1-4471-5122-7_3, © Springer-Verlag London 2013

of the variation of the function on this set. For example, as "small" sets we can take the sets $e_k = f^{-1}([y_{k-1}, y_k))$, where y_k $(k = 0, 1, \dots, n)$ is an increasing sequence (with $y_0 \leqslant \inf f$, $y_n > \sup f$). With this method of partitioning the interval, the definition of the integral sum should be modified: instead of the differences $x_k - x_{k-1}$, i.e., the lengths of the intervals $[x_{k-1}, x_k]$, we should consider the measures of the sets e_k. In this case, the integral sum takes the form $\sum_{k=1}^{n} f(\xi_k)\lambda(e_k)$, where $\xi_k \in e_k$ and λ is the Lebesgue measure. Postponing the discussion of the properties of these sums until the next chapter, we only note that the integral of f should be understood as their limit as $\max_k(y_k - y_{k-1}) \to 0$. However, in order to implement the new approach to the definition of the integral, we should fill a significant gap in our argument. Namely, we cannot be sure that the sets e_k are measurable (recall that not every set is Lebesgue measurable!) and hence we have no right to speak about their measures. Therefore, if we consider an arbitrary function, we cannot speak about any properties of the modified integral sums, since there is no guarantee that we can construct them. This is why, aiming at the implementation of the program suggested above, we will abandon attempts to define the integral for an arbitrary function and content ourselves with considering only functions for which the sets e_k constructed above are necessarily measurable. This chapter is devoted to the study of such functions, called measurable functions. The class of measurable functions is extremely wide and meets not only the demands of applications, but almost all needs of pure mathematics. At the same time, it is sufficiently tractable and, as we will see, in the case of functions defined in $\mathbb{R}^m$, is closely related to classes of simpler (e.g., continuous) functions.

3.1 Definition and Basic Properties of Measurable Functions

In what follows, we assume that there is a fixed set X and a σ-algebra $\mathfrak{A}$ of subsets of X. The pair $(X, \mathfrak{A})$ is called a *measurable space*, and the elements of the σ-algebra $\mathfrak{A}$ are called measurable sets.

As the reader will see below, it is convenient to consider real-valued functions not only with finite, but also with infinite values. Some technical complications related to arithmetic operations with such functions that arise at the first stages are well compensated for by the freedom we gain allowing ourselves to consider measurable functions with infinite values. We will see the first confirmation of this thesis in Theorem 3.1.4.

3.1.1 One can see from the remarks at the beginning of this chapter that it is crucial for the construction of the integral that the sets on which the oscillations of the function are small should be measurable. A key role here is played by sets on which the function is bounded from one side. Let us introduce the following important definition.

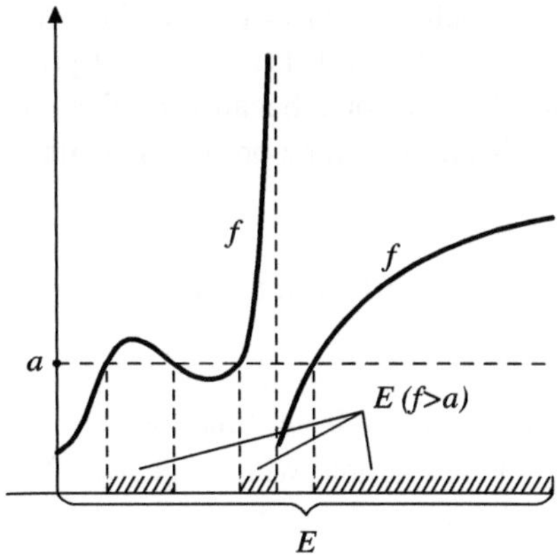

Fig. 3.1 Lesbegue set $E(f > a)$

Definition Let $f : E \to \overline{\mathbb{R}} = [-\infty, +\infty]$ be a function defined on a set $E \subset X$ and $a \in \mathbb{R}$. The sets

$$E(f < a) \equiv \{x \in E \mid f(x) < a\}, \qquad E(f \leqslant a) \equiv \{x \in E \mid f(x) \leqslant a\},$$
$$E(f > a) \equiv \{x \in E \mid f(x) > a\}, \qquad E(f \geqslant a) \equiv \{x \in E \mid f(x) \geqslant a\}$$

are called the *Lebesgue sets* of f (of the first, second, third, and fourth kind, respectively).

As follows from the definition, the Lebesgue sets are the inverse images of open and closed semi-axes, i.e., the sets

$$f^{-1}([-\infty, a)), \quad f^{-1}([-\infty, a]), \quad f^{-1}((a, +\infty]), \quad f^{-1}([a, +\infty]),$$

respectively (see Fig. 3.1).

As well as the notation $E(f < a)$, $E(f \leqslant a)$, and so on for the inverse images of semi-axes, we will also use similar notation for the inverse images of intervals, e.g., $E(a < f \leqslant b) = f^{-1}((a, b])$.

It turns out that the measurability of all Lebesgue sets of one kind implies the measurability of all Lebesgue sets of the other kinds. More precisely, the following theorem holds.

Theorem *Let E be a measurable set and $f : E \to \overline{\mathbb{R}}$. The following conditions are equivalent*:

(1) *the sets $E(f < a)$ are measurable for all a in $\mathbb{R}$*;
(2) *the sets $E(f \leqslant a)$ are measurable for all a in $\mathbb{R}$*;
(3) *the sets $E(f > a)$ are measurable for all a in $\mathbb{R}$*;
(4) *the sets $E(f \geqslant a)$ are measurable for all a in $\mathbb{R}$*.

Proof The proof follows the scheme (1) ⇒ (2) ⇒ (3) ⇒ (4) ⇒ (1).

Since $E(f \leqslant a) = \bigcap_{n\geqslant 1} E(f < a + 1/n)$, the first property implies the second one, which in turn implies the third one, because $E(f > a) = E \setminus E(f \leqslant a)$.

The remaining two implications can be proved in a similar way. We leave this to the reader. □

3.1.2 Now we introduce a class of functions that plays a key role in the theory of integration.

Definition Let $E \in \mathfrak{A}$ and $f : E \to \overline{\mathbb{R}}$. A function f is called *measurable* (more precisely, $\mathfrak{A}$-measurable, or measurable with respect to $\mathfrak{A}$) if its Lebesgue sets (of all four kinds) are measurable for any $a \in \mathbb{R}$.

If $E \subset \mathbb{R}^m$ and $\mathfrak{A} = \mathfrak{A}^m$ (or $\mathfrak{A} = \mathfrak{B}^m$), then measurable functions are also called Lebesgue (or Borel) measurable.

As follows from Theorem 3.1.1, for a function to be measurable it suffices that its Lebesgue sets of only one kind (the first, the second, etc.) be measurable for all $a \in \mathbb{R}$.

Remark 1 We emphasize that, when speaking about a measurable function, we always assume that it is defined on a measurable set.

Remark 2 Extending the definition, we say that a function $f : E \to \overline{\mathbb{R}}$ is measurable on a set E_0, $E_0 \subset E$, if the restriction $f|_{E_0}$ is measurable (of course, $E_0 \in \mathfrak{A}$).

Examples

(1) A constant function is measurable on every (measurable) set E. In particular, according to our definition, the function identically equal to $+\infty$ (or $-\infty$) on E is measurable.

(2) The *characteristic function* of a set $A \subset X$ is the function χ_A that is equal to one on A and zero outside A. As one can easily check by considering the Lebesgue sets of χ_A, this function is measurable if and only if the set A is measurable.

(3) Let $X = [0, 1] \times [0, 1]$, and let $\mathfrak{A}$ be the σ-algebra of sets of the form $e \times [0, 1]$, where $e \in \mathfrak{A}^1$, $e \subset [0, 1]$. Then the function $f(x, y) = y$ is not measurable with respect to $\mathfrak{A}$. However, it is obviously Lebesgue measurable.

Let us mention some simple properties of measurable functions.

(1) *The inverse images of one-point sets (including those of the points $+\infty$ and $-\infty$) are measurable.*

Indeed, if $a \in \mathbb{R}$, then $f^{-1}(\{a\}) = E(f \leqslant a) \cap E(f \geqslant a)$. Furthermore, $f^{-1}(\{+\infty\}) = \bigcap_{n\geqslant 1} E(f > n)$ and $f^{-1}(\{-\infty\}) = \bigcap_{n\geqslant 1} E(f < -n)$.

(2) *The inverse image of every interval Δ is measurable. In particular, the set on which the function takes finite values, i.e., $f^{-1}(\mathbb{R})$, is measurable.*

Indeed, by Property (1), we may assume that Δ is an open interval: $\Delta = (a,b)$. If $a, b \in \mathbb{R}$, then $f^{-1}(\Delta) = E(f < b) \setminus E(f \leqslant a) \in \mathfrak{A}$. If Δ is an infinite interval, then it can be exhausted by finite intervals: $\Delta = \bigcup_{n\geqslant 1}(a_n, b_n)$. Hence $f^{-1}(\Delta) = \bigcup_{n\geqslant 1} E(a_n < f < b_n) \in \mathfrak{A}$.

(3) *The absolute value of a measurable function is measurable*, since $E(|f| < a) = E(-a < f < a) \in \mathfrak{A}$ *for every* $a \in \mathbb{R}$.

(4) *If f and g are measurable functions, then the functions* $\varphi = \max\{f, g\}$ *and* $\psi = \min\{f, g\}$ *are also measurable. In particular, the functions* $f_+ = \max\{f, 0\}$ and $f_- = \min\{-f, 0\}$ *are measurable.*

To prove this, it suffices to observe that $E(\varphi < a) = E(f < a) \cap E(g < a)$ and $E(\psi > a) = E(f > a) \cap E(g > a)$ for every $a \in \mathbb{R}$.

(5) *The inverse image of an open set is measurable.*

By Theorem 1.1.7, a non-empty open subset G of $\mathbb{R}$ can be written in the form $G = \bigcup_{n\geqslant 1}[a_n, b_n)$. Hence the measurability of $f^{-1}(G)$ follows from the equality $f^{-1}(G) = \bigcup_{n\geqslant 1} E(a_n \leqslant f < b_n)$.

Using Theorem 1.6.1 on the inverse image of the Borel hull, we see that a more general result holds.

Proposition *For every measurable function, the inverse image of a Borel subset of the real line is measurable.*

Note that the proposition is no longer true if instead of Borel sets we consider Lebesgue measurable sets (see Exercise 5 in Sect. 2.3).

3.1.3 Let us continue to study the properties of measurable functions.

Theorem

(1) *The restriction of a measurable function to a measurable set is measurable.*

(2) *If* $E = \bigcup_{n\geqslant 1} E_n$ *and a function f is measurable on each* E_n, *then it is measurable on* E.

Proof (1) If f is defined on E and $E_0 \subset E$, then for every $a \in \mathbb{R}$, the set $E_0(f < a)$ can be written in the form $E_0(f < a) = E_0 \cap E(f < a)$ and, consequently, is measurable provided that E_0 is measurable.

(2) The measurability of f on E follows from the equality $E(f < a) = \bigcup_{n\geqslant 1} E_n(f < a)$. □

Corollary *Every measurable function f defined on E is the restriction to E of a measurable function defined on X.*

To prove this, it suffices to extend f by zero outside E. The measurability of the function obtained in this way follows from the theorem.

Remark In view of this corollary, when studying measurable functions, we may always assume that they are defined on the whole set X.

3.1.4 We proceed to the problem of passing to the limit in the class of measurable functions. We will prove that this class is closed under pointwise convergence, i.e., that the pointwise limit of a sequence of measurable functions is again a measurable function.

Recall that a function f is the pointwise limit of a sequence $\{f_n\}_{n\geqslant 1}$ on E if

$$f_n(x) \underset{n\to\infty}{\longrightarrow} f(x) \quad \text{for every point } x \text{ in } E.$$

Using Remark 3.1.3, in what follows we assume that all functions under consideration are defined on the whole set X.

Theorem *Let $\{f_n\}_{n\geqslant 1}$ be an arbitrary sequence of measurable functions, $g = \sup_n f_n$ and $h = \inf_n f_n$. Then:*

(1) *the functions g and h are measurable;*

(2) *the functions $\overline{\lim}_{n\to\infty} f_n$ and $\underline{\lim}_{n\to\infty} f_n$ are measurable; in particular, if the sequence $\{f_n\}_{n\geqslant 1}$ has a pointwise limit, then it is a measurable function.*

Since the definition of measurability allows one to consider functions with values in $\overline{\mathbb{R}}$, in the above theorem we need not make any assumptions on the finiteness of functions. In particular, for every monotone sequence of measurable functions, the limit function (possibly taking infinite values) is measurable.

Proof (1) For every $a \in \mathbb{R}$, we have

$$X(g > a) = \bigcup_{n\geqslant 1} X(f_n > a), \qquad X(h < a) = \bigcup_{n\geqslant 1} X(f_n < a);$$

the desired assertion follows by Theorem 3.1.1.

(2) It suffices to recall the formulas

$$\overline{\lim_{n\to\infty}}\, f_n(x) = \inf_{n\geqslant 1} \sup_{k\geqslant 1} f_{n+k}(x) \quad \text{and} \quad \underline{\lim_{n\to\infty}}\, f_n(x) = \sup_{n\geqslant 1} \inf_{k\geqslant 1} f_{n+k}(x)$$

known from the theory of limits and apply the first claim of the theorem. □

3.1.5 The following theorem shows the measurability of compositions.

Theorem *Let $f_1, \dots, f_n$ be measurable functions, and let $\varphi \in C(H)$, where $H \subset \mathbb{R}^n$. Assume that $(f_1(x), \dots, f_n(x)) \in H$ for every x. Then the function F defined by the formula $F(x) = \varphi(f_1(x), \dots, f_n(x))$ (for $x \in X$) is measurable.*

Proof We will use the fact that for every $a \in \mathbb{R}$, the set $H(\varphi < a)$ is relatively open in H by the continuity of f, i.e., $H(\varphi < a) = H \cap G_a$, where G_a is an open set in $\mathbb{R}^n$.

Consider the auxiliary map $U : X \to \mathbb{R}^n$ defined by the formula

$$U(x) = \big(f_1(x), \dots, f_n(x)\big) \quad (x \in X).$$

Let us check that for every open set G in $\mathbb{R}^n$, its inverse image $U^{-1}(G)$ is measurable. Indeed, the inverse image of every n-dimensional cell $P = \prod_{k=1}^{n}[a_k, b_k)$ is measurable, because

$$U^{-1}(P) = \{x \in X \mid a_k \leqslant f_k(x) < b_k \text{ for } k = 1, \ldots, n\} = \bigcap_{k=1}^{n} X(a_k \leqslant f_k < b_k).$$

Since G can be written as the union of a sequence of cells, $G = \bigcup_{j \geqslant 1} P_j$ (see Theorem 1.1.7), the set $U^{-1}(G) = \bigcup_{j \geqslant 1} U^{-1}(P_j)$ is measurable.

Thus the set

$$X(F < a) = \{x \in X \mid U(x) \in H(\varphi < a)\} = U^{-1}(H \cap G_a) = U^{-1}(G_a)$$

is also measurable. □

Using Theorem 3.1.4, we can slightly generalize the obtained result (for a further generalization, see Exercise 11).

Corollary *Theorem* 3.1.5 *remains valid if* φ *is the pointwise limit of a sequence of continuous functions* $\{\varphi_k\}_{k \geqslant 1}$.

To prove this corollary, it suffices to observe that $F = \varphi \circ U$ is the limit of the measurable functions $F_k = \varphi_k \circ U$, where U is the map defined in the proof of the theorem.

3.1.6 Now let us discuss the arithmetic operations on measurable functions. Since we allow measurable functions to take infinite values, we need to specify what we mean by the sum and the product of such functions. For the sum, this is necessary if the summands are infinities of opposite sign, and for the product, if one of the factors is infinite and the other one is equal to zero. To avoid repeatedly making stipulations, we extend the arithmetic operations to $\overline{\mathbb{R}}$ according to the following definition.

Definition

(1) If $x \in \overline{\mathbb{R}}$ and $x \neq 0$, then $x \cdot (\pm\infty) = (\pm\infty) \cdot x = \pm\infty$ for $x > 0$ and $x \cdot (\pm\infty) = (\pm\infty) \cdot x = \mp\infty$ for $x < 0$.
(2) For every $x \in \overline{\mathbb{R}}$, we set $0 \cdot x = x \cdot 0 = 0$.
(3) For every $x \in \overline{\mathbb{R}}$, we set $x/(\pm\infty) = 0$ (in particular, $(\pm\infty)/(\pm\infty) = 0$).
(4) For every $x \in \mathbb{R}$, we set $x + (+\infty) = x - (-\infty) = (+\infty) + x = +\infty$, $x + (-\infty) = x - (+\infty) = (-\infty) + x = -\infty$.
(5) $(+\infty) + (-\infty) = (-\infty) + (+\infty) = (+\infty) - (+\infty) = (-\infty) - (-\infty) = 0$.

As in $\mathbb{R}$, division by zero is not defined in $\overline{\mathbb{R}}$.

The first four conventions introduced above do not violate the associativity of the arithmetic operations. In view of the fifth convention, addition in $\overline{\mathbb{R}}$ is no longer associative. This will not cause considerable trouble, because we mainly deal with functions that take infinite values only on sets of zero measure, which, as will be seen from what follows, can be neglected (for more details, see Sect. 4.3).

Theorem *The following statements are true*:

(1) *The product and the sum of measurable functions are measurable.*
(2) *If a function f is measurable and a function φ is continuous, and their composition $\varphi \circ f$ is well defined, then it is measurable.*
(2′) *If $f \geqslant 0$ and $p > 0$, then the function f^p is measurable (in the case where f takes infinite values, we assume that $(+\infty)^p = +\infty$).*
(3) *The function $1/f$ is measurable on the set where $f \neq 0$.*

Proof Let f and g be functions defined on a set E.

(1) If f and g take only finite values, then the measurability of their product follows immediately from the previous theorem in which we put $\varphi(x, y) = xy$. If f and g may take infinite values, consider the sets

$$E_0(f) = E(f = 0), \quad E_1(f) = E(0 < f < +\infty), \quad E_2(f) = E(-\infty < f < 0),$$
$$E_3(f) = E(f = -\infty), \quad E_4(f) = E(f = +\infty)$$

and the similar sets $E_k(g)$. By the above, the product fg is measurable on $E_j(f) \cap E_k(g)$ for $j, k = 1, 2$ and constant on such intersections for other values of j, k ($j, k = 0, \ldots, 4$). Therefore, by Theorem 3.1.3, it is also measurable on the union of these sets, i.e., on E. The measurability of the sum can be proved in a similar way.

(2) This is a special case of Theorem 3.1.5.

(2′) The function f^p is measurable on the set $E(f < +\infty)$ by the previous claim of the theorem and constant on the set $E(f = +\infty)$. Therefore, it is also measurable on the union of these sets, i.e., on E.

(3) The set $\widetilde{E} = E(f \neq 0)$ is obviously measurable. Furthermore,

$$\widetilde{E}\left(\frac{1}{f} < a\right) = \begin{cases} E(f < 0) \cup E\left(f > \frac{1}{a}\right) & \text{for } a > 0, \\ E(-\infty < f < 0) & \text{for } a = 0, \\ E\left(\frac{1}{a} < f < 0\right) & \text{for } a < 0. \end{cases}$$

In all cases, the Lebesgue sets of the function $1/f$ are measurable. □

Corollary 1 *The product of a finite family of measurable functions is measurable.*

Corollary 2 *A positive integer power of a measurable function f is measurable*; *a negative integer power is measurable on the set where $f \neq 0$.*

Corollary 3 *A linear combination of measurable functions is measurable.*

3.1.7 In conclusion, we consider the question of the measurability of a function $f : E \to \mathbb{R}$ defined on a Lebesgue measurable subset E of the space $\mathbb{R}^m$. We denote the Lebesgue measure on $\mathbb{R}^m$ by λ without indicating the dimension.

Theorem *Let f be a function defined on a set E, $E \in \mathfrak{A}^m$, that takes only finite values. If for every $\varepsilon > 0$ there exists a measurable set $e \subset E$ such that*

$$\lambda(e) < \varepsilon \quad \text{and} \quad \text{the restriction of } f \text{ to } E \setminus e \text{ is continuous,} \tag{C}$$

then f is Lebesgue measurable. In particular, every function continuous on E is Lebesgue measurable.

Remark Condition (C) means that f will be continuous if we remove from its domain a set of arbitrarily small measure. It is this condition that Luzin called the C-property. He proved that it is not only sufficient, but also necessary for a function to be Lebesgue measurable, i.e., that every Lebesgue measurable function satisfies the C-property. We will return to this topic in Sect. 3.4.3.

It obviously follows from the last theorem that every function whose set of discontinuities has zero measure is measurable. However, the theorem allows one to establish the measurability of functions with "large" sets of discontinuities. An example of this kind is the Dirichlet function. As one can easily see, it is discontinuous at every point. However, its restriction to the set of irrational numbers is continuous (being identically zero). Hence the Dirichlet function satisfies condition (C) and, consequently, is measurable. On the other hand, its measurability is obvious without the theorem, since it is the characteristic function of the measurable set $\mathbb{Q}$.

Proof If f is continuous, then the Lebesgue set $E(f < a) = f^{-1}((-\infty, a))$ is relatively open in E. Hence it is the intersection of E with some set open in $\mathbb{R}^m$ and, consequently, is measurable as the intersection of measurable sets.

If f is an arbitrary function satisfying condition (C), consider sets $e_n \subset E$ (where $n \in \mathbb{N}$) such that

$$\lambda(e_n) < \frac{1}{n} \quad \text{and} \quad \text{the restriction of } f \text{ to } E_n \equiv E \setminus e_n \text{ is continuous.}$$

Put $E_0 = \bigcap_{n \geqslant 1} e_n$. Obviously, $\lambda(E_0) = 0$, and hence f is measurable on E_0 (since in the case of a complete measure, every function is measurable on a set of zero measure). Thus $E = \bigcup_{n \geqslant 0} E_n$, and, as we have already proved, f is measurable on each of the sets E_n. It remains to apply Theorem 3.1.3. □

EXERCISES

1. Establish the Lebesgue measurability of a monotone function φ defined on an arbitrary interval and of its composition $\varphi \circ f$ with every measurable function f (provided that this composition is well defined).
2. Let $\{f_n\}_{n \geqslant 1}$ be an arbitrary sequence of measurable functions. Establish the measurability of the sets

$$\Big\{x \in X \mid \exists \lim_{n \to \infty} f_n(x) \in \overline{\mathbb{R}}\Big\} \quad \text{and}$$

$$\big\{x \in X \mid \text{the sequence } \{f_n(x)\}_{n \geqslant 1} \text{ converges}\big\}.$$

3. Give an example of a (Lebesgue) measurable bounded function on $\mathbb{R}$ that is "so discontinuous" that one cannot make it continuous even at a single point by modifying it at a set of zero measure. *Hint*. Consider the characteristic function of the set from Exercise 9 in Sect. 2.1.

4. Show that the characteristic function of the set constructed in Exercise 8 of Sect. 2.1 satisfies condition (C).

5. Let K be a compact subset of $\mathbb{R}^{m+1} = \mathbb{R}^m \times \mathbb{R}$, P be the canonical projection of $\mathbb{R}^{m+1}$ to $\mathbb{R}^m$, and $Q = P(K)$. Show that there exists a function $f : \mathbb{R}^m \to \mathbb{R}$ such that the graph of its restriction to Q is contained in K and the set of its discontinuities has zero measure.

6. Using the result of Exercise 5 from Sect. 2.3, show that Theorem 3.1.5 is no longer true if instead of $\varphi \circ f$ one considers $f \circ \varphi$.

7. Let $f : \mathbb{R} \to \mathbb{R}$ be an arbitrary (possibly non-measurable) function. Show that the set of points where it is differentiable is measurable and the function $\overline{f'}$ defined by the formula $\overline{f'}(x) = \overline{\lim}_{y \to x} \frac{f(y)-f(x)}{y-x}$ is also measurable. Show that the function $\overline{f'}_+(x) = \overline{\lim}_{y \to x+0} \frac{f(y)-f(x)}{y-x}$ can be non-measurable.

8. The Rademacher functions r_n $(n \in \mathbb{N})$ are defined on $\mathbb{R}$ by the formula $r_n(x) = \operatorname{sign} \sin 2^n \pi x$ (see Sect. 6.4.5). Show that

$$\lambda\big(\{x \in (0,1) \,|\, r_{n_j}(x) < a_j \ \text{ for } \ j = 1, \dots, k\}\big)$$

$$= \prod_{j=1}^{k} \lambda\big(\{x \in (0,1) \,|\, r_{n_j}(x) < a_j\}\big)$$

for any $a_1, \dots, a_k \in \mathbb{R}$ and pairwise distinct $n_1, \dots, n_k \in \mathbb{N}$.
In probability theory, functions satisfying this condition are called statistically independent (see Sect. 6.4.4).

9. We say that a function f defined on $\mathbb{R}^m$ is radial if it is of the form $f(x) = f_0(\|x\|)$, where f_0 is a function defined on $\mathbb{R}_+$. Using Exercise 3 from Sect. 2.5, show that f is measurable if and only if f_0 is measurable.

10. Let $(X, \mathfrak{A})$ be a measurable space. A map $F : X \to \mathbb{R}^m$ is called measurable if at least one of the following conditions is satisfied:

(a) its coordinate functions are measurable;
(b) the inverse images of Borel sets are measurable;
(c) the inverse images of cells are measurable;
(d) the inverse image of every open set is measurable.

Show that conditions (a)–(d) are equivalent.

11. Let $F : X \to \mathbb{R}^m$ be a measurable map (see the previous exercise). Show that for every Borel measurable function $\varphi : \mathbb{R}^m \to \mathbb{R}$, the composition $\varphi \circ F$ is measurable.

12. Let E be an arbitrary subset of $\mathbb{R}^m$ and $f : E \to \mathbb{R}$ be an arbitrary function. Given $x \in \overline{E}$, put

$$g(x) = \lim_{r \to 0} \sup_{X \cap B(x,r)} f \quad \text{and} \quad h(x) = \lim_{r \to 0} \inf_{X \cap B(x,r)} f$$

(the functions g and h may take infinite values). Show that the sets $\overline{E}(g < a)$ and $\overline{E}(h > a)$ are relatively open in $\overline{E}$ and, therefore, the functions g and h are Borel measurable.

3.2 Simple Functions. The Approximation Theorem

As in the previous section, we assume that there is a fixed measurable space $(X, \mathfrak{A})$. All functions under consideration are defined on X.

3.2.1 We introduce a subclass of measurable functions, which will later be used systematically for the approximation.

Definition An $\mathbb{R}$-valued measurable function is called *simple* if the set of its values is finite.

If f is a simple function, there is a finite partition of X into measurable sets (we will call it *admissible* for f) such that f is constant on its elements. For instance, such a partition can be obtained as follows. Let $a_1, \dots, a_N$ be all pairwise distinct values of f. Put $e_k = f^{-1}(\{a_k\})$. Obviously, these sets are measurable and form a partition of X that is admissible for f.

In general, an admissible partition is not unique: splitting any of its elements into measurable parts, we will obtain a "finer" admissible partition. Thus f may take equal values on different elements of an admissible partition. Furthermore, we do not exclude the case where some of the elements are empty.

Example The characteristic function χ_E of a set E is simple if and only if E is measurable. In this case, the sets E, $X \setminus E$ form an admissible partition for χ_E. The family $\{E, X \setminus E, \varnothing\}$ is also an admissible partition for χ_E.

Let us mention some basic properties of simple functions.

(1) Every $\mathbb{R}$-valued function that is constant on the elements of some finite partition of X into measurable sets is simple.
Indeed, the set of values of such a function is finite, and the measurability follows, for example, from Theorem 3.1.3.

(2) Any two simple functions f and g have a common admissible partition.
Indeed, a desired partition consists, for example, of the sets $e_k \cap e'_j$, where $\{e_k\}_{k=1}^n$ and $\{e'_j\}_{j=1}^m$ are admissible partitions for f and g, respectively.

(3) The sum and the product of two simple functions is a simple function.
This fact follows immediately from the existence of a common admissible partition and Property (1).

(3′) A linear combination and the product of a finite family of simple functions are simple functions.

(4) The maximum and the minimum of a finite family of simple functions are simple functions.
To prove this, it suffices to consider a partition that is admissible for all functions of the given family.

3.2.2 The next theorem shows, in particular, that every measurable function is the pointwise limit of a sequence of simple functions. This result is not only an important technical tool which we will repeatedly use in what follows, it can be regarded as an alternative definition of a measurable function: a function is called measurable if it is the pointwise limit of a sequence of simple functions. In contrast to the purely descriptive definition given in the previous section, the new definition provides a method of constructing arbitrary measurable functions starting from the more tractable functions which we have called simple. In this sense, one may say that the new definition is constructive. The equivalence of these two definitions, which follows from the theorem proved below, is further evidence that the class of measurable functions is very natural. Indeed, in Theorem 3.1.4 we have shown that it is sufficiently wide to contain, together with every pointwise convergent sequence, the limit of this sequence. Accordingly, the question might arise whether the class of measurable functions is not too wide. Indeed, if we consider the space $\mathbb{R}^m$ with the Lebesgue measure, this class contains not only functions that are discontinuous at every point (for example, the Dirichlet function, i.e., the characteristic function of the set of rational points), but even functions that (in contrast to the Dirichlet function) cannot be made continuous by modifying them on a set of zero measure (see Exercises 3 and 4 in Sect. 3.1). However, it follows from the theorem proved below that if we assume that the class in question is closed under pointwise limits and contains the characteristic functions of measurable sets as well as their linear combinations, then no proper part of the class of all measurable functions will suffice: together with characteristic functions this class contains all simple functions, whose pointwise limits yield all measurable functions.

Theorem (Approximation by simple functions) *Every non-negative measurable function $f : X \to \overline{\mathbb{R}}$ is the pointwise limit of an increasing sequence of non-negative simple functions f_n. If f is bounded, then we may assume that the sequence $\{f_n\}_{n\geqslant 1}$ converges uniformly on X.*

Proof Fix a positive integer n and consider the intervals $\Delta_k = [k/n, (k+1)/n)$ for $k = 0, 1, \ldots, n^2 - 1$ and the interval $\Delta_{n^2} = [n, +\infty]$. Obviously, they form a partition of the set $[0, +\infty]$. Consider the sets $e_k = f^{-1}(\Delta_k)$ $(k = 0, 1, \ldots, n^2)$. They are measurable and form a partition of the set X. (It would be more accurate to write $\Delta_k^{(n)}$ and $e_k^{(n)}$, indicating that these sets depend not only on k, but also on n, but we will not do this.) Put

$$g_n(x) = \frac{k}{n} \quad \text{for } x \in e_k$$

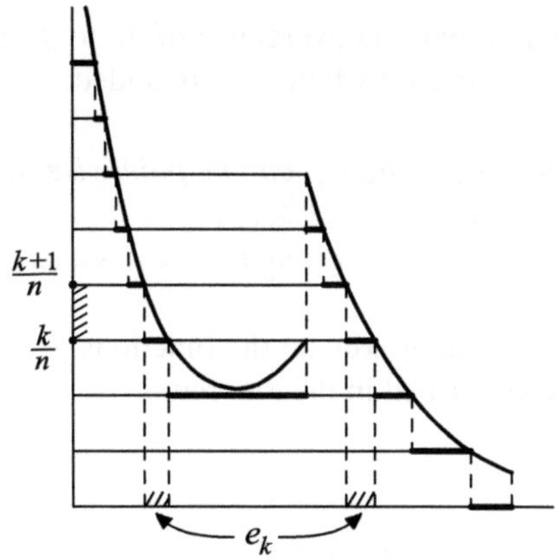

Fig. 3.2 Graphs of functions f and g_n

(the graph of this function is schematically shown in Fig. 3.2 by horizontal bold line segments). Obviously,

$$0 \leqslant g_n(x) \leqslant f(x) \quad \text{for every } x \in X. \tag{1}$$

Furthermore,

$$g_n(x) \leqslant f(x) \leqslant g_n(x) + \frac{1}{n}, \quad \text{if } x \notin e_{n^2}. \tag{2}$$

Now let us check that the constructed sequence $\{g_n\}_{n \geqslant 1}$ converges pointwise to f. Consider an arbitrary point $x \in X$. If $f(x) = +\infty$, then $x \in e_{n^2}$ for every n, whence

$$g_n(x) = n \underset{n \to \infty}{\longrightarrow} +\infty = f(x).$$

If $f(x) < +\infty$, then $x \notin e_{n^2}$ for

$$n > f(x). \tag{3}$$

Then, by (2),

$$0 \leqslant f(x) - g_n(x) \leqslant \frac{1}{n} \underset{n \to \infty}{\longrightarrow} 0. \tag{4}$$

If f is bounded and $f(x) \leqslant C$ for all $x \in X$, then, taking $n > C$, we see that inequalities (3) and hence (4) are satisfied simultaneously for all $x \in X$, which implies the uniform convergence.

Thus the constructed functions g_n have all the desired properties except for one. In general, they do not form an increasing sequence. Hence we need to slightly modify them. Put $f_n = \max\{g_1, \dots, g_n\}$. Obviously, the functions f_n are simple and $f_n \leqslant f_{n+1}$. In addition, it follows from (1) that

$$0 \leqslant g_n(x) \leqslant f_n(x) \leqslant f(x) \quad \text{for every } x \in X.$$

This guarantees both the pointwise convergence of f_n to f in the general case and the uniform convergence in the case where f is bounded. □

Corollary *Every measurable function f can be pointwise approximated by simple functions f_n satisfying the condition $|f_n| \leqslant |f|$.*

If f is bounded, then this approximation may be assumed uniform.

To prove this, it suffices to approximate the functions $f_+ = \max\{f, 0\}$ and $f_- = \max\{-f, 0\}$ separately as described in the theorem.

EXERCISES

1. Let $\{g_n\}_{n\geqslant 1}$ be the sequence constructed in the proof of Theorem 3.2.2, and let $h_k = g_{2^k}$. Show that the sequence $\{h_k\}_{k\geqslant 1}$ is increasing.

2. Show that every non-negative measurable function f on a set X can be written as the sum of a series $\sum_{n=1}^{\infty} \frac{1}{n}\chi_{A_n}$. *Hint*. Consider the sets

$$A_1 = \{x \in X \mid f(x) \geqslant 1\},$$

$$A_n = \left\{x \in X \,\middle|\, f(x) \geqslant \frac{1}{n} + \sum_{k=1}^{n-1} \frac{1}{k}\chi_{A_k}\right\} \quad \text{for } n \geqslant 2.$$

3.3 Convergence in Measure and Convergence Almost Everywhere

From a course in analysis the reader already knows two types of convergence of sequences of functions: pointwise and uniform. Now we will define two further types of convergence, which play an important role in the theory of integration and probability. Both of them apply to functions defined on a measure space.

We assume that a measure space $(X, \mathfrak{A}, \mu)$ is fixed. All sets we deal with are assumed measurable, i.e., they belong to the σ-algebra $\mathfrak{A}$. All functions are also assumed measurable, and furthermore we assume that they are finite almost everywhere, i.e., may take infinite values only on sets of zero measure. The class of all such functions on a set E will be denoted by $\mathcal{L}^0(E, \mu)$ or merely by $\mathcal{L}^0(E)$. Everywhere in this section (except for Sect. 3.3.7), we consider functions only from this class.

The pointwise convergence of a sequence $\{f_n\}_{n\geqslant 1}$ to a function f will be denoted, as usual, by a simple arrow, $f_n \underset{n\to\infty}{\longrightarrow} f$, and the uniform convergence will be denoted by a double arrow: $f_n \underset{n\to\infty}{\rightrightarrows} f$. Recall that χ_E stands for the characteristic function of a set E, and the set $\{x \in E \mid f(x) > a\}$ is also denoted by $E(f > a)$.

3.3.1 We introduce an important new type of convergence of functional sequences.

Definition 1 A sequence of functions $f_n \in \mathcal{L}^0(E,\mu)$ *converges to a function* $f \in \mathcal{L}^0(E,\mu)$ *in measure* (notation: $f_n \xrightarrow[n\to\infty]{\mu} f$) if

$$\mu\big(E(|f_n - f| > \varepsilon)\big) \underset{n\to\infty}{\longrightarrow} 0 \quad \text{for every positive } \varepsilon.$$

Thus $f_n \xrightarrow[n\to\infty]{\mu} f$ if for sufficiently large n each of the functions f_n is uniformly close to f on the set obtained from E by removing a subset of arbitrarily small measure. It is worth mentioning that, in general, the subset to be removed differs for each n and one cannot generally remove a single set outside of which all functions f_n with sufficiently large indices are uniformly close to the limit function.

Extending the definition, we say that a sequence $\{f_n\}_{n\geqslant 1}$ converges in measure on a set $\widetilde{E}$, $\widetilde{E} \subset E$, to a function $f \in \mathcal{L}^0(\widetilde{E})$ if the sequence $\widetilde{f}_n = f_n|_{\widetilde{E}}$ converges in measure to f. This is obviously equivalent to the condition that the sequence $\{f_n\chi_{\widetilde{E}}\}_{n\geqslant 1}$ converges in measure to the function f extended by zero from $\widetilde{E}$ to E. This observation allows us to assume, when discussing convergence in measure, that the functions under consideration are defined on the whole of X, since otherwise we can extend them to X by zero.

Let us discuss how convergence in measure is related to other types of convergence. Obviously, uniform convergence implies convergence in measure; indeed, in the case of uniform convergence, for every $\varepsilon > 0$ the set $E(|f_n - f| > \varepsilon)$ is empty for sufficiently large n. However, this is no longer true if we replace uniform convergence with pointwise convergence. To obtain a corresponding example, it suffices to consider the real line with the Lebesgue measure and the functions $\chi_{(n,+\infty)}$ or $\chi_{(n,n+1)}$, which converge to zero in $\mathbb{R}$ pointwise, but not in measure. The reader can easily check that these sequences have no limit in the sense of convergence in measure.

Of course, convergence in measure does not imply pointwise convergence. Indeed, if a sequence of functions f_n converges both pointwise and in measure (as is the case, for example, if the sequence converges uniformly), then we may break the pointwise convergence by modifying the values of f_n on sets of zero measure. However, this does not affect the convergence in measure, as follows from its definition. Hence it is natural to compare convergence in measure with "weakened pointwise convergence", which is insensitive to modifications of functions on sets of zero measure. We make the following definition.

Definition 2 A sequence of measurable functions $f_n : E \to \overline{\mathbb{R}}$ *converges to a function* f *almost everywhere on* E (notation: $f_n \xrightarrow[n\to\infty]{\text{a.e.}} f$) if there exists a set $e \subset E$ of zero measure such that $f_n \underset{n\to\infty}{\longrightarrow} f$ pointwise on $E \setminus e$.

In this definition (as well as in the previous one), we assume that there is a fixed measure μ. If we also consider other measures, then we speak about convergence μ-almost everywhere (respectively, convergence in measure with respect to μ).

By Theorem 3.1.4, the limit function f is measurable on the set $E \setminus e$. If the measure μ is complete, then f is measurable not only on $E \setminus e$, but also on E. If (in the case of a non-complete measure) f is not measurable on E, then, modifying it on a set of zero measure (for example, setting it equal to zero on e), we can obtain a measurable function that is the limit of the sequence $\{f_n\}_{n\geqslant 1}$ in the sense of almost everywhere convergence.

Formally speaking, we can drop the condition of measurability of f_n in Definition 2, but we will not need such a generalization.

There is a subtle relation between convergence in measure and almost everywhere convergence, see H. Lebesgue's and F. Riesz's theorems proved in this section. But we begin with a counterexample showing that almost everywhere convergence does not follow from convergence in measure.

Example Let $X = \mathbb{R}$ and $\mu = \lambda$ be the one-dimensional Lebesgue measure. For every positive integer k, consider the partition of the interval $[0, 1)$ into the subintervals $\Delta(k, p) = [\frac{p}{2^k}, \frac{p+1}{2^k})$, where $p = 0, 1, \dots, 2^k - 1$. To define a function f_n, write the index $n > 1$ in the form $n = 2^k + p$, where $0 \leqslant p < 2^k$ (such a representation is obviously unique, and k is just the integer part of $\log_2 n$), and set $f_n = \chi_{\Delta(k,p)}$. Since

$$X(f_n \neq 0) = \Delta(k, p) \quad \text{and} \quad \lambda\big(\Delta(k, p)\big) = \frac{1}{2^k} \leqslant \frac{2}{n} \underset{n\to\infty}{\longrightarrow} 0,$$

the constructed sequence converges in measure to zero. However, the numerical sequence $\{f_n(x)\}_{n\geqslant 1}$ has no limit for any $x \in [0, 1)$, since among the values $f_n(x)$ there are infinitely many ones and zeros.

3.3.2 As we have observed, convergence in measure does not follow from almost everywhere convergence. However, the situation changes dramatically if the set under consideration has finite measure.

Theorem (Lebesgue) *On a set of finite measure, almost everywhere convergence implies convergence in measure.*

Proof Let $f_n \overset{\text{a.e.}}{\underset{n\to\infty}{\longrightarrow}} f$ on E and $\mu(E) < +\infty$. Redefining the functions, if necessary, on a set of zero measure (for example, setting them equal to zero on this set), we assume that $f_n \underset{n\to\infty}{\longrightarrow} f$ everywhere on E.

For a monotone sequence $\{f_n\}_{n\geqslant 1}$ that converges pointwise to zero, the desired assertion is almost obvious. Indeed, in this case, for every $\varepsilon > 0$ the sets $E(|f_n| > \varepsilon)$ decrease as n grows and have an empty intersection. Since the measure is continuous from above, $\mu(E(|f_n| > \varepsilon)) \underset{n\to\infty}{\longrightarrow} 0$ (it is here that the condition $\mu(E) < +\infty$ is crucial). Thus we have established the convergence of $\{f_n\}_{n\geqslant 1}$ in measure in the special case under consideration.

In the general case, where $f_n(x) \underset{n\to\infty}{\longrightarrow} f(x)$ for all $x \in E$, we apply the result already proved to the functions $\varphi_n(x) = \sup_{k\geqslant n} |f_k(x) - f(x)|$. Clearly,

$\varphi_n(x) \underset{n\to\infty}{\longrightarrow} 0$ monotonically everywhere, and, by the above, $\mu(E(\varphi_n > \varepsilon)) \underset{n\to\infty}{\longrightarrow} 0$. It remains to use the inclusion $E(|f_n - f| > \varepsilon) \subset E(\varphi_n > \varepsilon)$, which follows from the inequality $|f_n - f| \leqslant \varphi_n$:

$$\mu\big(E(|f_n - f| > \varepsilon)\big) \leqslant \mu\big(E(\varphi_n > \varepsilon)\big) \underset{n\to\infty}{\longrightarrow} 0. \qquad \square$$

3.3.3 Before continuing to discuss the relations between convergence in measure and almost everywhere convergence, we prove a simple but important result often used in probability theory.

Lemma (Borel–Cantelli[1]) *Let $\{E_n\}_{n\geqslant 1}$ be a sequence of measurable sets and $E = \bigcap_{n=1}^{\infty}\bigcup_{k=n}^{\infty} E_k$, i.e.,*

$$E = \big\{x \in X \mid x \in E_n \text{ for infinitely many } n\big\}.$$

If $\sum_{n\geqslant 1}\mu(E_n) < +\infty$, then $\mu(E) = 0$.

Proof Since $E \subset \bigcup_{n\geqslant k} E_n$, we have $\mu(E) \leqslant \sum_{n\geqslant k}\mu(E_n) \underset{k\to\infty}{\longrightarrow} 0$. $\square$

This lemma implies a useful criterion for almost everywhere convergence.

Corollary *Let $\varepsilon_n > 0$, $\varepsilon_n \underset{n\to\infty}{\longrightarrow} 0$, $g_n \in \mathcal{L}^0(X,\mu)$, and $X_n = X(|g_n| > \varepsilon_n)$. If $\sum_{n\geqslant 1}\mu(X_n) < +\infty$, then $g_n \overset{\text{a.e.}}{\underset{n\to\infty}{\longrightarrow}} 0$. Furthermore, for every $\varepsilon > 0$ there exists a set e such that*

$$\mu(e) < \varepsilon \quad \text{and} \quad g_n(x) \underset{n\to\infty}{\rightrightarrows} 0 \text{ on } X \setminus e.$$

To prove the almost everywhere convergence, one should, given an arbitrary $\varepsilon > 0$, apply the Borel–Cantelli lemma to the sets $E_n = X(|g_n| > \varepsilon)$, taking into account that $E_n \subset X_n$ for sufficiently large n.

To prove the second claim of the corollary, choose N so large that

$$\sum_{n>N}\mu(X_n) < \varepsilon$$

and put $e = \bigcup_{n>N} X_n$. Then $|g_n(x)| < \varepsilon_n$ for $x \in X \setminus e$ and $n > N$.

3.3.4 Let us return to discussing the relations between almost everywhere convergence and convergence in measure. As we have seen, a sequence that converges in measure may be divergent at every point. However, the situation changes if we consider subsequences.

[1] Francesco Paolo Cantelli (1875–1966)—Italian mathematician.

Theorem (F. Riesz[2]) *Every sequence that converges in measure contains a subsequence that converges almost everywhere to the same limit.*

Note that, in contrast to Lebesgue's theorem, here we do not assume that the measure is finite.

Proof Let $f_n \xrightarrow[n\to\infty]{\mu} f$. Then

$$\mu\left(X\left(|f_n - f| > \frac{1}{k}\right)\right) \underset{n\to\infty}{\longrightarrow} 0$$

for every $k \in \mathbb{N}$. Hence there exists an increasing sequence of indices n_k such that

$$\mu\left(X\left(|f_n - f| > \frac{1}{k}\right)\right) < \frac{1}{2^k} \quad \text{for all } n \geqslant n_k.$$

The sequence $\{f_{n_k}\}_{k\geqslant 1}$ has the desired property. Indeed, applying the corollary of the Borel–Cantelli lemma to the functions $g_k = |f_{n_k} - f|$, we see that $g_k \xrightarrow[k\to\infty]{\text{a.e.}} 0$, i.e., $f_{n_k} \xrightarrow[k\to\infty]{\text{a.e.}} f$. □

Remark The subsequence constructed in the proof of Riesz's theorem, besides being almost everywhere convergent, has another useful (and stronger) property. Namely, for every $\varepsilon > 0$ there exists a set e such that

$$\mu(e) < \varepsilon \quad \text{and} \quad f_{n_k} \underset{k\to\infty}{\rightrightarrows} f \text{ on } X \setminus e.$$

To prove this, it suffices to apply the definition of the functions f_{n_k} and the corollary of the Borel–Cantelli lemma.

3.3.5 Using Riesz's theorem, one can reduce some questions about convergence in measure to similar questions about almost everywhere convergence. As examples, consider the problems related to the uniqueness of the limit and passing to the limit in inequalities.

Corollary 1 *If a sequence $\{f_n\}_{n\geqslant 1}$ converges in measure to functions f and g, then $f(x) = g(x)$ for almost all x.*

Proof By Riesz's theorem, there exists a subsequence $\{f_{n_k}\}_{k\geqslant 1}$ that converges to f almost everywhere. Since the subsequence $\{f_{n_k}\}_{k\geqslant 1}$, along with the original sequence, converges in measure to g, again applying Riesz's theorem, we can find a subsequence $\{f_{n_{k_j}}\}_{j\geqslant 1}$ that converges almost everywhere to g. Thus the functions

[2]Frigyes Riesz (1880–1956)—Hungarian mathematician.

f and g coincide almost everywhere as limits of the almost everywhere convergent sequence $\{f_{n_{k_j}}\}_{j\geqslant 1}$. □

Corollary 2 *If* $f_n \xrightarrow[n\to\infty]{\mu} f$ *and* $f_n \leqslant g$ *almost everywhere for every* n, *then* $f \leqslant g$ *almost everywhere on* E.

Proof Let f_{n_k} be a subsequence that converges to f almost everywhere. By our condition, $f_{n_k} \leqslant g$ outside of some set e_k of zero measure. Putting $e = \bigcup_{k=1}^{\infty} e_k$, we obtain a set of zero measure such that for any $x \notin e$ and $k \in \mathbb{N}$ the inequality $f_{n_k}(x) \leqslant g(x)$ holds. It remains to pass to the limit as $k \to \infty$. □

3.3.6 Almost everywhere convergence is closely related to a stronger type of convergence which we now define.

Definition We say that a sequence $\{f_n\}_{n\geqslant 1}$ converges to f *almost uniformly* on X if for every positive ε there exists a set A_ε such that

$$\mu(A_\varepsilon) < \varepsilon \quad \text{and} \quad f_n \underset{n\to\infty}{\rightrightarrows} f \text{ on } X \setminus A_\varepsilon.$$

Almost uniform convergence implies almost everywhere convergence. Indeed, the sequence $\{f_n\}_{n\geqslant 1}$ converges pointwise outside of each set $A_{1/k}$, and hence outside of their intersection $\bigcap_{k\geqslant 1} A_{1/k}$, which obviously has zero measure. As we observed after Riesz's theorem (see Remark 3.3.4), the sequence constructed in its proof converges not only almost everywhere, but almost uniformly.

Surprisingly, we have the following unexpected result: on a set of finite measure, almost uniform convergence is equivalent to almost everywhere convergence.

Theorem (Egorov[3]) *Let* $f_n, f \in \mathcal{L}^0(X,\mu)$, *and let* $f_n \xrightarrow[n\to\infty]{\text{a.e.}} f$. *If* $\mu(X) < +\infty$, *then* $f_n \underset{n\to\infty}{\longrightarrow} f$ *almost uniformly on* X.

Considering the sequence $\chi_{(n,n+1)}$ shows that this theorem cannot be extended to sets of infinite measure.

Proof Put $g_n(x) = \sup_{k\geqslant n} |f_k(x) - f(x)|$. Clearly, $g_n \xrightarrow[n\to\infty]{\text{a.e.}} 0$. By Lebesgue's theorem (see Sect. 3.3.2), $g_n \xrightarrow[n\to\infty]{\mu} 0$ (it is here that the finiteness of μ is crucial). Hence (cf. the proof of Riesz's theorem) there exists a subsequence $\{g_{n_k}\}_{k\geqslant 1}$ such that

$$\mu\left(X\left(g_{n_k} > \frac{1}{k}\right)\right) < \frac{1}{2^k}.$$

[3]Dmitri Fyodorovich Egorov (1869–1931)—Russian mathematician.

By the corollary of the Borel–Cantelli lemma, this subsequence converges to zero almost uniformly. Since $|f_n - f| \leqslant g_{n_k}$ for $n \geqslant n_k$, the sequence $\{f_n - f\}_{n\geqslant 1}$ also converges to zero almost uniformly. □

3.3.7 In conclusion, we establish another useful property of almost everywhere convergence.

Theorem (Diagonal sequence) *Let μ be a σ-finite measure, and let $f_k^{(n)} \in \mathcal{L}^0(X,\mu)$, $g_n \in \mathcal{L}^0(X,\mu)$ for $n,k \in \mathbb{N}$. If $f_k^{(n)} \xrightarrow[k\to\infty]{\text{a.e.}} g_n$ for every $n \in \mathbb{N}$ and $g_n \xrightarrow[n\to\infty]{\text{a.e.}} h$, then there exists a strictly increasing sequence of indices k_n such that $f_{k_n}^{(n)} \xrightarrow[n\to\infty]{\text{a.e.}} h$.*

Note that h, in contrast to $f_k^{(n)}$ and g_n, may take infinite values on sets of positive measure.

Proof First assume that the measure is finite. Then $f_k^{(n)} \xrightarrow[k\to\infty]{\mu} g_n$ by Lebesgue's theorem. This means that

$$\mu\big(X(|f_k^{(n)} - g_n| > \varepsilon)\big) \underset{k\to\infty}{\longrightarrow} 0 \quad \text{for every } n \in \mathbb{N} \text{ and every } \varepsilon > 0.$$

Hence for every n there exists an index k_n ($k_n > k_{n-1}$) such that

$$\mu\left(X\left(\left|f_{k_n}^{(n)} - g_n\right| > \frac{1}{n}\right)\right) < \frac{1}{2^n}.$$

By the corollary of the Borel–Cantelli lemma, $f_{k_n}^{(n)} - g_n \xrightarrow[n\to\infty]{\text{a.e.}} 0$. Thus

$$f_{k_n}^{(n)} = \big(f_{k_n}^{(n)} - g_n\big) + g_n \xrightarrow[n\to\infty]{\text{a.e.}} h,$$

which completes the proof of the theorem for a finite measure. □

The case of an infinite measure can be reduced to that considered above by the following lemma.

Lemma *If μ is a σ-finite measure, then there exists a finite measure ν such that $\nu(E) = 0$ if and only if $\mu(E) = 0$.*

Thus "almost everywhere" assertions for the measures μ and ν hold simultaneously. Hence we may assume without loss of generality that the measure μ in the diagonal sequence theorem is finite.

Proof of the Lemma Let $X = \bigcup_{n=1}^{\infty} X_n$, where $0 < \mu(X_n) < +\infty$. We will obtain a measure with the desired property by putting

$$\nu(E) = \sum_{n\geqslant 1} \frac{1}{2^n} \frac{\mu(E \cap X_n)}{\mu(X_n)}$$

for every measurable set E.

The reader can easily check that ν is a measure and that μ and ν vanish on the same sets. □

Remark The diagonal sequence theorem is no longer true if we replace almost everywhere convergence by pointwise convergence, see Exercise 6. Since pointwise convergence can also be interpreted as almost everywhere convergence with respect to the counting measure, this exercise also shows that in the diagonal sequence theorem one cannot drop the condition of σ-finiteness.

EXERCISES

1. Let $f_n \underset{n\to\infty}{\longrightarrow} f$ in measure. Show that if $\mu(X) < +\infty$ and $g \in \mathcal{L}^0(X)$, then $f_n g \underset{n\to\infty}{\longrightarrow} fg$ in measure. Is this true for an infinite measure?

2. Let $\{f_n\}_{n\geqslant 1}$ be the sequence constructed in the example of Sect. 3.3.1, and let $g_n = (-1)^k n\, f_n$ with $k = [\log_2 n]$. Show that the sequence $\{g_n\}$ converges to zero with respect to the Lebesgue measure, but

$$\varliminf_{n\to\infty} g_n(x) = -\infty, \qquad \varlimsup_{n\to\infty} g_n(x) = +\infty \quad \text{for all } x \in [0, 1).$$

3. Let $f, g, h \in \mathcal{L}^0([0, 1], \lambda)$, where λ is the Lebesgue measure and $g \leqslant f \leqslant h$. Show that there exists a sequence of functions $f_n \in \mathcal{L}^0([0, 1])$ that converges to f in measure and satisfies the following conditions:

$$\varlimsup_{n\to\infty} f_n(x) = h(x), \qquad \varliminf_{n\to\infty} f_n(x) = g(x) \quad \text{for every } x \in [0, 1].$$

4. Let $g \in \mathcal{L}^0(X, \mu)$, and let f_n be functions from $\mathcal{L}^0(X, \mu)$ such that $|f_n| \leqslant g$ almost everywhere on X for every n. Show that if $\mu(X(g > a)) < +\infty$ for every $a > 0$, then the almost everywhere convergence of the sequence $\{f_n\}_{n\geqslant 1}$ implies its convergence in measure.

5. Establish the following version of Riesz's theorem: if a measure is σ-finite and a sequence $\{f_n\}_{n\geqslant 1}$ converges to a function f in measure on every set of positive measure, then it contains a subsequence that converges to f almost everywhere.

6. Let $f_k^{(n)}(x) = \cos^{2k}(\pi n!\, x)$ $(x \in \mathbb{R})$. Show that:

(a) for every $x \in \mathbb{R}$, the limit $g_n(x) = \lim_{k\to\infty} f_k^{(n)}(x)$ exists;
(b) $g_n(x) \underset{n\to\infty}{\longrightarrow} \chi(x)$ everywhere on $\mathbb{R}$ (here χ is the Dirichlet function);

(c) there is no sequence of continuous functions (and, in particular, no diagonal sequence $\{f_{k_n}^{(n)}\}_{n\geqslant 1}$) that converges to the Dirichlet function pointwise on a non-degenerate interval.

7. Show that in the case of a σ-finite measure, for every sequence of functions $f_n \in \mathcal{L}^0(X,\mu)$ there exists a sequence of positive numbers c_n such that $c_n f_n(x) \xrightarrow[n\to\infty]{\text{a.e.}} 0$. *Hint*. Apply the diagonal sequence theorem to the functions $f_k^{(n)} = \frac{1}{k} f_n$.

8. Using the fact that the set of all numerical sequences has the cardinality of the continuum, show that the assertion of Exercise 7 is no longer true for the counting measure on $[0,1]$.

9. Assume that the measure under consideration is σ-finite and a sequence of measurable functions f_k converges to zero almost everywhere. Show that $c_k f_k \xrightarrow[n\to\infty]{\text{a.e.}} 0$ for some numerical sequence $c_k \to +\infty$ (stability of almost everywhere convergence). *Hint*. Assuming that the sequence $\{|f_k|\}_{k\geqslant 1}$ is decreasing, apply the diagonal sequence theorem to the functions $f_k^{(n)} = n f_k$.

10. Using the stability of almost everywhere convergence, show that if μ is a σ-finite measure, $f_k \in \mathcal{L}^0(X,\mu)$ $(k \in \mathbb{N})$, and $f_k \xrightarrow[k\to\infty]{\text{a.e.}} 0$, then there exists a function $g \in \mathcal{L}^0(X,\mu)$ and a sequence $c_k \to +\infty$ such that $|f_k(x)| \leqslant \frac{1}{c_k} g(x)$ for almost all $x \in X$ for every k (relatively uniform convergence, or convergence with a regulator). Prove Egorov's theorem using this result.

11. Let f be a function defined on the square $[0,1]^2$ and continuous in the first variable (for an arbitrary fixed second variable). Show that if $f(x,y) \underset{y\to 0}{\longrightarrow} 0$ for almost all $x \in [0,1]$, then the following version of Egorov's theorem holds: for every $\varepsilon > 0$ there exists a set $e \subset [0,1]$, $\lambda(e) < \varepsilon$, such that $f(x,y) \underset{y\to 0}{\longrightarrow} 0$ uniformly on $[0,1] \setminus e$. *Hint*. Consider the sets

$$G_n(\varepsilon) = \left\{(x,y) \,|\, 0 < x < 1, 0 < y < \frac{1}{n},\ \bigl|f(x,y)\bigr| > \varepsilon\right\}$$

and their projections to the x-axis.

12. Give an example of a Lebesgue measurable function f on the square $[0,1]^2$ with the following properties:

(a) for every $y \in [0,1]$, $f(x,y) \neq 0$ for at most one value $x \in [0,1]$;

(b) for every $x \in [0,1]$, $f(x,y) \neq 0$ for at most one value $y \in [0,1]$ (which implies that $f(x,y) \underset{y\to 0}{\longrightarrow} 0$ for every $x \in [0,1]$);

(c) there is no set $e \subset [0,1]$ of positive measure for which the convergence $f(x,y) \underset{y\to 0}{\longrightarrow} 0$ is uniform on e.

3.4 *Approximation of Measurable Functions by Continuous Functions. Luzin's Theorem

In this section, we discuss properties of measurable functions on $\mathbb{R}^m$. The measurability (of sets and functions) means their measurability with respect to the Lebesgue measure, which we denote by λ.

3.4.1 Let us first establish some auxiliary results. Recall the notion of the distance from a point to a set.

Definition Let $A \subset \mathbb{R}^m$ and $x \in \mathbb{R}^m$. The value

$$\operatorname{dist}(x, A) = \inf\bigl\{\|x - y\| \bigm| y \in A\bigr\}$$

is called the *distance from x to A*.

Clearly, $\operatorname{dist}(x, A) = 0$ only for points x lying in the closure of A. In particular, for a closed set A, the inequality $\operatorname{dist}(x, A) > 0$ holds everywhere outside A.

Lemma 1 *The function $x \mapsto \operatorname{dist}(x, A)$ is continuous on $\mathbb{R}^m$.*

Proof Let $y \in A$ and $x, x' \in \mathbb{R}^m$. Then $\|x - y\| \leqslant \|x' - y\| + \|x' - x\|$, whence

$$\operatorname{dist}(x, A) \leqslant \bigl\|x' - y\bigr\| + \bigl\|x' - x\bigr\|.$$

Taking the lower boundary in y of the right-hand side, we see that $\operatorname{dist}(x, A) \leqslant \operatorname{dist}(x', A) + \|x - x'\|$, i.e., $\operatorname{dist}(x, A) - \operatorname{dist}(x', A) \leqslant \|x - x'\|$. Since x and x' are interchangeable, it follows that

$$\bigl|\operatorname{dist}(x, A) - \operatorname{dist}\bigl(x', A\bigr)\bigr| \leqslant \bigl\|x - x'\bigr\|. \qquad \square$$

Lemma 2 *The characteristic function of a closed set $F \subset \mathbb{R}^m$ is the pointwise limit of a sequence of continuous functions.*

Proof Obviously, the set-theoretic difference $\mathbb{R}^m \setminus F$ can be exhausted by the closed sets $H_n = \{x \in \mathbb{R}^m \mid \operatorname{dist}(x, F) \geqslant 1/n\}$. Consider the following smoothings of the characteristic function of F:

$$f_n(x) = \frac{\operatorname{dist}(x, H_n)}{\operatorname{dist}(x, F) + \operatorname{dist}(x, H_n)} \quad \bigl(x \in \mathbb{R}^m\bigr).$$

These functions are continuous everywhere, since the denominator does not vanish. The reader can easily check that $f_n(x) \underset{n\to\infty}{\longrightarrow} \chi_F(x)$ for every $x \in \mathbb{R}^m$. $\square$

3.4.2 We prove that a measurable function can be arbitrarily well approximated in the sense of convergence almost everywhere by continuous functions.

Theorem (Fréchet[4]) *Every (Lebesgue) measurable function f on $\mathbb{R}^m$ is the limit of a sequence of continuous functions converging almost everywhere.*

Here we do not exclude the case where f takes infinite values on a set of positive measure.

Proof The proof will be split into several steps, with the function f becoming more and more complicated.

(1) Let f be the characteristic function of a measurable set E. By the regularity of the Lebesgue measure, $E = e \cup \bigcup_{n=1}^{\infty} K_n$, where $\lambda(e) = 0$ and K_n are compact sets that form an increasing sequence (see Corollary 2.3 in Sect. 2.2.2). Obviously, $\chi_{K_n} \to \chi_E$ almost everywhere. However, by Lemma 2, each of the characteristic functions χ_{K_n} is the limit of a sequence of continuous functions. Hence, by the diagonal sequence theorem, χ_E is also the limit of a sequence of continuous functions in the sense of almost everywhere convergence.
(2) If f is a simple function, i.e., it can be written in the form $f = \sum_{k=1}^{N} c_k \chi_{E_k}$, where E_k are measurable sets, then, in order to approximate f by continuous functions, it suffices to approximate the functions χ_{E_k}.
(3) In the general case, consider a sequence of simple functions f_n that converges to f pointwise (see Sect. 3.2.2, the corollary of the approximation theorem). It remains to approximate each function f_n by continuous functions and then apply the diagonal sequence theorem. □

3.4.3 We will use the Fréchet theorem to prove a result that gives deep insight into the structure of a measurable function on $\mathbb{R}^m$. It shows that Luzin's condition, which we used in Theorem 3.1.7, is not only sufficient, but also necessary for a function to be measurable. In other words, every measurable function on $\mathbb{R}^m$ can be transformed into a continuous function by removing from $\mathbb{R}^m$ a set of arbitrarily small measure.

Theorem (Luzin) *Every Lebesgue measurable function f on $\mathbb{R}^m$ that is finite almost everywhere satisfies the Luzin property, i.e., for every $\delta > 0$ there exists a set $e \subset \mathbb{R}^m$ such that*

$$\lambda(e) < \delta \quad \textit{and the restriction of } f \textit{ to } \mathbb{R}^m \setminus e \textit{ is continuous.}$$

Proof By the Fréchet theorem, there exists a sequence of continuous functions f_k that converges to f almost everywhere. According to Egorov's theorem, in every spherical layer $E_n = \{x \in \mathbb{R}^m \mid n-1 \leqslant \|x\| < n\}$ there is a subset e_n such that

$$\lambda(e_n) < \delta/2^n \quad \text{and} \quad f_k \rightrightarrows f \ \text{ on } E_n \setminus e_n.$$

Clearly, the restriction of f to $E_n \setminus e_n$ is continuous as the uniform limit of continuous functions. Put $e = \bigcup_{n=1}^{\infty}(e_n \cup S_n)$, where S_n is the sphere of radius n centered

[4]Maurice René Fréchet (1878–1973)—French mathematician.

at the origin. Then, obviously, $\lambda(e) < \delta$ and the restriction of f to $\mathbb{R}^m \setminus e$ is continuous. □

The result we have proved can be slightly strengthened by using the theorem on extension of continuous functions. The latter is formulated as follows.

Theorem *Every function continuous on a closed subset F of $\mathbb{R}^m$ is the restriction to F of a function continuous on $\mathbb{R}^m$.*

The proof of this theorem is given in Appendix 13.2. It allows us to state Luzin's theorem in the following form.

Theorem *Every Lebesgue measurable function f that is finite almost everywhere on $\mathbb{R}^m$ coincides with a function that is continuous on $\mathbb{R}^m$ except for a set of arbitrarily small measure. In other words, for every $\delta > 0$ there exists a function φ_δ continuous on $\mathbb{R}^m$ such that*

$$\lambda\left(\left\{x \in \mathbb{R}^m \mid f(x) \neq \varphi_\delta(x)\right\}\right) < \delta.$$

Proof Fix $\delta > 0$ and consider the set e from the statement of Luzin's theorem. By the regularity of the Lebesgue measure, there exists an open set G containing e whose measure is arbitrarily close to the measure of e. Hence we may assume that $\lambda(G) < \delta$. Let $F = \mathbb{R}^m \setminus G$, and let f_0 be the restriction of f to F. Now, to obtain φ_δ, it suffices to extend f_0 to a continuous function on $\mathbb{R}^m$. □

EXERCISES

1. Show that every function from $\mathcal{L}^0(\mathbb{R}^m)$ is the limit of a sequence of continuous functions with compact support that converges almost everywhere.
2. Show that a map $F : \mathbb{R}^m \to \mathbb{R}^m$ preserves Lebesgue measurability (i.e., sends measurable sets to measurable sets) if and only if it sends sets of zero measure to sets of zero measure.

Chapter 4
The Integral

At the beginning of the previous chapter, we briefly discussed the problem of constructing the integral of a bounded function defined on a finite interval $[a, b]$. As we noted, if the function f under consideration is discontinuous, Riemann sums of the form

$$\sum_{k=1}^{n} f(\xi_k)(x_k - x_{k-1}), \tag{1}$$

where $x_0 = a < x_1 < \cdots < x_n = b$, $\xi_k \in [x_{k-1}, x_k]$, are strongly affected by the choice of ξ_k, so we cannot hope that these sums will have a limit as the partition becomes finer. Hence, when constructing the integral of a discontinuous function, the idea is to replace the subintervals $[x_{k-1}, x_k]$ (on which, in spite of their "smallness", the oscillations of f may be quite large) by sets on which the oscillations of f can be controlled. More precisely, we replace (1) with the sums

$$\sum_{k=1}^{n} y_k \lambda(E_k), \tag{2}$$

where $y_0 < y_1 < \cdots < y_n$, $y_0 \leqslant \inf f$, $y_n > \sup f$ and $E_k = f^{-1}([y_{k-1}, y_k))$ $(k = 1, \ldots, n)$.

Lebesgue described the passage from (1) to (2) as follows:[1] the first approach "is comparable to a messy merchant who counts coins in the order they come to his hand whereas we act like a prudent merchant who says:

- I have mes E_1 coins à one crown, that is $1 \times$ mes E_1 crowns;
- mes E_2 coins à two crowns, that is $2 \times$ mes E_2 crowns;
- mes E_3 coins à five crowns, that is $5 \times$ mes E_3 crowns;
- ...

[1]The quotation is borrowed from [Lus, p. 499].

B. Makarov, A. Podkorytov, *Real Analysis: Measures, Integrals and Applications*, Universitext, DOI 10.1007/978-1-4471-5122-7_4, © Springer-Verlag London 2013

Therefore I have $1 \times \operatorname{mes} E_1 + 2 \times \operatorname{mes} E_2 + 5 \times \operatorname{mes} E_3 + \cdots$ crowns.

Both approaches—no matter how rich the merchant might be—lead to the same result since he only has to count a finite number of coins. But ... the difference between the approaches is essential."

We could give a definition of the integral based on the sums (2), as is done, for example, in [L, N], etc. However, we prefer a slightly different approach, first focusing on the (definition and) study of the basic properties of the integral of non-negative functions. This approach is based on a simple and clear geometric observation, essentially known to the ancient Greeks: the region lying under the graph of a non-negative function can be "exhausted" by the regions lying under the graphs of simple functions. Here the sums (2) are interpreted as the integrals of simple functions. The positivity of the integrand offers substantial technical advantages, making it possible to quickly derive all basic properties of the integral, which underlie the subsequent development (see Sect. 4.2.5).

4.1 Definition of the Integral

Everywhere in this section we consider a fixed measure space $(X, \mathfrak{A}, \mu)$. All sets and functions are assumed measurable. Unless otherwise stated, the values of all functions belong to the extended real line $\overline{\mathbb{R}} = [-\infty, +\infty]$.

4.1.1 Before proceeding to definitions, we prove a lemma.

Lemma *Let f be a non-negative simple function, $\{A_j\}_{j=1}^M$, $\{B_k\}_{k=1}^N$ be admissible partitions for f, and a_j, b_k be the values of f on A_j and B_k, respectively. Then*

$$\sum_{j=1}^{M} a_j \mu(A_j) = \sum_{k=1}^{N} b_k \mu(B_k).$$

Since the measures of the sets under consideration may be infinite, we recall our convention that $0 \cdot x = x \cdot 0 = 0$ for every $x \in \overline{\mathbb{R}}$ (see Sect. 3.1.6).

Proof It is clear that $C = \bigcup_{j=1}^M A_j \cap C = \bigcup_{k=1}^N B_k \cap C$ for every set $C \subset X$, and

$$\mu(C) = \sum_{j=1}^{M} \mu(A_j \cap C) = \sum_{k=1}^{N} \mu(B_k \cap C).$$

Furthermore, $a_j = b_k$ if $A_j \cap B_k \neq \varnothing$. Hence $a_j \mu(A_j \cap B_k) = b_k \mu(A_j \cap B_k)$ for all j, k. Therefore,

$$\sum_{j=1}^{M} a_j\mu(A_j) = \sum_{j=1}^{M}\sum_{k=1}^{N} a_j\mu(A_j \cap B_k) = \sum_{j=1}^{M}\sum_{k=1}^{N} b_k\mu(A_j \cap B_k)$$
$$= \sum_{k=1}^{N} b_k \sum_{j=1}^{M} \mu(A_j \cap B_k) = \sum_{k=1}^{N} b_k\mu(B_k).$$

Note that all equations remain valid in the case where some of the sets $A_1, \dots, A_M$ and $B_1, \dots, B_N$ have infinite measure. □

Replacing the set X by a subset E of X, we obtain an obvious generalization of the lemma.

Corollary *For every (measurable) set $E \subset X$,*

$$\sum_{j=1}^{M} a_j\mu(A_j \cap E) = \sum_{k=1}^{N} b_k\mu(B_k \cap E).$$

4.1.2 Now we are ready to define the integral of a non-negative function.

Definition 1 Let f be a non-negative simple function, $\{A_j\}_{j=1}^{M}$ be an arbitrary admissible partition for f, and a_j be the value of f on A_j. The *integral* of f over a set $E \subset X$ is defined as

$$\sum_{j=1}^{M} a_j\mu(E \cap A_j) \tag{1}$$

and is denoted by $\int_E f\,d\mu$.

The $\int$ symbol, which is the stylized first letter of the word *Summa*, was introduced by Leibniz[2] in a work published in 1686. In manuscripts, Leibniz started to employ it, instead of the original notation Omn, from 1675. The term "integral" first appeared in J. Bernoulli's[3] work published in 1690.

By the corollary, the sum (1) does not depend on the choice of an admissible partition. Hence Definition 1 is correct. Furthermore, the sum (1) does not depend on the values of f on $X \setminus E$, since if $E \cap A_j = \varnothing$, then $a_j\mu(E \cap A_j) = a_j \cdot 0 = 0$. In the case where f takes a single value C on the whole of E, by abuse of notation we denote the integral $\int_E f\,d\mu$ by $\int_E C\,d\mu$.

We now consider some properties of the integral.

---

[2]Gottfried Wilhelm Leibniz (1646–1716)—German philosopher and mathematician.

[3]Jacob Bernoulli (1654–1705)—Swiss mathematician.

(1) If C is a non-negative number, then $\int_E C\,d\mu = C\mu(E)$. In particular, the integral of the function identically equal to zero over an arbitrary set vanishes.

This property follows immediately from Definition 1.

(2) *Monotonicity*. If f and g are simple non-negative functions such that $f \leqslant g$ on E, then $\int_E f\,d\mu \leqslant \int_E g\,d\mu$.

Indeed, let $\{A_j\}_{j=1}^M$ be a common admissible partition for f and g, and $\{a_j\}_{j=1}^M$, $\{b_j\}_{j=1}^M$ be the corresponding values of these functions. Then $a_j \leqslant b_j$ if $A_j \cap E \neq \varnothing$, so that $a_j\mu(A_j \cap E) \leqslant b_j\mu(A_j \cap E)$ for all j, $1 \leqslant j \leqslant M$. Therefore,

$$\int_E f\,d\mu = \sum_{j=1}^M a_j\mu(A_j \cap E) \leqslant \sum_{j=1}^M b_j\mu(A_j \cap E) = \int_E g\,d\mu.$$

Definition 2 Let f be a non-negative measurable function on a set E. The integral of f over E is defined as

$$\int_E f\,d\mu = \sup\left\{\int_E g\,d\mu \,\middle|\, g \text{ is a non-negative simple function, } g \leqslant f \text{ on } E\right\}.$$

Remark 1 If f is a non-negative simple function, then its integrals over E in the sense of Definitions 1 and 2 coincide. This follows from the monotonicity of the integral of a simple function (Property (2)).

Remark 2 The integral of a non-negative (measurable) function is always defined and non-negative. It may take the value $+\infty$.

4.1.3 In order to define the integral of a signed measurable function f, we use the functions $f_+ = \max\{f, 0\}$ and $f_- = \max\{-f, 0\}$. They are obviously non-negative and, as we observed earlier (see Property 4 in Sect. 3.1.2), measurable. Furthermore, it is easy to check that

$$f_+ \cdot f_- = 0, \qquad f = f_+ - f_-, \qquad |f| = f_+ + f_-.$$

Definition Given an arbitrary measurable function f on a set E, we keep the notation introduced above and put

$$\int_E f\,d\mu = \int_E f_+\,d\mu - \int_E f_-\,d\mu$$

if at least one of the integrals $\int_E f_\pm\,d\mu$ is finite. In this case, the function f is said to be *integrable* on E (with respect to the measure μ). If both integrals $\int_E f_\pm\,d\mu$ are finite, then f is *summable* on E (with respect to the measure μ).

Remark If f is non-negative, then the integrals of f in the sense of the last definition and that of Definition 2 coincide, since in this case $f_+ = f$, $f_- = 0$, and $\int_E 0\,d\mu = 0$ (see Property (1)).

In conclusion, note that as well as the symbol $\int_E f\,d\mu$ we will also use the notation $\int_E f(x)\,d\mu(x)$, $\int_E f(y)\,d\mu(y)$, etc., which explicitly indicates the "variable of integration". This notation, which is formally superfluous, is very convenient when solving concrete problems, especially if the function f depends on parameters. For instance, the symbols $\int_{(0,1)} x^y\,d\mu(x)$ and $\int_{(0,1)} x^y\,d\mu(y)$ make it clear what function is being integrated, the power function $x \mapsto x^y$ in the first case, or the exponential function $y \mapsto x^y$ in the second case.

4.2 Properties of the Integral of Non-negative Functions

As in the previous section, hereafter we consider a fixed measure space $(X, \mathfrak{A}, \mu)$. All sets and functions are assumed measurable. The values of all functions belong to the extended real line and are non-negative, and every measurable function is defined on the whole set X (to satisfy the latter condition, we can extend a function by zero outside its domain, if necessary).

4.2.1 We now establish some simple properties of the integral.

(1) *Monotonicity*. If $f \leqslant g$ on E, then $\int_E f\,d\mu \leqslant \int_E g\,d\mu$.
For simple functions, this property has already been proved. In the general case, it follows immediately from Definition 2 of Sect. 4.1.2.

(2) If $\mu(E) = 0$, then $\int_E f\,d\mu = 0$ for every function f.
If f is simple, then it is bounded. Let $0 \leqslant f \leqslant C$. Then $0 \leqslant \int_E f\,d\mu \leqslant \int_E C\,d\mu = C\mu(E) = 0$. In the general case, the desired property follows immediately from Definition 2 of Sect. 4.1.2.

(3) $\int_E f\,d\mu = \int_X f\chi_E\,d\mu$.

This implies that the integral over E does not depend on the behavior of the integrand outside E.

If f is a simple function and $\{A_k\}_{1 \leqslant k \leqslant N}$ is an admissible partition for f, then $\{E \cap A_1, E \cap A_2, \ldots, E \cap A_N, X \setminus E\}$ is an admissible partition for $f\chi_E$. On the last element of this partition, the function $f\chi_E$ vanishes, and on the other elements, it takes the same values as f. Thus the desired equation follows immediately from Definition 1 of Sect. 4.1.2.

In the general case, consider arbitrary non-negative simple functions g and h such that

$$g \leqslant f \quad \text{on } E, \qquad h \leqslant f\chi_E \quad \text{on } X. \tag{1}$$

Then $h = h\chi_E$ and

$$\int_X h\,d\mu = \int_X h\chi_E\,d\mu = \int_E h\,d\mu \leqslant \int_E f\,d\mu \quad \text{and}$$

$$\int_E g\,d\mu = \int_X g\chi_E\,d\mu \leqslant \int_X f\chi_E\,d\mu$$

(both inequalities follow from the definition of the integral of a non-negative function). Taking the supremum of the left-hand sides of these inequalities over h and g satisfying conditions (1), we see (again using Definition 2 of Sect. 4.1.2) that $\int_X f\chi_E\,d\mu \leqslant \int_E f\,d\mu$ and $\int_E f\,d\mu \leqslant \int_X f\chi_E\,d\mu$.

Corollary *If (measurable non-negative) functions f and g coincide on a set A, then $\int_A f\,d\mu = \int_A g\,d\mu$, because $f\chi_A = g\chi_A$. In particular:*

(3′) *If $f(x) = C$ for all $x \in A$, then $\int_A f\,d\mu = C\,\mu(A)$.*

By abuse of notation, we denote the last integral by $\int_A C\,d\mu$.

Remark When proving various properties of the integral, Property (3) allows one to consider only the case where the domain of integration is the whole set X. In what follows, we will repeatedly use this observation.

(4) *Monotonicity with respect to the set*. If $A \subset B$ and $f \geqslant 0$ on B, then $\int_A f\,d\mu \leqslant \int_B f\,d\mu$.

Since $f\chi_A \leqslant f\chi_B$, this property follows from the previous ones.

4.2.2 Here we will prove one of the basic properties of the integral. According to the above remark, we consider only integrals over the whole set X. We do not assume them to be finite.

Theorem (B. Levi[4]) *Let $\{f_n\}_{n\geqslant 1}$ be a sequence of non-negative measurable functions that has a pointwise limit f on X. If*

$$f_n \leqslant f_{n+1} \quad \text{on } X \quad \text{for every } n \in \mathbb{N}, \tag{2}$$

then

$$\int_X f_n\,d\mu \underset{n\to\infty}{\longrightarrow} \int_X f\,d\mu.$$

Proof First of all, observe that f is measurable as the limit of measurable functions and $f_n \leqslant f$ for all $n \in \mathbb{N}$ in view of (2). By the monotonicity of the integral, we obtain

$$\int_X f_n\,d\mu \leqslant \int_X f_{n+1}\,d\mu \leqslant \int_X f\,d\mu.$$

Hence the limit $L = \lim_{n\to\infty} \int_X f_n\,d\mu$ exists and $L \leqslant \int_X f\,d\mu$.

The major part of the proof consists of verifying the reverse inequality

$$\int_X f\,d\mu \leqslant L.$$

---

[4]Beppo Levi (1875–1961)—Italian mathematician.

Let g be a simple function such that $0 \leqslant g \leqslant f$, $A_1, A_2, \dots, A_N$ be an admissible partition for g, and $a_1, a_2, \dots, a_N$ be the values of g on its elements. Fix an arbitrary number $\theta \in (0, 1)$ and put $X_n = X(f_n \geqslant \theta g)$. Note that

$$\begin{aligned}&\text{(a)}\quad X_n \subset X_{n+1} \qquad \text{and}\\ &\text{(b)}\quad \bigcup_{n\geqslant 1} X_n = X.\end{aligned} \tag{3}$$

Inclusion (3a) is obvious in view of (2). To prove (3b), consider an arbitrary point $x \in X$. If $g(x) = 0$, then $f_n(x) \geqslant 0 = \theta g(x)$, and hence $x \in X_n$ for every $n \in \mathbb{N}$. If $g(x) > 0$, then for sufficiently large n we have $f_n(x) > \theta g(x)$, because $f_n(x) \underset{n\to\infty}{\longrightarrow} f(x) \geqslant g(x) > \theta g(x)$ (it is here that we use the assumption $\theta < 1$). Therefore, $x \in \bigcup_{n\geqslant 1} X_n$, and (3b) is proved. It follows from (3) that for every set $A \subset X$ we have

$$(A \cap X_n) \subset (A \cap X_{n+1}), \quad A = \bigcup_{n\geqslant 1} A \cap X_n.$$

Hence, by the continuity of μ from below,

$$\mu(A \cap X_n) \underset{n\to\infty}{\longrightarrow} \mu(A). \tag{4}$$

Now we can estimate $\int_X f_n\, d\mu$ from below using the monotonicity of the integral (Properties (4) and (1)) and the definition of the integral of a simple function:

$$\int_X f_n\, d\mu \geqslant \int_{X_n} f_n\, d\mu \geqslant \int_{X_n} \theta g\, d\mu = \sum_{k=1}^{N} \theta a_k \mu(A_k \cap X_n).$$

In view of (4), passing to the limit as $n \to \infty$, we obtain the inequality

$$L \geqslant \sum_{k=1}^{N} \theta a_k \mu(A_k) = \theta \int_X g\, d\mu.$$

Passing to the limit as $\theta \to 1$, we conclude that $L \geqslant \int_X g\, d\mu$. Since g is an arbitrary function, it follows from the definition of $\int_X f\, d\mu$ that $L \geqslant \int_X f\, d\mu$, as required. □

4.2.3 Now we turn to the properties of the integral related to arithmetic operations.

(5) *Additivity*. If $f, g \geqslant 0$ on X, then

$$\int_X (f + g)\, d\mu = \int_X f\, d\mu + \int_X g\, d\mu. \tag{5}$$

First let f and g be simple functions, $C_1, C_2, \dots, C_N$ be a common admissible partition for f and g, and a_k, b_k be the values they take on C_k. Then

$$\int_X (f+g)\,d\mu = \sum_{k=1}^N (a_k+b_k)\,\mu(C_k) = \sum_{k=1}^N a_k\,\mu(C_k) + \sum_{k=1}^N b_k\,\mu(C_k)$$
$$= \int_X f\,d\mu + \int_X g\,d\mu.$$

The general case is proved by approximating the functions f and g by increasing sequences of simple functions f_n and g_n (see Theorem 3.2.2): since

$$\int_X (f_n+g_n)\,d\mu = \int_X f_n\,d\mu + \int_X g_n\,d\mu,$$

passing to the limit in this inequality according to Levi's theorem yields (5).

(6) *Positive homogeneity*. If a is a non-negative number, then $\int_X af\,d\mu = a\int_X f\,d\mu$.

The proof of this property goes along the same lines as that of the additivity. First we establish it for simple functions by direct computations, and then deduce the general case by passing to the limit. The details are left to the reader.

Corollary *By induction, Properties* (5) *and* (6) *immediately imply that*

$$\int_X \left(\sum_{k=1}^N a_k f_k\right) d\mu = \sum_{k=1}^N a_k \int_X f_k\,d\mu$$

for arbitrary numbers $a_k \geqslant 0$ *and functions* $f_k \geqslant 0$.

(7) *Additivity with respect to the set*. If $A, B \subset X$ and $A \cap B = \varnothing$, then

$$\int_{A\cup B} f\,d\mu = \int_A f\,d\mu + \int_B f\,d\mu.$$

Since $A\cap B = \varnothing$, we have $\chi_{A\cup B} = \chi_A + \chi_B$, whence $f\chi_{A\cup B} = f\chi_A + f\chi_B$. It remains to apply Properties (5) and (3).

Remark The last property means that the set function $A \mapsto \int_A f\,d\mu$ defined on $\mathfrak{A}$ is additive, i.e., it is a volume. Later we will prove (see Theorem 4.5.1) that it is in fact a measure.

In conclusion, we establish a useful inequality.

(8) *Strict positivity*. If $\mu(E) > 0$ and $f(x) > 0$ on E, then $\int_E f\,d\mu > 0$.

Let $E_n = E(f > \frac{1}{n})$ $(n \in \mathbb{N})$. Clearly, $\bigcup_{n\geqslant 1} E_n = E$ and, consequently, $\mu(E_n) > 0$ for some n. Therefore,

$$\int_E f\,d\mu \geqslant \int_{E_n} f\,d\mu \geqslant \int_{E_n} \frac{1}{n}\,d\mu = \frac{1}{n}\mu(E_n) > 0.$$

4.2.4 Now we will derive a formula for the integral with respect to a discrete measure (see Example 5 in Sect. 1.3.1). Let $\mathfrak{A}$ be a σ-algebra of subsets of X that contains all one-point sets, $\{\omega_x\}_{x\in X}$ be an arbitrary family of non-negative numbers, and μ be the corresponding discrete measure:

$$\mu(A) = \sum_{x\in A} \omega_x \quad (A \in \mathfrak{A}).$$

Let us verify that

$$\int_X f\,d\mu = \sum_{x\in X} f(x)\omega_x. \tag{6}$$

If f is a non-negative simple function that takes values $a_1,\dots,a_n$ on sets $A_1,\dots,A_n$ forming a partition of X, then, by the definition of the integral of a simple function,

$$\int_X f\,d\mu = \sum_{k=1}^{n} a_k\mu(A_k) = \sum_{k=1}^{n}\sum_{x\in A_k} a_k\omega_x = \sum_{x\in X} f(x)\omega_x$$

(the last equality follows from the additivity of the discrete measure corresponding to the family of numbers $\{f(x)\,\omega_x\}_{x\in X}$). Thus (6) holds for simple functions. Let us verify that it holds in the general case. Indeed, if $\int_X f\,d\mu = +\infty$, then, by Definition 2 of Sect. 4.1.2, for every $C > 0$ there exists a simple function g such that $0 \leqslant g \leqslant f$ and $\int_X g\,d\mu > C$. Using formula (6) for g, we see that $\sum_{x\in X} f(x)\,\omega_x \geqslant \sum_{x\in X} g(x)\,\omega_x = \int_X g\,d\mu > C$. Hence in the case under consideration, $\sum_{x\in X} f(x)\,\omega_x = +\infty$ and (6) holds.

If $\int_X f\,d\mu < +\infty$, then for every $\varepsilon > 0$ there exists a simple function g such that $0 \leqslant g \leqslant f$ and $\int_X f\,d\mu < \int_X g\,d\mu + \varepsilon$. For every finite set $E \subset X$, by the finite additivity of the integral, we have

$$\sum_{x\in E} f(x)\omega_x = \sum_{x\in E}\int_{\{x\}} f\,d\mu = \int_E f\,d\mu.$$

Hence

$$\sum_{x\in E} f(x)\omega_x = \int_E f\,d\mu \leqslant \int_X f\,d\mu < \int_X g\,d\mu + \varepsilon = \sum_{x\in X} g(x)\omega_x + \varepsilon$$
$$\leqslant \sum_{x\in X} f(x)\omega_x + \varepsilon.$$

Taking the supremum of the left-hand side over all finite subsets E, we obtain, by the definition of the sum of a family of numbers,

$$\sum_{x\in X} f(x)\omega_x \leqslant \int_X f\,d\mu \leqslant \sum_{x\in X} f(x)\omega_x + \varepsilon.$$

Since ε is arbitrary, (6) follows.

Along with (6), a more general formula holds:

$$\int_A f\,d\mu = \sum_{x\in A} f(x)\omega_x \quad (A \in \mathfrak{A}).$$

To prove it, we may repeat the above argument with A in place of X, or use formula (6) and the equalities

$$\int_A f\,d\mu = \int_X f\chi_A\,d\mu \quad \text{and} \quad \sum_{x\in X} f(x)\chi_A(x)\omega_x = \sum_{x\in A} f(x)\omega_x.$$

In particular, if $A = \{x_1, \dots, x_n, \dots\}$ is a countable set, then the integral $\int_A f\,d\mu$ is just the sum of a series:

$$\int_A f\,d\mu = \sum_{n=1}^{\infty} f(x_n)\,\omega_{x_n}. \tag{7}$$

4.2.5 *On the Axiomatic Definition of the Integral.* Among the properties of the integral established above, some are worth special mention. As we proved in Sects. 4.2.1–4.2.3, the integral has, in particular, the following properties: it is non-negative on non-negative functions (by definition), additive with respect to the set (Property (7)), positively homogeneous (Property (6)), and continuous with respect to increasing sequences (as follows from Levi's theorem). It turns out that these properties uniquely determine the integral. Let us consider this question in more detail.

Let $\mathcal{K}$ be the set (cone) of all non-negative measurable functions (which may take infinite values) defined on X. Restricting ourselves to non-negative functions, we may say that the integral is a map from $\mathcal{K} \times \mathfrak{A}$ to the extended real line: with each pair $(f, A) \in \mathcal{K} \times \mathfrak{A}$ it associates the value $\int_A f\,d\mu$. Usually, $\mathbb{R}$- and $\overline{\mathbb{R}}$-valued maps are called functions; however, the domain of our map (integral) is itself defined in terms of functions, so, to avoid overloading the term "function" with different meanings and causing ambiguities, we will call it a *functional*. Thus the integral is a functional defined on $\mathcal{K} \times \mathfrak{A}$.

When considering functionals on $\mathcal{K} \times \mathfrak{A}$, we do not fix a measure in advance, so now we assume that we are given not a measure space, but a measurable space $(X, \mathfrak{A})$. In this section, unless otherwise stated, all sets are measurable and all functions belong to $\mathcal{K}$; we denote the function identically equal to one on X by $\mathbb{I}$.

Assume that a functional $J : \mathcal{K} \times \mathfrak{A} \mapsto \overline{\mathbb{R}}$ enjoys the following properties:

(I) $J(f, A) \geqslant 0$ *for all* f *and* A;

(II) *if* $A \cap B = \varnothing$, *then* $J(f, A \cup B) = J(f, A) + J(f, B)$ (*additivity with respect to the set*);

(III) *if f takes the same value C at all points of A, then $J(f, A) = C\, J(\mathbb{I}, A)$;*

(IV) *if $\{f_n\}_{n\geqslant 1}$ is an increasing sequence of functions on X and $\lim_{n\to\infty} f_n(x) = g(x)$ for all $x \in X$, then $J(f_n, A) \underset{n\to\infty}{\longrightarrow} J(g, A)$ for every A.*

Let us show that these properties imply, for instance, the equation $J(f+g, A) = J(f, A) + J(g, A)$.

It follows immediately from the additivity with respect to the set that if $A_1, \dots, A_n$ are pairwise disjoint sets, then for every $n \in \mathbb{N}$,

$$J\left(f, \bigvee_{k=1}^{n} A_k\right) = \sum_{k=1}^{n} J(f, A_k). \tag{8}$$

If f, g are non-negative simple functions and a_k, b_k are their values at the elements of a common admissible partition $\{A_k\}_{k=1}^{n}$, then, using formula (8) and property (III), we obtain

$$\begin{aligned}
J(f+g, A) &= J\left(f+g, A \cap \bigvee_{k=1}^{n} A_k\right) = \sum_{k=1}^{n} J(f+g, A \cap A_k) \\
&= \sum_{k=1}^{n} (a_k + b_k) J(\mathbb{I}, A \cap A_k) \\
&= \sum_{k=1}^{n} a_k J(\mathbb{I}, A \cap A_k) + \sum_{k=1}^{n} b_k J(\mathbb{I}, A \cap A_k) = J(f, A) + J(g, A).
\end{aligned}$$

In the general case, one should approximate f and g by increasing sequences of simple functions f_n and g_n (see Theorem 3.2.2) and, using property (IV), pass to the limit in the equation $J(f_n + g_n, A) = J(f_n, A) + J(g_n, A)$.

In a similar way one can prove that $J(af, A) = aJ(f, A)$ for $a > 0$.

It is little wonder that the functional J has the last two properties, because, as we are going to prove now, every functional satisfying conditions (I)–(IV) is the integral with respect to some measure.

Theorem *Let $J : \mathcal{K} \times \mathfrak{A} \mapsto \overline{\mathbb{R}}$ be a functional satisfying conditions* (I)–(IV). *Then it has an integral representation, i.e.,*

$$J(f, A) = \int_A f\, d\mu \quad \textit{for all } (f, A) \in \mathcal{K} \times \mathfrak{A},$$

where μ is a measure defined on $\mathfrak{A}$.

It follows from the integral representation of J that $\mu(A) = \int_A \mathbb{I}\, d\mu = J(\mathbb{I}, A)$, so that μ is uniquely determined.

Proof The proof proceeds in several steps.

(1) $J(\chi_A, X) = J(\mathbb{I}, A)$.

Indeed, by (II) and (III),

$$J(\chi_A, X) = J(\chi_A, A) + J(\chi_A, X \setminus A) = 1 \cdot J(\mathbb{I}, A) + 0 \cdot J(\mathbb{I}, X \setminus A) = J(\mathbb{I}, A)$$

(recall that, according to our convention, the product $0 \cdot J(\mathbb{I}, X \setminus A)$ vanishes even if $J(\mathbb{I}, X \setminus A) = +\infty$).

(2) Now we set $\mu(A) = J(\mathbb{I}, A)$ and verify that μ is a measure. The additivity of μ follows from (II). Hence μ is a volume. To prove that it is countably additive, we verify that it is continuous from below (see Theorem 1.3.3).

Let $A_n \subset A_{n+1}$ $(n \in \mathbb{N})$, $\bigcup_{n=1}^{\infty} A_n = A$. Then $\chi_{A_n} \leqslant \chi_{A_{n+1}}$ and $\chi_{A_n} \underset{n\to\infty}{\longrightarrow} \chi_A$ pointwise on X. Hence, according to (IV), we have $J(\chi_{A_n}, X) \underset{n\to\infty}{\longrightarrow} J(\chi_A, X)$. It remains to observe that, in view of (1), $J(\chi_A, X) = J(\mathbb{I}, A) = \mu(A)$, and similar equations hold for A_n.

(3) Let us prove that J coincides with the integral with respect to μ on simple functions. Indeed, if f is a non-negative simple function and a_k are its values at the elements of an admissible partition $\{A_k\}_{k=1}^{N}$, then, using (8) and (III), we see that

$$J(f, A) = J\left(f, \bigcup_{k=1}^{N}(A \cap A_k)\right) = \sum_{k=1}^{N} J(f, A \cap A_k) = \sum_{k=1}^{N} a_k J(\mathbb{I}, A \cap A_k)$$
$$= \sum_{k=1}^{N} a_k \mu(A \cap A_k) = \int_A f\, d\mu.$$

(4) Finally, let f be an arbitrary function and A be an arbitrary set. Consider an increasing sequence of simple functions f_n that converges to f pointwise on X. Passing to the limit in the equality $J(f_n, A) = \int_A f_n\, d\mu$ (by (IV) on the left-hand side and by Levi's theorem on the right-hand side), we obtain the desired result: $J(f, A) = \int_A f\, d\mu$. □

This theorem allows us to declare that a functional J satisfying conditions (I)–(IV) is the integral with respect to the measure μ defined by the formula $\mu(A) = J(\mathbb{I}, A)$. All properties of the integral established in Sects. 4.2.1–4.2.3 can be deduced from these conditions. However, such an axiomatic approach leaves open the question of whether there exists a non-trivial (not identically equal to zero) functional satisfying conditions (I)–(IV), as well as the question of whether or not every measure can be obtained in this way. To resolve these questions, one produces a construction of a functional with the desired properties, just as we did at the very beginning.

EXERCISES In Exercises 1–7, μ is a measure defined on a σ-algebra $\mathfrak{A}$ of subsets of a set X and f is a measurable, non-negative, everywhere finite function on X.

1. Show that if the measure μ is finite, then the integral $\int_X f\,d\mu$ is finite if and only if either of the sums $\sum_{n=1}^{\infty} n\mu(X(n \leqslant f < n+1))$ and $\sum_{n=1}^{\infty} \mu(X(f \geqslant n))$ is finite.

2. Let $\mu(X) = 1$, and assume that every point of X belongs to at least k of the N measurable sets $E_1, \dots, E_N$. Show that $\mu(E_n) \geqslant \frac{k}{N}$ for some n.

3. For $p \geqslant 1$, set $I = \int_X f^p\,d\mu$ and

$$s = \sum_{n\in\mathbb{Z}} 2^n \big(\mu\big(X\big(2^n < f \leqslant 2^{n+1}\big)\big)\big)^{\frac{1}{p}}, \qquad S = \sum_{n\in\mathbb{Z}} 2^n \big(\mu\big(X\big(2^n < f\big)\big)\big)^{\frac{1}{p}}.$$

Show that $(s < +\infty) \Rightarrow (S < +\infty) \Rightarrow (I < +\infty)$.

4. Show that the integrals $\int_X f\,d\mu$ and $\int_X e^f\,d\mu$ are finite simultaneously for every nonnegative measurable function f on X if and only if the measure μ is finite and the set X cannot be divided into an infinite number of pieces of positive measure.

5. What can we say about a measure for which every non-negative measurable function (with finite values) is summable?

6. Prove the following version of Levi's theorem: if a sequence of measures $\{\mu_n\}$ defined on $\mathfrak{A}$ increases to μ, then $\int_X f\,d\mu_n \to \int_X f\,d\mu$.

7. Show that $\int_X f\,d\mu = \int_X f\,d\widetilde{\mu}$, where $\widetilde{\mu}$ is an arbitrary extension of the measure μ.

4.3 Properties of the Integral Related to the "Almost Everywhere" Notion

As in the previous sections, hereafter we fix a measure space $(X, \mathfrak{A}, \mu)$. All sets and ($\overline{\mathbb{R}}$-valued) functions under consideration are assumed measurable.

4.3.1 In the theory of functions, one often deals with propositions whose validity depends on a point $x \in X$. For example, "$f(x) > 0$", "the sequence $\{f_n(x)\}_{n\geqslant 1}$ is bounded", "the sequence $\{f_n(x)\}_{n\geqslant 1}$ converges", etc. The most important case is that of a proposition $P(x)$ which is valid for all x except for the points of a set of zero measure. Thus we introduce the following definition.

Definition A proposition $P(x)$ *is valid for almost all* x in a set $E \subset X$ (or *almost everywhere on* E) if there exists a set $e \subset E$ such that $\mu(e) = 0$ and $P(x)$ is valid for every point x in $E \setminus e$.

In Sect. 3.3.1 we already encountered a special case of this definition, when $P(x)$ is the proposition "the sequence $\{f_n(x)\}_{n\geqslant 1}$ converges" (almost everywhere convergence).

A set whose complement in X has zero measure is called a *set of full measure*. If a property $P(x)$ holds on a set of full measure, i.e., almost everywhere on X, then we say that it holds almost everywhere, omitting the reference to the set.

Remark One should bear in mind that when we consider several measures $\mu, \nu, \dots$, the fact that $P(x)$ holds almost everywhere with respect to one measure does not at all mean that it holds almost everywhere with respect to another measure. In such cases, to avoid ambiguity, we say that $P(x)$ holds μ-almost everywhere, ν-almost everywhere, etc.

In what follows, we will often use the following lemma.

Lemma *Let $\{P_n(x)\}_{n\geqslant 1}$ be a sequence of propositions and $P(x)$ be the proposition "all $P_n(x)$ hold at a point $x \in X$". If each $P_n(x)$ holds almost everywhere on a set $E \subset X$, then $P(x)$ also holds almost everywhere on E.*

Proof This follows from the fact that the union of a sequence of sets of zero measure is again a set of zero measure (see Corollary 1.3.2). The details are left to the reader. □

4.3.2 Now we establish a few properties of the integral related to the "almost everywhere" notion.

(1) If $\int_E |f|\,d\mu < +\infty$, then $|f(x)| < +\infty$ almost everywhere on E.
Let $E_0 = \{x \in E \mid |f(x)| = +\infty\}$. Then for every $t > 0$ we have $\int_E |f|\,d\mu \geqslant \int_{E_0} t\,d\mu = t\mu(E_0)$. Therefore,

$$\mu(E_0) \leqslant \frac{1}{t}\int_E |f|\,d\mu \underset{t\to\infty}{\longrightarrow} 0.$$

(2) If $\int_E |f|\,d\mu = 0$, then $f(x) = 0$ almost everywhere on E.
Indeed, if $\mu(E(|f| > 0)) > 0$, then, by Property (8) from Sect. 4.2.3, $\int_{E(|f|>0)} |f|\,d\mu > 0$, a contradiction.

(3) Let $E_0 \subset E$ such that $\mu(E \setminus E_0) = 0$. Then the integral $\int_E f\,d\mu$ exists if and only if $\int_{E_0} f\,d\mu$ exists; if either integral exists, they are equal.
Indeed, by the additivity of the integral and Property (2) from Sect. 4.2.1, we have

$$\int_E f_\pm\,d\mu = \int_{E_0} f_\pm\,d\mu + \int_{E\setminus E_0} f_\pm\,d\mu = \int_{E_0} f_\pm\,d\mu. \tag{1}$$

Thus the integrals $\int_E f_+\,d\mu$, $\int_{E_0} f_+\,d\mu$, as well as the integrals $\int_E f_-\,d\mu$, $\int_{E_0} f_-\,d\mu$, are finite or not simultaneously, which means, by definition, that the function f is integrable on E if and only if it is integrable on E_0. The fact that the integrals $\int_E f\,d\mu$ and $\int_{E_0} f\,d\mu$ are equal follows immediately from (1) and the definition.

(4) If measurable functions f and g coincide almost everywhere on E, then the integral $\int_E f\,d\mu$ exists if and only if $\int_E g\,d\mu$ exists; if either integral exists, they are equal.

Let $e = E(f \neq g)$. Since $f_\pm(x) = g_\pm(x)$ on $E \setminus e$, it follows from the previous property that

$$\int_E f_\pm \, d\mu = \int_{E \setminus e} f_\pm \, d\mu = \int_{E \setminus e} g_\pm \, d\mu = \int_E g_\pm \, d\mu,$$

which implies the desired assertion.

We see that in the framework of integration problems, functions that coincide almost everywhere can be treated as equal. It is convenient to introduce the following definition.

Definition Functions that coincide almost everywhere on X are called *equivalent* (with respect to the measure μ).

4.3.3 *Addendum to the Definition of the Integral.* Sometimes when dealing with functions measurable on some set E, we we have, for some natural reason, to consider also functions defined almost everywhere on E. This happens, for example, if we are interested in the limit of a sequence of measurable functions converging not everywhere, but only almost everywhere on E.

This situation arises often enough, and it is convenient to appropriately generalize the notions of a measurable function and the integral, to avoid the necessity of making repeated comments.

Definition A function f, defined and measurable on a set $E_0 \subset E$ such that $\mu(E \setminus E_0) = 0$, will be called *wide-sense measurable* on E; for such a function, by $\int_E f \, d\mu$ we will understand the integral $\int_{E_0} f \, d\mu$, if it exists. As before (see Definition 4.1.3), if the integrals $\int_E f_\pm \, d\mu$ are finite, then the function f is said to be summable on E.

Property (3) established above guarantees that this generalization of the notion of integral is well defined. It is clear that all properties of the integral proved in the last two sections remain valid for the integral understood in the wider sense.

We want to draw the reader's attention to the fact that a wide-sense measurable function may be defined everywhere on E, but be non-measurable on E (this may happen if the measure under consideration is not complete).

4.4 Properties of the Integral of Summable Functions

Everywhere in this section, we consider a fixed measure space $(X, \mathfrak{A}, \mu)$. Unless otherwise stated, all subsets of X are assumed measurable and all functions are assumed wide-sense measurable on X. According to Definition 4.3.3, a wide-sense measurable function f is summable on a set $E \in \mathfrak{A}$ with respect to the measure μ if the integrals $\int_E f_\pm \, d\mu$ are finite. The set of such functions is denoted by $\mathscr{L}(E, \mu)$, or $\mathscr{L}(E)$ for short if the measure is clear from the context. Studying the properties of the integral, we everywhere (except for Properties (2) and (8), for obvious reasons) consider integrals of summable functions over the whole set X. The corresponding properties of integrals over arbitrary measurable subsets of X can be obtained via the equality $\int_E f\,d\mu = \int_X f\chi_E\,d\mu$ (see Sect. 4.2, Property (3)); we leave the details to the reader.

4.4.1 *Properties of the Integral Expressed by Inequalities*

(1) A function f is summable on X if and only if $|f| \in \mathscr{L}(X)$. If $f \in \mathscr{L}(X)$, then $|\int_X f\,d\mu| \leqslant \int_X |f|\,d\mu$.

The summability of f means, by definition, that the integrals $\int_X f_+\,d\mu$ and $\int_X f_-\,d\mu$ are finite. This is equivalent to the summability of $|f|$, since $|f| = f_+ + f_-$. If f is summable, we have

$$\left|\int_X f\,d\mu\right| = \left|\int_X f_+\,d\mu - \int_X f_-\,d\mu\right| \leqslant \int_X f_+\,d\mu + \int_X f_-\,d\mu = \int_X |f|\,d\mu.$$

Corollary *Every function summable on E is finite almost everywhere on E.*

To prove this, it suffices to compare the property proved above and Property (1) from Sect. 4.3.2.

(2) Every function summable on E is summable on every (measurable) subset of E.

This follows immediately from Property (1) and the monotonicity of the integral over the set.

(3) Every bounded function f is summable on a set E of finite measure.

Indeed, let $|f| \leqslant C$ on E. Then

$$\int_E |f|\,d\mu \leqslant \int_E C\,d\mu = C\mu(E) < \infty,$$

and it remains to apply Property (1).

(4) *Monotonicity of the integral.* If $f, g \in \mathscr{L}(X)$ and $f \leqslant g$ almost everywhere, then $\int_X f\,d\mu \leqslant \int_X g\,d\mu$.

Since $f_+ - f_- \leqslant g_+ - g_-$, we have $f_+ + g_- \leqslant g_+ + f_-$. Hence, by the additivity and monotonicity of the integral of non-negative functions,

$$\int_X f_+\,d\mu + \int_X g_-\,d\mu \leqslant \int_X g_+\,d\mu + \int_X f_-\,d\mu.$$

Since all integrals are finite, the desired inequality follows:

$$\int_X f_+\,d\mu - \int_X f_-\,d\mu \leqslant \int_X g_+\,d\mu - \int_X g_-\,d\mu.$$

(5) If $|f| \leqslant g$ almost everywhere on X and $g \in \mathscr{L}(X)$, then $f \in \mathscr{L}(X)$.

The proof follows immediately from the monotonicity of the integral and Property (1).

4.4.2 *Properties of the Integral Expressed by Equalities*

(6) *Additivity*. If $f, g \in \mathscr{L}(X)$, then $f + g \in \mathscr{L}(X)$ and

$$\int_X (f+g)\,d\mu = \int_X f\,d\mu + \int_X g\,d\mu. \tag{1}$$

Let $h = f + g$. The functions f and g are finite almost everywhere, hence the function h is defined (and measurable) on a set of full measure. Since $|h| \leqslant |f| + |g|$ and $\int_X (|f| + |g|)\,d\mu = \int_X |f|\,d\mu + \int_X |g|\,d\mu$ by the additivity of the integral of non-negative functions, h is summable by Property (5). To prove (1), observe that

$$h_+ - h_- = f_+ - f_- + g_+ - g_-, \quad \text{i.e.,} \quad h_+ + f_- + g_- = f_+ + g_- + h_-.$$

Integrating the last equation and using the additivity of the integral of non-negative functions, we obtain

$$\int_X h_+\,d\mu + \int_X f_-\,d\mu + \int_X g_-\,d\mu = \int_X f_+\,d\mu + \int_X g_+\,d\mu + \int_X h_-\,d\mu.$$

All integrals here are finite, and hence

$$\int_X h_+\,d\mu - \int_X h_-\,d\mu = \int_X f_+\,d\mu - \int_X f_-\,d\mu + \int_X g_+\,d\mu - \int_X g_-\,d\mu.$$

(7) *Homogeneity*. If $f \in \mathscr{L}(X)$ and $\alpha \in \mathbb{R}$, then $\alpha f \in \mathscr{L}(X)$ and

$$\int_X \alpha f\,d\mu = \alpha \int_X f\,d\mu. \tag{2}$$

If $\alpha \geqslant 0$, then $(\alpha f)_+ = \alpha f_+$, $(\alpha f)_- = \alpha f_-$. By the definition of the integral (see Sect. 4.1.3) and the positive homogeneity, we have

$$\int_X \alpha f\,d\mu = \int_X \alpha f_+\,d\mu - \int_X \alpha f_-\,d\mu = \alpha \int_X f_+\,d\mu - \alpha \int_X f_-\,d\mu = \alpha \int_X f\,d\mu,$$

which proves both the summability of αf and formula (2).

For $\alpha = -1$ we have

$$(-f)_+ = \max\{-f, 0\} = f_-, \qquad (-f)_- = \max\big\{-(-f), 0\big\} = f_+.$$

Hence

$$\int_X (-f)\,d\mu = \int_X (-f)_+\,d\mu - \int_X (-f)_-\,d\mu = \int_X f_-\,d\mu - \int_X f_+\,d\mu$$

$$= -\int_X f\,d\mu.$$

The case $\alpha < 0$ follows from the above, due to the equality $\alpha = (-1)|\alpha|$.

Corollary (Linearity of the integral) *If* $f_1, \ldots, f_n \in \mathscr{L}(X)$, $\alpha_1, \ldots, \alpha_n \in \mathbb{R}$, *then* $(\alpha_1 f_1 + \cdots + \alpha_n f_n) \in \mathscr{L}(X)$ *and*

$$\int_X \sum_{k=1}^{n} \alpha_k f_k\,d\mu = \sum_{k=1}^{n} \alpha_k \int_X f_k\,d\mu.$$

For $n = 2$, this follows immediately from the additivity and homogeneity of the integral; the general case is proved by induction.

(8) *Additivity with respect to a set*. Let $E = \bigcup_{k=1}^{n} E_k$, and let f be a (wide-sense) measurable function on E. Then f is summable on E if and only if it is summable on each E_k. If $f \in \mathscr{L}(E)$ and the sets E_k are pairwise disjoint, then

$$\int_E f\,d\mu = \sum_{k=1}^{n} \int_{E_k} f\,d\mu. \tag{3}$$

Assuming that f is extended in an arbitrary way to the whole set X, observe that $|f|\chi_{E_k} \leqslant |f|\chi_E \leqslant |f|\chi_{E_1} + \cdots + |f|\chi_{E_n}$ for every $k = 1, \ldots, n$. Hence the inequality $\int_X |f|\chi_E\,d\mu < +\infty$, which is equivalent to the summability of f on E, holds if and only if all inequalities $\int_X |f|\chi_{E_k}\,d\mu < +\infty$ $(k = 1, \ldots, n)$ hold, i.e., f is summable on each E_k.

If the sets E_k are pairwise disjoint, then $f\chi_E = f\chi_{E_1} + \cdots + f\chi_{E_n}$. Integrating this equality, we arrive at the desired result.

Note that the summability of f on each set of an infinite family does not imply its summability on the union of these sets. A corresponding example can be obtained by considering the function identically equal to one and an arbitrary sequence of sets of finite measure whose union has infinite measure.

(9) *Integration with respect to a sum of measures*. If $\mu = \mu_1 + \mu_2$, then

$$\int_X f\,d\mu = \int_X f\,d\mu_1 + \int_X f\,d\mu_2 \tag{4}$$

for every non-negative function f. A (signed) function f is summable with respect to μ if and only if it is summable with respect to μ_1 and μ_2. In the latter case, (4) remains valid.

Since $\mu(E) = \int_X \chi_E \, d\mu$, formula (4) holds for characteristic functions, and hence for all non-negative simple functions. The general case can be obtained by passing to the limit (cf. the proof of Property (5) in Sect. 4.2.3). If f is a signed function, then the fact that the integrals $\int_X f\,d\mu$ and $\int_X f\,d_1\mu$, $\int_X f\,d_2\mu$ are finite or not simultaneously follows from formula (4) applied to $|f|$. Since (4) holds for $f_\pm$, we obtain it for f by subtraction (which is allowed, since the integrals are finite).

4.4.3 Now consider the integration of complex-valued functions. A complex-valued function f is called measurable if its real and imaginary parts, i.e., the functions $g = \mathcal{R}e(f)$ and $h = \mathcal{I}m(f)$, are measurable; the wide-sense measurability of f is understood in a similar way. A function f is called summable on a set E if g and h are summable on E. In this case, by definition,

$$\int_E f\,d\mu = \int_E g\,d\mu + i\int_E h\,d\mu.$$

This immediately implies a formula for integrating the conjugate function: $\int_E \overline{f}\,d\mu = \overline{\int_E f\,d\mu}$.

The equality properties (6)–(8) of the integral remain valid for complex-valued functions. An easy check is left to the reader.

Properties (1), (2), (3) and (5) (Property (4) no longer makes sense) also remain valid in the complex-valued case. Since Properties (2) and (5) easily follow from Property (1), we will prove only the latter.

Let f be a measurable complex-valued function. Keeping the notation introduced above, we see that $|f| = \sqrt{g^2 + h^2}$. Hence the function $|f|$ is also measurable. Furthermore,

$$|g|, |h| \leqslant |f| = \sqrt{g^2 + h^2} \leqslant |g| + |h|,$$

which implies that $|f|$ is summable if and only if both g and h are summable, i.e., f is summable.

Let us prove that if f is summable, then $|\int_E f\,d\mu| \leqslant \int_E |f|\,d\mu$. Obviously, $|\int_E f\,d\mu| = e^{i\alpha}\int_E f\,d\mu$ for some $\alpha \in \mathbb{R}$. Hence, by the homogeneity of the integral with respect to complex scalars, we have

$$\left|\int_E f\,d\mu\right| = e^{i\alpha}\int_E f\,d\mu = \int_E e^{i\alpha} f\,d\mu = \int_E \mathcal{R}e\big(e^{i\alpha} f\big)\,d\mu + i\int_E \mathcal{I}m\big(e^{i\alpha} f\big)\,d\mu.$$

Since this chain of equalities begins with a real number, it follows that $\int_E \mathcal{I}m(e^{i\alpha} f)\,d\mu = 0$. Therefore,

$$\left|\int_E f\,d\mu\right| = \int_E \mathcal{R}e\big(e^{i\alpha} f\big)\,d\mu \leqslant \int_E \big|\mathcal{R}e\big(e^{i\alpha} f\big)\big|\,d\mu \leqslant \int_E \big|e^{i\alpha} f\big|\,d\mu = \int_E |f|\,d\mu.$$

4.4.4 The remaining part of the section deals with important integral inequalities. The functions (in general, complex-valued) under consideration are assumed wide-sense measurable.

Theorem (Chebyshev's[5] inequality) *Let p and t be positive numbers. Given a function f defined on X, put $X_t = X(|f| \geqslant t)$. Then*

$$\mu(X_t) \leqslant \frac{1}{t^p} \int_X |f|^p \, d\mu.$$

Proof The proof is almost obvious:

$$\int_X |f|^p \, d\mu \geqslant \int_{X_t} |f|^p \, d\mu \geqslant \int_{X_t} t^p \, d\mu = t^p \mu(X_t). \qquad \square$$

4.4.5 The following inequality is a convenient tool for evaluating integrals.

Theorem (Hölder's[6] inequality) *Let $p, q > 1$ and $\frac{1}{p} + \frac{1}{q} = 1$. Then for any functions f and g,*

$$\int_X |f\, g| \, d\mu \leqslant \left(\int_X |f|^p \, d\mu \right)^{\frac{1}{p}} \cdot \left(\int_X |g|^q \, d\mu \right)^{\frac{1}{q}}.$$

Proof We may assume that f and g are non-negative (otherwise replace f with $|f|$ and g with $|g|$). If at least one of the integrals $\int_X f^p \, d\mu$ or $\int_X g^q \, d\mu$ vanishes, then the product fg vanishes almost everywhere and the inequality in question is obvious. The case where at least one of these integrals is infinite is also trivial. Hence in what follows we assume that

$$0 < A^p = \int_X f^p \, d\mu < +\infty, \qquad 0 < B^q = \int_X g^q \, d\mu < +\infty.$$

Let us use an auxiliary inequality to be proved a little later:

$$u\, v \leqslant \frac{u^p}{p} + \frac{v^q}{q} \quad \text{for } u, v \geqslant 0.$$

Substituting $u = \frac{f(x)}{A}$ and $v = \frac{g(x)}{B}$, we obtain

$$\frac{f(x)}{A} \cdot \frac{g(x)}{B} \leqslant \frac{1}{p} \cdot \frac{f^p(x)}{A^p} + \frac{1}{q} \cdot \frac{g^q(x)}{B^q}.$$

Integrating over X yields

$$\int_X \frac{fg}{AB} \, d\mu \leqslant \frac{1}{p} + \frac{1}{q} = 1,$$

which is equivalent to the inequality in question.

---

[5]Pafnuty L'vovich Chebyshev (1821–1894)—Russian mathematician.

[6]Ludwig Otto Hölder (1859–1937)—German mathematician.

Proceeding to the proof of the auxiliary inequality, we observe that the function $\varphi(u) = \frac{u^p}{p} + \frac{v^q}{q} - uv$ is convex on $[0, +\infty)$ for every $v \geqslant 0$. Since $\varphi'(u) = 0$ at the point $u_0 = v^{\frac{q}{p}}$, it follows that φ attains the minimum at this point. It is easy to calculate that $\varphi(u_0) = 0$, and the non-negativity of φ follows. □

Corollary 1 *If $\int_X |f|^p \, d\mu < +\infty$ and $\int_X |g|^q \, d\mu < +\infty$, then the function fg is summable and*

$$\left| \int_X fg \, d\mu \right| \leqslant \left(\int_X |f|^p \, d\mu \right)^{\frac{1}{p}} \cdot \left(\int_X |g|^q \, d\mu \right)^{\frac{1}{q}}$$

(*this is also called Hölder's inequality*).

The summability of fg follows immediately from Hölder's inequality, the right-hand side of which is finite.

Corollary 2 *If $p_1, \ldots, p_m$ are positive numbers such that $\frac{1}{p_1} + \cdots + \frac{1}{p_m} = 1$, then*

$$\int_X |f_1 \cdots f_m| \, d\mu \leqslant \prod_{j=1}^{m} \left(\int_X |f_j|^{p_j} \, d\mu \right)^{\frac{1}{p_j}}$$

for any measurable functions $f_1, \ldots, f_m$ on X.

The reader can easily prove this by induction.

We complement Corollary 1 with an inequality corresponding to the case $p = 1$. To this end, we introduce the notion of a "refined" upper boundary.

Definition The *essential supremum* of a function $f \in \mathscr{L}^0(X, \mu)$ is the value

$$\inf\{C \mid f \leqslant C \text{ almost everywhere on } X\}.$$

It is denoted by $\operatorname{esssup}_X f$.

Clearly, if $\operatorname{esssup}_X |f| < +\infty$, then we can make f bounded redefining it on a set of zero measure. Note also that in the definition of the essential supremum, the lower boundary can be replaced by the minimum, so that $f \leqslant \operatorname{esssup}_X f$ almost everywhere. Indeed, if $\operatorname{esssup}_X f = +\infty$, this is obvious, and if $\operatorname{esssup}_X f = C_0 < +\infty$, then $f \leqslant C_0 + \frac{1}{n}$ almost everywhere for every $n \in \mathbb{N}$, and the desired assertion follows by passing to the limit.

The set of functions f with $\operatorname{esssup}_X |f| < +\infty$ is denoted by $\mathscr{L}^\infty(X, \mu)$.

The monotonicity of the integral implies that if $f \in \mathscr{L}(X, \mu)$ and $g \in \mathscr{L}^\infty(X, \mu)$, then the function fg is summable and

$$\left| \int_X fg \, d\mu \right| \leqslant \operatorname*{esssup}_X |g| \cdot \int_X |f| \, d\mu.$$

Corollary 3 *Let $p > 1$, $\int_X |f|^p \, d\mu < +\infty$, and $\mu(X) < +\infty$. Then f is summable.*

Indeed, assuming that $\frac{1}{p} + \frac{1}{q} = 1$ and applying Hölder's inequality to the functions $|f|$ and 1, we see that

$$\int_X |f| \, d\mu = \int_X |f| \cdot 1 \, d\mu \leqslant \left(\int_X |f|^p \, d\mu \right)^{\frac{1}{p}} (\mu(X))^{\frac{1}{q}} < +\infty.$$

An important special case of Hölder's inequality is obtained for $p = q = 2$:

$$\int_X |f g| \, d\mu \leqslant \left(\int_X |f|^2 \, d\mu \right)^{\frac{1}{2}} \left(\int_X |g|^2 \, d\mu \right)^{\frac{1}{2}}.$$

This is usually called the Cauchy–Bunyakovsky[7] inequality.

Note also that if μ is the counting measure on a finite set $X_N = \{1, \ldots, N\}$, then, by the additivity of the integral, $\int_{X_N} f \, d\mu = \sum_{n=1}^N \int_{\{n\}} f \, d\mu = \sum_{n=1}^N f_n$, where $f_n = f(n)$. Hence in this case Hölder's inequality takes the form

$$\sum_{n=1}^N |f_n g_n| \leqslant \left(\sum_{n=1}^N |f_n|^p \right)^{\frac{1}{p}} \cdot \left(\sum_{n=1}^N |g_n|^q \right)^{\frac{1}{q}}.$$

Passing to the limit as $N \to \infty$, we obtain Hölder's inequality for series,

$$\sum_{n=1}^\infty |f_n g_n| \leqslant \left(\sum_{n=1}^\infty |f_n|^p \right)^{\frac{1}{p}} \cdot \left(\sum_{n=1}^\infty |g_n|^q \right)^{\frac{1}{q}},$$

which is just Hölder's inequality for integrals in the case where μ is the counting measure on $\mathbb{N}$ (see Example 4 in Sect. 1.3.1 and the example in Sect. 4.5.1 below).

The special case $p = q = 2$ yields the classical Cauchy inequality:

$$\sum_{n=1}^\infty |f_n g_n| \leqslant \left(\sum_{n=1}^\infty |f_n|^2 \right)^{\frac{1}{2}} \cdot \left(\sum_{n=1}^\infty |g_n|^2 \right)^{\frac{1}{2}}.$$

4.4.6 The following inequality can be viewed as a generalization of the triangle inequality for the function.

Theorem (Minkowski's inequality) *Let $p \geqslant 1$, and let f and g be functions that are finite almost everywhere on X. Then*

$$\left(\int_X |f + g|^p \, d\mu \right)^{\frac{1}{p}} \leqslant \left(\int_X |f|^p \, d\mu \right)^{\frac{1}{p}} + \left(\int_X |g|^p \, d\mu \right)^{\frac{1}{p}}.$$

---

[7] Viktor Yakovlevich Bunyakovsky (1804–1889)—Russian mathematician.

Proof Since $|f+g| \leqslant |f|+|g|$, Minkowski's inequality for $p=1$ is obvious. Hence in what follows we assume that $p>1$. Set

$$A^p = \int_X |f|^p \, d\mu, \qquad B^p = \int_X |g|^p \, d\mu, \qquad C^p = \int_X |f+g|^p \, d\mu.$$

Clearly, the inequality needs to be proved only if A and B are finite. Let us show that in this case $C < +\infty$. Indeed, since

$$|f+g|^p \leqslant \bigl(2\max\{|f|,|g|\}\bigr)^p \leqslant 2^p\bigl(|f|^p + |g|^p\bigr),$$

we have $C^p \leqslant 2^p(A^p + B^p) < +\infty$. Obviously,

$$\begin{aligned} C^p &\leqslant \int_X \bigl(|f|+|g|\bigr)\bigl(|f|+|g|\bigr)^{p-1} \, d\mu \\ &= \int_X |f|\bigl(|f|+|g|\bigr)^{p-1} \, d\mu + \int_X |g|\bigl(|f|+|g|\bigr)^{p-1} \, d\mu. \end{aligned} \tag{4}$$

Applying Hölder's inequality (see Sect. 4.4.5) with $q = \frac{p}{p-1} > 1$ to the first integral on the right-hand side, we obtain

$$\int_X |f|\bigl(|f|+|g|\bigr)^{p-1} \, d\mu \leqslant \left(\int_X |f|^p \, d\mu\right)^{\frac{1}{p}} \cdot \left(\int_X |f+g|^{(p-1)q} \, d\mu\right)^{\frac{1}{q}} = A \cdot C^{\frac{p}{q}}.$$

Analogously,

$$\int_X |g|\,|f+g|^{p-1} \, d\mu \leqslant B \cdot C^{\frac{p}{q}}.$$

Together with (4) this yields

$$C^p \leqslant A \cdot C^{\frac{p}{q}} + B \cdot C^{\frac{p}{q}} = (A+B)C^{\frac{p}{q}}.$$

For $C>0$, dividing both sides by $C^{\frac{p}{q}}$ yields the desired result, since $p - \frac{p}{q} = 1$. For $C=0$, the inequality being proved is obvious. □

We also mention the version of Minkowski's inequality for sums:

$$\left(\sum_{n=1}^{\infty} |f_n + g_n|^p\right)^{\frac{1}{p}} \leqslant \left(\sum_{n=1}^{\infty} |f_n|^p\right)^{\frac{1}{p}} + \left(\sum_{n=1}^{\infty} |g_n|^p\right)^{\frac{1}{p}}.$$

4.5 The Integral as a Set Function

In this section, as in the previous one, we consider a fixed measure space $(X, \mathfrak{A}, \mu)$. We assume that all subsets of X under consideration are measurable and all function are defined at least almost everywhere on X and are wide-sense measurable.

4.5.1 We establish one of the most important integral properties.

Theorem (Countable additivity of the integral) *Let f be a non-negative function on a set $A \subset X$ and $A = \bigvee_{k=1}^{\infty} A_k$. Then*

$$\int_A f\,d\mu = \sum_{k=1}^{\infty} \int_{A_k} f\,d\mu.$$

Note that we do not assume the integrals to be finite.

Proof Since $\{A_n\}_{n \geqslant 1}$ is a partition of A, we have

$$f\chi_A = \sum_{k=1}^{\infty} f\chi_{A_k}.$$

Let S_n be the nth partial sum of the series on the right-hand side. Clearly, $0 \leqslant S_n \leqslant S_{n+1}$ and $S_n \underset{n\to\infty}{\longrightarrow} f\chi_A$. By Levi's theorem,

$$\int_X f\chi_A\,d\mu = \lim_{n\to\infty} \int_X S_n\,d\mu = \lim_{n\to\infty} \sum_{k=1}^{n} \int_X f\chi_{A_k}\,d\mu.$$

Thus

$$\int_A f\,d\mu = \lim_{n\to\infty} \sum_{k=1}^{n} \int_{A_k} f\,d\mu = \sum_{k=1}^{\infty} \int_{A_k} f\,d\mu. \qquad \square$$

Remark The theorem can be restated as follows: if f is a non-negative function on X, then the set function $A \mapsto \int_A f\,d\mu$ is a measure on $\mathfrak{A}$ (cf. the remark after Property (7) in Sect. 4.2.3).

Corollary 1 *The theorem remains valid if f is a signed summable function.*

To prove this, it remains to apply the theorem to the functions $f_{\pm}$.

The next result shows that the integral of a summable function has the same continuity properties as a finite measure (see Theorems 1.3.3 and 1.3.4).

Corollary 2 *The integral of a summable function f is continuous from below and from above. More explicitly, if*

$$A = \bigcup_{n \geqslant 1} A_n, \quad A_n \subset A_{n+1}, \quad \textit{or} \quad A = \bigcap_{n \geqslant 1} A_n, \quad A_n \supset A_{n+1},$$

then $\int_{A_n} f\,d\mu \underset{n\to\infty}{\longrightarrow} \int_A f\,d\mu$.

In particular, if $A_n \supset A_{n+1}$ and $\bigcap_{n \geqslant 1} A_n = \varnothing$, then $\int_{A_n} f\,d\mu \underset{n\to\infty}{\longrightarrow} 0$.

Proof This follows directly from the fact that the finite measures $\nu_\pm$, where $\nu_+(A) = \int_A f_+ \, d\mu$ and $\nu_-(A) = \int_A f_- \, d\mu$, are continuous from below and from above. □

If f is a function summable on a set X of infinite measure, then the integral $\int_X f \, d\mu$ is "essentially concentrated" on a set of finite measure. More precisely, this means the following.

Corollary 3 *If $f \in \mathscr{L}(X)$, then for every $\varepsilon > 0$ there exists a set A of finite measure such that $\int_{X \setminus A} |f| \, d\mu < \varepsilon$.*

Proof Let us verify that A can be taken equal to $X(|f| \geqslant \frac{1}{n})$ for sufficiently large n. To this end, observe that the sets $A_n = X(|f| < \frac{1}{n})$ decrease and their intersection coincides with $X(f = 0)$. Since the integral is continuous from above, $\int_{A_n} |f| \, d\mu \underset{n\to\infty}{\longrightarrow} \int_{X(f=0)} |f| \, d\mu = 0$. Hence $\int_{X \setminus A} |f| \, d\mu = \int_{A_n} |f| \, d\mu < \varepsilon$ provided that n is sufficiently large. It remains to observe that $\mu(A) < +\infty$, since

$$\frac{1}{n}\mu\left(X\left(|f| \geqslant \frac{1}{n}\right)\right) \leqslant \int_A |f| \, d\mu \leqslant \int_X |f| \, d\mu < +\infty.$$

□

Example If μ is the counting measure defined on the algebra of all subsets of $\mathbb{N}$, then every sequence $f = \{f_n\}_{n\in\mathbb{N}}$ is a measurable function. By the countable additivity of the integral,

$$\int_{\mathbb{N}} |f| \, d\mu = \sum_{n=1}^{\infty} \int_{\{n\}} |f| \, d\mu = \sum_{n=1}^{\infty} |f_n|.$$

Thus the summability of f means the absolute convergence of the series $\sum_{n=1}^{\infty} f_n$, and the sum of this series is the integral of f with respect to the counting measure. The comparison theorems, rearrangement property, and other properties of absolutely convergent series are just special cases of the corresponding properties of the integral of summable functions.

More generally, if μ is the discrete measure corresponding to a family of point masses $\{\omega_x\}_{x\in X}$ and the set $X_0 = \{x \in X \mid \omega_x > 0\}$ is finite or countable ($X_0 = \{x_1, x_2, \ldots\}$), then for $f \geqslant 0$ we have

$$\int_X f \, d\mu = \sum_{n\geqslant 1} f(x_n)\omega_{x_n},$$

and this equality holds for every (possibly complex-valued) summable function.

4.5.2 We now establish another important property of the integral.

Theorem (Absolute continuity of the integral) *Let $f \in \mathscr{L}(X)$. Then for every $\varepsilon > 0$ there exists a $\delta > 0$ such that $\int_e |f| \, d\mu < \varepsilon$ if $\mu(e) < \delta$.*

Proof By the definition of the integral,

$$\int_X |f|\,d\mu = \sup\left\{\int_X g\,d\mu \mid 0 \leqslant g \leqslant |f| \text{ on } X,\ g \text{ is a simple function}\right\},$$

hence there exists a simple function g_ε such that

$$0 \leqslant g_\varepsilon \leqslant |f|, \qquad \int_X |f|\,d\mu < \int_X g_\varepsilon\,d\mu + \frac{\varepsilon}{2}. \tag{1}$$

It is clear that the function g_ε is bounded. Let $g_\varepsilon \leqslant C_\varepsilon$ on X. We will show that it suffices to put $\delta = \frac{\varepsilon}{2C_\varepsilon}$. Indeed, if $\mu(e) < \delta$, then, using (1) and the monotonicity of the integral with respect to the set, we see that

$$\int_e |f|\,d\mu = \int_e \bigl(|f| - g_\varepsilon\bigr)\,d\mu + \int_e g_\varepsilon\,d\mu$$
$$\leqslant \int_X \bigl(|f| - g_\varepsilon\bigr)\,d\mu + \int_e C_\varepsilon\,d\mu < \frac{\varepsilon}{2} + C_\varepsilon \mu(e) < \varepsilon,$$

as required. □

The theorem immediately implies the following result.

Corollary *Let* $\{e_n\}_{n\geqslant 1}$ *be a sequence of sets such that* $\mu(e_n) \to 0$. *If* f *is a summable function, then*

$$\int_{e_n} |f|\,d\mu \underset{n\to\infty}{\longrightarrow} 0.$$

4.5.3 We consider the problem of calculating the integral with respect to a measure of special form.

Definition Let ν be a measure defined on the same σ-algebra $\mathfrak{A}$ as μ. If there exists a non-negative function ω such that $\nu(A) = \int_A \omega\,d\mu$ for all $A \in \mathfrak{A}$, then ω is called the *density* (or the *weight*) of ν with respect to μ.

Let us find a formula relating the integrals with respect to μ and ν.

Theorem *If* ν *has a density* ω *with respect to* μ, *then for any non-negative function* f,

$$\int_X f\,d\nu = \int_X f\omega\,d\mu. \tag{2}$$

A (signed) function f *is summable with respect to* ν *if and only if the product* $f\omega$ *is summable with respect to* μ. *In the latter case,* (2) *remains valid.*

In view of this result, the fact that ω is the density of ν with respect to μ is often denoted as follows: $d\nu = \omega\, d\mu$.

Proof If f is a characteristic function, then (2) follows immediately from the definition of ν. Therefore, this formula is also valid for all non-negative simple functions. To obtain the general case, it suffices to approximate f by simple functions (see Sect. 3.2.2) and apply Levi's theorem.

The summability condition for f can be obtained from (2) by replacing f with $|f|$. The fact that (2) is valid for a signed summable function easily follows from the equalities $\int_X f_\pm\, d\nu = \int_X f_\pm \omega\, d\mu$. □

Example Let ν be the discrete measure (see Sect. 1.3.1) defined on the σ-algebra of all subsets of X that corresponds to a family $\omega = \{\omega(x)\}_{x\in X}$. Clearly, ω is the density of ν with respect to the counting measure.

4.5.4 It is obvious that two densities that coincide μ-almost everywhere generate the same measure. We will prove that the converse is also true, i.e., that a function is determined up to equivalence by the values of its integrals.

Theorem *Let f and g be summable functions. If*

$$\int_A f\, d\mu = \int_A g\, d\mu \quad \textit{for all } A \in \mathfrak{A},$$

then $f(x) = g(x)$ for almost all $x \in X$.

Proof Let $h = f - g$. Obviously, $\int_A h\, d\mu = 0$ for every $A \in \mathfrak{A}$. In particular, for $A = A_\pm$, where $A_+ = \{x \in X \mid h(x) \geqslant 0\}$ and $A_- = \{x \in X \mid h(x) < 0\}$, we have

$$\int_{A_+} |h|\, d\mu = \int_{A_+} h\, d\mu = 0, \qquad \int_{A_-} |h|\, d\mu = -\int_{A_-} h\, d\mu = 0.$$

Since the sets A_+ and A_- form a partition of X, it follows that $\int_X |h|\, d\mu = \int_{A_+} |h|\, d\mu + \int_{A_-} |h|\, d\mu = 0$. Therefore, $h(x) = 0$ almost everywhere on X (see Property (2) in Sect. 4.3.2). □

Corollary *Let f be a function summable with respect to the Lebesgue measure on $\mathbb{R}^m$. If $\int_P f\, d\lambda_m = 0$ for every cell P, then $f(x) = 0$ almost everywhere.*

Proof By assumption, the measures $\nu_\pm(A) = \int_A f_\pm\, d\lambda_m$ coincide on the semiring $\mathscr{P}^m$. Hence, by the uniqueness of the extension 1.5.1, they coincide on the whole σ-algebra $\mathfrak{A}^m$, i.e., $\int_A f\, d\lambda_m = 0$ for all $A \in \mathfrak{A}^m$. It remains to apply the theorem. □

EXERCISES

1. Show that if μ is a σ-finite measure, then Theorem 4.5.4 remains valid for all non-negative (not necessarily summable) functions f and g.

2. Show that a measure is σ-finite if and only if there exists a positive summable function.

4.6 The Lebesgue Integral of a Function of One Variable

In this section, λ stands for the one-dimensional Lebesgue measure. The integral $\int_E f\,d\lambda$, where $E \subset \mathbb{R}$ is a Lebesgue measurable set, is called the Lebesgue integral (of the function f over the set E). Recall that a function summable on E may be defined not everywhere, but only almost everywhere on E. Here we will consider only the simplest sets, namely, intervals (possibly infinite). Note that the type of an interval is irrelevant, since the Lebesgue measure of a one-point set is equal to zero. Hence the integrals over (a,b), $[a,b]$, $[a,b)$ and $(a,b]$ coincide. An arbitrary interval with endpoints a and b will be denoted by $\langle a,b\rangle$.

Note that every (measurable) function that is bounded on a finite interval is summable on this interval. In particular, a function that is continuous on a closed interval is summable.

4.6.1 First let us study the properties of the function $t \mapsto \int_{(a,t)} f\,d\lambda$, which is associated in a natural way with every summable function f on $\langle a,b\rangle$.

In the theorem below we consider a function F defined on a non-degenerate closed interval $[a,b]$ contained in the extended real line $\overline{\mathbb{R}}$. Observe that we do not exclude the cases $a=-\infty$ or $b=+\infty$. The continuity of F at the points $\pm\infty$ means that $F(\pm\infty)=\lim_{t\to\pm\infty}F(t)$. In other words, the continuity on $[a,b]$ is understood in the sense of the topological space $\overline{\mathbb{R}}$.

Theorem *Let f be a summable function on an interval $\langle a,b\rangle$, $-\infty \leqslant a < b \leqslant +\infty$, and $F(t)=\int_{(a,t)} f\,d\lambda$ for $t\in[a,b]$. Then:*

(1) *if $f \geqslant 0$, then F is non-decreasing;*

(2) *F is bounded and continuous on $[a,b]$; in particular, if $b=+\infty$ ($a=-\infty$), then*

$$F(t) \underset{t\to+\infty}{\longrightarrow} \int_{(a,+\infty)} f\,d\lambda \quad \Big(F(t) \underset{t\to-\infty}{\longrightarrow} 0\Big); \tag{1}$$

(3) *if f is continuous at a point $t_0 \in \langle a,b\rangle$, then F is differentiable at t_0 and $F'(t_0)=f(t_0)$.*

Claim (3), which establishes a link between integral and differential calculus, was essentially known to Barrow.[8]

---

[8]Isaac Barrow (1630–1677)—English mathematician.

Proof (1) The fact that F is non-decreasing follows from the inequality

$$F(t)-F(s)=\int_{(s,t)} f\,d\lambda \quad \text{for } s\leqslant t,\ s,t\in[a,b], \tag{2}$$

whose right-hand side is non-negative for $f\geqslant 0$.

(2) The boundedness of F is obvious, since

$$\bigl|F(t)\bigr|\leqslant \int_{(a,t)}|f|\,d\lambda\leqslant \int_{(a,b)}|f|\,d\lambda<+\infty.$$

Equation (2) shows that the continuity of F at a point $s\in\mathbb{R}$ is a consequence of the absolute continuity of the integral (see Sect. 4.5.2).

To prove (1), observe that, by the definition of F,

$$F(a)=\int_{\varnothing} f\,d\lambda=0 \quad \text{and} \quad F(b)=\int_{(a,b)} f\,d\lambda.$$

Since $F(t)=F(b)-\int_{(t,b)} f\,d\lambda$, in the case $b=+\infty$ ($a=-\infty$) it remains to check that

$$\int_{(t,+\infty)}|f|\,d\lambda\underset{t\to+\infty}{\longrightarrow}0 \quad \left(\text{respectively, } \int_{(-\infty,t)}|f|\,d\lambda\underset{t\to-\infty}{\longrightarrow}0\right),$$

which follows immediately from the continuity of the integral from above.

(3) Let us prove the existence of the right derivative of F at a point t_0, $t_0<b$. We will assume that f is defined everywhere on $\langle a,b\rangle$ (otherwise extend f to $\langle a,b\rangle$ by setting it equal to $f(t_0)$ at a set of zero measure; this affects neither the value of the integral nor the continuity of f at t_0).

Taking Eq. (2) with $s=t_0$, dividing it by $t-t_0$, and subtracting the equation $f(t_0)=\frac{1}{t-t_0}\int_{(t_0,t)} f(t_0)\,d\lambda$ from the result, we see that

$$\frac{F(t)-F(t_0)}{t-t_0}-f(t_0)=\frac{1}{t-t_0}\int_{(t_0,t)}\bigl(f-f(t_0)\bigr)\,d\lambda.$$

Hence

$$\left|\frac{F(t)-F(t_0)}{t-t_0}-f(t_0)\right|\leqslant\frac{1}{t-t_0}\int_{(t_0,t)}\bigl|f-f(t_0)\bigr|\,d\lambda\leqslant\sup_{x\in[t_0,t]}\bigl|f(x)-f(t_0)\bigr|.$$

The right-hand side tends to zero as $t\to t_0$, since f is continuous at t_0. Thus we have proved that F is differentiable at t_0 from the right and $F'_+(t_0)=f(t_0)$. The fact that F is differentiable at t_0 from the left and $F'_-(t_0)=f(t_0)$ can be proved in a similar way. □

Corollary 1 *Every continuous function f on an interval $\langle a,b\rangle$ has an antiderivative.*

Proof The function f is summable on every closed interval contained in $\langle a, b\rangle$. Assume that the interval $\langle a, b\rangle$ is closed from the left and put

$$F(t) = \int_{(a,t)} f\,d\lambda \quad \text{for } t \in [a, b\rangle. \tag{3}$$

It follows from the theorem that F is an antiderivative for f. In the case where the interval $\langle a, b\rangle$ is closed from the right, one should put

$$F(t) = -\int_{(t,b)} f\,d\lambda \quad \text{for } t \in \langle a, b].$$

In the case of an arbitrary interval, fix a point $c \in (a, b)$ and put

$$F(t) = \begin{cases} -\int_{(t,c)} f\,d\lambda & \text{for } t \in \langle a, c),\\ \int_{(c,t)} f\,d\lambda & \text{for } t \in [c, b\rangle. \end{cases}$$

We leave the reader to check that the constructed function is indeed an antiderivative for f on $\langle a, b\rangle$. □

Corollary 2 (Fundamental theorem of calculus) *If Φ is an antiderivative of a continuous function f on an interval $[a, b]$, then*

$$\int_{[a,b]} f\,d\lambda = \Phi(b) - \Phi(a), \quad \textit{i.e.}, \quad \int_{[a,b]} \Phi'\,d\lambda = \Phi(b) - \Phi(a).$$

Proof Indeed, let F be the antiderivative of f defined by (3). Then $F(a) = 0$. Since the difference of two antiderivatives is constant, it follows that $\Phi(b) - F(b) = \Phi(a) - F(a) = \Phi(a)$. Hence

$$\Phi(b) - \Phi(a) = F(b) = \int_{[a,b]} f\,d\lambda.$$ □

The difference $\Phi(b) - \Phi(a)$ is often denoted by $\Phi(x)|_{x=a}^{x=b}$ or, in short, $\Phi|_a^b$, so that the fundamental theorem of calculus can be rewritten as

$$\int_{[a,b]} f\,d\lambda = \Phi(x)|_{x=a}^{x=b}.$$

The reader familiar with other definitions of integral may conclude from the fundamental theorem of calculus that for continuous functions, the integral over an interval in the sense of each of these definitions coincides with the integral with respect to the Lebesgue measure. With this in mind, for the integral over $\langle a, b\rangle$ of a function f (continuous or just integrable with respect to the Lebesgue measure) we use the traditional notation $\int_a^b f(x)\,dx$, calling a and b the lower and the upper limit of integration,[9] respectively (of course, x can be replaced by any other letter). With this notation, the function F considered in the theorem can be written as $F(t) = \int_a^t f(x)\,dx$; that is why it is often called an "integral with a variable upper limit".

We complement the new notation for the Lebesgue integral over an interval $[a, b]$ with the following convenient convention: by definition, we set

$$\int_b^a f(x)\,dx = -\int_a^b f(x)\,dx.$$

Obviously, the fundamental theorem of calculus remains valid: swapping a and b results in changing the sign of both sides of the formula.

Remark The fundamental theorem of calculus shows that the increment of a smooth function F over an interval is equal to the integral of its derivative. As we will see later, this is also true for functions from wider classes, for instance, for functions satisfying the Lipschitz condition (see Sect. 11.4.1). Now we are going to verify that it is true if F is continuous and convex on $[a, b]$.

As is well known, the derivative of a convex function exists at all but at most countably many points and is increasing (see Sect. 13.4.3). Hence it suffices to prove the fundamental theorem of calculus under the assumption that F' is of constant sign (otherwise we may divide the interval $[a, b]$ into two parts on which this condition is satisfied). We assume without loss of generality that $F' \geqslant 0$ and divide $[a, b]$ into equal parts of length $h = (b - a)/n$ by the points $x_k = a + kh$ $(k = 0, 1, \dots, n)$. It follows from the three chords lemma (see Sect. 13.4.3) that for $k = 0, \dots, n - 1$,

$$F'_+(x_k)\,h \leqslant F(x_{k+1}) - F(x_k) \leqslant F'_-(x_{k+1})h.$$

Since F' is increasing, for $k = 1, \dots, n - 2$ we also have the estimates

$$\int_{x_{k-1}}^{x_k} F'(x)\,dx \leqslant F(x_{k+1}) - F(x_k) \leqslant \int_{x_{k+1}}^{x_{k+2}} F'(x)\,dx.$$

Summing these inequalities, we see that

$$\int_a^{x_{n-2}} F'(x)\,dx \leqslant F(x_{n-1}) - F(x_1) \leqslant \int_{x_2}^b F'(x)\,dx,$$

that is,

$$\int_a^{b-2h} F'(x)\,dx \leqslant F(b - h) - F(a + h) \leqslant \int_{a+2h}^b F'(x)\,dx.$$

---

[9] Following tradition, we often denote the integral over an infinite interval $(a, +\infty)$ by $\int_a^\infty f(x)\,dx$, omitting the plus sign in front of the symbol ∞.

Passing to the limit in this double inequality, we obtain the desired formula. In particular, we see that F' is a summable function, since the left-hand side has a finite limit.

4.6.2 Let us discuss two important methods of computing integrals.

Proposition 1 (Integration by parts) *Let u and v be continuously differentiable functions on an interval $[a,b]$. Then*

$$\int_a^b u(x)v'(x)\,dx = u(x)v(x)\Big|_{x=a}^{x=b} - \int_a^b u'(x)v(x)\,dx.$$

This formula is often written in the form

$$\int_a^b u\,dv = uv\Big|_a^b - \int_a^b v\,du.$$

Proof Integrating the equation $u'v + uv' = (uv)'$ and using the fundamental theorem of calculus, we obtain

$$\int_a^b u'(x)v(x)\,dx + \int_a^b u(x)v'(x)\,dx = \int_a^b \big(u(x)v(x)\big)'\,dx$$
$$= u(b)v(b) - u(a)v(a).$$ □

Various generalizations of Proposition 1 can be found in Sects. 4.6.4, 4.10.6, 4.11.4 and Exercise 9.

Example 1 Let us compute the integrals

$$W_n = \int_0^{\frac{\pi}{2}} \cos^n x\,dx \quad (n = 0, 1, 2, \ldots).$$

It is clear that $W_0 = \frac{\pi}{2}$ and $W_1 = 1$. Assuming that $n \geqslant 2$ and applying integration by parts, we obtain

$$W_n = \int_0^{\frac{\pi}{2}} \cos^{n-1} x\,d\sin x = \sin x \cos^{n-1} x\Big|_{x=0}^{x=\frac{\pi}{2}} - \int_0^{\frac{\pi}{2}} \sin x\,d\cos^{n-1} x$$
$$= (n-1)\int_0^{\frac{\pi}{2}} \sin^2 x \cos^{n-2} x\,dx = (n-1)\int_0^{\frac{\pi}{2}} \big(\cos^{n-2} x - \cos^n x\big)\,dx$$
$$= (n-1)(W_{n-2} - W_n).$$

Hence the integrals W_n satisfy the recurrence relation

$$W_n = \frac{n-1}{n} W_{n-2} \quad (n = 2, 3, \ldots).$$

For an even n, repeatedly using this relation, we obtain

$$W_{2k} = \frac{2k-1}{2k}\,W_{2(k-1)} = \frac{2k-1}{2k}\,\frac{2k-3}{2k-2}\,W_{2(k-2)} = \cdots$$
$$= \frac{(2k-1)(2k-3)\cdots 3\cdot 1}{(2k)(2k-2)\cdots 2}\,W_0.$$

Since $W_0 = \frac{\pi}{2}$, it follows that[10] $W_{2k} = \frac{(2k-1)!!}{(2k)!!}\,\frac{\pi}{2}$. In a similar way we can prove that $W_{2k+1} = \frac{(2k)!!}{(2k+1)!!}$. Thus

$$W_n = \frac{(n-1)!!}{n!!}\,v_n, \quad \text{where } v_n = \begin{cases} 1 & \text{for odd } n, \\ \frac{\pi}{2} & \text{for even } n. \end{cases}$$

This result leads to Wallis'[11] famous formula, which is historically the first example of a representation of π as the limit of a sequence of rational numbers. Indeed, since $v_n v_{n-1} \equiv \frac{\pi}{2}$, we have

$$W_n W_{n-1} = \frac{\pi}{2}\,\frac{(n-1)!!}{n!!}\,\frac{(n-2)!!}{(n-1)!!} = \frac{\pi}{2n}.$$

The obvious inequalities $W_n < W_{n-1} < W_{n-2} = \frac{n}{n-1}W_n$ imply that $W_n \sim W_{n-1}$. Hence

$$W_n^2 \sim \frac{\pi}{2n} \tag{4}$$

and, consequently, $4k\,W_{2k+1}^2 \to \pi$. This is an abbreviated form of Wallis' formula; in expanded form, it reads as follows:

$$\pi = \lim_{k\to\infty} \frac{1}{k}\left(\frac{2\cdot 4\cdots(2k)}{3\cdot 5\cdots(2k-1)}\right)^2.$$

Example 2 Let us establish a famous result due to Euler:[12]

$$\sum_{k=1}^{\infty} \frac{1}{k^2} = \frac{\pi^2}{6}.$$

The ingenious trick described below is borrowed from [M] (for other methods, based on Fourier series, see Sects. 10.2.1, 10.3.5).

---

[10] Recall that $n!!$ stands for the product of all positive integers less than or equal to n and having the same parity as n.

[11] John Wallis (1616–1703)—English mathematician.

[12] Leonhard Euler (1707–1783)—Swiss mathematician.

First, using integration by parts, we obtain a recurrence formula for the integrals $J_n = \int_0^{\frac{\pi}{2}} x^2 \cos^n x\,dx$:

$$J_n = \int_0^{\frac{\pi}{2}} x^2 \cos^{n-1} x\,d\sin x = -\int_0^{\frac{\pi}{2}} \sin x\,d\big(x^2\cos^{n-1} x\big)$$

$$= (n-1)\int_0^{\frac{\pi}{2}} x^2\big(\cos^{n-2} x - \cos^n x\big)dx + \frac{2}{n}\int_0^{\frac{\pi}{2}} x\,d\cos^n x$$

$$= (n-1)(J_{n-2} - J_n) - \frac{2}{n}W_n$$

(by W_n we denote the integral computed in the previous example). Therefore,

$$\frac{2}{n}W_n = (n-1)J_{n-2} - nJ_n, \quad \text{i.e.,} \quad \frac{2}{n^2} = \frac{n-1}{n}\frac{J_{n-2}}{W_n} - \frac{J_n}{W_n}.$$

For even n, the latter equation takes the form

$$\frac{1}{2k^2} = \frac{2k-1}{2k}\frac{J_{2(k-1)}}{W_{2k}} - \frac{J_{2k}}{W_{2k}} = \frac{J_{2(k-1)}}{W_{2(k-1)}} - \frac{J_{2k}}{W_{2k}}.$$

Hence

$$\frac{1}{2}\sum_{k=1}^{n}\frac{1}{k^2} = \frac{J_0}{W_0} - \frac{J_{2n}}{W_{2n}}.$$

Since $\frac{J_0}{W_0} = \frac{\pi^2}{12}$, it remains to verify that the ratio J_{2n}/W_{2n} tends to zero. Indeed,

$$\frac{J_{2n}}{W_{2n}} = \frac{1}{W_{2n}}\int_0^{\frac{\pi}{2}} x^2\cos^{2n} x\,dx < \frac{1}{W_{2n}}\int_0^{\frac{\pi}{2}}\left(\frac{\pi}{2}\sin x\right)^2\cos^{2n} x\,dx$$

$$= \frac{\pi^2}{4W_{2n}}\big(W_{2n} - W_{2(n+1)}\big) = \frac{\pi^2}{4}\left(1 - \frac{2n+1}{2n+2}\right) \underset{n\to\infty}{\longrightarrow} 0.$$

Proposition 2 (Integration by substitution) *Let f be a continuous function on $\langle a,b\rangle$ and φ be a continuously differentiable function on $[p,q]$. If $\varphi([p,q]) \subset \langle a,b\rangle$, then*

$$\int_p^q f\big(\varphi(x)\big)\,\varphi'(x)\,dx = \int_{\varphi(p)}^{\varphi(q)} f(y)\,dy.$$

One says that these integrals are related by the substitution $y = \varphi(x)$. To emphasize this, one sometimes writes the left-hand side in the form $\int_p^q f(\varphi(x))\,d\varphi(x)$. Note that φ is not required to be one-to-one or monotone, so that $\varphi(p)$ may be less or greater than (or equal to) $\varphi(q)$. Later (see Sect. 6.2) we will see that if $\varphi'(x) \neq 0$ on (p,q) (and, consequently, φ is strictly monotone), then the substitution rule is valid not only for continuous, but also for arbitrary summable functions f.

Proof Let F be an antiderivative of f on $\langle a,b\rangle$. Put $H = F(\varphi)$. Clearly, $H' = F'(\varphi)\,\varphi' = f(\varphi)\,\varphi'$. Hence H is an antiderivative of $f(\varphi)\,\varphi'$ on $[p,q]$. Applying the fundamental theorem of calculus twice, we obtain

$$\int_p^q f\big(\varphi(x)\big)\varphi'(x)\,dx = H(q) - H(p) = F\big(\varphi(q)\big) - F\big(\varphi(p)\big)$$

$$= \int_{\varphi(p)}^{\varphi(q)} f(y)\,dy. \qquad \square$$

The substitution rule is of great importance for the computation and study of integrals. To extend its range of applicability, we now generalize it to the case where φ is defined not on a closed, but only on an open (possibly infinite) interval. However, we assume additionally that it is monotone.

Proposition 3 *Let f be a non-negative continuous function on $\langle a,b\rangle$ and φ be a continuously differentiable and monotone function on (p,q). If $\varphi((p,q)) \subset \langle a,b\rangle$, then*

$$\int_p^q f\big(\varphi(x)\big)\varphi'(x)\,dx = \int_A^B f(y)\,dy,$$

where $A = \lim_{x\to p+0}\varphi(x)$ and $B = \lim_{x\to q-0}\varphi(x)$.

This formula is also valid for every continuous summable function f on (a,b).

Proof Let $a < s < t < b$. Then, by Proposition 2,

$$\int_s^t f\big(\varphi(x)\big)\varphi'(x)\,dx = \int_{\varphi(s)}^{\varphi(t)} f(y)\,dy.$$

It remains to pass to the limit as $s \to a$ and $t \to b$.

If f is an arbitrary continuous summable function, then it suffices to apply the obtained result to the non-negative functions $f_+ = \max\{f,0\}$ and $f_- = \max\{-f,0\}$. $\square$

Corollary *If a continuous function f is summable on a symmetric interval $(-a,a)$, where $0 < a \leqslant +\infty$, then $\int_{-a}^a f(x)\,dx = \int_0^a (f(x) + f(-x))\,dx$. In particular, if f is even (odd), then $\int_{-a}^a f(x)\,dx = 2\int_0^a f(x)\,dx$ (respectively, $\int_{-a}^a f(x)\,dx = 0$).*

Proof To prove this, it suffices to write the integral $\int_{-a}^a f(x)\,dx$ in the form $\int_{-a}^0 f(y)\,dy + \int_0^a f(x)\,dx$ and make the substitution $y = -x$ in the first term. $\square$

4.6.3 Let us give some important examples of summable functions. The first three of them serve as a kind of reference function; comparing with them often helps one to establish the summability of many other functions.

Example 1 Let $a > 0$ and $f(x) = e^{-ax}$ for $x \geqslant 0$. Obviously, the antiderivative $F(x) = -\frac{1}{a}e^{-ax}$ tends to zero as $x \to +\infty$. Therefore,

$$\int_0^{\infty} e^{-ax}\,dx = \lim_{t\to+\infty}\int_0^t e^{-ax}\,dx = \lim_{t\to+\infty}\bigl(F(t) - F(0)\bigr) = -F(0) = \frac{1}{a} < +\infty.$$

So, the function e^{-ax} is summable on the half-line $[0, +\infty)$.

Example 2 Let $f(x) = x^{-p}$ for $1 \leqslant x < +\infty$. An antiderivative of this function is equal to $\frac{1}{1-p}x^{1-p}$ for $p \neq 1$ and $\ln x$ for $p = 1$. If $p \leqslant 1$, it tends to infinity as $x \to +\infty$. Hence for such p the function f is not summable. If $p > 1$, then

$$\int_1^{\infty} \frac{1}{x^p}\,dx = \lim_{t\to+\infty}\int_0^t \frac{1}{x^p}\,dx = \lim_{t\to+\infty}\frac{t^{1-p}}{1-p} - \frac{1}{1-p} = \frac{1}{p-1} < +\infty.$$

Thus the function x^{-p} is summable on $[1, +\infty)$ only for $p > 1$.

Example 3 Let $f(x) = x^{-p}$ for $0 < x \leqslant 1$. Arguing as in the previous example, we arrive at the conclusion that the function x^{-p} is summable on $(0, 1]$ only for $p < 1$.

As one can easily see, a similar result holds for the integrals $\int_a^b \frac{dx}{(x-a)^p}$ and $\int_a^b \frac{dx}{(b-x)^p}$, where (a, b) is an arbitrary finite interval.

It follows from Examples 2 and 3 that the function x^{-p} is not summable on $(0, +\infty)$ for any p.

Example 4 Consider the *beta function* (the Euler integral of the first kind) introduced by Euler:

$$B(s, t) = \int_0^1 x^{s-1}(1 - x)^{t-1}\,dx.$$

As follows from the result of the previous example, $B(s, t) < +\infty$ only for $s, t > 0$. Making the substitution $x = \frac{y}{1+y}$, we can write the beta function in the form

$$B(s, t) = \int_0^{\infty} \frac{y^{s-1}}{(1 + y)^{s+t}}\,dy.$$

As we will see later, this function happens to be useful for computing many integrals.

Now consider the gamma function, which plays an important role in various fields of mathematics.

Example 5 The *gamma function* (the Euler integral of the second kind) is defined by the formula

$$\Gamma(t) = \int_0^{\infty} x^{t-1}e^{-x}\,dx.$$

Since the integrand does not exceed $Ce^{-x/2}$ for $x \geqslant 1$, the integral over $[1, +\infty)$ is finite. Hence the integral $\Gamma(t)$ is finite if and only if $\int_0^1 x^{t-1} e^{-x}\,dx$ is finite and, consequently, if and only if $\int_0^1 x^{t-1}\,dx$ is finite, i.e., if $t > 0$. Thus the function Γ is well defined on the positive half-line. We now consider its basic properties (it will be studied in more detail in Sect. 7.2).

The function Γ satisfies the functional equation

$$\Gamma(t+1) = t\Gamma(t) \quad \text{for } t > 0.$$

Indeed, using the remark to Proposition 1, we obtain

$$\Gamma(t+1) = \int_0^\infty x^t e^{-x}\,dx = -\int_0^\infty x^t\,de^{-x} = t\int_0^\infty x^{t-1}e^{-x}\,dx = t\,\Gamma(t)$$

(we have used the fact that the limit $L = \lim_{x\to+\infty} x^t e^{-x}$ is obviously equal to zero).

The functional equation reveals a close relationship between the gamma function and the factorial:

$$\Gamma(n) = (n-1)! \quad \text{for } n \in \mathbb{N}$$

(recall that, by definition, $0! = 1$).

This formula can be proved by induction. The base $\Gamma(1) = 1$ is obvious, and the inductive step relies on the functional equation:

$$\Gamma(n+1) = n\Gamma(n) = n \cdot (n-1)! = n!.$$

In a similar way, the computation of $\Gamma(n+a)$, where $0 < a < 1$, reduces to the computation of $\Gamma(a)$. One can write $\Gamma(a)$ in terms of known constants only for $a = \frac{1}{2}$, but this is not easy. To solve this problem, we need one "non-elementary integral" (in what follows, we will compute it in several different ways).

Theorem (Euler–Poisson[13] integral)

$$\int_{-\infty}^{\infty} e^{-x^2}\,dx = \sqrt{\pi}.$$

Proof Since $e^u \geqslant 1 + u$, we have $1 - x^2 \leqslant e^{-x^2} \leqslant \frac{1}{1+x^2}$ for all $x \in \mathbb{R}$. Hence for every $k \in \mathbb{N}$,

$$(1-x^2)^k \leqslant e^{-kx^2} \quad \text{for } |x| \leqslant 1 \quad \text{and} \quad e^{-kx^2} \leqslant \frac{1}{(1+x^2)^k} \quad \text{for } x \in \mathbb{R}.$$

[13] Siméon Denis Poisson (1781–1840)—French mathematician.

Integrating these inequalities, we obtain

$$\int_{-1}^{1}(1-x^2)^k\,dx \leqslant \int_{-\infty}^{\infty} e^{-kx^2}\,dx \leqslant \int_{-\infty}^{\infty}\frac{dx}{(1+x^2)^k}.$$

In the left integral, make the substitution $x=\sin t$; in the middle integral, $x=\frac{t}{\sqrt{k}}$; and in the right integral, $x=\tan t$. This yields the two-sided bound

$$\int_{-\frac{\pi}{2}}^{\frac{\pi}{2}}\cos^{2k+1}t\,dt \leqslant \frac{1}{\sqrt{k}}\int_{-\infty}^{\infty} e^{-t^2}\,dt \leqslant \int_{-\frac{\pi}{2}}^{\frac{\pi}{2}}\cos^{2k-2}t\,dt.$$

Using the notation introduced in Example 1 of Sect. 4.6.2, we can rewrite this in the form

$$2\sqrt{k}\,W_{2k+1} \leqslant \int_{-\infty}^{\infty} e^{-t^2}\,dt \leqslant 2\sqrt{k}\,W_{2k-2}.$$

It remains to observe that, in view of (4), both the left-hand side and the right-hand side of this inequality tend to the common limit $\sqrt{\pi}$ as $k\to\infty$. □

Corollary 1 $\Gamma(\frac{1}{2})=\sqrt{\pi}$.

Proof Making the substitution $x=y^2$ in the integral $\int_0^\infty x^{-\frac{1}{2}}e^{-x}\,dx=\Gamma(\frac{1}{2})$, we obtain

$$\Gamma\left(\frac{1}{2}\right)=2\int_0^\infty e^{-y^2}\,dy=\int_{-\infty}^{\infty} e^{-y^2}\,dy=\sqrt{\pi}.$$ □

Corollary 2 $\Gamma(n+\frac{1}{2})=\frac{(2n-1)!!}{2^n}\sqrt{\pi}$ *for every* $n\in\mathbb{N}$.

Proof The proof is an almost verbatim reproduction of the computation of the gamma function at integer points and is left to the reader. □

4.6.4 The remaining part of this section is devoted to so-called improper integrals. Our main purpose here is to formulate the conditions under which an improper integral over an interval coincides with the corresponding Lebesgue integral. For additional information on improper integrals and some important examples, see Sect. 7.4. In what follows, all functions under consideration may be either real- or complex-valued.

Definition Let f be a measurable function on an interval $\langle a,b\rangle$ $(-\infty\leqslant a<b\leqslant+\infty)$. We say f is admissible from the left on $\langle a,b\rangle$ if it is summable on every interval (a,t), where $a<t<b$. If the limit $\lim_{t\to b}\int_a^t f(x)\,dx$ exists, it is called the *improper integral* of the function f over the interval $\langle a,b\rangle$ and is denoted by $\int_a^{\to b} f(x)\,dx$. If an improper integral is finite, then we say that it *converges*, and in the remaining cases (i.e., if the limit does not exist or is infinite), we say that it *diverges*.

In a similar way we define a function admissible from the right and the improper integral of such a function. In what follows, we study the improper integrals of functions admissible from the left, leaving it to the reader to extend the obtained results to the case of functions admissible from the right.

It is clear that every function summable on an interval is admissible from the left (as well as from the right), and Theorem 4.6.1 implies that the improper integral of such a function converges and coincides with the Lebesgue integral. With this in mind, outside this subsection we usually denote improper integrals in the ordinary way, employing the notation $\int_a^{\to b} f(x)\,dx$ only in exceptional cases. A point near which a function f is not summable is sometimes called a singular point of f.

For improper integrals, the substitution rule stated in Proposition 3 of Sect. 4.6.2 remains valid (the assumption that the integrand is summable should be replaced by the assumption that the improper integral converges). Integration by parts is also available, provided that at least one of the integrals under consideration converges and the limit $L = \lim_{x\to b} u(x)v(x)$ exists and is finite (see Exercise 9).

Note that for a function f admissible from the left on $\langle a, b\rangle$, the convergence of the integral $\int_a^{\to b} f(x)\,dx$ is equivalent to the convergence of the integral $\int_c^{\to b} f(x)\,dx$, where c is an arbitrary point from (a, b). It is also obvious that for a non-negative function admissible from the left, the improper integral always exists and coincides with the Lebesgue integral. However, for signed functions, this is no longer the case.

Example Let us show that the *Fresnel*[14] *integral* $\int_0^\infty e^{ix^2}\,dx$ converges (it will be computed in Sect. 7.4.8). To do this, we use integration by parts; this trick does not only underly the convergence criteria established below, but can be successfully used (as in the case under consideration) beyond the framework of these criteria. Clearly,

$$\int_1^t e^{ix^2}\,dx = \int_1^t \frac{1}{2ix}\,d\big(e^{ix^2}\big) = \frac{1}{2ix}\,e^{ix^2}\Big|_1^t + \frac{1}{2i}\int_1^t \frac{1}{x^2}\,e^{ix^2}\,dx.$$

The first term on the right-hand side has a finite limit as $t \to +\infty$, and the function $\frac{1}{x^2}\,e^{ix^2}$ is summable on $[1, +\infty)$; the convergence of the Fresnel integral follows.

At the same time, obviously, the function e^{ix^2} is not summable on $(0, +\infty)$. Moreover, its real and imaginary parts are not summable either. For example,

$$\int_0^N \left|\cos x^2\right| dx = \int_0^{N^2} \frac{|\cos y|}{2\sqrt{y}}\,dy > \int_0^{N^2} \frac{\cos^2 y}{2N}\,dy = \frac{1}{4N}\int_0^{N^2} (1+\cos 2y)\,dy$$
$$= \frac{N}{4} + o(1) \underset{N\to\infty}{\longrightarrow} +\infty.$$

The integration by parts formula can be extended to improper integrals. Here we confine ourselves to its simplest version.

---

[14] Augustin-Jean Fresnel (1788–1827)—French physicist.

Proposition *If u and v are continuously differentiable functions on $[a, b)$ $(-\infty < a < b \leqslant +\infty)$ such that there exists a finite limit $L = \lim_{x\to b} u(x)\, v(x)$ and the integral $\int_a^b u'(x)\, v(x)\, dx$ converges, then the integral $\int_a^b u(x)\, v'(x)\, dx$ also converges and*

$$\int_a^b u(x)\, v'(x)\, dx = L - u(a)\, v(a) - \int_a^b u'(x)\, v(x)\, dx.$$

By analogy with the integration by parts formula obtained in Proposition 1, one also writes the last equation in the form

$$\int_a^b u(x)\, v'(x)\, dx = u(x)\, v(x)\Big|_a^b - \int_a^b u'(x)\, v(x)\, dx.$$

To prove the desired equation, it suffices to apply the integration by parts formula to the interval $[a, t]$ and to pass to the limit as $t \to b$.

For other generalizations of Proposition 1, which allow one to consider non-smooth functions, see Sects. 4.10.6, 4.11.4.

Proposition 3 of Sect. 4.6.2 can be extended to improper integrals in a similar way. We encourage the reader to formulate this generalization as an exercise.

4.6.5 To establish a relation between the summability of a function and the existence of the corresponding improper integral, one uses the notion of the absolute convergence of an improper integral.

Definition We say that an improper integral of a (measurable) function f *converges absolutely* if the improper integral of the function $|f|$ converges.

An improper integral that does converge but does not absolutely converge is sometimes said to *converge conditionally.*

Theorem *An improper integral $\int_a^{\to b} f(x)\, dx$ converges absolutely if and only if the function f is summable on (a, b).*

Thus an absolutely convergent improper integral is just the integral of a summable function.

Proof We have already observed that the summability of a function implies the convergence of the improper integral. Since the function $|f|$ is summable simultaneously with f, the summability of f guarantees the absolute convergence of the integral.

If the integral $\int_a^{\to b} f(x)\, dx$ converges absolutely, then, since the integral of a non-negative function is continuous from below, we see that f is summable:

$$\int_a^b \big|f(x)\big|\, dx = \lim_{t\to b} \int_a^t \big|f(x)\big|\, dx = \int_a^{\to b} \big|f(x)\big|\, dx < +\infty. \qquad \square$$

For a real-valued function f, the conditional convergence of the improper integral $\int_a^b f(x)\,dx$ implies that both functions $f_+ = \max\{f, 0\}$ and $f_- = \max\{-f, 0\}$ are not "small" (more exactly, not summable):

$$\int_a^b f_+(x)\,dx = \int_a^b f_-(x)\,dx = +\infty.$$

Indeed, these integrals cannot be finite simultaneously, since the last theorem implies that the function f is not summable; and if only one of them were finite, this would cause the divergence of the improper integral. At the same time, the integral

$$\int_a^t f(x)\,dx = \int_a^t f_+(x)\,dx - \int_a^t f_-(x)\,dx$$

has a finite limit as $t \to b$, and, consequently, the integrals on the right-hand side, each growing unboundedly with t, must nearly cancel. Thus the conditional convergence of an improper integral may occur only in the case where the integrand f oscillates strongly enough in the vicinity of b, taking both positive and negative values (this is clearly seen by considering the real or imaginary part of the Fresnel integral).

4.6.6 It is crucial to have easy-to-check conditions that guarantee the convergence of an improper integral even in the case where there is no absolute convergence. We will consider two such results (convergence tests for improper integrals). The reader familiar with the theory of numerical series will notice that these are analogs of Dirichlet's[15] and Abel's[16] tests, which allow one to establish the convergence of a numerical series even in the absence of absolute convergence. This is why the corresponding results on convergence of improper integrals are also named after these mathematicians. Here we will consider only simplified statements containing some superfluous assumptions. Less restrictive conditions will be formulated later, see Sect. 7.4.6.

Theorem (Dirichlet's test for improper integrals) *Let $f \in C([a, b))$, $g \in C^1([a, b))$, where $-\infty < a < b \leqslant +\infty$. If an antiderivative F of f is bounded on $[a, b)$, the function g is decreasing, and $\lim_{x\to b} g(x) = 0$, then the improper integral $\int_a^b f(x)\,g(x)\,dx$ converges.*

Here f may be either real- or complex-valued (while g is, of course, real-valued).

---

[15] Johann Peter Gustav Lejeune Dirichlet (1805–1859)—German mathematician.

[16] Niels Henrik Abel (1802–1829)—Norwegian mathematician.

Proof Integrating by parts on an interval $[a,t]$ $(a<t<b)$, we obtain

$$\int_a^t f(x)g(x)\,dx=\int_a^t g(x)\,dF(x)=g(t)F(t)-g(a)F(a)-\int_a^t F(x)g'(x)\,dx. \tag{5}$$

By assumption, $g(t)F(t)\underset{t\to b}{\longrightarrow}0$. Furthermore, the function Fg' is summable on (a,b), since

$$\int_a^b \left|F(x)g'(x)\right|dx\leqslant \sup_{[a,b)}|F|\int_a^b\bigl(-g'(x)\bigr)\,dx=g(a)\sup_{[a,b)}|F|<+\infty.$$

Hence the right-hand side in (5) has a finite limit (as $t\to b$) and

$$\int_a^b f(x)g(x)\,dx=-g(a)F(a)-\int_a^b F(x)g'(x)\,dx. \tag{5'}$$

□

Curiously enough, formula (5′) relates an improper integral on the left-hand side which does not in general absolutely converge to an absolutely convergent integral on the right-hand side.

The above test is often used when studying integrals of the form

$$\int_a^{\infty} g(x)\,e^{i\omega x}\,dx\qquad(\omega\in\mathbb{R}).$$

If g is continuously differentiable on $[a,+\infty)$ and decreases to zero at infinity, then this integral converges by Dirichlet's test for $\omega\neq 0$, notwithstanding that g may be not summable.

Example 1 Let $p>0$. The improper integral

$$\int_1^{\infty}\frac{1}{x^p}\,e^{ix}\,dx$$

converges, the convergence being absolute only for $p>1$.

For $p>1$, the integrand is obviously summable. We will show that for $p\leqslant 1$, both real and imaginary parts of the integrand are not summable. Consider, for instance, the imaginary part. Since $\frac{1}{x^p}\geqslant\frac{1}{x}$ for $x\geqslant 1$, it suffices to show that $\int_1^{\infty}\frac{|\sin x|}{x}\,dx=+\infty$. Let us consider in more detail the integrals $I_n=\int_0^{\pi n}\frac{|\sin x|}{x}\,dx$, which will be useful in what follows; we will not only prove that they grow unboundedly, but also find the growth rate. Obviously,

$$I_n=\sum_{k=1}^{n}\int_{(k-1)\pi}^{k\pi}\frac{|\sin x|}{x}\,dx=\sum_{k=1}^{n}\int_0^{\pi}\frac{\sin x}{x+\pi(k-1)}\,dx.$$

Since $\pi(k-1) \leqslant x + \pi(k-1) \leqslant \pi k$, the value of the kth integral lies between $\frac{2}{\pi k}$ and $\frac{2}{\pi(k-1)}$. Taking into account that the first integral is less than π, we obtain

$$\frac{2}{\pi}\left(1 + \frac{1}{2} + \cdots + \frac{1}{n}\right) < I_n < \pi + \frac{2}{\pi}\left(1 + \frac{1}{2} + \cdots + \frac{1}{n-1}\right).$$

Since

$$\frac{1}{2} + \cdots + \frac{1}{n} < \int_1^n \frac{dx}{x} < 1 + \frac{1}{2} + \cdots + \frac{1}{n-1},$$

the two-sided bound on I_n implies that

$$\frac{2}{\pi} \ln n < I_n < \pi + \frac{2}{\pi}(1 + \ln n) < 4 + \frac{2}{\pi} \ln n.$$

In particular, $I_n \sim \frac{2}{\pi} \ln n$ as $n \to \infty$.

Example 2 Let f be a convex function on $(0, +\infty)$ summable near the origin and such that $f(x) \underset{x \to +\infty}{\longrightarrow} 0$. Then the integral $C(y) = \int_0^\infty f(x) \cos yx \, dx$ converges and is non-negative for every $y > 0$.

We may assume that $y = 1$ (otherwise make the substitution $x \mapsto yx$). The product $f(x) \cos x$ is summable near the origin, since f is. The improper integral over $[1, +\infty)$ converges by Dirichlet's test, since f decreases on $(0, +\infty)$. Indeed, by the convexity of f, the difference $f(x) - f(x + t)$ for $t \geqslant 0$ decreases with x: $f(x) - f(x + t) \geqslant f(x') - f(x' + t)$ for every $x' \geqslant x$. Passing to the limit as $t \to +\infty$, we see that $f(x) \geqslant f(x')$.

To verify that $C(1) \geqslant 0$, we will prove that $\int_{2\pi k}^{2\pi(k+1)} f(x) \cos x \, dx \geqslant 0$ for $k = 0, 1, \ldots$. The substitution $x \mapsto x + 2\pi k$ reduces this problem to the case $k = 0$. We have

$$\int_0^{2\pi} f(x) \cos x \, dx = \int_0^{\frac{\pi}{2}} \big(f(x) - f(\pi - x) - f(\pi + x) + f(2\pi - x)\big) \cos x \, dx.$$

It remains to observe that $f(x) - f(\pi - x) - f(\pi + x) + f(2\pi - x) \geqslant 0$ (this is a special case of the inequality mentioned above with $x' = \pi - x$ and $t = \pi$).

Dirichlet's test easily implies another convergence criterion for improper integrals, which shows that multiplying the integrand of a convergent improper integral by a bounded monotone function does not affect the convergence.

Corollary (Abel's test) *Let $f \in C([a, b))$ and $g \in C^1([a, b))$. If the improper integral $\int_a^b f(x) \, dx$ converges and g is a bounded monotone function on $[a, b)$, then the integral $\int_a^b f(x) g(x) \, dx$ also converges.*

Proof Let $L = \lim_{x \to b} g(x)$. We may assume without loss of generality that g is decreasing. Since the improper integral converges, the antiderivative $F(t) =$

$\int_a^t f(x)\,dx$ $(a \leqslant t < b)$ is bounded. Obviously,

$$f(x)g(x) = f(x)\big(g(x) - L\big) + Lf(x).$$

Each of the integrals $\int_a^b f(x)(g(x) - L)\,dx$, $\int_a^b L\,f(x)\,dx$ converges (the first one, by Dirichlet's test). Hence the integral $\int_a^b f(x)g(x)\,dx$ also converges. □

EXERCISES

1. Compute the integral $\int_0^\infty e^{-a(x^2+1/x^2)}\,dx$ $(a > 0)$.

2. Show that Propositions 1 and 3 of Sect. 4.6.2 remain valid for improper integrals.

3. For which $a, b, c \in \mathbb{R}_+$ is the function $\frac{\sin^a(x^b)}{x^c}$ summable on $(0, 1)$?

4. For which real a is the integral $\int_1^\infty x^a e^{-x^3 \sin^2 x}\,dx$ finite?

5. Let f be bounded and decrease to zero at $(a, +\infty)$. Show that if the product $f(x)\sin x$ is summable on $(a, +\infty)$, then f is also summable. This result cannot be extended to the two-dimensional case (see Exercise 3 of the next section).

6. Let $p > 1$, f be a non-negative summable function on $\mathbb{R}$, and $\{x_n\}_{n\in\mathbb{Z}}$ be a two-sided sequence of real numbers such that $\inf_n(x_{n+1} - x_n) > 0$. Show that

$$\sum_{n\in\mathbb{Z}} \int_{x_{n+1}}^\infty \frac{f(x)}{(x - x_n)^p}\,dx < +\infty.$$

7. Compute the integral $I = \int_0^{\frac{\pi}{2}} \ln\sin x\,dx$, originally found by Euler, by making the substitution $x = 2y$ in the integral $2I = \int_0^\pi \ln\sin x\,dx$.

8. Compute the Euler–Poisson integral once again, by replacing the function e^{-x^2} on the interval $[0, \sqrt{n}]$ with the polynomial $(1 - \frac{x^2}{n})^n$. *Hint.* To estimate the error caused by this approximation, prove the inequality $0 \leqslant e^{-y} - (1 - \frac{y}{n})^n \leqslant \frac{3}{n} y^2 e^{-y}$ for $0 \leqslant y \leqslant n$. Reduce the integral $\int_0^{\sqrt{n}} (1 - \frac{x^2}{n})^n\,dx$ to W_{2n+1} and use (4).

9. Let $u, v \in C^1([a, b))$. Show that if two of the limits

$$\lim_{t\to b} \int_a^t u(x)v'(x)\,dx, \qquad \lim_{t\to b} \int_a^t u'(x)v(x)\,dx, \qquad \lim_{t\to b} u(t)v(t)$$

exist and are finite, then the third one also exists and the integration by parts formula holds true.

10. For which $a \in \mathbb{Z}$, $b, c \in \mathbb{R}$ does the integral $\int_0^1 \frac{\sin^a(x^b)}{x^c}\,dx$ converge absolutely (conditionally)?

11. Considering the function $f(x) = \frac{\sin x}{\sqrt{x}}$, show that the convergence on $[0, +\infty)$ of the improper integral of a function f that tends to zero at infinity is not sufficient for the integral of f^2 to converge. The same example demonstrates that we cannot drop the assumption on the monotonicity of g in Abel's test (even if the integral of g converges).

12. Verify that the integral $\int_0^\infty \frac{1}{x^p} \sin x^3\,dx$ converges not only for positive p. Can $\sin x^3$ be replaced by $\sin^3 x$?

13. Show that the integral $\int_a^\infty e^{i\,P(x)}\,dx$, where P is a real polynomial of degree greater than 1, converges.

14. Does the convergence of the improper integral $\int_1^\infty f(x)\,dx$ imply the summability of the function $\frac{f(x)}{x^3}$?

15. Compute Frullani's[17] integral $\int_0^\infty (f(ax) - f(bx))\frac{dx}{x}$, where $a, b > 0$ and f is a continuous function on $[0, +\infty)$ that satisfies one of the following conditions:

(a) the improper integral $\int_1^\infty f(x)\frac{dx}{x}$ converges;
(b) $f(x+T) = f(x)$ for some $T > 0$ and arbitrary $x \geqslant 0$;
(c) the limit $L = \lim_{x\to+\infty} f(x)$ exists and is finite.

16. Compute the integral $\int_0^\infty (c_1 \cos\frac{x}{a_1} + \cdots + c_n \cos\frac{x}{a_n})\frac{dx}{x}$ under the assumption that $c_1 + \cdots + c_n = 0$.

17. Show that the integral $\int_0^\infty f(x)\sin yx\,dx$ for $y > 0$ is non-negative provided that f is a function decreasing to zero on $(0, +\infty)$ such that the product $xf(x)$ is summable near the origin.

18. Let φ be a continuous 2π-periodic function. Show that the integral $\int_\pi^\infty \frac{\varphi(x)\,dx}{\ln x+\cos x}$ converges only if φ is odd.

19. Show that the integral $\int_\pi^\infty \frac{\sin x\,dx}{\ln x+\cos x+\sin x}$ diverges and the integral $\int_\pi^\infty \frac{\sin x\,dx}{\ln x+\cos x^2}$ converges.

20. Show that for every $\varepsilon > 0$ and every measurable non-negative function f on $(0, +\infty)$, the following inequality holds:

$$\int_1^2 \sum_{n=1}^\infty f(n\varepsilon x)\,dx \leqslant \frac{1}{\varepsilon}\int_0^\infty f(x)\,dx.$$

21. Let $x_n > 0$, $x_n \underset{n\to\infty}{\longrightarrow} 0$, $f(t) = \mathrm{card}\{n \in \mathbb{N} \mid x_n > t\}$. Show that $\int_0^\infty f(t)\,dt = \sum_{n=1}^\infty x_n$.

22. Show that $|\int_E e^{ix}\,dx| \leqslant 2\sin\frac{\lambda_1(E)}{2}$ for every measurable set $E \subset [0, 2\pi]$.

4.7 The Multiple Lebesgue Integral

In this section, we consider a few properties of the integral with respect to the Lebesgue measure on a multi-dimensional space. As in the previous section, the integral with respect to the Lebesgue measure is called the Lebesgue integral and is denoted by $\int_E f(x)\,dx$ by analogy with the one-dimensional case. The Lebesgue measure itself is usually denoted by λ, without indicating the dimension.

[17] Giuliano Frullani (1795–1834)—Italian mathematician.

Note that the integrals with respect to the planar, three-dimensional, and m-dimensional Lebesgue measures are usually called the double, triple, and m-multiple integrals, respectively, and are often conveniently denoted by the symbols $\iint$, $\iiint$, and $\int\cdots\int$.

4.7.1 The theorem below is a generalization of the results of Examples 2 and 3 of Sect. 4.6.3. It deals with a power of the norm, which in many cases serves as a reference function with which one compares other functions when studying their summability.

Theorem *Let B be a ball in $\mathbb{R}^m$ of radius r centered at a point a. Given $p > 0$, set $f(x) = \|x - a\|^{-p}$ for $x \in \mathbb{R}^m$. Then:*

(1) f is summable on B if and only if $p < m$;
(2) f is summable on $\mathbb{R}^m \setminus B$ if and only if $p > m$.

Proof First recall that the volume (m-dimensional Lebesgue measure) of an m-dimensional ball of radius R is equal to $\alpha_m R^m$, where α_m is the volume of a ball of unit radius (see Corollary 2 in Sect. 2.5.2). Hence the volume of the spherical layer $E(R) = \{x \in \mathbb{R}^m \mid \frac{R}{2} \leqslant \|x - a\| < R\}$ is, obviously, equal to

$$\lambda\big(E(R)\big) = \alpha_m R^m - \alpha_m \left(\frac{R}{2}\right)^m = \alpha_m (2^m - 1)\left(\frac{R}{2}\right)^m = \beta_m R^m,$$

where $\beta_m = \alpha_m(1 - 2^{-m})$.

Now divide the ball B into the spherical layers $E_k = E(\frac{r}{2^k})$: $B = \{0\} \vee \bigvee_{k\geqslant 1} E_k$. Then

$$\lambda(E_k) = \beta_m \left(\frac{r}{2^k}\right)^m \qquad \text{for all } k \in \mathbb{N}.$$

Furthermore,

$$\left(\frac{2^{k-1}}{r}\right)^p \leqslant f(x) \leqslant \left(\frac{2^k}{r}\right)^p \qquad \text{for } x \in E_k.$$

Integrating this inequality, we see that

$$\left(\frac{2^{k-1}}{r}\right)^p \cdot \beta_m \left(\frac{r}{2^k}\right)^m \leqslant \int_{E_k} f(x)\,dx \leqslant \left(\frac{2^k}{r}\right)^p \cdot \beta_m \left(\frac{r}{2^k}\right)^m,$$

i.e.,

$$A 2^{k(p-m)} \leqslant \int_{E_k} f(x)\,dx \leqslant B 2^{k(p-m)},$$

where A and B are positive coefficients that do not depend on k. The obtained two-sided bound on the integrals $\int_{E_k} f(x)\,dx$ implies that if either of the series

$$\sum_{k=1}^{\infty} 2^{k(p-m)} \quad \text{and} \quad \sum_{k=1}^{\infty} \int_{E_k} f(x)\,dx$$

converges, then the other one also converges. But it is obvious that the first series has a finite sum only for $p < m$, and the sum of the second series, by the countable additivity of the integral, is equal to $\int_B f(x)\,dx$; the first claim of the theorem follows.

The proof of the second claim is entirely similar (one should consider the spherical layers $E(2^k r)$), and is left to the reader. □

4.7.2 The mean value theorem, known for the integral over the interval (see Sect. 13.1.2), is also valid for multiple integrals.

Theorem (Mean value theorem) *Let $E \subset \mathbb{R}^m$ be a connected set of finite measure. If f is a continuous summable (in particular, continuous bounded) function on E, then there exists a point $c \in E$ such that*

$$\int_E f(x)\,dx = f(c)\lambda(E).$$

Proof We may assume that $\lambda(E) \neq 0$, since otherwise any point of E can be taken as c. Let $A = \inf_E f$ and $B = \sup_E f$ $(A, B \in \overline{\mathbb{R}})$. Integrating the inequality $A \leqslant f \leqslant B$ and dividing the result by $\lambda(E)$, we obtain

$$A \leqslant C = \frac{1}{\lambda(E)} \int_E f(x)\,dx \leqslant B. \tag{1}$$

It remains to prove that C is a value of f. If $A < C < B$, this follows from the intermediate value theorem, which states that the set of values of f contains the interval (A, B). If, however, $C = A$ (or $C = B$), then f is equal to A (respectively, B) almost everywhere on E. Indeed, in the case $C = A$ it follows from (1) that $\int_E (f(x) - A)\,dx = 0$. Since the integrand is non-negative, this in turn implies that $f(x) - A = 0$ almost everywhere on E (see Property (2) in Sect. 4.3.2). Thus almost every point of E can be taken as c. □

Remark As one can easily see, the proof of the mean value theorem does not use any properties of the Lebesgue measure except for the finiteness of $\lambda(E)$. Hence the theorem remains valid for every Borel measure μ such that $\mu(E) < +\infty$. Here E may be assumed to be a connected subset of an arbitrary Hausdorff topological space.

4.7.3 *The Integral as the Limit of Riemann Sums*

Definition Let $\tau = \{e_k\}_{k=1}^N$ be a partition of a set $E \subset \mathbb{R}^m$ into measurable subsets. The value $r(\tau) = \max_{1 \leqslant k \leqslant N} \operatorname{diam}(e_k)$ is called the *mesh* of τ.

If a point (tag) ξ_k is fixed in each set $E \cap \overline{e_k}$, then τ together with the family $\xi \equiv \{\xi_k\}_{k=1}^N$ of tags is called a *tagged partition*.

Now let f be a function defined on E. With each tagged partition (τ, ξ) we can associate the following sum:

$$\sigma(f, \tau, \xi) = \sum_{k=1}^{N} f(\xi_k)\,\lambda(e_k)$$

(where λ is the Lebesgue measure on $\mathbb{R}^m$). It is called the *Riemann sum* of the function f with respect to the tagged partition (τ, ξ).

Theorem (On the limit of the Riemann sums) *If E is a compact set and f is a continuous function on E, then $\sigma(f, \tau, \xi) \underset{r(\tau)\to 0}{\longrightarrow} \int_E f(x)\,dx$. In more detail, this means that for every $\varepsilon > 0$ there exists a $\delta > 0$ such that whatever family of tags ξ one chooses,*

$$\left| \sigma(f, \tau, \xi) - \int_E f(x)\,dx \right| < \varepsilon$$

as soon as $r(\tau) < \delta$.

Proof Let ω be the modulus of continuity of f:

$$\omega(t) = \sup\bigl\{ \bigl| f(x) - f(y) \bigr| \;\big|\; \|x - y\| \leqslant t,\ x, y \in E \bigr\}.$$

In particular, $|f(x) - f(y)| \leqslant \omega(\|x - y\|)$ for $x, y \in E$. Hence for every point $\xi_k \in \overline{e_k}$,

$$\left| \int_{e_k} f(x)\,dx - f(\xi_k)\lambda(e_k) \right| = \left| \int_{e_k} \bigl(f(x) - f(\xi_k)\bigr)\,dx \right| \leqslant \int_{e_k} \bigl| f(x) - f(\xi_k) \bigr|\,dx$$
$$\leqslant \omega\bigl(\operatorname{diam}(\overline{e_k})\bigr)\lambda(e_k).$$

Since $\operatorname{diam}(\overline{e_k}) = \operatorname{diam}(e_k)$, we have

$$\left| \int_E f(x)\,dx - \sigma(f, \tau, \xi) \right| \leqslant \sum_{k=1}^{N} \left| \int_{e_k} f(x)\,dx - f(\xi_k)\lambda(e_k) \right|$$
$$\leqslant \sum_{k=1}^{N} \omega\bigl(\operatorname{diam}(e_k)\bigr)\lambda(e_k) \leqslant \omega\bigl(r(\tau)\bigr)\lambda(E).$$

The theorem follows, because $\omega(t) \underset{t\to 0}{\longrightarrow} 0$ (here we use Cantor's uniform continuity theorem). □

Remark The above proof, as well as the proof of Theorem 4.7.2, does not use special properties of the Lebesgue measure. The reader can easily check that the proof remains valid in a much more general situation. In particular, we may assume that E is a compact metric space and λ is an arbitrary finite measure defined on a σ-algebra containing all open sets (the latter condition is needed to guarantee the measurability of a continuous function).

EXERCISES

1. Let f be a function summable on every ball $B(x,r) \subset \mathbb{R}^m$. Show that the function $(x,r) \mapsto \int_{B(x,r)} |f(y)|\,dy$ is continuous on $\mathbb{R}^m \times \mathbb{R}_+$.
2. Show that the integral of a bounded monotone function over an interval is the limit of the Riemann sums.
3. Show that $\int_0^\infty\int_0^\infty \frac{|\sin x|\,|\sin y|}{e^{xy\ln(x+y+2)}}\,dx\,dy < +\infty$, although $\int_0^\infty\int_0^\infty \frac{dx\,dy}{e^{xy\ln(x+y+2)}} = +\infty$.

4.8 Interchange of Limits and Integration

Here we will prove several important results that allow us to justify the formula $\lim_{n\to\infty}\int_X f_n\,d\mu = \int_X f\,d\mu$ provided that the sequence f_n converges, in some sense or other, to f. Thus our aim is to obtain conditions under which one can interchange limits and integration.

Everywhere in this section, μ stands for a measure defined on a σ-algebra of subsets of a set X and the functions under consideration are assumed to be defined at least almost everywhere on X.

4.8.1 We begin with an easy theorem, which is probably familiar, in some form or other, to the reader. To simplify the statement, we assume that the functions under consideration are defined everywhere on X.

Theorem *Let $\mu(X) < +\infty$, and let $\{f_n\}_{n\geqslant 1}$ be a sequence of summable functions that converges to a limit function f uniformly on X. Then f is summable and $\int_X f_n\,d\mu \to \int_X f\,d\mu$ as $n \to \infty$.*

Proof The function f is measurable as the limit of a sequence of measurable functions. Since $\mu(X) < +\infty$ and $|f_n - f| < 1$ everywhere on X for sufficiently large n, the difference $f_n - f$ is summable. Therefore, the limit function f is also summable. The convergence $\int_X f_n\,d\mu \to \int_X f\,d\mu$ is obvious, since

$$\left|\int_X f_n\,d\mu - \int_X f\,d\mu\right| \leqslant \int_X |f_n - f|\,d\mu \leqslant \mu(X)\,\sup_X |f_n - f| \underset{n\to\infty}{\longrightarrow} 0. \qquad \square$$

4.8.2 Theorem 4.8.1 is sufficient for solving simple problems related to interchanging limits and integration. However, in many cases its conditions turn out to be too restrictive, so that we need more general results. The first of them will be obtained by slightly generalizing one of the most important theorems on the interchange of limits and integration proved in Sect. 4.2.2.

Theorem (B. Levi) *Let f_n be a sequence of measurable functions that converges to a function f almost everywhere on X. If for every $n \in \mathbb{N}$,*

$$0 \leqslant f_n(x) \leqslant f_{n+1}(x) \quad \textit{for almost all } x \in X,$$

then

$$\int_X f_n \, d\mu \underset{n\to\infty}{\longrightarrow} \int_X f \, d\mu.$$

Proof Since the countable union of sets of zero measure is again a set of zero measure, there is a set $X_0 \subset X$ of full measure on which all assumptions of B. Levi's theorem 4.2.2 are satisfied. Therefore, $\int_{X_0} f_n \, d\mu \underset{n\to\infty}{\longrightarrow} \int_{X_0} f \, d\mu$. This is just what we wanted to show, because the integrals over X and X_0 coincide. □

Corollary 1 *A series of almost everywhere non-negative measurable functions can be integrated term by term.*

Proof It suffices to apply B. Levi's theorem to the partial sums of the series under consideration. □

Note that in Corollary 1 we impose no assumptions on the convergence of the series. This proves useful in the next result.

Corollary 2 *If the number series $\sum_{n\geqslant 1} \int_X |f_n| \, d\mu$ converges, then the function series $\sum_{n\geqslant 1} f_n(x)$ converges absolutely almost everywhere.*

Proof Let $S = \sum_{n\geqslant 1} |f_n|$. It follows from Corollary 1 (regardless of the convergence of the series $\sum_{n\geqslant 1} |f_n|$) that

$$\int_X S \, d\mu = \sum_{n\geqslant 1} \int_X |f_n| \, d\mu < +\infty.$$

Thus the function S is summable on X. Hence it is finite almost everywhere, which is equivalent to the required assertion. □

Corollary 2 provides a useful method of proving the almost everywhere convergence of various function series.

Example Let $\{x_n\}_{n\geqslant 1}$ be an arbitrary sequence of numbers (for instance, the set of rational numbers arranged in an arbitrary order). If the series $\sum_{n\geqslant 1} a_n$ converges absolutely, then the series $\sum_{n\geqslant 1} \frac{a_n}{\sqrt{|x-x_n|}}$ converges absolutely almost everywhere (with respect to the Lebesgue measure) on $\mathbb{R}$.

To prove this, it suffices to check that the series under consideration converges absolutely almost everywhere on an arbitrary interval $(-A, A)$. Obviously,

$$\int_{-A}^{A}\left|\frac{a_n}{\sqrt{|x-x_n|}}\right|dx = |a_n|\int_{-A-x_n}^{A-x_n}\frac{dx}{\sqrt{|x|}} \leqslant |a_n|\int_{-A}^{A}\frac{dx}{\sqrt{|x|}} = 4\sqrt{A}\,|a_n|.$$

Hence the series $\sum_{n\geqslant 1}\int_{-A}^{A}\frac{|a_n|}{\sqrt{|x-x_n|}}\,dx$ converges, and it remains to apply Corollary 2.

4.8.3 B. Levi's theorem applies only to increasing sequences of non-negative functions and cannot be used if these conditions are not satisfied. The following two important *dominated convergence* theorems fill this gap, providing convenient sufficient conditions for the interchange of limits and integration for arbitrary sequences of functions (either real- or complex-valued).

It is intuitively clear that if the integral $\int_X |f-g|\,d\mu$ is small, then the functions f and g are "close" on a set of "sufficiently large" measure. If we want to obtain conditions under which $\int_X |f_n - f|\,d\mu \underset{n\to\infty}{\longrightarrow} 0$, then we would expect the functions f_n to be close to f on sets of ever increasing measure. To obtain a precise formulation of this condition, we will use the notion of convergence in measure (see Sect. 3.3). Recall that $X(f > a) = \{x \in X \mid f(x) > a\}$.

Theorem (Lebesgue) *Let $\{f_n\}_{n\geqslant 1}$ be a sequence of measurable functions that converges in measure to a function f on X. If*

$$\begin{cases} \text{(a)} \quad |f_n(x)| \leqslant g(x) & \textit{almost everywhere on } X \textit{ for every } n \in \mathbb{N}, \\ \text{(b)} \quad g \textit{ is summable on } X, \end{cases} \tag{L}$$

then the functions f_n and f are summable,

$$\int_X |f_n - f|\,d\mu \underset{n\to\infty}{\longrightarrow} 0, \quad \textit{and, consequently,} \quad \int_X f_n\,d\mu \underset{n\to\infty}{\longrightarrow} \int_X f\,d\mu.$$

Note that the convergence of f_n to f in measure is a necessary condition for the conclusion of the theorem to be true, since $\mu(X(|f-f_n|>\varepsilon)) \leqslant \frac{1}{\varepsilon}\int_X |f-f_n|\,d\mu$ by Chebyshev's inequality (see Sect. 4.4.4).

Proof The summability of f_n is guaranteed by condition (L). The function f is measurable by the definition of convergence in measure. Passing to the limit in inequality (a) (see Corollary 2 in Sect. 3.3.5), we see that $|f(x)| \leqslant g(x)$ almost everywhere on X, which implies the summability of f.

Since

$$\left|\int_X f_n\,d\mu - \int_X f\,d\mu\right| \leqslant \int_X |f_n - f|\,d\mu,$$

it suffices to establish the first of the relations to be proved.

First assume that $\mu(X) < +\infty$. Fix an arbitrary $\varepsilon > 0$ and set $X_n(\varepsilon) = X(|f_n - f| > \varepsilon)$. Obviously,

$$\int_X |f_n - f|\,d\mu = \int_{X_n(\varepsilon)} \cdots + \int_{X\setminus X_n(\varepsilon)} \cdots \leqslant \int_{X_n(\varepsilon)} 2g\,d\mu + \int_X \varepsilon\,d\mu.$$

Since $\mu(X_n(\varepsilon)) \underset{n\to\infty}{\longrightarrow} 0$, we have $\int_{X_n(\varepsilon)} g\,d\mu \underset{n\to\infty}{\longrightarrow} 0$ by the absolute continuity of the integral. Hence $\int_{X_n(\varepsilon)} g\,d\mu < \frac{\varepsilon}{2}$ for sufficiently large n, and, consequently,

$$\int_X |f_n - f|\,d\mu < \varepsilon + \mu(X)\varepsilon,$$

which proves the theorem in the case under consideration.

If $\mu(X) = +\infty$, then fix $\varepsilon > 0$ and consider a set A of finite measure such that $\int_{X\setminus A} g\,d\mu < \varepsilon$ (see Corollary 3 in Sect. 4.5.1). Then

$$\int_X |f_n - f|\,d\mu \leqslant \int_A |f_n - f|\,d\mu + \int_{X\setminus A} 2g\,d\mu < \int_A |f_n - f|\,d\mu + 2\varepsilon.$$

Since $\mu(A) < +\infty$, it follows from the above that $\int_A |f_n - f|\,d\mu \underset{n\to\infty}{\longrightarrow} 0$, and hence $\int_X |f_n - f|\,d\mu < 3\varepsilon$ for sufficiently large n. □

Remark As one can see from the proof, in the case where μ is an infinite but σ-finite measure, the theorem remains valid if we replace the convergence in measure on X with the weaker assumption that $f_n \underset{n\to\infty}{\longrightarrow} f$ in measure on every set of finite measure (for a more general result, see Exercise 8). Since for a finite measure, convergence in measure follows from almost everywhere convergence (see Theorem 3.3.2), Lebesgue's theorem remains valid in the case of a σ-finite measure if we replace convergence in measure with almost everywhere convergence. We will prove that this is in fact true for an arbitrary measure.

4.8.4 Theorem 4.8.3 remains valid even if the convergence in measure is replaced with the convergence almost everywhere.

Theorem (Lebesgue) *Let $\{f_n\}_{n\geqslant 1}$ be a sequence of measurable functions that converges to a function f almost everywhere on X. If condition* (L) *is satisfied, then the functions f_n and f are summable,*

$$\int_X |f_n - f|\,d\mu \underset{n\to\infty}{\longrightarrow} 0, \quad \textit{and, consequently,} \quad \int_X f_n\,d\mu \underset{n\to\infty}{\longrightarrow} \int_X f\,d\mu.$$

Proof The summability of f_n and f can be proved in exactly the same way as in Theorem 4.8.3. Set

$$h_n = \sup\{|f_n - f|, |f_{n+1} - f|, \dots\}.$$

Obviously, $\lim_{n\to\infty} h_n(x) = \overline{\lim}_{n\to\infty} |f_n(x) - f(x)| = 0$ almost everywhere on X. Moreover, $h_{n+1} \leqslant h_n \leqslant 2g$ for every $n \in \mathbb{N}$. Applying B. Levi's theorem to the increasing sequence $\{2g - h_n\}_{n\geqslant 1}$, we see that

$$\int_X (2g - h_n)\,d\mu \underset{n\to\infty}{\longrightarrow} \int_X 2g\,d\mu.$$

Hence $\int_X h_n\,d\mu \underset{n\to\infty}{\longrightarrow} 0$, and, consequently,

$$\int_X |f_n - f|\,d\mu \leqslant \int_X h_n\,d\mu \underset{n\to\infty}{\longrightarrow} 0. \qquad \square$$

Condition (L) is not necessary for the interchange of limits and integration. One can see this from the following example. Let μ be the Lebesgue measure on $\mathbb{R}$ and consider the functions f_n defined by the formula $f_n(x) = c_n > 0$ for $\frac{1}{n+1} < x < \frac{1}{n}$ and $f_n(x) = 0$ for the other values of x. Obviously, $f_n(x) \underset{n\to\infty}{\longrightarrow} 0$ everywhere. Furthermore, $\int_{\mathbb{R}} f_n(x)\,dx = \frac{c_n}{n(n+1)} \underset{n\to\infty}{\longrightarrow} 0$ provided that $c_n = o(n^2)$. However, the functions f_n are not necessarily dominated by a summable function. Indeed, such a function is, obviously, not less than the sum $\sum_{n\geqslant 1} f_n(x)$, the integral of which is equal to $\sum_{n\geqslant 1} \frac{c_n}{n(n+1)}$. If, for example, $c_n = n$, the latter series diverges.

Example 1 Let μ be a finite Borel measure on $[0, +\infty)$. Let us find the limit of the sequence of integrals

$$I_n = \int_{[0,+\infty)} \varphi(x^n)\,d\mu(x),$$

where φ is a continuous function that has a finite limit C at infinity.

The pointwise convergence obviously holds:

$$\varphi(x^n) \underset{n\to\infty}{\longrightarrow} f(x) = \begin{cases} \varphi(0) & \text{if } 0 \leqslant x < 1, \\ \varphi(1) & \text{if } x = 1, \\ C & \text{if } x > 1. \end{cases}$$

Since the function φ is bounded on $[0, +\infty)$ and the measure μ is finite, condition (L) is satisfied (the sequence is dominated by the constant function equal to $\sup|\varphi|$ everywhere). Hence we may apply Lebesgue's theorem:

$$\lim_{n\to\infty} I_n = \int_{[0,+\infty)} f(x)\,d\mu(x) = \varphi(0)\,\mu\big([0,1)\big) + \varphi(1)\,\mu\big(\{1\}\big) + C\,\mu\big((1,+\infty)\big).$$

In particular, if μ is a discrete measure generated by point masses ω_k at integer points, and $\sum_{k\geqslant 0} \omega_k < +\infty$, then

$$I_n = \sum_{k\geqslant 0} \varphi(k^n)\omega_k \underset{n\to\infty}{\longrightarrow} \varphi(0)\omega_0 + \varphi(1)\omega_1 + C\sum_{k\geqslant 2} \omega_k.$$

In some cases, not only the integrand f_n, but also the domain of integration depends on the index n. However, extending f_n by zero outside of this set, we can reduce such a situation to the standard one (where the domain of integration is constant).

Example 2 Let $a > 0$. We will prove that the integrals

$$I_n = \int_0^n x^{a-1}\left(1 - \frac{x}{n}\right)^n dx$$

tend to $\int_0^\infty x^{a-1}e^{-x}\,dx = \Gamma(a)$ as $n \to \infty$.

To this end, set $f_n(x) = x^{a-1}(1 - \frac{x}{n})^n$ for $x \in (0, n]$ and $f_n(x) = 0$ for $x > n$. Clearly, $f_n(x) \underset{n\to\infty}{\longrightarrow} f(x) = x^{a-1}e^{-x}$ for every $x > 0$. To prove that $\lim_{n\to\infty} I_n = \Gamma(a)$, we will check that the functions f_n satisfy condition (L) of Lebesgue's theorem. Indeed, since $1 - \frac{x}{n} \leqslant e^{-x/n}$, we have $(1 - \frac{x}{n})^n \leqslant e^{-x}$ for $0 < x \leqslant n$, whence $0 \leqslant f_n(x) \leqslant x^{a-1}e^{-x}$ for all $x > 0$. Thus the functions f_n are dominated by a summable function on $(0, +\infty)$.

4.8.5 The next application of Lebesgue's theorem is of a more general nature; we preface it with an auxiliary result.

Let f be a function defined on a bounded set $E \subset \mathbb{R}^m$. With every tagged partition (τ, ξ) of E, which consists of sets $e_1, \dots, e_n$ and tags $\xi_1, \dots, \xi_n$ (by construction, $\xi_k \in E \cap \overline{e_k}$), we associate the simple function

$$f_\tau = \sum_{k=1}^{n} f(\xi_k)\chi_{e_k}.$$

Thus $f_\tau(x) = f(\xi_k)$ for $x \in e_k$. Recall that the mesh of τ is equal to $r(\tau) = \max_{1\leqslant k\leqslant n} \operatorname{diam}(e_k)$ (see Sect. 4.7.3).

Lemma *If $r(\tau) \to 0$, then $f_\tau(x) \to f(x)$ at all points of continuity of f. More precisely: if x is a point of continuity of f, then for every $\varepsilon > 0$ there exists a $\delta > 0$ such that $|f_\tau(x) - f(x)| < \varepsilon$ as soon as $r(\tau) < \delta$.*

Proof It suffices, given ε, to choose a $\delta > 0$ such that $|f(x) - f(y)| < \varepsilon$ for $\|x - y\| < \delta$ and $y \in E$. In this case, if $r(\tau) < \delta$ and $x \in e_k$, then $\|\xi_k - x\| < \delta$, whence $|f(x) - f_\tau(x)| = |f(x) - f(\xi_k)| < \varepsilon$. □

The following theorem generalizes Theorem 4.7.3.

Theorem *Let E be a bounded (measurable) subset of $\mathbb{R}^m$. If f is a bounded function defined on E and the set of discontinuities of f is of zero measure, then the integral $\int_E f(x)\,dx$ is the limit of Riemann sums (in the same sense as in Theorem 4.7.3).*

Proof First observe that, by Theorem 3.1.7, the function f is measurable. Let $\tau = \{e_k\}_{k=1}^n$ be a partition of E and $\xi = \{\xi_k\}_{k=1}^n$, where $\xi_k \in E \cap \overline{e_k}$, be a family of tags for τ. By definition, the Riemann sum $S(f, \tau, \xi)$ corresponding to the tagged partition (τ, ξ) is equal to $S(f, \tau, \xi) = \sum_{k=1}^n f(\xi_k)\lambda_m(e_k)$. In the notation of the lemma, this formula can be rewritten in the form

$$S(f, \tau, \xi) = \int_E f_\tau(x)\,dx.$$

Since $f_\tau(x) \to f(x)$ as $r(\tau) \to 0$ at all points of continuity of f, we see that $f_{\tau_n} \underset{n\to\infty}{\longrightarrow} f$ almost everywhere for every sequence of partitions τ_n such that $r(\tau_n) \underset{n\to\infty}{\longrightarrow} 0$. Furthermore, it is obvious that the functions f_τ are uniformly bounded. Hence, by Lebesgue's theorem, $\int_E f_{\tau_n}(x)\,dx \underset{n\to\infty}{\longrightarrow} \int_E f(x)\,dx$, which is equivalent to the required assertion. □

By tradition, for functions defined on an interval $[a, b]$, the integral is defined as the limit of the Riemann sums corresponding to partitions of $[a, b]$ into finer and finer subintervals. This definition was suggested by Riemann,[18] so the integral understood in this way is called the Riemann integral. It is worth mentioning that such sums and their limits were earlier considered by Cauchy, but only for continuous functions. As follows from the theorem proved above, if f is bounded and continuous almost everywhere on $[a, b]$, then the corresponding Riemann integral exists and coincides with the integral of f with respect to the Lebesgue measure. We leave the reader to prove that the assumptions made in the theorem (that f is bounded and the set of discontinuities of f has zero measure) are not only sufficient, but also necessary for the integral $\int_E f(x)\,dx$ to coincide with the limit of the Riemann sums (see Exercises 10–12). Thus the Riemann integral of a bounded function over a finite interval exists if and only if it is continuous almost everywhere.

4.8.6 The next theorem is not exactly a result on the interchange of limits and integration, but it shows that in a wide class of cases one can pass to the limit in an inequality. More precisely, the integral of non-negative functions has an important property: it is lower semicontinuous with respect to almost everywhere convergence. This property is often used in the cases where one has to establish the summability of a limit function.

Theorem 1 (Fatou[19]) *Let $\{f_n\}_{n\geqslant 1}$ be a sequence of non-negative measurable functions that converges to f almost everywhere on X. If for some $C > 0$,*

$$\int_X f_n\,d\mu \leqslant C \quad \textit{for every } n \in \mathbb{N}, \tag{1}$$

then $\int_X f\,d\mu \leqslant C$.

[18] Georg Friedrich Bernhard Riemann (1826–1866)—German mathematician.

[19] Pierre Joseph Louis Fatou (1878–1929)—French mathematician.

Remark Even if the integrals of all functions f_n are equal, the integral of the limit function may be strictly less than their common value. To obtain a corresponding example, assume that our measure space is the interval (0, 1) with Lebesgue measure and f_n is the function defined by the formula

$$f_n(x) = \begin{cases} n & \text{for } 0 < x < \frac{1}{n}, \\ 0 & \text{for } \frac{1}{n} \leqslant x < 1. \end{cases}$$

Obviously, $f_n(x) \underset{n\to\infty}{\longrightarrow} 0$ pointwise on (0, 1) and $\int_0^1 f_n(x)\,dx = 1$, while the integral of the limit function vanishes.

The same example shows that Fatou's theorem is no longer true if we reverse the inequalities in condition (1) and in the conclusion of the theorem; that is, the integral, while being lower semicontinuous, is not upper semicontinuous.

Changing the signs of f_n in the above example, we see that Fatou's theorem is not true without the assumption that the functions under consideration are non-negative.

Proof Let $g_n(x) = \inf\{f_n(x), f_{n+1}(x), \dots, f_{n+k}(x), \dots\}$ $(x \in X)$. Clearly, $g_n \leqslant g_{n+1}$, $g_n \underset{n\to\infty}{\longrightarrow} f$ almost everywhere on X, and

$$\int_X g_n\,d\mu \leqslant \int_X f_n\,d\mu \leqslant C \quad \text{for all } n \in \mathbb{N}.$$

Therefore, by B. Levi's theorem,

$$\int_X f\,d\mu = \lim_{n\to\infty} \int_X g_n\,d\mu \leqslant C.$$

□

Since the sequence $\{g_n\}_{n\geqslant 1}$ is monotone, we can drop the assumption that the sequence $\{f_n\}_{n\geqslant 1}$ converges and use the equation $\lim_{n\to\infty} g_n = \underline{\lim}_{n\to\infty} f_n$ to prove a formally stronger version of Fatou's theorem:

for every sequence of non-negative measurable functions $\{f_n\}_{n\geqslant 1}$,

$$\int_X \underline{\lim}_{n\to\infty} f_n\,d\mu \leqslant \underline{\lim}_{n\to\infty} \int_X f_n\,d\mu. \tag{2}$$

The reader has probably encountered situations where a more general result is much less important than a central special case. In our opinion, Fatou's theorem and its generalization provide such an example. For another example, see Exercise 4, which generalizes B. Levi's theorem.

Theorem 1 remains valid if we replace almost everywhere convergence with convergence in measure.

Theorem 1′ (Fatou) *Let* $\{f_n\}_{n\geqslant 1}$ *be a sequence of non-negative measurable functions that converges in measure to a function* f. *If for some* $C > 0$,

$$\int_X f_n\,d\mu \leqslant C \quad \textit{for every } n \in \mathbb{N},$$

then $\int_X f\,d\mu \leqslant C$.

Proof Using Riesz' theorem (see Sect. 3.3.4), choose a subsequence $\{f_{n_k}\}_{k\geqslant 1}$ that converges to f almost everywhere. Applying Theorem 1 to this subsequence, we obtain the desired result. □

Note that in the case of a finite measure Theorem 1′ is stronger than Theorem 1. In addition, the result obtained in the former does not follow from (2), because the lower limit $\underline{\lim}_{n\to\infty} f_n$ can be substantially less than f (see Sect. 3.3, Exercises 2, 3).

4.8.7 As we have seen, the existence of a summable dominating function is not a necessary condition for the interchange of limits and integration; however, in Lebesgue's theorem it is essential and cannot be dispensed with. But an analysis of the proof shows that this condition can be weakened. Indeed, what we in fact need is not the existence of a summable dominating function, but the smallness of the integrals $\int_e |f_n|\,d\mu$ for sets e of sufficiently small measure implied by this assumption. So we introduce the following definition.

Definition We say that functions f_α $(\alpha \in A)$ have *absolutely equicontinuous* integrals if they are summable and

$$\forall \varepsilon > 0\, \exists \delta > 0 : \big(\mu(e) < \delta\big) \quad \Rightarrow \quad \left(\forall \alpha \in A \int_e |f_\alpha|\,d\mu < \varepsilon\right). \tag{3}$$

If a family $\{f_\alpha\}_{\alpha\in A}$ is dominated by a summable function, i.e., there exists a summable function g such that $|f_\alpha| \leqslant g$ almost everywhere for every α, then $\int_e |f_\alpha|\,d\mu \leqslant \int_e g\,d\mu$, and condition (3) is satisfied by the absolute continuity of the integral of g.

It turns out that the absolute equicontinuity of the integrals of f_n $(n \in \mathbb{N})$ is a necessary condition for the integrals $\int_E f_n\,d\mu$ to have a finite limit for every measurable set E and, in particular, for the convergence $\int_X |f_n - f|\,d\mu \underset{n\to\infty}{\longrightarrow} 0$. The proof of this theorem, due to Vitali, is rather involved (see, for example, [Bo, Vol. I]), so we do not reproduce it here, but establish a much easier result that the absolute equicontinuity is sufficient for the interchange of limits and integration. Note that the proof of this result provides a typical application of Fatou's theorem, which is used to estimate the integral of the limit function.

Theorem (Vitali) *Let $\{f_n\}_{n\geqslant 1}$ be a sequence of measurable functions that converges to a function f in measure on X. If $\mu(X) < \infty$ and the functions f_n have absolutely equicontinuous integrals, then f is summable and $\int_X |f_n - f|\,d\mu \underset{n\to\infty}{\longrightarrow} 0$.*

Proof Fix an arbitrary $\varepsilon > 0$, and let δ be a number such that condition (3) is satisfied. Set $e_n = X(|f - f_n| > \varepsilon)$. Since $\mu(e_n) \underset{n\to\infty}{\longrightarrow} 0$ by assumption, it follows that $\mu(e_n) < \delta$ for sufficiently large n; by condition (3), for such n and for all k we have

$\int_{e_n} |f_k|\,d\mu < \varepsilon$. By Fatou's theorem, it follows that $\int_{e_n} |f|\,d\mu \leqslant \varepsilon$. Therefore,

$$\begin{aligned}\int_X |f - f_n|\,d\mu &= \int_{X\setminus e_n} |f - f_n|\,d\mu + \int_{e_n} |f - f_n|\,d\mu \\ &\leqslant \int_{X\setminus e_n} \varepsilon\,d\mu + \int_{e_n} |f|\,d\mu + \int_{e_n} |f_n|\,d\mu \leqslant \varepsilon\,\mu(X) + \varepsilon + \varepsilon \\ &= (\mu(X) + 2)\varepsilon.\end{aligned}$$

Since this inequality holds for sufficiently large n, it follows that $\int_X |f - f_n|\,d\mu \underset{n\to\infty}{\longrightarrow} 0$. Furthermore, f is summable, because $f = f_n + (f - f_n)$, with both terms on the right-hand side being summable. □

Vitali's theorem implies a useful corollary.

Corollary *Let $\mu(X) < +\infty$, and let $\{f_n\}_{n\geqslant 1}$ be a sequence of measurable functions that converges in measure to a function f. If there exist $p > 1$ and $C > 0$ such that*

$$\int_X |f_n|^p\,d\mu \leqslant C \quad \textit{for all } n, \tag{V}$$

then the functions f_n and f are summable and $\int_X |f_n - f|\,d\mu \underset{n\to\infty}{\longrightarrow} 0$.

Proof To apply Vitali's theorem, we should check the summability of f_n and the absolute equicontinuity of the integrals of f_n. Both these facts follow from Hölder's inequality. Indeed, assuming that $\frac{1}{p} + \frac{1}{q} = 1$, for every set e we have

$$\int_e |f_n|\,d\mu \leqslant \left(\int_e |f_n|^p\,d\mu\right)^{\frac{1}{p}} \big(\mu(e)\big)^{\frac{1}{q}} \leqslant C^{\frac{1}{p}} \big(\mu(e)\big)^{\frac{1}{q}}.$$

This implies both the summability of f_n (for $e = X$) and condition (3), since the integrals $\int_e |f_n|\,d\mu$ are arbitrarily small for all n provided that the measure of e is sufficiently small. □

Following the same scheme, we can use Vitali's theorem to deduce a more general result, whose proof we leave to the reader.

Theorem (de La Vallée Poussin[20]) *Let $\mu(X) < \infty$, and let $\{f_n\}_{n\geqslant 1}$ be a sequence of measurable functions that converges in measure to a function f. If there exists*

[20]Charles-Jean Étienne Gustave Nicolas de La Vallée Poussin (1866–1962)—Belgian mathematician.

a non-negative function Φ that grows unboundedly on $[0,+\infty)$ and satisfies the condition

$$\sup_n \int_X |f_n|\Phi(|f_n|)\,d\mu < +\infty,$$

then the functions f_n and f are summable and $\int_X |f_n - f|\,d\mu \underset{n\to\infty}{\longrightarrow} 0$.

EXERCISES

1. Give an example of a sequence of positive continuous functions f_n on $[a,b]$ such that $\int_a^b f_n(x)\,dx \underset{n\to\infty}{\longrightarrow} 0$ and $\sup_n f_n(x) = +\infty$ at every point $x \in [a,b]$.

2. Show, by an example, that B. Levi's theorem is no longer true if we drop the assumption that the functions under consideration are non-negative.

3. Let $a_n \geqslant 0$ and $\sum_{n=1}^{\infty} a_n < +\infty$. Show that:

(a) if $\sum_{n=1}^{\infty} a_n \ln n < +\infty$, then the series $\sum_{n=1}^{\infty} \frac{a_n}{|x-x_n|}$ converges almost everywhere on $\mathbb{R}$ (with respect to the Lebesgue measure) for every sequence $\{x_n\} \subset \mathbb{R}$;

(b) if $\sum_{n=1}^{\infty} a_n \ln n = +\infty$ and X is a countable dense subset of the interval $(0,1)$, then, depending on the numbering $\{x_n\}$ of X, the series in question may converge almost everywhere on $(0,1)$ as well as diverge almost everywhere on $(0,1)$ (and even at every point of $(0,1)$).

4. Prove the following generalization of B. Levi's theorem. Let $\{f_n\}_{n\geqslant 1}$ be a sequence of non-negative measurable functions that converges to f almost everywhere on X. If $f_n \leqslant f$ almost everywhere for every n, then $\int_X f_n\,d\mu \underset{n\to\infty}{\longrightarrow} \int_X f\,d\mu$.

5. Let $\{f_n\}_{n\geqslant 1}$ be a sequence of non-negative measurable functions that converges to a summable function f almost everywhere on X. If $\int_X f_n\,d\mu \underset{n\to\infty}{\longrightarrow} \int_X f\,d\mu$, then $\int_E f_n\,d\mu \underset{n\to\infty}{\longrightarrow} \int_E f\,d\mu$ for every measurable set $E \subset X$. Moreover, $\int_X |f_n - f|\,d\mu \underset{n\to\infty}{\longrightarrow} 0$. *Hint.* To prove the first claim, apply Fatou's theorem; to prove the second claim, use the identity $|f_n - f| = f_n - f + 2(f_n - f)_-$ and Lebesgue's theorem.

6. Is the sequence of functions $\frac{1}{n}(\frac{\sin nx}{x})^2$, which pointwise converges to zero, dominated by a summable function on $(0,\pi)$?

7. Show that if μ is a finite measure, then $f_n \underset{n\to\infty}{\longrightarrow} f$ in measure if and only if $\int_X \frac{|f_n - f|}{1+|f_n - f|}\,d\mu \underset{n\to\infty}{\longrightarrow} 0$.

8. Let μ be a measure such that $\mu(A) = \sup\{\mu(E) \mid E \subset A,\ \mu(E) < +\infty\}$ for every measurable set A. Show that Theorem 4.8.3 remains valid if we replace the convergence of f_n to f in measure on X with convergence in measure on every set of finite measure. The latter condition is obviously satisfied if $f_n \underset{n\to\infty}{\longrightarrow} f$ almost everywhere.

9. Is the sequence of functions $f_n(x) = \frac{n}{n-(n-1)e^{ix}}$ dominated by a summable function on $(-\pi,\pi)$? What is the limit $\lim_{n\to\infty} \int_{-\pi}^{\pi} |f_n(x)|\,dx$?

10. Show that if a function f defined on a ball is not bounded, then the corresponding Riemann sums $S(f, \tau, \xi)$ cannot have a finite limit as $r(\tau) \to 0$.

11. Let f be a measurable function defined on a bounded subset e of $\mathbb{R}^m$ such that the set of discontinuities of f is of positive measure. Show that the Riemann sums of f corresponding to finer and finer partitions have no limit (even if we restrict ourselves to partitions of E into sets of the form $E \cap P$, where P is a cell). *Hint.* Consider a set of positive measure $K \subset E$ such that

$$\overline{\lim_{y \to x}}\, f(y) - \underline{\lim_{y \to x}}\, f(y) \geqslant \varepsilon > 0 \quad \text{for all } x \in K.$$

Verify that there exist a partition τ of arbitrarily small mesh and families of tags ξ and ξ' for τ such that $S(f, \tau, \xi) - S(f, \tau, \xi') > \frac{\varepsilon}{2}\lambda_m(K)$.

12. Show that Theorem 4.8.5 and the result of Exercise 11 remain valid for every finite Borel measure. The result of Exercise 10 also remains valid under the additional assumption that every non-empty open set has positive measure.

13. Show that in the definition of absolute equicontinuity, the integrals $\int_e |f_\alpha|\, d\mu$ may be replaced by $|\int_e f_\alpha\, d\mu|$.

4.9 ⋆The Maximal Function and Differentiation of the Integral with Respect to a Set

In this section, we study the following problem: to what degree can Barrow's theorem on differentiation of the integral of a continuous function with respect to a variable upper limit (see Sect. 4.6.1) be extended to summable functions? As one can easily see, there is no difficulty in generalizing it to the case of multiple integrals keeping the assumption that the integrand is continuous. However, an attempt to extend the class of functions under consideration encounters major difficulties even in the one-dimensional case. If the integrand is only summable, we cannot expect the derivative with respect to a variable upper limit to exist at every point. It is also clear that even if this derivative exists, it does not necessarily coincide with the corresponding value of the integrand (since we can modify the latter at a set of zero measure in an arbitrary way without affecting the integral). Hence we should adjust the statement of the problem. Obviously, we can hope for the derivative to coincide with the integrand only almost everywhere. It is extremely important to find out whether or not the derivative does indeed exist almost everywhere. More precisely, we formulate the question as follows. Given a summable function on $\mathbb{R}^m$, is it true that the limit of the average values of f over balls shrinking to a point, i.e., $\lim_{r \to 0} \frac{1}{v(r)} \int_{B(x,r)} f(y)\, dy$, exists almost everywhere? We will see that the answer to this question is positive and, moreover, the above limit coincides with $f(x)$ almost everywhere.

By λ we denote the Lebesgue measure on $\mathbb{R}^m$ and by λ^* the corresponding outer measure; $v(r) = \lambda(B(0, r))$. Let $\mathscr{L}(\mathbb{R}^m)$ be the set of functions summable on $\mathbb{R}^m$ with respect to the Lebesgue measure.

4.9.1 The problem of differentiating the integral with a variable upper limit deals in fact with the behavior of average values of the form $\frac{1}{h}\int_x^{x+h} f(y)\,dy$. In various estimates related to average values of a function (of one or several variables), it is often useful to employ a function dominating these averages. The most convenient of such functions was introduced by Hardy[21] and Littlewood[22]; it is the smallest function dominating the averages of f over balls. Here is the corresponding definition.

Definition Let $f \in \mathscr{L}(\mathbb{R}^m)$. The function M_f defined by the formula

$$M_f(x) = \sup_{r>0} \frac{1}{v(r)} \int_{B(x,r)} |f(y)|\,dy \quad (x \in \mathbb{R}^m)$$

is called the *maximal function* (for f).

Note that the maximal function is measurable. Indeed, as follows from the absolute continuity of the integral, the function $(x,r) \mapsto \frac{1}{v(r)}\int_{B(x,r)} |f(y)|\,dy$ is continuous. Hence the supremum in the definition of M_f can be taken only over the rational values of r. Thus the maximal function is measurable as the supremum of a countable family of measurable functions. If $I = \int_{\mathbb{R}^m} |f(x)|\,dx > 0$, then M_f is not summable. Indeed, if the norm $\|x\|$ is sufficiently large, then

$$M_f(x) \geqslant \frac{1}{v(2\|x\|)} \int_{B(x,2\|x\|)} |f(y)|\,dy \geqslant \frac{\text{const}}{\|x\|^m} I.$$

One can show that the maximal function is not necessarily summable even on sets of finite measure (see Exercise 1). However, it is finite almost everywhere, as the following important theorem implies.

Theorem *Let $f \in \mathscr{L}(\mathbb{R}^m)$ and $E_t = \{x \in \mathbb{R}^m \mid M_f(x) > t\}$ for $t > 0$. Then*

$$\lambda(E_t) \leqslant \frac{5^m}{t} \int_{\mathbb{R}^m} |f(x)|\,dx. \tag{1}$$

Since $\{x \in \mathbb{R}^m \mid M_f(x) = +\infty\} \subset E_t$ for every $t > 0$, it follows that the function M_f is finite almost everywhere.

Proof To estimate the measure of the set E_t, we use Theorem 2.7.1. Since $M_f(x) = \sup_{r>0} \frac{1}{v(r)} \int_{B(x,r)} |f(y)|\,dy > t$ for $x \in E_t$, for every point $x \in E_t$ there exists a ball $B(x, r_x)$ such that

$$\frac{1}{v(r_x)} \int_{B(x,r_x)} |f(y)|\,dy > t.$$

[21] Godfrey Harold Hardy (1877–1947)—English mathematician.

[22] John Edensor Littlewood (1885–1977)—English mathematician.

This inequality can be rewritten as

$$v(r_x) < \frac{1}{t} \int_{B(x,r_x)} |f(y)|\,dy. \tag{2}$$

It follows that $v(r_x) \leqslant \frac{1}{t} \int_{\mathbb{R}^m} |f(y)|\,dy$, and hence the radii of the balls are uniformly bounded. To apply Theorem 2.7.1, instead of the whole set E_t, which may be unbounded, consider an arbitrary bounded part E_t^0 of E_t. Then, according to this theorem, in the family $\{B(x, r_x)\}_{x \in E_t^0}$ there is a (possibly finite) sequence of pairwise disjoint balls $B_k = B(x_k, r_{x_k})$ such that $E_t^0 \subset \bigcup_{k \geqslant 1} B_k^*$, where $B_k^* = B(x_k, 5r_{x_k})$. Hence, using (2), we obtain

$$\lambda(E_t^0) \leqslant \sum_{k=1}^{\infty} \lambda(B_k^*) = 5^m \sum_{k=1}^{\infty} \lambda(B_k) \leqslant \frac{5^m}{t} \sum_{k=1}^{\infty} \int_{B_k} |f(y)|\,dy$$

$$= \frac{5^m}{t} \int_{\bigvee_{k \geqslant 1} B_k} |f(y)|\,dy \leqslant \frac{5^m}{t} \int_{\mathbb{R}^m} |f(y)|\,dy.$$

Since E_t^0 is arbitrary, this inequality holds for E_t as well. □

4.9.2 Now we turn to the main problem of this section: is it true that the limit $\lim_{n\to\infty} \frac{1}{\lambda(E_n(x))} \int_{E_n(x)} f(y)\,dy$, where $E_n(x)$ are sets of positive measure shrinking to a point x, exists almost everywhere and coincides with $f(x)$? Keeping in mind the analogy with the one-dimensional case, where $E_n(x)$ are intervals shrinking to x, it is natural to interpret our question as asking about the derivative of the integral with respect to the system of sets $\{E_n(x)\}$.

It is obvious that the behavior of f at points "far" from x (in particular, the summability of f on the whole space $\mathbb{R}^m$) does not affect the existence and the value of the limit in question. So it makes sense to introduce a wider class of functions than $\mathscr{L}(\mathbb{R}^m)$; this class of measurable functions often appears in function theory as well as in other areas of mathematics.

Definition A measurable function f in $\mathbb{R}^m$ is called *locally summable* in $\mathbb{R}^m$ if it is summable on every bounded set, i.e.,

$$\int_{B(0,R)} |f(x)|\,dx < +\infty \quad \text{for every } R > 0.$$

The set of all such functions will be denoted by $\mathscr{L}_{\mathrm{loc}}(\mathbb{R}^m)$. It is clear that every locally summable function is finite almost everywhere and the class $\mathscr{L}_{\mathrm{loc}}(\mathbb{R}^m)$ contains both continuous and simple functions.

First we consider the case of differentiating the integral with respect to a family of concentric balls. Our main goal is to prove the following important result.

Theorem (Lebesgue) *If* $f \in \mathscr{L}_{\mathrm{loc}}(\mathbb{R}^m)$, *then*

$$\frac{1}{v(r)} \int_{B(x,r)} |f(y) - f(x)|\, dy \underset{r\to 0}{\longrightarrow} 0 \tag{3}$$

for almost all x. *In particular,*

$$\frac{1}{v(r)} \int_{B(x,r)} f(y)\, dy \underset{r\to 0}{\longrightarrow} f(x) \quad \textit{almost everywhere.} \tag{3'}$$

A point x at which (3) holds is called a *Lebesgue point* of f.[23] Thus the Lebesgue differentiation theorem can also be stated as follows:

almost every point of a locally summable function f *is a Lebesgue point of* f.

Of course, every continuity point of f is a Lebesgue point of f, since

$$\frac{1}{v(r)} \int_{B(x,r)} |f(y) - f(x)|\, dy \leqslant \sup_{y\in B(x,r)} |f(y) - f(x)| \underset{r\to 0}{\longrightarrow} 0.$$

We preface the proof of the theorem with a useful lemma.

Lemma *A function* f *from the class* $\mathscr{L}(X, \mu)$ *can be approximated by simple functions in the following sense: for every* $\varepsilon > 0$ *there exists a simple function* g *such that*

$$\int_X |f - g|\, d\mu < \varepsilon.$$

Proof If f is non-negative, then this claim follows immediately from the definition of the integral. Indeed,

$$\int_X f\, d\mu = \sup\left\{\int_X h\, d\mu \,|\, 0 \leqslant h \leqslant f,\ h \text{ is a simple function}\right\},$$

and hence there exists a simple function g such that $0 \leqslant g \leqslant f$ and $\int_X f\, d\mu < \int_X g\, d\mu + \varepsilon$. It provides the desired approximation:

$$\int_X |f - g|\, d\mu = \int_X (f - g)\, d\mu = \int_X f\, d\mu - \int_X g\, d\mu < \varepsilon.$$

In the general case, it suffices to approximate the functions f_+ and f_-. □

Proof of the theorem We assume without loss of generality that f is real-valued. It suffices to show that for every $R > 0$ almost all points of the ball $B(0, R)$ are Lebesgue points of f. To prove this, fix a radius R and observe that for $\|x\| < R$ the

[23] In some books, the term "Lebesgue set" refers to the set of Lebesgue points of a function. We draw the reader's attention to this terminological ambiguity.

validity of (3) does not depend on the values of f outside of $B(0, R)$. This allows us to assume that $f \in \mathscr{L}(\mathbb{R}^m)$ (it suffices to redefine f by zero outside of the ball).

In the subsequent argument, we will consider functions f of more and more complicated structure. First assume that $f = \chi_E$ is the characteristic function of a measurable set E. Then

$$|f(y) - f(x)| = |\chi_E(y) - \chi_E(x)| = \begin{cases} 1 - \chi_E(y) & \text{if } x \in E, \\ \chi_E(y) & \text{if } x \notin E. \end{cases}$$

Hence

$$\frac{1}{v(r)} \int_{B(x,r)} |f(y) - f(x)|\, dy = \begin{cases} 1 - \frac{\lambda(E \cap B(x,r))}{v(r)}, & \text{if } x \in E, \\ \frac{\lambda(E \cap B(x,r))}{v(r)}, & \text{if } x \notin E. \end{cases}$$

By Corollary 1 of Vitali's theorem (see Sect. 2.7.3), almost every point of E is a density point of this set, which implies that the right-hand side tends to zero almost everywhere as $r \to 0$.

It is clear that (3) remains valid for every linear combination of characteristic functions, i.e., for every simple function.

Now we turn to the main case, where f is an arbitrary summable function. We will show that for an arbitrary $a > 0$, the measure of the set

$$E_a(f) = \left\{ x \in \mathbb{R}^m \;\middle|\; \overline{\lim_{r \to 0}} \frac{1}{v(r)} \int_{B(x,r)} |f(y) - f(x)|\, dy > a \right\}$$

vanishes. This will complete the proof of the theorem, since points at which (3) does not hold are contained in the union $\bigcup_{n=1}^{\infty} E_{1/n}(f)$.

Fix $a > 0$; we will estimate the outer measure of the set $E_a(f)$ (leaving aside the question of its measurability). Obviously,

$$\begin{aligned} \overline{\lim_{r \to 0}} \frac{1}{v(r)} \int_{B(x,r)} |f(y) - f(x)|\, dy &\leqslant \overline{\lim_{r \to 0}} \frac{1}{v(r)} \int_{B(x,r)} |f(y)|\, dy + |f(x)| \\ &\leqslant M_f(x) + |f(x)|, \end{aligned}$$

whence

$$E_a(f) \subset \left\{ x \in \mathbb{R}^m \;\middle|\; M_f(x) > \frac{a}{2} \right\} \cup \left\{ x \in \mathbb{R}^m \;\middle|\; |f(x)| > \frac{a}{2} \right\}.$$

But

$$\lambda\left(\left\{ x \in \mathbb{R}^m \;\middle|\; |f(x)| > \frac{a}{2} \right\} \right) \leqslant \frac{2}{a} \int_{\mathbb{R}^m} |f(x)|\, dx \quad \text{(by Chebyshev's inequality)},$$

$$\lambda\left(\left\{ x \in \mathbb{R}^m \;\middle|\; M_f(x) > \frac{a}{2} \right\} \right) \leqslant 2 \frac{5^m}{a} \int_{\mathbb{R}^m} |f(x)|\, dx \quad \text{(by Theorem 4.9.1)}.$$

Therefore,

$$\lambda^*\big(E_a(f)\big) \leqslant \frac{C}{a}\int_{\mathbb{R}^m} \big|f(x)\big|\,dx, \tag{4}$$

where C is a coefficient that depends only on the dimension.

To complete the proof of the theorem, we will show that $\lambda^*(E_a(f)) = 0$. As we have already established, (3) holds almost everywhere for a simple function. Hence, taking an arbitrary simple function g, averaging the inequality

$$\begin{aligned}\big|f(y)-f(x)\big| - \big|g(y)-g(x)\big| &\leqslant \big|\big(f(y)-g(y)\big) - \big(f(x)-g(x)\big)\big| \\ &\leqslant \big|f(y)-f(x)\big| + \big|g(y)-g(x)\big|\end{aligned}$$

over the ball $B(x,r)$, and taking the limit superior as $r \to 0$, we see that $\lambda^*(E_a(f)) = \lambda^*(E_a(f-g))$. Thus inequality (4) can be substantially strengthened: for an arbitrary simple function g,

$$\lambda^*\big(E_a(f)\big) = \lambda^*\big(E_a(f-g)\big) \leqslant \frac{C}{a}\int_{\mathbb{R}^m} \big|f(x)-g(x)\big|\,dx. \tag{4'}$$

As follows from the lemma, the right-hand side can be made arbitrarily small by the choice of g. Thus $\lambda^*(E_a(f)) = 0$. □

Remark 1 Since equality (3) holds for every continuous function g, it follows from the above argument that inequality (4′) also holds for such g. As we will show in Chap. 9, Lemma 4.9.2 remains valid if we replace simple functions with continuous ones. Hence we could prove the theorem using continuous rather than simple functions and applying Theorem 9.2.3 instead of the lemma.

Remark 2 The theorem can easily be extended to the (formally more general) case where a function is defined only on a subset of $\mathbb{R}^m$. Let us say that a function f is locally summable on an open set $\mathcal{O} \subset \mathbb{R}^m$, or on an arbitrary interval $\Delta \subset \mathbb{R}$, if it is summable on every compact subset. In this case, (3) holds for almost all points of $\mathcal{O}$.

Indeed, $\mathcal{O}$ can be exhausted by a sequence of closed cubes contained in it. Hence it suffices to prove (3) for almost all points of every such cube Q. The corresponding assertion follows immediately by applying the theorem to the function f_Q that coincides with f on Q and vanishes outside of Q (note that $f_Q \in \mathscr{L}(\mathbb{R}^m) \subset \mathscr{L}_{\mathrm{loc}}(\mathbb{R}^m)$, since $\int_Q |f(x)|\,dx < +\infty$).

4.9.3 Now we turn to the famous Lebesgue theorem, which provides the generalization of Barrow's theorem that we discussed at the beginning of this section. It concerns functions that can be written as integrals with a variable upper limit.

Definition A function F defined on an interval Δ, $\Delta \subset \mathbb{R}$, is called *absolutely continuous* on Δ if it can be written in the form

$$F(x) = F(c) + \int_c^x f(y)\,dy \quad (x \in \Delta), \tag{5}$$

where $c \in \Delta$ and f is locally summable on Δ.

We draw the reader's attention to the fact that if the interval Δ is closed, then f is summable on Δ. Otherwise this is not necessarily so (see Exercise 4).

It follows from Theorem 4.6.1 that every absolutely continuous function is continuous. The converse is not true even for monotone functions (see Exercises 4, 5). The simplest examples of absolutely continuous functions are C^1 functions. Clearly, the functions $|x|$, $\sqrt{|x|}$ are absolutely continuous on $\mathbb{R}$. As follows from the remark after the fundamental theorem of calculus (Sect. 4.6.1), a convex continuous function on an interval is absolutely continuous on this interval.

In the one-dimensional case, Theorem 4.9.2 shows that if F is the function defined by (5), then the limit of the ratio $\frac{F(x+h)-F(x-h)}{2h} = \frac{1}{2h}\int_{x-h}^{x+h} f(y)\,dy$ as $h \to 0$ exists almost everywhere and coincides with $f(x)$. One can strengthen this result by showing that the function F is almost everywhere differentiable in the classical sense.

Theorem (Lebesgue) *If F is a function that is absolutely continuous on an interval Δ, then it is differentiable almost everywhere, its derivative is locally summable, and $F(y) - F(x) = \int_x^y F'(t)\,dt$ for any $x, y \in \Delta$.*

Proof Let F be a function satisfying (5). We will prove that for almost all x the right derivative of F at x exists and coincides with $f(x)$. Indeed, if $h > 0$, then

$$\frac{F(x+h) - F(x)}{h} - f(x) = \frac{1}{h}\int_x^{x+h} f(y)\,dy - f(x) = \frac{1}{h}\int_x^{x+h} \big(f(y) - f(x)\big)\,dy.$$

Therefore,

$$\left|\frac{F(x+h) - F(x)}{h} - f(x)\right| \leqslant \frac{1}{h}\int_x^{x+h} \big|f(y) - f(x)\big|\,dy$$
$$\leqslant \frac{1}{h}\int_{x-h}^{x+h} \big|f(y) - f(x)\big|\,dy,$$

and the right-hand side tends to zero as $h \to 0$ almost everywhere on Δ by Theorem 4.9.2. It follows that the right derivative exists. Clearly, the existence of the left derivative can be proved in a similar way, and then the desired formula is obvious. □

The theorem shows that absolutely continuous functions admit the following description: a function F is absolutely continuous on an interval Δ if the derivative F'

exists almost everywhere, is locally summable, and F can be recovered from F' by the formula

$$F(x) = F(c) + \int_c^x F'(y)\,dy \quad (c,\, x \in \Delta).$$

Note that the local summability of the derivative F', which exists almost everywhere, is only necessary, but not sufficient for this equality to hold (see Exercise 5).

4.9.4 One may differentiate the integral with respect to other families of sets instead of concentric balls. Using the notion of a regular cover (see Sect. 2.7.4), we can easily obtain the following corollary of Theorem 4.9.2.

Corollary *If f is a locally summable function on an open subset $\mathcal{O}$ of $\mathbb{R}^m$ and $\{E_n(x)\}_{x\in X, n\in\mathbb{N}}$ is a regular cover of $X \subset \mathcal{O}$, then*

$$\lim_{n\to\infty} \frac{1}{\lambda(E_n(x))} \int_{E_n(x)} \left| f(y) - f(x) \right| dy \underset{n\to\infty}{\longrightarrow} 0 \quad \textit{almost everywhere on } X.$$

Note that we do not assume the set X to be measurable.

Proof For every x in X, let

$$E_n(x) \subset B\big(x, r_n(x)\big), \quad r_n(x) \underset{n\to\infty}{\longrightarrow} 0, \quad \text{and} \quad \inf_n \frac{\lambda(E_n(x))}{v(r_n(x))} = \theta(x).$$

Then the desired assertion follows from the inequality

$$\frac{1}{\lambda(E_n(x))} \int_{E_n(x)} \left| f(y) - f(x) \right| dy \leqslant \frac{1}{\theta(x) v(r_n(x))} \int_{B(x, r_n(x))} \left| f(y) - f(x) \right| dy,$$

whose right-hand side tends to zero almost everywhere by Theorem 4.9.2. □

EXERCISES

1. Let $f(x) = \frac{1}{x \ln^2 x}$ for $x \in (0, \frac{1}{2})$ and $f(x) = 0$ at the other points. Show that $M_f(x) \geqslant \frac{1}{|x \ln x|}$ for $x \in (0, \frac{1}{4})$ and, consequently, the maximal function is not summable in any neighborhood of the origin.
2. Give an example of a function f in $\mathscr{L}(\mathbb{R})$ such that the maximal function M_f is not summable on any non-empty interval.
3. Let $f(x) = \sin \frac{1}{x}$ for $x \neq 0$, $f(0) = 0$ and $F(x) = \int_0^x f(y)\,dy$. Show that 0 is not a Lebesgue point of f, but nevertheless the derivative $F'(0)$ does exist and is equal to zero.
4. Show that the function $f(x) = \int_0^x \sin \frac{1}{t} \frac{dt}{t}$ $(x \in [0, 1])$ is continuous but not absolutely continuous on the closed interval $[0, 1]$, though it is absolutely continuous on the semi-open interval $(0, 1]$.
5. Show that the Cantor function (see Sect. 2.3.2) is not absolutely continuous (while having zero derivative almost everywhere).

4.10 *The Lebesgue–Stieltjes Measure and Integral

Here we consider an important class of measures generated by increasing functions. The one-dimensional Lebesgue measure of a subset of the real line can be interpreted as the mass of this subset provided that the mass is distributed with a constant density. Dropping the latter condition, we arrive at the notion of the Lebesgue–Stieltjes[24] measure.

4.10.1 We now proceed to precise definitions. Let Δ be a non-empty open interval (finite or not), and let g be an increasing function defined on Δ. Given $c \in \Delta$, by $g(c-0)$ and $g(c+0)$ we denote the one-sided limits $\lim_{x \to c-0} g(x)$ and $\lim_{x \to c+0} g(x)$, respectively. These limits are finite, $g(c-0) \leqslant g(c+0)$, and g has a discontinuity at a point c if and only if $g(c-0) < g(c+0)$. Since g is increasing, it follows that $g(c+0) \leqslant g(c'-0)$ for $c < c'$ ($c, c' \in \Delta$). Hence the intervals $(g(c-0), g(c+0))$ corresponding to different points of discontinuity are disjoint. Since every such interval contains a rational number, the set of discontinuities of a monotone function is at most countable.

Now consider the semiring $\mathscr{P}(\Delta)$ of all semi-open finite intervals $[a, b)$ whose closures are contained in Δ. We define a volume μ_g on $\mathscr{P}(\Delta)$ by the formula

$$\mu_g\big([a, b)\big) = g(b-0) - g(a-0) \quad (a, b \in \Delta,\ a \leqslant b).$$

We leave the reader to check that the function μ_g thus defined is indeed a volume, i.e., that it is non-negative and additive. One may ask why we did not define μ_g by the simpler formula $\nu_g([a, b)) = g(b) - g(a)$ (see Example (3) in Sect. 1.2.2). Of course, if g is continuous, or at least left-continuous, both formulas give the same result. The reason why we have to use the more complicated formula is that the volume μ_g, as we will soon prove, is always a measure, while the function ν_g (being a volume) is not a measure in the case where g is not left-continuous (see Example (2) in Sect. 1.3.1).

Since $g(u) \leqslant g(v-0) \leqslant g(v)$ for $u < v$ ($u,\ v \in \Delta$), it follows that $\lim_{x \to c-0} g(x-0) = g(c-0)$ for $c \in \Delta$. This immediately implies the following property of the volume μ_g, which will be useful when proving its countable additivity: if $[a, b] \subset \Delta$, then

$$\mu_g\big([a, b)\big) = \lim_{s \to a-0} \mu_g\big([s, b)\big) = \lim_{t \to b-0} \mu_g\big([a, t)\big). \tag{1}$$

4.10.2 First we establish that volume μ_g is countably additive.

Theorem *The volume μ_g is a σ-finite measure.*

---

[24] Thomas Joannes Stieltjes (1856–1894)—Dutch mathematician.

Proof [25] We need to prove only the countable additivity of the volume μ_g, its σ-finiteness being obvious. As we know (see Theorem 1.3.2), it suffices to verify that μ_g is countably subadditive: if $P, P_n \in \mathscr{P}(\Delta)$, $P \subset \bigcup_{n=1}^{\infty} P_n$, then

$$\mu_g(P) \leqslant \sum_{n=1}^{\infty} \mu_g(P_n). \tag{2}$$

We will prove inequality (2) up to ε, where ε is an arbitrary positive number. Let $P = [a,b) \neq \varnothing$ and $P_n = [a_n, b_n)$. Using (1), find $s_n \in \Delta$ such that $s_n < a_n$ and

$$\mu_g\big([s_n, b_n)\big) < \mu_g\big([a_n, b_n)\big) + \frac{\varepsilon}{2^n} \quad (n \in \mathbb{N}). \tag{3}$$

Let us estimate the volume $\mu_g([a,t))$ from above for an arbitrary $t \in (a,b)$. Clearly,

$$[a,t] \subset P \subset \bigcup_{n=1}^{\infty} P_n \subset \bigcup_{n=1}^{\infty} (s_n, b_n).$$

Since the interval $[a,t]$ is compact, for sufficiently large N we have $[a,t] \subset \bigcup_{n=1}^{N} (s_n, b_n)$. Then *a fortiori* $[a,t) \subset \bigcup_{n=1}^{N} [s_n, b_n)$. Since the volume μ_g is subadditive, the inequalities (3) yield the bound

$$\mu_g\big([a,t)\big) \leqslant \sum_{n=1}^{N} \mu_g\big([s_n,b_n)\big) < \sum_{n=1}^{N} \left(\mu_g\big([a_n,b_n)\big) + \frac{\varepsilon}{2^n}\right) < \sum_{n=1}^{\infty} \mu_g\big([a_n,b_n)\big) + \varepsilon.$$

Applying (1) once again, we see that

$$\mu_g\big([a,b)\big) = \lim_{t \to b-0} \mu_g\big([a,t)\big) \leqslant \sum_{n=1}^{\infty} \mu_g\big([a_n,b_n)\big) + \varepsilon.$$

Since ε is arbitrary, this implies (2). □

4.10.3 Now we can introduce the main notion of this section.

Definition The *Lebesgue–Stieltjes measure* generated by an increasing function g is the Carathéodory extension of the volume μ_g.

For this measure, we keep the notation μ_g; the σ-algebra of subsets of the interval Δ on which it is defined will be denoted by $\mathfrak{A}_g(\Delta)$. The Lebesgue measure is a special case of the Lebesgue–Stieltjes measure, corresponding to $\Delta = \mathbb{R}$ and $g(x) \equiv x$.

---

[25] It is instructive to compare this proof with that of the countable additivity of the ordinary volume (Theorem 2.1.1).

Note that the σ-algebra $\mathfrak{A}_g(\Delta)$ contains all subintervals of Δ and hence all open and Borel subsets of Δ.

Let us compute the measure μ_g of a one-point set. Let $c \in \Delta$, and let $c_n \in \Delta$ be points of continuity of g such that $c_n > c_{n+1}$, $c_n \underset{n\to\infty}{\longrightarrow} c$. Put $P_n = [c, c_n)$. Then $P_n \supset P_{n+1}$ and $\bigcap_{n\geqslant 1} P_n = \{c\}$. Since every measure is conditionally continuous from above, $\mu_g(P_n) \underset{n\to\infty}{\longrightarrow} \mu_g(\{c\})$. Furthermore,

$$\mu_g(P_n) = g(c_n) - g(c-0) \underset{n\to\infty}{\longrightarrow} g(c+0) - g(c-0),$$

whence $\mu_g(\{c\}) = g(c+0) - g(c-0)$. Thus $\mu_g(\{c\}) > 0$ if and only if c is a point of discontinuity of g; the measure concentrated at c is equal to the jump of g at this point.

Knowing the measure of one-point sets, one can easily compute the measure of an arbitrary interval contained in Δ. For example, if $[a, b] \subset \Delta$, then

$$\mu_g([a,b]) = \mu_g([a,b) \cup \{b\}) = \mu_g([a,b)) + \mu_g(\{b\}) = g(b+0) - g(a-0).$$

We leave the reader to find the measure of intervals of other types.

In general, the σ-algebras $\mathfrak{A}_g$ for different functions g do not coincide. For example, if an increasing function g is constant on $(a, b) \subset \Delta$, then $\mu_g((a, b)) = 0$ and, by the completeness of μ_g, the σ-algebra $\mathfrak{A}_g(\Delta)$ contains *all* subsets of this interval. At the same time we know that every non-degenerate interval (a, b) contains sets that are not Lebesgue measurable (see Sect. 2.1.3).

In order to deal with measures defined on the same σ-algebra, one often considers Lebesgue–Stieltjes measures only on Borel subsets. The restriction of μ_g to the σ-algebra of Borel sets is called the *Borel–Stieltjes measure*.

Up to now we have only considered the case where the function g, which generates a Lebesgue–Stieltjes measure, is defined on an open interval Δ. If Δ has the form $\Delta = [p, q)$, then we define μ_g on semi-open subintervals of Δ of the form $[a, b)$ in the same way as above, with the only difference that $g(p-0)$ should now be understood as $g(p)$. Thus the mass concentrated at the point p will be equal to the jump of g at p. If Δ is a right-closed interval, then we should assume that the mass concentrated at the point q is equal to $g(q) - g(q-0)$. One may say that if $\Delta = \langle p, q\rangle$ and $p \in \Delta$ (or $q \in \Delta$), we extend the function g by assuming it constant on the half-line $(-\infty, p]$ (respectively, $[q, +\infty)$), and then consider the measure generated by the extended function only on subsets of the original interval.

It is clear that if the difference of two increasing functions is constant, then they generate the same Lebesgue–Stieltjes measure. However, this may also happen in other cases, because the volume, and hence the measure, μ_g does not depend on the values of g at points of discontinuity. Replacing $g(x)$ at each point of discontinuity x by the value $\widetilde{g}(x) = g(x-0)$, we obtain a "corrected" function, which generates the same volume as g but is left-continuous at every point. Thus we may assume without loss of generality that the volume μ_g is generated by a left-continuous function; this is sometimes technically convenient. For a description of all functions generating the same measure, see Exercise 6.

Remark We have introduced a class of Borel measures defined on subsets of a given interval Δ. These measures are finite on compact subsets of Δ. A natural question is whether there exist other Borel measures having this property. We will show that the answer to this question is negative, assuming, to avoid some minor complications, that Δ is an open interval.

Consider a Borel measure ν that is finite on $\mathscr{P}(\Delta)$, fix an arbitrary interior point $p \in \Delta$, and define a function g on Δ by the formula

$$g(x) = \begin{cases} \nu([p,x)) & \text{for } x \geqslant p, \\ -\nu([x,p)) & \text{for } x < p. \end{cases}$$

We leave the reader to show that if $[a,b] \subset \Delta$, then $g(b) - g(a) = \nu([a,b))$, and that g is increasing and left-continuous. Thus the measures ν and μ_g coincide on $\mathscr{P}(\Delta)$ and hence, by the uniqueness theorem, on all Borel subsets of Δ.

To complete our discussion of the definition of the Lebesgue–Stieltjes measure, observe that if g_1 and g_2 are increasing functions defined on Δ, then for Borel subsets of Δ we have $\mu_{g_1+g_2}(A) = \mu_{g_1}(A) + \mu_{g_2}(A)$, i.e., $\mu_{g_1+g_2} = \mu_{g_1} + \mu_{g_2}$ for any Borel–Stieltjes measures. However, for Lebesgue–Stieltjes measures, this is not generally the case, since, as we have already mentioned, these measures may be defined on different σ-algebras.

4.10.4 Consider two classes of increasing functions generating Stieltjes measures of different types.

Let g be an increasing function on an interval Δ, Δ_0 be the set of points of discontinuity of g, and ω_x be the (possibly zero) jump of g at a point $x \in \Delta$. Note that if $a, b \in \Delta$, $a < b$, then the increment of g over the interval $[a,b]$ is not less than the sum of its jumps corresponding to the points of discontinuity in (a,b). Indeed,

$$\sum_{x\in(a,b)} \omega_x = \sum_{x\in(a,b)\cap\Delta_0} \omega_x \leqslant \mu_g\bigl((a,b)\bigr) = g(b-0) - g(a+0) \leqslant g(b) - g(a).$$

This implies, in particular, that the sum $\sum_{x\in[a,b]} \omega_x$ is finite.

Definition An increasing function g on an interval Δ is called a *jump function* if its increment corresponding to any two points of continuity is equal to the sum of the jumps between them, i.e., for any two points of continuity $a, b \in \Delta$, $a < b$, the equality $g(b) - g(a) = \sum_{x\in(a,b)\cap\Delta_0} \omega_x$ $(= \sum_{x\in(a,b)} \omega_x)$ holds.

One of the simplest examples of a jump function is $[x]$, the integer part of x. However, there are more complicated cases; for instance, the set of points of discontinuity of a jump function may be dense in Δ.

Example A jump function can be constructed as follows. Let $\{x_1, x_2, \ldots\}$ be an arbitrary countable subset of an interval Δ and $\omega_1, \omega_2, \ldots$ be positive numbers with

$\sum_{n=1}^{\infty} \omega_n < +\infty$. Set

$$h(x) = \sum_{x_n < x} \omega_n = \sum_{n=1}^{\infty} \omega_n \chi_+(x - x_n),$$

where χ_+ is the characteristic function of the half-line $(0, +\infty)$. Since the series defining h uniformly converges, this function is continuous (from both sides) at all points $x \neq x_n$. Further, for every m we have

$$h(x) = \omega_m \chi_+(x - x_m) + \sum_{n \neq m}^{\infty} \omega_n \chi_+(x - x_n),$$

and the sum of the series is continuous at x_m, which implies that the function h, as well as $\chi_+(x - x_m)$, is left-continuous at x_m, with the jump at x_m equal to ω_m. At the same time, if a and b are points of continuity and $a < b$, then $h(b) - h(a) = \sum_{a < x_n < b} \omega_n$, so that h is a jump function.

By increasing the values of h at points of discontinuity in an appropriate way, we again obtain a jump function with the given jumps from the left and from the right.

Note that the condition $\sum_{n=1}^{\infty} \omega_n < +\infty$ can be weakened. The reader can easily check that all arguments used in the construction of h remain valid if we replace it with a weaker condition: $\sum_{x_n \in [a,b]} \omega_n < +\infty$ for any $a, b \in \Delta$, $a < b$.

Let us find the measure generated by the jump function g. As above, let Δ_0 be the set of points of discontinuity of g and ω_x be the (possibly zero) jump of g at a point $x \in \Delta$. If $a, b \in \Delta$, $a < b$, then

$$\mu_g\big([a, b)\big) = g(b - 0) - g(a - 0) = \sum_{x \in [a,b) \cap \Delta_0} \omega_x = \mu_g\big([a, b) \cap \Delta_0\big). \quad (4)$$

If a and b are points of continuity of g, then the middle equality holds by the definition of a jump function; in the general case, it can be proved by passing to the limit. It follows from (4) that $\mu_g(\Delta) = \mu_g(\Delta_0)$ and, consequently, $\mu_g(\Delta \setminus \Delta_0) = 0$. Since the measure μ_g is complete, the σ-algebra $\mathfrak{A}_g(\Delta)$ coincides with the algebra of all subsets of the interval Δ.

Equation (4) shows that on the semiring $\mathscr{P}(\Delta)$ the measure μ_g coincides with the discrete measure generated by the masses $\{\omega_x\}_{x \in \Delta}$ (see Sect. 1.3.1). By the uniqueness theorem (Sect. 1.5.1), these measures are identical. Thus if g is a jump function, then the measure μ_g is just the discrete measure generated by the family of jumps of g.

Now consider a situation that is in a sense opposite to the previous one; namely, the case where the function g not only has no jumps, i.e., is continuous, but is absolutely continuous (see Sect. 4.9.3). By Theorem 4.9.3, g is then differentiable almost everywhere, and g is increasing if and only if g' is non-negative.

We will prove that in this case the measure μ_g has a density (see Sect. 4.5.3) with respect to the Lebesgue measure.

Lemma *Let g be an increasing function absolutely continuous on an interval Δ. Then $\mu_g(A) = \int_A g'(x)\,dx$ for every Lebesgue measurable set $A \subset \Delta$.*

Proof Consider the measure ν defined on the σ-algebra $\mathfrak{A}(\Delta)$ of Lebesgue measurable subsets of Δ by the formula

$$\nu(A) = \int_A g'(x)\,dx \quad \bigl(A \in \mathfrak{A}(\Delta)\bigr).$$

Since the measures ν and μ_g coincide on the semiring $\mathscr{P}(\Delta)$, it follows from the uniqueness theorem (see Sect. 1.5.1) that they coincide on all Borel sets, and hence, by the completeness of μ_g, on the whole σ-algebra $\mathfrak{A}(\Delta)$. Thus

$$\mathfrak{A}(\Delta) \subset \mathfrak{A}_g(\Delta) \quad \text{and} \quad \mu_g(A) = \nu(A) \quad \text{for } A \in \mathfrak{A}(\Delta). \qquad \square$$

Remark Applying Theorem 4.5.3 to the measure μ_g generated by a function g satisfying the assumptions of the lemma, we see that for every Lebesgue measurable non-negative function f,

$$\int_\Delta f\,d\mu_g = \int_\Delta f g'\,d\lambda, \tag{5}$$

where λ is the one-dimensional Lebesgue measure.

4.10.5 Bearing in mind that Lebesgue–Stieltjes measures may be defined on different σ-algebras, in this subsection, when speaking about the sum of measures, we mean Borel–Stieltjes measures, i.e. we consider only the measures of Borel sets.

Let g be an increasing function defined on an interval Δ (to avoid obvious minor technicalities, we assume it open), $\{\omega_x\}_{x\in\Delta}$ be the family of jumps of g, and $\Delta_0 = \{x \in \Delta \mid \omega_x > 0\}$ be the set of points of discontinuity of g. Fix an arbitrary point $p \in \Delta$ and put

$$h(x) = \begin{cases} \sum_{t\in[p,x)} \omega_t & \text{for } x > p, \\ 0 & \text{for } x = p, \\ -\sum_{t\in[x,p)} \omega_t & \text{for } x < p \end{cases}$$

(cf. the formula from the remark in Sect. 4.10.3). Let $\Delta_0 = \{x_1, x_2, \ldots\}$ and $h_n \equiv \omega_{x_n}$; then h coincides with the function considered in Example 4.10.4. As we have mentioned, it is not necessary to assume that the family of masses is summable; in our case, it is summable on every closed subinterval of Δ, which suffices to construct h. As we have shown in Example 4.10.4, the function h is increasing, has the same points of discontinuity and the same jumps as g, and is a jump function. Modifying, if necessary, the values of h at points of discontinuity, we can make it have the same jumps from the left and from the right as g. Assuming that h has this property, we see that the difference $g_c = g - h$ is a continuous function. It is increasing. Indeed, let $a, b \in \Delta$, $a < b$. When proving the inequality $g_c(b) - g_c(a) \geqslant 0$, we may assume, by the continuity of g_c, that a and b are points of continuity of g (and h). In this case,

$$g_c(b) - g_c(a) = g(b) - g(a) - \big(h(b) - h(a)\big) = g(b) - g(a) - \sum_{x \in (a,b)} \omega_x \geqslant 0,$$

since the increment of an increasing function over the interval $[a,b]$ is not less than the sum of its jumps corresponding to the points of discontinuity lying in (a,b).

So, every increasing function can be written as the sum of a jump function and a continuous increasing function: $g = h + g_c$.

Now consider the (Borel) measures μ_g, μ_{g_c} and μ_h corresponding to these functions. It is clear that if $a, b \in \Delta$, $a < b$, then

$$g(b-0) - g(a-0) = h(b-0) - h(a-0) + g_c(b) - g_c(a),$$

i.e.,

$$\mu_g\big([a,b)\big) = \mu_h\big([a,b)\big) + \mu_{g_c}\big([a,b)\big).$$

Thus on the semiring $\mathscr{P}(\Delta)$ the measure μ_g coincides with the sum $\mu_h + \mu_{g_c}$. By the uniqueness theorem, these measures coincide on the Borel hull of the semiring $\mathscr{P}(\Delta)$, i.e., on all Borel subsets of Δ. Therefore (see Sect. 4.4.2, Property (9)), for every non-negative (measurable) function f,

$$\int_\Delta f\, d\mu_g = \int_\Delta f\, d\mu_h + \int_\Delta f\, d\mu_{g_c}.$$

Since a jump function generates a discrete measure, the integral with respect to μ_h can be computed according to the general formula (see Sect. 4.2.4):

$$\int_\Delta f\, d\mu_h = \sum_{x \in \Delta_0} f(x)\omega_x.$$

Computing the integral with respect to μ_{g_c} may be rather difficult (see Exercises 7 and 8). It simplifies substantially if the function g_c is absolutely continuous. In this case, by (5),

$$\int_\Delta f\, d\mu_{g_c} = \int_\Delta f g_c'\, d\lambda.$$

In conclusion, note that the integral with respect to the measure μ_g is called the Lebesgue–Stieltjes integral, or simply the Stieltjes integral. To denote it, along with the symbols $\int_A f\, d\mu_g$, $\int_A f(x)\, d\mu_g(x)$, the shorter classical notation $\int_A f\, dg$, $\int_A f(x)\, dg(x)$ is also used; in what follows, we will usually employ the latter notation.

4.10.6 In this section, we obtain a generalization of the integration by parts formula to Stieltjes integrals.

Theorem *Let g be a non-decreasing function and F an absolutely continuous function on $[a,b]$. Then*

$$\int_{[a,b]} F(x)\,dg(x) = F(x)\,g(x)\Big|_a^b - \int_a^b F'(x)g(x)\,dx.$$

Proof Since $F' = (F')_+ - (F')_-$, with $(F')_\pm \geqslant 0$, it suffices to prove the desired formula in the case where $F' \geqslant 0$, so in what follows we assume that this condition is satisfied.

Let τ be an arbitrary partition of the interval $[a,b]$ formed by points $x_0 = a < x_1 < \cdots < x_n = b$. Since g is increasing, for $k = 0, 1, \ldots, n-1$ we have

$$\begin{aligned}\big(F(x_{k+1}) - F(x_k)\big)g(x_k) &= g(x_k)\int_{x_k}^{x_{k+1}} F'(x)\,dx \leqslant \int_{x_k}^{x_{k+1}} F'(x)\,g(x)\,dx \\ &\leqslant g(x_{k+1})\int_{x_k}^{x_{k+1}} F'(x)\,dx = \big(F(x_{k+1}) - F(x_k)\big)g(x_{k+1}).\end{aligned}$$

Summing these inequalities, we obtain

$$\begin{aligned}\sum_{k=0}^{n-1}\big(F(x_{k+1}) - F(x_k)\big)g(x_k) &\leqslant \int_a^b F'(x)\,g(x)\,dx \\ &\leqslant \sum_{k=0}^{n-1}\big(F(x_{k+1}) - F(x_k)\big)g(x_{k+1}).\end{aligned} \tag{6}$$

Let us transform the sum on the left-hand side:

$$\begin{aligned}&\sum_{k=0}^{n-1}\big(F(x_{k+1}) - F(x_k)\big)g(x_k) \\ &\quad = \sum_{k=1}^{n} F(x_k)g(x_{k-1}) - \sum_{k=0}^{n-1} F(x_k)g(x_k) \\ &\quad = F(b)g(b) - F(a)g(a) - \sum_{k=1}^{n} F(x_k)\big(g(x_k) - g(x_{k-1})\big) \\ &\quad = F(x)g(x)\Big|_{x=a}^{x=b} - S_\tau.\end{aligned}$$

Transforming the sum on the right-hand side of (6) in a similar way, we obtain

$$\sum_{k=0}^{n-1}\big(F(x_{k+1}) - F(x_k)\big)g(x_{k+1}) = F(x)g(x)\Big|_{x=a}^{x=b} - S'_\tau,$$

where $S'_\tau = \sum_{k=0}^{n-1} F(x_{k+1})(g(x_k) - g(x_{k-1}))$. Thus

$$F(x)g(x)\Big|_{x=a}^{x=b} - S_\tau \leqslant \int_a^b F'(x)\,g(x)\,dx \leqslant F(x)g(x)\Big|_{x=a}^{x=b} - S'_\tau.$$

Now assume that the partition points lying inside $[a,b]$ are points of continuity of g. In this case, the sums S_τ and S'_τ turn into Riemann sums for the integral $\int_{[a,b]} F(x)\,dg(x)$. Hence, refining the partition, passing to the limit, and using the remark to Theorem 4.7.3, we obtain the equation

$$F(x)g(x)\Big|_{x=a}^{x=b} - \int_{[a,b]} F(x)\,dg(x) = \int_a^b F'(x)g(x)\,dx,$$

which is equivalent to the desired one. □

For another proof of this theorem, see Corollary 3 in Sect. 5.3.4.

One should bear in mind that the integration by parts formula proved above is valid in the case where g is defined on the closed interval $[a,b]$, so that, by definition, the measure μ_g assigns the masses $g(a+0) - g(a)$ and $g(b) - g(b-0)$ to the points a and b, respectively. If the measure μ_g is generated by a function defined on an interval containing $[a,b]$, then the equation

$$\int_{[a,b]} F(x)\,dg(x) = F(x)g(x)\Big|_{x=a}^{x=b} - \int_a^b F'(x)g(x)\,dx$$

may no longer be true, since the measures of the one-point sets $\{a\}$ and $\{b\}$ may differ from the above one-sided jumps. However, the integration by parts formula clearly remains true if the measure has no masses at a and b, i.e., if they are points of continuity of g. In the case where the function g is left-continuous, the integration by parts formula always holds when integrating over an interval closed from the left and open from the right:

$$\int_{[a,b)} F(x)\,dg(x) = F(x)g(x)\Big|_{x=a}^{x=b} - \int_a^b F'(x)g(x)\,dx.$$

EXERCISES

1. Compute the integral $\int_{[\frac{1}{\sqrt{15}},2]} x\,dg(x)$, where $g(x) = x - [\frac{1}{x}]$ (the symbol $[a]$ stands for the integer part of a).
2. Let $g(x) = \sum_{n=1}^{\infty} 2^{-n}\chi_+(x - \frac{1}{n})$ $(x \in \mathbb{R})$, where χ_+ is the characteristic function of the half-line $(0,+\infty)$. Do the integrals $\int_\delta x^2\,dg(x)$ over the intervals $\delta = (\frac{2}{3},1)$ and $\delta = [\frac{2}{3},1]$ differ (and, if so, what is the difference)? Consider the same questions for the intervals $(\frac{1}{3},\frac{1}{2})$, $(\frac{1}{3},\frac{1}{2}]$ and $[\frac{1}{3},\frac{1}{2})$.
3. Compute the integrals $\int_{\frac{1}{2}}^{2} g\,dg$ for the functions g from Exercises 1 and 2. Are these integrals equal to the limits of the corresponding Riemann sums?

4. Show that if an increasing function g is continuous on $[a,b]$, then the formula

$$\int_{[a,b]} g^\sigma \, dg = \frac{g^{\sigma+1}(b) - g^{\sigma+1}(a)}{\sigma+1}$$

holds for every $\sigma > 0$. Is this true if we drop the condition that g is continuous?

5. Show that the measure generated on $(0,+\infty)$ by the function $g(x) = \ln x$ is defined on Lebesgue measurable sets and invariant under multiplication by a positive number c (i.e., the sets $A \subset (0,+\infty)$ and $cA = \{cx \mid x \in A\}$ have the same measure provided that they are measurable).

6. Show that if two increasing functions generate the same Lebesgue–Stieltjes measure, then their difference is constant on the set of (common) points of continuity.

7. Compute the integral $\int_0^1 x \, d\varphi(x)$, where φ is the Cantor function (see Sect. 2.3.2).

8. Show that the integral $F(y) = \int_0^1 e^{iyx} d\varphi(x)$ $(y \in \mathbb{R})$, where φ is the Cantor function, is equal to $e^{iy/2} \prod_{k=1}^\infty \cos \frac{y}{3^k}$. Verify that $F(y) \not\to 0$ as $|y| \to +\infty$.

9. Show that if f is a continuous function and g is an increasing function on $[a,b]$, then the Lebesgue–Stieltjes integral $\int_{[a,b]} f \, d\mu_g$ coincides with the limit of the classical Riemann sums $S_\tau(f,\xi) = \sum_{k=0}^{n-1} f(\xi_k)(g(x_{k+1}) - g(x_k))$ as the mesh of τ tends to zero.

10. Let $f \in C([-1,1])$, φ be the Cantor function, and $a_\varepsilon = 2\sum_{k=1}^n \frac{\varepsilon_k}{3^k}$, where $\varepsilon = (\varepsilon_1, \dots, \varepsilon_n)$, $\varepsilon_k = 0$ or 1 (a_ε are the left endpoints of segments of the nth rank appearing in the construction of the Cantor set). Show that as $n \to \infty$,

$$\frac{1}{2^n} \sum_\varepsilon f(x - a_\varepsilon) \rightrightarrows \int_0^1 f(x-y)\, d\varphi(y) \quad \text{on } [0,1].$$

11. One says that two measure spaces $(X, \mathfrak{A}, \mu)$ and $(Y, \mathfrak{B}, \nu)$ are isomorphic if there exist sets $e \subset X$ and $e' \subset Y$ of zero measure and a bijection $\Phi : X \setminus e \to Y \setminus e'$ such that the set $A \subset X \setminus e$ is measurable if and only if the set $\Phi(A)$ is measurable and, in the latter case, the measures of these sets coincide. Show that if we replace the Lebesgue measure on the interval $[0,1]$ by the measure corresponding to the Cantor function φ, then we will obtain an isomorphic measure space. *Hint*. Use the equality $\varphi(\mathcal{C}) = [0,1]$.

4.11 *Functions of Bounded Variation

4.11.1 Consider a function f defined on a closed interval $[a,b]$. Given an arbitrary partition τ of $[a,b]$ formed by points $x_0 = a < x_1 < \dots < x_n = b$, set

$$S_\tau = \sum_{k=0}^{n-1} \bigl| f(x_{k+1}) - f(x_k) \bigr|.$$

Obviously, when new partition points are added to τ, the sum S_τ may only increase.

Definition The value $\sup_\tau S_\tau$ is called the *total variation* of the function f on the interval $[a,b]$ and is denoted by $\mathbf{V}_a^b(f)$. If $\mathbf{V}_a^b(f)$ is finite, f is called a function *of bounded variation*.

It is clear that if f satisfies the Lipschitz condition on the interval $[a,b]$, then it is of finite variation. However, one should bear in mind that if f satisfies the Lipschitz condition of order $\alpha < 1$, then its variation may be infinite not only on the interval $[a,b]$, but on every (non-degenerate) subinterval (see Exercise 4).

Let us mention a few properties of the total variation.

(1) $\mathbf{V}_a^b(f) \geqslant |f(b) - f(a)|$.
(2) A monotone function f is of bounded variation, with $\mathbf{V}_a^b(f) = |f(b) - f(a)|$.
(3) A linear combination of functions of bounded variation is again a function of bounded variation, with

$$\mathbf{V}_a^b(f+g) \leqslant \mathbf{V}_a^b(f) + \mathbf{V}_a^b(g) \quad \text{and} \quad \mathbf{V}_a^b(\alpha f) = |\alpha| \mathbf{V}_a^b(f) \quad \text{for } \alpha \in \mathbb{R}.$$

Now we establish a less obvious property of the total variation, namely, its additivity.

Theorem *If $a < c < b$, then $\mathbf{V}_a^b(f) = \mathbf{V}_a^c(f) + \mathbf{V}_c^b(f)$.*

The theorem applies both to the case of bounded and unbounded variation.

Proof Let τ be an arbitrary partition of the interval $[a,b]$ formed by points $x_0, \dots, x_n$. Assume that one of these points, say x_m, coincides with c. Then the points $x_0, \dots, x_m$ and $x_m, \dots, x_n$ form partitions of the intervals $[a,c]$ and $[c,b]$, respectively. Hence

$$S_\tau = \sum_{k=0}^{m-1} \left| f(x_{k+1}) - f(x_k) \right| + \sum_{k=m}^{n-1} \left| f(x_{k+1}) - f(x_k) \right| \leqslant \mathbf{V}_a^c(f) + \mathbf{V}_c^b(f).$$

This inequality remains valid in the case where c is not a partition point, since adding it to the set of partition points does not decrease the sum S_τ. Therefore,

$$\mathbf{V}_a^b(f) \leqslant \mathbf{V}_a^c(f) + \mathbf{V}_c^b(f).$$

On the other hand, if τ' and τ'' are arbitrary partitions of the intervals $[a,c]$ and $[c,b]$ formed by points $y_0, \dots, y_p$ and $z_0, \dots, z_q$, respectively, then $z_0 = y_p$ and the points $y_0, \dots, y_p, z_1, \dots, z_q$ form a partition τ of the interval $[a,b]$, with

$$S_{\tau'} + S_{\tau''} = S_\tau \leqslant \mathbf{V}_a^b(f).$$

Taking the supremum first over τ' and then over τ'', we see that

$$\mathbf{V}_a^c(f) + \mathbf{V}_c^b(f) \leqslant \mathbf{V}_a^b(f).$$

Together with the reverse inequality obtained above, this proves the theorem. □

As one can see from Properties (2) and (3), the difference of increasing functions is a function of bounded variation. It follows from the last theorem that the converse is also true.

Corollary *A function of bounded variation can be written as the difference of increasing functions.*

Proof Indeed, it is clear that the function $V(x) = \mathbf{V}_a^x(f)$ is increasing. Furthermore, it follows from Property (1) and the above theorem that the difference $W(x) = \mathbf{V}_a^x(f) - f(x)$ is also increasing: if $x, y \in [a, b]$, $x < y$, then

$$W(y) - W(x) = \big(\mathbf{V}_a^y(f) - \mathbf{V}_a^x(f)\big) - \big(f(y) - f(x)\big) \geqslant \mathbf{V}_x^y(f) - \big|f(y) - f(x)\big| \geqslant 0.$$

Hence

$$f = V - W \tag{1}$$

is a representation of f in the desired form. □

This corollary implies, in particular, that the set of points of discontinuity of a function of bounded variation is at most countable.

4.11.2 It turns out that the continuity of the function is stored in the transition to variation. More precisely, the following statement is valid.

Theorem *Let f be a function of bounded variation on $[a, b]$ and $V(x) = \mathbf{V}_a^x(f)$ for $a < x \leqslant b$, $V(a) = 0$. If f is continuous at a point $c \in [a, b]$, then V is also continuous at this point.*

Proof We will prove that V is right-continuous (the left-continuity can be proved in a similar way). Let $a \leqslant c < b$. By the corollary of Theorem 4.11.1, f can be written as the difference of increasing functions: $f = g - h$. Hence for $x \in (c, b)$ we have

$$\begin{aligned} 0 \leqslant V(x) - V(c) &= \mathbf{V}_c^x(f) = \mathbf{V}_c^x(g - h) \leqslant \mathbf{V}_c^x(g) + \mathbf{V}_c^x(h) \\ &= \big(g(x) - g(c)\big) + \big(h(x) - h(c)\big). \end{aligned}$$

The right-hand side tends to zero as $x \to c$ if the functions g and h are continuous at c. Let us verify that we may assume this without loss of generality. Indeed, if these functions are discontinuous at c, then their jumps at this point are equal, since the difference $g - h$ is continuous. Modify g and h by decreasing them at the interval $(c, b]$ by the jump at c and setting their values at c equal to their right limits at c. As one can easily check, the modified functions are increasing, continuous at c, and their difference coincides with $g - h$, i.e., with f. □

In view of the representation (1), the above theorem implies that a function of bounded variation can be written as the difference of increasing functions that are continuous at the same points as f.

4.11.3 As one can see from the following theorem, on transition to the variation not only continuity but also the absolute continuity persist.

Theorem *If a function f is absolutely continuous on $[a,b]$, then it is of bounded variation, with*

$$\mathbf{V}_a^b(f) = \int_a^b |f'(t)|\,dt. \tag{2}$$

Proof By assumption, $f(x) = f(a) + \int_a^x \omega(t)\,dt$ for $x \in [a,b]$, where the function ω (which coincides with f' almost everywhere by Theorem 4.9.3) is summable on $[a,b]$. Therefore, for every partition $x_0 = a < x_1 < \cdots < x_n = b$,

$$\sum_{k=0}^{n-1} |f(x_{k+1}) - f(x_k)| = \sum_{k=0}^{n-1} \left| \int_{x_k}^{x_{k+1}} f'(t)\,dt \right| \leqslant \sum_{k=0}^{n-1} \int_{x_k}^{x_{k+1}} |f'(t)|\,dt$$
$$= \int_a^b |f'(t)|\,dt.$$

Hence f is of bounded variation and

$$\mathbf{V}_a^b(f) \leqslant \int_a^b |f'(t)|\,dt. \tag{3}$$

We will prove the reverse inequality up to an arbitrary $\varepsilon > 0$. For this, using the absolute continuity of the integral and the regularity of the Lebesgue measure, we find closed sets $Q_+ \subset \{x \in [a,b] \mid f'(x) \geqslant 0\}$ and $Q_- \subset \{x \in [a,b] \mid f'(x) < 0\}$ such that

$$\int_{[a,b]\setminus Q} |f'(x)|\,dx < \varepsilon, \quad \text{where } Q = Q_+ \cup Q_-. \tag{4}$$

Since the sets $Q_\pm$ are disjoint and compact, they are separated, that is, there exists a $\delta > 0$ such that $|x - y| \geqslant \delta$ for any $x \in Q_+$ and $y \in Q_-$.

Now consider a partition τ of the interval $[a,b]$ formed by points $x_0 = a < x_1 < \cdots < x_n = b$ such that $x_{k+1} - x_k < \delta$ for all k. Then every interval $\Delta_k = [x_k, x_{k+1}]$ may have a non-empty intersection with at most one of the sets $Q_\pm$. Hence the function f' does not change sign at the intersection $\Delta_k \cap Q$, and, consequently,

$$|f(x_{k+1}) - f(x_k)| = \left| \int_{\Delta_k} f'(x)\,dx \right| \geqslant \left| \int_{\Delta_k \cap Q} f'(x)\,dx \right| - \int_{\Delta_k \setminus Q} |f'(x)|\,dx$$
$$= \int_{\Delta_k \cap Q} |f'(x)|\,dx - \int_{\Delta_k \setminus Q} |f'(x)|\,dx.$$

Summing these inequalities, we obtain a lower bound on the sum $S_\tau = \sum_{k=0}^{n-1} |f(x_{k+1}) - f(x_k)|$:

$$S_\tau \geqslant \int_{[a,b]\cap Q} |f'(x)|\,dx - \int_{[a,b]\setminus Q} |f'(x)|\,dx$$
$$= \int_a^b |f'(x)|\,dx - 2\int_{[a,b]\setminus Q} |f'(x)|\,dx.$$

In view of (4), this implies the inequality $\mathbf{V}_a^b(f) \geqslant S_\tau > \int_a^b |f'(x)|\,dx - 2\varepsilon$. Since ε is arbitrary, we have $\mathbf{V}_a^b(f) \geqslant \int_a^b |f'(x)|\,dx$, which together with (3) implies (2). □

Later (see Theorem 11.1.6) we will use another idea to obtain a more general result.

Note that a function f absolutely continuous on $[a,b]$ can be written as the difference of absolutely continuous increasing functions, since

$$f(x) - f(a) = \int_a^x f'(y)\,dy = \int_a^x (f'(y))_+\,dy - \int_a^x (f'(y))_-\,dy.$$

4.11.4 Starting from the definition of the Stieltjes integral, we can introduce the notion of the integral with respect to a function of bounded variation, which is useful in some cases. First we make a preliminary observation: if increasing functions g, h, g_1, h_1 on an interval $[a,b]$ satisfy the condition $g - h = g_1 - h_1$, then for every bounded Borel measurable function φ we have $\int_a^b \varphi\,dg - \int_a^b \varphi\,dh = \int_a^b \varphi\,dg_1 - \int_a^b \varphi\,dh_1$. Indeed, by assumption, $g + h_1 = g_1 + h$, and the corresponding equality holds also for the Borel–Stieltjes measures: $\mu_g + \mu_{h_1} = \mu_{g_1} + \mu_g$. Hence (see Sect. 4.4.2, Property (9))

$$\int_a^b \varphi\,d\mu_g + \int_a^b \varphi\,d\mu_{h_1} = \int_a^b \varphi\,d\mu_{g_1} + \int_a^b \varphi\,d\mu_h,$$

and our claim follows.

Definition Let f be a function of bounded variation on $[a,b]$ and φ be a Borel measurable bounded function on $[a,b]$. The *integral of φ with respect to f over* $[a,b]$, denoted by $\int_a^b \varphi\,df$, is the difference $\int_a^b \varphi\,dg - \int_a^b \varphi\,dh$, where g and h are increasing functions such that $g - h = f$.

The remark made before the definition shows that this integral is well defined: the difference $\int_a^b \varphi\,dg - \int_a^b \varphi\,dh$ does not depend on the choice of increasing functions g and h satisfying the condition $g - h = f$.

It is clear that the integral with respect to a function of bounded variation is linear, since this is true for Stieltjes integrals. For the same reason, the integral with respect to a function of bounded variation satisfies the integration by parts formula (cf. Theorem 4.10.6):

$$\int_a^b \varphi(x)\,df(x) = \varphi(x)\,f(x)\Big|_a^b - \int_a^b \varphi'(x) f(x)\,dx,$$

where φ is absolutely continuous and f is of bounded variation on $[a, b]$.

If f is absolutely continuous on $[a, b]$, then there is a formula generalizing formula (5) from the previous section:

$$\int_a^b \varphi\, df = \int_a^b \varphi f'\, d\lambda,$$

where λ is the one-dimensional Lebesgue measure.

Let us establish another property of the integral with respect to a function of bounded variation.

Theorem *If φ is continuous and f is of bounded variation on* $[a, b]$, *then*

$$\left| \int_a^b \varphi\, df \right| \leqslant \sup_{[a,b]} |\varphi| \cdot \mathbf{V}_a^b(f).$$

As we will show later (see Theorem 11.1.8), this inequality holds not only for continuous, but also for any Borel measurable bounded functions φ.

Proof We use the fact that for the integral with respect to a function of bounded variation, as for the Stieltjes integral, Theorem 4.7.3 holds, i.e., the integral is the limit of the Riemann sums.

Consider an arbitrary partition $\tau = \{x_0, \ldots, x_n\}$ of $[a, b]$ and the corresponding sum:

$$S_\tau = \sum_{k=0}^{n-1} \varphi(\xi_k)\big(f(x_{k+1}) - f(x_k)\big),$$

where $\xi_k \in [x_k, x_{k+1})$. Obviously,

$$|S_\tau| \leqslant \sum_{k=0}^{n-1} |\varphi(\xi_k)| \big|f(x_{k+1}) - f(x_k)\big| \leqslant M \sum_{k=0}^{n-1} \big|f(x_{k+1}) - f(x_k)\big| \leqslant M \cdot \mathbf{V}_a^b(f), \tag{4}$$

where $M = \sup_{[a,b]} |\varphi|$.

The function f can be written as the difference $f = g - h$, where g and h are increasing functions (see the corollary of Theorem 4.11.1). Hence

$$S_\tau = S'_\tau - S''_\tau,$$

where

$$S'_\tau = \sum_{k=0}^{n-1} \varphi(\xi_k)\big(g(x_{k+1}) - g(x_k)\big), \qquad S''_\tau = \sum_{k=0}^{n-1} \varphi(\xi_k)\big(h(x_{k+1}) - h(x_k)\big).$$

If we assume (and we may do this without loss of generality) that all interior partition points are points of continuity of g and h, then the sums S'_τ and S''_τ turn into Riemann sums. Therefore, by Theorem 4.7.3, these sums, and hence the sum S_τ, tend to the corresponding integrals as the mesh of τ tends to zero. To complete the proof, it suffices to pass to the limit in inequality (4). □

EXERCISES

1. The product of two functions of bounded variation is again a function of bounded variation; the quotient of two functions of bounded variation is again a function of bounded variation provided that the denominator is bounded away from zero.
2. Let f and g be functions of bounded variation defined on $[a, b]$. Show that the integration by parts formula for $\int_a^b f\,dg$ may be false. Is it true under the additional assumption that at least one of the functions is continuous on $[a, b]$?
3. Using the function $x^2 \sin \frac{1}{x^2}$, show that a differentiable function (unlike a smooth one) may have unbounded variation on a closed interval.
4. Show that the function $x \mapsto f(x) = (\ln x)^{-1} \sin(\ln x)$ for $0 < x \leqslant 1$, $f(0) = 0$, is of unbounded variation and satisfies the Lipschitz condition of an arbitrary order less than one. Using a series of the form $\sum_{n=1}^{\infty} a_n f(x - x_n)$, construct a function that satisfies the Lipschitz condition of an arbitrary order less than one and is of unbounded variation on every subinterval.

Chapter 5
The Product Measure

5.1 Definition of the Product Measure

Given two measures on the subsets of the sets X and Y, our goal is to construct a new measure (the so-called product measure) defined on subsets of the Cartesian product $X \times Y$. The definition of the product measure relies on Theorem 1.4.5 on the standard extension of measures and on Theorem 5.1.2. When proving the latter, we will use the properties of the integral. There exists yet another approach to the proof, which is independent of the notion of the integral. Technically it is more complicated than the one we present but it is of independent interest because it allows us to give an alternative definition of the integral. We will discuss this in more detail in Sect. 5.5.

All measures in this chapter are assumed to be σ-finite.

5.1.1 We leave the proof of the following lemma to the reader.

Lemma *Let A, $A' \subset X$, B, $B' \subset Y$, and let $\{B_\omega\}_{\omega\in\Omega}$ be a family of subsets of the set Y. Then:*

(1) $A \times B \subset A' \times B'$ *if and only if either* $A \subset A'$ *and* $B \subset B'$, *or* $A \times B = \varnothing$;
(2) $(A \times B) \cap (A' \times B') = (A \cap A') \times (B \cap B')$;
(3) $A \times (B \setminus B') = (A \times B) \setminus (A \times B')$;
(4) $A \times \bigcup_{\omega\in\Omega} B_\omega = \bigcup_{\omega\in\Omega}(A \times B_\omega)$;
(5) $A \times \bigcap_{\omega\in\Omega} B_\omega = \bigcap_{\omega\in\Omega}(A \times B_\omega)$.

The same properties hold with the roles of the first and the second factors exchanged.

5.1.2 Now we turn to the construction of the product measure.

Let $(X, \mathfrak{A}, \mu)$ and $(Y, \mathfrak{B}, \nu)$ be two measure spaces with σ-finite measures. Put

$$\mathscr{P} = \big\{A \times B \mid A \in \mathfrak{A},\ \mu(A) < +\infty,\ \ B \in \mathfrak{B},\ \nu(B) < +\infty\big\}.$$

We will call the sets $A \times B \in \mathscr{P}$ *measurable rectangles*.

B. Makarov, A. Podkorytov, *Real Analysis: Measures, Integrals and Applications*, Universitext, DOI 10.1007/978-1-4471-5122-7_5, © Springer-Verlag London 2013

Define the function m_0 on $\mathscr{P}$ by

$$m_0(A \times B) = \mu(A)\,\nu(B).$$

Theorem *The collection $\mathscr{P}$ of all measurable rectangles is a semiring. The function m_0 is a σ-finite measure on $\mathscr{P}$.*

Proof Since the collections of sets $\{A \in \mathfrak{A} \mid \mu(A) < +\infty\}$ and $\{B \in \mathfrak{B} \mid \nu(B) < +\infty\}$ are, obviously, semirings, the first statement of the theorem is a special case of Theorem 1.1.5.

To prove the second statement, we will show first that the function m_0 is countably additive. Note that if $A \subset X$, $B \subset Y$, then

$$\chi_{A\times B}(x, y) = \chi_A(x)\,\chi_B(y) \quad \text{for all } x \in X,\ y \in Y.$$

Assume that the measurable rectangles $P_k = A_k \times B_k$, $k \in \mathbb{N}$, are pairwise disjoint and their union $P \equiv \bigcup_{k\geqslant 1} P_k$ belongs to the semiring $\mathscr{P}$. Then $P = A \times B$, where $A \in \mathfrak{A}$, $B \in \mathfrak{B}$, and $\chi_P = \sum_{k\geqslant 1} \chi_{P_k}$, i.e.,

$$\chi_A(x)\chi_B(y) = \sum_{k\geqslant 1} \chi_{A_k}(x)\,\chi_{B_k}(y) \quad \text{for all } x \in X,\ y \in Y.$$

Integrating this non-negative series termwise with respect to the measure ν (which is possible by Levy's theorem, see Sect. 4.8.2), we get the equality

$$\chi_A(x)\,\nu(B) = \sum_{k\geqslant 1} \chi_{A_k}(x)\,\nu(B_k) \quad \text{for all } x \in X.$$

Integrating termwise again (this time with respect to the measure μ), we obtain

$$\mu(A)\,\nu(B) = \sum_{k\geqslant 1} \mu(A_k)\nu(B_k), \quad \text{that is,} \quad m_0(P) = \sum_{k\geqslant 1} m_0(P_k).$$

Thus, the countable additivity of the function m_0 is proved.

Since the measures μ and ν are σ-finite, the sets X and Y can be represented as

$$X = \bigcup_{k\geqslant 1} X_k, \qquad Y = \bigcup_{k\geqslant 1} Y_k, \quad \text{where } \mu(X_k) < +\infty,\ \nu(Y_k) < +\infty \text{ for all } k \in \mathbb{N}.$$

So, the σ-finiteness of the measure m_0 follows from the identity

$$X \times Y = \bigcup_{k,n\geqslant 1} X_k \times Y_n.$$

□

Remark It is clear from the proof of the theorem that we have not used the σ-finiteness of the measures μ and ν when proving the countable additivity of m_0. The σ-finiteness of these measures implies the σ-finiteness of m_0, which, in turn, guarantees the uniqueness of the extension of m_0.

5.1.3 The theorem we just proved allows us to introduce the following

Definition Let $(X, \mathfrak{A}, \mu)$ and $(Y, \mathfrak{B}, \nu)$ be measure spaces with σ-finite measures. The measure obtained by the standard extension of the measure m_0 described in Theorem 5.1.2 from the semiring $\mathscr{P}$ is called *the product measure* of the measures μ and ν. It is denoted by $\mu \times \nu$, and the σ-algebra on which it is defined is denoted by $\mathfrak{A} \otimes \mathfrak{B}$. The measure space $(X \times Y, \mathfrak{A} \otimes \mathfrak{B}, \mu \times \nu)$ is called *the product of the measure spaces* $(X, \mathfrak{A}, \mu)$ and $(Y, \mathfrak{B}, \nu)$.

Remarks

(1) A very simple example of a product measure is the product of two one-dimensional Lebesgue measures, which, as we shall prove in Sect. 5.4, is simply the Lebesgue measure on the plane. Similarly, the Lebesgue measure on $\mathbb{R}^3$ is the product of the planar and the one-dimensional Lebesgue measures.

(2) The Cartesian product of measurable sets is measurable. If $\mu(e) = 0$, then $(\mu \times \nu)(e \times Y) = 0$.

Indeed, let $A \in \mathfrak{A}$, $B \in \mathfrak{B}$. If the measures of these sets are finite, then their product $A \times B$ is measurable by the definition of the product measure. In the general case, each of the sets A and B can be represented as a union of sets A_k and B_n of finite measure respectively ($k, n \in \mathbb{N}$). Thereby, the set

$$A \times B = \bigcup_{k \geqslant 1} (A_k \times B) = \bigcup_{k \geqslant 1} \bigcup_{n \geqslant 1} (A_k \times B_n)$$

is measurable as a countable union of measurable sets.

Let $\mu(e) = 0$. Since $Y = \bigcup_{k \geqslant 1} Y_k$ where $\nu(Y_k) < +\infty$, we have $e \times Y = \bigcup_{k \geqslant 1} (e \times Y_k)$ and $(\mu \times \nu)(e \times Y_k) = \mu(e) \cdot \nu(Y_k) = 0$. Thus

$$(\mu \times \nu)(e \times Y) \leqslant \sum_{k \geqslant 1} (\mu \times \nu)(e \times Y_k) = \sum_{k \geqslant 1} 0 = 0.$$

The definition of the product of two measure spaces can be generalized naturally to the case of an arbitrary number of factors. For instance, if

$$(X_1, \mathfrak{A}_1, \mu_1), \qquad (X_2, \mathfrak{A}_2, \mu_2), \qquad (X_3, \mathfrak{A}_3, \mu_3)$$

are three measure spaces with σ-finite measures and $\mathcal{R}_0$ is the collection of the "measurable parallelepipeds", i.e., of the sets of the form $A \times B \times C$ where $A \subset X_1$, $B \subset X_2$, $C \subset X_3$ are measurable sets of finite measure, then we can define the function ν_0 on $\mathcal{R}_0$ by

$$\nu_0(A \times B \times C) = \mu_1(A)\, \mu_2(B)\, \mu_3(C).$$

Repeating the arguments used in the proof of Theorem 5.1.2 with some necessary modifications, we can show that $\mathcal{R}_0$ is a semiring and ν_0 is a σ-finite measure. The product measure $\mu_1 \times \mu_2 \times \mu_3$ is then the standard extension of the measure ν_0. The product measure operation defined in this way is associative: $(\mu_1 \times \mu_2) \times \mu_3 = \mu_1 \times (\mu_2 \times \mu_3) = \mu_1 \times \mu_2 \times \mu_3$. We leave it to the reader to verify this claim by himself (see Exercise 1). Similarly, one can define the product measure for every finite family of measures.

EXERCISES

1. Let $(X_1, \mathfrak{A}_1, \mu_1)$, $(X_2, \mathfrak{A}_2, \mu_2)$, $(X_3, \mathfrak{A}_3, \mu_3)$ be three measure spaces with σ-finite measures. Identifying the sets $(X_1 \times X_2) \times X_3$, $X_1 \times (X_2 \times X_3)$ and $X_1 \times X_2 \times X_3$ in the canonical way, show that the product operation is associative, i.e., that $(\mu_1 \times \mu_2) \times \mu_3 = \mu_1 \times (\mu_2 \times \mu_3) = \mu_1 \times \mu_2 \times \mu_3$. *Hint.* Using the uniqueness of the extension of measures (Theorem 1.5.1) and the completeness of the standard extension, show that these measures are defined on the same σ-algebra.

5.2 The Computation of the Measure of a Set via the Measures of Its Cross Sections. The Integral as the Measure of the Subgraph

Let us remind the reader that the function f defined almost everywhere on the measure space $(X, \mathfrak{A}, \mu)$ is called measurable in the wide sense if it is measurable on some subset $X_0 \subset X$ of full measure. In this case, it coincides with a function measurable on X almost everywhere. The integral $\int_X f\,d\mu$ is then defined as $\int_{X_0} f\,d\mu$ (see Sect. 4.3.3).

5.2.1 Let X and Y be two arbitrary sets and $C \subset X \times Y$. Put

$$C_x = \{y \in Y \mid (x, y) \in C\}, \qquad C^y = \{x \in X \mid (x, y) \in C\}.$$

Definition We will call the sets C_x and C^y *cross sections* of the set C of the *first* and the *second kind* respectively.

It is worth emphasizing that the cross sections of the first and the second kind are subsets of the sets Y and X respectively. Let us exhibit some properties of cross sections.

Lemma *Let $\{C_\omega\}_{\omega\in\Omega}$ be a family of subsets of the Cartesian product $X \times Y$. Then*

$$\Big(\bigcup_{\omega\in\Omega} C_\omega\Big)_x = \bigcup_{\omega\in\Omega} (C_\omega)_x \quad \text{and} \quad \Big(\bigcap_{\omega\in\Omega} C_\omega\Big)_x = \bigcap_{\omega\in\Omega} (C_\omega)_x.$$

Also, $(C \setminus C')_x = C_x \setminus C'_x$ for all sets C, $C' \subset X \times Y$, and $C_x \cap C'_x = \varnothing$ when $C \cap C' = \varnothing$.

We leave the proof of this lemma to the reader.

5.2.2 The following theorem shows that the measure of a set $C \subset X \times Y$ is completely determined by the measures of its cross sections. This is a far-reaching generalization of the famous Cavalieri[1] principle, about which we will reveal more later (see the end of Sect. 5.4.1).

Theorem *Let* $(X, \mathfrak{A}, \mu)$ *and* $(Y, \mathfrak{B}, \nu)$ *be measure spaces with* σ*-finite complete measures. Let* $m = \mu \times \nu$. *If* $C \in \mathfrak{A} \otimes \mathfrak{B}$, *then*:

(1) $C_x \in \mathfrak{B}$ *for almost every* $x \in X$;
(2) *the function* $x \mapsto \nu(C_x)$ *is measurable on* X *in the wide sense*;
(3) $m(C) = \int_X \nu(C_x)\, d\mu(x)$.

The analogous statements also hold for cross sections of the second kind.

Note that we do not exclude the case when the function in (2) takes infinite values.

One should keep in mind that the measurability of the cross sections of the set C (of both the first and the second kind) by no means guarantees that C is measurable even if condition (2) of the theorem holds as well. It follows, for instance, from the existence of a Lebesgue non-measurable set on the plane whose intersection with every line consists of at most two points. An example of such a set, constructed by Sierpinski,[2] can be found in [GO], p. 142.

Proof We will carry out the proof in several steps. For the first three steps, we will assume that the measures μ and ν are finite.

(1) We start by proving the statements of the theorem for the sets in the Borel hull of the semiring $\mathscr{P}$. Here, as in the previous section, $\mathscr{P}$ is the semiring of the measurable rectangles, i.e., of the sets of the form $A \times B$ where $A \in \mathfrak{A}$ and $B \in \mathfrak{B}$. Note that in the case under consideration, we have $X \times Y \in \mathscr{P}$.

Consider the collection $\mathcal{E}$ of all sets $E \subset X \times Y$ satisfying the following conditions:

(I) $E_x \in \mathfrak{B}$ for all $x \in X$;
(II) the function $x \mapsto \nu(E_x)$ is measurable on X.

Every set E belongs or does not belong to $\mathcal{E}$ simultaneously with its complement E^c because $(E^c)_x = Y \setminus E_x$ and $\nu((E^c)_x) = \nu(Y) - \nu(C_x)$ (we need the finiteness of the measure ν to derive the last equality).

---

[1]Francesco Bonaventura Cavalieri (1598–1647)—Italian mathematician.

[2]Wacław Franciszek Sierpiński (1882–1969)—Polish mathematician.

Every union of an increasing sequence of sets in $\mathcal{E}$ also belongs to $\mathcal{E}$. Indeed, assume that

$$E = \bigcup_{n=1}^{\infty} E_n, \quad \text{where } E_1 \subset E_2 \subset \cdots \text{ and } E_n \in \mathcal{E} \text{ for all } n \in \mathbb{N}.$$

Then $E_x \in \mathfrak{B}$ for all $x \in X$ because, by the lemma, $E_x = \bigcup_{n \geqslant 1} (E_n)_x$. In addition, by the theorem on the continuity from below of measure, one has $\nu((E_n)_x) \to \nu(E_x)$. So the function $x \mapsto \nu(E_x)$ is measurable as a limit of measurable functions. Thus, the collection $\mathcal{E}$ is a monotone class. Let us note one more of its properties: if the sets $A, B \in \mathcal{E}$ are disjoint, then $A \vee B \in \mathcal{E}$. This property follows from the identities

$$(A \vee B)_x = A_x \vee B_x, \qquad \nu\big((A \vee B)_x\big) = \nu(A_x) + \nu(B_x).$$

Clearly, the collection $\mathcal{E}$ contains the semiring $\mathscr{P}$. Moreover, it contains all finite unions of sets from $\mathscr{P}$ because, by the theorem on properties of semirings (see Sect. 1.1.4), each such union can be represented as a union of pairwise disjoint sets from $\mathscr{P}$. Since $X \times Y \in \mathcal{E}$, by the corollary to that theorem, the collection $\mathcal{E}$ contains the algebra generated by $\mathscr{P}$. Therefore $\mathcal{E}$ satisfies the assumptions of the monotone class theorem (see Sect. 1.6.3) and, thereby, contains the entire Borel hull $\mathfrak{B}(\mathscr{P})$ of the semiring $\mathscr{P}$. In particular, the statements (1) and (2) of the theorem hold for all sets in $\mathfrak{B}(\mathscr{P})$.

Let us show now that the equality (3) also holds for all sets in $\mathfrak{B}(\mathscr{P})$. Consider the function $E \mapsto \int_X \nu(E_x)\, d\mu(x)$ on this σ-algebra. It follows from Lemma 5.2.1 and the countable additivity of the integral that this function is a measure. The reader can easily check that it coincides with m on the sets from the semiring $\mathscr{P}$. So the equality (3) follows from the uniqueness of the extension of a measure.

Thus, the theorem has been proved for all sets in $\mathfrak{B}(\mathscr{P})$.

(2) Consider now the case when C is a set from $\mathfrak{A} \otimes \mathfrak{B}$ and $m(C) = 0$. Let $\widetilde{C}$ be a set in $\mathfrak{B}(\mathscr{P})$ of zero measure containing C (the existence of such a set has been proved in the corollary to Theorem 1.5.2). Then

$$\int_X \nu(\widetilde{C}_x)\, d\mu(x) = m(\widetilde{C}) = 0.$$

Therefore $\nu(\widetilde{C}_x) = 0$ for almost all $x \in X$. The inclusion $C_x \subset \widetilde{C}_x$ and the completeness of the measure ν imply that the set C_x is measurable whenever $\nu(\widetilde{C}_x) = 0$, i.e., for almost all $x \in X$. The remaining statements of the theorem for the set C now become obvious.

(3) Let us turn to the general case. Again, using the corollary to Theorem 1.5.2, we can represent C as $C = \widetilde{C} \setminus e$ where $\widetilde{C}$ is a set in $\mathfrak{B}(\mathscr{P})$ and $m(e) = 0$. Therefore the set $C_x = \widetilde{C}_x \setminus e_x$ is measurable for almost all $x \in X$ together with the set e_x. It follows that the values of the function $x \mapsto \nu(C_x) = \nu(\widetilde{C}_x) - \nu(e_x)$ (defined almost

everywhere) coincide with $\nu(\widetilde{C}_x)$ almost everywhere, which implies its measurability on a set of full measure and the equality

$$m(C) = m(\widetilde{C}) = \int_X \nu(\widetilde{C}_x)\,d\mu(x) = \int_X \nu(C_x)\,d\mu(x).$$

Thus, the theorem is proved for the case when the measures μ and ν are finite.

As can be seen from the above arguments, one can relax the boundedness condition imposed on the measures μ and ν, assuming instead that the set C is contained in a measurable rectangle.

(4) Let us turn to the case when the measures μ and ν are infinite. Then the sets X and Y can be represented as disjoint unions $X = \bigvee_{n\geqslant 1} X_n$ and $Y = \bigvee_{n\geqslant 1} Y_n$, where X_n, Y_n are sets of finite measure. Consider a measurable set $C \subset X \times Y$. It is clear that

$$m(C) = \sum_{k=1}^{\infty}\sum_{n=1}^{\infty} m(C_{k,n}), \quad \text{where } C_{k,n} = C \cap (X_k \times Y_n).$$

Applying the part of the theorem proved above to each of the sets $C_{k,n} \subset X_k \times Y_n$, we see that, for every $k, n \in \mathbb{N}$, one has

$$m(C_{k,n}) = \int_{X_k} \nu(Y_n \cap C_x)\,d\mu(x).$$

Since $C_x = \bigvee_{n=1}^{\infty}(Y_n \cap C_x)$, we have $\nu(C_x) = \sum_{n=1}^{\infty} \nu(Y_n \cap C_x)$. Therefore,

$$\int_X \nu(C_x)\,d\mu(x) = \sum_{k=1}^{\infty} \int_{X_k} \nu(C_x)\,d\mu(x)$$
$$= \sum_{k=1}^{\infty}\sum_{n=1}^{\infty} \int_{X_k} \nu(Y_n \cap C_x)\,d\mu(x) = \sum_{k=1}^{\infty}\sum_{n=1}^{\infty} m(C_{k,n}) = m(C).$$

The concluding part of the proof is, of course, also valid in the case when only one of the measures is infinite. For example, if $\nu(Y) < +\infty$, we can just consider the sets $X_k \times Y$ instead of $X_k \times Y_n$. □

Remark We would like to draw the reader's attention to the fact that at the first step of the proof we established that the sections C_x of a set C in $\mathfrak{B}(\mathcal{P})$ are measurable for all (rather than almost all) $x \in X$. Moreover, the proof of this result did not use the completeness of the measures, so that it holds for arbitrary measures, not necessarily complete. (We retain our assumption that all measures in question are σ-finite.)

One can see from the proof that, if only the cross sections of the first kind are considered, then the theorem remains valid if only the completeness of the measure ν is assumed. We shall use this observation in the next theorem (on the measure of the subgraph).

Corollary *Let*

$$P_1(C)=\{x\in X\,|\,C_x\neq\varnothing\},\qquad P_2(C)=\{y\in Y\,|\,C^y\neq\varnothing\}$$

be the canonical projections of the subset $C\subset X\times Y$ to the sets X and Y. If the projection $P_1(C)$ ($P_2(C)$) is measurable, then $m(C)=\int_{P_1(C)}\nu(C_x)\,d\mu(x)$ (respectively, $m(C)=\int_{P_2(C)}\mu(C^y)\,d\nu(y)$).

Proof This equality follows from the theorem directly because $C_x=\varnothing$ and $\nu(C_x)=0$ when $x\notin P_1(C)$. □

Note that we cannot drop the assumption that the projection is measurable because the projection of a measurable set may be non-measurable. For example, if E is a non-measurable subset of X and F is a non-empty subset of Y of measure 0, then $E\times F$ is measurable but its projection to X is not.

5.2.3 Now we shall discuss the "geometric meaning" of the integral. We will fix a measure space $(X,\mathfrak{A},\mu)$ with σ-finite measure and a function f on X with values in $\overline{\mathbb{R}}$. Throughout the rest of this section, the symbol m will denote the product measure of the measure μ and the one-dimensional Lebesgue measure λ.

Definition Given a non-negative function f, we will call the set

$$\mathscr{P}_f(E)=\{(x,y)\in X\times\mathbb{R}\,|\,x\in E,\ 0\leqslant y\leqslant f(x)\}$$

the *subgraph* of f over the set $E\subset X$.

We will call the set

$$\Gamma_f(E)=\{(x,y)\in X\times\mathbb{R}\,|\,x\in E,\ y=f(x)\}$$

the *graph* of the function $f\ E\to\overline{\mathbb{R}}$. Note that the function f may take infinite values. Nevertheless, even in this case, the sets $\mathscr{P}_f(E)$, $\Gamma_f(E)$ are contained in $X\times\mathbb{R}$, not in $X\times\overline{\mathbb{R}}$, according to our definition. In the case when $E=X$, we will just call these sets the subgraph and the graph of the function f and denote them by $\mathscr{P}_f$ and Γ_f respectively.

First of all, let us check the following claim, some special cases of which we have already met (see Corollary 2.3.1 and Exercise 1 in Sect. 2.3).

Lemma *If a real-valued function f is measurable on the set E, then $m(\Gamma_f(E))=0$. If a non-negative function f is measurable in the wide sense, then its subgraph is measurable.*

Proof Since the measure μ is σ-finite, we may restrict ourselves to the case $\mu(E)<+\infty$. Fix an arbitrarily small $\varepsilon>0$ and put

$$e_k=\{x\in E\,|\,k\varepsilon\leqslant f(x)<(k+1)\varepsilon\},\quad \text{where } k\in\mathbb{Z}.$$

Obviously, the sets e_k are pairwise disjoint (they exhaust E if the function f takes only finite values). In addition, $\Gamma_f(e_k) \subset e_k \times [k\varepsilon, (k+1)\varepsilon)$ and, thereby,

$$\Gamma_f(E) \subset \bigcup_{k\in\mathbb{Z}} e_k \times \big[k\varepsilon, (k+1)\varepsilon\big) = H_\varepsilon.$$

The set H_ε is measurable and

$$m(H_\varepsilon) = \sum_{k\in\mathbb{Z}} \varepsilon\mu(e_k) \leqslant \varepsilon\mu(E).$$

Thus, the graph can be covered by a set of arbitrarily small measure. Taking into account that the measure m is complete, we conclude from this that the graph is measurable and has zero measure (see Lemma 1.5.3).

Let us turn to the proof of the measurability of the subgraph of a measurable function. First, consider the case when the function f is simple. Let $\{E_k\}_{k=1}^N$ be an admissible partition for f and let $\{a_k\}_{k=1}^N$ be the corresponding values of the function. It is clear that

$$\mathscr{P}_f(E_k) = E_k \times [0, a_k] \quad \text{and} \quad \mathscr{P}_f = \bigcup_{k=1}^N E_k \times [0, a_k].$$

One can see from this that the subgraph of a simple function is measurable as a union of measurable rectangles.

A general non-negative function f measurable on X can be approximated by a pointwise increasing sequence of non-negative simple functions $\{f_n\}_{n\geqslant 1}$ (see Theorem 3.2.2). The reader can easily verify the inclusions

$$\mathscr{P}_f \setminus \Gamma_f \subset \bigcup_{n\geqslant 1} \mathscr{P}_{f_n} \subset \mathscr{P}_f.$$

Since, as we have proved, $m(\Gamma_f) = 0$, these inclusions imply that the subgraph differs from the union of a sequence of measurable sets just by a set of zero measure and, thereby, is itself measurable. This implies the measurability of $\mathscr{P}_f(E)$ too because $\mathscr{P}_f(E) = \mathscr{P}_f \cap (E \times \mathbb{R})$. The subgraph $\mathscr{P}_f(E)$ is measurable for every function f measurable on E because we can view f as a restriction of a function measurable on the entire set X.

Lastly, assume that the function f is measurable in the wide sense, i.e., it is measurable on some subset X_0 of full measure. It is clear that

$$\mathscr{P}_f = \mathscr{P}_f(X_0) \cup \mathscr{P}_f(e),$$

where $e = X \setminus X_0$, $\mu(e) = 0$. The set $\mathscr{P}_f(X_0)$ is measurable according to what we have proved above, and the subgraph $\mathscr{P}_f(e)$ is measurable due to the completeness of the measure m because

$$\mathscr{P}_f(e) \subset e \times \mathbb{R} \quad \text{and} \quad m(e \times \mathbb{R}) = 0. \qquad \square$$

Now, let us turn to the main result of this section.

Theorem (On the measure of the subgraph) *Let λ be the Lebesgue measure on $\mathbb{R}$, let $(X, \mathfrak{A}, \mu)$ be a measure space with σ-finite measure, let $m = \mu \times \lambda$, and let f be a non-negative function defined on X. The function f is measurable in the wide sense if and only if its subgraph is measurable. In this case,*

$$\int_X f\,d\mu = m(\mathscr{P}_f). \tag{1}$$

Proof The measurability of the subgraph of a non-negative measurable function has been established in the lemma.

Assume now that the subgraph of the function f is measurable. Obviously, the cross section $(\mathscr{P}_f)_x$ coincides with the closed interval $[0, f(x)]$ when $f(x) < +\infty$ and with $[0, +\infty)$ when $f(x) = +\infty$. By Theorem 5.2.2 (see also the remark to it, where it has been pointed out that if only the cross sections of the first kind are considered, the assumption about the completeness of the measure μ may be dropped), we obtain that the function $x \mapsto \lambda((\mathscr{P}_f)_x) = f(x)$ is measurable in the wide sense and the equality (1) holds. □

Remarks

(1) The theorem just proved confirms once more that the definition of a measurable function we accepted is reasonable: the non-negative measurable in the wide sense functions are exactly the functions to whose subgraphs one can assign a measure in a natural way. If the product measure is constructed without using the notion of the integral, then the equality (1) can be taken as the definition of the integral of a non-negative measurable function. In this case, some properties of integrals become obvious. For example, Levy's theorem follows directly from the continuity from below of the measure $\mu \times \lambda$. We shall return to the discussion of such a definition in Sect. 5.5.2.

(2) For a non-positive function f, one can introduce an analog of the subgraph: the set

$$\widetilde{\mathscr{P}}_f(E) = \{(x, y) \in E \times \mathbb{R} \mid x \in E,\ f(x) \leqslant y \leqslant 0\}.$$

Approximating the function $(-f)$ by simple functions, one can easily check that Theorem 5.2.3 remains valid for non-positive functions if one replaces the subgraph by the set $\widetilde{\mathscr{P}}_f(E)$, and the equality (1) by $m(\widetilde{\mathscr{P}}_f(E)) = \int_E |f|\,d\mu = -\int_E f\,d\mu$. Thus, for every integrable function f, the equality

$$\int_X f\,d\mu = m\bigl(\mathscr{P}_f(E_+)\bigr) - m\bigl(\widetilde{\mathscr{P}}_f(E_-)\bigr)$$

holds where $E_\pm = E(\pm f > 0)$. If μ is the Lebesgue measure and the sets $\mathscr{P}_f(E_+)$, $\widetilde{\mathscr{P}}_f(E_-)$ are congruent (in particular, if the set E is symmetric with respect to the origin and f is odd), then $\int_E f\,d\mu = 0$.

EXERCISES

1. Let f and g be two measurable almost everywhere finite functions defined on the measure space $(X, \mathfrak{A}, \mu)$. Prove that, if $g \leqslant f$, then the set

$$Q = \{(x, y) \in X \times \mathbb{R} \,|\, x \in X,\ g(x) \leqslant y \leqslant f(x)\}$$

is measurable in $X \times \mathbb{R}$ and $m(Q) = \int_X (f - g)\, d\mu$ where $m = \mu \times \lambda$, and λ is the Lebesgue measure on $\mathbb{R}$.

2. Prove that if pairwise disjoint disks contained in a square cover it up to a set of measure 0, then the sum of the lengths of their boundary circumferences is infinite.

5.3 Double and Iterated Integrals

Our goal is to reduce the computation of the integral with respect to the product measure $\mu \times \nu$ to the computation of integrals with respect to the measures μ and ν. We shall consider only real-valued functions here, although all results we will obtain below can be generalized to the complex-valued case in the obvious way.

5.3.1 With every function f defined on the set $C \subset X \times Y$, one can associate two families of functions obtained by "fixing one of the variables". More precisely, this means that on every non-empty cross section C_x, one can define the function f_x by the rule $f_x(y) = f(x, y)$. Similarly on every cross section C^y, one can define the function f^y by $f^y(x) = f(x, y)$. This notation will frequently be used later.

Passing to the study of the connection between the integral with respect to the product measure $\mu \times \nu$ and the integrals with respect to the measures μ and ν, consider first the case when the function to integrate is non-negative. The following important theorem holds.

Theorem (Tonelli[3]) *Let $(X, \mathfrak{A}, \mu)$ and $(Y, \mathfrak{B}, \nu)$ be two measure spaces with σ-finite complete measures. Let $m = \mu \times \nu$. Let f be a non-negative function defined on $X \times Y$ that is measurable with respect to the σ-algebra $\mathfrak{A} \otimes \mathfrak{B}$. Then:*

(1) *for almost all $x \in X$, the function f_x is measurable on Y;*
(1′) *for almost all $y \in Y$, the function f^y is measurable on X;*
(2) *the function*

$$x \mapsto \varphi(x) \equiv \int_Y f_x \, d\nu = \int_Y f(x, y)\, d\nu(y)$$

is measurable on X in the wide sense;

[3] Leonida Tonelli (1885–1946)—Italian mathematician.

(2′) *the function*

$$y \mapsto \psi(y) \equiv \int_X f^y \, d\mu = \int_X f(x, y)\, d\mu(x)$$

is measurable on Y in the wide sense;

(3) *the equalities*

$$\int_{X \times Y} f \, dm = \int_X \varphi \, d\mu = \int_Y \psi \, d\nu \tag{1}$$

hold.

Remark The last equality can be rewritten as

$$\int_{X \times Y} f(x, y)\, dm(x, y) = \int_X \left(\int_Y f(x, y)\, d\nu(y) \right) d\mu(x)$$

$$= \int_Y \left(\int_X f(x, y)\, d\mu(x) \right) d\nu(y).$$

The integral on the left-hand side of this equality is called a double integral, and the other two integrals are called repeated integrals. Let us emphasize that this equality of the repeated integrals, often referred to as "the validity of changing the order of integration", is used very frequently when computing double integrals (see, in particular, Examples 1 and 2 below).

Proof Consider three cases corresponding to more and more general functions f.

(1) Let $f = \chi_C$ be the characteristic function of a measurable set $C \subset X \times Y$. Then, for all x in X and y in Y,

$$f_x(y) = \chi_C(x, y) = \begin{cases} 1, & \text{when } (x, y) \in C, \\ 0, & \text{when } (x, y) \notin C, \end{cases}$$

$$= \begin{cases} 1, & \text{when } y \in C_x, \\ 0, & \text{when } y \notin C_x, \end{cases} = \chi_{C_x}(y).$$

Thus, $f_x = \chi_{C_x}$. Since, by Theorem 5.2.2, the sets C_x are measurable for almost all x, the function f_x is measurable as well. Integrating the equality $f_x = \chi_{C_x}$, we see that

$$\varphi(x) = \int_Y f_x \, d\nu = \nu(C_x).$$

By Theorem 5.2.2, the function φ is measurable in the wide sense. Finally, integrating the last equality and using Theorem 5.2.2 again, we get the desired equality

$$\int_X \varphi(x)\, d\mu(x) = \int_X \nu(C_x)\, d\mu(x) = m(C) = \int_{X \times Y} f \, dm.$$

(2) Now let f be a simple function. Then $f = \sum_{k=1}^{N} c_k \chi_{C_k}$ where $c_k \geqslant 0$. It follows that $f_x = \sum_{k=1}^{N} c_k (\chi_{C_k})_x$ and $\varphi(x) = \sum_{k=1}^{N} c_k \nu((C_k)_x)$, which implies the statements (1)–(3).

(3) In the general case, approximate f by an increasing sequence of simple functions f_n. Then $f_x = \lim_{n\to\infty}(f_n)_x$, which guarantees the measurability of f_x for almost all x in X. Since $(f_n)_x \leqslant (f_{n+1})_x$, Levy's theorem yields

$$\varphi(x) = \int_Y f_x(y)\,d\nu(y) = \lim_{n\to\infty} \varphi_n(x),$$

where the function φ_n is defined by $\varphi_n(x) = \int_Y (f_n)_x(y)\,d\nu(y)$. Obviously, $\varphi_n \leqslant \varphi_{n+1}$ almost everywhere. Using Levy's theorem again, we get

$$\int_X \varphi(x)\,d\mu(x) = \lim_{n\to\infty} \int_X \varphi_n(x)\,d\mu(x) = \lim_{n\to\infty} \int_{X\times Y} f_n\,dm = \int_{X\times Y} f\,dm.$$

The statements (1′), (2′) and the second of the equalities (1) can be proved similarly. □

Corollary 1 *Let f be a non-negative measurable function defined on a (measurable) set $C \subset X \times Y$. If the projection $P_1(C)$ is measurable, then*

$$\int_C f\,dm = \int_{P_1(C)} \left(\int_{C_x} f(x, y)\,d\nu(y) \right) d\mu(x). \tag{1′}$$

Proof To prove the corollary, it is enough to extend the function f by zero outside the set C and to use statement (3) of the theorem. □

A similar equality holds when the projection $P_2(C)$ is measurable. In that case

$$\int_C f\,dm = \int_{P_2(C)} \left(\int_{C^y} f(x, y)\,d\mu(x) \right) d\nu(y). \tag{1″}$$

Corollary 2 *If the function f is measurable on $X \times Y$, then:*

(1) *for almost all $x \in X$, the function f_x is measurable on Y;*
(2) *if $\int_Y |f_x(y)|\,d\nu(y) < +\infty$ for almost all $x \in X$, then the function $x \mapsto \int_Y f(x, y)\,d\nu(y)$ is measurable on X in the wide sense.*

Similar statements hold for the function f^y.

Proof The first statement follows from the equality $f_x = (f_+)_x - (f_-)_x$ and the measurability of the functions $(f_\pm)_x$ (see Tonelli's theorem). To prove the second statement, it suffices to note that (again, by Tonelli's theorem) the functions $x \mapsto \int_Y (f_\pm)_x(y)\,d\nu(y)$ are measurable in the wide sense. They are finite almost everywhere, so their difference

$$\int_Y f_+(x,y)\,d\nu(y) - \int_Y f_-(x,y)\,d\nu(y) = \int_Y f(x,y)\,d\nu(y)$$

is well-defined and measurable on a set of full measure. □

5.3.2 Let us consider a few examples demonstrating applications of Tonelli's theorem. Note that in all cases we shall use only the equality of the repeated integrals and we will not be interested in the product measure itself. The only related fact that we will really need is the measurability of a function defined and continuous on an open subset of the space $\mathbb{R}^2$ with respect to the product measure of the one-dimensional Lebesgue measures λ_1. This is obvious because the measure $\lambda_1 \times \lambda_1$ is defined on the two-dimensional rectangles and, thereby, on all open sets as well. (As we shall see in Sect. 5.4, the measure $\lambda_1 \times \lambda_1$ is just the planar Lebesgue measure, but we do not need this fact right now.)

Example 1 We will use Tonelli's theorem to compute the Euler–Poisson integral $I = \int_{-\infty}^{\infty} e^{-x^2}\,dx$ again (see also Sect. 4.6.3).

It is clear that

$$I^2 = \left(2\int_0^\infty e^{-x^2}\,dx\right)\left(2\int_0^\infty e^{-y^2}\,dy\right)dx = 4\int_0^\infty e^{-x^2}\left(\int_0^\infty e^{-y^2}\,dy\right)dx.$$

Make the change of variable $y = xu$ in the inner integral:

$$I^2 = 4\int_0^\infty e^{-x^2}\left(\int_0^\infty e^{-x^2u^2}\,x\,du\right)dx.$$

Taking into account that the integrand $(x,u) \mapsto x\,e^{-x^2(1+u^2)}$ is measurable and non-negative, we can change the order of integration using Tonelli's theorem:

$$I^2 = 4\int_0^\infty \left(\int_0^\infty xe^{-(1+u^2)x^2}\,dx\right)du.$$

The inner integral can be computed using an explicit antiderivative:

$$\int_0^\infty xe^{-(1+u^2)x^2}\,dx = -\frac{1}{2(1+u^2)}e^{-(1+u^2)x^2}\Big|_0^\infty = \frac{1}{2(1+u^2)}.$$

Therefore

$$I^2 = 2\int_0^\infty \frac{1}{1+u^2}\,du = \pi.$$

Thus, $I = \sqrt{\pi}$.

Example 2 Let us use Tonelli's theorem to derive an important formula relating the functions B and Γ, which is due to Euler (see Sect. 4.6.3):

$$B(s,t)=\int_0^1 x^{s-1}(1-x)^{t-1}\,dx=\frac{\Gamma(s)\Gamma(t)}{\Gamma(s+t)}\quad\text{for all } s,t>0.$$

To prove it, write the product $\Gamma(s)\Gamma(t)$ as a repeated integral with the outer integration taken with respect to x and make the change of variable $y=u-x$ in the inner integral:

$$\begin{aligned}\Gamma(s)\Gamma(t)&=\int_0^\infty x^{s-1}e^{-x}\biggl(\int_0^\infty y^{t-1}e^{-y}\,dy\biggr)dx\\&=\int_0^\infty x^{s-1}\biggl(\int_x^\infty (u-x)^{t-1}e^{-u}\,du\biggr)dx.\end{aligned}$$

The resulting repeated integral equals the double integral over the angle $C=\{(x,u)\,|\,0<x<u\}$. It is clear that $C_x=(x,+\infty)$ and $C^u=(0,u)$. Changing the order of integration and using the formula (1″), we get

$$\Gamma(s)\Gamma(t)=\int_0^\infty\biggl(\int_0^u x^{s-1}(u-x)^{t-1}e^{-u}\,dx\biggr)du,$$

which, after one more change of variable $x=uv$, yields the identity

$$\Gamma(s)\Gamma(t)=\int_0^\infty u^{s+t-1}e^{-u}\biggl(\int_0^1 v^{s-1}(1-v)^{t-1}\,dv\biggr)du.$$

It remains to note that the inner integral equals $B(s,t)$.

Putting $s=t=\frac{1}{2}$ in the Euler formula and computing the integral $\int_0^1\frac{dx}{\sqrt{x(1-x)}}$, we again arrive at the identity $\Gamma(\frac{1}{2})=\sqrt{\pi}$, which we obtained in Sect. 4.6.3.

The Euler formula also allows one to express the frequently encountered integrals $\int_0^{\frac{\pi}{2}}\sin^p\varphi\cos^q\varphi\,d\varphi$ $(p,q>-1)$ in terms of the Γ-function. Indeed,

$$\begin{aligned}\int_0^{\frac{\pi}{2}}\sin^p\varphi\cos^q\varphi\,d\varphi&=\frac{1}{2}\int_0^{\frac{\pi}{2}}\sin^{p-1}\varphi\cos^{q-1}\varphi\,d\sin^2\varphi\\&=\frac{1}{2}\int_0^1 x^{\frac{p-1}{2}}(1-x)^{\frac{q-1}{2}}\,dx\\&=\frac{1}{2}B\biggl(\frac{p+1}{2},\frac{q+1}{2}\biggr)=\frac{\Gamma(\frac{p+1}{2})\Gamma(\frac{q+1}{2})}{2\Gamma(\frac{p+q}{2}+1)}.\end{aligned}$$

5.3.3 Tonelli's theorem remains valid for sign-changing functions if one replaces the assumption "measurable in the wide sense" by "summable". Let us discuss this important observation in more detail.

Theorem (Fubini[4]) *Let $(X, \mathfrak{A}, \mu)$ and $(Y, \mathfrak{B}, \nu)$ be two measure space with σ-finite complete measures and let $m = \mu \times \nu$. If a (real or complex-valued) function f is summable on $X \times Y$ with respect to the measure m, then:*

(1) *for almost all $x \in X$, the function f_x is summable on Y;*
(1′) *for almost all $y \in Y$, the function f^y is summable on X;*
(2) *the function*

$$x \mapsto \varphi(x) \equiv \int_Y f_x \, d\nu = \int_Y f(x, y) \, d\nu(y)$$

is summable on X;
(2′) *the function*

$$y \mapsto \psi(y) \equiv \int_X f^y \, d\mu = \int_X f(x, y) \, d\mu(x)$$

is summable on Y;
(3) *the equalities*

$$\int_{X \times Y} f \, dm = \int_X \varphi \, d\mu = \int_Y \psi \, d\nu \tag{2}$$

hold.

Proof Obviously, we can restrict ourselves to the case when the function f is real-valued. Due to the symmetry between X and Y, it suffices to prove the statements (1) and (2) and the first of the equalities (2). Let $f_\pm = \max\{\pm f, 0\}$. By Tonelli's theorem,

$$\int_{X \times Y} f_\pm \, dm = \int_X \left(\int_Y f_\pm(x, y) \, d\nu(y) \right) d\mu(x) < +\infty. \tag{3}$$

By the same theorem, the functions $(f_\pm)_x$ are measurable for almost all x, and the functions

$$x \mapsto \varphi_1(x) \equiv \int_Y f_+(x, y) \, d\nu(y), \qquad x \mapsto \varphi_2(x) \equiv \int_Y f_-(x, y) \, d\nu(y)$$

are measurable in the wide sense. The inequalities (3) show that the functions φ_1 and φ_2 are summable and, thereby, finite almost everywhere. The latter means that the functions $(f_\pm)_x$ are summable on Y for almost all $x \in X$. Now, to prove statements (1) and (2) of the theorem, it remains to note that

$$f_x = (f_+)_x - (f_-)_x, \qquad \varphi = \varphi_1 - \varphi_2. \tag{4}$$

To prove the first of the equalities (2), one should just take the difference of the equalities (3) and use the relations (4). □

---

[4] Guido Fubini (1879–1943)—Italian mathematician.

Note that if the function f is summable on a (measurable) set $C \subset X \times Y$ and the projection $P_1(C)$ is measurable, then formula (1′) remains valid:

$$\int_C f\,dm = \int_{P_1(C)} \left(\int_{C_x} f(x, y)\,d\nu(y) \right) d\mu(x). \tag{2′}$$

For the proof, it is enough to extend the function f by zero outside the set C and to use statement (3) of Fubini's theorem.

Needless to say, in the case when the projection $P_2(C)$ is measurable, a similar equality is valid (see (1″)).

Remark Both the Tonelli and the Fubini theorems require the assumption that the function f under consideration is measurable on $X \times Y$, or, as is often said, "as a function of two variables". This assumption is stronger than the assumption that f is "measurable in each variable separately", i.e., that the functions f_x and f^y are measurable. On the other hand, if the functions g, h are measurable on X, Y respectively, then the functions $\widetilde{g}, \widetilde{h}$ defined on $X \times Y$ by $\widetilde{g}(x, y) = g(x)$, $\widetilde{h}(x, y) = h(y)$ are measurable on $X \times Y$. To check this, it suffices to consider only the function g, assuming it real-valued. Then it is clear that the Lebesgue sets of the function $\widetilde{g}$ are of the form $E \times Y$ where $E \in \mathfrak{A}$. Therefore they are measurable by Remark (2) from Sect. 5.1.3.

The measurability of the functions $\widetilde{g}$ and $\widetilde{h}$ on $X \times Y$ implies the measurability of their product $f = \widetilde{g} \cdot \widetilde{h}$, which is sometimes denoted by the symbol $g \otimes h$.

5.3.4 Let us point out some useful formulae implied by Fubini's theorem.

Corollary 1 *Assume that the functions g and h are summable on the measure spaces $(X, \mathfrak{A}, \mu)$ and $(Y, \mathfrak{B}, \nu)$ with σ-finite measures respectively. Then the function $f = g \otimes h$ is summable on $X \times Y$ with respect to the measure $m = \mu \times \nu$ and*

$$\int_{X \times Y} f(x, y)\,dm(x, y) = \int_X g(x)\,d\mu(x) \cdot \int_Y h(y)\,d\nu(y).$$

Proof Assuming for the time being that the measures μ and ν are complete, we check that the function f is summable using Tonelli's theorem. The measurability of the function f is established in the remark in Sect. 5.3.3. Let us check that it is summable using Tonelli's theorem. Indeed,

$$\begin{aligned}
\int_{X \times Y} |f(x, y)|\,dm(x, y) &= \int_X \left(\int_Y |g(x) h(y)|\,d\nu(y) \right) d\mu(x) \\
&= \int_X |g(x)| \left(\int_Y |h(y)|\,d\nu(y) \right) d\mu(x) \\
&= \left(\int_Y |h(y)|\,d\nu(y) \right) \cdot \left(\int_X |g(x)|\,d\mu(x) \right) < +\infty.
\end{aligned}$$

Now, when the summability of the function f is established, the desired equality follows from Fubini's theorem.

In the case where the measures are not complete, one should write down the equality in question for their standard extensions and use the fact that the integral over the extended measures remains the same (see Exercise 7 from Sect. 4.2). □

The argument we just presented is very typical. When computing the integral, we rely on Fubini's theorem but first we need to check that the integrand is summable, which can be done using Tonelli's theorem.

In Corollary 2, we will show that the integration by parts formula obtained earlier for smooth functions (see Sect. 4.6.2) is valid under less restrictive assumptions as well. Let us remind the reader (see Sect. 4.9.3) that a function f is called absolutely continuous on a closed interval $[a,b]$ if it can be represented as $f(x) = f(a) + \int_a^x \varphi(t)\,dt$ where the function φ is summable on $[a,b]$. By Theorem 4.9.3, one has $\varphi = f'$ almost everywhere.

Corollary 2 *Let the functions f and g be absolutely continuous on a closed interval $[a,b]$. Then*

$$\int_a^b f(x)\,g'(x)\,dx = f(x)\,g(x)\Big|_{x=a}^{x=b} - \int_a^b f'(x)\,g(x)\,dx.$$

Proof First, let us prove this formula under the additional assumption $f(a) = g(b) = 0$. Then the substitution term vanishes and our task can be reduced to the change of the order of integration. Indeed, since the functions f' and g' are summable on $[a,b]$, Corollary 1 implies that the function $(x,y) \mapsto f'(x)\,g'(y)$ is summable on the square $[a,b]^2$ and, thereby, on the triangle $C = \{(x,y) \in [a,b]^2 \mid a \leqslant y \leqslant x \leqslant b\}$ as well. It is easy to check that its cross sections for $x,\ y \in [a,b]$ are

$$C_x = [a,x], \qquad C^y = [y,b].$$

Since $f(x) = \int_a^x f'(y)\,dy$, Formula (2′) implies

$$\int_a^b f(x)\,g'(x)\,dx = \int_a^b g'(x)\left(\int_a^x f'(y)\,dy\right)dx = \iint_C f'(y)\,g'(x)\,dx\,dy$$
$$= \int_a^b f'(y)\left(\int_y^b g'(x)\,dx\right)dy = -\int_a^b f'(y)\,g(y)\,dy,$$

which establishes the desired formula in the special case under consideration. To prove it in the general case, one should merely apply the result just obtained to the functions $f(x) - f(a)$, $g(x) - g(b)$. □

Let us generalize this corollary and obtain the integration by parts formula for the integral with respect to the Lebesgue–Stieltjes measure (another proof of this formula is given in Sect. 4.10.6).

Corollary 3 *Let g be a non-decreasing function on the closed interval $[a,b]$. If the function f is absolutely continuous on $[a,b]$, then*

$$\int_{[a,b]} f(x)\,dg(x) = f(x)g(x)\Big|_{x=a}^{x=b} - \int_a^b f'(x)g(x)\,dx.$$

Proof As in the proof of Corollary 2, it suffices to consider the case $f(a) = g(b) = 0$. Applying Fubini's theorem to the product measure $\mu_g \times \lambda$ and changing the order of integration, we get

$$\int_{[a,b]} f(x)\,dg(x) = \int_{[a,b]} \left(\int_a^x f'(u)\,du\right) dg(x) = \int_a^b f'(u)\left(\int_{[u,b]} dg(x)\right) du.$$

When $u > a$, the inner integral on the right-hand side equals $g(b) - g(u-0)$ and, therefore, coincides with $-g(u)$ almost everywhere (with respect to the Lebesgue measure). Therefore,

$$\int_{[a,b]} f(x)\,dg(x) = -\int_a^b f'(u)\,g(u)\,du.$$

□

The formula just obtained remains valid in the case when g is a function of bounded variation as well.

5.3.5 The summability of the functions f_x, f^y, φ and ψ considered in Fubini's theorem does not guarantee the equality of the repeated integrals, much less the summability of the function f with respect to the measure $\mu \times \nu$ even in the case when the measures are finite and the repeated integrals are equal. We will demonstrate this using the following two examples. In both, we assume that the measures μ and ν coincide with the one-dimensional Lebesgue measure on $[-1,1]$.

Consider the functions $f(x,y) = \frac{x^2-y^2}{(x^2+y^2)^2}$ and $g(x,y) = \frac{2xy}{(x^2+y^2)^2}$ for $x^2+y^2 > 0$. It is clear that the functions f_x, f^y, g_x, g^y are summable on $[-1,1]$ for all $x, y \neq 0$ in $[-1,1]$. Obviously,

$$\int_{-1}^1 g(x,y)\,dy = \int_{-1}^1 g(x,y)\,dx = 0.$$

The reader will easily establish the identity

$$\int_{-1}^1 f(x,y)\,dy = \int_{-1}^1 d\left(\frac{y}{x^2+y^2}\right) = \frac{2}{1+x^2} \quad (x \neq 0),$$

which, in view of the fact that f is antisymmetric, yields

$$\int_{-1}^1 f(x,y)\,dx = -\frac{2}{1+y^2} \quad (y \neq 0).$$

Therefore,

$$\int_{-1}^{1}\left(\int_{-1}^{1}\frac{x^2-y^2}{(x^2+y^2)^2}\,dy\right)dx=\pi,\qquad \int_{-1}^{1}\left(\int_{-1}^{1}\frac{x^2-y^2}{(x^2+y^2)^2}\,dx\right)dy=-\pi.$$

Thus, the repeated integrals associated with the function f are finite but have opposite signs, which implies, in particular, that this function is not summable on $[-1,1]^2$.

The repeated integrals associated with the function g give the same value (zero). Despite this, the function g is not summable. Indeed,

$$\begin{aligned}\int_{-1}^{1}\left(\int_{-1}^{1}|g(x,y)|\,dy\right)dx &= 4\int_{0}^{1}\left(\int_{0}^{1}\frac{2xy}{(x^2+y^2)^2}\,dy\right)dx\\ &= 4\int_{0}^{1}\left(-\frac{x}{x^2+y^2}\Big|_{y=0}^{y=1}\right)dx\\ &= 4\int_{0}^{1}\left(\frac{1}{x}-\frac{x}{1+x^2}\right)dx=+\infty.\end{aligned}$$

We leave it to the reader to construct examples of functions such that one of the repeated integrals is finite and the other one either does not exist or exists but is infinite.

EXERCISES

1. Let f be a non-negative function on $X\times Y$ and let $x\in X$. Prove that the subgraph $\mathscr{P}_{f_x}$ of the function f_x coincides with the cross section $(\mathscr{P}_f)_x$ of the subgraph $\mathscr{P}_f$ of f.

2. Prove Tonelli's theorem using Theorem 5.2.3 on the measure of the subgraph and Exercise 1.

3. Let μ be any finite Borel measure on $\mathbb{R}^m$. Prove that, for every $0<p<m$, the integral $\int_{\mathbb{R}^m}\frac{d\mu(x)}{\|x-y\|^p}$ is finite for almost all (with respect to the Lebesgue measure) $y\in\mathbb{R}^m$.

4. If a measurable function f is positive on a set E and $\mu(E)<+\infty$, then $\int_E f\,d\mu\cdot\int_E\frac{1}{f}\,d\mu\geqslant\mu^2(E)$. *Hint.* Use the inequality $\frac{f(x)}{f(y)}+\frac{f(y)}{f(x)}\geqslant 2$.

5. Let μ be a Borel measure on the closed interval $[a,b]$ such that $\mu([a,b])=1$. Prove that for all increasing (or decreasing) functions f and g on $[a,b]$ the Chebyshev inequality

$$\int_a^b fg\,d\mu\geqslant\int_a^b f\,d\mu\cdot\int_a^b g\,d\mu$$

holds. If one of the functions is increasing and the other one is decreasing then the inequality sign should be reversed. *Hint.* Use the fact that the product $(f(x)-f(y))(g(x)-g(y))$ does not change sign.

6. Let φ be the Cantor function. For which $p > 0$ are the integrals

$$\iint_{[0,1]^2} \frac{d\varphi(x)\,d\varphi(y)}{(x^2+y^2)^{\frac{p}{2}}}, \qquad \iint_{[0,1]^2} \frac{d\varphi(x)\,d\varphi(y)}{|x-y|^p}$$

finite?

5.4 Lebesgue Measure as a Product Measure

Our goal is to relate the Lebesgue measure λ_{p+q} on the space $\mathbb{R}^{p+q}$ to the Lebesgue measures λ_p and λ_q on the spaces $\mathbb{R}^p$ and $\mathbb{R}^q$ respectively. We shall identify the space $\mathbb{R}^{p+q}$ with the Cartesian product $\mathbb{R}^p \times \mathbb{R}^q$, assuming that the pair (x, y), where $x = (x_1, \dots, x_p) \in \mathbb{R}^p$ and $y = (y_1, \dots, y_q) \in \mathbb{R}^q$, coincides with the point $(x_1, \dots, x_p, y_1, \dots, y_q)$ in $\mathbb{R}^{p+q}$.

Let us remind the reader that the symbol $\mathscr{P}^m$ denotes the semiring of cells in $\mathbb{R}^m$.

5.4.1 We proceed directly to the main statement of this Section.

Theorem $\lambda_{p+q} = \lambda_p \times \lambda_q$.

This implies, in particular, that the product measure operation is associative on the class of Lebesgue measures:

$$(\lambda_p \times \lambda_q) \times \lambda_r = \lambda_p \times (\lambda_q \times \lambda_r) = \lambda_{p+q+r}.$$

Proof Let $\mathscr{P}$ be a semiring of all sets of the form $A \times B$, where A and B are measurable subsets of finite measure of the spaces $\mathbb{R}^p$ and $\mathbb{R}^q$ respectively. Every cell from $\mathscr{P}^{p+q}$ is, obviously, a product of two cells from $\mathscr{P}^p$ and $\mathscr{P}^q$. Thus, $\mathscr{P}^{p+q} \subset \mathscr{P}$.

The measures λ_{p+q} and $\lambda_p \times \lambda_q$ have been obtained as the standard extensions of the measures l_{p+q} (the classical volume defined on $\mathscr{P}^{p+q}$) and m_0 (the measure defined on $\mathscr{P}$—see Sect. 5.1.2) respectively. To prove that the measures λ_{p+q} and $\lambda_p \times \lambda_q$ coincide, it suffices to show that the measures l_{p+q} and m_0 generate the same outer measures: $l^*_{p+q} = m^*_0$. Since m_0 extends l_{p+q} from the semiring $\mathscr{P}^{p+q}$ to the wider semiring $\mathscr{P}$, the definition of the outer measure generated by a measure immediately implies the inequality $m^*_0 \leqslant l^*_{p+q}$. It remains to check the opposite inequality. It suffices to prove that $l^*_{p+q}(E) < m^*_0(E) + \varepsilon$ for every set $E \subset \mathbb{R}^{p+q}$ such that $m^*_0(E) < +\infty$, and every $\varepsilon > 0$. By the definition of m^*_0, there exist measurable subsets $A_j \subset \mathbb{R}^p$ and $B_j \subset \mathbb{R}^q$ of finite measure ($j \in \mathbb{N}$) such that $E \subset \bigcup_{j \geqslant 1} A_j \times B_j$ and

$$\sum_{j \geqslant 1} \lambda_p(A_j)\lambda_q(B_j) < m^*_0(E) + \varepsilon.$$

Due to the regularity of the Lebesgue measure, the sets A_j, B_j can be covered by open sets G_j, H_j (in the respective spaces) so close to them in measure that if we replace A_j by G_j and B_j by H_j, the last inequality will still hold. As a result, we shall obtain the inclusion $E \subset \bigcup_{j\geqslant 1} G_j \times H_j$ and the inequality

$$\sum_{j\geqslant 1} \lambda_p(G_j)\lambda_q(H_j) < m_0^*(E) + \varepsilon.$$

Since the measures λ_{p+q} and $\lambda_p \times \lambda_q$ coincide on $\mathscr{P}^{p+q}$, they coincide on all open sets in $\mathbb{R}^{p+q}$. The sets $G_j \times H_j$ are open, so

$$l_{p+q}^*(G_j \times H_j) = \lambda_{p+q}(G_j \times H_j) = \lambda_p(G_j)\lambda_q(H_j) \quad (j \in \mathbb{N}).$$

Now the desired estimate follows from the countable subadditivity of l_{p+q}^*:

$$l_{p+q}^*(E) \leqslant \sum_{j\geqslant 1} l_{p+q}^*(G_j \times H_j) = \sum_{j\geqslant 1} \lambda_p(G_j)\lambda_q(H_j) < m_0^*(E) + \varepsilon.$$

□

Remark The integrals with respect to the planar, the three-dimensional, and the m-dimensional Lebesgue measures (over a subset E of the corresponding space) are called double, triple, and m-fold integrals and are often denoted by the symbols

$$\iint_E f(x,y)\,dx\,dy, \qquad \iiint_E f(x,y,z)\,dx\,dy\,dz \quad \text{and}$$

$$\int \cdots \int_E f(x_1,\ldots,x_m)\,dx_1 \cdots dx_m.$$

Since for summable and arbitrary non-negative functions, the integral with respect to the product measure equals the repeated integral, this notation does not lead to any confusion.

5.4.2 According to the classical Cavalieri principle, if two bodies can be positioned in space so that each plane parallel to a given one intersects the two bodies by planar domains of equal areas, then the volumes of these bodies are equal. Since, as we have established above, the measure λ_{p+q} is the product measure of the measures λ_p and λ_q, Theorem 5.2.2 implies the following assertion, which we will refer to as the Cavalieri principle throughout the rest of the book:

If two measurable sets contained in $\mathbb{R}^{p+q} \equiv \mathbb{R}^p \times \mathbb{R}^q$ can be positioned so that the Lebesgue measures of all their cross sections of the first (or the second) kind are equal, then their $(p+q)$-dimensional Lebesgue measures are equal.

Now we will consider some applications of Theorem 5.2.2, the Tonelli theorem, and the Cavalieri principle. By volume, we shall mean the m-dimensional Lebesgue measure.

First, we will compute the volume of a cone in several dimensions.

Example 1 We will call the set $C = \{(t, y) \in \mathbb{R}^m \mid t \in [0, H],\ y \in \frac{t}{H}E\}$ a cone with altitude H and base E, $E \subset \mathbb{R}^{m-1}$. The cone with a measurable base E is measurable because it is the image of the cylinder $[0, H] \times E$ under the smooth mapping $(t, y) \mapsto (t, \frac{t}{H}y)$. For fixed t, the cone cross section C_t is either empty (if $t \notin [0, H]$), or the set $\frac{t}{H}E$, whose measure equals $\lambda_{m-1}(E)\,(\frac{t}{H})^{m-1}$. By Theorem 5.2.2,

$$\lambda_m(C) = \int_{\mathbb{R}} \lambda_m(C_t)\,dt = \int_0^H \lambda_{m-1}(E)\left(\frac{t}{H}\right)^{m-1} dt = \frac{1}{m}H\lambda_{m-1}(E),$$

when $m = 2$ and $m = 3$ this implies the well-known school formulae for the area of a triangle and the volumes of a pyramid and a circular cone.

In the next example, we shall obtain an important result: the formula for the volume of a multi-dimensional ball.

Example 2 When studying the change of the Lebesgue measure under linear transformations (see Sect. 2.5.2), we established that the volume of any m-dimensional ball of radius R equals $\alpha_m R^m$ where α_m is the volume of the unit ball. Obviously, $\alpha_1 = 2$ and $\alpha_2 = \int_{-1}^{1} 2\sqrt{1-t^2}\,dt = \pi$.

To compute α_m for $m > 2$, we will identify the space $\mathbb{R}^m$ with the Cartesian product $\mathbb{R}^{m-1} \times \mathbb{R}$. By definition, the cross section $(B^m)^y$ of the open unit ball B^m is the set

$$\left\{x \in \mathbb{R}^{m-1} \mid (x, y) \in B^m\right\} = \left\{x \in \mathbb{R}^{m-1} \mid \|x\|^2 < 1 - y^2\right\}.$$

For $|y| \geqslant 1$, it is empty, and for $|y| < 1$ it is an $(m-1)$-dimensional ball of radius $\sqrt{1-y^2}$. The $(m-1)$-dimensional volume of the latter equals $\alpha_{m-1}(1-y^2)^{\frac{m-1}{2}}$, so, by Theorem 5.2.2, $\alpha_m = \int_{-1}^{1} \alpha_{m-1}(1-y^2)^{\frac{m-1}{2}}\,dy$. The change of variable $y = \sin u$ gives the recurrence relation

$$\alpha_m = 2\alpha_{m-1}\int_0^{\frac{\pi}{2}} \cos^m u\,du.$$

We computed the last integral in Sect. 4.6.2. It equals $\frac{(m-1)!!}{m!!}v_m$ where $v_m = \frac{\pi}{2}$ for even m and $v_m = 1$ for odd m. Obviously, $v_m v_{m-1} \equiv \frac{\pi}{2}$. Applying the obtained recurrence relation twice, we arrive at the formula

$$\alpha_m = 2\alpha_{m-1}\frac{(m-1)!!}{m!!}v_m = 4\alpha_{m-2}\frac{(m-2)!!}{(m-1)!!}v_{m-1}\frac{(m-1)!!}{m!!}v_m = \frac{2\pi}{m}\alpha_{m-2}.$$

Since we know the initial values $\alpha_1 = 2$ and $\alpha_2 = \pi$, this formula yields

$$\alpha_{2k} = \frac{\pi^k}{k!}, \qquad \alpha_{2k+1} = 2\frac{(2\pi)^k}{(2k+1)!!} \qquad \text{for all } k \in \mathbb{N}.$$

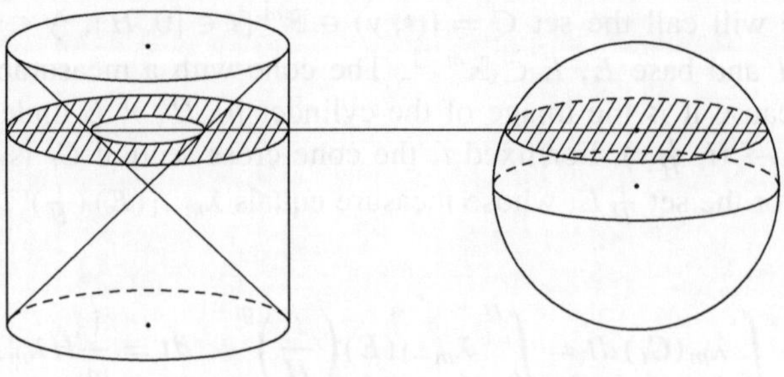

Fig. 5.1 Horizontal cross sections of equal area

The Γ-function allows us to cover both the odd and the even cases in one common formula. Indeed, $k! = \Gamma(k+1)$ and $\sqrt{\pi}(2k+1)!! = 2^{k+1}\Gamma(k+\frac{3}{2})$ (see Sect. 4.6.3). Plugging these values of factorials into the formulae for α_{2k} and α_{2k+1}, we see that, for every $m \in \mathbb{N}$, the equality

$$\alpha_m = \frac{\pi^{\frac{m}{2}}}{\Gamma(\frac{m}{2}+1)}$$

holds.

In relation to Examples 1 and 2, let us remind the reader of the following discovery of Archimedes,[5] which he was very proud of: the ball fills 2/3 of the volume of its circumscribed cylinder (Cicero claimed that he had found Archimedes' grave in an abandoned cemetery by a small column with the engraving of a ball and a cylinder above an accompanying verse).

To obtain this beautiful result, one should compare the ball and the body obtained by removing from the cylinder two cones with vertex at the center of the ball and bases at each end of the cylinder. It is easy to see from the Fig. 5.1 that the ball and this body have horizontal cross sections of equal area (compare with Exercise 11).

Similarly, one can find the volume of the four-dimensional ball avoiding any integration. Indeed, it is clear that the volume of the Cartesian product of two unit disks equals π^2. Identifying the point (x, y, u, v) with the pair (ξ, η) where $\xi = (x, y)$, $\eta = (u, v)$, we will split the product $C = B^2 \times B^2$ into two parts as follows:

$$K = \{(\xi, \eta) \in C \mid \|\xi\| \leqslant \|\eta\|\}, \qquad K' = \{(\xi, \eta) \in C \mid \|\xi\| \geqslant \|\eta\|\}$$

(their two-dimensional analogs $\{(s, t) \mid |s| \leqslant |t| \leqslant 1\}$ and $\{(s, t) \mid 1 \geqslant |s| \geqslant |t|\}$ are formed by two pairs of vertical triangles tiling the square $[-1, 1] \times [-1, 1]$). It is clear that these sets are congruent and $\lambda_4(K \cap K') = 0$. Therefore

$$\lambda_4(K) = \frac{1}{2}\lambda_4(C) = \frac{\pi^2}{2}.$$

[5] Archimedes ($'A\rho\chi\iota\mu\acute{\eta}\delta\eta\varsigma$, circa 287 – 212 BC)—Greek mathematician and inventor.

Let us find the area of the cross section K_ξ of the body K for $\|\xi\| < 1$ (otherwise it is empty). Since

$$K_\xi = \{\eta \in \mathbb{R}^2 \mid (\xi, \eta) \in K\} = \{\eta \in \mathbb{R}^2 \mid \|\xi\| \leqslant \|\eta\| < 1\},$$

this cross section is an annulus whose area equals $\pi(1 - \|\xi\|^2)$. An easy computation shows that the two-dimensional cross section $(B^4)_\xi$ of the four-dimensional unit ball has exactly the same area. Thus, according to the Cavalieri principle, its volume is equal to that of K, i.e.,

$$\lambda_4(B^4) = \lambda_4(K) = \frac{\pi^2}{2}.$$

Example 3 Let us compute the integral $I_m(a) = \int_{\mathbb{R}^m} e^{-a\|x\|^2}\,dx$ $(a > 0)$. In the one-dimensional case, this reduces to the Euler–Poisson integral:

$$I_1(a) = \int_{-\infty}^{\infty} e^{-ax^2}\,dx = \frac{1}{\sqrt{a}}\int_{-\infty}^{\infty} e^{-u^2}\,du = \sqrt{\frac{\pi}{a}}.$$

Representing the m-dimensional Lebesgue measure as the product measure of the $(m-1)$-dimensional and the one-dimensional Lebesgue measures and using Tonelli's theorem, we get the recurrence relation $I_m(a) = I_{m-1}(a) \cdot I_1(a)$, which immediately implies that $I_m(a) = (\frac{\pi}{a})^{\frac{m}{2}}$.

Example 4 Let us generalize the result obtained in Example 2 and find the volume $V_P(R)$ of the set

$$W_P(R) = \{(x_1, \dots, x_m) \in \mathbb{R}^m \bigm| |x_1|^{p_1} + \dots + |x_m|^{p_m} < R\} \quad (R > 0),$$

where $P = (p_1, \dots, p_m) \in \mathbb{R}^m_+$.

First, note that the linear change of variable $x_j = R^{1/p_j} u_j$ $(j = 1, \dots, m)$ maps the set $W_P(R)$ to $W_P(1)$. Therefore (see Sect. 2.5.2) $V_P(R) = R^q V_P(1)$ where $q = \frac{1}{p_1} + \dots + \frac{1}{p_m}$. Thus it suffices to compute $V_P(1)$. To this end, we will use Theorem 5.2.2. Assuming that $m > 1$, put $P' = (p_1, \dots, p_{m-1})$ and $q' = \frac{1}{p_1} + \dots + \frac{1}{p_{m-1}}$. Since the cross section of the set $W_P(1)$ corresponding to the fixed coordinate x_m is, obviously, $W_{P'}(1 - |x_m|^{p_m})$, we obtain

$$V_P(1) = \int_{-1}^{1} V_{P'}\bigl(1 - |x_m|^{p_m}\bigr)\,dx_m = 2V_{P'}(1)\int_0^1 \bigl(1 - x_m^{p_m}\bigr)^{q'}\,dx_m.$$

After the change of variable $u = x_m^{p_m}$, this identity becomes

$$V_P(1) = \frac{2}{p_m} V_{P'}(1) \int_0^1 (1-u)^{q'} u^{\frac{1}{p_m}-1}\,du.$$

The last integral can be expressed in terms of values of the Γ-function (see Example 2 in Sect. 5.3.2) and we obtain the dimension reduction formula

$$V_P(1) = \frac{2}{p_m} V_{P'}(1) \frac{\Gamma(1+q')\Gamma(\frac{1}{p_m})}{\Gamma(1+q)} = 2V_{P'}(1) \frac{\Gamma(1+q')\Gamma(1+\frac{1}{p_m})}{\Gamma(1+q)}.$$

It follows easily from this that

$$V_P(1) = \frac{2^m}{\Gamma(1+\frac{1}{p_1}+\cdots+\frac{1}{p_m})} \prod_{j=1}^m \Gamma\left(1+\frac{1}{p_j}\right).$$

When $p_1 = \cdots = p_m = p$, this yields the formula for the volume of the set $W_p = \{(x_1, \ldots, x_m) \in \mathbb{R}^m \mid |x_1|^p + \cdots + |x_m|^p < 1\}$:

$$\lambda_m(W_p) = \frac{2^m \Gamma^m(1+\frac{1}{p})}{\Gamma(1+\frac{m}{p})}.$$

When $p = 2$, we get the formula for the volume of the ball once more.

5.4.3 Let us mention a nice formula relating the double and the repeated integrals. As a preliminary step, we establish a lemma that will also be of use for us later. In this lemma, we will identify the space $\mathbb{R}^{2m}$ with the Cartesian product $\mathbb{R}^m \times \mathbb{R}^m$ (see the beginning of Sect. 5.4).

Lemma *Let f be a measurable function defined on $\mathbb{R}^m$. Then the functions $(x, y) \mapsto f(x-y)$ and $(x, y) \mapsto f(x+y)$ are measurable on the space $\mathbb{R}^{2m}$.*

Proof It suffices to prove the result for the function $(x, y) \mapsto F(x, y) = f(x-y)$, which we may also assume real-valued (the argument for the second function is similar). Let $E = \{x \in \mathbb{R}^m \mid f(x) < a\}$. Then

$$\begin{aligned} &\{(x, y) \in \mathbb{R}^{2m} \mid F(x, y) = f(x-y) < a\} \\ &= \{(x, y) \in \mathbb{R}^{2m} \mid x - y \in E\} = T^{-1}(E \times \mathbb{R}^m), \end{aligned}$$

where $T : \mathbb{R}^{2m} \to \mathbb{R}^{2m}$ is the linear mapping defined by $T(x, y) = (x - y, y)$. The mapping T is, obviously, invertible. Therefore, the Lebesgue set of the function F is measurable as the image of the measurable set $E \times \mathbb{R}^m$. □

The following example essentially repeats the derivation of Euler's formula relating the functions B and Γ (see Sect. 5.3.2, Example 2).

Example (Liouville's identity[6]) Let f be a non-negative measurable function on $\mathbb{R}_+$. Then, for all positive numbers p and q the identity

$$\iint_{\mathbb{R}^2_+} f(x+y)x^{p-1}y^{q-1}\,dx\,dy = B(p,q)\int_0^\infty f(t)t^{p+q-1}\,dt$$

holds where $B(p,q)=\int_0^1 s^{p-1}(1-s)^{q-1}\,ds$ is the Euler B-function.

Indeed, we can extend f to the negative semi-axis by zero. Then, according to the lemma, the function $(x,y)\mapsto f(x+y)$ is measurable on $\mathbb{R}^2_+$. Using Tonelli's theorem, replace the double integral by the repeated integral with the outer integration with respect to x and make the change of variable $y=t-x$ in the inner integral:

$$\iint_{\mathbb{R}^2_+} f(x+y)x^{p-1}y^{q-1}\,dx\,dy = \int_0^\infty x^{p-1}\biggl(\int_x^\infty f(t)(t-x)^{q-1}\,dt\biggr)\,dx.$$

The repeated integral on the right-hand side of this equality equals the double integral over the angle $C=\{(x,t)\,|\,0<x<t\}$. Clearly, $C_x=(x,+\infty)$ and $C^t=(0,t)$. Changing the order of integration, we see that

$$\iint_{\mathbb{R}^2_+} f(x+y)x^{p-1}y^{q-1}\,dx\,dy = \int_0^\infty f(t)\biggl(\int_0^t x^{p-1}(t-x)^{q-1}\,dx\biggr)\,dt.$$

To obtain the desired result, it remains to make the change of variable $x=ts$ in the inner integral.

5.4.4 In the conclusion of this section, we will, relying on the representation of the double integral as a repeated one, prove an inequality that plays an important role in mathematical physics. It concerns the domination of an integral of a certain power of function of class $C_0^1(\mathbb{R}^m)$ (i.e., a smooth compactly supported function) by the integral of the appropriate power of norm of its gradient. In the one-dimensional case, we obviously have $f(x)=\int_{-\infty}^x f'(t)\,dt=-\int_x^\infty f'(t)\,dt$, so

$$|f(x)|\leqslant \frac{1}{2}\int_{-\infty}^{\infty}|f'(t)|\,dt. \tag{1}$$

This estimate can be generalized for functions of several variables in the following way.

---

[6] Joseph Liouville (1809–1882)—French mathematician.

Theorem (The Gagliardo[7]–Nirenberg[8]–Sobolev[9] inequality) *Let* $1 \leqslant p < m$, $q = \frac{mp}{m-p}$ *and* $C = p\frac{m-1}{m-p}$. *Then, for every function* $f \in C_0^1(\mathbb{R}^m)$, *the inequality*

$$\left(\int_{\mathbb{R}^m} |f(x)|^q \, dx\right)^{\frac{1}{q}} \leqslant \frac{C}{2}\left(\int_{\mathbb{R}^m} \|\operatorname{grad} f(x)\|^p \, dx\right)^{\frac{1}{p}} \tag{2}$$

holds.

To begin with, we will establish a nice inequality, which strengthens (2) somewhat in the case $p = 1$.

Lemma *Let* $q = \frac{m}{m-1}$ *and let* $f \in C_0^1(\mathbb{R}^m)$. *Then*

$$\left(\int_{\mathbb{R}^m} |f(x)|^q \, dx\right)^{\frac{1}{q}} \leqslant \frac{1}{2}\left(\int_{\mathbb{R}^m} |f'_{x_1}(x)| \, dx \cdots \int_{\mathbb{R}^m} |f'_{x_m}(x)| \, dx\right)^{\frac{1}{m}}.$$

For $q = +\infty$ (i.e., in the case $m = 1$), the left-hand side should be understood as $\sup_{\mathbb{R}^m} |f|$, so the statement of the lemma coincides with the inequality (1).

Proof We will carry out the proof by induction on m. Since for $m = 1$ the desired result reduces to (1), it remains to prove the inductive step. Let $m > 1$. Assume that the statement of the lemma is true for functions of $m - 1$ variables. Writing the vector $x \in \mathbb{R}^m$ as (s, t) where $s \in \mathbb{R}^{m-1}$ and $t \in \mathbb{R}$, put

$$I_j(t) = \int_{\mathbb{R}^{m-1}} |f'_{x_j}(s, t)| \, ds \quad \text{for } j = 1, \ldots, m-1 \qquad \text{and}$$

$$I_m(s) = \int_{\mathbb{R}} |f'_{x_m}(s, t)| \, dt.$$

In addition to the exponent $q = \frac{m}{m-1}$, corresponding to the dimension m, we shall need the exponent $r = \frac{m-1}{m-2}$, corresponding to the dimension $m - 1$. By the induction assumption,

$$\left(\int_{\mathbb{R}^{m-1}} |f(s, t)|^r \, ds\right)^{\frac{1}{r}} \leqslant \frac{1}{2}\big(I_1(t) \cdots I_{m-1}(t)\big)^{\frac{1}{m-1}}. \tag{3}$$

Note also that $|f(s, t)| \leqslant \frac{1}{2} I_m(s)$ (this is nothing but the inequality (1)) and, therefore, $|f(s, t)|^q \leqslant 2^{1-q} |f(s, t)| \, I_m^{\frac{1}{m-1}}(s)$. The Hölder inequality with the exponent r

[7]Emilio Gagliardo (1930–2008)—Italian mathematician.

[8]Louis Nirenberg (born 1925)—American mathematician.

[9]Sergey L'vovich Sobolev (1908–1989)—Russian mathematician.

yields

$$\int_{\mathbb{R}^{m-1}} |f(s,t)|^q\,ds \leqslant 2^{1-q}\int_{\mathbb{R}^{m-1}} |f(s,t)| I_m^{\frac{1}{m-1}}(s)\,ds$$

$$\leqslant 2^{1-q}\Big(\int_{\mathbb{R}^{m-1}} |f(s,t)|^r\,ds\Big)^{\frac{1}{r}}\Big(\int_{\mathbb{R}^{m-1}} I_m(s)\,ds\Big)^{\frac{1}{m-1}}.$$

Taking the inequality (3) into account, we see that

$$\int_{\mathbb{R}^{m-1}} |f(s,t)|^q\,ds \leqslant 2^{-q}\big(I_1(t)\cdots I_{m-1}(t)\big)^{\frac{1}{m-1}}\cdot\Big(\int_{\mathbb{R}^{m-1}} I_m(s)\,ds\Big)^{\frac{1}{m-1}}.$$

Integrating this inequality with respect to t, we obtain

$$\int_{\mathbb{R}^m} |f(x)|^q\,dx \leqslant 2^{-q}\int_{\mathbb{R}}\big(I_1(t)\cdots I_{m-1}(t)\big)^{\frac{1}{m-1}}\,dt\cdot\Big(\int_{\mathbb{R}^{m-1}} I_m(s)\,ds\Big)^{\frac{1}{m-1}}.$$

Estimating the first integral on the right by Hölder's inequality for several functions (see Corollary 2 in Sect. 4.4.5 with $p_k = m-1$), we get the inequality

$$\int_{\mathbb{R}^m} |f(x)|^q\,dx \leqslant 2^{-q}\Big(\int_{\mathbb{R}} I_1(t)\,dt\cdots\int_{\mathbb{R}} I_{m-1}(t)\,dt\Big)^{\frac{1}{m-1}}dt$$

$$\cdot\Big(\int_{\mathbb{R}^{m-1}} I_m(s)\,ds\Big)^{\frac{1}{m-1}},$$

which is, obviously, equivalent to the one we set out to prove. □

Proof of the theorem For $p = 1$, the inequality (2) with the coefficient $C = 1$ follows from the lemma immediately because $|f'_{x_k}(x)| \leqslant \|\operatorname{grad} f(x)\|$ for all k and x.

Now let $p > 1$. Then $C > 1$, and an easy computation shows that $q = C\frac{m}{m-1} = (C-1)\frac{p}{p-1}$. Introduce the auxiliary function $\varphi = |f|^C$. Since $C > 1$, φ is smooth and $\|\operatorname{grad}\varphi\| = C|f|^{C-1}\|\operatorname{grad} f\|$. Applying the inequality (2) with $p = 1$ to φ, we obtain:

$$\Big(\int_{\mathbb{R}^m} \varphi^{\frac{m}{m-1}}(x)\,dx\Big)^{\frac{m-1}{m}} \leqslant \frac{1}{2}\int_{\mathbb{R}^m} \|\operatorname{grad}\varphi(x)\|\,dx,$$

i.e.,

$$\Big(\int_{\mathbb{R}^m} |f(x)|^q\,dx\Big)^{\frac{m-1}{m}} \leqslant \frac{C}{2}\int_{\mathbb{R}^m} |f(x)|^{C-1}\|\operatorname{grad} f(x)\|\,dx. \tag{4}$$

Estimating the last integral by Hölder's inequality with exponent p and taking into account that $(C-1)\frac{p}{p-1}=q$, we see that

$$\int_{\mathbb{R}^m} |f(x)|^{C-1} \|\text{grad}\, f(x)\|\, dx \leqslant \left(\int_{\mathbb{R}^m} |f(x)|^q\, dx\right)^{\frac{p-1}{p}} \left(\int_{\mathbb{R}^m} \|\text{grad}\, f(x)\|^p\, dx\right)^{\frac{1}{p}}.$$

Together with (4), this yields the desired result because $\frac{m-1}{m} - \frac{p-1}{p} = \frac{1}{p} - \frac{1}{m} = \frac{1}{q}$. □

EXERCISES

1. Let $E \subset \mathbb{R}_+$ be a measurable set. Prove that the "annulus" $A = \{(x,y) \in \mathbb{R}^2 \mid \sqrt{x^2+y^2} \in E\}$ is measurable and $\lambda_2(A) = 2\pi \int_E t\, dt$.

2. Assume that the set $E \subset \mathbb{R}^2$, contained in the half-plane $y > 0$, is measurable. Prove that the volume of the body $T = \{(x,y,z) \in \mathbb{R}^3 \mid (x, \sqrt{y^2+z^2}) \in E\}$, which is obtained by the revolution of E around the x-axis, equals $2\pi \iint_E y\, dx\, dy$.

3. Prove by induction that for every vector $a \in \mathbb{R}^m_+$, the volume of the simplex $S(a) = \{x \in \mathbb{R}^m_+ \mid \frac{x_1}{a_1} + \cdots + \frac{x_m}{a_m} \leqslant 1\}$ equals $\frac{a_1 \cdots a_m}{m!}$.

4. Prove that the volume of the regular m-dimensional simplex Σ with edges of unit length equals $\frac{\sqrt{m+1}}{m! 2^{m/2}}$. Find the ellipsoid E of maximal volume for Σ. Investigate the growth of the quantity $(\frac{\lambda_m(\Sigma)}{\lambda_m(E)})^{\frac{1}{m}}$ (volume ratio for Σ) as the dimension increases.

5. Let $1 \leqslant p < +\infty$, $V_p = \{(x_1, \ldots, x_m) \mid \sum_{k=1}^m |x_k|^p \leqslant 1\}$. Find the ellipsoid E_p of maximal volume for V_p. For which C does the inclusion $V_p \subset C\, E_p$ hold? Investigate the growth of the quantity $(\frac{\lambda_m(V_p)}{\lambda_m(E_p)})^{\frac{1}{m}}$ as the dimension increases. For which p is it bounded?

6. Let $A \subset \mathbb{R}^m$ and $B \subset \mathbb{R}^n$ be two convex origin-symmetric compact bodies, let $C \subset \mathbb{R}^{m+n}$ be the convex hull of the union $(A \times \{0\}) \cup (\{0\} \times B)$. Prove that

$$\lambda_{m+n}(C) = \frac{m!n!}{(m+n)!} \lambda_m(A) \lambda_n(B).$$

7. Let K be an arbitrary convex body in $\mathbb{R}^m$ and $V = \lambda_m(K)$. Prove that if the $(m-1)$-dimensional volume of the projection of K to every hyperplane is at least S, then $\text{diam}(K) \leqslant \frac{mV}{S}$.

8. Prove that a non-zero polynomial of several variables (either algebraic or trigonometric) takes non-zero values almost everywhere.

9. Let $E_1, \ldots, E_n \subset [0,1)$ and $S = \lambda(E_1) + \cdots + \lambda(E_n)$. Prove that there exist translations of the sets E_j modulo 1 (see Exercise 6 in Sect. 2.4) whose union covers $[0,1)$ almost entirely: the measure of the difference $[0,1) \setminus \bigcup_{j=1}^n \{x_j + E_j\}$ is less than e^{-S}. Generalize this statement to the multi-dimensional case.
Hint. Consider the integral

$$\int_0^1 \cdots \int_0^1 \left(\int_0^1 \chi_1(\{t-x_1\}) \cdots \chi_n(\{t-x_n\})\, dt\right) dx_1 \cdots dx_n,$$

where χ_j is the characteristic function of the set $[0, 1) \setminus E_j$ $(j = 1, \ldots, n)$.

10. Applying the method used in the proof of Theorem 5.4.4, prove the following generalization of Lemma 5.4.4:

$$\left(\int_{\mathbb{R}^m} |f(x)|^q\,dx\right)^{\frac{1}{q}} \leqslant \frac{C}{2}\left(\int_{\mathbb{R}^m} |f'_{x_1}(x)|^p\,dx \cdots \int_{\mathbb{R}^m} |f'_{x_m}(x)|^p\,dx\right)^{\frac{1}{mp}}.$$

11. Let f be a function that is summable on the square $(0, 1)^2$ and satisfies the condition $|\iint_{A\times B} f(x, y)\,dx\,dy| \leqslant 1$ for any measurable sets $A, B \subset (0, 1)$. Show that the integral $\iint_{(0,1)^2} |f(x, y)|\,dx\,dy$ can be arbitrarily large (one possible example is given in Exercise 9 of Sect. 10.2).

12. In three-dimensional space, consider the ball inscribed into a cube and the tetrahedron that is the convex hull of two non-coplanar diagonals of opposite faces of this cube (say, horizontal for the sake of definiteness). Prove that the ratio of the areas of the horizontal cross sections of the ball and the tetrahedron is constant and find the volume of the ball using the Cavalieri principle.[10]

13. Using the Cavalieri principle obtain the formula for the volume of a cone (see Example 1 in Sect. 5.4.2) without employing integration. *Hint*. Verify that the measure $E \mapsto \lambda_m(C_E)$, where $E \in \mathfrak{A}^{m-1}$ and $C_E = \{(t, ty) \mid t \in [0, 1],\ y \in E\}$, is translation invariant, so it is a multiple of the Lebesgue measure.

14. Let

$$E \subset \mathbb{R}^{m-2}, \qquad K_E = \left\{(tw, tx) \in \mathbb{R}^2 \times \mathbb{R}^{m-2} \mid 0 \leqslant t \leqslant 1,\ (w, x) \in S^1 \times E\right\}$$

be the cone with vertex at the origin and "cylindrical base" $S^1 \times E$. Using the Cavalieri principle, prove that the measure $E \mapsto \lambda_m(K_E)$ defined on $\mathfrak{A}^{m-2}$ is proportional to $\lambda_{m-2}(E)$.

Representing the polydisk $B^2 \times \cdots \times B^2$ (k factors) as the union of k congruent cones with cylindrical bases and the common vertex at the origin, find the proportionality coefficient and derive the formula $\lambda_m(K_E) = \frac{2\pi}{m}\lambda_{m-2}(E)$ for even m.

15. Taking the Cartesian product of the polydisk and $[-1, 1]$ and refining the argument from the previous exercise, prove that the formula $\lambda_m(K_E) = \frac{2\pi}{m}\lambda_{m-2}(E)$ obtained there remains valid for odd m.

16. Using the Cavalieri principle alone, derive the recurrent formula for the volume of the m-dimensional ball: $\alpha_m = \frac{2\pi}{m}\alpha_{m-2}$ $(m \geqslant 3)$. *Hint*. Use the results of the two previous exercises with $E = B^{m-2}$.

17. Prove that the interval $[0, 1]$ and the square $[0, 1]^2$ endowed with the corresponding Lebesgue measures are isomorphic as measure spaces (the definition of an isomorphism of measure spaces was given in Exercise 11, Sect. 4.10). Generalizing this result, prove that the measure spaces $\mathbb{R}^m$ and $\mathbb{R}^n$ with the

---

[10] This problem was proposed by A. Andzans in a slightly different formulation (see "Kvant", 1990, No. 3, p. 27, Problem M1211). The authors are grateful to A.N. Petrov for drawing their attention to this result.

corresponding Lebesgue measures are isomorphic. *Hint*. Using the binary representations of numbers $x \in [0, 1]$, consider the mapping

$$x = \sum_{k=1}^{\infty} \varepsilon_k 2^{-k} \longmapsto \Phi(x) = \left(\sum_{k=1}^{\infty} \varepsilon_{2k-1} 2^{-k}, \sum_{k=1}^{\infty} \varepsilon_{2k} 2^{-k} \right) \in [0, 1]^2.$$

5.5 ⋆An Alternative Approach to the Definition of the Product Measure and the Integral

In this section, we shall give an alternative proof of Theorem 5.1.2 on the countable additivity of the product measure that does not use the notion of the integral. This allows us to define the integral of a non-negative function as the measure of its subgraph. As we shall see, this approach to the construction of the integral is equivalent to the one in Chap. 4.

5.5.1 As in Sect. 5.1, let $(X, \mathfrak{A}, \mu)$ and $(Y, \mathfrak{B}, \nu)$ be two measure spaces with σ-finite measures, let

$$\mathscr{P} = \{ A \times B \mid A \in \mathfrak{A},\ \mu(A) < +\infty,\ B \in \mathfrak{B},\ \nu(B) < +\infty \}$$

be the semiring of measurable rectangles, and let m_0 be the product of the measures μ and ν, defined on $\mathscr{P}$ by

$$m_0(A \times B) = \mu(A)\, \nu(B). \tag{1}$$

It was shown in Theorem 1.2.4 that m_0 is a volume. We now want to prove its countable additivity.

Assume first that the measures μ and ν are finite. Then $X \times Y \in \mathscr{P}$ and, by the remark in Sect. 1.2.3, we may assume that the volume m_0 has been extended to the algebra $\mathfrak{C}$ of all sets representable as finite unions of measurable rectangles. We will use the same notation m_0 for this extended volume.

As a preliminary step, let us prove the following lemma, which is a substantially weakened version of Theorem 5.2.2. We shall need it for estimating the volumes of sets from the algebra $\mathfrak{C}$. The notions of the cross section C_x and the canonical projection $P_1(C)$ used in the lemma are defined in Sects. 5.2.1 and 5.2.2.

Lemma *If C is a set from the algebra $\mathfrak{C}$ such that $\nu(C_x) \leqslant \delta$ for all $x \in X$, then $m_0(C) \leqslant \delta \cdot \mu(P_1(C))$. In particular, $m_0(C) \leqslant \delta \cdot \mu(X)$.*

Proof By the definition of the algebra $\mathfrak{C}$, all its elements are representable as unions of finitely many measurable rectangles. We will carry out the proof by induction on the number of rectangles comprising the set C. The induction base (C is a measurable rectangle) is obvious. Now we will assume that the statement of the lemma holds for all unions of at most $n-1$ measurable rectangles and will prove it for the set $C=\bigcup_{k=1}^{n}(A_k\times B_k)$, where $A_k\in\mathfrak{A}$, $B_k\in\mathfrak{B}$.

Put $U=\bigcup_{k=1}^{n-1}A_k$ and split the set C into three parts D, E and F so that $P_1(D)=A_n\setminus U$, $P_1(E)=U\setminus A_n$ and $P_1(F)=U\cap A_n$ (the sets D, E and F are disjoint because their projections to X do not overlap). Since D, E and F are subsets of C, each of their cross sections is contained in the corresponding cross section of the set C. Therefore, $\nu(D_x),\nu(E_x),\nu(F_x)\leqslant\delta$ for all x in X. To apply the induction assumption to the sets D, E and F, let us check that each of them is a union of at most $n-1$ measurable rectangles. This is obvious for the sets D and E because

$$D=(A_n\setminus U)\times B_n \quad\text{and}\quad E=\bigcup_{k=1}^{n-1}(A_k\setminus A_n)\times B_k.$$

It follows directly from the definition of the set F that, if $x\in U\cap A_n$, then

$$F_x=C_x=\left(\bigcup_{k=1}^{n-1}(A_k\cap A_n)\times B_k\right)_x\cup B_n=\left(\bigcup_{k=1}^{n-1}(A_k\cap A_n)\times (B_k\cup B_n)\right)_x,$$

and, therefore, $F=\bigcup_{k=1}^{n-1}(A_k\cap A_n)\times(B_k\cup B_n)$. One can see from this that the induction assumption can also be applied to F. Using the additivity of m_0, we obtain the desired inequality:

$$\begin{aligned}m_0(C)&=m_0(D)+m_0(E)+m_0(F)\leqslant\delta\mu\big(P_1(D)\big)+\delta\mu\big(P_1(E)\big)+\delta\mu\big(P_1(F)\big)\\&=\delta\big(\mu(U\setminus A_n)+\mu(U\cap A_n)+\mu(A_n\setminus U)\big)=\delta\mu(U\cup A_n)\\&=\delta\mu\big(P_1(C)\big).\end{aligned}$$ □

It can be seen from the proof that we have used only the finite additivity of the measures μ and ν, not the countable additivity, so the lemma is valid not only for measures, but also for volumes.

Now we can prove that the product of measures is a measure.

Theorem *The volume m_0 is countably additive.*

Proof Assume first that the measures μ and ν are normalized, i.e., $\mu(X)=\nu(Y)=1$, and that the volume m_0 has already been extended to the algebra $\mathfrak{C}$ consisting of all sets representable as finite unions of measurable rectangles.

Let us prove that this volume is continuous from above on the empty set, which, by Theorem 1.3.4, implies its countable additivity. So, let the sets C_n from $\mathfrak{C}$ satisfy the conditions

$$C_n\supset C_{n+1}\quad\text{for } n\in\mathbb{N},\quad \bigcap_{n=1}^{\infty}C_n=\varnothing.$$

We have to prove that $m_0(C_n)\to 0$ as $n\to\infty$.

Assume the contrary. Then for some $\delta > 0$,

$$m_0(C_n) > \delta \quad \text{for all } n.$$

Consider the set E_n consisting of those points x for which the cross section $(C_n)_x$ has "large" measure. More precisely, put

$$E_n = \left\{ x \in X \,\middle|\, \nu\big((C_n)_x\big) > \frac{\delta}{2} \right\}.$$

It is easy to check that the function $x \mapsto \nu((C_n)_x)$ is simple and the set E_n is measurable with respect to $\mathfrak{A}$. Clearly, $C_n \subset (E_n \times Y) \cup C_n'$ where $C_n' = C_n \setminus (E_n \times Y)$. Therefore,

$$\delta < m_0(C_n) \leqslant m_0(E_n \times Y) + m_0(C_n').$$

Also, $\nu((C_n')_x) \leqslant \frac{\delta}{2}$. Using the lemma proved above to estimate $m_0(C_n')$, we obtain

$$\delta < m_0(C_n) \leqslant m_0(E_n \times Y) + \frac{\delta}{2} = \mu(E_n) + \frac{\delta}{2}$$

(recall that $\mu(X) = \nu(Y) = 1$). Therefore, $\mu(E_n) > \frac{\delta}{2}$. Thus, the measures of the sets E_n do not tend to zero. Since the sets E_n form a decreasing sequence, their intersection cannot be empty. Let $x_0 \in \bigcap_{n=1}^{\infty} E_n$. Then $\nu((C_n)_{x_0}) > \frac{\delta}{2}$ for each n. Since the sets $(C_n)_{x_0}$ form a decreasing sequence, their intersection is not empty. Let $y_0 \in \bigcap_{n=1}^{\infty} (C_n)_{x_0}$. Then the point (x_0, y_0) belongs to each of the sets C_n, which is impossible by our assumptions, and we get the contradiction sought.

Once we have established the statement of the theorem for normalized measures, we immediately get it for arbitrary finite measures as well. Consider now the case when μ and ν are arbitrary σ-finite measures. Assume that $C = A \times B \in \mathscr{P}$, $C = \bigcup_{n=1}^{\infty} C_n$, and the sets C_n from $\mathscr{P}$ are pairwise disjoint. Then $\mu(A) < +\infty$ and $\nu(B) < +\infty$ by the definition of the semiring $\mathscr{P}$. Therefore we can replace X by A, and Y by B, consider the restriction of m_0 to the semiring of those measurable rectangles that are contained in $A \times B$, and then just refer to the already considered case of finite measures. □

Now we can justifiably define (as in Sect. 5.1.1) the product measure $\mu \times \nu$ as the standard extension of the measure m_0.

5.5.2 Let us sketch an alternative approach to the definition of the integral of a non-negative measurable function (for measurable functions of arbitrary sign, we will preserve the definition from Sect. 4.1.3). Let us remind the reader that, as it has been proved in Lemma 5.2.3 (without using the integral), the subgraph of a non-negative measurable function (in the wide sense) is measurable.

Definition Let $(X, \mathfrak{A}, \mu)$ be a measure space with σ-finite measure, let $m = \mu \times \lambda$ where λ is the one-dimensional Lebesgue measure. The integral of a non-negative measurable function f over a set $A \in \mathfrak{A}$ is the measure of its subgraph $\mathscr{P}_f(A)$ over A.

To distinguish this integral from the integral introduced in Chap. 4, we will denote it by the symbol $I(f, A)$. Thus, I is a functional (with values in $[0, +\infty]$) defined on the set $\mathcal{K} \times \mathfrak{A}$, where $\mathcal{K}$ is the cone of non-negative functions that are measurable on X.

It is easy to check that the functional I satisfies the conditions (I)–(IV) from Sect. 4.2.5. Indeed, condition (I) is obvious. Condition (II) follows from the identity $\mathscr{P}_f(A \vee B) = \mathscr{P}_f(A) \vee \mathscr{P}_f(B)$ and the additivity of the measure m.

If $f(x) = c$ for all x in A, then $\mathscr{P}_f(A) = A \times [0, c]$ and, therefore, $I(f, A) = c\mu(A) = c\,I(\mathbb{I}, A)$, which means that condition (III) is satisfied.

Finally, condition (IV) is also satisfied. Indeed, if $\{f_n\}_{n\geqslant 1}$ is an increasing sequence of non-negative measurable functions that converges to f pointwise, then the inclusions

$$\mathscr{P}_f \setminus \Gamma_f \subset \bigcup_{n\geqslant 1} \mathscr{P}_{f_n} \subset \mathscr{P}_f$$

hold. In addition, we have $m(\Gamma_f) = 0$ and $\mathscr{P}_{f_n} \subset \mathscr{P}_{f_{n+1}}$. Therefore, $m(\mathscr{P}_{f_n}) \to m(\mathscr{P}_f)$ by the continuity from below of the measure. This means that $I(f_n, X) \to I(f, X)$, which coincides with the statement of condition (IV).

As we have already pointed out in Sect. 4.2.5, all other properties of the integral obtained in Sect. 4.2 follow from (I)–(IV).

EXERCISES

1. Let $(X, \mathfrak{A}, \mu)$ be a measure space with σ-finite complete measure. We will call a non-negative function f measurable if its subgraph is measurable with respect to the algebra $\mathfrak{A} \otimes \mathfrak{A}^1$. Prove that this definition is equivalent to the definition of measurability using Lebesgue sets.

5.6 *Infinite Products of Measures

5.6.1 Now we will define the product measure of an infinite sequence of measures. Let us remind the reader that the product measure operation is associative for finite families of measures (see Sect. 5.1.3), so, in particular, $\mu_1 \times \mu_2 \times \cdots \times \mu_n = \mu_1 \times (\mu_2 \times \cdots \times \mu_n)$.

Let $(X_n, \mathfrak{A}_n, \mu_n)$ $(n \in \mathbb{N})$ be measure spaces with normalized measures, i.e., with measures satisfying the condition $\mu_n(X_n) = 1$ (such measures are also called probability measures). Put

$$Y = \prod_{k=1}^{\infty} X_k, \qquad Y_n = \prod_{k=n+1}^{\infty} X_k \quad (n = 1, 2, \ldots).$$

If all the sets X_k coincide with X, we will denote their product by the symbol $X^{\mathbb{N}}$.

A set $A \subset Y$ will be called a *cylindrical* subset of rank n if it is representable as $A = B \times Y_n$, where the set B (which we will call the *base* of the set A) belongs to the σ-algebra on which the product measure $\mu_1 \times \cdots \times \mu_n$ is defined. Obviously, every cylindrical set of rank n with base B is simultaneously a cylindrical set of rank $n+1$ with base $B \times X_{n+1}$.

We leave it to the reader to check that the cylindrical sets of all possible ranks form an algebra. For every cylindrical set A of rank n with base B, put

$$\nu(A) = (\mu_1 \times \cdots \times \mu_n)(B).$$

This definition is self-consistent because

$$(\mu_1 \times \cdots \times \mu_n)(B) = (\mu_1 \times \cdots \times \mu_n \times \mu_{n+1})(B \times X_{n+1}) = \cdots.$$

Let us verify that the function ν is additive, i.e., that it is a volume. Indeed, let A and A' be cylindrical sets. Obviously, without loss of generality, we may assume that they have the same rank. Then $A = B \times Y_n$ and $A' = B' \times Y_n$. If A and A' are disjoint, then so are their bases, and, since $A \cup A' = (B \cup B') \times Y_n$, we have

$$\begin{aligned} \nu\big(A \cup A'\big) &= (\mu_1 \times \cdots \times \mu_n)\big(B \cup B'\big) \\ &= (\mu_1 \times \cdots \times \mu_n)(B) + (\mu_1 \times \cdots \times \mu_n)\big(B'\big) = \nu(A) + \nu\big(A'\big). \end{aligned}$$

We will call the volume ν the *product* of the measures $\mu_1, \mu_2, \ldots$.

Note also that, for almost all $x_1 \in X_1$, the cross sections of the cylindrical set $A = B \times Y_n$ of rank n are cylindrical sets (of rank $n-1$) in Y_1. This follows from the identity

$$\begin{aligned} A_{x_1} &= \big\{(x_2, \ldots, x_n, \ldots) \in Y_1 \mid (x_1, x_2, \ldots, x_n, \ldots) \in A\big\} \\ &= \big\{(x_2, \ldots, x_n) \in X_2 \times \cdots \times X_n \mid (x_1, x_2, \ldots, x_n) \in B\big\} \times Y_n = B_{x_1} \times Y_n \end{aligned}$$

and Theorem 5.2.2, which guarantees the measurability of B_{x_1} for almost all $x_1 \in X_1$.

5.6.2 Let us prove the countable additivity of the volume ν following the idea used in the proof of Theorem 5.5.1.

Theorem *The infinite product of measures is a measure.*

Proof Since the collection of all cylindrical sets is an algebra and ν is a finite volume, to prove the countable additivity of the latter, it suffices to check that it is continuous from above on the empty set (see Theorem 1.3.4). Let A^k be cylindrical sets, $A^k \supset A^{k+1}$, $\bigcap_{k=1}^{\infty} A^k = \varnothing$. We will prove that $\nu(A^k) \underset{k\to\infty}{\longrightarrow} 0$. Arguing by contradiction, assume that for some $\delta > 0$, we have

$$\nu\big(A^k\big) \geqslant \delta > 0 \quad \text{for all } k. \tag{1}$$

We shall derive from this that there exists a $c_1 \in X_1$ such that for the cross sections $A^k_{c_1}$, the inequalities

$$\nu_1\bigl(A^k_{c_1}\bigr) \geqslant \frac{\delta}{2} \quad \text{hold for all } k, \tag{1'}$$

where ν_1 is the product of the measures $\mu_2, \mu_3, \ldots$. Let $A^k = B^k \times Y_{n_k}$ be a cylindrical set of rank n_k and let $\lambda = \mu_2 \times \cdots \times \mu_{n_k}$. Put

$$E_k = \left\{x_1 \in X_1 \;\middle|\; \nu_1\bigl(A^k_{x_1}\bigr) = \lambda\bigl(B^k_{x_1}\bigr) \geqslant \frac{\delta}{2}\right\}.$$

Then

$$\begin{aligned}\delta \leqslant \nu\bigl(A^k\bigr) &= (\mu_1 \times \mu_2 \times \cdots \times \mu_{n_k})\bigl(B^k\bigr) = (\mu_1 \times \lambda)\bigl(B^k\bigr)\\ &= \int_{E_k} \lambda\bigl(B^k_{x_1}\bigr)\,d\mu_1(x_1) + \int_{X_1\setminus E_k} \lambda\bigl(B^k_{x_1}\bigr)\,d\mu_1(x_1) \leqslant \mu_1(E_k) + \frac{\delta}{2}\end{aligned}$$

and, therefore, $\mu_1(E_k) \geqslant \frac{\delta}{2}$. Since the sets E_k decrease, we have $\mu_1(\bigcap_{k=1}^{\infty} E_k) > 0$. Obviously, the inequalities (1′) hold for all points $c_1 \in \bigcap_{k=1}^{\infty} E_k$ for which the cross sections are measurable.

Replacing (1) with (1′), and ν with ν_1 and repeating the above argument, we will find a point $c_2 \in X_2$ such that

$$\nu_2\bigl(A^k_{(c_1,c_2)}\bigr) \geqslant \frac{\delta}{4} \quad \text{for all } k,$$

where ν_2 is the product of the measures $\mu_3, \mu_4, \ldots$.

Continuing this process by induction, we will get a sequence of points $c_j \in X_j$ such that for all j and k, the cross sections $A^k_{(c_1,\ldots,c_j)}$ have positive volumes (products of measures $\mu_{j+1}, \mu_{j+2}, \ldots$) and, thereby, are non-empty. This is the crux of the argument: contrary to our assumptions, the point $c = (c_1, c_2, \ldots) \in X$ belongs to all sets A^k. Indeed, for $j = n_k$, the statement that the cross section $A^k_{(c_1,\ldots,c_{n_k})}$ is non-empty means that it coincides with Y_{n_k}. Therefore, A^k contains all points of the form $(c_1, \ldots, c_{n_k}, x_{n_k+1}, x_{n_k+2}, \ldots)$. In particular, A^k contains the point c. Since k is arbitrary, we obtain the sought contradiction. □

The infinite product of measures $\mu_1, \mu_2, \ldots$ we have constructed is defined on the algebra of cylindrical sets, which is usually not a σ-algebra. Extending it in the standard way (see Sect. 1.4), we obtain a measure defined on a σ-algebra. We will call this extension the product measure of the measures $\mu_1, \mu_2, \ldots$ and denote it by the symbol $\mu_1 \times \mu_2 \times \cdots$.

In conclusion, note that some properties of the infinite product of measures may seem unusual. For instance, a set $A \subset X^{\mathbb{N}}$ may have zero measure even though for each its points $x = (x_1, x_2, \ldots)$, all "cross sections"

$$A_n(x) = \bigl\{y \mid (x_1, \ldots, x_{n-1}, y, x_{n+1}, \ldots) \in A\bigr\} \quad (n \in \mathbb{N})$$

coincide with X (A can be taken to be the set of all bounded sequences, see Exercise 3).

EXERCISES

1. Prove that the closed interval $[0, 1]$ with the Lebesgue measure λ is isomorphic, as a measure space (see the definition in Exercise 11, Sect. 5.4), to the measure space $(X, \mathfrak{A}, \mu)$, where $X = [0, 1]^{\mathbb{N}}$, and $\mu = \lambda \times \lambda \times \cdots$.
2. Give an example of a sequence of non-negative functions $f_n \in \mathscr{L}(X, \mu)$ such that their integrals are bounded and that

$$\text{for every subsequence } \{n_k\}, \quad \sup_k \left| f_{n_k}(x) \right| = +\infty \text{ almost everywhere.}$$

Hint. Consider the measure μ from Exercise 1 and the functions $f_n(x) = \frac{1}{\sqrt{x_n}}$, where $x = (x_1, x_2, \ldots) \in (0, 1)^{\mathbb{N}}$.
3. Let γ be a probability measure on $\mathbb{R}$ with density $\frac{1}{\sqrt{\pi}} e^{-t^2}$, $\mu = \gamma \times \gamma \times \cdots$. Prove that every infinite-dimensional cube and the set of all bounded sequences have zero measure but, for sufficiently large $a > 0$, the μ-measure of the set

$$P(a) = \left\{ (x_1, x_2, \ldots) \mid |x_n| \leqslant a\sqrt{\ln(n+1)},\ n \in \mathbb{N} \right\}$$

is arbitrarily close to one.
4. Let μ be the measure on $\mathbb{R}^{\mathbb{N}}$ defined in the previous exercise. Put

$$E_a = \left\{ x = (x_1, x_2, \ldots) \in \mathbb{R}^{\mathbb{N}} \,\middle|\, \overline{\lim_{n\to\infty}} \frac{|x_n|}{\sqrt{\ln n}} < a \right\}.$$

Prove that $\mu(E_a) = 0$ for $a = 1$ and $\mu(E_a) = 1$ when $a > 1$. Derive from this that $\mu(H) = 1$ where H is the set of all points $x \in \mathbb{R}^{\mathbb{N}}$ such that $\overline{\lim}_{n\to\infty} \frac{|x_n|}{\sqrt{\ln n}} = 1$.

Chapter 6
Change of Variables in an Integral

6.1 Integration over a Weighted Image of a Measure

6.1.1 Our main goal in this chapter is to learn how to change variables in an integral with respect to Lebesgue measure. As often happens, it is useful to begin with a more general question: is it possible to use a "parametrization" $\Phi : X \to Y$ of a set Y to reduce the integration with respect to a measure given on Y to the integration with respect to a measure given on X? More precisely, given measure spaces $(X, \mathfrak{A}, \mu)$ and $(Y, \mathfrak{B}, \nu)$, a map $\Phi : X \to Y$, and a function f defined on Y, it is extremely important to know conditions under which we can establish a relation between the integral of f with respect to ν and the integral of $f \circ \Phi$ with respect to μ. Of course, to make it possible, we must assume that the measures μ and ν are somehow compatible. We describe this compatibility by introducing the notion of a weighted image of a measure.

Definition Let $(X, \mathfrak{A}, \mu)$ be a measure space, let $\mathfrak{B}$ be an arbitrary σ-algebra of subsets of Y, and let $\Phi : X \to Y$ be a mapping satisfying the condition

$$\Phi^{-1}(B) \in \mathfrak{A} \quad \text{for every set } B \text{ in } \mathfrak{B}.$$

For a non-negative measurable function ω on X, we define the function $\nu : \mathfrak{B} \to \overline{\mathbb{R}}$ as follows:

$$\nu(B) = \int_{\Phi^{-1}(B)} \omega(x)\, d\mu(x) \quad (B \in \mathfrak{B}). \tag{1}$$

Obviously, ν is a measure on $\mathfrak{B}$. We call it a *weighted image* (more precisely, the *ω-weighted Φ-image*) of μ. We call the function ω a *weight* or a *weight function.*

We note that here we do not assume that the map Φ is one-to-one or surjective.

The following theorem demonstrates a connection between the integrals with respect to the measures ν and μ.

B. Makarov, A. Podkorytov, *Real Analysis: Measures, Integrals and Applications*, Universitext, DOI 10.1007/978-1-4471-5122-7_6, © Springer-Verlag London 2013

Theorem *Let ν be an ω-weighted image of a measure μ under a map $\Phi : X \to Y$. Then, for every non-negative measurable function f on Y, the composition $f \circ \Phi$ is also measurable, and the following holds:*

$$\int_Y f(y)\,d\nu(y) = \int_X f\big(\Phi(x)\big)\omega(x)\,d\mu(x). \tag{2}$$

The above relation is also valid for every summable function f on Y.

Proof The fact that the composition $g = f \circ \Phi$ is measurable follows from the definition of a weighted image of a measure. Indeed, $X(g < a) = \Phi^{-1}(Y(f < a)) \in \mathfrak{A}$ since the inequality $g(x) = f(\Phi(x)) < a$ is equivalent to the inclusion $\Phi(x) \in Y(f < a)$ for every real a.

We verify Eq. (2) by successively complicating the function f. If $f = \chi_B$ is the characteristic function of B, $B \in \mathfrak{B}$, then

$$(f \circ \Phi)(x) = \begin{cases} 1 & \text{if } \Phi(x) \in B, \\ 0 & \text{if } \Phi(x) \notin B \end{cases}$$

$$= \begin{cases} 1 & \text{if } x \in \Phi^{-1}(B), \\ 0 & \text{if } x \notin \Phi^{-1}(B) \end{cases} = \chi_{\Phi^{-1}(B)}(x).$$

Thus, $f \circ \Phi = \chi_{\Phi^{-1}(B)}$. In this case, Eq. (2) follows directly from the definition of ν. For a non-negative simple function f, Eq. (2) follows from the linearity of the integral.

In the case where f is an arbitrary non-negative measurable function, we consider an increasing sequence of non-negative simple functions f_n that converges pointwise to f. Then

$$\int_Y f_n(y)\,d\nu(y) = \int_X f_n\big(\Phi(x)\big)\omega(x)\,d\mu(x).$$

Passing to the limit (this is possible by Levi's theorem), we obtain Eq. (2), which completes the proof of the theorem for $f \geqslant 0$.

As we proved, the relation

$$\int_Y \big|f(y)\big|\,d\nu(y) = \int_X \big|f\big(\Phi(x)\big)\big|\,\omega(x)\,d\mu(x)$$

is valid for every measurable function f on Y. Therefore, the functions f and $(f \circ \Phi)\,\omega$ are simultaneously summable with respect to the measures ν and μ, respectively. If f is summable, we write Eq. (2) for the functions $f_+ = \max\{0, f\}$ and $f_- = \max\{0, -f\}$. Subtracting the equations obtained, we see that Eq. (2) is also valid for a real function f. The complex case is obvious. □

Equation (2) can be represented formally in a more general form.

Corollary *Let $B \in \mathfrak{B}$. Then*

$$\int_B f(y)\,d\nu(y) = \int_{\Phi^{-1}(B)} f\big(\Phi(x)\big)\omega(x)\,d\mu(x).$$

For the proof, it is sufficient to apply the theorem to the function $f \cdot \chi_B$.

6.1.2 We consider two important specific cases of a weighted image of a measure.

First, we consider the case where $\omega \equiv 1$. Then Eq. (1) takes the form $\nu(B) = \mu(\Phi^{-1}(B))$. The measure ν is called the Φ-*image* of μ and is denoted by $\Phi(\mu)$. For more details concerning integration over the image of a measure in the case where $Y = \mathbb{R}$, see Sect. 6.4.

The second case is obtained by putting $Y = X$, $\mathfrak{B} = \mathfrak{A}$ and $\Phi = \mathrm{Id}$. Now, Eq. (1) takes the form

$$\nu(B) = \int_B \omega\,d\mu \quad (B \in \mathfrak{A}), \tag{1'}$$

and, by (2), we have

$$\int_X f(x)\,d\nu(x) = \int_X f(x)\omega(x)\,d\mu(x) \tag{2'}$$

for every non-negative function.

We already know this result (see Sect. 4.5.3). In this specific case, we called the function ω the density of the measure ν with respect to μ. Equation (2′) suggests the following symbolic notation for this situation: $d\nu = \omega\,d\mu$.

From Theorem 4.5.4, it follows that the density of a measure ν is determined uniquely up to equivalence if the measure is finite.

The same is true for a σ-finite measure (see Exercise 1, Sect. 4.5). Using the notion of image of a measure, we can say that the ω-weighted Φ-image of μ is the Φ-image of the measure having density ω with respect to μ: $\nu = \Phi(\mu_1)$, where $d\mu_1 = \omega\,d\mu$.

To make Eq. (2′) easy-to-use, it is desirable to have convenient criteria for ω to be the density of a given measure with respect to another one. Now, we establish one such simple and important criterion.

Theorem *Let μ and ν be measures defined on a σ-algebra $\mathfrak{A}$ of subsets of a set X. In order that a non-negative function ω be the density of ν with respect to μ it is necessary and sufficient that the following two-sided estimate*[1] *be valid for every set A in $\mathfrak{A}$:*

$$\mu(A)\inf_A \omega \leqslant \nu(A) \leqslant \mu(A)\sup_A \omega.$$

[1] As usual, we assume that the products $0 \cdot (+\infty)$ and $(+\infty) \cdot 0$ are zero.

Proof The necessity is obvious, so we proceed to prove sufficiency, i.e., to prove Eq. (1′). We may assume that $\omega > 0$ on B since we obviously have $\nu(e) = 0 = \int_e \omega\, d\mu$ for $e = \{x \in B \mid \omega(x) = 0\}$. Assuming that the function ω is positive, we fix an arbitrary number q in the interval $(0, 1)$ and consider the sets

$$B_j = \{x \in B \mid q^j \leqslant \omega(x) < q^{j-1}\} \quad (j \in \mathbb{Z}).$$

These sets are measurable and form a partition of B. From the two-sided estimate, it follows immediately that

$$q^j \mu(B_j) \leqslant \nu(B_j) \leqslant q^{j-1} \mu(B_j).$$

Similar inequalities,

$$q^j \mu(B_j) \leqslant \int_{B_j} \omega(x)\, d\mu(x) \leqslant q^{j-1} \mu(B_j),$$

are valid for the integrals over the sets B_j.

Consequently,

$$q \int_B \omega(x)\, d\mu(x) \leqslant \sum_j q^j \mu(B_j) \leqslant \nu(B) \leqslant \frac{1}{q} \sum_j q^j \mu(B_j) \leqslant \frac{1}{q} \int_B \omega(x)\, d\mu(x).$$

Thus,

$$q \int_B \omega(x)\, d\mu(x) \leqslant \nu(B) \leqslant \frac{1}{q} \int_B \omega(x)\, d\mu(x)$$

for every q in the interval $(0, 1)$. Passing to the limit as $q \to 1$, we obtain the required relation. □

6.1.3 We give an example showing how to use the measure conservation condition. Let $(X, \mathfrak{A}, \mu)$ be a finite measure space, and let $T : X \to X$ be a measure-preserving map. Then the following theorem of Poincaré[2] holds.

Theorem (Poincaré recurrence theorem) *Let $\mu(X) < \infty$, and let $T : X \to X$ be a measure-preserving map. Under the map T, almost every point of an arbitrary measurable set $A \subset X$ returns to A infinitely many times, i.e., for almost all x in A, we have $T^n(x) \in A$ for infinitely many n.*

Proof First, we verify that almost every point of A returns to A at least once, i.e., that, for almost every point x of A, there exists a power T^n of T such that $T^n(x) \in A$. Indeed, the points whose T-images do not belong to A form the set $A \cap T^{-1}(X \setminus A)$. Similarly, the points whose images do not belong to A after n-fold

[2]Jules Henri Poincaré (1854–1912)—French mathematician.

action of T form the set $A \cap T^{-n}(X \setminus A)$. Therefore, the points that never return to A form the set

$$B = A \cap T^{-1}(X \setminus A) \cap \cdots \cap T^{-n}(X \setminus A) \cap \cdots.$$

The sets B, $T^{-1}(B)$, $T^{-2}(B), \ldots$ are disjoint. Indeed, if $x_0 \in T^{-k}(B)$ and $x_0 \in T^{-(k+l)}(B)$ $(l > 0)$, then, by the definition of a preimage, we have $y_0 = T^k(x_0) \in B$, and so $T^{k+l}(x_0) = T^l y_0 \in B$. This means that the point y_0 of B returns to B, a contradiction. Since the sets B, $T^{-1}(B)$, $T^{-2}(B), \ldots$ are disjoint, have that same measure, and $\mu(X) < \infty$, we obtain $\mu(B) = 0$. Thus, all points of A except those of the set B of measure zero return to A.

Applying this result to the maps $T^2, T^3, \ldots$ and using the fact that the union of a sequence of sets of measure zero has measure zero, we see that, for almost every point of A, there exist arbitrarily large powers of T that return the point to A. The theorem is proved. □

EXERCISES

1. Let ν' be the image and ν be the ω-weighted image of a measure μ under a bijective map Φ. Prove that $d\nu = \omega \circ \Phi^{-1}\, d\nu'$.
2. Define the map $\Phi : [0, 1) \to [0, 1) \times [0, 1)$ as follows: if the binary expansion of x has the form $x = 0, \alpha_1\alpha_2\alpha_3 \ldots$, then $\Phi(x) = (y_1, y_2)$, where $y_1 = 0, \alpha_1\alpha_3 \ldots$, and $y_2 = 0, \alpha_2\alpha_4 \ldots$ (we arbitrarily fix one of the binary expansions if x has more than one such expansion). Prove that the set $A \subset [0, 1) \times [0, 1)$ is Lebesgue measurable if and only if its preimage $\Phi^{-1}(A)$ is Lebesgue measurable. Find the Φ-image of Lebesgue measure.
3. Let λ be Lebesgue measure on $[0, 1)$, and let $\{x\}$ be the fractional part of x. Consider the map $\varphi(x) = \{\frac{1}{x}\}$ from $[0, 1)$ to itself (by definition, $\varphi(0) = 0$). Prove that $\omega(x) = \sum_{k=0}^{\infty} \frac{1}{(k+x)^2}$ is the density of the measure $\varphi(\lambda)$, i.e.,

$$\lambda\bigl(\varphi^{-1}(A)\bigr) = \int_A \omega(x)\, dx$$

for every measurable set A lying in $[0, 1)$. We note that by formula (9) from Sect. 7.2.6, we have $\omega(x) = (\ln \Gamma(x))''$.
4. Prove that the measure defined on $(0, 1)$ and having density $\frac{1}{1+x}$ with respect to the Lebesgue measure is invariant under the map $\varphi(x) = \{\frac{1}{x}\}$, i.e.,

$$\int_{\varphi^{-1}(A)} \frac{dx}{1+x} = \int_A \frac{dx}{1+x} \quad \bigl(A \subset (0, 1)\bigr).$$

5. Let φ and ψ be non-decreasing functions bounded on $\mathbb{R}$, and let

$$g(x) = \int_{\mathbb{R}} \varphi(x - t)\, d\psi(t) \quad (x \in \mathbb{R}).$$

Prove that the measure μ_g is the image of the measure $\mu_\varphi \times \mu_\psi$ under the map $(x, y) \mapsto x + y$ and $\mu_g(A) = \int_{\mathbb{R}} \mu_\varphi(-t + A)\, d\psi(t)$ for every Borel set A. Prove that the function g is continuous if at least one of the functions φ or ψ is continuous.

6. Prove that the function g from the previous exercise is strictly increasing on $[0, 2]$ if $\varphi = \psi$ is the Cantor function (from the left and from the right of $[0, 1]$ the values of φ are equal to 0 and 1, respectively). *Hint*. On every interval of the form $\Delta_k = [2t_k, 2t_k + 2 \cdot 3^{-n}]$, where $t_k = k \cdot 3^{-n}$ ($n \in \mathbb{N}$, $k = 0, 1, \ldots, 3^n - 1$), the increment of g is positive since the strip $\{(x, y) \in \mathbb{R}^2 \mid x + y \in \Delta_k\}$ contains a square whose sides are segments of rank n arising in the construction of the Cantor set.

7. Prove that the function g in Exercise 6 is not absolutely continuous. *Hint*. Verify that, for each n, at least half of the measure μ_g is concentrated on the intervals Δ_k for which the ternary expansion of t_k contains at least $n/2$ ones; prove that the total length of these intervals is arbitrarily small for large n.

6.2 Change of Variable in a Multiple Integral

We want to concretize the general scheme developed in Sect. 6.1 and find a relation between the integrals over open subsets $\mathcal{O}$ and $\mathcal{O}'$ of the space $\mathbb{R}^m$ in the case where the first set is mapped onto the second one by a diffeomorphism. In this section, by measurable sets we mean Lebesgue measurable sets and the integrals are regarded only with respect to Lebesgue measure on $\mathbb{R}^m$, which is denoted by the letter λ without indicating the dimension.

In what follows, $\Phi'(x)$ is the Jacobi[3] matrix of a smooth map Φ at a point x (the matrix corresponding to the linear map $d_x\Phi$ in the canonical basis of the space $\mathbb{R}^m$); the determinant of this matrix (*the Jacobian* of Φ) is denoted by $J_\Phi(x)$ ($x \in \mathcal{O}$).

We recall (see Sect. 13.7.3) that a *diffeomorphism* is a bijective smooth map of an open subset of $\mathbb{R}^m$ to an open subset of $\mathbb{R}^m$ with smooth inverse. As proved in Theorem 2.3.1, the image of a measurable set under a smooth map is measurable and the image of a set of measure zero has measure zero.

6.2.1 Before applying Theorem 6.1.1 to our situation, it is necessary to find out how Lebesgue measure transforms under a diffeomorphism. It is convenient to state this question as a problem on the calculation of the measure ν defined on the σ-algebra of measurable subsets of $\mathcal{O}$ by the equation $\nu(A) = \lambda(\Phi(A))$. More specifically, we want to find out whether the measure ν has a density with respect to the Lebesgue measure and find the density if it exists.

In search of a hypothetic density at an arbitrary point a, $a \in \mathcal{O}$, the key point is the fact that, in the vicinity of this point, the diffeomorphism Φ is well approximated

[3] Carl Gustav Jacob Jacobi (1804–1851)—German mathematician.

by the affine map $x \mapsto \widetilde{\Phi}(x) = \Phi(a) + d_a\Phi(x-a)$ the influence of which on the Lebesgue measure is well known (see Theorems 2.4.1 and 2.5.2):

$$\lambda\big(\widetilde{\Phi}(A)\big) = \lambda\big(d_a\Phi(A)\big) = |\det d_a\Phi|\lambda(A) = \big|J_\Phi(a)\big|\lambda(A).$$

Therefore, it is natural to assume that the following approximate equation is valid for a measurable set A lying in a small neighborhood of a:

$$\lambda\big(\Phi(A)\big) \approx \lambda\big(\widetilde{\Phi}(A)\big) = \big|J_\Phi(a)\big|\lambda(A).$$

At the same time, it follows from the mean value theorem that $|J_\Phi(a)|\lambda(A) \approx \int_A |J_\Phi(x)|\,dx$, from which we obtain

$$\nu(A) = \lambda\big(\Phi(A)\big) \approx \int_A \big|J_\Phi(x)\big|\,dx.$$

The last relation makes it very probable that the function $|J_\Phi|$ might be the density of the measure ν with respect to the Lebesgue measure.

Now, we give a precise statement and a formal proof of this fact.

Theorem *Let Φ be a diffeomorphism defined on an open set $\mathcal{O}$, $\mathcal{O} \subset \mathbb{R}^m$. Then the following relation is valid for every measurable set A, $A \subset \mathcal{O}$:*

$$\lambda\big(\Phi(A)\big) = \int_A \big|J_\Phi(x)\big|\,dx. \tag{1}$$

Proof On the σ-algebra of measurable sets contained in $\mathcal{O}$, we define a measure ν by the equation

$$\nu(A) = \lambda\big(\Phi(A)\big) \quad (A \subset \mathcal{O})$$

and verify that the measure satisfies the condition

$$\inf_A |J_\Phi|\lambda(A) \leqslant \nu(A) \leqslant \sup_A |J_\Phi|\lambda(A). \tag{2}$$

As stated in Theorem 6.1.2, this implies the relation $\nu(A) = \int_A |J_\Phi(x)|\,dx$, which proves the theorem.

Proceeding to prove inequality (2), we note that it is sufficient to verify the right-hand inequality since, applying it to the map Φ^{-1} and to the set $\Phi(A)$, we obtain the left-hand inequality (recall that $J_\Phi(x) \cdot J_{\Phi^{-1}}(y) = 1$ for $y = \Phi(x)$ and $x \in \mathcal{O}$).

As the first and most difficult step, we prove by contradiction the right-hand inequality (2) for an arbitrary cubic cell whose closure lies in $\mathcal{O}$. We assume that $\lambda(Q)\sup_Q |J_\Phi| < \nu(Q)$ for a cubic cell Q such that $\overline{Q} \subset \mathcal{O}$. Then $C\lambda(Q) < \nu(Q)$ for some $C > \sup_Q |J_\Phi|$. We divide Q into 2^m cells the edges of which are two times smaller than those of Q. Among the new cells, there is a cell, call it Q_1, such that $C\lambda(Q_1) < \nu(Q_1)$. Repeating the above construction, we inductively construct a sequence of embedded cubic cells $\{Q_n\}$ such that $\operatorname{diam}(Q_n) \to 0$ and

$$C\lambda(Q_n) < \nu(Q_n) \quad \text{for all } n.$$

Let $a \in \bigcap_n \overline{Q}_n$ and $L = d_a\Phi$. By assumption, L is an invertible linear map, and since $a \in \overline{Q}$, we have $|\det L| = |J_\Phi(a)| < C$. We consider the auxiliary map

$$\Psi(x) = a + L^{-1}\big(\Phi(x) - \Phi(a)\big).$$

Near the point a, this map is close to the identity, $\Psi(x) = x + o(x-a)$. Therefore, for every $\varepsilon > 0$, there is a small ball B centered at a such that

$$\|\Psi(x) - x\| \leqslant \frac{\varepsilon}{\sqrt{m}}\|x - a\| \quad \text{for all } x \text{ in } B.$$

By construction, we have $a \in \overline{Q}_n$ and $\overline{Q}_n \subset B$ for sufficiently large n. Let h be the length of an edge of the cube Q_n and $x \in Q_n$. Since $\|x - a\| \leqslant \sqrt{m}h$, we obtain $\|\Psi(x) - x\| \leqslant \varepsilon h$, and a similar inequality is valid for all coordinates of the difference $\Psi(x) - x$. Therefore, the vector $\Psi(x)$ belongs to a cube whose edge is at most $(1 + 2\varepsilon)h$. Consequently,

$$\lambda\big(\Psi(Q_n)\big) \leqslant (1 + 2\varepsilon)^m h^m = (1 + 2\varepsilon)^m \lambda(Q_n).$$

Using Theorem 2.5.2 and the fact that the Lebesgue measure is translation invariant (see Sect. 2.4.1), we obtain

$$\lambda\big(\Psi(Q_n)\big) = \lambda\big(L^{-1} \circ \Phi(Q_n)\big) = \big|\det L^{-1}\big| \cdot \lambda\big(\Phi(Q_n)\big) = \frac{\nu(Q_n)}{|\det L|}.$$

Thus,

$$C\lambda(Q_n) < \nu(Q_n) = |\det L| \cdot \lambda\big(\Psi(Q_n)\big) \leqslant (1 + 2\varepsilon)^m |\det L| \cdot \lambda(Q_n).$$

Therefore, $C < (1 + 2\varepsilon)^m |\det L|$ for all $\varepsilon > 0$, i.e., $C \leqslant |\det L| = |J_\Phi(a)|$. However, this is impossible since $C > \sup_Q |J_\Phi|$ and $a \in \overline{Q}$. The contradiction obtained proves that our assumption is false and the inequality $\nu(Q) \leqslant \lambda(Q) \sup_Q |J_\Phi|$ is valid for each cubic cell Q such that $\overline{Q} \subset \mathcal{O}$.

We note that the estimate from above in (2) is valid for a set A if it is valid for the sets of some at most countable partition of A. From this it follows immediately that the estimate is valid for every open set G, $G \subset \mathcal{O}$ (it is sufficient to divide G into cubic cells with closures in G, see Theorem 1.1.7). Moreover, we can assume in what follows that A is a bounded set whose closure is contained in the set $\mathcal{O}$. For such a set, the right-hand side of inequality (2) can be obtained using the regularity of the Lebesgue measure:

$$\nu(A) \leqslant \inf_{\substack{A \subset G \subset \mathcal{O} \\ G \text{ is open}}} \nu(G) \leqslant \inf_{\substack{A \subset G \subset \mathcal{O} \\ G \text{ is open}}} \Big(\lambda(G) \cdot \sup_G |J_\Phi|\Big) = \lambda(A) \cdot \sup_A |J_\Phi|.$$

This completes the proof of (2) and the theorem. □

From the continuity of J_Φ and formula (1), it follows that

$$\left|J_\Phi(a)\right| = \lim \frac{\lambda(\Phi(A))}{\lambda(A)}, \tag{3}$$

where the limit is calculated under the assumption that $\lambda(A) > 0$ and the sets A "shrink" to the point a (i.e., $A \subset B(a, r),\ r \to 0$). Thus, as we surmised from the very beginning, "in the small", the number $|J_\Phi(a)|$ can be regarded as the measure distortion coefficient under the map Φ (in much the same way as in the case of a linear map, the absolute value of the determinant is a "global" measure distortion coefficient).

As follows from Theorem 8.8.1, the assertion of Theorem 6.2.1 remains valid if instead of the smoothness of Φ we assume that it is a homeomorphism such that both Φ and its inverse satisfy the Lipschitz condition. For a generalization of the theorem to maps that are not one-to-one, see, for example, [EG].

6.2.2 Now we have everything we need to obtain the main result of the present section, the change of variable formula for multiple integrals.

Theorem *Let Φ be a diffeomorphism defined on an open set $\mathcal{O}$, $\mathcal{O} \subset \mathbb{R}^m$. Then, for every measurable non-negative function f defined on $\mathcal{O}' = \Phi(\mathcal{O})$, we have*

$$\int_{\mathcal{O}'} f(y)\,dy = \int_{\mathcal{O}} f\big(\Phi(x)\big)\big|J_\Phi(x)\big|\,dx. \tag{4}$$

The above equation is valid for every summable function f on $\mathcal{O}'$

Proof By the previous theorem, this is a special case of Theorem 6.1.1, where $X = \mathcal{O}$, $Y = \mathcal{O}'$, $\omega = |J_\Phi|$, and μ and ν are the Lebesgue measures on the σ-algebras of measurable subsets of $\mathcal{O}$ and $\mathcal{O}'$, respectively. The fact that the set $\Phi^{-1}(B)$ is measurable follows from the smoothness of Φ^{-1}, and the equation $\lambda(B) = \int_{\Phi^{-1}(B)} |J_\Phi(x)|\,dx$ required by Definition 6.1.1 is equivalent to the statement of Theorem 6.2.1. □

As in Sect. 6.1 (see Corollary 6.1.1), the formula proved above is valid in a more general form. Namely, for every measurable set A lying in $\mathcal{O}$, we have

$$\int_{\Phi(A)} f(y)\,d\lambda(y) = \int_A f\big(\Phi(x)\big)\big|J_\Phi(x)\big|\,d\lambda(x).$$

The function f is summable on $\Phi(A)$ if and only if the function $(f \circ \Phi)|J_\Phi|$ is summable on A.

Remark The conditions of Theorem 6.2.2 can be weakened slightly by allowing the function Φ to "worsen" on a "negligible" set. We describe this in more detail. Let $X \subset \mathbb{R}^m$, $\Phi : X \to \mathbb{R}^m$ and $Y = \Phi(X)$. If the restriction of Φ to an open subset $\mathcal{O}$

of X is a diffeomorphism and both the difference $e = X \setminus \mathcal{O}$ and its image $\Phi(e)$ have zero measure, then the conclusion of Theorem 6.2.2 remains valid, and the equation

$$\int_Y f(y)\,dy = \int_X f(\Phi(x))\,|J_\Phi(x)|\,dx \tag{4'}$$

is valid for every function f summable on Y.

Indeed, since e and $Y \setminus \Phi(\mathcal{O}) \subset \Phi(e)$ are sets of measure zero, we have

$$\int_Y f(y)\,dy = \int_{\Phi(\mathcal{O})} f(y)\,dy = \int_{\mathcal{O}} f\big(\Phi(x)\big)\,|J_\Phi(x)|\,dx = \int_X f\big(\Phi(x)\big)\,|J_\Phi(x)|\,dx.$$

We note that the map Φ, which is one-to-one on $\mathcal{O}$, need not satisfy this condition on X and may be not only non-smooth, but even discontinuous on e.

We consider the simplest special case of the theorem. Let $m = 1$, $\Phi \in C^1([a,b])$, and let $\Phi'(x) \neq 0$ for $x \in (a,b)$. By the last condition, the function Φ' preserves sign on (a,b) and the function Φ is strictly monotonic. By Theorem 6.2.2, we obtain that the equation

$$\int_{[p,q]} f(y)\,dy = \int_{[a,b]} f\big(\Phi(x)\big)\,|\Phi'(x)|\,dx$$

is valid for every measurable non-negative function f on $[p,q] = \Phi([a,b])$.

Considering the cases $\Phi' > 0$ and $\Phi' < 0$, the reader can easily verify that, in both cases, the above equation implies the formula

$$\int_a^b f\big(\Phi(x)\big)\Phi'(x)\,dx = \int_{\Phi(a)}^{\Phi(b)} f(y)\,dy$$

obtained in Proposition 2, Sect. 4.6.2 only for a continuous function f on (p,q) (however, under some weaker assumptions on Φ).

We mention two simple specific cases of Theorem 6.2.2 that will be used repeatedly in the sequel.

TRANSLATION. For every vector $v \in \mathbb{R}^m$, we have the equation

$$\int_{\mathbb{R}^m} f(y)\,dy = \int_{\mathbb{R}^m} f(v+x)\,dx = \int_{\mathbb{R}^m} f(v-x)\,dx.$$

For the proof, it is sufficient to observe that a translation, as well as a translation followed by a reflection, is a diffeomorphism of the space $\mathbb{R}^m$ the absolute value of the Jacobian of which is equal to 1 everywhere.

LINEAR CHANGE. Let $L : \mathbb{R}^m \to \mathbb{R}^m$ be an invertible linear map. Then

$$\int_{\mathbb{R}^m} f(y)\,dy = |\det L| \int_{\mathbb{R}^m} f\big(L(x)\big)\,dx.$$

In particular, the equation

$$\int_{\mathbb{R}^m} f(y)\,dy = |c|^m \int_{\mathbb{R}^m} f(cx)\,dx$$

is valid for every non-zero coefficient c.

In both cases, for simplicity, we consider integration over the entire space $\mathbb{R}^m$. From this, the formulas for integration over a part of $\mathbb{R}^m$ can easily be obtained.

6.2.3 If $\Phi : \mathcal{O} \to \mathcal{O}'$ is a diffeomorphism, then the position of a point y in $\mathcal{O}'$ is completely determined by the point $x = \Phi^{-1}(y)$, and, therefore, the Cartesian coordinates of x are often called the curvilinear coordinates of y. It is convenient to think of $\mathcal{O}$ and $\mathcal{O}'$ as sets lying in different spaces $\mathbb{R}^m$ by considering two copies of this space (it is natural to denote the coordinates of the points in these spaces by different letters). A subset of the set $\mathcal{O}'$ on which the curvilinear coordinate with a given index k is constant is called a coordinate surface (a coordinate line in the two-dimensional case). A coordinate surface is the image of the intersection of $\mathcal{O}$ and a plane $x_k = \text{const}$. This surface can also be regarded as the level surface for the kth coordinate function of the map Φ^{-1}. Thus, $\mathcal{O}'$ is "foliated" into the coordinate surfaces $x_k = \text{const}$, which are obviously disjoint. Such a foliation can be produced in m ways, depending on the index of a coordinate. Every point in $\mathcal{O}'$ lies in the intersection of m coordinate surfaces.

Fixing all coordinates of a point $a = (a_1, \ldots, a_m) \in \mathcal{O}$ except the kth one and changing this coordinate in the vicinity of a_k, we obtain a path parametrizing a curve passing through the point $\Phi(a)$. The corresponding curve is called a *coordinate line*. The tangent vector to it at the point $\Phi(a)$ is simply the kth column of the Jacobi matrix $\Phi'(a)$; we denote this vector by τ_k. It is well known (see Sect. 2.5.2) that the number $|J_\Phi(a)|$ has a simple geometric interpretation as the volume of the parallelepiped spanned by the vectors $\tau_1, \ldots, \tau_m$. Sometimes, especially in the cases where the curvilinear coordinates have a simple geometric interpretation, the situation in question can be described without mentioning the diffeomorphism Φ. Instead, we say that the set $\mathcal{O}'$ "is equipped with curvilinear coordinates" and give the dependencies $y_k = \varphi_k(x_1, \ldots, x_m)$ of the Cartesian coordinates of a point in $\mathcal{O}'$ on the curvilinear coordinates, i.e., the coordinate functions of the diffeomorphism Φ. Since the diffeomorphism Φ is not given explicitly, instead of the determinant $J_\Phi(x) = \det d_x\Phi$ one uses the functional determinant $\frac{D(\varphi_1,\ldots,\varphi_m)}{D(x_1,\ldots,x_m)} = \det \| \frac{\partial \varphi_k}{\partial x_j} \|$ corresponding to the system of functions $\varphi_1, \ldots, \varphi_m$.

Sometimes it is possible to calculate the absolute value of the Jacobian without using its definition directly but applying Eq. (3) to sets A of one form or another. Let, for example, A be a cell $\prod_{k=1}^m [a_k, a_k + h_k)$ lying in $\mathcal{O}$, where $h = (h_1, \ldots, h_m) \in \mathbb{R}_+^m$. Its image is a "curvilinear parallelepiped" bounded by the corresponding coordinate surfaces. The "edges" of the parallelepiped lie on the coordinate lines and are close to the tangent vectors $h_k\tau_k$ for small h.

Quite often, the principal part of the volume of the curvilinear parallelepiped can be found directly, using the geometric interpretation of curvilinear coordinates, which makes it possible to calculate $|J_\Phi(a)|$ as well.

Example We calculate the area S of the curvilinear quadrangle

$$M = \left\{ (x, y) \in \mathbb{R}_+^2 \;\middle|\; a^2 \leqslant xy \leqslant b^2,\ \alpha \leqslant \frac{y}{x} \leqslant \beta \right\}$$

(a, b, α and β are positive parameters, $a < b$ and $\alpha < \beta$). To this end, we introduce curvilinear coordinates u and v in $\mathbb{R}^2_+$ by the equations

$$u = xy \quad \text{and} \quad v = \frac{y}{x}.$$

The corresponding coordinate lines are hyperbolas and rays. Since the curvilinear coordinates on M can take, respectively, the values from a^2 to b^2 and from α to β independently of one another, the points (u, v) corresponding to the points (x, y) in M "on the $u\,v$-plane" fill the rectangle $[a^2, b^2] \times [\alpha, \beta]$. As a rule, such a simplification of a given domain is one of the main goals when changing variables. It is easy to prove that $\frac{D(u,v)}{D(x,y)} = 2\frac{y}{x} = 2v$. Consequently, the Jacobian of the map $(u, v) \mapsto (x, y)$ is equal to $\frac{1}{2v}$ (we call the reader's attention to the fact that here it was easier to first find the Jacobian of the map inverse to $(u, v) \mapsto (x, y)$). Therefore, the required area is equal to

$$S = \iint_M 1\, dx\, dy = \int_{a^2}^{b^2} \int_{\alpha}^{\beta} \frac{du\, dv}{2v} = \frac{b^2 - a^2}{2} \ln \frac{\beta}{\alpha}.$$

6.2.4 *Polar Coordinates*. Besides the Cartesian coordinates x and y, there are other numerical parameters that can be used to locate points in the plane. For example, the distance r from a point to the origin O (of a Cartesian coordinate system) and the polar angle φ, i.e., the angle formed by a fixed ray from O and the radius-vector of the point. The numbers r and φ are called the *polar coordinates* of the point. Introducing Cartesian coordinates so that the polar angle is counted anticlockwise from the positive x-axis towards the positive y-axis, we see that the Cartesian and polar coordinates are connected by the formulas

$$x = r\cos\varphi, \qquad y = r\sin\varphi.$$

Formally speaking, these equations define a smooth map

$$(r, \varphi) \mapsto \Phi(r, \varphi) = (r\cos\varphi, r\sin\varphi),$$

taking the r, φ plane into the $x,\ y$ plane. However, taking into account the geometric meaning of the parameter r (the distance from the origin), we assume that the map Φ is defined in the half-plane $r \geqslant 0$. Obviously, the map Φ is not one-to-one. To make it one-to-one, we must assume that the angle φ changes in an open interval the length of which does not exceed 2π.

As the reader can easily verify, the restriction of Φ to a strip of the form $P_\alpha = (0, +\infty) \times (\alpha, \alpha + 2\pi)$ is one-to-one, and its image is the plane with the ray $L_\alpha = \{(r\cos\alpha, r\sin\alpha) \mid r \geqslant 0\}$ removed, or, as one says, the plane cut along the ray L_α. It is obvious that $L_\alpha = \Phi(\partial P_\alpha)$, and so $\Phi(\overline{P}_\alpha) = \mathbb{R}^2$. Since the map Φ is not one-to-one, it is necessary to specify the range of the polar angle when passing from Cartesian to polar coordinates. As a rule, one uses the intervals $(0, 2\pi)$ and $(-\pi, \pi)$ (corresponding to $\alpha = 0$ and $\alpha = -\pi$).

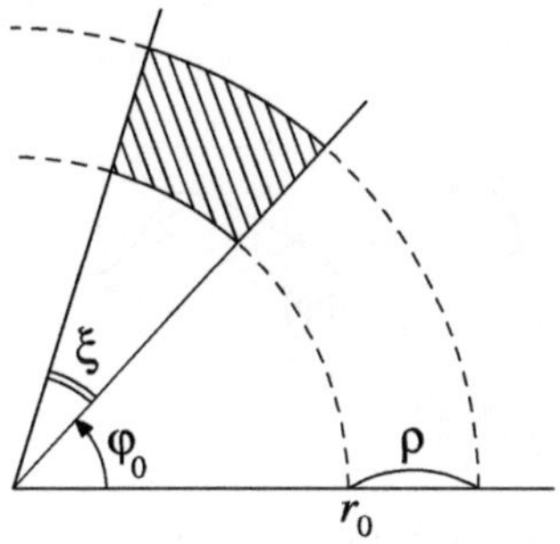

Fig. 6.1 Increment of a circular sector

The coordinate lines, i.e., the lines $r = \text{const}$ and $\varphi = \text{const}$, are circles (centered at the origin O) and rays (from O), respectively. The rectangle $[r_0, r_0 + \rho] \times [\varphi_0, \varphi_0 + \xi]$ transforms into the curvilinear quadrangle bounded by the circles $r = r_0$ and $r = r_0 + \rho$ and by the rays $\varphi = \varphi_0$ and $\varphi = \varphi_0 + \xi$ (see Fig. 6.1).

For small ρ and ξ, this curvilinear quadrangle is almost a rectangle with sides $r_0\xi$ and ρ. Therefore, up to higher order infinitesimals, the area is equal to $r_0\rho\xi$. Recalling that the value of the Jacobian J_Φ at (r_0, φ_0) is the area distortion coefficient, we come to the conclusion that $J_\Phi(r_0, \varphi_0) = r_0$. The reader can easily obtain this result by calculating the second order functional determinant. By the remark following Theorem 6.2.2, the general change of variables formula (4′) takes the following form in the case of transition to polar coordinates:

$$\iint_A f(x, y)\,dx\,dy = \iint_{\Phi_\alpha^{-1}(A)} f(r\cos\varphi, r\sin\varphi)r\,dr\,d\varphi,$$

where $A \subset \mathbb{R}^2$ and Φ_α is the restriction of Φ to $\overline{P}_\alpha$. In particular,

$$\iint_{\mathbb{R}^2} f(x, y)\,dx\,dy = \int_\alpha^{\alpha+2\pi} \left(\int_0^\infty f(r\cos\varphi, r\sin\varphi)r\,dr \right) d\varphi.$$

Example 1 Using polar coordinates, we can easily find the area of the "curvilinear triangle" (see Fig. 6.2)

$$T = \{(r\cos\varphi, r\sin\varphi) \in \mathbb{R}^2 \mid \varphi \in \Delta,\ 0 \leqslant r \leqslant \rho(\varphi)\},$$

where $\Delta \subset \mathbb{R}$ is an interval of length less than or equal to 2π and ρ is a non-negative function measurable on Δ.

Putting $f = \chi_T$ in the last formula, we obtain the required result

$$\lambda_2(T) = \iint_T 1\,dx\,dy = \int_\Delta \left(\int_0^{\rho(\varphi)} r\,dr \right) d\varphi = \frac{1}{2} \int_\Delta \rho^2(\varphi)\,d\varphi.$$

Example 2 The use of polar coordinates gives us one more way of calculating the Euler–Poisson integral $I = \int_{-\infty}^{\infty} e^{-x^2}\,dx$ (cf. Sect. 5.3.2, Example 1). As before, we

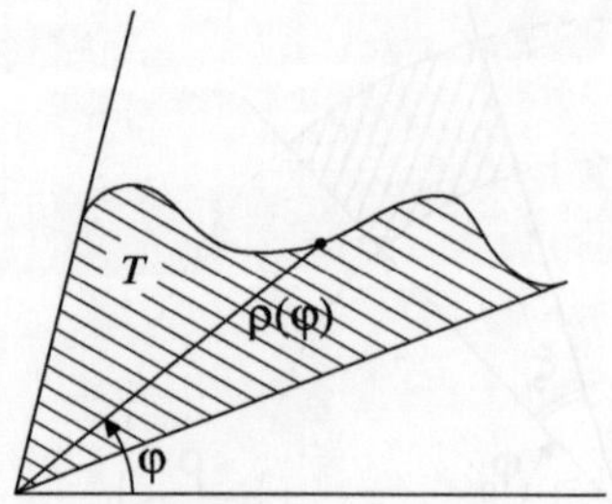

Fig. 6.2 Curvilinear triangle

transform I^2 by Fubini's theorem,

$$I^2 = \int_{-\infty}^{\infty} e^{-x^2}\,dx \cdot \int_{-\infty}^{\infty} e^{-y^2}\,dy = \iint_{\mathbb{R}^2} e^{-(x^2+y^2)}\,dx\,dy.$$

Now, passing to polar coordinates, we obtain

$$I^2 = \iint_{\mathbb{R}^2} e^{-(x^2+y^2)}\,dx\,dy = \int_0^{2\pi} \left(\int_0^{\infty} e^{-r^2} r\,dr \right) d\varphi = \pi \int_0^{\infty} e^{-r^2}\,d\bigl(r^2\bigr) = \pi.$$

Therefore, $I = \sqrt{\pi}$.

6.2.5 *Spherical Coordinates.* Spherical coordinates in three-dimensional space are an analog of polar coordinates in a plane. The location of a point (x, y, z) can be determined by the following three numerical parameters: the distance r from the point to the origin, the polar angle φ corresponding to the projection of the point on the x, y plane, and the angle θ between the radius-vector of the point and the positive z-axis.

The spherical and Cartesian coordinates are connected by the formulas

$$x = r\cos\varphi\sin\theta, \qquad y = r\sin\varphi\sin\theta, \qquad z = r\cos\theta.$$

Formally speaking, these equations define a smooth map

$$(r, \varphi, \theta) \mapsto \Phi(r, \varphi, \theta) = (r\cos\varphi\sin\theta, r\sin\varphi\sin\theta, r\cos\theta)$$

taking the r, φ, θ space to the x, y, z space. However, taking into account the geometric meaning of the parameter r (the distance from the origin), we assume that the map Φ is defined on the half-space $r \geqslant 0$. Obviously, the map Φ is not one-to-one. To make it one-to-one, we must restrict the ranges of the angles φ and θ. As to θ, we will always assume that $0 \leqslant \theta \leqslant \pi$. We also assume that the angle φ changes from 0 to 2π (sometimes, it is convenient to change these bounds to $-\pi$ and π, respectively). As the reader can easily verify, the restriction of Φ to an infinite parallelepiped of the form $P = (0, +\infty) \times (0, 2\pi) \times (0, \pi)$ is one-to-one and its image is the entire space $\mathbb{R}^3$ with the half-plane $L_0 = \{(r\sin\theta, 0, r\cos\theta) \mid r \geqslant 0,\ 0 \leqslant \theta \leqslant \pi\}$

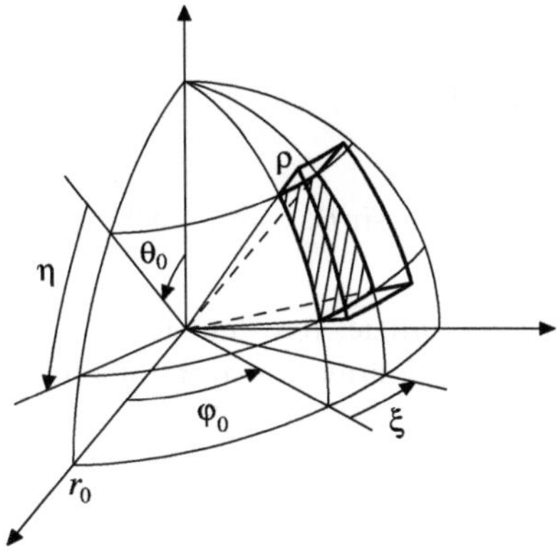

Fig. 6.3 Curvilinear parallelepiped corresponding to increments of spherical coordinates

removed. Obviously, $L_0 = \Phi(\partial P)$, and so $\Phi(\overline{P}) = \mathbb{R}^3$. In what follows, we assume that Φ is defined on $\overline{P}$.

The coordinate surfaces, i.e., the surfaces $r = \text{const}$, $\varphi = \text{const}$, and $\theta = \text{const}$ are spheres (centered at the origin O), half-planes bounded by the z-axis, and circular cones with vertex at O that are symmetric with respect to the z-axis. The intersections of the sphere with the half-planes and cones forms a grid of meridians and parallels (this is why the angle $\theta' = \pi/2 - \theta$ (the "latitude") is sometimes considered instead of the angle θ).

The map Φ transforms the parallelepiped $[r_0, r_0+\rho] \times [\varphi_0, \varphi_0+\xi] \times [\theta_0, \theta_0+\eta]$ into the curvilinear parallelepiped bounded by the spheres $r = r_0$ and $r = r_0+\rho$, the half-planes $\varphi = \varphi_0$ and $\varphi = \varphi_0 + \xi$, and the conical surfaces $\theta = \theta_0$ and $\theta = \theta_0 + \eta$ (see Fig. 6.3).

For small ρ, ξ, and η, this parallelepiped is almost rectangular. Its base lying on the sphere $r = r_0$ is bounded by the arcs of meridians and parallels. This base is almost a rectangle with length of sides equal to $r_0\eta$ and $r_0 \sin\theta_0 \xi$, respectively. Therefore, up to higher order infinitesimals, the volume of the curvilinear parallelepiped is equal to $(r_0^2 \sin\theta_0)\rho\xi\eta$. Recalling that the value of the Jacobian J_Φ at $(r_0, \varphi_0, \theta_0)$ is the volume distortion coefficient, we come to the conclusion that $J_\Phi(r_0, \varphi_0, \theta_0) = r_0^2 \sin\theta_0$. The reader can easily obtain the same result by performing all necessary formal calculation. In the case of transition to spherical coordinates, we can take into account the remark following Theorem 6.2.2 and represent the general change of variables formula in the integral as follows:

$$\iiint_A f(x, y, z)\,dx\,dy\,dz$$
$$= \iiint_{\Phi^{-1}(A)} f(r\cos\varphi\sin\theta, r\sin\varphi\sin\theta, r\cos\theta)\,r^2\sin\theta\,dr\,d\varphi\,d\theta,$$

where $A \subset \mathbb{R}^3$. In particular,

$$\iiint_{\mathbb{R}^3} f(x,y,z)\,dx\,dy\,dz$$
$$= \int_0^{\pi}\int_0^{2\pi}\int_0^{\infty} f(r\cos\varphi\sin\theta, r\sin\varphi\sin\theta, r\cos\theta)\,r^2\sin\theta\,dr\,d\varphi\,d\theta.$$

Example We use spherical coordinates to calculate the Fourier transform of a radial function. In the general case, the Fourier transform of a function f summable on $\mathbb{R}^m$ is defined by the equation

$$\widehat{f}(y) = \int_{\mathbb{R}^m} f(x)e^{-2\pi i\langle x,y\rangle}\,dx.$$

Let f be a measurable radial function of three variables, i.e., a function of the form $f(x) = f_0(\|x\|)$, where f_0 is a measurable function on $\mathbb{R}_+$. Converting to spherical coordinates, we see that $\int_{\mathbb{R}^3}|f(x)|\,dx = 4\pi\int_0^{\infty}|f_0(r)|r^2\,dr$. Therefore, the function f is summable in $\mathbb{R}^3$ if and only if the inequality $\int_0^{\infty}|f_0(r)|r^2\,dr < +\infty$ is valid. In this case, the calculation of $\widehat{f}$ can be reduced to the calculation of the integral over the semi-axis $\mathbb{R}_+$.

As $y \neq 0$, we make an orthogonal change of variables $x \mapsto u$ in the integral $\widehat{f}(y)$ that takes the unit vector $y/\|y\|$ to $(0,0,1)$. Then

$$\widehat{f}(y) = \int_{\mathbb{R}^3} f_0(\|x\|)e^{-2\pi i\langle x,y\rangle}\,dx = \int_{\mathbb{R}^3} f_0\big(\|u\|\big)e^{-2\pi i\|y\|u_3}\,du.$$

Converting the last integral to spherical coordinates, we obtain

$$\widehat{f}(y) = \int_0^{\infty} f_0(r)r^2\left(\int_0^{\pi}\left(\int_0^{2\pi} e^{-2\pi i r\|y\|\cos\theta}\sin\theta\,d\varphi\right)d\theta\right)dr$$
$$= 2\pi\int_0^{\infty} f_0(r)r^2\left(\int_0^{\pi} e^{-2\pi i r\|y\|\cos\theta}\sin\theta\,d\theta\right)dr.$$

The integral with respect to θ can easily be calculated, and we obtain the required formula

$$\widehat{f}(y) = \frac{2}{\|y\|}\int_0^{\infty} f_0(r)r\sin\big(2\pi r\|y\|\big)\,dr.$$

We see that the Fourier transform of a radial function is a radial function.

6.2.6 We consider the question of the change of volume under diffeomorphisms generated by a system of differential equations

$$\frac{dx_1}{dt} = f_1(x_1,\dots,x_m),$$
$$\vdots$$
$$\frac{dx_m}{dt} = f_m(x_1,\dots,x_m),$$

where $f_1, \dots, f_m$ are smooth functions defined on the entire space $\mathbb{R}^m$ and the variable t is regarded as time.

This system can be written in the concise form

$$\frac{dx}{dt} = f(x), \tag{5}$$

where $f : \mathbb{R}^m \to \mathbb{R}^m$ denotes a smooth map with coordinate functions $f_1, \dots, f_m$ called a *direction field.*

In the theory of ordinary differential equations, it is proved that, for all initial conditions, system (5) has a unique solution defined for all t close to the initial moment t_0. We assume that all these solutions are defined for all $t \in \mathbb{R}$. Assuming that the initial conditions correspond to the moment $t = 0$, we obtain that, for every t, there is a unique solution $x(t)$ corresponding to the given initial condition $x = x(0)$. This gives rise to the map $S_t : \mathbb{R}^m \to \mathbb{R}^m$ taking the initial point $x(0)$ to the point $x(t)$ ($S_0 = id$). In the theory of differential equations, it is proved that the map $(x, t) \mapsto S_t(x)$ is smooth (see, e.g., P. Hartman "Ordinary Differential Equations"). Since the solution satisfying given initial conditions is unique, we obtain the equation $S_{t+\tau} = S_t \circ S_\tau$ valid for all $t, \tau \in \mathbb{R}$. In particular, the map S_t is invertible since $S_t \circ S_{-t} = S_{-t} \circ S_t = S_0 = id$, and, consequently, $(S_t)^{-1} = S_{-t}$. Thus, S_t is a diffeomorphism. The family of diffeomorphisms $\{S_t\}_{t \in \mathbb{R}}$ is called a *flow*. Our goal is clarify how the volume (i.e., Lebesgue measure on $\mathbb{R}^m$) changes under the action of diffeomorphisms forming the flow.

From (5) it follows that

$$x(t) = x(0) + \int_0^t f\bigl(x(u)\bigr)\,du, \quad \text{i.e.,} \quad S_t(x) = x + \int_0^t f\bigl(S_u(x)\bigr)\,du. \tag{6}$$

First, we prove the following formula describing the derivative of a diffeomorphism $S_t : \mathbb{R}^m \to \mathbb{R}^m$ for small t:

$$S_t'(x) = id + tf'(x) + \alpha(t, x), \tag{7}$$

where $\frac{1}{t}\alpha(t, x) \rightrightarrows 0$ as $t \to 0$ if x belongs to a bounded set.

Differentiating Eq. (6) with respect to x (for justification of differentiation under the integral sign, see Sect. 7.1), we obtain the relation

$$S_t'(x) = id + \int_0^t f'\bigl(S_u(x)\bigr)S_u'(x)\,du.$$

From this equation we obtain that $S_t'(x) \to id$ as $t \to 0$. Continuing the last equation, we obtain

$$S_t'(x) = id + tf'(x) + \int_0^t \bigl(f'\bigl(S_u(x)\bigr)S_u'(x) - f'(x)\bigr)\,du. \tag{8}$$

It is clear that the difference $f'(S_u(x))S_u'(x) - f'(x)$ tends to zero as $u \to 0$, and the convergence is uniform if x is taken from a bounded set. Therefore, the last term on the right-hand side of Eq. (8) is $o(t)$ (uniformly with respect to x), which proves (7). Consequently, as $t \to 0$, we have

$$\det S_t'(x) = \det\bigl(id + tf'(x) + o(t,x)\bigr) = 1 + t\,\text{trace}\, f'(x) + o(t,x), \tag{9}$$

where trace $f'(x)$ is the trace of the matrix $f'(x)$, which is also called the *divergence* of the direction field and is denoted by $\text{div}\, f(x)$: $\text{div}\, f(x) = \frac{\partial f_1}{\partial x_1}(x) + \cdots + \frac{\partial f_m}{\partial x_m}(x)$. We leave it as an exercise (connected with the calculation of a determinant) for the reader to check the second equality in (9).

Let A be a bounded measurable set, $A_t = S_t(A)$ and $V(t) = \lambda_m(A_t)$. By Theorem 6.2.1, we obtain

$$V(t) = \int_A \bigl|\det S_t'(x)\bigr|\,dx.$$

Using Eq. (9) for sufficiently small t and taking into account the fact that the o-term is uniformly small on A, we see that

$$V(t) = \int_A \bigl(1 + t\,\text{trace}\, f'(x) + o(t)\bigr)\,dx = V(0) + t\int_A \text{div}\, f(x)\,dx + o(t).$$

Consequently,

$$V'(0) = \int_A \text{div}\, f(x)\,dx. \tag{10}$$

Since $S_{\tau+t} = S_t \circ S_\tau$, we obtain $V(\tau + t) = \lambda_m(S_t(A_\tau))$. Therefore, replacing A by A_τ and applying formula (10), we obtain that the relation

$$V'(\tau) = \int_{A_\tau} \text{div}\, f(x)\,dx$$

is valid for all $\tau \in \mathbb{R}$. This result is well known as Liouville's theorem. The theorem implies the following statement.

Corollary *If* $\text{div}\, f(x) \equiv 0$, *then the flow preserves the volume.*

Example The motion of material particles of mass m and charge q in a stationary electromagnetic field is described by the Lorentz equation

$$m\frac{dv}{dt} = q\left(E + \frac{1}{c}v \times B\right),$$

where $v = \frac{dx}{dt}$ is the velocity of the particle, c is the speed of light in vacuum, and $E = E(x)$ and $B = B(x)$ are certain smooth vector functions (the intensity and the inductance of the field); the symbol $\times$ denotes the vector product.

The Lorentz equation takes the form (5) for the vectors $w = (x, v)$ in the six-dimensional space if we rewrite it as $m\frac{dw}{dt} = f(w)$, where the right-hand side is $f(w) = (mv, q(E + \frac{1}{c} v \times B)) = (v_1, v_2, v_3, V_1, V_2, V_3)$, where v_1, v_2, v_3 and V_1, V_2, V_3 are the coordinates of the vectors $m\,v$ and $q(E + \frac{1}{c} v \times B)$, respectively. It is easy to verify that V_i does not depend on v_i $(i = 1, 2, 3)$. Therefore,

$$\operatorname{div} f(w) = \frac{\partial v_1}{\partial x_1} + \frac{\partial v_2}{\partial x_2} + \frac{\partial v_3}{\partial x_3} + \frac{\partial V_1}{\partial v_1} + \frac{\partial V_2}{\partial v_2} + \frac{\partial V_3}{\partial v_3} \equiv 0.$$

Thus, the Lebesgue measure λ_6 is invariant under the flow corresponding to the Lorentz equation.

We observe that to describe the properties of a material particle motion it is helpful to use the measure on a six-dimensional space.

Remark The reader can verify that the diffeomorphisms S_t are volume-preserving if and only if $\operatorname{div} f(x) \equiv 0$.

EXERCISES

1. Calculate the integral $\iint_{\mathbb{R}^2} |ax + by| e^{-(x^2+y^2)}\, dx\, dy$.
2. Calculate the integral $\iiiint_{x^2+y^2+u^2+v^2 \leqslant 1} e^{x^2+y^2-u^2-v^2}\, dx\, dy\, du\, dv$.
3. Calculate the integral $\int_{\langle Ax,x\rangle \leqslant 1} e^{\langle Ax,x\rangle}\, dx$, where A is a positive definite $m \times m$ matrix.
4. Let $E = \{(x_1, x_2, x_3, x_4) \in \mathbb{R}^4 \mid \sqrt{x_2^2 + x_3^2 + x_4^2} \leqslant x_1\}$. For which values of $t \in \mathbb{R}^4$ is the integral $\int_E e^{-\langle x,t\rangle}\, dx$ finite? Calculate the integral.
5. Making an appropriate orthogonal transformation, calculate the integral $\int_{\|x\|<r} |\langle a, x\rangle|^p\, dx$ over the m-dimensional ball for $p > -1$.
6. Calculate the integral $\int_{\mathbb{R}^m} e^{-Q(x)}\, dx$, where $Q(x) = \sum_{1 \leqslant j \leqslant k \leqslant m} x_j x_k$.
7. For which values of a and b is the integral

$$\int_{(0,1)^m} \bigl(\min(x_1, \dots, x_m)\bigr)^a \bigl(\max(x_1, \dots, x_m)\bigr)^b\, dx,$$

   finite, where $m \geqslant 2$? Express it in terms of the beta function.
8. Prove that, for every non-negative measurable function f on $\mathbb{R}$ and all $a, b \in \mathbb{R}$, the relation

$$\frac{1}{\pi} \iint_{\mathbb{R}^2} \frac{f(ax + by)}{(1 + x^2)(1 + y^2)}\, dx\, dy = \int_{-\infty}^{\infty} \frac{f(cu)}{1 + u^2}\, du, \quad \text{where } c = |a| + |b|$$

   holds.
9. Using the previous problem and induction, prove that, for $p \in (-1, 1)$ and $a = (a_1, \dots, a_m) \in \mathbb{R}^m$, the relation

$$\frac{1}{\pi^m} \int_{\mathbb{R}^m} \frac{|\langle x, a\rangle|^p\, dx}{(1 + x_1^2)(1 + x_2^2) \cdots (1 + x_m^2)} = C_p \left(\sum_{k=1}^m |a_k| \right)^p,$$

   where $C_p = \frac{2}{\pi} \int_0^\infty \frac{t^p}{1+t^2}\, dt$, holds.

10. Regarding the plane $\mathbb{R}^2$ as the set of complex numbers, find a function $\omega > 0$ such that the measure ν with density $\omega > 0$ ($d\nu = \omega\, d\lambda_2$) is invariant under multiplication, i.e., such that the image of ν under the map $z \mapsto az$ coincides with ν for all $a \neq 0$.

11. Let $p > 0$, $E \subset \mathbb{R}^m$, and $\lambda_m(E) = \lambda_m(B(0,r))$. Prove that

$$\int_E \frac{dy}{\|x-y\|^p} \leqslant \int_{B(0,r)} \frac{dz}{\|z\|^p} \quad \text{for every } x \text{ in } \mathbb{R}^m.$$

12. Prove that the inequality

$$\left| \iint_E \frac{dx\,dy}{x+iy} \right| \leqslant \sqrt{\pi\, \lambda_2(E)}$$

holds for every set $E \subset \mathbb{R}^2$ of finite measure.
Hint. By a rotation, reduce the left-hand side of the inequality to an integral of the function $\mathcal{R}e\frac{1}{z}$ and verify that, for a given area of the integration region, this integral is maximal if the integration is performed over an appropriate Lebesgue set of the integrand.

13. Let $f(x, y)$ be the number of points (k, j) with integer coordinates satisfying the condition $k^2 + j^2 < x^2 + y^2$, and let $S = \sum_{n\in\mathbb{Z}} e^{-n^2}$. Prove that

$$\iint_{\mathbb{R}^2} f(x, y) e^{-(x^2+y^2)}\, dx\, dy = \pi S^2.$$

Hint. Converting to polar coordinates, use integration by parts by means of the functions $F(r) = f(r, 0)$.

14. Let $A = \{(z_1, z_2) \in \mathbb{C}^2 \mid 0 < |z_1| < |z_2| < 1\}$, and let

$$\Phi(z_1, z_2) = \left(z_1, z_2\sqrt{1 - \frac{|z_1|^2}{|z_2|^2}} \right) \quad ((z_1, z_2) \in A).$$

Prove that Φ is a diffeomorphism preserving the four-dimensional Lebesgue measure. Find $\lambda_4(A)$ and $\Phi(A)$.

6.3 Integral Representation of Additive Functions

Since its inception, the integral calculus proved to be a very successful tool in solving applied problems of mechanics and physics. Among them are the problems associated with additive quantities such as calculation of mass, statical moments, energy, etc. In the present section, we consider a general scheme that allows us to evaluate and estimate such quantities in a wide range of cases.

Turning to applications, we set ourselves a restricted task. We are concerned only with *evaluation* of quantities based on their given properties. As a rule, these properties are quite obvious intuitively, and justifying the use of them, we apply only the simplest plausible considerations, leaving a more thorough justification to other branches of science.

6.3.1 Modifications of Theorem 6.1.2 allow us to obtain integral representations of various additive physical and mechanical quantities. We consider one of these modifications.

Proposition *Let* $(X, \mathfrak{A}, \mu)$ *be a finite measure space, and let* φ *be an additive function defined on a* σ*-algebra* $\mathfrak{A}$. *If there exists a bounded measurable function* f *such that*

$$\mu(A) \inf_A f \leqslant \varphi(A) \leqslant \mu(A) \sup_A f \quad \textit{for every } A \textit{ in } \mathfrak{A},$$

then $\varphi(A) = \int_A f\, d\mu$ $(A \in \mathfrak{A})$.

Generalizing the definition from Sect. 6.1.2, we call the function f the *density of the additive function* φ. As follows from Theorem 4.5.4, the density is determined uniquely up to equivalence.

We note that we did not assume in the proposition that the additive function φ is countably additive. This weakening of the conditions imposed on φ is compensated by the assumptions that the measure μ is finite and the density f is bounded.

Proof We fix an arbitrary $\varepsilon > 0$ and consider the sets

$$A_k = \{x \in A \mid k\varepsilon \leqslant f(x) < (k+1)\varepsilon\} \quad (k \in \mathbb{Z}).$$

These sets are measurable and constitute a finite partition of the set A (if the quantity $|k|$ is sufficiently small, then $A_k = \varnothing$ since f is bounded). Summing the inequalities $\varepsilon k\mu(A_k) \leqslant \varphi(A_k) \leqslant \varepsilon(k+1)\mu(A_k)$, which follow from the two-sided estimate, we see that $\varphi(A)$ is closely approximated by the sum $S = \varepsilon \sum_{k \in \mathbb{Z}} k\mu(A_k)$,

$$S \leqslant \varphi(A) \leqslant S + \varepsilon\mu(A).$$

In the same way, from the inequalities $\varepsilon k\mu(A_k) \leqslant \int_{A_k} f(x)\, d\mu(x) \leqslant \varepsilon(k+1)\mu(A_k)$, it follows that

$$S \leqslant \int_A f(x)\, d\mu(x) \leqslant S + \varepsilon\mu(A).$$

Thus, $|\varphi(A) - \int_A f(x)\, d\mu(x)| \leqslant \varepsilon\mu(A)$, which is equivalent to the required statement since ε is arbitrary. □

6.3.2 We use Proposition 6.3.1 to calculate the attractive force between a material particle with mass μ_0 and a compact set $A \subset \mathbb{R}^3$ on which the mass μ is distributed. We assume that the particle lies outside the set A.

Not to deal with a vector quantity, we consider the projection of the attractive force $\vec{F}(A)$ in a fixed direction corresponding to a unit vector $\vec{l}$, i.e., the inner product $F_l(A) = \langle \vec{F}(A), \vec{l} \rangle$.

Obviously, $F_l(A)$ is an additive set function. Without loss of generality, we may assume that the point mass is concentrated at the origin. If the set A degenerates to a point $w_0 \neq 0$, then, by the law of gravitation, we have

$$F_l(A) = \gamma \mu_0 \mu(A) \frac{\langle w_0, \vec{l} \rangle}{r^3},$$

where $r = \|w_0\|$ and γ is a proportionality coefficient (the gravitational constant). It is natural to assume that the following estimates are valid:

$$\gamma \mu_0 \mu(A) \inf_{w \in A} \frac{\langle w, \vec{l} \rangle}{\|w\|^3} \leqslant F_l(A) \leqslant \gamma \mu_0 \mu(A) \sup_{w \in A} \frac{\langle w, \vec{l} \rangle}{\|w\|^3}.$$

By Proposition 6.3.1, we obtain that

$$F_l(A) = \gamma \mu_0 \int_A \frac{\langle w, \vec{l} \rangle}{\|w\|^3} \, d\mu(w).$$

Changing variables, we easily obtain that if the mass μ_0 is concentrated at a point w_0 with coordinates a, b, c, then the force components are calculated by the formulas

$$F_x = \gamma \mu_0 \int_A \frac{x - a}{\|r\|^3} \, d\mu(w),$$

$$F_y = \gamma \mu_0 \int_A \frac{y - b}{\|r\|^3} \, d\mu(w),$$

$$F_z = \gamma \mu_0 \int_A \frac{z - c}{\|r\|^3} \, d\mu(w),$$

where $w = (x, y, z)$ and $r = \|w - w_0\|$.

Example We calculate the force $\vec{F}$ that the uniform ball of radius R exerts on a particle of unit mass (we assume that the particle lies outside the ball).

We assume that the center of the ball coincides with the origin, the particle has the coordinates $(0, 0, c)$, $c > R$, and the mass is distributed in the ball with (constant) density ρ. Using the formulas for the components of the attractive force, we obtain

$$F_x = \gamma \iiint_{B(R)} \frac{\rho x}{(x^2 + y^2 + (z - c)^2)^{3/2}} \, dx\, dy\, dz,$$

$$F_y = \gamma \iiint_{B(R)} \frac{\rho y}{(x^2 + y^2 + (z - c)^2)^{3/2}} \, dx\, dy\, dz,$$

$$F_z = \gamma \iiint_{B(R)} \frac{\rho (z - c)}{(x^2 + y^2 + (z - c)^2)^{3/2}} \, dx\, dy\, dz.$$

From symmetry considerations, it is clear that $F_x = F_y = 0$, which, of course, follows easily from the fact that the integrands are odd functions. Converting to polar coordinates, we see that

$$
\begin{aligned}
F_z &= \gamma \iiint_{B(R)} \frac{\rho\,(z-c)}{(x^2+y^2+(z-c)^2)^{3/2}}\,dx\,dy\,dz \\
&= \gamma \int_0^{2\pi}\int_0^{\pi}\int_0^{R} \frac{\rho\,(r\cos\theta - c)r^2\sin\theta}{(r^2-2rc\cos\theta+c^2)^{3/2}}\,dr\,d\theta\,d\varphi \\
&= 2\pi\gamma\rho \int_0^R r^2\left(\int_0^{\pi} \frac{(r\cos\theta - c)\sin\theta}{(r^2-2rc\cos\theta+c^2)^{3/2}}\,d\theta\right)dr.
\end{aligned}
$$

We leave it as an exercise for the reader to verify that the resulting integral with respect to θ is equal to $-\frac{2}{c^2}$. Therefore,

$$
F_z = 2\pi\gamma\rho \int_0^R r^2\left(-\frac{2}{c^2}\right)dr = -\gamma\rho\frac{4\pi}{3}R^3\frac{1}{c^2} = -\gamma\frac{\mu(B(R))}{c^2}.
$$

Thus, a material particle is attracted by a uniform ball as if all the ball's mass were concentrated at its center.

6.3.3 We consider one more application of Proposition 6.3.1.

Let P be a fixed plane in the space $\mathbb{R}^m$. The plane divides the space $\mathbb{R}^m$ into two half-spaces one of which will be called the (+)-half-space and the other one the (−)-half-space. By the *arm* $p(x)$ of a point x with respect to the plane P, we mean the distance from x to P taken with the plus sign if the point belongs to the (+)-half-space and with the minus sign otherwise. If $P = P_k$ is the coordinate plane $x_k = 0$, then by the (+)-half-space, we mean the half-space $x_k \geqslant 0$. Then the arm of x with respect to P_k is just the kth coordinate of x. By a mass distributed on a set $A \subset \mathbb{R}^m$, we mean a Borel measure μ concentrated on A ($\mu(\mathbb{R}^m \setminus A) = 0$). In particular, by a point mass μ_0 concentrated at a point x we mean the measure generated by the load μ_0 at x (see Sect. 1.2.2, Example (4)).

It is well known from theoretical mechanics that the statical moment of a distributed mass of a set A with respect to a plane P is the physical quantity $M_P(A)$ characterizing the "disequilibrium degree". It has the following properties.

(1) Additivity:

$$
M_P(A \cup B) = M_P(A) + M_P(B), \quad \text{if } A \cap B = \varnothing
$$

(here and below, we assume that all sets under consideration are Borel sets).

(2) The moment satisfies the inequality

$$
\mu(A)\inf_{x\in A} p(x) \leqslant M_P(A) \leqslant \mu(A)\sup_{x\in A} p(x),
$$

where $\mu(A)$ is the mass of A.

If $A = \{x_0\}$ is a singleton and μ_0 is a mass concentrated at a point x_0, then Property (2) implies that $M_P(A) = \mu_0\, p(x_0)$.

We point out that condition (2) is natural. Indeed, if a set A lies in the (+)-half-space and we concentrate all mass distributed in A at a point that is farther from the plane P than the points of the set A, then we obtain a system with "even less equilibrium" than before. This corresponds to the right-hand inequality in property (2).

Property (2) implies that the moment is positive in the sense that the moment of a set lying in the (+)-half-space is non-negative. Since the moment is additive and positive, it is monotonic for the sets lying in the (+)-half-subspace: if $A \subset B$, then $M_P(A) \leqslant M_P(B)$. However, there is no need to dwell on these properties because they follow from the integral representation of the moment. Since the moment is an additive set function satisfying the two-sided estimate, we can use Proposition 6.3.1. The direct application of this proposition shows that the following statement is valid.

Proposition *Let μ be a finite mass distributed on a bounded set A. Then*

$$M_P(A) = \int_A p(x)\, d\mu(x).$$

Definition The *center of mass* of a set A with mass distributed on it is a point C such that the moment of A with respect to any plane passing through C is equal to zero.

We prove that the center of mass always exists.

First, we find necessary conditions for a point to be a center of mass. Let μ be non-zero mass distributed on A, and let $C = (c_1, \ldots, c_m)$ be a center of mass. Let P be a plane that passes through C and is defined by the equation $x_k - c_k = 0$. Obviously, the arm of a point $x = (x_1, \ldots, x_m)$ with respect to this plane coincides (depending on the choice of (+)-half-subspace) either with $x_k - c_k$ or with $c_k - x_k$. In any case, we have

$$0 = M_P(A) = \int_A p(x)\, d\mu(x) = \int_A (x_k - c_k)\, d\mu(x) = M_k(A) - c_k \mu(A),$$

where $M_k(A)$ is the moment with respect to the plane $x_k = 0$. Thus, only the point C with coordinates $c_k = M_k(A)/\mu(A)$ $(k = 1, \ldots, m)$ can be a center of mass.

Now, we prove that this point is indeed the center of mass. Let P be an arbitrary plane passing through C. This plane is given by an equation of the form

$$\sum_{k=1}^{m} a_k (x_k - c_k) = 0.$$

Without loss of generality, we may assume that $\sum_{k=1}^{m} a_k^2 = 1$. Taking the half-subspace defined by the inequality $\sum_{k=1}^{m} a_k (x_k - c_k) > 0$ as the (+)-half-space, we obtain that the arm of the point $x = (x_1, \dots, x_m)$ coincides with the sum $\sum_{k=1}^m a_k(x_k - c_k)$. Therefore,

$$M_P(A) = \int_A \sum_{k=1}^m a_k(x_k - c_k)\, d\mu = \sum_{k=1}^m a_k\big(M_k(A) - c_k\mu(A)\big) = 0,$$

which proves the statement.

As well as a proof of the existence of a center of mass, we obtain the following formulas for its coordinates:

$$c_k = \frac{1}{\mu(A)} \int_A x_k\, d\mu(x) \quad (k = 1, \dots, m).$$

We note that if the set A in question is finite, $A = \{a_1, \dots, a_N\}$, and the mass μ_k is concentrated at a_k, then the above formulas imply that the center of mass of such a system is a convex combination of the points a_k,

$$C = \frac{\mu_1 a_1 + \cdots + \mu_N a_N}{\mu_1 + \cdots + \mu_N}.$$

The coefficients of this convex combination are proportional to the masses concentrated at the corresponding points.

Example We find the center of mass C of the uniform set $B_+^m = B(0,1) \cap \mathbb{R}_+^m$ (the set $\mathbb{R}_+^m$ consists of the points with positive coordinates). We may assume that the density of the mass distribution is equal to 1, i.e., $\mu = \lambda_m$. Then the mass is equal to the volume of B_+^m, $\mu(B_+^m) = \frac{\alpha_m}{2^m}$ (recall that $\alpha_m = \lambda_m(B(0,1))$).

By symmetry, the coordinates of the vector C are equal, $C = (c, \dots, c)$. By the formulas for the coordinates of the center of mass, we obtain

$$c = \frac{1}{\mu(B_+^m)} \int_{B_+^m} x_m\, dx = \frac{2^m}{\alpha_m} \int_{B_+^m} x_m\, dx.$$

To evaluate this integral, we represent the vector x from B_+^m in the form $x = (y, t)$, where $y \in B_+^{m-1}$, $t \in (0,1)$ and $\|y\|^2 + t^2 < 1$. Then

$$\begin{aligned} c &= \frac{2^m}{\alpha_m} \int_0^1 t\lambda_{m-1}\big(\sqrt{1-t^2}\, B_+^{m-1}\big)\, dt \\ &= \frac{2^m}{\alpha_m} \int_0^1 t \cdot \frac{\alpha_{m-1}}{2^{m-1}} \cdot \big(1 - t^2\big)^{\frac{m-1}{2}}\, dt = \frac{2}{m+1}\, \frac{\alpha_{m-1}}{\alpha_m}. \end{aligned}$$

Since $\alpha_m = \pi^{\frac{m}{2}} / \Gamma(1 + \frac{m}{2})$ (see Sect. 5.4.2), we have

$$c = \frac{2}{m+1}\, \frac{\Gamma(\frac{m+2}{2})}{\sqrt{\pi}\, \Gamma(\frac{m+1}{2})} = \frac{1}{\sqrt{\pi}}\, \frac{\Gamma(\frac{m+2}{2})}{\Gamma(\frac{m+3}{2})}.$$

For $m = 3$, we obtain $C = (\frac{3}{8}, \frac{3}{8}, \frac{3}{8})$. We observe that Stirling's formula (see Sect. 7.2.6) implies that the coordinates of the point C tend to zero like $\sqrt{\frac{2}{\pi m}}$ as $m \to \infty$.

Therefore, the norm $\|C\|$ tends to $\sqrt{\frac{2}{\pi}}$. A more detailed study shows that $\|C\|$ increases with the dimension, which means that an increasingly larger portion of volume of the set B^m_+ is concentrated near the spherical part of its boundary.

EXERCISES

1. Find the force with which the uniform spherical layer $B_{r,R} = \{w \in \mathbb{R}^3 | r \leqslant \|w\| \leqslant R\}$ attracts a material particle w_0, $w_0 \notin B_{r,R}$. Consider the cases $\|w_0\| > R$ and $\|w_0\| < r$.

2. Assume that a set A lies in a plane on one side of a line ℓ. Use the result of Exercise 2, Sect. 5.4 to prove Guldin's theorem:[4] *the volume of a solid of revolution obtained by rotating the set A about the line ℓ is equal to the product of the area of A and the distance traveled by the center of mass of A* (it is assumed that the mass is distributed on A with constant density).

Assume that a finite mass μ is distributed on a bounded set $A \subset \mathbb{R}^3$. The moment of inertia $I_\ell(A)$ of a set A with respect to an axis ℓ is a physical quantity characterizing the kinetic energy of a body rotating about this axis. More precisely, the kinetic energy is equal to $\frac{1}{2} I_\ell(A)\omega^2$, where ω is the angular velocity. For a point mass μ_0 located at distance r from the axis of rotation, the kinetic energy E is calculated by the formula $E = \frac{\mu_0 v^2}{2} = \frac{\mu_0 r^2}{2}\omega^2$. Thus, in this case, the moment of inertia is equal to $\mu_0 r^2$.

It is clear from physical considerations that the moment of inertia with respect to a fixed axis is an additive set function that does not decrease as the distance between the body and the axes increases. Thus, if we concentrate all mass at a point of the body farthest from the axis, then the moment of inertia can only increase. Respectively, if we concentrate all mass at a point closest to the axis, then the moment of inertia can only decrease. This means that the following two-sided estimates are valid for $I_\ell(A)$:

$$\mu(A) \inf_{x \in A} \operatorname{dist}^2(x, \ell) \leqslant I_\ell(A) \leqslant \mu(A) \sup_{x \in A} \operatorname{dist}^2(x, \ell).$$

This allows us to use Proposition 6.3.1 in the calculation of moments of inertia.

3. Find the moments of inertia of a uniform ball with respect to its diameter and a tangent line.

4. Find the moments of inertia of a uniform right circular cylinder with respect to the axis of symmetry, a generatrix, and a diameter of the base.

5. Find the moment of inertia of a ball with respect to its diameter if the mass distribution density is inversely proportional to the distance from the origin.

---

[4]Paul Guldin (1577–1643)—Swiss mathematician.

6. For which of the lines parallel to each other is the moment of inertia of a body minimal?
7. Assume that mass is uniformly distributed on a measurable cone (see, e.g., Sect. 5.4.2, Example 1). Prove that the distance from the center of mass to the plane containing the base of the cone is proportional to the hight of the cone. Prove that the proportionality coefficient depends only on the dimension and find this coefficient.
8. Assume that mass is uniformly distributed on a convex body $K \subset \mathbb{R}^m$ and that the center of mass coincides with the origin. Prove that $-K \subset mK$. *Hint*. Prove that each chord passing through the center of mass is divided by the center of mass into segments with length at least $\frac{1}{m+1}$ of the length of the chord.
9. Verify that the moment of inertia of a uniform cube (of arbitrary dimension) with respect to a line passing through the center of cube does not depend on the line. For which mass distribution does this property remain valid? Is it true that the sum of the squares of the distances from the vertices to a line passing through the center of the cube is the same whichever line we take?

6.4 *Distribution Functions. Independent Functions

6.4.1 We consider an important specific case of the weighted image of a measure. As in Sect. 6.1, let $(X, \mathfrak{A}, \mu)$ be a measure space. Unless otherwise stated, we assume that all functions in question are measurable.

Let $Y = \mathbb{R}$, and let $\mathfrak{B} = \mathfrak{B}(\mathbb{R})$ be the σ-algebra of Borel sets. Further, let h be a measurable almost everywhere finite function on X. It is well known (see Proposition 3.1.2) that the preimage $h^{-1}(B)$ is measurable for every Borel set $B \subset \mathbb{R}$. Therefore, we can define the measure $\nu = h(\mu)$ on $\mathfrak{B}$, which is the image of μ with respect to h. We assume in addition that the measure ν is finite on intervals. Then ν is a Borel–Stieltjes measure and, consequently, is generated by a non-decreasing function. To specify this function, we introduce the following definition.

Definition Let h be a measurable almost everywhere finite function on X. We assume that the set

$$X(h < t) = \{x \in X \mid h(x) < t\}$$

has a finite measure for every $t \in \mathbb{R}$ and put $H(t) = \mu(X(h < t))$. The function H is called the *distribution function* of the function h (with respect to the measure μ or in measure μ).

It is obvious that a distribution function is non-decreasing. From the lower continuity of measure, it follows that a distribution function is left-continuous. We note that the function $t \mapsto \mu(X(h \leqslant t))$ coincides with H at all points of continuity. If the measure is finite, then every measurable almost everywhere finite function has a distribution function.

Proposition *Under the assumption of the definition, $h(\mu)$ coincides with the Borel–Stieltjes measure generated by H.*

For the definition of a Borel–Stieltjes measure, see Sect. 4.10.3.

Proof By the uniqueness theorem 1.5.1, it is sufficient to check that the measures in question coincide on the right-open semi-intervals, which, in turn, follows from the definition of the functions H. □

In our specific case, the general theorem proved in Sect. 6.1.1 turns into the theorem stated below. We notice that a function f considered in the above-mentioned general theorem must be measurable with respect to the σ-algebra $\mathfrak{B}$, which now is the σ-algebra of Borel subsets of the real line. Such functions are called Borel measurable. It is obvious that all continuous functions are Borel measurable.

Theorem *Let f be a non-negative Borel measurable function defined on $\mathbb{R}$, let h be an almost everywhere finite measurable function on X, and let H be the distribution function of h. Then*

$$\int_X f\big(h(x)\big)\,d\mu(x) = \int_{\mathbb{R}} f(t)\,dH(t). \tag{1}$$

This relation remains valid for functions f taking values of an arbitrary sign provided the composition $f \circ h$ is summable.

The above theorem is obtained from Theorem 6.1.1 by putting $(Y, \mathfrak{B}, \nu) = (\mathbb{R}, \mathfrak{B}(\mathbb{R}), h(\mu))$, $\Phi = h$ and $\omega \equiv 1$. We note that the condition in Definition 6.1.1 (the preimage $h^{-1}(B)$ of a Borel set B is measurable) is fulfilled by Proposition 3.1.2.

Remark Specific cases of Eq. (1) are the formulas

$$\int_X h\,d\mu = \int_{-\infty}^{\infty} t\,dH(t), \qquad \int_X |h|^p\,d\mu = \int_{-\infty}^{\infty} |t|^p\,dH(t),$$

which are frequently used in probability theory. The reader familiar with probability theory will recognize these formulas as those for the mean and the absolute moments of a random variable h.

6.4.2 We give several examples.

Example 1 We consider the integral $\int_{\mathbb{R}^m} f(\|x\|)\,dx$, where f is a non-negative function measurable on the semi-axis $(0, +\infty)$.

Let h be a function defined by the equation $h(x) = \|x\|$ for $x \in \mathbb{R}^m$. Its distribution function is as follows: $H(t) = 0$ if $t \leqslant 0$ and $H(t) = \alpha_m t^m$ if $t > 0$, where α_m

is the volume of the unit ball. Therefore,

$$\int_{\mathbb{R}^m} f(\|x\|)\,dx = \int_0^\infty f(t)\,dH(t) = m\alpha_m \int_0^\infty f(t)t^{m-1}\,dt$$

(the last equality is a consequence of formula (5) of Remark 4.10.4 and the fact that H is smooth on $(0, +\infty)$).

Example 2 The formula from Example 1 provides a new way to calculate the Euler–Poisson integral $I = \int_{-\infty}^{\infty} e^{-x^2}\,dx$, the value of which is already known (see Sect. 5.3.2, Example 1 and Sect. 6.2.4, Example 2).

As before, we transform I^2 by Fubini's theorem,

$$I^2 = \int_{-\infty}^{\infty} e^{-x^2}\,dx \cdot \int_{-\infty}^{\infty} e^{-y^2}\,dy = \iint_{\mathbb{R}^2} e^{-(x^2+y^2)}\,dx\,dy.$$

Now, using the formula from Example 1 for $m = 2$ and taking $f(t) = e^{-t^2}$, we obtain

$$\iint_{\mathbb{R}^2} e^{-(x^2+y^2)}\,dx\,dy = \int_0^\infty e^{-t^2}\,d(\pi t^2) = (-\pi)e^{-t^2}\Big|_0^\infty = \pi.$$

Thus, $I = \sqrt{\pi}$.

Example 3 Generalizing the method used in the previous example, we find the volume V of the set

$$W = \{(x_1, \dots, x_m) \in \mathbb{R}^m \mid |x_1|^{p_1} + \dots + |x_m|^{p_m} \leqslant 1\},$$

where $p_1, \dots, p_m$ are positive numbers (in Sect. 5.4.2, Example 4, this problem is solved without using the distribution function). To this end, we calculate the integral

$$I = \int_{\mathbb{R}^m} \exp\left(-\sum_{j=1}^m |x_j|^{p_j}\right) dx$$

in two different ways.

On the one hand, we use Fubini's theorem and obtain

$$I = \prod_{j=1}^m \int_{-\infty}^\infty e^{-t^{p_j}}\,dt = 2^m \prod_{j=1}^m \Gamma\left(1 + \frac{1}{p_j}\right).$$

On the other hand, we can use formula (1), with $f(t) = e^{-t}$ and $h(x) = |x_1|^{p_1} + \dots + |x_m|^{p_m}$. The corresponding distribution function $H(t)$ for $t > 0$ can be calculated by the linear change of variables $x_j = t^{1/p_j} u_j$ ($j = 1, \dots, m$):

$$\begin{aligned} H(t) &= \lambda_m(\{(x_1, \dots, x_m) \mid |x_1|^{p_1} + \dots + |x_m|^{p_m} \leqslant t\}) \\ &= t^q \lambda_m(\{(u_1, \dots, u_m) \mid |u_1|^{p_1} + \dots + |u_m|^{p_m} \leqslant 1\}) = t^q \lambda_m(W) = t^q V, \end{aligned}$$

where $q = \frac{1}{p_1} + \cdots + \frac{1}{p_m}$. Therefore, formula (1) yields the relation

$$I = \int_0^\infty f(t)\,dH(t) = \int_0^\infty e^{-t}\,d\big(Vt^q\big) = V\Gamma(1+q).$$

Thus,

$$V = \frac{I}{\Gamma(1+q)} = \frac{2^m}{\Gamma(1+\frac{1}{p_1}+\cdots+\frac{1}{p_m})}\prod_{j=1}^m \Gamma\Big(1+\frac{1}{p_j}\Big).$$

In the case where $p_1 = \cdots = p_m = 2$, we once again obtain the formula for the volume of the m-dimensional unit ball B^m (see Sect. 5.4.2),

$$\lambda_m\big(B^m\big) = \frac{2^m\Gamma^m(1+\frac{1}{2})}{\Gamma(1+\frac{m}{2})} = \frac{\pi^{\frac{m}{2}}}{\Gamma(1+\frac{m}{2})}.$$

In conclusion, we present a more general result. We use a distribution function to estimate the ratio of the volumes of the compact sets $K \subset \mathbb{R}^m$ and $\Delta K = \{x - y \mid x, y \in K\}$. In the general case, such an estimate is impossible (for example, if $K \subset \mathbb{R}^2$ consists of two non-parallel intervals, then $\lambda_2(K) = 0$ but $\lambda_2(\Delta K) > 0$). However, the following statement is proved in [RS].

Theorem *Let $K \subset \mathbb{R}^m$ be a convex body. Then*

$$2^m\lambda_m(K) \leqslant \lambda_m(\Delta K) \leqslant C_{2m}^m\lambda_m(K).$$

Proof The estimate from below is easily obtained from the Brunn–Minkowski inequality. Indeed, since $\lambda_m(-K) = \lambda_m(K)$ and $\Delta K = K + (-K)$, we have $2\lambda_m^{\frac{1}{m}}(K) = \lambda_m^{\frac{1}{m}}(K) + \lambda_m^{\frac{1}{m}}(-K) \leqslant \lambda_m^{\frac{1}{m}}(K + (-K)) = \lambda_m^{\frac{1}{m}}(\Delta K)$.

The estimate from above is harder to prove. It is obvious that

$$\lambda_m^2(K) = \int_{\mathbb{R}^m} \chi_K(x)\Big(\int_{\mathbb{R}^m} \chi_K(y)\,dy\Big)dx = \int_{\mathbb{R}^m} \chi_K(x)\Big(\int_{\mathbb{R}^m} \chi_K(x-z)\,dz\Big)dx$$

$$= \int_{\mathbb{R}^m}\Big(\int_{\mathbb{R}^m} \chi_K(x)\chi_K(x-z)\,dx\Big)dz = \int_{\mathbb{R}^m} \lambda_m\big(K \cap (z+K)\big)\,dz.$$

If $z \notin \Delta K$, then $K \cap (z+K) = \varnothing$. Indeed, in the representation $z = x - (x - z)$, we have either $x \notin K$ or $x - z \notin K$. Consequently, $\lambda_m(K \cap (z+K)) = 0$ for $z \notin \Delta K$, and so,

$$\lambda_m^2(K) = \int_{\Delta K} \lambda_m\big(K \cap (z+K)\big)\,dz.$$

To estimate $\lambda_m(K \cap (z+K))$ from below, we take $z \in \Delta K$, $z \neq 0$, and find $h = h(z) \in (0,1]$ such that $\frac{z}{h} \in \partial\Delta K$. Let $\frac{z}{h} = a - b$, where $a, b \in K$. We prove that

$$ha + (1-h)K \subset K \cap (z+K).$$

The inclusion $ha + (1-h)K \subset K$ is obvious, and the inclusion $ha + (1-h)K \subset z + K$ follows from the fact that $ha = hb + h$, and, therefore,

$$ha + (1-h)K = z + hb + (1-h)K \subset z + K.$$

Thus,

$$\lambda_m\bigl(K \cap (z+K)\bigr) \geqslant \lambda_m\bigl(ha + (1-h)K\bigr) = (1-h)^m \lambda_m(K).$$

Consequently,

$$\lambda_m^2(K) \geqslant \lambda_m(K) \int_{\Delta K} \bigl(1 - h(z)\bigr)^m \, dz.$$

To calculate the last integral, we consider the distribution function for h:

$$H(t) = \lambda_m\bigl(\{z \in \Delta K \mid h(z) < t\}\bigr) = \lambda_m(t\,\Delta K) = t^m \lambda_m(\Delta K) \quad \text{if } 0 < t < 1,$$
$$H(t) = 0 \quad \text{if } t \leqslant 0,$$
$$H(t) = \lambda_m(\Delta K) \quad \text{if } t \geqslant 1.$$

By Theorem 6.4.1, we have

$$\int_{\Delta K} \bigl(1 - h(z)\bigr)^m \, dz = \int_0^1 (1-t)^m \, dH(t) = m\lambda_m(\Delta K) \int_0^1 t^{m-1}(1-t)^m \, dt$$
$$= m\mathrm{B}(m, m+1)\lambda_m(\Delta K) = \frac{m!\,m!}{(2m)!}\lambda_m(\Delta K).$$

Thus,

$$\lambda_m^2(K) \geqslant \frac{m!m!}{(2m)!}\lambda_m(\Delta K)\lambda_m(K),$$

which is equivalent to the required inequality. □

Remark Obviously, $\Delta K = 2K$ for a centrally symmetric convex body K and, consequently, $\lambda_m(\Delta K) = 2^m \lambda_m(K)$. Therefore, the estimate from below for the volume $\lambda_m(\Delta K)$ given in the theorem is sharp. The estimate from above is also sharp; it becomes an equality if K is a simplex since, in this case, we have $K \cap (z+K) = ha + (1-h)K$. We leave it to the reader to verify this equation. It is convenient to verify it for the simplex $K = \{(x_1, \dots, x_m) \in \mathbb{R}_+^m | x_1 + \cdots + x_m \leqslant 1\}$ by proving that $h(x) = \max\{\sum_{k=1}^m (x_k)_+, \sum_{k=1}^m (-x_k)_+\}$.

6.4.3 As we said before, if a measure μ is finite, then the distribution function always exists, but, in the case of an infinite measure, this is not the case. For example, every positive function summable with respect to an infinite measure does not have a distribution function. Therefore, it is often useful to change the definition of a distribution function.

Definition Let h be a non-negative measurable function on X such that, for all $t > 0$, the set

$$X(h > t) = \{x \in X \mid h(x) > t\}$$

has a finite measure. We put $\widetilde{H}(t) = \mu(X(h > t))$ and call the function $\widetilde{H}$ the *decreasing distribution function* for h.

To avoid ambiguity, we sometimes call the distribution function defined in Sect. 6.4.1 the *increasing* distribution function.

From the continuity of a measure, it follows that $\widetilde{H}(t) \underset{t\to+\infty}{\longrightarrow} 0$ if and only if $h(x) < +\infty$ almost everywhere on X. As well as the increasing distribution function, the function $\widetilde{H}$ is also left-continuous. The sets $X(h \leqslant t)$ and $X(h > t)$ both have a finite measure only if the measure μ is finite. In this case, we obtain $H(t+0) + \widetilde{H}(t) = \mu(X)$. We note that a non-negative measurable function h certainly has a decreasing distribution function if $\int_X h^p \, d\mu < +\infty$ for some $p > 0$. This follows directly from Chebyshev's inequality (see Theorem 4.4.4):

$$\widetilde{H}(t) = \mu\big(X(h > t)\big) \leqslant \frac{1}{t^p} \int_X h^p \, d\mu < +\infty \quad \text{for all } t > 0.$$

We do not state an analog of Theorem 6.4.1 for decreasing distribution functions, contenting ourselves instead with a more specific statement.

Proposition *Let $p > 0$, and let h be a non-negative measurable function with a decreasing distribution function $\widetilde{H}$. Then*

$$\int_X h^p \, d\mu = p \int_0^\infty t^{p-1} \widetilde{H}(t) \, dt.$$

Proof We transform the integral $\int_X h^p \, d\mu$ as follows:

$$\int_X h^p \, d\mu = p \int_X \left(\int_0^{h(x)} t^{p-1} \, dt \right) d\mu(x).$$

The repeated integral on the right-hand side is equal to the double integral of the function $(x,t) \mapsto t^{p-1}$ over the subgraph of the function $\mathscr{P} = \mathscr{P}_h(X)$ of h. To change the order of integration, we observe that, for $t > 0$, the cross section $\mathscr{P}^t$ of the subgraph is the set $X(h \geqslant t)$ (see Fig. 6.4). Therefore, changing the order of integration, we obtain

$$\int_X h^p \, d\mu = p \int_0^\infty t^{p-1} \left(\int_{X(h \geqslant t)} 1 \, d\mu \right) dt = p \int_0^\infty t^{p-1} \mu\big(X(h \geqslant t)\big) \, dt.$$

It remains to observe that $\mu(X(h \geqslant t)) = \widetilde{H}(t)$ almost everywhere, namely, at the points of continuity of the function $\widetilde{H}$. □

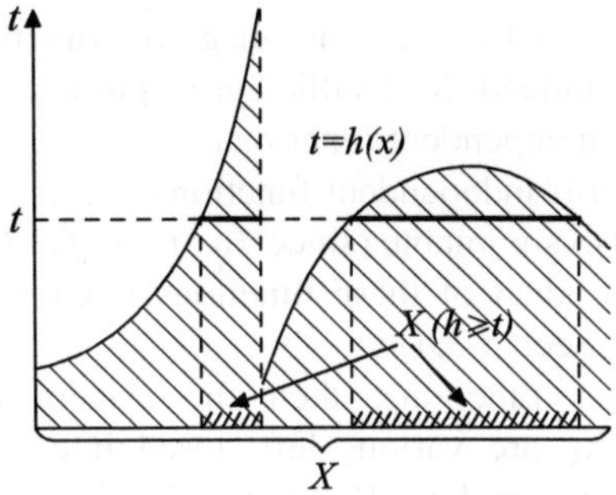

Fig. 6.4 Cross section of the subgraph of h on a level t

6.4.4 Throughout this section, we assume that all functions in question are defined on a fixed normalized measure space $(X, \mathfrak{A}, \mu)$ $(\mu(X) = 1)$. Let $f_1, \dots, f_n$ be measurable almost everywhere finite real functions. For each function f_k, there exists a Borel measure ν_k that is the image of μ with respect to f_k; this measure is called the *distribution* of f_k. We also consider the map $\Phi : X \to \mathbb{R}^n$ with coordinate functions $f_1, \dots, f_n$, and put $\nu = \Phi(\mu)$. The measure ν is called the *simultaneous distribution* of the functions $f_1, \dots, f_n$. We introduce the notion of independent functions, which play a fundamental role in probability theory.

Definition Functions $f_1, \dots, f_n$ are called *independent* if ν coincides with the measure $\nu_1 \times \cdots \times \nu_n$ (the product of the measures $\nu_1, \dots, \nu_n$). Functions of an infinite family are called independent if the functions in each finite subfamily are independent.

The uniqueness of measure extensions implies that to prove that the measures ν and $\nu_1 \times \cdots \times \nu_n$ coincide it is sufficient to prove that they coincide on the cells, i.e., that for every cell $P = \prod_{k=1}^{n} [a_k, b_k)$ the equation

$$\mu\big(\Phi^{-1}(P)\big) = \prod_{k=1}^{n} \mu\big(f_k^{-1}\big([a_k, b_k)\big)\big)$$

is valid. Since the set $\Phi^{-1}(P)$ coincides with $\bigcap_{k=1}^{n} f_k^{-1}([a_k, b_k))$, the last equation can be represented in the form

$$\mu\left(\bigcap_{k=1}^{n} f_k^{-1}\big([a_k, b_k)\big)\right) = \prod_{k=1}^{n} \mu\big(f_k^{-1}\big([a_k, b_k)\big)\big). \tag{2}$$

If the functions $f_1, \dots, f_n$ are independent and a non-negative function h defined on $\mathbb{R}^n$ is Borel measurable, then Theorem 6.1.1 implies

$$\int_X h\big(f_1(x), \dots, f_n(x)\big)\, d\mu(x) = \int_{\mathbb{R}^n} h(t_1, \dots, t_n)\, d\nu_1(t_1) \cdots d\nu_n(t_n), \tag{3}$$

and this equation remains valid if the function h is summable.

In turn, if Eq. (3) is valid for every non-negative function h, then, in particular, Eq. (2) is also valid. Indeed, it is sufficient to put $h = \chi_P$. Thus, Eq. (3) is a characteristic property of independent functions.

It follows from (3) that if independent functions $f_1, \dots, f_n$ are summable, then the product $f_1 \cdots f_n$ is also summable (since $\int_X |f_1 \cdots f_n| \, d\mu = \prod_{k=1}^n \int_X |f_k| \, d\mu$), and the integral of the product of these functions is equal to the product of the integrals (cf. Corollary 1, Sect. 5.3.4).

If $f_1, \dots, f_n$ are real functions, then the system of sets of the form $\bigcap_{k=1}^n f_k^{-1}(\Delta_k)$, where Δ_k are various left-closed intervals, is a semiring; we denote it by $\mathscr{P}(f_1, \dots, f_n)$ and its Borel hull by $\mathfrak{A}(f_1, \dots, f_n)$. It is obvious that the functions $f_1, \dots, f_n$ are measurable with respect to $\mathfrak{A}(f_1, \dots, f_n)$ and that $\mathfrak{A}(f_1, \dots, f_n)$ is the minimal σ-algebra with respect to which all functions $f_1, \dots, f_n$ are measurable. Similarly, we denote by $\mathfrak{A}(\{f_n\}_{n\geqslant 1})$ the Borel hull of the union $\bigcup_{n=1}^\infty \mathscr{P}(f_1, \dots, f_n)$. This is the minimal σ-algebra with respect to which all functions of the sequence $\{f_n\}_{n\geqslant 1}$ are measurable.

Lemma *Let the functions $f_1, \dots, f_n, g_1, \dots, g_m$ be independent. Then the algebras $\mathfrak{A}(f_1, \dots, f_n)$ and $\mathfrak{A}(g_1, \dots, g_m)$ are independent in the sense that*

$$\mu(A \cap B) = \mu(A)\,\mu(B) \quad \text{for all } A \in \mathfrak{A}(f_1, \dots, f_n),\ B \in \mathfrak{A}(g_1, \dots, g_m). \tag{4}$$

Proof If $A \in \mathscr{P}(f_1, \dots, f_n)$ and $B \in \mathscr{P}(g_1, \dots, g_m)$, then Eq. (4) is valid by the definition of independent functions. We fix a set $Q \in \mathscr{P}(g_1, \dots, g_m)$ and consider the following two measures defined on the σ-algebra $\mathfrak{A}(f_1, \dots, f_n)$:

$$\mu_1(A) = \mu(A \cap Q) \quad \text{and} \quad \mu_2(A) = \mu(A)\,\mu(Q) \quad \big(A \in \mathfrak{A}(f_1, \dots, f_n)\big).$$

These measures coincide on the semiring $\mathscr{P}(f_1, \dots, f_n)$, and, by the uniqueness theorem, they coincide on $\mathfrak{A}(f_1, \dots, f_n)$. Now, we fix an arbitrary set $A \in \mathfrak{A}(f_1, \dots, f_n)$ and consider the following two measures defined on the σ-algebra $\mathfrak{A}(g_1, \dots, g_m)$:

$$\nu_1(B) = \mu(A \cap B) \quad \text{and} \quad \nu_2(B) = \mu(A)\,\mu(B) \quad \big(B \in \mathfrak{A}(g_1, \dots, g_m)\big).$$

If follows from the above that ν_1 and ν_2 coincide on $\mathscr{P}(g_1, \dots, g_m)$, and, by the uniqueness theorem, they also coincide on $\mathfrak{A}(g_1, \dots, g_m)$, which completes the proof. □

Remark Equation (4) also remains valid in the case of infinite families of independent functions since this equation is valid for $A \in \bigcup_{n=1}^\infty \mathscr{P}(f_1, \dots, f_n)$ and $B \in \bigcup_{n=1}^\infty \mathscr{P}(g_1, \dots, g_n)$.

Corollary *Let the functions $f_1, \dots, f_n, g_1, \dots, g_m$ be independent. If functions F and G are measurable with respect to the σ-algebras $\mathfrak{A}(f_1, \dots, f_n)$ and $\mathfrak{A}(g_1, \dots, g_m)$, respectively, then they are independent.*

Proof The fact that F and G are independent follows from Eq. (4) applied to the sets $A = F^{-1}(\Delta)$ and $B = G^{-1}(\Delta')$, where Δ and Δ' are arbitrary intervals. □

Let $\{f_n\}_{n\geqslant 1}$ be a sequence of independent functions, and let $\mathfrak{A}_n = \mathfrak{A}(f_n, f_{n+1}, \dots)$ be the minimal σ-algebra with respect to which all functions $f_n, f_{n+1}, \dots$ are measurable. Obviously, the σ-algebras $\mathfrak{A}_n$ decrease as n increases. The intersection of all $\mathfrak{A}_n$ contains sets that "do not depend on any number of the initial functions $f_1, \dots, f_m$". Such a set is, for example, the convergence set of the series $\sum_{k=1}^{\infty} f_k$, since the series $\sum_{k=1}^{\infty} f_k$ and $\sum_{k=n}^{\infty} f_k$ converge simultaneously.

It turns out that the following statement is valid for the sets belonging to the intersection $\bigcap_{n=1}^{\infty} \mathfrak{A}_n$.

Theorem (Zero-one law) *If $A \in \bigcap_{n=1}^{\infty} \mathfrak{A}_n$, then either $\mu(A) = 0$ or $\mu(A) = 1$.*

Proof We verify that if $B \in \mathfrak{A}_1$, then

$$\mu(A \cap B) = \mu(A)\,\mu(B). \tag{4'}$$

Indeed, by the remark to the lemma, the algebras $\mathfrak{A}(f_1, \dots, f_n)$ and $\mathfrak{A}_{n+1}$ are independent for every n, and, therefore, Eq. (4′) is valid for every set B in the semiring $\bigcup_{n=1}^{\infty} \mathscr{P}(f_1, \dots, f_n)$. By the uniqueness theorem, the measures $B \mapsto \mu(A \cap B)$ and $B \mapsto \mu(A)\,\mu(B)$ also coincide on the Borel hull of this semiring, i.e., on $\mathfrak{A}_1$, which proves Eq. (4′). For $B = A$, Eq. (4′) turns into $\mu(A) = (\mu(A))^2$ which holds only if $\mu(A) = 0$ or $\mu(A) = 1$. □

Corollary *If the functions $f_1, f_2, \dots$ are independent, then the series $\sum_{n=1}^{\infty} f_n$ either converges almost everywhere or diverges almost everywhere.*

Proof This is a special case of the zero-one law since, as noted above, the convergence set of the series belongs to the intersection $\bigcap_{n=1}^{\infty} \mathfrak{A}_n$. □

Using an infinite product of measures, we can easily verify that there exists a sequence of independent functions with arbitrarily prescribed distributions ν_n ($n = 1, 2, \dots$). Indeed, it is sufficient to consider the measure $\mu = \nu_1 \times \nu_2 \times \cdots$ in the infinite product $\mathbb{R}^{\mathbb{N}} = \mathbb{R} \times \mathbb{R} \times \cdots$ and put $h_n(t)$ equal to the nth coordinate of the point $t \in \mathbb{R}^{\mathbb{N}}$.

6.4.5 An important example of independent functions is provided by the *Rademacher functions*[5] r_n ($n = 1, 2, \dots$). The function r_n is defined as follows. We divide the interval $(0, 1)$ into equal parts by the points $k2^{-n}$, and put $r_n(x) = (-1)^k$ in the interval $\Delta_{n,k} = (k2^{-n}, (k+1)2^{-n})$ ($k = 0, 1, \dots, 2^n - 1$). Furthermore, we assume

[5] Hans Adolph Rademacher (1892–1969)—German mathematician.

that $r_n(k2^{-n}) = 0$ for $k = 0, 1, \dots, 2^n$ and that the function r_n is 1-periodic. It can easily be seen that

$$r_n(x) = r_1(2^{n-1}x) = \operatorname{sign}\sin(2^n \pi x).$$

The reader can easily verify that the values of the Rademacher functions at a point $x \in (0, 1)$ are closely related to the digits of the binary expansion of x, i.e., if x is not a dyadic fraction and $x = \sum_{n=1}^{\infty} \frac{\varepsilon_n(x)}{2^n}$, where $\varepsilon_n(x) = 0$ or 1, then $r_n(x) = 1 - 2\varepsilon_n(x)$.

It is clear that $\lambda_1(\{x \in (0,1) \mid r_n(x) = 1\}) = \lambda_1(\{x \in (0,1) \mid r_n(x) = -1\}) = \frac{1}{2}$, and, therefore, all Rademacher functions have the same increasing distribution function F,

$$F(t) = \begin{cases} 0 & t \leqslant -1, \\ \frac{1}{2} & -1 < t \leqslant 1, \\ 1 & t > 1. \end{cases}$$

The function F gives rise to the measure ν on $\mathbb{R}$ generated by the loads $\frac{1}{2}$ at the points ± 1. Obviously,

$$\int_0^1 h(r_n(x))\,dx = \frac{h(-1) + h(1)}{2} = \int_{\mathbb{R}} h(t)\,dF(t).$$

Being regarded as functions on the interval $(0, 1)$ with Lebesgue measure, the Rademacher functions form an independent system. To prove this, it is sufficient to check Eq. (3). In our case, this means that

$$\int_0^1 h(r_1(x), \dots, r_n(x))\,dx = \int_{\mathbb{R}^n} h(t_1, \dots, t_n)\,dF(t_1) \cdots dF(t_n)$$

for all n. We calculate the left-hand and right-hand sides of this equation separately. Since the values of the functions $r_1, \dots, r_n$ are constant on each interval $\Delta_{n,k}$, we see that the family $\{r_k(x)\}_{k=1}^n$ is a sequence of ± 1 for $x \in \Delta_{n,k}$. The reader can easily prove by induction that distinct intervals $\Delta_{n,k}$ give rise to distinct sequences. Since the number of intervals $\Delta_{n,k}$ as well as the number of n-tuples $\varepsilon = (\varepsilon_1, \dots, \varepsilon_n)$ with $\varepsilon_k = \pm 1$ is equal to 2^n, there is a one-to-one correspondence between them. Therefore,

$$\begin{aligned} \int_0^1 h(r_1(x), \dots, r_n(x))\,dx &= \sum_{k=1}^{2^n - 1} \int_{\Delta_{n,k}} h(r_1(x), \dots, r_n(x))\,dx \\ &= \frac{1}{2^n} \sum_{\varepsilon \in \{-1,1\}^n} h(\varepsilon_1, \dots, \varepsilon_n). \end{aligned}$$

On the other hand, since F gives rise to the measure ν generated by the loads $\frac{1}{2}$ at the points ± 1, we see that $\nu \times \cdots \times \nu$ (n times) is the measure generated by the loads 2^{-n} in the vertices of the cube $[-1,1]^n$, i.e., at all points of the form $\varepsilon=(\varepsilon_1,\dots,\varepsilon_n)$, where $\varepsilon_k=\pm1$. Therefore,

$$\int_{\mathbb{R}^n} h(t_1,\dots,t_n)\,dF(t_1)\cdots dF(t_n)=\sum_{\varepsilon\in\{-1,1\}^n} h(\varepsilon_1,\dots,\varepsilon_n)2^{-n}.$$

Since the right-hand side of this equation coincides with the right-hand side of the previous equation, we see that (3) is valid for the Rademacher functions. Similarly, it is easy to prove that the digits of the binary (decimal, p-ary) expansion of $x\in(0,1)$ are independent functions.

The use of distribution functions is helpful not only in the calculation of integrals but also in proofs of integral inequalities. In the remainder of this section, we consider such an example related to an important inequality.

We estimate the decreasing distribution function of the function $|S|$, where $S=\sum_{j=1}^n a_j r_j$ (here, $a_1,\dots,a_n\in\mathbb{R}$ and $r_1,\dots,r_n$ are Rademacher functions), i.e., the measure of the set $E_t=\{x\in(0,1)\mid |S(x)|>t\}$. It is important to obtain an estimate that depends not on the number n of summands but on the total value of the coefficients $a_1,\dots,a_n$. More precisely, our goal is to estimate the measure $\widetilde{F}(t)=\lambda_1(E_t)$ by the parameter $A=(\sum a_j^2)^{\frac{1}{2}}$.

We start with the inequality $1\leqslant e^{u(|S(x)|-t)}$, valid for $x\in E_t$ and all $u>0$ (we will use the freedom in the choice of the parameter u later). Obviously,

$$\widetilde{F}(t)=\lambda_1(E_t)\leqslant\int_{E_t} e^{u(|S(x)|-t)}\,dx\leqslant e^{-ut}\int_0^1\left(e^{uS(x)}+e^{-uS(x)}\right)dx.$$

Since the Rademacher functions are independent, the integrals $\int_0^1 e^{\pm uS(x)}\,dx$ split into the product of integrals

$$\int_0^1 e^{\pm uS(x)}\,dx=\prod_{j=1}^n\int_0^1 e^{\pm ua_jr_j(x)}\,dx=\prod_{j=1}^n\cosh(ua_j).$$

Therefore, $\widetilde{F}(t)\leqslant 2e^{-ut}\prod_{j=1}^n\cosh(ua_j)$. Using the inequality $\cosh x\leqslant e^{x^2/2}$, which can easily be proved by comparing the coefficients of the Taylor expansions, we can find an upper bound for the latter product. We obtain

$$\widetilde{F}(t)\leqslant 2e^{-ut}\prod_{j=1}^n e^{u^2a_j^2/2}=2e^{-ut+u^2A^2/2}.$$

Now, we choose a u so that the right-hand side of the inequality is minimal. To this end, we put $u=t/A^2$. As a result, we come to the required estimate

$$\widetilde{F}(t)=\lambda(E_t)\leqslant 2e^{-t^2/(2A^2)}.$$

This estimate enables us to obtain the *Khintchine*[6] *inequality*, which says that

$$\left(\int_0^1 |S(x)|^p\,dx\right)^{1/p} \leqslant C_p\big(a_1^2+\cdots+a_n^2\big)^{1/2}, \tag{5}$$

for every $p>0$ (the constant C_p depends only on p). Since $\int_0^1 |S(x)|^2\,dx = a_1^2+\cdots+a_n^2$, Hölder's inequality implies that we can take $C_p=1$ if $p\in(0,2]$. We also notice that the Khintchine inequality is sharp in order of magnitude: it can be supplemented by a similar estimate from below (see Exercise 7 in Sect. 9.1).

For the proof, we use Proposition 6.4.3. Applying the estimate obtained above for the distribution function, we have

$$\int_0^1 |S(x)|^p\,dx = p\int_0^\infty t^{p-1}\widetilde{F}(t)\,dt \leqslant 2p\int_0^\infty t^{p-1}e^{-t^2/(2A^2)}\,dt.$$

It remains to express the latter integral in terms of the function Γ (see Sect. 4.6.3, Example 5). Using the change of variables $s=t^2/(2A^2)$ in the latter integral, we see that

$$\int_0^\infty t^{p-1}e^{-t^2/(2A^2)}\,dt = p2^{p/2}A^p\int_0^\infty s^{\frac{p}{2}-1}e^{-s}\,ds = p2^{p/2}\Gamma(p/2)A^p.$$

This gives inequality (5) with $C_p=\sqrt{2}(p\Gamma(p/2))^{1/p}$. By Stirling's formula (see Sect. 7.2.6), we can easily prove that $C_p\sim\sqrt{p/e}$ as $p\to+\infty$.

EXERCISES

1. Let f be a non-negative Lebesgue measurable function defined on $\mathbb{R}_+$, and let $\mathbb{R}_+^m=\{x=(x_1,x_2,\ldots,x_m)\mid x_1>0,x_2>0,\ldots,x_m>0\}$. Prove the relations:

(a) $$\int_{\mathbb{R}_+^m} f(x_1+x_2+\cdots+x_m)\,dx = \frac{1}{(m-1)!}\int_0^\infty t^{m-1}f(t)\,dt;$$

(b) $$\int_{\mathbb{R}_+^m} f\big(\max\{x_1,x_2,\ldots,x_m\}\big)\,dx = m\int_0^\infty t^{m-1}f(t)\,dt.$$

2. Prove the following Catalan[7] formula: if f is a non-negative Borel measurable function on $\mathbb{R}$, then

$$\int_X g(x)f\big(h(x)\big)\,d\mu(x) = \int_{\mathbb{R}} f(t)\,dH(t).$$

Here, the function $g\geqslant 0$ is summable on X, h is measurable, and H is the increasing distribution function of h in a measure with density g with respect to μ (i.e., $H(t)=\int_{X(h<t)} g\,d\mu$).

---

[6] Aleksandr Yakovlevich Khintchine (1894–1959)—Russian mathematician.

[7] Eugène Charles Catalan (1814–1894)—Belgian mathematician.

3. Prove the following generalization of Proposition 6.4.3. Let φ be a continuously differentiable increasing function on $[0, +\infty)$ such that $\varphi(0) = 0$. Then

$$\int_X \varphi(h)\,d\mu = \int_0^\infty \varphi'(t)\widetilde{H}(t)\,dt,$$

where the function h is non-negative and $\widetilde{H}$ is the decreasing distribution function of h.

4. If μ is a non-zero finite measure on X, $\mu(X) \neq 1$, then there are no pairs of independent functions defined on X.

5. Let $\rho(x)$ be the distance from a point x to a convex bounded set $A \subset \mathbb{R}^2$. Calculate the increasing distribution function of the function ρ.

6. Let $f(x) = \frac{a_1}{x-c_1} + \cdots + \frac{a_n}{x-c_n}$, where $a_1, \ldots, a_n$ are positive and $c_1, \ldots, c_n$ are arbitrary real numbers. Prove Boole's[8] equations:

$$\lambda\big(\{x \in \mathbb{R} \mid f(x) > t\}\big) = \lambda\big(\{x \in \mathbb{R} \mid f(x) < -t\}\big) = \frac{a_1 + \cdots + a_n}{t} \quad \text{for all } t > 0.$$

6.5 *Computation of Multiple Integrals by Integrating over the Sphere

In this section, we will use the following notation:

$\mathbb{R}^m_\pm = \{x = (x_1, \ldots, x_m) \in \mathbb{R}^m \mid \pm x_m > 0\}$;
$S^{m-1}_\pm = S^{m-1} \cap \mathbb{R}^m_\pm$;
$B^m = \{x \in \mathbb{R}^m \mid \|x\| < 1\}$ is the unit ball in the space $\mathbb{R}^m$;
P is the orthogonal projection of $\mathbb{R}^m$ onto $\mathbb{R}^{m-1}$: if $x = (x_1, \ldots, x_{m-1}, x_m)$, then $P(x) = (x_1, \ldots, x_{m-1}) \in \mathbb{R}^{m-1}$ $(m \geqslant 2)$.

6.5.1 In Chap. 8, we define a measure on smooth surfaces (the "surface area"). Here, running a few steps forward, we note only that this measure σ $(= \sigma_{m-1})$ is constructed on the sphere S^{m-1} in such a way that a set $A \subset S^{m-1}_+$ is measurable if and only if its projection $P(A)$ is Lebesgue measurable in the space $\mathbb{R}^{m-1}$, and if $P(A)$ is measurable, then the measure $\sigma(A)$ is calculated by the formula

$$\sigma(A) = \int_{P(A)} \frac{du}{\sqrt{1 - \|u\|^2}}.$$

[8]George Boole (1815–1864)—British mathematician.

We remark that the restriction of P to S_+^{m-1} is the map inverse to the map $T: B^{m-1} \to S_+^{m-1}$ defined by the equation

$$T(u) = \left(u, \sqrt{1 - \|u\|^2}\right), \quad \text{where } u \in B^{m-1}$$

(throughout this section, we identify a pair (u, t), where $u = (u_1, \dots, u_{m-1}) \in \mathbb{R}^{m-1}$, $t \in \mathbb{R}$, with the point $(u_1, \dots, u_{m-1}, t) \in \mathbb{R}^m$).

Thus, the restriction of the measure σ to the σ-algebra of measurable subsets of the upper hemisphere is the ω-weighted image of the measure λ_{m-1} in the unit ball under the map T, where $\omega(u) = \frac{1}{\sqrt{1-\|u\|^2}}$. From the definition of the measure σ given above, it follows that a function g defined on S_+^{m-1} and the composition $g \circ T$ are measurable simultaneously. Using Theorem 6.1.1, we see that

$$\int_{S_+^{m-1}} g(\xi)\, d\sigma(\xi) = \int_{B^{m-1}} g(T(u)) \frac{du}{\sqrt{1 - \|u\|^2}} \tag{1}$$

for every non-negative measurable function g on S_+^{m-1}.

Similar facts also hold for the lower hemisphere.

We will find the total area of the sphere S^{m-1} $(m > 1)$ right away. It is clear that

$$\sigma\left(S^{m-1}\right) = 2\sigma\left(S_+^{m-1}\right) = 2\int_{B^{m-1}} \frac{du}{\sqrt{1 - \|u\|^2}}.$$

By the formula obtained in Example 1 of Sect. 6.4.2, we have

$$\begin{aligned} 2\int_{B^{m-1}} \frac{du}{\sqrt{1 - \|u\|^2}} &= 2(m-1)\alpha_{m-1} \int_0^1 \frac{t^{m-2}}{\sqrt{1-t^2}}\, dt \\ &= (m-1)\alpha_{m-1} \int_0^1 \frac{t^{m-3}}{\sqrt{1-t^2}}\, d(t^2) \\ &= (m-1)\alpha_{m-1} \int_0^1 s^{\frac{m-1}{2}-1}(1-s)^{\frac{1}{2}-1}\, ds, \end{aligned}$$

where α_{m-1} is the $(m-1)$-dimensional volume of the ball B^{m-1}. As proved in Sect. 5.4.2 (Example 2) and in Sect. 5.3.2 (Example 2),

$$\alpha_{m-1} = \frac{\pi^{(m-1)/2}}{\Gamma((m+1)/2)}, \qquad \int_0^1 s^{\frac{m-1}{2}-1}(1-s)^{\frac{1}{2}-1}\, ds = \frac{\Gamma((m-1)/2)\Gamma(1/2)}{\Gamma(m/2)}.$$

Substituting these expressions into the previous equation, we obtain

$$\sigma\left(S^{m-1}\right) = (m-1)\frac{\pi^{(m-1)/2}}{\Gamma((m+1)/2)} \frac{\Gamma((m-1)/2)\Gamma(1/2)}{\Gamma(m/2)} = \frac{2\pi^{m/2}}{\Gamma(m/2)}.$$

The formula for the area of a higher-dimensional sphere was found for the first time by Jacobi.

6.5.2 After giving the above information needed in the sequel, we now turn to the question we are presently interested in. Our goal is to generalize the results of Sects. 6.2.4 and 6.2.5 (the calculation of integrals with the help of polar and spherical coordinates) to higher dimensional spaces.

Theorem *For every non-negative Lebesgue measurable function f defined on $\mathbb{R}^m$, the following relation holds*:

$$\int_{\mathbb{R}^m} f(y)\,dy = \int_0^\infty t^{m-1}\left(\int_{S^{m-1}} f(t\xi)\,d\sigma(\xi)\right)dt. \tag{2}$$

Here, the function $\xi \mapsto f(t\xi)$ is measurable on S^{m-1} for almost all $t > 0$ and the internal integral on the right-hand side of (2) *is a measurable function of t.*

This statement can obviously be carried over to the functions summable on an arbitrary ball $B(0,r)$,

$$\int_{B(0,r)} f(y)\,dy = \int_0^r t^{m-1}\left(\int_{S^{m-1}} f(t\xi)\,d\sigma(\xi)\right)dt. \tag{2'}$$

Proof We prove a formula similar to (2), where the space $\mathbb{R}^m$ and the sphere S^{m-1} are replaced by the half-space $\mathbb{R}^m_+$ and the hemisphere S^{m-1}_+,

$$\int_{\mathbb{R}^m_+} f(y)\,dy = \int_0^\infty t^{m-1}\left(\int_{S^{m-1}_+} f(t\xi)\,d\sigma(\xi)\right)dt.$$

It can easily be proved that a similar equation holds for the half-space $\mathbb{R}^m_-$ and the hemi-sphere S^{m-1}_-. Therefore, Eq. (2) is also valid.

By (1), we have

$$\int_{S^{m-1}_+} f(t\xi)\,d\sigma(\xi) = \int_{B^{m-1}} f\bigl(tT(u)\bigr)\frac{du}{\sqrt{1-\|u\|^2}},$$

where $T(u) = (u, \sqrt{1-\|u\|^2})$. Therefore, it is sufficient to verify the equation

$$\int_{\mathbb{R}^m_+} f(y)\,dy = \int_0^\infty t^{m-1}\left(\int_{B^{m-1}} f\bigl(tT(u)\bigr)\frac{du}{\sqrt{1-\|u\|^2}}\right)dt \tag{3}$$

the proof to which we now proceed.

We put $G = B^{m-1} \times \mathbb{R}_+$ and define the map $\Phi : G \to \mathbb{R}^m_+$ by the equation

$$\Phi(x) = tT(u), \quad \text{where } x = (u,t) \in G.$$

This map is, obviously, smooth. We leave it to the reader to verify that the map

$$y \mapsto \Psi(y) = \left(P\left(\frac{y}{\|y\|}\right), \|y\|\right) \in B^{m-1} \times \mathbb{R}_+ \quad \left(y \in \mathbb{R}^m_+\right)$$

(which is also smooth) is inverse to Φ. Thus, Φ is a diffeomorphism.

Since the composition $f \circ \Phi$ is measurable, we can use Tonelli's theorem and represent the right-hand side of Eq. (3) as follows:

$$\int_0^\infty t^{m-1}\bigg(\int_{B^{m-1}} f\big(tT(u)\big)\frac{du}{\sqrt{1-\|u\|^2}}\bigg)\,dt = \int_G f\big(\Phi(x)\big)\frac{t^{m-1}}{\sqrt{1-\|u\|^2}}\,dx.$$

We notice that, by Tonelli's theorem, the function $u \mapsto f(tT(u))$ is measurable for almost all $t > 0$. By the definition of the measure on the sphere, we obtain that the function $\xi \mapsto f(t\xi)$, where $\xi \in S_+^{m-1}$ is also measurable for almost all $t > 0$.

Thus, Eq. (3) is equivalent to the equation

$$\int_{\mathbb{R}_+^m} f(y)\,dy = \int_G f\big(\Phi(x)\big)\frac{t^{m-1}}{\sqrt{1-\|u\|^2}}\,dx.$$

We show that the last equation follows from the change of variables formula for multiple integrals and smooth maps (see Theorem 6.2.2). To this end, we prove that the Jacobian of $J_\Phi(x)$ of Φ at a point $x = (u,t)$ ($x \in G$) is equal to $\frac{t^{m-1}}{\sqrt{1-\|u\|^2}}$. Since $\Phi(x) = (tu, t\sqrt{1-\|u\|^2})$, we have

$$J_\Phi(x) = \begin{vmatrix} t & 0 & \dots & 0 & u_1 \\ 0 & t & \dots & 0 & u_2 \\ \vdots & \vdots & \ddots & \vdots & \vdots \\ 0 & 0 & \dots & t & u_{m-1} \\ -tcu_1 & -tcu_2 & \dots & -tcu_{m-1} & 1/c \end{vmatrix},$$

where, for brevity, we put $c = \frac{1}{\sqrt{1-\|u\|^2}}$. Multiplying the kth row ($1 \leqslant k < m$) by cu_k and adding all rows to the last row, we obtain

$$J_\Phi(x) = \begin{vmatrix} t & 0 & \dots & 0 & u_1 \\ 0 & t & \dots & 0 & u_2 \\ \vdots & \vdots & \ddots & \vdots & \vdots \\ 0 & 0 & \dots & t & u_{m-1} \\ 0 & 0 & \dots & 0 & v \end{vmatrix} = t^{m-1}v,$$

where

$$v = \frac{1}{c} + c\|u\|^2 = \sqrt{1-\|u\|^2} + \frac{\|u\|^2}{\sqrt{1-\|u\|^2}} = \frac{1}{\sqrt{1-\|u\|^2}}.$$

Thus, the equation

$$\big|J_\Phi(x)\big| = \frac{t^{m-1}}{\sqrt{1-\|u\|^2}} \quad \text{for } x = (u,t) \in G.$$

is proved, which completes the proof of the theorem. □

If the function $f(x)$ in Eq. (2) is a product of two functions one of which depends only on $t = \|x\|$ and the other only on the direction of the vector x, i.e., only on $\xi = x/\|x\|$, then the integral on the right-hand side of (2) splits into the product of two integrals (over the positive semi-axis and over the unit sphere). This allows us to reduce the calculation of the integral over the sphere to the calculation of a multiple integral. Here is a typical example of such a situation.

Example 1 We calculate the integral

$$J = \int_{S^{m-1}} |\xi_1|^{p_1} \cdots |\xi_m|^{p_m}\, d\sigma(\xi) \quad (p_1, \ldots, p_m \in \mathbb{R}).$$

As will be seen, this integral is finite only if all p_j are larger than -1.

To this end, we consider the auxiliary integral

$$I = \int_{\mathbb{R}^m} |x_1|^{p_1} \cdots |x_m|^{p_m} e^{-\|x\|^2}\, dx$$

(as usual, $\|x\|$ is the Euclidean norm of $x = (x_1, \ldots, x_m)$).

On the one hand, we have $I = I_1 \cdots I_m$, where

$$I_j = \int_{-\infty}^{\infty} |u|^{p_j} e^{-u^2}\, du = \begin{cases} \Gamma(\frac{p_j+1}{2}) & \text{for } p_j > -1, \\ +\infty & \text{for } p_j \leqslant -1. \end{cases}$$

It follows, in particular, that the integral I is finite only if $p_1, \ldots, p_m > -1$.

On the other hand, formula (2) gives

$$I = J \int_0^{\infty} t^{m+p-1} e^{-t^2}\, dt = \frac{J}{2} \Gamma\left(\frac{m+p}{2}\right),$$

where $p = p_1 + \cdots + p_m$. Thus, if all p_j are larger than -1, then

$$J = \frac{2I}{\Gamma(\frac{m+p}{2})} = \frac{2}{\Gamma(\frac{m+p_1+\cdots+p_m}{2})} \prod_{j=1}^{m} \Gamma\left(\frac{1+p_j}{2}\right),$$

and $J = +\infty$ otherwise.

Example 2 Let q and $p_1, \ldots, p_m$ be real numbers. For $x = (x_1, \ldots, x_m)$, $x \in \mathbb{R}^m$, we put

$$f(x) = \frac{|x_1|^{p_1} \cdots |x_m|^{p_m}}{(1 + \|x\|^2)^q}.$$

We find the conditions under which f is summable on the space $\mathbb{R}^m$. Using the notation introduced in the previous example, we obtain by (2) that

$$\int_{\mathbb{R}^m} f(x)\, dx = J \int_0^{\infty} \frac{t^{p+m-1}}{(1+t^2)^q}\, dt, \quad \text{where } p = p_1 + \cdots + p_m.$$

Taking into account the result of Example 1, we see that f is summable if and only if the inequalities $q > \frac{p+m}{2}$ and $p_1, \ldots, p_m > -1$ are valid.

We note also that the change $t = \operatorname{tg}\varphi$ reduces the integral on the right-hand side of the last formula to an integral expressible in terms of the function Γ (see the end of Sect. 5.3.2).

Remark The above theorem can be restated as follows. We consider the measure $\mu = \lambda_1 \times \sigma$ on the direct product $\mathbb{R}_+ \times S^{m-1}$. The map $\Theta : \mathbb{R}_+ \times S^{m-1} \to \mathbb{R}^m \setminus \{0\}$ defined by the equation $\Theta(t, \xi) = t\xi$ is, obviously a homeomorphism. If f is the characteristic function of a set $A \subset \mathbb{R}^m$, then Theorem 6.5.2 implies that

$$\lambda_m(A) = \int_{\mathbb{R}_+ \times S^{m-1}} t^{m-1} f(t\xi)\, d\mu(t, \xi) = \int_{\Theta^{-1}(A)} t^{m-1}\, d\mu(t, \xi).$$

Thus, the Lebesgue measure λ_m is the ω-weighted image of the measure $\mu = \lambda_1 \times \sigma$ with respect to the map Θ with $\omega(t, \xi) = t^{m-1}$. We leave it to the reader to verify that the converse is also true, i.e., that Theorem 6.5.2 follows from this statement.

6.5.3 We present some corollaries of Theorem 6.5.2. The first of them is a direct generalization of the formula for the area in polar coordinates (see Sect. 6.2.4).

Corollary 1 *The volume of the set* $V = \{r\xi \mid \xi \in S^{m-1},\ 0 \leqslant r \leqslant \rho(\xi)\}$, *where* ρ *is a non-negative measurable function on the sphere* S^{m-1}, *can be calculated by the formula*

$$\lambda_m(V) = \frac{1}{m} \int_{S^{m-1}} \rho^m(\xi)\, d\sigma(\xi). \tag{4}$$

In particular, for $\rho \equiv 1$, *we again obtain the Jacobi formula for the surface area of a sphere.*

Proof For the proof, we apply formula (2) to the characteristic function of V, changing the order of integration on the right-hand side. We have

$$\lambda_m(V) = \int_{\mathbb{R}^m} \chi_V(x)\, dx = \int_{S^{m-1}} \left(\int_0^\infty r^{m-1} \chi_V(r\xi)\, dr \right) d\sigma(\xi).$$

Since $\chi_V(r\xi) = 0$ for $r > \rho(\xi)$ and $\chi_V(r\xi) = 1$ for $r \leqslant \rho(\xi)$, we obtain the required equation

$$\lambda_m(V) = \int_{S^{m-1}} \left(\int_0^{\rho(\xi)} r^{m-1}\, dr \right) d\sigma(\xi) = \frac{1}{m} \int_{S^{m-1}} \rho^m(\xi)\, d\sigma(\xi). \qquad \square$$

To obtain one more result by Jacobi, we represent Eq. (4) in the following form: if f is a non-negative measurable function on $\mathbb{R}^m$ satisfying the conditions $f(x) > 0$

for $x \neq 0$ and $f(tx) = tf(x)$ for $t \geqslant 0$, $x \in \mathbb{R}^m$, then

$$\lambda_m\big(\{x \in \mathbb{R}^m \mid f(x) \leqslant 1\}\big) = \frac{1}{m}\int_{S^{m-1}} \frac{d\sigma(\xi)}{f^m(\xi)}. \tag{4'}$$

To prove this, it is sufficient to apply formula (4) to the function $\rho(\xi) = 1/f(\xi)$, since the inequalities $f(r\xi) \leqslant 1$ and $r \leqslant \rho(\xi)$ are equivalent.

Corollary 2 *Let A be a positive definite $m \times m$ matrix. Then the following Jacobi equation holds*:

$$\int_{S^{m-1}} \frac{d\sigma(\xi)}{\langle A\xi, \xi\rangle^{\frac{m}{2}}} = \frac{m\alpha_m}{\sqrt{\det A}}.$$

Proof Indeed, representing A in the form $A = A_1^2$, where A_1 is a positive definite matrix, we put $f(x) = \|A_1(x)\|$. Then $\sqrt{\langle A\xi, \xi\rangle} = \|A_1(\xi)\| = f(\xi)$, and the set $V = \{x \in \mathbb{R}^m \mid f(x) \leqslant 1\}$ is $A_1^{-1}(B^m)$. Therefore, by formula (4′), we obtain

$$\int_{S^{m-1}} \frac{d\sigma_{m-1}(\xi)}{\langle A\xi, \xi\rangle^{\frac{m}{2}}} = m\lambda_m(V) = m \det\big(A_1^{-1}\big)\lambda_m\big(B^m\big) = \frac{m\alpha_m}{\sqrt{\det A}}.$$ □

We finish this section with a result obtained earlier by a different method (see Sect. 6.4.2, Example 1).

Corollary 3 *Let φ be a non-negative (Lebesgue) measurable function defined on $\mathbb{R}_+$. Then*

$$\int_{\mathbb{R}^m} \varphi\big(\|x\|\big)\,dx = m\alpha_m \int_0^\infty t^{m-1}\varphi(t)\,dt.$$

Proof This follows from Theorem 6.5.2 applied to the radial function $f(x) = \varphi(\|x\|)$. □

6.5.4 From the formula for the volume of the m-dimensional unit ball, it immediately follows that, for large m, the majority of its volume is concentrated near the boundary sphere. For example, the volume of the ball of radius $1 - \frac{1}{\sqrt{m}}$ is negligibly small in comparison with the volume of B^m. In other words, for large m, the thin spherical layer $\{x \in \mathbb{R}^m \mid 1 - \frac{1}{\sqrt{m}} < \|x\| < 1\}$ almost exhausts the unit ball. Therefore, Alice finding herself in a 1000-dimensional space would be unable to regale herself with watermelon. Even if the thickness of the rind of a watermelon is incredibly small and makes up 1 % of its radius, the rind makes up 99.99 % of the watermelon.

This phenomenon leads to the following result, unexpected at first sight: for a function varying sufficiently regularly in $\mathbb{R}^m$ and for large m, the mean values of the function in the ball and on the boundary sphere almost coincide. More precisely, let f be a function defined on the unit ball $\overline{B}^m$ and satisfying the Lipschitz condition

$|f(x)-f(y)| \leqslant L\|x-y\|$, where L is a constant, and let f_B and f_S be the mean values of f in the ball $\overline{B}^m$ and on the sphere S^{m-1}, respectively,

$$f_B = \frac{1}{\alpha_m}\int_{\|x\|\leqslant 1} f(x)\,dx, \qquad f_S = \frac{1}{s_m}\int_{\|\xi\|=1} f(\xi)\,d\sigma(\xi)$$

(here α_m is the volume of $\overline{B}^m$ and s_m is the surface area of S^{m-1}). Then

$$|f_B - f_S| \leqslant \frac{L}{m}.$$

For the proof, we write the integral over the ball in spherical coordinates (see Eq. (2′)),

$$\int_{\|x\|\leqslant 1} f(x)\,dx = \int_{\|\xi\|=1}\left(\int_0^1 f(t\xi)\,t^{m-1}\,dt\right)d\sigma(\xi).$$

Taking into account the equation $s_m = m\alpha_m$, we obtain

$$\begin{aligned} f_S - f_B &= \frac{1}{s_m}\int_{\|\xi\|=1}\left(f(\xi) - m\int_0^1 f(t\xi)\,t^{m-1}\,dt\right)d\sigma(\xi) \\ &= \frac{m}{s_m}\int_{\|\xi\|=1}\left(\int_0^1 \big(f(\xi)-f(t\xi)\big)\,t^{m-1}\,dt\right)d\sigma(\xi). \end{aligned}$$

Therefore,

$$|f_S - f_B| \leqslant \frac{m}{s_m}\int_{\|\xi\|=1}\left(\int_0^1 L(1-t)\,t^{m-1}\,dt\right)d\sigma(\xi) = \frac{L}{m+1}.$$

EXERCISES

1. Let f be a continuous function on $\mathbb{R}^m$ ($m \geqslant 2$), and let $I(r) = \int_{\|x\|\leqslant r} f(x)\,dx$ for $r \geqslant 0$. Prove that

$$I'(r) = r^{m-1}\int_{S^{m-1}} f(r\xi)\,d\sigma(\xi).$$

2. For which numbers $p_1 > 0, \ldots, p_m > 0$ are the following integrals finite?

(a) $$\int_{B^m} \frac{dx}{|x_1|^{p_1}+\cdots+|x_m|^{p_m}},$$

(b) $$\int_{\mathbb{R}^m\setminus B^m} \frac{dx}{|x_1|^{p_1}+\cdots+|x_m|^{p_m}}.$$

3. For which a, p and q is the function $\frac{x^p y^q}{(1+x^2+y^2)^a}$ summable in the angles $\{(x,y)\in\mathbb{R}^2 \mid 0<y<x\}$ and $\{(x,y)\in\mathbb{R}^2 \mid 0<y<x<2y\}$? Compare the result obtained with the summability condition from Example 2 in Sect. 6.5.2.

4. Using the geometric interpretation of the Jacobian, prove that it is equal to $\|x\|^{-2m}$ for the inversion with respect to the unit sphere (i.e., for the map $x \mapsto x/\|x\|^2$).

5. Generalize the result of Sect. 6.5.4, by estimating the difference $f_S - f_B$ in terms of the modulus of continuity of f.

6.6 *Some Geometric Applications

In this section, we give analytic proofs of interesting geometric results connected with Brouwer's fixed point theorem and vector fields on a sphere.

6.6.1 Brouwer's theorem stating that every continuous map of a closed ball into itself has a fixed point plays an important role in the theory of non-linear equations and topology. It can be deduced from the theorem stating that there is no a smooth retraction of a ball to its boundary or from the theorem stating that a non-degenerate smooth tangent vector field does not exist on an even-dimensional sphere. We prove these important theorems following the papers [Mi] and [R].

As usual, let $\overline{B}$ be the closure of the unit ball B in $\mathbb{R}^m$, I be the identity matrix, Φ' be the Jacobi matrix of the smooth map Φ, and J_Φ be the determinant of Φ' (Jacobian). We say that a map defined on a subset of the space $\mathbb{R}^m$ is smooth if it is a restriction of a smooth map defined on a neighborhood of this subset.

Lemma *Let $\mathcal{O} \subset \mathbb{R}^m$ be an open set, and let K be a compact subset of $\mathcal{O}$. Let $\Psi \in C^1(\mathcal{O}, \mathbb{R}^m)$ and $\Phi_t(x) = x + t\Psi(x)$, where $x \in \mathcal{O}$ and $t \in \mathbb{R}$. Then, for sufficiently small $t > 0$, we have:*

(1) *Φ_t is one-to-one on K;*

(2)
$$J_{\Phi_t}(x) > 0 \quad \text{on } K. \tag{1}$$

Moreover, $J_{\Phi_t}(x)$ is a polynomial in t (with coefficients depending on x).

Proof First, we verify that Ψ satisfies the Lipschitz condition on K, i.e., that, for some $L > 0$ the inequality

$$\big\|\Psi(x) - \Psi(y)\big\| \leqslant L\|x - y\| \tag{2}$$

is valid for all x and y in K. We know that this is the case if K is a convex compact set (see Theorem 13.7.2). If K is not convex, then it can be covered by a finite number of open balls $B_1, \ldots, B_N$ the closures of which are contained in $\mathcal{O}$. Let L' be the maximal Lipschitz constant for these balls. If points $x, y \in K$ belong to one of the balls, then $\|\Psi(x) - \Psi(y)\| \leqslant L'\|x - y\|$. Otherwise, the pair (x, y) belongs to the compact set

$$(K \times K) \setminus \bigcup_{n=1}^{N} (B_n \times B_n),$$

and, therefore, the quantity $\|x-y\|$ is separated from zero by a number $\eta > 0$. Consequently, in this case, we have

$$\|\Psi(x)-\Psi(y)\| \leqslant 2M \leqslant \frac{2M}{\eta}\|x-y\|,$$

where $M = \max_{z\in K}\|\Psi(z)\|$. Thus, if $L = \max\{L', 2M/\eta\}$, then inequality (2) is valid for all $x, y \in K$. We see that

$$\|\Phi_t(x)-\Phi_t(y)\| \geqslant \|x-y\| - t\|\Psi(x)-\Psi(y)\| \geqslant (1-tL)\|x-y\| > 0$$

for $x, y \in K,\ x \neq y$, and $0 < t < 1/L$, which proves the first statement of the lemma.

The last statement of the lemma follows from the properties of determinants and the fact that $\Phi_t'(x) = I + t\Psi'(x)$. The second statement is obtained by continuity considerations if we take into account that $J_{\Phi_0}(x) = \det(I) = 1 > 0$. □

6.6.2 Now we are ready to turn to the retraction theorem.

Definition Let $A \subset X \subset \mathbb{R}^m$. A continuous map $\Phi : X \to \mathbb{R}^m$ is called a *retraction of X to A* if

$$\Phi(X) \subset A \quad \text{and} \quad \Phi(x) = x \quad \text{for all } x \in A.$$

Theorem (Retraction theorem) *There is no retraction of the ball $\overline{B}$ to its boundary.*

Proof We confine ourselves to the proof of the non-existence of a smooth retraction. The general case can be obtained from the smooth one by approximation (see Exercise 1). We assume the contrary. Let $\mathcal{O}$ be an open set containing $\overline{B}$, and let $\Phi = (\varphi_1, \ldots, \varphi_m) : \mathcal{O} \to \mathbb{R}^m$ be a smooth map whose restriction to $\overline{B}$ is a retraction of $\overline{B}$ to the sphere S^{m-1}. Since the image of the unit ball has no interior points, it follows from the open mapping theorem 13.7.3 that

$$J_\Phi(x) = \det(\Phi'(x)) = 0 \quad \text{for all } x \in \overline{B}. \tag{3}$$

We consider the family $\{\Phi_t\}_{0\leqslant t\leqslant 1}$ of maps, where $\Phi_t : \mathcal{O} \to \mathbb{R}^m$ is defined by the equation

$$\Phi_t(x) = x + t(\Phi(x) - x) = (1-t)x + t\Phi(x) \quad \text{for } x \in \mathcal{O}\ (0 \leqslant t \leqslant 1).$$

It is clear that $\Phi_t(\overline{B}) \subset \overline{B}$ and, moreover,

$$\Phi_t(B) \subset B \quad \text{for } 0 \leqslant t < 1 \tag{4}$$

since

$$\|\Phi_t(x)\| \leqslant (1-t)\|x\| + t\|\Phi(x)\| < (1-t) + t = 1 \quad \text{for all } x \in B \text{ and } 0 \leqslant t < 1.$$

On the sphere S^{m-1}, the map Φ_t is the identity. Indeed,

$$\Phi_t(x) = (1-t)x + t\Phi(x) = (1-t)x + tx = x \quad \text{for } x \in S^{m-1}. \tag{5}$$

We assume that inequality (1) is valid for $0 < t < \delta$. By the open mapping theorem (see Sect. 13.7.3), it follows from (1) that

$$\text{the set } \Phi_t(B) \text{ is open for } 0 < t < \delta. \tag{6}$$

We verify that relations (4), (5) and (6) imply that

$$\Phi_t(B) = B \quad \text{for } 0 < t < \delta. \tag{7}$$

By (4), it is sufficient to show that the set $\Phi_t(B)$ is not only open but also (relatively) closed in B. Then Eq. (7) will follow from the fact that B is connected. So, let $\{y_n\}_{n\geqslant 1} \subset \Phi_t(B)$, $y_n \underset{n\to\infty}{\longrightarrow} y_0 \in B$. We choose an $x_n \in B$ such that $y_n = \Phi_t(x_n)$ for $n \in \mathbb{N}$. Without loss of generality, we may assume that the sequence $\{x_n\}_{n\geqslant 1}$ converges (otherwise, it can be replaced by a convergent subsequence). Let $x_n \underset{n\to\infty}{\longrightarrow} x_0$. If $\|x_0\| = 1$, then $\|\Phi_t(x_n)\| \underset{n\to\infty}{\longrightarrow} \|\Phi_t(x_0)\| = \|x_0\| = 1$, which is impossible since $\|\Phi_t(x_n)\| = \|y_n\| \underset{n\to\infty}{\longrightarrow} \|y_0\| < 1$. Therefore, $\|x_0\| < 1$. Consequently, $x_0 \in B$ and

$$y_0 = \lim_{n\to\infty} y_n = \lim_{n\to\infty} \Phi(x_n) = \Phi(x_0) \in \Phi(B),$$

which proves that $\Phi(B)$ is closed in B along with Eq. (7). As stated in the lemma, the map Φ_t is one-to-one for small t. Thus, it is a diffeomorphism of B onto itself. Using the theorem on a smooth change of variables in a multiple integral and taking into account (1), we obtain that

$$\lambda_m(B) = \int_B J_{\Phi_t}(x)\,dx \quad \text{for sufficiently small } t > 0. \tag{8}$$

By the lemma, $J_{\Phi_t}(x)$ is a polynomial in t. Consequently, the right-hand side of Eq. (8) is also a polynomial in t. Since this polynomial is constant for small t, it is identically constant. Therefore, Eq. (8) is valid not only for small but for all $t \in [0, 1]$ and, in particular, for $t = 1$. Since $\Phi_1 \equiv \Phi$, it follows from (8) that

$$\lambda_m(B) = \int_B J_\Phi(x)\,dx.$$

However, this is impossible since the right-hand side of the last equation is zero by (3).

Thus, the assumption of the existence of a smooth retraction leads to a contradiction. □

6.6.3 Now, we show how to deduce Brouwer's theorem from the retraction theorem. We recall that a *fixed point of a map* $f : X \to X$ is a point $x \in X$ such that $f(x) = x$.

Theorem (Brouwer[9] fixed point theorem) *Every continuous map of a ball $\overline{B}$ into itself has a fixed point.*

Proof First, we prove that the theorem is valid for smooth maps. Assuming the contrary, let $f : \overline{B} \to \overline{B}$ be a smooth map without a fixed point. We use f to construct a smooth retraction Φ of the ball to its boundary.

Since $y = f(x) \neq x$ for $x \in \overline{B}$, the points x and y uniquely determine the ray $\ell_x = \{y + t(x-y) \mid t \geqslant 0\}$ that has origin at y and passes through x. Since the points x and y lie in $\overline{B}$, the open ray $\ell_x \setminus \{y\}$ meets the sphere S^{m-1} at a single point (sketch it!). We take this point as $\Phi(x)$. Analytically, this means that the equation $\|y + t(x-y)\|^2 = 1$ quadratic in t has a unique positive root. We can represent the equation in the form

$$\|x-y\|^2 t^2 + 2\langle y, x-y\rangle t + \|y\|^2 - 1 = 0.$$

The unique positive root of this equation, which we denote by t^*, is calculated by the formula

$$t^* = \frac{-\langle y, x-y\rangle + \sqrt{\langle y, x-y\rangle^2 + \|x-y\|^2(1-\|y\|^2)}}{\|x-y\|^2} \tag{9}$$

(note that if $\|y\| = 1$, then $\langle y, x-y\rangle = \langle y, x\rangle - 1 < 0$, since, otherwise, $\langle y, x\rangle = 1$, which is possible only if the vectors x and y coincide). The map Φ can be given by the formula

$$\Phi(x) = y + t^*(x-y) \quad \text{for } x \in \overline{B}, \tag{10}$$

where the number t^* is defined by Eq. (9). Since $\|y\| \leqslant 1$ and $x \neq y$ for $x \in \overline{B}$, the denominator and the expression under the root sign in (9) do not vanish not only in the ball $\overline{B}$ but also in its neighborhood. Therefore, the right-hand side of Eq. (10) is defined and is smooth in a neighborhood of $\overline{B}$. Thus, the map Φ is smooth on $\overline{B}$. By the definition of t^*, we obtain that $\|\Phi(x)\| = 1$ for all $x \in \overline{B}$. Moreover, if $\|x\| = 1$, then the point at which the open ray $\ell_x \setminus \{y\}$ meets the sphere coincides with x (it can easily be seen that this is equivalent to the equation $t^* = 1$). Consequently, the map Φ is the identity on the sphere, and, therefore, is a smooth retraction, which contradicts the previous theorem.

Now, we prove by contradiction that the theorem is valid for an arbitrary continuous map. Let $f : \overline{B} \to \overline{B}$ be a continuous map without fixed points. Then the difference $x - f(x)$ does not vanish on $\overline{B}$, and so there is a positive δ such that

$$\|x - f(x)\| > 2\delta$$

for all $x \in \overline{B}$. We construct a smooth map of the ball to itself without fixed points.

---

[9]Luitzen Egbertus Jan Brouwer (1881–1966)—Dutch mathematician.

Let $f_1, \dots, f_m$ be the coordinate functions of f. By the Weierstrass approximation theorem (see Sect. 7.6.4), there exist polynomials $P_1, \dots, P_m$ such that

$$\left|f_k(x) - P_k(x)\right| < \frac{\delta}{m} \quad \text{for all } x \in \overline{B} \text{ and } k,\ 1 \leqslant k \leqslant m.$$

Let $F : \mathbb{R}^m \to \mathbb{R}^m$ be the map with coordinate functions $P_1, \dots, P_m$. It is clear that

$$\left\|f(x) - F(x)\right\|^2 = \sum_{k=1}^{m} \left|f_k(x) - P_k(x)\right|^2 < \delta^2 \tag{11}$$

for all $x \in \overline{B}$. The image of $\overline{B}$ under F need not belong to the ball. Therefore, we consider the map $\varphi = (1+\delta)^{-1} F$. Obviously, $\varphi \in C^\infty(\mathbb{R}^m, \mathbb{R}^m)$. We verify that $\varphi(\overline{B}) \subset B$. Indeed, for $x \in \overline{B}$, we have

$$\left\|\varphi(x)\right\| = \frac{1}{1+\delta}\left\|F(x)\right\| \leqslant \frac{1}{1+\delta}\left(\left\|f(x)\right\| + \left\|F(x) - f(x)\right\|\right) < \frac{1}{1+\delta}(1+\delta) = 1.$$

We prove that the map φ has no fixed point in $\overline{B}$. Indeed, by (10) and (11), we obtain

$$\begin{aligned}\left\|x - \varphi(x)\right\| &\geqslant \left\|x - f(x)\right\| - \left\|f(x) - \varphi(x)\right\| \geqslant 2\delta - \left\|\frac{1+\delta}{1+\delta} f(x) - \frac{1}{1+\delta} F(x)\right\| \\ &\geqslant 2\delta - \frac{\delta}{1+\delta}\left\|f(x)\right\| - \frac{1}{1+\delta}\left\|f(x) - F(x)\right\| \geqslant 2\delta - \frac{2\delta}{1+\delta} > 0\end{aligned}$$

for $x \in \overline{B}$. Thus, if there is a continuous map of the ball to itself without fixed points, then there is a smooth map with the same property, which, as proved above, is impossible. □

For another proof of the theorem, see Exercise 2.

Corollary *Let $K \subset \mathbb{R}^m$ be a compact set homeomorphic to a ball $\overline{B}$. Every continuous map of K to itself has a fixed point.*

Proof The proof of the corollary is left as an exercise to the reader. □

The general case of the retraction theorem (i.e., for arbitrary continuous maps) can be obtained from the case of smooth maps by approximation considerations as in the proof of Brouwer's theorem. However, the retraction theorem can be deduced from Brouwer's theorem directly. Indeed, if $f : \overline{B} \to S^{m-1}$ is a retraction of the ball $\overline{B}$ to its boundary, then the map $-f$ cannot have fixed points, which contradicts Brouwer's theorem.

6.6.4 Now, we turn our attention to questions concerning vector fields on spheres. By a vector field, we mean a "continuous map on $\mathbb{R}^m$". More precisely, by a *vector field* on a set $X \subset \mathbb{R}^m$, we mean a continuous map $V : X \to \mathbb{R}^m$. A vector field is called *normalized* if $\|V(x)\| = 1$ for all $x \in X$. We say that a vector field V on the sphere S^{m-1} consists of *tangent vectors* if $V(x) \perp x$ (i.e., if $\langle x, V(x)\rangle = 0$) for $x \in S^{m-1}$. Such a vector field is called a *tangent vector field*.

An example of a normalized tangent vector field on the sphere $S^{2n-1} \subset \mathbb{R}^{2n}$ can be obtained as follows. We put

$$V(x) = (x_2, -x_1, \dots, x_{2n}, -x_{2n-1}), \qquad \text{for } x = (x_1, x_2, \dots, x_{2n-1}, x_{2n}) \in S^{2n-1}.$$

It is clear that $V(x) \perp x$ since

$$\big\langle x, V(x)\big\rangle = x_1x_2 - x_2x_1 + \dots + x_{2n-1}x_{2n} - x_{2n}x_{2n-1} = 0.$$

However, there are no nowhere-zero tangent vector fields on even-dimensional spheres. This statement is also known as the hairy ball theorem: you cannot comb a hairy ball flat without creating a cowlick. Here is a precise formulation.

Theorem *Let $m > 1$ be an even integer. Then there are no non-vanishing tangent vector fields on the sphere $S^{m-1} \subset \mathbb{R}^m$.*

Proof We assume the contrary, considering first the smooth case. Let V be a smooth nowhere-zero tangent vector field on S^{m-1} (as in Sect. 6.6.1, the smoothness on the sphere means that V is the restriction of a map smooth in a neighborhood of the sphere). Without loss of generality, we will assume that V is normalized (otherwise, we replace it by the field $\frac{1}{\|V(x)\|}V(x)$). We extend V to $\mathbb{R}^m \setminus \{0\}$ as follows: $\Psi(x) = \|x\|\, V(x/\|x\|)$. It is obvious that Ψ is a smooth map.

We consider the map $\Phi_t(x) = x + t\Psi(x)$, where t is a positive number. Since $x \perp \Psi(x)$ and $\|\Psi(x)\| = \|x\|$, we see that Φ_t takes the sphere of radius r to the sphere of radius $r\sqrt{1+t^2}$. Now, we fix a spherical layer $G_0 = \{x \in \mathbb{R}^m \mid a < \|x\| < b\}$, where $0 < a < b$. It is clear that its image $\Phi_t(G_0)$ is contained in the spherical layer $G_t = \{x \in \mathbb{R}^m \mid a\sqrt{1+t^2} < \|x\| < b\sqrt{1+t^2}\}$. We verify that the map is onto for sufficiently small $t > 0$, i.e., that

$$\Phi_t(G_0) = G_t. \tag{12}$$

We notice that, by Lemma 6.6.1 (for $K = \overline{G_0}$) and the smoothness of the inverse map, we see that Φ_t is a diffeomorphism for sufficiently small $t > 0$. To prove (12), it is sufficient to show that, for $a < r < b$, the image of the sphere $S(r)$ of radius r coincides with the sphere $\widetilde{S}(r) = S(r\sqrt{1+t^2})$. As noted above, $\Phi_t(S(r)) \subset \widetilde{S}(r)$. The set $\Phi_t(S(r))$ is, obviously, closed. At the same time, $\Phi_t(S(r))$ is relatively open in $\widetilde{S}(r)$, which follows from the equation

$$\Phi_t\big(S(r)\big) = \Phi_t\big(S(r) \cap G_0\big) = \widetilde{S}(r) \cap \Phi_t(G_0)$$

since the set $\Phi_t(G_0)$ is open. Since the sphere is connected, the set $\widetilde{S}(r)$ must coincide with its (obviously, non-empty) closed-open subset, and, therefore, $\Phi_t(S(r)) = \widetilde{S}(r)$. By arbitrariness of r, we obtain Eq. (12).

Now, using Eq. (12), we calculate the volume of the set G_t in two ways. On the one hand, we obviously have

$$\lambda_m(G_t) = \lambda_m\big(B\big(b\sqrt{1+t^2}\big) \setminus B\big(a\sqrt{1+t^2}\big)\big) = \alpha_m\big(b^m - a^m\big)\big(1+t^2\big)^{m/2}. \quad (13)$$

On the other hand, assuming that $t > 0$ is sufficiently small and using the formula for the image of a measure under a diffeomorphism, we obtain

$$\lambda_m(G_t) = \int_{G_0} J_{\Phi_t}(x)\,dx.$$

By Lemma 6.6.1, the right-hand side of this equation is a polynomial in t. Taking into account (13), we see that the function $(1+t^2)^{m/2}$ is a polynomial for sufficiently small positive t, which is impossible if m is odd because, for such m, infinitely many derivatives of this function would be non-zero.

Thus, we have proved that there are no smooth non-degenerate tangent vector fields on an even-dimensional sphere. Now, we consider the case of non-smooth fields, which (as in the proof of Brouwer's theorem) is settled by approximation.

Let V be a field of non-zero tangent vectors on the sphere S^{m-1}. As in the smooth case, we may assume that the field is normalized. We use this field to construct a smooth field of non-zero tangent vectors, which will contradict the part of the theorem proved above.

Let $f_1, \dots, f_m$ be the coordinate functions of the field V. By the Weierstrass theorem (see Sect. 7.6.4), there exist polynomials $P_1, \dots, P_m$ such that

$$\big|f_k(x) - P_k(x)\big| < \frac{1}{2m} \quad \text{for all } x \in S^{m-1} \text{ and } k = 1, \dots, m.$$

Let $F : \mathbb{R}^m \to \mathbb{R}^m$ be the map with coordinate functions $P_1, \dots, P_m$. Clearly,

$$\big\|V(x) - F(x)\big\|^2 = \sum_{k=1}^{m} \big|f_k(x) - P_k(x)\big|^2 < \frac{1}{4}$$

for all $x \in S^{m-1}$. In general, the vectors $F(x)$ are not tangent to the sphere but, being close to $V(x)$, they have a non-zero "tangent component". This enables us to modify the vectors to obtain a non-degenerate smooth field of tangent vectors.

Subtracting the radial component from the vector $F(x)$, we put

$$W(x) = F(x) - \big\langle x, F(x)\big\rangle x \quad \text{for } x \in S^{m-1}.$$

It is clear that W is a smooth vector field. Moreover,

$$\begin{aligned}\big\|W(x)\big\| &\geqslant \big\|F(x)\big\| - \big|\big\langle x, F(x)\big\rangle\big| \\ &\geqslant \big\|V(x)\big\| - \big\|V(x) - F(x)\big\| - \big|\big\langle x, V(x) - F(x)\big\rangle\big| \\ &\geqslant 1 - 2\big\|V(x) - F(x)\big\| > 0\end{aligned}$$

since $\langle x, V(x)\rangle = 0$. We prove that the field W consists of tangent vectors. Indeed,

$$\langle x, W(x)\rangle = \langle x, F(x) - \langle x, F(x)\rangle x\rangle = \langle x, F(x)\rangle - \langle x, F(x)\rangle \|x\|^2 = 0$$

for all $x \in S^{m-1}$. Thus, the smooth field W consists of non-zero tangent vectors, which, as proved above, is impossible. □

EXERCISES

1. Complete the proof of the retraction theorem in the general case without using Brouwer's theorem. *Hint.* If Φ is an arbitrary retraction, then the difference $\Phi(x) - x$ is small in a neighborhood of the sphere. Therefore, in the ball $\overline{B}$, the function $\Phi(x) - x$ can be approximated up to an accuracy of $1/2$ by a smooth map Ψ that vanishes on the sphere. Then a smooth retraction Φ_1 can be obtained by putting $\Phi_1(x) = \frac{x+\Psi(x)}{\|x+\Psi(x)\|}$, which leads to a contradiction.

2. Give another proof of Brouwer's theorem by verifying that if a continuous map $f : \overline{B} \to \overline{B}$ has no fixed points, then the map $\Phi(x) = g(x)/\|g(x)\|$, where

$$g(x) = x - \frac{1 - \|x\|^2}{1 - \langle x, f(x)\rangle} f(x) \quad (x \in \overline{B}),$$

is a retraction of the ball to its boundary.

3. Prove the following sharpening of Brouwer's theorem: if the map $\Phi : \overline{B} \to \mathbb{R}^m$ is continuous and $\Phi(S^{m-1}) \subset \overline{B}$, then Φ has a fixed point. *Hint.* Verify that the map $x \mapsto \Phi(x)/\max\{1, \|\Phi(x)\|\}$ has the same fixed points as Φ.

6.7 *Some Geometric Applications (Continued)

6.7.1 In this section, we discuss an interesting geometric problem connected with the calculation of the measure of a set by the measures of its cross sections. Let A and B be measurable sets in the space $\mathbb{R}^m$. Is it possible to compare their measures (volumes) if we know only the measures (areas) of some of their intersections with subspaces of smaller dimension? Cavalieri's principle implies that $\lambda_m(A) < \lambda_m(B)$ if $\lambda_{m-1}(A \cap H) < \lambda_{m-1}(B \cap H)$ for all planes H perpendicular to a fixed direction.

It turns out that the situation changes drastically if instead of parallel planes we consider planes passing through a fixed point.

In 1956 Busemann[10] and Petty[11] considered the following question. Let A and B be convex sets symmetric with respect to the origin. Is it true that $\lambda_m(A) < \lambda_m(B)$ if $\lambda_{m-1}(A \cap H) < \lambda_{m-1}(B \cap H)$ for every plane H passing through the origin? It is clear that certain geometric restrictions (convexity, symmetry) of the set are necessary since otherwise the answer is negative (see Exercise 1).

---

[10] Herbert Busemann (1905–1994)—American mathematician.

[11] Clinton Petty (born 1923)—American mathematician.

The Busemann–Petty question, a positive answer to which is clear in the two-dimensional case (it is enough to use polar coordinates), is not so simple in the spaces of higher dimension. It was not until almost 20 years later, in 1975, that the following unexpected fact was established: for large m the answer is negative. Ten years later, Ball[12] proved that if the dimension is sufficiently large (more precisely, if $m \geqslant 10$), then counterexamples are given by a ball and a cube.

Following [NP], we prove a beautiful result of Ball: the area of a plane cross section of a cube takes its maximum value for a cross section that passes through a diagonal of a two-dimensional face and is perpendicular to this face. More precisely, the area of an arbitrary plane cross section of the cube $Q = [-\frac{1}{2}, \frac{1}{2}]^m$ does not exceed $\sqrt{2}$, and equality is attained only for the planes of the form $x_k = \pm x_j$, $k \neq j$.

Knowing the estimate for the areas of the plane cross sections of the cube, we can easily answer the Busemann–Petty question if the dimension m is large. It is sufficient to compare the unit cube Q and the ball $B(r_m)$ in $\mathbb{R}^m$ having Lebesgue measure 1. Indeed, since $1 = \lambda_m(B(r_m)) = \alpha_m\, r_m^m$ (as usual, α_m is the volume of the unit ball in $\mathbb{R}^m$), then $r_m = 1/\sqrt[m]{\alpha_m}$. Consequently, the measure s_m of the central cross section of $B(r_m)$ is equal to

$$s_m = \alpha_{m-1}\, r_m^{m-1} = \frac{\alpha_{m-1}}{\alpha_m^{1-\frac{1}{m}}}.$$

Taking into account the equation $\alpha_m = \pi^{\frac{m}{2}} / \Gamma(1+\frac{m}{2})$ (see Sect. 5.4.2) and Stirling's formula, we obtain $s_m \to \sqrt{e}$. Therefore, $s_m > \sqrt{2}$ if the dimension m is sufficiently large. Thus, we come to the following paradoxical result: for large m, the area of an arbitrary central cross section of the ball is greater than the area of an arbitrary cross section of the unit cube but their volumes are equal.

We verify that $s_m > \sqrt{2}$ for $m \geqslant 10$. By direct calculation, we obtain

$$s_{10} = \frac{(120)^{\frac{9}{10}}}{\frac{945}{32}\sqrt{\pi}} = 1.420\ldots > \sqrt{2} \quad \text{and}$$

$$s_{11} = \frac{1}{120}\left(\frac{10395}{64}\sqrt{\pi}\right)^{\frac{10}{11}} = 1.433\ldots > \sqrt{2}.$$

Therefore, it is sufficient to prove that the inequality $s_{m-2} < s_m$ is valid for $m > 10$. Since

$$s_m = \frac{\alpha_{m-1}}{\alpha_m^{\frac{m-1}{m}}} = \frac{\Gamma^{\frac{m-1}{m}}(\frac{m+2}{2})}{\Gamma(\frac{m+1}{2})} = s_{m-2}\frac{(2m^{m-3})^{\frac{1}{m-2}}}{m-1}\Gamma^{\frac{2}{m(m-2)}}\left(\frac{m+2}{2}\right),$$

---

[12] Keith Martin Ball (born 1960)—British mathematician.

this is equivalent to the inequality

$$\Gamma\Big(\frac{m+2}{2}\Big) > \frac{(m-1)^{\frac{m(m-2)}{2}}}{(2m^{m-3})^{\frac{m}{2}}} = \frac{m^{\frac{m}{2}}}{2^{\frac{m}{2}}}\Big(\Big(1-\frac{1}{m}\Big)^m\Big)^{\frac{m-2}{2}}.$$

Taking into account that $1 - \frac{1}{m} < e^{-\frac{1}{m}}$, we must show that

$$\Gamma\Big(\frac{m+2}{2}\Big) \geqslant e\Big(\frac{m}{2e}\Big)^{\frac{m}{2}}.$$

This immediately follows from Stirling's formula (see Eq. (8″) in Sect. 7.2.6) but can also be obtained by induction (with the inductive step from m to $m+2$), which we leave as an exercise to the reader.

We note that it is known today that the answer to the Busemann–Petty question is affirmative only if $m \leqslant 4$. For a more detailed history of the Busemann–Petty problem, see [Ko].

6.7.2 Now, we estimate the area of a cross section of $Q = [-\frac{1}{2}, \frac{1}{2}]^m$. To obtain an estimate from above, we first find an integral representation for this area. Although it is natural to look for the cross sections with maximal area among the cross sections passing through the center of the cube, we also need other cross sections.

The volume of a set A, $A \subset \mathbb{R}^m$, can be represented in terms of the areas of its cross sections

$$A(\omega, r) = \{x \in A \mid \langle x, \omega\rangle = r\} \quad (r \in \mathbb{R}^m)$$

by the planes perpendicular to the vector $\omega \neq 0$. Indeed, Cavalieri's principle implies the equation

$$\lambda_m(A) = \int_{\mathbb{R}} \lambda_{m-1}\big(A(\omega, r)\big)\, dr, \quad \text{if } \|\omega\| = 1.$$

We prove that

$$\lambda_{m-1}\big(Q(\omega, r)\big) = \frac{2}{\pi}\|\omega\| \int_0^\infty \cos(2rt) \prod_{j=1}^m \frac{\sin \omega_j t}{\omega_j t}\, dt \tag{1}$$

(if $\omega_j = 0$, then the quotient $\frac{\sin \omega_j t}{\omega_j t}$ must be replaced by 1).

If ω is proportional to a vector from the canonical basis, then Eq. (1) is valid for all $r \neq \pm\frac{1}{2}\|\omega\|$. This immediately follows from the formula $\int_0^\infty \frac{\sin ct}{t}\, dt = \frac{\pi}{2}\,\mathrm{sign}\, c$ (see Example 2 in Sect. 7.1.6). It can easily be seen that the above formula also gives the required result in the two-dimensional case. Therefore, we will assume that $m > 2$ and that at least two of the coordinates $\omega_1, \ldots, \omega_m$ of ω are non-zero. Then (1) is valid for all r. We give two proofs of this formula. The first proof, though completely elementary, is more technically involved and uses induction on dimension. The second, less cumbersome proof assumes an acquaintance with Fourier transforms (see Sect. 10.5).

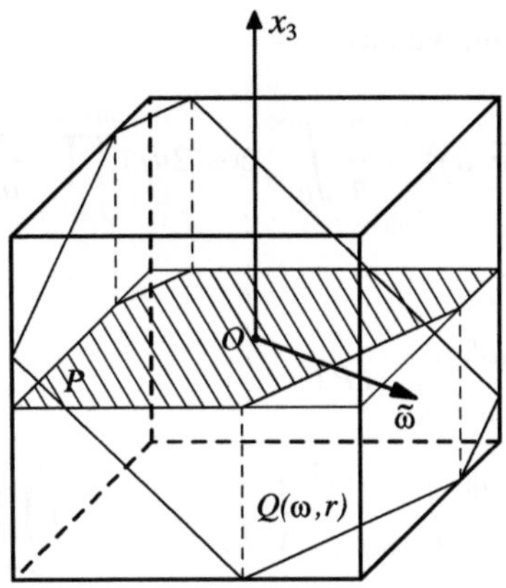

Fig. 6.5 Cross section of the cube by an oblique plane and its projection

We note that it is sufficient to prove formula (1) for a single vector proportional to ω since $Q(a\omega, r) = Q(\omega, \frac{r}{a})$.

Assuming that formula (1) is valid for the cross sections of a cube of dimension $m-1$, we prove it for the cross sections of the m-dimensional cube Q. To this end, for $\omega_m \neq 0$, we consider the projection P of a cross section $Q(\omega, r)$ on the plane $x_m = 0$. Assuming that ω_m is positive, we see that

$$\lambda_{m-1}\big(Q(\omega, r)\big) = \frac{\|\omega\|}{\omega_m}\lambda_{m-1}(P).$$

To calculate $\lambda_{m-1}(P)$, we put $\widetilde{\omega} = (\omega_1, \dots, \omega_{m-1})$. This is a non-zero vector since at least two coordinates of the vector ω are non-zero. As seen in Fig. 6.5, non-empty cross sections (in $\mathbb{R}^{m-1}$) of P by planes perpendicular to $\widetilde{\omega}$ coincide with the corresponding cross sections of the cube $\widetilde{Q} = [-\frac{1}{2}, \frac{1}{2}]^{m-1}$.

Indeed, since

$$\begin{aligned} P &= \left\{\widetilde{x} \in \widetilde{Q} \mid \exists x_m \in \left[-\frac{1}{2}, \frac{1}{2}\right] : \langle\widetilde{\omega}, \widetilde{x}\rangle + \omega_m x_m = r\right\} \\ &= \left\{\widetilde{x} \in \widetilde{Q} \mid \langle\widetilde{\omega}, \widetilde{x}\rangle \in \left[r - \frac{1}{2}\omega_m, r + \frac{1}{2}\omega_m\right]\right\}, \end{aligned}$$

the cross section $P(\widetilde{\omega}, u)$ coincides with $\widetilde{Q}(\widetilde{\omega}, u)$ for $u \in [r - \omega_m/2, r + \omega_m/2]$ and is empty for the remaining u.

Replacing, if necessary, the vector ω by a proportional vector, we may assume that $\|\widetilde{\omega}\| = 1$. Then, as noted above, Cavalieri's principle implies the equation

$$\lambda_{m-1}(P) = \int_{\mathbb{R}} \lambda_{m-2}\big(P(\widetilde{\omega}, u)\big)\,du = \int_{r-\omega_m/2}^{r+\omega_m/2} \lambda_{m-2}\big(\widetilde{Q}(\widetilde{\omega}, u)\big)\,du.$$

By the induction assumption, we have

$$\lambda_{m-2}\big(\widetilde{Q}(\widetilde{\omega}, u)\big) = \frac{2}{\pi}\int_0^\infty \cos(2ut)\prod_{j=1}^{m-1}\frac{\sin\omega_j t}{\omega_j t}\,dt.$$

Therefore,

$$\begin{aligned}
\lambda_{m-1}\big(Q(\omega, r)\big) &= \frac{\|\omega\|}{\omega_m}\lambda_{m-1}(P) \\
&= \frac{\|\omega\|}{\omega_m}\int_{r-\omega_m/2}^{r+\omega_m/2}\frac{2}{\pi}\int_0^\infty \cos(2ut)\prod_{j=1}^{m-1}\frac{\sin\omega_j t}{\omega_j t}\,dt\,du \\
&= \frac{2}{\pi}\frac{\|\omega\|}{\omega_m}\int_0^\infty\left(\int_{r-\omega_m/2}^{r+\omega_m/2}\cos(2ut)\,du\right)\prod_{j=1}^{m-1}\frac{\sin\omega_j t}{\omega_j t}\,dt \\
&= \frac{2}{\pi}\|\omega\|\int_0^\infty \cos(2rt)\frac{\sin\omega_m t}{\omega_m t}\prod_{j=1}^{m-1}\frac{\sin\omega_j t}{\omega_j t}\,dt,
\end{aligned}$$

which completes the proof.

Proceeding to the proof of formula (1) based on the Fourier transform, we fix a unit vector ω (with at least two non-zero coordinates) and find the Fourier transform of the function

$$r \mapsto s(r) = \lambda_{m-1}\big(Q(\omega, r)\big) \quad (r \in \mathbb{R}).$$

To this end, we calculate the value the Fourier transform $\widehat{\chi}$ of the characteristic function χ of Q at the point $t\omega$. By definition, we have

$$\widehat{\chi}(t\omega) = \int_Q e^{-2\pi i\langle t\omega, x\rangle}\,dx = \prod_{j=1}^m\int_{-\frac{1}{2}}^{\frac{1}{2}} e^{-2\pi i t\omega_j x_j}\,dx_j = \prod_{j=1}^m\frac{\sin(\pi\omega_j t)}{\pi\omega_j t}.$$

Now, we consider an orthogonal transformation L in $\mathbb{R}^m$ taking the first vector of the canonical basis to the vector ω. By Fubini's theorem, we obtain

$$\begin{aligned}
\widehat{\chi}(t\omega) &= \int_{\mathbb{R}^m}\chi(x)e^{-2\pi i t\langle\omega, x\rangle}\,dx = \int_{\mathbb{R}^m}\chi(Ly)e^{-2\pi i t y_1}\,dy \\
&= \int_{-\infty}^\infty s(y_1)e^{-2\pi i t y_1}\,dy_1 = \widehat{s}(t),
\end{aligned}$$

where $\widehat{s}$ is the Fourier transform of s. Comparing the two equations obtained, we see that

$$\widehat{s}(t) = \prod_{j=1}^m\frac{\sin(\pi\omega_j t)}{\pi\omega_j t}.$$

Since the function $\widehat{s}$ is summable on $\mathbb{R}$, we can find $s(r)$ by the inversion formula (see Theorem 10.5.4):

$$s(r) = \int_{-\infty}^{\infty} \widehat{s}(t)e^{2\pi irt}\,dt = 2\int_0^{\infty} \cos 2\pi rt \prod_{j=1}^{m} \frac{\sin(\pi\omega_j t)}{\pi\omega_j t}\,dt.$$

It remains to make the change of variables $\pi t \mapsto t$.

6.7.3 In the proof of the inequality $\lambda_{m-1}(Q(\omega,r)) \leqslant \sqrt{2}$, we may assume without loss of generality that $\|\omega\| = 1$ and all coordinates of the vector ω are positive. If at least one of the coordinates, e.g., ω_m is large, $\omega_m \geqslant \frac{1}{\sqrt{2}}$, then the inequality $\lambda_{m-1}(Q(\omega,r)) \leqslant \sqrt{2}$ is obvious (since the measure of the projection of the cross section on the plane $\omega_m = 0$ is at most 1). Now, we assume that $0 < \omega_j < \frac{1}{\sqrt{2}}$ for all $j = 1,\dots,m$. From Eq. (1) and Hölder's inequality (see Sect. 4.4.5, Corollary 2) with exponents $1/\omega_j^2$, we obtain

$$\lambda_{m-1}\big(Q(\omega,r)\big) \leqslant \frac{2}{\pi}\int_0^{\infty} \prod_{j=1}^{m} \left|\frac{\sin\omega_j t}{\omega_j t}\right| dt \leqslant \frac{2}{\pi}\prod_{j=1}^{m}\left(\int_0^{\infty}\left|\frac{\sin\omega_j t}{\omega_j t}\right|^{1/\omega_j^2} dt\right)^{\omega_j^2}$$

$$= \frac{2}{\pi}\prod_{j=1}^{m}\left(\frac{1}{\omega_j}\int_0^{\infty}\left|\frac{\sin t}{t}\right|^{1/\omega_j^2} dt\right)^{\omega_j^2}.$$

Now, we use Ball's integral inequality (whose proof we postpone until the next section)

$$\int_0^{\infty}\left|\frac{\sin x}{x}\right|^p dx < \frac{\pi}{\sqrt{2p}} \qquad \text{for } p > 2.$$

Setting p equal to $\frac{1}{\omega_j^2} > 2$ $(j = 1,\dots,m)$ and taking into account the relation $\omega_1^2 + \cdots + \omega_m^2 = 1$, we see that

$$\lambda_{m-1}\big(Q(\omega,r)\big) < \frac{2}{\pi}\prod_{j=1}^{m}\left(\frac{\pi}{\sqrt{2}}\right)^{\omega_j^2} = \sqrt{2},$$

as required.

The above proof shows that the area of a cross section is equal to $\sqrt{2}$ only if all coordinates of the vector ω, $\|\omega\| = 1$, are either zero or $\pm 1/\sqrt{2}$. Therefore, only the cross sections by the planes $x_j = \pm x_k$ $(k \neq j)$ are extremal.

We note also that Eq. (1) gives the following expression for the area S_m of the central cross section of the unit cube by the plane orthogonal to its main diagonal (i.e., by the plane $x_1 + \cdots + x_m = 0$):

$$S_m = \frac{2}{\pi}\sqrt{m}\int_0^{\infty}\left(\frac{\sin t}{t}\right)^m dt.$$

Ball's integral inequality implies that these areas are not maximal for $m > 2$. The fact that they are not maximal for sufficiently large m follows from the asymptotic Laplace formula (see Sect. 7.3.3, Example 3) since, by this formula, we have $S_m \to \sqrt{\frac{6}{\pi}} < \sqrt{2}$.

6.7.4 Despite the deceptive simplicity of Ball's inequality, its proof is technically involved. We obtain it from an integral inequality interesting in itself and connected with decreasing distribution functions.

Let $(X, \mathfrak{A}, \mu)$ be a measure space, and let f be a non-negative measurable almost everywhere finite function on X. The integral $I_p(f) = \int_X f^p\, d\mu$ $(0 < p < +\infty)$ contains a great deal of information about the function f and is used in various problems. Often it is important to compare the values of these integrals for different p. In the case of a normalized measure (i.e., if $\mu(X) = 1$), the behavior of $I_p(f)$ is quite simple: it follows from Hölder's inequality (see Sect. 4.4.5) that the quantities $I_p^{1/p}(f)$ increase with p. It is more complicated to compare their growth for two distinct functions. In particular, the following question is of some interest: under which conditions is the inequality $I_p(g) \leqslant I_p(f)$ valid for all $p > q$ if we know that it is valid at the "initial point", i.e., at $p = q$? The answer is given by the following statement.

Proposition *Let $(X, \mathfrak{A}, \mu)$ be a measure space, and let non-negative measurable almost everywhere finite functions f and g on X have finite decreasing distribution functions F and G,*

$$F(t) = \mu\big(X(f > t)\big), \qquad G(t) = \mu\big(X(g > t)\big) \quad (t > 0).$$

If at some point $t_0 > 0$, the difference $F - G$ changes its sign from minus to plus (i.e., $F(t) \leqslant G(t)$ for $t \in (0, t_0)$ and $F(t) \geqslant G(t)$ for $t > t_0$), then for $p > q > 0$

$$\text{the inequality } \int_X g^q\, d\mu \leqslant \int_X f^q\, d\mu \quad \text{implies the inequality } \int_X g^p\, d\mu \leqslant \int_X f^p\, d\mu.$$

The latter inequality becomes an equality only in the two trivial cases: $F \equiv G$ or $\int_X g^p\, d\mu = \int_X f^p\, d\mu = +\infty$.

Proof Assuming that $\int_X f^p\, d\mu < +\infty$ (otherwise everything is obvious), we put

$$\Delta(p) = \int_X f^p\, d\mu - \int_X g^p\, d\mu.$$

By assumption $\Delta(q) \geqslant 0$, and we must prove that $\Delta(p) \geqslant 0$ for $p > q$. To this end, we must verify that the difference $R = A\Delta(p) - B\Delta(q)$ is non-negative for some positive A and B. We will see that, for our purposes, it is convenient to take A and B equal, respectively, to the derivatives of t^q and t^p at the point t_0.

With the help of distribution functions, we represent the integrals $\int_X f^p\,d\mu$ and $\int_X g^p\,d\mu$ in the form (see Proposition 6.4.3)

$$\Delta(p)=\int_X f^p\,d\mu-\int_X g^p\,d\mu=\int_0^\infty \big(F(t)-G(t)\big)pt^{p-1}\,dt.$$

Taking $A=qt_0^{q-1}$ and $B=pt_0^{p-1}$, we obtain

$$R=\int_0^\infty \big(F(t)-G(t)\big)\big(qt_0^{q-1}pt^{p-1}-pt_0^{p-1}qt^{q-1}\big)\,dt.$$

The difference

$$pqt_0^{q-1}t^{p-1}-pqt_0^{p-1}t^{q-1}=pq(t_0t)^{q-1}\big(t^{p-q}-t_0^{p-q}\big)$$

has the same sign as $F(t)-G(t)$. Therefore, we have a non-negative function under the last integral. Consequently, $R\geqslant 0$. Moreover, $R>0$ if $F\not\equiv G$. □

Remark The above proof suggest the following slightly stronger result:

$$I(p)=\int_0^\infty \big(F(t)-G(t)\big)\Big(\frac{t}{t_0}\Big)^{p-1}dt=\frac{1}{pt_0^{p-1}}\int_X \big(f^p-g^p\big)\,d\mu$$

is non-decreasing.

6.7.5 We finish this section with a proof of Ball's integral inequality

$$\int_0^\infty \left|\frac{\sin x}{x}\right|^p dx<\frac{\pi}{\sqrt{2p}}\qquad \text{for } p>2.$$

If $p=2$, then the inequality becomes an equality. This can easily be obtained by integrating by parts the equation $\int_0^\infty \frac{\sin x}{x}\,dx=\frac{\pi}{2}$ (see Sect. 7.1.6, Example 2). For $1<p<2$, the sign in Ball's inequality must be replaced with its opposite (see Exercise 2).

We represent the inequality in question in the form

$$\int_0^\infty g^p(x)\,dx<\int_0^\infty f^p(x)\,dx\quad (p>2),$$

where $g(x)=|\frac{\sin x}{x}|$ and $f(x)=e^{-\frac{x^2}{2\pi}}$. Since the inequality becomes an equality for $p=2$, it is sufficient to prove that the difference $F-G$ of decreasing distribution functions at some point t_0 changes its sign from minus to plus. Since each of the functions f and g is at most 1, we have $F(t)=G(t)=0$ for $t\geqslant 1$. Therefore, below we may assume that $t\in(0,1)$. Obviously, $F(t)=f^{-1}(t)=\sqrt{2\pi\ln\frac{1}{t}}$. It is more complicated to find the values of the function G, and to estimate them, we need the quantities $t_m=\max_{(\pi m,\pi m+\pi)} g$, $m\in\mathbb{N}$. It is clear that $\frac{1}{\pi(m+\frac{1}{2})}<t_m<\frac{1}{\pi m}$.

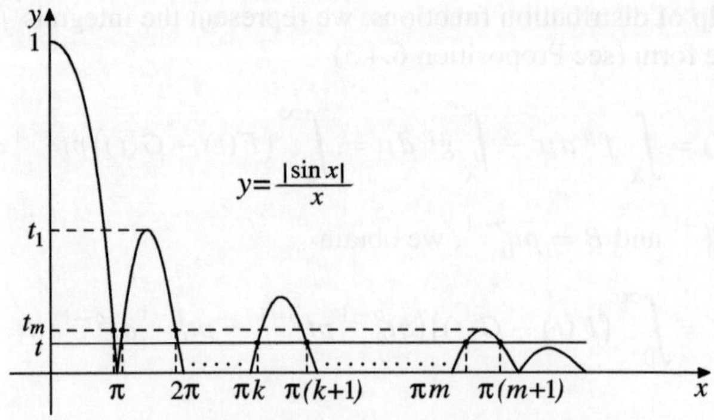

Fig. 6.6 Roots of the equation $\frac{|\sin x|}{x} = t$

Using the expansion of sine into an infinite product (see Sect. 7.2.5, formula (7)), we obtain that the function g decreases on the interval $(0, 1)$ and does not exceed $e^{-x^2/6}$,

$$g(x) = \prod_{k=1}^{\infty}\left(1 - \frac{x^2}{\pi^2 k^2}\right) \leqslant \prod_{k=1}^{\infty} e^{-x^2/(\pi k)^2} = e^{-x^2/6}$$

(at the end, we used the relation $\sum_{k=1}^{\infty} \frac{1}{k^2} = \frac{\pi^2}{6}$ proved in Example 2 of Sect. 4.6.2). Therefore, for $t \in (t_1, 1)$, we obtain

$$G(t) = (g|_{(0,1)})^{-1}(t) \leqslant \sqrt{6 \ln \frac{1}{t}} < \sqrt{2\pi \ln \frac{1}{t}} = F(t)$$

and, consequently, the difference $F - G$ is positive on $(t_1, 1)$. At the same time, the difference changes its sign since

$$2\int_0^{\infty} t\big(F(t) - G(t)\big)\,dt = \int_0^{\infty} \big(f^2(x) - g^2(x)\big)\,dx = 0.$$

To prove that the change of the sign takes place only once, it is sufficient to prove that $F - G$ increases on $(0, t_1)$. To this end, we prove that $|G'(t)| > |F'(t)|$ for $t \in (0, t_1)$, $t \neq t_m$. It is clear that the function G is everywhere continuous and differentiable at the points distinct from t_m $(m \in \mathbb{N})$. Moreover,

$$\big|G'(t)\big| = -G'(t) = \sum_{\substack{x>0:\\ g(x)=t}} \frac{1}{|g'(x)|}.$$

Let $t \in (t_{m+1}, t_m)$. For such t, the equation $g(x) = t$ has a unique root on $(0, \pi)$ and two roots on the intervals $(\pi k, \pi k + \pi)$ for $k = 1, \ldots, m$ (see Fig. 6.6).

We estimate $|g'(x)|$ from above at these points. If $x \in (0, \pi)$, then

$$\big|g'(x)\big| = \frac{\sin x - x\cos x}{x^2} \leqslant \frac{1}{2}$$

(the inequality $\sin x - x\cos x \leqslant x^2/2$ can easily be proved by differentiation). If $x \in (\pi k, \pi k + \pi)$ for $k \geqslant 1$, then

$$|g'(x)| = \frac{1}{x}\left|\cos x - \frac{\sin x}{x}\right| \leqslant \frac{1}{x}\left(1 + \frac{|\sin(x-\pi k)|}{\pi k}\right) \leqslant \frac{1}{x}\left(1 + \frac{x-\pi k}{\pi k}\right) = \frac{1}{\pi k}.$$

Consequently, for $t \in (t_{m+1}, t_m)$, we have

$$|G'(t)| \geqslant 2 + 2\sum_{k=1}^{m} \pi k = 2 + \pi m(m+1) > \pi\left(m + \frac{3}{2}\right) > \frac{1}{t_{m+1}}.$$

Thus,

$$\left|\frac{G'(t)}{F'(t)}\right| = |G'(t)|t\sqrt{\frac{2}{\pi}\ln\frac{1}{t}} \geqslant \frac{1}{t_{m+1}}\sqrt{\frac{2}{\pi}t^2\ln\frac{1}{t}}.$$

Since $t_1 < \frac{1}{\pi}$, the product $t^2 \ln\frac{1}{t}$ increases on $(0, t_1)$, and, therefore, for $t > t_{m+1}$, we obtain

$$\left|\frac{G'(t)}{F'(t)}\right| > \frac{1}{t_{m+1}}\sqrt{\frac{2}{\pi}t_{m+1}^2\ln\frac{1}{t_{m+1}}} \geqslant \sqrt{\frac{2}{\pi}\ln\frac{1}{t_2}} > \sqrt{\frac{2}{\pi}\ln 2\pi}.$$

It remains to observe that the right-hand side of the inequality is greater than 1 since $\ln 4x > x$ on $[1, 2]$ by the concavity of the logarithm.

EXERCISES

1. Prove that a spherical layer can have an arbitrarily large volume whereas the area of its cross section by every plane is arbitrarily small.
2. Prove that $\int_0^\infty |\frac{\sin x}{x}|^p\,dx > \frac{\pi}{\sqrt{2p}}$ for $0 < p < 2$. *Hint.* Use Remark 6.7.4.

Chapter 7
Integrals Dependent on a Parameter

7.1 Basic Theorems

When dealing with functions of "two variables", i.e., with functions defined on the direct product of two sets, the reader has probably encountered the situation in which it is required to decide whether it is possible to perform an operation (passage to the limit, differentiation, integration) for one variable independently of the operations for the other variable. In other words, do the operations for different variables commute? Speaking of differentiation, we should mention the well-known theorem on the equality of mixed partial derivatives. The reader probably also knows the theorem on equality of iterated limits which says that under certain conditions the two limiting passages commute. We will study this question in the case where one of the operations is integration.

Our goal in this section is to study the properties of an "integral dependent on a parameter", i.e., a function J defined by an equation of the form

$$J(y) = \int_X f(x, y)\, d\mu(x) \quad (y \in Y).$$

Here μ is a measure defined on a σ-algebra of subsets of a set X, the function $x \mapsto f(x, y)$ is summable on X for every $y \in Y$, and Y is a subset of a metrizable topological space $\widetilde{Y}$. If X is a topological space, then we always assume that the measure μ is defined on all Borel sets (and, consequently, all continuous functions on X are measurable). We do not exclude the case where μ is the counting measure; therefore the results below are valid, in particular, for absolutely convergent series.

First of all, we are interested in the continuity and (in the case where $Y \subset \mathbb{R}^m$) in the differentiability of the function J. Actually, this is a question about the validity of interchanging the integration with respect to the first variable with other analytical operations (passage to the limit, differentiation) with respect to the second variable (see the Theorems in Sects. 7.1.2 and 7.1.5). We encountered such a situation in Sect. 4.8, where we discussed the passage to the limit under the integral sign and the index of a function played the role of a parameter. These results will serve as a basis for subsequent reasoning.

B. Makarov, A. Podkorytov, *Real Analysis: Measures, Integrals and Applications*, Universitext, DOI 10.1007/978-1-4471-5122-7_7,

It is also natural to study the problem of integration with respect to a parameter. However, there is no need to touch on this here, since to a great extent the problem is solved by Fubini's theorem.

7.1.1 In this section and the next, we restate three theorems from Sect. 4.8 for the case of a "continuous parameter". In all three statements, a is a limit point[1] of the set Y in the space $\widetilde{Y}$ and $\varphi(x) = \lim_{y \to a} f(x, y)$, where f and φ are functions (in general, complex-valued) defined on $X \times Y$ and X, respectively. We present conditions under which the following relation holds:

$$J(y) = \int_X f(x, y)\, d\mu(x) \underset{y \to a}{\longrightarrow} \int_X \varphi(x)\, d\mu(x), \tag{1}$$

i.e.,

$$\lim_{y \to a} \int_X f(x, y)\, d\mu(x) = \int_X \Big(\lim_{y \to a} f(x, y) \Big)\, d\mu(x).$$

Theorem *If $\mu(X) < +\infty$ and the convergence $f(x, y) \underset{y \to a}{\longrightarrow} \varphi(x)$ is uniform with respect to $x \in X$, then the function φ is summable on X and relation* (1) *holds.*

Proof Since the space $\widetilde{Y}$ is metrizable, we can argue "in terms of sequences". We must prove that

$$J(y_n) = \int_X f(x, y_n)\, d\mu(x) \underset{n \to \infty}{\longrightarrow} \int_X \varphi(x)\, d\mu(x)$$

for every sequence $\{y_n\}$ of points $y_n \in Y \setminus \{a\}$, $n \in \mathbb{N}$, converging to a. This fact and the summability of φ follows directly from Theorem 4.8.1 all conditions of which are fulfilled with $f_n(x) = f(x, y_n)$. □

7.1.2 For convenience of reference, we present here modifications of Lebesgue's theorems (see Sects. 4.8.3 and 4.8.4) and the corollary to Vitali's theorem (see Sect. 4.8.7) for the case of a "continuous parameter".

Theorem 1 *Let $\varphi(x) = \lim_{y \to a} f(x, y)$ for almost all $x \in X$. Assume that there exist a neighborhood U of a and a function $g : X \to \mathbb{R}$ such that the following conditions hold*:

$$\left.\begin{array}{ll} \text{(a)} & \textit{for almost all } x \in X \textit{ and every } y \in (Y \cap U) \setminus \{a\} \\ & \textit{the inequality } \big|f(x, y)\big| \leqslant g(x) \textit{ holds,} \\ \text{(b)} & \textit{the function } g \textit{ is summable on } X. \end{array}\right\} \qquad (L_{\mathrm{loc}})$$

Then the function φ is summable on X and relation (1) *holds.*

---

[1]In particular, if $\widetilde{Y} = [-\infty, +\infty]$, then the cases $a = \pm\infty$ are possible.

Condition (L_{loc}) can be weakened by requiring that the inequality $|f(x, y)| \leqslant g(x)$ be valid for each $y \in Y$ only on a set of full measure possibly depending on y. The above proof of the theorem remains valid for this generalization of condition (L_{loc}). However, in the sequel, (see Theorems 7.1.5 and 7.1.7) we need the exact formulation of condition (L_{loc}) given in Theorem 1.

Proof As in Theorem 7.1.1, we consider the sequence of functions $f_n(x) = f(x, y_n)$, where $y_n \to a$, $y_n \in (Y \cap U) \setminus \{a\}$ and apply Lebesgue's theorem 4.8.4. □

In the case where $X = \mathbb{N}$ and μ is the counting measure, the integral $\int_X f(x, y)\, d\mu(x)$ becomes the sum of the (absolutely convergent) series $\sum_{n=1}^{\infty} f(n, y)$, and condition ($L_{\mathrm{loc}}$) coincides with the condition in the Weierstrass M-test for uniform convergence of a series in a neighborhood of a. It follows from Theorem 1 that the limit of the sum can be found termwise.

If $\mu(X) < +\infty$ and the function f is bounded, then condition (L_{loc}) obviously holds for every limit point of Y.

In the case of finite measure, condition (L_{loc}) can be replaced by a modification of condition (V) of Corollary 4.8.7.

Theorem 2 *Let $\mu(X) < +\infty$ and $\varphi(x) = \lim_{y \to a} f(x, y)$ for almost all $x \in X$. If there exists a neighborhood U of a and numbers $s > 1$ and $C > 0$ such that*

$$\int_X |f(x, y)|^s \, d\mu(x) \leqslant C \quad \text{for all } y \in (Y \cap U) \setminus \{a\}, \qquad (V_{\mathrm{loc}})$$

then the function φ is summable on X and relation (1) *holds.*

Proof The proof of this theorem is the same as the proof of the preceding one, the only difference being that now we refer to Corollary 4.8.7 instead of Lebesgue's theorem. □

In some cases, condition (V_{loc}) is a useful alternative to condition (L_{loc}). For example, if $X = Y$ is a ball in $\mathbb{R}^m$, μ is Lebesgue measure, and $f(x, y) = \frac{1}{\|x-y\|^p}$, where $p < m$, then condition (V_{loc}) (for $1 < s < m/p$) is fulfilled at an arbitrary point $a \in Y$, and, therefore, the function J is continuous on Y. At the same time, condition (L_{loc}) cannot hold at any $a \in Y$, since we have

$$\sup_{y \in U \setminus \{x\}} f(x, y) = +\infty \quad \text{for all } x \in U$$

in each neighborhood U of a.

7.1.3 The following theorem is obviously a special case of Theorem 1 of Sect. 7.1.2.

Theorem *If a function f satisfies condition (L_{loc}) at a point $y_0 \in Y$ and is continuous with respect to the second variable at almost all $x \in X$, i.e.,*

$$f(x, y) \underset{y \to y_0}{\longrightarrow} f(x, y_0) \quad \text{for almost all } x \in X, \qquad (2)$$

then the function J is continuous at y_0:

$$J(y) = \int_X f(x, y)\, d\mu(x) \underset{y \to y_0}{\longrightarrow} J(y_0) = \int_X f(x, y_0)\, d\mu(x).$$

We remark that condition (2) is certainly fulfilled if X is a topological space and the function f is continuous on $X \times Y$.

Corollary *If X is a compact space with a finite measure and $Y \subset \mathbb{R}$ is an arbitrary interval, then the continuity of f on $X \times Y$ implies the continuity of the integral $J(y) = \int_X f(x, y)\, d\mu(x)$ on this interval.*

Proof Indeed, every point of the interval has a relative neighborhood U whose closure is a compact set contained in the interval. By the Weierstrass theorem, the function f is bounded on the product $X \times U$, which guarantees the fulfillment of condition (L_{loc}). □

It is clear that the corollary is valid not only for an interval but for every locally compact space Y; in particular, it is valid if Y is an open or closed subset of a Euclidean space.

7.1.4 We consider two examples. We prove that the functions H and K defined by the equations

$$H(y) = \int_0^\infty \frac{\cos xy}{1 + x^2}\, dx \quad \text{for } y \in \mathbb{R},$$

$$K(y) = \int_0^\infty e^{-xy} \sin x\, dx \quad \text{for } y \in (0, +\infty)$$

are continuous.

In the first case, we have $f(x, y) = \frac{\cos xy}{1+x^2}$. Since

$$\left|\frac{\cos xy}{1 + x^2}\right| \leqslant \frac{1}{1 + x^2} \quad \text{for all } x, y \in \mathbb{R} \quad \text{and} \quad \int_0^\infty \frac{dx}{1 + x^2} < +\infty,$$

we see that the function f satisfies condition (L_{loc}) at every point $y \in \mathbb{R}$. It remains to refer to Theorem 1 of Sect. 7.1.2.

In the second case, we have $f(x, y) = e^{-xy} \sin x$. In contrast to the preceding example, there is now no majorant g_0 common for all $y \in Y$ and summable on $(0, +\infty)$ for which the inequality $|f(x, y)| \leqslant g_0(x)$ holds for all $x, y > 0$. Nevertheless, condition (L_{loc}) still holds for every point $y \in (0, +\infty)$, but now, for every $y > 0$, we must choose a neighborhood and a summable majorant depending on the neighborhood. Indeed, let $y_0 > 0$ and $U = (y_0/2, +\infty)$. Then

$$\left|e^{-xy} \sin x\right| \leqslant e^{-\frac{xy_0}{2}} \quad \text{for all } x > 0,\ y \in U, \quad \text{and} \quad \int_0^\infty e^{-\frac{xy_0}{2}}\, dx < +\infty.$$

The second of the above examples can be handled in a different way by direct calculation. Indeed, integrating by parts twice, we obtain

$$K(y) = -e^{-xy}\cos x\Big|_0^\infty - y\int_0^\infty e^{-xy}\cos x\,dx$$
$$= 1 - y\left(e^{-xy}\sin x\Big|_0^\infty + y\int_0^\infty e^{-xy}\sin x\,dx\right) = 1 - y^2K(y).$$

Consequently, $K(y) = 1/(1+y^2)$ for every $y > 0$.

The first solution, based on the general scheme, is presented here for two reasons. First, it is typical for such problems. For example, in the same way, we can prove that the integral $\int_0^\infty e^{-xy}h(x)\,dx$ is continuous with respect to the parameter y for every bounded function h. Secondly, even if we know how to calculate the integral, we must sometimes check condition (L_{loc}) for the integrand (see Example 2 of Sect. 7.1.6 below).

7.1.5 Theorem 1 of Sect. 7.1.2 allows us easily obtain conditions not only for the continuity but also for the differentiability of an integral depending on a parameter.

Theorem *Let $Y \subset \mathbb{R}$ be an arbitrary interval. Assume that:*

(a) *the derivative*

$$f'_y(x,y) = \lim_{h\to 0}\frac{f(x,y+h) - f(x,y)}{h}$$

exists for almost all $x \in X$ and every $y \in Y$;

(b) *the function f'_y satisfies condition (L_{loc}) at a point $y_0 \in Y$.*

Then the function J is differentiable at y_0 and

$$J'(y_0) = \int_X f'_y(x,y_0)\,d\mu(x). \tag{3}$$

This formula is called the *Leibniz rule*.

Proof Let $x \in X$, $y_0 + h \in Y$, $h \neq 0$, and

$$F(x,h) = \frac{f(x,y_0+h) - f(x,y_0)}{h}.$$

Since

$$\frac{J(y_0+h) - J(y_0)}{h} = \int_X \frac{f(x,y_0+h) - f(x,y_0)}{h}\,d\mu(x) = \int_X F(x,h)\,d\mu(x), \tag{4}$$

we see that the existence of a finite derivative $J'(y_0)$ and Eq. (3) is immediately obtained by passing to the limit as $h \to 0$ under the integral sign in Eq. (4). We can justify the passage to the limit by Theorem 1 of Sect. 7.1.2 if we prove that the function F satisfies condition (L_{loc}) at the point $h = 0$. Let us check this. Since the function f'_y satisfies condition (L_{loc}), there exist a positive number δ and a function g summable on X such that

$$\left|f'_y(x, y)\right| \leqslant g(x) \quad \text{for almost all } x \in X \text{ and for } y \in Y,\ 0 < |y - y_0| < \delta.$$

The Lagrange mean value theorem applied to the function $y \mapsto f(x, y)$ on the interval with endpoints y_0 and $y_0 + h$ gives the relation $F(x, h) = f'_y(x, y_0 + \theta h)$, where θ is a number in the interval $(0, 1)$. Therefore, $|F(x, h)| \leqslant g(x)$ for almost all $x \in X$ and $0 < |h| < \delta$. Consequently, condition (L_{loc}) is fulfilled for F. □

Usually, when using the theorem proved above, there is no doubt as to the existence of the partial derivative f'_y and it only remains to check that it satisfies condition (L_{loc}). The situation is even simpler in the case where $X = [p, q]$, $Y = \langle a, b\rangle$, and the functions f and f'_y are continuous in the rectangle $X \times Y$. Then the function $J(y) = \int_p^q f(x, y)\,dx$ is continuously differentiable on $\langle a, b\rangle$ and $J'(y) = \int_p^q f'_y(x, y)\,dx$.

Remark Theorem 7.1.5 obviously also remains valid in the more general setting where Y is a subset of a multi-dimensional space and the derivative $J'(y)$ is replaced by the partial derivative with respect to one of the coordinates.

7.1.6 We consider some applications of the results obtained. First of all, we apply them to calculate two important integrals of functions whose primitives cannot be expressed in terms of elementary functions.

Example 1 Calculate the integral

$$J(y) = \int_0^{\infty} e^{-x^2} \cos yx\, dx \quad \text{for } y \in \mathbb{R}.$$

By the theorem on differentiation of an integral with respect to a parameter (all conditions of this theorem are obviously met), this is a smooth function and

$$J'(y) = -\int_0^{\infty} x e^{-x^2} \sin yx\, dx.$$

Integrating by parts, we obtain

$$J'(y) = \frac{1}{2} e^{-x^2} \sin yx \Big|_0^{\infty} - \frac{y}{2}\int_0^{\infty} e^{-x^2} \cos yx\, dx = -\frac{y}{2}\int_0^{\infty} e^{-x^2} \cos yx\, dx.$$

Therefore, $J'(y) + \frac{y}{2} J(y) = 0$. Consequently, $(e^{y^2/4} J(y))' = 0$. Thus, $J(y) = Ce^{-y^2/4}$. Since $C = J(0) = \frac{\sqrt{\pi}}{2}$ (see Sect. 4.6.3), we come to the required result,

$$J(y) = \int_0^{\infty} e^{-x^2} \cos yx\, dx = \frac{\sqrt{\pi}}{2} e^{-y^2/4}.$$

Example 2 We consider the integral

$$J(y) = \int_0^\infty e^{-xy} \frac{\sin x}{x}\, dx \quad \text{for } y \in (0, +\infty).$$

We prove that the function J is differentiable and use this fact to find $J(y)$. It is clear that, in our case, we have $f'_y(x, y) = -e^{-xy} \sin x$ for all $x,\ y > 0$. As proved in Sect. 7.1.4, the function f'_y satisfies condition (L_{loc}) at every point of the semi-axis $(0, +\infty)$. Therefore, we may use the Leibniz rule,

$$J'(y) = -\int_0^\infty e^{-xy} \sin x\, dx \quad \text{for } y > 0.$$

The last integral was calculated in Example 7.1.4 (we remark that the knowledge of this integral does not spare us the necessity of using condition (L_{loc}) for justification of the above equation). Consequently,

$$J'(y) = -\frac{1}{1+y^2} \quad \text{and} \quad J(y) = C - \arctan y \quad \text{for all } y > 0,$$

where C is a constant. To determine the constant, we observe that $J(y) \underset{y\to+\infty}{\longrightarrow} 0$ since $|J(y)| \leqslant \int_0^\infty e^{-xy}\, dx = \frac{1}{y}$. Therefore, $C = \frac{\pi}{2}$, and thus

$$J(y) = \frac{\pi}{2} - \arctan y \quad \text{for all } y > 0. \tag{5}$$

Up to now, we have considered the integral $J(y)$ only for $y > 0$. However, the integrand also makes sense for $y = 0$. Moreover, we know (see Example 1 of Sect. 4.6.6) that, although the function $f(x, 0) = \frac{\sin x}{x}$ is not summable, the improper integral $\int_0^\infty \frac{\sin x}{x}\, dx$ nevertheless converges. Therefore, it is natural to define the integral $J(y)$ also for $y = 0$ by $J(0) = \int_0^\infty \frac{\sin x}{x}\, dx$. This naturally raises the question of whether the integral $J(y)$ thus defined is continuous at zero. It is clear that

$$e^{-xy} \frac{\sin x}{x} \underset{y\to 0}{\longrightarrow} \frac{\sin x}{x} \quad \text{for all } x > 0.$$

The justification of the passage to the limit $J(y) \to J(0)$ is complicated by the fact that the integrand $J(0)$ is not summable. Therefore, we cannot use Theorem 1 of Sect. 7.1.2 here, the conditions of which guarantee the summability of the limiting function. In Sect. 7.4, we obtain general theorems allowing us to verify the continuity of an improper integral depending on a parameter, but now we prove that the function J is continuous at zero directly. We verify that the difference

$$J(y) - J(0) = \int_0^\infty \left(e^{-yx} - 1\right) \frac{\sin x}{x}\, dx$$

tends to zero as $y \to 0$. To this end, we estimate the integral over the intervals $[0, t]$ and $[t, +\infty)$ separately; here $t > 0$ is an auxiliary parameter which will be specified later. The integral over the interval $[0, t]$ can be coarsely estimated: since $0 \leqslant 1 - e^{-yx} \leqslant yx$, we have

$$\left| \int_0^t (e^{-yx} - 1) \frac{\sin x}{x} \, dx \right| \leqslant \int_0^t xy \frac{1}{x} \, dx = yt.$$

Integrating by parts in the second integral, we obtain

$$\int_t^\infty (e^{-yx} - 1) \frac{\sin x}{x} \, dx = \int_t^\infty (e^{-yx} - 1) \frac{d(-\cos x)}{x}$$

$$= -(1 - e^{-ty}) \frac{\cos t}{t} + \int_t^\infty \cos x \left(\frac{e^{-yx} - 1}{x} \right)_x' dx.$$

Consequently,

$$\left| \int_t^\infty (e^{-yx} - 1) \frac{\sin x}{x} \, dx \right| \leqslant \frac{1}{t} + \int_t^\infty \left(\frac{1}{x^2} + \frac{y}{x} e^{-yx} \right) dx$$

$$\leqslant \frac{2}{t} + \frac{y}{t} \int_t^\infty e^{-yx} \, dx < \frac{3}{t}.$$

Thus, $|J(y) - J(0)| \leqslant yt + \frac{3}{t}$ for all positive y and t. Putting $t = \frac{1}{\sqrt{y}}$, we see that $|J(y) - J(0)| \leqslant 4\sqrt{y}$, which implies the continuity of $J(y)$ as $y \to 0$.

Taking into account (5), we obtain the value of the important integral

$$\int_0^\infty \frac{\sin x}{x} \, dx = \frac{\pi}{2}.$$

7.1.7 Theorem 7.1.5 also remains valid in the case of differentiability with respect to a complex parameter.

Theorem *Let Y be an open subset of the complex plane. If the conditions*:

(a) *the function $y \mapsto f(x, y)$ is holomorphic in Y for almost all $x \in X$*;
(b) *the partial derivative f_y' satisfies condition (L_{loc}) at a point $y_0 \in Y$, are fulfilled, then the integral $J(y) = \int_X f(x, y) \, d\mu(x)$ is differentiable at y_0 and*

$$J'(y_0) = \int_X f_y'(x, y_0) \, d\mu(x).$$

Proof The proof of Theorem 7.1.5 can be repeated verbatim, the only difference being that now, in the case where the disk $\overline{B}(y_0, |h|)$ lies in Y, we use the estimate

$$|F(x, h)| = \left| \int_0^1 f_y'(x, y_0 + th) \, dt \right| \leqslant \max_{0 \leqslant t \leqslant 1} |f_y'(x, y_0 + th)|$$

instead of the Lagrange mean value theorem. By condition (L_{loc}), for h sufficiently small in absolute value, the right-hand side of the above inequality has a summable majorant independent of h. Knowing this, we can finish the proof as in the case of a real parameter. □

It follows from the above theorem that if the function φ is summable on a finite interval $[a, b]$, then the function

$$F(z) = \int_a^b \varphi(t)e^{zt}\,dt$$

is holomorphic on the entire complex plane. Thus, the Laplace and Fourier transforms of a summable function with compact support, i.e., the integrals

$$\mathcal{L}(z) = \int_{\mathbb{R}_+} \varphi(t)e^{-zt}\,dt \quad \text{and} \quad \mathcal{F}(z) = \int_{\mathbb{R}} \varphi(t)e^{-izt}\,dt$$

are entire functions.

Example 1 We find the Laplace transform of a power function. Let $a > 0$, $z \in \mathbb{C}$, $x = \mathcal{R}e(z) > 0$, and

$$\mathcal{L}(z) = \int_0^\infty t^{a-1}e^{-zt}\,dt.$$

Obviously, $|f(t, z)| = |t^{a-1}e^{-zt}| = t^{a-1}e^{-xt}$, and, therefore, the integrand is summable for every z, $\mathcal{R}e(z) > 0$. The derivative f'_z satisfies the condition (L_{loc}) at every point in the right half-plane. Therefore,

$$\mathcal{L}'(z) = -\int_0^\infty t^a e^{-zt}\,dt = \frac{1}{z}t^a e^{-tz}\Big|_{t=0}^{\infty} - \frac{a}{z}\int_0^\infty t^{a-1}e^{-zt}\,dt = -\frac{a}{z}\mathcal{L}(z).$$

This equation can be represented in the form $(z^a\mathcal{L}(z))' \equiv 0$, which implies that $z^a\mathcal{L}(z) \equiv \text{const}$. We will assume that z^a is the branch of the power function equal to 1 at $z = 1$. Then $\mathcal{L}(z) = \frac{\mathcal{L}(1)}{z^a}$, and it remains to recall the definition of the gamma function (see Sect. 4.6.3) to complete the calculation,

$$\mathcal{L}(1) = \int_0^\infty t^{a-1}e^{-t}\,dt = \Gamma(a).$$

Thus, $\mathcal{L}(z) = \frac{\Gamma(a)}{z^a}$.

Example 2 Let X be a closed subset of the complex plane, let G be the complement of X, and let h be a function summable on X with respect to the measure μ (we recall that according to our agreement at the beginning of the section, a measure on a topological space is defined at least for all Borel subsets). We define a function J on G by the equation

$$J(z) = \int_X \frac{h(\zeta)}{\zeta - z}\,d\mu(\zeta) \quad (z \in G).$$

The function J is called an integral of Cauchy type.

We verify that this function is holomorphic in G and its derivatives can be calculated by differentiation with respect to the parameter under the integral sign, i.e.,

$$J^{(n)}(z) = n! \int_X \frac{h(\zeta)}{(\zeta - z)^{n+1}}\, d\mu(\zeta) \quad \text{for all } z \in G,\ n \in \mathbb{N}.$$

In our case, we have $f(\zeta, z) = h(\zeta)/(\zeta - z)$ and $f'_z(\zeta, z) = h(\zeta)/(\zeta - z)^2$ for $\zeta \in X$, $z \in G$. In a neighborhood of z_0, the denominator $\zeta - z$ is separated from zero. Indeed, if the disk $B(z_0, 2r)$ is contained in G, then the inequality $|\zeta - z| \geqslant r$ holds for $|z - z_0| < r$ and $\zeta \in X$. Therefore, the function $\zeta \mapsto f(\zeta, z)$ is summable on X for every $z \in G$ and

$$\left|f'_z(\zeta, z)\right| = \left|\frac{h(\zeta)}{(\zeta - z)^2}\right| \leqslant \frac{|h(\zeta)|}{r^2} \quad \text{for all } \zeta \in X,\ |z - z_0| < r.$$

The last estimate shows that the function f'_z satisfies condition (L_{loc}) at z_0. Since z_0 is arbitrary, we obtain by Theorem 7.1.7 that the function J is holomorphic in G and

$$J'(z) = \int_X \frac{h(\zeta)}{(\zeta - z)^2}\, d\mu(\zeta) \quad \text{for all } z \in G.$$

The higher order derivatives are calculated similarly.

EXERCISES

1. For the family of functions $\{\ln(1 - 2r\cos x + r^2)\}_{0<r<1}$, find a majorant summable on $(0, 2\pi)$.

2. Does the family $\{1/|1 - re^{ix}|\}_{0<r<1}$ have a majorant summable on $(0, 2\pi)$?

3. Prove that

$$\int_0^{2\pi} \frac{dx}{|1 - re^{ix}|} \underset{r\to 1-0}{\sim} 2\ln\frac{1}{1-r};$$

$$\int_0^{2\pi} \frac{dx}{|1 - re^{ix}|^p} \underset{r\to 1-0}{\sim} \frac{C_p}{(1-r)^{p-1}} \quad (p > 1),$$

where $C_p = 2\int_0^\infty \frac{dt}{(1+t^2)^{p/2}}$.

4. Let $E \subset \mathbb{R}^m$ be a bounded measurable set. Prove that the function $y \mapsto \int_E \frac{dx}{\|x-y\|^p}$ is continuous in the space $\mathbb{R}^m$ for $p < m$.

5. Calculate the integral $\int_0^{\frac{\pi}{2}} \frac{x}{\tan x}\, dx$. *Hint*. Consider the integral $\int_0^{\frac{\pi}{2}} \frac{\arctan(y\tan x)}{\tan x}\, dx$ as $y \geqslant 0$.

7.2 *The Gamma Function

In the present section, we consider an important example of an integral depending on a parameter. Here we are talking about the gamma function introduced by Euler, or the Euler integral of the second kind, which is of the same fundamental significance as the elementary functions. We have already encountered it episodically (see Sects. 4.6.3 and 5.3.2). In particular, we used the gamma function in Sect. 5.4.2 to calculate the volume of the m-dimensional ball.

7.2.1 We recall that the gamma function is defined for $x > 0$ by the formula

$$\Gamma(x) = \int_0^{\infty} t^{x-1} e^{-t}\, dt. \tag{1}$$

We leave it to the reader to verify that the derivative f'_x of the integrand $f(t,x) = t^{x-1}e^{-t}$ satisfies condition (L_{loc}) in a neighborhood of every point $x_0 > 0$. By Theorem 7.1.5, the gamma function is differentiable and

$$\Gamma'(x) = \int_0^{\infty} t^{x-1} e^{-t} \ln t\, dt.$$

Similarly, we can prove that the gamma function has a derivative of an arbitrary order and find a formula for it. In particular, $\Gamma''(x) = \int_0^{\infty} t^{x-1} e^{-t} \ln^2 t\, dt > 0$. Therefore, the gamma function is a convex function of class $C^{\infty}((0,+\infty))$.

Integrating by parts, we can easily verify that Γ satisfies the functional equation

$$\Gamma(x+1) = x\Gamma(x) \quad \text{for } x > 0. \tag{2}$$

We evaluate Γ for positive integers. It is clear that $\Gamma(1) = 1$. By Eq. (2) and induction, we obtain $\Gamma(n+1) = n!$ for all $n \in \mathbb{N}$. Thus, the gamma function is a continuation of the function $n!$ to the positive real axis (at first sight, the function $n!$ is intimately connected only with positive integers).

By the change of variable $t = u^2$, the integral $\int_0^{\infty} t^{-1/2} e^{-t}\, dt = \Gamma(1/2)$ can be reduced to the Euler–Poisson integral $I = \int_{-\infty}^{\infty} e^{-u^2}\, du$, which we calculated repeatedly (see, e.g., Sect. 6.2.4). Thus, $\Gamma(1/2) = I = \sqrt{\pi}$. Based on this result and functional equation (2), we can find the values of Γ at half-integers,

$$\Gamma\left(n + \frac{1}{2}\right) = \frac{(2n-1)!!}{2^n}\sqrt{\pi} \quad (n \in \mathbb{N}).$$

Equation (2) enables us to study the behavior of Γ in the vicinity of zero,

$$\Gamma(x) = \frac{1}{x}\Gamma(x+1) \sim \frac{1}{x} \quad \text{as } x \to +0.$$

For large x, the values $\Gamma(x)$ are large, since

$$\Gamma(1+x) = \int_0^{\infty} t^x e^{-t}\, dt \geqslant \int_x^{\infty} t^x e^{-t}\, dt \geqslant x^x \int_x^{\infty} e^{-t}\, dt = \left(\frac{x}{e}\right)^x.$$

This simple estimate describes well the growth of Γ at infinity. Below (see Sect. 7.2.6), we obtain the precise asymptotic behavior of $\Gamma(x)$ as $x \to +\infty$.

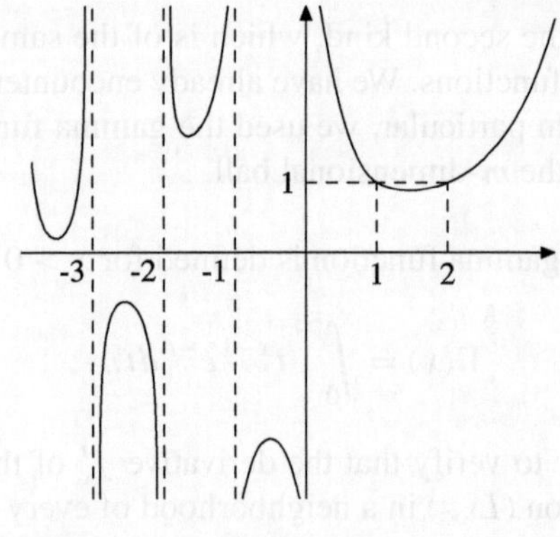

Fig. 7.1 Graph of the gamma function

The functional equation (2) suggests a natural continuation of Γ to the negative semi-axis. Indeed, we should take the formula $\Gamma(x) = \frac{1}{x}\Gamma(x+1)$ as the definition of Γ on the interval $(-1, 0)$. Then the values of Γ on $(-1, 0)$ are negative and the one-sided limits at the points 0 and -1 are infinite. Using the definition of Γ on $(-1, 0)$, we can define it on the interval $(-2, -1)$. Proceeding in this way, we define $\Gamma(x)$ for all $x < 0, x \neq -1, -2, \ldots$. We see that $(-1)^n\Gamma(x) > 0$ if $x \in (-n, -n+1)$, and $|\Gamma(x)| \underset{x\to -n}{\longrightarrow} +\infty$ $(n = 1, 2, \ldots)$. Now it is clear that Eq. (2) can be generalized as follows:

$$\Gamma(x+1) = x\,\Gamma(x) \quad \text{for } x \in \mathbb{R} \setminus \{0, -1, -2, \ldots\}. \tag{2'}$$

The properties of the gamma function obtained above allows us to sketch the graph of Γ (see Fig. 7.1). We remark that since $\Gamma(2) = 1 = \Gamma(1)$, Rolle's theorem implies that there is a (unique) critical point of Γ in the interval $(1, 2)$. At this point the function assumes a local minimum. Moreover, every interval $(-n, -n+1)$, $n \in \mathbb{N}$, contains a unique critical point of Γ (see Exercise 8).

Replacing x by a complex number z in Eq. (1) (and regarding t^{z-1} as $e^{(z-1)\ln t}$), we see that this equation allows us to define Γ not only at $x > 0$ but also at complex z provided $\mathcal{R}e(z) > 0$, i.e., in the right complex half-plane. It follows from Theorem 7.1.7 that Γ is holomorphic in this half-plane. Moreover, the identity $\Gamma(z+1) = z\Gamma(z)$ remains valid and can be used to define Γ in the entire complex plane except at the points $0, -1, -2, \ldots$ in the same way as for Γ on the semi-axis $(-\infty, 0)$. However, we content ourselves with the study of the gamma function only on the real axis.

7.2.2 In this section and the next, we obtain very important formulas for the gamma function.

First of all, we recall the formula connecting the functions B and Γ. The function B is defined (see Sect. 4.6.3) by the formula $\mathrm{B}(x, y) = \int_0^1 t^{x-1}(1-t)^{y-1}\,dt$, where

$x, y > 0$. As proved in Sect. 5.3.2, we have $\mathrm{B}(x, y) = \frac{\Gamma(x)\Gamma(y)}{\Gamma(x+y)}$, i.e.,

$$\int_0^1 t^{x-1}(1-t)^{y-1}\,dt = \frac{\Gamma(x)\Gamma(y)}{\Gamma(x+y)}. \tag{3}$$

From this equation, we derive the following asymptotic relation (see also Exercise 9):

$$\Gamma(x+a) \sim x^a \Gamma(x) \quad \text{for } x \to +\infty. \tag{4}$$

By virtue of the functional equation for Γ it is sufficient to prove this for $a > 0$. By (3), we obtain

$$\frac{\Gamma(x)\Gamma(a)}{\Gamma(x+a)} = \int_0^1 t^{a-1}(1-t)^{x-1}\,dt \quad (a, x > 0).$$

For convenience, we replace x with $x + 1$. Using the change of variables $t = u/x$, we obtain

$$\frac{\Gamma(x+1)\Gamma(a)}{\Gamma(x+a+1)} = \frac{1}{x^a}\int_0^x u^{a-1}\left(1-\frac{u}{x}\right)^x du.$$

Since $1 - t \leqslant e^{-t}$, we see that $1 - \frac{u}{x} \leqslant e^{-u/x}$ and $(1 - \frac{u}{x})^x \leqslant e^{-u}$ for $0 \leqslant u \leqslant x$. Consequently, for every x, the integrand in the last integral (we assume that this function is zero for $u > x$) has the majorant $u^{a-1}e^{-u}$ summable on $(0, +\infty)$. Therefore, Theorem 1 of Sect. 7.1.2 implies

$$x^a \frac{\Gamma(x+1)\Gamma(a)}{\Gamma(x+a+1)} = \int_0^x u^{a-1}\left(1-\frac{u}{x}\right)^x du \underset{x\to+\infty}{\longrightarrow} \int_0^\infty u^{a-1}e^{-u}\,du = \Gamma(a)$$

(the passage to the limit is actually justified in Example 2 of Sect. 4.8.4). Dividing by $\Gamma(a)$, we can represent this in a form equivalent to (4):

$$x^a \frac{x\Gamma(x)}{(x+a)\Gamma(x+a)} \underset{x\to+\infty}{\longrightarrow} 1.$$

For a sharpening of this relation, see Exercise 9 and Example 1 of Sect. 7.3.5.

7.2.3 The following formula makes it possible to find the values of Γ without integration:

$$\Gamma(x) = \lim_{n\to\infty} \frac{n^x\, n!}{x(x+1)\cdots(x+n-1)(x+n)} \qquad \text{for } x \in \mathbb{R},\ x \neq 0, -1, -2, \ldots$$

This formula is similar to Euler's definition of Γ (see Exercise 2) and is known as the *Euler–Gauss formula.*[2]

[2]Carl Friedrich Gauss (1777–1855)—German mathematician.

For the proof we observe that $\Gamma(x+n) = (x+n-1)\cdots(x+1)x\Gamma(x)$. Therefore,

$$\frac{n^x n!}{x(x+1)\cdots(x+n)} = \frac{n}{x+n}\cdot\frac{n^x(n-1)!}{x(x+1)\cdots(x+n-1)} = \frac{n}{x+n}\cdot\Gamma(x)\cdot\frac{n^x\Gamma(n)}{\Gamma(x+n)}.$$

It remains to use relation (4).

For $x=\frac{1}{2}$, the Euler–Gauss formula essentially coincides with the Wallis formula (see Sect. 4.6.2). Indeed, for $x=\frac{1}{2}$ we obtain

$$\sqrt{\pi} = \Gamma\left(\frac{1}{2}\right) = \lim_{n\to\infty}\frac{\sqrt{n}\,n!}{\frac{1}{2}\,\frac{3}{2}\cdots(\frac{1}{2}+n)} = 2\lim_{n\to\infty}\sqrt{n}\frac{(2n)!!}{(2n+1)!!},$$

which is equivalent to the Wallis formula.

To obtain one more famous formula connected with the gamma function, we recall the asymptotic behavior of the partial sums of the harmonic series: there exists a γ *(the Euler constant)* such that

$$1+\frac{1}{2}+\frac{1}{3}+\cdots+\frac{1}{n} = \ln n + \gamma + o(1).$$

This follows from the convergence of the series $\sum_{k=1}^{\infty}(\frac{1}{k}-\ln(1+\frac{1}{k}))$, since its nth partial sum is equal to $1+\frac{1}{2}+\cdots+\frac{1}{n}-\ln(n+1)$.

We will use this result to obtain a beautiful expansion of the function $1/\Gamma$ in an infinite product. We recall that by the infinite product of a numerical sequence $a_1, a_2, \ldots$, we mean the limit $\lim_{n\to\infty}\prod_{k=1}^{n} a_k$, which is denoted by $\prod_{k=1}^{\infty} a_k$.

We prove that

$$\frac{1}{\Gamma(x)} = xe^{\gamma x}\prod_{k=1}^{\infty}\left(1+\frac{x}{k}\right)e^{-\frac{x}{k}} \quad (x\in\mathbb{R}) \tag{5}$$

(since $|\Gamma(x)|\to+\infty$ as $x\to 0,-1,-2,\ldots$, it is natural to assume that the quotient $1/\Gamma$ is zero at these points). The relation obtained is called the *Weierstrass formula.*[3]

For the proof, we rewrite the Euler–Gauss formula as

$$\frac{1}{\Gamma(x)} = \lim_{n\to\infty} n^{-x}x(1+x)\cdots\left(1+\frac{x}{n}\right).$$

Now, after elementary transformations, we obtain

$$\frac{1}{\Gamma(x)} = x\lim_{n\to\infty} n^{-x}\prod_{k=1}^{n}\left(1+\frac{x}{k}\right) = x\lim_{n\to\infty} e^{x(1+\frac{1}{2}+\cdots+\frac{1}{n}-\ln n)}\prod_{k=1}^{n}\left(1+\frac{x}{k}\right)e^{-\frac{x}{k}}.$$

Since $1+\frac{1}{2}+\cdots+\frac{1}{n}-\ln n\to\gamma$, we see that the limit $\lim_{n\to\infty}\prod_{k=1}^{n}(1+\frac{x}{k})e^{-\frac{x}{k}}$ exists and the Weierstrass formula is valid.

---

[3] Karl Theodor Wilhelm Weierstrass (1815–1897)—German mathematician.

7.2.4 Equation (3) enables us to obtain *Legendre's (duplication) formula*,[4] also simply called the *duplication formula*:

$$\Gamma(x)\Gamma\left(x+\frac{1}{2}\right)=\frac{\sqrt{\pi}}{2^{2x-1}}\Gamma(2x) \quad (x>0).$$

To this end, we transform the right-hand side of the equation

$$\frac{\Gamma^2(x)}{\Gamma(2x)}=\int_0^1 t^{x-1}(1-t)^{x-1}\,dt.$$

We have

$$\frac{\Gamma^2(x)}{\Gamma(2x)}=\int_0^1 \left(t-t^2\right)^{x-1}dt=\int_0^1\left(\frac{1}{4}-\left(\frac{1}{2}-t\right)^2\right)^{x-1}dt=2\int_0^{\frac{1}{2}}\left(\frac{1}{4}-s^2\right)^{x-1}ds.$$

Substituting $u=4s^2$, we obtain by (3)

$$\frac{\Gamma^2(x)}{\Gamma(2x)}=2^{1-2x}\int_0^1 u^{-\frac{1}{2}}(1-u)^{x-1}\,du=2^{1-2x}\,\frac{\Gamma(\frac{1}{2})\Gamma(x)}{\Gamma(x+\frac{1}{2})}.$$

Since $\Gamma(\frac{1}{2})=\sqrt{\pi}$, we come to the required formula.

As follows from (2′), the formula proved above is valid not only for positive x but also for all real x such that $2x\neq 0,-1,-2,\dots$.

7.2.5 Now we obtain one of the most important formulas connected with the gamma function. This is *Euler's reflection formula*

$$\Gamma(x)\Gamma(1-x)=\frac{\pi}{\sin\pi x} \quad \text{for } x\in\mathbb{R}\setminus\mathbb{Z}.$$

Our elegant proof of this formula follows the proof in the book [Ar].

We prove that the product $\theta(x)=\frac{\sin\pi x}{\pi}\Gamma(x)\Gamma(1-x)$ is constant on $\mathbb{R}\setminus\mathbb{Z}$. It follows from Eq. (2′) that the function θ has period 1. Indeed,

$$\theta(x+1)=-\frac{\sin(\pi x)}{\pi}\Gamma(x+1)\Gamma(-x)=-\frac{\sin(\pi x)}{\pi}x\Gamma(x)\frac{\Gamma(1-x)}{-x}=\theta(x).$$

Moreover,

$$\theta(x)=\frac{\sin(\pi x)}{\pi x}\Gamma(x+1)\Gamma(1-x).$$

Hence, extending θ by the formula $\theta(n)=1$ for $n\in\mathbb{Z}$, we obtain a 1-periodic function infinitely differentiable in a neighborhood of zero and, consequently, on the entire real axis. It is clear that $\theta>0$ on $\mathbb{R}$.

---

[4] Adrien-Marie Legendre (1752–1833)—French mathematician.

For $x > 0$, Legendre's formula implies (as the reader can easily verify) the relation

$$\theta\left(\frac{x}{2}\right)\theta\left(\frac{1+x}{2}\right) = \theta(x).$$

Taking logarithms, we see that

$$g\left(\frac{x}{2}\right) + g\left(\frac{x+1}{2}\right) = g(x), \tag{6}$$

where $g = \ln\theta$. Consequently, the continuous and 1-periodic function g'' satisfies the identity

$$g''\left(\frac{x}{2}\right) + g''\left(\frac{1+x}{2}\right) = 4g''(x).$$

For $M = \max|g''|$, we obtain that $2M \geqslant 4M$. Since $0 \leqslant M < +\infty$, this means that $M = 0$, i.e., $g = \ln\theta$ is a linear function. Taking into account that $g(0) = g(1) = 0$, we obtain $g \equiv 0$, i.e., $\theta \equiv 1$.

The reflection formula can be used to obtain Euler's famous factorization of the sine function into "simple factors" just as polynomials can be represented in a similar form.

Since the sine function has infinitely many zeros, we have to use infinite products. Euler's result is as follows:

$$\sin\pi x = \pi x \prod_{n=1}^{\infty}\left(1 - \frac{x^2}{n^2}\right) \quad \text{for each } x \in \mathbb{R}. \tag{7}$$

Rejecting the trivial case, we may assume that $x \notin \mathbb{Z}$. Multiplying the Weierstrass formulas for $\Gamma(x)$ and $\Gamma(-x)$, we obtain

$$\frac{1}{\Gamma(x)\Gamma(-x)} = -x^2 \prod_{n=1}^{\infty}\left(1 - \frac{x^2}{n^2}\right).$$

It remains to apply the reflection formula,

$$\sin\pi x = \frac{\pi}{\Gamma(x)\Gamma(1-x)} = \frac{\pi}{(-x)\Gamma(x)\Gamma(-x)} = \pi x \prod_{n=1}^{\infty}\left(1 - \frac{x^2}{n^2}\right).$$

We remark that, as seen from the above proof, the reflection formula can in turn be derived from the Weierstrass formula and Eq. (7).

7.2.6 Now we turn to a more substantial study of the asymptotic behavior of $\Gamma(x)$ as $x \to +\infty$. The asymptotic behavior is described by *Stirling's formula*[5]

$$\Gamma(x) \underset{x\to+\infty}{\sim} \sqrt{2\pi}\, x^{x-\frac{1}{2}} e^{-x}. \tag{8}$$

In Sect. 7.3 we obtain this result as a particular case of a more general statement, but now we use a different approach based on our knowledge of the gamma function and allowing us to obtain a sharpening of asymptotic formula (8).

First of all, we replace the rapidly decreasing gamma function by its logarithm. The next step is to find the asymptotic behavior of the second derivative of $\ln\Gamma(x)$.

Taking logarithms in Eq. (5) for $x > 0$, we obtain

$$-\ln\Gamma(x) = \ln x + \gamma x + \sum_{n=1}^{\infty}\left(\ln\left(1+\frac{x}{n}\right) - \frac{x}{n}\right).$$

Differentiating twice, we obtain

$$\big(\ln\Gamma(x)\big)'' = \sum_{n=0}^{\infty}\frac{1}{(x+n)^2}. \tag{9}$$

The termwise differentiation is legal since the series obtained converges uniformly on every closed interval lying in $(0,+\infty)$.

The general method that, in particular, makes it possible to obtain arbitrarily precise asymptotic representation of the sum of series (9) as $x \to +\infty$ is provided by the Euler–Maclaurin formula (see [F], vol. II, [Bou]). However, we will not use it, instead obtaining the first several terms of the asymptotic of $(\ln\Gamma(x))''$ directly. The principal term of the asymptotic can easily be found since the sum of series (9) is close to the integral. We obtain

$$\frac{1}{x} = \int_0^{\infty}\frac{dt}{(x+t)^2} \leqslant \sum_{n=0}^{\infty}\frac{1}{(x+n)^2} \leqslant \frac{1}{x^2} + \int_0^{\infty}\frac{dt}{(x+t)^2} = \frac{1}{x} + \frac{1}{x^2}.$$

Thus,

$$\big(\ln\Gamma(x)\big)'' = \frac{1}{x} + O\left(\frac{1}{x^2}\right)$$

(from here to the end of this section, we assume that $x > 0$ and that the symbol O refers to $x \to +\infty$ without saying it explicitly). The trick that we will use here is as follows. We will successively sharpen the asymptotic formula obtained, extracting the principal parts by series whose sums can easily be found. First, we represent $\frac{1}{x}$

[5] James Stirling (1692–1770)— Scottish mathematician.

in the form

$$\frac{1}{x} = \sum_{n=0}^{\infty}\left(\frac{1}{x+n} - \frac{1}{x+n+1}\right) = \sum_{n=0}^{\infty} \frac{1}{(x+n)(x+n+1)}$$

and subtract it from (9). We obtain

$$\begin{aligned}(\ln\Gamma(x))'' - \frac{1}{x} &= \sum_{n=0}^{\infty}\left(\frac{1}{(x+n)^2} - \frac{1}{(x+n)(x+n+1)}\right)\\ &= \sum_{n=0}^{\infty} \frac{1}{(x+n)^2(x+n+1)}.\end{aligned} \tag{10}$$

Again, comparing the series obtained with the corresponding integral, we see that

$$\begin{aligned}\frac{1}{2(x+1)^2} = \int_0^{\infty} \frac{dt}{(x+t+1)^3} &\leqslant (\ln\Gamma(x))'' - \frac{1}{x} = \sum_{n=0}^{\infty} \frac{1}{(x+n)^2(x+n+1)}\\ &\leqslant \frac{1}{x^3} + \int_0^{\infty} \frac{dt}{(x+t)^3} = \frac{1}{2x^2} + \frac{1}{x^3}.\end{aligned}$$

Consequently,

$$(\ln\Gamma(x))'' - \frac{1}{x} = \frac{1}{2x^2} + h(x), \quad \text{where } h(x) = O\left(\frac{1}{x^3}\right). \tag{11}$$

This result (for another proof of which, see Exercise 13) is already sufficient to prove (8). Indeed, it is clear that

$$\int_1^x h(t)\,dt = \int_1^{\infty} h(t)\,dt - \int_x^{\infty} h(t)\,dt = \text{const} + O\left(\frac{1}{x^2}\right).$$

Therefore, integrating expansion (11) from 1 to x, we obtain the equation

$$(\ln\Gamma(x))' = A + \ln x - \frac{1}{2x} + O\left(\frac{1}{x^2}\right).$$

One more integration gives the relation

$$\ln\Gamma(x) = B + Ax + x\ln x - x - \frac{1}{2}\ln x + O\left(\frac{1}{x}\right).$$

To find A and B, it is convenient to write this equation (slightly coarsening it) as the equivalence

$$\Gamma(x) \underset{x\to+\infty}{\sim} Cx^{x-\frac{1}{2}}e^{(A-1)x},$$

where $C = e^B$. To determine A, we use the functional equation, which implies that

$$\Gamma(x) = \frac{\Gamma(x+1)}{x} \underset{x\to+\infty}{\sim} \frac{C}{x}(x+1)^{x+\frac{1}{2}} e^{(A-1)(x+1)}.$$

Taking the ratio of the right-hand sides of these equivalencies, we obtain

$$\left(1+\frac{1}{x}\right)^{x+\frac{1}{2}} e^{A-1} \underset{x\to+\infty}{\longrightarrow} 1,$$

which is possible only if $A = 0$. The constant B can be found similarly by means of Legendre's formula, which implies that

$$C^2 x^{x-\frac{1}{2}} e^{-x} \left(x+\frac{1}{2}\right)^x e^{-x-\frac{1}{2}} \underset{x\to+\infty}{\sim} \frac{\sqrt{\pi}}{2^{2x-1}} C (2x)^{2x-\frac{1}{2}} e^{-2x}.$$

Dividing by $Cx^{2x-\frac{1}{2}} e^{-2x}$, we see that $C(1+\frac{1}{2x})^x e^{-\frac{1}{2}} \underset{x\to+\infty}{\longrightarrow} \sqrt{2\pi}$, which implies the equality $C = \sqrt{2\pi}$. Thus,

$$\ln \Gamma(x) = \left(x - \frac{1}{2}\right) \ln x - x + \frac{1}{2} \ln(2\pi) + O\left(\frac{1}{x}\right),$$

i.e.,

$$\Gamma(x) = \sqrt{2\pi} x^{x-\frac{1}{2}} e^{-x} \left(1 + O\left(\frac{1}{x}\right)\right). \tag{8'}$$

The above relations, as well as Eq. (8), are also called Stirling's formulas.

To sharpen the asymptotic, we represent $\frac{1}{x^2}$ in the form

$$\frac{1}{x^2} = \sum_{n=0}^{\infty} \left(\frac{1}{(x+n)^2} - \frac{1}{(x+n+1)^2} \right) = \sum_{n=0}^{\infty} \frac{2(x+n)+1}{(x+n)^2(x+n+1)^2}$$

and, multiplying by $\frac{1}{2}$, we subtract it from (10). We obtain

$$\big(\ln \Gamma(x)\big)'' - \frac{1}{x} - \frac{1}{2x^2} = \frac{1}{2} \sum_{n=0}^{\infty} \frac{1}{(x+n)^2(x+n+1)^2}. \tag{12}$$

Since

$$\begin{aligned} \frac{1}{6(x+1)^3} &= \frac{1}{2} \int_0^{\infty} \frac{dt}{(x+t+1)^4} \\ &\leqslant \big(\ln \Gamma(x)\big)'' - \frac{1}{x} - \frac{1}{2x^2} = \frac{1}{2} \sum_{n=0}^{\infty} \frac{1}{(x+n)^2(x+n+1)^2} \\ &\leqslant \frac{1}{2x^4} + \frac{1}{2} \int_0^{\infty} \frac{dt}{(x+t)^4} = \frac{1}{2x^4} + \frac{1}{6x^3}, \end{aligned}$$

we see that

$$(\ln\Gamma(x))'' - \frac{1}{x} - \frac{1}{2x^2} = \frac{1}{6x^3} + O\left(\frac{1}{x^4}\right).$$

The further sharpening of the asymptotic can be performed repeatedly, but we make only one more step. Applying the trick used twice, we represent $\frac{1}{x^3}$ in the form

$$\frac{1}{x^3} = \sum_{n=0}^{\infty}\left(\frac{1}{(x+n)^3} - \frac{1}{(x+n+1)^3}\right) = \sum_{n=0}^{\infty}\frac{3(x+n)^2+3(x+n)+1}{(x+n)^3(x+n+1)^3}.$$

Multiplying by $\frac{1}{6}$ and subtracting from (12), we obtain

$$(\ln\Gamma(x))'' - \frac{1}{x} - \frac{1}{2x^2} - \frac{1}{6x^3} = -\frac{1}{6}\sum_{n=0}^{\infty}\frac{1}{(x+n)^3(x+n+1)^3} \equiv -\frac{1}{6}s(x),$$

where

$$\frac{1}{5(x+1)^5} = \int_0^{\infty}\frac{dt}{(x+t+1)^6} \leqslant s(x) = \sum_{n=0}^{\infty}\frac{1}{(x+n)^3(x+n+1)^3}$$

$$< \frac{1}{x^6} + \int_0^{\infty}\frac{dt}{(x+t)^6} = \frac{1}{5x^5} + \frac{1}{x^6}.$$

It can easily be verified that $\frac{1}{5x^5} - \frac{1}{x^6} < \frac{1}{5(x+1)^5}$. Therefore, $|s(x) - \frac{1}{5x^5}| < \frac{1}{x^6}$. Thus,

$$(\ln\Gamma(x))'' = \frac{1}{x} + \frac{1}{2x^2} + \frac{1}{6x^3} - \frac{1}{30x^5} + \frac{\theta}{6x^6}, \quad |\theta| < 1.$$

After integration we obtain the following sharpening of formula (8′):

$$\Gamma(x) = \sqrt{2\pi}\, x^{x-\frac{1}{2}} e^{-x} e^{\frac{1}{12x} - \frac{1}{360x^3} + \frac{\theta}{120x^4}}, \quad |\theta| < 1. \qquad (8'')$$

7.2.7 We generalize Legendre's formula and verify that the relation (the Gauss multiplication theorem)

$$\Gamma(x)\Gamma\left(x+\frac{1}{p}\right)\cdots\Gamma\left(x+\frac{p-1}{p}\right) = \frac{(2\pi)^{\frac{p-1}{2}}}{p^{px-\frac{1}{2}}}\Gamma(px) \quad (px \neq 0, -1, -2, \ldots)$$

is valid for every $p = 2, 3, 4, \ldots$. For the proof, we use the Euler–Gauss formula and represent the left-hand side in the form

$$\prod_{k=0}^{p-1}\Gamma\left(x+\frac{k}{p}\right) = \lim_{n\to\infty}\prod_{k=0}^{p-1}\frac{n!\,n^{x+\frac{k}{p}}}{\prod_{j=0}^{n}(x+\frac{k}{p}+j)} = \lim_{n\to\infty}\frac{(n!)^p n^{px+\frac{p-1}{2}} p^{(n+1)p}}{\prod_{k=0}^{p-1}\prod_{j=0}^{n}(px+pj+k)}.$$

It can easily be seen that the arising product is equal to the product of factors of the form $px + i$ for $0 \leqslant i \leqslant pn + p - 1$. Replacing the last $p - 1$ factors by the equivalent quantities pn (as $n \to \infty$), we obtain that the product is equivalent to

$$(pn)^{p-1} \prod_{i=0}^{pn} (px + i).$$

Therefore,

$$\begin{aligned}\prod_{k=0}^{p-1} \Gamma\left(x + \frac{k}{p}\right) &= \lim_{n\to\infty} \frac{(n!)^p n^{px+\frac{p-1}{2}} p^{(n+1)p}}{(pn)^{p-1} \prod_{i=0}^{pn}(px+i)} \\ &= p^{-px} \lim_{n\to\infty} \frac{(n!)^p p^{np+1}}{(np)! n^{\frac{p-1}{2}}} \cdot \lim_{n\to\infty} \frac{(np)^{px}(np)!}{px(px+1)\cdots(px+pn)}.\end{aligned}$$

By the Gauss formula, the second limit is $\Gamma(px)$. It remains to observe that the first limit (independent of x) is equal to $(2\pi)^{\frac{p-1}{2}} \sqrt{p}$. This can easily be proved by Stirling's formula and is left to the reader. Thus, we arrive at the required result.

7.2.8 We now pause to discuss one more property of the gamma function. It will be shown that this property along with functional equation (2) characterizes the gamma function up to a constant factor. We speak of the logarithmic convexity. A positive function f is called *logarithmically convex* if $\ln f$ is a convex function.

The convexity of $\ln \Gamma$ certainly follows from formula (9) demonstrating that $(\ln \Gamma)'' > 0$. However, the logarithmic convexity of Γ can be proved directly from the definition of Γ. Indeed, the logarithmic convexity of Γ is obviously equivalent to the fact that $\Gamma(\alpha x + (1-\alpha)y) \leqslant \Gamma^\alpha(x)\Gamma^{1-\alpha}(y)$ for all positive x, y and $\alpha \in (0, 1)$. The last inequality follows from Hölder's inequality (see Theorem 4.4.5 for $p = 1/\alpha$). Indeed,

$$\begin{aligned}\Gamma\big(\alpha x + (1-\alpha)y\big) &= \int_0^\infty \left(t^{x-1}e^{-t}\right)^\alpha \left(t^{y-1}e^{-t}\right)^{1-\alpha} dt \\ &\leqslant \left(\int_0^\infty t^{x-1}e^{-t}\,dt\right)^\alpha \left(\int_0^\infty t^{y-1}e^{-t}\,dt\right)^{1-\alpha} = \Gamma^\alpha(x)\,\Gamma^{1-\alpha}(y).\end{aligned}$$

The gamma function is not a unique solution of the functional equation $f(x+1) = xf(x)$. For example, other solutions can be obtained by multiplying the gamma function by 1-periodic functions. Thus, this equation does not determine the gamma function uniquely. The state of affairs changes radically if we seek solutions in the class of logarithmically convex functions. In this class the equation in question has a unique (up to a positive coefficient) solution.

In other words, the following statement is true.[6]

---

[6] To the best of our knowledge, this was first published by H. Bohr and J. Mollerup in "Laerebog i Matematisk Analyse" in 1922 (see e.g. [LO]).

Theorem *If a logarithmically convex function f on $(0,+\infty)$ satisfies the functional equation $f(x+1)=xf(x)$, then $f(x)=f(1)\Gamma(x)$.*

Proof We verify that the quotient f/Γ is constant. To this end, we consider the function $M=\ln(f/\Gamma)$, which is continuous on $(0,+\infty)$ as the difference of two convex functions. Moreover, M is one-sided 1-periodic, i.e., $M(x+1)=M(x)$ for all $x>0$. Assuming that M is not constant, we consider a point $x_0\in(1,2]$ at which M attains its maximum value. In this case, for some $h\in(0,1)$, the second difference $\Delta_h^2 M(x)=M(x+h)-2M(x)+M(x-h)$ is negative, $\Delta_h^2 M(x_0)=\delta<0$. At the same time, $\Delta_h^2(\ln f(x))\geqslant 0$ since f is logarithmically convex. However, for each n, the one-sided periodicity implies

$$0\leqslant \Delta_h^2\big(\ln f(x_0+n)\big)=\Delta_h^2 M(x_0+n)+\Delta_h^2\big(\ln\Gamma(x_0+n)\big)=\delta+\Delta_h^2\big(\ln\Gamma(x_0+n)\big).$$

It follows from (4) that $\Delta_h^2(\ln\Gamma(x))\to 0$ for $x\to+\infty$. Therefore, passing to the limit as $n\to\infty$ in the inequality $0\leqslant\delta+\Delta_h^2(\ln\Gamma(x_0+n))$, we obtain $0\leqslant\delta<0$, a contradiction. □

EXERCISES

1. Express the following integrals in terms of Γ:

(a) $$\int_0^1 x^{a-1}\big(1-x^b\big)^c\,dx \quad (a,b,c>0);$$

(b) $$\int_0^\infty \frac{x^{a-1}}{(1+x^b)^c}\,dx \quad (a,b,c>0,\ bc>a).$$

2. Prove the relation

$$\Gamma(x)=\frac{1}{x}\prod_{n=1}^{\infty}\frac{(1+\frac{1}{n})^x}{1+\frac{x}{n}} \quad \text{for } x\neq 0,-1,-2,\dots$$

used by Euler as a definition of Γ.

3. If the numbers $a_1,b_1,\dots,a_k,b_k$ in $\mathbb{R}\setminus\mathbb{N}$ are such that $a_1+\dots+a_k=b_1+\dots+b_k$, then

$$\prod_{n=1}^{\infty}\frac{(n-a_1)\cdots(n-a_k)}{(n-b_1)\cdots(n-b_k)}=\frac{\Gamma(1-b_1)\cdots\Gamma(1-b_k)}{\Gamma(1-a_1)\cdots\Gamma(1-a_k)}.$$

4. Let $S(t)=\prod_{n=1}^{\infty}(1+\frac{t^2}{n^2})=\frac{\operatorname{sh}\pi t}{\pi t}$. Prove that

$$S\bigg(\frac{a}{\sqrt{x^2-a^2}}\bigg)\leqslant\frac{\Gamma(x-a)\Gamma(x+a)}{\Gamma^2(x)}\leqslant S(a) \quad \text{for } x>1+|a|.$$

5. Differentiating the series expansion of $\ln\Gamma$ (see Sect. 7.2.6), prove that

$$\Gamma'(1) = -\gamma \quad \text{and} \quad \Gamma'(n+1) = n!\left(-\gamma + 1 + \frac{1}{2} + \cdots + \frac{1}{n}\right).$$

6. Calculate $\Gamma'(n+\frac{1}{2})$ for $n = 0, 1, 2, \ldots$

7. Find the limit $\lim_{x\to -n}(x+n)\Gamma(x)$ for $n \in \mathbb{N}$.

8. Use the logarithmic convexity of Γ on $(0, +\infty)$ and Eq. (2′) to prove that the function $|\Gamma(x)|$ is logarithmically convex (and, consequently, convex) on each interval $(-n, -n+1)$, $n \in \mathbb{N}$.

9. By sharpening relation (4), prove that the inequality

$$x^a \frac{x}{x+a}\Gamma(x) \leqslant \Gamma(x+a) \leqslant x^a\Gamma(x)$$

holds for $0 < a < 1$ and $x > 0$ (use the logarithmic convexity of Γ).

10. Use the preceding exercise to prove the inequality $(\ln\Gamma(x))' \leqslant \ln x$ for $x > 0$.

11. Verify that Theorem 7.2.8 remains valid in the class of positive functions $f \in C((0, +\infty))$ satisfying the condition

$$\varliminf_{x\to+\infty} \Delta_h^2(\ln f(x)) \geqslant 0$$

("the logarithmic convexity of f at infinity") for sufficiently small $h > 0$.

12. By Stirling's formula (8) (do not use formula (8″)), prove that

$$\sqrt{2\pi n}n^n e^{-n} e^{\frac{1}{12n+1}} < n! < \sqrt{2\pi n}n^n e^{-n} e^{\frac{1}{12n}} \quad (n \in \mathbb{N}),$$

(verify that the ratios of the left-hand and right-hand sides of the inequality to $n!$ are monotonic).

13. Prove relation (11) by comparing series (9) with the sum

$$\sum_{n=0}^{\infty} \frac{1}{(x+n)^2 - \frac{1}{4}} = \sum_{n=0}^{\infty}\left(\frac{1}{x+n-\frac{1}{2}} - \frac{1}{x+n+\frac{1}{2}}\right).$$

14. Supplementing the estimate for s obtained in the proof of Stirling's formula, prove that $0 < s(x) < \frac{1}{5x^5}$ and obtain the inequality

$$\sqrt{2\pi}\, x^{x-\frac{1}{2}} e^{-x} e^{\frac{1}{12x} - \frac{1}{360x^3}} < \Gamma(x) < \sqrt{2\pi}\, x^{x-\frac{1}{2}} e^{-x} e^{\frac{1}{12x}}.$$

15. Use the identity $\frac{1}{y^2} = \frac{1}{y(y+1)} + \frac{1}{y(y+1)(y+2)} + \cdots + \frac{(j-1)!}{y(y+1)\cdots(y+j)} \cdots$ for $y = x$, $x+1, x+2, \ldots$ to derive from (11) the equation

$$(\ln\Gamma(x))'' = \frac{1}{x} + \sum_{n=1}^{\infty} \frac{n!}{n+1} \cdot \frac{1}{x(x+1)\cdots(x+n)} \qquad (x > 0),$$

which can be used to sharpen relation (8″).

7.3 *The Laplace Method

This section is devoted to the study of the asymptotic behavior of an important class of integrals depending on a parameter in a specific way, namely, the integrals

$$\Phi(x) = \int_T f(t)\varphi^x(t)\,d\mu(t),$$

where the function φ is non-negative and bounded. The interest in the asymptotic behavior of such integrals as $x \to +\infty$ comes from problems of classical analysis, mathematical physics, probability theory, etc. A systematic study of this problem was first made by Laplace[7] to substantiate the law of large numbers. Throughout this section, when we speak of the asymptotic behavior of the integrals $\Phi(x)$, we refer to the asymptotic as $x \to +\infty$ without saying it explicitly.

7.3.1 We start the study of the integrals in question from the simplest and, at the same time, the most important case where the set T is an interval (possibly infinite) and μ is Lebesgue measure. We will assume that the function φ is positive, bounded and piecewise monotonic, but the function f is summable on (a, b). Thus, the integral

$$\Phi(x) = \int_a^b f(t)\,\varphi^x(t)\,dt \tag{1}$$

is finite for all $x \geqslant 0$.

Instead of the summability of f, we may assume that only the product $f\varphi^{x_0}$ is summable for some $x_0 > 0$ and consider the integral $\Phi(x)$ for $x \geqslant x_0$. Obviously, we can reduce this case to the preceding one by replacing f with $f\varphi^{x_0}$.

The Laplace method is based on the fact that the main contribution to the integral $\Phi(x)$ comes from the integrals over the neighborhoods of the points at which the function φ attains its maximum value. This is well illustrated on the graph of φ^x, which, for large x, has "humps" in neighborhoods of such points; the larger the x, the more pronounced the hump (see Fig. 7.2 illustrating the case $\max\varphi = 1$).

Usually, such sharp oscillations of the integrand complicate the calculation of the integral, but, in the case in question, they simplify the determination of the asymptotic. Two hundred years ago, in the preface to his famous "Analytical Theory of Probability", Laplace wrote with enthusiasm that the method discovered by him "is the more required, the more exact".

Dividing, if necessary, the interval of integration into several parts, we may assume that the function φ is monotonic. Clearly, it is sufficient to consider only the case where φ decreases, since the case where φ increases can be reduced the preceding one by a change of variable. We assume that φ is decreasing on $[a, b)$, where $-\infty < a < b \leqslant +\infty$, and verify, first of all, that the integrals of the form (1) demonstrate the localization phenomenon, i.e., their asymptotic depends on the behavior

---

[7] Pierre-Simon Laplace (1749–1827)—French mathematician.

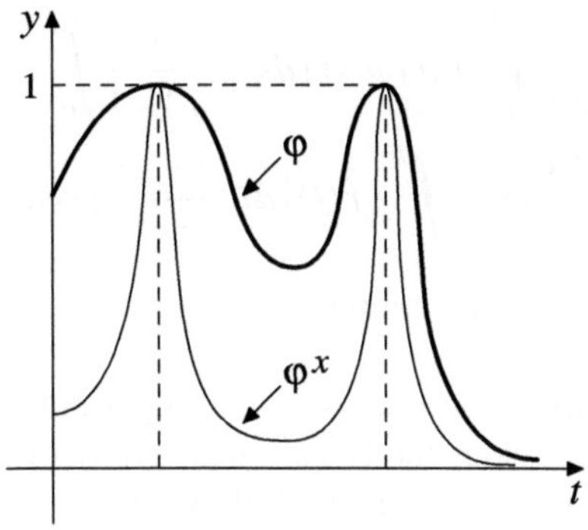

Fig. 7.2 Graphs of the functions φ and φ^x for large x

of the integrand only in an arbitrarily small neighborhood of the point a. More precisely, the following simple but important assertion concerning localization is valid.

Lemma *Assume that φ decreases, f is summable on $[a,b)$, and that the following conditions hold*:

(1) $0 < \varphi(t) < \varphi(a) = \lim_{u\to a} \varphi(u)$ *for* $t \in (a,b)$;
(2) f *preserves sign in a neighborhood of the point* a *and does not vanish at the points close to* a, *i.e.*,

$$I_c = \int_a^c f(t)\,dt \neq 0 \quad \textit{for all } c \textit{ sufficiently close to } a.$$

Then the asymptotic expansions

$$\Phi(x) \sim \int_a^c f(t)\,\varphi^x(t)\,dt \quad \textit{and} \quad \varphi^x(c) = o\big(\Phi(x)\big) \quad \textit{as } x \to +\infty$$

are valid for all $c \in (a,b)$.

Thus, the main contribution to $\Phi(x)$ comes from the integral over an arbitrary small interval (a,c), and the contribution of the integral over the remaining interval is negligibly small.

Proof From the inequality $|\int_c^b f(t)\varphi^x(t)\,dt| \leqslant \varphi^x(c)\int_c^b |f(t)|\,dt$, it follows that $\Phi(x) = \int_a^c f(t)\varphi^x(t)\,dt + O(\varphi^x(c))$. Therefore, we need to prove only the relation $\varphi^x(c) = o(\Phi(x))$. Since the function φ decreases, the point c can be chosen arbitrarily close to a. Without loss of generality, we may assume that the function f is non-negative on the interval $[a,c]$.

By assumption, there exists a point $\overline{c} \in (a,c)$ such that $\varphi(c) < \varphi(\overline{c})$. Then $I_{\overline{c}} > 0$ and

$$\frac{\Phi(x)}{\varphi^x(c)} \geqslant \frac{1}{\varphi^x(c)} \int_a^{\overline{c}} f(t)\,\varphi^x(t)\,dt + \frac{1}{\varphi^x(c)} \int_c^b f(t)\,\varphi^x(t)\,dt$$

$$\geqslant \frac{\varphi^x(\overline{c})}{\varphi^x(c)} I_{\overline{c}} - \int_c^b \bigl|f(t)\bigr|\,dt \underset{x\to+\infty}{\longrightarrow} +\infty,$$

which completes the proof of the lemma. □

The localization property proved above is an important qualitative characteristic of the integrals $\Phi(x)$. It provides the basis for the study of these integrals for large values of the parameter x. In particular, it allows us to compare the behavior of the integrals $\Phi(x) = \int_a^b f(t)\varphi^x(t)\,dt$ and $\Psi(x) = \int_a^b g(t)\varphi^x(t)\,dt$ as $x \to +\infty$ if we take into account the information on the behavior of the quotient $g(t)/f(t)$ as $t \to a$. We state this technically useful result in more detail.

Corollary *Assume that functions φ and f satisfy the conditions of the lemma. Assume that a function g is summable on $[a, b)$ and $\Psi(x) = \int_a^b g(t)\,\varphi^x(t)\,dt$. Then:*

(a) *if $g(t) = O(f(t))$ as $t \to a$, then $\Psi(x) = O(\Phi(x))$ as $x \to +\infty$;*
(b) *if $g(t) = o(f(t))$ as $t \to a$, then $\Psi(x) = o(\Phi(x))$ as $x \to +\infty$;*
(c) *if $g(t) \sim f(t)$ as $t \to a$, then $\Psi(x) \sim \Phi(x)$ as $x \to +\infty$.*

Proof (a) By assumption, there exists a coefficient $C > 0$ and a point $c \in (a, b)$ such that $|g(t)| \leqslant C|f(t)|$ on (a, c). We can choose c so close to a that the function f preserves the sign on (a, c). Then

$$\bigl|\Psi(x)\bigr| \leqslant \int_a^c \bigl|g(t)\bigr|\varphi^x(t)\,dt + \int_c^b \bigl|g(t)\bigr|\varphi^x(t)\,dt$$

$$\leqslant C \int_a^c \bigl|f(t)\bigr|\varphi^x(t)\,dt + O\bigl(\varphi^x(c)\bigr).$$

Since f preserves the sign on (a, c), the lemma implies that

$$\bigl|\Psi(x)\bigr| \leqslant C\left|\int_a^c f(t)\,\varphi^x(t)\,dt\right| + o\bigl(\Phi(x)\bigr) = C\bigl|\Phi(x)\bigr| + o\bigl(\Phi(x)\bigr),$$

which completes the proof of statement (a).

The same reasoning proves statement (b), since the coefficient C can be taken arbitrarily small. Finally, to prove statement (c), we apply (b) to the difference $f - g$. □

7.3.2 The study of the integrals $\Phi(x)$ is based on the following simple idea: we use localization to approximate the functions f and φ in the vicinity of a by functions generating an easily computable integral.

Obviously, the behavior of $\Phi(x)$ as $x \to +\infty$ is determined to a great extent by the rate of decrease of the maximum value of $\varphi(t)$ when the argument t moves away from the point a. In other words, in the problem in question, the infinitesimal $\varphi(a)-\varphi(t)$ (as $t \to a$) plays a decisive role. Therefore, its rate of change should be taken into account in the first place when making a choice of an approximation.

For smooth φ, the following two cases are of greatest importance:

$$\text{(a)}\quad \varphi(t)-\varphi(a) \underset{t\to a}{\sim} \varphi'(a)(t-a), \quad \text{where } \varphi'(a)<0;$$

$$\text{(b)}\quad \varphi(t)-\varphi(a) \underset{t\to a}{\sim} \frac{1}{2}\varphi''(a)(t-a)^2, \quad \text{where } \varphi''(a)<0 \ \big(\varphi'(a)=0\big).$$

In these cases, if $f(t) \underset{t\to a}{\longrightarrow} L \neq 0$, then the following *Laplace asymptotic formulas*

$$\begin{aligned} &\text{(a)}\quad \Phi(x) \underset{x\to+\infty}{\sim} L\frac{\varphi(a)}{|\varphi'(a)|x}\varphi^x(a), \\ &\text{(b)}\quad \Phi(x) \underset{x\to+\infty}{\sim} L\sqrt{\frac{\pi\varphi(a)}{2|\varphi''(a)|x}}\varphi^x(a) \end{aligned} \tag{2}$$

are valid.

We obtain a more general result provided the difference $\varphi(a)-\varphi(t)$ is a power-type infinitesimal, i.e., $\varphi(a)-\varphi(t) \underset{t\to a}{\sim} C(t-a)^p$, where $C, p>0$. Representing the function φ in the form $\varphi(t)=e^{-S(t)}$, we see that the above condition is equivalent to $S(t)-S(a) \underset{t\to a}{\sim} \frac{C}{\varphi(a)}(t-a)^p$.

The proof of the main result for $\varphi(a)=1$ essentially reduces to the justification of the natural idea of replacing $\varphi(t)$ by a "similar" function $e^{-C(t-a)^p}$ in the vicinity of the point a. We fulfill this idea in three steps.

First of all, we consider the integral $\int_0^\infty e^{-xt^p}\,dt$, which will serve as a standard model and can easily be calculated by the gamma function,

$$\int_0^\infty e^{-xt^p}\,dt = \int_0^\infty e^{-u}\,d\left(\frac{u}{x}\right)^{\frac{1}{p}} = \frac{1}{p}\Gamma\left(\frac{1}{p}\right)x^{-\frac{1}{p}}.$$

In the next step, we use the localization property to verify that the replacement of t^p by an equivalent function does not change the asymptotic of the integral. The following statement holds.

Lemma *Let φ be a function defined on $[0,b)$ and satisfying the conditions of the localization lemma. If $\ln\varphi(t) \underset{t\to+0}{\sim} -t^p$ for some $p>0$, then*

$$\Phi(x)=\int_0^b \varphi^x(t)\,dt \underset{x\to+\infty}{\sim} \int_0^\infty e^{-xt^p}\,dt = C_p x^{-\frac{1}{p}}, \quad \textit{where } C_p=\frac{1}{p}\Gamma\left(\frac{1}{p}\right).$$

Proof We put $S(t)=-\ln\varphi(t)$. Fixing an arbitrary number $\theta\in(0,1)$, we find a $c\in(0,b)$ such that $(\theta t)^p<S(t)<(\frac{t}{\theta})^p$ for $0<t<c$. By the localization lemma,

we have

$$\Phi(x) \underset{x\to+\infty}{\sim} \Psi(x) = \int_0^c e^{-xS(t)}\,dt.$$

We estimate the integral Ψ. Obviously,

$$\int_0^c e^{-x(\frac{t}{\theta})^p}\,dt < \Psi(x) < \int_0^c e^{-x(\theta t)^p}\,dt.$$

As $x \to +\infty$, the integrals on the right-hand and left-hand sides of the above inequality are equivalent to $C_p\theta x^{-\frac{1}{p}}$ and $\frac{C_p}{\theta}x^{-\frac{1}{p}}$, respectively. Slightly changing the coefficients, we obtain

$$C_p\theta^2 < x^{\frac{1}{p}}\Psi(x) < \frac{C_p}{\theta^2}$$

for sufficiently large x. Since $x^{1/p}\Phi(x) \underset{x\to+\infty}{\sim} x^{1/p}\Psi(x)$, we see that (again for sufficiently large x)

$$C_p\theta^3 < x^{\frac{1}{p}}\Phi(x) < \frac{C_p}{\theta^3}.$$

Since $\theta \in (0,1)$ is arbitrary, we obtain that $x^{\frac{1}{p}}\Phi(x) \underset{x\to+\infty}{\longrightarrow} C_p$. □

Now we are ready to turn to the concluding step and obtain the main result.

Theorem *Let φ be an decreasing positive function, and let f be a summable function on $[a,b)$. Assume that:*

(a) *there exist positive numbers C and p such that $\varphi(a) - \varphi(t) \underset{t\to a}{\sim} C(t-a)^p$;*
(b) *there exist numbers $L \neq 0$ and $q > -1$ such that $f(t) \underset{t\to a}{\sim} L(t-a)^q$.*

Then

$$\Phi(x) = \int_a^b f(t)\varphi^x(t)\,dt \underset{x\to+\infty}{\sim} \frac{L}{p}\Gamma\left(\frac{q+1}{p}\right)\left(\frac{\varphi(a)}{Cx}\right)^{\frac{q+1}{p}}\varphi^x(a). \tag{3}$$

In particular, for $q = 0$ and $p = 1$ or 2, we obtain the Laplace formulas (2).

As has already been pointed out, the case where the function φ increases can be reduced to what has just been considered by a change of variable. Therefore, if φ is a non-negative non-decreasing function on the interval $(a,b]$ (here $-\infty \leqslant a < b < +\infty$), $\varphi(b) - \varphi(t) \sim C(b-t)^p$, and $f(t) \sim L(b-t)^q$ as $t \to b$, then relation (3) remains valid (with a replaced by b).

If the function φ attains its maximum value at a point t_0 of the interval (a,b), increases from the left and decreases from the right of it, and $\varphi(t_0) - \varphi(t) \sim C|t-t_0|^p$

and $f(t) \sim L|t-t_0|^q$ as $t \to t_0$, then, applying the formula to each of the intervals $(a, t_0]$ and $[t_0, b)$ separately, we see that the right-hand side must be doubled,

$$\Phi(x) = \int_a^b f(t)\varphi^x(t)\,dt \underset{x\to+\infty}{\sim} \frac{2L}{p}\Gamma\left(\frac{q+1}{p}\right)\left(\frac{\varphi(t_0)}{Cx}\right)^{\frac{q+1}{p}} \varphi^x(t_0). \qquad (3')$$

Proof Replacing φ by $\varphi/\varphi(a)$, we may assume that $\varphi(a) = 1$. Moreover, we will assume that $a = 0$ (which can be achieved by the change of variable $t \mapsto t - a$). Thus, we must prove that

$$\Phi(x) = \int_0^b f(t)\varphi^x(t)\,dt \underset{x\to+\infty}{\sim} \frac{L}{p}\Gamma\left(\frac{q+1}{p}\right)(Cx)^{-\frac{q+1}{p}}$$

if $1 - \varphi(t) \underset{t\to 0}{\sim} Ct^p$. By virtue of the localization lemma, we may change the function f by an equivalent, not changing the asymptotic. We have

$$\begin{aligned}\Phi(x) \underset{x\to+\infty}{\sim} \int_0^b Lt^q\varphi^x(t)\,dt &= \frac{L}{q+1}\int_0^{b^{q+1}} \varphi^x\left(u^{\frac{1}{q+1}}\right)du \\ &= \frac{L}{q+1}\int_0^B e^{-CxS(u)}\,du,\end{aligned}$$

where $S(u) = -\frac{1}{C}\ln\varphi(u^{\frac{1}{q+1}})$ and $B = b^{q+1}$. As $u \to 0$, we again obtain, by assumption, that

$$S(u) \sim \frac{1}{C}\left(1 - \varphi\left(u^{\frac{1}{q+1}}\right)\right) \sim u^{\frac{p}{q+1}}.$$

This allows us to apply the lemma (with Cx in place of x and $\frac{p}{q+1}$ in place of p) to the integral over the interval $[0, B)$ and obtain

$$\Phi(x) \underset{x\to+\infty}{\sim} \frac{L}{q+1}\,\frac{q+1}{p}\Gamma\left(\frac{q+1}{p}\right)(Cx)^{-\frac{q+1}{p}} = \frac{L}{p}\Gamma\left(\frac{q+1}{p}\right)(Cx)^{-\frac{q+1}{p}}. \qquad \square$$

Remark A negligibly small contribution to $\Phi(x)$ comes not only from the point of the interval (c, b) with a fixed $c \in (a, b)$. We can choose the parameter c dependent on x and tending (not very rapidly) to the point a. Under the conditions of the theorem, the following sharpening of the localization property is valid: if $\alpha(x) \to 0$ and $x^{1/p}\alpha(x) \to +\infty$ as $x \to +\infty$, then

$$\int_a^b f(t)\varphi^x(t)\,dt \underset{x\to+\infty}{\sim} \int_a^{a+\alpha(x)} f(t)\varphi^x(t)\,dt.$$

Indeed, as in the proof of the theorem, we assume that $a = 0$ and $\varphi(a) = 1$. Since the integral over the interval $[c, b)$ is exponentially small for a fixed $c \in (a, b)$, it is sufficient to estimate the integral over the interval $[\alpha(x), c]$. We choose a number c

so small that $|f(t)| \leqslant 2|L|t^q$ and $1-\varphi(t) \geqslant \frac{C}{2}t^p$ for $t \in [0,c]$. Then $\varphi(t) \leqslant e^{-Ct^p/2}$ and, therefore,

$$\left|\int_{\alpha(x)}^{c} f(t)\varphi^x(t)\,dt\right| \leqslant 2|L| \int_{\alpha(x)}^{\infty} t^q e^{-\frac{C}{2}xt^p}\,dt = \frac{2|L|}{x^{\frac{q+1}{p}}} \int_{x^{\frac{1}{p}}\alpha(x)}^{\infty} u^q e^{-\frac{C}{2}u^p}\,du$$

$$= o\big(\Phi(x)\big).$$

This estimate implies that if $(\frac{x}{\ln x})^{\frac{1}{p}}\alpha(x) \to +\infty$ as $x \to +\infty$, the relative error of the approximate equality $\Phi(x) \approx \int_a^{a+\alpha(x)} f(t)\varphi^x(t)\,dt$ decreases "overpowerly", i.e., faster that every negative power of x.

In Theorem 7.3.2, we assumed that the difference $\varphi(a) - \varphi(t)$ and the function $f(t)$ have power type principal parts as $t \to a$. In Sect. 7.3.4, we consider examples where this condition is violated. In our study of these examples, an important role is played by the choice of a neighborhood of the point a that shrinks as x grows. It should be noted that if the infinitesimal $\varphi(a) - \varphi(t)$ is not of power type, then, replacing it by an equivalent function, we can change the asymptotic of the integral in question (see Exercise 10). Additional restrictions allowing us to make this change are given in Exercise 11.

7.3.3 We consider several examples of application of the Laplace formula.

Example 1 We find the asymptotic of the integral $\int_0^{\frac{\pi}{2}} \cos^x t\,dt$. To this end, we use Laplace formula (2b) with functions $\varphi(t) = \cos t$ and $f(t) \equiv 1$, which immediately leads to the relation

$$\int_0^{\frac{\pi}{2}} \cos^x t\,dt \underset{x\to+\infty}{\sim} \sqrt{\frac{\pi}{2x}}.$$

We recall that, if $x = n$ is an integer, then the integral is equal to $v_n \frac{(n-1)!!}{n!!}$, where $v_n = 1$ for odd n and $v_n = \frac{\pi}{2}$ for even n. Therefore, for such x, the asymptotic can be obtained by Stirling's formula.

Example 2 The asymptotic of the gamma function (Stirling's formula). The integrand in the integral $\Gamma(x+1) = \int_0^\infty t^x e^{-t}\,dt$ attains its maximum value at $t = x$, and in a neighborhood of this point the graph has a sharply pronounced "peak", which suggests that we may use the Laplace method. However, this method cannot be applied directly since the local maximum of the integrand changes with the parameter. To represent the integrand in the form considered in Theorem 7.3.2, it is necessary to use a sliding peak, which can be achieved by the substitution $t = xu$. This gives us the relation

$$\Gamma(x+1) = x^{x+1} \int_0^\infty \varphi^x(u)\,du,$$

where $\varphi(u) = ue^{-u}$. Obviously, the function φ increases on $[0, 1]$, decreases on $[1, +\infty)$ and $\varphi''(1) = -e^{-1}$. Taking into account that the maximum value $\varphi(1) = e^{-1}$ is attained at an interior point of the interval, we obtain by (3′) that

$$\int_0^{\infty} \varphi^x(u)\,du \underset{x\to+\infty}{\sim} \sqrt{\frac{2\pi}{x}}e^{-x}.$$

Consequently,

$$\Gamma(x) = \frac{1}{x}\Gamma(x+1) = x^x \int_0^{\infty} \varphi^x(u)\,du \underset{x\to+\infty}{\sim} \sqrt{2\pi}x^{x-\frac{1}{2}}e^{-x}.$$

Previously (see Sect. 7.2.6), this result was obtained by a different method.

Example 3 In Sect. 6.7.3, we discussed the cross sections of the cube $[-\frac{1}{2}, \frac{1}{2}]^n$ and incidentally obtained a formula for the area S_n of the cross section created by the plane that passes through the center of the cube and is orthogonal to its main diagonal,

$$S_n = \frac{2}{\pi}\sqrt{n}\int_0^{\infty} \left(\frac{\sin t}{t}\right)^n dt.$$

Let us trace the behavior of the quantities S_n as n increases unboundedly. This behavior is determined by the asymptotic of the integrals

$$I_n = \int_0^{\infty} \left(\frac{\sin t}{t}\right)^n dt.$$

A direct application of the Laplace formula is complicated by the fact that the function $\frac{\sin t}{t}$ changes its sign as $t \geqslant 0$. However, we can easily overcome this difficulty since, as $n \to \infty$, the integral over the interval $[\pi, +\infty)$ is exponentially small,

$$\int_\pi^{\infty} \left|\frac{\sin t}{t}\right|^n dt \leqslant \int_\pi^{\infty} \frac{1}{t^n}\,dt \leqslant \frac{1}{\pi^{n-1}} \quad (n \geqslant 2).$$

At the same time, the Laplace formula can be applied to the integral over the interval $[0, \pi]$. Indeed, since $\frac{1}{t}\sin t = 1 - \frac{1}{6}t^2 + o(t^2)$ as $t \to 0$, we obtain

$$\int_0^{\pi} \left(\frac{\sin t}{t}\right)^n dt \underset{n\to\infty}{\sim} \frac{1}{2}\sqrt{\frac{6\pi}{n}}.$$

Therefore, $I_n \sim \frac{1}{2}\sqrt{\frac{6\pi}{n}}$ and, consequently, $S_n \to \sqrt{\frac{6}{\pi}}$. This result will be supplemented below in Example 2 of Sect. 7.3.5.

Example 4 We find the asymptotic of the sums

$$S_n = 1 + n + \frac{n^2}{2!} + \cdots + \frac{n^k}{k!} + \cdots + \frac{n^n}{n!}.$$

Obviously, S_n is the value of the nth Taylor polynomial of the function e^x calculated at the point n. Using the integral representation of the remainder in the Taylor formula, we obtain

$$e^x = \sum_{k=0}^{n} \frac{x^k}{k!} + \frac{1}{n!}\int_0^x (x-t)^n e^t\,dt.$$

For $x = n$, the substitution $t = ns$ gives

$$e^n - S_n = \frac{1}{n!}\int_0^n (n-t)^n e^t\,dt = \frac{n^{n+1}}{n!}\int_0^1 \left((1-s)e^s\right)^n ds.$$

Since $(1-s)e^s = 1 - \frac{1}{2}s^2 + o(s^2)$ as $s \to 0$, the Laplace formula gives

$$e^n - S_n \underset{n\to\infty}{\sim} \frac{n^{n+1}}{n!}\sqrt{\frac{\pi}{2n}}.$$

Now, to obtain the final result, it remains to use Stirling's formula, which implies that $e^n - S_n \sim \frac{1}{2}e^n$, i.e., $S_n \sim \frac{1}{2}e^n$.

Remark 1 If we weaken conditions (a) and (b) in Theorem 7.3.2 by replacing them with the two-sided estimates $\varphi(a) - \varphi(t) \underset{t\to a}{\asymp} (t-a)^p$ and $0 \leqslant f(t) \underset{t\to a}{\asymp} (t-a)^q$, then, by Lemma 7.3.1, we can obtain the two-sided estimate $\Phi(x) \underset{x\to+\infty}{\asymp} x^{-\frac{q+1}{p}}$. For example, it can easily be verified that the Cantor function φ satisfies the identity $\varphi(t)+\varphi(1-t) = 1$ and the double inequality $(t/2)^p \leqslant \varphi(t) \leqslant t^p$, where $p = \log_3 2$, on the interval $[0, 1]$. Therefore, the integral $\Phi(x) = \int_0^1 \varphi^x(t)\,dt$ satisfies the two-sided estimate $\Phi(x) \underset{x\to+\infty}{\asymp} x^{-\log_2 3}$. It is considerably harder to describe its asymptotic behavior (see Exercise 14).

Remark 2 It is worth noting that the assumption that the function φ is monotonic has been used only in the proof of the localization lemma. In this lemma and, consequently, in Theorem 7.3.2 this assumption can be replaced by the following less restrictive assumption:

$$\lim_{t\to a} \varphi(t) > \sup_{t>c} \varphi(t) \quad \text{for all } c \text{ in the interval } (a, b).$$

This condition is fulfilled, for example, if the function φ is continuous on $[a, b]$ and $\varphi(a) > \varphi(t)$ for $t \neq a$.

7.3.4 Applications of the Laplace method are not confined to the justification of formulas (2), (3) and (3′). This method is widely used in many different situations. The main idea of the method, localization and replacing the integrand by its Taylor expansion in a small neighborhood of a point, also turns out to be quite effective in the cases where the conditions of Theorem 7.3.2 are violated. The main difficulty is to choose a neighborhood. On the one hand, the neighborhood should not be too large, since otherwise the error caused by the application of the Taylor formula will manifest itself. On the other hand, to compensate the error arising in replacing the given integral with the integral over a neighborhood, we cannot take the neighborhood to be too small. A successful choice of a neighborhood allowing us to make both of the above-mentioned errors negligible is the core of the method.

Here, having no intention of making general statements and wanting only to give an idea of how to find the asymptotic in such situations, we follow Newton, who said that "in studies of science, examples are more useful than rules", and confine ourselves to considering two specific problems (see also Exercises 9, 10 and 12).

Example 1 Let

$$\Phi(x) = \int_0^b |\ln t|^r e^{-xt^p}\,dt,$$

where $p > 0$ and r is an arbitrary real number (for $r \leqslant -1$, we assume that $0 < b < 1$ to guarantee that the non-summable singularity, at the point $t = 1$, of the integrand does not belong to the interval of integration). We remark that, in this example, the function $f(t) = |\ln t|^r$ does not satisfy condition (b) of Theorem 7.3.2 as $t \to 0$ (and is not even summable on $(0, +\infty)$ if $b = +\infty$). To find the asymptotic, we choose a neighborhood by the method described in the remark to this theorem.

Since the function $\varphi(t) = e^{-t^p}$ attains its maximum value at $t = 0$, the main contribution to $\Phi(x)$ comes from the integral over a neighborhood of zero. We have a great freedom in the choice of such a neighborhood. We put $\alpha(x) = x^{-1/(2p)}$ and study the behavior of the integrals $\Phi_1(x) = \int_0^{\alpha(x)} \cdots$ and $\Phi_2(x) = \int_{\alpha(x)}^b \cdots$, the sum of which is $\Phi(x)$. We make the change of variables $u = x^{1/p}t$ in the first integral and then, after elementary transformations, we obtain

$$\Phi_1(x) = x^{-1/p} \ln^r\left(x^{1/p}\right) \int_0^{x^{1/(2p)}} \left|1 - p\frac{\ln u}{\ln x}\right|^r e^{-u^p}\,du.$$

For $0 < u < x^{1/(2p)}$ and sufficiently large x, the following estimates are valid:

$$\frac{1}{2} \leqslant \left|1 - p\frac{\ln u}{\ln x}\right| \leqslant 1 + p|\ln u|.$$

Consequently, for every r (positive or negative), we have

$$\left|1 - p\frac{\ln u}{\ln x}\right|^r \leqslant 2^{|r|} + \left(1 + p|\ln u|\right)^r, \quad \text{if } 0 < u < x^{1/(2p)}.$$

This inequality enables us to use Lebesgue's theorem, and we obtain

$$\int_0^{x^{1/(2p)}} \left|1 - p\frac{\ln u}{\ln x}\right|^r e^{-u^p}\,du \underset{x\to+\infty}{\longrightarrow} \int_0^\infty e^{-u^p}\,du = \Gamma\left(1 + \frac{1}{p}\right).$$

Therefore, $\Phi_1(x) \sim \Gamma(1+\frac{1}{p})x^{-1/p}\ln^r(x^{1/p})$. Now, we verify that $\Phi_2(x) = o(\Phi_1(x))$. Indeed,

$$\Phi_2(x) \leqslant \int_{\alpha(x)}^b |\ln t|^r e^{-xt^p}\, dt \leqslant e^{-(x-1)\alpha^p(x)} \int_{\alpha(x)}^b |\ln t|^r e^{-t^p}\, dt \leqslant \frac{\text{const}}{e^{\sqrt{x}}}$$
$$= o(\Phi_1(x)).$$

Thus,

$$\int_0^b |\ln t|^r e^{-xt^p}\, dt \underset{x\to+\infty}{\sim} \Gamma\left(1+\frac{1}{p}\right)\left(\frac{\ln x}{p}\right)^r x^{-1/p}.$$

This integral, as well as the integral $\int_0^\infty e^{-xt^p}\, dt$ considered in Theorem 7.3.2, may serve as a standard model in the study of integrals of the form $\int_a^b f(t)\varphi^x(t)\, dt$, where $f(t) \underset{t\to+0}{\sim} L(t-a)^q|\ln(t-a)|^r$ (see Exercises 5 and 6).

Example 2 We find the asymptotic of the integral

$$\Phi(x) = \int_0^{1/e} \left(1+\frac{1}{\ln t}\right)^x dt.$$

The function $\varphi(t) = 1+\frac{1}{\ln t}$ (we assume that $\varphi(0)=1$) strictly decreases on the interval of integration. Therefore, the main contribution to the integral comes from the points t close to zero. We cannot apply formula (3) since the infinitesimal $\varphi(0)-\varphi(t) = -1/\ln t$ is not of power type. To overcome this difficulty, it is worthwhile to make the change of variables $t = e^{-u}$, which leads to the relation

$$\Phi(x) = \int_1^\infty \left(1-\frac{1}{u}\right)^x e^{-u}\, du.$$

For a fixed x, the integrand attains its maximum value at the point $u_x = \frac{1}{2}(1+\sqrt{1+4x}) \approx \sqrt{x}$. Its value at $u=\sqrt{x}$ is equal to

$$\left(1-\frac{1}{\sqrt{x}}\right)^x e^{-\sqrt{x}} = e^{-2\sqrt{x}+O(1)}.$$

The integrand is considerably smaller at the points $u \leqslant \sqrt{x}/3$ and does not exceed $e^{-u-\frac{x}{u}} \leqslant e^{-\frac{10}{3}\sqrt{x}}$. Therefore, the contribution of these points to the integral $\Phi(x)$ is small; it admits the estimate $o(e^{-3\sqrt{x}})$.

Now, we consider the integral $\widetilde{\Phi}(x)$ over the remaining interval $(\sqrt{x}/3, +\infty)$, where $\sqrt{x}/3 > 1$. This is just the "small" neighborhood of the point u_x in which it will be possible to approximate the integrand by its Taylor expansion,

$$\left(1-\frac{1}{u}\right)^x e^{-u} = \exp\left(-u - x\left(\frac{1}{u}+\frac{1}{2u^2}+O\left(\frac{1}{u^3}\right)\right)\right) \quad \text{for } u \geqslant \sqrt{x}/3.$$

As we have already noted, the maximum value of the integrand is attained at the point $u_x \approx \sqrt{x}$ changing with x. To decrease this dependence and reduce the situation to the case considered in Theorem 7.3.2, we make the change of variable $u = \sqrt{x}v$. Then

$$\widetilde{\Phi}(x) = \sqrt{x}\int_{\frac{1}{3}}^{\infty} \exp\Big(-\sqrt{x}\Big(v+\frac{1}{v}\Big) - \frac{1}{2v^2} + O\Big(\frac{1}{\sqrt{x}\,v^3}\Big)\Big)\,dv$$
$$= \sqrt{x}\int_{\frac{1}{3}}^{\infty} \exp\Big(-\sqrt{x}\Big(v+\frac{1}{v}\Big) - \frac{1}{2v^2}\Big)\Big(1 + O\Big(\frac{1}{\sqrt{x}}\Big)\Big)\,dv.$$

It is clear that

$$\widetilde{\Phi}(x) \underset{x\to+\infty}{\sim} \sqrt{x}\int_{\frac{1}{3}}^{\infty} \exp\Big(-\sqrt{x}\Big(v+\frac{1}{v}\Big) - \frac{1}{2v^2}\Big)\,dv.$$

We can apply Theorem 7.3.2 (with parameter $\sqrt{x}$ instead of x) to the integral obtained. Taking into account that the function $e^{-(v+\frac{1}{v})}$ attains its maximum value e^{-2} at the interior point $v = 1$ of the interval of integration, we obtain by (3′) that

$$\widetilde{\Phi}(x) \underset{x\to+\infty}{\sim} \sqrt{\frac{\pi}{e}}\,\sqrt[4]{x}e^{-2\sqrt{x}}.$$

The given integral $\Phi(x)$ has the same asymptotic since, as we already know, $\Phi(x) - \widetilde{\Phi}(x) = o(e^{-3\sqrt{x}})$ as $x \to +\infty$.

We call the reader's attention to the fact that, under the conditions of Theorem 7.3.2, for $\varphi(a) = 1$, the integral $\Phi(x)$ decreases like a power function. However, in the example in question, the function $\Phi(x)$ decreases faster than any negative power of the parameter. The reason is that the function $1 + \frac{1}{\ln t}$ has a "supersharp peak", i.e., as the argument t increases, the function loses its maximum values faster than any function of the form $1 - t^p$ $(p > 0)$ (see also Exercise 10).

7.3.5 Up to this point, we have used the Laplace method only for extracting the principal part of an integral of the form (1) and have said nothing about the further sharpening of an asymptotic obtained. Turning to this question, we will assume throughout this section that the functions φ and f satisfy the conditions of Theorem 7.3.2 with $a = 0$. Certainly, to sharpen the result obtained in this Theorem, we need additional information about the behavior of the function φ and f in the vicinity of zero. Usually, one proceeds from local asymptotic expansions of these functions, generalizing in some sense the Taylor expansion. We recall that a series $\sum_{j=0}^{\infty} a_j t^{q_j}$ (or $\sum_{j=0}^{\infty} a_j t^{-q_j}$), where $q_0 < q_1 < \cdots$ and $q_j \to +\infty$, is called an asymptotic expansion of F as $t \to 0$ (as $t \to +\infty$) if the relation $F(t) = \sum_{j=0}^{n} a_j t^{q_j} + o(t^{q_n})$ (respectively, $F(t) = \sum_{j=0}^{n} a_j t^{-q_j} + o(t^{-q_n})$) is valid for each n. In particular, if F has derivatives of all orders at zero, then the Taylor formula implies the asymptotic expansion $\sum_{j=0}^{\infty} \frac{F^{(j)}(0)}{j!} t^j$ as $t \to 0$ (regardless of whether the Taylor series converges for some $t \neq 0$ or not).

In what follows, we will repeatedly use the formula

$$\int_0^\infty t^q e^{-xt^p}\,dt = \int_0^\infty \left(\frac{u}{x}\right)^{\frac{q}{p}} e^{-u}\,d\left(\frac{u}{x}\right)^{\frac{1}{p}} = \frac{1}{p}\Gamma\left(\frac{q+1}{p}\right)x^{-\frac{q+1}{p}}. \tag{4}$$

We remark that the error that arises when replacing the infinite interval of integration on the right-hand side of this equation by a finite one is exponentially small, more precisely,

$$\int_0^b t^q e^{-xt^p}\,dt = \frac{1}{p}\Gamma\left(\frac{1+q}{p}\right)x^{-\frac{1+q}{p}} + O\left(e^{-xb^p}\right) \quad \text{as } x \to +\infty. \tag{4'}$$

Indeed, since $t^q e^{-xt^p} \leqslant \frac{C}{t^2}e^{-xb^p}$ for $t \geqslant b$, $x > 0$, we have

$$\int_0^\infty t^q e^{-xt^p}\,dt - \int_0^b t^q e^{-xt^p}\,dt = \int_b^\infty t^q e^{-xt^p}\,dt \leqslant \int_b^\infty \frac{C}{t^2}e^{-xb^p}\,dt = \frac{C}{b}e^{-xb^p}.$$

The following lemma describes the asymptotic behavior of the integral $\Phi(x)$ in the simplest and, at the same time, the most important case where $\varphi(t) = e^{-t^p}$.

Lemma (Watson[8]) *Let f be a summable function on an interval* $[0, b)$ $(0 < b \leqslant +\infty)$, *and let* $\Phi(x) = \int_0^b f(t)e^{-xt^p}\,dt$. *If*

$$f(t) = a_1 t^{q_1} + \cdots + a_n t^{q_n} + o\left(t^{q_n}\right) \quad \textit{as } t \to 0,$$

where $-1 < q_1 < \cdots < q_n$, *then*

$$\Phi(x) = \int_0^b f(t)\,e^{-xt^p}\,dt = \frac{1}{p}\sum_{j=1}^n a_j \Gamma\left(\frac{1+q_j}{p}\right)x^{-\frac{1+q_j}{p}} + o\left(x^{-\frac{1+q_n}{p}}\right) \tag{5}$$

as $x \to +\infty$.

Proof The required statement is obtained directly from relation (4′) since, by Corollary to Lemma 7.3.1, we have

$$\int_0^b o\left(t^{q_n}\right)e^{-xt^p}\,dt = o\left(\int_0^\infty t^{q_n}e^{-xt^p}\,dt\right) = o\left(x^{-\frac{1+q_n}{p}}\right). \qquad \square$$

From Watson's lemma and the definition of an asymptotic expansion, we immediately obtain the following statement.

---

[8] George Neville Watson (1886–1965)—British mathematician.

Corollary *If, for $0 \leqslant t < \delta$, the function f can be represented in the form $f(t) = t^q g(t)$, where $g \in C^\infty([0,\delta))$, then the asymptotic expansion*

$$\Phi(x) = \frac{1}{p}\sum_{j=0}^{n} \frac{g^{(j)}(0)}{j!}\Gamma\left(\frac{q+j+1}{p}\right)x^{-\frac{q+j+1}{p}} + o\left(x^{-\frac{q+n+1}{p}}\right) \quad \textit{as } x \to +\infty \quad (6)$$

is valid for every n.

It is interesting to note that, though asymptotic equalities can break down under formal differentiation (e.g., an infinitesimal can have an unbounded derivative), under the conditions of Watson's lemma, the asymptotic equality (5) (and, consequently, also (6)) remains valid under repeated differentiation.

Indeed, applying the Leibniz rule (Theorem 7.1.5), we see that the function $\Phi(x)$ belongs to the class C^∞ and

$$\Phi'(x) = \int_0^b f(t)\left(e^{-xt^p}\right)'_x dt = -\int_0^b \widetilde{f}(t)e^{-xt^p}\,dt,$$

where

$$\widetilde{f}(t) = t^p f(t) = a_1 t^{p+q_1} + \cdots + a_n t^{p+q_n} + o\left(t^{p+q_n}\right) \quad \text{as } t \to 0.$$

We may assume that $b < +\infty$ (otherwise, making an exponentially small error, we can integrate over a smaller interval). Then the function $\widetilde{f}$ is summable on $(0, b)$. Applying Watson's lemma to it, we obtain

$$\Phi'(x) = -\int_0^b \widetilde{f}(t)e^{-xt^p}\,dt = -\frac{1}{p}\sum_{j=1}^{n} a_j\Gamma\left(\frac{1+p+q_j}{p}\right)x^{-\frac{1+p+q_j}{p}} + o\left(x^{-\frac{1+p+q_n}{p}}\right)$$

$$= \left(\frac{1}{p}\sum_{j=1}^{n} a_j\Gamma\left(\frac{1+q_j}{p}\right)x^{-\frac{1+q_j}{p}}\right)'_x + o\left(x^{-\frac{1+q_n}{p}-1}\right),$$

which guarantees the validity of differentiating relation (5) termwise.

How to treat the case where the function φ does not coincide with e^{-t^p} and is more complicated? Without going into details, we outline two possible approaches. The first approach is straightforward and consists of reducing the problem to the preceding one. Assuming that φ is a smooth function defined on $(0, b)$ and strictly decreasing from 1 to zero, we make the change of variable $u = -\ln\varphi(t)$ in the integral $\Phi(x) = \int_0^b f(t)\varphi^x(t)\,dt$. Then the integral takes the form

$$\Phi(x) = \int_0^\infty g(u)e^{-xu}\,du,$$

where $g(u) = f(\psi(u))\psi'(u)$, ψ is a function inverse to $-\ln\varphi$. Knowing asymptotic expansions of the functions f, φ and φ', we can find an asymptotic expansion of g, which makes it possible to use Watson's lemma. If $f \in C^\infty$, then Lagrange's formula for the power series expansion of the inverse function can be useful.

The second approach is to use additional information about the behavior of the difference $\varphi(0) - \varphi(t)$ to represent φ^x as the product of e^{-Cxt^p} and the sum of a series in powers of $xs(t)$, where the function $s(t)$ decreases rapidly as $t \to 0$.

We consider this approach in more detail. As before, to simplify the notation, we assume that $\varphi(0) = 1$. Then the function $S = -\ln\varphi$ can be represented in the form $S(t) = Ct^p + s(t)$, where $s(t) = o(t^p)$ as $t \to 0$. We will assume that $s(t) = O(t^r)$ as $t \to 0$, where $r > p$. Moreover, without loss of generality, we may assume that the function s is bounded on $[0, b)$. Otherwise, we can make the interval of integration smaller, since the replacement of $[0, b)$ by $[0, c)$ gives an exponentially small error. Since

$$\varphi^x(t) = e^{-xCt^p - xs(t)} = e^{-xCt^p}\left(1 - xs(t) + \cdots + \frac{(-xs(t))^{n-1}}{(n-1)!} + O\big((xt^r)^n\big)\right)$$

and $f(t) = O(t^q)$, we have

$$\Phi(x) = \sum_{j=0}^{n-1} \frac{(-x)^j}{j!} \int_0^b f(t)s^j(t)e^{-xCt^p}\,dt + O\left(x^n \int_0^b t^{q+nr} e^{-xCt^p}\,dt\right). \tag{7}$$

Every term can be estimated by formula (4′). In particular, the O-term has order $x^{-\frac{q+1+n(r-p)}{p}}$. If we know more about the behavior of the functions f and s as $t \to 0$, we can make more precise the asymptotic of each term by Watson's lemma. Of course, this sharpening makes sense so long as the precision guaranteed by the remainder is not exceeded.

The following two examples are intended to illustrate this approach. The first example is connected with the gamma function.

Example 1 Applying the method described above, we will sharpen the relation $\frac{x^a\Gamma(x)}{\Gamma(x+a)} \underset{x\to+\infty}{\longrightarrow} 1$ obtained in Sect. 7.2.2. To this end, we use the well-known formula connecting the functions B and Γ,

$$\frac{\Gamma(a)\Gamma(x+1)}{\Gamma(x+a+1)} = \mathrm{B}(a, x+1) = \int_0^1 t^{a-1}(1-t)^x\,dx$$

(here $a > 0$; to simplify the subsequent formulas, we consider $\mathrm{B}(a, x+1)$ instead of $\mathrm{B}(a, x)$). Obviously,

$$(1-t)^x = e^{-xt}\big((1-t)e^t\big)^x = e^{-xt}e^{-xs(t)},$$

where $s(t) = -\ln(1-t) - t = \frac{t^2}{2} + \frac{t^3}{3} + \cdots = O(t^2)$, for $t \in [0, 1/2]$. Therefore,

$$\int_0^1 t^{a-1}(1-t)^x\,dx = \int_0^{1/2} t^{a-1}e^{-xt}\Big(1-\frac{x}{2}t^2+O\big(xt^3+x^2t^4\big)\Big)\,dt + O\Big(\frac{1}{2^x}\Big)$$
$$= \frac{\Gamma(a)}{x^a} - \frac{x}{2}\frac{\Gamma(a+2)}{x^{a+2}} - \frac{x}{3}\frac{\Gamma(a+3)}{x^{a+3}} + x^2 O\Big(\frac{1}{x^{a+4}}\Big)$$
$$= \frac{\Gamma(a)}{x^a}\Big(1-\frac{(a+1)a}{2x}+O\Big(\frac{1}{x^2}\Big)\Big).$$

Dividing both sides of the equation

$$\frac{\Gamma(a)\Gamma(x+1)}{\Gamma(x+a+1)} = \frac{\Gamma(a)x\Gamma(x)}{(x+a)\Gamma(x+a)}$$

by $\Gamma(a)$, we obtain

$$\frac{x^a\Gamma(x)}{\Gamma(x+a)} = \Big(1+\frac{a}{x}\Big)\Big(1-\frac{(a+1)a}{2x}+O\Big(\frac{1}{x^2}\Big)\Big) = 1+\frac{a(1-a)}{2x}+O\Big(\frac{1}{x^2}\Big).$$

The next example has already been considered in the above-mentioned "Analytical theory of probability" by Laplace (Book 1, No 42). Here we are dealing with the integral

$$\int_0^{\pi/2}\Big(\frac{\sin t}{t}\Big)^x \cos yt\,dt$$

in which x and the real parameter y may vary independently. The result becomes especially apparent if we represent the parameter in the form $y = r\sqrt{x}$ ($r \geqslant 0$).

Example 2 We find the asymptotic of the integral

$$I_r(x) = \int_0^{\pi/2}\Big(\frac{\sin t}{t}\Big)^x \cos yt\,dt \quad \text{as } x \to +\infty,$$

where $y = r\sqrt{x}$. In this case $\varphi(t) = \frac{\sin t}{t}$ and

$$S(t) = -\ln\varphi(t) = \frac{1}{6}t^2 + s(t), \quad \text{where } s(t) = \frac{1}{180}t^4 + O\big(t^6\big) \quad \text{as } t \to 0.$$

Putting $n = 2$ in formula (7), we obtain

$$I_r(x) = \int_0^{\pi/2}\big(1-xs(t)\big)e^{-\frac{x}{6}t^2}\cos yt\,dt + O\Big(x^2\int_0^{\pi/2} t^8|\cos yt|e^{-\frac{x}{6}t^2}\,dt\Big).$$

We replace (with exponentially small error) the integration over the interval $[0,\pi/2]$ by the integration over $[0,+\infty)$ and $|\cos yt|$ in the O-term by 1. Then we come to the relation

$$I_r(x) = \int_0^{\infty}\Big(1-\frac{x}{180}t^4\Big)e^{-\frac{x}{6}t^2}\cos yt\,dt + O\Big(\int_0^{\infty}\big(xt^6+x^2t^8\big)e^{-\frac{x}{6}t^2}\,dt\Big).$$

It can easily be seen that the O-term has order of smallness $O(x^{-5/2})$ (with an absolute constant). Thus,

$$I_r(x) = \int_0^\infty \left(1 - \frac{x}{180}t^4\right) e^{-\frac{x}{6}t^2} \cos(\sqrt{x}\,rt)\,dt + O\big(x^{-\frac{5}{2}}\big).$$

Now, making the change of variable $t = \sqrt{\frac{6}{x}}\,u$ in the integral, we obtain

$$I_r(x) = \sqrt{\frac{6}{x}} \int_0^\infty \left(1 - \frac{u^4}{5x}\right) e^{-u^2} \cos(\sqrt{6}\,ru)\,du + O\big(x^{-\frac{5}{2}}\big).$$

To calculate the last integral, we use the following relation obtained in Example 1 of Sect. 7.1.6:

$$\int_0^\infty e^{-u^2} \cos(\sqrt{6}\,ru)\,du = \frac{\sqrt{\pi}}{2} e^{-\frac{3}{2}r^2}.$$

Differentiating this identity with respect to the parameter r four times, we can also find the integral $\int_0^\infty u^4 e^{-u^2} \cos(\sqrt{6}\,ru)\,du$. A simple calculation leads to the required formula,

$$I_r(x) = \sqrt{\frac{3\pi}{2x}} \left(1 - \frac{3}{20x}\big(1 - 6r^2 + 3r^4\big)\right) e^{-\frac{3}{2}r^2} + O\big(x^{-\frac{5}{2}}\big).$$

Of particular interest is the case where $x = m \in \mathbb{N}$. Then the power $(\frac{\sin t}{t})^m$ is defined for all t. Replacing (with exponentially small error) the integral $I_{2r}(m)$ by the integral over the interval $(0, +\infty)$, we obtain

$$\int_0^\infty \left(\frac{\sin t}{t}\right)^m \cos(2\sqrt{m}\,rt)\,dt = \sqrt{\frac{3\pi}{2m}} \left(1 - \frac{3}{20m}\big(1 - 24r^2 + 48r^4\big)\right) e^{-6r^2} + O\big(m^{-\frac{5}{2}}\big).$$

We recall that (see Eq. (1) of Sect. 6.7.2 for $\omega = (1, \dots, 1)$), up to a factor, the integral on the left-hand side coincides with the area $S_m(r)$ of the cross section of the m-dimensional unit cube by the plane perpendicular to the main diagonal of the cube and lying at distance r from its center,

$$S_m(r) = \frac{2}{\pi}\sqrt{m} \int_0^\infty \left(\frac{\sin t}{t}\right)^m \cos(2\sqrt{m}\,rt)\,dt.$$

Therefore,

$$S_m(r) = \sqrt{\frac{6}{\pi}} \left(1 - \frac{3}{20m}\big(1 - 24r^2 + 48r^4\big)\right) e^{-6r^2} + O\big(m^{-2}\big)$$

(the constant in the O-term is absolute).

For $r = 0$, i.e., for the central cross sections, this gives the following sharpening of the asymptotic formula obtained in Example 3 of Sect. 7.3.3:

$$S_m(0) = \sqrt{\frac{6}{\pi}}\left(1 - \frac{3}{20m}\right) + O\left(m^{-2}\right).$$

A more precise calculation (formula (7) with $n = 3$ must be applied to the integral $I_0(x)$) leads to the asymptotic relation

$$S_m(0) = \sqrt{\frac{6}{\pi}}\left(1 - \frac{3}{20m} - \frac{13}{1120m^2}\right) + O\left(m^{-3}\right).$$

7.3.6 We discuss the Laplace method in the general situation mentioned at the beginning of this section. Let $(T, \mathfrak{A}, \mu)$ be a measure space, and let φ and f be non-negative measurable functions on T. We assume that φ is bounded and f is summable. Under these assumptions, the integrals

$$\Phi(x) = \int_T f(t)\varphi^x(t)\,d\mu(t)$$

are finite for all $x \geqslant 0$. The question of their behavior as $x \to +\infty$ is a direct generalization of the problem considered in the preceding subsections of the present section. We certainly exclude the trivial case where the product $f\varphi$ vanishes almost everywhere on T. Replacing the measure, we can reduce the problem to the case of a finite measure and $f \equiv 1$. Indeed, it follows from Eq. (2′) of Sect. 6.1.2 that

$$\Phi(x) = \int_T \varphi^x(t)\,d\nu(t), \quad \text{where } \nu(A) = \int_A f(t)\,d\mu(t) \quad (A \in \mathfrak{A})$$

and $\nu(T) = \int_T f\,d\mu < +\infty$. To avoid additional constraints, we will assume in this section that the functions f and φ are everywhere positive on T (this can be achieved by replacing, if necessary, T by the set $\{t \in T \mid f(t)\varphi(t) > 0\}$). Then the inequalities $\nu(A) > 0$ and $\mu(A) > 0$ are equivalent, and, therefore, the conditions "almost everywhere with respect to the measure μ" and "almost everywhere with respect to the measure ν" have the same meaning.

It appears at first sight that dropping the assumptions concerning the nature of the monotonicity of the function φ (its piecewise monotonicity) made in the one-dimensional case, we are unable to give even a qualitative description of the behavior of $\Phi(x)$. However, in the new situation the following principle underlying all preceding reasoning is preserved: the contribution to the integral $\Phi(x)$ coming from the points at which the value of the function φ is below a certain level, say, $\varphi < h$, is negligibly small in comparison with that coming from the points where $\varphi \geqslant h$ (of course, under the assumption that the set $T_h = \{t \in T \mid \varphi(t) \geqslant h\}$ is of positive measure). By abuse of language, we may say that the main contribution to the integral $\Phi(x)$ comes from the points $t \in T$ at which the values $\varphi(t)$ are "almost maximal". To make this statement more precise, we use the notion of the genuine supremum of a function f introduced in Sect. 4.4.5. We denote it by H, and, by definition, obtain $H = \inf\{C \in \mathbb{R} \mid \varphi \leqslant C$ almost everywhere on $T\}$. It can easily be seen that

$$H = \sup\{h \in \mathbb{R} \mid \nu(T_h) > 0\},$$

and that, under our assumptions, we have $0 < H < +\infty$.

The above equation implies that $\nu(T_h) > 0$ for $h < H$ and $\nu(T_h) = 0$ for $h > H$. If the set T_H has positive measure, then the asymptotic of the integral $\Phi(x)$ is obvious: since the fraction $(\varphi/H)^x$ tends to zero as $x \to +\infty$ outside T_H and is equal to 1 almost everywhere on T_H, we obtain by Theorem 1 of Sect. 7.1.2 that

$$\frac{\Phi(x)}{H^x} = \int_T \left(\frac{\varphi}{H}\right)^x d\nu \underset{x\to+\infty}{\longrightarrow} \nu(T_H).$$

Thus, in the simplest case in question, we have $\Phi(x) \sim H^x \nu(T_H)$.

Now we discuss a more interesting case where $\nu(T_H) = 0$. Then $\varphi(t) < H$ almost everywhere on T, and, for all $h < H$, the sets T_h are of positive measure. Obviously,

$$\int_{T_h} \varphi^x \, d\nu \geqslant h^x \nu(T_h) \quad \text{and} \quad \int_{T\setminus T_h} \varphi^x \, d\nu = h^x \int_{T\setminus T_h} \left(\frac{\varphi}{h}\right)^x d\nu = o\big(h^x\big).$$

Therefore, the integral over $T \setminus T_h$ is negligibly small in comparison with the integral over T_h as $x \to +\infty$. Hence we obtain the following counterpart of the localization principle in the case in question: the main contribution to $\Phi(x)$ comes from the integral over the set T_h with parameter h arbitrarily close to H. Therefore, to find the asymptotic of the integral $\Phi(x)$, it is important to know the rate of decrease of the measure of T_h as $h \to H$. In other words, the asymptotic is determined by the decreasing distribution function $\widetilde{f}(h) = \nu(T_h)$ (see Sect. 6.4.3). By Proposition 6.4.3, the integral $\Phi(x)$ can be represented in the form

$$\Phi(x) = x \int_0^{\infty} h^{x-1} \widetilde{f}(h) \, dh = x \int_0^H h^{x-1} \widetilde{f}(h) \, dh \tag{8}$$

(we have taken into account that $\widetilde{f}(h) = 0$ for $h > H$). Thus, the Laplace "abstract" integrals are reduced to the classical ones studied above with the difference that now the integrand $\widetilde{f}$ is given not directly but by the equation

$$\widetilde{f}(h) = \nu(T_h) = \int_{T_h} f \, d\mu.$$

It follows from the above that conceptually there is nothing new in the general case in comparison with the classical one. However, some technical difficulties connected with estimating the function $\widetilde{f}(h)$ must be overcome to find the asymptotic. For example, in contrast to the one-dimensional case, now it is quite natural to consider the situation where the sets T_h shrink not to a point but to a set of measure zero, say, to a surface, as $h \to H$ (see Example 2 below).

Example 1 For $t = (t_1, \dots, t_m) \in \mathbb{R}^m$, we put $\|t\|_p = (|t_1|^p + \dots + |t_m|^p)^{\frac{1}{p}}$ $(p > 0)$. We find the asymptotic of the integral

$$\Phi(x) = \int_{S^{m-1}} \|t\|_p^x \, d\sigma,$$

where σ is the area on the $(m-1)$-dimensional unit sphere S^{m-1}.

The maximum value of $\|t\|_p$ on the sphere depends on the parameter p. The case where $p = 2$ is trivial. If $p > 2$, then $H = \max_{t \in S^{m-1}} \|t\|_p = 1$ (this value is attained at the $2m$ points $\pm e_1, \dots, \pm e_m$, where $e_1, \dots, e_m$ are the vectors of the canonical basis). If $p \in (0, 2)$, then $H = \max_{t \in S^{m-1}} \|t\|_p = m^{\frac{1}{p} - \frac{1}{2}}$ (this value is attained at the 2^m points whose coordinates have absolute values equal to $1/\sqrt{m}$).

We consider the case where $p > 2$ in more detail. We will find the asymptotic of $\Phi(x)$ with the help of Eq. (8) containing the distribution function $\widetilde{f}(h)$, which is the area of the set $T_h = \{t \in S^{m-1} \mid \|t\|_p \geqslant h\}$. We need to estimate the area for h close to $H = 1$. For such h, this set is partitioned into $2m$ congruent parts. It is sufficient to estimate the area of one of them, say, of the part lying close to e_m. As $h \to 1$, the area of this part is equivalent to the area of its projection on the subspace $t_m = 0$. The projection is formed by the points $t' = (t_1, \dots, t_{m-1})$ the coordinates of which satisfy the inequality

$$|t_1|^p + \dots + |t_{m-1}|^p + \left(1 - \|t'\|^2\right)^{\frac{p}{2}} \geqslant h^p$$

($\|t'\|$ is the Euclidean norm of the vector t'). As $h \to 1$, the projection shrinks to the origin, and, therefore, is formed by the points satisfying the relation

$$1 - \frac{p}{2}\|t'\|^2 + o\left(\|t'\|^2\right) \geqslant h^p.$$

It contains an $(m-1)$-dimensional ball of radius $(1-\varepsilon)\sqrt{\frac{2}{p}(1-h^p)}$ and is contained in a ball of radius $(1+\varepsilon)\sqrt{\frac{2}{p}(1-h^p)}$, where ε tends to zero as $h \to 1$. Consequently, the area of the projection is equivalent to $\alpha_{m-1}(\frac{2}{p}(1-h^p))^{\frac{m-1}{2}}$, where α_{m-1} is the volume of the unit ball B^{m-1}. Thus,

$$\widetilde{f}(h) = \sigma(\varphi \geqslant h) \underset{h \to 1}{\sim} 2m\alpha_{m-1}\left(\frac{2}{p}\left(1 - h^p\right)\right)^{\frac{m-1}{2}} \underset{h \to 1}{\sim} 2m\alpha_{m-1}\left(2(1-h)\right)^{\frac{m-1}{2}}.$$

We obtain

$$\Phi(x) = x\int_0^1 h^{x-1}\widetilde{f}(h)\, dh \underset{x \to +\infty}{\sim} m\alpha_{m-1} 2^{\frac{m+1}{2}} x \int_0^1 h^{x-1}(1-h)^{\frac{m-1}{2}}\, dh.$$

After simple calculations, we obtain that

$$\Phi(x) = \int_{S^{m-1}} \left(|t_1|^p + \dots + |t_m|^p\right)^{\frac{x}{p}} d\sigma \underset{x \to +\infty}{\sim} 2m\left(\frac{2\pi}{x}\right)^{\frac{m-1}{2}}$$

for $p > 2$. The case where $0 < p < 2$ is considered similarly. We leave it for the reader as an exercise (see Exercise 18). We will return to this example in Sect. 7.3.8, illustrating one more method of studying the integrals $\Phi(x)$.

The following example shows that sometimes it is helpful to take into account the specific features of the problem and use modifications of the general method.

Example 2 Let $r, p_1, \ldots, p_m$ be positive numbers. We find the asymptotic of the integral

$$\Phi(x) = \int_{\mathbb{R}^m} \varphi^x(t)\, dt,$$

where

$$\varphi(t) = e^{-||t_1|^{p_1}+\cdots+|t_m|^{p_m}-1|^r} \quad (t = (t_1, \ldots, t_m) \in \mathbb{R}^m).$$

Obviously, the genuine supremum of φ coincides with its maximum value and is equal to 1. The maximum is attained at all points of the surface

$$\{t \in \mathbb{R}^m \mid |t_1|^{p_1} + \cdots + |t_m|^{p_m} = 1\}.$$

We use the distribution function F of the sum $\sum_{j=1}^m |t_j|^{p_j}$ (see Example 3 of Sect. 6.4.2),

$$F(u) = V\, u^s \ (u > 0), \quad \text{where } s = \frac{1}{p_1} + \cdots + \frac{1}{p_m} \text{ and}$$

$$V = \frac{2^m}{\Gamma(1+s)} \prod_{j=1}^m \Gamma\Big(1 + \frac{1}{p_j}\Big).$$

Then

$$\Phi(x) = \int_0^\infty e^{-x|u-1|^r}\, dF(u) = s\, V \int_0^\infty e^{-x|u-1|^r}\, u^{s-1}\, du.$$

To find the asymptotic of $\Phi(x)$, it remains to apply formula (3′) for $t_0 = 1$, $p = r$, $q = 0$, and $C = L = 1$. We obtain

$$\int_{\mathbb{R}^m} e^{-x||t_1|^{p_1}+\cdots+|t_m|^{p_m}-1|^r}\, dt \underset{x\to+\infty}{\sim} 2s\, V\, \Gamma\Big(1 + \frac{1}{r}\Big) x^{-\frac{1}{r}}.$$

7.3.7 Here, we consider a multi-dimensional version of the Laplace integral

$$\Phi(x) = \int_T f(t)\, \varphi^x(t)\, dt,$$

where T is a Lebesgue measurable subset of the space $\mathbb{R}^m$. Our goal is to obtain a version of Theorem 7.3.2 in the situation where the sets $T_h = \{t \in T \mid \varphi(t) \geqslant h\}$

shrink to a single point as h increases to $\max\varphi$. This is a replacement of the monotonicity assumption of φ in the one-dimensional case. This assumption taken together with other conditions allows us to obtain a multi-dimensional generalization of Theorem 7.3.2 without using the distribution function. Additional restrictions imposed on the functions f and φ below are similar to the conditions of Theorem 7.3.2. They can conveniently be described in terms of spherical coordinates (see Sect. 6.5.2). We write σ for the area on the unit sphere S^{m-1}.

Theorem *Let $T \subset \mathbb{R}^m$ and $a \in \mathrm{Int}(T)$. Let f be a summable function and φ be a measurable function defined on T, and let $0 \leqslant \varphi(t) \leqslant \varphi(a)$ and $\mathrm{diam}(T_h) \to 0$ as $h \to \varphi(a) - 0$. Assume that there exist numbers $p > 0$, $q > -m$ and $c > 0$ and non-negative functions F and G on the unit sphere S^{m-1} such that the following conditions are fulfilled for almost all $\xi \in S^{m-1}$:*

(a) *the limits $C(\xi) = \lim_{r\to 0} \frac{\varphi(a)-\varphi(a+r\xi)}{r^p}$ and $L(\xi) = \lim_{r\to 0} \frac{f(a+r\xi)}{r^q}$ exist;*
(b) $\frac{\varphi(a)-\varphi(a+r\xi)}{r^p} \geqslant F(\xi)$ *and* $|\frac{f(a+r\xi)}{r^q}| \leqslant G(\xi)$ *for* $0 < r < c$.

If, in addition, the function $GF^{-\frac{q+m}{p}}$ is summable on S^{m-1}, then

$$\Phi(x) = \frac{I + o(1)}{p}\Gamma\left(\frac{q+m}{p}\right)\left(\frac{\varphi(a)}{x}\right)^{\frac{q+m}{p}} \varphi^x(a) \quad \text{as } x \to +\infty,$$

where

$$I = \int_{S^{m-1}} L(\xi) C^{-\frac{q+m}{p}}(\xi)\, d\sigma(\xi).$$

In particular, if $I \neq 0$, then

$$\Phi(x) \underset{x\to+\infty}{\sim} \frac{I}{p}\Gamma\left(\frac{q+m}{p}\right)\left(\frac{\varphi(a)}{x}\right)^{\frac{q+m}{p}} \varphi^x(a). \tag{9}$$

Obviously, formula (9) is a multi-dimensional version of (3′). Formula (3) corresponds to the situation in which a is a boundary point of the set T. Putting $f = 0$ outside T, we can reduce this case to the case considered in the theorem. In particular, the theorem remains valid if the intersection of T with a ball $B(a,\rho)$ coincides with the intersection of the ball with the cone $\{a + r\xi \mid \xi \in E \subset S^{m-1},\ 0 \leqslant r < +\infty\}$ and φ and f satisfy the conditions of the theorem for almost all $\xi \in E$. In this case, we can use relation (9), assuming that the function L is zero outside E.

If the function C is separated from zero and the limit relations (a) are fulfilled uniformly on S^{m-1}, then the theorem can be proved similarly to Theorem 7.3.2 by dropping condition (b) and assuming only that the function L is summable (we advise the reader to verify this). For a more general result, we need a more powerful tool, namely Lebesgue's dominated convergence theorem (Theorem 7.1.2). The application of this theorem in the proof of Theorem 7.3.2 could simplify the reasoning somewhat, but we preferred to manage with comparatively elementary tools.

Proof Without loss of generality, we may assume that $a = 0$ and $\varphi(0) = 1$. Since $\operatorname{diam}(T_h) \underset{h\to 1-0}{\longrightarrow} 0$, we see that, for h sufficiently close to 1, the set T_h lies in the ball $B = B(0, c)$ in which condition (b) is fulfilled. Consequently, $\varphi < h$ outside the ball B, which implies that the integral $\int_{T\setminus B} f(t)\varphi^x(t)\,dt$ is exponentially small. Therefore, it is sufficient to consider the case where $T = B$. Introducing spherical coordinates and making the change of variable $r = x^{-1/p}u$, we obtain

$$x^{\frac{q+m}{p}}\Phi(x) = x^{\frac{q+m}{p}}\int_{S^{m-1}}\left(\int_0^c r^{m-1} f(r\xi)\varphi^x(r\xi)\,dr\right)d\sigma(\xi)$$

$$= \int_{S^{m-1}}\left(\int_0^{cx^{\frac{1}{p}}} u^{m-1}x^{\frac{q}{p}} f\big(x^{-\frac{1}{p}}u\xi\big)\varphi^x\big(x^{-\frac{1}{p}}u\xi\big)\,dr\right)d\sigma(\xi).$$

By condition (a), the integrand (we assume that the integrand is defined on the product $P = S^{m-1}\times(0,+\infty)$ and is equal to zero if $u > cx^{\frac{1}{p}}$) converges pointwise to the limit function

$$u^{q+m-1}L(\xi)e^{-u^pC(\xi)}$$

as $x\to+\infty$. The passage to the limit on the right-hand side of the last equation (we will justify it later) gives

$$x^{\frac{q+m}{p}}\Phi(x) \underset{x\to+\infty}{\longrightarrow} J = \iint_P u^{q+m-1}L(\xi)e^{-u^pC(\xi)}\,d\sigma(\xi)\,du.$$

The integral J can easily be calculated,

$$J = \int_{S^{m-1}} L(\xi)\left(\int_0^\infty u^{q+m-1}e^{-u^pC(\xi)}\,du\right)d\sigma(\xi) = \frac{I}{p}\Gamma\left(\frac{q+m}{p}\right).$$

To justify the passage to the limit, we can use Theorem 1 of Sect. 7.1.2. Indeed, since condition (b) implies the inequalities

$$x^{\frac{q}{p}}\big|f\big(x^{-\frac{1}{p}}u\xi\big)\big| \leqslant u^qG(\xi) \quad\text{and}\quad \varphi^x\big(x^{-\frac{1}{p}}u\xi\big) \leqslant \left(1-\frac{u^p}{x}F(\xi)\right)^x \leqslant e^{-u^pF(\xi)},$$

we see that the integrand is dominated by $u^{q+m-1}G(\xi)e^{-u^pF(\xi)}$ on the set P. Up to notation, the proof of its summability based on Tonelli's theorem coincides with the calculation of the integral J. □

Of most importance is the specific case in which the smooth function φ attains its maximum value at an interior point a, the second differential $d_a^2\varphi$ is a negative definite quadratic form, and there exists a finite non-zero limit $L_0 = \lim_{t\to a} f(t)$. Then $p = 2$, $q = 0$, $C = -\frac{1}{2}d_a^2\varphi$, and $L \equiv L_0$. Therefore,

$$I = L_0\int_{S^{m-1}}\left(-\frac{1}{2}d_a^2\varphi\right)^{-\frac{m}{2}}(\xi)\,d\sigma(\xi).$$

This integral can be expressed (see Corollary 2 of Sect. 6.5.3) in terms of the determinant of the second derivative matrix (the determinant of the *Hesse matrix*[9]) of φ,

$$\int_{S^{m-1}} \left(-d_a^2\varphi\right)^{-\frac{m}{2}}(\xi)\,d\sigma(\xi) = \frac{2\pi^{\frac{m}{2}}}{\Gamma(\frac{m}{2})\sqrt{|\Delta|}}, \qquad \text{where } \Delta = \det\left(\frac{\partial^2\varphi}{\partial t_k \partial t_j}(a)\right)_{1\leqslant k,j\leqslant m}.$$

Thus, we come to the following multi-dimensional version of Laplace's formula (2b):

$$\Phi(x) \underset{x\to+\infty}{\sim} \frac{L_0}{\sqrt{|\Delta|}}\left(\frac{2\pi\varphi(a)}{x}\right)^{\frac{m}{2}} \varphi^x(a). \tag{10}$$

7.3.8 Completing the study of the Laplace method, we consider some applications of Theorem 7.3.7.

Example 1 We find the asymptotic of the integral

$$\Phi(x) = \iint_{[-1,1]^2} (\cos t_1 + \cos t_2)^x\,dt$$

as $x \to +\infty$. In this case, $\varphi(t) = \cos t_1 + \cos t_2$, $\max\varphi = 2$ and $\varphi(t) = 2 - \frac{1}{2}\|t\|^2 + o(\|t\|^2)$ as $t \to 0$. Therefore, $\Delta = \det(\frac{\partial^2\varphi}{\partial t_k \partial t_j}(0))_{1\leqslant k,j\leqslant 2} = 1$, and, by formula (10), we obtain

$$\iint_{[-1,1]^2} (\cos t_1 + \cos t_2)^x\,dt \underset{x\to+\infty}{\sim} \frac{4\pi}{x}2^x.$$

Example 2 We return to Example 1 of Sect. 7.3.6 and find the asymptotic of the integral $\Phi(x) = \int_{S^{m-1}} \|t\|_p^x\,d\sigma(t)$ in the case where $p \in (0,2)$.

We reduce the problem to the case in which it is possible to apply formula (10). To this end, we use a method which is effective in many similar situations. The idea of the method, to use the positive homogeneity of the function $t \mapsto \|t\|_p$ and replace the integral over the sphere by the integral over the entire space, has already been exploited in Example 1 of Sect. 6.5.2.

We consider the integral

$$\widetilde{\Phi}(x) = \int_{\mathbb{R}^m} \|t\|_p^x e^{-\frac{x}{2m}\|t\|^2}\,dt$$

(here, as usual, $\|t\| = \|t\|_2$ is the Euclidean norm of the vector t; the coefficient $\frac{1}{2m}$ in the exponent is introduced to simplify the calculation). Using spherical coordinates (see Sect. 6.5.2), we can easily obtain the following relation between the

[9]Ludwig Otto Hesse (1811–1874)—German mathematician.

integrals $\Phi(x)$ and $\widetilde{\Phi}(x)$:

$$\widetilde{\Phi}(x) = \int_{S^{m-1}} \int_0^\infty r^{m+x-1} \,\|\xi\|_p^x\, e^{-\frac{x}{2m}r^2}\, d\sigma(\xi)\, dr = \frac{\Phi(x)}{2}\left(\frac{2m}{x}\right)^{\frac{x+m}{2}} \Gamma\left(\frac{x+m}{2}\right).$$

Thus, the problem is reduced to the study of the integral

$$\widetilde{\Phi}(x) = 2^m \int_{\mathbb{R}_+^m} \varphi^x(t)\, dt, \quad \text{where } \varphi(t) = \|t\|_p e^{-\|t\|^2/(2m)}.$$

Since $\|t\|_p \leqslant m^{\frac{1}{p}-\frac{1}{2}}\|t\|$ for $p \in (0,2)$, we obtain $\varphi(t) \leqslant m^{\frac{1}{p}-\frac{1}{2}}\|t\|e^{-\|t\|^2/(2m)} \leqslant m^{\frac{1}{p}}/\sqrt{e}$. At the point $a = (1,\dots,1)$ (and only at this point) the inequalities turn into equalities, i.e., $\varphi(a) = m^{\frac{1}{p}}/\sqrt{e}$ is the strict maximum of φ. Now, we can use formula (10). We leave it to the reader to carry out the necessary calculations and verify that the determinant of the Hesse matrix of the function φ at the point a is equal to $\frac{2}{p-2}(\frac{2-p}{m}\varphi(a))^m$. Therefore, the following asymptotic relation is valid for $p \in (0,2)$:

$$\int_{S^{m-1}} \big(|t_1|^p + \cdots + |t_m|^p\big)^{\frac{x}{p}}\, d\sigma(t) \underset{x\to+\infty}{\sim} 2\left(\frac{8\pi}{(2-p)x}\right)^{\frac{m-1}{2}} m^{(\frac{1}{p}-\frac{1}{2})x}.$$

In the theorem, we considered the case where the increment $\varphi(a) - \varphi(a + r\xi)$ tends to zero as $r \to 0$ whose order does not depend on the direction of $\xi \in S^{m-1}$. However, this assumption is often violated. In many situations the smooth function φ attains its maximum value at the point a but the second differential $d_a^2\varphi$ is degenerate and negative semi-definite (possibly, $d_a^2\varphi \equiv 0$). Then, in the vicinity of a, the increment $\varphi(a) - \varphi(t)$ is described by a non-negative polynomial of higher degree. For example, we have the asymptotic relation

$$\varphi(a) - \varphi(t) = c_1(t_1 - a_1)^{2n_1} + \cdots + c_m(t_m - a_m)^{2n_m} + o\big(\|t - a\|^n\big), \quad \text{as } t \to a,$$

where the coefficients $c_1,\dots,c_m$ are positive and $n = \max(n_1,\dots,n_m)$. By the change of variables $u_j = (t_j - a_j)^{n_j}$ $(j = 1,\dots,m)$, we can reduce this situation to the case described by the theorem. We clarify this by the following example.

Example 3 We find the asymptotic of the integral

$$\Phi(x) = \iint_{\mathbb{R}^2} \big(u^2 + v^2\big)^{\beta/2} e^{-x(|u|+v^4)}\, du\, dv.$$

It is clear that the integral is finite for $\beta > -2$ and is equal to $4\iint_{\mathbb{R}_+^2} \dots$. We make the change of variables $u = t^2$ and $v = s^{\frac{1}{2}}$ in the integral over the set $\mathbb{R}_+^2$. As a result, we obtain

$$\Phi(x) = \iint_{\mathbb{R}^2} \big(t^4 + |s|\big)^{\beta/2} \frac{|t|}{\sqrt{|s|}} e^{-x(t^2+s^2)}\, dt\, ds. \tag{11}$$

Now the conditions of Theorem 7.3.7 are fulfilled with the functions

$$\varphi(t,s)=e^{-(t^2+s^2)} \quad \text{and} \quad f(t,s)=\left(t^4+|s|\right)^{\frac{\beta}{2}}\frac{|t|}{\sqrt{|s|}}=r^{\frac{\beta+1}{2}}\left(r^3\xi_1^4+|\xi_2|\right)^{\frac{\beta}{2}}\frac{|\xi_1|}{\sqrt{|\xi_2|}},$$

where $r=\sqrt{t^2+s^2}$, $\xi_1=t/r$, $\xi_2=s/r$. We have $p=2$, $C(\xi)\equiv 1$, $q=\frac{\beta+1}{2}$ and $L(\xi)=\lim_{r\to 0} r^{-\frac{\beta+1}{2}} f(r\xi_1,r\xi_2)=|\xi_1||\xi_2|^{\frac{\beta-1}{2}}$. Obviously, the function L is summable on the circle for $\beta>-1$. This is not the case if $\beta\leqslant -1$, and the theorem is not applicable. In this case, we have to invoke some additional considerations (see Example 4). As to the case where $\beta>-1$, the reader can easily verify that the conditions of the theorem are fulfilled. Since

$$I=4\int_0^{\pi/2}\cos\theta\,\sin^{\frac{\beta-1}{2}}\theta\,d\theta=\frac{8}{\beta+1},$$

formula (9) implies the following relation for $m=2$, $p=2$, $q=(\beta+1)/2$, $a=0$ and $\varphi(0)=1$:

$$\Phi(x)\underset{x\to+\infty}{\sim}\frac{I}{p}\Gamma\left(\frac{q+2}{p}\right)x^{-\frac{q+2}{p}}=\Gamma\left(\frac{\beta+1}{4}\right)x^{-\frac{\beta+5}{4}}.$$

Now we consider the problem in the situation where the conditions of the theorem do not hold.

Example 4 We find the asymptotic of the integral from Example 3 for $-2<\beta<-1$ (we invite the reader to investigate the case $\beta=-1$ independently, see Exercise 17).

Passing to polar coordinates in Eq. (11), we can represent the integral in the form

$$\begin{aligned}\Phi(x)&=4\int_0^{\infty} r^{\frac{\beta+3}{2}}e^{-xr^2}\left(\int_0^{\pi/2}\left(r^3\cos^4\theta+\sin\theta\right)^{\beta/2}\frac{\cos\theta}{\sqrt{\sin\theta}}\,d\theta\right)dr\\&=4\int_0^{\infty} r^{\frac{\beta+3}{2}}e^{-xr^2}\left(\int_0^{1}\left(r^3\left(1-u^2\right)^2+u\right)^{\beta/2}\frac{du}{\sqrt{u}}\right)dr.\end{aligned}$$

Since $\beta<-1$, the inner integral tends to infinity as $r\to 0$. To be able to apply Theorem 7.3.2, we must find its asymptotic behavior. It can easily be verified (the reader is invited to prove this independently) that

$$\int_0^1\left(r^3\left(1-u^2\right)^2+u\right)^{\beta/2}\frac{du}{\sqrt{u}}\underset{r\to 0}{\sim} g(r)=\int_0^1\left(r^3+u\right)^{\beta/2}\frac{du}{\sqrt{u}}.$$

Making the change of variables $u=r^3v$ in the last integral, we see that

$$g(r)=r^{\frac{3}{2}(\beta+1)}\int_0^{1/r^3}(1+v)^{\beta/2}\frac{dv}{\sqrt{v}}\underset{r\to 0}{\sim} r^{\frac{3}{2}(\beta+1)}\int_0^{\infty}(1+v)^{\beta/2}\frac{dv}{\sqrt{v}}.$$

Transforming the integral obtained by the change of variable $v = z/(1-z)$, we obtain

$$\int_0^{\infty} (1+v)^{\beta/2} \frac{dv}{\sqrt{v}} = \int_0^1 z^{-1/2}(1-z)^{-\frac{\beta+1}{2}-1}\, dz = \frac{\Gamma(1/2)\Gamma(|\beta+1|/2)}{\Gamma(|\beta|/2)}.$$

Thus,

$$\Phi(x) \underset{x\to+\infty}{\sim} 4\int_0^{\infty} r^{\frac{\beta+3}{2}} g(r) e^{-xr^2}\, dr,$$

where

$$g(r) \underset{r\to 0}{\sim} \frac{\Gamma(1/2)\Gamma(|\beta+1|/2)}{\Gamma(|\beta|/2)} r^{\frac{3}{2}(\beta+1)}.$$

By Theorem 7.3.2 with $p=2$ and $q = \frac{\beta+3}{2} + \frac{3}{2}(\beta+1) = 2\beta+3$, we obtain

$$\Phi(x) \underset{x\to+\infty}{\sim} 2\frac{\Gamma(1/2)\Gamma(|\beta+1|/2)}{\Gamma(|\beta|/2)}\Gamma(\beta+2)\, x^{-\beta-2}.$$

Using Legendre's formula and the reflection formula for the Gamma function (see Sects. 7.2.4 and 7.2.5), we can simplify the coefficient on the right-hand side of the equation and obtain

$$\Phi(x) \underset{x\to+\infty}{\sim} \frac{2^{3+\beta}\pi^2}{\Gamma^2(\frac{|\beta|}{2})\sin\pi\beta} x^{-\beta-2}.$$

EXERCISES

1. Let f and φ be non-negative functions on (a,b) and $\varphi(t) \to 1$ as $t \to a$. Prove that if $\int_a^c f(t)\,dt > 0$ for $a < c < b$, then the integral $\Phi(x) = \int_a^b f(t)\varphi^x(t)\,dt$ cannot decrease exponentially, i.e., that $q^x = o(\Phi(x))$ as $x \to +\infty$ for every $q \in (0,1)$.
2. Prove that $\int_0^{\infty} \frac{\cos^2 1/t}{t^{1/2}} e^{-xt}\,dt \underset{x\to+\infty}{\sim} \frac{1}{2}\sqrt{\frac{\pi}{x}}$.
3. Prove that $\int_0^{\infty} \frac{\cos 1/t}{t^{2/3}} e^{-xt}\,dt = o(x^{-N})$ as $x \to +\infty$ for every $N > 0$.
   This example together with Exercise 2 shows that the non-negativity condition in the corollary to Lemma 7.3.1 cannot be dropped.
4. Use the representation $\frac{\sin u}{u} = \int_0^1 \cos ut\,dt$ to find the asymptotic of the derivatives $(\frac{\sin u}{u})^{(n)}$ as $n \to \infty$.
5. Use the result of Example 1 of Sect. 7.3.4 to generalize Lemma 7.3.2 by proving that

$$\int_0^b |\ln t|^r \varphi^x(t)\,dt \underset{x\to+\infty}{\sim} \Gamma\left(1+\frac{1}{p}\right)\left(\frac{\ln x}{p}\right)^r x^{-\frac{1}{p}}.$$

6. Use the result of the preceding exercise and follow the proof of Theorem 7.3.2 to verify that replacing condition (b) of the theorem by

$$f(t) \underset{t\to a}{\sim} L(t-a)^q \left|\ln(t-a)\right|^r, \quad \text{where } r \in \mathbb{R},$$

implies the asymptotic relation

$$\Phi(x) \underset{x\to+\infty}{\sim} \frac{L}{p}\Gamma\left(\frac{q+1}{p}\right)\left(\frac{\ln x}{p}\right)^r \left(\frac{\varphi(a)}{Cx}\right)^{\frac{q+1}{p}} \varphi^x(a).$$

7. Prove that

$$\Gamma'(x) = \Gamma(x)\left(\ln x + O\left(\frac{1}{x}\right)\right) \quad \text{and} \quad \Gamma''(x) = \Gamma(x)\left(\ln^2 x + O\left(\frac{\ln x}{x}\right)\right)$$

as $x \to +\infty$.

8. Let $\varphi(a) = 1$ in Theorem 7.3.2. Prove that replacing condition (b) with $\frac{1-\varphi(t)}{(t-a)^p} \to +\infty$ (or $\frac{1-\varphi(t)}{(t-a)^p} \to 0$) as $t \to a$ leads to the relation $x^{\frac{q+1}{p}}\Phi(x) \to 0$ (respectively, $x^{\frac{q+1}{p}}\Phi(x) \to +\infty$).

9. Find the asymptotic of the integrals $\int_0^{\frac{1}{2}} t^{xt}\,dt$ and $\int_0^1 t^{xt}\,dt$.

10. Prove that $\int_0^{\frac{1}{e}} e^{x/\ln t}\,dt \underset{x\to+\infty}{\sim} \sqrt{\pi}\frac{\sqrt[4]{x}}{e^{2\sqrt{x}}}$.
It is instructive to compare this result with Example 2 of Sect. 7.3.4. Although the functions $\varphi(t) = 1 + \frac{1}{\ln t}$ and $\psi(t) = e^{1/\ln t}$ are very close to each other in the vicinity of zero ($\varphi(t) - 1 \sim \psi(t) - 1$ as $t \to 0$), the corresponding integrals, nevertheless, have distinct asymptotics.

11. Let a non-negative function f be summable and functions S and R be continuous and strictly increasing on $[a, b)$. Let, in addition, $S(a) = R(a) = 0$ and $S(t) - R(t) = O((t-a)^2)$ as $t \to a$. Prove that the integrals

$$\Phi(x) = \int_a^b f(t)e^{-xS(t)}\,dt \quad \text{and} \quad \Psi(x) = \int_a^b f(t)e^{-xR(t)}\,dt$$

are equivalent if one of them, say, $\Psi(x)$ decreases not too fast to zero, i.e., if $\ln\Psi(x) = o(\sqrt{x})$ as $x \to +\infty$.
Comparing Exercise 10 with Example 2 of Sect. 7.3.4, we see that the last condition cannot be weakened and replaced by $\ln\Psi(x) = O(\sqrt{x})$.

12. Let $p > 0$. Prove that

$$\int_0^\infty e^{-xt^p - t^{-p}}\,dt \underset{x\to+\infty}{\sim} \frac{\sqrt{\pi}}{p} x^{-\frac{2+p}{4p}} e^{-2\sqrt{x}}.$$

13. Sharpen the result of Example 4 of Sect. 7.3.3 by proving that

$$S_n = \frac{e^n}{2}\left(1 + \frac{2}{3}\sqrt{\frac{2}{\pi n}} + O\left(\frac{1}{n^{\frac{3}{2}}}\right)\right).$$

14. Prove that the following asymptotic relation is valid for the integral $\Phi(x) = \int_0^1 \varphi^x(t)\,dt$, where φ is the Cantor function:

$$\Phi(x) \underset{x\to+\infty}{\sim} \frac{\theta(\log_2 x)}{x^{\log_2 3}}, \quad \text{where } \theta(u) = \frac{1}{2}\sum_{k\in\mathbb{Z}} \frac{3^{k+u}}{\sinh(2^{k+u})}.$$

In particular, $\Phi(2^n) \sim \theta(0)\, 3^{-n}$ $(n \in \mathbb{N})$.

It can be proved that the function θ is "almost constant": $1.9964 < \theta(u) < 1.997$.

15. Let H be the genuine supremum of a positive measurable function φ. By generalizing Exercise 1, prove that the limit relation $\frac{1}{x}\ln\Phi(x) \underset{x\to+\infty}{\longrightarrow} \ln H$ is valid for the integral $\Phi(x) = \int_T \varphi^x(t)\,d\nu(t)$.

16. Let B be the unit ball in $\mathbb{R}^3$. Prove that

$$\iiint_B \sqrt{s^2+t^2+u^2}\left(1 - \frac{|s|+t^2+u^4}{2}\right)^x ds\,dt\,du \underset{x\to+\infty}{\sim} \frac{16}{3x^2}.$$

17. Use the result of Exercise 6 to prove that the function Φ considered in Examples 3 and 4 of Sect. 7.3.8 has the asymptotic

$$\Phi(x) \underset{x\to+\infty}{\sim} \frac{3}{4}\frac{\ln x}{x}$$

if $\beta = -1$.

18. Reasoning as in the case $p > 2$ in Example 1 of Sect. 7.3.6, find the asymptotic of the integral $\int_{S^{m-1}} \|t\|_p^x\,d\sigma(t)$ for $p \in (0,2)$.

19. Verify that, without substantial changes in the proof, Theorem 7.3.7 can be generalized to the case in which the function f satisfies the following conditions: for some $s \geqslant 0$ and almost all $\xi \in S^{m-1}$, the limit $L(\xi) = \lim_{r\to 0} \frac{f(a+r\xi)}{r^q|\ln r|^s}$ exists and $\frac{|f(a+r\xi)|}{r^q|\ln r|^s} \leqslant G(\xi)$ $(0 < r < c)$. Then, keeping the other conditions and notation in the theorem unchanged, we have the relation

$$\Phi(x) = \frac{I+o(1)}{p}\,\Gamma\left(\frac{q+m}{p}\right)\left(\frac{\ln x}{p}\right)^s\left(\frac{\varphi(a)}{x}\right)^{\frac{q+m}{p}}\varphi^x(a) \quad \text{as } x\to+\infty.$$

The above relation is also valid for $s < 0$ if the quotients $(\varphi(a) - \varphi(a+r\xi))/r^p$ are separated from zero.

20. Show that if $s < 0$, then the result of the preceding exercise is not valid without additional assumptions; namely, the product $x^{\frac{q+m}{p}}\Phi(x)$ (for simplicity, we assume that $\varphi(a) = 1$) can tend to zero arbitrarily slowly. *Hint.* For $a = 0$, consider the functions $f(r\xi) = r^q \ln^s \frac{1}{r}$ and $\varphi(r\xi) = e^{-r^p(C(\xi)+\theta(r))}$, where θ is a continuous slowly increasing function on $[0,1]$, $\theta(0) = 0$, and C is a non-negative measurable function on the sphere S^{m-1} such that the integral $\int_{S^{m-1}} C^{-\frac{q+m}{p}}(\xi)\,d\sigma(\xi)$ is finite.

7.4 *Improper Integrals Dependent on a Parameter

7.4.1 Up to now we have investigated the integrals $J(y) = \int_X f(x, y)\,d\mu(x)$ dependent on a parameter under the assumption that the integrand is summable for every y. However, sometimes this requirement turns out to be too burdensome, as we have seen in Example 2 of Sect. 7.1.6, where we had to appeal to the concept of improper integral introduced in Sect. 4.6.4. Now we consider this problem more systematically. Here, we of course have to reduce the generality considerably, replacing a measure space by an interval with Lebesgue measure.

Thus, we will assume that $X = \langle a, b\rangle$, the function f is defined on the product $\langle a, b\rangle \times Y$, and the summability requirement for the function $x \mapsto f(x, y)$ for every $y \in Y$ is weakened and replaced with the convergence requirement for the improper integral $\int_a^{\to b} f(x, y)\,dx$. We recall that, by definition, the latter requirement means that the function $x \mapsto f(x, y)$ is summable on each interval (a, t), where $a < t < b$, (we called such functions left admissible, see Definition 4.6.4) and the limit $J(y) = \lim_{t\to b} \int_a^t f(x, y)\,dx$ exists and is finite. In the absence of absolute convergence the properties of such integrals cannot be studied by the previous methods that assume the summability of the integrand. Therefore, to extend the results of Sect. 7.1 to improper integrals, we need a new notion, the uniform convergence of an improper integral, which replaces the condition (L_{loc}) used in Sect. 7.1. We motivate this notion with two examples.

Let us consider the integrals

$$J(y) = \int_0^\infty \left(1 - e^{-xy}\right)\frac{\sin x}{x}\,dx \quad \text{and} \quad I(y) = \int_0^\infty \frac{\sin xy}{x}\,dx \quad (y \geqslant 0).$$

Obviously, $J(0) = I(0) = 0$. For $y > 0$, we already know the value of each of the integrals (see Sect. 7.1.6): $J(y) = \arctan y$ and $I(y) = I(1) = \frac{\pi}{2}$. Hence we see that

$$J(y) \underset{y\to 0}{\longrightarrow} J(0) = 0, \qquad I(y) \underset{y\to 0}{\not\to} I(0).$$

That is all we need to know about these integrals in the sequel. What is the reason why the former function is continuous at zero and the latter is not? After all, in both cases the integrand tends to zero as $y \to 0$. It is obvious that if the integrals J and I were calculated over an arbitrary finite interval $[0, t]$, both integrals would converge to zero. Thus, the behavior of the integrals J and I is determined by the behavior of the "remainder integrals"

$$j_t(y) = \int_t^\infty \left(1 - e^{-xy}\right)\frac{\sin x}{x}\,dx \quad \text{and} \quad i_t(y) = \int_t^\infty \frac{\sin xy}{x}\,dx.$$

In the proof that the integral J is continuous at zero, we actually proved (see Sect. 7.1.6) that the inequality $|j_t(y)| \leqslant 3/t$ holds for all $y > 0$. Consequently,

$$|J(y)| = \left|\int_0^t \left(1 - e^{-xy}\right)\frac{\sin x}{x}\,dx + j_t(y)\right| \leqslant \left|\int_0^t \left(1 - e^{-xy}\right)\frac{\sin x}{x}\,dx\right| + \frac{3}{t}.$$

Therefore, we can first make the second summand arbitrarily small (for all $y > 0$ at once) by choosing an appropriate t, and then, fixing a t, make the first summand small for small $y > 0$.

We see a different picture in the second case. Splitting the integral $I(y)$ as before into two summands,

$$I(y) = \int_0^t \frac{\sin xy}{x}\,dx + i_t(y),$$

we use the change of variables $z = xy$ to verify that

$$i_t(y) = \int_{yt}^{\infty} \frac{\sin z}{z}\,dz.$$

Thus, for an arbitrarily large parameter t, the integral $i_t(y)$ is close to $I(1) \neq 0$ for small y and does not tend to zero as $y \to 0$.

The above reasoning shows that the fact that the former integral is continuous and the latter discontinuous is caused by the distinct behavior of the remainder integrals j_t and i_t. The continuity of J follows from that fact that the remainder integral j_t can be made small for sufficiently large t and for all values of the parameter at once. This is the property that underlies the definition of the uniform convergence of improper integrals.

7.4.2 By an improper integral dependent on a parameter, we mean a function J defined on a set Y by the formula

$$J(y) = \int_a^{\to b} f(x, y)\,dx \quad (y \in Y), \tag{1}$$

where f is a function (in general, complex-valued) defined on $\langle a, b\rangle \times Y$. We always assume that improper integral (1) converges for each value of y in Y (see Sect. 4.6.4). We do not exclude the possibility that the integrand is summable for some values of y.

Naturally, in the same way we can also define an improper integral with a singularity at the left-endpoint of the interval of integration. This case, as well as the more general case of an integral with several singularities, can be investigated in much the same way. Therefore, in the sequel we will confine ourselves to considering improper integrals of the above-mentioned form.

In the present section, the following important concept plays a decisive role.

Definition We say that the improper integral (1) *converges uniformly* on Y (or with respect to $y \in Y$) if

$$\int_a^t f(x, y)\,dx \underset{t \to b}{\longrightarrow} J(y) \quad \text{uniformly on } Y.$$

Since it is obvious that

$$J(y) - \int_a^t f(x,y)\,dx = \int_t^{\to b} f(x,y)\,dx,$$

the definition can be rewritten in the following form:

$$\sup_{y\in Y} \left| \int_t^{\to b} f(x,y)\,dx \right| \longrightarrow 0 \quad \text{as } t \to b. \tag{2}$$

This condition is certainly fulfilled in the simplest situation where the family of functions $\{x \mapsto f(x,y)\}_{y\in Y}$ has a summable majorant, i.e., where there exists a summable function F on the interval (a,b) such that $|f(x,y)| \leqslant F(x)$ for all $x \in (a,b)$ and $y \in Y$. Indeed, in this case, we have $\sup_{y\in Y} |\int_t^b f(x,y)\,dx| \leqslant \int_t^b F(x)\,dx$, and it remains to observe that $\int_t^b F(x)\,dx \to 0$ as $t \to b$ since the function F is summable on (a,b). However, we are not presently interested in this situation since the existence of a summable majorant implies the conditions investigated in Sect. 7.1. At the same time the uniform convergence may be useful in the case where all functions $x \mapsto f(x,y)$ are summable but do not have a summable majorant (see Exercise 4).

7.4.3 Our first result concerning the behavior of the improper integral $J(y)$ is as follows.

Theorem *Let Y be a subset of a metrizable topological space $\widetilde{Y}$ and $y_0 \in \widetilde{Y}$ be a limit point of Y. Assume that a function $f : X \times Y \mapsto \mathbb{C}$ satisfies the following conditions*:

(a) *for almost all x in (a,b), the limit $f_0(x) = \lim_{y\to y_0} f(x,y)$ exists*;
(b) *the function f_0 is summable on each interval (a,t) $(a < t < b)$ and*

$$\int_a^t f(x,y)\,dx \to \int_a^t f_0(x)\,dx \quad \textit{as } y \to y_0;$$

(c) *the improper integral* (1) *converges uniformly with respect to $y \in Y$.*

Then the improper integral $J_0 = \int_a^{\to b} f_0(x)\,dx$ converges and

$$J(y) = \int_a^{\to b} f(x,y)\,dx \underset{y\to y_0}{\longrightarrow} J_0.$$

The following statement is a direct consequence of the theorem.

Corollary *If $y_0 \in Y$ and conditions* (b) *and* (c) *of the theorem are preserved but condition* (a) *is replaced with the condition*

(a′) *for all x in (a,b) the function $y \mapsto f(x,y)$ is continuous at y_0, then the integral J is continuous at y_0.*

Before passing to the proof of the theorem, we make two remarks.

Remark 1 Condition (b) is fulfilled if there exists a left-admissible function g such that $|f(x, y)| \leqslant g(x)$ almost everywhere on (a, b) for each $y \in Y$ (see Theorem 1 of Sect. 7.1.2).

Remark 2 Since the existence of a limit is a local property, there is no need to assume that the integral in the theorem (and in the corollary) converges uniformly on the entire set Y. It is sufficient to require the uniform convergence only on the intersection $Y \cap U(y_0)$, where $U(y_0)$ is a neighborhood of y_0.

To prove the theorem we simply apply the well-known statement about interchanging the order of integration (see, for example, [Z] v. II, Chap. XVI, Sect. 3.2; [Fi] v. II, No. 436).

Proposition *Let T and Y be subsets of metrizable topological spaces $\widetilde{T}$ and $\widetilde{Y}$, respectively, and let $t_0 \in \widetilde{T}$ and $y_0 \in \widetilde{Y}$ be their limit points. Assume that F is a function defined on the product $T \times Y$ and satisfying the following conditions*:

(I) *for every $t \in T$, the limit $L(t) = \lim_{y \to y_0} F(t, y)$ exists and is finite*;
(II) *for every $y \in Y$, the limit $J(y) = \lim_{t \to t_0} F(t, y)$ exists and is finite.*

If in at least one of these cases the convergence is uniform (on T or Y), then the functions J and L have equal finite limits: $\lim_{t \to t_0} L(t) = \lim_{y \to y_0} J(y)$.

In other words, the conditions of the proposition guarantee that the iterated limits exist, are finite, and coincide,

$$\lim_{t \to t_0} \lim_{y \to y_0} F(t, y) = \lim_{y \to y_0} \lim_{t \to t_0} F(t, y).$$

Proof of the theorem Turning to the proof of the theorem, we assume that $T = (a, b)$, $T_0 = \overline{\mathbb{R}}$, $t_0 = b$ and $F(t, y) = \int_a^t f(x, y)\, dx$. Then the statement in question reduces to the proposition given above. Indeed, it follows from condition (b) that F satisfies assumption I with $L(t) = \int_a^t f_0(x)\, dx$. Assumption II is also fulfilled since the uniform convergence of the integral (i.e., condition (c)) is the uniform convergence of $F(t, y)$ to $J(y)$ with respect to $y \in Y$ as $t \to b$. Therefore, the theorem on the change of order of integration guarantees that the limit $\lim_{t \to b} L(t)$ exists and is finite, i.e. that the improper integral $\int_a^{\to b} f_0(x)\, dx$ converges and coincides with the limit $\lim_{y \to y_0} J(y)$. □

7.4.4 After conditions for the continuity of an improper integral dependent on a parameter have been found, it is natural to obtain a counterpart of Theorem 7.1.5, the Leibniz rule. However, we postpone this topic until the next section since it is convenient to be able to integrate with respect to a parameter. We did not discuss results of this type in Sect. 7.1 since, for summable functions, the problem of integration with respect to a parameter is covered by Fubini's theorem. Now we will discuss its generalization to improper integrals.

Let μ be a complete measure defined on a σ-algebra of subsets of a set Y. We shall show that, under natural additional assumptions, the change in the order of integration in improper integrals is legal (for an alternative condition, see Exercise 5).

Theorem *Let a function f be summable with respect to the measure*[10] $\lambda_1 \times \mu$ *on every set $(a,t)\times Y, a<t<b$, and let $I(x)=\int_Y f(x,y)\,d\mu(y)$. If $\mu(Y)<+\infty$ and the improper integral* (1) *converges uniformly on Y, then the function J is summable on Y, the improper integral $\int_a^{\to b} I(x)\,dx$ converges, and the following equation holds*:

$$\int_Y J(y)\,d\mu(y) = \int_a^{\to b} I(x)\,dx. \tag{3}$$

Proof Let $J_t(y)=\int_a^t f(x,y)\,dx$ for $t\in(a,b)$ and $y\in Y$. By Fubini's theorem, the function J_t is summable on Y. Since the uniform convergence of the integral $J(y)$ is equivalent to relation (2), we see that for a t sufficiently close to b the following inequality holds:

$$\bigl|J(y)-J_t(y)\bigr| = \left|\int_t^{\to b} f(x,y)\,dx\right| \leqslant 1 \quad \text{for all } y\in Y.$$

Since the measure μ is finite, we obtain that the function $J-J_t$ is summable and, consequently, the function J is also summable.

By Fubini's theorem, the function I is summable on the interval (a,t) and

$$\begin{aligned}\int_a^t I(x)\,dx &= \int_a^t\left(\int_Y f(x,y)\,d\mu(y)\right)dx = \int_Y\left(\int_a^t f(x,y)\,dx\right)d\mu(y)\\ &= \int_Y J(y)\,d\mu(y) - \int_Y\left(\int_t^{\to b} f(x,y)\,dx\right)d\mu(y).\end{aligned}$$

Thus,

$$\begin{aligned}\left|\int_a^t I(x)\,dx - \int_Y J(y)\,d\mu(y)\right| &\leqslant \int_Y\left|\int_t^{\to b} f(x,y)\,dx\right| d\mu(y)\\ &\leqslant \mu(Y)\sup_{y\in Y}\left|\int_t^{\to b} f(x,y)\,dx\right| \underset{t\to b}{\longrightarrow} 0.\end{aligned}$$

This means that the improper integral $\int_a^{\to b} I(x)\,dx$ converges and is equal to $\int_Y J(y)\,d\mu(y)$. $\square$

Corollary *If Y is a compact subset of the space $\mathbb{R}^m$, $\mu=\lambda_m$, and the function f is continuous on $[a,b)\times Y$, then Eq.* (3) *is valid under the single condition that the improper integral $J(y)$ converges uniformly on Y.*

---

[10] We recall that by λ_1 we denote the one-dimensional Lebesgue measure.

7.4.5 We are now ready to turn to the proof of the Leibniz rule for differentiating improper integrals with respect to a parameter, which is an important tool for studying such integrals.

Theorem *Let f be a continuous function defined on the set $[a,b)\times\langle c,d\rangle$, and let integral* (1) *converge for each $y\in\langle c,d\rangle$. Assume that:*

(a) *for each $x\in[a,b)$, $y\in\langle c,d\rangle$, the partial derivative $f'_y(x,y)$ exists and is continuous on $[a,b)\times\langle c,d\rangle$;*
(b) *the integral $I(y)=\int_a^{\to b} f'_y(x,y)\,dx$ converges uniformly on Y.*

Then $J\in C^1(\langle c,d\rangle)$ and $J'(y)=I(y)$, i.e.,

$$\left(\int_a^{\to b} f(x,y)\,dx\right)'_y=\int_a^{\to b} f'_y(x,y)\,dx.$$

Proof First of all, we note that the integral I depends continuously on y by the corollary to Theorem 7.4.3. Applying to I the theorem on integrating with respect to a parameter on an arbitrary interval with endpoints $s_0,\ s\in\langle c,d\rangle$, we obtain

$$\int_{s_0}^{s} I(y)\,dy=\int_{s_0}^{s}\left(\int_a^{\to b} f'_y(x,y)\,dx\right)dy=\int_a^{\to b}\left(\int_{s_0}^{s} f'_y(x,y)\,dy\right)dx$$
$$=\int_a^{\to b}\big(f(x,s)-f(x,s_0)\big)\,dx=J(s)-J(s_0).$$

Since the integral on the left-hand side of the equation is differentiable, so is J. Barrow's theorem implies that $J'(s)=I(s)$. □

7.4.6 After basic properties of an integral dependent on a parameter have been established and we have become convinced of the usefulness of the concept of uniform convergence, it is desirable to have convenient and easy-to-use uniform convergence tests. We prove two such tests similar to the Dirichlet and Abel convergence tests for improper integrals (see Sect. 4.6.6), but we first generalize them somewhat, dropping superfluous smoothness requirements, and obtain some estimates. In these statements, the function f is, in general, complex-valued.

Lemma *Let a function f be left admissible on an interval $[a,b)$, $-\infty<a<b\leqslant+\infty$, and let g be a function that tends monotonically to zero as $x\to b$. Assume that the function $t\mapsto\int_a^t f(x)\,dx$ $(a<t<b)$ is bounded. Then the improper integral $\int_a^{\to b} f(x)g(x)\,dx$ converges and the following inequality holds:*

$$\left|\int_a^{\to b} f(x)g(x)\,dx\right|\leqslant|g(a)|\sup_{t\in(a,b)}\left|\int_a^t f(x)\,dx\right|. \tag{4}$$

Proof To a great extent, the proof repeats the proof of the Dirichlet test. Both proofs are based on the integration by parts formula, but this time g is not necessarily smooth, and we use a version of this formula obtained in Corollary 3 of Sect. 5.3.4. The equation proved there implies (without loss of generality, we assume that the function g is increasing) that

$$\int_a^t f(x)g(x)\,dx = F(t)g(t) - \int_{(a,t]} F(x)\,dg(x) \quad (a < t < b),$$

where $F(t) = \int_a^t f(x)\,dx$. Obviously, the measure μ generated by the function g is finite and the bounded function F is summable with respect to μ. Therefore, the integral $\int_{(a,t]} F(x)\,dg(x)$ tends to the finite limit $\int_{(a,b)} F(x)\,dg(x)$ as $t \to b$. Since $F(t)g(t) \underset{t\to b}{\longrightarrow} 0$, we obtain from the above equation that the improper integral $\int_a^{\to b} f(x)g(x)\,dx$ converges and the equation

$$\int_a^{\to b} f(x)g(x)\,dx = -\int_{(a,b)} F(x)\,dg(x)$$

holds. This immediately implies estimate (4):

$$\left|\int_a^{\to b} f(x)g(x)\,dx\right| \leqslant \mu\big((a,b)\big) \sup_{t\in(a,b)} \big|F(t)\big| \leqslant \big|g(a)\big| \sup_{t\in(a,b)} \big|F(t)\big|. \qquad \square$$

If the limit $\lim_{t\to b} F(t)$ exists and is finite, i.e., the improper integral $\int_a^{\to b} f(x)\,dx$ converges, then the condition $g(t) \underset{t\to b}{\longrightarrow} 0$ can be dropped, and we obtain the following version of the lemma.

Corollary *Assume that an integral $\int_a^{\to b} f(x)\,dx$ converges and a function g is monotonic and bounded on $[a,b)$. Then the integral $\int_a^{\to b} f(x)g(x)\,dx$ converges and the following inequality holds*:

$$\left|\int_a^{\to b} f(x)g(x)\,dx\right| \leqslant 5 \sup_{t\in(a,b)} \big|g(t)\big| \sup_{t\in(a,b)} \left|\int_t^{\to b} f(x)\,dx\right|.$$

Proof By the lemma, the integral $\int_a^{\to b} f(x)(g(x)-L)\,dx$, where $L = \lim_{x\to b} g(x)$, converges. Representing the product $f(x)g(x)$ in the form $Lf(x)+f(x)(g(x)-L)$, we obtain the convergence of the integral $\int_a^{\to b} f(x)g(x)\,dx$ as well as the following consequence of inequality (4):

$$\left|\int_a^{\to b} f(x)g(x)\,dx\right| \leqslant |L| \left|\int_a^{\to b} f(x)\,dx\right| + \big|g(a+0)-L\big| \sup_{t\in(a,b)} \left|\int_a^t f(x)\,dx\right|.$$

Since the numbers $|L|$ and $|g(a+0)|$ do not exceed $\sup_{(a,b)} |g|$, we can complete the proof via the obvious estimate

$$\left|\int_a^t f(x)\,dx\right| = \left|\int_a^{\to b} f(x)\,dx - \int_t^{\to b} f(x)\,dx\right| \leqslant 2 \sup_{s\in(a,b)} \left|\int_s^{\to b} f(x)\,dx\right|. \qquad \square$$

Before passing to our main goal in this section, uniform convergence tests for an improper integral dependent on a parameter, we make a convention about terminology. We will have to consider functions $x \mapsto f(x, y)$ (for a fixed $y \in Y$) and $y \mapsto f(x, y)$ (for a fixed $x \in X$). In Sect. 5.3.1, these functions were denoted by f^y and f_x, respectively. In what follows, we say, by abuse of language, that a function f has a certain property (is measurable, continuous, etc.) for a given y if the function f^y does; likewise for the first variable. For example, the statement "for a given y, a function f is monotonic with respect to the first variable" means that f^y is monotonic.

Theorem (Dirichlet uniform convergence test for an improper integral) *Let $-\infty < a < b \leqslant +\infty$, and let Y be an arbitrary set. Let f and g be functions defined on $[a, b) \times Y$ and satisfying the following conditions:*

(1) *for each $y \in Y$, the function f is left admissible as a function of x and the function g is monotonic with respect to x on $[a, b)$;*
(2) *the function $(t, y) \mapsto F(t, y) = \int_a^t f(x, y)\,dx$ is bounded on $(a, b) \times Y$;*
(3) *$g(x, y) \underset{x \to b}{\longrightarrow} 0$ converges uniformly with respect to $y \in Y$.*

Then the improper integral

$$J(y) = \int_a^{\to b} f(x, y) g(x, y)\,dx$$

converges uniformly on Y.

Proof The convergence of $J(y)$ for each y in Y follows from the lemma. As noted in Sect. 7.4.2, the uniform convergence is equivalent to the relation

$$\sup_{y \in Y} \left| \int_s^{\to b} f(x, y) g(x, y)\,dx \right| \underset{s \to b}{\longrightarrow} 0. \tag{2'}$$

By condition (2), there exists a number C such that $|F(t, y)| \leqslant C$ for all t in (a, b) and y in Y. Replacing the interval $[a, b)$ by $[s, b)$ and the integral $\int_a^t f(x)\,dx$ by

$$\int_s^t f(x, y)\,dx = F(t, y) - F(s, y)$$

in inequality (4), we obtain that

$$\left| \int_s^{\to b} f(x, y) g(x, y)\,dx \right| \leqslant 2C \left| g(s, y) \right| \quad (a < s < b).$$

To verify relation (2′), it now remains for us only to use condition (3). □

Theorem (Abel uniform convergence test for an improper integral) *If an improper integral $\int_a^{\to b} f(x, y)\,dy$ converges uniformly on a set Y and a function g is bounded* *on a set* $(a,b)\times Y$ *and monotonic with respect to the first variable for each* $y\in Y$, *then the integral* $\int_a^{\to b} f(x,y)g(x,y)\,dx$ *also converges uniformly on* Y.

For the proof it suffices to refer the inequality proved in the corollary to the lemma.

7.4.7 The following two particular cases are especially convenient.

Corollary 1 *If a function* φ *is defined on an interval* $[a,+\infty)$ *and tends monotonically to zero as* $x\to+\infty$, *then the integral*

$$J(y)=\int_a^{\infty}\varphi(x)e^{ixy}\,dx \tag{5}$$

converges uniformly on every set $\mathbb{R}\setminus(-\delta,\delta)$, *where* $\delta>0$.

This statement follows from the Dirichlet test applied to the functions $f(x,y)=e^{ixy}$ and $g(x,y)=\varphi(x)$. In this case, we obviously have

$$\left|F(t,y)\right|=\left|\int_a^t e^{ixy}dx\right|=\left|\frac{e^{ity}-e^{iay}}{iy}\right|\leqslant\frac{2}{|y|}\leqslant\frac{2}{\delta}.$$

It is useful to note that the monotonicity of $\varphi(x)$ is essential only for large x since the uniform convergence is connected with the behavior of the integrals $\int_a^t\varphi(x)e^{ixy}\,dx$ as $t\to+\infty$. For an arbitrary interval $[a,c]$ the summability of φ is sufficient.

Since the integrand is continuous with respect to y, we obtain by the corollary to Theorem 7.4.3 that $J\in C(\mathbb{R}\setminus\{0\})$. Moreover, the parameter y may take complex values provided that $\mathcal{Im}\,y\geqslant 0$, $y\neq 0$. In this case, the estimate $|F(t,y)|\leqslant\frac{2}{|y|}$ remains valid, and therefore, the same reasoning gives the continuity of the integral J in the entire half-plane $\mathcal{Im}\,y\geqslant 0$ except at the origin. We note also that Theorem 7.1.7 implies that at the interior points (i.e., if $\mathcal{Im}\,y>0$) the function J is holomorphic.

Corollary 2 *If an improper integral* $I=\int_0^\infty f(x)\,dx$ *converges and a function* Φ *is monotonic and bounded on* $[0,+\infty)$, *then the integral*

$$J(y)=\int_0^{\infty}\Phi(yx)f(x)\,dx$$

converges uniformly with respect to $y>0$ *and* $J(y)\underset{y\to 0}{\longrightarrow}\Phi(+0)I$, *where* $\Phi(+0)=\lim_{y\to 0}\Phi(y)$.

This is a direct consequence of the Abel test. It is frequently used in the calculation of conditionally convergent improper integrals. For example, if the function f is bounded, then, taking $\Phi(x)=e^{-x}$, we can represent the required integral I as the limit of the absolutely convergent integral $J(y)$ as $y \to +0$, which is usually easier to calculate than the given integral I. We have already encountered this situation in Example 2 of Sect. 7.1.6. There we first used differentiation with respect to a parameter to calculate the integral $J(y) = \int_0^\infty \frac{\sin x}{x} e^{-xy}\, dx$ for $y > 0$, and then proved its continuity at $y = 0$ "by hand". Now this passage to the limit can be justified by Corollary 2.

7.4.8 We will apply the results obtained to calculate important integrals of functions whose primitives cannot be expressed in terms of elementary functions.

Example 1 As proved in Example 1 of Sect. 7.1.7, for $a > 0$ and $\mathcal{R}e(z) > 0$, the integral $\mathcal{L}(z) = \int_0^\infty x^{a-1} e^{-zx}\, dx$ is equal to $\Gamma(a)/z^a$, where z^a is the branch of the power function such that $z^a = 1$ at $z = 1$. At the same time, if $0 < a < 1$, then the integral $\mathcal{L}(z)$ also converges for purely imaginary $z \neq 0$, and so, for such a, the function $\mathcal{L}$ is defined on the entire half-plane $\mathcal{R}e\, z \geqslant 0$ except at zero. The uniform convergence on the set $|z| \geqslant \delta > 0$, $\mathcal{R}e\, z \geqslant 0$ follows easily from the Dirichlet test. Therefore, the function $\mathcal{L}(z)$ is continuous for $\mathcal{R}e\, z \geqslant 0$, $z \neq 0$, and the equation $\mathcal{L}(z) = \Gamma(a)/z^a$ remains valid also for purely imaginary z. In particular, for $z = i$, we obtain

$$\int_0^\infty x^{a-1} e^{-ix}\, dx = \frac{\Gamma(a)}{i^a} = \Gamma(a) e^{-i\frac{\pi a}{2}} \quad (0 < a < 1).$$

Separating the real and imaginary parts, we can represent this equation in the form

$$\text{(C)} \qquad \int_0^\infty \frac{\cos x}{x^{1-a}}\, dx = \Gamma(a) \cos\frac{\pi a}{2},$$

$$\text{(S)} \qquad \int_0^\infty \frac{\sin x}{x^{1-a}}\, dx = \Gamma(a) \sin\frac{\pi a}{2}.$$

Equation (S) is also valid for $-1 < a < 0$ (it is sufficient to integrate by parts equation (C)). Moreover, passing to the limit in (S) as $a \to 0$, we obtain the value of the required important integral $\int_0^\infty \frac{\sin x}{x}\, dx = \frac{\pi}{2}$ once again (see Exercise 2).

We also mention the special case corresponding to the value $a = \frac{1}{2}$,

$$\int_0^\infty \frac{1}{\sqrt{x}} e^{-ix}\, dx = \Gamma\left(\frac{1}{2}\right) e^{-i\frac{\pi}{4}} = (1 - i)\sqrt{\frac{\pi}{2}}.$$

By the substitution $x = t^2$, the integral on the left-hand side of this equation can be reduced to the Fresnel integral (see Sect. 4.6.4) $\int_0^\infty \frac{1}{\sqrt{x}} e^{-ix}\, dx = 2\int_0^\infty e^{-it^2}\, dt$. Thus,

$$\int_0^\infty e^{-it^2}\, dt = \frac{1-i}{2}\sqrt{\frac{\pi}{2}}, \quad \text{and so} \quad \int_0^\infty \cos t^2\, dt = \int_0^\infty \sin t^2\, dt = \frac{1}{2}\sqrt{\frac{\pi}{2}}.$$

Example 2 Here we compute certain integrals that were first found by Laplace:

$$C(y) = \int_0^\infty \frac{\cos yx}{1+x^2}\,dx \quad \text{and} \quad S(y) = \int_0^\infty \frac{x \sin yx}{1+x^2}\,dx \quad (y \in \mathbb{R}).$$

The function C is continuous and bounded on $\mathbb{R}$ since the integrand has the summable majorant $\frac{1}{1+x^2}$.

The integrals $C(y)$ and $S(y)$ are closely connected with each other: using the Leibniz rule, we obtain that $C'(y) = -S(y)$ for $y > 0$. To justify the applicability of the Leibniz rule we have only to refer (see Theorem 7.4.5) to uniform convergence of the integral $S(y)$ near an arbitrary point $y_0 > 0$, which follows immediately from the uniform convergence of the integrals of the form (5) for $\varphi(x) = x/(1+x^2)$.

The Leibniz rule cannot be applied to the integral $S(y)$ directly since the improper integral obtained by formal differentiation with respect to the parameter diverges. Here the following artificial device will be helpful: before the differentiation is performed, we extract from $S(y)$ a "slowly converging" part, the integral $\int_0^\infty \frac{\sin yx}{x}\,dx$, which is known (see Example 2 of Sect. 7.1.6),

$$S(y) = \int_0^\infty \left(\frac{x}{1+x^2} - \frac{1}{x}\right) \sin yx\,dx + \int_0^\infty \frac{\sin yx}{x}\,dx = -\int_0^\infty \frac{\sin yx}{x(1+x^2)}\,dx + \frac{\pi}{2}.$$

Now the Leibniz rule can obviously be applied to the arising integral, and we obtain the relation $S'(y) = -C(y)$. Consequently, $C''(y) = C(y)$. It is known from the theory of ordinary differential equations that the general solution of this equation has the form $C(y) = Ae^y + Be^{-y}$ $(A, B \in \mathbb{R})$. Since the function C is bounded, the coefficient A is zero, i.e., $C(y) = Be^{-y}$ for $y > 0$. To find the coefficient B, we use the continuity of C at zero,

$$B = \lim_{y\to 0} C(y) = C(0) = \int_0^\infty \frac{dx}{1+x^2} = \frac{\pi}{2}.$$

Thus, $C(y) = \frac{\pi}{2}e^{-y}$ and, consequently, $S(y) = \frac{\pi}{2}e^{-y}$ for $y > 0$. Taking into account the fact that the former function is even and the latter is odd, we obtain

$$C(y) = \frac{\pi}{2}e^{-|y|} \quad \text{and} \quad S(y) = \frac{\pi}{2}e^{-|y|}\operatorname{sign} y \quad \text{for } y \in \mathbb{R}.$$

We remark that for the calculation of the integral $C(y) = \int_0^\infty \frac{\cos xy}{1+x^2}\,dx$, where the integrand is summable, it was convenient to go outside the class of summable functions and use the theory developed for improper integrals (see also Exercise 3).

7.4.9 In conclusion, we discuss the asymptotic of integrals similar to those considered in Sect. 7.3 in the course of deriving the Laplace formula. We mean the integrals of the form

$$I(x) = \int_{\mathbb{R}^m} f(t)e^{ix\varphi(t)}\,dt \quad (x \in \mathbb{R}), \tag{6}$$

which play an important role in the stationary phase method used in the study of wave processes. From physical considerations, which we will not dwell on here, it is natural to call the function f the amplitude and the function φ the phase function. Imposing certain restrictions on these functions, we find the rate of change of the integral $I(x)$ as $x \to \pm\infty$. It should be noted that the reason for this decrease is fundamentally different from that which determined the asymptotic in Sect. 7.3. While there the smallness of the integral was a consequence of the smallness of the integrand, here the integral $I(x)$ is small for large x because the real and imaginary parts of the exponential factor frequently change their signs (we will return to this phenomenon in Sect. 9.2.5 in the course of the proof of the Riemann–Lebesgue theorem).

So that technical details will not befog the main idea, we do not seek for maximum generality, instead confining ourselves to the case of infinitely smooth functions f and φ, assuming everywhere that the phase function is real-valued and the amplitude has a compact support. The latter condition guarantees, in particular, the summability of the integrand.

Our first result concerns the case where φ has no critical points on the support of f.

Theorem *If $f, \varphi \in C^\infty(\mathbb{R}^m)$ and* $\operatorname{grad}\varphi \neq 0$ *on* $\operatorname{supp} f$, *then integral* (6) *decreases "overpowerly" as $x \to \infty$, i.e., for each $n \in \mathbb{N}$*

$$I(x) = O\left(\frac{1}{x^n}\right) \quad \text{as } x \to \pm\infty.$$

Proof Using the partition of unity subordinate to the covering of the support $K = \operatorname{supp} f$ by the sets $\{t \in \mathbb{R}^m \mid \frac{\partial\varphi}{\partial t_j}(t) \neq 0\}$ (see Sect. 8.1.8), we may assume that $\frac{\partial\varphi}{\partial t_1}(t) \neq 0$ on K. Then the Jacobian J_Φ of the map

$$t = (t_1, t_2, \dots, t_m) \mapsto \Phi(t) = \big(\varphi(t), t_2, \dots, t_m\big)$$

is separated from zero on K, and, by the theorem on local invertibility, Φ is a diffeomorphism in a neighborhood of each point of K. Again using the partition of unity, if required, we may assume that Φ is a diffeomorphism (of class C^∞) in a neighborhood G of the support of f. Using the change of variable $u = \Phi(t)$, we obtain

$$I(x) = \int_G \frac{f(t)}{|J_\Phi(t)|} e^{ix\varphi(t)} \big|J_\Phi(t)\big|\, dt = \int_{\Phi(G)} \frac{f(\Phi^{-1}(u))}{|J_\Phi(\Phi^{-1}(u))|} e^{ixu_1}\, du$$

$$= \int_{\mathbb{R}^m} g(u) e^{ixu_1} du,$$

where $g = \frac{f(\Phi^{-1})}{|J_\Phi(\Phi^{-1})|} \in C_0^\infty$. Integrating by parts the right-hand side of this equation n times with respect to the first coordinate, we see that

$$\left|I(x)\right| = \left|\frac{1}{(ix)^n}\int_{\mathbb{R}^m}\frac{\partial^n g}{\partial u_1^n}(u)e^{ixu_1}\,du\right| \leqslant \frac{1}{|x|^n}\int_{\mathbb{R}^m}\left|\frac{\partial^n g}{\partial u_1^n}(u)\right|du. \qquad \square$$

7.4.10 We begin the investigation of the integral $I(x)$ in the case where the function φ has critical points lying in $\operatorname{supp} f$ with the most important specific case where φ is a non-degenerate quadratic form. The general case can be reduced to this one provided that the Hesse matrix of the phase function is invertible at the critical points (see Sect. 7.4.11 below).

Thus, let

$$I(x) = \int_{\mathbb{R}^m} f(t)e^{ixQ(t)}\,dt,$$

where Q is a non-degenerate real quadratic form and $f \in C_0^\infty(\mathbb{R}^m)$. Since the form Q can be diagonalized by an orthogonal transformation, from now on we may assume that

$$Q(t) = \sum_{j=1}^{m} a_j t_j^2 \quad \bigl(t = (t_1, \dots, t_m) \in \mathbb{R}^m\bigr). \tag{7}$$

It will be convenient to use a special case of this formula in which $|a_1| = \dots = |a_m| = 1$. Up to renumbering the coordinates this means that

$$Q(t) = \sum_{j=1}^{p} t_j^2 - \sum_{j=1}^{q} t_{p+j}^2, \tag{7'}$$

where $p + q = m$ (if $p = 0$, then the first sum in Eq. (7′) must be replaced by zero, and if $q = 0$, then the second sum must be replaced by zero).

We begin with an estimate of integral $I(1)$ in this special case.

Lemma *If a quadratic form Q has the form (7′) and a function f belonging to the class $C_0^\infty(\mathbb{R}^m)$ is such that the inequality $|\frac{\partial^\alpha f}{\partial t^\alpha}| \leqslant M$ is valid everywhere for $0 \leqslant |\alpha| \leqslant 2m$, then $|I(1)| \leqslant 8^m M$.*

Here $\alpha = (\alpha_1, \dots, \alpha_m) \in \mathbb{Z}_+^m$ is a multi-index and $|\alpha| = \alpha_1 + \dots + \alpha_m$.

Proof We use induction on the number of variables. For $m = 1$, we estimate the integrals of $f(t)e^{it^2}$ over each of the intervals $[1, +\infty)$, $(-\infty, -1]$ and $[-1, 1]$ separately. Integrating by parts two times, we obtain

$$\int_1^\infty f(t)e^{it^2}\,dt = \frac{1}{2i}\int_1^\infty \frac{f(t)}{t}\,d\big(e^{it^2}\big)$$

$$= -\frac{f(1)}{2i}e^i - \frac{1}{(2i)^2}\int_1^\infty \frac{1}{t}\left(\frac{f(t)}{t}\right)' d\big(e^{it^2}\big)$$

$$= -\frac{f(1)}{2i}e^i - \frac{f'(1)-f(1)}{4}e^i$$

$$- \int_1^\infty \frac{t^2 f''(t) - 3tf'(t) + 3f(t)}{4t^4}e^{it^2}\,dt.$$

Therefore, the integral over the interval $[1,+\infty)$ does not exceed $\frac{M}{2} + \frac{M+M}{4} + \frac{7M}{4}\int_1^\infty \frac{dt}{t^2} \leqslant 3M$. The same estimate is also valid for the integral $\int_{-\infty}^{-1} f(t)e^{it^2}\,dt$. Consequently,

$$\left|\int_{-\infty}^{\infty} f(t)e^{it^2}\,dt\right| \leqslant \left|\int_{-\infty}^{-1}\cdots\right| + \left|\int_{-1}^{1}\cdots\right| + \left|\int_{1}^{\infty}\cdots\right| \leqslant 3M + 2M + 3M = 8M.$$

Obviously, this estimate is also valid for the integral $\int_{-\infty}^{\infty} f(t)e^{-it^2}\,dt$.

Now assume that the assertion of the lemma is valid for the functions of $m-1$ variables. For brevity we put $u = (t_1,\dots,t_{m-1})$ and represent $Q(t)$ in the form $Q(t) = \widetilde{Q}(u) \pm t_m^2$. Then

$$I(1) = \int_{-\infty}^{\infty} h(t_m)e^{\pm it_m^2}\,dt_m, \quad \text{where } h(t_m) = \int_{\mathbb{R}^{m-1}} f(u,t_m)e^{i\widetilde{Q}(u)}\,du.$$

Since, by the induction assumption applied to the functions f, $\frac{\partial f}{\partial t_m}$ and $\frac{\partial^2 f}{\partial t_m^2}$, the functions $|h|$, $|h'|$, and $|h''|$ do not exceed $8^{m-1}M$, to complete the proof it remains to use the fact that the statement is valid for $m = 1$. □

Theorem *Let Q be a non-degenerate quadratic form, $f \in C_0^\infty(\mathbb{R}^m)$, and $I(x) = \int_{\mathbb{R}^m} f(t)e^{ixQ(t)}\,dt$. Then*

$$I(x) = \frac{f(0)}{\sqrt{|\det(A)|}}\left(\frac{\pi}{x}\right)^{\frac{m}{2}} e^{i\frac{\pi}{4}S} + O\left(\frac{1}{x^{\frac{m}{2}+1}}\right), \quad \text{as } x \to +\infty$$

where A is the matrix of Q and S is its signature (the difference between the number of positive and negative eigenvalues of A).

Proof First we consider the case where $f(0) = 0$. We will assume that $x > 1$. By Hadamard's lemma (see Sect. 13.7.8) we have $f(t) = t_1g_1(t) + \cdots + t_mg_m(t)$, where $g_1,\dots,g_m \in C_0^\infty(\mathbb{R}^m)$. Therefore, it is sufficient to consider a function f of the form $f(t) = t_kg(t)$ with g belonging to the class $C_0^\infty(\mathbb{R}^m)$. Integrating by parts with respect to the kth coordinate and making the change of variables $t = \frac{u}{\sqrt{x}}$, we obtain

$$I(x) = \pm\frac{1}{2ix}\int_{\mathbb{R}^m}\frac{\partial g}{\partial t_k}(t)e^{ixQ(t)}\,dt = \pm\frac{1}{2ix}\frac{1}{x^{\frac{m}{2}}}\int_{\mathbb{R}^m}\frac{\partial g}{\partial t_k}\left(\frac{u}{\sqrt{x}}\right)e^{iQ(u)}\,du$$

(the choice of signs in this formula is determined by the sign with which the term t_k^2 appears in Q). Since, for $x > 1$, the function $\frac{\partial g}{\partial t_k}(\frac{u}{\sqrt{x}})$ has uniformly bounded derivatives up to order $2m$ inclusive, it follows from the lemma that the last integral is bounded by a constant independent of x. This implies the statement in the case $f(0) = 0$.

Now let $f(0) \neq 0$. Making an orthogonal change of variables, if necessary, we may assume that the matrix A is diagonal and Q has the form (7). Then $\det(A) = a_1 \cdots a_m$. We transform the integral $I(x)$ by dilations in the directions of coordinate axes with coefficients $\sqrt{|a_j|}$:

$$I(x) = \frac{1}{\sqrt{|\det(A)|}}I_1(x), \qquad I_1(x) = \int_{\mathbb{R}^m} f_1(u)e^{ixQ_1(u)}\,du.$$

Here $f_1(0) = f(0)$ and Q_1 is a quadratic form of the form (7′), where p is the number of positive and q is the number of negative eigenvalues of A. Thus, in what follows, we may and will assume without loss of generality that Q has the form (7′).

We consider a function $\theta \in C_0^\infty(\mathbb{R})$ such that $\theta(u) = 1$ in a neighborhood of zero and put

$$K_\pm(x) = \int_{\mathbb{R}} \theta(u)e^{\pm ixu^2}\,du.$$

It is clear that the product $\Theta(t) = \theta(t_1)\cdots\theta(t_m)$ belongs to $C_0^\infty(\mathbb{R}^m)$ and

$$\int_{\mathbb{R}^m} \Theta(t)e^{ixQ(t)}\,dt = K_+^p(x)K_-^q(x). \tag{8}$$

Since the difference $\widetilde{f} = f - f(0)\Theta$ is an infinitely differentiable function with compact support and $\widetilde{f}(0) = 0$, we obtain

$$I(x) - f(0)K_+^p(x)K_-^q(x) = \int_{\mathbb{R}^m} \widetilde{f}(t)e^{ixQ(t)}\,dt = O\left(\frac{1}{x^{\frac{m}{2}+1}}\right). \tag{9}$$

Thus, to complete the proof it remains only for us to find the asymptotic of the integrals $K_\pm(x)$. Obviously,

$$K_\pm(x) = \int_{-\infty}^{\infty} e^{\pm ixu^2}\,du + \int_{-\infty}^{\infty}\bigl(\theta(u) - 1\bigr)e^{\pm ixu^2}\,du.$$

The first of the integrals is reduced to the Fresnel integral calculated in Example 1 of the preceding section,

$$\int_{-\infty}^{\infty} e^{\pm ixu^2}\,du = \frac{1}{\sqrt{x}}\int_{-\infty}^{\infty} e^{\pm it^2}\,dt = \sqrt{\frac{\pi}{x}}e^{\pm i\frac{\pi}{4}}.$$

The second integral converges rapidly to zero. Indeed, since $\theta \in C_0^\infty(\mathbb{R})$ and $\theta(u) = 1$ in a neighborhood of zero, we see that the function $(\theta(u) - 1)/u$ is infinitely smooth on $\mathbb{R}$. Moreover, this function tends to zero at infinity together with all its derivatives. Therefore, representing the integral in question in the form $\frac{\pm 1}{2ix}\int_{-\infty}^{\infty} \frac{\theta(u)-1}{u}\, d(e^{\pm ixu^2})$ and integrating by parts two times, we see that the integral decreases at least as x^{-2}. Thus,

$$K_\pm(x) = \sqrt{\frac{\pi}{x}} e^{\pm i\frac{\pi}{4}} + O(x^{-2}).$$

Taking into account (8), (9) and the equations $p + q = m$ and $p - q = S$, we can complete the proof as follows:

$$\begin{aligned} I(x) &= f(0)\left(\sqrt{\frac{\pi}{x}} e^{i\frac{\pi}{4}} + O\left(\frac{1}{x^2}\right)\right)^p \left(\sqrt{\frac{\pi}{x}} e^{-i\frac{\pi}{4}} + O\left(\frac{1}{x^2}\right)\right)^q + O\left(\frac{1}{x^{\frac{m}{2}+1}}\right) \\ &= f(0)\left(\frac{\pi}{x}\right)^{\frac{m}{2}} e^{i\frac{\pi}{4}(p-q)} + O\left(\frac{1}{x^{\frac{m}{2}+1}}\right) = f(0)\left(\frac{\pi}{x}\right)^{\frac{m}{2}} e^{i\frac{\pi}{4}S} + O\left(\frac{1}{x^{\frac{m}{2}+1}}\right). \quad \square \end{aligned}$$

7.4.11 We generalize the result obtained by replacing the quadratic form Q by a smooth function φ. As follows from Theorem 7.4.9, the contribution to integral (6) that comes from the complement of a neighborhood of the set of critical points is small. Therefore, everything reduces to the calculation of the contribution that comes from (arbitrarily small) neighborhoods of the critical points. We carry out this calculation, assuming that the critical points are non-degenerate. We recall that a critical point p of a function φ is called non-degenerate if its Hesse matrix $H(p) = (\frac{\partial^2\varphi}{\partial t_j\, \partial t_k}(p))_{1\leqslant j,k\leqslant m}$ is invertible.

We denote by $S(p)$ the signature of the second differential of a function φ at a point p.

Theorem *Let $f, \varphi \in C^\infty(\mathbb{R}^m)$, where f has a compact support and φ is a real-valued function having a finite number of critical points $p_1, \ldots, p_n$ in* supp(f) *all of which are non-degenerate. Then as $x \to +\infty$,*

$$\begin{aligned} I(x) &= \int_{\mathbb{R}^m} f(t) e^{ix\varphi(t)}\, dt \\ &= \left(\frac{2\pi}{x}\right)^{\frac{m}{2}} \sum_{j=1}^{n} \frac{f(p_j)}{\sqrt{|\det(H(p_j))|}} e^{ix\varphi(p_j)} e^{i\frac{\pi}{4}S(p_j)} + O\left(\frac{1}{x^{\frac{m}{2}+1}}\right). \end{aligned}$$

Proof We begin with a basic particular case where $n = 1$ and $p_1 = 0$ (if $p_1 \neq 0$ it is necessary to use a shift). We verify that the support of f can be assumed arbitrarily small. To this end, we consider the function $\theta \in C_0^\infty(\mathbb{R}^m)$ that is zero outside the ball $B(\rho)$ and is equal to 1 near the origin. The choice of a radius ρ will be made precise later (it depends only on the properties of φ). Since the product $f \cdot (1 - \theta)$

satisfies the conditions of Theorem 7.4.9, the replacement of f by $f\theta$ leads to a small change of $I(x)$ (the error caused by the change converges to zero overpowerly). This allows us to assume in the sequel that $\operatorname{supp}(f) \subset \overline{B}(\rho)$. By the Morse lemma (see Sect. 13.7.8), for a sufficiently small ρ, there exists a diffeomorphism $\Phi \in C^\infty(B(\rho), \mathbb{R}^m)$ such that $\Phi(0) = 0$, $J_\Phi(0) = 1$, and the following relation is valid for $u = \Phi(t)$:

$$\varphi(t) - \varphi(0) = Q(u) = \sum_{j=1}^{m} a_j u_j^2.$$

The change of variable $u = \Phi(t)$ reduces the integral $I(x)$ to the following form considered in the theorem of the preceding section:

$$I(x) = \int_{\Phi(B(\rho))} \frac{f(\Phi^{-1}(u))}{|J_\Phi(\Phi^{-1}(u))|} e^{ix(\varphi(0)+Q(u))}\, du = e^{ix\varphi(0)} \int_{\mathbb{R}^m} \widetilde{f}(u) e^{ixQ(u)}\, du,$$

where $\widetilde{f} = \frac{f(\Phi^{-1})}{|J_\Phi(\Phi^{-1})|}$ on $\Phi(B(\rho))$ and $\widetilde{f} = 0$ outside this set. Moreover, $\widetilde{f}(0) = f(0)$ since $J_\Phi(0) = 1$. As $\det(H(0)) = 2^m a_1 \cdots a_m$, it only remains to refer to Theorem 7.4.10.

In the general case, we construct disjoint balls with centers at the points $p_1, \dots, p_n$ and the functions $\theta_1, \dots, \theta_n$ with the properties described above. Since φ has no critical points outside the balls, it follows from Theorem 7.4.9 that the replacement of f by $(\theta_1 + \dots + \theta_n) f$ changes the integral $I(x)$ by an amount that decreases overpowerly at infinity. The integral of the function $(\theta_1 + \dots + \theta_n) f$ splits into a sum of integrals of the types considered above. □

EXERCISES

1. Prove that $\int_0^\infty \frac{1-e^{-x}}{x} \cos ax\, dx = \frac{1}{2}\ln(1 + \frac{1}{a^2})$ for $a \in \mathbb{R} \setminus \{0\}$. *Hint.* Check that the integral is equal to the limit of the integral $I(y) = \int_0^\infty e^{-xy} \frac{1-e^{-x}}{x} \cos ax\, dx$ as $y \to +0$. Use the Leibniz rule to calculate $I(y)$.

2. Prove that the integral on the right-hand side of Eq. (S) of Sect. 7.4.8 converges uniformly in a neighborhood of $a = 0$. Passing to the limit as $a \to 0$, find once again the integral $\int_0^\infty \frac{\sin x}{x}\, dx$.

3. By analogy with the calculation of the Laplace integrals, find the integral $J(y) = \int_0^\infty \frac{\cos yx}{1+x^4}\, dx$ $(y \in \mathbb{R})$.

4. Verify that the integral $\int_0^\infty \frac{\sin x}{x} \frac{dx}{\ln^2(2+xy)}$ converges uniformly on $(0, 1)$ but does not satisfy condition (L_{loc}) in any neighborhood of zero.

5. Preserving the notation of Theorem 7.4.4 on integration with respect to a parameter, prove that the requirement that the measure be finite and the integral be uniformly convergent can be replaced by the following condition: there exists a summable function φ on Y such that $|\int_a^t f(x, y)\, dx| \leqslant \varphi(y)$ for $t \in (a, b)$ and $y \in Y$.

6. Use the preceding exercise to justify a reversal in the order of integration in the iterated integrals

$$\int_0^{\infty}\left(\int_0^{\infty} e^{-xy}\sin x\,dy\right)dx \quad\text{and}\quad \int_0^{\infty}\left(\int_0^{\infty} e^{-xy^2}\sin x\,dy\right)dx.$$

Use this to find once again the integrals $\int_0^{\infty}\frac{\sin x}{x}\,dx$ and $\int_0^{\infty}\frac{\sin x}{\sqrt{x}}\,dx$.

7. In the notation of Theorem 7.4.5 on differentiation with respect to a parameter prove the following sharpening of this theorem: if

(a) for some point y_0 in Y the integral $J(y_0)$ converges;
(b) for almost all $x \in (a,b)$ and each $y \in Y$, the partial derivative $f'_y(x,y)$ exists and satisfies the inequality $|f'_y(x,y)| \leqslant g(x)$, where g is a left-admissible function on the interval (a,b);
(c) the integral $I(y) = \int_a^{\to b} f'_y(x,y)\,dx$ converges uniformly on Y, then the improper integral $J(y)$ converges for each $y \in Y$, the function J is differentiable on Y and $J'(y) = I(y)$.

8. Prove that $\int_{-\infty}^{\infty}\frac{\sin x}{1+y^2\sin^2 x}\frac{dx}{x} = \frac{\pi}{\sqrt{1+y^2}}$ for $y \in \mathbb{R}$. *Hint.* Use the following partial fraction expansion of $\frac{1}{\sin x}$:

$$\frac{1}{\sin x} = \frac{1}{x} + 2x\sum_{n=1}^{\infty}\frac{(-1)^n}{x^2-(\pi n)^2}$$

(see Example 2 of Sect. 10.3.5).

9. Use the result of Exercise 8 to find the value of the integral $\int_0^{\infty}\frac{\arctan(y\sin x)}{x}\,dx$ $(y \in \mathbb{R})$.

10. Let f be a function defined almost everywhere on $\mathbb{R}^m$ and summable in every ball $B(R)$. We will say that an improper integral of f converges if the limit

$$\lim_{R\to+\infty}\int_{\|x\|\leqslant R} f(x)\,dx$$

exists and is finite (in which case, it will be denoted, as before, by $\int_{\mathbb{R}^m} f(x)\,dx$). Prove that if the integral converges, then the following multi-dimensional version of Corollary 2 of Sect. 7.4.7 is valid: if Φ is a bounded monotonic function, then the improper integral $J(y) = \int_{\mathbb{R}^m}\Phi(y\|x\|)\,f(x)\,dx$ on $(0,+\infty)$ converges for all $y > 0$ and

$$J(y) \underset{y\to 0}{\longrightarrow} \Phi(+0)\int_{\mathbb{R}^m} f(x)\,dx.$$

7.5 Existence Conditions and Basic Properties of Convolution

We will assume that all functions considered in the present section are, in general, complex-valued and measurable on $\mathbb{R}^m$ (in the wide sense; see Sect. 4.3.3), and a measure will mean Lebesgue measure. As before, let $B(r)$ be the ball of radius r with center at the origin.

7.5.1 We introduce the main concept to which this Section is devoted.

Definition Let f and g be functions measurable on $\mathbb{R}^m$. If

$$\int_{\mathbb{R}^m} \big|f(x-y)g(y)\big|\,dy < +\infty \quad \text{for almost all } x \in \mathbb{R}^m, \tag{1}$$

then the function h defined almost everywhere by the equation

$$h(x) = \int_{\mathbb{R}^m} f(x-y)g(y)\,dy \tag{2}$$

is called the *convolution* of f and g and is denoted by $f * g$.

Condition (1) will be called the *convolution existence condition.* By the change of variable $y \to z = x - y$, we can easily verify that the above condition is equivalent to the condition

$$\int_{\mathbb{R}^m} \big|f(z)g(x-z)\big|\,dz < +\infty \quad \text{for almost all } x \in \mathbb{R}^m,$$

in which case equation (2) implies $h(x) = \int_{\mathbb{R}^m} f(z)g(x-z)\,dz$. Therefore, the convolutions $f * g$ and $g * f$ exist simultaneously and are equal. Thus, convolution is commutative, i.e., $f * g = g * f$ (if at least one of the convolutions exists).

We see that the properties of convolution are similar to those of multiplication of numbers. The convolution is not only commutative, but, obviously, also distributive, i.e., $f * (g_1 + g_2) = f * g_1 + f * g_2$. Without going deeply into this analogy (see Exercise 1), we will use the terminology invoked by this association. In particular, we call the functions f and g the convolution factors.

We also mention that convolution commutes with a shift: if f_h is a shift of f, i.e., $f_h(x) = f(x-h)$, then it follows directly from the definition of convolution that $(f * g)_h = f_h * g = f * g_h$.

Besides pure mathematical questions (among them, as we will see in the next section, are approximation problems) the concept of convolution has its origins in applied problems. For example, convolution arises as a natural mathematical model of a real device that transforms incoming signals. Let us discuss it in more detail. Suppose we have a device ("black box") reacting to signals, which will be regarded as functions of time with compact support. It is natural to assume that the reaction of the device (its "response") to the signal f_h coming with a delay h differs from its response to the signal f only in the corresponding delay in time. In other words, the transformation performed by the device that takes an incoming signal f to its response $\widetilde{f}$ commutes with the shift in time: $\widetilde{f_h} = (\widetilde{f})_h$. Furthermore, we assume that the device takes a linear combination of signals to a linear combination of the responses. The main characteristic of such a device (or, as is often said, the system function) is its reaction to a pulse action δ_α, which can be regarded as a function with unit integral (the "pulse energy") constant on a very small interval $\Delta_\alpha = [0, \alpha)$ and equal zero outside it. In other words, $\delta_\alpha = \frac{1}{\alpha}\chi_\alpha$, where χ_α is the characteristic function of the interval Δ_α. For sufficiently small α, the reaction of the device to the signals δ_α does not practically depend on α. Therefore, replacing δ_α by the "limit function", we can regard an instantaneous unit pulse as the Dirac function[11] δ (the conventionality of this term will be discussed in Sect. 7.6.1), which has the following properties:

$$\delta(t) = 0 \quad \text{if } t \neq 0, \qquad \delta(0) = +\infty, \qquad \int_{-\infty}^{\infty} \delta(t)\,dt = 1.$$

The response E to a signal $\delta \approx \delta_\alpha$ is called the system function of the device. Representing an arbitrary signal f as a linear combination of step functions constant on the intervals $[n\alpha, (n+1)\alpha)$ with required accuracy, we obtain that

$$f(t) \approx \sum_n f(n\alpha)\chi_\alpha(t - n\alpha) \approx \alpha \sum_n f(n\alpha)\delta_\alpha(t - n\alpha).$$

Because the transformation performed by the device is linear and commutes with a shift, we obtain

$$\widetilde{f}(s) \approx \alpha \sum_n f(n\alpha)\,E(s - n\alpha).$$

This sum is nothing but an integral sum for the integral

$$\int_{-\infty}^{\infty} f(t)E(s - t)\,dt.$$

Taking into account the fact that the above approximation becomes arbitrarily accurate for sufficiently small α, we may assume that $\widetilde{f}(s) = \int_{-\infty}^{\infty} f(u)E(s-u)\,du$. Thus, the response of the device to a signal f coincides with the convolution of f and the system function of the device. For that reason convolution is of essential importance in the theoretical foundations of optics and radio engineering.

7.5.2 First, we establish an auxiliary statement.

Lemma *If f and g are measurable functions on $\mathbb{R}^m$ satisfying condition* (1), *then their convolution $f * g$ is also measurable on $\mathbb{R}^m$.*

Proof It is sufficient to proof the theorem for real-valued functions. In this case, we can use Lemma 5.4.3, which implies that the integrand in Eq. (2) is not only measurable for almost all $x \in \mathbb{R}^m$ as a function of y, but also measurable with respect to the "totality" of the variables x and y (i.e., the function $(x, y) \mapsto F(x, y) = f(x - y)g(y)$ is measurable on $\mathbb{R}^m \times \mathbb{R}^m$). Therefore, to prove the lemma, it remains to refer to Corollary 2 of Tonelli's theorem (see Sect. 5.3.1). □

[11] Paul Adrien Maurice Dirac (1902–1984)—British physicist.

Theorem *The convolution of functions f and g summable on $\mathbb{R}^m$ is defined almost everywhere on $\mathbb{R}^m$, is summable, and satisfies the inequality*

$$\int_{\mathbb{R}^m} |(f*g)(x)|\,dx \leqslant \int_{\mathbb{R}^m} |f(x)|\,dx \int_{\mathbb{R}^m} |g(x)|\,dx.$$

Proof We put $H(x) = \int_{\mathbb{R}^m} |f(x-y)g(y)|\,dy$. Since (by Lemma 5.4.3) the function $(x, y) \mapsto f(x-y)g(y)$ is measurable on $\mathbb{R}^m \times \mathbb{R}^m$, it follows from Tonelli's theorem that

$$\int_{\mathbb{R}^m} H(x)\,dx = \int_{\mathbb{R}^m} \left(\int_{\mathbb{R}^m} |f(x-y)|\,dx \right) |g(y)|\,dy.$$

The change of variable $x \mapsto x - y$ shows that the inner integral is equal to $\int_{\mathbb{R}^m} |f(x)|\,dx$ for every y. Therefore,

$$\int_{\mathbb{R}^m} H(x)\,dx = \int_{\mathbb{R}^m} |f(x)|\,dx \int_{\mathbb{R}^m} |g(y)|\,dy < +\infty.$$

Consequently, the function H summable and, therefore, $H(x) < +\infty$ almost everywhere. Thus, condition (1) is fulfilled, and the convolution $(f*g)(x)$ exists. Its measurability is established in the lemma, and the summability follows from the inequality $|(f*g)(x)| \leqslant H(x)$. Moreover,

$$\int_{\mathbb{R}^m} |(f*g)(x)|\,dx \leqslant \int_{\mathbb{R}^m} H(x)\,dx = \int_{\mathbb{R}^m} |f(x)|\,dx \int_{\mathbb{R}^m} |g(y)|\,dy. \qquad \square$$

7.5.3 We supplement Theorem 7.5.2 and obtain alternative conditions sufficient for the existence of a convolution. We consider a wider class of functions than $\mathscr{L}(\mathbb{R}^m)$, the class of measurable functions, which is frequently encountered in function theory and in other branches of mathematics.

Definition A measurable function f in $\mathbb{R}^m$ is called *locally summable* in $\mathbb{R}^m$ if it is summable on every bounded set, i.e., if

$$\int_{B(R)} |f(x)|\,dx < +\infty \quad \text{for every } R > 0.$$

The set of all functions locally summable in $\mathbb{R}^m$ will be denoted by $\mathscr{L}_{\text{loc}}(\mathbb{R}^m)$. Obviously, every locally summable function is almost everywhere finite and $C(\mathbb{R}^m) \subset \mathscr{L}_{\text{loc}}(\mathbb{R}^m)$.

We remind the reader that the closure of the set $\{x \in \mathbb{R}^m \mid f(x) \neq 0\}$ is called the *support* of a function f and is denoted by $\operatorname{supp}(f)$. By $A + B$, where $A, B \subset \mathbb{R}^m$, we denote the set $\{a + b \mid a \in A,\ b \in B\}$.

Theorem *If $f \in \mathscr{L}_{\mathrm{loc}}(\mathbb{R}^m)$ and g is a summable function with compact support, then the convolution $f * g$ exists and*

$$\operatorname{supp}(f * g) \subset \operatorname{supp}(f) + \operatorname{supp}(g). \tag{3}$$

Proof Let $\operatorname{supp}(g) \subset B(r)$. As in Theorem 7.5.2, we put

$$H(x) = \int_{\mathbb{R}^m} \big|f(x-y)g(y)\big|\,dy = \int_{B(r)} \big|f(x-y)g(y)\big|\,dy$$

and prove that $H(x) < +\infty$ almost everywhere. For this, we check that $H \in \mathscr{L}_{\mathrm{loc}}(\mathbb{R}^m)$, i.e., that $\int_{B(R)} H(x)\,dx < +\infty$ for every $R > 0$. Indeed, since $\operatorname{supp}(g) \subset B(r)$, we have

$$\begin{aligned}
\int_{B(R)} H(x)\,dx &= \int_{B(R)} \left(\int_{B(r)} \big|f(x-y)g(y)\big|\,dy \right) dx \\
&= \int_{B(r)} \big|g(y)\big| \left(\int_{B(R)} \big|f(x-y)\big|\,dx \right) dy \\
&\leqslant \int_{B(r)} \big|g(y)\big| \left(\int_{B(r+R)} \big|f(u)\big|\,du \right) dy \\
&= \int_{B(r+R)} \big|f(u)\big|\,du \cdot \int_{B(r)} \big|g(y)\big|\,dy < +\infty.
\end{aligned}$$

The last inequality is valid since f is locally summable. Thus, the function H is finite almost everywhere in the ball $B(R)$, and, consequently, almost everywhere on $\mathbb{R}^m$. Therefore, condition (1) is fulfilled, and the convolution $f * g$ exists.

To prove inclusion (3), we remark that if $f(x-y)g(y) \neq 0$, then $x - y \in \operatorname{supp}(f)$ and $y \in \operatorname{supp}(g)$, and so, $x = (x-y) + y \in \operatorname{supp}(f) + \operatorname{supp}(g)$. Therefore, $f(x-y)g(y) \equiv 0$ in the case where $x \notin \operatorname{supp}(f) + \operatorname{supp}(g)$. Consequently, $f * g = 0$ outside the set $\operatorname{supp}(f) + \operatorname{supp}(g)$, i.e.,

$$\big\{x \in \mathbb{R}^m \mid (f * g)(x) \neq 0\big\} \subset \operatorname{supp}(f) + \operatorname{supp}(g).$$

Since $\operatorname{supp}(g)$ is compact, the set on the right-hand side of this inclusion is closed (we leave it to the reader to prove this independently), which implies that

$$\operatorname{supp}(f * g) = \overline{\big\{x \in \mathbb{R}^m \mid (f * g)(x) \neq 0\big\}} \subset \operatorname{supp}(f) + \operatorname{supp}(g). \qquad \square$$

Corollary *The convolution of two summable functions with compact supports has a compact support.*

7.5.4 We now discuss differential properties of convolution. First, we prove an auxiliary result.

Lemma (Truncation lemma) *Let $f, \widetilde{f} \in \mathscr{L}_{\mathrm{loc}}(\mathbb{R}^m)$, and let a function g be bounded and satisfy the inclusion* $\mathrm{supp}(g) \subset B(r)$. *If f coincides with $\widetilde{f}$ in the ball $B(R+r)$, then the convolutions $f * g$ and $\widetilde{f} * g$ coincide in the ball $B(R)$.*

Proof Let $\|x\| < R$. Then $\|x-y\| < R+r$ for $\|y\| < r$. Therefore,

$$(f * g)(x) = \int_{B(r)} f(x-y)g(y)\,dy = \int_{B(r)} \widetilde{f}(x-y)g(y)\,dy = (\widetilde{f} * g)(x). \qquad \square$$

Theorem *Let $f \in \mathscr{L}_{\mathrm{loc}}(\mathbb{R}^m)$, and let g be a bounded function with compact support. Then*:

(1) *if at least one of the functions f or g is continuous, then the convolution $f * g$ is continuous*;

(2) *if at least one of the functions f or g is continuously differentiable, then the convolution is continuously differentiable and its derivatives can be calculated by the formula* ($k = 1, \dots, m$)

$$\frac{\partial (f * g)}{\partial x_k}(x) = \begin{cases} (f * \frac{\partial g}{\partial x_k})(x) & \text{if } g \in C^1(\mathbb{R}^m); \\ (\frac{\partial f}{\partial x_k} * g)(x) & \text{if } f \in C^1(\mathbb{R}^m). \end{cases} \tag{4}$$

Remark The first assertion of the theorem admits an essential sharpening. As we will see in the sequel (see Corollary 9.3.2), the convolution of a locally summable function and a bounded summable function is continuous without any additional assumptions.

Proof We will assume that $\mathrm{supp}(g) \subset B(r)$.

(1) If g is continuous and f is summable, then the integral $\int_{\mathbb{R}^m} f(y)g(x-y)\,dy$ is continuous with respect to the parameter by Theorem 7.1.3. If f is not summable, then we use the obvious fact that it is sufficient to prove the continuity of the convolution in an arbitrary ball $B(R)$. We can also use the truncation lemma and replace f by a summable function $\widetilde{f}$ that has a compact support and coincides with f on a ball $B(R+r)$. The same method can be applied if f is continuous because, in this case, we may assume that $\widetilde{f}$ is continuous.

(2) Turning to the proof of the smoothness of convolution, we first assume that the function g is smooth. It is obvious that if $x_0 \in \mathbb{R}^m$ and $\|x - x_0\| < 1$, then

$$(f * g)(x) = \int_{\mathbb{R}^m} f(y)g(x-y)\,dy = \int_{B(x_0, r+1)} f(y)g(x-y)\,dy.$$

Applying the Leibniz rule (see Theorem 7.1.5) to the right-hand side of this equation, we immediately obtain the required result,

$$\frac{\partial (f * g)}{\partial x_k}(x) = \int_{B(x_0, r+1)} f(y)\frac{\partial g}{\partial x_k}(x-y)\,dy = \left(f * \frac{\partial g}{\partial x_k}\right)(x).$$

We verify that, in the case in question, the application of the Leibniz rule is legal. For this, we must check that the partial derivative $\frac{\partial}{\partial x_k}(f(y)g(x-y)) = f(y)\frac{\partial g}{\partial x_k}(x-y)$ satisfies condition (L_{loc}) at x_0.

This is indeed the case because

$$\left| f(y)\frac{\partial g}{\partial x_k}(x-y) \right| \leqslant M \left| f(y) \right| \chi_{B(x_0,r+1)}(y) \quad \text{for all } x \in B(x_0, r+1),$$

where $M = \max_x |\frac{\partial g(x)}{\partial x_k}|$.

Now assume that f is continuously differentiable. To prove that the convolution is differentiable on $B(R)$, we should replace f by a function $\widetilde{f}$ that has a compact support and coincides with f on a sufficiently large ball, as we did in the proof of the continuity of convolution, the only difference being that the function $\widetilde{f}$ must now be smooth. For example, we can multiply f by a smooth function that has a compact support and is equal to 1 on a ball $B(R+r)$. Then, interchanging the roles of g and $\widetilde{f}$ and using the formula proved above, we find that

$$\frac{\partial (f*g)}{\partial x_k}(x) = \frac{\partial (g*\widetilde{f})}{\partial x_k}(x) = \left(g * \frac{\partial \widetilde{f}}{\partial x_k} \right)(x) = \left(g * \frac{\partial f}{\partial x_k} \right)(x) = \left(\frac{\partial f}{\partial x_k} * g \right)(x)$$

for $\|x\| < R$. Since R is arbitrary, this proves the theorem. □

Corollary *The convolution of a locally summable function f and a bounded function φ with compact support is infinitely differentiable if at least one of the functions f or φ is infinitely differentiable.*

Proof The assertion should be proved by induction using (4). □

In particular, it follows from the corollary that a linear differential operator with constant coefficients commutes with convolution.

7.5.5 The concept of convolution has different generalizations and modifications. We mention some of them.

In the case where the functions in question are periodic on the real line, the convolution is defined in the same way as above with the only difference that now the integral over $\mathbb{R}$ is replaced by the integral over an interval with length equal to the period (no matter which interval is used). For definiteness, we will assume that the period is 2π. It is clear that the convolution of periodic functions is also periodic. We leave it to the reader to verify independently that an analog of Theorem 7.5.2 is valid for the convolution of periodic functions. One simply repeats the proof of this theorem, changing the domain of integration appropriately.

The above applies in full to functions defined on $\mathbb{R}^m$ and 2π-periodic with respect to each variable. Their convolution is defined by the equation

$$(f*g)(x) = \int_{(-\pi,\pi)^m} f(x-y)g(y)\,dy.$$

One more version of the definition of convolution can be obtained as follows. If a function g is summable and non-negative, then the integral $\int_{\mathbb{R}^m} f(x-y)g(y)\,dy$ can be regarded as an integral with respect to the measure ν having density g with respect to Lebesgue measure, $(f*g)(x) = \int_{\mathbb{R}^m} f(x-y)\,d\nu(y)$. The right-hand side of this equation will be used as the definition of the convolution of a function and a measure. To guarantee the existence of the convolution, we will assume that all measures are finite and all functions are bounded.

Definition Let ν be a finite Borel measure on $\mathbb{R}^m$, and let f be a measurable bounded function on $\mathbb{R}^m$. The convolution $f*\nu$ is defined by the equation

$$(f*\nu)(x) = \int_{\mathbb{R}^m} f(x-y)\,d\nu(y) \quad \bigl(x \in \mathbb{R}^m\bigr).$$

One more version of convolution can be considered if μ is the counting measure defined on the integer lattice $\mathbb{Z}^m$. In this case, instead of functions, we speak of multiple sequences. By analogy with (2), the convolution of such sequences $f = \{f_n\}_{n\in\mathbb{Z}^m}$ and $g = \{g_n\}_{n\in\mathbb{Z}^m}$ is defined by the formula

$$(f*g)_k = \int_{\mathbb{Z}^m} f(k-n)g(n)\,d\mu(n) = \sum_{n\in\mathbb{Z}^m} f_{k-n}\,g_n \quad \bigl(k \in \mathbb{Z}^m\bigr).$$

We invite the reader to state and prove a counterpart of Theorem 7.5.2 for this case.

EXERCISE

1. Prove the associativity of convolution: if $f, g, h \in \mathscr{L}(\mathbb{R}^m)$, the functions $(f*g)*h$ and $f*(g*h)$ coincide almost everywhere.
2. Prove that the convolution of two functions of class C^r one of which has a compact support is a function of class C^{2r} $(r = 0, 1, \ldots)$.
3. Verify that the measure μ defined on the semi-axis $\mathbb{R}_+ = (0, +\infty)$ by the equation $d\mu = \frac{dx}{x}$ is invariant with respect to multiplication, i.e., for every measurable set $E \subset \mathbb{R}_+$ and every $a > 0$, the relation $\mu(E) = \mu(aE)$ is valid, where $aE = \{ax \mid x \in E\}$. This fact makes it possible to define a convolution on the semi-axis as follows:

$$(f*g)(x) = \int_0^\infty f\left(\frac{x}{y}\right) g(y)\frac{dy}{y}.$$

   Verify that all theorems proved in this Sect. 7.5 remain valid for the convolution thus defined (by definition, a function belongs to the class $\mathscr{L}_{\mathrm{loc}}(\mathbb{R}_+)$ if it is summable on every compact set lying in $\mathbb{R}_+$).
4. Prove that the convolution of a Borel measure finite on compact sets and a function of class C^r with compact support is again a function of the same class.

7.6 Approximate Identities

7.6.1 If one of the convolution factors is non-negative and its integral is 1, then the convolution can be regarded as the mean value of the other factor. Indeed, if $g \geqslant 0$ and $\int_{\mathbb{R}^m} g(y)\,dy = 1$, then $\inf_{\mathbb{R}^m} f \leqslant (f * g)(x) \leqslant \sup_{\mathbb{R}^m} f$ for $x \in \mathbb{R}^m$. If the support of g is contained in a ball $B(r)$, then the estimate can be sharpened as follows:

$$\inf_{B(x,r)} f \leqslant (f * g)(x) \leqslant \sup_{B(x,r)} f \quad (x \in \mathbb{R}^m).$$

Therefore, if f is continuous, then the convolution must be close to f for a small r. At the same time, the convolution often has higher degree of smoothness than the function f itself. In particular, as we will prove in Example 1 of Sect. 7.6.2, the convolution of an arbitrary locally summable function and the characteristic function of an arbitrary ball is continuous. Thus, we may hope that convolution can be used to obtain a method of approximating functions by smoother ones.

Since the convolution of a locally summable function and a characteristic function of a ball is continuous, we obtain that there is no locally summable function playing the role of identity for convolution; in other words, there is no locally summable function the convolution with which would not change the other convolution factor (even if this factor is a continuous function with compact support, see Exercise 1). At the same time, a convolution with measure δ_0 generated by a unit point mass concentrated at zero has this property,

$$(f * \delta_0)(x) = \int_{\mathbb{R}^m} f(x-y)\,d\delta_0(y) = f(x) \quad \text{for all } x \text{ in } \mathbb{R}^m.$$

The measure δ_0 certainly does not have a density with respect to Lebesgue measure. However, avoiding integration with respect to the measure δ_0, the famous physicist Paul Dirac actually suggested to assume that such a density nevertheless exists. He introduced a "function" δ (now known as the Dirac delta function) having the following properties:

$$\text{I.} \quad \delta(x) = \begin{cases} 0 & \text{for } x \neq 0, \\ +\infty & \text{for } x = 0, \end{cases} \quad \text{and}$$

$$\text{II.} \quad \int_{\mathbb{R}^m} \delta(x)\,dx = 1.$$

From this he concluded that, for every continuous function f on $\mathbb{R}^m$, the relation $f(x) = \int_{\mathbb{R}^m} f(x-y)\delta(y)\,dy$ is valid, i.e., that δ is an identity for convolution in the class of continuous functions. It is this fact that plays a crucial role. Properties I and II characterizing the Dirac delta function are clearly incompatible. However, if we regard the integral $\int_{\mathbb{R}^m} f(x-y)\delta(y)\,dy$ simply as a new notation for the integral $\int_{\mathbb{R}^m} f(x-y)\,d\delta_0(y)$, then the calculation involving the function δ becomes legal.

As has already been said, the measure δ_0 has no density with respect to Lebesgue measure, and a function satisfying properties I and II does not exist. In this connection there is a problem of approximating δ_0 by measures having densities, i.e., by measures of the form $\omega(x)\,dx$. In a wide range of cases, we will see that, on the one hand, a convolution with a measure of this form causes little change in the function (since the measure is close to δ_0) and, on the other hand, it results in a function smoother than the initial one. This opens possibilities for approximating arbitrary functions by smooth ones. The character of an approximation may be different and requires clarification. In the present and the next section, we obtain results connected mainly with pointwise and uniform approximation. A different approach to this problem will be considered in Chap. 9.

First of all we define a family of functions by which the measure δ_0 is approximated.

Definition Let $T \subset (0, +\infty)$, and let t_0 be a limit point of T $(0 \leqslant t_0 \leqslant +\infty)$. A family of functions $\{\omega_t\}_{t\in T}$ defined on $\mathbb{R}^m$ is called an *approximate identity* in $\mathbb{R}^m$ (as $t \to t_0$) if

(a) $\omega_t \geqslant 0$,

(b) $\displaystyle\int_{\mathbb{R}^m} \omega_t(x)\,dx = 1$,

(c) $\displaystyle\int_{\|x\|>\delta} \omega_t(x)\,dx \underset{t\to t_0}{\longrightarrow} 0$ for every $\delta > 0$.

Remarks

(1) Taking into account equation (b), we can restate condition (c) in the following form:

$$\int_{\|x\|<\delta} \omega_t(x)\,dx \underset{t\to t_0}{\longrightarrow} 1 \quad \text{for every } \delta > 0.$$

Thus, the main contribution to the integral $\int_{\mathbb{R}^m} \omega_t(x)\,dx$ comes from the integral over an arbitrarily small neighborhood of zero. This property of an approximate identity is sometimes called the *localization property*. It says that, for t close to t_0, the graph of ω_t can schematically be displayed as a "narrow and tall hump". Such functions are sometimes called δ-images.

(2) Sometimes the positivity condition for ω_t is lifted and condition a) is replaced by the less restrictive assumption

$$(\text{a}') \quad \int_{\mathbb{R}^m} \bigl|\omega_t(x)\bigr|\,dx \leqslant C \quad \text{for some } C > 0 \text{ and all } t \in T$$

(and the function ω_t in condition (c) is replaced by $|\omega_t|$).

Because of equation (b), condition (a′) is automatically fulfilled for non-negative functions. Many of the results obtained below also remain valid in a more general setting, but we will not dwell on this.

7.6.2 We consider some examples of approximate identities. In all cases the families under consideration are approximate identities as $t \to +0$, $T = (0, +\infty)$, and the convolution factors are assumed to be locally summable on $\mathbb{R}^m$.

Example 1 (Steklov[12] averages) Let $\omega_t = \frac{1}{v(t)}\chi_{B(t)}$, where $v(t)$ is the volume (Lebesgue measure) of a ball $B(t)$ in $\mathbb{R}^m$. Obviously, this family is an approximate identity. The value of the convolution $f * \omega_t$ at a point x is the average of f over the ball $B(x,t)$:

$$(f * \omega_t)(x) = \int_{\mathbb{R}^m} f(y)\frac{1}{v(t)}\chi_{B(t)}(x-y)\,dy = \frac{1}{v(t)}\int_{B(x,t)} f(y)\,dy.$$

This average has systematically been used by Steklov, and the convolutions $f_t = f * \omega_t$ are called *Steklov averages* of f. They are continuous if the function is locally summable (in the sequel, we will establish a more general result, see Sect. 9.3.2). Indeed, assume that $\|x - x_0\| < 1$. For such x, the symmetric difference e_x of the balls $B(x_0,t)$ and $B(x,t)$ lies in the ball $B(x_0, 1+t)$. Since the function f is summable on $B(x_0, 1+t)$, and $\lambda(e_x) \to 0$ as $x \to x_0$, it follows from the absolute continuity of the integral that

$$\left|f_t(x) - f_t(x_0)\right| \leqslant \frac{1}{v(t)}\int_{e_x} \left|f(y)\right| dy \underset{x\to x_0}{\longrightarrow} 0.$$

Example 2 The example considered above fits into a general scheme allowing one to construct different approximate identities. The scheme is as follows.

Let ψ be a non-negative summable function on $\mathbb{R}^m$, and let

$$C = \int_{\mathbb{R}^m} \psi(x)\,dx > 0.$$

We put

$$\omega_t(x) = \frac{1}{Ct^m}\psi\left(\frac{x}{t}\right) \quad (x \in \mathbb{R}^m).$$

The family $\{\omega_t\}_{t>0}$ is an approximate identity as $t \to 0$. Condition (a) in the definition of an approximate identity is obviously fulfilled, and the fact that condition (b) is valid can be verified by the change of variable $y = x/t$:

$$\int_{\mathbb{R}^m} \omega_t(x)\,dx = \frac{1}{C}\int_{\mathbb{R}^m} \psi(y)\,dy = 1.$$

At the same time, condition (c) is also fulfilled since we have

$$\int_{\|x\|<\delta} \omega_t(x)\,dx = \frac{1}{C}\int_{\|y\|<\delta/t} \psi(y)\,dy \underset{t\to 0}{\longrightarrow} 1$$

for every $\delta > 0$.

[12] Vladimir Andreevich Steklov (1863–1926)—Russian mathematician.

In Example 1, the characteristic function of the ball $B(1)$ plays the role of ψ.

It is especially convenient to use approximate identities obtained by the method described above in the case where ψ is a function of class C^∞ and its support lies in the unit ball. Such approximate identities were first systematically used by Sobolev, and we call them *Sobolev approximate identities.*

7.6.3 Now we state the main result concerning approximate identities. We will return to this question in Sect. 9.3.

Theorem *Let f be a bounded measurable function on $\mathbb{R}^m$, and let $\{\omega_t\}_{t\in T}$ be an approximate identity as $t \to t_0$, $f_t = f * \omega_t$. Then:*

(1) *if the limit $L = \lim_{x\to x_0} f(x)$ exists and is finite for a point x_0 in $\mathbb{R}^m$, then $f_t(x_0) \underset{t\to t_0}{\longrightarrow} L$;*
(2) *if $f \in C(\mathbb{R}^m)$, then $f_t \underset{t\to t_0}{\rightrightarrows} f$ on every bounded set.*

Proof By definition

$$f_t(x_0) = \int_{\mathbb{R}^m} f(x_0 - y)\,\omega_t(y)\,dy.$$

Multiplying the equation

$$1 = \int_{\mathbb{R}^m} \omega_t(y)\,dy$$

by L and subtracting the equation obtained from the preceding one, we obtain

$$f_t(x_0) - L = \int_{\mathbb{R}^m} \big(f(x_0 - y) - L\big)\omega_t(y)\,dy.$$

We prove that the right-hand side of this relation tends to zero as $t \to t_0$. By assumption, we have $|f| \leqslant C$ everywhere. Therefore, the inequality

$$\begin{aligned}\big|f_t(x_0) - L\big| &\leqslant \int_{\mathbb{R}^m} \big|f(x_0 - y) - L\big|\omega_t(y)\,dy = \int_{\|y\|<\delta} \cdots + \int_{\|y\|>\delta} \cdots \\ &\leqslant \sup_{0<\|z-x_0\|<\delta} \big|f(z) - L\big| \int_{\|y\|<\delta} \omega_t(y)\,dy + 2C \int_{\|y\|>\delta} \omega_t(y)\,dy \quad (1)\end{aligned}$$

holds for every $\delta > 0$. Since $\int_{\|y\|<\delta} \omega_t(y)\,dy \leqslant \int_{\mathbb{R}^m} \omega_t(y)\,dy = 1$, it follows that

$$\big|f_t(x_0) - L\big| \leqslant \sup_{0<\|z-x_0\|<\delta} \big|f(z) - L\big| + 2C \int_{\|y\|>\delta} \omega_t(y)\,dy.$$

Now, we can make the first summand on the right-hand side of the inequality arbitrarily small by an appropriate choice of δ, and then, fixing δ, we can make the second summand small by condition (c).

The proof of the second assertion of the theorem will repeat the proof of the first one if we replace x_0 by x, L by $f(x)$, and take into account that, for every bounded set E, we can choose the same δ for all $x \in E$ since f is uniformly continuous on every bounded set. □

Remark It can be seen from the proof of the theorem that if f is uniformly continuous on the entire space, then $f_t \underset{t \to t_0}{\rightrightarrows} f$ on $\mathbb{R}^m$.

Corollary *If g is a bounded function continuous at zero, then*

$$\int_{\mathbb{R}^m} g(y)\,\omega_t(y)\,dy \underset{t \to t_0}{\longrightarrow} g(0).$$

Proof This is a particular case of the statement of the theorem where $x_0 = 0$, $f(x) = g(-x)$ and $L = g(0)$. □

The corollary reinforces our motivation to introduce approximate identities. It follows from the corollary that the measures ν_t with densities ω_t converge to the measure δ_0 generated by the unit load concentrated at zero in the sense that, for every bounded continuous function g, we have

$$\int_{\mathbb{R}^m} g(x)\,d\nu_t(x) \underset{t \to t_0}{\longrightarrow} g(0) = \int_{\mathbb{R}^m} g(x)\,d\delta_0(x).$$

If we also assume that ω_t are functions with compact supports contracting to zero, then this statement is valid for every (possibly unbounded) continuous function (see Exercise 2).

7.6.4 We consider an important application of approximate identities and prove Weierstrass' famous approximation theorem stating that every continuous function on a closed bounded interval can be approximated by a polynomial as closely as desired. The method of proof we use here is that we first replace the given function, with small error, by the convolution with some "nice" function, and then construct a polynomial approximation for the convolution. This method works equally well for functions of one variable and for functions of several variables.[13]

Following Weierstrass, we will consider the convolutions of a given function and functions of the form

$$W_t(x) = \frac{1}{t^m} e^{-\pi \frac{\|x\|^2}{t^2}} \quad \bigl(x \in \mathbb{R}^m, t > 0\bigr).$$

This family is an approximate identity as $t \to +0$. Condition (a) of the definition of an approximate identity is obviously fulfilled; using the value of the multidimensional Euler integral found in Sect. 5.4.2, we can easily verify condition (b). We leave the verification of the localization property to the reader.

[13]A brief and clear account of the idea of the method (as well as a clever parody of a formal and pseudo-scientific style of exposition) can be found in the remarkable book [Li], Sect. 11.

Theorem 1 (Weierstrass approximation theorem) *Let $f \in C(\mathbb{R}^m)$. Then for any $R > 0$ and $\varepsilon > 0$ there exists a polynomial P in m variables such that*

$$\left|f(x) - P(x)\right| < \varepsilon \quad \textit{for all } x \textit{ in } \overline{B}(R).$$

Proof First we assume that the support of f is a compact set lying in the ball $\overline{B} \equiv \overline{B}(R)$ (otherwise we can increase the radius R). We put $f_t = f * W_t$. As pointed out in the remark to Theorem 7.6.3, $f_t \rightrightarrows f$ as $t \to 0$. We fix a t such that

$$\left|f(x) - f_t(x)\right| < \varepsilon \quad \text{for each } x \text{ in } \mathbb{R}^m. \tag{2}$$

Now we show that every function f_t can be uniformly approximated by a polynomial in the ball $\overline{B}$. Since f is zero outside $\overline{B}$, we obtain

$$f_t(x) = \int_{B(R)} f(y) W_t(x-y)\,dy.$$

We assume that $x \in \overline{B}$, which implies that $x - y \in B(2R)$ in the last integral.

The next idea is to find a good polynomial approximation for the function W_t in the ball $B(2R)$ and use the fact that the convolution of a function with compact support and a polynomial is again a polynomial. To verify the last assertion, we consider an arbitrary polynomial Q. It is clear that $Q(x-y)$ is also a polynomial in the coordinates of x with coefficients dependent on y. After multiplying by the function $f(y)$ with compact support, we obtain that the coefficients become summable. Integrating them, we obtain certain numbers, and, therefore, the convolution is a polynomial.

Now we turn our attention to approximating the function W_t by a polynomial. By Taylor's formula (with the Lagrange form of the remainder) we obtain $e^{-u} = T_{n-1}(u) + r_n(u)$, where T_{n-1} is a polynomial of degree $n-1$, $r_n(u) = \frac{1}{n!}e^{-\theta u}(-u)^n$, $0 < \theta < 1$. It is clear that $|r_n(u)| \leqslant u^n/n!$ for $u \geqslant 0$. By the definition of W_t, we obtain

$$W_t(x) = \frac{1}{t^m} T_{n-1}\left(\pi \frac{\|x\|^2}{t^2}\right) + \frac{1}{t^m} r_n\left(\pi \frac{\|x\|^2}{t^2}\right) = P_n(x) + \rho_n(x),$$

where P_n is a polynomial (as the composition of T_{n-1} and the polynomial $\pi \frac{\|x\|^2}{t^2}$) and ρ_n satisfies the estimate

$$\left|\rho_n(x)\right| = \left|\frac{1}{t^m} r_n\left(\pi \frac{\|x\|^2}{t^2}\right)\right| \leqslant \frac{1}{t^m n!}\left(\pi \frac{\|x\|^2}{t^2}\right)^n \leqslant \frac{1}{t^m n!}\left(\frac{4\pi R^2}{t^2}\right)^n \tag{3}$$

for $\|x\| \leqslant 2R$. Since $f(x) = 0$ outside the ball $\overline{B}$, we see that the convolution $f * \rho_n$ satisfies the inequality

$$\left|(f * \rho_n)(x)\right| = \left|\int_{B(R)} f(y)\rho_n(x-y)\,dy\right| \leqslant M \int_{B(R)} \left|\rho_n(x-y)\right| dy,$$

where $M = \max_x |f(x)|$. Since the inequality $\|x - y\| \leqslant 2R$ is valid for $\|x\| \leqslant R$, we can use (3) to estimate the integral on the right-hand side of the above inequality. We obtain

$$\left|(f * \rho_n)(x)\right| \leqslant Mv(R)\frac{1}{t^m n!}\left(\frac{4\pi R^2}{t^2}\right)^n,$$

where $v(R)$ is the m-dimensional volume of the ball $B(R)$. Now we fix an n so that the right-hand side of the last inequality is less than ε. Then we obviously obtain

$$\left|f_t(x) - (f * P_n)(x)\right| = \left|(f * \rho_n)(x)\right| < \varepsilon \tag{4}$$

for $\|x\| \leqslant R$. This inequality together with (2) shows that $|f(x) - (f * P_n)(x)| < 2\varepsilon$ for $x \in \overline{B}$. This completes the proof of the theorem for a function with compact support because, as noted above, the convolution $f * P_n$ is a polynomial.

In the general case, it is sufficient to replace f by a continuous function f_1 that has a compact support and coincides with f in the ball $\overline{B}$. Constructing a polynomial that approximates f_1 in $\overline{B}$, we also find an approximation for f. □

Corollary 1 *Let f be a continuous function on a compact set $K \subset \mathbb{R}^m$. Then for every $\varepsilon > 0$ there exists a polynomial P such that $|f(x) - P(x)| < \varepsilon$ for all $x \in K$.*

Proof By the Tietze–Urysohn theorem (see Sect. 13.2.2), every continuous function on a closed subset of the space $\mathbb{R}^m$ can be extended to a continuous function defined on the entire space. Therefore, it is sufficient to apply the theorem to the extended function, assuming that R is so large that $K \subset B(R)$. □

We shall mention one more consequence of the Weierstrass approximation theorem.

Corollary 2 *Let f be a continuous function on $\mathbb{R}^m$. Assume that f has a compact support. Then, for every $\varepsilon > 0$, there exists an infinitely differentiable function g with a compact support such that $|f(x) - g(x)| < \varepsilon$ for all $x \in \mathbb{R}^m$.*

Proof Assume that f vanishes outside the ball $B(R)$, and let P be a polynomial approximating f with accuracy ε in the ball $B(R+1)$. We obtain the required function g if we multiply P by a function φ of class C^∞ such that $0 \leqslant \varphi \leqslant 1$, $\varphi(x) = 1$ for $x \in B(R)$, and φ vanishes outside $B(R+1)$. □

Remark If the function f in Corollary 2 is non-negative, then we may assume that the function g is also non-negative.

Indeed, otherwise, the function g can be replaced by $\varphi \cdot (g + \varepsilon)$, which, obviously, is non-negative and approximates f with accuracy 2ε.

Generalizing Theorem 1, we prove that a smooth function together with its derivatives can be approximated by a polynomial. In the next theorem, the letter

k denotes a multi-index ($k \in \mathbb{Z}_+^m$), and the symbol $D^k f$, where $k = (k_1, \ldots, k_m)$, denotes the derivative of f of order $|k| = k_1 + \cdots + k_m$ such that the differentiation with respect to the jth coordinate is carried out k_j times.

Theorem 2 *Let $f \in C^r(\mathbb{R}^m)$ ($r \in \mathbb{N}$). Then, for all $R > 0$ and $\varepsilon > 0$, there exists a polynomial P in m variables such that*

$$\left|D^k f(x) - D^k P(x)\right| < \varepsilon \quad \text{for all } x \text{ in } \overline{B}(R) \text{ and all } k,\ 0 \leqslant |k| \leqslant r.$$

Proof As in Theorem 1, we may assume without loss of generality that $\operatorname{supp}(f) \subset \overline{B}(R)$. Since, by properties of convolution, we have $D^k(f_t) = (D^k f) * W_t$, we can choose the parameter $t > 0$ so that inequality (2) and similar inequalities for $D^k f$ are valid for $|k| \leqslant r$ and all $x \in \mathbb{R}^m$. We put $M = \max_{x, |k| \leqslant r} |D^k f(x)|$. Then, for an appropriate choice of n, inequality (4) turns out to be valid not only for the function f, but also for all its derivatives up to order r inclusive. □

7.6.5 Here, relying on the concept of the convolution of periodic functions (see Sect. 7.5.5), we define a periodic approximate identity and prove Weierstrass' theorem on approximation by trigonometric polynomials. We can easily change a period by contraction. Therefore, we may assume without loss of generality that all functions considered in the present section are 2π-periodic with respect to each variable (and only such functions will be called periodic).

For the case of periodic functions, the definition of an approximate identity from Sect. 7.6.1 can easily be modified as follows (below, $Q = [-\pi, \pi]^m$): a family of periodic functions $\{\omega_t\}_{t \in T}$ is called a *periodic approximate identity* (as $t \to t_0$) if:

(a) $\omega_t \geqslant 0$,

(b) $\displaystyle\int_Q \omega_t(x)\,dx = 1$,

(c) $\displaystyle\int_{Q \setminus B(\delta)} \omega_t(x)\,dx \underset{t \to t_0}{\longrightarrow} 0$ for each $\delta \in (0, \pi)$.

We also introduce the following strong version of the localization property:

(c′) $\omega_t(x)\,dx \underset{t \to t_0}{\rightrightarrows} 0$ on $Q \setminus B(\delta)$ for each $\delta \in (0, \pi)$.

An almost verbatim repetition of the proof of Theorem 7.6.3 verifies the following approximative properties for the periodic convolution $f_t = f * \omega_t$.

Theorem *Let a periodic function f be measurable and bounded on the cube Q. Then:*

(a) *if the limit $L = \lim_{x \to x_0} f(x)$ exists and is finite at a point x_0, $x_0 \in \mathbb{R}^m$, then $f_t(x_0) \underset{t \to t_0}{\longrightarrow} L$;*

(b) *if $f \in C(\mathbb{R}^m)$, then $f_t \underset{t \to t_0}{\rightrightarrows} f$ on $\mathbb{R}^m$.*

Remark If an approximate identity satisfies condition (c′), then assertion (a) remains valid for every periodic function summable on Q. For the proof, we replace inequality (1) by the inequality

$$\left|f_t(x_0)-L\right| \leqslant \int_Q \left|f(x_0-y)-L\right|\omega_t(y)\,dy = \int_{B(\delta)} \ldots + \int_{Q\setminus B(\delta)} \ldots$$

$$\leqslant \sup_{y\in B(\delta)} \left|f(x_0-y)-L\right| + \sup_{y\in Q\setminus B(\delta)} \omega_t(y) \int_Q \left|f(x_0-y)-L\right| dy,$$

after which the proof can be completed as in Theorem 7.6.3: first, by a choice of δ, we make the first summand on the right-hand side of the inequality small, and then make the second summand small with the help of condition (c′).

Leaning on the last theorem, we now obtain a periodic version of the Weierstrass approximation theorem (see Sect. 7.6.4). Since the convolution of a summable function with a trigonometric polynomial is again a trigonometric polynomial, it is sufficient for us to construct an approximate identity consisting of such polynomials.

We begin with the one-dimensional case and consider the trigonometric polynomial

$$\Theta_n(x) = \frac{1}{c_n}\cos^{2n}\frac{x}{2} = \frac{1}{c_n}\left(\frac{1+\cos x}{2}\right)^n,$$

where the coefficient c_n is such that $\int_{-\pi}^{\pi}\Theta_n(x)\,dx = 1$, i.e., $c_n = \int_{-\pi}^{\pi}\cos^{2n}\frac{x}{2}\,dx$. These integrals were calculated in Example of Sect. 4.6.2, but in what follows it is important only that they tend to zero not too fast,

$$c_n = 4\int_0^{\frac{\pi}{2}}\cos^{2n} y\,dy \geqslant 4\int_0^{\frac{\pi}{2}}\sin y\cos^{2n} y\,dy = \frac{4}{2n+1} > \frac{1}{n}.$$

It follows that the sequence of functions Θ_n have the strong localization property (c′) stated at the beginning of the present section,

$$\sup_{\delta<|x|<\pi}\Theta_n(x) \leqslant n\cos^{2n}\frac{\delta}{2} \underset{n\to\infty}{\longrightarrow} 0 \quad \text{for each } \delta\in(0,\pi).$$

Using the periodic approximate identity Θ_n of one variable constructed above, we can easily construct its multi-dimensional counterpart,

$$\omega_n(x) = \Theta_n(x_1)\cdots\Theta_n(x_m) \quad \text{for } x=(x_1,\ldots,x_m)\in\mathbb{R}^m.$$

This sequence of trigonometric polynomials satisfies conditions (a)–(c) and, therefore, the theorem is valid for the sequence with $T=\mathbb{N}$ and $t_0=+\infty$. Since $f_n = f*\omega_n$ is a trigonometric polynomial, we come to a periodic version of the Weierstrass approximation theorem 7.6.4.

Corollary (Weierstrass) *Let f be everywhere continuous periodic function. Then there exists a sequence of trigonometric polynomials converging to f uniformly on $\mathbb{R}^m$.*

EXERCISES

1. Prove that there is no identity for convolution, i.e., there is no locally summable function g such that $f * g = f$ for each continuous function f with compact support.
2. Let the supports of the functions ω_t forming an approximate identity in $\mathbb{R}^m$ "contract to zero" as $t \to 0$, i.e., satisfy the condition $\text{supp}(\omega_t) \subset B(r_t)$, $r_t \to 0$ as $t \to 0$. Prove that $f * \omega_t \underset{t\to 0}{\longrightarrow} f$ pointwise for every continuous function f on $\mathbb{R}^m$ and that the convergence is uniform on each compact set.
3. Prove that the first assertion of Theorem 7.6.3 also remains valid in the case where $L = +\infty$.
4. Supplement the first assertion of Theorem 7.6.3 in the one-dimensional case by proving that if all functions ω_t are even, then the relation

$$(f * \omega_t)(x_0) \underset{t\to t_0}{\longrightarrow} \frac{f(x_0 - 0) + f(x_0 + 0)}{2}$$

holds for every bounded function f having finite one-sided limits $f(x_0 - 0)$ and $f(x_0 + 0)$ at a point x_0.
5. On the real line find an approximate identity $\{\omega_t\}_{t>0}$ such that $(f * \omega_t)(x_0) \underset{t\to 0}{\longrightarrow} f(x_0 + 0)$ if the one-sided limit $f(x_0 + 0)$ exists.
6. Supplementing Corollary 2 of Sect. 7.6.4, prove that every continuous function on $\mathbb{R}^m$ can be approximated by a function of class C^∞ uniformly on $\mathbb{R}^m$.
7. Prove that assertion (a) of Theorem 7.6.5 is not valid for an approximate identity violating condition (c$'$).

Chapter 8
Surface Integrals

In this chapter, our main aim is to give an exact meaning to the notion of the area of a smooth surface and to develop a means of its calculation. Undoubtedly, everybody has an intuitive idea of the area of a curved surface that one uses in everyday life (for instance, when one estimates the amount of paint consumption). At the same time, the evaluation of a curved surface area is quite a difficult problem compared with the analogous problem in the case of a plane figure. It is possible to reduce the first problem to the second one by elementary means only in the cases of conic and cylindrical surfaces: it is sufficient to "unroll" them. At school, one adds to this the calculation of the area of a sphere or its parts as a result of some unobvious argumentation.

Before we proceed to the computational side of the problem, it is necessary to overcome the principal difficulty, that is, to define the area of a surface.

We do not restrict ourselves to two-dimensional surfaces, so in what follows we discuss the construction of the measure (surface area) on smooth manifolds of arbitrary dimension.[1] For this purpose, we need some results from the theory of smooth maps. For the convenience of the reader, the required material is collected in the auxiliary first section of the chapter.

8.1 Auxiliary Notions

Here we remind the reader of the principal notions and facts of the theory of smooth manifolds and fix the related notation and terminology.

We denote by the symbol $C^r(\mathcal{O}, \mathbb{R}^m)$ $(1 \leqslant r \leqslant +\infty)$ the set of r times continuously differentiable maps defined on the set $\mathcal{O} \subset \mathbb{R}^k$, which is always assumed to be open, taking values in $\mathbb{R}^m$. We call C^1-maps *smooth* maps. If we talk about a map which is smooth on an arbitrary (non-open) set, we will always mean that it is

[1] The reader familiar with the theory of manifolds will note that, except in the case of dimension one, we consider only manifolds without a boundary.

B. Makarov, A. Podkorytov, *Real Analysis: Measures, Integrals and Applications*, Universitext, DOI 10.1007/978-1-4471-5122-7_8, © Springer-Verlag London 2013

defined and continuously differentiable on some neighborhood of this set, i.e., on a wider open set.

We denote by $d_a\Phi$ the differential of the map $\Phi \in C^1(\mathcal{O}, \mathbb{R}^m)$ at a point $a \in \mathcal{O}$, and by $\Phi'(a)$ the corresponding matrix (in the canonical bases of the spaces $\mathbb{R}^k$ and $\mathbb{R}^m$), i.e., the Jacobian matrix. This $m \times k$ matrix (with m rows and k columns) is formed, as is well known, by the partial derivatives $\frac{\partial \varphi_j}{\partial t_i}$ $(1 \leqslant j \leqslant m,\ 1 \leqslant i \leqslant k)$ of the coordinate functions $\varphi_1, \ldots, \varphi_m$ of the map Φ. If $m = k$, then the Jacobian matrix is square. Its determinant $\det \| \frac{\partial \varphi_j}{\partial t_i} \|$ (called the Jacobian of the map Φ) is also denoted by $\frac{D(\varphi_1,\ldots,\varphi_k)}{D(t_1,\ldots,t_k)}$.

8.1.1 We introduce a concept that is essential in what follows.

Definition A set M, $M \subset \mathbb{R}^m$, is called a *simple k-dimensional manifold* if it is homeomorphic to an open subset $\mathcal{O}$ of the set $\mathbb{R}^k$ $(k \leqslant m)$. The homeomorphism $\Phi : \mathcal{O} \xrightarrow{\text{on}} M$ is called a *parametrization* of the manifold M. If for some $r = 1, 2, \ldots, +\infty$

$$\Phi \in C^r(\mathcal{O}, \mathbb{R}^m) \quad \text{and} \quad \operatorname{rank} d_a\Phi = k \text{ at every point } a \in \mathcal{O},$$

then the parametrization Φ is said to be *smooth* of class C^r. A simple manifold that has such a parametrization is also called smooth (of class C^r).

We emphasize that, by definition, the domain of the parametrization is always an open set.

Since the position of a point $p = \Phi(t)$ on a manifold is uniquely determined by the parameter t, its coordinates $t_1, \ldots, t_k$ are often called the curvilinear coordinates of the point p. In particular cases they often have a simple geometrical meaning that simplifies the solution of the problem.

The simplest example of a k-dimensional manifold is a k-dimensional vector subspace. Its parametrization can be obtained, for example, in the following way. Fix an arbitrary basis $\tau_1, \ldots, \tau_k$ in the subspace and set

$$\Phi(t) = t_1\tau_1 + \cdots + t_k\tau_k \quad \bigl(t = (t_1, \ldots, t_k) \in \mathbb{R}^k\bigr).$$

Clearly, the map Φ satisfies all the requirements. The "curvilinear coordinates" of a vector in the subspace are simply its coordinates in the basis $\tau_1, \ldots, \tau_k$.

Definition (The first definition of a smooth manifold) A set M, $M \subset \mathbb{R}^m$, is called a *k-dimensional manifold of class C^r* if every point p from M has a neighborhood U such that the intersection $U \cap M$ is a simple k-dimensional manifold of class C^r. Its parametrization is called a *local parametrization* of the manifold M in the vicinity of the point p.

A k-dimensional manifold of class C^0 is defined in an analogous way: it is a set M which is, locally, a simple k-dimensional manifold (without any smoothness conditions). Each of its points has a neighborhood U such that the intersection $U \cap M$

is a simple k-dimensional manifold. The number k is called *the dimension of the manifold* M and is denoted by the symbol $\dim M$. The difference $m - \dim M$ is called *the codimension* of the manifold. A manifold of codimension 1 is called a *surface*.

Contrary to local parametrization, a parametrization of a simple manifold is also called *a global parametrization*.

If the value of the parameter $r \geqslant 1$ is not important (in most cases, it is sufficient to take $r = 1$), we call the set M a smooth k-dimensional manifold, or a smooth manifold, and sometimes simply a manifold since otherwise the character of the manifold is indicated explicitly.

If a point p belongs to a manifold $M \subset \mathbb{R}^m$, then by its M-neighborhood, or relative neighborhood, we mean the intersection of the neighborhood of p in $\mathbb{R}^m$ with the manifold M. It is clear that every point of the manifold has a base of M-neighborhoods whose closures lie in M. A *coordinate neighborhood* is a relative neighborhood which is a simple manifold, i.e., it admits a parametrization (and, consequently, curvilinear coordinates can be introduced).

In simple and important cases we come across examples of "almost smooth" manifolds (consider, for example, the boundary of a square, or a cube, etc.). Therefore, we expand the definition of a smooth manifold as follows: a *piecewise smooth* k-dimensional manifold is a union of a smooth k-dimensional manifold (possibly non-connected) and a set of zero k-dimensional Hausdorff measure. It is clear that the boundaries of polyhedral bodies are piecewise smooth surfaces according to this definition.

Speaking formally, when we consider a smooth manifold in $\mathbb{R}^m$, we do not exclude the possibility that $\dim M$ equals m. In this case, as follows from the definition, M is simply an open subset of $\mathbb{R}^m$. The problem of the definition of a measure on such a set has been resolved in Chap. 2, where the Lebesgue measure has been constructed. Therefore, in what follows, we consider only manifolds whose dimension is less than the dimension of the enveloping space unless otherwise stated. At the same time, we admit the possibility that $\dim M = 1$. In this case we use the term "curve" instead of the term "manifold". A connected simple curve is also called a simple arc.

Another definition of a smooth manifold will be of use (it is equivalent to the first definition, as is proved in Sect. 13.7.7).

Definition (The second definition of a smooth manifold) A set $M \subset \mathbb{R}^m$ is called a k-dimensional ($1 \leqslant k < m$) manifold of class C^r if for every point $p \in M$ there exist a neighborhood U and functions $F_1, \dots, F_{m-k}$ of class C^r defined on it such that:

(1) $x \in M \cap U$ if and only if

$$F_1(x) = 0, \quad \dots, \quad F_{m-k}(x) = 0, \tag{1}$$

and

(2) the vectors

$$\operatorname{grad} F_1(p), \quad \ldots, \quad \operatorname{grad} F_{m-k}(p) \tag{2}$$

are linearly independent.

In particular, a smooth surface (a manifold of codimension 1) is locally a level set of some smooth function with non-zero gradient. As is seen from the latter definition, locally every manifold lies in some surface.

8.1.2 Related to the notion of a smooth manifold are the important notions of tangent vector and tangent space. Recall that a *path* (in $\mathbb{R}^m$) is any continuous map from some segment into $\mathbb{R}^m$. A path is called *smooth* if its coordinate functions are smooth, and *piecewise smooth* if it is defined on a union $\bigcup_{j=0}^{n-1}[c_j, c_{j+1}]$ and its restrictions to the segments $[c_j, c_{j+1}]$ are smooth paths.

Definition Let M be a smooth manifold in $\mathbb{R}^m$. A vector $\tau \in \mathbb{R}^m$ is called a *tangent vector* to M at a point p, $p \in M$, if there exists a smooth path $\gamma : [a, b] \mapsto \mathbb{R}^m$ such that $\gamma(t) \in M$ for $t \in [a, b]$, and for some $c \in (a, b)$ we have $\gamma(c) = p$ and $\gamma'(c) = \tau$.

If some M-neighborhood of a point $p = \gamma(c)$ lies in a level set of a smooth function F, then $F(\gamma(t)) \equiv \text{const}$ for t close to c. Therefore, $\langle \operatorname{grad} F(p), \gamma'(c)\rangle = 0$, i.e., the tangent vector at the point p is orthogonal to the vector $\operatorname{grad} F(p)$.

Let Φ be a local parametrization of the manifold M in the vicinity of a point $p = \Phi(a)$. "Freezing" all coordinates of a point $a = (a_1, \ldots, a_k)$, except the j-th one, and making the latter change in the vicinity of a_j, we get a path that parametrizes the curve that passes through the point p. This curve is called a *coordinate line*. The vector tangent to this curve at the point p that corresponds to the mentioned parametrization is the j-th column of the matrix $\Phi'(a)$; we denote it by $D_j\Phi(a)$ or $\tau_j = \tau_j(a)$. Since $\operatorname{rank} d_a\Phi = k$, the vectors $\tau_1, \ldots, \tau_k$ are linearly independent. It is clear that $\tau_j(a) = d_a\Phi(e_j)$ (where the vectors $e_1, \ldots, e_k$ form the canonical basis in $\mathbb{R}^k$). We call them the *canonical tangent vectors* related to the parametrization Φ. The set of all vectors tangent to the manifold M at the point p is called the *tangent space* and is denoted by $T_p(M)$, or T_p for short. Note that this term needs validation, that is, one must check that T_p is actually a vector space.

Lemma *T_p is a k-dimensional subspace of the space $\mathbb{R}^m$.*

In the case where $k = m - 1$, we also call the subspace T_p a *tangent plane*.

Proof Assume that in the vicinity of the point p the manifold M is given by the Eqs. (1) and that the vectors (2) are linearly independent.

We check that, along with any two vectors τ_1, τ_2, the set T_p contains their linear combination $\tau = \alpha_1\tau_1 + \alpha_2\tau_2$. We may assume that $\tau \neq 0$, otherwise it suffices to take a constant path. Augmenting the vector system (2), which is orthogonal to the vector τ, with vectors $h_1, \ldots, h_{k-1}$ to a basis in the orthogonal complement to τ, we consider the system of equations

$$F_1(x) = 0, \quad \ldots, \quad F_{m-k}(x) = 0, \quad \langle x - p, h_1 \rangle = 0, \quad \ldots, \\ \langle x - p, h_{k-1} \rangle = 0. \tag{3}$$

According to the second definition of a smooth manifold, this system defines a smooth one-dimensional manifold in the vicinity of the point p, i.e., a smooth curve that obviously lies in M and passes through p.

Let γ be some parametrization of this curve in the vicinity of the point p. Without loss of generality, we may assume that $p = \gamma(0)$. Then the (non-zero!) vector $\gamma'(0)$ is orthogonal to the gradients (at the point p) of all functions appearing in system (3). Therefore, it is proportional to the vector τ.

For an appropriate choice of the coefficient θ, the vector tangent to the path $\widetilde{\gamma}(t) = \gamma(\theta t)$, $|t| \leqslant \delta$, at $t = 0$ coincides with τ, i.e., $\tau \in T_p$.

Thus, we have proved that T_p is a vector subspace of the space $\mathbb{R}^m$. Its dimension does not exceed k since all vectors in it are orthogonal to the vectors of system (2). Moreover, it contains k linearly independent vectors that are tangent to the coordinate lines. Therefore, $\dim T_p = k$. □

Remark If the surface M is defined by the equation $F(x) = 0$ in the vicinity of the point p and $\operatorname{grad} F(p) \neq 0$, then, as noted before the lemma, the vectors tangent to it at the point p are orthogonal to the vector $\operatorname{grad} F(p)$. Therefore, the tangent space to M at the point p is the plane that consists of the vectors orthogonal to $\operatorname{grad} F(p)$, i.e., it is defined by the equation $\langle x, \operatorname{grad} F(p) \rangle = 0$.

We note that since the canonical tangent vectors corresponding to the parametrization Φ of the manifold M are linearly independent, they form a basis in the tangent space. The linearity of the map $d_a\Phi : \mathbb{R}^k \to \mathbb{R}^m$ implies that for every vector $t = (t_1, \ldots, t_k)$ in $\mathbb{R}^k$ the equality

$$d_a\Phi(t) = \sum_{j=1}^{k} t_j \tau_j$$

holds. Therefore, the differential of the parametrization maps $\mathbb{R}^k$ onto the tangent space isomorphically.

Sometimes it is more geometrically clear to consider the *affine tangent space* L_p instead of the tangent space T_p; it is the shift of T_p by the vector p: $L_p = p + T_p$. Since $p + d\Phi_a(t - a) \in L_p$, where $p = \Phi(a)$, and

$$\Phi(t) = p + d\Phi_a(t - a) + o\big(\|t - a\|\big) \quad \text{as } t \to a,$$

the point $x = \Phi(t)$ satisfies the relation

$$\begin{aligned} \operatorname{dist}(x, L_p) &= \operatorname{dist}\big(\Phi(t), L_p\big) \leqslant \big\|\Phi(t) - \big(p + d\Phi_a(t - a)\big)\big\| \\ &= o\big(\|t - a\|\big) \quad \text{as } t \to a. \end{aligned}$$

By Corollary 1 from Sect. 8.1.4, the map Φ^{-1} satisfies the Lipschitz condition

$$\|t-a\| = \left\|\Phi^{-1}(x) - \Phi^{-1}(p)\right\| \leqslant C\|x-p\|$$

in the vicinity of the point p, and therefore

$$\operatorname{dist}(x, L_p) = o\big(\|x-p\|\big) \quad \text{as } x \to p,\ x \in M.$$

Thus, when we substitute points in the subspace L_p for points in the manifold, the relative error tends to zero, i.e., the manifold M is "almost flat" in the small. The latter relation is a formalization of our intuitive idea of the tangent space as the space "tight-fitting" to the manifold. It can be proved that this property of the affine tangent space uniquely determines it (see Exercise 2).

8.1.3 We give some examples.

Example 1 An important example of a surface is the graph of a smooth function f defined on an open subset of the space $\mathbb{R}^{m-1}$. By definition of the graph, it is the set

$$\Gamma_f = \big\{(x_1, \ldots, x_{m-1}, y) \in \mathbb{R}^m \mid (x_1, \ldots, x_{m-1}) \in \mathcal{O},\ y = f(x_1, \ldots, x_{m-1})\big\}.$$

The map

$$\mathcal{O} \ni x = (x_1, \ldots, x_{m-1}) \mapsto \Phi(x) = \big(x_1, \ldots, x_{m-1}, f(x)\big)$$

is, obviously, a global parametrization of the graph. We call this parametrization *canonical*.

The graph of the function f may be considered as a zero level set of the function $F(x_1, \ldots, x_{m-1}, y) = y - f(x_1, \ldots, x_{m-1})$ defined on the set $\mathcal{O}' = \mathcal{O} \times \mathbb{R}$. We note that $\operatorname{grad} F \neq 0$ everywhere in the set $\mathcal{O}'$ and, in particular, in Γ_f. As follows from the remark after the proof of Lemma 8.1.2, the affine tangent plane at the point $p = (a_1, \ldots, a_{m-1}, f(a))$, where $a = (a_1, \ldots, a_{m-1}) \in \mathcal{O}$, is given by the equation

$$y - f(a) = \big\langle \operatorname{grad} f(a), x - a\big\rangle = \sum_{j=1}^{m-1} f'_{x_j}(a)(x_j - a_j).$$

A set that can be obtained from the graph by changing the order of coordinates (so that the "dependent" coordinate does not occur in the last position) is also called a graph, or, more precisely, a graph in a wider sense. Clearly, a set $M \subset \mathbb{R}^m$ such that $M \cap U$ is a graph (in this wide sense) for some neighborhood U of each of its points is a surface.

Using the implicit function theorem (see Sect. 13.7.6), one can prove the reverse: every surface in $\mathbb{R}^m$ is (locally) a graph of a smooth function (see Exercise 4).

Example 2 Consider the sphere

$$S(R) = \{(x_1, \dots, x_m) \in \mathbb{R}^m \mid x_1^2 + \dots + x_m^2 = R^2\}$$

in $\mathbb{R}^m$. We check that it is a surface. For every point $p = (p_1, \dots, p_m)$ in $S(R)$, at least one coordinate is non-zero. Assume, for the sake of definiteness, that $p_m > 0$. Then the point p belongs to the upper hemisphere $S_+(R) = \{x \in S(R) \mid x_m > 0\}$ which is simply the graph of the function $f(x_1, \dots, x_{m-1}) = \sqrt{R^2 - x_1^2 - \dots - x_{m-1}^2}$ defined in a ball of the space $\mathbb{R}^{m-1}$. This function is of class C^∞. Therefore, the hemisphere $S_+(R)$, and thus all the sphere $S(R)$, are C^∞-surfaces.

It is intuitively clear that the sphere has no global parametrization. At the same time, one can easily give a map that parametrizes almost all of the sphere. We restrict ourselves to the most obvious particular case of the two-dimensional sphere in $\mathbb{R}^3$ (see the discussion of the general case in Exercise 5). Recall the geographical coordinates, the longitude φ and the latitude θ of a point on the surface of the Earth. For $\varphi \in [-\pi, \pi]$ and $\theta \in [-\frac{\pi}{2}, \frac{\pi}{2}]$, we set

$$\Phi(\varphi, \theta) = (R\cos\varphi\cos\theta, R\sin\varphi\cos\theta, R\sin\theta) \tag{4}$$

(the corresponding coordinate lines are parallels and meridians; the eastern hemisphere corresponds to the positive values of φ, the western to the negative values, the northern hemisphere is determined by the inequality $\theta > 0$, whereas the southern hemisphere corresponds to $\theta < 0$). In the given example we deal with a rather typical situation. Speaking formally, the map Φ is defined for any φ and θ, but we are interested only in its restrictions to some subsets that are convenient for our considerations. It is clear that Φ is an infinitely differentiable map, but it is not bijective since $\Phi(\varphi, \pm\frac{\pi}{2}) = (0, 0, \pm R)$ for all values of φ (there is no natural way to ascribe a longitude to the north or south pole). Moreover, for any θ we lose injectivity for $\varphi = \pm\pi$ since the angles $\varphi = \pi$ and $\varphi = -\pi$ correspond to the same point on the sphere. These values of the parameter φ correspond to the meridian on the Earth, called the International Date Line, where the date changes as a ship or aeroplane travels east or west across it.[2] Deleting it (together with the poles), we get the "cut sphere", i.e., the C^∞-surface that has a global parametrization (4) defined on an open rectangle $|\varphi| < \pi$, $|\theta| < \frac{\pi}{2}$. The condition $\operatorname{rank} d\Phi \equiv 2$, i.e., the linear independence of the tangent vectors

$$\tau_1 = D_1\Phi(\varphi, \theta) = (-R\sin\varphi\cos\theta, R\cos\varphi\cos\theta, 0),$$

$$\tau_2 = D_2\Phi(\varphi, \theta) = (-R\cos\varphi\sin\theta, -R\sin\varphi\sin\theta, R\cos\theta)$$

is a consequence of their orthogonality (τ_1 is tangent to a parallel and τ_2 to a meridian), since $\|\tau_1\| = R\cos\theta \neq 0$ and $\|\tau_2\| = R$.

[2] In reality, the International Date Line is determined by special agreements and does not coincide with this meridian completely.

As we will see later, the deletion of a meridian is inessential when one integrates over a sphere.

It should also be noted that in analysis the angle θ' between the radius vector and the positive direction of the OZ axis is often used instead of the latitude. It varies in the interval $[0, \pi]$ and is related to the latitude θ by the relation $\theta + \theta' = \pi/2$.

Example 3 Consider the torus, the surface in $\mathbb{R}^3$ created by rotating the circle $(R-x)^2 + z^2 = r^2$ $(0 < r < R)$ around the axis OZ. As can easily be seen, the torus may be defined by the equation $(R - \sqrt{x^2+y^2})^2 + z^2 = r^2$. No global parametrization of the torus exists (we do not dwell on the proof of this fact). The position of a point on this surface is determined by two angles φ and θ (the analogs of the longitude and the latitude on the sphere) by the relations

$$x = (R + r\cos\theta)\cos\varphi, \qquad y = (R + r\cos\theta)\sin\varphi, \qquad z = r\sin\theta.$$

The infinitely differentiable map (defined in $\mathbb{R}^2$)

$$\Phi(\varphi, \theta) = \big((R + r\cos\theta)\cos\varphi, (R + r\cos\theta)\sin\varphi, r\sin\theta\big)$$

maps the square $[-\pi, \pi]^2$ onto the torus. It is not bijective due to the 2π-periodicity of trigonometric functions. Deleting two circles corresponding to the angles $\varphi = \pm\pi$ and $\theta = \pm\pi$, we obtain "the torus with two cuts", a surface of the class C^∞, for which the restriction of Φ to the square $(-\pi, \pi)^2$ is a global parametrization. The condition $\operatorname{rank} d\Phi \equiv 2$ is fulfilled because the tangent vectors

$$\tau_1 = D_1\Phi(\varphi, \theta) = \big(-(R + r\cos\theta)\sin\varphi, (R + r\cos\theta)\cos\varphi, 0\big),$$
$$\tau_2 = D_2\Phi(\varphi, \theta) = (-r\cos\varphi\sin\theta, -r\sin\varphi\sin\theta, r\cos\theta)$$

are linearly independent: they are orthogonal and $\|\tau_1\| = R + r\cos\theta > 0$, $\|\tau_2\| = r > 0$.

In order to check that not only the torus with the cuts, but also the torus in the whole, is a smooth surface, we need to show that every point $p = \Phi(\varphi_0, \theta_0)$ has a neighborhood in the torus that admits a global parametrization. This parametrization can be obtained if we change the definition domain of the mapping Φ. We leave it to the reader to check that the square $(\varphi_0 - \pi, \varphi_0 + \pi) \times (\theta_0 - \pi, \theta_0 + \pi)$ may be regarded as such a domain. The corresponding neighborhood of the point p on the torus is the torus with the cuts along the circles $\varphi = \varphi_0 \pm \pi$ and $\theta = \theta_0 \pm \pi$.

We note that in the limit case where $r = R$, we rotate the circle $(R - x)^2 + z^2 = R^2$ around the OZ-axis. The set M thus obtained is not a smooth surface since there is no M-neighborhood of the origin that is a simple surface. The reader can check however that the set $M \setminus \{0\}$ is a smooth surface of class C^∞, and so M is a piecewise smooth surface.

Example 4 Consider a manifold of minimal dimension, i.e., a curve. Its parametrization in a vicinity of an arbitrary point is a smooth vector function defined on an interval of a real line. It is a homeomorphism with non-zero derivative. It is clear that the graph of a function of one variable defined on an interval is a smooth flat curve, i.e., a curve in $\mathbb{R}^2$.

Another well-known curve is a circle. To get a more general example, recall that according to the second definition of a smooth manifold, the level set of a smooth function of two variables with non-zero gradient is a smooth curve. The example of the lemniscate of Bernoulli, a plane set consisting of the points (x, y) such that

$$\left(x^2+y^2\right)^2-\left(x^2-y^2\right)=0,$$

demonstrates that the hypothesis about the gradient is important: the point $(0,0)$, where the gradient of the function $F(x,y)=(x^2+y^2)^2-(x^2-y^2)$ is equal to zero, has no relative neighborhood which is homeomorphic to an interval. Near the origin, the lemniscate may be viewed as a union of two graphs, i.e., "a self-intersecting curve". Note that if we delete the origin from this set, we get a (disconnected) smooth curve. This shows that the lemniscate is a piecewise smooth curve.

Example 5 Consider the group $O(n)$ of orthogonal $n\times n$ matrices. We regard it as a subset of the n^2-dimensional Euclidean space which we identify with the set of all $n\times n$ matrices $U=\{u_{i,j}\}_{i,j=1}^n$ with elements u_{ij}. This subset is defined by the system of equations

$$\begin{aligned} u_{i,1}^2+\cdots+u_{i,n}^2&=1, && 1\leqslant i\leqslant n,\\ u_{i,1}u_{k,1}+\cdots+u_{i,n}u_{k,n}&=0, && 1\leqslant i<k\leqslant n. \end{aligned}$$

The gradients of the functions $F_i(U)=u_{i,1}^2+\cdots+u_{i,n}^2$ and $F_{ik}(U)=u_{i,1}u_{k,1}+\cdots+u_{i,n}u_{k,n}$ evaluated at the points of $O(n)$ are linearly independent. To convince ourselves that this is true, represent these gradients as the matrices made up of the derivatives over $u_{i,j}$ placed at the intersection of the ith row and the jth column. Then each row of a matrix which represents a linear combination of the gradients contains only a linear combination of (pairwise orthogonal) rows of the matrix U, whence the needed property easily follows. Thus, $O(n)$ is a smooth manifold of dimension $n^2-n-n(n-1)/2=n(n-1)/2$. It is natural to call the map $U\mapsto U_0U$ (or $U\mapsto UU_0$), where $U\in\mathbb{R}^{n^2}$ and U_0 is a certain element in $O(n)$, the left (correspondingly, the right) shift in the set of all matrices. The shift preserves the Euclidean distance between matrices since, as is easily seen, the Euclidean norms of the matrices U and U_0U (UU_0), considered as elements of the space $\mathbb{R}^{n^2}$, are the same. Therefore, the shift in $O(n)$ is an isometry relative to the metric in $O(n)$ induced from the enveloping n^2-dimensional space.

8.1.4 In what follows, it is important that a parametrization of a k-dimensional manifold in $\mathbb{R}^m$ can be viewed, locally, as a restriction to the subspace $\mathbb{R}^k$ of a diffeomorphism defined on an open subspace of the space $\mathbb{R}^m$. More precisely, we can consider a canonical embedding of $\mathbb{R}^k$ in $\mathbb{R}^m$ where the vectors $(x_1,\dots,x_k)$ in $\mathbb{R}^k$ are identified with the vectors $(x_1,\dots,x_k,0,\dots,0)$ in $\mathbb{R}^m$. Then the following statement about the extension of a parametrization to a diffeomorphism is true.

Lemma *Let $\mathcal{O}$ be an open subset of the space $\mathbb{R}^k$ and a be a point in $\mathcal{O}$. For a smooth parametrization Φ of the set $\Phi(\mathcal{O}) \subset \mathbb{R}^m$, a neighborhood $V \subset \mathbb{R}^m$ of the point $a = (a_1, \dots, a_k, 0, \dots, 0)$ and a diffeomorphism F defined on it can be given such that Φ and F coincide on $V \cap \mathbb{R}^k$.*

Proof Since the rank of the Jacobian matrix $\Phi'(a)$ is k, it has a $k \times k$ non-zero minor. Without loss of generality, we can assume that it is formed by the first rows of the matrix. Then $\frac{D(\varphi_1,\dots,\varphi_k)}{D(t_1,\dots,t_k)}(a) \neq 0$, where $\varphi_1, \dots, \varphi_k$ are the coordinate functions of the map Φ.

We consider the map Θ from $\mathcal{O} \times \mathbb{R}^{m-k}$ to $\mathbb{R}^m$ defined by the formula

$$\Theta(t_1, \dots, t_k, t_{k+1}, \dots, t_m) = \Phi(t_1, \dots, t_k) + (0, \dots, 0, t_{k+1}, \dots, t_m),$$

where $(t_1, \dots, t_k) \in \mathcal{O}$ and $(t_{k+1}, \dots, t_m) \in \mathbb{R}^{m-k}$. It is clear that Θ is a smooth map and extends Φ to $\mathcal{O} \times \mathbb{R}^{m-k}$. Moreover, $\operatorname{rank} d_a\Theta = m$ since $\det \Theta'(a) = \frac{D(\varphi_1,\dots,\varphi_k)}{D(t_1,\dots,t_k)}(a) \neq 0$.

By the local invertibility theorem, the restriction of Θ to a (sufficiently small) neighborhood V of the point a is a diffeomorphism (see Sect. 13.7.5). This restriction should be taken for F. □

If Φ is a local parametrization of a k-dimensional manifold M in the vicinity of a point p and F is the diffeomorphism described in the above lemma, then Φ^{-1} and F^{-1} coincide on some M-neighborhood of the point p, more precisely, on the set $\Phi(V_0)$, where $V_0 = V \cap \mathbb{R}^k$. Thus, the next two statements follow from this lemma.

Corollary 1 *In a sufficiently small M-neighborhood of a point p, the map Φ^{-1} satisfies the Lipschitz condition, i.e.,*

$$\left\|\Phi^{-1}(x) - \Phi^{-1}(y)\right\| \leqslant C\|x - y\| \quad \textit{for } x, y \in M$$

and some C.

To prove the result, it suffices to note that the smooth map F^{-1} satisfies the Lipschitz condition in every closed ball in the domain of the map (see Theorem 13.7.2).

Corollary 2 *Let $\mathcal{O}$ and $\mathcal{O}'$ be open sets in $\mathbb{R}^k$ and $\Phi \in C^1(\mathcal{O}, \mathbb{R}^m)$ be a parametrization of the manifold M, $M \subset \mathbb{R}^m$. If $\Psi \in C^1(\mathcal{O}', \mathbb{R}^m)$ and $\Psi(\mathcal{O}') \subset M$, then the composition $\Phi^{-1} \circ \Psi$ is a smooth map.*

Indeed, for any point $t_0 \in \mathcal{O}'$ the map Φ^{-1} coincides with the smooth map F^{-1} in some M-neighborhood of the point $\Psi(t_0)$. Therefore, in a sufficiently small neighborhood of the point t_0, the map $\Phi^{-1} \circ \Psi = F^{-1} \circ \Psi$ is a composition of smooth maps.

8.1.5 We will use a simple geometrical fact based upon the following observation.

Every open subset G of the space $\mathbb{R}^m$ is a union of balls in G for which the radii and the coordinates of the centers are rational numbers.

Since every point x of the set G may be regarded as the center of a ball $B(x,r)$ in G, it suffices to note that $x \in B(y,\rho) \subset B(x,r)$ for $0 < \rho < r/2$ and $\|x-y\| < \rho$. Clearly, the number ρ and the coordinates of the vector y may be chosen to be rational.

Theorem (Lindelöf[3]) *For any family $\{G_\alpha\}_{\alpha\in A}$ of sets that are open in $\mathbb{R}^m$ there exists an at most countable subfamily $\{G_\alpha\}_{\alpha\in A_0}$ (the set A_0, $A_0 \subset A$, is at most countable) with the same union*:

$$\bigcup_{\alpha\in A} G_\alpha = \bigcup_{\alpha\in A_0} G_\alpha.$$

Proof Consider arbitrary balls, with rational radii and rational center coordinates, that are contained in at least one of the sets G_α. The collection of such balls is countable. Let $\{B_n\}_{n\in\mathbb{N}}$ be an enumeration of this collection. By the choice of these balls, for any $n \in \mathbb{N}$ there exists an index $\alpha_n \in A$ such that $B_n \subset G_{\alpha_n}$. Moreover, each set G_α is exhausted by the balls chosen this way:

$$G_\alpha \subset \bigcup_{n\in\mathbb{N}} B_n \quad \text{for any index } \alpha \in A.$$

Consequently,

$$\bigcup_{\alpha\in A} G_\alpha \subset \bigcup_{n\in\mathbb{N}} B_n \subset \bigcup_{n\in\mathbb{N}} G_{\alpha_n}.$$

Due to the evident inclusion $\bigcup_{n\in\mathbb{N}} G_{\alpha_n} \subset \bigcup_{\alpha\in A} G_\alpha$, one can take the set A_0 as the set of indices α_n. □

Corollary 1 *A smooth manifold can be represented as a union of an at most countable family of simple manifolds.*

Since the range of curvilinear coordinates is a countable union of compact sets, the following statement is true.

Corollary 2 *A smooth manifold is an at most countable union of compact sets such that each such set is a subset of a simple manifold.*

The following corollary is an immediate consequence of the preceding one.

Corollary 3 *A smooth manifold in $\mathbb{R}^m$ is a Borel subset of this space.*

[3] Ernest Leonhard Lindelöf (1870–1946)—Finnish mathematician.

Since a smooth surface is locally a graph (in the broad sense) of a smooth function, we also have the following.

Corollary 4 *A smooth surface is an at most countable union of graphs of smooth functions.*

8.1.6 Now we prove a useful fact that allows us to represent a smooth function as a sum of smooth functions with small supports. It often leads to important technical simplifications due to the "localization" of the problem (see Sect. 8.6.5). Recall that the support of a function $\varphi : \mathbb{R}^m \to \mathbb{R}$, denoted by the symbol $\operatorname{supp}(\varphi)$, is the closure of the set $\{x \mid \varphi(x) \neq 0\}$.

Theorem (On a smooth partition of unity) *For every* $\varepsilon > 0$ *there exists a non-negative function* φ_ε *of class* $C^\infty(\mathbb{R}^m)$ *such that* $\operatorname{supp}(\varphi_\varepsilon) = [-\varepsilon, \varepsilon]^m$ *and*

$$\sum_{n \in \mathbb{Z}^m} \varphi_\varepsilon(x - \varepsilon n) = 1 \quad \textit{for any } x \textit{ in } \mathbb{R}^m.$$

Note that the number of non-zero summands of this sum is finite near every point $a \in \mathbb{R}^m$. More precisely, if $x \in a + (-\varepsilon, \varepsilon)^m$ and $\varphi_\varepsilon(x - \varepsilon n) \neq 0$, then $n \in \frac{1}{\varepsilon} a + (-2, 2)^m$.

Proof We use the following well-known example of a function of class $C^\infty(\mathbb{R})$:

$$\Psi(t) = \begin{cases} 0 & \text{if } t \leqslant 0, \\ e^{-1/t} & \text{if } t > 0. \end{cases}$$

The existence of its derivatives of all orders is evident for non-zero t, and at zero is a consequence of the easy-to-check representation of $\Psi^{(n)}(t)$, for $t > 0$, in the form $\Psi^{(n)}(t) = P_n(1/t)e^{-1/t}$, where P_n is a polynomial.

Set $\psi(x) = \prod_{k=1}^m \Psi(1 - x_k^2)$ where $x = (x_1, \dots, x_m)$. Clearly, ψ is a class $C^\infty(\mathbb{R}^m)$ function which is positive in the cube $(-1, 1)^m$ and equal to zero outside it. Therefore, each function $x \mapsto \psi(x - n)$ is positive in the shifted cube $n + (-1, 1)^m$ ($n \in \mathbb{Z}^m$). Since every point x belongs to at least one such cube, the sum

$$\Phi(x) = \sum_{n \in \mathbb{Z}^m} \psi(x - n)$$

is positive. It is the sum of only a finite number of infinitely-differentiable functions (see the remark that follows the statement of the theorem). Consequently, $\Phi \in C^\infty(\mathbb{R}^m)$. Take $\varphi_1(x) = \frac{\psi(x)}{\Phi(x)}$. It is clear that this function satisfies the hypothesis of the theorem for $\varepsilon = 1$. To construct the function φ_ε for arbitrary ε, it suffices, via a scaling, to set $\varphi_\varepsilon(x) = \varphi_1(\frac{1}{\varepsilon} x)$. □

8.1.7 We show how one can construct a smooth approximation of characteristic functions using a partition of unity. It is intuitively clear that outside the set E, the values of its characteristic function can be altered gradually, without sudden jumps, decreasing to zero. It is also plausible that such a "descent" can be effectuated in the vicinity of E, without overstepping the limits of its arbitrarily small ε-neighborhood. Recall that the ε-neighborhood of the set E is the set

$$E_\varepsilon = \{y \in \mathbb{R}^m \mid \operatorname{dist}(y, E) < \varepsilon\} = \bigcup_{x \in E} B(x, \varepsilon).$$

We also show that this smoothing can be made without a steep drop of the smoothing function, i.e., we can control the norm of its gradient so that, under the circumstances considered, it is of the smallest possible order.

Theorem (On a smooth descent) *For every set $E \subset \mathbb{R}^m$ and every $\varepsilon > 0$ there exists a function θ_ε of class $C^\infty(\mathbb{R}^m)$ such that:*

(a) $0 \leqslant \theta_\varepsilon \leqslant 1$ *on* $\mathbb{R}^m$;
(b) $\theta_\varepsilon(x) = 1$ *if* $x \in E$;
(c) $\theta_\varepsilon(x) = 0$ *outside* E_ε;
(d) $\|\operatorname{grad} \theta_\varepsilon\| \leqslant \frac{c_m}{\varepsilon}$ *on* $\mathbb{R}^m$, *where c_m is a coefficient that depends only on the dimension.*

Proof Take $\delta = \varepsilon/(2\sqrt{m})$ and let φ_δ be the function constructed in Theorem 8.1.6, $Q = (-1, 1)^m$. We keep only those summands in the sum

$$\sum_{n \in \mathbb{Z}^m} \varphi_\delta(x - \delta n) = 1$$

for which the cube $\delta(n + Q)$ intersects E, and set

$$\theta_\varepsilon(x) = \sum_{\substack{n \in \mathbb{Z}^m: \\ \delta(n+Q) \cap E \neq \varnothing}} \varphi_\delta(x - \delta n).$$

It is clear that $0 \leqslant \theta_\varepsilon(x) \leqslant 1$ everywhere and $\theta_\varepsilon(x) = 1$ if $x \in E$. Moreover, since $\operatorname{diam}(Q) = 2\sqrt{m}$, we have $\theta_\varepsilon(x) = 0$ outside E_ε. Therefore, the function θ_ε satisfies conditions (a)–(c). Now we check the condition (d). Since the sum that defines θ_ε consists of a finite number of summands near every point (as was mentioned after the statement of Theorem 8.1.6), it can be differentiated termwise. It is clear that

$$\|\operatorname{grad} \theta_\varepsilon(x)\| \leqslant \sum_{n \in \mathbb{Z}^m} \|\operatorname{grad} \varphi_\delta(x - \delta n)\|.$$

Since every point x belongs to at most 2^m cubes $\delta(n + Q)$, it follows that

$$\begin{aligned}
\|\operatorname{grad} \theta_\varepsilon(x)\| &\leqslant 2^m \max_x \|\operatorname{grad} \varphi_\delta(x)\| = \frac{2^m}{\delta} \max_x \left\| \operatorname{grad} \varphi_1\left(\frac{1}{\delta} x\right) \right\| \\
&= \frac{2^m}{\delta} L = \frac{2^{m+1}\sqrt{m}}{\varepsilon} L,
\end{aligned}$$

where $L = \max_y \|\operatorname{grad} \varphi_1(y)\|$ does not depend on ε (but does depend on m). □

8.1.8 We conclude this section with a modification of the theorem on the partition of unity. First, we prove a useful geometric fact.

Lemma *Let K be a compact set in the space $\mathbb{R}^m$ and let $\{G_\alpha\}_{\alpha\in A}$ be its open cover. Then there exists a number $\delta > 0$ such that every set e that intersects K and has the property* $\operatorname{diam}(e) < \delta$ *is contained in at least one set belonging to the cover.*

Proof Assume that the statement of the lemma is false. Then for every $n \in \mathbb{N}$ there exists a set e_n that is not contained in any set G_α and at the same time

$$e_n \cap K \neq \varnothing, \qquad \operatorname{diam}(e_n) < \frac{1}{n}.$$

Fix a point x_n in each $e_n \cap K$. Without loss of generality, one can consider that $x_n \to x_0$ for some $x_0 \in K$ (otherwise, one may pass to a subsequence). The point x_0 belongs to some set in the family G_α, say, to G_{α_0}. Therefore, $B(x_0, r) \subset G_{\alpha_0}$ for some $r > 0$. If n is large enough, then $\|x_n - x_0\| < r/2$ and $\operatorname{diam}(e_n) < r/2$, whence $e_n \subset B(x_0, r) \subset G_{\alpha_0}$. This shows that the sets e_n with large indices are contained in the set G_{α_0}, contrary to the choice of e_n. □

It is convenient to use the following theorem in those situations where one needs to replace an arbitrary function by functions with supports lying in the prescribed sets (see, for instance, Theorems 8.4.2 and 8.6.5).

Theorem (On a partition of unity subordinate to a cover) *Let K be a compact subset of the space $\mathbb{R}^m$ and let $\{G_\alpha\}_{\alpha\in A}$ be its open cover. Then there exists a finite family of non-negative finitary functions $\psi_1, \dots, \psi_N$ of class $C^\infty(\mathbb{R}^m)$ such that*

$$\sum_{j=1}^{N} \psi_j \leqslant 1 \quad \text{on } \mathbb{R}^m, \qquad \sum_{j=1}^{N} \psi_j(x) = 1 \quad \text{if } x \in K$$

and the support of ψ_j is contained in one of the sets that make up the cover for all j.

The family of functions $\psi_1, \dots, \psi_N$ is called a partition of unity for K *subordinate to the cover* $\{G_\alpha\}_{\alpha\in A}$.

Proof Let δ be a number from the lemma above corresponding to the given cover. Consider the partition of unity $1 = \sum_{n\in\mathbb{Z}^m} \varphi_\varepsilon(x - \varepsilon n)$ constructed in Theorem 8.1.6 taking ε so small that $\operatorname{diam}(\operatorname{supp}(\varphi_\varepsilon)) < \delta$. Keeping only those summands $\varphi_\varepsilon(x - \varepsilon n)$ in the partition of unity whose supports intersect K, we obviously get a finite family, as required, and it only remains to enumerate it. □

EXERCISES

1. Let $M = \{(x, y, u, v) \in \mathbb{R}^4 \mid x^2 + y^2 = 1,\ u^2 + v^2 = 1\}$. Prove that:

(a) M is a two-dimensional C^∞-smooth compact manifold homeomorphic to a Cartesian product of two circles;

(b) the mapping $(x, y, u, v) \mapsto ((R+ru)x, (R+ru)y, rv)$ is a homeomorphism of M and the torus considered in Example 3 of Sect. 8.1.3.

2. Let p be a point of a smooth manifold M. Prove that the tangent subspace is a unique among affine subspaces L ($\dim(L) = \dim(M)$) that pass through the point p and satisfy the property

$$\operatorname{dist}(x, L) = o\bigl(\|x - p\|\bigr) \quad \text{as } x \to p,\ x \in M.$$

3. Let S^2 be a two-dimensional sphere in $\mathbb{R}^3$ with center at zero and $N = (0, 0, 1)$ its north pole. Consider a mapping of the set $S^2 \setminus \{N\}$ to the equatorial plane defined as follows: a point p in the sphere maps to the point where the equatorial plane intersects the line through p and N. This map is called the stereographic projection. Prove that:

(a) the stereographic projection is a one-to-one mapping of the set $S^2 \setminus \{N\}$ to the plane;

(b) the map that is the inverse to the stereographic projection is a C^∞-smooth parametrization of the set $S^2 \setminus \{N\}$;

(c) the angle between two intersecting curves on the sphere S^2 is the same as the angle between their images under the stereographic projection (this property is called the conformality of the stereographic projection).

4. Prove that every smooth surface in $\mathbb{R}^m$ is, locally, the graph of a smooth function.

5. Let $R > 0$. Consider the map Φ that sends each point $\varphi = (\varphi_1, \dots, \varphi_{m-1})$ in $\mathbb{R}^{m-1}$ ($m \geqslant 3$) to the vector $x = (x_1, \dots, x_m)$ in $\mathbb{R}^m$ according to the rule

$$\begin{aligned}
x_1 &= R\cos\varphi_1,\\
x_2 &= R\sin\varphi_1\cos\varphi_2,\\
x_3 &= R\sin\varphi_1\sin\varphi_2\cos\varphi_3,\\
&\ \ \vdots\\
x_{m-1} &= R\sin\varphi_1\sin\varphi_2\cdots\sin\varphi_{m-2}\cos\varphi_{m-1},\\
x_m &= R\sin\varphi_1\sin\varphi_2\cdots\sin\varphi_{m-2}\sin\varphi_{m-1}.
\end{aligned}$$

Prove that Φ sends the rectangular parallelepiped $[0, \pi]^{m-2} \times [0, 2\pi]$ onto the sphere of radius R (with center at zero) and the restriction of Φ to the open parallelepiped $(0, \pi)^{m-2} \times (0, 2\pi)$ maps the latter in a one-to-one manner onto the "cut sphere" (this cut is a compact subset of an $m - 2$-dimensional sphere). The numbers $\varphi_1, \dots, \varphi_{m-1}$ are called the spherical coordinates of a point in the boundary of a ball.

6. Let T_p be the tangent space to a smooth manifold M at a point p and let P be the orthogonal projection to T_p. Prove that for any $\varepsilon > 0$, in a sufficiently small M-neighborhood U of the point p the following inequality holds:

$$(1-\varepsilon)\|x-y\| \leqslant \big\|P(x)-P(y)\big\| \leqslant \|x-y\| \quad (x, y \in U).$$

7. It follows from the solution of the preceding problem that the restriction of the projection P to a sufficiently small M-neighborhood U of the point p is invertible. Prove that $(P|_U)^{-1}$ is a smooth map.

8. Let $M \subset \mathbb{R}^m$ be a smooth manifold with $\dim M < m$. Using the fact that the graph of a smooth function has zero measure (see Corollary 2.3.1), prove that $\lambda_m(M) = 0$.

8.2 Surface Area

8.2.1 By a k-dimensional area in $\mathbb{R}^m$ ($1 \leqslant k \leqslant m$) we will understand a Borel measure (see Sect. 2.2.3) satisfying properties similar to the properties of the Lebesgue measure λ_k. In particular, on subsets of k-dimensional affine subspaces, this measure must coincide with the Lebesgue measure. This allows us to speak about the area of sets consisting of planar parts, in particular, in the case $k = m - 1$, the faces of polyhedra. However, this does not, of course, suffice for a reasonable definition of the area of "curvilinear figures", and we must specify some property of area that would allow us to compare its values on non-planar sets (i.e., sets that do not lie in k-dimensional affine subspaces). In our approach, the role of such a condition is played by the following intuitively clear requirement: the area does not increase under a weak contraction (see Sect. 2.6.2, Property (6)). Since the image of a Borel set under a weak contraction (and even under a projection) may not be a Borel set, we will assume that the latter condition applies only to compact sets. So, we adopt the following definition.

Definition Let $k, m \in \mathbb{N}$, $1 \leqslant k \leqslant m$. A measure σ_k defined on the σ-algebra $\mathfrak{B}^m$ of all Borel subsets of $\mathbb{R}^m$ is called a *k-dimensional area* (in $\mathbb{R}^m$) if it satisfies the following two axioms:

(I) on every k-dimensional affine subspace L of $\mathbb{R}^m$, the measure σ_k coincides with the restriction of the Lebesgue measure λ_k to the σ-algebra of Borel subsets of L;

(II) on compact sets, σ_k does not increase under weak contractions: if Φ is a weak contraction of a compact set Q, then

$$\sigma_k\big(\Phi(Q)\big) \leqslant \sigma_k(Q).$$

We have agreed to call manifolds of codimension 1 surfaces. Hence it is natural to call an $(m-1)$-dimensional area a surface area. However, by abuse of language, we will use this term for a k-dimensional area for arbitrary k. As we will soon see, a k-dimensional area in $\mathbb{R}^m$ exists for all $k = 1, \dots, m$.

As for every Borel measure, a surface area enjoys the regularity property stated in Corollary 2.2.3:

$$\text{if } \sigma_k(E) < +\infty, \quad \text{then } \sigma_k(E) = \sup\{\sigma_k(Q) \mid Q \text{ is a compact set, } Q \subset E\}. \quad (1)$$

By axiom (I), σ_m coincides with the Lebesgue measure λ_m (more precisely, with its restriction to $\mathfrak{B}^m$). Property (II) holds for the Lebesgue measure by Corollary 2.6.4.

Note that axiom (II) and condition (1) imply that a surface area is invariant under any isometry, since both an isometry and the map inverse to an isometry are weak contractions. In particular, it is invariant under translations and rotations, so that the areas of congruent sets are equal. Under an orthogonal projection (which is, obviously, a weak contraction), the area of a compact set does not increase.

Let us establish another important property of σ_k.

Theorem *The area of a Borel set of finite area does not decrease under an expanding map.*

Proof Let $E \subset \mathbb{R}^m$ be a Borel set of finite area and Θ be an expanding map on E. By Proposition 2.3.3, the image $E' = \Theta(E)$ is again a Borel set. If E is a compact set, we can apply axiom (II) to the map Θ^{-1} (since it is a weak contraction), whence $\sigma_k(E) \leqslant \sigma_k(E')$. In the general case, use condition (1). □

We complement the theorem with a simple, but important result. It provides a two-sided bound on the area of a set that has a Lipschitz parametrization. This property will be repeatedly used in what follows.

Lemma *Let $E \subset \mathbb{R}^m$ be a Borel set and Ψ be a map from E to $\mathbb{R}^k$. If there exists a $C > 1$ such that*

$$\frac{1}{C}\|x - y\| \leqslant \|\Psi(x) - \Psi(y)\| \leqslant C\|x - y\| \quad \textit{for } x, y \in E,$$

then

$$\frac{1}{C^k}\lambda_k\big(\Psi(E)\big) \leqslant \sigma_k(E) \leqslant C^k\lambda_k\big(\Psi(E)\big).$$

Proof Obviously, it suffices to prove the desired inequality for bounded sets. Thus we assume that the set E and, consequently, its image E' are bounded. It follows from the assumptions of the lemma that $H = C\Psi$ and $\Theta : u \mapsto \Psi^{-1}(Cu)$ are expanding maps on E and $\frac{1}{C}E'$, respectively. As we have established in Proposition 2.3.3, each of these maps sends a Borel set to a Borel set. Since the weak contraction H^{-1} is uniformly continuous, we may assume that it is defined on the compact set $Q = C\overline{E'}$, whose area (coinciding with the Lebesgue measure) is finite. Therefore,

$$\sigma_k(E) = \sigma_k\big(H^{-1}(CE')\big) \leqslant \sigma_k\big(H^{-1}(Q)\big) \leqslant \sigma_k(Q) = \lambda_k(Q) < +\infty.$$

This allows us to apply the theorem to the expanding map H and obtain the required upper bound:

$$\sigma_k(E) \leqslant \sigma_k\big(H(E)\big) = \lambda_k\big(C\Psi(E)\big) = C^k\,\lambda_k\big(\Psi(E)\big).$$

On the other hand,

$$\sigma_k(E) = \sigma_k\left(\Theta\left(\frac{1}{C}\Psi(E)\right)\right) \geqslant \sigma_k\left(\frac{1}{C}\Psi(E)\right) = \lambda_k\left(\frac{1}{C}\Psi(E)\right) = \frac{1}{C^k}\lambda_k\big(\Psi(E)\big),$$

which yields the lower bound. □

Remark As can be seen from the proof, the lemma remains valid if the codomain of Ψ is not $\mathbb{R}^k$, but a k-dimensional (linear or affine) subspace of a space of arbitrary dimension, where a k-dimensional area coincides with the Lebesgue measure by axiom (I).

8.2.2 The question of the existence of a surface area has been essentially solved. Indeed, the Hausdorff measure μ_k satisfies axiom (II) (see Property (6) in Sect. 2.6.2) and is proportional to the Lebesgue measure on k-dimensional subspaces. As we have established in Sect. 2.6.5, the proportionality coefficient is equal to the volume α_k of the unit ball in $\mathbb{R}^k$. Therefore, the function $\alpha_k\mu_k$ regarded on all Borel subsets of $\mathbb{R}^m$ is a k-dimensional area. Thus the following theorem holds.

Theorem *For every positive integer k, $1 \leqslant k \leqslant m$, there is a k-dimensional area in $\mathbb{R}^m$.*

One can show (see [F]) that a Borel measure satisfying conditions (I) and (II) is not unique. The discussion of related subtle results is beyond the scope of this book. Note, however, that the non-uniqueness of area may manifest itself only on quite complicated sets. We will soon see that the area of Borel sets satisfying some natural geometric conditions is uniquely determined.

8.2.3 Now we turn to the one-dimensional case and consider the problem of computing the measure σ_1 on *simple arcs*, i.e., homeomorphic images of intervals. For brevity, we will also use the term “arc” as a synonym. Of course, it is natural to call $\sigma_1(L)$ the length of an arc L. However, using this term now may cause a certain ambiguity. Indeed, way back in school the reader learned, in the example of a circle, the definition of the length of a curve as the limit of the lengths of inscribed polygonal chains. A natural generalization of this definition leads to the classical definition of arc length. Thus it is desirable to ensure that the measure σ_1 agrees with this definition. Before proceeding to the solution of this problem, we first introduce the notion of the length of a path.

Consider an arbitrary path $\gamma : [a, b] \to \mathbb{R}^m$. Given a partition τ of the interval $[a, b]$ formed by points $t_0 = a < t_1 < \cdots < t_n = b$, set

$$S_\tau = \sum_{i=0}^{n-1} \|\gamma(t_{i+1}) - \gamma(t_i)\|.$$

By definition, the *length* of γ is equal to $s(\gamma) = \sup_\tau S_\tau$. A path is called *rectifiable* if it has finite length. Note that $s(\gamma) \geqslant \|\gamma(b) - \gamma(a)\|$, since $S_\tau \geqslant \|\gamma(b) - \gamma(a)\|$ by the triangle inequality.

If $\varphi_1, \ldots, \varphi_m$ are the coordinate functions of γ, then, obviously, for any $i = 1, \ldots, m$ and $k = 1, \ldots, n$,

$$|\varphi_i(t_{k+1}) - \varphi_i(t_k)| \leqslant \|\gamma(t_{k+1}) - \gamma(t_k)\| \leqslant \sum_{j=1}^{m} |\varphi_j(t_{k+1}) - \varphi_j(t_k)|.$$

Hence, the variations of the functions φ_k (see definition in Sect. 4.11.1) satisfy the inequality

$$\mathbf{V}_a^b(\varphi_i) \leqslant s(\gamma) \leqslant \sum_{j=1}^{m} \mathbf{V}_a^b(\varphi_j). \tag{1}$$

Thus a path is rectifiable if and only if all its coordinate functions are of bounded variation. Reproducing the proof of Theorem 4.11.1, one can see that the path length is additive, i.e., $s(\gamma) = s(\gamma_1) + s(\gamma_2)$, where γ_1, γ_2 are the restrictions of γ to the intervals $[a, c]$ and $[c, b]$, respectively ($a < c < b$).

To define the classical arc length, we need an easy auxiliary result. For a path, which is a homeomorphism of a line segment onto an arc, we keep the term "parametrization", even though it is defined not on an open interval, as it should be according to Definition 8.1.1, but on a closed interval.

Lemma *The lengths of two parametrizations of a simple arc coincide.*

Proof Let $\gamma : [a, b] \to \mathbb{R}^m$, $\gamma_1 : [p, q] \to \mathbb{R}^m$ be two parametrizations of a simple arc L. Set $\omega(x) = \gamma_1^{-1}(\gamma(x))$ ($a \leqslant x \leqslant b$). Then the function ω is continuous, one-to-one, and, consequently, strictly monotone. Therefore, every partition $\tau = \{x_0, \ldots, x_n\}$ of the interval $[a, b]$ gives rise to a partition of the interval $[p, q]$, which is formed by the points $\omega(x_0), \ldots, \omega(x_n)$ if ω is increasing, and by the points $\omega(x_n), \ldots, \omega(x_0)$ if ω is decreasing. Furthermore, $\gamma(x) = \gamma_1(\omega(x))$. Hence

$$S_\tau = \sum_{k=0}^{n-1} \|\gamma(x_{k+1}) - \gamma(x_k)\| = \sum_{k=0}^{n-1} \|\gamma_1(\omega(x_{k+1})) - \gamma_1(\omega(x_k))\| \leqslant s(\gamma_1).$$

But τ is arbitrary, whence $s(\gamma) \leqslant s(\gamma_1)$. Since the parametrizations γ and γ_1 are interchangeable, this means that $s(\gamma) = s(\gamma_1)$. □

Let us define the *length of an arc* as the common value of the lengths of all its parametrizations. The length of an arc L will be denoted by $s(L)$. Thus $s(L) = s(\gamma)$ if γ is a parametrization of L (using the symbol s both for the length of an arc and the length of its parametrization does not lead to a contradiction). An arc is called *rectifiable* if it has finite length. Since the length of a path is not less than the distance between its endpoints, the arc length satisfies the clear geometric principle "a line segment is the shortest arc connecting two given points":

$$s(L) \geqslant \|B - A\| \quad \text{if } L \text{ contains } A \text{ and } B. \tag{2}$$

It follows from axiom (II) that this principle can be extended to (any) measure σ_1:

$$\sigma_1(L) \geqslant \|B - A\| \quad \text{if } L \text{ contains } A \text{ and } B \tag{2'}$$

(since the projection of L to the line passing through A and B contains the whole segment connecting them).

Note also that if a path γ is rectifiable, then the function $x \mapsto \theta(x) = s(\gamma|_{[a,x]})$ is continuous. Indeed, if $a \leqslant x < y \leqslant b$, then, in view of (1) (hereafter $\varphi_1, \ldots, \varphi_m$ are the coordinate functions of the parametrization γ),

$$\left|\theta(y) - \theta(x)\right| = s(\gamma|_{[x,y]}) \leqslant \sum_{j=1}^{m} \mathbf{V}_x^y(\varphi_j),$$

where $\mathbf{V}_x^y(\varphi_j) \to 0$ as $x - y \to 0$ by Theorem 4.11.2.

The continuity of θ allows us to introduce a new parametrization of a rectifiable arc L. Since the set of values of θ coincides with $[0, S]$ where $S = s(L)$, the map $u \mapsto \delta(u) = \gamma(\theta^{-1}(u))$ is defined on $[0, S]$, continuous, one-to-one, and satisfies $\delta([0, S]) \subset L$. The reader can easily check that $\delta([0, S]) = L$. Thus δ is a parametrization of L. It follows from the definition of θ that for $0 \leqslant u \leqslant S$ we always have $s(\delta([0, u])) = u$. Furthermore, by the additivity of length, this parametrization also has the following property: if $0 \leqslant u_1 < u_2 \leqslant S$, then $s(\delta([u_1, u_2])) = u_2 - u_1$. Thus the parameter u in δ has a simple geometric interpretation: the difference of two values u_1, u_2 (where $u_1 < u_2$) is equal to the length of the arc corresponding to the interval $[u_1, u_2]$. This parametrization of a simple arc is called *natural*. It is a weak contraction on $[0, S]$, since $u_2 - u_1 = s(\delta([u_1, u_2])) \geqslant \|\delta(u_2) - \delta(u_1)\|$ by (2).

Theorem *For every simple arc L, $\sigma_1(L) = s(L)$.*

Proof First we check that $s(L) \leqslant \sigma_1(L)$. Let γ be a parametrization of L defined on $[a, b]$. Consider an arbitrary partition τ of $[a, b]$ formed by points $t_0 = a < t_1 <$

$\cdots < t_n = b$ and set $L_k = \gamma([t_k, t_{k+1}])$ $(0 \leqslant k < n)$. Since $\|\gamma(t_{k+1}) - \gamma(t_k)\| \leqslant \sigma_1(L_k)$ by (2′), it follows that

$$S_\tau \equiv \sum_{k=0}^{n-1} \|\gamma(t_{k+1}) - \gamma(t_k)\| \leqslant \sum_{k=0}^{n-1} \sigma_1(L_k) = \sigma_1(L).$$

But τ is arbitrary, whence $s(L) \leqslant \sigma_1(L)$.

When proving the reverse inequality, we may assume that L is rectifiable, i.e., $S = s(L) < +\infty$. As we have already noticed, the natural parametrization of L is a weak contraction on $[0, S]$. Hence $\sigma_1(L) \leqslant \sigma_1([0, S]) = S = s(L)$. □

This theorem allows us to call σ_1 the length.

8.2.4 As we have seen in the previous subsection, the σ_1 measure of any simple arc L is equal to the supremum of the lengths of polygonal chains inscribed into L. "Common sense" suggests that in order to compute the area of a curved surface M, we should apply a similar procedure: consider polyhedral surfaces with vertices on M (polyhedra inscribed in M), compute the sum of the areas of their faces, and then take the limit as the faces get smaller and smaller. However, simple analysis shows that this approach cannot lead to a reasonable result even for a cylinder. We briefly sketch the construction of the corresponding classical counterexample, the so-called "Schwarz[4] lantern".

Consider a right circular cylinder of radius R and height H and inscribe into it a polyhedral surface constructed as follows. Cut the cylinder into m equal small cylinders by planes perpendicular to its axis. Into the top and the bottom of each small cylinder inscribe a regular n-gon so that the vertices of the top polygon lie above the middles of the arcs subtended by the sides of the bottom polygon. In other words, the top polygon is rotated by $\frac{\pi}{n}$ with respect to the bottom one. Join each vertex of each polygon with the closest vertices of the polygons one level up or down by line segments. The pair of such segments going from a given vertex to a neighboring polygon, along with the corresponding side of this polygon, forms an isosceles triangle. These triangles together form a polyhedral surface that resembles a Chinese lantern (see Fig. 8.1).

Between two neighboring cross sections there are $2n$ triangular faces (half of them are based on the bottom n-gon, and the other half, on the top n-gon). Thus our polyhedral surface consists of $2mn$ equal triangular faces. Clearly, the lengths of the sides of these faces tend to zero as $m, n \to \infty$. Observe that the planes of the faces are almost perpendicular to the axis of the cylinder provided that the height of the levels is small compared to the side length of the polygons.

The area s_{mn} of each face is easy to compute:

$$s_{mn} = R \sin \frac{\pi}{n} \sqrt{\left(2R \sin^2 \frac{\pi}{2n}\right)^2 + \left(\frac{H}{m}\right)^2}.$$

[4] Karl Hermann Amandus Schwarz (1843–1921)—German mathematician.

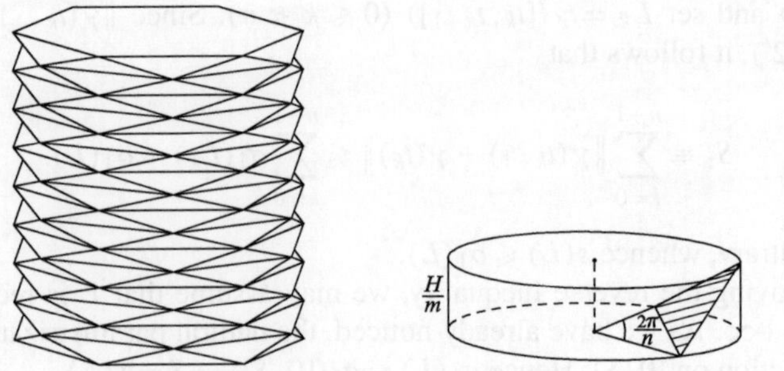

Fig. 8.1 "Schwarz lantern"

Discarding the second term under the square root sign, we have (taking into account that $\sin\varphi \geqslant \frac{2}{\pi}\varphi$ for $\varphi \in [0, \frac{2}{\pi}]$)

$$s_{mn} \geqslant R\sin\frac{\pi}{n}\cdot 2R\sin^2\frac{\pi}{2n} \geqslant \frac{4R^2}{n^3}.$$

Hence the total area of the polyhedral surface, i.e., $2mns_{mn}$, is not less than $8R^2\frac{m}{n^2}$. Therefore, we can inscribe into the cylinder a polyhedral surface of arbitrarily large area (taking $m \gg n^2$), though its faces are triangles with arbitrarily small sides. Note that the limit is equal to the area of the cylinder, i.e., $2\pi RH$, only if $\frac{m}{n^2} \to 0$.

The above construction makes it clear why the approach which is effective when computing arc lengths fails if we need to compute areas. The reason is simple: the segments of a polygonal chain inscribed into a smooth curve are almost tangent to the curve provided that their lengths are sufficiently small. For surfaces, the situation is quite different: arbitrarily small faces of a polyhedron inscribed into a curved surface may be almost orthogonal to the surface (the inscribed surface may be "rugged"). Thus, when computing the area of a surface, we should abandon the naive approach related to inscribed polyhedra.

EXERCISES

1. Show that any two parametrizations of a simple arc can be obtained from each other by a strictly monotone change of variables. Deduce (without using Theorem 8.2.3) that the classical length of a simple arc does not depend on the parametrization.
2. Using only the definition of the classical length, but not Theorem 8.2.3, show that the length is additive: $s(L) = s(L_1) + s(L_2)$, where L is a simple arc, $L_1 = \gamma([a, c])$, $L_2 = \gamma([c, b])$, $a < c < b$, and γ is an arbitrary parametrization of L.
3. Let L be a smooth simple arc and denote by $L_{x,y}$ the subarc of L with endpoints x, y. Show that for every point $p \in L$,

$$\lim_{x,y\to p}\frac{s(L_{x,y})}{\|x-y\|} = 1,$$

i.e., the length of an arc shrinking to a point is equivalent to the length of the corresponding chord.

4. Show that the length of the boundary of a planar domain is not less than the length of the boundary of its convex hull. Does a similar result hold in the three-dimensional case?

5. Show that σ_1 is uniquely determined on Borel subsets of rectifiable arcs.

6. Let $f \in C([a,b])$. Prove that if f is monotone, then the length of its graph does not exceed $b - a + |f(b) - f(a)|$. Show that the inequality is strict if f (where $f \not\equiv \text{const}$) satisfies the Lipschitz condition.

7. What is the length of the graph of the Cantor function?

8. Show that the interval $[0, 1)$, regarded as a measure space with Lebesgue measure, is isomorphic (for the definition, see Exercise 11 of Sect. 4.10) to the unit circle with measure $\frac{1}{2\pi}\sigma_1$.

8.3 Properties of the Surface Area of a Smooth Manifold

In this section, all manifolds and parametrizations are assumed smooth by default. Manifolds of codimension one are called surfaces.

8.3.1 Our immediate goal is to obtain a formula for computing the area of Borel subsets of a simple smooth k-dimensional manifold. Then, by the countable additivity of the area, we will be able to compute the areas of countable unions of such sets, in particular, the areas of subsets of arbitrary smooth manifolds.

Let us discuss geometric considerations that suggest what form the desired formula should have. Let $\Phi \in C^1(\mathcal{O}, \mathbb{R}^m)$ be a parametrization of a simple manifold M with $\dim M = k$, and let $\widetilde{\Phi}(t) = \Phi(a) + d_a\Phi(t - a)$ be the linearization of Φ at a point $a \in \mathcal{O}$. The set of values of $\widetilde{\Phi}$ is the k-dimensional affine tangent space, on which the Lebesgue measure λ_k is defined. Consider a cubic cell $Q_h \subset \mathcal{O}$ with a vertex at a and edge length h. Its image under Φ is a "curved parallelotope" $R_h = \Phi(Q_h)$. For small h, it is almost isometric to the corresponding "scale" $\widetilde{R}_h$, the image of Q_h under the linearized map $\widetilde{\Phi}$ (see the lemma in the next section). Hence we should expect that the area of the "curved parallelotope" R_h is close to the Lebesgue measure of the set $\widetilde{R}_h$. It can be obtained by a translation from the parallelotope $d_a\Phi([0,h)^k)$ lying in the tangent space. Therefore, $\lambda_k(\widetilde{R}_h) = \lambda_k(d_a\Phi([0,h)^k)) = h^k\lambda_k(d_a\Phi([0,1)^k))$. We will call $C_a = d_a\Phi([0,1)^k)$ the *accompanying* parallelotope. Thus $\sigma_k(R_h)$ must be close to $h^k\lambda_k(C_a)$, in the sense that their ratio tends to one as h decreases. This suggests that the surface area of a simple smooth manifold M is just a weighted image (see Sect. 6.1.1) of the k-dimensional Lebesgue measure[5] under Φ, with the weight $\omega_\Phi : t \mapsto \lambda_k(C_t)$ equal to the volume of the accompanying parallelotope.

[5]More precisely, of its restriction to $\mathfrak{B}^k$.

8.3.2 In order to verify that our heuristic considerations are correct, we first prove a two-sided bound on the deviation of a point of a manifold from the tangent subspace. One of the possible "straightening" maps, which is of a simple geometric nature, could be obtained by orthogonally projecting a sufficiently small neighborhood of the tangency point to the tangent space (see Exercise 6 in Sect. 8.1). However, it is technically more convenient to associate a separate straightening map (close to a projection) with every parametrization.

Lemma *Let $\Phi \in C^1(\mathcal{O}, \mathbb{R}^m)$ be a local parametrization of a manifold M and $\widetilde{\Phi}(t) = \Phi(a) + d_a\Phi(t-a)$ be its linearization at a point $a \in \mathcal{O}$. Then:*

(1) *the map $\Psi = \widetilde{\Phi} \circ \Phi^{-1}$ is almost isometric in the vicinity of the point $p = \Phi(a)$: for every $C > 1$, in a sufficiently small M-neighborhood U of p,*

$$\frac{1}{C}\|x-y\| \leqslant \big\|\Psi(x)-\Psi(y)\big\| \leqslant C\|x-y\| \quad (x, y \in U);$$

(2) *for every set $A \in \mathfrak{A}^k$,*

$$\lambda_k\big(\widetilde{\Phi}(A)\big) = \omega_\Phi(a)\,\lambda_k(A). \tag{1}$$

Proof It suffices to check that in the vicinity of p the inequality

$$\big\|(x-y)-\big(\Psi(x)-\Psi(y)\big)\big\| \leqslant \left(1-\frac{1}{C}\right)\|x-y\|$$

holds. Since the map Φ^{-1} locally satisfies the Lipschitz condition (see Corollary 1 in Sect. 8.1.4), it suffices to check that for every $\varepsilon > 0$ there exists a small M-neighborhood U of the point p such that

$$\big\|(x-y)-\big(\Psi(x)-\Psi(y)\big)\big\| \leqslant \varepsilon\big\|\Phi^{-1}(x)-\Phi^{-1}(y)\big\| \quad (x, y \in U).$$

Setting $s = \Phi^{-1}(x)$ and $t = \Phi^{-1}(y)$, we see that this inequality is equivalent to the condition

$$\big\|\big(\Phi(s)-\Phi(t)\big)-d_a\Phi(s-t)\big\| \leqslant \varepsilon\|s-t\| \quad \text{in the vicinity of } a. \tag{2}$$

The latter follows from the smoothness of Φ. Indeed, let r be so small that $\|d_u\Phi - d_a\Phi\| \leqslant \varepsilon$ for every u from the k-dimensional ball $B(a, r)$. Then, by Corollary from Lagrange's inequality (see Sect. 13.7.2),

$$\big\|\big(\Phi(s)-\Phi(t)\big)-d_a\Phi(s-t)\big\| \leqslant \sup_{u \in B(a,r)} \|d_u\Phi - d_a\Phi\|\,\|s-t\|,$$

which implies (2). Thus, as a desired M-neighborhood U of p, we can take $\Phi(B(a, r))$ provided that the radius r is sufficiently small.

To prove (1), observe that the measure $A \mapsto \lambda_k(\widetilde{\Phi}(A))$ is translation-invariant and hence (see Sect. 2.4.2) proportional to λ_k. Since $\lambda_k(\widetilde{\Phi}([0,1)^k)) = \lambda_k(C_a) = \omega_\Phi(a)$, the proportionality coefficient is equal to $\omega_\Phi(a)$. □

Now we are in a position to prove a formula for computing the area of a set lying on a smooth manifold. The idea of the proof is the same as in Theorem 6.2.1.

Theorem *For every Borel set E contained in a simple smooth manifold M,*

$$\sigma_k(E) = \int_{\Phi^{-1}(E)} \omega_\Phi(t)\,dt, \tag{3}$$

where Φ is an arbitrary smooth parametrization of M.

As we mentioned in Sect. 8.2.2, the axioms of area do not uniquely determine it on all Borel sets. In contrast, the above theorem shows that on sufficiently "good" sets—smooth manifolds and their Borel subsets—all areas coincide. In Sect. 8.8 it is shown that the same is true for subsets of Lipschitz (in particular, convex) surfaces. Thus the difference between various areas may manifest itself only on Borel sets of quite a complicated nature.

Proof Let $\mathcal{O}$ be the open set on which the parametrization Φ is defined, and consider the measure $\nu(A) = \sigma_k(\Phi(A))$ on Borel subsets of $\mathcal{O}$. We will verify that it satisfies the condition

$$\inf_A \omega_\Phi\, \lambda_k(A) \leqslant \nu(A) \leqslant \sup_A \omega_\Phi \lambda_k(A). \tag{4}$$

As established in Theorem 6.1.2, this implies that $\nu(A) = \int_A \omega_\Phi(t)\,dt$, which is equivalent to the desired assertion.

If these inequalities hold for sets forming an increasing sequence, then they obviously hold for the union of these sets. Hence it suffices to prove (4) assuming that A is a bounded set whose closure is contained in $\mathcal{O}$. Both inequalities (4) are proved in the same way, so we will prove only the upper bound, leaving the reader to carry out a similar argument for the lower bound.

If the right inequality in (4) is false, then for some $C_0 > 1$ we have

$$\nu(A) > C_0 \sup_A \omega_\Phi \lambda_k(A). \tag{5}$$

Divide A into finitely many parts with the diameter of each part at most $\operatorname{diam}(A)/2$. Then (5) must hold for one of these parts, which we denote by A_1. Replacing A by A_1 and repeating the argument, we obtain a set A_2, etc. By induction, we will construct a sequence of nested sets A_n with diameters tending to zero. Take a point a in the intersection $\bigcap_{n=1}^{\infty} \overline{A}_n$. By the construction of the sets A_n, they satisfy (5):

$$\nu(A_n) > C_0 \sup_{A_n} \omega_\Phi \lambda_k(A_n). \tag{6}$$

Let us show that this leads to a contradiction. By the lemma, for every $C > 1$ (to be specified later) there exists a neighborhood V of a such that for $x, y \in U = \Phi(V)$

we have

$$\frac{1}{C}\|x-y\| \leqslant \|\Psi(x)-\Psi(y)\| \leqslant C\|x-y\|,$$

where, as in the lemma, $\Psi = \widetilde{\Phi} \circ \Phi^{-1}$ and $\widetilde{\Phi}$ is the linearization of Φ (i.e., $\widetilde{\Phi}(t) = \Phi(a) + d_a\Phi(t-a)$). If n is so large that $A_n \subset V$, then $\Phi(A_n) \subset U$. Hence, by Lemma 8.2.1, for the set $E_n = \Phi(A_n)$ we have $\sigma_k(E_n) \leqslant C^k \lambda_k(\Psi(E_n))$. Since $\Psi(E_n) = \widetilde{\Phi}(A_n)$ and, according to (1), $\lambda_k(\widetilde{\Phi}(A_n)) = \omega_\Phi(a)\lambda_k(A_n)$, it follows that

$$\nu(A_n) = \sigma_k(\Phi(A_n)) = \sigma_k(E_n) \leqslant C^k \lambda_k(\Psi(E_n)) = C^k \lambda_k(\widetilde{\Phi}(A_n))$$
$$\leqslant C^k \sup_{A_n} \omega_\Phi \lambda_k(A_n).$$

From (6) and the last inequality we see that $1 < C_0 \leqslant C^k$. However, this is not possible if C is chosen sufficiently close to 1. Therefore, our assumption is false, and the theorem follows. □

A special case of this result (for $k = m$) is Theorem 6.2.1 on the behavior of the Lebesgue measure under a diffeomorphism, since in this case $\omega_\Phi = J_\Phi \equiv |\det(\Phi')|$. The proofs of these theorems are similar, but here we have used the properties of area, which has allowed us not to keep track of the measures of the images of small cubic cells.

8.3.3 Here we will discuss the basic properties of the area σ_k in the space $\mathbb{R}^m$, always assuming that $k < m$. For brevity, we will call it just the area, omitting the reference to the dimension.

It is clear that the properties of σ_k substantially differ in some respects from the familiar properties of the Lebesgue measure. For example, since every m-dimensional cube contains a continuum of congruent pairwise disjoint k-dimensional cubes, in $\mathbb{R}^m$ there are compact sets of infinite area. It is also clear that every non-empty open set has infinite area. This fact immediately implies that the area is not σ-finite and cannot be a regular measure. Thus, when studying the properties of σ_k in more detail, we will rather consider not the area "as a whole", but its restrictions to subsets contained in a fixed manifold. To be more precise, we introduce the following notation related to a smooth k-dimensional manifold $M \subset \mathbb{R}^m$ (with $k < m$). By $\mathfrak{B}_M$ we denote the system of all Borel sets contained in M, and by σ_M the restriction of the k-dimensional area to $\mathfrak{B}_M$. Since M itself is a Borel set, $\mathfrak{B}_M$ is a σ-algebra and σ_M is a measure.

Formula (3) allows us to compute the area of a Borel subset of a simple manifold. It is obviously valid also for subsets of a coordinate neighborhood of an arbitrary, not necessarily simple, manifold provided that Φ is the corresponding parametrization.

Let us establish several important properties of the area.

(1) *The area of a compact subset of a smooth manifold is finite.*

For a compact subset of a coordinate neighborhood, this follows from (3), since its inverse image under every parametrization is a compact set and the weight ω_Φ

is continuous. An arbitrary compact subset of a manifold can be covered by finitely many coordinate neighborhoods of finite area.

(2) *The measure σ_M is σ-finite.*

This property follows from the previous one and Corollary 2 from Sect. 8.1.5.

Since Borel sets of zero measure may have non-Borel subsets, the measure σ_M is not complete. To obtain a complete measure, we should consider its Carathéodory extension. The σ-algebra on which it is defined will be denoted by $\mathfrak{A}_M$, and the elements of $\mathfrak{A}_M$ will be called *Lebesgue measurable* or, in short, *measurable*. By Theorem 1.5.1, the extension of σ_M to $\mathfrak{A}_M$ is unique. For this extension we keep the old notation σ_M and still call it the area (of the manifold M). By Corollary 1.5.2, every measurable set can be approximated from the inside and from the outside by Borel sets of the same measure.

If Φ is a parametrization of a simple manifold M and $E \subset M$ is an arbitrary Lebesgue measurable set, then formula (3), established for Borel sets, remains valid. Thus:

(3) *the measure σ_M on a simple k-dimensional manifold M is a weighted image of the k-dimensional Lebesgue measure with respect to an arbitrary parametrization Φ. A subset of a simple manifold is measurable if and only if its inverse image is measurable.*

Indeed, a measurable set E can be written in the form $E = A \cup e$, where A is a Borel set and $\sigma_M(e) = 0$. Besides, $e \subset e'$, where e' is a Borel set of zero area. Hence $\Phi^{-1}(e) \subset \Phi^{-1}(e')$ and, moreover, $\lambda_k(\Phi^{-1}(e')) = 0$. The latter holds, since $0 = \sigma_M(e') = \int_{\Phi^{-1}(e')} \omega_\Phi(t)\,dt$ and $\omega_\Phi > 0$. By the completeness of the Lebesgue measure, the set $\Phi^{-1}(e)$ is measurable (and has zero measure). Therefore,

$$\sigma_M(E) = \sigma_M(A) = \int_{\Phi^{-1}(A)} \omega_\Phi(t)\,dt = \int_{\Phi^{-1}(E)} \omega_\Phi(t)\,dt.$$

In a similar way one can show that the measurability of $\Phi^{-1}(E)$ implies the measurability of E.

(4) *The measure σ_M is regular*, i.e. (see Sect. 2.2.2),

$$\sigma_M(E) = \inf_{\substack{E \subset G \subset M \\ G \text{ is open in } M}} \sigma_M(G) = \sup_{\substack{E \supset F \\ F \text{ is compact set}}} \sigma_M(F)$$

for every measurable set E, $E \subset M$.

If M is a simple manifold, then this property is an immediate consequence of the regularity of the Lebesgue measure and formula (3). We leave the reader to prove it in the case of an arbitrary manifold.

(5) *Let L be an arbitrary smooth manifold*, $\dim L < k$. *Then $\sigma_k(L) = 0$.*

Let us show that every point from L has a neighborhood U such that the intersection $L \cap U$ has zero area (this suffices because, by Lindelöf's theorem, L can be covered by countably many such neighborhoods).

According to the second definition of a smooth manifold, U can be chosen in such a way that $L \cap U$ is contained in a simple smooth manifold M of dimension k. Moreover, we may assume without loss of generality that $\dim L = k-1$.

Let Φ be a parametrization of M. Then $\Phi^{-1}(L)$ is a smooth surface in $\mathbb{R}^k$, which can be written as a countable union of graphs of smooth functions (see Corollary 4 in Sect. 8.1.5). Since the k-dimensional volume of every such graph vanishes by Corollary 2.3.1, we have $\lambda_k(\Phi^{-1}(L)) = 0$. Thus $\sigma_k(L) = \int_{\Phi^{-1}(L)} \omega_\Phi(t)\,dt = 0$.

It follows, for example, that when computing the area of a subset of a sphere, we may discard manifolds of smaller dimension. This allows us to assume without loss of generality that the set under consideration is contained in the "cut" sphere and use the corresponding parametrization and formula (3).

(6) *Under the homothety with ratio $a > 0$, the measure of a set contained in a k-dimensional manifold M is multiplied by a^k:* $\sigma_k(aE) = a^k \sigma_k(E)$ if $E \in \mathfrak{A}_M$.

Indeed, the area is proportional to the Hausdorff measure, which has the desired property (see Property (8) in Sect. 2.6.2).

Note that in the case of an arbitrary linear transformation of a manifold M, the measures on M and on its image do not have such a simple relation. To see this, it suffices to consider compressing a circle: a simple calculation shows that the length of an arc of an ellipse with eccentricity $\varepsilon \neq 1$ can be expressed via the elliptic integral $\int_0^\varphi \sqrt{1-\varepsilon^2 \sin^2 t}\,dt$, which is not an elementary function.

In conclusion, we state the property of the area mentioned immediately after Definition 8.2.1.

(7) *The area is invariant under isometries.*

In particular, the area on a sphere is rotation-invariant.

8.3.4 In order to compute the areas of manifolds via formula (3), we need explicit expressions for the weight $\omega_\Phi(t)$ equal to the measure of the accompanying parallelotope C_t. For $k = 1$, C_t is just the line segment with endpoints 0 and $\Phi'(t)$. Hence $\omega_\Phi(t) = \lambda_1(C_t) = \|\Phi'(t)\|$, so that, in order to compute, for example, the length of an arc $L = \Phi([a,b])$, we should integrate the length of the tangent vector: $\sigma_1(L) = \int_a^b \|\Phi'(t)\|\,dt$.

In the general case, the parallelotope C_t is spanned by the canonical tangent vectors $\tau_1 = \tau_1(t), \ldots, \tau_k = \tau_k(t)$ corresponding to the parametrization Φ. Since they are linearly independent, the volume of C_t is positive, i.e., we always have $\omega_\Phi(t) > 0$. The value $\omega_\Phi(t)$ can be computed via the Gram determinant (see Sect. 2.5.3):

$$\omega_\Phi(t) = \lambda_k(C_t) = \sqrt{\Gamma(\tau_1, \ldots, \tau_k)} = \sqrt{\det\bigl(\bigl(\langle \tau_i, \tau_j\rangle\bigr)_{i,j=1}^k\bigr)}$$

$$= \sqrt{\det\bigl[\bigl(\Phi'(t)\bigr)^T \Phi'(t)\bigr]}.$$

It follows from the Binet–Cauchy formula that

$$\omega_\Phi^2(t) = \det\big[\big(\Phi'(t)\big)^T\, \Phi'(t)\big] = \sum_{j_1<j_2<\cdots<j_k} \left(\frac{D(\varphi_{j_1}, \varphi_{j_2}, \dots, \varphi_{j_k})}{D(t_1, t_2, \dots, t_k)}(t)\right)^2,$$

where $\varphi_1, \varphi_2, \dots, \varphi_m$ are the coordinate functions of Φ.

For a surface, i.e., in the case $k = m - 1$, the expression for $\omega_\Phi(t)$ provided by the Binet–Cauchy formula simplifies:

$$\omega_\Phi^2(t) = \sum_{j=1}^{m} \left(\frac{D(\varphi_1, \dots, \widehat{\varphi}_j, \dots, \varphi_m)}{D(t_1, \dots, t_{m-1})}(t)\right)^2$$

(the symbol $\widehat{\ }$ indicates that the corresponding function is omitted).

The right-hand side has a simple geometric interpretation. Let $e_1, \dots, e_m$ be the canonical basis in $\mathbb{R}^m$. Consider the vector

$$N_\Phi(t) = \sum_{j=1}^{m} (-1)^{j+1} \frac{D(\varphi_1, \dots, \widehat{\varphi}_j, \dots, \varphi_m)}{D(t_1, \dots, t_{m-1})}(t) \cdot e_j.$$

It can be written via the symbolic determinant

$$N_\Phi(t) = \begin{vmatrix} e_1 & e_2 & \dots & e_m \\ \frac{\partial \varphi_1(t)}{\partial t_1} & \frac{\partial \varphi_2(t)}{\partial t_1} & \dots & \frac{\partial \varphi_m(t)}{\partial t_1} \\ \frac{\partial \varphi_1(t)}{\partial t_2} & \frac{\partial \varphi_2(t)}{\partial t_2} & \dots & \frac{\partial \varphi_m(t)}{\partial t_2} \\ \vdots & \vdots & \ddots & \vdots \\ \frac{\partial \varphi_1(t)}{\partial t_{m-1}} & \frac{\partial \varphi_2(t)}{\partial t_{m-1}} & \dots & \frac{\partial \varphi_m(t)}{\partial t_{m-1}} \end{vmatrix},$$

whose rows, except for the first one, consist of tangent vectors.

The vector $N_\Phi(t)$ is orthogonal to each tangent vector $\tau_j(t)$, since the inner product $\langle N_\Phi(t), \tau_j(t)\rangle$ can be written as a determinant with two equal rows. Thus $N_\Phi(t)$ is a normal to M at the point $\Phi(t)$. It will be called the *normal corresponding to the parametrization* Φ. Obviously, the length of $N_\Phi(t)$ is equal to $\omega_\Phi(t)$.

For $m = 3$, $k = 2$, we see that $N_\Phi(t)$ is simply the vector product of the tangent vectors: $N_\Phi(t) = \tau_1(t) \times \tau_2(t)$. Clearly,

$$\omega_\Phi^2 = \begin{vmatrix} \langle \tau_1, \tau_1\rangle & \langle \tau_1, \tau_2\rangle \\ \langle \tau_2, \tau_1\rangle & \langle \tau_2, \tau_2\rangle \end{vmatrix} = EG - F^2,$$

where E, F, G are the coefficients of the first fundamental form of the surface, i.e., $E = \|\tau_1\|^2$, $G = \|\tau_2\|^2$ and $F = \langle \tau_1, \tau_2\rangle$. If the tangent vectors τ_1 and τ_2 are orthogonal, then $\omega_\Phi = \|\tau_1\| \cdot \|\tau_2\|$.

If $M = \Gamma_f$ is the graph of a function f from $C^1(\mathcal{O}, \mathbb{R})$, then the map

$$\mathcal{O} \ni x = (x_1, \dots, x_{m-1}) \mapsto \Phi(x) = \big(x, f(x)\big)$$

is its canonical parametrization, and

$$\frac{D(\varphi_1, \ldots, \widehat{\varphi}_j, \ldots, \varphi_m)}{D(x_1, \ldots, x_{m-1})}(x) = (-1)^{m+j+1} \frac{\partial f}{\partial x_j}(x)$$

for $1 \leqslant j \leqslant m-1$. If $j = m$, then this determinant is equal to one. Hence $N_\Phi(x) = (-1)^m (f'_{x_1}(x), \ldots, f'_{x_{m-1}}(x), -1)$ and $\omega_\Phi(x) = \sqrt{1 + \|\text{grad} f(x)\|^2}$.

The density ω_Φ corresponding to the canonical parametrization can easily be computed directly from geometric considerations, without using the general formula. Indeed, the tangent vectors corresponding to the canonical parametrization of the graph are $\tau_j = (0, \ldots, 0, 1, 0, \ldots, 0, f'_{x_j}(x))$. The projection to $\mathbb{R}^{m-1}$ sends the accompanying parallelotope C_x spanned by these vectors to the unit cube $[0, 1]^{m-1}$. Hence (see Sect. 2.4.6) $\lambda_{m-1}(C_x) = \frac{1}{|\cos\theta(x)|}$, where $\theta(x)$ is the angle between the last coordinate axis and an arbitrary normal to the tangent plane $T_{(x, f(x))}$. As is well known (see Example 1 in Sect. 8.1.3), one such normal is the vector $\nu(x) = (-f'_{x_1}(x), \ldots, -f'_{x_{m-1}}(x), 1)$ (observe that the normals $N_\Phi(x)$ and $\nu(x)$ coincide for odd m and are opposite for even m). Therefore,

$$\left|\cos\theta(x)\right| = \frac{\langle \nu(x), e_m \rangle}{\|\nu(x)\| \, \|e_m\|} = \frac{1}{\|\nu(x)\|}.$$

Hence

$$\omega_\Phi(x) = \lambda_{m-1}(C_x) = \frac{1}{|\cos\theta(x)|} = \left\|\nu(x)\right\| = \sqrt{1 + \left\|\text{grad} f(x)\right\|^2}.$$

Thus the area of every set E contained in the graph Γ_f can be computed by the formula

$$\sigma_{\Gamma_f}(E) = \int_{\Phi^{-1}(E)} \frac{dx}{|\cos\theta(x)|} = \int_{P(E)} \sqrt{1 + \left\|\text{grad} f(x)\right\|^2} \, dx, \tag{7}$$

where $P(E)$ is the orthogonal projection of E to $\mathbb{R}^{m-1}$.

In particular, if $f(x) = \sqrt{R^2 - \|x\|^2}$ for $x \in B^{m-1}(R)$, then Γ_f is the upper hemisphere $S_+^{m-1}(R)$. Since the radius vector of a point on the sphere $S^{m-1}(R)$ is a normal to $S^{m-1}(R)$, we have $\cos\theta = f(x)/R$. Hence for the area of a measurable set E lying on the hemisphere $S_+^{m-1}(R)$ we obtain the formula

$$\sigma_{\Gamma_f}(E) = \int_{P(E)} \frac{1}{|\cos\theta(x)|} dx = \int_{P(E)} \frac{R}{\sqrt{R^2 - \|x\|^2}} dx, \tag{7'}$$

which we already used in Sect. 6.5 for $R = 1$.

8.3.5 Now we consider several examples.

Example 1 (The area of subsets of a two-dimensional sphere) According to Property (5) from Sect. 8.3.3, when computing the areas of subsets of the sphere

$S^2(R) = \{(x, y, z) \mid x^2 + y^2 + z^2 = R^2\}$, one may discard smooth curves. We introduce spherical coordinates φ, θ, i.e., consider the map

$$(\varphi, \theta) \mapsto \Phi(\varphi, \theta) = (R\cos\varphi\cos\theta,\ R\sin\varphi\cos\theta,\ R\sin\theta) \in S^2$$
$$\left(|\varphi| < \pi,\ |\theta| < \frac{\pi}{2}\right).$$

It provides a parametrization of the sphere cut along the meridian $\varphi = \pm\pi$. The coordinate lines for this parametrization are meridians and parallels. This allows one to easily compute the approximate area of the small quadrilateral bounded by the coordinate lines corresponding to the angles $\varphi, \varphi + h$ and $\theta, \theta + h$ ($h > 0$). Since the meridians and the parallels are orthogonal, the accompanying parallelotope is a rectangle whose side lengths for small h are almost equal to the lengths of the arcs bounding the curved quadrilateral. The latter are circular arcs of radius R (along the meridian) and $R\cos\theta$ (along the parallel). Hence the area of the accompanying parallelogram is approximately equal to $(R^2\cos\theta)\,h^2$. This suggests that the weight ω_Φ corresponding to the parametrization in question is equal to $R^2\cos\theta$. We leave the reader to check that the obtained result is correct by finding the tangent vectors and computing the corresponding Gram determinant.

Knowing the weight, one can easily find the area of a set lying on the sphere. For simplicity, consider a spherical quadrilateral Q bounded by parallels and meridians:

$$Q = \left\{\Phi(\varphi, \theta) \,\Big|\, -\pi \leqslant \alpha_1 < \varphi < \alpha_2 \leqslant \pi,\ -\frac{\pi}{2} \leqslant \beta_1 < \theta < \beta_2 \leqslant \frac{\pi}{2}\right\}.$$

Obviously,

$$\sigma_2(Q) = \int_{\alpha_1}^{\alpha_2}\int_{\beta_1}^{\beta_2} R^2\cos\theta\,d\varphi\,d\theta = R^2(\sin\beta_2 - \sin\beta_1)(\alpha_2 - \alpha_1).$$

For the extreme values of $\alpha_1, \alpha_2, \beta_1, \beta_2$, we obtain the area of the whole sphere: $\sigma_2(S^2(R)) = 4\pi R^2$.

Example 2 (The area of subsets of a two-dimensional torus) According to Property (5) from Sect. 8.3.3, when computing the areas of subsets of the torus $T^2 = \{(x, y, z) \mid (R - \sqrt{x^2 + y^2})^2 + z^2 = r^2\}$, we may discard smooth curves. Let us find the weight corresponding to the parametrization of the cut torus considered in Example 3 from Sect. 8.1.3. Recall that this parametrization Φ is as follows:

$$\Phi(\varphi, \theta) = \big((R + r\cos\theta)\cos\varphi,\ (R + r\cos\theta)\sin\varphi,\ r\sin\theta\big),$$

where $\varphi,\ \theta \in (-\pi, \pi)$.

As in the case of a sphere, the coordinate lines (the analogs of parallels and meridians) form two families of orthogonal circles. Computing, as in Example 1, the area of a small curved quadrilateral bounded by coordinate lines, we arrive at the plausible conclusion that the weight corresponding to the chosen parametrization has the form $\omega_\Phi(\varphi,\theta)=r(R+r\cos\theta)$. The reader can easily perform all necessary formal calculations.

It is clear that the area of the curved quadrilateral on the torus bounded by "meridians" $\varphi=\alpha_1$, $\varphi=\alpha_2$ and "parallels" $\theta=\beta_1$, $\theta=\beta_2$ is equal to

$$\int_{\alpha_1}^{\alpha_2}\int_{\beta_1}^{\beta_2} r(R+r\cos\theta)\,d\varphi\,d\theta = r\big(R(\beta_2-\beta_1)+r(\sin\beta_2-\sin\beta_1)\big)(\alpha_2-\alpha_1).$$

For the extreme values of $\alpha_1,\alpha_2,\beta_1,\beta_2$, we obtain the area of the whole torus: $\sigma_2(T^2)=4\pi^2 rR$.

Example 3 (The area of subsets of a conic surface) Let us find out how the area of a set E lying on the conic surface $\{(x,y)\,|\,x\in\mathbb{R}^{m-1},\ y=c\|x\|\}$ is related to the area of its projection $P(E)$ to the plane $y=0$. If $0\notin E$, then E lies on a smooth surface, the graph of the function $f(x)=c\|x\|$ defined on $\mathbb{R}^{m-1}\setminus\{0\}$. One can easily see that $\|\operatorname{grad} f(x)\|=|c|$ everywhere (the angle between the tangent plane and the plane $y=0$ is constant). Hence

$$\sigma_{m-1}(E)=\int_{P(E)}\sqrt{1+\|\operatorname{grad} f(x)\|^2}\,dx=\sqrt{1+c^2}\,\lambda_{m-1}\big(P(E)\big).$$

We now consider several examples related to the multi-dimensional sphere.

Example 4 (The area of a multi-dimensional sphere) To compute the area of a sphere $S^{m-1}(R)$ of arbitrary radius, it suffices to compute the area of the unit sphere S^{m-1}, since $S^{m-1}(R)=RS^{m-1}$ and, consequently (see Property (6) in Sect. 8.3.3), $\sigma_{m-1}(S^{m-1}(R))=R^{m-1}\sigma_{m-1}(S^{m-1})$. The area of the unit sphere has already been computed in Sect. 6.5.1 using the fact that it consists of two hemispheres each of which is the graph of a smooth function. We will not reproduce these calculations, but only write down the formula they lead to:

$$\sigma_{m-1}\big(S^{m-1}(R)\big)=\frac{2\pi^{m/2}}{\Gamma(m/2)}R^{m-1}.$$

The right-hand side is equal to $m\alpha_m R^{m-1}=(\alpha_m R^m)'_R$. Hence the area of a sphere is equal to the derivative (with respect to the radius) of the volume of the corresponding ball. Later (see Remark (3) in Sect. 8.4.3 and Theorem 13.4.7) we will discuss this question in more detail.

Let us also compute the area of the spherical "cap" cut from the sphere $S^{m-1}(R)$ by a plane at distance H $(0\leqslant H<R)$ from its center (for $H=0$, we thus obtain a hemisphere). Since the area is rotation-invariant, we may say that the set in question is

$$S_H(R)=\big\{(x,y)=(x_1,\dots,x_{m-1},y)\in S^{m-1}(R)\,|\,y>H\big\}.$$

It is obviously the part of the graph of the function $x \mapsto f(x) = \sqrt{R^2 - \|x\|^2}$ that lies above the ball $B^{m-1}(\sqrt{R^2 - H^2})$. Formula (7′) yields

$$\sigma_{m-1}\big(S_H(R)\big) = \int_{B^{m-1}(\sqrt{R^2-H^2})} \frac{R}{\sqrt{R^2 - \|x\|^2}}\, dx.$$

Now we use the formula for computing the integral of a radial function (see Example 1 in Sect. 6.4.2 or Corollary 3 in Section 6.5.3 in which m should be replaced by $m-1$):

$$\sigma_{m-1}\big(S_H(R)\big) = (m-1)\alpha_{m-1} R \int_0^{\sqrt{R^2-H^2}} \frac{u^{m-2}\, du}{\sqrt{R^2 - u^2}}$$

$$= (m-1)\alpha_{m-1} R^{m-1} \int_0^{\sqrt{R^2-H^2}/R} \frac{v^{m-2}\, dv}{\sqrt{1 - v^2}}. \tag{8}$$

Here $\alpha_{m-1} = \lambda_{m-1}(B^{m-1}) = \pi^{(m-1)/2}/\Gamma((m+1)/2)$. Setting $H = \delta R$, we can rewrite (8) as

$$\sigma_{m-1}\big(S_{\delta R}(R)\big) = (m-1)\alpha_{m-1} R^{m-1} \int_0^{\sqrt{1-\delta^2}} \frac{v^{m-2}\, dv}{\sqrt{1 - v^2}}$$

$$= (m-1)\alpha_{m-1} R^{m-1} \int_{\arcsin\delta}^{\pi/2} \cos^{m-2} t\, dt.$$

Now we find out what part of the multi-dimensional sphere falls into the spherical cap as the dimension m increases indefinitely while the distance from the plane cutting off the cap to the center of the sphere remains constant (to simplify the formulas, we consider spheres in $\mathbb{R}^{m+1}$ rather than in $\mathbb{R}^m$). It is particularly instructive to consider this question in the two cases where the sphere radius is equal to one (in all dimensions) and where it is proportional to $\sqrt{m}$.

First, we consider the case of the unit sphere.

Example 5 In a space of large dimension, we have the "concentration of measure" phenomenon: almost all of the area of a sphere is concentrated in an arbitrarily narrow zone near the "equator". More precisely, for the caps $S_\delta = S_\delta(1)$, the ratio $\sigma_m(S_\delta)/\sigma_m(S_0)$ decays rapidly as the dimension grows:

$$\frac{\sigma_m(S_\delta)}{\sigma_m(S_0)} = \frac{\int_{\arcsin\delta}^{\pi/2} \cos^{m-1} t\, dt}{\int_0^{\pi/2} \cos^{m-1} t\, dt} < 3e^{-\frac{m\delta^2}{2}}. \tag{9}$$

One of the consequences of this surprising phenomenon is described in Exercise 7.

To prove (9), let us first estimate the denominator. As is well known (see Example 1 in Sect. 4.6.2),

$$W_m \equiv \int_0^{\pi/2} \cos^m t\, dt = \frac{(m-1)!!}{m!!} v_m,$$

where v_m is equal to 1 or $\pi/2$ depending on the parity of m. Hence $W_m W_{m+1} = \frac{\pi}{2m+2}$. Since the integrals W_m decrease, it follows that

$$\sqrt{\frac{\pi}{2m+2}} < W_m < \sqrt{\frac{\pi}{2m}}.$$

To estimate the numerator, apply the inequalities $\delta \leqslant \arcsin\delta$ and $\cos t \leqslant e^{-t^2/2}$:

$$W_m(\delta) \equiv \int_{\arcsin\delta}^{\pi/2} \cos^m t\, dt \leqslant \int_{\delta}^{\pi/2} e^{-\frac{m}{2}t^2}\, dt < \int_0^{\infty} e^{-\frac{m}{2}(t+\delta)^2}\, dt$$
$$\leqslant \sqrt{\frac{\pi}{2m}} e^{-\frac{m}{2}\delta^2}.$$

Thus for $m > 1$ we have

$$\frac{\sigma_m(S_\delta)}{\sigma_m(S_0)} = \frac{W_{m-1}(\delta)}{W_{m-1}} < \sqrt{\frac{m}{m-1}} e^{-\frac{m-1}{2}\delta^2} \leqslant \sqrt{\frac{me}{m-1}} e^{-\frac{m}{2}\delta^2} < 3e^{-\frac{m\delta^2}{2}}.$$

Observe that, having replaced $\arcsin\delta$ with δ, we have estimated the area of the cap determined by the inequality $y \geqslant \sin\delta$, which is a little larger than S_δ. Such a set appears naturally if we replace the Euclidean metric on the sphere by the stronger geodesic metric. In the latter metric, the distance between two points of the sphere is equal to the angle between their radius vectors. It is clear that the larger cap consists of the points for which the deviation from the equator (the intersection of the sphere with the plane $y = 0$) is at least δ in the geodesic metric.

Example 6 Now let us discuss the second of the questions posed above: as m grows, how does the ratio of the area of the cap $S_H(R)$ to the area of the ambient sphere $S^m(R)$ behave under the condition that $R = R_m = \theta\sqrt{m}$, where $\theta > 0$. Set $P_m(H) = \sigma_m(S_H(R))/\sigma_m(S^m(R))$. This value may be regarded as the probability that a point "picked at random" from the sphere falls into the cap. One may also say that $P_m(H)$ is the probability that the last coordinate of a point picked at random from the sphere is greater than H.

It follows from (8) (with m replaced by $m+1$) that

$$P_m(H) = \frac{m\alpha_m}{(m+1)\alpha_{m+1}} \int_0^{\sqrt{1-H^2/(m\theta^2)}} \frac{t^{m-1}\, dt}{\sqrt{1-t^2}}.$$

Since $\Gamma(x+\frac{1}{2}) \sim \sqrt{x}\,\Gamma(x)$ as $x \to +\infty$ (see formula (4) in Sect. 7.2.2), we have

$$\frac{m\alpha_m}{(m+1)\alpha_{m+1}} = \frac{m}{m+1} \frac{\pi^{m/2}}{\Gamma(\frac{m+2}{2})} \frac{\Gamma(\frac{m+3}{2})}{\pi^{\frac{m+1}{2}}} \underset{m\to\infty}{\sim} \sqrt{\frac{m}{2\pi}},$$

whence

$$P_m(H) \underset{m\to\infty}{\sim} \sqrt{\frac{m}{2\pi}} \int_0^{\sqrt{1-H^2/(m\theta^2)}} \frac{t^{m-1}}{\sqrt{1-t^2}}\, dt.$$

Let us see how the last integral behaves, setting for brevity $\delta = H/\theta$. Making the substitution $v = \sqrt{m}\sqrt{1-t^2}$, we obtain

$$\sqrt{m}\int_0^{\sqrt{1-\delta^2/m}} \frac{t^{m-1}}{\sqrt{1-t^2}}\,dt = \int_\delta^{\sqrt{m}} \left(1-\frac{v^2}{m}\right)^{\frac{m-2}{2}} dv \underset{m\to\infty}{\longrightarrow} \int_\delta^\infty e^{-v^2/2}\,dv.$$

The passage to the limit can be justified in exactly the same way as in Example 2 from Sect. 4.8.4. Therefore,

$$P_m(H) \underset{m\to\infty}{\longrightarrow} \frac{1}{\sqrt{2\pi}}\int_{H/\theta}^\infty e^{-v^2/2}\,dv = \frac{1}{\theta\sqrt{2\pi}}\int_H^\infty e^{-t^2/(2\theta^2)}\,dt.$$

This limit may be interpreted as the probability that the Gaussian random variable with density $\frac{1}{\theta\sqrt{2\pi}}e^{-t^2/(2\theta^2)}$ takes a value greater than H. This result, sometimes called Poincaré's or Maxwell's[6] lemma, means that with respect to the normalized surface area of the m-dimensional sphere of radius $\theta\sqrt{m}$, the distribution of the coordinates is "almost Gaussian".

8.3.6 Now we are going to discuss a more special question, namely, the behavior of the area under a bending. By a bending of a manifold M one usually means a transformation under which the lengths of curves lying on M do not change. For our purposes, this sense is too wide, since under such a map a smooth manifold may transform into a set that is not a manifold. For instance, an interval of the real axis can be bent into a "figure-of-eight" (the continuous map $t \mapsto ((1-\cos t)\,\mathrm{sign}\,t, \sin t)$ transforms the interval $(-2\pi, 2\pi)$ into a pair of touching circles; it is one-to-one, but not homeomorphic). In addition, we continue to restrict ourselves to smooth maps. Thus it would be wise to impose additional conditions on bending transformations.

Definition A *bending* of a smooth manifold M lying in $\mathbb{R}^m$ is a smooth map $\Theta : M \to \mathbb{R}^d$ satisfying the following conditions:

(1) Θ is a homeomorphism between M and $\Theta(M)$;
(2) Θ preserves the lengths of smooth curves: $\sigma_1(L) = \sigma_1(\Theta(L))$ for every smooth curve L contained in M.

Recall that the smoothness of Θ on M means that this map is smooth on an open set containing M.

It is intuitively clear that a bending does not change the area of a set lying on the surface. This observation underlies, for example, the computation of the areas of a cone and a cylinder known from school. Let us discuss the first of these examples in more detail (for the second one, see Exercise 5).

[6]James Clerk Maxwell (1831–1879)—Scottish physicist.

Example Consider a cone K in $\mathbb{R}^3$ formed by rays starting at the origin (the vertex of the cone). Such a cone is uniquely determined by its intersection with the unit sphere, i.e., the set $\ell = K \cap S^2$. Obviously, the smoothness of the surface $K \setminus \{0\}$ means that ℓ is a smooth curve. Adopting this assumption, we also assume that the length Δ of ℓ is finite and Ψ is the natural parametrization defined on $(0, \Delta)$. Let us "unfold" the cone in such a way that the curve ℓ turns into an arc of the unit circle of the same length; more precisely, a point $\Psi(s)$ maps to $z(s) = (\cos s, \sin s)$, and the ray passing through $\Psi(s)$ maps to the ray passing through $z(s)$ (all rays are assumed to start at the origin). We also assume that $\Delta < 2\pi$ (otherwise ℓ should be divided into several parts).

To formally verify that the described map is a bending, it is convenient to use the inverse map Θ. We will assume that it is defined in an angle C with the vertex at the origin whose points are determined by their polar angles lying in the interval $(0, \Delta)$. To obtain an analytic expression for Θ, consider the smooth map $P : C \to (0, \Delta) \times (0, +\infty)$ that associates with each point z of C its polar coordinates $\varphi(z)$ and $r(z)$. Then $\Theta(z) = r(z)\Psi(\varphi(z))$. Let us check that Θ is indeed a bending.

Let $\gamma(t)$ ($t \in (a, b)$) be a parametrization of a smooth curve L lying in C. It generates the parametrization $\Phi = \Theta \circ \gamma$ of the curve $\Theta(L) \subset K$. We must show that the lengths $\sigma_1(L)$ and $\sigma_1(\Theta(L))$ of these curves coincide. Set $P(\gamma(t)) = (\omega(t), \rho(t))$. Note that $\sigma_1(L) = \int_a^b \sqrt{(\rho'(t))^2 + \rho^2(t) \cdot (\omega'(t))^2}\, dt$. Since $\Phi(t) = \rho(t)\Psi(\omega(t))$, the tangent vector Φ' breaks into two terms: $\Phi' = \rho'\Psi + \rho\,\omega'\,\Psi'$. Since $\|\Psi\| = \|\Psi'\| = 1$ and $\Psi' \perp \Psi$, it follows that $\|\Phi'\|^2 = (\rho')^2 + \rho^2 \cdot (\omega')^2$. Hence

$$\sigma_1\big(\Theta(L)\big) = \int_a^b \|\Phi'(t)\|\, dt = \int_a^b \sqrt{\big(\rho'(t)\big)^2 + \rho^2(t) \cdot \big(\omega'(t)\big)^2}\, dt = \sigma_1(L),$$

as required.

Thus Θ is a bending. By Theorem 8.3.6 (see below), it preserves the area. In particular, the area of the part of the cone K lying in the ball of radius R centered at the vertex of the cone is equal to the area of the circular sector $C \cap B(0, R)$.

Now let us study under what conditions a smooth map is a bending.

Lemma *Let $M \subset \mathbb{R}^m$ be a smooth manifold and $\Theta : M \to \mathbb{R}^d$ be a smooth homeomorphism. It preserves the lengths of curves lying in M if and only if for every point p in M, the map $d_p\Theta$ is an isometry of the tangent space $T_p(M)$ onto $\mathbb{R}^d$.*

Proof Let L ($L \subset M$) be a smooth curve passing through a point p, and let $t \mapsto \gamma(t)$ ($t \in (\alpha, \beta)$) be a (smooth) parametrization of L with $p = \gamma(t_0)$. Then $\delta = \Theta \circ \gamma$ is a parametrization of the curve $\widetilde{L} = \Theta(L)$. It is clear that $\delta'(t) = d_{\gamma(t)}\Theta(\gamma'(t))$, and the lengths of the arcs $\ell = \gamma((t_0, t))$, $\widetilde{\ell} = \delta((t_0, t))$ are given by the formulas

$$\sigma_1(\ell) = \int_{t_0}^{t} \|\gamma'(u)\|\, du, \qquad \sigma_1(\widetilde{\ell}) = \int_{t_0}^{t} \|\delta'(u)\|\, du. \tag{10}$$

If Θ is a bending, then these lengths are equal. Differentiating with respect to t, we see that $\|\delta'(t_0)\| = \|\gamma'(t_0)\|$. This means that $\|d_p\Theta(\gamma'(t_0))\| = \|\gamma'(t_0)\|$, i.e., $\|d_p\Theta(x)\| = \|x\|$ for every vector x that can be written in the form $x = \gamma'(t_0)$. By the definition of the tangent space, every vector from $T_p(M)$ can be written in this form. Thus it follows from the preservation of the lengths of curves that $\|d_p\Theta(x)\| = \|x\|$ for $x \in T_p(M)$, i.e., $d_p\Theta$ is an isometry on the tangent space $T_p(M)$. If this condition is satisfied for every $p \in M$, then $\|\delta'(t)\| = \|\gamma'(t)\|$ for every t, so that the right-hand sides of equalities (10) coincide, which implies that Θ is a bending. □

It follows from the lemma that if Θ is a bending, then $\operatorname{rank}\Theta' = \dim M$ and, consequently, the set $\widetilde{M} = \Theta(M)$ is a smooth manifold of the same dimension as M.

Now we can easily prove that a bending preserves the area.

Theorem *Let M be a smooth k-dimensional manifold, Θ be a bending of M, and $\widetilde{M} = \Theta(M)$. Then for every set $E \in \mathfrak{A}_M$, its image $\widetilde{E} = \Theta(E)$ has the same area:*

$$\sigma_k(E) = \sigma_k(\widetilde{E}).$$

Proof Obviously, it suffices to prove the assertion of the theorem for a set lying in some coordinate neighborhood $U \subset M$. Let Φ be a parametrization of U and $\widetilde{\Phi} = \Theta \circ \Phi$. The differential $d_{\Phi(t)}\Theta$ is an isometry of the accompanying parallelotope corresponding to Φ onto the parallelotope corresponding to $\widetilde{\Phi}$. Since isometries preserve Lebesgue measure, the measures of these parallelotopes coincide, i.e., the weights ω_Φ and $\omega_{\widetilde{\Phi}}$ are equal. Hence the areas of sets contained in U do not change. □

Note that a bending may also be an expanding map; for instance, the "straightening" of a circular arc, the "unfolding" of a cylinder into a plane, etc. These examples show that under an expanding map that strictly increases the distance between some points, the length and the area do not always strictly increase.

8.3.7 We have already observed that the area of a sphere is rotation-invariant (see Property (7) in Sect. 8.3.3). Let us discuss another example of an invariant measure. Consider the measure $\sigma = \sigma_{n(n-1)/2}$ on the group $O(n)$ of orthogonal $n \times n$ matrices with the metric induced from $\mathbb{R}^{n^2}$ (see Sect. 8.1.3, Example 5). Since this set is compact, the measure σ is finite. As we have established in Sect. 8.1.3, a translation on the group $O(n)$ (the multiplication on the left or on the right by a fixed element U_0 from $O(n)$) maps $O(n)$ isometrically onto itself. Since the area is invariant under isometries, the measure σ is invariant under translations on $O(n)$. In particular, for every summable function f on $O(n)$ and every V in $O(n)$, we have (see formula (2′) from Sect. 6.1.2)

$$\int_{O(n)} f(UV)\,d\sigma(U) = \int_{O(n)} f(VU)\,d\sigma(U) = \int_{O(n)} f(U)\,d\sigma(U). \qquad (11)$$

Using the existence of an invariant measure on the group $O(n)$, one can prove both the uniqueness of such a measure and the uniqueness of a rotation-invariant measure on the sphere. More precisely, the following theorem holds.

Theorem

(1) *A finite Borel rotation-invariant measure on S^{m-1} is unique up to a multiplicative constant.*

(2) *A finite Borel measure on $O(n)$ invariant under an arbitrary right or left translation (i.e., under left or right multiplication by an element of the group $O(n)$) is unique up to a multiplicative constant.*

Proof (1) Let ν be a Borel rotation-invariant measure on S^{m-1} and σ be the area on $O(m)$, which we know to be translation-invariant. Assume that $\nu(S^{m-1}) = \sigma_{m-1}(S^{m-1})$; we are going to prove that the measures ν and σ_{m-1} coincide.

Consider the measure μ on $O(m)$ obtained by normalizing the measure σ (i.e., $\mu = \frac{1}{\sigma(O(m))}\sigma$), and let E be a Borel subset of the sphere S^{m-1}. First let us show that the value $\int_{O(m)} \chi_E(Ux_0)\,d\mu(U)$ does not depend on the choice of a point x_0 from S^{m-1}. Indeed, for every vector $x \in S^{m-1}$ there is an orthogonal transformation V such that $x = Vx_0$. Hence, by (11) with $f(U) = \chi_E(Ux_0)$,

$$\int_{O(m)} \chi_E(Ux)\,d\mu(U) = \int_{O(m)} \chi_E(UVx_0)\,d\mu(U) = \int_{O(m)} \chi_E(Ux_0)\,d\mu(U),$$

as required. Since, by the invariance of ν,

$$\nu(E) = \int_{S^{m-1}} \chi_E(x)\,d\nu(x) = \int_{S^{m-1}} \chi(Ux)\,d\nu(x)$$

for every U in $O(m)$, integrating this equality with respect to the (normalized) measure μ and changing the order of integration yields

$$\begin{aligned}\nu(E) &= \int_{O(m)} \nu(E)\,d\mu(U) = \int_{O(m)} \left(\int_{S^{m-1}} \chi_E(Ux)\,d\nu(x)\right) d\mu(U)\\ &= \int_{S^{m-1}} \left(\int_{O(m)} \chi_E(Ux)\,d\mu(U)\right) d\nu(x) = \nu\big(S^{m-1}\big) \int_{O(m)} \chi_E(Ux)\,d\mu(U),\end{aligned}$$

where the right-hand side, as we have established above, does not depend on x. Obviously, a similar equality can be written with ν replaced by σ_{m-1}. The right-hand sides of these equalities are equal. Therefore, the left-hand sides also coincide, as required.

When changing the order of integration (and, consequently, using Tonelli's theorem), we have assumed that the function $(x, U) \mapsto \varphi(x, U) \equiv \chi_E(Ux)$ is measurable on $S^{m-1} \times O(m)$. This is indeed the case, since φ is the characteristic function of the set $\{(x, U) \mid Ux \in E\} = \Psi^{-1}(E)$, where $\Psi(x, U) = Ux$. Since the map Ψ is obviously continuous and E is a Borel set, its inverse image $\Psi^{-1}(E)$ is also a Borel set (see Corollary 2 in Sect. 1.6.2).

The proof of Claim (2) is completely analogous. □

EXERCISES

1. What fraction of the area of a sphere centered at the origin is occupied by points (x, y, z) satisfying the inequalities $0 \leqslant y \leqslant \sqrt{3}x$ and $0 \leqslant z \leqslant \sqrt{2}x$?

2. Find the area of the surface obtained by rotating about the OX axis the graph of a smooth function defined on an interval (a, b). Prove Guldin's theorem: for a function of fixed sign, this area is equal to the product of the length of the graph and the length of the circle described by its center of mass (under the assumption that the mass is distributed over the graph with constant density).

3. Consider the bounded part of the right circular cone with an angle 2α at the apex cut off by a plane making an angle β $(0 < \beta \leqslant \pi/2)$ to the axis of the cone. Show that the ratio of the area of this part to the area of the ellipse obtained in the cross section is equal to $\sin\beta/\sin\alpha$.

4. Find the three-dimensional area of the "bodily torus" M:

$$M = \{(x, y, u, v) \in \mathbb{R}^4 \mid x^2 + y^2 = r^2,\ u^2 + v^2 < R^2\}.$$

5. Show that the cylindric surface $C = \ell \times \mathbb{R} \subset \mathbb{R}^3$, where $\ell \subset \mathbb{R}^2$ is a simple smooth curve of finite length S (it is called the directrix of C), can be obtained by bending the strip $(0, S) \times \mathbb{R}$.

6. Refine the result of Lemma 8.3.6 by proving that the differential of a smooth expanding map Θ on a smooth manifold M expands the tangent subspace T_p at an arbitrary point $p \in M$, i.e., $\|d_p\Theta(x)\| \geqslant \|x\|$ for all $x \in T_p$.

7. Using (9), show that as $m \to \infty$, the sphere S^m is almost entirely (with respect to the area) contained in an infinitesimal cube: the area of the set-theoretic difference $S^m \setminus (-\delta, \delta)^{m+1}$ for $\delta = \delta_m = 2\sqrt{\frac{\ln m}{m}}$ is less than $\frac{6}{m}\sigma_m(S^m)$.

8. To what result does the argument from Example 6 of Sect. 8.3.5 lead in the case of spheres of unit area?

9. Using only the invariance of the area under rotations, show that the mean values of the functions $x_j^2, x_j^4, x_j^2x_k^2$ (here $j, k = 1, \dots, m,\ j \neq k$) on the sphere S^{m-1} are equal to $\frac{1}{m}$, $\frac{3}{m(m+2)}$ and $\frac{1}{m(m+2)}$, respectively. *Hint*. One can obtain simple equations relating the mean values x_j^4 and $x_j^2x_k^2$ by integrating the functions $((x_j + x_k)/\sqrt{2})^4$ and $(x_1^2 + \cdots + x_m^2)^2$ over the sphere.

10. For $m = m_1 + m_2$, identify $\mathbb{R}^m$ with $\mathbb{R}^{m_1} \times \mathbb{R}^{m_2}$. Let M_1 and M_2 be smooth manifolds in $\mathbb{R}^{m_1}$ and $\mathbb{R}^{m_2}$, respectively, and $M = M_1 \times M_2$. Show that the measure σ_M is the product of the measures σ_{M_1} and σ_{M_2}.

11. The Grassmanian G_m^k is the set of all k-dimensional subspaces of $\mathbb{R}^m$. The distance between two elements L_1 and L_2 of G_m^k is defined as the norm $\|P_1 - P_2\|$, where P_1 and P_2 are the orthogonal projections from $\mathbb{R}^m$ to L_1 and L_2, respectively. The group $O(m)$ of orthogonal matrices can be canonically mapped onto G_m^k by associating with every such matrix U the linear hull of the first k rows of U. We define a measure on G_m^k as the image of the area in $O(m)$ under this map. Show that the measure on G_m^k obtained in this way is "invariant under rotations", i.e., under transformations of the form $G_m^k \ni L \mapsto U(L)$, where U is an arbitrary element of $O(m)$, and that a finite Borel measure on G_m^k invariant under rotations is unique up to a multiplicative constant.

12. Denote by $\widetilde{\sigma}_n$ the n-dimensional area normalized so that $\widetilde{\sigma}_n(S^n) = 1$. Using the uniqueness of a rotation-invariant measure, show that for $1 < k < m$

$$\int_{S^{m-1}} f(x)\, d\widetilde{\sigma}_{m-1}(x) = \int_{G_m^k} \left(\int_{L \cap S^{m-1}} f(x)\, d\widetilde{\sigma}_{k-1}(x) \right) d\nu(L),$$

where $f \in C(S^{m-1})$ and ν is the normalized measure on the manifold G_m^k constructed in Exercise 11.

8.4 Integration over a Smooth Manifold

8.4.1 The computation of an integral with respect to the surface area of a smooth manifold, or, in short, of a surface integral, can be reduced to the computation of a multiple integral with respect to the Lebesgue measure. This transition does not require additional efforts, since, by Property (3) from Sect. 8.3.3, the area of a simple manifold is a weighted image of the Lebesgue measure. Hence we may apply the general Theorem 6.1.1 on the computation of an integral with respect to a weighted image of a measure, which in the case under consideration leads to the following result.

Theorem *Let M be a simple smooth manifold in $\mathbb{R}^m$, $\dim M = k$, and f be a non-negative measurable function on M. Then*

$$\int_M f(x)\, d\sigma_k(x) = \int_{\Phi^{-1}(M)} f\big(\Phi(t)\big)\omega_\Phi(t)\, dt$$

for every parametrization Φ of M.

This formula also holds for every summable function on M.

Recall that $\omega_\Phi(t)$ has a simple geometric interpretation: this is the volume of the accompanying parallelotope. For $k = 1$, it is equal to the length of the tangent vector $\Phi'(t)$, and for $k = m - 1$, to the length of the normal $N_\Phi(t)$ corresponding to the parametrization Φ (see Sect. 8.3.4).

As in the general situation (see Corollary 6.1.1), a similar formula holds for functions defined not on the whole manifold M, but only on a measurable subset of M. If the manifold M is not simple, then an integral over M can be computed by considering a partition of M into at most countably many sets each of which is contained in a coordinate neighborhood.

In the case $\dim M = m$, the assertion of the theorem is just the change of variables formula for a diffeomorphism (see Sect. 6.2.2).

Observe the important special case when the manifold is the graph of a smooth function φ defined on an open subset $\mathcal{O}$ of $\mathbb{R}^{m-1}$. Consider the canonical parametrization of the graph $\mathcal{O} \ni x \mapsto \Phi(x) = (x, \varphi(x))$. Then (see Sect. 8.3.4) $\omega_\Phi(x) = \sqrt{1 + \|\operatorname{grad}\varphi(x)\|^2}$ and $\Phi^{-1}(E) = P(E)$, where P is the orthogonal projection to $\mathbb{R}^{m-1}$. Hence for every measurable set $E \subset M = \Gamma_\varphi$,

$$\int_E f\, d\sigma_{m-1} = \int_{P(E)} f\big(x, \varphi(x)\big)\sqrt{1 + \big\|\operatorname{grad}\varphi(x)\big\|^2}\, dx. \tag{1}$$

Example 1 Let $\Sigma_m = S^{m-1} \cap \mathbb{R}^m_+$ be the part of the unit sphere S^{m-1} of $\mathbb{R}^m$ lying in the "first octant". Assuming that the sphere is homogeneous, we are going to find the center of mass C of the surface Σ_m. By symmetry, all coordinates of this vector are equal: $C = (c, \dots, c)$. As we have established in Sect. 6.3.3, they are given by the formula

$$c = \frac{1}{\sigma_{m-1}(\Sigma_m)} \int_{\Sigma_m} x_m\, d\sigma_{m-1}(x) = \frac{2^m}{m\alpha_m} \int_{\Sigma_m} x_m\, d\sigma_{m-1}(x).$$

To compute this integral, observe that Σ_m is a subset of the graph of the function $\varphi(t) = \sqrt{1 - \|t\|^2}$ defined on the unit ball B^{m-1} and the projection Σ_m coincides with the intersection $A = B^{m-1} \cap \mathbb{R}^{m-1}_+$. Applying (1) with $f(x) = x_m$, we obtain

$$c = \frac{2^m}{m\alpha_m} \int_A \varphi(t)\sqrt{1 + \big\|\operatorname{grad}\varphi(t)\big\|^2}\, dt = \frac{2^m}{m\alpha_m} \int_A 1\, dt = \frac{2^m}{m\alpha_m}\lambda_{m-1}(A)$$
$$= \frac{2\alpha_{m-1}}{m\alpha_m} = \frac{\Gamma(\frac{m}{2})}{\sqrt{\pi}\,\Gamma(\frac{m+1}{2})}.$$

In particular,

$$\|C\| = \frac{\sqrt{m}\,\Gamma(\frac{m}{2})}{\sqrt{\pi}\,\Gamma(\frac{m+1}{2})}.$$

As $m \to \infty$, these norms tend to $\sqrt{\frac{2}{\pi}}$. One can show that they decrease. Note that the center of mass C' of the part of the unit ball lying in $\mathbb{R}^m_+$ (see Sect. 6.3.3) has the coordinates $c' = \frac{1}{\sqrt{\pi}} \frac{\Gamma(\frac{m}{2})}{\Gamma(\frac{m+1}{2})} \frac{m}{m+1} = \frac{m}{m+1} c$ and $\|C'\|$ tends to the same limit as $\|C\|$, but increases rather than decreases.

Example 2 Let M be a smooth k-dimensional manifold in $\mathbb{R}^m$, $\sigma_k(M) < +\infty$, and $x_0 \in M$. Let us find out for which $p > 0$ the integral

$$I_0 = \int_M \frac{d\sigma_k(x)}{\|x - x_0\|^p}$$

is finite. First we will find a necessary condition. Consider a parametrization Φ of an M-neighborhood of the point $x_0 = \Phi(a)$. In some ball $B(a, \rho) \subset \mathbb{R}^k$, Φ satisfies the Lipschitz condition: $\|\Phi(t) - \Phi(a)\| \leqslant L\|t - a\|$, where L is a fixed positive number. If ρ is sufficiently small, then $\omega_\Phi(t) \geqslant \frac{1}{2}\omega_\Phi(a)$ for $t \in B(a, \rho)$. Hence

$$I_0 \geqslant \int_{B(a,\rho)} \frac{\omega_\Phi(t)\,dt}{\|\Phi(t) - \Phi(a)\|^p} \geqslant \frac{\omega_\Phi(a)}{2L^p} \int_{B(a,\rho)} \frac{dt}{\|t - a\|^p}.$$

If I_0 is finite, then the integral on the right-hand side is also finite, which is possible only for $p < k$ (see Theorem 4.7.1).

Now we will show that the condition $p < k$ is not only necessary, but also sufficient for I_0 to be finite. In order to prove at once a somewhat stronger result, we introduce the integral

$$I(y) = \int_M \frac{d\sigma_k(x)}{\|x - y\|^p} \quad (y \in \mathbb{R}^m).$$

Obviously, $I_0 = I(x_0)$. We will prove that for $p < k$, the integral I is bounded in the vicinity of x_0. Note that in general the condition $y \notin M$ is not sufficient for $I(y)$ to be finite if $y \in \overline{M} \setminus M$ (see Exercise 2).

We still assume that Φ is a parametrization of an M-neighborhood of x_0 and $x_0 = \Phi(a)$. Recall that in the vicinity of a the parametrization Φ is the restriction of some diffeomorphism P (see Lemma 8.1.4; we assume that the space $\mathbb{R}^k$, on a subset of which Φ is defined, is canonically embedded into $\mathbb{R}^m$). In a sufficiently small ball $B(x_0, r)$, the map F^{-1} satisfies the Lipschitz condition with some constant C:

$$\big\|F^{-1}(x) - F^{-1}(y)\big\| \leqslant C\|x - y\| \quad \text{for } x, y \in B(x_0, r). \tag{2}$$

Assuming that r is so small that $\omega_\Phi(\Phi^{-1}(x)) \leqslant 2\omega_\Phi(a)$ for all x in $M_r = M \cap B(x_0, r)$, we will prove that the integral I is bounded on the ball $B(x_0, r)$.

Taking an arbitrary point y from this ball, write the integral $I(y)$ in the form

$$I(y) = \int_{M_r} \frac{d\sigma_k(x)}{\|x - y\|^p} + \int_{M \setminus M_r} \frac{d\sigma_k(x)}{\|x - y\|^p} = I_1(y) + I_2(y).$$

It is clear that

$$I_2(y) \leqslant \frac{1}{r^p} \int_{M \setminus M_r} d\sigma_k(x) \leqslant \frac{1}{r^p}\sigma_k(M) < +\infty.$$

It remains to estimate the integral $I_1(y)$ over the simple manifold M_r:

$$I_1(y) = \int_{\Phi^{-1}(M_r)} \frac{\omega_\Phi(t)\,dt}{\|\Phi(t) - y\|^p} \leqslant 2\omega_\Phi(a) \int_{\Phi^{-1}(M_r)} \frac{dt}{\|\Phi(t) - y\|^p}.$$

Let us estimate the norm $\|\Phi(t) - y\|$ from below. Let $x = \Phi(t)$ and $s = F^{-1}(y)$. It follows from (2) that for $x, y \in B(x_0, r)$,

$$\big\|\Phi(t) - y\big\| = \|x - y\| \geqslant \frac{1}{C}\|t - s\| \geqslant \frac{1}{C}\|t - u\|,$$

where u is the projection of s to $\mathbb{R}^k$. Since the points $x = \Phi(t)$ and y lie in the ball $B(x_0, r)$, we have $\|\Phi(t) - y\| < 2r$ and, consequently, $\|t - u\| \leqslant C\|\Phi(t) - y\| < 2Cr$. Hence

$$\int_{\Phi^{-1}(M_r)} \frac{dt}{\|\Phi(t) - y\|^p} \leqslant \int_{\|t-u\|<2Cr} \left(\frac{C}{\|t-u\|}\right)^p dt = C^p \int_{\|v\|<2Cr} \frac{dv}{\|v\|^p}$$
$$= C^p \frac{k\alpha_k}{k-p}(2Cr)^{k-p}.$$

So, for $y \in B(x_0, r)$,

$$I(y) \leqslant 2C^k \omega_\Phi(a) \frac{k\alpha_k}{k-p}(2r)^{k-p} + \frac{1}{r^p}\sigma_k(M)$$

(the parameters C and r depend on the manifold M and the point x_0, but not on the exponent p).

Example 3 (Integrals similar to a simple-layer potential) Let M be a smooth k-dimensional manifold in $\mathbb{R}^m$, E be a compact subset of M, and $w \in C(E)$. Let us check that for $p < k$, the function

$$y \mapsto I(y) = \int_E \frac{w(x)}{\|x - y\|^p} d\sigma_k(x) \quad \left(y \in \mathbb{R}^m\right)$$

is continuous in the whole space and infinitely differentiable in $\mathbb{R}^m \setminus E$.

The smoothness of I outside E follows from the fact that for $y_0 \notin E$ the norm $\|y - x\|$ is bounded away from zero if $x \in E$ and y lies in a sufficiently small neighborhood of y_0. Hence the integrand, as well as all its partial derivatives with respect to the coordinates $y_1, \ldots, y_m$, are bounded in the vicinity of y_0. Thus at y_0 the condition (L_{loc}) is satisfied and we can apply the Leibniz rule.

To prove that I is continuous at a point $y_0 \in E$, we use Theorem 2 of Sect. 7.1.2. Fix a number $s > 1$ such that $sp < k$ and put $C = \max_E |w|$. As we have established in the previous example, the integral $\widetilde{I}(y) = \int_E \frac{C\, d\sigma_k(x)}{\|x-y\|^{sp}}$ is bounded in the vicinity of y_0, and this, by Theorem 2 of Sect. 7.1.2, suffices for I to be continuous at this point.

8.4.2 In this section we will obtain a generalization of Fubini's theorem to the case where an open subset of $\mathbb{R}^m$ ($m \geqslant 2$) stratifies not into affine subspaces, but into the level surfaces of a smooth function. In the special case where the level surfaces are concentric spheres, we have essentially solved this problem in Theorem 6.5.2. Indeed, in this theorem it is proved that for every function f summable in the ball $B(0, r) \subset \mathbb{R}^m$,

$$\int_{B(0,r)} f(x)\,dx = \int_0^r t^{m-1}\left(\int_{S^{m-1}} f(t\xi)\,d\sigma_{m-1}(\xi)\right) dt.$$

Since both the area σ_{m-1} and the volume λ_m are translation-invariant, we may consider spheres with arbitrary center. Furthermore, using the equality $\sigma_{m-1}(tE) = t^{m-1}\sigma_{m-1}(E)$ for $t > 0$ (see Property (6) in Sect. 8.3.3) and the change of variables theorem 6.1.1, we can rewrite the assertion of Theorem 6.5.2 as follows:

$$\int_{B(a,r)} f(x)\,dx = \int_0^r \left(\int_{S(a,t)} f(x)\,d\sigma_{m-1}(x) \right) dt. \tag{3}$$

The theorem we are going to consider next is a far-reaching generalization of this result.

Theorem (Kronrod[7]–Federer[8]) *Let $\mathcal{O}$ be an open subset of $\mathbb{R}^m$, $F \in C^1(\mathcal{O})$ and* $\operatorname{grad} F \neq 0$ *in $\mathcal{O}$. Then for every function f summable in $\mathcal{O}$,*

$$\int_{\mathcal{O}} f(x)\,dx = \int_{-\infty}^{\infty} \left(\int_{M(t)} \frac{f(x)}{\|\operatorname{grad} F(x)\|}\,d\sigma_{m-1}(x) \right) dt, \tag{4}$$

where $M(t) = \{x \in \mathcal{O} \mid F(x) = t\}$.

Proof Let us first prove a local version of this theorem: every point $x_0 \in \mathcal{O}$ has a small neighborhood U such that (4) holds for every function f vanishing outside U.

We assume without loss of generality that $F(x_0) = 0$. Moreover, applying if necessary a translation and an orthogonal transformation, we may assume that $x_0 = 0$ and that the tangent plane to $M(0)$ at the origin coincides with the coordinate subspace $x_m = 0$. To simplify formulas, denote by u the projection of x to this subspace, and let v be the last coordinate x_m, so that $x = (u, v)$, $u = (u_1, \dots, u_{m-1}) \in \mathbb{R}^{m-1}$, $v \in \mathbb{R}$. Then $F'_{u_k}(0) = 0$ for $1 \leqslant k < m$ and $F'_v(0) \neq 0$. Consider the map $T : \mathcal{O} \to \mathbb{R}^m$ that "straightens" the level surfaces: $T(x) = (u, F(x))$. It transforms the level surfaces into planes parallel to the subspace $\mathbb{R}^{m-1}$. The Jacobian J_T of this map at $x = 0$ does not vanish, since $J_T(0) = F'_v(0) \neq 0$. Hence the restriction of T to some neighborhood U of the origin is a diffeomorphism. Let us assume that U is projected into a ball of radius δ and lies between level surfaces $M(-\varepsilon)$ and $M(\varepsilon)$, i.e.,

$$U = \{x = (u, v) \mid \|u\| < \delta, |F(x)| < \varepsilon\},$$

where δ and ε are sufficiently small positive numbers. Then for $|t| < \varepsilon$ the set $T(M(t) \cap U)$ is contained in the affine subspace $v = t$, and $T(U)$ coincides with the Cartesian product $W = B^{m-1}(0, \delta) \times (-\varepsilon, \varepsilon)$. Clearly, the map Φ inverse to the restriction of T to U, as well as T itself, affects only the last coordinate of the argument, so that it has the form

$$\Phi(u, t) = \big(u, \varphi(u, t)\big), \quad \text{where } \|u\| < \delta,\ |t| < \varepsilon$$

(φ is the last coordinate function of the map Φ, $\varphi \in C^1(W)$).

---

[7] Alexandr Semenovich Kronrod (1921–1986)—Russian mathematician.

[8] Herbert Federer (1920–2010)—American mathematician.

Thus, for $|t| < \varepsilon$, the part of $M(t)$ lying in U is just the graph of the smooth function $u \mapsto \varphi_t(u) \equiv \varphi(u,t)$. Since $F(u,\varphi_t(u)) \equiv t$, it is easy to establish a relation between the gradients of the functions φ_t and F:

$$F'_{u_j}\big(u,\varphi_t(u)\big) + F'_v\big(u,\varphi_t(u)\big)\frac{\partial\varphi(u,t)}{\partial u_j} \equiv 0 \quad \text{for } 1 \leqslant j < m.$$

Hence

$$\frac{1}{|F'_v|} = \frac{\sqrt{1+\|\operatorname{grad}\varphi_t\|^2}}{\|\operatorname{grad} F\|}. \tag{5}$$

Moreover, using the identity $F'_v(u,\varphi(u,t))\,\frac{\partial\varphi(u,t)}{\partial t} \equiv 1$, we can compute the Jacobian of Φ:

$$J_\Phi(u,t) = \frac{\partial\varphi(u,t)}{\partial t} = \frac{1}{F'_v(\Phi(u,t))}.$$

Assuming that f vanishes outside U, we obtain, making a substitution, that

$$\int_U f(x)\,dx = \int_W \frac{f(\Phi(u,t))}{|F'_v(\Phi(u,t))|}\,du\,dt = \int_{-\varepsilon}^{\varepsilon}\left(\int_{\|u\|<\delta}\frac{f(\Phi(u,t))}{|F'_v(\Phi(u,t))|}\,du\right)dt.$$

Let us write the inner integral as an integral over the graph of φ_t (see (1)). In view of (5), we have

$$\begin{aligned}\int_{\|u\|<\delta}\frac{f(\Phi(u,t))}{|F'_v(\Phi(u,t))|}\,du &= \int_{\|u\|<\delta}\frac{f(u,\varphi_t(u))}{\|\operatorname{grad} F(u,\varphi_t(u))\|}\sqrt{1+\big\|\operatorname{grad}\varphi_t(u)\big\|^2}\,du\\ &= \int_{M'(t)}\frac{f(x)}{\|\operatorname{grad} F(x)\|}\,d\sigma_{m-1}(x),\end{aligned} \tag{6}$$

where $M'(t)$ is the graph of φ_t, i.e., the part of $M(t)$ contained in U. Thus

$$\begin{aligned}\int_U f(x)\,dx &= \int_{-\varepsilon}^{\varepsilon}\left(\int_{M(t)\cap U}\frac{f(x)}{\|\operatorname{grad} F(x)\|}\,d\sigma_{m-1}(x)\right)dt\\ &= \int_{-\infty}^{\infty}\left(\int_{M(t)\cap U}\frac{f(x)}{\|\operatorname{grad} F(x)\|}\,d\sigma_{m-1}(x)\right)dt.\end{aligned}$$

Since $f = 0$ outside U, it follows that (4) holds for functions that do not vanish only in a sufficiently small neighborhood of x_0.

It follows from the obtained local version of the theorem that (4) holds for a summable function f supported by a compact subset of $\mathcal{O}$. Indeed, for each point $x \in \mathcal{O}$ choose a neighborhood $U_x \subset \mathcal{O}$ such that (4) holds for functions vanishing outside U_x. By Theorem 8.1.8, there exists a partition of unity $\varphi_1, \ldots, \varphi_N$ on the set $\operatorname{supp}(f)$ subordinate to the family $\{U_x\}_{x\in\mathcal{O}}$. Write (4) for $f\varphi_k$:

$$\int_{\mathcal{O}} f(x)\varphi_k(x)\,dx = \int_{-\infty}^{\infty}\left(\int_{M(t)} f(x)\varphi_k(x)\,d\sigma_{m-1}(x)\right)dt.$$

Adding these equalities, we obtain the desired result for compactly supported functions. To prove it for a non-negative function with arbitrary support (obviously, this suffices for proving the theorem in full strength), we should exhaust the set $\mathcal{O}$ by an increasing sequence of compact subsets K_n and apply Levi's theorem to both sides of (4) with f replaced by $f\chi_{K_n}$. □

Remark If f is continuous, then the function

$$t \mapsto \int_{M(t)} \frac{f(x)}{\|\mathrm{grad}\, F(x)\|}\, d\sigma_{m-1}(x)$$

is also continuous. If the support of f is small, this follows from (6). The general case can be proved using a partition of unity.

Observe also that the theorem remains valid if the smoothness of F is violated at a closed set E satisfying the conditions

$$\lambda_m(E) = 0, \qquad \sigma_{m-1}\big(M(t) \cap E\big) = 0 \quad \text{for every } t.$$

To prove this, it suffices to apply the theorem to $\mathcal{O} \setminus E$ replacing $M(t)$ with $M(t) \setminus E$.
A deeper generalization of Theorem 8.4.2 can be found in [F] or [EG].

8.4.3 We now make a few more remarks about the obtained result.

(1) Obviously, formula (3) follows from the above theorem with $\mathcal{O} = \mathbb{R}^m \setminus \{a\}$ and $F(x) = \|x - a\|$ (note that $\|\mathrm{grad}\, F(x)\| \equiv 1$). We encourage the reader to compare the proof of the theorem with the arguments from Sect. 6.5.2, where, due to the special form of F, we did not need local considerations.

(2) Replacing f with $f\,\|\mathrm{grad}\, F\|$, we can rewrite (4) in the form

$$\int_{\mathcal{O}} f(x)\|\mathrm{grad}\, F(x)\|\, dx = \int_{-\infty}^{\infty} \left(\int_{M(t)} f(x)\, d\sigma_{m-1}(x) \right) dt. \tag{4'}$$

If F is sufficiently smooth, the condition $\mathrm{grad}\, F \neq 0$ can be dropped, since, by Sard's theorem 13.5.2, the set of critical values of the function $F \in C^m(\mathcal{O})$ has zero measure. Indeed, let $\widetilde{\mathcal{O}} = \{x \in \mathcal{O} \mid \mathrm{grad}\, F(x) \neq 0\}$, and let $E \subset \mathbb{R}$ be the set of critical values of F, so that $\lambda_1(E) = 0$. If $t \notin E$, then F does not take the value t on $\mathcal{O} \setminus \widetilde{\mathcal{O}}$, and, consequently, $M(t) \cap \widetilde{\mathcal{O}} = M(t)$. Thus we can obtain the desired result by applying (4′) to $\widetilde{\mathcal{O}}$.

(3) Let $V(u) = \lambda_m(\mathcal{O}(F < u))$ $(u \in \mathbb{R})$. Assuming that $V(u) < +\infty$, $f \equiv 1$ and applying (4) to $\mathcal{O}(F < u)$, we have (taking into account that $M(t) = \varnothing$ for $t > u$)

$$V(u) = \int_{-\infty}^{u} \left(\int_{M(t)} \frac{d\sigma_{m-1}(x)}{\|\mathrm{grad}\, F(x)\|} \right) dt.$$

Since the function $t \mapsto \int_{M(t)} \frac{d\sigma_{m-1}(x)}{\|\operatorname{grad} F(x)\|}$ is continuous (see Remark 8.4.2), differentiating the last equality, we arrive at the following result:

$$V'(u) = \int_{M(u)} \frac{d\sigma_{m-1}(x)}{\|\operatorname{grad} F(x)\|}. \tag{7}$$

In the special case where $F(x) = \|x\|$ (and, correspondingly, $\|\operatorname{grad} F(x)\| \equiv 1$), the obtained formula leads to the equality $(\lambda_m(B(u)))' = \sigma_{m-1}(S^{m-1}(u))$, which we have already encountered (see Sect. 8.3.5, Example 4).

8.4.4 Now we use formula (7) to relate the area of a surface and its Minkowski area (see Sect. 2.8.2). One can prove that for a compact set A contained in a smooth surface $S \subset \mathbb{R}^m$,

$$\sigma_{m-1}(A) = \lim_{\varepsilon \to 0} \frac{\lambda_m(A_\varepsilon \setminus A)}{2\varepsilon},$$

where A_ε is the ε-neighborhood of A (see [F, Theorem 3.2.39]). We will prove a similar formula not for an arbitrary compact subset of a smooth surface, but for the boundary of a compact Lebesgue set of a smooth function.

Theorem *Let $\mathcal{O}$ be an open subset of $\mathbb{R}^m$, $F \in C^2(\mathcal{O})$, $K = \mathcal{O}(F \leqslant C)$ and $M = \partial K$. If the set K is compact and* $\operatorname{grad} F \neq 0$ *on M, then the area of M coincides with the Minkowski area, i.e.,*

$$\sigma_{m-1}(M) = \lim_{\varepsilon \to 0} \frac{\lambda_m(K_\varepsilon \setminus K)}{\varepsilon}.$$

Proof First assume that $\|\operatorname{grad} F\| \equiv 1$ on M; we may also assume without loss of generality that $C = 0$. Fix $\delta > 0$ such that $\overline{K_\delta} \subset \mathcal{O}$. Let ω be the modulus of continuity of the map $x \mapsto \operatorname{grad} F(x)$ on the set $K_\delta \setminus K$. We will assume that δ is so small that $\omega(\delta) < 1/2$. Let us show that for small $\varepsilon > 0$ the sets $V(\varepsilon) = \{x \in K_\delta \mid F(x) \leqslant \varepsilon\}$ are close to the ε-neighborhoods of K. For this, keeping the above notation, we will show that the following lemma holds.

Lemma *Let $\varepsilon < \delta/2$, $\varepsilon' = \varepsilon(1 - \omega(2\varepsilon))$ and $\varepsilon'' = \varepsilon(1 + \omega(2\varepsilon))$; then*

$$V(\varepsilon') \subset K_\varepsilon \subset V(\varepsilon'').$$

Proof of the lemma If $x \in K_\delta \setminus K$ and x_0 is the point of M that is closest to x, then

$$\begin{aligned} F(x) = F(x) - F(x_0) &\leqslant \max_{z \in [x, x_0]} \|\operatorname{grad} f(z)\| \|x - x_0\| \\ &\leqslant \bigl(1 + \omega(\|x - x_0\|)\bigr) \|x - x_0\|. \end{aligned}$$

Hence for every point x in K_ε, for $\varepsilon < \delta$ we have $F(x) \leqslant (1 + \omega(\varepsilon))\varepsilon \leqslant \varepsilon''$ and, consequently, $K_\varepsilon \subset V(\varepsilon'')$.

Now we will prove that $V(\varepsilon') \subset K_\varepsilon$. Let $x \in V(\varepsilon) \setminus K$, and let x_0 be the point of M that is closest to x. By the definition of $V(\varepsilon)$, we have $\|x - x_0\| < \delta$. One can easily see that the vectors $x - x_0$ and $\operatorname{grad} F(x_0)$ are proportional: $x - x_0 = \|x - x_0\| \operatorname{grad} F(x_0)$. Hence for some $z \in [x_0, x]$ we have

$$\begin{aligned}\varepsilon \geqslant F(x) - F(x_0) &= \langle \operatorname{grad} F(z), x - x_0\rangle = \langle \operatorname{grad} F(x_0), x - x_0\rangle \\ &\quad + \langle \operatorname{grad} F(z) - \operatorname{grad} F(x_0), x - x_0\rangle \\ &\geqslant \|x - x_0\| - \omega\big(\|x - x_0\|\big)\|x - x_0\|. \end{aligned} \tag{8}$$

Since $\|x - x_0\| < \delta$, we obtain $\omega(\|x - x_0\|) \leqslant \omega(\delta) \leqslant 1/2$, whence $\|x - x_0\| \leqslant 2\varepsilon$. Returning to (8), we see that

$$\varepsilon \geqslant \|x - x_0\| - \omega\big(\|x - x_0\|\big)\|x - x_0\| \geqslant \|x - x_0\| - \omega(2\varepsilon)\|x - x_0\|,$$

i.e., $\|x - x_0\| \leqslant t = \varepsilon/(1 - \omega(2\varepsilon))$. It follows that $V(\varepsilon) \subset K_t$. Since $t > \varepsilon$, we have $\omega(2\varepsilon) \leqslant \omega(2t)$, whence $\varepsilon = t(1 - \omega(2\varepsilon)) > t(1 - \omega(2t))$. Therefore,

$$V\big(t\big(1 - \omega(2t)\big)\big) \subset V(\varepsilon) \subset K_t,$$

which, replacing t by ε, can be rewritten in the form

$$V\big(\varepsilon'\big) = V\big(\varepsilon\big(1 - \omega(2\varepsilon)\big)\big) \subset K_\varepsilon. \qquad \square$$

Let us return to the proof of the theorem. It follows from the lemma that

$$\big(1 - \omega(2\varepsilon)\big)\frac{\lambda_m(V(\varepsilon') \setminus K)}{\varepsilon'} \leqslant \frac{\lambda_m(K_\varepsilon \setminus K)}{\varepsilon} \leqslant \big(1 + \omega(2\varepsilon)\big)\frac{\lambda_m(V(\varepsilon'') \setminus K)}{\varepsilon''}.$$

By (7), as $\varepsilon \to 0$, the outermost parts of this inequality tend to $\int_M \frac{d\sigma_{m-1}(x)}{\|\operatorname{grad} F(x)\|} = \sigma_{m-1}(M)$, which completes the proof of the theorem under the additional assumption made above. Note that at this stage of the proof, we haven't yet used the C^2-smoothness of F, but have used only the C^1-smoothness.

In the general case (still assuming that $C = 0$), we introduce an auxiliary function H by the formula

$$H(x) = \frac{F(x)}{\sqrt{\|\operatorname{grad} F(x)\|^2 + F^2(x)}}.$$

It is obvious that $H \in C^1(\mathcal{O})$, $K = \mathcal{O}(H \leqslant 0)$, and

$$\operatorname{grad} H(x) = \frac{\operatorname{grad} F(x)}{\sqrt{\|\operatorname{grad} F(x)\|^2 + F^2(x)}} + F(x) \operatorname{grad} \frac{1}{\sqrt{\|\operatorname{grad} F(x)\|^2 + F^2(x)}}.$$

Hence $\|\operatorname{grad} H(x)\| = 1$ for $x \in M$, and the desired equality holds by the first part of the proof. $\square$

8.4.5 Using the isoperimetric inequality (see Sect. 2.8.2) and Theorem 8.4.4, we can obtain the main special case of the Gagliardo–Nirenberg–Sobolev inequality. For $p = 1$, it takes the form (see Sect. 5.4.4)

$$\left(\int_{\mathbb{R}^m} |F(x)|^{\frac{m}{m-1}}\, dx\right)^{\frac{m-1}{m}} \leqslant \frac{1}{2}\int_{\mathbb{R}^m} \|\operatorname{grad} F(x)\|\, dx,$$

where $F \in C_0^1(\mathbb{R}^m)$. We will prove it and, in passing, reduce the coefficient on the right-hand side. Since every smooth function, along with all its derivatives, can be uniformly approximated by functions of the class C_0^∞ (see Theorem 2 in Sect. 7.6.4), in what follows we assume that $F \in C_0^\infty(\mathbb{R}^m)$. Then formula (4′) with $f \equiv 1$ yields

$$\int_{\mathbb{R}^m} \|\operatorname{grad} F(x)\|\, dx = \int_{-\infty}^{\infty} \sigma_{m-1}\big(M(t)\big)\, dt, \tag{9}$$

where $M(t)$ is the boundary of the set $V(t) = \{x \in \mathbb{R}^m \mid F(x) \geqslant t\}$. Since, by Sard's theorem 13.5.2, the set of critical values of the function F has zero measure, this equality obviously remains valid if we integrate $\|\operatorname{grad} F(x)\|$ not over the whole space $\mathbb{R}^m$, but only over the set $\mathcal{O} = \{x \in \mathbb{R}^m \mid F(x) \neq 0,\ \operatorname{grad} F(x) \neq 0\}$. In this case we may assume that $F \geqslant 0$, since otherwise F can be replaced by $|F|$.

By Theorem 8.4.4, for non-critical values $t \in \mathbb{R}$,

$$\sigma_{m-1}\big(M(t)\big) = \lim_{\varepsilon\to 0} \frac{1}{\varepsilon}\lambda_m\big(\big(V(t)\big)_\varepsilon \setminus V(t)\big);$$

by the isoperimetric inequality, the right-hand side is not less than $m\alpha_m^{\frac{1}{m}}\lambda_m^{\frac{m-1}{m}}(V(t))$, where α_m is the volume of the unit ball. Thus (9) implies that

$$\int_{\mathbb{R}^m} \|\operatorname{grad} F(x)\|\, dx \geqslant m\alpha_m^{\frac{1}{m}} \int_0^{\infty} \lambda_m^{\frac{m-1}{m}}\big(V(t)\big)\, dt. \tag{10}$$

To estimate the last integral, we need the following lemma.

Lemma *If a non-negative function* ψ *does not increase on* $[0, +\infty)$, *then for any* $r > 1$ *and* $s > 0$,

$$\left(\int_0^s \psi^r(t)\, dt^r\right)^{\frac{1}{r}} \leqslant \int_0^s \psi(t)\, dt.$$

Proof of the lemma Denote the left- and right-hand sides of the inequality by $I(s)$ and $J(s)$, respectively. Since the function ψ does not increase, $I(s) \geqslant \psi(s)(\int_0^s dt^r)^{\frac{1}{r}} = s\psi(s)$. Hence for almost all $s > 0$ we have

$$I'(s) = \frac{1}{r} I^{1-r}(s)\, r\, s^{r-1}\psi^r(s) \leqslant \big(s\psi(s)\big)^{1-r} s^{r-1}\psi^r(s) = \psi(s) = J'(s).$$

The lemma follows, since the functions I, J are absolutely continuous and $I(0) = J(0)$ $(=0)$. □

Applying the lemma to the function $\psi(t) = \lambda_m^{\frac{1}{r}}(V(t))$, we obtain

$$\int_0^\infty \lambda_m^{\frac{1}{r}}\big(V(t)\big)\,dt \geqslant \left(r\int_0^\infty t^{r-1}\lambda_m\big(V(t)\big)\,dt\right)^{\frac{1}{r}} = \left(\int_{\mathbb{R}^m} |F(x)|^r\,dx\right)^{\frac{1}{r}}$$

(at the end, we have used Proposition 6.4.3). For $r = \frac{m}{m-1}$, this inequality together with (10) yields the desired bound:

$$\left(\int_{\mathbb{R}^m} |F(x)|^{\frac{m}{m-1}}\,dx\right)^{\frac{m-1}{m}} \leqslant \frac{1}{m\alpha_m^{1/m}}\int_{\mathbb{R}^m} \|\operatorname{grad} F(x)\|\,dx.$$

Note that $m\alpha_m^{1/m} \geqslant 2\sqrt{m}$, because

$$\alpha_m^{\frac{1}{m}} = \lambda_m^{\frac{1}{m}}\big(B(1)\big) \geqslant \lambda_m^{\frac{1}{m}}\left(\left(-\frac{1}{\sqrt{m}}, \frac{1}{\sqrt{m}}\right)^m\right) = \frac{2}{\sqrt{m}}.$$

EXERCISES

1. Let $M = \{(u, v, w) \in \mathbb{R}^3 \mid u^2 + v^2 + w^2 = R^2,\ u^2 + v^2 > R|u|\}$ be the part of the sphere $S^2(R)$ lying outside the cylinders $u^2 + v^2 = \pm Ru$. For which α is the integral $\int_M \frac{d\sigma_2(x)}{\|x-x_0\|^\alpha}$, where $x_0 = (0, 0, R)$, finite?

2. Show that for $\alpha < 1$, the graph of the function $\varphi(x) = x\sin\frac{1}{x^\alpha}$ $(x \in (0, 1))$ is a rectifiable curve. For which p is the integral $\int_{\Gamma_\varphi} \frac{d\sigma_1(x)}{\|x\|^p}$ finite?

3. Let E be a compact subset of a smooth k-dimensional manifold and $w \in C(E)$. Show that the function $I(y) = \int_E w(x)\ln\|x - y\|\,d\sigma_k(x)$ is infinitely differentiable on $\mathbb{R}^m \setminus E$ and has $(k-1)$ continuous derivatives on E (cf. Example 3 in Sect. 8.4.1).

4. Let $f \in C(B(a, 2r))$. Set $g(x) = \int_{B(r)} f(x + y)\,dy$ for $x \in B(a, r)$. Then $g \in C^1(B(a, r))$ and

$$\frac{\partial g}{\partial e}(x) = \frac{1}{r}\int_{S(r)} f(x + y)\langle y, e\rangle\,d\sigma(y) \quad \text{for every vector } e \neq 0.$$

5. Let $f \in C([-1, 1])$ and e be a unit vector in $\mathbb{R}^m$ $(m > 1)$. Prove Poisson's formula

$$\int_{S^{m-1}} f\big(\langle x, e\rangle\big)\,d\sigma_{m-1}(x) = 2(m-1)\frac{\pi^{\frac{m-1}{2}}}{\Gamma(\frac{m-1}{2})}\int_{-1}^1 f(t)\big(1-t^2\big)^{\frac{m-3}{2}}\,dt.$$

6. Let f be a positive continuous function on $\mathbb{R}^m$ such that $f(tx) = t^m f(x)$ for $t > 0$. Show that for every non-degenerate linear transformation A in $\mathbb{R}^m$,

$$\int_{S^{m-1}} \frac{1}{f(A(x))}\,d\sigma_{m-1}(x) = \frac{1}{|\det(A)|}\int_{S^{m-1}} \frac{1}{f(x)}\,d\sigma_{m-1}(x).$$

Hint. Use formula (4′) from Sect. 6.5.3.

7. Let $v \in \mathbb{R}^m$, $A : \mathbb{R}^m \to \mathbb{R}^m$ be a non-degenerate linear transformation and $F \in C(\mathbb{R})$. Show that

$$\int_{S^{m-1}} F\left(\frac{\langle x, v\rangle}{\|A(x)\|}\right)\frac{d\sigma_{m-1}(x)}{\|A(x)\|^m} = \frac{s_{m-1}}{|\det(A)|}\int_{-1}^{1} F(ct)\big(1-t^2\big)^{m-\frac{3}{2}}\,dt,$$

where $c = \|(A^{-1})^*(v)\|$ and s_{m-1} is the area of the unit sphere.

8. Let θ be the angle between non-zero vectors $a, b \in \mathbb{R}^m$ $(0 \leqslant \theta \leqslant \pi)$. Show that

$$\int_{S^{m-1}} \operatorname{sign}\big(\langle a, x\rangle\big)\operatorname{sign}\big(\langle b, x\rangle\big)\,d\sigma_{m-1}(x) = \left(1 - \frac{2}{\pi}\theta\right)s_{m-1}$$

(where s_{m-1} is the area of the unit sphere).

8.5 Integration of Vector Fields

8.5.1 In problems of mechanics and physics, one often encounters integrals of the form

$$\int_M \big\langle V(x), \theta(x)\big\rangle\,d\sigma(x),$$

where M is a smooth manifold, $V(x)$ are vectors corresponding to the problem under consideration, $\theta(x)$ is a unit vector related only to M, and σ is the surface area of M. Note that for the integrand to be summable it suffices that all coordinates of the vector $V(x)$ be summable on M, which is equivalent to the condition $\int_M \|V(x)\|\,d\sigma(x) < +\infty$. In what follows, we assume that this condition is satisfied.

We will restrict ourselves to the discussion of two extreme cases, which are of special interest.

(I) M is a one-dimensional manifold and $\theta(x)$ is a unit tangent vector to M at x.
(II) M is a manifold of codimension 1 (surface) and $\theta(x)$ is a unit normal to M at x.

We introduce several terms that will allow us to clarify the physical interpretation of the arising integrals.

Let us regard a continuous map $V : E \to \mathbb{R}^m$, where $E \subset \mathbb{R}^m$, as the family of vectors $\{V(x)\}_{x\in E}$ and call it a *vector field on* E. As a rule, we assume that the set E is open and the field is smooth (the latter means that the map V is C^1-smooth).

We can interpret $V(x)$ as the force applied at the point x and speak about a force field. We may also imagine that in the set E there is a steady-state flow of matter (fluid or gas) such that the velocity of the particle that at time t is at position $x \in E$ does not depend on the time and is equal to $V(x)$. In this case, one says that in E there is a stationary flow and V is its velocity field. We will abide by these mechanical interpretations. However, one should bear in mind that in applications an important role is also played by vector fields of another nature, for instance, the electric or magnetic fields appearing in Maxwell's equations.

8.5.2 *Integration over an Oriented Curve.* Let us first discuss case I. For simplicity, we assume that the vector field V is defined in a domain $\mathcal{O} \subset \mathbb{R}^m$. It makes sense to change the notation, in order to emphasize the one-dimensional nature of the problem under consideration. A one-dimensional manifold will be called a curve and denoted by L. The measure $\sigma = \sigma_1$ will be called the length, as usual.

Denote a unit tangent vector to L at a point x by $\tau(x)$. Clearly, there exist only two such vectors: $\tau(x)$ and $-\tau(x)$. An *orientation* on a smooth curve L is a continuous family of unit tangent vectors defined on L. In other words, a continuous family $\tau = \{\tau(x)\}_{x\in L}$ is an orientation on L if $\|\tau(x)\| = 1$ and $\tau(x)$ is a tangent vector to L at x for all $x \in L$. A curve equipped with an orientation, i.e., the pair (L, τ), is called an *oriented curve*.

Using the coordinate functions $V_1, \dots, V_m$ of the field V, the line integral of $\langle V, \tau\rangle$ can be written in the form

$$\int_L \bigl\langle V(x), \tau(x)\bigr\rangle\, d\sigma(x) = \int_{(L,\tau)} V_1(x)\, dx_1 + \dots + V_m(x)\, dx_m. \tag{1}$$

It is also denoted by $\int_L V_1(x)\, dx_1 + \dots + V_m(x)\, dx_m$; the latter notation does not explicitly indicate the orientation (which is assumed to be given).

Clearly, reversing the orientation from $\tau = \{\tau(x)\}_{x\in L}$ to $\{-\tau(x)\}_{x\in L}$ changes the sign of the line integral.

On a connected curve there are only two opposite orientations. Indeed, if $\{\widetilde{\tau}(x)\}_{x\in L}$ is an orientation on L, then the function $x \mapsto \langle \tau(x), \widetilde{\tau}(x)\rangle$ is continuous on L and takes only the values ± 1. By the connectedness, this function is constant on L, which implies that $\widetilde{\tau}$ coincides either with τ or with the opposite orientation. Note that in order to define an orientation on a connected curve, it suffices to define a tangent vector only at one point.

Using a smooth parametrization $\gamma : (a, b) \mapsto \mathbb{R}^m$ of a simple curve L, one can easily construct an orientation on L which we will call the *orientation corresponding to* γ. It is defined by the formula

$$\tau(x) = \frac{\gamma'(\gamma^{-1}(x))}{\|\gamma'(\gamma^{-1}(x))\|} \quad (x \in L).$$

This implies (see Theorem 8.4.1) a formula for computing the integral (1):

$$\int_{(L,\tau)} V_1(x)\, dx_1 + \dots + V_m(x)\, dx_m = \int_a^b \bigl\langle V\bigl(\gamma(t)\bigr), \gamma'(t)\bigr\rangle\, dt.$$

This leads to a useful generalization. Let γ be a piecewise smooth path in $\mathcal{O}$ defined on $[a, b]$. The *integral over* γ of a vector field V is the integral on the right-hand side of the last formula. It is denoted by $\int_\gamma V_1(x)\, dx_1 + \dots + V_m(x)\, dx_m$.

Now we will explain how one can interpret the integral (1) over an oriented curve (L,τ). Assume that V is a force field and L is a curve contained in $\mathcal{O}$. Consider a small L-neighborhood U of a point $x\in L$. In view of the smallness of U, we may assume that this piece of the curve is almost straight and the field V is almost constant on it. Hence the work done by the force V along U must be approximately equal to the work done by the constant force $V(x)$ in moving the particle by the vector $\sigma(U)\tau(x)$. The latter work is equal to $\langle V(x),\tau(x)\rangle\,\sigma(U)$. Thus it is natural to assume that the work $A(e,\tau)$ done by the force V along an arbitrary segment e of the oriented curve satisfies the estimates

$$\inf_{x\in e}\bigl\langle V(x),\tau(x)\bigr\rangle\sigma(e)\leqslant A(e,\tau)\leqslant\sup_{x\in e}\bigl\langle V(x),\tau(x)\bigr\rangle\sigma(e).$$

In addition, $A(e,\tau)$ depends additively on e. Under these assumptions, using the general scheme considered in Sect. 6.3, we see that the work done by the force V in moving a particle along the oriented curve (L,τ) is given by the integral (1). It is clear that the integral over a piecewise smooth path lying in $\mathcal{O}$ has the same interpretation.

Definition A vector field $V=(V_1,\dots,V_m)$ defined in a domain $\mathcal{O}$ is called *potential* if there exists a smooth function F on $\mathcal{O}$ (a *potential of* V) such that $V(x)=\operatorname{grad}F(x)$ for all points $x\in\mathcal{O}$.

In the case of a potential field, the integral $\int_\gamma V_1(x)\,dx_1+\dots+V_m(x)\,dx_m$ satisfies the so-called gradient theorem, or the fundamental theorem of calculus for line integrals.

Proposition 1 *Let F be a potential of a vector field $V=(V_1,\dots,V_m)$ defined in a domain $\mathcal{O}$, and let γ be a piecewise smooth path in $\mathcal{O}$ starting at A and ending at B. Then*

$$\int_\gamma V_1(x)\,dx_1+\dots+V_m(x)\,dx_m=F(B)-F(A).$$

Proof It suffices to prove the assertion only for a smooth path γ. We assume that it is defined on an interval $[a,b]$, so that $A=\gamma(a)$ and $B=\gamma(b)$. It is easy to check that $\langle V(\gamma),\gamma'\rangle=(F(\gamma))'$. Hence

$$\begin{aligned}\int_\gamma V_1(x)\,dx_1+\dots+V_m(x)\,dx_m&=\int_a^b\bigl\langle V\bigl(\gamma(t)\bigr),\gamma'(t)\bigr\rangle\,dt=\int_a^b\bigl(F\bigl(\gamma(t)\bigr)\bigr)'\,dt\\&=F\bigl(\gamma(b)\bigr)-F\bigl(\gamma(a)\bigr)=F(B)-F(A).\qquad\square\end{aligned}$$

Thus the work done by a potential field along a path depends only on the values of the potential at its endpoints and is equal to the increment of the potential. In this case, one says that the integral is path-independent. Crucially, the converse is also true.

Proposition 2 *If a line integral is path-independent, then the corresponding vector field is potential.*

Proof Note that any two points A and B of a domain $\mathcal{O}$ can be joined by a piecewise smooth path lying in $\mathcal{O}$. Let V be a vector field in $\mathcal{O}$ for which all integrals along such paths coincide. Denote their common value by $\int_A^B V_1(z)\,dz_1 + \cdots + V_m(z)\,dz_m$. Fixing a point $A \in \mathcal{O}$, consider the "integral with variable upper limit"

$$F(x) = \int_A^x V_1(z)\,dz_1 + \cdots + V_m(z)\,dz_m \quad (x \in \mathcal{O}).$$

It is easy to see that $F(y) - F(x) = \int_x^y V_1(z)\,dz_1 + \cdots + V_m(z)\,dz_m$ (to check this, write $F(y)$ as the integral over a path passing through x). Let us show that F is a potential of V. Fix x and consider an arbitrary vector e_j of the canonical basis. For a sufficiently small real t, we have

$$\begin{aligned} F(x + te_j) - F(x) &= \int_x^{x+te_j} V_1(z)\,dz_1 + \cdots + V_m(z)\,dz_m \\ &= \int_0^t \langle V(x + se_j), e_j \rangle\,ds \\ &= \int_0^t V_j(x + se_j)\,ds \\ &= t \int_0^1 V_j(x + tue_j)\,du. \end{aligned}$$

Therefore, $F(x + te_j) - F(x) = t(V_j(x) + o(1))$ as $t \to 0$, i.e., $\frac{\partial F}{\partial x_j}(x) = V_j(x)$. □

In the case of path-independence, the integral over a closed path vanishes (recall that a path is called closed if both its endpoints coincide). However, in the general case, this is not true.

Example Consider the force field $V(x, y) = (-\frac{y}{x^2+y^2}, \frac{x}{x^2+y^2})$ defined in the "punctured" plane $\mathbb{R}^2 \setminus \{0\}$. Let us compute its work A along a circle, or, more precisely, along the closed path $\gamma(t) = (\cos t, \sin t)$, where $t \in [0, 2\pi]$:

$$A = \int_\gamma -\frac{y}{x^2 + y^2}\,dx + \frac{x}{x^2 + y^2}\,dy = \int_0^{2\pi} dt = 2\pi.$$

This example shows that the work done by a non-potential field along a closed path may be non-zero. Note that the restriction of the field under consideration to every half-plane that does not contain the origin is potential. In particular, its restriction to the half-plane $x > 0$ is the gradient field of the function $F(x, y) = \arctan \frac{y}{x}$.

For smooth fields, one can easily establish a simple and important necessary condition for potentiality. Indeed, if F is a potential of a smooth field $V = (V_1, \dots, V_m)$ in a domain $\mathcal{O}$, then, by the mixed derivatives (or Clairaut's) theorem, we have

$$\frac{\partial V_k}{\partial x_j}(x) = \frac{\partial^2 F}{\partial x_j \partial x_k}(x) = \frac{\partial^2 F}{\partial x_k \partial x_j}(x) = \frac{\partial V_j}{\partial x_k}(x).$$

Thus the equalities

$$\frac{\partial V_k}{\partial x_j}(x) = \frac{\partial V_j}{\partial x_k}(x) \quad (x \in \mathcal{O},\ j, k = 1, \dots, m) \tag{2}$$

are necessary conditions for the field V to be potential. As the above example shows, in the general case, these conditions are not sufficient. However, in "good" domains, they are. Leaving aside the thorough investigation of this problem, we restrict ourselves to a special case of the result known as the Poincaré lemma.

Proposition 3 *A smooth field $V = (V_1, \dots, V_m)$ defined in a convex domain $\mathcal{O}$ and satisfying condition* (2) *is potential.*

Proof To simplify formulas, we assume that $\mathcal{O}$ contains the origin. Then for every point $x = (x_1, \dots, x_m)$ in $\mathcal{O}$, the straight path $\gamma_x(t) = tx$, $t \in [0, 1]$, lies in $\mathcal{O}$. Set $F(x) = \int_{\gamma_x} V_1(z)\, dz_1 + \dots + V_m(z)\, dz_m$; we will show that F is a potential of V. Indeed,

$$F(x) = \int_0^1 \langle V(\gamma_x(t)), \gamma_x'(t) \rangle\, dt = \sum_{k=1}^m x_k \int_0^1 V_k(tx)\, dt.$$

Differentiating with respect to x_j and using (2), we obtain

$$\begin{aligned}
\frac{\partial F}{\partial x_j}(x) &= \int_0^1 V_j(tx)\, dt + \sum_{k=1}^m x_k \int_0^1 t \frac{\partial V_k}{\partial x_j}(tx)\, dt \\
&= \int_0^1 \left(V_j(tx) + t \sum_{k=1}^m x_k \frac{\partial V_j}{\partial x_k}(tx) \right) dt \\
&= \int_0^1 \big(t V_j(tx)\big)_t'\, dt = V_j(x).
\end{aligned}$$

□

Remark The proof does not use the convexity of the domain in full strength. In particular, it remains valid for star domains ($\mathcal{O}$ is a star domain if there is a point $x_0 \in \mathcal{O}$ such that the line segment $\{x_0 + t(x - x_0) \mid t \in [0, 1]\}$ lies in $\mathcal{O}$ for every $x \in \mathcal{O}$).

We will say that a vector field defined in a domain $\mathcal{O}$ is *locally potential* if every point of $\mathcal{O}$ has a neighborhood in which the field has a potential. Proposition 3 implies an obvious but useful corollary.

Corollary *A smooth field is locally potential if and only if it satisfies condition* (2).

The above example shows that a locally potential field may not be globally potential, and the integral of a locally potential field over a closed path may be non-zero. Later, in Sect. 8.6.7, we will return to this question.

8.5.3 *Side of a Surface and the Flow of a Vector Field.* Now we proceed to case II. Consider integrals of the form $\int_M \langle V(x), \nu(x)\rangle \, d\sigma(x)$, which often appear in problems of physics and mechanics. Here M is a smooth surface, $\nu(x)$ is a unit normal to M at a point x, and $V(x)$ is a vector corresponding to the problem under investigation.

Recall that a normal to a smooth surface $M \subset \mathbb{R}^m$ at a point $x \in M$ is a non-zero vector orthogonal to the tangent space T_x. A unit normal is a normal of unit length. At every point of a surface there exist only two (opposite) normals.

A *side* of a smooth surface M is a continuous family of unit normals defined on M. In other words, a continuous family $\{\nu(x)\}_{x\in M}$ is a side of M if $\|\nu(x)\| = 1$ and $\nu(x)$ is a normal to M at x for every $x \in M$.

Using our intuitive notion of the surface area as a value proportional to the amount of paint needed to paint it (as mentioned at the beginning of this chapter), we may now say that a side of a surface may be thought of as the surface together with a coat of paint, or as the collection of all positions of the paintbrush. There is also a more widely used everyday interpretation, that of the "visible side". This is determined by the part of the surface that is "visible" to an observer, or, more exactly, by the part on which the normals are oriented "towards" the visual ray (making an obtuse angle with it).

Now we proceed from an informal discussion to necessary elaborations related to the notion of a side of a surface. If $\nu = \{\nu(x)\}_{x\in M}$ is a side of a surface M, then, obviously, the opposite family $\{-\nu(x)\}_{x\in M}$ is also a side of M. On a connected surface, there are no other sides (to prove this, it suffices to reproduce almost literally the argument used when considering an orientation on a curve). With this in mind, surfaces on which there is a side are called *two-sided.* To indicate a side of a connected surface, it suffices to define a normal at least at one point.

Clearly, if a smooth surface M has a global parametrization Φ, then one can easily construct a side of M using the vector N_Φ (see Sect. 8.3.4):

$$\nu(x) = \frac{N_\Phi(\Phi^{-1}(x))}{\|N_\Phi(\Phi^{-1}(x))\|} \quad (x \in M).$$

We say that this side is generated by Φ, or *corresponds* to Φ.

The graph Γ_φ of a smooth function φ is a two-sided surface. Its canonical parametrization generates the side

$$x = \big(u, \varphi(u)\big) \mapsto \nu(x) = \frac{(-\operatorname{grad}\varphi(u), 1)}{\sqrt{1 + \|\operatorname{grad}\varphi(u)\|^2}}. \tag{3}$$

Note that all vectors of this side make acute angles with the x_m axis. Hence we will say that it is the *upper* side of the graph and the opposite side is the *lower* side.

Another important example of a two-sided surface, the boundary of a "sufficiently good" compact set, will be considered in the next section.

We see from the above that every sufficiently small M-neighborhood of every point of a smooth surface M has a side. However, this does not mean that the whole surface M also has a side. A counterexample is the surface called the Möbius[9] strip, which can be obtained by "giving a half-twist" to a rectangle and then "gluing together" its opposite sides. Speaking more formally, given a rectangle $[-a, a] \times (-b, b)$, we identify the centrally symmetric points lying at the vertical sides (note that by identifying the points symmetric with respect to the y axis, we will obtain an ordinary cylindric surface, which is obviously two-sided). It can be proved that one cannot define a side on the Möbius strip. We encourage the reader to experiment by painting the surface obtained by gluing a twisted narrow rectangular strip of paper. Smooth surfaces on which one cannot define a side are called one-sided.

Having chosen a side $\{\nu(x)\}_{x \in M}$ of a two-sided surface lying in a domain where a vector field V is defined, we can consider the surface integral

$$\int_M \langle V(x), \nu(x) \rangle \, d\sigma(x) \tag{4}$$

(reversing the side obviously changes the sign of the integral). If the chosen side is generated by a parametrization $\Phi \in C^1(G)$, then the computation of this integral reduces to the computation of a multiple integral (see Theorem 8.4.1 and the formula for N_Φ in Sect. 8.3.4):

$$\int_M \langle V(x), \nu(x) \rangle \, d\sigma(x) = \int_G \langle V\big(\Phi(u)\big), N_\Phi(u) \rangle \, du$$

$$= \int_G \begin{vmatrix} V_1(\Phi(u)) & \dots & V_m(\Phi(u)) \\ \frac{\partial \varphi_1(u)}{\partial u_1} & \dots & \frac{\partial \varphi_m(u)}{\partial u_1} \\ \vdots & \ddots & \vdots \\ \frac{\partial \varphi_1(u)}{\partial u_{m-1}} & \dots & \frac{\partial \varphi_m(u)}{\partial u_{m-1}} \end{vmatrix} \, du.$$

Now we turn to a physical problem leading to the integral (4). Assume that in a domain $\mathcal{O} \subset \mathbb{R}^m$ there is a vector field V, which we regard as the velocity field of a stationary fluid flow. How can one compute the amount of fluid flowing through a smooth two-sided surface $M \subset \mathcal{O}$ per unit time? When solving this problem, one should bear in mind that particles of fluid can traverse the surface in different directions "moving from one side to the other". If the surface bounds a body, this means that the fluid may flow out of it as well as into it. Hence, to make our problem more definite, we fix a side $\{\nu(x)\}_{x \in M}$ of M.

Consider a small M-neighborhood U of a point $x \in M$. Then we may assume that this piece of the surface is almost planar and the velocity V is almost constant on it. Hence the fluid flowing through U per unit time fills a curved parallelotope

[9] August Ferdinand Möbius (1790–1868)—German mathematician.

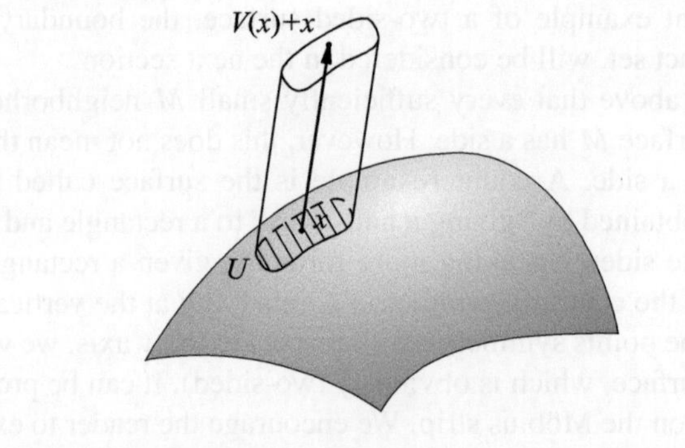

Fig. 8.2 The parallelotope with base U and edges equal to $V(x)$

close to the parallelotope with base U and edges equal to $V(x)$. The volume of the latter is equal to $\sigma(U)|\langle V(x), \nu(x)\rangle|$ (see Fig. 8.2).

The inner product $\langle V(x), \nu(x)\rangle$ is positive if the vectors $V(x)$ and $\nu(x)$ make an acute angle, i.e., if the fluid traverses M "in the direction $\nu(x)$", and is negative otherwise. Hence the absolute value of the integral $\int_U \langle V(x), \nu(x)\rangle\, d\sigma(x)$ is equal to the amount of fluid flowing through U per unit time. Its sign depends on the choice of a side of the surface and characterizes the direction of the fluid motion. In view of these considerations, the integral (4) is called the *flow* of the vector field V through M in the given direction.[10]

Example Let f be a smooth function on a domain $\mathcal{O} \subset \mathbb{R}^m$ that has no critical points. Set

$$\nu(x) = \frac{1}{\|\operatorname{grad} f(x)\|}\operatorname{grad} f(x), \qquad V(x) = \frac{1}{\|\operatorname{grad} f(x)\|}\nu(x) \quad (x \in \mathcal{O}).$$

It is clear that the family $\{\nu(x)\}_{x\in M_C}$ is a side of the level surface $M_C = \{x \in \mathcal{O} \mid f(x) = C\}$. The flow of ν in this direction is just the area of M_C. The flow of V through M_C also has a simple geometric interpretation: it is the derivative at $u = C$ of the volume of the set $\mathcal{O}_u = \{x \in \mathcal{O} \mid f(x) \leqslant u\}$ (see Remark (3) in Sect. 8.4.3).

8.5.4 Having discussed integration of vector fields over manifolds of minimal (Sect. 8.5.2) and maximal (Sect. 8.5.3) dimension, a few words are in order regarding integration over plane curves. In this situation, the maximal and the minimal dimensions coincide (both are equal to 1). Hence a plane curve L has not only a direction, but also a side. Formally, we obtain two types of line integrals of a vector field $V = (V_1, V_2)$ over L. First, the integral over an oriented curve

$$\int_{(L,\tau)} V_1(x,y)\,dx + V_2(x,y)\,dy = \int_L \bigl\langle V(x,y), \tau(x,y)\bigr\rangle\, d\sigma_1(x,y);$$

---

[10] We leave the reader to deduce this formula using the intuitively clear properties of the flow and the general scheme considered in Sect. 6.3.

second, the integral over L corresponding to a side $\nu = \{\nu(x, y)\}_{(x,y)\in L}$:

$$\int_L \big\langle V(x, y), \nu(x, y)\big\rangle\, d\sigma_1(x, y).$$

One can easily see that there is a close relation between these two integrals. To make it precise, consider the orthogonal transformation $z = (x, y) \mapsto U(z) = (-y, x)$ that rotates a vector z by $\pi/2$ "counterclockwise" (identifying $\mathbb{R}^2$ with the set of complex numbers $\mathbb{C}$, we can write it simply as $U(z) = iz$). Since by rotating a normal by a right angle we obtain a tangent vector, every side ν of L gives rise to an orientation $\tau = U(\nu)$. Conversely, every orientation τ on L gives rise to the side $\nu = U^{-1}(\tau)$. Given an orientation and a side related by the formula $\tau = U(\nu)$, we say that they agree with each other. Obviously, $\langle V, \nu\rangle = \langle \overline{V}, \tau\rangle$, where $\overline{V} = U(V)$, and hence the flow of V in the direction ν is equal to the integral of the field $\overline{V} = U(V)$ over the oriented curve (L, τ), where $\tau = U(\nu)$:

$$\int_L \langle V, \nu\rangle\, d\sigma_1 = \int_L \langle \overline{V}, \tau\rangle\, d\sigma_1.$$

This equality can be rewritten in the form

$$\int_L \langle V, \nu\rangle\, d\sigma_1 = \int_{(L,\tau)} -V_2(x, y)\, dx + V_1(x, y)\, dy. \tag{5}$$

We will use this in the next section when discussing Green's formula (Sect. 8.6.7).

EXERCISES

1. Find a potential F of the vector field $V(x) = -\frac{x}{\|x\|^m}$ defined in $\mathbb{R}^m \setminus \{0\}$. In the three-dimensional case, V is proportional to the gravitational field created by a point mass at the origin. Hence F is called the Newton potential. What is the work done by V along a path that starts at a point $x \neq 0$ and moves away to infinity?

2. A vector field V is called central if there exists a continuous function f on $(0, +\infty)$ such that $V(x) = f(\|x\|)x$ for $x \neq 0$. Show that such a field is potential. What is its potential?

3. Let M be the part of the boundary of the ellipsoid $\frac{x_1^2}{a_1^2} + \cdots + \frac{x_m^2}{a_m^2} < 1$ lying in the "first octant" $\mathbb{R}^m_+$. Find the flow of the vector field $V(x) = (\frac{c_1}{x_1}, \ldots, \frac{c_m}{x_m})$ through the side of M invisible from the origin (i.e., the upper side).

8.6 The Gauss–Ostrogradski Formula

8.6.1 The classical integral calculus is based on the Newton–Leibniz formula

$$\int_a^b f'(x)\, dx = f(b) - f(a),$$

which expresses the integral of the derivative in terms of the values of the function at the endpoints of the interval of integration.

What is an analog of this formula in the multi-dimensional case? It is natural to assume that, for a function of several variables, one should replace f' by a partial derivative, the interval by a compact set K, and consider the integral $\int_K \frac{\partial f}{\partial x_j}\, d\lambda_m$. Is it possible to express it by a formula including the values of f on the boundary of the integration domain only? The goal of this section is to show that the answer to this question is affirmative under very wide assumptions on the structure of the boundary of the set K.

The simplest version of the formula we seek can be obtained using Fubini's theorem. Integrating the partial derivative $\frac{\partial f}{\partial x_m}$ of the function f that is smooth on the parallelepiped $P = Q \times [a, b]$, where $Q \subset \mathbb{R}^{m-1}$, we get

$$\int_P \frac{\partial f}{\partial x_m}(x)\, dx = \int_Q \left(\int_a^b \frac{\partial f}{\partial x_m}(u, v)\, dv \right) du = \int_Q f(u, b)\, du - \int_Q f(u, a)\, du$$

(we identify a point x from P with the pair (u, v), $u \in Q$, $v \in [a, b]$). The integrals on the right-hand side are simply the integrals of f over the top and the bottom bases of the parallelepiped P. Denoting these parts of the boundary of P by the symbols ∂P_+ and ∂P_-, one can, obviously, write

$$\int_P \frac{\partial f}{\partial x_m}(x)\, dx = \int_{\partial P_+} f(x)\, d\sigma_{m-1}(x) - \int_{\partial P_-} f(x)\, d\sigma_{m-1}(x). \tag{1}$$

The next step is crucial for our argument. In the situation that arose above, one should ponder over the fact that we have to consider the difference and not, say, the sum of the integrals over ∂P_+ and ∂P_-. It is desirable to find an explanation of this phenomenon that would allow one to get rid of the "asymmetry" between the integrals over ∂P_+ and ∂P_-. This can be done using the notion of the outer normal, which will enable us to rewrite the right-hand side of the equality (1) as an integral over the boundary of the parallelepiped P.

To do this, consider the outer normal ν on ∂P. We postpone the precise definition of this notion until the next subsection. Still, using intuitive considerations, one can say that the outer normal coincides with the vector e_m on ∂P_+, coincides with $(-e_m)$ on ∂P_-, and is orthogonal to e_m on the rest of ∂P. Thus, the formula (1) can be rewritten as

$$\begin{aligned}
\int_P \frac{\partial f}{\partial x_m}(x)\, dx &= \int_{\partial P_+} f(x)\langle \nu(x), e_m\rangle\, d\sigma_{m-1}(x) + \int_{\partial P_-} f(x)\langle \nu(x), e_m\rangle\, d\sigma_{m-1}(x) \\
&= \int_{\partial P_+ \cup \partial P_-} f(x)\langle \nu(x), e_m\rangle\, d\sigma_{m-1}(x) \\
&= \int_{\partial P} f(x)\langle \nu(x), e_m\rangle\, d\sigma_{m-1}(x).
\end{aligned}$$

Taking into account that the partial derivative $\frac{\partial f}{\partial x_m}$ is the directional derivative in the direction e_m, it is useful to transform this equality into

$$\int_P \frac{\partial f}{\partial e_m}(x)\,dx = \int_{\partial P} f(x)\langle \nu(x), e_m\rangle\, d\sigma_{m-1}(x), \tag{2}$$

emphasizing the connection of the integrand on the right with the direction of the differentiation on the left.

Thus, we have obtained the simplest version of the classical Gauss–Ostrogradski[11] formula, which is precisely the generalization of the Newton–Leibniz formula we are aiming at.

Note that in the one-dimensional case, the Newton–Leibniz formula can be interpreted as a special case of the formula (2) if one considers the interval $[a,b]$ as a parallelepiped P, and defines the measure σ_0 on its boundary as the sum of two unit point masses at the points a and b and the "unit normals" at these points as the vectors $-e$ and e correspondingly where e is the unit vector on the real line.

Even now, the reader can easily check that in the equality (2), one can replace $\frac{\partial f}{\partial e_m}$ by the partial derivative with respect to any other coordinate or, more generally, the directional derivative in any direction. It is much harder to prove that the formula we obtained is valid not only for parallelepipeds, but also for more general compact sets. The description of such sets together with the verification of the corresponding equality are the main topics of this section. The final result will be obtained as the outcome of the process of the gradual extension of the class of admissible sets.

Everywhere in this section, we assume that $m > 1$. The surface area and the Lebesgue measure λ_m will be denoted by the letters σ and λ respectively without specifying the dimension explicitly.

8.6.2 Let $A \subset \mathbb{R}^m$ and let $p \in \partial A$. If near the point p the boundary ∂A coincides with a smooth surface M, then a normal $N(p)$ to M at the point p is called a normal to the boundary of A.

The normal $N(p)$ is called an *outer normal* to ∂A at the point p if $p + tN(p) \notin A$ and $p - tN(p) \in \operatorname{Int} A$ for all sufficiently small positive t. The side of M consisting of outer normals is called the outer side of M, and the opposite side is called the inner side. In the case when the set A can be defined by an inequality near the point $p \in \partial A$, i.e., $A \cap B(p,\delta) = \{x \in B(p,\delta) \mid F(x) \leqslant 0\}$ where F is a function smooth in the ball $B(p,\delta)$ with non-vanishing gradient, the corresponding part of the boundary ∂A is nothing but the zero level set of the function F. In this case $\operatorname{grad} F(p)$ is an outer normal to ∂A because the function F strictly increases in the direction of the gradient.

Later we shall need some special sets closely related to the subgraph of a smooth function. To define them, we will identify the space $\mathbb{R}^m$ with the Cartesian product $\mathbb{R}^{m-1} \times \mathbb{R}$ and will write a point x from $\mathbb{R}^m$ as $x = (u, v)$ where $u \in \mathbb{R}^{m-1}$ and

[11] Mikhail Vasil'evich Ostrogradski (1801–1862)—Russian mathematician.

$v \in \mathbb{R}$. Let φ be a function smooth on the closed cube $Q \subset \mathbb{R}^{m-1}$. The sets

$$\{(u,v) \mid u \in Q,\ c \leqslant v \leqslant \varphi(u)\} \quad \text{and} \quad \{(u,v) \mid u \in Q,\ \varphi(u) \leqslant v \leqslant d\},$$

where $c < \varphi$ and, respectively, $\varphi < d$, will be called the *lower* and the *upper beams*. The sets obtained from them by reenumerations of the coordinates will be called *beams.*

The points (u, v) of the graph of the function φ whose projections u lie in the interior of the cube Q form the *non-trivial part* of the beam boundary. The remaining points on the boundary form its *trivial part* (for the lower beam, this is contained in the boundary of the infinite parallelepiped $Q \times [c, +\infty)$).

If the function φ is not constant, the non-trivial part of the beam boundary is determined uniquely. Otherwise the non-trivial part should be specified explicitly (for example, every face of a cubic beam can be its non-trivial part).

It is clear that the beam boundary consists of finitely many compact subsets of smooth surfaces. The outer normal $\nu(x)$ to the boundary exists at every point x of smoothness, i.e., almost everywhere with respect to σ.

Completing the definition from Sect. 8.5.3, we will call the family of outer normals $\{\nu(x)\}_{x \in \partial\mathcal{B}}$ *the outer side of the boundary of the beam* $\mathcal{B}$. Note that, according to this definition, the outer side is defined not everywhere but only almost everywhere on $\partial\mathcal{B}$. Also, for every vector $e \in \mathbb{R}^m$, the function $x \mapsto \langle \nu(x), e\rangle$ is continuous almost everywhere on $\partial\mathcal{B}$ and, therefore, is measurable.

Since, near each point of the non-trivial part of the boundary, the lower beam is described by the inequality $F(u,v) = v - \varphi(u) \leqslant 0$, the gradient $\operatorname{grad} F(u,v) = (-\operatorname{grad}\varphi(u), 1)$ is an outer normal. In other words, the outer side of the non-trivial part of the lower beam is the upper side of the graph Γ_φ, and the unit outer normal at the point $x = (u, \varphi(u))$ belonging to this part of the boundary is equal to

$$\nu(x) = \frac{(-\operatorname{grad}\varphi(u), 1)}{\sqrt{1 + \|\operatorname{grad}\varphi(u)\|^2}}. \tag{3}$$

The outer side of the non-trivial part of the boundary of the upper beam is the lower side of the graph consisting of the opposite normals.

8.6.3 The next theorem constitutes the first step in the generalization of the formula (2). The symbol ν will denote the outer side of the beam boundary.

Theorem *Assume that a function f smooth on the beam $\mathcal{B} \subset \mathbb{R}^m$ vanishes on the trivial part of its boundary. Then for every unit vector $e \in \mathbb{R}^m$, the equality*

$$\int_{\mathcal{B}} \frac{\partial f}{\partial e}(x)\,dx = \int_{\partial\mathcal{B}} f(x)\langle \nu(x), e\rangle\,d\sigma(x) \tag{4}$$

holds.

Proof Changing the enumeration of the coordinates, if necessary, we may assume that $\mathcal{B}$ is either a lower beam, or an upper one. Since the arguments are essentially the same in these two cases, we will restrict ourselves to the consideration of the lower beam. By definition, it is a set of the form

$$\mathcal{B} = \{(u, v) \mid u \in Q, c \leqslant v \leqslant \varphi(u)\},$$

where φ is a smooth function on the closed cube $Q \subset \mathbb{R}^{m-1}$ satisfying the condition $\varphi > c$. Note that, since the directional derivative is a linear combination of partial derivatives, it suffices to prove the equality (4) for the case when e is one of the vectors $e_1, \ldots, e_m$ in the canonical basis of $\mathbb{R}^m$.

First, consider the case $e = e_m$. By Fubini's theorem, we have

$$\int_{\mathcal{B}} \frac{\partial f}{\partial e_m}(x)\,dx = \int_Q \left(\int_c^{\varphi(u)} \frac{\partial f}{\partial v}(u, v)\,dv\right) du = \int_Q \big(f\big(u, \varphi(u)\big) - f(u, c)\big)\,du.$$

In addition, $f(u, c) = 0$ because the function f vanishes on the trivial part of the beam boundary. Therefore,

$$\begin{aligned}\int_{\mathcal{B}} \frac{\partial f}{\partial e_m}(x)\,dx &= \int_Q f\big(u, \varphi(u)\big)\,du \\ &= \int_Q \frac{f(u, \varphi(u))}{\sqrt{1 + \|\operatorname{grad}\varphi(u)\|^2}} \sqrt{1 + \|\operatorname{grad}\varphi(u)\|^2}\,du.\end{aligned}$$

In view of equality (3), this means that

$$\int_{\mathcal{B}} \frac{\partial f}{\partial e_m}(x)\,dx = \int_{\Gamma_\varphi} f(x)\langle \nu(x), e_m\rangle\,d\sigma(x) = \int_{\partial\mathcal{B}} f(x)\langle \nu(x), e_m\rangle\,d\sigma(x)$$

(in the last equality, we again used the assumption $f \equiv 0$ on $\partial\mathcal{B} \setminus \Gamma_\varphi$).

Let now $e = e_k$, $1 \leqslant k < m$. The proof is the same for all such k, so we may consider $k = m - 1$ only. We may assume that $m \geqslant 3$. In the two-dimensional case, the argument, which the reader can easily check himself, is much simpler.

Represent the cube Q as the product $Q = R \times [a, b]$, where R is a cube in $\mathbb{R}^{m-2}$. We will write a point u from Q as $u = (s, t)$ where $s \in R$ and $a \leqslant t \leqslant b$. Using this notation, we get

$$\int_{\mathcal{B}} \frac{\partial f}{\partial e_{m-1}}(x)\,dx = \int_R \left(\int_a^b \left(\int_c^{\varphi(s,t)} \frac{\partial f}{\partial t}(s, t, v)\,dv\right) dt\right) ds \tag{5}$$

by Fubini's theorem. To transform the inner integral (to swap the integration with respect to v and the differentiation with respect to t), we shall need a generalization of the Leibniz rule for differentiation of an integral depending on a parameter. This generalization is given by the following lemma.

Lemma *Let $\psi \in C^1([a, b])$ and $c < \psi(t)$ for $a \leqslant t \leqslant b$. If the function f is smooth in a neighborhood of the curvilinear trapezoid*

$$T = \{(t, v) \in \mathbb{R}^2 \mid t \in [a, b], c \leqslant v \leqslant \psi(t)\},$$

then

$$\int_c^{\psi(t)} \frac{\partial f}{\partial t}(t,v)\,dv = \frac{d}{dt}\left(\int_c^{\psi(t)} f(t,v)\,dv\right) - f\big(t,\psi(t)\big)\psi'(t).$$

Proof of Lemma Put $F(t,\theta) = \int_c^\theta f(t,v)\,dv$ for (t,θ) in a sufficiently small neighborhood of the trapezoid T. Since $F'_\theta(t,\theta) = f(t,\theta)$ and, by the Leibniz rule, $F'_t(t,\theta) = \int_c^\theta f'_t(t,v)\,dv$, we can differentiate the composition $F(t,\psi(t))$ to get

$$\frac{d}{dt}\left(\int_c^{\psi(t)} f(t,v)\,dv\right) = \big(F\big(t,\psi(t)\big)\big)'_t = F'_t\big(t,\psi(t)\big) + F'_\theta\big(t,\psi(t)\big)\psi'(t)$$

$$= \int_c^{\psi(t)} \frac{\partial f}{\partial t}(t,v)\,dv + f\big(t,\psi(t)\big)\psi'(t),$$

which is equivalent to the equality we sought to prove. □

Let us return to the proof of the theorem. Take $\psi(t) = \varphi(s,t)$ and apply the lemma to the inner integral on the right-hand side of Eq. (5):

$$\int_c^{\varphi(s,t)} \frac{\partial f}{\partial t}(s,t,v)\,dv = \frac{\partial}{\partial t}\left(\int_c^{\varphi(s,t)} f(s,t,v)\,dv\right) - f\big(s,t,\varphi(s,t)\big)\,\frac{\partial \varphi}{\partial t}(s,t).$$

Integrating this equality with respect to t, we obtain

$$\int_a^b \left(\int_c^{\varphi(s,t)} \frac{\partial f}{\partial t}(s,t,v)\,dv\right) dt$$

$$= \int_c^{\varphi(s,t)} f(s,t,v)\,dv\Big|_{t=a}^{t=b} - \int_a^b f\big(s,t,\varphi(s,t)\big)\frac{\partial \varphi}{\partial t}(s,t)\,dt.$$

Since the points (s,a,v) and (s,b,v) belong to the trivial part of the beam boundary on which $f \equiv 0$, the substitution term vanishes. This allows us to rewrite Eq. (5) as

$$\int_{\mathcal{B}} \frac{\partial f}{\partial e_{m-1}}(x)\,dx = -\int_R \left(\int_a^b f\big(s,t,\varphi(s,t)\big)\frac{\partial \varphi}{\partial t}(s,t)\,dt\right) ds$$

$$= \int_Q f\big(u,\varphi(u)\big)\big\langle\big(-\operatorname{grad}\varphi(u),1\big), e_{m-1}\big\rangle\,du.$$

Taking Eq. (3) into account, we can represent the resulting integral as an integral with respect to the measure σ:

$$\int_{\mathcal{B}} \frac{\partial f}{\partial e_{m-1}}(x)\,dx = \int_Q f\big(u,\varphi(u)\big)\big\langle \nu\big(u,\varphi(u)\big), e_{m-1}\big\rangle\sqrt{1+\|\operatorname{grad}\varphi(u)\|^2}\,du$$

$$= \int_{\Gamma_\varphi} f(x)\big\langle \nu(x), e_{m-1}\big\rangle\,d\sigma(x) = \int_{\partial\mathcal{B}} f(x)\big\langle \nu(x), e_{m-1}\big\rangle\,d\sigma(x)$$

(here we used the condition $f \equiv 0$ on $\partial\mathcal{B} \setminus \Gamma_\varphi$ again). Thus, the Gauss–Ostrogradski formula for the beam $\mathcal{B}$ and the unit vectors $e_1, \ldots, e_{m-1}$ is proved, which also proves Eq. (4). □

The assumption that $f \equiv 0$ on the trivial part of $\partial\mathcal{B}$ is, of course, superfluous. It has been made only to simplify the proof of the preliminary version of the Gauss–Ostrogradski formula. The final version (see Sect. 8.6.5) contains no such assumption.

8.6.4 *Standard Compact Sets.* We will now introduce the compact sets that will be used in the general Gauss–Ostrogradski formula. We shall obtain the formula for compacts sets with smooth boundary without using the results of this subsection (see the first stage of the proof of Theorem 8.6.5). Since our goal is to prove the Gauss–Ostrogradski formula for more general compact sets, not only for compact sets with smooth boundaries (on which all notions of surface area coincide), we will now abandon the consideration of arbitrary surface areas and instead use only the area proportional to the Hausdorff measure μ_{m-1}. This area will still be denoted by the letter σ.

Definition A compact set $K \subset \mathbb{R}^m$ is called a *standard compact set* if its boundary can be represented as $\partial K = M \cup E$ where:

(a) for every point $p \in M$, there exists a ball B_p centered at p and a function $F \in C^1(B_p)$ such that $F > 0$ on $B_p \setminus K$, $F \leqslant 0$ on $B_p \cap K$, and $\operatorname{grad} F(p) \neq 0$;
(b) $\sigma(M) < +\infty$;
(c) E is a compact set and $\sigma(E) = 0$.

Condition (a) implies that M is a smooth surface. We shall call M the *regular part of the boundary* of the compactum K, and E its singular part. Condition (c) allows us to ignore the integral over the singular part when integrating over ∂K because it vanishes.

It is obvious that a beam is a standard compact set. As a rule (see, however, Exercise 10), compacta bounded by one or several smooth surfaces (e.g., a ball, a torus, or an m-dimensional annulus) are also standard compact sets. All bounded domains studied in school geometry (a polyhedron, a truncated cone, etc.) are also standard compact sets.

It is clear that the function F from condition (a) of the definition vanishes on $B_p \cap M$. As has already been pointed out in Sect. 8.6.2, $\operatorname{grad} F(p)$ is an outer normal to ∂K at the point $p \in M$. Therefore an outer normal exists at every point of M. The mapping that sends an arbitrary point of M to the unit outer normal at that point is continuous on M because, locally in a neighborhood of a point $p \in M$, the unit outer normals coincide with the normalized gradients of the function F. Thus, the family of unit outer normals forms a side of the surface M, which, according to the definition of Sect. 8.6.2, is an outer side of M. Since in a neighborhood of the point p, the level set $F(x) = 0$ coincides with the graph of some smooth function, there exists a sufficiently small open parallelepiped $P \subset B_p$ containing p for which the intersection $\overline{P} \cap K = \mathcal{B}_p$ is a beam. Obviously, the non-trivial part of its boundary coincides with $M \cap P$, and on this part, the outer normals to M are also outer normals to $\partial \mathcal{B}_p$.

It is important to have some sufficiently simple conditions that would allow us to check the equality $\sigma(E) = 0$ in condition (c). In particular, this is so if E is a subset of a smooth manifold L of codimension greater than 1 because, by property (5) from Sect. 8.3.3, we have $\sigma(L) = 0$. In what follows, we shall also use another condition that ensures the equality $\sigma(E) = 0$. To state it, let us remind the reader that the ε-neighborhood of a set E is the open set $E_\varepsilon = \bigcup_{x \in E} B(x, \varepsilon)$ (it consists of all points $y \in \mathbb{R}^m$ for which $\operatorname{dist}(y, E) < \varepsilon$).

Definition A set $E \subset \mathbb{R}^m$ is called *negligible* in $\mathbb{R}^m$ if the volume of its ε-neighborhood E_ε satisfies $\lambda(E_\varepsilon) = o(\varepsilon)$ as $\varepsilon \to 0$.

It is obvious that every negligible set is bounded. Since a set and its closure have the same ε-neighborhoods, the closure of every negligible set is negligible.

On the line, only the empty set is negligible. On the plane every finite set is negligible, but not every discrete set (see Example 6). The reader can easily check that the union of a finite family of negligible sets is negligible. In the space $\mathbb{R}^m$, $m \geqslant 3$, every bounded subset of an affine subspace L is negligible if $\dim L \leqslant m - 2$. The next proposition is useful when verifying condition (c) in the definition of a standard compact set.

Proposition *Every negligible subset of the space $\mathbb{R}^m$ has zero area.*

Proof Let $E \subset \mathbb{R}^m$ be a negligible set. As we have already mentioned, it is bounded. Let us check that $\sigma(E) = \alpha_{m-1}\mu_{m-1}(E) = 0$.

Fix an arbitrary $\varepsilon > 0$. We will call the points x and y ε-distinguishable if $\|x - y\| \geqslant \varepsilon$. Obviously, a bounded set can contain only finitely many pairwise ε-distinguishable points. Consider a set A consisting of the maximal possible number of ε-distinguishable points belonging to E. We have

$$E \subset \bigcup_{x \in A} B(x, \varepsilon)$$

because otherwise the set A could be augmented by a point from $E \setminus \bigcup_{x \in A} B(x, \varepsilon)$, which would contradict its maximality. Furthermore, the balls $B(x, \varepsilon/2)$ and $B(y, \varepsilon/2)$ centered at two different points of A are disjoint because the points in this set have distances at least ε between them. Since the balls $B(x, \varepsilon)$ $(x \in A)$ form a cover of the set E, according to the definition of $\mu_{m-1}(E, \varepsilon)$ (see Sect. 2.6.1), we get

$$\mu_{m-1}(E,\varepsilon) \leqslant \sum_{x\in A} \varepsilon^{m-1} = \frac{2^m}{\alpha_m \varepsilon} \sum_{x\in A} \lambda\big(B(x,\varepsilon/2)\big) = \frac{2^m}{\alpha_m \varepsilon} \lambda\Big(\bigcup_{x\in A} B(x,\varepsilon/2)\Big)$$

$$\leqslant \frac{2^m}{\alpha_m \varepsilon} \lambda(E_\varepsilon) \underset{\varepsilon\to 0}{\longrightarrow} 0.$$

Hence $\mu_{m-1}(E) = \lim_{\varepsilon\to 0} \mu_{m-1}(E,\varepsilon) = 0$. □

For compact subsets of smooth surfaces, the converse statement also holds.

Lemma *If a compact subset E of a smooth surface M has zero area, it is negligible.*

Proof Every point of the surface has an M-neighborhood whose closure is contained in the graph of some smooth function. It is clear that the set E can be covered by finitely many such neighborhoods U_n: $E \subset \bigcup_{n=1}^{N} U_n$. Put

$$E_n = E \cap \overline{U}_n \quad (n = 1, \dots, N).$$

Obviously, the sets E_n are compact and $E = \bigcup_{n=1}^{N} E_n$. Therefore, it suffices to prove the statement of the lemma for the sets E_n, which allows us to assume in what follows that M is the graph of a smooth function $\varphi \in C^1(G)$ where G is an open subset of the space $\mathbb{R}^{m-1}$.

As before, we will represent a point x of the space $\mathbb{R}^m$ as $x = (u, v)$ where $u \in \mathbb{R}^{m-1}$, $v \in \mathbb{R}$, identifying $\mathbb{R}^{m-1}$ with the plane $v = 0$. Let H be the projection of the set E to $\mathbb{R}^{m-1}$. By the δ-neighborhood of H, we will mean the δ-neighborhood in the space $\mathbb{R}^{m-1}$, preserving the notation H_δ for it. Choose $\delta > 0$ so small that H_δ is contained in G together with its closure $\overline{H_\delta}$, and put $L = \max_{u\in\overline{H_\delta}} \|\operatorname{grad}\varphi(u)\|$.

Since the canonical parametrization Φ of the graph Γ_φ is an expansion and $E = \Phi(H)$, we have $\lambda_{m-1}(H) \leqslant \sigma(\Phi(H)) = \sigma(E) = 0$. Therefore, $\lambda_{m-1}(H) = 0$. Since $\bigcap_{\varepsilon>0} H_\varepsilon = H$, the upper semicontinuity of measure implies that

$$\lambda_{m-1}(H_\varepsilon) \underset{\varepsilon\to 0}{\longrightarrow} 0. \tag{6}$$

Consider the layer

$$A(\varepsilon) = \big\{(u,v) \in \mathbb{R}^m \mid u \in H_\varepsilon, \big|v - \varphi(u)\big| < (L+1)\varepsilon\big\}$$

around the graph of φ over H_ε with $0 < \varepsilon < \delta$. Since $\lambda(A(\varepsilon)) = 2(L+1)\varepsilon\lambda_{m-1}(H_\varepsilon)$, by (6), we have $\lambda(A(\varepsilon)) = o(\varepsilon)$ as $\varepsilon \to 0$. Hence, to prove that E is negligible, it suffices to show that $E_\varepsilon \subset A(\varepsilon)$. Let $x = (u,v) \in E_\varepsilon$. Let us check that $x \in A(\varepsilon)$, that is, that $u \in H_\varepsilon$ and $|v - \varphi(u)| < (L+1)\varepsilon$. By the definition of the ε-neighborhood, there exists a point $x' = (u',v') \in E \subset \Gamma_\varphi$ such that $\|x - x'\| < \varepsilon$. Since $\|u - u'\| \leqslant \|x - x'\| < \varepsilon$ and $u' \in H$, we have $u \in H_\varepsilon$. Furthermore, $v' = \varphi(u')$ and $|v - v'| \leqslant \|x - x'\|$. Thus

$$\big|v - \varphi(u)\big| \leqslant \big|v - v'\big| + \big|\varphi(u') - \varphi(u)\big| \leqslant \big\|x - x'\big\| + L\big\|u - u'\big\| < (L+1)\varepsilon,$$

whence $x \in A(\varepsilon)$. □

Note that one cannot relax the conditions of the lemma by replacing compactness with boundedness (see Exercise 7). The lemma also fails if one assumes that E is contained not in one but in the union of two smooth surfaces (see Exercise 8).

We will now present a simple but useful corollary to this lemma.

Corollary *A compact subset E of a smooth manifold of codimension greater than* 1 *is negligible.*

Proof Indeed, the area of such a manifold equals zero (see property (5) in Sect. 8.3.3). Moreover, locally it is contained in a manifold of codimension 1, i.e., in a surface (see the end of Sect. 8.1.1). Thus E can be covered by finitely many compact sets, each of which is contained in a smooth surface and has zero area. It remains to use the lemma. □

8.6.5 In this subsection, we will generalize the preliminary version of the Gauss–Ostrogradski formula obtained in Sect. 8.6.3 replacing beams by an arbitrary standard compact set. We will also call the outer side ν of the surface M the outer side of ∂K. Thus, the outer side is defined and continuous almost everywhere on ∂K. Fixing an arbitrary vector $e \in \mathbb{R}^m$, we conclude that the function $x \mapsto \langle \nu(x), e\rangle$ is continuous almost everywhere on ∂K (with respect to the measure σ) and, thereby, measurable.

Theorem (The Gauss–Ostrogradski formula) *Let f be a function smooth on a standard compact set $K \subset \mathbb{R}^m$. Then for every unit vector $e \in \mathbb{R}^m$, one has*

$$\int_K \frac{\partial f}{\partial e}(x)\,dx = \int_{\partial K} f(x)\langle \nu(x), e\rangle\,d\sigma(x).$$

Before we start the proof, let us note that it will be carried out in three stages. For compacta with smooth boundaries (a ball, a torus, etc.) the result that will be obtained at the first stage is enough. If the boundary of the compactum contains a singular part, in most cases, it is a negligible set (as it is for a polyhedron, a half-ball, a cone, etc.). This case will be covered at the second stage of the proof.

Proof By the definition of a standard compact set, $\partial K = E \cup M$, where M is the regular part of ∂K, and $\mu_{m-1}(E) = 0$.

(I) Assume that $f \equiv 0$ on an open set G containing E (this assumption on the function f is vacuous in the smooth boundary case, i.e., when $E = \varnothing$). We will construct a cover of K of a special form. For each point $x \in \operatorname{Int} K$, choose an open cube $Q_x \subset \operatorname{Int} K$ centered at x. For each point $p \in M$, choose an open parallelepiped R_p containing p such that the intersection $\overline{R_p} \cap K$ is a beam lying in the interior of K except for the closure of the non-trivial part of its boundary, which is contained in M. Such a parallelepiped exists by the definition of a standard compact set.

The sets G, $\{Q_x\}_{x\in \operatorname{Int} K}$ and $\{R_p\}_{p\in M}$ form an open cover of the compactum K. Let G, $Q_{x_1}, \dots, Q_{x_J}$ and $R_{p_1}, \dots, R_{p_N}$ be a finite subcover. Consider the partition of unity subordinate to this subcover (see Theorem 8.1.8). It consist of the smooth functions ω, $\psi_1, \dots, \psi_J$ and $\theta_1, \dots, \theta_N$ ($\omega \equiv 0$ outside G, $\psi_j \equiv 0$ outside Q_{x_j} and $\theta_n \equiv 0$ outside R_{p_n} for $j = 1, \dots, J$, $n = 1, \dots, N$). We have

$$1 = \omega(x) + \sum_{j=1}^{J} \psi_j(x) + \sum_{n=1}^{N} \theta_n(x) \quad \text{for all } x \in K.$$

Due to the condition $f \equiv 0$ on G, it follows that

$$f(x) = \sum_{j=1}^{J} \psi_j(x) f(x) + \sum_{n=1}^{N} \theta_n(x) f(x) \quad \text{for all } x \in K. \tag{7}$$

Hence

$$\int_K \frac{\partial f}{\partial e}(x)\,dx = \sum_{j=1}^{J} \int_K \frac{\partial(\psi_j f)}{\partial e}(x)\,dx + \sum_{n=1}^{N} \int_K \frac{\partial(\theta_n f)}{\partial e}(x)\,dx.$$

Taking into account that $\psi_j \equiv 0$ outside Q_{x_j} and $\theta_n \equiv 0$ outside R_{p_n}, we obtain

$$\int_K \frac{\partial f}{\partial e}(x)\,dx = \sum_{j=1}^{J} \int_{Q_{x_j}} \frac{\partial(\psi_j f)}{\partial e}(x)\,dx + \sum_{n=1}^{N} \int_{K \cap R_{pn}} \frac{\partial(\theta_n f)}{\partial e}(x)\,dx. \tag{8}$$

By Theorem 8.6.3, all terms in the first sum are equal to zero (because $\psi_j \equiv 0$ on the entire boundary of the cube Q_{x_j}). Let us transform the integrals in the second sum. To this end, note that $\theta_n \equiv 0$ on ∂R_{p_n} and, therefore, the function $\theta_n f$ vanishes on the trivial part of the boundary of the beam $K \cap \overline{R}_{p_n}$. Thus we can apply the Gauss–Ostrogradski formula for beams to these integrals too (see Theorem 8.6.3):

$$\int_{K \cap R_{pn}} \frac{\partial(\theta_n f)}{\partial e}(x)\,dx = \int_{\partial(K \cap R_{pn})} \theta_n(x) f(x) \langle \nu_n(x), e \rangle\,d\sigma(x)$$

where ν_n is the unit outer normal to $\partial(K \cap R_{p_n})$. Since $\theta_n(x) \neq 0$ only on the non-trivial part of the boundary of the beam $K \cap \overline{R}_{p_n}$, i.e., on $M \cap R_{p_n}$, and since on that part ν_n coincides with the unit outer normal ν to M, we have

$$\begin{aligned}\int_{K \cap R_{pn}} \frac{\partial(\theta_n f)}{\partial e}(x)\,dx &= \int_{M \cap R_{pn}} \theta_n(x) f(x) \langle \nu_n(x), e \rangle\,d\sigma(x) \\ &= \int_M \theta_n(x) f(x) \langle \nu(x), e \rangle\,d\sigma(x)\end{aligned}$$

(in the end, we have taken into account that $\theta_n \equiv 0$ outside R_{p_n}). Thus, Eq. (8) implies that

$$\begin{aligned}\int_K \frac{\partial f}{\partial e}(x)\,dx &= \sum_{n=1}^{N} \int_M \theta_n(x) f(x) \langle \nu(x), e \rangle\,d\sigma(x) \\ &= \int_M f(x) \sum_{n=1}^{N} \theta_n(x) \langle \nu(x), e \rangle\,d\sigma(x).\end{aligned}$$

Since the functions $\psi_1, \ldots, \psi_J$ vanish on M, if follows from Eq. (7) that $f(x)\sum_{n=1}^{N}\theta_n(x) = f(x)$ for $x \in M$. Thus,

$$\int_K \frac{\partial f}{\partial e}(x)\,dx = \int_M f(x)\langle \nu(x), e\rangle\,d\sigma(x) = \int_{\partial K} f(x)\langle \nu(x), e\rangle\,d\sigma(x).$$

(II) Let us now turn to the case where the singular part E is negligible. We will verify that the difference

$$\Delta = \int_K \frac{\partial f}{\partial e}(x)\,dx - \int_{\partial K} f(x)\langle \nu(x), e\rangle\,d\sigma(x)$$

between the left- and the right-hand sides of the formula we wish to prove is arbitrarily small. To this end, fix an arbitrary positive number ε and apply Theorem 8.1.7 on a smooth descent to the set E_ε (it is easy to see that its ε-neighborhood coincides with $E_{2\varepsilon}$). We conclude that there exists a function $\theta \in C^\infty(\mathbb{R}^m)$ such that:

(a) $0 \leqslant \theta \leqslant 1$ on the entire space $\mathbb{R}^m$;
(b) $\theta = 1$ on E_ε;
(c) $\theta = 0$ outside $E_{2\varepsilon}$;
(d) $\|\operatorname{grad}\theta\| \leqslant \frac{C}{\varepsilon}$ everywhere on $\mathbb{R}^m$, where C is some constant depending only on the dimension m.

Put $L_0 = \max_K |f|$ and $L_1 = \max_K |\frac{\partial f}{\partial e}|$.

Since the function $(1-\theta)f$ vanishes on E_ε, we can apply to it the already proven part of the theorem. Therefore

$$\begin{aligned}\int_K \frac{\partial f}{\partial e}(x)\,dx &= \int_K \frac{\partial(\theta f)}{\partial e}(x)\,dx + \int_K \frac{\partial(f-\theta f)}{\partial e}(x)\,dx \\ &= \int_K \frac{\partial(\theta f)}{\partial e}(x)\,dx + \int_{\partial K} \big(1-\theta(x)\big) f(x)\langle \nu(x), e\rangle\,d\sigma(x).\end{aligned}$$

Hence,

$$\Delta = \int_K \frac{\partial(\theta f)}{\partial e}(x)\,dx - \int_{\partial K} \theta(x) f(x)\langle \nu(x), e\rangle\,d\sigma(x).$$

Due to property (c), one can reduce the sets of integration in both integrals to their intersections with $E_{2\varepsilon}$. This gives us the inequality

$$|\Delta| \leqslant \int_{K\cap E_{2\varepsilon}} \left(\left|\frac{\partial\theta}{\partial e}(x)\right| |f(x)| + |\theta(x)| \left|\frac{\partial f}{\partial e}(x)\right|\right) dx + \int_{E_{2\varepsilon}\cap\partial K} |\theta(x) f(x)|\,d\sigma(x).$$

So,

$$\begin{aligned}|\Delta| &\leqslant \int_{K\cap E_{2\varepsilon}} \left(\frac{C}{\varepsilon}L_0 + L_1\right) dx + \int_{E_{2\varepsilon}\cap\partial K} L_0\,d\sigma(x) \\ &\leqslant \left(\frac{C}{\varepsilon}L_0 + L_1\right)\lambda(E_{2\varepsilon}) + L_0\sigma(E_{2\varepsilon}\cap\partial K).\end{aligned}$$

Since E is a negligible set, the term $(\frac{C}{\varepsilon}L_0 + L_1)\lambda(E_{2\varepsilon})$ gets arbitrarily small as $\varepsilon \to 0$. The same can be said about the second term. Indeed,

$$\sigma(E_{2\varepsilon} \cap \partial K) \to \sigma(E) \quad \text{as } \varepsilon \to 0$$

due to the upper semicontinuity of the measure σ (it is here that we use the finiteness of the area of the boundary of a standard compact set). It remains to recall that $\sigma(E) = 0$. Thus, $|\Delta| = 0$.

(III) Consider now the general case. We will carry out the proof as followings. We will start by improving the set K somewhat by expanding it and applying the Gauss–Ostrogradski formula to the expanded set, and then we will pass to the limit contracting the auxiliary set back to K.

Since $\mu_{m-1}(E) = 0$, one can fix an arbitrarily small positive number ε and choose the balls $B_j = B(x_j, r_j)$ so that

$$E \subset \bigcup_{j=1}^{\infty} B_j, \qquad \sum_{j=1}^{\infty} r_j^{m-1} < \varepsilon^{m-1}.$$

We will assume ε to be so small that the function f is continuously differentiable in $K_{2\varepsilon}$. Taking into account the compactness of the set E, one can assume that $E \subset \bigcup_{j=1}^{N} B_j \subset E_{2\varepsilon}$. Note also that, when intersecting a surface of finite area with concentric spheres, we will get sets of zero areas except, perhaps, for a countable set of radii because the family $\{\sigma(M \cap \partial B(a,r))\}_{r>0}$ is summable (see Sect. 1.2.2). Thus, without loss of generality, we may assume that

$$\sigma(M \cap \partial B_j) = 0 \quad (j = 1, \ldots, N).$$

Now, introduce the set $K(\varepsilon) = K \cup \bigcup_{j=1}^{N} \overline{B}_j$. Obviously, its boundary is disjoint with E and consists only of points belonging to the regular part of ∂K or to the spheres $\partial B_1, \ldots, \partial B_N$. The boundary of $K(\varepsilon)$ can lose its smoothness only on the intersections of spheres or on the intersection of the set M with the spheres. Therefore the area of the singular part of the boundary of $K(\varepsilon)$ equals zero. This part consists of finitely many compact sets, each of which is negligible (by Lemma 8.6.4). Thus, their union is negligible too. Therefore, the set $K(\varepsilon)$ is a standard compact set whose boundary has a negligible singular part, so the Gauss–Ostrogradski formula is valid for this set:

$$\int_{K(\varepsilon)} \frac{\partial f}{\partial e}(x)\, dx = \int_{\partial K(\varepsilon)} f(x) \langle \nu(x), e \rangle \, d\sigma(x). \tag{9}$$

Putting

$$M'(\varepsilon) = \partial K \setminus \bigcup_{j=1}^{N} B_j, \qquad M''(\varepsilon) = \partial K(\varepsilon) \setminus M'(\varepsilon),$$

and separating the integrals over K and M in Eq. (9) from the rest, we can rewrite it in the following form:

$$\int_K \frac{\partial f}{\partial e}(x)\,dx + \int_{K(\varepsilon)\setminus K} \frac{\partial f}{\partial e}(x)\,dx$$

$$= \int_{M'(\varepsilon)} f(x)\langle\nu(x), e\rangle\,d\sigma(x) + \int_{M''(\varepsilon)} f(x)\langle\nu(x), e\rangle\,d\sigma(x)$$

$$= \int_M \ldots - \int_{M\setminus M'(\varepsilon)} \ldots + \int_{M''(\varepsilon)} \ldots. \tag{10}$$

Since $K(\varepsilon)\setminus K \subset E_{2\varepsilon}$ and $M\setminus M'(\varepsilon) \subset M\cap E_{2\varepsilon}$, we have

$$\lambda_m\big(K(\varepsilon)\setminus K\big) \leqslant \lambda_m(E_{2\varepsilon}), \qquad \sigma\big(M\setminus M'(\varepsilon)\big) \leqslant \sigma(M\cap E_{2\varepsilon}).$$

Furthermore,

$$\sigma\big(M''(\varepsilon)\big) \leqslant \sum_{j=1}^{N} \sigma(\partial B_j) = m\alpha_m \sum_{j=1}^{N} r_j^{m-1} < m\alpha_m \varepsilon^{m-1}.$$

The right-hand sides of these three inequalities tend to zero as $\varepsilon \to 0$. Thus, passing to the limit as $\varepsilon \to 0$ in Eq. (10), we obtain the desired formula. □

Note that the theorem proved admits various generalizations in the direction of extending the class of standard compact sets (see Sect. 8.8.4) as well as in the direction of relaxing the smoothness properties of the function f (see Sect. 9.3.5).

Example The Gauss–Ostrogradski formula allows one to express the volume of a body as an integral over its boundary. For instance, applying this formula to the function $f(x) = \langle x, e\rangle$, we get

$$\lambda(K) = \int_{\partial K} \langle x, e\rangle\langle e, \nu(x)\rangle\,d\sigma(x).$$

In other words, the volume of the body K is equal to the flux of the vector field $V(x) = \langle x, e\rangle e$ through its boundary "outwards".

This result can be generalized as follows. Let L be a subspace of $\mathbb{R}^m$, and let P be the orthogonal projection to L. Then $\dim L \cdot \lambda(K) = \int_{\partial K} \langle P(x), \nu(x)\rangle\,d\sigma(x)$, i.e., the flux of the projection P through the outer side is proportional to the volume of the compactum, the proportionality coefficient being equal to the dimension of the subspace to which one projects. In particular, for $L = \mathbb{R}^m$, we get $\lambda(K) = \frac{1}{m}\int_{\partial K} \langle x, \nu(x)\rangle\,d\sigma(x)$.

8.6.6 Let us transform the Gauss–Ostrogradski formula to clarify its physical meaning.

Let $\mathcal{O}$ be an open set in $\mathbb{R}^m$. Let $\{V(x)\}_{x\in\mathcal{O}}$ be a smooth vector field with the coordinate functions $V_1, \dots, V_m$. According to the Gauss–Ostrogradski formula,

$$\int_{\partial K} \langle V(x), \nu(x)\rangle \, d\sigma(x) = \sum_{j=1}^{m} \int_{\partial K} V_j(x)\langle e_j, \nu(x)\rangle \, d\sigma(x)$$

$$= \sum_{j=1}^{m} \int_K \frac{\partial V_j}{\partial x_j}(x)\, dx = \int_K \left(\sum_{j=1}^{m} \frac{\partial V_j}{\partial x_j}(x) \right) dx,$$

where ν is the outer side of the standard compact set $K \subset \mathcal{O}$. The left-hand side of this equation is the flux of the vector field V through the outer side of the boundary ∂K. The integrand $\sum_{j=1}^{m} \frac{\partial V_j}{\partial x_j}$ on the right-hand side is called the *divergence* of the vector field V and denoted $\operatorname{div} V$. Using this notation, the obtained formula can be rewritten in the following form (the so-called "vector form" of the Gauss–Ostrogradski formula, or the divergence formula):

$$\int_K \operatorname{div} V(x)\, dx = \int_{\partial K} \langle V(x), \nu(x)\rangle \, d\sigma(x). \tag{11}$$

Note that $\operatorname{div} V(x)$ is simply the trace of the Jacobian matrix $(\frac{\partial V_j}{\partial x_k}(x))_{j,k=1}^{m}$, i.e., the trace of the operator $d_x V$. Since the trace does not depend on the choice of the basis, when computing the divergence, one can use any orthonormal coordinate system, not only the canonical one.

The last result can also be established in another way. According to the mean value theorem, for every $a \in \mathcal{O}$ and for every sufficiently small $\varepsilon > 0$, one has

$$\frac{1}{\alpha_m \varepsilon^m} \int_{B(a,\varepsilon)} \operatorname{div} V(x)\, dx = \operatorname{div} V(x_\varepsilon), \quad \text{where } x_\varepsilon \in B(a,\varepsilon).$$

Therefore,

$$\operatorname{div} V(a) = \lim_{\varepsilon\to 0} \frac{1}{\alpha_m \varepsilon^m} \int_{B(a,\varepsilon)} \operatorname{div} V(x)\, dx$$

$$= \lim_{\varepsilon\to 0} \frac{1}{\alpha_m \varepsilon^m} \int_{\|x-a\|=\varepsilon} \langle V(x), \nu(x)\rangle \, d\sigma(x).$$

It can be seen from this that the value $\operatorname{div} V(a)$ does not depend on the choice of the coordinate system.

If one views V as an incompressible fluid velocity field, then the flux through the boundary of a body can be non-zero only if the body contains some sources (if the flux is positive) or sinks (if the flux is negative). The quantity $\frac{1}{\alpha_m \varepsilon^m} \int_{\|x-a\|=\varepsilon} \langle V(x), \nu(x)\rangle \, d\sigma(x)$ on the right-hand side of the last equality characterizes the average intensity of the sources (sinks) in the ball $B(a,\varepsilon)$, and its limit $\operatorname{div} V(a)$ can be interpreted as the intensity of the source (sink) at the point a.

Example (The law of Archimedes) Let us show how the Gauss–Ostrogradski formula can be used to derive the law of Archimedes from Pascal's law.[12] Let us remind the reader that, according to Pascal's law, the pressure a liquid exerts on a submersed flat area is directed along the normal to the area and is equal to the weight of the pillar of the liquid whose base is congruent to the submersed area and whose height is the submersion depth. Let us compute the Archimedes force acting on a body $K \subset \mathbb{R}^3$ submersed into a liquid. To this end, introduce the Cartesian coordinates for which the OXY-plane coincides with the liquid surface and the axis OZ is directed downward. At each point $(x, y, z) \in \partial K$, the body K is subjected to the pressure force $F(x, y, z) = -g\rho z\nu(x, y, z)$, where $\nu(x, y, z)$ is the unit outer normal to ∂K, ρ is liquid density and g is acceleration of gravity. The resultant, i.e., the Archimedean force $\iint_{\partial K}(-g\rho z)\nu(x, y, z)\,d\sigma(x, y, z)$ has the coordinates

$$F_x = -g\rho \iint_{\partial K} z\langle \nu(x, y, z), e_1\rangle\, d\sigma(x, y, z),$$

$$F_y = -g\rho \iint_{\partial K} z\langle \nu(x, y, z), e_2\rangle\, d\sigma(x, y, z),$$

$$F_z = -g\rho \iint_{\partial K} z\langle \nu(x, y, z), e_3\rangle\, d\sigma(x, y, z).$$

Rewriting the first of these equalities as $F_x = \iint_{\partial K}\langle V, \nu\rangle\, d\sigma$, where $V(x, y, z) = (-g\rho\, z, 0, 0)$, and, using the Gauss–Ostrogradski formula, we obtain

$$F_x = \iiint_K \operatorname{div} V(x, y, z)\, dx\, dy\, dz = \iiint_K 0\, dx\, dy\, dz = 0.$$

Similarly, it can be shown that $F_y = 0$. The vertical component of the Archimedean force can be expressed in terms of the divergence of the vector field $\widetilde{V}(x, y, z) = (0, 0, -g\rho\, z)$, so it equals

$$F_z = \iint_{\partial K} \langle \widetilde{V}(x, y, z), \nu(x, y, z)\rangle\, d\sigma(x, y, z) = \iiint_K \operatorname{div} \widetilde{V}(x, y, z)\, dx\, dy\, dz$$

$$= \iiint_K (-g\rho)\, dx\, dy\, dz = -g\rho\lambda_3(K).$$

Thus, a buoyancy force numerically equal to the weight of the fluid forced out by the body acts on this body in a vertical direction.

8.6.7 *Green's Formula.* Let us single out the two-dimensional case of the Gauss–Ostrogradski formula. Let K be a standard compact set in $\mathbb{R}^2$ and let ν be its outer side. On the regular part L of the boundary ∂K, the side ν agrees with the direction $\tau = U(\nu)$ (see Sect. 8.5.4). The pair $(\partial K, \tau)$ will be called an *oriented boundary* of the planar standard compact set K and denoted by the symbol $\partial^+ K$.

---

[12]Blaise Pascal (1623–1662)—French philosopher, mathematician, and physicist.

For a vector field $V = (V_1, V_2)$ that is smooth in some neighborhood of the compactum K, the vector form (11) of the Gauss–Ostrogradski formula yields

$$\iint_K \operatorname{div} V(x, y)\, dx\, dy = \int_{\partial K} \big\langle V(x, y), \nu(x, y)\big\rangle\, d\sigma_1(x, y).$$

By Eq. (5) from Sect. 8.5.4, this equality can be rewritten as

$$\iint_K \left(\frac{\partial V_1}{\partial x}(x, y) + \frac{\partial V_2}{\partial y}(x, y)\right) dx\, dy = \int_{\partial^+ K} -V_2(x, y)\, dx + V_1(x, y)\, dy.$$

Putting $P = -V_2$ and $Q = V_1$, we arrive at an important result known as Green's[13] formula:

$$\iint_K \left(\frac{\partial Q}{\partial x} - \frac{\partial P}{\partial y}\right) dx\, dy = \int_{\partial^+ K} P(x, y)\, dx + Q(x, y)\, dy.$$

In particular, Green's formula allows one to express the area of a standard compact set as an integral over its boundary: taking the functions $P(x, y) \equiv 0$, $Q(x, y) \equiv x$ or $P(x, y) \equiv -y$, $Q(x, y) \equiv 0$, we obtain

$$\lambda_2(K) = \int_{\partial^+ K} x\, dy = -\int_{\partial^+ K} y\, dx = \frac{1}{2} \int_{\partial^+ K} -y\, dx + x\, dy.$$

In Sect. 8.5, it was noted (for an example, see Sect. 8.5.2) that the integral of a locally potential field over a closed oriented curve can be non-zero. On the other hand, it is obvious that Green's formula implies the following.

Corollary 1 *Let $V = (P, Q)$ be a locally potential vector field that is smooth in some domain $\mathcal{O} \subset \mathbb{R}^2$. Let $K \subset \mathcal{O}$ be a standard compact set with oriented boundary. Then*

$$\int_{\partial^+ K} P(x, y)\, dx + Q(x, y)\, dy = 0.$$

This corollary gives a simple geometric condition for the integral of a locally potential vector field over a closed curve to vanish. This is the case if the curve "bounds a set in $\mathcal{O}$", i.e., coincides with the boundary of some standard compact set *contained in $\mathcal{O}$*. Otherwise (for example, if the curve "surrounds" a point that does not belong to the domain) it is easy to find a smooth locally potential vector field in $\mathcal{O}$ that has a non-zero integral over this curve (for an example, see Sect. 8.5.2).

Let us point out one important special case of Corollary 1 related to holomorphic functions. Let $L \subset \mathbb{C}$ be a piecewise smooth oriented curve that lies in the domain of a continuous complex-valued function f. Let $g = \mathcal{Re}\, f$ and $h = \mathcal{Im}\, f$. Guided by the formal multiplication identity $f(z)\, dz = (g + ih)(dx + i\, dy) =$

[13]George Green (1793–1841)—English mathematician and physicist.

$(g\,dx - h\,dy) + i(h\,dx + g\,dy)$, we define the integral $\int_L f(z)\,dz$ as the sum $\int_L g\,dx - h\,dy + i\int_L h\,dx + g\,dy$. It is easy to see that $\int_L f(z)\,dz = \int_a^b f(\gamma(t))\gamma'(t)\,dt$ for every smooth parametrization γ that agrees with the orientation of the curve L.

Corollary 2 (The Cauchy theorem) *If a function f has a continuous derivative $\frac{df}{dz}$ in a domain $\mathcal{O} \subset \mathbb{C}$ and $K \subset \mathcal{O}$ is a standard compact set, then $\int_{\partial^+ K} f(z)\,dz = 0$.*

Proof By our assumptions, the functions $g = \mathcal{R}e\, f$ and $h = \mathcal{I}m\, f$ belong to the class $C^1(\mathcal{O})$. Moreover, the Cauchy–Riemann conditions $g'_x = h'_y$, $h'_x = -g'_y$ hold, which ensures that the vector fields $(g, -h)$ and (h, g) are locally potential (see Sect. 8.5.2, the corollary to Proposition 3). Thus, the equalities

$$\int_{\partial^+ K} g\,dx - h\,dy = 0 \quad \text{and} \quad \int_{\partial^+ K} h\,dx + g\,dy = 0$$

follow from Corollary 1. □

EXERCISES Let K be a standard compact set in $\mathbb{R}^m$ and let ν be the outer side of its boundary.

1. Prove that $\int_{\partial K} \nu(x)\,d\sigma(x) = 0$ (this equality should be understood coordinate-wise).
2. As noted in the example in Sect. 8.6.5, $\lambda(K) = \frac{1}{m}\int_{\partial K} \langle x, \nu(x)\rangle\,d\sigma(x)$. Generalizing this result, prove that

$$\int_{\partial K} L\big(x, \nu(x)\big)\,d\sigma(x) = \lambda(K)\sum_{j=1}^{m} L(e_j, e_j)$$

for every bilinear form L defined in $\mathbb{R}^m \times \mathbb{R}^m$. In particular, for $m = 3$, this implies the equality $\int_{\partial K} x \times \nu(x)\,d\sigma(x) = 0$ (the symbol $x \times y$ denotes the vector (cross) product of the vectors x and y).
3. Prove the following version of the integration by parts formula for functions of several variables:

$$\int_K \frac{\partial f}{\partial e}(x)\cdot g(x)\,dx = \int_{\partial K} f(x)g(x)\big\langle \nu(x), e\big\rangle\,d\sigma(x) - \int_K f(x)\cdot\frac{\partial g}{\partial e}(x)\,dx$$

(here the functions f and g are continuously differentiable in some neighborhood of K and e is an arbitrary unit vector in $\mathbb{R}^m$).
4. Let f be continuously differentiable in some neighborhood of K and let $y \in \mathbb{R}^m$. Prove that

$$\int_K f(x)e^{i\langle x,y\rangle}\,dx = \frac{1}{i\,\|y\|^2}\bigg(\int_{\partial K} f(x)\big\langle \nu(x), y\big\rangle e^{i\langle x,y\rangle}\,d\sigma(x) - \int_K \big\langle \operatorname{grad} f(x), y\big\rangle e^{i\langle x,y\rangle}\,dx\bigg).$$

In particular, it follows that $\int_K f(x)e^{i\langle x,y\rangle}dx = O(\frac{1}{1+\|y\|})$.

5. A submersible that has the shape of an ellipsoid whose vertical semi-axis equals c is lowered to the bottom of a sea of depth $H > 2c$. If it partially sinks into the sea floor, the sunk part of the surface of the submersible is not subjected to the water pressure. Therefore, the expelling force reduces (a submersible that is half-sunk in the sea floor is no longer expelled from the water but, on the contrary, is pushed by the water towards the sea bottom). Assuming that the average density θ of the submersible is less than 1 (water density), estimate the sinking depth h for which the submersible loses its floatability. Show that this depth is almost inversely proportional to the sea depth. More precisely, prove that $\frac{2}{3}(1-\theta)\frac{c^2}{H} < h < \frac{2}{3}(1-\theta)\frac{c^2}{H-c}$.

6. Prove that the discrete set consisting of the points $(\frac{1}{n}, \sin\sqrt{n})$ $(n \in \mathbb{N})$, is not negligible on the plane.

7. Give an example of a smooth curve in $\mathbb{R}^3$ that is bounded but not negligible.

8. Prove that the conclusion of Lemma 8.6.4 may fail if K is contained in the union of two smooth curves. *Hint.* Consider in the plane $\mathbb{R}^2$ the union of the y-axis L and the graph of the function $x \mapsto f(x) = \sin\frac{1}{x}$ $(x > 0)$. Let $K = ([0,1] \times \mathcal{C}) \cap (L \cup \Gamma_f)$, where $\mathcal{C}$ is the Cantor set. Verify that the set K is not negligible despite $\sigma_1(K) = 0$.

9. Prove that a curve of finite length in $\mathbb{R}^3$ is negligible if it is connected and that one cannot drop the connectedness assumption in general.

10. Show that the subgraph of a function infinitely smooth on an interval may fail to be a standard compact set. *Hint.* Consider a non-negative function that vanishes on a set of Cantor type of positive measure.

8.7 *Harmonic Functions

Everywhere in this section, $\mathcal{O}$ is a domain in $\mathbb{R}^m$, K is a standard compact set contained in $\mathcal{O}$ and ν is the outer side of its boundary. As in the previous section, σ is the surface area proportional to the Hausdorff measure μ_{m-1}. The symbols $B(a,r)$ and $S(a,r)$ stand respectively for the open ball and the sphere in $\mathbb{R}^m$ of radius r centered at a (for brevity, we do not specify the dimension explicitly). Lastly, put

$$B = B(0,1), \qquad S = S(0,1), \qquad s(r) = \sigma\big(S(0,r)\big), \qquad v(r) = \lambda_m\big(B(0,r)\big).$$

8.7.1 The mapping $U \mapsto \sum_{k=1}^{m} \frac{\partial^2 U}{\partial x_k^2}$, defined on $C^2(\mathcal{O})$ is denoted by the symbol Δ and is called the *Laplace operator*. It is clear that $\Delta U = \operatorname{div} \operatorname{grad} U$.

Since the Laplace operator is a composition of the gradient and the divergence operators, it does not depend on the choice of the coordinate system (contrary to the first impression that one can get looking at its definition). We can also prove this in the following way.

Let $U \in C^2(\mathcal{O})$, $B(a, R) \subset \mathcal{O}$. According to the Taylor formula,

$$U(x) = U(a) + d_a U(x-a) + \frac{1}{2} d_a^2 U(x-a) + o\big(\|x-a\|^2\big).$$

Integrate this equality over the sphere $S(a, r)$ with $0 < r < R$. Since

$$\int_{S(a,r)} (x_j - a_j)\, d\sigma(x) = 0 \quad \text{and}$$

$$\int_{S(a,r)} (x_j - a_j)(x_k - a_k)\, d\sigma(x) = 0 \quad \text{for all } j, k,\ j \neq k,$$

we obtain

$$\int_{S(a,r)} U(x)\, d\sigma(x) = s(r) U(a) + \frac{1}{2} \sum_{j=1}^{m} U''_{x_j^2}(a) \int_{S(a,r)} (x_j - a_j)^2\, d\sigma(x) + o\big(r^{m+1}\big).$$

Since

$$\int_{S(a,r)} (x_j - a_j)^2\, d\sigma(x) = \frac{1}{m} \int_{S(a,r)} \|x - a\|^2\, d\sigma(x) = \frac{r^2}{m} s(r),$$

we have

$$\frac{1}{s(r)} \int_{S(a,r)} U(x)\, d\sigma(x) = U(a) + \frac{r^2}{2m} \Delta U(a) + o\big(r^2\big)$$

and, therefore, the value of ΔU at the point a can be found from the average values of the function over the spheres centered at this point:

$$\Delta U(a) = \lim_{r \to 0} \frac{2m}{r^2} \left(\frac{1}{s(r)} \int_{S(a,r)} U(x)\, d\sigma(x) - U(a) \right).$$

Definition A function U, $U \in C^2(\mathcal{O})$, is called *harmonic on* $\mathcal{O}$ if $\Delta U(x) = 0$ for all $x \in \mathcal{O}$.

We will consider real-valued harmonic functions only because the real and the complex parts of a complex-valued harmonic function are again harmonic.

Obviously, the functions harmonic on $\mathcal{O}$ form a vector space. In the one-dimensional case, it is just the set of all linear polynomials. So we will assume that $m > 1$. In this case, the class of harmonic functions is very wide and plays an important role in mathematics as well as in its applications. For instance, the stationary temperature of a body that has no parts emitting or absorbing heat is a harmonic function. If the velocity field of a homogeneous incompressible fluid is a gradient of some function, it follows from the Gauss–Ostrogradski formula that this function must be harmonic.

Our first example of a harmonic function is the *point mass potential* at a point a. This function, which will be denoted by $\mathcal{N}_a$, is defined in a space of three or more dimensions as follows:

$$\mathcal{N}_a(x) = \frac{1}{\|x-a\|^{m-2}} \quad \big(x \in \mathbb{R}^m,\ x \neq a\big).$$

In the two-dimensional case, instead of $\mathcal{N}_a$ one uses the *logarithmic potential*: $x \mapsto \ln \frac{1}{\|x-a\|}$. The reader can easily verify that these potentials are indeed harmonic functions.

The point mass potentials, their linear combinations, and, especially the convolutions of these potentials and their partial derivatives with various measures play an extremely important role in many problems related to harmonic functions. One particular example is the integral $\int_E \frac{d\mu(y)}{\|x-y\|^{m-2}}$, which is called the Newton potential corresponding to the measure μ concentrated on the set E.

Let us mention the identity

$$\operatorname{grad} \mathcal{N}_a(x) = -(m-2)\,\frac{x-a}{\|x-a\|^m}, \tag{1}$$

which will be useful to us later. For $m = 3$, it shows that the vector $\operatorname{grad} \mathcal{N}_a$ coincides, up to a constant factor, with the intensity of the gravitational or the electrostatic field generated by a mass or a charge concentrated at the point a.

If L is a rigid motion or a homothety in $\mathbb{R}^m$, then it is easy to check that, for every harmonic function U, the composition $U \circ L$ is also harmonic (this is not true for an arbitrary linear transformation).

8.7.2 Here we will derive some important corollaries of the Gauss–Ostrogradski formula that are valid for all functions in the class C^2. As we shall see, applying them to harmonic functions, one can obtain remarkable results. The first of these corollaries is also known as Green's theorem (just as the formula in Sect. 8.6.7 is).

Below, we denote by $\frac{\partial U}{\partial \nu}(x)$ the quantity $\langle \operatorname{grad} U(x), \nu(x)\rangle$, i.e., the directional derivative of the function U at the point $x \in \partial K$ in the direction of the outer normal $\nu(x)$.

Theorem 1 (Green) *Let $U, V \in C^2(\mathcal{O})$. Then*

$$\begin{aligned}&\int_K \big(U(x)\Delta V(x) - V(x)\Delta U(x)\big)\,dx\\&\quad= \int_{\partial K}\left(U(x)\frac{\partial V}{\partial \nu}(x) - V(x)\frac{\partial U}{\partial \nu}(x)\right) d\sigma(x).\end{aligned} \tag{2}$$

Taking $V \equiv 1$, we get a useful equality

$$\int_K \Delta U(x)\,dx = \int_{\partial K} \frac{\partial U}{\partial \nu}(x)\,d\sigma(x). \tag{2'}$$

Proof It suffices to apply the Gauss–Ostrogradski formula (see Eq. (11) in Sect. 8.6.6) to the vector field $U \operatorname{grad} V - V \operatorname{grad} U$, whose divergence coincides with $U\Delta V - V\Delta U$. □

The next formula, being valid for all $m \geqslant 2$, gives nothing new for $m = 2$ because in this case it reduces to Eq. (2′). Its meaningful analog in the two-dimensional case can be obtained if one replaces the point mass potential by the logarithmic one (see Exercise 3).

Theorem 2 *Let $U \in C^2(\mathcal{O})$. Then, for every point $x \in \operatorname{Int}(K)$ the equality*

$$(m-2)s(1)U(x) = -\int_K \frac{\Delta U(y)}{\|y-x\|^{m-2}}dy + \int_{\partial K} \frac{\partial U}{\partial \nu}(y)\frac{d\sigma(y)}{\|y-x\|^{m-2}} + (m-2)\int_{\partial K} U(y)\frac{\langle y-x, \nu(y)\rangle}{\|y-x\|^m}\,d\sigma(y) \tag{3}$$

holds.

Remark In mathematical physics, integrals of the form

$$\int_K \frac{\omega(y)\,dy}{\|y-x\|^{m-2}}, \qquad \int_{\partial K} \frac{\omega(y)\,d\sigma(y)}{\|y-x\|^{m-2}} \quad \text{and}$$

$$\int_{\partial K} \omega(y)\frac{\langle y-x, \nu(y)\rangle}{\|y-x\|^m}\,d\sigma(y)$$

play an important role. They are called the volume potential, the single layer potential and the double layer potential respectively.

Proof To apply formula (2) to the function $V = \mathcal{N}_x$, we shall modify the compactum K a little. Since x is an interior point, one has $\overline{B}(x,r) \subset \operatorname{Int}(K)$ for sufficiently small $r > 0$. The set $K_r = K \setminus B(x,r)$ is also a standard compact set. Its boundary is the union of ∂K and the sphere $S(x,r) = \partial B(x,r)$. Note that at each point y on that sphere, the unit outer normal $\nu(y)$ to ∂K_r is opposite to the outer normal to the ball $\overline{B}(x,r)$, which, obviously, coincides with $(y-x)/\|y-x\|$. Hence, on the sphere $S(x,r)$, the directional derivative of $\mathcal{N}_x$ with respect to the outer normal to K_r coincides with $(m-2)/\|y-x\|^{m-1}$ (see (1)). Applying formula (2) to the compactum K_r and using the harmonicity of the function $\mathcal{N}_x$, we get

$$\begin{aligned}
&-\int_{K_r} \frac{\Delta U(y)}{\|y-x\|^{m-2}}dy \\
&= \int_{\partial K}\Big(U(y)\frac{\partial \mathcal{N}_x}{\partial \nu}(y) - \mathcal{N}_x(y)\frac{\partial U}{\partial \nu}(y)\Big)\,d\sigma(y) \\
&\quad + \int_{S(x,r)}\Big(U(y)\frac{m-2}{\|y-x\|^{m-1}} - \frac{1}{\|y-x\|^{m-2}}\frac{\partial U}{\partial \nu}(y)\Big)\,d\sigma(y)
\end{aligned}$$

$$= \int_{\partial K} \left(U(y) \frac{\partial \mathcal{N}_x}{\partial \nu}(y) - \mathcal{N}_x(y) \frac{\partial U}{\partial \nu}(y) \right) d\sigma(y)$$

$$+ \frac{m-2}{r^{m-1}} \int_{S(x,r)} U(y)\, d\sigma(y) - \frac{1}{r^{m-2}} \int_{S(x,r)} \frac{\partial U}{\partial \nu}(y)\, d\sigma(y).$$

As $r \to 0$, the last term is, obviously, $O(r)$, and the middle term tends to $(m-2)s(1)U(x)$ by the mean value theorem (see Sect. 4.7.2). On the other hand, the integral over K_r on the left-hand side of this equality tends to the integral over the whole compactum K. So, passing to the limit as $r \to 0$, we get

$$- \int_K \frac{\Delta U(y)}{\|y-x\|^{m-2}}\, dy = \int_{\partial K} \left(U(y) \frac{\partial \mathcal{N}_x}{\partial \nu}(y) - \mathcal{N}_x(y) \frac{\partial U}{\partial \nu}(y) \right) d\sigma(y)$$
$$+ (m-2)s(1)U(x),$$

which, in view of (1), is equivalent to Eq. (3). □

8.7.3 Let us point out several corollaries related to harmonic functions that directly follow from Theorems 1 and 2 of preceding subsection.

Note, first of all, a necessary condition for harmonicity that is a special case of Eq. (2′): if a function U is harmonic on $\mathcal{O}$, then

$$\int_{\partial K} \frac{\partial U}{\partial \nu}(y)\, d\sigma(y) = 0 \qquad (2'')$$

for every standard compact set $K \subset \mathcal{O}$. It turns out that this condition is not only necessary but also sufficient for the harmonicity of a C^2-smooth function, even in a relaxed form.

Proposition *A function $U \in C^2(\mathcal{O})$ is harmonic on $\mathcal{O}$ if $\int_{\partial \overline{B}} \frac{\partial U}{\partial \nu}(y)\, d\sigma(y) = 0$ for every closed ball $\overline{B}$ contained in $\mathcal{O}$.*

In other words, a function is harmonic if the "outward" flux of its gradient vanishes for every ball contained in $\mathcal{O}$.

Proof Equation (2′) implies that the function ΔU has zero integral over every ball $\overline{B}(x,r) \subset \mathcal{O}$. Hence, $\Delta U(y_r) = 0$ for some point y_r in $\overline{B}(x,r)$. Therefore, $\Delta U(x) = \lim_{r\to 0} \Delta U(y_r) = 0$, which completes the proof due to the arbitrary choice of the point x. □

The next theorem is devoted to a remarkable property of harmonic functions. It turns out that one can determine their values at interior points from their values at points near the boundary. This result, also known as the integral representation of harmonic functions, plays the fundamental role in the theory of harmonic functions.

Theorem *If a function U is harmonic on $\mathcal{O}$, then*

$$\frac{1}{s(1)}\int_{\partial K}\left(U(y)\frac{\langle y-x,\nu(y)\rangle}{\|y-x\|^m}+\frac{1}{m-2}\frac{\partial U(y)}{\partial\nu}\frac{1}{\|y-x\|^{m-2}}\right)d\sigma(y)$$

$$=\begin{cases} U(x), & \text{if } x\in \operatorname{Int}(K),\\ 0, & \text{if } x\notin K.\end{cases} \tag{4}$$

Proof For interior points x, this is a direct corollary of (3), and for outer points it follows from Eq. (2) applied to the harmonic function $V=\mathcal{N}_x$ if one notes that $\frac{\partial\mathcal{N}_x}{\partial\nu}(y)=-(m-2)\frac{\langle y-x,\nu(y)\rangle}{\|y-x\|^m}$ by (1). □

The functions $x\mapsto\langle y-x,\nu(y)\rangle/\|y-x\|^m$ and $x\mapsto 1/\|y-x\|^{m-2}$ are differentiable infinitely many times in $\operatorname{Int}(K)$ for all $y\in\partial K$. This leads to an important conclusion.

Corollary *Every harmonic function is differentiable infinitely many times.*

Proof Since differentiability is a local property, to prove this statement, it suffices to consider an arbitrary point $a\in\mathcal{O}$ and to apply formula (4) to the closed ball $K=\overline{B}(a,r)$. The differentiability of the integral in the ball (and, thereby, the infinite smoothness) follows from Theorem 7.1.5 and the remark to it. □

Let us point out an interesting special case of formula (4). Taking $U\equiv 1$, we obtain an integral allowing one to compute the area of a sphere:

$$\int_{\partial K}\frac{\langle y-x,\nu(y)\rangle}{\|y-x\|^m}\,d\sigma(y)=\begin{cases} s(1), & \text{if } x\in\operatorname{Int}(K),\\ 0, & \text{if } x\notin K.\end{cases} \tag{4'}$$

The formulae (4) and (4′) have two-dimensional analogs in which the potential $\mathcal{N}_x$ is replaced by the logarithmic potential (see Exercise 3).

Remark The integral representation (4) allows one to complement Eq. (4′) in the following way (see Exercise 5): if C is a cone with the vertex at the origin whose boundary is so good that $K\cap C$ is a standard compact set, then

$$\int_{C\cap\partial K}\frac{\langle y,\nu(y)\rangle}{\|y\|^m}\,d\sigma(y)=\begin{cases}\sigma(S\cap C), & \text{if } 0\in\operatorname{Int}(K),\\ 0, & \text{if } 0\notin K.\end{cases}$$

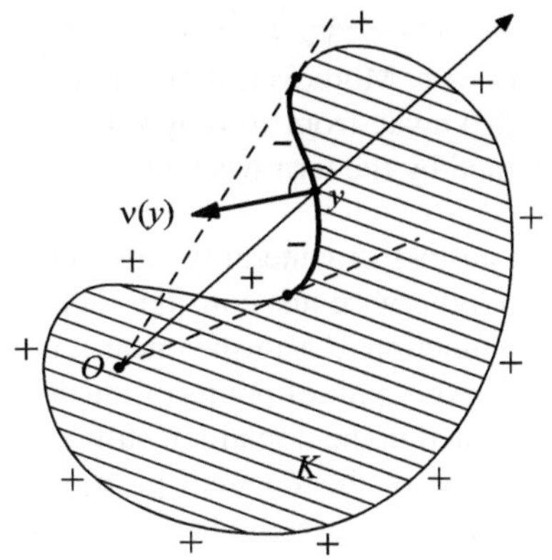

Fig. 8.3 Parts of boundary seen under acute and obtuse angles

Thus, the integral on the left-hand side of this equation equals the aperture of the solid angle at which the part of the boundary of the compactum K contained in C is seen from the origin. This was considered by Gauss in his study of surface curvature. Speaking of apertures of solid angles, one has to keep in mind that the area of the central projection of a part of the boundary is taken with the plus sign if it is the inner part of the boundary, which is seen from the origin (i.e., if the vision rays form acute angles with the outer normals to ∂K), and with the minus sign otherwise (see Fig. 8.3 corresponding to the two-dimensional case).

8.7.4 For harmonic functions the important theorem of uniqueness is valid.

Theorem *If two functions harmonic on a domain coincide on some ball, then they coincide everywhere.*

The proof of this theorem is based on the real analyticity of harmonic functions. It follows from the Poisson formula (14) proved in Sect. 8.7.10. Here we will restrict ourselves to proving a weaker version of this property which is still sufficient for our purposes, the real analyticity of harmonic functions "along line segments".

Lemma *If a function U is harmonic on $\mathcal{O}$, then, for every point $a \in \mathcal{O}$ and for every vector $e \in \mathbb{R}^m$, the function $\varphi : t \mapsto U(a + te)$ can be decomposed into a power series on a sufficiently small interval $(-\delta, \delta) \subset \mathbb{R}$.*

Proof Let $K \subset \mathcal{O}$ be a standard compact set (e.g., a ball) for which a is an interior point. To simplify the formulae, we will assume that $a = 0$. Then the integral representation (4) implies the equality

$$\varphi(t) = \frac{1}{s(1)} \int_{\partial K} \left(U(y) \frac{\langle y - te, \nu(y) \rangle}{\|y - te\|^m} + \frac{1}{m-2} \frac{\partial U(y)}{\partial \nu} \frac{1}{\|y - te\|^{m-2}} \right) d\sigma(y).$$

For every $y \in \partial K$, the functions $t \mapsto \frac{\langle y-te, \nu(y)\rangle}{\|y-te\|^m}$ and $t \mapsto \frac{1}{\|y-te\|^{m-2}}$ can be expanded into power series in powers of t. Moreover, their radii of convergence are at least $\|y\|/\|e\|$, so they are bounded away from zero by some positive quantity. Thus, the result we need can be obtained by the termwise integration of these series. □

Proof of the theorem Without loss of generality, one may assume that one of these functions is identically zero (otherwise just consider their difference). Thus we need to prove that if a harmonic function U vanishes near a point a, then $U(x) = 0$ for every x in the domain. Since every pair of points of the domain can be connected by a piecewise linear path contained in the domain, it suffices to prove that the function vanishes in some neighborhood of the segment $[a, b] \subset \mathcal{O}$ if it vanishes near the point a. For every point x sufficiently close to $[a, b]$ the line segment with endpoints a and x is contained in $\mathcal{O}$. By the lemma, the function $\varphi(t) = U(a + t(x - a))$ is real analytic on $[0, 1]$. Since it vanishes for small t, the uniqueness theorem for real analytic functions yields $\varphi \equiv 0$. In particular, $U(x) = \varphi(1) = 0$. □

8.7.5 As it follows from the integral representation, the values of a harmonic function at the interior points of a standard compact set are determined by the values of the function and the values of its normal derivative on the compactum boundary. This implies several fundamental properties of harmonic functions. The first of them is given by the following theorem.

Theorem (Mean value theorem for harmonic functions) *Assume that a function U is harmonic on $\mathcal{O}$. Then it has the mean value property: for every closed ball $\overline{B}(x, r)$, contained in $\mathcal{O}$, the equality*

$$U(x) = \frac{1}{s(r)} \int_{S(x,r)} U(y)\, d\sigma(y) \tag{5}$$

holds.

Proof Apply formula (4) to the ball $\overline{B}(x, r)$ (in the two-dimensional case, one should use the result of Exercise 3 instead of (4)). Then $\|y - x\| = r$ and $\nu(y) = (y - x)/r$. Hence

$$U(x) = \frac{1}{s(1)} \int_{S(x,r)} \left(U(y) \frac{\langle y - x, (y - x)/r\rangle}{r^m} + \frac{1}{m-2} \frac{\partial U(y)}{\partial \nu} \frac{1}{r^{m-2}} \right) d\sigma(y).$$

Since the integral of the normal derivative vanishes (see (2″)), it follows that

$$U(x) = \frac{1}{s(1)} \int_{S(x,r)} U(y) \frac{d\sigma(y)}{r^{m-1}} = \frac{1}{s(r)} \int_{S(x,r)} U(y)\, d\sigma(y).$$ □

Corollary *Under the assumptions of the lemma, the equality*

$$U(x) = \frac{1}{v(r)} \int_{B(x,r)} U(y)\, dy \tag{5′}$$

holds.

Proof For the proof, it suffices to note that $s(\rho)U(x) = \int_{S(x,\rho)} U(z)\,d\sigma(z)$ for $0 < \rho < r$ and, thereby,

$$v(r)U(x) = \int_0^r s(\rho)U(x)\,d\rho = \int_0^r \Big(\int_{S(x,\rho)} U(z)\,d\sigma(z)\Big)\,d\rho = \int_{B(x,r)} U(z)\,dz$$

(the last equality holds according to formula (3) of Sect. 8.4.2). □

Let us point out one important property of functions harmonic on the entire space, which is known as Liouville's theorem.

Theorem *If a harmonic function U on $\mathbb{R}^m$ is bounded, then it is constant.*

Proof Let us show that $U(x) = U(0)$ for all $x \in \mathbb{R}^m$. Let $C = \sup|U|$. By the corollary to the mean value theorem, for every $r > 0$, we have

$$U(0) = \frac{1}{v(r)} \int_{B(0,r)} U(y)\,dy, \qquad U(x) = \frac{1}{v(r)} \int_{B(x,r)} U(y)\,dy.$$

Hence,

$$\big|U(0) - U(x)\big| \leqslant \frac{1}{v(r)} \int_{E_r} \big|U(y)\big|\,dy \leqslant C\frac{\lambda_m(E_r)}{v(r)},$$

where E_r is the set of all points contained in exactly one of the balls $B(0,r)$ and $B(x,r)$. E_r is contained in the annulus $\{y \mid r - \|x\| \leqslant \|y\| \leqslant r + \|x\|\}$, whose volume is $O(r^{m-1})$. Therefore, $\lambda_m(E_r) = O(r^{m-1})$ and $|U(0) - U(x)| = O(\frac{1}{r})$ as $r \to +\infty$. Thus, $U(x) = U(0)$. □

As a matter of fact, we have proved a bit more than what was claimed in the statement of the theorem: a harmonic on $\mathbb{R}^m$ function U is constant if $U(x) = o(\|x\|)$ as $\|x\| \to +\infty$. The example of a linear function shows that the condition $U(x) = o(\|x\|)$ cannot be relaxed to $U(x) = O(\|x\|)$. One can prove (see Exercise 11) that among all harmonic functions on $\mathbb{R}^m$, only polynomials satisfy the power growth bound $U(x) = O(\|x\|^p)$ as $\|x\| \to +\infty$.

The boundedness assumption of the theorem can be relaxed to the one-sided boundedness (see Corollary 8.7.11).

8.7.6 It turns out that the mean value property completely characterizes harmonic functions. More precisely, the following theorem holds.

Theorem *If a function U locally summable in $\mathcal{O}$ satisfies Eq. (5′) for every closed ball $\overline{B}(x,r) \subset \mathcal{O}$, then it is infinitely smooth and harmonic on $\mathcal{O}$.*

Note that we shall prove the infinite smoothness of the function under consideration without using the integral representation (4).

Proof The harmonicity of a C^2-smooth function satisfying Eq. (5′) is easy to prove. Indeed, in this case, repeated differentiation of the equality

$$U(x) = \frac{1}{v(1)} \int_B U(x + ry)\,dy$$

with respect to r yields:

$$0 = \int_B \sum_{j,k=1}^m \frac{\partial^2 U(x+ry)}{\partial x_j\,\partial x_k} y_j y_k\,dy.$$

Passing to the limit as $r \to 0$, we obtain

$$0 = \sum_{j,k=1}^m \frac{\partial^2 U(x)}{\partial x_j\,\partial x_k} \int_B y_j y_k\,dy = \sum_{k=1}^m \frac{\partial^2 U(x)}{\partial x_k^2} \int_B y_k^2\,dy.$$

The integrals $\int_B y_k^2\,dy$ are, obviously, equal. Hence $\Delta U(x) = 0$ at every point $x \in \mathcal{O}$.

Now let us prove the smoothness of the function U. To this end, we will use the standard trick for such problems of mollifying U by a convolution with a compactly supported smooth function.

Since the smoothness is a local property, we may assume in the proof that U is locally summable in the entire space (otherwise one can replace $\mathcal{O}$ by a sufficiently small ball and extend U by zero outside that ball).

We will use the infinite smoothness of the convolution $U * \varphi$ for $\varphi \in C_0^\infty(\mathbb{R}^m)$ (see Corollary 7.5.4). Choose a function φ of radial type: $\varphi(y) = \psi(\|y\|)$ where $\psi \in C^\infty([0,\infty))$ and $\psi(t) = 0$ for all $t \geqslant r$. Assume that $a \in \mathcal{O}$ and $B(a, 2r) \subset \mathcal{O}$. Then for $x \in B(a,r)$, Fubini's theorem and Eq. (5′) imply

$$\begin{aligned} U * \varphi(x) &= \int_{\mathbb{R}^m} U(x-y)\psi(\|y\|)\,dy = -\int_{\mathbb{R}^m} \left(\int_{\|y\|}^\infty U(x-y)\psi'(t)\,dt \right) dy \\ &= -\int_0^r \psi'(t) \left(\int_{\|y\|<t} U(x-y)\,dy \right) dt = -U(x) \int_0^r \psi'(t) v(t)\,dt. \end{aligned}$$

If ψ is non-increasing on $(0,r)$, then $\int_0^r \psi'(t)v(t)\,dt \neq 0$. Thus the function U is proportional to the infinitely smooth convolution $U * \varphi$ on $B(a,r)$. □

Corollary *If a sequence of functions harmonic on the domain $\mathcal{O}$ converges to a function U uniformly on every compact set contained in $\mathcal{O}$, then U is a harmonic function.*

Proof Indeed, the function U is continuous and has the mean value property. □

8.7.7 An important property of harmonic functions follows from the mean value theorem.

Theorem (Maximum principle for harmonic functions) *If a function harmonic on a domain is not constant, then it has no local extrema.*

It follows from here immediately that a harmonic function is constant if its absolute value attains its maximum.

Proof Obviously, it suffices to consider the case when a function U harmonic on a domain has a local maximum. Then there exists a closed ball centered at the point a such that $U(a) \geqslant U(y)$ for all y in that ball. The average value of U over it equals $U(a)$ by the corollary to the mean value theorem. This is possible only if $U(y) \equiv U(a)$ in the entire ball. But then the uniqueness theorem implies that $U(y) \equiv U(a)$ in the whole domain, which contradicts our assumptions. □

Corollary *Q be a compact set and $U \in C(Q)$. If the function U is harmonic on* $\mathrm{Int}(Q)$, *then*

$$\max_{x\in Q} U(x) = \max_{x\in\partial Q} U(x) \quad \textit{and} \quad \min_{x\in Q} U(x) = \min_{x\in\partial Q} U(x).$$

Proof It suffices to prove the first of these equalities only assuming that the set $\mathrm{Int}(Q)$ is connected. If U is not constant in $\mathrm{Int}(Q)$, then, according to the maximum principle, it cannot attain its maximal value there. The remaining case $U \equiv \mathrm{const}$ is obvious. □

Note that if one interprets a harmonic function as the stationary temperature in a body that contains no parts emitting or absorbing heat, then, from a physicist's standpoint, the maximum principle is completely obvious. Indeed, if under these assumptions the temperature had a local maximum at some point, then the heat would flow away from a neighborhood of that point to the nearby regions lowering the temperature at the point, which contradicts the stationarity assumption.

8.7.8 Until this point, when talking of harmonic functions, as a rule, we considered the case $m > 2$, leaving the derivation of the analogous results for the two-dimensional case to the reader (see Exercises 2, 3 and 6).

However, the two-dimensional case has one specific feature, which we will discuss now. Namely, we will talk about the notion of a harmonic conjugate function.

Definition Let U be a function that is harmonic on a domain $\mathcal{O} \subset \mathbb{R}^2$. The function $V \in C^2(\mathcal{O})$ is called a *harmonic conjugate of U* if $U'_x = V'_y$ and $U'_y = -V'_x$.

By definition, the gradients of the functions U and V are orthogonal at every point of $\mathcal{O}$, so that the level sets of U and V are mutually orthogonal at their intersection points. In the class of harmonic functions vanishing at some fixed point, the harmonic conjugate function is unique and repeated conjugation leads, as one can easily see, to the function $-U$. Note that every harmonic conjugate function is harmonic itself, because $\Delta V = -(U'_y)'_x + (U'_x)'_y = 0$. A reader familiar with complex analysis will notice that a function V is a harmonic conjugate of a function U if and only if the function $U + iV$ is holomorphic.

Proposition *Each function U harmonic on a convex planar domain has a harmonic conjugate function.*

Proof Consider the vector field $(-U'_y, U'_x)$. Since $\frac{\partial}{\partial x}(U'_x) = -\frac{\partial}{\partial y}(U'_y)$ by the harmonicity of U, the Poincaré lemma (see Sect. 8.5.2) implies that this field is a potential one. The corresponding potential is a harmonic conjugate function to U. □

If the domain is not convex, a harmonic conjugate function may fail to exist (even though the proposition implies that it exists locally). This can be seen if one considers the harmonic function $U(x, y) = \ln(x^2 + y^2)$ in $\mathbb{R}^2 \setminus \{0\}$ (see the example in Sect. 8.5.2).

8.7.9 *The Dirichlet Problem.* This classical problem about harmonic functions is as follows. One needs to find a function that is continuous in the closure of a given domain $\mathcal{O}$ and harmonic on $\mathcal{O}$ with the prescribed values on the boundary of the domain. In other words, one needs to find a function $U \in C(\overline{\mathcal{O}}) \cap C^2(\mathcal{O})$, satisfying the following conditions:

$$(1)\quad \Delta U(x) = 0 \quad \text{when } x \in \mathcal{O}$$

(this equation is called the *Laplace equation*) and

$$(2)\quad U(x) = f(x) \quad \text{when } x \in \partial\mathcal{O},$$

where f is a given function defined and continuous on $\partial\mathcal{O}$. This function is called the boundary function.

We will restrict ourselves to the case $m \geqslant 3$ here. The corollary to the maximum principle implies that in a bounded domain, the solution of the Dirichlet problem is unique. To outline an approach that can lead to finding the solution, assume that the closure $\overline{\mathcal{O}}$ is a standard compact set. If U is a solution of the Dirichlet problem that is smooth in some neighborhood of $\overline{\mathcal{O}}$, then, according to the integral representation formula (see Theorem 8.7.3), for every $x \in \mathcal{O}$, one has

$$U(x) = \frac{1}{s(1)} \int_{\partial\mathcal{O}} \left(f(y) \frac{\langle y - x, \nu(y) \rangle}{\|x - y\|^m} + \frac{1}{m-2} \mathcal{N}_x(y) \frac{\partial U}{\partial \nu}(y) \right) d\sigma(y). \quad (6)$$

The right-hand side of this formula contains an unknown function $\frac{\partial U}{\partial \nu}$. To eliminate it, we will do the following. Fix a point $x \in \mathcal{O}$ and consider a function W_x harmonic on $\mathcal{O}$ whose boundary values are the same as those of the potential $\mathcal{N}_x$. If such a function exists and is sufficiently smooth in a neighborhood of the set $\overline{\mathcal{O}}$, then Green's formula (2) applied to $V = W_x$ yields

$$0 = \frac{1}{s(1)} \int_{\partial\mathcal{O}} \left(f(y) \frac{\partial W_x}{\partial \nu}(y) - \mathcal{N}_x(y) \frac{\partial U}{\partial \nu}(y) \right) d\sigma(y).$$

Dividing this equality by $(m-2)$ and adding the result to (6), we obtain

$$U(x)=\frac{1}{s(1)}\int_{\partial\mathcal{O}}\left(\frac{\langle y-x,\nu(y)\rangle}{\|x-y\|^m}+\frac{1}{m-2}\frac{\partial W_x}{\partial\nu}(y)\right)f(y)\,d\sigma(y). \tag{7}$$

Thus, the solution to the Dirichlet problem with a given boundary function f can be expressed in terms of this boundary function using the function

$$\frac{1}{s(1)}\left(\frac{\langle y-x,\nu(y)\rangle}{\|x-y\|^m}+\frac{1}{m-2}\frac{\partial W_x}{\partial\nu}(y)\right)=\frac{\partial G}{\partial\nu}(x,y)\quad(x\in\mathcal{O},\ y\in\overline{\mathcal{O}},\ x\neq y),$$

where

$$G(x,y)=\frac{1}{(m-2)s(1)}\left(W_x(y)-\frac{1}{\|x-y\|^{m-2}}\right). \tag{8}$$

The function G is called the *Green function* for the domain $\mathcal{O}$. When using the symbol $\frac{\partial G}{\partial\nu}$, we will always mean the derivative with respect to the second argument. Using the normal derivative of the Green function, Eq. (7) can be rewritten as

$$U(x)=\int_{\partial\mathcal{O}}\frac{\partial G}{\partial\nu}(x,y)f(y)\,d\sigma(y). \tag{7$'$}$$

The reader can find the proof of the existence of Green's function and the investigation of its properties for a wide class of domains in Sect. 29 of the book [V]. We will restrict ourselves to the most important special cases: the construction of Green's function and the solution of the Dirichlet problem for a ball and for a half-space.

8.7.10 *The Dirichlet Problem for a Ball.* Since the harmonicity property is translation and dilation invariant, it suffices to construct the solution for the unit ball centered at the origin. In this case, Green's function can be obtained using the so-called spherically symmetric points. The heuristics leading to its construction comes from the following theorem of Kelvin.

Definition Let $x\neq 0$. The point x' lying on the same ray as x and satisfying the condition $\|x\|\cdot\|x'\|=1$ (i.e., the point $x'=\frac{x}{\|x\|^2}$) is called the spherically symmetric point to the point x.

It is clear that the spherically symmetric point to the point x' is x. The points on the unit sphere are spherically symmetric to themselves. If $x\notin S$, then the points x and x' are separated by the sphere S. The spherically symmetric points have one useful geometric property: their distances to the points on the sphere S are proportional. More precisely,

$$\|x\|\cdot\|y-x'\|=\|y-x\|,\quad\text{when } \|y\|=1. \tag{9}$$

Indeed,

$$\|x\| \cdot \|y - x'\| = \left\| \|x\| y - \frac{x}{\|x\|} \right\| = \sqrt{\|x\|^2 - 2\langle y, x\rangle + 1} = \|x - y\|.$$

Theorem (Kelvin[14]) *Assume that the function U is harmonic on the domain $\mathcal{O}$, $0 \notin \mathcal{O}$. Let $\mathcal{O}'$ be the domain spherically symmetric to $\mathcal{O}$ with respect to the sphere S. Then the function V defined by the equality $V(x) = \frac{1}{\|x\|^{m-2}} U(\frac{x}{\|x\|^2})$ is harmonic on $\mathcal{O}'$.*

Since we will never refer to this theorem formally, we leave its proof (the technical details of which are rather cumbersome) to the reader (see Exercise 9).

Now, we shall turn to the construction of Green's function for the unit ball B. Assume that $\|x\| < 1$. To obtain a function harmonic on B and taking at the points $y \in S$ the values $\mathcal{N}_x(y) = \frac{1}{\|x-y\|^{m-2}}$, we will use the harmonicity of the function $\mathcal{N}_x$ outside B and "transplant" it to B using the spherical symmetry. More precisely, put

$$W(x, y) = \begin{cases} \frac{1}{\|x\|^{m-2}} \frac{1}{\|y-x'\|^{m-2}} & \text{if } x \neq 0, \ \|x\| \|y\| < 1, \\ 1 & \text{if } x = 0, \ y \in \mathbb{R}^m. \end{cases}$$

The harmonicity of the function W as a function of x (for $x \neq 0$) follows from Kelvin's theorem. However we will establish this fact directly. Obviously, $W(x, y) = (1 - 2\langle x, y\rangle + \|x\|^2 \|y\|^2)^{(2-m)/2}$ when $\|x\| \, \|y\| < 1$ and, therefore, W is a symmetric function of its arguments. For a fixed $x \neq 0$, the function $y \mapsto W(x, y)$ is proportional to the point mass potential at the point x', so it is harmonic. Due to the symmetry, the function $x \mapsto W(x, y)$ is harmonic for every fixed y as well. In particular, if $\|y\| = 1$, then this function is harmonic on the unit ball. This allows us to avoid referring to Kelvin's theorem.

Now, take $W_x(y) = W(x, y)$. Motivated by the formula (8), put

$$G(x, y) = \frac{1}{(m-2)s(1)} \left(W(x, y) - \frac{1}{\|x - y\|^{m-2}} \right)$$

for $\|x\| < 1$ and $\|y\| = 1$.

Since the unit outer normal to the sphere at a point $y \in S$ is y, (1) and (9) imply that, for any fixed $x \in B$, one has

$$\frac{\partial W}{\partial \nu}(x, y) = (m-2) \frac{\langle x, y\rangle - \|x\|^2}{\|x - y\|^m} \quad \text{for all } \ y \in S.$$

Therefore, for all $y \in S$, we have

$$\frac{\partial G}{\partial \nu}(x, y) = \frac{1}{s(1)} \frac{1 - \|x\|^2}{\|x - y\|^m}.$$

[14] William Thomson, Lord Kelvin (1824–1907)—English physicist and mathematician.

In our case, formula (7′) shows that the solution of the Dirichlet problem for the unit ball with the boundary function f should be of the form

$$U(x)=\int_S f(y)\frac{\partial G}{\partial \nu}(x,y)\,d\sigma(y)=\frac{1}{s(1)}\int_S \frac{1-\|x\|^2}{\|x-y\|^m}f(y)\,d\sigma(y) \quad \bigl(\|x\|<1\bigr).$$

Let us check that this formula does indeed give a solution of the Dirichlet problem. Put

$$P(x,y)=\frac{\partial G}{\partial \nu}(x,y)=\frac{1}{s(1)}\frac{1-\|x\|^2}{\|x-y\|^m} \quad \text{for } (x,y)\in B\times S, \tag{10}$$

where the derivative is taken with respect to the outer normal to the unit sphere at the point y. This function is called the *Poisson kernel* (for the ball). Let us now establish its main properties.

Lemma

(1) *The Poisson kernel is positive and, for any fixed $y\in S$, is harmonic on B as a function of x.*
(2) *For every x in B, we have*

$$\int_S P(x,y)\,d\sigma(y)=1. \tag{11}$$

(3) *If $a\in S$, $x\in B$ and $\|x-a\|<\delta$, then*

$$\int_{S\setminus B(a,\delta)} P(x,y)\,d\sigma(y)\leqslant \frac{2\|x-a\|}{(\delta-\|x-a\|)^m}.$$

Proof (1) The inequality $P>0$ is obvious. As has already been mentioned, for $\|y\|=1$, the function $x\mapsto W(x,y)$ is harmonic on the unit ball, so the function $x\mapsto G(x,y)$ is harmonic as well. Since the partial derivatives of a harmonic function are again harmonic, it remains to refer to Eq. (10).

(2) Since for $x=0$, the equality (11) is obvious, we may assume that $0<\|x\|<1$. Write the Gauss formula (4′) with $K=\overline{B}$ for the interior point x and for the exterior point x'. In the first case, we have

$$1=\frac{1}{s(1)}\int_S \frac{\langle y-x,y\rangle}{\|y-x\|^m}\,d\sigma(y). \tag{12}$$

In the second case,

$$0=\frac{1}{s(1)}\int_S \frac{\langle y-x',y\rangle}{\|y-x'\|^m}\,d\sigma(y).$$

By (9), it follows that

$$0=\frac{1}{s(1)}\int_S \frac{\langle y-x',y\rangle}{\|y-x\|^m}\,d\sigma(y).$$

Multiplying this equality by $\|x\|^2$ and subtracting it from (12), we obtain

$$1 = \frac{1}{s(1)} \int_S \langle y - x - \|x\|^2(y - x'), y\rangle \frac{d\sigma(y)}{\|y - x\|^m}.$$

To arrive at the final result, it remains just to transform the scalar product:

$$\big\langle y - x - \|x\|^2(y - x'), y\big\rangle = \big\langle (1 - \|x\|^2)y, y\big\rangle = 1 - \|x\|^2.$$

(3) This inequality follows from (10) because $1 - \|x\|^2 < 2(1 - \|x\|) \leqslant 2\|x - a\|$ and $\|x - y\| > \delta - \|x - a\|$ for $y \notin B(a, \delta)$. □

Now we are ready to consider the Dirichlet problem for the ball. For its solution, it is important that, as one can see from the lemma just proved, the Poisson kernel has properties analogous to those of an approximate identity. The only difference is that instead of integration with respect to Lebesgue measure we use integration with respect to the area measure on the sphere. The formula (13) shows that one can obtain a solution of the Dirichlet problem considering the generalized convolution of the boundary function and the Poisson kernel.

Theorem *The solution of the Dirichlet problem in the ball B with a boundary function $f \in C(S)$ exists and is unique. For $x \in B$ this solution U is given by*

$$U(x) = \int_S P(x, y) f(y)\, d\sigma(y). \tag{13}$$

Proof As already mentioned, the uniqueness follows from the maximum principle. The harmonicity of the function U in the ball B follows from the harmonicity of the Poisson kernel and the validity of differentiating under the integral sign. Put $U(x) = f(x)$ at the points $x \in S$. To finish the proof of the theorem, it remains to check that the function U is continuous at the boundary points of the ball. To this end, estimate the difference $U(x) - U(a)$ between the values at the points $a \in S$ and $x \in B$. Multiplying the equality (11) by $f(a)$ and subtracting the result from (13), we see that

$$U(x) - U(a) = \int_S \big(f(y) - f(a)\big) P(x, y)\, d\sigma(y).$$

Let ω be the modulus of continuity of f. Let $C = \max_S |f|$. For every $\delta > 0$ and $x \in B$ with $\|x - a\| < \delta$, the lemma implies

$$\begin{aligned}\big|U(x) - U(a)\big| &\leqslant \int_S \big|f(y) - f(a)\big| P(x, y)\, d\sigma(y) = \int_{S\cap B(a,\delta)} \cdots + \int_{S\setminus B(a,\delta)} \cdots \\ &\leqslant \int_{S\cap B(a,\delta)} \omega(\delta) P(x, y)\, d\sigma(y) + \int_{S\setminus B(a,\delta)} 2C P(x, y)\, d\sigma(y) \\ &\leqslant \omega(\delta) + 4C \frac{\|x - a\|}{(\delta - \|x - a\|)^m}.\end{aligned}$$

Now we can make the term $\omega(\delta)$ as small as we wish by choosing a sufficiently small δ, after which the second term can also be made small by taking the point x sufficiently close to a. □

A simple computation shows that the solution of the Dirichlet problem in an arbitrary ball $B(a, R)$ is given by

$$U(x) = \frac{1}{s(1)R}\int_{S(a,R)} \frac{R^2 - \|x-a\|^2}{\|y-x\|^m} f(y)\,d\sigma(y) \quad \big(x \in B(a, R)\big). \tag{14}$$

In particular, if the function U is harmonic on some domain containing $\overline{B}(a, R)$, or, at least, is continuous in $\overline{B}(a, R)$ and harmonic on $B(a, R)$, then, due to the uniqueness of the solution of the Dirichlet problem, for every $\|x - a\| < R$, one has the *Poisson formula*

$$U(x) = \frac{1}{s(1)R}\int_{S(a,R)} \frac{R^2 - \|x-a\|^2}{\|y-x\|^m} U(y)\,d\sigma(y).$$

8.7.11 The Poisson formula allows one to complement the mean value theorem for a harmonic function and to estimate the deviations of its values in the ball from its value at the center. We shall state the corresponding result for a ball centered at the origin.

Theorem (Harnack's[15] inequality) *Assume that a non-negative function U is harmonic on the m-dimensional ball $B(0, R)$. Then, at every point x with $\|x\| < R$, the two-sided inequality*

$$\left(1 - \frac{\|x\|}{R}\right)\left(\frac{R}{R+\|x\|}\right)^{m-1} U(0) \leqslant U(x) \leqslant \left(1 + \frac{\|x\|}{R}\right)\left(\frac{R}{R-\|x\|}\right)^{m-1} U(0)$$

holds.

Proof Take a number r in the interval $(\|x\|, R)$. According to the Poisson formula, we have

$$U(x) = \frac{1}{s(1)r}\int_{S(0,r)} \frac{r^2 - \|x\|^2}{\|y-x\|^m} U(y)\,d\sigma(y).$$

Hence

$$U(x) \leqslant \frac{r^2 - \|x\|^2}{r(r-\|x\|)^m}\frac{1}{s(1)}\int_{S(0,r)} U(y)\,d\sigma(y) = \frac{(r+\|x\|)r^{m-2}}{(r-\|x\|)^{m-1}} U(0).$$

Passing to the limit as $r \to R$, we get the upper bound for $U(x)$. The lower bound is proved similarly. □

[15]Carl Gustav Axel Harnack (1851–1888)—German mathematician.

Using Harnack's inequality, one can easily obtain a refinement of Liouville's theorem (Sect. 8.7.5) (some other implications of this inequality are mentioned in Exercises 12–15).

Corollary *If a harmonic function U on $\mathbb{R}^m$ is bounded either from above, or from below, then it is constant.*

Proof Without loss of generality, we may assume that $U \geqslant 0$. Passing to the limit as $R \to +\infty$ in Harnack's inequality, we obtain that $U(0) \leqslant U(x) \leqslant U(0)$ for every point $x \in \mathbb{R}^m$. □

The Poisson formula implies yet another important estimate. It turns out that the size of the gradient of a function harmonic on a ball can be estimated by the maximum of the function itself on the ball boundary (cf. Exercise 17).

Theorem *Assume that a function U is continuous in $\overline{B}(0,R)$ and harmonic on $B(0,R)$. Then, at every point $x \in B(0,R)$, the inequality*

$$\|\operatorname{grad} U(x)\| \leqslant \frac{\sqrt{m}}{R-\|x\|} \max_{\|y\|=R} |U(y)|$$

holds.

Proof Put $C = \max_{\|y\|=R} |U(y)|$. It suffices to check that $|\frac{\partial U}{\partial e}(x)| \leqslant \frac{C\sqrt{m}}{R-\|x\|}$ for every unit vector $e \in \mathbb{R}^m$. By the Poisson formula,

$$U(x) = \frac{1}{s(1)R} \int_{S(0,R)} \frac{R^2 - \|x\|^2}{\|y-x\|^m} U(y)\, d\sigma(y).$$

Let us first estimate the derivative at the center of the ball. Differentiating under the integral sign and making the change of variable $y = Ru$, we obtain

$$\begin{aligned}\frac{\partial U}{\partial e}(0) &= \frac{1}{s(1)R} \int_{S(0,R)} mR^2 \frac{\langle y, e\rangle}{\|y\|^{m+2}} U(y)\, d\sigma(y) \\ &= \frac{m}{s(1)R} \int_S \langle u, e\rangle U(Ru)\, d\sigma(u).\end{aligned}$$

Hence

$$\left|\frac{\partial U}{\partial e}(0)\right| \leqslant \frac{mC}{R} \frac{1}{s(1)} \int_S \big|\langle u, e\rangle\big|\, d\sigma(u) \leqslant \frac{mC}{R} \sqrt{\frac{1}{s(1)} \int_S \big|\langle u, e\rangle\big|^2\, d\sigma(u)}$$

(in the second inequality, we used the Cauchy–Bunyakovsky inequality, see Sect. 4.4.5). It is clear that

$$\int_S \big|\langle u, e\rangle\big|^2\, d\sigma(u) = \int_S u_1^2\, d\sigma(u) = \int_S \frac{u_1^2 + \cdots + u_m^2}{m}\, d\sigma(u) = \frac{s(1)}{m}.$$

Thus, $|\frac{\partial U}{\partial e}(0)| \leqslant \frac{C\sqrt{m}}{R}$. Obviously, the estimate obtained is valid for a ball centered at an arbitrary point. When $0 < \|x\| < R$, one just needs to apply it to the ball $\overline{B}(x, R - \|x\|)$ and to use the fact that, on the boundary of that ball, the function does not exceed C in absolute value due to the maximum principle (see Sect. 8.7.7). □

8.7.12 The solvability of the Dirichlet problem and the uniqueness of the solution allow us to obtain an important "singularity removal principle". It is natural to ask under what conditions one can guarantee the harmonicity of a function on the entire domain $\mathcal{O}$ if it is known that it is harmonic on $\mathcal{O}$ outside some "small" set. If that set has no interior points, then, obviously, the C^2-smoothness of the function implies its harmonicity everywhere in $\mathcal{O}$. To what extent can this smoothness assumption be relaxed? The example of the function $x \mapsto |x_m|$, which is harmonic for $x_m \neq 0$ but not in the entire space $\mathbb{R}^m$, shows that the continuity on the exceptional set alone is, generally speaking, not enough. However, if the exceptional set is contained in some hyperplane, one can give a useful and easy to verify condition that is formally not related to the differentiation but still ensures the harmonicity of the function on the entire domain.

Since the harmonicity is preserved under rigid motions, we shall assume that the set where the harmonicity may be violated is contained in the hyperplane $x_m = 0$. Let us introduce some notation. For an arbitrary domain $\mathcal{O}$, put

$$\mathcal{O}_+ = \mathcal{O} \cap \{x = (x_1, \dots, x_m) \mid x_m > 0\},$$

$$\mathcal{O}_- = \mathcal{O} \cap \{x = (x_1, \dots, x_m) \mid x_m < 0\},$$

$$\mathcal{O}_0 = \mathcal{O} \cap \{x = (x_1, \dots, x_m) \mid x_m = 0\}.$$

It turns out that if a continuous function is odd with respect to the last coordinate, then the singularity removal occurs automatically without any additional smoothness assumptions. Some other singularity removal conditions can be derived from this result (see Exercises 18 and 19).

Theorem (The symmetry principle) *Assume that a function V is continuous in a domain $\mathcal{O}$ symmetric with respect to the hyperplane $x_m = 0$, and is odd with respect to the last coordinate, i.e., $V(x_1, \dots, x_{m-1}, -x_m) = -V(x_1, \dots, x_{m-1}, x_m)$ whenever $x = (x_1, \dots, x_m) \in \mathcal{O}$. If V is harmonic on $\mathcal{O}_+$, then it is harmonic on the entire domain $\mathcal{O}$.*

Proof It is obvious that V is harmonic on $\mathcal{O}_-$ as well. Thus it remains to prove that it is harmonic on a neighborhood of every point in $\mathcal{O}_0$. Let $a \in \mathcal{O}_0$, $\overline{B}(a, R) \subset \mathcal{O}$, and let f be the restriction of V to $S(a, R)$. It is obvious that the function f is continuous and odd with respect to the last coordinate. Let U be the solution of the Dirichlet problem in the ball $B(a, R)$ with the boundary function f. At the interior points of the ball, this solution is given by formula (14), which immediately implies (we leave it to the reader to check this step) that U is odd with respect to the last coordinate. The function U, like V, vanishes on the set $\mathcal{O}_0 \cap \overline{B}(a, R)$. Thus, the functions V and U coincide on the boundary of the upper half-ball $\mathcal{O}_+ \cap B(a,R)$ and, thereby, coincide on the entire half-ball due to the uniqueness of the solution of the Dirichlet problem. The same can be said about the lower half of the ball $B(a,R)$. Thus the function V coincides with the harmonic function U on the entire ball, proving the statement. □

8.7.13 In conclusion, let us discuss the Dirichlet problem in an unbounded domain. We shall restrict ourselves to the case when this domain is the "upper" half-space. For technical reasons, we will consider the $(m+1)$-dimensional half-space $\mathbb{R}^{m+1}_+$ consisting of all points of the kind $\xi = (x_1, \dots, x_m, t)$, where $t > 0$. In this case the solution is, generally speaking, not unique. For example, the Dirichlet problem with zero boundary function has, beside the trivial solution ($U \equiv 0$), another solution $U(\xi) = t$. However one can restore uniqueness if one narrows the class of considered functions by demanding that they possess some additional properties. For the half-space, such an additional property ensuring uniqueness is the boundedness of the solution. More precisely, the following theorem holds.

Proposition *Let U and V be two continuous bounded functions on the upper half-space $\overline{\mathbb{R}^{m+1}_+}$ that are harmonic on $\mathbb{R}^{m+1}_+$. If they coincide on $\partial\mathbb{R}^{m+1}_+$, then they are identical.*

Proof Obviously, the function $F = U - V$ is bounded and vanishes on the hyperplane $\partial\mathbb{R}^{m+1}_+$. Extend it to the whole space as an odd function with respect to the last variable by putting $F(x_1, \dots, x_m, -t) = -F(x_1, \dots, x_m, t)$ for $t > 0$. It is clear that this extension satisfies all the assumptions of the symmetry principle and, thereby, is harmonic on the entire space. Since it is bounded, it must be constant by Liouville's theorem (see Sect. 8.7.5), whence it is identically zero. □

Now let us turn to the proof of the existence of the solution of the Dirichlet problem for the half-space in the case when the boundary function is continuous and bounded. Following the general scheme, fix a point $\xi = (x_1, \dots, x_m, t) \in \mathbb{R}^{m+1}_+$ and construct a harmonic function W_ξ on $\mathbb{R}^{m+1}_+$ that has the same boundary values as $\mathcal{N}_\xi$. Obviously, one can take $W_\xi = \mathcal{N}_{\xi'}$ where $\xi' = (x_1, \dots, x_m, -t)$. Therefore, the Green function for the half-space must be of the form

$$G(\xi,\eta) = \frac{1}{(m-1)\sigma(1)}\left(\frac{1}{\|\xi' - \eta\|^{m-1}} - \frac{1}{\|\xi - \eta\|^{m-1}}\right),$$

where $\sigma(1)$ is the area of the unit sphere in $\mathbb{R}^{m+1}$ and $\eta = (y_1, \dots, y_m, \tau)$. Since in the case under consideration, one has $\nu = -e_{m+1}$, we get $\frac{\partial}{\partial\nu} = -\frac{\partial}{\partial\tau}$, whence, according to (7′), the solution of the Dirichlet problem with boundary function f must be of the form

$$U(\xi) = -\int_{\mathbb{R}^m} f(\eta)\frac{\partial G}{\partial\tau}(\xi,\eta)\,d\eta.$$

Let us check that this formula does indeed yield a solution of the Dirichlet problem. Obviously,

$$-\frac{\partial G}{\partial \tau}(\xi,\eta)=\frac{1}{\sigma(1)}\left(\frac{t+\tau}{\|\xi'-\eta\|^{m+1}}+\frac{t-\tau}{\|\xi-\eta\|^{m+1}}\right).$$

Since $\|\xi'-\eta\|=\|\xi-\eta\|$ for $\tau=0$, we get

$$-\frac{\partial G}{\partial \tau}(\xi,\eta)=\frac{1}{\sigma(1)}\frac{2t}{\|\xi-\eta\|^{m+1}}$$

for all $\eta\in\partial\mathbb{R}^{m+1}_+$.

Put

$$P_t(x)=\frac{1}{\sigma(1)}\frac{2t}{\|\xi\|^{m+1}}=\frac{1}{\sigma(1)}\frac{2t}{(\|x\|^2+t^2)^{\frac{m+1}{2}}}\qquad \big(\xi=(x,t),\ x\in\mathbb{R}^m,\ t>0\big).$$

The function $x\mapsto P_t(x)$ is called the *Poisson kernel* (for the half-space). We will also use this definition for the case $m=1$.

According to the general scheme, the solution of the Dirichlet problem for the half-space $\mathbb{R}^{m+1}_+$ with boundary function f should be of the form

$$U(x,t)=\int_{\mathbb{R}^m} f(y)P_t(x-y)\,dy.$$

In other words the solution of the Dirichlet problem can be represented as the convolution of the boundary function with the Poisson kernel. To prove it, let us establish the main properties of the Poisson kernel.

Lemma

(1) *The Poisson kernel for the half-space is positive and harmonic on* $\mathbb{R}^{m+1}_+$ *(i.e., as a function of the point* $\xi=(x,t)$, $x\in\mathbb{R}^m$, $t>0$*).*

(2) *For every* $t>0$*, one has*

$$\int_{\mathbb{R}^m} P_t(x)\,dx=1. \tag{16}$$

(3) *For every* $\delta>0$*, one has*

$$\int_{\|x\|>\delta} P_t(x)\,dx \underset{t\to 0}{\longrightarrow} 0.$$

This lemma is valid for every m, starting with $m=1$. In particular, it shows that the family $\{P_t\}_{t>0}$ is an approximate identity in $\mathbb{R}^m$ as $t\to 0$ (see Sect. 7.6.1).

Proof (1) It is obvious that $P_t(x)$ is positive. For $m>1$, its harmonicity follows from the fact that, up to a constant factor, the Poisson kernel is the derivative of a point mass potential $P_t(x) = \frac{1}{s(1)}\frac{\partial \mathcal{N}_0}{\partial t}(\xi)$, where $\xi = (x,t)$. For $m = 1$, one should replace the potential $\mathcal{N}_0$ by the logarithmic potential.

(2) Equation (16) can be verified by the direct computation

$$\int_{\mathbb{R}^m} \frac{2t}{(\|x\|^2+t^2)^{\frac{m+1}{2}}}\,dx = m\alpha_m \int_0^\infty \frac{2tr^{m-1}}{(r^2+t^2)^{\frac{m+1}{2}}}\,dr = m\alpha_m \int_0^\infty \frac{u^{\frac{m}{2}-1}}{(u+1)^{\frac{m+1}{2}}}\,du$$

(in the second equality, we made the change of variable $u = r^2/t^2$). Now the desired result follows from the formula

$$\int_0^\infty \frac{u^{a-1}}{(u+1)^{a+b}}\,du = B(a,b) = \frac{\Gamma(a)\Gamma(b)}{\Gamma(a+b)}$$

in Example 4 of Sect. 4.6.3 and the equality $\sigma(1) = (m+1)\alpha_{m+1}$:

$$\int_{\mathbb{R}^m} P_t(x)\,dx = \frac{1}{\sigma(1)}\int_{\mathbb{R}^m} \frac{2t}{(\|x\|^2+t^2)^{\frac{m+1}{2}}}\,dx = \frac{m\alpha_m}{(m+1)\alpha_{m+1}}\,\frac{\Gamma(\frac{m}{2})\Gamma(\frac{1}{2})}{\Gamma(\frac{m+1}{2})} = 1.$$

(3) It suffices to use the obvious inequality $P_t(x) \leqslant \frac{\text{const}}{\|x\|^{m+1}}\,t$. □

Theorem *A bounded solution of the Dirichlet problem in the half-space* $\mathbb{R}^{m+1}_+$ *with a bounded boundary function* $f \in C(\mathbb{R}^m)$ *exists and is unique. For* $\xi = (x,t) \in \mathbb{R}^{m+1}_+$*, this solution* U *is given by*

$$U(\xi) = \int_{\mathbb{R}^m} P_t(x-y)f(y)\,dy. \tag{17}$$

Proof The uniqueness has already been established in the proposition in the beginning of this section. Let us prove the existence. Put $C = \sup_{y\in\mathbb{R}^m}|f(y)|$. Let us check, first of all, that the function U is bounded. Indeed, for every $\xi = (x,t) \in \mathbb{R}^{m+1}_+$, we have

$$\big|U(\xi)\big| \leqslant C\int_{\mathbb{R}^m} P_t(x-y)\,dy = C.$$

The harmonicity of U follows from the first statement of the lemma and the Leibniz rule.

Defining $U(\xi) = f(\xi)$ for $\xi \in \partial\mathbb{R}^{m+1}_+$, let us check the continuity of U at an arbitrary boundary point $\xi_0 = (a,0) \in \partial\mathbb{R}^{m+1}_+$. Since $U(\xi) = (P_t * f)(x)$, we have

$$\big|U(\xi) - U(\xi_0)\big| = \big|(P_t * f)(x) - f(a)\big| \leqslant \big|(P_t * f)(x) - f(x)\big| + \big|f(x) - f(a)\big|.$$

It is clear that the quantity $|f(x) - f(a)|$ becomes arbitrarily small if x is restricted to a sufficiently small ball $B(a,\delta)$. By Theorem 7.6.3, $(P_t * f)(x) \underset{t\to 0}{\rightrightarrows} f(x)$ on every bounded set. Thus, t can be chosen to be so small that the first term on the right-hand side of the last inequality is as small as we wish for all $x \in B(a,\delta)$. Therefore, $U(\xi) \to U(\xi_0)$ as $\xi \to \xi_0$, which is exactly what we need. □

EXERCISES

1. Prove that if $U, V \in C^2(\mathcal{O})$, then

$$\int_K \langle \operatorname{grad} U, \operatorname{grad} V\rangle\, dx + \int_K V\Delta U\, dx = \int_{\partial K} V\frac{\partial U}{\partial \nu}\, d\sigma.$$

In particular, for every function U harmonic on $\mathcal{O}$, we have

$$\int_K \|\operatorname{grad} U(x)\|^2\, dx = \int_{\partial K} U\frac{\partial U}{\partial \nu}\, d\sigma.$$

2. For every compactly supported function $\varphi \in C^2(\mathbb{R}^m)$ $(m \geqslant 3)$ the equality

$$(\mathcal{N}_0 * \Delta\varphi)(x) = \int_{\mathbb{R}^m} \mathcal{N}_0(x-y)\Delta\varphi(y)\, dy = -(m-2)s(1)\varphi(x)$$

holds. Thus, using the convolution with $\mathcal{N}_0$, one can recover a compactly supported function from its Laplacian. This fact provides the grounds for calling $-\frac{1}{(m-2)s(1)}\mathcal{N}_0$ the fundamental solution of the Laplace equation.
Replacing the potential $\mathcal{N}_0$ by the logarithmic potential, obtain an analogous result for functions of two variables.

3. Prove the two-dimensional analogs of Eqs. (3) and (4) (Theorems 8.7.2 and 8.7.3), replacing the potential $\mathcal{N}_x$ by the logarithmic potential.

4. Complementing the Gauss formula (see formula (4′) for $x = 0$), prove that

$$\int_{\partial K} \frac{\langle y, \nu(x)\rangle}{\|y\|^m}\, d\sigma(y) = \frac{1}{2}s(1),$$

provided that the origin belongs to the regular part of ∂K.

5. Prove the statement of Remark 8.7.3. *Hint.* Use the equality $\langle x, \nu(x)\rangle = 0$ for the points x lying on the boundary of the cone C.

6.

(a) Prove the mean value theorem for harmonic functions of two variables (see Sect. 8.7.5).

(b) By the mean value theorem, calculate the integral over the circle

$$\int_{|z|=r} \ln\left|\frac{1-z^2}{2}\right| ds(z) \quad \text{for } r < 1.$$

Justify the passage to the limit as $r \to 1$ and use it to calculate the Euler integral $\int_0^{2\pi} \ln|\sin t|\, dt$.

7. For $f \in C^2(\mathcal{O})$ and $a \in \mathcal{O}$ put $F(r) = \frac{1}{s(r)}\cdot\int_{S(a,r)} f(x)\, d\sigma(x)$, when $r > 0$ and $\overline{B}(a,r) \subset \mathcal{O}$, and put $F(0) = f(a)$. Prove that $F \in C^2([0,R))$ and $F''(0) = \frac{1}{m}\Delta f(a)$.

8. Using Theorem 8.7.6, prove that in Proposition 8.7.3, the condition $U \in C^2(\mathcal{O})$ can be relaxed to the assumption that U is merely continuously differentiable. Based on this, prove the following "singularity removal principle": if a function from $C^1(\mathcal{O})$ is harmonic on $\mathcal{O} \setminus L$ where L is some hyperplane, then it is harmonic on $\mathcal{O}$.

9. Prove Kelvin's theorem: if a function U is harmonic on some domain, then the function $V(x) = \frac{1}{\|x\|^{m-2}} U(\frac{x}{\|x\|^2})$ is also harmonic (in the corresponding domain). *Hint.* Since the harmonicity of a function is preserved under rotations, it is enough to compute ΔV at the points of the form $(t, 0, \ldots, 0)$.

10. Prove that a homogeneous sphere in $\mathbb{R}^3$ attracts an outer point as if the entire mass of the sphere were concentrated at its center, and that the interior points are in zero gravity. Generalize these results to the m-dimensional case assuming that the gravitational attraction force between two point masses is proportional to $\frac{1}{r^{m-1}}$, where r is the distance between the points. *Hint.* When computing the arising integrals, argue similarly to the proof of Eq. (11).

11. Assume that for some $p > 0$, a harmonic function U on $\mathbb{R}^m$ satisfies the condition $|U(x)| = O(\|x\|^p)$ as $\|x\| \to +\infty$. Using the gradient bound (see Sect. 8.7.11), prove that U is a polynomial of degree at most $[p]$.

12. Assume that a sequence of functions continuous in a closed ball and harmonic on its interior is uniformly bounded and converges pointwise on the boundary sphere. Prove that it also converges pointwise inside the ball and the limit function is harmonic.

13. Assume that a sequence of functions harmonic on a domain $\mathcal{O}$ converges pointwise to some function. Using Harnack's inequality, prove that if this sequence is monotone, then the limiting function is harmonic on $\mathcal{O}$ (Harnack's theorem).

14. Prove that if a series of non-negative functions harmonic on some domain converges at some point, then it converges uniformly on any compact set contained in the domain.

15. Using Harnack's inequality, prove that if a non-constant function harmonic on a ball and continuous in its closure attains its extremum at some boundary point, then it cannot happen that the normal derivative at that point is zero.

16. Prove that for every non-negative function U harmonic on the m-dimensional ball $B(a, R)$, the inequality $\|\operatorname{grad} U(a)\| \leqslant \frac{m}{R} U(a)$ holds.

17. Refine the gradient bound obtained in Sect. 8.7.11 by proving that

$$\|\operatorname{grad} U(x)\| \leqslant \frac{c_m}{R - \|x\|} \max_{\|y\|=R} |U(y)|, \quad \text{where } c_m = 2\frac{\alpha_{m-1}}{\alpha_m}.$$

Prove that the coefficient c_m cannot be improved.

18. Prove that the symmetry principle remains valid for a function $U \in C(\mathcal{O})$ that is even with respect to the last coordinate under the assumption that $U'_{x_m} \in C(\mathcal{O})$. *Hint.* Apply the symmetry principle to the derivative U'_{x_m}.

19. Using the result of the previous exercise, prove the following refinement of the "singularity removal principle" (Exercise 8): if a function from $C(\mathcal{O})$ is continuously differentiable in $\mathcal{O}$ with respect to the last coordinate and is harmonic for $x_m \neq 0$, then it is harmonic on $\mathcal{O}$.

20. Prove the following point singularity removal principle for harmonic functions. If a function U is harmonic on the punctured m-dimensional ($m \geqslant 3$) ball $B(a,r) \setminus \{a\}$ and $U(x) = o(\frac{1}{\|x-a\|^{m-2}})$ as $x \to a$, then one can assign some value to this function at the point a so that the resulting function is harmonic on the entire ball $B(a,r)$. What is the two-dimensional analog of this statement?

21. Find the dimension of the linear space of all degree n homogeneous harmonic polynomials of two variables.

22. Present a basis for the linear space of all degree four homogeneous harmonic polynomials of three variables.

8.8 *Area on Lipschitz Manifolds

8.8.1 In this section, by area, we understand a k-dimensional area in the sense of Definition 8.2.1 (we do not assume that it is generated by the Hausdorff measure μ_k). We will denote this area by the letter σ, and the k-dimensional Lebesgue measure by the letter λ (without specifying the dimension explicitly).

Let us remind the reader that Theorem 8.3.2 answers the question about the uniqueness of the area on the (Borel) subsets of smooth manifolds. Our goal is to generalize this result. Let us mention in this connection that Lebesgue himself was interested in the definition and the properties of the area on non-smooth surfaces and, moreover, devoted one of his first works to this topic. Denjoy in his memoirs mentions that Lebesgue, explaining his interest in this problem, pointed at a crumpled napkin as an example of a non-smooth surface for which the existence and the uniqueness of the area are as much beyond doubt as for a smooth one. We will show here that the area is unique not only on smooth manifolds but also on the much wider class of so-called Lipschitz manifolds. Let us give the rigorous definition of this notion.

A homeomorphism Φ is called a *bi-Lipschitz mapping* if the Lipschitz condition is satisfied for Φ as well as for Φ^{-1}. A *simple Lipschitz manifold* is a manifold that has a bi-Lipschitz parametrization (defined on some open subset of the space $\mathbb{R}^k$ where k is the dimension of the manifold). If one can find such a parametrization near every point of a manifold, the manifold is called *Lipschitz*. In particular, this terminology applies to surfaces (manifolds of codimension 1).

It is obvious that the canonical parametrization of the graph of a function is a bi-Lipschitz mapping if and only if the function itself is Lipschitz (see also Exercise 1). However, in contrast to smooth surfaces, Lipschitz manifolds do not need to be graphs of Lipschitz functions even locally (see Exercise 2).

Beyond the σ-algebra generated by the Borel subsets of k-dimensional Lipschitz manifolds, the area is not defined uniquely. We will not discuss this issue, which is outside the scope of this book. We note only that the condition that a set belongs to this σ-algebra is not only sufficient, but also in a certain sense necessary for the area to be uniquely defined on this set. The reader can find additional information in [F, Sect. 3.3], and [BZ, Sect. III.2].

In the derivation of the formula for the area on a simple Lipschitz manifold, a crucial role is played by the Rademacher theorem 11.4.2, according to which the functions satisfying the Lipschitz condition are differentiable almost everywhere. It implies that the coordinate functions of a Lipschitz parametrization Φ are differentiable almost everywhere. Hence, the accompanying parallelepiped $C_t = d_t\Phi([0, 1)^k)$ and the density $\omega_\Phi(t) = \lambda(C_t)$ are defined almost everywhere. The next theorem, whose proof is given in Sect. 8.8.3, extends the result of Theorem 8.3.2 to Lipschitz manifolds.

Theorem *For every Borel set E contained in a simple Lipschitz manifold M, one has*

$$\sigma(E) = \int_{\Phi^{-1}(E)} \omega_\Phi(t)\,dt,$$

where Φ is an arbitrary bi-Lipschitz parametrization of M.

In particular, the theorem remains valid if the dimension of the manifold coincides with the dimension of the ambient Euclidean space. In this case we obtain a generalization of Theorem 6.2.1 with a bi-Lipschitz homeomorphism in place of a diffeomorphism.

As we shall see in Appendix 13.4, the almost everywhere differentiability of convex functions can be established without referring to the Rademacher theorem. Therefore the Rademacher theorem is not needed for the proof of the uniqueness of the area on convex surfaces.

This theorem immediately implies the following corollary that will be used later.

Corollary *Let f be Lipschitz on an open subset $\mathcal{O}$ of the space $\mathbb{R}^m$. Then for every Borel set E contained in the graph of the function f, the equality*

$$\sigma(E) = \int_{P(E)} \sqrt{1 + \|\operatorname{grad} f(x)\|^2}\,dx$$

holds.

Here $P(E)$ denotes the orthogonal projection of the set E to $\mathbb{R}^m$.

8.8.2 Before proving the theorem, we will state and prove the following lemma, which is an improvement upon Lemma 8.2.1 and gives estimates for the area of a subset of an "almost affine" manifold.

Lemma *Let Φ be a bi-Lipschitz parametrization of a simple manifold M ($M \subset \mathbb{R}^m$) defined on an open set $\mathcal{O} \subset \mathbb{R}^k$, and let $\varkappa$ be the Lipschitz constant for Φ^{-1}. Let, further, $A \subset \mathcal{O}$ be a Borel set, and let $L : \mathbb{R}^k \to \mathbb{R}^m$ be a linear mapping. If for some $\varepsilon \in (0, 1/\varkappa)$, the inequality*

$$\|\Phi(t) - \Phi(s) - L(t-s)\| \leqslant \varepsilon\|t-s\| \quad \textit{holds for all } t, s \in A, \qquad (1)$$

then

$$(1-\varkappa\varepsilon)^k\,\lambda\big(L(A)\big)\leqslant\sigma\big(\Phi(A)\big)\leqslant\frac{1}{(1-\varkappa\varepsilon)^k}\,\lambda\big(L(A)\big).$$

Proof Taking $x, y \in E = \Phi(A)$ and putting $s = \Phi^{-1}(x)$, $t = \Phi^{-1}(y)$, we obtain from (1) that the "straightening" mapping $\Psi = L\circ\Phi^{-1}$ satisfies the inequality

$$\big\|y-x-\big(\Psi(y)-\Psi(x)\big)\big\|\leqslant\varepsilon\|t-s\|\leqslant\varepsilon\varkappa\|y-x\|.$$

Hence Ψ is an almost isometry for small ε:

$$(1-\varepsilon\varkappa)\|y-x\|\leqslant\big\|\Psi(y)-\Psi(x)\big\|\leqslant(1+\varepsilon\varkappa)\|y-x\|\quad\text{for } x,y\in E.$$

Applying Lemma 8.2.1 with $C=(1-\varepsilon\varkappa)^{-1}$ and taking into account the remark after that lemma, we get the two-sided estimate

$$(1-\varepsilon\varkappa)^k\lambda\big(\Psi(E)\big)\leqslant\sigma(E)\leqslant\frac{1}{(1-\varepsilon\varkappa)^k}\lambda\big(\Psi(E)\big),$$

which is equivalent to the inequality we sought to prove, because $\Psi(E)=L(A)$. □

8.8.3 *Proof of Theorem 8.8.1.* Without loss of generality, we may assume that the parametrization Φ is defined on an open set $\mathcal{O}\subset\mathbb{R}^k$ of finite measure. Consider the measure $\nu(A)=\sigma(\Phi(A))$ on Borel sets $A\subset\mathcal{O}$. We will check that it satisfies the condition

$$\inf_{t\in A}\omega_\Phi(t)\,\lambda(A)\leqslant\nu(A)\leqslant\sup_{t\in A}\omega_\Phi(t)\,\lambda(A).\tag{2}$$

As was established in Theorem 6.1.2, it follows that $\nu(A)=\int_A\omega_\Phi(t)\,dt$, which is equivalent to the statement of the theorem.

Since both the parametrization Φ and the inverse mapping Φ^{-1} are Lipschitz, we conclude that $\nu(A)=0$ if and only if $\lambda(A)=0$ (see Lemma 8.2.1). This allows us to neglect zero measure subsets of the set A when establishing the inequalities (2).

Let D be the set of points in $\mathcal{O}$ for which all coordinate functions of the mapping Φ are differentiable. By the Rademacher theorem $\lambda(\mathcal{O}\setminus D)=0$. Replacing, if needed, the set D by a Borel subset of the same measure (see Corollary 5 in Sect. 2.2.2), we may assume without loss of generality that D is Borel. Since $\nu(\mathcal{O}\setminus D)=\lambda(\mathcal{O}\setminus D)=0$, it suffices to prove the inequalities (2) for all sets contained in D, to which we will restrict our attention from now on.

Both inequalities (2) are established in the same way. We will prove the upper bound only, leaving the analogous argument for the lower bound to the reader.

Suppose that the right inequality (2) fails for a set $A\subset D$. Then, for some $C>1$, we have

$$\nu(A)>C\sup_{t\in A}\omega_\Phi(t)\,\lambda(A).\tag{3}$$

Note that, if the inequality (3) were violated for either an increasing sequence of sets whose union equals A, or for sets forming a countable partition of A, then it would also be violated for A itself.

Now fix a positive number ε to be chosen later, and consider, for every $r > 0$, the sets

$$D^r = \{s \in D \mid B(s,r) \subset \mathcal{O} \text{ and } \|\Phi(t) - \Phi(s) - d_s\Phi(t-s)\| \leqslant \varepsilon \|t-s\| \text{ for all } t \in B(s,r)\}.$$

Obviously, they expand and exhaust the entire set D as r decreases to 0. Thus, the inequality (3) is valid for some set $A \cap D^r$ as well. Replacing A by such an intersection if needed, we may assume that $A \subset D^r$ for some $r > 0$. If we split the set A into countably many parts of diameter less than r, the inequality (3) will hold for at least one of them. So, we may assume without loss of generality that $\operatorname{diam}(A) < r$. In this case, on the set A, we have the inequality

$$\|\Phi(t) - \Phi(s) - d_s\Phi(t-s)\| \leqslant \varepsilon \|t-s\| \quad (s,t \in A). \tag{4}$$

Since the partial derivatives of the coordinate functions of the mapping Φ are measurable and bounded, the set A can be partitioned into finitely many (Borel) parts so that on each part the oscillations of all these partial derivatives will be arbitrarily small. Then, obviously, the oscillation of the differential mapping of the mapping Φ will be arbitrarily small as well. Construct this partition in such a way that for each of its elements A', the inequality

$$\|d_s\Phi - d_t\Phi\| < \varepsilon \quad \text{for } s,t \in A' \tag{5}$$

holds. The inequality (3) will hold for at least one element of this partition. Replacing A by that element, if needed, we may assume without loss of generality that the set A satisfies both conditions (4) and (5). Thus, defining the linear mapping L as the differential of Φ at some point $a \in A$, we will conclude that the mapping Φ is almost affine:

$$\begin{aligned}&\|\Phi(t) - \Phi(s) - L(t-s)\| \\ &\quad \leqslant \|\Phi(t) - \Phi(s) - d_s\Phi(t-s)\| + \|d_s\Phi(t-s) - d_a\Phi(t-s)\| \leqslant 2\varepsilon \|t-s\|\end{aligned}$$

for all t,s in A. So, condition (1) of Lemma 8.8.2 is satisfied (with 2ε in place of ε). Assuming that $\varepsilon < 1/(2\varkappa)$ where $\varkappa$ is the Lipschitz constant for the inverse mapping Φ^{-1}, we obtain the inequality

$$\nu(A) = \sigma(\Phi(A)) \leqslant \frac{\lambda(L(A))}{(1-2\varepsilon\varkappa)^k} = \frac{\omega_\Phi(a)\lambda(A)}{(1-2\varepsilon\varkappa)^k} \leqslant \frac{1}{(1-2\varepsilon\varkappa)^k} \sup_{t\in A} \omega_\Phi(t)\lambda(A).$$

Together with (3), this implies

$$C \sup_{t\in A} \omega_\Phi(t)\lambda(A) < \nu(A) \leqslant \frac{1}{(1-2\varepsilon\varkappa)^k} \sup_{t\in A} \omega_\Phi(t)\lambda(A).$$

Hence,

$$1 < C \leqslant \frac{1}{(1-2\varepsilon\varkappa)^k}.$$

This gives the contradiction sought if ε is chosen small enough. Thus, our initial assumption was false. The theorem is proved.

Corollary *The restriction of a k-dimensional area to the σ-algebra of Borel subsets of a k-dimensional Lipschitz manifold is a regular measure finite on compact sets.*

This follows from the formula proved in the theorem if one takes into account the boundedness of the function ω_Φ and the regularity of the Lebesgue measure.

8.8.4 Once we are able to compute the area on Lipschitz surfaces, we can expand the class of compact sets for which the Gauss–Ostrogradski formula (Sect. 8.6.5) is valid. To this end, replace the smooth function in the definition of a beam (Sect. 8.6.2) with a Lipschitz function φ. Since it is differentiable almost everywhere, the outer normal is defined almost everywhere on the non-trivial part of the beam boundary. Moreover, the area of a set contained in the graph of a Lipschitz function is computed in the same way as in the case of a smooth function (see Corollary 8.8.1). Since the increment of a Lipschitz function over each coordinate can be represented as the integral of the corresponding partial derivative (see Theorem 11.4.1), both the statement and the proof of Theorem 8.6.3 remain the same for this more general case.

Moreover, Theorem 8.6.3 remains valid not only for beams corresponding to Lipschitz functions, but also for sets obtained from them by a rigid motion (provided that the integrand vanishes on the image of the trivial part of the beam boundary), which we leave the reader to verify (using the invariance of the Lebesgue measure and the surface area).

Generalizing the notion of a standard compact set (Definition 8.6.4), we shall assume, as before, that $\partial K = M \cup E$, where M and E satisfy the conditions (b) and (c) of that definition, replacing the condition (a) by the following one:

(a′) for every point $p \in M$ there exist a rigid motion W and an open parallelepiped R_p such that $p = W(p) \in R_p$ and the intersection $W(K) \cap \overline{R_p}$ is a beam corresponding to a Lipschitz function which lies in the interior of $W(K)$ except for the closure of the non-trivial part of its boundary lying in $W(M)$.

Repeating the proof of the Gauss–Ostrogradski theorem, we see that it is valid for such compacts sets. In particular, it is valid for every convex set, since condition (a′) is satisfied at each point of its boundary. The latter follows from the fact that a sufficiently small part of the boundary of a convex body coincides, up to a rotation, with the graph of a convex function (see Proposition 13.4.5).

For a further weakening of the conditions under which the Gauss–Ostrogradski theorem is valid, see [EG].

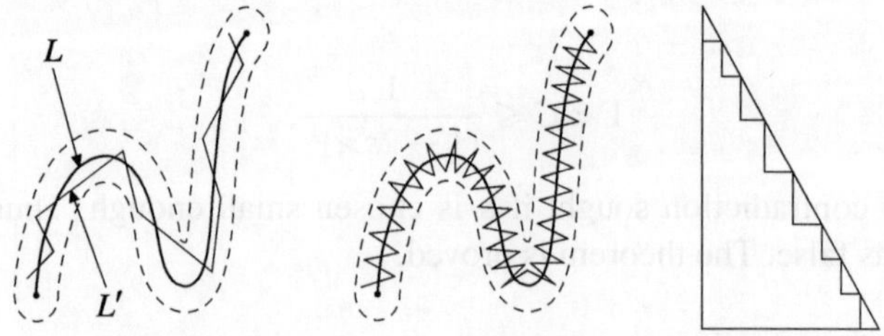

Fig. 8.4 Approximation of curves by polygonal lines

8.8.5 Let us now discuss the question of whether the areas of "close" sets are close, or, in other words whether the area is continuous. To formalize the notion of closeness, we will introduce a numeric quantity characterizing the deviation of two sets from each other. Let us remind the reader (see Sect. 8.1.7) that the symbol E_ε stands for the ε-neighborhood of the set E. For bounded sets A and B, put

$$\rho(A, B) = \inf\{\varepsilon > 0 \mid A \subset B_\varepsilon,\ B \subset A_\varepsilon\}.$$

Obviously, thus defined the function ρ is non-negative and symmetric. Moreover, $\rho(A, B) = 0$ if and only if $\overline{A} = \overline{B}$. We leave it to the reader to check that the function ρ satisfies the triangle inequality and, therefore, is a metric (or, more precisely, pseudometric). It is called the *Hausdorff metric*. On the class of compact sets, it is a true metric. If $\rho(A, B) < \varepsilon$, the sets A, B are ε-close in the sense that $A \subset B_\varepsilon$ and $B \subset A_\varepsilon$. Conversely, if A, B are ε-close, then $\rho(A, B) \leqslant \varepsilon$.

If L is a simple planar arc and the number $\varepsilon > 0$ is small, then every curve L' that is ε-close to L lies in the ε-neighborhood of the curve L and "mainly" follows its bends (see Fig. 8.4(a)). As Fig. 8.4(b) demonstrates, the length of the curve L' can be arbitrarily large for an arbitrarily small ε. It is also easy to construct a sequence of piecewise linear paths (see Fig. 8.4(c)) approximating the hypotenuse of a right triangle such that the length of each path is equal to the sum of the leg lengths.

Thus, already in the two-dimensional case, there is no hope that the length would be continuous with respect to the Hausdorff metric even if one only considers smooth curves. However, the examples leading to this negative conclusion also allow us to make an important observation. Indeed, the length of the curve L' can differ from the length of the curve L as much as one wants but only by being much larger! The pictures we have considered show that the curve L' cannot be much shorter than the curve L (for not only does L' lie in the ε-neighborhood of L, but L also lies in the ε-neighborhood of L'). It is this property, which is called the lower semicontinuity of the length, that we shall discuss. It is, of course, very important that the curve L is approximated by curves and not just by arbitrary sets. It is clear that we can always construct a countable set ε-close to L. The length of this set is zero (due to its countability). So, beyond the set of curves, even the semicontinuity of the length fails. The phenomenon that we have just discovered for curves also occurs for surfaces. This, in particular, is illustrated by the Schwartz example in which the area of the approximating polyhedral surfaces can noticeably differ

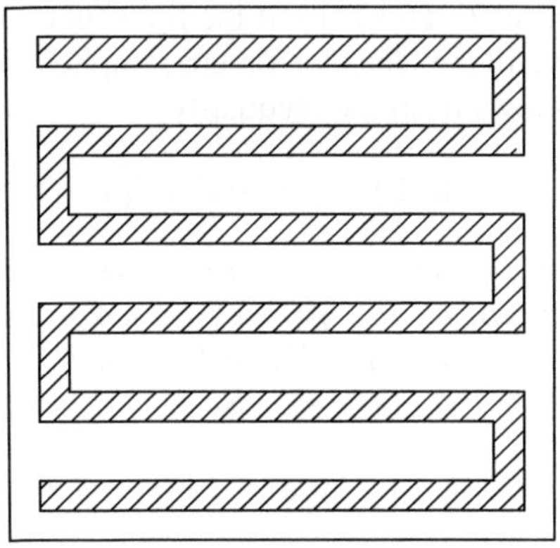

Fig. 8.5 Approximation of a square by a figure of small area

from the area of the cylinder only by being much larger. However, the situation for surfaces is more complicated than that for the curves. It turns out that (unlike what we have observed for curves) the ε-closeness of two surfaces does not imply that the approximating surface has a sufficiently large area. This can be seen in Fig. 8.5 where the approximated surface is just the unit square and the approximating one is a narrow snakelike strip that passes very close to each point in the square. Although this strip is ε-close to the square, its area can be arbitrarily small.

The way out is that in the multi-dimensional case, in addition to the ε-closeness of the surfaces themselves, one should also assume the ε-closeness of their boundaries. For general manifolds, this requires the introduction of a rigorous notion of a boundary and leads to additional topological difficulties to overcome (see [Bol, pp. 88–141]). To avoid this, we will consider below only the graphs of continuous functions instead of dealing with arbitrary manifolds.

Let us make the notion of semicontinuity more precise for this case. Recall that the symbol Γ_f stands for the graph of the function f. All considered functions are assumed to be defined in some fixed bounded open set $\mathcal{O} \subset \mathbb{R}^m$. Below, the letters σ and λ will stand for an arbitrary m-dimensional area (in $\mathbb{R}^{m+1}$) and the Lebesgue measure on $\mathbb{R}^m$ respectively. The points of the space $\mathbb{R}^{m+1}$ will be written as (x, y) where $x \in \mathbb{R}^m$ and $y \in \mathbb{R}$.

Definition Assume that the function $f \in C(\mathcal{O})$ is bounded and that $\sigma(\Gamma_f) < +\infty$. We will say that the area is *lower semicontinuous* on Γ_f if for every number $\varepsilon > 0$, there exists a number $\delta > 0$ such that $\sigma(\Gamma_g) > \sigma(\Gamma_f) - \varepsilon$ whenever $g \in C(\mathcal{O})$ and $\rho(\Gamma_f, \Gamma_g) < \delta$.

This definition can be restated as follows: the area is lower semicontinuous on Γ_f if for every sequence of functions f_n continuous in $\mathcal{O}$, the convergence $\Gamma_{f_n} \underset{n\to\infty}{\longrightarrow} \Gamma_f$ in the Hausdorff metric implies that $\underline{\lim}_{n\to\infty} \sigma(\Gamma_{f_n}) \geqslant \sigma(\Gamma_f)$.

Note that the uniform convergence of f_n to f on $\mathcal{O}$ implies the convergence of the corresponding graphs in the Hausdorff metric. These types of convergence are not equivalent (see Exercise 3). However, if the limit function is Lipschitz, then the convergence of the graphs in the Hausdorff metric implies the uniform convergence of the functions. This follows from the inequality

$$\left|f(x)-g(x)\right| \leqslant (C+1)\rho(\Gamma_f,\Gamma_g) \quad (x\in\mathcal{O}), \tag{6}$$

which is valid if at least one of the functions, say f, satisfies the Lipschitz condition with the constant C. Indeed, let $r>\rho(\Gamma_f,\Gamma_g)$, $x\in\mathcal{O}$. Since $(x,g(x))\in\Gamma_g\subset(\Gamma_f)_r$, there exists a point $(x_0,f(x_0))\in\Gamma_f$ such that $\|(x,g(x))-(x_0,f(x_0))\|<r$. Then $\|x-x_0\|<r$ and $|g(x)-f(x_0)|<r$. So,

$$\left|g(x)-f(x)\right| \leqslant \left|g(x)-f(x_0)\right| + \left|f(x_0)-f(x)\right| < r + C\|x-x_0\| < (C+1)r.$$

This implies the required result because r can be taken arbitrarily close to $\rho(\Gamma_f,\Gamma_g)$.

8.8.6 Before turning to the proof of the lower semicontinuity of the area, let us establish one auxiliary result.

Lemma *Let a function f be continuous in the ball $B(a,r)$. Assume that for some ε, $0<\varepsilon<r$, its graph Γ_f is contained in the ε-neighborhood of a plane L. Then the orthogonal projection of the graph to this plane contains all its points lying above the ball $B(a,r-\varepsilon)$, i.e., all points of L of the form (x,y) where $x\in B(a,r-\varepsilon)$.*

Proof It is clear that L is not parallel to the last coordinate axis. Fix a point $p=(x,y)\in L$ lying above $B(a,r-\varepsilon)$ and check that the line $\ell=\{p+tv\mid t\in\mathbb{R}\}$, where $v=(v',\alpha)$ is the unit normal to L, crosses Γ_f. By our assumption, $\alpha\neq 0$. Assume for definiteness that $\alpha>0$. The line ℓ crosses the boundary of the ε-neighborhood of L at the points $p\pm\varepsilon v=(x\pm\varepsilon v', y\pm\varepsilon\alpha)$. Moreover, $x\pm\varepsilon v'\in B(a,r)$ because $x\in B(a,r-\varepsilon)$ and $\|\varepsilon v'\|\leqslant\|\varepsilon v\|=\varepsilon$. Since Γ_f lies in the ε-neighborhood of L, we have $f(x+\varepsilon v')<y+\varepsilon\alpha$ and $f(x-\varepsilon v')>y-\varepsilon\alpha$. Hence the difference $(y+t\alpha)-f(x+tv')$ takes values of opposite signs at the endpoints of the interval $[-\varepsilon,\varepsilon]$. Therefore, $f(x+\tau v')=y+\tau\alpha$ for some $\tau\in(-\varepsilon,\varepsilon)$. Thus, the point $p+\tau v=(x+\tau v', f(x+\tau v'))$ belongs to Γ_f and is mapped to the point p under the orthogonal projection to L because v is a normal to L. □

Now we are ready to turn to the main result of this section and to present a condition guaranteeing the lower semicontinuity of the area in the sense of Definition 8.8.5.

Theorem *The area is lower semicontinuous on the graph of a Lipschitz function.*

Proof Let the function f satisfy the Lipschitz condition with constant C on a bounded open set $\mathcal{O}\subset\mathbb{R}^m$ and let D be the set of points at which f is differentiable. Obviously, $\|\operatorname{grad} f(x)\|\leqslant C$ on D. Put $\omega(x)=\sqrt{1+\|\operatorname{grad} f(x)\|^2}$ and $\Omega=\sqrt{1+C^2}$, so that $\omega(x)\leqslant\Omega$ and $\sigma(\Gamma_f)\leqslant\Omega\lambda(\mathcal{O})<+\infty$.

Fix an arbitrary number $\varepsilon \in (0,1)$. It is clear that the set D is exhausted by the sets

$$D^t = \Big\{a \in D \mid B(a,t) \subset \mathcal{O} \text{ and } \big|f(x) - f(a) - \langle \operatorname{grad} f(a), x-a\rangle\big| \leqslant \frac{\varepsilon}{3}\|x-a\| \text{ for } x \in B(a,t)\Big\},$$

which expand as t decreases. Since, by Rademacher's theorem, $\lambda(\mathcal{O} \setminus D) = 0$, we can fix a $t > 0$ so small that $\lambda(\mathcal{O} \setminus D^t) < \varepsilon$. Let us construct balls whose union almost coincides with D^t. To this end, note that by the corollary to Vitali's theorem (see Corollary 1 in Sect. 2.7.3), almost every point of D^t is its density point. Let E^t be the set of such points:

$$E^t = \Big\{x \in D_t \;\Big|\; \lim_{r\to 0} \frac{\lambda(D^t \cap B(x,r))}{\lambda(B(x,r))} = 1\Big\}.$$

Then $\lambda(E^t \cap B(x,r)) = \lambda(D_t \cap B(x,r)) > (1-\varepsilon)\lambda(B(x,r))$ for $x \in E^t$ and $0 < r < r(x)$, so

$$\lambda\big(B(x,r) \setminus E^t\big) < \varepsilon\lambda\big(B(x,r)\big). \tag{7}$$

We may assume that $r(x) < t/2$ for $x \in E^t$. The collection of the balls $B(x,r)$ where $x \in E^t$ and $0 < r < r(x)$ is a Vitali cover of the set E^t. By Vitali's theorem, we can find a subcollection of pairwise disjoint balls B_i such that $\lambda(E^t \setminus \bigcup_i B_i) = 0$.

Let B be one of the balls B_i, and let r be its radius. Choose a point a in $B \cap E^t$ so that $\omega(a) + \varepsilon\Omega \geqslant \sup_{B\cap E^t} \omega$. Since, according to (7), $\lambda(B \setminus E^t) < \varepsilon\lambda(B)$, the area of the graph of f over B satisfies the estimate

$$\begin{aligned}\sigma\big(\Gamma(f,B)\big) &= \int_B \omega(x)\,dx = \int_{B\cap E^t} \cdots + \int_{B\setminus E^t} \cdots \\ &\leqslant \big(\omega(a) + \varepsilon\Omega\big)\lambda\big(B \cap E^t\big) + \Omega\lambda\big(B \setminus E^t\big) \leqslant \big(\omega(a) + 2\varepsilon\Omega\big)\lambda(B). \end{aligned}\tag{8}$$

(By $\Gamma(f,B)$, we denote the part of the graph of the function f lying above the set B.)

The inequality (8) allows one to compare $\sigma(\Gamma(f,B))$ with the area of the graph of a function close to f in the ball B. Indeed, let $g \in C(\mathcal{O})$ and $|f(x) - g(x)| < \frac{\varepsilon}{3}r$ for all $x \in B$. Consider the equation $y = h(x) = f(a) + \langle x-a, \operatorname{grad} f(a)\rangle$ of the affine tangent plane L to the graph Γ_f at the point $(a, f(a))$. Since $a \in B$, we have $\|x-a\| < 2r < t$ for every point x in this ball. Taking into account the inclusion $a \in D^t$, we obtain the inequality $|f(x) - h(x)| \leqslant \frac{\varepsilon}{3}\|x-a\| < \frac{2}{3}\varepsilon r$. Hence $|g(x) - h(x)| < \eta = \varepsilon r$ and, thereby, the graph $\Gamma(g,B)$ lies in the η-neighborhood of the plane L. To use the lemma, consider the set $B^\eta = \{x \in B \mid B(x,\eta) \subset B\}$. Clearly, it is a ball of radius $r - \eta = (1-\varepsilon)r$. Therefore $\lambda(B^\eta) = (1-\varepsilon)^m\lambda(B) > (1-m\varepsilon)\lambda(B)$. By the lemma, the orthogonal projection Γ' of the graph $\Gamma(g,B)$

to L contains the graph $\Gamma(h, B^\eta)$. Since the areas of compact sets do not increase under an orthogonal projection and since $\Gamma(g, B)$ is a countable union of compact sets, we have $\sigma(\Gamma(g, B)) \geqslant \sigma(\Gamma')$, so $\sigma(\Gamma(g, B)) \geqslant \sigma(\Gamma(h, B^\eta))$. Hence,

$$\sigma\big(\Gamma(g, B)\big) \geqslant \sigma\big(\Gamma\big(h, B^\eta\big)\big) = \omega(a)\,\lambda\big(B^\eta\big) \geqslant (1 - m\varepsilon)\omega(a)\lambda(B)$$
$$\geqslant \big(\omega(a) - m\Omega\varepsilon\big)\lambda(B).$$

Together with inequality (8), this yields

$$\sigma\big(\Gamma(f, B)\big) \leqslant \sigma\big(\Gamma(g, B)\big) + (m+2)\varepsilon\Omega\lambda(B), \tag{9}$$

provided that $|f(x) - g(x)| < \frac{\varepsilon}{3} r$ for all $x \in B$.

To conclude the proof, we will estimate the area of Γ_g from below assuming that $\rho(\Gamma_f, \Gamma_g) < \delta$ and δ is small enough. Fix an index N so large that $\lambda(\bigcup_{i>N} B_i) < \varepsilon$. Put $\delta = \frac{\varepsilon}{3(C+1)} \min_{1\leqslant i\leqslant N} r_i$ where r_i is the radius of the ball B_i. If $\rho(\Gamma_f, \Gamma_g) < \delta$, then $|f(x) - g(x)| < (C+1)\delta$ for all $x \in \mathcal{O}$ due to the inequality (6), and, thereby, $|f(x) - g(x)| < \frac{\varepsilon}{3} r_i$ in the ball B_i for every $i = 1, \dots, N$. Hence, the inequalities (9) are valid for $B = B_i$, $i = 1, \dots, N$. Adding them, we obtain

$$\sigma\left(\Gamma\left(f, \bigcup_{i=1}^N B_i\right)\right) = \sum_{i=1}^N \sigma\big(\Gamma(f, B_i)\big) \leqslant \sigma\left(\Gamma\left(g, \bigcup_{i=1}^N B_i\right)\right) + \widetilde{\Omega}\varepsilon \leqslant \sigma(\Gamma_g) + \widetilde{\Omega}\varepsilon,$$

where $\widetilde{\Omega} = (m+2)\Omega\lambda(\mathcal{O})$. Thus,

$$\sigma(\Gamma_g) \geqslant \int_{\bigcup_{i=1}^N B_i} \omega(x)\,dx - \widetilde{\Omega}\varepsilon = \int_{\bigcup_{i=1}^\infty B_i} \cdots - \int_{\bigcup_{i>N} B_i} \cdots - \widetilde{\Omega}\varepsilon$$
$$> \int_{D_t} \omega(x)\,dx - (\Omega + \widetilde{\Omega})\varepsilon > \int_{\mathcal{O}} \omega(x)\,dx - \Omega\varepsilon - (\Omega + \widetilde{\Omega})\varepsilon$$
$$= \sigma(\Gamma_f) - (2\Omega + \widetilde{\Omega})\varepsilon.$$

Since ε was arbitrary, this proves the lower semicontinuity of the area of the graph Γ_f. □

Corollary *If the function f is Lipschitz and a sequence of continuous functions g_n converges to f uniformly, then*

$$\sigma(\Gamma_f) \leqslant \varliminf_{n\to\infty} \sigma(\Gamma_{g_n}).$$

To prove this, it suffices to note that $\rho(\Gamma_{g_n}, \Gamma_f) \leqslant \sup_{\mathcal{O}} |g_n - f| \underset{n\to\infty}{\longrightarrow} 0$, and, therefore, for $\varepsilon > 0$, the inequality $\sigma(\Gamma_{g_n}) < \sigma(\Gamma_f) - \varepsilon$ may hold only for a finite number of indices n.

This corollary can be generalized somewhat by replacing the uniform convergence in its assumption by convergence almost everywhere or in measure. Let us show that it holds for convergence almost everywhere.

Indeed, let $f_n \underset{n\to\infty}{\longrightarrow} f$ almost everywhere on a (bounded) set $\mathcal{O}$. Fix an arbitrarily small number $\varepsilon > 0$ and, applying Egorov's theorem 3.3.6, find a subset $e \subset \mathcal{O}$ such that

$$f_n \rightrightarrows f \quad \text{on } \mathcal{O}\setminus e \text{ and } \lambda(e) < \varepsilon.$$

Put $t_n = \sup_{\mathcal{O}\setminus e} |f_n - f|$ and correct the functions f_n, "truncating" them at places where they differ from f "too much". To this end, introduce the functions

$$g_n(x) = \begin{cases} f(x) + t_n, & \text{when } f_n(x) \geqslant f(x) + t_n, \\ f(x) - t_n, & \text{when } f_n(x) \leqslant f(x) - t_n, \\ f_n(x), & \text{when } |f_n(x) - f(x)| < t_n. \end{cases}$$

They are continuous and converge to f uniformly on $\mathcal{O}$ because $|f - g_n| \leqslant t_n \underset{n\to\infty}{\longrightarrow} 0$. Note that the graph of g_n above the set $E_n = \mathcal{O}(|g_n - f| \geqslant t_n)$ consists of two parts lying on the graphs of the Lipschitz functions $f \pm t_n$. Also, if $t_n < \varepsilon$, then the set E_n is contained in e. Thus, for all sufficiently large indices, we have (recall that $\omega \leqslant \sqrt{1 + C^2} = \Omega$)

$$\begin{aligned} \sigma(\Gamma_{f_n}) &\geqslant \sigma\big(\Gamma(f_n, \mathcal{O}\setminus E_n)\big) = \sigma\big(\Gamma(g_n, \mathcal{O}\setminus E_n)\big) \\ &= \sigma(\Gamma_{g_n}) - \int_{E_n} \omega(x)\,dx \geqslant \sigma(\Gamma_{g_n}) - \Omega\lambda(e) > \sigma(\Gamma_{g_n}) - \Omega\varepsilon. \end{aligned}$$

Passing to the lower limit in this inequality, we obtain

$$\varliminf_{n\to\infty} \sigma(\Gamma_{f_n}) \geqslant \varliminf_{n\to\infty} \sigma(\Gamma_{g_n}) - \Omega\varepsilon.$$

Since ε was arbitrary and since $\varliminf_{n\to\infty} \sigma(\Gamma_{g_n}) \geqslant \sigma(\Gamma_f)$ according to the corollary, we arrive at the required result.

The case of convergence in measure can be reduced to the one just considered using Riesz's theorem 3.3.4.

EXERCISES

1. Show by example that the graph of a function can be a Lipschitz manifold even when the function does not satisfy the Lipschitz condition on any (non-empty) interval. *Hint.* Consider an increasing expanding map whose derivative is unbounded on every interval.
2. Let L_0 be the graph of the function $f(x) = 2x^2 \sin\frac{2\pi}{x}$ for $0 < x < 1/2$, $f(0) = 0$, and let L be the union of L_0 and the set obtained from it by a $\pi/2$ rotation. Prove that the curve L admits a bi-Lipschitz parametrization but its intersection with every neighborhood of the point $(0, 0)$ is not congruent to the graph of any function.
3. Assume that continuous bounded functions f, f_n $(n = 1, 2, \ldots)$ are defined on a bounded set $\mathcal{O}$. Prove that:

(a) if $f_n \rightrightarrows f$ on $\mathcal{O}$, then $\rho(\Gamma_{f_n}, \Gamma_f) \underset{n\to\infty}{\longrightarrow} 0$;

(b) if $\rho(\Gamma_{f_n}, \Gamma_f) \underset{n\to\infty}{\longrightarrow} 0$, then $f_n \underset{n\to\infty}{\longrightarrow} f$ pointwise in $\mathcal{O}$;

(c) if $\rho(\Gamma_{f_n}, \Gamma_f) \underset{n\to\infty}{\longrightarrow} 0$ and the function f is uniformly continuous, then $f_n \rightrightarrows f$ on $\mathcal{O}$.

Give an example showing that statement (c) fails without the assumption that f is uniformly continuous.

Chapter 9
Approximation and Convolution in the Spaces $\mathscr{L}^p$

9.1 The Spaces $\mathscr{L}^p$

In the solution of various problems, it is important to be able to approximate the functions of a certain class by functions with better properties. For example, measurable functions can be approximated by simple functions (see Theorem 3.2.2) and continuous functions can be approximated by smooth functions. Here the way we understand the closeness between two functions, or what is taken as a "dissimilarity" measure, is of great importance. The reader is probably familiar with the uniform, or Chebyshev, deviation. We recall that the uniform deviation of a function f from a function g on a set X is defined as $\sup_X |f - g|$. It is clear that the Chebyshev deviation of f_n from g tends to zero if and only if $f_n \rightrightarrows g$ on X. If functions are defined on a measure space, then as well as the classical uniform deviation it is also useful to consider its modification, the uniform deviation on a set of full measure. For functions f and g, this is defined by the notion of essential supremum (see Sect. 4.4.5) as $\operatorname{esssup}_X |f - g|$.

However, the conditions of the problem under consideration often exclude in advance the possibility of uniform approximation. This happens, in particular, when approximating unbounded functions by bounded functions or, which is especially important, when approximating discontinuous functions by continuous functions. In such cases, it is necessary to use other characteristics of the deviation between functions. For summable functions f and g, this can be done by the so-called *deviation in mean*, by which we mean the integral $\int_X |f - g|\, d\mu$. If X is a subset of the space $\mathbb{R}^m$ and μ is the m-dimensional Lebesgue measure, then the deviation in mean has a simple geometric meaning, namely, it is the $(m+1)$-dimensional volume of the set confined between the graphs of the functions in question. The deviation in mean is essentially different from the uniform deviation. The latter is already large when the functions differ greatly at a single point, but the deviation in mean takes into account the behavior of the functions on the entire set of integration. It is easy find examples for which the deviation in mean can be arbitrarily small even if the Chebyshev deviation is arbitrarily large.

B. Makarov, A. Podkorytov, *Real Analysis: Measures, Integrals and Applications*, Universitext, DOI 10.1007/978-1-4471-5122-7_9, © Springer-Verlag London 2013

It is natural to consider the deviation in mean on the set $\mathscr{L}(X,\mu)$ of all summable functions, whereas the modified uniform deviation is naturally defined for the functions in the set

$$\mathscr{L}^\infty(X,\mu) = \{f \in \mathscr{L}^0(X,\mu) \mid \operatorname*{esssup}_X |f| < +\infty\}.$$

To increase the number of possible applications, it is also useful to introduce sets "intermediate" between $\mathscr{L}(X,\mu)$ and $\mathscr{L}^\infty(X,\mu)$.

9.1.1 In what follows, we will use the usual notation $(X,\mathfrak{A},\mu)$ for a space X with an arbitrary (non-zero) measure μ, and $\mathscr{L}^0(X,\mu)$ for the set of measurable (real or complex) functions that are finite almost everywhere on X. In what follows, all functions will be taken from this set.

We fix an arbitrary number p, $1 < p < +\infty$, and put

$$\mathscr{L}^p(X,\mu) = \left\{ f \in \mathscr{L}^0(X,\mu) \,\middle|\, \int_X |f|^p \, d\mu < +\infty \right\}.$$

For uniformity, we will assume that $\mathscr{L}^1(X,\mu) = \mathscr{L}(X,\mu)$ (the set of all summable functions). Since

$$\begin{aligned}|f+g|^p &\leqslant \big(|f|+|g|\big)^p \leqslant \big(2\max\{|f|,|g|\}\big)^p = 2^p \max\{|f|^p,|g|^p\}\\ &\leqslant 2^p\big(|f|^p+|g|^p\big),\end{aligned}$$

we see that, along with arbitrary functions f and g, the set $\mathscr{L}^p(X,\mu)$ also contains their sum and, consequently their linear combinations. Thus, $\mathscr{L}^p(X,\mu)$ is a vector space. The set $\mathscr{L}^p(X,\mu)$ is often called the set of pth power summable functions. More precisely, these are functions for which the pth power of the absolute value is summable. It may happen that the function itself is not summable. For example, the function $x \mapsto \frac{1}{x+1}$ is not summable with respect to Lebesgue measure λ on $\mathbb{R}_+$ but belongs to $\mathscr{L}^2(\mathbb{R}_+,\lambda)$ or, in other words, is square-summable on $(0,+\infty)$.

However, if the measure μ is finite, then $\mathscr{L}^p(X,\mu) \subset \mathscr{L}^1(X,\mu)$. Moreover, if $1 \leqslant r < p \leqslant +\infty$, then $\mathscr{L}^p(X,\mu) \subset \mathscr{L}^r(X,\mu)$. This is obvious for $p = +\infty$. If $p < +\infty$, then, putting $s = p/r$, $1/s + 1/s' = 1$, and applying Hölder's inequality with exponent s to the functions $|f|^r$ and 1, where $f \in \mathscr{L}^p(X,\mu)$, we see that

$$\begin{aligned}\int_X |f|^r \, d\mu &= \int_X |f|^r \cdot 1 \, d\mu \leqslant \left(\int_X |f|^{rs} \, d\mu\right)^{\frac{r}{p}} (\mu(X))^{\frac{1}{s'}}\\ &= \left(\int_X |f|^p \, d\mu\right)^{\frac{r}{p}} (\mu(X))^{\frac{1}{s'}} < +\infty. \qquad (1)\end{aligned}$$

Thus, in the case of a finite measure, the sets $\mathscr{L}^p(X,\mu)$ decrease when p increases. In particular, $\mathscr{L}^1(X,\mu)$ is the largest among them and $\mathscr{L}^\infty(X,\mu)$ is the smallest.

It is convenient to introduce a generalized deviation in mean by using a norm.

Definition Let $f \in \mathscr{L}^p(X,\mu)$, $1 \leqslant p \leqslant +\infty$. The *norm*[1] (more precisely, the *$\mathscr{L}^p$-norm*) of a function f is defined by the equation

$$\|f\|_p = \begin{cases} (\int_X |f|^p \, d\mu)^{1/p} & \text{if } 1 \leqslant p < +\infty; \\ \operatorname{esssup}_X |f| & \text{if } p = +\infty. \end{cases}$$

If $\mu(X) = 1$, then it can be seen from (1) that the $\mathscr{L}^p$-norm increases with p. Moreover, it can be proved (see Exercise 5) that $\|f\|_p \underset{p\to+\infty}{\longrightarrow} \|f\|_\infty$ for f in $\mathscr{L}^\infty(X,\mu)$. This limit relation can serve as an extra motivation of the notation $\|f\|_\infty$ for $\operatorname{esssup}_X |f|$.

We note the basic properties of a norm,

(1) $\|f\|_p \geqslant 0$;
(2) $\|af\|_p = |a|\|f\|_p$;
(3) $\|f+g\|_p \leqslant \|f\|_p + \|g\|_p$

for all $f, g \in \mathscr{L}^p(X,\mu)$ and every scalar a. Properties (1)–(2) are obvious. We also remark that $\|f-g\|_p = 0$ if and only if the functions f and g are equivalent, i.e., coincide almost everywhere. Inequality (3) is called the *triangle inequality*. For finite p this is simply the Minkowski inequality established in Sect. 4.4.6. We leave it to the reader to verify that the inequality is valid in the case where $p = +\infty$.

The triangle inequality implies the following useful inequality:

$$\big| \|f\|_p - \|g\|_p \big| \leqslant \|f-g\|_p.$$

Indeed, $\|f\|_p = \|g + (f-g)\|_p \leqslant \|g\|_p + \|f-g\|_p$, i.e., $\|f\|_p - \|g\|_p \leqslant \|f-g\|_p$. By the symmetry of the functions f and g, this gives the required inequality.

Obviously, the deviation in mean of f from g is just the $\mathscr{L}^1$-norm of their difference. Therefore, if $\|f_n - f\|_1 \underset{n\to\infty}{\longrightarrow} 0$, then we say that the sequence $\{f_n\}_{n\geqslant 1}$ *converges to f in mean*. If $p \geqslant 1$, then, by abuse of language, we will also say that f_n converges to f in mean, more precisely, *in mean with exponent p* if $\|f_n - f\|_p \underset{n\to\infty}{\longrightarrow} 0$ (the *convergence in $\mathscr{L}^p$-norm*). Using the triangle inequality, one can easily prove that, with respect to convergence in the $\mathcal{L}^p$-norm, the limit is unique up to equivalence. Indeed, if $\|f_n - f\|_p \underset{n\to\infty}{\longrightarrow} 0$ and $\|f_n - g\|_p \underset{n\to\infty}{\longrightarrow} 0$, then

$$\|f-g\|_p \leqslant \|f-f_n\|_p + \|f_n - g\|_p \underset{n\to\infty}{\longrightarrow} 0,$$

and so $\|f-g\|_p = 0$.

The $\mathscr{L}^p$-norm is continuous with respect to convergence in mean, i.e.,

$$\text{if } \|f_n - f\|_p \underset{n\to\infty}{\longrightarrow} 0, \quad \text{then } \|f_n\|_p \underset{n\to\infty}{\longrightarrow} \|f\|_p.$$

This follows from the inequality $| \|f_n\|_p - \|f\|_p | \leqslant \|f_n - f\|_p$ just proved.

[1] This terminology is not in full agreement with the conventional terminology; usually, a function satisfying the properties (1)–(3) listed below is called a seminorm.

The definitions of the set $\mathscr{L}^p$ and the $\mathscr{L}^p$-norm given above can be extended to the case where $0 < p < 1$. However, in this case, the "norm" does not satisfy the triangle inequality (see Exercise 14), and the set $\mathscr{L}^p(\mathbb{R})$ contains not only non-summable but also locally non-summable functions (see Exercise 15), which narrows down the range of possible applications considerably. Therefore, we confine ourselves to the study of the properties of $\mathscr{L}^p$ only for $p \geqslant 1$. In the sequel, we will tacitly assume that this condition holds.

9.1.2 Let us discuss the necessary conditions as well as the sufficient conditions for convergence in the $\mathscr{L}^p$-norm.

Theorem *Let $1 \leqslant p < +\infty$ and $f_n \in \mathscr{L}^p(X, \mu)$ for all $n \in \mathbb{N}$.*

(a) *If $f \in \mathscr{L}^p(X, \mu)$ and $\| f_n - f \|_p \underset{n\to\infty}{\longrightarrow} 0$, then $f_n \underset{n\to\infty}{\longrightarrow} f$ in measure.*

(b) *If $f_n \underset{n\to\infty}{\longrightarrow} f$ in measure or almost everywhere and $|f_n(x)| \leqslant g(x)$ almost everywhere for all n, where $g \in \mathscr{L}^p(X, \mu)$, then $f \in \mathscr{L}^p(X, \mu)$ and $\| f - f_n \|_p \underset{n\to\infty}{\longrightarrow} 0$.*

Proof (a) We fix an arbitrary positive number ε and put

$$X_n(\varepsilon) = \{x \in X \mid |f(x) - f_n(x)| \geqslant \varepsilon\}.$$

Then

$$\mu(X_n(\varepsilon)) \leqslant \frac{1}{\varepsilon^p} \int_{X_n(\varepsilon)} |f - f_n|^p \, d\mu \leqslant \frac{1}{\varepsilon^p} \| f - f_n \|_p^p \underset{n\to\infty}{\longrightarrow} 0,$$

as required.

(b) Since $|f_n| \leqslant g$, we have $|f| \leqslant g$ (in the case of convergence in measure, this is established in Corollary 2 of Sect. 3.3.5) and $|f_n - f|^p \leqslant (2g)^p \in \mathscr{L}^1(X, \mu)$. Therefore, $\| f_n - f \|_p^p = \int_X |f_n - f|^p \, d\mu \underset{n\to\infty}{\longrightarrow} 0$ by Lebesgue's theorem (see Sects. 4.8.3–4.8.4). □

Remark It can easily be seen that the convergence of a sequence in the space $\mathscr{L}^\infty(X, \mu)$ is equivalent to the uniform convergence of this sequence on a subset of full measure.

9.1.3 Here we establish an important property of the space $\mathscr{L}^p(X, \mu)$.

Definition A sequence $\{f_n\}_{n\geqslant 1} \subset \mathscr{L}L^p(X, \mu)$ is called *fundamental* in $\mathscr{L}^p(X, \mu)$ if $\| f_n - f_k \|_p \to 0$ for $k, n \to \infty$, i.e.,

$$\forall \varepsilon > 0 \; \exists N : \| f_n - f_k \|_p < \varepsilon \quad \text{for } k, n > N.$$

Every convergent sequence is fundamental because if $f_n \underset{n\to\infty}{\longrightarrow} f$, then

$$\| f_n - f_k \|_p \leqslant \| f_n - f \|_p + \| f - f_k \|_p \underset{k,n\to\infty}{\longrightarrow} 0$$

by the triangle inequality. It turns out that the converse is also true. Now we prove this property, called the completeness of the space $\mathscr{L}^p$.

Theorem *Every sequence fundamental in the* $\mathscr{L}^p$*-norm* ($1 \leqslant p \leqslant +\infty$), *has a limit.*

Proof We leave it to the reader to consider the case $p = +\infty$ (it reduces to the completeness of the space of bounded functions with respect to uniform convergence). In the sequel, we assume that $p < +\infty$.

First, we show that every fundamental sequence $\{f_n\}_{n\geqslant 1}$ has a subsequence that converges almost everywhere. For this, we use the fact that the sequence $\{f_n\}_{n\geqslant 1}$ is fundamental and extract from it a subsequence $\{f_{n_j}\}_{j\geqslant 1}$ such that

$$\sum_{j=1}^{\infty} \|f_{n_{j+1}} - f_{n_j}\|_p \leqslant 1.$$

We verify that the sequence $\{f_{n_j}\}_{j\geqslant 1}$ converges almost everywhere. We consider the series

$$\sum_{j=1}^{\infty} |f_{n_{j+1}} - f_{n_j}|. \tag{2}$$

Let S and S_k be its sum and its kth partial sum, respectively. By the triangle inequality, we obtain $\|S_k\|_p \leqslant \sum_{j=1}^{\infty} \|f_{n_{j+1}} - f_{n_j}\|_p \leqslant 1$, and, therefore,

$$\int_X S_k^p \, d\mu \leqslant 1 \quad \text{for all } k.$$

Since $S_k \underset{k\to\infty}{\longrightarrow} S$ pointwise, Fatou's theorem implies that $\int_X S^p \, d\mu \leqslant 1$. Since S^p is summable, we obtain that $S(x) < +\infty$ almost everywhere, which just means that series (2) converges almost everywhere. Now, we consider the series

$$f_{n_1} + \sum_{j=1}^{\infty} (f_{n_{j+1}} - f_{n_j}).$$

Like (2), it converges almost everywhere and its partial sums are the functions f_{n_k}. Thus, $f_{n_k}(x) \underset{k\to\infty}{\longrightarrow} f(x)$ almost everywhere, where f is the sum of the last series.

Now, we prove that f is the limit of the sequence $\{f_n\}_{n\geqslant 1}$ in the sense of convergence in mean, i.e., that $f \in \mathscr{L}^p(X,\mu)$ and $\|f_n - f\|_p \underset{n\to\infty}{\longrightarrow} 0$. We fix an arbitrary number $\varepsilon > 0$ and, by the definition of a fundamental sequence, we find an N such that

$$\int_X |f_n - f_l|^p \, d\mu < \varepsilon^p \quad \text{for } l, n > N.$$

Substituting $l = n_k > N$ in this inequality, we see that

$$\int_X |f_n - f_{n_k}|^p \, d\mu < \varepsilon^p.$$

Passing to the limit with respect to k (for a fixed $n > N$) in this inequality and using Fatou's theorem, we obtain

$$\int_X |f_n - f|^p \, d\mu \leqslant \varepsilon^p.$$

Thus, the function $f_n - f$, and along with it the function f (since $f = (f - f_n) + f_n$), belongs to $\mathscr{L}^p(X, \mu)$, and the last inequality can be represented in the form

$$\|f_n - f\|_p \leqslant \varepsilon \quad \text{for } n > N. \qquad \square$$

9.1.4 In conclusion of this section, we discuss a generalization of the important estimate of the maximal function obtained in Theorem 4.9.1. First of all, we generalize the concept of the maximal function, dropping the summability requirement.

In the sequel, we will denote the space $\mathscr{L}^p(\mathbb{R}^m, \lambda_m)$ by $\mathscr{L}^p(\mathbb{R}^m)$. It is clear that $\mathscr{L}^p(\mathbb{R}^m) \subset \mathscr{L}_{\text{loc}}(\mathbb{R}^m)$ (for the definition of $\mathscr{L}_{\text{loc}}(\mathbb{R}^m)$, see Sect. 4.9.2).

Definition Let $f \in \mathscr{L}_{\text{loc}}(\mathbb{R}^m)$. The function M_f defined by the formula

$$M_f(x) = \sup_{r>0} \frac{1}{v(r)} \int_{B(x,r)} |f(y)| \, dy \quad (x \in \mathbb{R}^m)$$

is called the *maximal function* (for f).

Here, as in Sect. 4.9, $v(r)$ is the volume of a ball of radius r.

Repeating the argument given in Sect. 4.9.1, we can convince ourselves that the maximal function is measurable. We mention two more obvious properties of the maximal function (in the sequel, they will be used without any special reference),

$$M_{f+g} \leqslant M_f + M_g; \qquad \text{if } |f| \leqslant C, \quad \text{then } M_f \leqslant C.$$

If $f \in \mathscr{L}_{\text{loc}}(\mathbb{R}^m)$ and $|f(x)|$ increases unboundedly as $\|x\| \to +\infty$, the maximal function has the value $+\infty$ everywhere. If the function f is summable, as proved in Theorem 4.9.1, the maximal function is finite almost everywhere. Moreover, in this theorem, we obtained the estimate $F(t) \leqslant \frac{5^m}{t}\|f\|_1$, where $F(t) = \lambda_m(\{x \in \mathbb{R}^m \mid M_f(x) > t\})$ is the decreasing distribution function for M_f. However, the summability may not be preserved when passing to the maximal function (see Exercise 1, Sect. 4.9). In contrast to this, if some function belongs to a class $\mathscr{L}^p(\mathbb{R}^m)$, where $p > 1$, then its maximal function belongs to the same class. To prove this result, we first sharpen somewhat the estimate obtained in Theorem 4.9.1 for a summable function.

Lemma *Let $f \in \mathscr{L}^1(\mathbb{R}^m)$, $E_t = \{x \in \mathbb{R}^m \mid |f(x)| > t\}$, and let F be the decreasing distribution function for M_f. Then*

$$F(t) \leqslant 2\frac{5^m}{t}\int_{E_{t/2}} |f(x)|\,dx \tag{3}$$

for all $t > 0$.

Proof To estimate $F(t)$, we represent f in the form of the sum of two functions the choice of which depends on t. This important idea was first used by Marcinkiewicz[2] in the proof of his interpolation theorem, a particular case of which is the theorem proved below. Thus, we put $g = f \cdot \chi_{E_{t/2}}$ and $h = f - g = f \cdot (1 - \chi_{E_{t/2}})$. Then $\|g\|_1 = \int_{E_{t/2}} |f|\,d\mu$ and $|h| \leqslant t/2$. Since $f = g + h$, we have $M_f \leqslant M_g + M_h$. We estimate F in terms of the distribution functions M_g and M_h, which we denote by G and H, respectively. Obviously,

$$\{x \in \mathbb{R}^m \mid M_f(x) > t\} \subset \{x \in \mathbb{R}^m \mid M_g(x) > t/2\} \cup \{x \in \mathbb{R}^m \mid M_h(x) > t/2\},$$

and, therefore, $F(t) \leqslant G(t/2) + H(t/2)$. By Theorem 4.9.1, we have $G(t) \leqslant \frac{5^m}{t}\|g\|_1$. Since $|h| \leqslant t/2$, we have also $M_h \leqslant t/2$, and so $H(t/2) = 0$. Consequently,

$$F(t) \leqslant G(t/2) \leqslant \frac{5^m}{t/2}\|g\|_1 = 2\frac{5^m}{t}\int_{E_{t/2}} |f(x)|\,dx.$$

□

Now, we proceed to the main result of this section.

Theorem *If $1 < p < +\infty$ and $f \in \mathscr{L}^p(\mathbb{R}^m)$, then $M_f \in \mathscr{L}^p(\mathbb{R}^m)$ and*

$$\|M_f\|_p \leqslant 2\left(\frac{p}{p-1}\right)^{\frac{1}{p}} 5^{\frac{m}{p}} \|f\|_p.$$

Proof Let F be the decreasing distribution function for M_f. By Proposition 6.4.3, we have

$$I \equiv \int_{\mathbb{R}^m} M_f^p(x)\,dx = p\int_0^\infty t^{p-1}F(t)\,dt.$$

Estimating $F(t)$ by inequality (3), we obtain

$$\begin{aligned} I &\leqslant 2p5^m \int_0^\infty t^{p-2}\left(\int_{E_{t/2}} |f(x)|\,dx\right)dt \\ &= 2p5^m \int_0^\infty t^{p-2}\left(\int_{\mathbb{R}^m} |f(x)|\chi_+(|f(x)| - t/2)\,dx\right)dt, \end{aligned}$$

[2] Józef Marcinkiewicz (1910–1940)—Polish mathematician.

where χ_+ is the characteristic function of the semi-axis $(0, +\infty)$. Changing the order of integration (the verification that the integrand is measurable jointly in the variables x, t is left to the reader), we obtain

$$I \leqslant 2p5^m \int_{\mathbb{R}^m} |f(x)| \left(\int_0^{\infty} t^{p-2} \chi_+\big(|f(x)| - t/2\big)\, dt \right) dx$$

$$= 2p5^m \int_{\mathbb{R}^m} |f(x)| \left(\int_0^{2|f(x)|} t^{p-2}\, dt \right) dx$$

$$= 2p5^m \int_{\mathbb{R}^m} |f(x)| \frac{1}{p-1} |2f(x)|^{p-1}\, dx = 2^p 5^m \frac{p}{p-1} \|f\|_p^p,$$

which is equivalent to the assertion of the theorem. □

EXERCISES

1. Verify that neither of the sets $\mathscr{L}^1(\mathbb{R})$, $\mathscr{L}^2(\mathbb{R})$ is contained in the other. Give an example of a function in $\mathscr{L}^2(\mathbb{R})$ that does not belong to any space $\mathscr{L}^p(\mathbb{R})$ for $p \neq 2$.
2. Verify that $\mathscr{L}^\infty([0,1], \lambda) \neq \bigcap_{p<+\infty} \mathscr{L}^p([0,1], \lambda)$ (λ is Lebesgue measure).
3. Prove that the inclusion

$$\mathscr{L}^p(X, \mu) \subset \mathscr{L}^1(X, \mu) + \mathscr{L}^\infty(X, \mu) \equiv \left\{ f + g \mid f \in \mathscr{L}^1(X, \mu),\ g \in \mathscr{L}^\infty(X, \mu) \right\}$$

holds for $1 < p < +\infty$.
4. Let $\mu(X) < +\infty$, $f_n \in \mathscr{L}^p(X, \mu)$ $(n \in \mathbb{N})$, and $f_n \rightrightarrows f$ on X. Prove that $f \in \mathscr{L}^p(X, \mu)$ and $\|f_n - f\|_p \underset{n\to\infty}{\longrightarrow} 0$.
5. Prove that, for $0 < r < p < s \leqslant +\infty$, the intersection $\mathscr{L}^r(X, \mu) \cap \mathscr{L}^s(X, \mu)$ is contained in $\mathscr{L}^p(X, \mu)$. Moreover, $\|f\|_\infty = \lim_{p\to+\infty} \|f\|_p$ for each function f in $\mathscr{L}^r(X, \mu) \cap \mathscr{L}^\infty(X, \mu)$.
6. Let f be a measurable function and $I(p) = \int_X |f|^p\, d\mu > 0$. Prove that the set $\{p > 0 \mid I(p) < +\infty\}$ is an interval. Verify that if the interval is non-degenerate, then the function $p \mapsto I(p)$ is logarithmically convex.
7. Verify that if $0 < r < p < s \leqslant +\infty$ and $\|f\|_s \leqslant C\|f\|_p$, then $\|f\|_p \leqslant K\|f\|_r$, where K depends on C and on p, r, and s but not on f. Use this result to prove the following supplement to the Khintchine inequality (see Sect. 6.4.5):

$$B_p\big(a_1^2 + \cdots + a_n^2\big)^{1/2} \leqslant \left(\int_0^1 |a_1 r_1(t) + \cdots + a_n r_n(t)|^p\, dt \right)^{1/p}$$

for all $p > 0$, $n \in \mathbb{N}$ and $a_1, \ldots, a_n \in \mathbb{R}$, where $r_1, \ldots, r_n$ are Rademacher functions and $B_p > 0$ is a constant depending only on p.
8. Prove that the set $B = \{f \in \mathscr{L}^p(X, \mu) \mid \|f\|_p \leqslant R\}$ is closed in $\mathscr{L}^p(X, \mu)$ with respect to convergence in measure: if $\{f_n\}_{n\geqslant 1} \subset B$ and $f_n \underset{n\to\infty}{\longrightarrow} f$ in measure, then $f \in B$.

9. Verify that the convergence of the series $\sum_{n=1}^{\infty} \|f_n - f\|_p$ implies the convergence of the functions f_n to f almost everywhere.

10. Describe the measures for which

(a) $\mathscr{L}^1(X,\mu) = \mathscr{L}^2(X,\mu)$;
(b) $\mathscr{L}^1(X,\mu) \not\subset \mathscr{L}^2(X,\mu)$ and $\mathscr{L}^2(X,\mu) \not\subset \mathscr{L}^1(X,\mu)$.

11. Find a sequence of functions f_n in $\mathscr{L}^1([0,1],\lambda)$ that converges in mean to zero and is such that $\overline{\lim}_{n\to\infty} f_n(x) = +\infty$ and $\underline{\lim}_{n\to\infty} f_n(x) = -\infty$ everywhere.

12. Let $\mathscr{L}^p(X,\mu) = \{f \in \mathscr{L}^0(X,\mu) \mid \|f\|_p < +\infty\}$, $0 < p < 1$. Prove that this set is a vector space and verify that

$$\|f+g\|_p^p \leqslant \|f\|_p^p + \|g\|_p^p \quad \text{and} \quad \|f+g\|_p \leqslant 2^{\frac{1}{p}-1}\big(\|f\|_p + \|g\|_p\big).$$

In particular, the function $\rho(f,g) = \|f-g\|_p^p$ is a metric in $\mathscr{L}^p(X,\mu)$ (with the proviso that the equation $\rho(f,g)=0$ implies that f and g coincide only almost everywhere).

13. Verify that the theorems of the present section are valid for all $p > 0$.

14. Verify by example that if $0 < p < 1$, then the triangle inequality fails for $\|f\|_p = (\int_X |f|^p \, d\mu)^{1/p}$.

15. Give an example of a function f such that $\int_0^1 \sqrt{|f(x)|}\, dx < +\infty$ and $\int_a^b |f(x)|\, dx = +\infty$ for all a, b, $0 \leqslant a < b \leqslant 1$.

9.2 *Approximation in the Spaces $\mathscr{L}^p$

Beginning from Sect. 9.2.2, X is a Lebesgue measurable subset of the space $\mathbb{R}^m$ and $\mathscr{L}^p(X)$ is a shorthand for $\mathscr{L}^p(X,\lambda_m)$, $p \geqslant 1$. As usual, χ_E is the characteristic function of a set E.

9.2.1 Our first result forms the basis for subsequent theorems on the approximation of functions with respect to the $\mathscr{L}^p$-norm. It says that simple functions are densely scattered everywhere in the space $\mathscr{L}^p(X,\mu)$ just as the rational numbers are everywhere dense in the real numbers (a special case of this statement is established in Lemma 4.9.2). Since we do not confine ourselves to consideration of real functions only, by a simple function we mean a measurable function with a finite number of values (real or complex), i.e., a complex linear combination of the characteristic functions of measurable sets.

Theorem *Let $(X, \mathfrak{A}, \mu)$ be an arbitrary measure space. For every function f in $\mathscr{L}^p(X,\mu)$, $1 \leqslant p \leqslant +\infty$, and every $\varepsilon > 0$, there is a simple function g such that $\|f-g\|_p < \varepsilon$.*

Proof Without loss of generality, we will assume that f is real. For $p = +\infty$, the assertion of the theorem is established in the corollary to Theorem 3.2.2. Let $p < +\infty$.

By the same corollary, there exist simple functions g_n $(n = 1, 2, \ldots)$ such that

$$g_n(x) \underset{n\to\infty}{\longrightarrow} f(x) \quad \text{and} \quad |g_n(x)| \leqslant |f(x)| \quad \text{for all } x \in X \text{ and } n \in \mathbb{N}.$$

Consequently, we have $\|f - g_n\|_p \underset{n\to\infty}{\longrightarrow} 0$ by Theorem 9.1.2. Thus, as the required function g, we can take an arbitrary function g_n with sufficiently large index n. □

9.2.2 One of our goals is to show that the functions in $\mathscr{L}^p(X)$ can be approximated (in $\mathscr{L}^p$-norm) as closely as desired by smooth functions. We begin with approximation by simple functions of a special form.

Definition A linear combination of the characteristic functions of cells is called a *step function*.

It is clear that each step function belongs to $\mathscr{L}^p(\mathbb{R}^m)$ for every p.

Theorem *For $1 \leqslant p < +\infty$, every function f in $\mathscr{L}^p(X)$ can be approximated (in the $\mathscr{L}^p$-norm) as closely as desired by a step function.*

Proof Extending f by zero outside X, we assume that $X = \mathbb{R}^m$. We fix an arbitrary $\varepsilon > 0$. Now we divide the proof into several steps, gradually complicating the function f.

(1) Let $f = \chi_E$ be the characteristic function of a set E of finite measure. By the definition of Lebesgue measure, we have

$$\lambda_m(E) = \inf\left\{\sum_{k=1}^{\infty} \lambda_m(P_k) \,\middle|\, E \subset \bigcup_{k=1}^{\infty} P_k,\ P_k \in \mathscr{P}^m \text{ for } k = 1, 2, \ldots\right\}.$$

Let $\{P_k\}_{k\geqslant 1}$ be a sequence of cells such that

$$E \subset \bigcup_{k=1}^{\infty} P_k, \qquad \sum_{k=1}^{\infty} \lambda_m(P_k) < \lambda_m(E) + \varepsilon.$$

We fix a number N so large that $\sum_{k>N} \lambda_m(P_k) < \varepsilon$ and put

$$A = \bigcup_{k=1}^{\infty} P_k, \qquad B = \bigcup_{k=1}^{N} P_k.$$

By the theorem on the properties of semirings, the set B can be represented as the union of pairwise disjoint cells. Without loss of generality, we will assume that the sets $P_1, \ldots, P_N$ are pairwise disjoint. Then $\chi_B = \chi_{P_1} + \cdots + \chi_{P_N}$ and, thus, $g = \chi_B$ is a step function. We estimate $\|f - g\|_p$. By the triangle inequality, we obtain

$$\|f - g\|_p = \|\chi_E - \chi_B\|_p \leqslant \|\chi_E - \chi_A\|_p + \|\chi_A - \chi_B\|_p.$$

We consider each summand separately. It is obvious that

$$\|\chi_E - \chi_A\|_p^p = \int_{\mathbb{R}^m} (\chi_A - \chi_E)^p \, d\lambda_m = \int_{A\setminus E} 1 \, d\lambda_m$$
$$= \lambda_m(A) - \lambda_m(E) \leqslant \sum_{k=1}^{\infty} \lambda_m(P_k) - \lambda_m(E) < \varepsilon$$

(the last inequality is valid by the choice of the sequence $\{P_k\}$). Further,

$$\|\chi_A - \chi_B\|_p^p = \int_{A\setminus B} 1 \, d\lambda_m = \lambda_m(A \setminus B) \leqslant \lambda_m\Big(\bigcup_{k>N} P_k\Big) \leqslant \sum_{k>N} \lambda_m(P_k) < \varepsilon$$

(the last inequality is valid by the choice of N). Consequently, $\|\chi_E - g\|_p \leqslant 2\varepsilon^{1/p}$. We see that the function χ_E can be approximated as closely as desired by a step function.

(2) If f is a simple function, i.e., a linear combination of the characteristic functions of sets E_k of finite measure, then the assertion of the theorem is valid since, by what was just proved, each function χ_{E_k} can be approximated as closely as desired.

(3) In the general case, we use Theorem 9.2.1 to approximate a function f by simple functions h so that $\|f - h\|_p < \varepsilon$. By what has been proved, we can find a step function g such that $\|h - g\|_p < \varepsilon$. Then $\|f - g\|_p \leqslant \|f - h\|_p + \|h - g\|_p < 2\varepsilon$, which completes the proof. □

The theorem just proved is valid not only for Lebesgue measure but also for many other σ-finite measures (see Exercise 2).

9.2.3 Now we turn to the problem of approximation of summable functions by smooth functions. We recall that the closure of the set $\{x \in \mathbb{R}^m \mid \varphi(x) \neq 0\}$ is called the support of φ and is denoted by $\operatorname{supp}(\varphi)$, and a function with compact support is called a compactly supported function. In what follows, $C_0^\infty(\mathbb{R}^m)$ is the class of infinitely differentiable compactly supported functions on $\mathbb{R}^m$.

Our goal is to prove that, for a finite p, every function in $\mathscr{L}^p(X)$ can be approximated in mean as closely as desired by a function in $C_0^\infty(\mathbb{R}^m)$. We note that not all functions in $\mathscr{L}^\infty(X)$ admit such approximations (see Exercise 1).

Theorem *For $1 \leqslant p < +\infty$, every function f in $\mathscr{L}^p(X)$ can be approximated (in the $\mathscr{L}^p$-norm) as closely as desired by a function in $C_0^\infty(\mathbb{R}^m)$.*

Proof As in the proof of Theorem 9.2.2, we will assume that $X = \mathbb{R}^m$. By this theorem, every function in $\mathscr{L}^p(\mathbb{R}^m)$ can be approximated as closely as desired by step functions. Therefore, it is sufficient to prove the theorem for the characteristic functions of cells. For this, we use the theorem on a smooth descent (see Sect. 8.1.7), which says that, for an arbitrary cell P and every positive ε, there is a function φ in $C_0^\infty(\mathbb{R}^m)$ such that

$$0 \leqslant \varphi \leqslant 1, \qquad \operatorname{supp}(\varphi) \subset P_\varepsilon, \quad \text{and} \quad \varphi(x) = 1 \quad \text{for } x \in P,$$

where P_ε is an ε-neighborhood of P.

Since $\chi_P - \varphi = 0$ on P and outside P_ε, we have

$$\|\chi_P - \varphi\|_p^p = \int_{P_\varepsilon \setminus P} \varphi^p \, d\lambda_m \leqslant \lambda_m(P_\varepsilon \setminus P).$$

Obviously, the right-hand side of this inequality is arbitrarily small along with ε. Thus, the existence of the required approximation for χ_P has been proved, which completes the proof of the theorem. □

Corollary *Let $X \subset \mathbb{R}^m$ be a bounded measurable set, $1 \leqslant p < +\infty$, and $f \in \mathscr{L}^p(X)$. For every $\varepsilon > 0$, there is a polynomial P such that $\|f - P\|_p < \varepsilon$.*

Proof By the theorem, there exist a function $\varphi \in C_0^\infty(\mathbb{R}^m)$ such that $\|f - \varphi\|_p < \varepsilon$. By the Weierstrass approximation theorem (see Corollary 1 of Sect. 7.6.4), the function φ can be uniformly approximated by a polynomial P on the closure of X, and so $|\varphi(x) - P(x)| < \varepsilon$ on X. Then

$$\|f - P\|_p \leqslant \|f - \varphi\|_p + \|\varphi - P\|_p < \varepsilon + \varepsilon\big(\lambda_m(X)\big)^{1/p}.$$

Thus, the function f can be approximated in the $\mathscr{L}^p$-norm as closely as desired by a polynomial. □

9.2.4 Here we establish one unexpected and interesting feature of the functions in $\mathscr{L}^p(\mathbb{R}^m)$ with finite p. Although such functions can be discontinuous everywhere, some characteristics of their global behavior make it possible to speak of "continuity in mean". For the precise formulation of this statement, we need the concept of a shift of a function.

Definition Let $f \in \mathscr{L}^0(\mathbb{R}^m)$ and $h \in \mathbb{R}^m$. By the *shift* of a function f by a vector h, we mean the function f_h defined by the formula

$$f_h(x) = f(x - h) \quad \big(x \in \mathbb{R}^m\big).$$

Since Lebesgue measure is shift invariant, it is clear that $f_h \in \mathscr{L}^p(\mathbb{R}^m)$ along with f, and $\|f_h\|_p = \|f\|_p$.

Theorem (On continuity in the mean) *Let $1 \leqslant p < +\infty$ and $f \in \mathscr{L}^p(\mathbb{R}^m)$. Then*

$$\|f - f_h\|_p = \left(\int_{\mathbb{R}^m} |f(x) - f(x-h)|^p \, dx\right)^{1/p} \underset{h \to 0}{\longrightarrow} 0.$$

Thus, the map $h \mapsto f_h$ from $\mathbb{R}^m$ to $\mathscr{L}^p(\mathbb{R}^m)$ is continuous.

Proof By Theorem 9.2.2, the function f can be approximated in $\mathscr{L}^p(\mathbb{R}^m)$ as closely as desired by a step function g. Obviously,

$$\|f - f_h\|_p \leqslant \|f - g\|_p + \|g - g_h\|_p + \|f_h - g_h\|_p = 2\|f - g\|_p + \|g - g_h\|_p.$$

Therefore, it is sufficient to prove the theorem for step functions. Since every such function is a linear combination of the characteristic functions of cells, it remains for us to verify the assertion of the theorem for an arbitrary function f of the form $f = \chi_P$, where P is a cell. As the reader can easily verify, this assertion follows from Lebesgue's theorem. However, it is possible to dispense with the reference to this theorem. It is clear that f_h is simply the characteristic function of the shifted cell $P_h = \{x + h \mid x \in P\}$. Since $|f - f_h| = 0$ outside the union $(P \setminus P_h) \cup (P_h \setminus P)$ and $|f - f_h| = 1$ on $(P \setminus P_h) \cup (P_h \setminus P)$, we see that

$$\|f - f_h\|_p^p = \int_{P\setminus P_h} 1\, d\lambda_m + \int_{P_h\setminus P} 1\, d\lambda_m = \lambda_m(P \setminus P_h) + \lambda_m(P_h \setminus P).$$

The right-hand side of this equation tends to zero as $h \to 0$. □

As the example of the function $f = \chi_{(0,1)}$ shows, the theorem is valid only for finite p.

We also point out the following statement concerning the continuity in mean of periodic functions.

Corollary *Let $1 \leqslant p < +\infty$, and let f be a measurable function defined on $\mathbb{R}^m$ and 2π-periodic in each variable. If $\int_{(-\pi,\pi)^m} |f(x)|^p\, dx < +\infty$, then*

$$\int_{(-\pi,\pi)^m} |f(x) - f(x-h)|^p\, dx \underset{h\to 0}{\longrightarrow} 0.$$

Proof Let g coincide with f on the cube $(-2\pi, 2\pi)^m$ and equal zero outside the cube. Then $g \in \mathscr{L}^p(\mathbb{R}^m)$ and $f(x) - f(x-h) = g(x) - g(x-h)$ for $x \in (-\pi, \pi)^m$ and $\|h\| < \pi$. Therefore,

$$\begin{aligned}\int_{(-\pi,\pi)^m} |f(x) - f(x-h)|^p\, dx &= \int_{(-\pi,\pi)^m} |g(x) - g(x-h)|^p\, dx \\ &\leqslant \|g - g_h\|_p^p \underset{h\to 0}{\longrightarrow} 0.\end{aligned}$$

□

9.2.5 Now, we use approximation to establish a result which plays an important role in harmonic analysis.

Theorem (Riemann–Lebesgue) *Let $X \subset \mathbb{R}^m$ and $f \in \mathscr{L}^1(X)$. Then*

$$I_f(y) = \int_X f(x)e^{i\langle x,y\rangle}\, dx \underset{\|y\|\to+\infty}{\longrightarrow} 0$$

(the symbol $\langle x, y\rangle$ denotes the scalar product of vectors x and y in $\mathbb{R}^m$).

If f is a function summable on a finite interval $[a, b]$, then the theorem says that

$$\int_a^b f(x)e^{ixy}\, dx \underset{|y|\to+\infty}{\longrightarrow} 0.$$

The reason the integral converges to zero, of course, is that, for large y, the real and imaginary parts of e^{ixy} oscillate rapidly near zero. If f is continuous, then the result to be proved is intuitively absolutely clear: the integral over $[a, b]$ can be split into a sum of integrals over intervals of length $2\pi/|y|$. For large $|y|$, the function f is almost constant on each such interval, and on the left and right halves of each interval the oscillating factor e^{ixy} assumes values of opposite sign. Therefore, the integrals over these intervals "almost cancel each other out", which leads to the result to be proved. It is surprising, however, that $I_f(y) \underset{|y|\to+\infty}{\longrightarrow} 0$ not only for continuous functions but for all summable functions, which may not have points of continuity at all.

We give two proofs of the Riemann–Lebesgue theorem. In these proofs, despite the intuitive clearness of the reasoning, we do not use the concept of continuity. It turns out that it is technically more convenient to rely on the result proved above that each summable function can be approximated by step functions. In the second proof, we establish not only the purely qualitative result formulated in the theorem but also obtain an estimate for the integral $I_f(y)$.

Proof I As in the proof of Theorems 9.2.2 and 9.2.3, we can extend f by zero outside X. Therefore, without loss of generality, we will assume that $X = \mathbb{R}^m$.

We divide the proof into several steps, gradually complicating the function f.

(1) Let f be the characteristic function of a cell P of the form $P = [a_1, b_1) \times \cdots \times [a_m, b_m)$. Then, for a vector $y = (y_1, \ldots, y_m)$ in $\mathbb{R}^m$, we have

$$I_f(y) = \prod_{k=1}^{m} \int_{a_k}^{b_k} e^{ix_k y_k}\, dx_k = \prod_{k=1}^{m} \frac{\exp(ib_k y_k) - \exp(ia_k y_k)}{iy_k}.$$

It is clear that all factors on the right are bounded, and if $\|y\| \to +\infty$, then the right-hand side of this equation tends to zero since the absolute value of at least one of the denominators is not less than $\|y\|/m$. Thus, the assertion has been proved for characteristic functions of cells.

(2) Let f be a step function. In this case, the statement is obvious since such an f is a linear combination of the characteristic functions of cells.

(3) Now, let f be an arbitrary function in $\mathscr{L}^1(\mathbb{R}^m)$. Since $I_f(y) = I_g(y) + I_{f-g}(y)$ and $|I_{f-g}(y)| \leqslant \|f - g\|_1$ for every function g, we obtain

$$|I_f(y)| \leqslant |I_g(y)| + |I_{f-g}(y)| \leqslant |I_g(y)| + \|f - g\|_1.$$

By Theorem 9.2.2, the summand $\|f - g\|_1$ can be made arbitrarily small by the choice of a step function g. Then, fixing g, we can make the summand $|I_g(y)|$ arbitrarily small for all vectors y with a sufficiently large norm. □

Since the shift in the space $\mathscr{L}^1(\mathbb{R}^m)$ is continuous, we can give one more (very short but somewhat formal) proof of the Riemann–Lebesgue theorem. This proof is based on a method often used in harmonic analysis.

Proof II We assume that $X = \mathbb{R}^m$. Let $h = \pi y/\|y\|^2$. It is clear that $\|h\| = \pi/\|y\| \to 0$ as $\|y\| \to +\infty$. After the change of variable $x \mapsto x - h$, we obtain

$$\int_{\mathbb{R}^m} f(x)e^{i\langle x,y\rangle}\,dx = \int_{\mathbb{R}^m} f(x-h)e^{-i\pi+i\langle x,y\rangle}\,dx = -\int_{\mathbb{R}^m} f_h(x)e^{i\langle x,y\rangle}\,dx.$$

Consequently,

$$2\int_{\mathbb{R}^m} f(x)e^{i\langle x,y\rangle}\,dx = \int_{\mathbb{R}^m} \big(f(x) - f_h(x)\big)e^{i\langle x,y\rangle}\,dx.$$

Therefore,

$$2\left|\int_{\mathbb{R}^m} f(x)e^{i\langle x,y\rangle}\,dx\right| \leqslant \int_{\mathbb{R}^m} \big|f(x) - f_h(x)\big|\,dx.$$

The integral on the right-hand side of this inequality tends to zero as $h \to 0$ since the function f is continuous in mean. □

From the second proof of the theorem, it is clear that the fast convergence to zero of the norm $\|f - f_h\|_1$ as $h \to 0$ implies the fast decrease of the integral $I_f(y)$ as $\|y\| \to +\infty$. For a smooth function f with compact support we, obviously, have $\|f - f_h\|_1 = O(\|h\|)$ and so $I_f(y) = O(1/\|y\|)$. In the one-dimensional case where $X = [a, b]$, the estimate $I_f(y) = O(1/|y|)$ is valid not only for a smooth but also for an absolutely continuous function f, which can easily be verified by integration by parts. However, in some problems (see Example 2 of Sect. 10.5.2) we need estimates for I_f valid under less restrictive assumptions. In the following example, we obtain one such result.

Example Let us find the rate at which the integral $I_f(y)$ tends to zero as $y \to +\infty$ if the function f is defined on the interval $X = [0, 1)$ and has the form $f(x) = \frac{F(x)}{\sqrt{1-x^2}}$, where $F \in C^1([0, 1])$.

We represent f in the form $f(x) = \frac{F(1)}{\sqrt{2(1-x)}} + g(x)$, where the function $g(x) = \frac{1}{\sqrt{1-x}}\big(\frac{F(x)}{\sqrt{1+x}} - \frac{F(1)}{\sqrt{2}}\big)$ is absolutely continuous on $[0, 1)$. Therefore, $I_g(y) = O(1/y)$, and we obtain that

$$\begin{aligned} I_f(y) &= \frac{F(1)}{\sqrt{2}}\int_0^1 \frac{e^{iyx}}{\sqrt{1-x}}\,dx + O\Big(\frac{1}{y}\Big) = \frac{F(1)}{\sqrt{2}}\int_0^1 e^{iy(1-t)}\frac{dt}{\sqrt{t}} + O\Big(\frac{1}{y}\Big) \\ &= \frac{F(1)}{\sqrt{2y}}e^{iy}\int_0^y e^{-iu}\frac{du}{\sqrt{u}} + O\Big(\frac{1}{y}\Big). \end{aligned}$$

Since the integral $\int_0^y e^{-iu}\frac{du}{\sqrt{u}}$ tends to the Fresnel integral $\int_0^\infty e^{-iu}\frac{du}{\sqrt{u}}$ as $y \to +\infty$, which is equal to $(1-i)\sqrt{\frac{\pi}{2}}$ (see Example 1 of Sect. 7.4.8), we have

$$I_f(y) = \frac{F(1)}{\sqrt{2y}}e^{iy}\bigg((1-i)\sqrt{\frac{\pi}{2}} - \int_y^\infty e^{-iu}\frac{du}{\sqrt{u}}\bigg) + O\Big(\frac{1}{y}\Big)$$

$$= \frac{F(1)}{2} e^{iy}(1-i)\sqrt{\frac{\pi}{y}} + O\left(\frac{1}{y}\right). \tag{1}$$

We use this formula to find the asymptotic of the integrals

$$C(y) = \int_{-1}^{1} \frac{e^{-iyx}}{\sqrt{1-x^2}}\,dx \quad \text{and} \quad S(y) = \int_{-1}^{1} \sqrt{1-x^2}e^{-iyx}\,dx.$$

Obviously,

$$C(y) = 2\int_0^1 \frac{\cos xy}{\sqrt{1-x^2}}\,dx = 2\mathcal{R}e \int_0^1 \frac{e^{ixy}}{\sqrt{1-x^2}}\,dx,$$

$$S(y) = -\frac{1}{iy}\int_{-1}^{1} \frac{x}{\sqrt{1-x^2}}e^{-ixy} = \frac{2}{y}\mathcal{I}m \int_0^1 \frac{x}{\sqrt{1-x^2}}e^{ixy}\,dx.$$

Applying formula (1) (in both cases $F(1) = 1$), we obtain by direct calculation that

$$C(y) = \sqrt{\frac{\pi}{y}}(\sin y + \cos y) + O\left(\frac{1}{y}\right),$$

$$S(y) = \frac{1}{y}\sqrt{\frac{\pi}{y}}(\sin y - \cos y) + O\left(\frac{1}{y^2}\right)$$

as $y \to +\infty$.

EXERCISES

1. Verify that the condition $p < +\infty$ in Theorems 9.2.2 and 9.2.3 cannot be dropped.
In problems 2–6, we assume that $1 \leqslant p < +\infty$.
2. Prove that Theorems 9.2.2 and 9.2.3 remain valid for every Borel measure finite on the cells.
3. Let μ be the standard extension of a measure from a semiring $\mathscr{P}$ of subsets of a set X. Repeating the reasoning in the proof of Theorem 9.2.2, prove that every function in $\mathscr{L}^p(X,\mu)$ can be approximated in mean as closely as desired by linear combinations of the characteristic functions of sets belonging to $\mathscr{P}$.
4. Let $(X, \mathfrak{A}, \mu)$ and $(Y, \mathfrak{B}, \nu)$ be spaces with σ-finite measures. Use the previous exercise to prove that every function in $\mathscr{L}^p(X \times Y, \mu \times \nu)$ can be approximated as closely as desired by linear combinations of products of the form $\varphi(x)\psi(y)$, where $\varphi \in \mathscr{L}^p(X,\mu)$ and $\psi \in \mathscr{L}^p(Y,\nu)$.
5. Prove that every function in $\mathscr{L}^p(\mathbb{R}^m,\mu)$, where μ is an arbitrary Borel measure finite on the cells, can be approximated as closely as desired by linear combinations of products of the form $\psi_1(x_1)\cdots\psi_m(x_m)$, where $\psi_1,\ldots,\psi_m \in C_0^\infty(\mathbb{R})$.
6. Prove that every function in $\mathscr{L}^p(\mathbb{R}^m)$ can be approximated as closely as desired by rational functions (i.e., by quotients of the form P/Q, where P and Q are algebraic polynomials in m variables and $Q \neq 0$).

7. Prove that $\frac{1}{uv} + \frac{1}{u-v}(\frac{1}{u}e^{iu} - \frac{1}{v}e^{iv}) \to 0$ as $u^2 + v^2 \to +\infty$. *Hint*. Apply the Riemann–Lebesgue theorem to the characteristic function of the triangle with vertices $(0,0)$, $(1,0)$ and $(0,1)$.

8. Let f be a function in $\mathscr{L}^0(\mathbb{R}^m)$ such that the norms $\|f_h - f\|_p$ $(h \in \mathbb{R}^m)$ are bounded for some $p \in [1, +\infty)$. Prove that f can be represented in the form $f = \varphi + \text{const}$, where $\varphi \in \mathscr{L}^p(\mathbb{R}^m)$. *Hint*. Verifying that $f \in \mathscr{L}_{\text{loc}}(\mathbb{R}^m)$, prove that the mean value $\lim_{R\to+\infty} \frac{1}{(2R)^m}\int_{[-R,R]^m} f(x)\,dx$ exists and is finite.

9. Let $0 < p < 1$, $F \in C^1([0,1])$. By analogy with Example 9.2.5, find the asymptotics of the integral $\int_0^1 \frac{F(x)}{(1-x^2)^p} e^{ixy}\,dx$.

10. Let $f \in \mathscr{L}^1(X)$, $X \subset \mathbb{R}^m$. Show that $\int_X f(x)e^{it\|x\|}\,dx \to 0$ as $t \to \pm\infty$.

9.3 ⋆Convolution and Approximate Identities in the Spaces $\mathscr{L}^p$

In this section, we supplement the information on convolution obtained in Sects. 7.5–7.6. All functions under consideration are assumed to be measurable (and, in general, complex-valued).

In the periodic and non-periodic cases, the properties of convolution are similar. Therefore, we consider only the non-periodic case in detail. In the periodic case, we confine ourselves to the statements of results, which we give for convenience of reference.

9.3.1 First, we generalize Theorem 7.5.2 on the existence of the convolution of two summable functions.

Theorem *Let $f \in \mathscr{L}^p(\mathbb{R}^m)$ and $g \in \mathscr{L}^1(\mathbb{R}^m)$, $1 \leqslant p \leqslant +\infty$. Then the convolution $f * g$ exists, belongs to $\mathscr{L}^p(\mathbb{R}^m)$, and*

$$\|f * g\|_p \leqslant \|g\|_1 \cdot \|f\|_p.$$

Proof Following the scheme of the proof of Theorem 7.5.2, we first prove the existence of the convolution, i.e., that the function $H(x) = \int_{\mathbb{R}^m} |f(x-y)g(y)|\,dy$ is almost everywhere finite. Moreover, we verify that $H \in \mathscr{L}^p(\mathbb{R}^m)$.

This is obvious for $p = +\infty$ since

$$H(x) \leqslant \|f\|_\infty \int_{\mathbb{R}^m} |g(y)|\,dx = \|f\|_\infty \|g\|_1.$$

If $1 < p < +\infty$, then, assuming that $\frac{1}{p} + \frac{1}{q} = 1$, we obtain by Hölder's inequality that

$$H(x) = \int_{\mathbb{R}^m} \left(|f(x-y)||g(y)|^{\frac{1}{p}}\right)|g(y)|^{\frac{1}{q}}\,dy$$
$$\leqslant \left(\int_{\mathbb{R}^m} |f(x-y)|^p |g(y)|\,dy\right)^{\frac{1}{p}} \left(\int_{\mathbb{R}^m} |g(y)|\,dy\right)^{\frac{1}{q}}$$

(for $p=1$ and $q=+\infty$, this inequality becomes equality). Consequently,

$$H^p(x) \leqslant \|g\|_1^{p/q} \int_{\mathbb{R}^m} |f(x-y)|^p |g(y)|\,dy = \|g\|_1^{p/q} \big(|f|^p * |g|\big)(x).$$

Thus, the function H^p is dominated (with the coefficient $\|g\|_1^{p/q}$) by the convolution of the summable functions $|f|^p$ and $|g|$. By Theorem 7.5.2, this function is summable and the following estimate is valid:

$$\int_{\mathbb{R}^m} H^p(x)\,dx \leqslant \|g\|_1^{p/q} \big\| |f|^p \big\|_1 \|g\|_1 = \|g\|_1^{1+p/q} \|f\|_p^p.$$

In particular, it follows that $H(x) < +\infty$ almost everywhere, and so the existence of the convolution is proved. Moreover, since $|(f*g)(x)| \leqslant H(x)$, we have $f*g \in \mathscr{L}^p(\mathbb{R}^m)$ and $\|f*g\|_p \leqslant \|H\|_p \leqslant \|g\|_1^{\frac{1}{p}+\frac{1}{q}} \cdot \|f\|_p = \|g\|_1 \cdot \|f\|_p$, which is what was to be proved. □

9.3.2 As we saw in Chap. 7, the degree of smoothness of a function can only increase under convolution. Now, we use the theorem on continuity in the mean to supplement this result and to prove that, in a wide range of cases, (in general) the convolution of discontinuous functions is continuous.

Theorem *Let $f \in \mathscr{L}^p(\mathbb{R}^m)$ and $g \in \mathscr{L}^q(\mathbb{R}^m)$, where $1 \leqslant p, q \leqslant +\infty$ and $\frac{1}{p}+\frac{1}{q}=1$. Then the convolution $f*g$ exists, is uniformly continuous on $\mathbb{R}^m$, and the inequality*

$$\big|(f*g)(x)\big| \leqslant \|f\|_p \, \|g\|_q \tag{1}$$

holds for every x in $\mathbb{R}^m$.

Proof By the symmetry of f and g, we may assume that $p < +\infty$. To verify that the convolution exists, we prove that the integral $H(x)$ is finite (this notation was introduced in the proof of Theorem 9.3.1). For $1 < p < +\infty$, Hölder's inequality yields

$$\begin{aligned} H(x) &\leqslant \left(\int_{\mathbb{R}^m} |f(x-y)|^p\,dy\right)^{\frac{1}{p}} \left(\int_{\mathbb{R}^m} |g(y)|^q\,dy\right)^{\frac{1}{q}} \\ &= \left(\int_{\mathbb{R}^m} |f(y)|^p\,dy\right)^{\frac{1}{p}} \left(\int_{\mathbb{R}^m} |g(y)|^q\,dy\right)^{\frac{1}{q}}; \end{aligned}$$

for $p=1$, this inequality takes the form $H(x) \leqslant \|f\|_1 \operatorname{esssup}|g|$. Thus, $H(x) \leqslant \|f\|_p \, \|g\|_q < +\infty$, which proves that the convolution exists. Since $|(f*g)(x)| \leqslant H(x)$, we obtain inequality (1). It remains to verify that the function $u = f*g$

is uniformly continuous. For this, we estimate the difference $u(x-h)-u(x) = u_h(x)-u(x) = (f_h - f) * g(x)$ by means of inequality (1) as follows:

$$\left|u(x-h)-u(x)\right| = \left|(f_h - f) * g(x)\right| \leqslant \|f_h - f\|_p \|g\|_q.$$

The right-hand side of this inequality does not depend on x and tends to zero as $h \to 0$, since the function f is continuous in mean. □

Corollary *If $f \in \mathscr{L}_{\mathrm{loc}}(\mathbb{R}^m)$ and the function g is bounded and compactly supported, then the convolution $f * g$ is continuous.*

Proof If f is summable, then the continuity of the convolution is established in the theorem. In the general case, the continuity of $f * g$ in an arbitrary ball $B(0, R)$ follows from the fact that the convolutions $f * g$ and $f_1 * g$ coincide in this ball. Here, f_1 is the function equal to zero outside the ball $B(0, 2R)$ and coinciding with f in this ball (see the truncation lemma of Sect. 7.5.4). □

Theorems 9.3.1 and 9.3.2 are the main special cases of a more general statement known as *Young's inequality*.[3] It consists of the following.

Let $f \in \mathscr{L}^p(\mathbb{R}^m)$ and $g \in \mathscr{L}^q(\mathbb{R}^m)$, where $p, q \geqslant 1$. We assume that $\frac{1}{p}+\frac{1}{q} \geqslant 1$ and consider an r, $1 \leqslant r \leqslant +\infty$, such that

$$\frac{1}{p}+\frac{1}{q} = 1+\frac{1}{r}.$$

Then the convolution $f * q$ exists and $\|f * g\|_r \leqslant \|f\|_p \cdot \|g\|_q$ (see also Exercise 10).

It follows from $\frac{1}{p}+\frac{1}{q} = 1+\frac{1}{r}$ that $p, q \leqslant r$. We assume that $1 < p$, $q < r < \infty$ (the remaining cases correspond to Theorems 9.3.1 and 9.3.2). In addition, we assume that the functions f and g are non-negative since otherwise they can be replaced by $|f|$ and $|g|$.

The idea of the remaining calculation is to majorize the convolution $f * g$ (the existence of which has to be proved) by the convolution of the summable functions $F = f^p$ and $G = g^q$. To this end, we represent the product $f(y)g(x-y)$ as a product of three factors,

$$f(y)g(x-y) = \big(F(y)G(x-y)\big)^{\frac{1}{r}} \cdot F^{\frac{1}{p}-\frac{1}{r}}(y) \cdot G^{\frac{1}{q}-\frac{1}{r}}(x-y),$$

and integrate this identity for a fixed x. Since $\frac{1}{r}+(\frac{1}{p}-\frac{1}{r})+(\frac{1}{q}-\frac{1}{r}) = 1$ by the definition of r, Hölder's inequality for three functions (see Corollary 2, Sect. 4.4.5) with exponents r, $(\frac{1}{p}-\frac{1}{r})^{-1}$ and $(\frac{1}{q}-\frac{1}{r})^{-1}$ gives us the following estimate from above:

$$\int_{\mathbb{R}^m} f(y)g(x-y)\,dy \leqslant \big((F * G)(x)\big)^{\frac{1}{r}} \cdot \|F\|_1^{\frac{1}{p}-\frac{1}{r}} \cdot \|G\|_1^{\frac{1}{q}-\frac{1}{r}}.$$

[3] William Henry Young (1863–1942)—English mathematician.

Since the functions F and G are summable, their convolution is finite almost everywhere by Theorem 9.3.1. Therefore, the left-hand side of the latter inequality is finite for almost all x, proving the existence of the convolution $f * g$. The convolution is measurable by Lemma 7.5.2, and, as we have just proved, satisfies the inequality

$$0 \leqslant (f * g)(x) \leqslant \|F\|_1^{\frac{1}{p}-\frac{1}{r}} \cdot \|G\|_1^{\frac{1}{q}-\frac{1}{r}} \cdot \big((F * G)(x)\big)^{\frac{1}{r}}.$$

Consequently,

$$\begin{aligned}\|f * g\|_r^r &\leqslant \|F\|_1^{\frac{r}{p}-1} \cdot \|G\|_1^{\frac{r}{q}-1} \|F * G\|_1 \leqslant \|F\|_1^{\frac{r}{p}-1} \cdot \|G\|_1^{\frac{r}{q}-1} \cdot \|F\|_1 \|G\|_1 \\ &= \|F\|_1^{\frac{r}{p}} \cdot \|G\|_1^{\frac{r}{q}}\end{aligned}$$

(here we applied the inequality from Theorem 9.3.1). Thus,

$$\|f * g\|_r \leqslant \|F\|_1^{\frac{1}{p}} \cdot \|G\|_1^{\frac{1}{q}} = \|f\|_p \cdot \|g\|_q.$$

9.3.3 In this section, we consider the properties of the convolution of a function in $\mathscr{L}^p(\mathbb{R}^m)$ with an approximate identity. We recall (see Sect. 7.6.1) that an approximate identity in $\mathbb{R}^m$ (as $t \to t_0$) is a family of functions $\{\omega_t\}_{t>0}$ satisfying the following conditions:

$$\text{(a)}\quad \omega_t \geqslant 0,$$

$$\text{(b)}\quad \int_{\mathbb{R}^m} \omega_t(x)\,dx = 1,$$

$$\text{(c)}\quad \int_{\|x\|>\delta} \omega_t(x)\,dx \underset{t\to t_0}{\longrightarrow} 0 \quad \text{for every } \delta > 0.$$

From Theorem 7.6.3 it follows, in particular, that if a bounded function f is continuous on $\mathbb{R}^m$, then the convolutions $f * \omega_t$ converge pointwise to f as $t \to t_0$ (the convergence is uniform if the function f is uniformly continuous on $\mathbb{R}^m$). For the functions in $\mathscr{L}^p(\mathbb{R}^m)$, this statement is modified as follows:

Theorem *Let $\{\omega_t\}_{t>0}$ be an approximate identity in $\mathbb{R}^m$ as $t \to t_0$ and let $f \in \mathscr{L}^p(\mathbb{R}^m)$, $1 \leqslant p < +\infty$. Then the functions $f_t = f * \omega_t$ converge to f as $t \to t_0$ in the $\mathscr{L}^p$-norm.*

Proof It is clear that

$$\begin{aligned}\big|f_t(x) - f(x)\big| &\leqslant \int_{\mathbb{R}^m} \big|f(x-y) - f(x)\big|\omega_t(y)\,dy \\ &= \int_{\mathbb{R}^m} \big|f(x-y) - f(x)\big|\omega_t^{\frac{1}{p}}(y)\omega_t^{\frac{1}{q}}(y)\,dy.\end{aligned}$$

By Hölder's inequality, we obtain

$$
\begin{aligned}
|f_t(x) - f(x)| &\leqslant \left(\int_{\mathbb{R}^m} |f(x-y) - f(x)|^p \omega_t(y)\,dy\right)^{\frac{1}{p}} \left(\int_{\mathbb{R}^m} \omega_t(y)\,dy\right)^{\frac{1}{q}} \\
&= \left(\int_{\mathbb{R}^m} |f(x-y) - f(x)|^p \omega_t(y)\,dy\right)^{\frac{1}{p}}.
\end{aligned}
$$

Raising to the pth power and changing the order of integration, we obtain

$$
\begin{aligned}
\int_{\mathbb{R}^m} |f_t(x) - f(x)|^p\,dx &\leqslant \int_{\mathbb{R}^m} \left(\int_{\mathbb{R}^m} |f(x-y) - f(x)|^p \omega_t(y)\,dy\right) dx \\
&= \int_{\mathbb{R}^m} \omega_t(y) \int_{\mathbb{R}^m} |f(x-y) - f(x)|^p\,dx\,dy \\
&= \int_{\mathbb{R}^m} g(y)\omega_t(y)\,dy,
\end{aligned}
$$

where $g(y) = \int_{\mathbb{R}^m} |f(x-y) - f(x)|^p\,dx \underset{y\to 0}{\longrightarrow} 0$ since f is continuous in mean. By the Corollary of Theorem 7.6.3, the right-hand side of the last inequality tends to zero as $t \to t_0$. □

The result obtained allows us to give one more proof of the important Theorem 9.2.3.

Corollary *Let* $1 \leqslant p < +\infty$, $f \in \mathscr{L}^p(\mathbb{R}^m)$, *and* $\varepsilon > 0$. *Then there is a function* $\varphi \in C_0^\infty(\mathbb{R}^m)$ *such that* $\|f - \varphi\|_p < \varepsilon$.

Proof Let $\{\omega_t\}_{t>0}$ be a Sobolev approximate identity in $\mathbb{R}^m$. By the corollary of Theorem 7.5.4, the convolution $f * \omega_t$ is infinitely differentiable. By the theorem just proved, we have $\|f - f * \omega_t\|_p \underset{t\to t_0}{\longrightarrow} 0$. We fix a t such that $\|f - f * \omega_t\|_p < \varepsilon/2$ and put $g = f * \omega_t$.

To complete the proof, it remains to approximate g by a function of class $C_0^\infty(\mathbb{R}^m)$ within $\varepsilon/2$. This can be done by multiplying g by a function obtained by smoothing the characteristic function of a ball of sufficiently large radius. We leave it to the reader to fill in the details. □

9.3.4 Here, we supplement the results of the previous section by the study of the convergence of the convolutions $f_t = f * \omega_t$ to the function f in $\mathscr{L}^p(\mathbb{R}^m)$ not with respect to the norm of this space but almost everywhere. To obtain the required result, we have to impose an additional restriction on the approximate identity and require that the functions ω_t have sufficiently a nice ("hump-shaped") majorant. More precisely, we assume that the estimates

$$
\omega_t(x) \leqslant \psi_t(\|x\|), \qquad \int_{\mathbb{R}^m} \psi_t(\|x\|)\,dx \leqslant C, \tag{2}
$$

where ψ_t are decreasing functions on $(0, +\infty)$ and C is a constant, are valid for all $t > 0$. In particular, if ψ_t has the form $\psi_t(r) = t^{-m}\Psi(r/t)$, where Ψ is a decreasing function on $(0, +\infty)$, then the second condition in (2) reduces to the inequality $\int_0^\infty r^{m-1}\Psi(r)\,dr < +\infty$.

Theorem *Let $\{\omega_t\}_{t>0}$ be an approximate identity in $\mathbb{R}^m$ as $t \to t_0$ satisfying condition (2). Then if $f \in \mathcal{L}^p(\mathbb{R}^m)$, $1 \leqslant p \leqslant +\infty$, then the convolutions $f_t = f * \omega_t$ converge to f as $t \to t_0$ at each Lebesgue point of f, and, consequently, almost everywhere.*

Proof For the characteristic function of a ball, the assertion of the theorem follows immediately from properties (b) and (c) of an approximate identity. Therefore, we may assume, without loss of generality, that x is a Lebesgue point of f and $f(x) = 0$ (otherwise, it is necessary to replace the function f by the difference $f - f(x)\chi$, where χ is the characteristic function of an arbitrary ball centered at x). Thus, we will prove that $f_t(x) \to 0$ as $t \to t_0$ if

$$\frac{1}{r^m}\int_{B(r)} \left|f(x-y)\right| dy \underset{r\to 0}{\longrightarrow} 0 \tag{3}$$

(as usual, $B(r)$ is the ball with radius r and center zero).

Since $f_t(x) = \int_{\mathbb{R}^m} f(x-y)\omega_t(y)\,dy$, we have

$$\left|f_t(x)\right| \leqslant \int_{\mathbb{R}^m} \left|f(x-y)\right|\omega_t(y)\,dy = \int_{B(\varepsilon)} \cdots + \int_{\mathbb{R}^m\setminus B(\varepsilon)} \cdots = I(\varepsilon) + J(\varepsilon)$$

for every $\varepsilon > 0$ (the freedom in the choice of this parameter will be used later). We estimate the integrals $I(\varepsilon)$ and $J(\varepsilon)$ separately. We represent the first integral as the sum of integrals over the spherical layers $S_j = B(\varepsilon_{j-1}) \setminus B(\varepsilon_j)$, where $\varepsilon_j = \varepsilon/2^j$ for $j \in \mathbb{N}$ and $\varepsilon_0 = \varepsilon$. We note that $\lambda_m(S_j) = \beta_m\varepsilon_j^m$, where the coefficient β_m depends only on m. Since $\omega_t(y) \leqslant \psi_t(\|y\|) \leqslant \psi_t(\varepsilon_j)$ on the layer S_j, we obtain

$$\int_{S_j} \left|f(x-y)\right|\omega_t(y)\,dy \leqslant \psi_t(\varepsilon_j)\int_{B(\varepsilon_{j-1})} \left|f(x-y)\right| dy.$$

We put

$$\Delta(\varepsilon) = \sup_{r\leqslant\varepsilon}\frac{1}{r^m}\int_{B(r)} \left|f(x-y)\right| dy.$$

Then

$$\int_{S_j} \left|f(x-y)\right|\omega_t(y)\,dy \leqslant \varepsilon_{j-1}^m\Delta(\varepsilon_{j-1})\psi_t(\varepsilon_j) \leqslant (4\varepsilon_{j+1})^m\Delta(\varepsilon)\psi_t(\varepsilon_j)$$

$$\leqslant \frac{4^m}{\beta_m}\Delta(\varepsilon)\int_{S_{j+1}} \psi_t\bigl(\|y\|\bigr)\,dy.$$

Adding these inequalities and taking into account (2), we obtain

$$I(\varepsilon) = \sum_{j=1}^{\infty} \int_{S_j} \left|f(x-y)\right| \omega_t(y)\, dy \leqslant \frac{4^m}{\beta_m} \Delta(\varepsilon) \int_{B(\varepsilon)} \psi_t\big(\|y\|\big)\, dy \leqslant \frac{4^m}{\beta_m} C \Delta(\varepsilon).$$

It follows from relation (3) that $\Delta(\varepsilon)$, and along with it $I(\varepsilon)$, tends to zero as $\varepsilon \to 0$.

Now we estimate the integral $J(\varepsilon)$. If $p = +\infty$, then everything is very simple: $J(\varepsilon) \leqslant \|f\|_\infty \int_{\mathbb{R}^m \setminus B(\varepsilon)} \omega_t(y)\, dy$ and, therefore,

$$\left|f_t(x)\right| \leqslant I(\varepsilon) + J(\varepsilon) \leqslant \frac{4^m}{\beta_m} C \Delta(\varepsilon) + \|f\|_\infty \int_{\mathbb{R}^m \setminus B(\varepsilon)} \omega_t(y)\, dy.$$

The first summand can be made arbitrarily small by an appropriate choice of ε. For a fixed ε, the second summand tends to zero as $t \to t_0$ by property (c) of an approximate identity.

It is harder to estimate the integral $J(\varepsilon)$ if $1 \leqslant p < +\infty$ because the values of $|f(x-y)|$ can be arbitrarily large on some part of the set $\mathbb{R}^m \setminus B(\varepsilon)$. We consider this part and put $E_R = \{y \in \mathbb{R}^m \mid |f(x-y)| > R\}$, where R is a large numerical parameter. From Chebyshev's inequality (see Sect. 4.4.4), it follows that the measure of E_R is infinitesimal as $R \to +\infty$. By absolute continuity, the integral $\int_{E_R} |f(x-y)|^p\, dy$ is also small. From the first inequality in (2) and the definition of the set E_R with $R > 1$, we obtain the estimate

$$\left|f(x-y)\right| \omega_t(y) \leqslant \begin{cases} \psi_t(\varepsilon)|f(x-y)|^p & \text{for } y \in E_R,\ \|y\| > \varepsilon, \\ R\omega_t(y) & \text{for } y \notin E_R. \end{cases}$$

Moreover, by the second inequality in (2), we have

$$C \geqslant \int_{B(\varepsilon)} \psi_t\big(\|y\|\big)\, dy \geqslant \psi_t(\varepsilon) \lambda_m\big(B(\varepsilon)\big)$$

and, consequently, $\psi_t(\varepsilon) \leqslant A(\varepsilon) = C/\lambda_m(B(\varepsilon))$ for all t. Therefore,

$$J(\varepsilon) \leqslant A(\varepsilon) \int_{E_R} \left|f(x-y)\right|^p dy + R \int_{\mathbb{R}^m \setminus B(\varepsilon)} \omega_t(y)\, dy.$$

Thus,

$$\begin{aligned} \left|f_t(x)\right| &\leqslant I(\varepsilon) + J(\varepsilon) \\ &\leqslant \frac{4^m}{\beta_m} C \Delta(\varepsilon) + A(\varepsilon) \int_{E_R} \left|f(x-y)\right|^p dy + R \int_{\mathbb{R}^m \setminus B(\varepsilon)} \omega_t(y)\, dy. \end{aligned}$$

Now, we can make the first summand arbitrarily small by the choice of a small ε. Then, for a fixed ε, we can make the second summand small by the choice of an R. It remains to observe that, for fixed R and ε, the third summand tends to zero as $t \to t_0$ by property (c) of an approximate identity. □

Remark The assertion of the theorem is not valid for an arbitrary approximate identity (see Exercise 4). At the same time, the assumptions concerning the functions ψ_t can be weakened somewhat. It can be seen from the proof that the second inequality in (2) can be replaced by the following less restrictive requirement: for some positive C and ρ and all $t > 0$, the inequality $\int_{B(\rho)} \psi_t(\|x\|)\, dx \leqslant C$ is valid.

9.3.5 In this section and the next, we consider two results in the proofs of which we use approximate identities. In both cases, the idea of the proof is that the required statement is obtained by a passage to the limit, based on the fact that the statement is valid for a "smoothed" function constructed by a convolution. The first of these results is connected to the Gauss–Ostrogradski formula (see Sect. 8.6.5). Without striving for maximal generality, we will assume that the function being integrated satisfies the Lipschitz condition. Here, as in Sect. 8.6.5, by the $(m-1)$-dimensional area σ we mean an area proportional to the Hausdorff measure μ_{m-1}.

Theorem *Let f be a function defined on a standard compact set $K \subset \mathbb{R}^m$ and satisfying the Lipschitz condition. Then, for every unit vector $e \in \mathbb{R}^m$, we have*

$$\int_K \frac{\partial f}{\partial e}(x)\, dx = \int_{\partial K} f(x)\langle \nu(x), e\rangle\, d\sigma(x)$$

(here ν is the outer normal to ∂K).

Proof Without loss of generality, we may assume that f is bounded and satisfies the Lipschitz condition on the entire space $\mathbb{R}^m$ (see Theorem 13.2.3). As will be proved in Sect. 11.4, a function f satisfying the Lipschitz condition is differentiable almost everywhere, and, for every smooth function φ, the following equality is valid:

$$\int_{\mathbb{R}^m} f'_{x_j}(x)\varphi(x)\, dx = -\int_{\mathbb{R}^m} f(x)\varphi'_{x_j}(x)\, dx \tag{4}$$

(see Eq. (1) in Sect. 11.4.2).

We consider the convolution f_t of the function f with a Sobolev approximate identity. It follows from Eq. (4) that $(f_t)'_{x_j} = (f'_{x_j})_t$. Moreover, $\int_K |f'_{x_j} - (f'_{x_j})_t|\, dx \underset{t\to 0}{\longrightarrow} 0$ by Theorem 9.3.3, and $f_t(x) \underset{t\to 0}{\longrightarrow} f(x)$ for every x by Theorem 7.6.3. Since $|f_t| \leqslant \|f\|_\infty$, we obtain by Lebesgue's theorem that $\int_{\partial K} |f_t - f|\, d\sigma \underset{t\to 0}{\longrightarrow} 0$. Thus, to complete the proof, it remains to use the Gauss–Ostrogradski formula for f_t and pass to the limit as $t \to 0$. □

9.3.6 We give one more application of Theorem 9.3.3. The following statement, known as Lagrange's lemma,[4] plays an important role in the theory of generalized functions and in the calculus of variations.

[4]Joseph Louis Lagrange (1736–1813)—French mathematician.

Theorem (Lagrange) *Let $\mathcal{O}$ be an open subset of the space $\mathbb{R}^m$, and let f be a function defined on $\mathcal{O}$ and summable on each compact set lying in $\mathcal{O}$. If*

$$\int_{\mathcal{O}} f(x)\varphi(x)\,dx = 0 \textit{ for every function } \varphi \in C_0^\infty(\mathbb{R}^m) \quad \textit{such that} \quad \operatorname{supp}(\varphi) \subset \mathcal{O},$$

then $f(x) = 0$ for almost all x in $\mathcal{O}$.

Proof We divide the proof into several steps. Without loss of generality, we will assume that f is real.

(1) First, let $\mathcal{O} = \mathbb{R}^m$, and let the function f be continuous. If we suppose that $f(x_0) \neq 0$, then $f(x) \neq 0$ in a ball $B(x_0, r)$. We consider a non-negative function $\varphi \not\equiv 0$ in $C_0^\infty(\mathbb{R}^m)$ such that $\operatorname{supp}(\varphi) \subset B(x_0, r)$. Then $f\varphi$ preserves the sign, and, therefore, $\int_{\mathbb{R}^m} f(x)\varphi(x)\,dx \neq 0$, which contradicts the condition of the theorem.

(2) Now, we assume, as in the previous step, that $\mathcal{O} = \mathbb{R}^m$ and that the function f is summable on $\mathbb{R}^m$, and use the convolutions of f with the even functions ω_t forming a Sobolev approximate identity. If $f_t = f * \omega_t$ and $\varphi \in C_0^\infty(\mathbb{R}^m)$, then

$$\begin{aligned}\int_{\mathbb{R}^m} f_t(x)\varphi(x)\,dx &= \int_{\mathbb{R}^m} \varphi(x)\left(\int_{\mathbb{R}^m} f(y)\omega_t(x-y)\,dy\right)dx \\ &= \int_{\mathbb{R}^m} f(y)\left(\int_{\mathbb{R}^m} \varphi(x)\omega_t(x-y)\,dx\right)dy = \int_{\mathbb{R}^m} f(y)\varphi_t(y)\,dy,\end{aligned}$$

where $\varphi_t = \varphi * \omega_t$. By Corollaries 7.5.3 and 7.5.4, we have $\varphi_t \in C_0^\infty(\mathbb{R}^m)$. Therefore,

$$\int_{\mathbb{R}^m} f_t(x)\varphi(x)\,dx = \int_{\mathbb{R}^m} f(y)\varphi_t(y)\,dy = 0.$$

Since this equation is valid for an arbitrary function φ in $C_0^\infty(\mathbb{R}^m)$ and f_t is continuous, we know from the previous step that $f_t \equiv 0$. At the same time, $f_t \underset{t\to 0}{\longrightarrow} f$ in mean, and, therefore, $\|f\|_1 = \|f - f_t\|_1 \underset{t\to 0}{\longrightarrow} 0$. Thus, $\|f\|_1 = \int_{\mathbb{R}^m} |f(x)|\,dx = 0$, which is equivalent to the conclusion of the theorem.

(3) Turning to the general situation, we note that since $\mathcal{O}$ can be represented as the union of a sequence of compact sets, it is sufficient to verify that $f(x) = 0$ is valid almost everywhere on each compact set $K \subset \mathcal{O}$. Fixing such a set K, we consider a function $\varphi_0 \in C_0^\infty(\mathbb{R}^m)$ with the properties

$$\varphi_0 = 1 \quad \text{on } K, \qquad \operatorname{supp}(\varphi_0) \subset \mathcal{O}, \quad \text{and} \quad 0 \leqslant \varphi_0 \leqslant 1$$

(see Theorem 8.1.7 on a smooth descent). Let f_1 be the function equal to $f\varphi_0$ in $\mathcal{O}$ and equal to zero outside $\mathcal{O}$. This function is summable on $\mathbb{R}^m$ since

$$\int_{\mathbb{R}^m} |f_1(x)|\,dx = \int_{\operatorname{supp}(\varphi_0)} |f(x)|\varphi_0(x)\,dx \leqslant \int_{\operatorname{supp}(\varphi_0)} |f(x)|\,dx < +\infty.$$

We note also that $\operatorname{supp}(\varphi_0\varphi) \subset \operatorname{supp}(\varphi_0) \subset \mathcal{O}$ for every function φ in $C_0^\infty(\mathbb{R}^m)$. Therefore,

$$\int_{\mathbb{R}^m} f_1(x)\varphi(x)\,dx = \int_{\mathcal{O}} f(x)\big(\varphi_0(x)\varphi(x)\big)\,dx = 0.$$

By what was proved in the previous step, we have $f_1(x) = 0$ almost everywhere on $\mathbb{R}^m$, and, in particular, almost everywhere on K where f_1 coincides with f. Thus, $f(x) = 0$ almost everywhere on K, as required. □

9.3.7 In conclusion of this section, we turn to the properties of the convolution in the periodic case. Here, by a periodic function (of several variables), we mean a function that is 2π-periodic in each variable. By $\widetilde{\mathscr{L}}^p(\mathbb{R}^m)$ $(1 \leqslant p < +\infty)$, we denote the space of periodic functions summable on the cube $Q = (-\pi,\pi)^m$ with degree $p \geqslant 1$. As p increases, these spaces, obviously, decrease. By $\|f\|_p$, where $f \in \widetilde{\mathscr{L}}^p(\mathbb{R}^m)$, we mean the $\mathscr{L}^p$-norm of the restriction of f to the cube Q.

As mentioned in Sect. 7.5.5, the convolution of periodic measurable functions f and g on $\mathbb{R}^m$ is defined by the formula

$$(f * g)(x) = \int_Q f(x-y)g(y)\,dy \quad \big(x \in \mathbb{R}^m\big).$$

It exists and is summable if the functions f and g are summable. Thus, in the periodic case, the convolution of f and g belonging to $\widetilde{\mathscr{L}}^p(\mathbb{R}^m)$ and $\widetilde{\mathscr{L}}^q(\mathbb{R}^m)$ exists since these spaces consist of summable functions. Analogs of Theorems 9.3.1–9.3.4 are valid for a periodic approximate identity (see the definition in Sect. 7.6.5). Their proofs differ from the proofs in the non-periodic case only in the replacement of the integration over the entire space $\mathbb{R}^m$ by integration over the cube Q.

We present some statements for convenience of reference.

Theorem 1 *If $f \in \widetilde{\mathscr{L}}^p(\mathbb{R}^m)$ and $g \in \widetilde{\mathscr{L}}^1(\mathbb{R}^m)$, then the convolution $f * g$ exists, belongs to $\widetilde{\mathscr{L}}^p(\mathbb{R}^m)$, and $\|f * g\|_p \leqslant \|f\|_p\|g\|_1$.*

Theorem 2 *Let $\{\omega_t\}_{t>0}$ be a periodic approximate identity in $\mathbb{R}^m$ as $t \to t_0$, and let $f \in \widetilde{\mathscr{L}}^p(\mathbb{R}^m)$, $1 \leqslant p < +\infty$. Then the functions $f_t = f * \omega_t$ converge to f as $t \to t_0$ in the $\mathscr{L}^p$-norm.*

Theorem 3 *Let $\{\omega_t\}_{t>0}$ be a periodic approximate identity $\mathbb{R}^m$ as $t \to t_0$ such that the functions $\psi_t(x) = \sup_{\|x\| \leqslant \|y\| \leqslant \rho} \omega_t(y)$ satisfy the condition*

$$\sup_{t>0} \int_{B(\rho)} \psi_t(x)\,dx < +\infty$$

for some ρ, $0 < \rho \leqslant \pi$. Then, for every function f in $\widetilde{\mathscr{L}}^p(\mathbb{R}^m)$, $1 \leqslant p \leqslant +\infty$, the convolutions $f_t = f * \omega_t$ converge to f as $t \to t_0$ almost everywhere.

By Theorem 2, it is easy to obtain a periodic analog of Corollary 9.2.3 on polynomial approximation with respect to the $\mathscr{L}^p$-norm.

Theorem 4 *Let $1 \leqslant p < +\infty$, $f \in \widetilde{\mathscr{L}}^p(\mathbb{R}^m)$ and $\varepsilon > 0$. Then there is a trigonometric polynomial T such that $\|f - T\|_p < \varepsilon$.*

Proof We consider the approximate identity Θ_n constructed in the proof of Corollary 7.6.5. It consists of trigonometric polynomials, and, therefore, the convolutions $f * \Theta_n$ are also trigonometric polynomials. Since $\|f * \Theta_n - f\|_p \underset{n\to\infty}{\longrightarrow} 0$ by Theorem 2, we see that every convolution $f * \Theta_n$ with a sufficiently large index n can be taken as T. □

EXERCISES

1. Let f_E be the averaging of a function $f \in \mathscr{L}^p(\mathbb{R}^m)$ over a set $E \subset \mathbb{R}^m$ of positive measure, $f_E(x) = \frac{1}{\lambda_m(E)} \int_E f(x+y)\,dy$ $(x \in \mathbb{R}^m)$. Prove that $\|f_E\|_p \leqslant \|f\|_p$.
2. Verify that the Poisson kernel for a half-space, i.e., the family of the functions $\{P_t\}_{t>0}$, where

$$P_t(x) = C\frac{t}{(t^2 + \|x\|^2)^{\frac{m+1}{2}}}, \qquad C = \pi^{-\frac{m+1}{2}}\Gamma\Big(\frac{m+1}{2}\Big) \quad (x \in \mathbb{R}^m),$$

   forms an approximate identity as $t \to +0$ satisfying condition (2) (use Lemma 8.7.13).
3. Prove that if $f \in \mathscr{L}^p(\mathbb{R}^m)$ and $U(x,t) = (f * P_t)(x)$, then the function U is harmonic on the half-space $t > 0$ and is a solution of the Dirichlet problem for the half-space in the sense that $\|U(\cdot,t) - f\|_p \to 0$ and $U(x,t) \to f(x)$ almost everywhere as $t \to +0$.
4. Let $\Delta_n = (\frac{1}{n}, \frac{1}{n} + \frac{1}{n^2})$ and $\omega_n = n^2\chi_{\Delta_n}$. Prove that the functions ω_n form an approximate identity for which the assertion of Theorem 9.3.4 is not valid. *Hint.* Verify that 0 is a Lebesgue point of the function $f(x) = \chi_E(-x)$, where $E = \bigcup_{k=1}^\infty \Delta_{2^k}$, however, $(f * \omega_n)(0) \not\to f(0)$.
5. Let f and g be locally summable functions on $\mathbb{R}^m$. The function g is called a generalized derivative of f with respect to the kth coordinate if

$$\int_{\mathbb{R}^m} g(x)\varphi(x)\,dx = -\int_{\mathbb{R}^m} f\varphi'_{x_k}(x)\,dx$$

   for every function φ of the class C_0^∞. Use Lagrange's theorem (see Sect. 9.3.6) to prove that the generalized derivative is unique up to equivalence.
6. Let $p \geqslant 1$, and let f and g be functions in $\mathscr{L}^p(\mathbb{R}^m)$ such that

$$\int_{\mathbb{R}^m} \left|\frac{f(x + te_k) - f(x)}{t} - g(x)\right|^p dx \underset{t\to 0}{\longrightarrow} 0,$$

   where e_k is a vector of the canonical basis of $\mathbb{R}^m$. Prove that g is the generalized derivative of f with respect to the kth coordinate.

7. Consider the function defined on the space $\mathbb{R}^m$ by the formula $x \mapsto f_\alpha(x) = \frac{1}{\|x\|^{m-\alpha}}$, where $\alpha > 0$ (this function does not belong to any space $\mathscr{L}^p(\mathbb{R}^m)$ for $p > 0$). Prove that if $\alpha + \beta < m$, then the convolution $f_\alpha * f_\beta$ exists and is equal to $Cf_{\alpha+\beta}$. Calculate the coefficient C. Verify that, for an appropriate choice of the coefficient C_α, the function $g_\alpha = C_\alpha f_\alpha$ has the property $g_\alpha * g_\beta = g_{\alpha+\beta}$ ($\alpha > 0,\ \beta > 0,\ \alpha + \beta < m$). *Hint.* To calculate C, use the formula

$$\frac{\Gamma(p/2)}{\|x\|^p} = \int_0^\infty t^{\frac{p}{2}-1} e^{-\|x\|^2 t}\, dt.$$

8. Let $Q = [-\pi, \pi]^m$ and let μ be a finite Borel measure on Q. Prove that, for finite p, trigonometric polynomials are dense in the space $\mathscr{L}^p(Q, \mu)$ if one of the possible products of half-open intervals with endpoints $\pm\pi$ (for example, the cell $[-\pi, \pi)^m$) has full measure. In the one-dimensional case, this condition is not only sufficient but also necessary. For necessary and sufficient conditions in the multi-dimensional case, see Exercise 7 of Sect. 11.2.

9. Prove a periodic analog of Theorem 9.3.2.

10. As a supplement to Young's inequality (see Sect. 9.3.2) show by example that the convolution of functions $f \in \mathscr{L}^p(\mathbb{R})$ and $g \in \mathscr{L}^q(\mathbb{R})$ may not exist if the condition $\frac{1}{p} + \frac{1}{q} \geqslant 1$ is violated (the integral $\int_{-\infty}^{\infty} f(y)g(x-y)\, dy$ can identically be equal to $+\infty$).

Chapter 10
Fourier Series and the Fourier Transform

10.1 Orthogonal Systems in the Space $\mathscr{L}^2(X, \mu)$

In the present section, we consider only the norm in the space $\mathscr{L}^2(X, \mu)$. For brevity, we denote it by $\|\cdot\|$ without index.

10.1.1 The norm in the space $\mathscr{L}^2(X, \mu)$ has an important specific feature: just like a norm in a finite dimensional Euclidean space, it is generated by a scalar product. The scalar product of functions f and g belonging to the (in general, complex) space $\mathscr{L}^2(X, \mu)$ is defined by the formula

$$\langle f, g\rangle = \int_X f\overline{g}\,d\mu$$

(the product $f\,\overline{g}$ is summable since $2|f\,\overline{g}| \leqslant |f|^2 + |g|^2$).

Obviously, $\langle g, f\rangle = \overline{\langle f, g\rangle}$ and $\langle f, f\rangle = \|f\|^2$. Moreover, by the Cauchy–Bunyakovsky inequality, we have $|\langle f, g\rangle| \leqslant \|f\|\,\|g\|$, which implies the continuity of the scalar product with respect to convergence in norm. Indeed, if $f_n \underset{n\to\infty}{\longrightarrow} f$ and $g_n \underset{n\to\infty}{\longrightarrow} g$, then

$$\begin{aligned} \big|\langle f_n, g_n\rangle - \langle f, g\rangle\big| &\leqslant \big|\langle f_n - f, g_n\rangle\big| + \big|\langle f, g_n - g\rangle\big| \\ &\leqslant \big\|f_n - f\big\|\,\|g_n\| + \|f\|\,\|g_n - g\| \underset{n\to\infty}{\longrightarrow} 0. \end{aligned}$$

From the continuity of the scalar product, it follows that the scalar multiplication of a series convergent in norm by a function can be carried out termwise, $\langle\sum_{n=1}^\infty f_n, g\rangle = \sum_{n=1}^\infty\langle f_n, g\rangle$. To verify this, it is sufficient to pass to the limit in the equation $\langle\sum_{n=1}^k f_n, g\rangle = \sum_{n=1}^k\langle f_n, g\rangle$ (the limit on the left-hand side of the equation exists since the series converges and the scalar product is continuous).

We point out one more property of the norm in $\mathscr{L}^2(X, \mu)$, the so-called parallelogram identity

$$\|f+g\|^2 + \|f-g\|^2 = 2\big(\|f\|^2 + \|g\|^2\big) \quad \big(f, g \in \mathscr{L}^2(X, \mu)\big),$$

which is connected with the fact that the norm is generated by a scalar product.

B. Makarov, A. Podkorytov, *Real Analysis: Measures, Integrals and Applications*, Universitext, DOI 10.1007/978-1-4471-5122-7_10, © Springer-Verlag London 2013

The reader can easily verify that if a measure is non-degenerate (more precisely, if there exist two disjoint sets of positive finite measure), then in each space $\mathscr{L}^p(X,\mu)$ with $p \neq 2$ the parallelogram identity is violated.

10.1.2 In the presence of a scalar product, as in a finite-dimensional Euclidean space, we can introduce the notion of the angle between vectors. We are not going to do this in the general setting, instead restricting ourselves to the most important case where the angle is $\pi/2$. We introduce the following definition.

Definition Functions $f, g \in \mathscr{L}^2(X,\mu)$ are called *orthogonal* if $\langle f, g\rangle = 0$.

We remark that if $\langle f, g\rangle = 0$, then also $\langle g, f\rangle = \overline{\langle g, f\rangle} = 0$, and so the orthogonality relation is symmetric. We denote it by $f \perp g$. A function that is zero almost everywhere is orthogonal to every function in $\mathscr{L}^2(X,\mu)$ and, obviously, the converse is also true. For orthogonal functions the *Pythagorean*[1] *theorem* is valid: if $f \perp g$, then $\|f+g\|^2 = \|f\|^2 + \|g\|^2$. This result remains valid for an arbitrary number of pairwise orthogonal summands: if $f_j \perp f_k$ for $j \neq k$ $(j,k = 1,\dots,n)$, then

$$\|f_1 + \cdots + f_n\|^2 = \|f_1\|^2 + \cdots + \|f_n\|^2. \tag{1}$$

Indeed, since $\langle f_j, f_k\rangle = 0$ for $j \neq k$, we have

$$\|f_1 + \cdots + f_n\|^2 = \langle f_1 + \cdots + f_n, f_1 + \cdots + f_n\rangle = \sum_{j,k=1}^{n} \langle f_j, f_k\rangle = \sum_{k=1}^{n} \|f_k\|^2.$$

The Pythagorean theorem is also valid for an "infinite number of summands". If functions $f_1, f_2, \dots$ are pairwise orthogonal and the series $\sum_{k=1}^{\infty} f_k$ converges, then

$$\Big\|\sum_{k=1}^{\infty} f_k\Big\|^2 = \sum_{k=1}^{\infty} \|f_k\|^2. \tag{1'}$$

For the proof, it remains only to pass to the limit in Eq. (1).

Due to the scalar product, every n-dimensional space L contained in $\mathscr{L}^2(X,\mu)$ is isomorphic (as a Euclidean space) to $\mathbb{R}^n$ or $\mathbb{C}^n$ (depending on the field of scalars under consideration). Therefore, we can speak of the orthogonal projection of a function f onto a subspace L. In particular, the projection of f onto the one-dimensional subspace generated by the unit vector e, is $\langle f, e\rangle e$.

In the space $\mathscr{L}^2(X,\mu)$, the families of pairwise orthogonal functions play a role similar to that of the orthogonal bases in finite dimensional Euclidean spaces.

[1] Pythagoras (Πυθαγόρας) (circa 570–500 BC)—Greek philosopher and mathematician.

Definition A family of functions $\{e_\alpha\}_{\alpha\in A}$ is called an *orthogonal system* (briefly, OS) if $e_\alpha \perp e_{\alpha'}$ for $\alpha \neq \alpha'$ and $\|e_\alpha\| \neq 0$ for every $\alpha \in A$. An orthogonal system is called *orthonormal* if $\|e_\alpha\| = 1$ for every $\alpha \in A$.

It follows immediately from the Pythagorean theorem (1) that the functions from an OS are linearly independent. Obviously, dividing each element of an orthogonal system by its norm, we obtain an orthonormal system.

Let the functions $e_1, \dots, e_n$ form an OS, and let L be the subspace generated by $e_1, \dots, e_n$ (i.e., the set of all linear combinations of these functions). It is important to know how to find the best approximation to a given function f by elements of L. The following theorem gives a solution of this extremal problem.

Theorem *The minimum value of the norm* $\|f - \sum_{k=1}^n a_k e_k\|$ *is attained if and only if* $a_k = c_k(f)$, *where*

$$c_k(f) = \frac{\langle f, e_k\rangle}{\|e_k\|^2} \quad (k = 1, \dots, n). \tag{2}$$

The function $f - \sum_{k=1}^n c_k(f) e_k$ *is orthogonal to every element of* L.

Thus, the function $\sum_{k=1}^n c_k(f)e_k$ is the best approximation for f in the set L. The above-stated theorem can be regarded as a generalization of the following well-known fact of school geometry: "the perpendicular dropped from a point f to L", i.e., the difference $f - \sum_{k=1}^n c_k(f)e_k$, is shorter than any "slant" $f - \sum_{k=1}^n a_k e_k$.

Proof We begin with the second assertion of the theorem. We put $S_n = \sum_{k=1}^n c_k(f)e_k$ and verify that $f - S_n \perp \sum_{k=1}^n a_k e_k$. It is sufficient to prove that $f - S_n \perp e_m$ for all $m = 1, \dots, n$. Indeed,

$$\begin{aligned}\langle f - S_n, e_m\rangle &= \langle f, e_m\rangle - \langle S_n, e_m\rangle = \langle f, e_m\rangle - \sum_{k=1}^n c_k(f)\langle e_k, e_m\rangle \\ &= \langle f, e_m\rangle - c_m(f)\|e_m\|^2 = 0.\end{aligned}$$

The last equality holds by the definition of $c_m(f)$.

Now, the extremal property of the sum S_n follows from the Pythagorean theorem. Indeed, if $g = \sum_{k=1}^n a_k e_k$ is an arbitrary function L, then $S_n - g \in L$, and, consequently, $f - S_n \perp S_n - g$. Therefore, by the Pythagorean theorem, we obtain

$$\begin{aligned}\|f - g\|^2 &= \big\|(f - S_n) + (S_n - g)\big\|^2 = \|f - S_n\|^2 + \|S_n - g\|^2 \\ &= \|f - S_n\|^2 + \sum_{k=1}^n \big|a_k - c_k(f)\big|^2 \|e_k\|^2.\end{aligned} \tag{3}$$

From this it follows that

$$\left\| f - \sum_{k=1}^{n} a_k e_k \right\|^2 \geqslant \left\| f - \sum_{k=1}^{n} c_k(f) e_k \right\|^2,$$

and the equality holds only in the case where $a_k = c_k(f)$ for all k. □

For $g = 0$ Eq. (3) takes the form

$$\|f\|^2 = \left\| f - \sum_{k=1}^{n} c_k(f) e_k \right\|^2 + \sum_{k=1}^{n} |c_k(f)|^2 \|e_k\|^2,$$

and, therefore, the *Bessel*[2] *inequality*

$$\sum_{k=1}^{n} |c_k(f)|^2 \|e_k\|^2 \leqslant \|f\|^2 \tag{4}$$

holds.

10.1.3 Let $\{e_n\}_{n\in\mathbb{N}}$ be an OS in the space $\mathscr{L}^2(X, \mu)$. Obviously, there are functions in $\mathscr{L}^2(X, \mu)$ that cannot be represented as linear combinations of functions e_n. Therefore, the question naturally arises, what are the conditions under which a function $f \in \mathscr{L}^2(X, \mu)$ is the sum of a series of the form $\sum_{n=1}^{\infty} a_n e_n$. From the theorem proved above it follows that such a series can converge to f only if it coincides with the series $\sum_{n=1}^{\infty} c_n(f) e_n$ whose coefficients are calculated by formula (2). Indeed, Eq. (3) shows that if $a_m \neq c_m(f)$ and $n \geqslant m$, then

$$\left\| f - \sum_{k=1}^{n} a_k e_k \right\| \geqslant \left| a_m - c_m(f) \right| \|e_m\|^2 > 0,$$

and, therefore, the series $\sum_{n=1}^{\infty} a_k e_k$ cannot converge to f.

The series with coefficients calculated by formula (2) play an important role, which justifies the following definition.

Definition Let $\{e_n\}_{n\in\mathbb{N}}$ be an orthogonal system, and let $f \in \mathscr{L}^2(X, \mu)$. The numbers $c_n(f)$ obtained by formula (2) are called the *Fourier*[3] *coefficients*, and the series $\sum_{n=1}^{\infty} c_n(f) e_n$ is called the *Fourier series* of f with respect to the given OS.

As we will see, the Fourier series of an arbitrary function $f \in \mathscr{L}^2(X, \mu)$ converges in the norm $\| \cdot \|$ (but not necessarily to f).

---

[2] Friedrich Wilhelm Bessel (1784–1846)—German mathematician.

[3] Jean Baptiste Joseph Fourier (1768–1830)—French mathematician.

In the case of an orthonormal system, formula (2) becomes simpler and takes the form $c_n(f)=\langle f,e_n\rangle$. If an orthogonal system $\{e_n\}_{n\in\mathbb{N}}$ is not orthonormal, then we can pass to the system $\widetilde{e}_n=e_n/\|e_n\|$ (to "normalize" the given system). The Fourier coefficients, obviously, can change, but the terms of the Fourier series do not change as the following relation shows:

$$c_n(f)e_n=\left\langle f,\frac{e_n}{\|e_n\|}\right\rangle\frac{e_n}{\|e_n\|}=\langle f,\widetilde{e}_n\rangle\widetilde{e}_n.$$

Thus, the terms of the Fourier series of a function f are simply the projections of f onto the lines generated by the elements of the orthogonal system.

Passing to the limit in Bessel inequality (4) as $n\to\infty$, we obtain the estimate

$$\sum_{k=1}^{\infty}\left|c_k(f)\right|^2\|e_k\|^2\leqslant\|f\|^2 \tag{4'}$$

also called *Bessel's inequality*. As follows from (1′), inequality (4′) becomes an equality if $f=\sum_{n=1}^{\infty}c_n(f)e_n$.

10.1.4 We do not yet know whether a Fourier series converges or, in the case of convergence, what its sum is. The following important theorem establishes that the sum of a Fourier series always exists. As a preliminary, we prove the following lemma.

Lemma *Let $\{e_n\}_{n\in\mathbb{N}}$ be an orthogonal system. A series*

$$\sum_{n=1}^{\infty}a_ne_n \tag{5}$$

converges in norm if and only if

$$\sum_{n=1}^{\infty}|a_n|^2\|e_n\|^2<+\infty. \tag{5'}$$

In the case of convergence, series (5) *is the Fourier series of its sum.*

Proof Let S_n and T_n be the partial sums of series (5) and (5′), respectively. Then, for all $n,p\in\mathbb{N}$, we have

$$\|S_{n+p}-S_n\|^2=\left\|\sum_{k=n+1}^{n+p}a_ke_k\right\|^2=\sum_{k=n+1}^{n+p}|a_k|^2\|e_k\|^2=T_{n+p}-T_n.$$

It follows that the partial sums of series (5) and (5′) are fundamental simultaneously. Since the space $\mathscr{L}^2(X,\mu)$ is complete (see Theorem 9.1.3), we obtain the first assertion of the lemma. The concluding assertion follows from the fact that scalar multiplication of a convergent series by a function can be performed termwise, i.e., if S is the sum of series (5), then the relation

$$\langle S, e_m\rangle = \sum_{n=1}^{\infty} a_n \langle e_n, e_m\rangle = a_m \|e_m\|^2$$

is valid for every $m \in \mathbb{N}$. Thus, $a_m = c_m(S)$ for all m, i.e., series (5) is the Fourier series of its sum. □

Theorem (Riesz–Fischer[4]) *For every orthogonal system* $\{e_n\}_{n\in\mathbb{N}}$, *the Fourier series of a function* $f \in \mathscr{L}^2(X, \mu)$ *converges in norm and*

$$f = \sum_{n=1}^{\infty} c_n(f) e_n + h, \quad \textit{where } h \perp e_n \textit{ for all } n \in \mathbb{N}. \tag{6}$$

Proof By Bessel's inequality, we obtain $\sum_{n=1}^{\infty} |c_n(f)|^2 \|e_n\|^2 \leqslant \|f\|^2 < +\infty$, and so the series $\sum_{n=1}^{\infty} c_n(f) e_n$ converges by the lemma. Let S be its sum. By the second assertion of the lemma, we have $c_n(f) \equiv c_n(S)$. Therefore, the Fourier coefficients of the difference $h = f - S$ are zero, i.e., $h \perp e_n$ for all n. □

10.1.5 Obviously, the sum of the Fourier series may not coincide with the function generating this series. For example, if we replace an OS $e_1, e_2, \ldots$ by the system $e_2, e_3, \ldots$ obtained by deleting the first vector, then the Fourier coefficients of the function e_1 with respect to the new system are zeros, and e_1 is not equal to the sum of its Fourier series (with respect to the new system).

Definition An orthogonal system $\{e_n\}_{n\in\mathbb{N}}$ is called a *basis* if every function in $\mathscr{L}^2(X, \mu)$ coincides with the sum of its Fourier series almost everywhere.

If $\{e_n\}_{n\in\mathbb{N}}$ is a basis, then, by (1′), the relation $f = \sum_{n=1}^{\infty} c_n(f) e_n$ implies that $\|f\|^2 = \sum_{n=1}^{\infty} |c_n(f)|^2 \|e_n\|^2$. Thus, for a basis, the Bessel inequality becomes an equality. We will prove that this property characterizes a basis.

We remark that if $\{e_n\}_{n\in\mathbb{N}}$ is a basis, then the scalar product of two functions can be calculated by their Fourier coefficients since

$$\langle f, g\rangle = \left\langle \sum_{n=1}^{\infty} c_n(f) e_n, g \right\rangle = \sum_{n=1}^{\infty} c_n(f) \langle e_n, g\rangle = \sum_{n=1}^{\infty} c_n(f) \overline{c_n(g)} \|e_n\|^2.$$

This relation (as well as the special case where $g = f$) is called *Parseval's*[5] *identity*.

We introduce one more important property which, like Parseval's identity, is characteristic for a basis.

[4] Ernest Sigismund Fisher (1875–1954)—German mathematician.

[5] Marc-Antoine Parseval (1755–1836)—French mathematician.

Definition A family of functions $\{f_\alpha\}_{\alpha\in A}$ in $\mathscr{L}^2(X,\mu)$ is called *complete* if the condition

$$f \in \mathscr{L}^2(X,\mu) \quad \text{and} \quad f \perp f_\alpha \quad \text{for every } \alpha \in A$$

implies that $f = 0$ almost everywhere, i.e., $\|f\| = 0$.

Lemma *A family $\{f_\alpha\}_{\alpha\in A}$ is complete if the set of all linear combinations of functions contained in this family is everywhere dense, i.e., if, for every function $f \in \mathscr{L}^2(X,\mu)$ and every $\varepsilon > 0$, there exists a linear combination $g = \sum_{k=1}^n c_k f_{\alpha_k}$ such that $\|f - g\| < \varepsilon$.*

Proof Let $f \perp f_\alpha$ for each α. If $\|f\| \neq 0$, then there is a function $g = \sum_{k=1}^n c_k f_{\alpha_k}$ such that $\|f - g\| < \|f\|$. Since $f \perp g$, we obtain a contradiction:

$$\|f\|^2 > \|f-g\|^2 = \|f\|^2 + \|g\|^2 \geqslant \|f\|^2. \qquad \square$$

Theorem (On the characterization of bases) *Let $\{e_n\}_{n\in\mathbb{N}}$ be an orthogonal system. The following conditions are equivalent:*

(1) *the system $\{e_n\}_{n\in\mathbb{N}}$ is a basis;*
(2) *for every function $f \in \mathscr{L}^2(X,\mu)$, Parseval's identity $\sum_{n=1}^\infty |c_n(f)|^2\|e_n\|^2 = \|f\|^2$ holds;*
(3) *the system $\{e_n\}_{n\in\mathbb{N}}$ is complete.*

Proof We prove the chain of implications (1) $\Rightarrow$ (2) $\Rightarrow$ (3) $\Rightarrow$ (1).

(1) $\Rightarrow$ (2) This implication was proved just after the definition of a basis.

(2) $\Rightarrow$ (3) Assume that $f \perp e_n$, i.e., $c_n(f) = 0$ for all $n = 1, 2, \ldots$. By hypothesis, $\|f\|^2 = \sum_{n=1}^\infty |c_n(f)|^2\|e_n\|^2 = 0$, which means that the system $\{e_n\}_{n\in\mathbb{N}}$ is complete.

(3) $\Rightarrow$ (1) Let $f \in \mathscr{L}^2(X,\mu)$. By the Riesz–Fischer theorem, $f = g + h$, where $g = \sum_{n=1}^\infty c_n(f)e_n$ and $h \perp e_n$ for all n. Since the system is complete, we obtain that $h = 0$ almost everywhere. Taking account of the arbitrariness of f, we obtain that the OS in question is a basis. $\square$

Comparing the theorem with the preceding lemma, we see that the following statement is valid.

Corollary *An orthogonal system $\{e_n\}_{n\in\mathbb{N}}$ is complete if and only if the set of all linear combinations of the functions contained in this system is everywhere dense.*

10.1.6 We will see in the next section (see also Sect. 10.2) that it is often convenient to label naturally arising orthogonal systems not by positive integers but by some other indices. Therefore, it is useful to generalize the definition of the Fourier series and coefficients. Let $\{e_\alpha\}_{\alpha\in A}$ be an arbitrary OS in the space $\mathscr{L}^2(X,\mu)$, and let $f \in \mathscr{L}^2(X,\mu)$. As above, the numbers $c_\alpha(f) = \frac{\langle f, e_\alpha\rangle}{\|e_\alpha\|^2}$ will be called the *Fourier* *coefficients* of the function f with respect to the given OS. Since Bessel's inequality $\sum_{k=1}^{n} |c_{\alpha_k}(f)|^2 \|e_{\alpha_k}\|^2 \leqslant \|f\|^2$ is valid for every finite set of indices $\alpha_1, \ldots, \alpha_n$, the family $\{|c_\alpha(f)|^2 \|e_\alpha\|^2\}_{\alpha \in A}$ is summable (see Sect. 1.2.2). Therefore, the set A_f of indices of the non-zero coefficients $c_\alpha(f)$ is at most countable (see Sect. 1.2.2), which, after enumeration, can be written in the form $\{\alpha_1, \alpha_2, \ldots\}$. By the Riesz–Fischer theorem, the series $\sum_{k=1}^{\infty} c_{\alpha_k}(f) e_{\alpha_k}$ converges, and its sum will also be called the sum of the Fourier series of f with respect to $\{e_\alpha\}_{\alpha \in A}$. To verify that the sum is well-defined, we must prove that different enumerations of the set A_f give the same sum. A change of enumeration of the set A_f results in a series obtained by rearranging the terms of the series $\sum_{k=1}^{\infty} c_{\alpha_k}(f) e_{\alpha_k}$. Therefore, it is sufficient to prove the following auxiliary statement.

Lemma *Let $\{e_n\}_{n \in \mathbb{N}}$ be an orthogonal system and $\omega : \mathbb{N} \to \mathbb{N}$ be a bijection. Then the series*

$$\text{(a)} \quad \sum_{n=1}^{\infty} a_n e_n \quad \textit{and}$$

$$\text{(b)} \quad \sum_{k=1}^{\infty} a_{\omega(k)} e_{\omega(k)}$$

converge simultaneously and, in the case of convergence, their sums are equal.

Proof As established in Lemma 10.1.4, series (a) and (b) converge simultaneously with the series $\sum_{n=1}^{\infty} |a_n|^2 \|e_n\|^2$ and $\sum_{k=1}^{\infty} |a_{\omega(k)}|^2 \|e_{\omega(k)}\|^2$, respectively. The last two series converge simultaneously because the sum of a positive series is independent of any rearrangement of the terms. This proves that series (a) and (b) converge simultaneously. Now, let series (a) and (b) converge and S_n be a partial sum of (a). By the Pythagorean theorem (see Eq. (1′)), we obtain

$$\left\| \sum_{k=1}^{\infty} a_{\omega(k)} e_{\omega(k)} - S_n \right\|^2 = \sum_{\omega(k)>n}^{\infty} |a_{\omega(k)}|^2 \|e_{\omega(k)}\|^2 = \sum_{j=n+1}^{\infty} |a_j|^2 \|e_j\|^2 \underset{n\to\infty}{\longrightarrow} 0,$$

which implies that the sums of series (a) and (b) coincide. □

As in the case of sequences, a family $\{e_\alpha\}_{\alpha \in A}$ is called a basis if every function is the sum of its Fourier series. It can easily be seen that the theorem on the characterization of bases and its corollary remain valid in the more general setting in question.

10.1.7 Let $\{e_k\}_{k \in \mathbb{N}}$ and $\{g_n\}_{n \in \mathbb{N}}$ be orthogonal systems in the spaces $\mathscr{L}^2(X, \mu)$ and $\mathscr{L}^2(Y, \nu)$, respectively. We use these systems to construct an OS $\{h_{k,n}\}_{k,n \in \mathbb{N}}$ in the space $\mathscr{L}^2(X \times Y, \mu \times \nu)$ by putting

$$h_{k,n}(x, y) = e_k(x) g_n(y) \quad (x \in X, y \in Y).$$

Using Fubini's theorem, we can easily verify that the functions $h_{k,n}$ are square-summable and pairwise orthogonal. We will prove that the above construction preserves completeness.

Theorem *If orthogonal systems* $\{e_k\}_{k\in\mathbb{N}}$ *and* $\{g_n\}_{n\in\mathbb{N}}$ *are complete, then the system* $\{h_{k,n}\}_{k,n\in\mathbb{N}}$ *is also complete.*

Proof Let $f \perp h_{k,n}$ for all $k, n \in \mathbb{N}$. This means that

$$\int_{X\times Y} f(x,y)\overline{e_k(x)}\,\overline{g_n(y)}\,d(\mu\times\nu)(x,y)$$
$$= \int_X\left(\int_Y f(x,y)\overline{g_n(y)}\,d\nu(y)\right)\overline{e_k(x)}\,d\mu(x) = 0 \tag{7}$$

for all $k, n \in \mathbb{N}$. We fix an arbitrary n and consider the function

$$x \mapsto \varphi_n(x) = \int_Y f(x,y)\overline{g_n(y)}\,d\nu(y).$$

This function is measurable by Corollary 2 to Tonelli's theorem. Moreover, $\varphi_n \in \mathscr{L}^2(X, \mu)$ since

$$|\varphi_n(x)| \leqslant \left(\int_Y |f(x,y)|^2\,d\nu(y)\right)^{1/2}\|g_n\|,$$

and, therefore,

$$\int_X |\varphi_n(x)|^2\,d\mu(x) \leqslant \int_X\left(\int_Y |f(x,y)|^2\,d\nu(y)\right)d\mu(x)\|g_n\|^2 < +\infty.$$

Equation (7) means that the Fourier coefficients of φ_n with respect to the system $\{e_k\}_{k\in\mathbb{N}}$ are zero. Since the system is complete, we have $\varphi_n(x) = 0$ almost everywhere. Since this is true for all indices n, we have

$$\sum_{n=1}^{\infty} |\varphi_n(x)|^2 = 0 \quad \text{almost everywhere on } X. \tag{8}$$

Since $\int_X\int_Y |f(x,y)|^2\,d\nu(y)\,d\mu(x) < +\infty$, Fubini's theorem implies that $\int_Y |f(x,y)|^2\,d\nu(y) < +\infty$ almost everywhere. In other words, the function $y \mapsto f_x(y) = f(x,y)$ is square-summable for almost all x. The numbers $\varphi_n(x)$ are simply the Fourier coefficients of this function with respect to the system $\{g_n\}_{n\in\mathbb{N}}$. Since the system $\{g_n\}_{n\in\mathbb{N}}$ is complete, Eq. (8) means that

$$\int_Y |f(x,y)|^2\,d\nu(y) = \|f_x\|^2 = \sum_{n=1}^{\infty} |\varphi_n(x)|^2 = 0 \quad \text{almost everywhere on } X.$$

Integrating the above equation over X, we obtain

$$0 = \int_X \int_Y |f(x,y)|^2 \, d\nu(y) \, d\mu(x) = \|f\|^2.$$

Consequently, $f = 0$ almost everywhere, which proves that the system $\{h_{k,n}\}_{k,n\in\mathbb{N}}$ is complete. □

By induction, the statement just proved can obviously be carried over to the case of more than two orthogonal systems.

10.1.8 Lemma 10.1.4 shows that, for a given orthonormal system, an arbitrary sequence $\{a_n\}_{n\geqslant 1}$ satisfying the condition $\sum_{n=1}^{\infty} |a_n|^2 < +\infty$ can serve as the sequence of Fourier coefficients of a square-summable function. It is natural to assume that the smaller the class of functions in question, the greater, in general, the rate of decrease of the Fourier coefficients. In Sect. 10.3, we will find more evidence for this conjecture. However, if, instead of square-summable functions, we consider arbitrary bounded functions (assuming, naturally, that they belong to $\mathscr{L}^2(X,\mu)$, i.e., that the measure μ is finite) then our conjecture is false: the Fourier coefficients of bounded functions tend to zero "no faster" than the Fourier coefficients of arbitrary functions from $\mathscr{L}^2$. A more precise formulation of this result of F.L. Nazarov[6] [Na] is as follows.

Theorem *Let $\{e_n\}_{n\in\mathbb{N}}$ be an orthonormal system in $\mathscr{L}^2(X,\mu)$, $\mu(X) < +\infty$, such that $\int_X |e_n| \, d\mu \geqslant \beta > 0$, where β does not depend on n. Then, for every series $\sum_{n=1}^{\infty} a_n^2 = 1$ ($a_n > 0$), there exists a measurable function F_a such that $|F_a| \leqslant 1$ and $|c_n(F_a)| \geqslant \theta\, a_n$ for all n (the coefficient $\theta > 0$ depends only on $\mu(X)$ and β).*

We note that the condition $\int_X |e_n| \, d\mu \geqslant \beta > 0$ is certainly fulfilled if the orthogonal system consists of uniformly bounded functions since $1 = \int_X |e_n|^2 \, d\mu \leqslant \|e_n\|_\infty \int_X |e_n| \, d\mu$.

Proof We consider only the real case, leaving the complex case to the reader (see Exercises 6 and 7).

For an arbitrary sequence of signs $\varepsilon = \{\varepsilon_n\}$, where $\varepsilon_n = \pm 1$, we construct the sum

$$f_\varepsilon = \sum_{n=1}^{\infty} \varepsilon_n a_n e_n$$

(the series on the right-hand side converges by Lemma 10.1.4). Let A be the set formed by all functions f_ε. This set is compact as a continuous image of the Cantor set (the reader can verify independently the continuity of the mapping

[6] Fedor L'vovich Nazarov (born 1967)—Russian mathematician.

that takes a number $\sum_{n=1}^{\infty} t_n 3^{-n}$ ($t_n = 0$ or 2) from the Cantor set to the point $\sum_{n=1}^{\infty}(t_n-1)a_n e_n$ of the set A).

Now, we consider the function Φ of class $C^2(\mathbb{R})$ such that $|\Phi'|, |\Phi''| \leqslant 1$ (the choice of Φ will be specified later). Since $|\Phi(u)| \leqslant |\Phi(0)| + |u|$ and the measure is finite, the integral $I(f) = \int_X \Phi(f)\,d\mu$ is finite for every function $f \in \mathscr{L}^2(X,\mu)$. Obviously, the integral continuously depends on f and so, by the Weierstrass extreme value theorem, it assumes its maximum value on A: there exists a sequence of signs $\varepsilon = \{\varepsilon_n\}$ such that $I(f_\varepsilon) \geqslant I(f)$ for every function f in A. We show that the required function has the form $F_a = \Phi'(f_\varepsilon)$ for an appropriate choice of Φ. Since $|F_a| \leqslant \sup|\Phi'| \leqslant 1$, it only remains for us to estimate the Fourier coefficients $c_n(F_a)$. To this end, we use the fact that the replacement of ε_n by $-\varepsilon_n$ leaves a function in the class A and, therefore, does not increase the integral I,

$$\int_X \big(\Phi(f_\varepsilon) - \Phi(f_\varepsilon - 2\varepsilon_n a_n e_n)\big)\,d\mu \geqslant 0.$$

The application of the Taylor formula to the integrand leads to the inequality

$$\int_X \left(2\varepsilon_n a_n e_n \Phi'(f_\varepsilon) - \frac{1}{2}(2\varepsilon_n a_n e_n)^2 \Phi''(g_n)\right) d\mu \geqslant 0, \tag{9}$$

where g_n is a function whose values lie between f_ε and $f_\varepsilon - 2\varepsilon_n a_n e_n$.

Dividing both sides of inequality (9) by $2a_n$, we obtain the required estimate for the Fourier coefficients of the function $F_a = \Phi'(f_\varepsilon)$,

$$\left|c_n(F_a)\right| \geqslant \varepsilon_n \int_X e_n \Phi'(f_\varepsilon)\,d\mu \geqslant a_n \int_X e_n^2 \Phi''(g_n)\,d\mu.$$

Now, it is necessary to choose Φ so that the integrals $J_n = \int_X e_n^2 \Phi''(g_n)\,d\mu$ be separated from zero. If we take an antiderivative of $\frac{2}{\pi}\arctan u$ as Φ, then

$$J_n = \frac{2}{\pi}\int_X \frac{e_n^2}{1+g_n^2}\,d\mu.$$

To estimate this integral, we use the Cauchy–Bunyakovsky inequality,

$$\beta \leqslant \int_X |e_n|\,d\mu = \int_X \frac{|e_n|}{\sqrt{1+g_n^2}} \cdot \sqrt{1+g_n^2}\,d\mu \leqslant \sqrt{\frac{\pi}{2} J_n} \cdot \sqrt{\int_X (1+g_n^2)\,d\mu}.$$

Since $|g_n| \leqslant |f_\varepsilon| + |f_\varepsilon - 2\varepsilon_n a_n e_n|$, we obtain $\int_X g_n^2\,d\mu \leqslant 2(\|f_\varepsilon\|^2 + \|f_\varepsilon - 2\varepsilon_n a_n e_n\|^2) = 4$. Therefore, $J_n \geqslant \theta = \frac{2\beta^2}{\pi(\mu(X)+4)}$ and $|c_n(F_a)| \geqslant \theta\, a_n$ for all n. □

EXERCISES

1. Supplement Lemma 10.1.4 by the following statement: if a system (in general, nonorthogonal) $\{e_n\}_{n\in\mathbb{N}}$ is such that the inequality $\|a_1e_1+\dots+a_ne_n\|^2 \leqslant |a_1|^2 + \dots + |a_n|^2$ is valid for all n and all scalars $a_1,\dots,a_n$, then the series $\sum_{n=1}^{\infty} a_n e_n$ converges as soon as $\sum_{n=1}^{\infty}|a_n|^2 < +\infty$.

2. Let an orthonormal system $\{e_n\}_{n\in\mathbb{N}}$ in $\mathscr{L}^2(X,\mu)$ be uniformly bounded. Prove that $\int_X f\overline{e}_n\,d\mu \underset{n\to\infty}{\longrightarrow} 0$ for every function f not only from $\mathscr{L}^2(X,\mu)$ but also from $\mathscr{L}^1(X,\mu)$.

3. Let $\{e_n\}_{n\in\mathbb{N}}$ be an orthonormal basis in $\mathscr{L}^2(X,\mu)$, and let $E\subset X$ be such that $0<\mu(E)<+\infty$. Prove that $\sum_{n=1}^{\infty}\int_E |e_n|^2\,d\mu \geqslant 1$.

4. Supplement the previous exercise by proving that $\sum_{n=1}^{\infty}|e_n(x)|^2=+\infty$ is valid almost everywhere if the σ-finite measure μ is such that every set of positive measure can be partitioned into two sets of positive measure. Can this additional condition be dropped?

5. Let $\{\varphi_n\}$ be an orthonormal basis. Prove that the system of functions $\{\psi_n\}$ is complete if $\sum_n \|\varphi_n-\psi_n\|^2<1$. If, in addition, we know that $\{\psi_n\}$ is an orthonormal system, then it is complete if $\sum_n \|\varphi_n-\psi_n\|^2<2$. *Hint*. Assuming that a function $f=\sum c_n\varphi_n$ is orthogonal to all functions ψ_n, estimate the norm of the difference $f-\sum_n c_n\psi_n$ from above and from below.

6. Verify that Theorem 10.1.8 remains valid in the real case if the orthonormality condition is replaced by the condition from Exercise 1 (the quantities $\langle F_a, e_n\rangle$ are estimated instead of the Fourier coefficients).

7. Generalize the result of the previous exercise to complex systems.

10.2 *Examples of Orthogonal Systems

Throughout this section, we consider the convergence of Fourier series only with respect to the $\mathscr{L}^2$-norm, which is denoted by $\|\cdot\|$. Instead of $\mathscr{L}^2(X,\lambda_m)$, where $X\subset\mathbb{R}^m$, we will write briefly $\mathscr{L}^2(X)$, omitting the indication of a measure.

10.2.1 *Trigonometric Systems*. The most important orthogonal systems are the following real and complex *trigonometric systems* in the space $\mathscr{L}^2((a,a+2\ell))$:

$$1,\quad \cos\frac{\pi x}{\ell},\quad \sin\frac{\pi x}{\ell},\quad \dots,\quad \cos\frac{\pi n x}{\ell},\quad \sin\frac{\pi n x}{\ell},\quad \dots\quad \text{and}\quad \{e^{\frac{\pi i n x}{\ell}}\}_{n\in\mathbb{Z}}.$$

We leave the simple verification of the orthogonality to the reader. The Fourier series with respect to these systems have, respectively, the form

$$A(f)+\sum_{n=1}^{\infty}\Big(a_n(f)\cos\frac{\pi n x}{\ell}+b_n(f)\sin\frac{\pi n x}{\ell}\Big)\quad\text{and}\quad \sum_{n=-\infty}^{\infty} c_n(f)e^{\frac{\pi i n x}{\ell}},$$

where the Fourier coefficients are calculated by the formulas

$$A(f)=\frac{1}{2\ell}\int_a^{a+2\ell} f(x)\,dx,$$

$$a_n(f)=\frac{1}{\ell}\int_a^{a+2\ell} f(x)\cos\frac{\pi n x}{\ell}\,dx,$$

$$b_n(f) = \frac{1}{\ell}\int_a^{a+2\ell} f(x)\sin\frac{\pi n x}{\ell}\,dx \quad (n \in \mathbb{N});$$

$$c_n(f) = \frac{1}{2\ell}\int_a^{a+2\ell} f(x)e^{-\frac{\pi i n x}{\ell}}\,dx \quad (n \in \mathbb{Z}).$$

In the study of Fourier series, we may assume that the functions are defined on the intervals of the form $(0, 2\ell)$, because the general case can be reduced to the case $a = 0$ by a translation. It is often convenient to use a symmetric interval $(-\ell, \ell)$.

The study of Fourier series with respect to a trigonometric system with some period can be reduced to the study of Fourier series with a different period. Following tradition, we will consider (with rare exceptions) only the Fourier series

$$A(f) + \sum_{n=1}^{\infty}\bigl(a_n(f)\cos nx + b_n(f)\sin nx\bigr) \quad \text{and} \quad \sum_{n=-\infty}^{\infty} c_n(f)e^{inx}$$

with respect to more natural and convenient 2π-periodic systems

$$1, \quad \cos x, \quad \sin x, \quad \dots, \quad \cos nx, \quad \sin nx, \quad \dots \quad \text{and} \quad \{e^{inx}\}_{n\in\mathbb{Z}}. \qquad \text{(T)}$$

In this case, the Fourier coefficient $c_n(f)$ will also be denoted by the symbol $\widehat{f}(n)$. Thus,

$$\widehat{f}(n) = \frac{1}{2\pi}\int_0^{2\pi} f(x)e^{-inx}\,dx \quad (n \in \mathbb{Z}).$$

The transition from the expansion in one system to the expansion in a different system proceeds as follows. For a function $f \in \mathscr{L}^2((0, 2\ell))$, we define a function g by putting $g(y) = f(\frac{\ell}{\pi}y)$, where $y \in (0, 2\pi)$. It is clear that $g \in \mathscr{L}^2((0, 2\pi))$. There is an obvious relation connecting the Fourier coefficients of these functions (with respect to the corresponding systems):

$$c_k(f) = \frac{1}{2\ell}\int_0^{2\ell} f(x)e^{-\frac{\pi i k x}{\ell}}\,dx = \frac{1}{2\pi}\int_0^{2\pi} f\Bigl(\frac{\ell}{\pi}y\Bigr)e^{-iky}\,dy = \widehat{g}(k)$$

for each $k \in \mathbb{Z}$. Consequently,

$$\sum_{|k|\leqslant n} c_k(f)e^{\frac{\pi i k x}{\ell}} = \sum_{|k|\leqslant n} \widehat{g}(k)e^{\frac{\pi i k x}{\ell}} = \sum_{|k|\leqslant n} \widehat{g}(k)e^{iky},$$

i.e., the partial sums of the Fourier series of the functions f and g at the corresponding points coincide. From this, it follows, in particular, that both series converge simultaneously and their sums coincide (or do not coincide) simultaneously with the values of the functions f and g. Thus, the transition from f to g makes it possible to reduce the study of a Fourier series in a system with an arbitrary period to the study of a Fourier series in the 2π-periodic system.

By Euler's formula, the systems (T) are tightly connected with each other: their linear spans coincide (the functions from these spans are called *trigonometric polynomials*), and the Fourier coefficients in one system are expressed in terms of the Fourier coefficients in the other one by the following formula:

$$\widehat{f}(\pm n) = \frac{1}{2\pi}\int_0^{2\pi} f(x)(\cos nx \mp i \sin nx)\, dx = \frac{a_n(f) \mp i b_n(f)}{2} \quad (n \in \mathbb{N})$$

and

$$a_n(f) = \widehat{f}(n) + \widehat{f}(-n) \quad \text{and} \quad b_n(f) = i\big(\widehat{f}(n) - \widehat{f}(-n)\big) \quad (n \in \mathbb{N}).$$

It follows that the Fourier series in systems (T) essentially coincide. More precisely, the relation

$$A(f) + \sum_{k=1}^{n}\big(a_k(f)\cos kx + b_k(f)\sin kx\big) = \sum_{k=-n}^{n} \widehat{f}(k)e^{ikx},$$

showing that the partial sums of the Fourier series in the real system (T) coincides with symmetric partial sums of the Fourier series in the complex system, is valid for each n.

In the following theorem, we establish one of the most important properties of the systems (T).

Theorem *The real and complex trigonometric systems form bases in* $\mathscr{L}^2((0, 2\pi))$.

Proof The assertion of the theorem follows immediately from Corollary 10.1.5 the assumptions of which are fulfilled by Theorem 4 of Sect. 9.3.7. □

Since each of the systems (T) is a basis, it satisfies Parseval's identity: if $f, g \in \mathscr{L}^2((0, 2\pi))$, then

$$\frac{1}{2\pi}\int_0^{2\pi} f(x)\overline{g(x)}\, dx = A(f)\overline{A(g)} + \frac{1}{2}\sum_{n=1}^{\infty}\big(a_n(f)\overline{a_n(g)} + b_n(f)\overline{b_n(g)}\big)$$

$$= \sum_{n=-\infty}^{\infty} \widehat{f}(n)\overline{\widehat{g}(n)}$$

in particular, every function f in $\mathscr{L}^2((0, 2\pi))$ satisfies the equation

$$\frac{1}{2\pi}\int_0^{2\pi} |f(x)|^2\, dx = |A(f)|^2 + \frac{1}{2}\sum_{n=1}^{\infty}\big(|a_n(f)|^2 + |b_n(f)|^2\big) = \sum_{n=-\infty}^{\infty} |\widehat{f}(n)|^2,$$

which is often called the closeness relation. As we have already noted, in these formulas and in the theorem, the interval $(0, 2\pi)$ can be replaced by an arbitrary interval of length 2π, in particular, by $(-\pi, \pi)$.

We will now give several examples that illustrate the importance of this formula.

Example 1 Let $f(x) = x$ for $x \in (-\pi, \pi)$. The Fourier series of this function has the form $\sum_{n=1}^{\infty}(-1)^{n-1}\frac{2}{n}\sin nx$. By Parseval's identity, we have

$$\frac{1}{\pi}\int_{-\pi}^{\pi} x^2\,dx = \sum_{n=1}^{\infty}|b_n(f)|^2 = 4\sum_{n=1}^{\infty}\frac{1}{n^2}.$$

Thereby we have arrived at the following result first obtained by Euler: $\sum_{n=1}^{\infty}\frac{1}{n^2} = \frac{\pi^2}{6}$. The same reasoning applied to the function $f(x) = x^2$ ($|x| \leqslant \pi$) gives another result of Euler's: $\sum_{n=1}^{\infty}\frac{1}{n^4} = \frac{\pi^4}{90}$.

Example 2 As we have seen (see Corollary 9.2.4), 2π-periodic functions in $\widetilde{\mathscr{L}}^2$, i.e., square integrable functions on $(-\pi, \pi)$ are continuous in mean. By the closeness equation, we can obtain an exact value for the deviation of a function from its translation.

We will assume that a function $f \in \mathscr{L}^2((-\pi, \pi))$ is extended by periodicity from $[-\pi, \pi]$ to $\mathbb{R}$. Let $h \in \mathbb{R}$, and let f_h be the corresponding translation of f, i.e., $f_h(x) = f(x-h)$ for $x \in \mathbb{R}$. It can easily be verified that $\widehat{f_h}(k) = e^{-ikh}\widehat{f}(k)$. Therefore, by Parseval's identity, we obtain

$$\|f_h - f\|^2 = 2\pi\sum_{k=-\infty}^{\infty}|\widehat{f}(k)|^2|e^{-ikh} - 1|^2 = 8\pi\sum_{k=-\infty}^{\infty}|\widehat{f}(k)|^2\sin^2\frac{kh}{2}.$$

From this formula, the continuity in the mean, $f_h \underset{h\to 0}{\longrightarrow} f$, follows directly.

Example 3 We apply Parseval's identity to prove an elegant inequality (see [EF]), which, in some cases, makes it possible to estimate from above the mean value of a function on an interval by its mean value on a smaller interval.

Let the Fourier coefficients of a function φ in $\mathscr{L}^2((-\pi, \pi))$ be non-negative. Then the inequality

$$\frac{1}{2\pi}\int_{-\pi}^{\pi}|\varphi(t)|^2\,dt \leqslant \frac{3}{2\alpha}\int_{-\alpha}^{\alpha}|\varphi(t)|^2\,dt$$

is valid for every $\alpha \in (0, \pi)$.

Since the function $h(t) = (1 - \frac{|t|}{\alpha})_+$ does not exceed 1, it is sufficient for us to estimate the integral $I = \int_{-\pi}^{\pi}|\varphi(t)h(t)|^2\,dt$ from below. The product $f = \varphi h$, obviously, belongs to $\mathscr{L}^2((-\pi, \pi))$. We calculate its Fourier coefficients (in what follows, $e_n(t) = e^{int}$),

$$\widehat{f}(n) = \frac{1}{2\pi}\langle\varphi h, e_n\rangle = \frac{1}{2\pi}\langle\varphi, he_n\rangle = \frac{1}{2\pi}\sum_{k=-\infty}^{\infty}\widehat{\varphi}(k)\langle e_k, he_n\rangle$$

$$= \sum_{k=-\infty}^{\infty}\widehat{\varphi}(k)\widehat{h}(n-k) = \sum_{k+j=n}\widehat{\varphi}(k)\widehat{h}(j).$$

Now, by Parseval's identity we obtain

$$I = 2\pi \sum_{n=-\infty}^{\infty} \left|\widehat{f}(n)\right|^2 = 2\pi \sum_{n=-\infty}^{\infty} \left| \sum_{k+j=n} \widehat{\varphi}(k)\,\widehat{h}(j)\right|^2.$$

A direct calculation shows that $\widehat{h}(j) \geqslant 0$ for all $j \in \mathbb{Z}$ (this also follows from the result of Example 2 of Sect. 4.6.6 since the function h is convex on $(0, \pi)$). Therefore, replacing the square of the sum by the sum of squares (here we use the inequalities $\widehat{\varphi}(k) \geqslant 0$), we obtain

$$I \geqslant 2\pi \sum_{n=-\infty}^{\infty} \sum_{k+j=n} \widehat{\varphi}^2(k)\,\widehat{h}^2(j) = 2\pi \sum_{k=-\infty}^{\infty} \widehat{\varphi}^2(k) \sum_{j=-\infty}^{\infty} \widehat{h}^2(j)$$

$$= \frac{1}{2\pi} \int_{-\pi}^{\pi} \left|\varphi(t)\right|^2 dt \int_{-\pi}^{\pi} h^2(t)\,dt = \frac{\alpha}{3\pi} \int_{-\pi}^{\pi} \left|\varphi(t)\right|^2 dt.$$

Thus,

$$\int_{-\alpha}^{\alpha} \left|\varphi(t)\right|^2 dt \geqslant I \geqslant \frac{\alpha}{3\pi} \int_{-\pi}^{\pi} \left|\varphi(t)\right|^2 dt.$$

Example 4 Hurwitz[7] found an unexpected application of trigonometric series. It turns out that they can be used to obtain a very simple proof of the classical *isoperimetric inequality* connected with the problem of determining a closed plane curve that has a given circumference L and bounds a figure of the largest area. This inequality has the form

$$4\pi S \leqslant L^2,$$

where S is the area of the figure. The equality is attained only in the case where the curve is a circle (the multi-dimensional case of the isoperimetric inequality is considered in Sects. 2.8.2 and 13.4.7).

The proof given by Hurwitz is analytic. It uses only the closeness equation and the formula for the area in terms of a curvilinear integral.

Let $K \subset \mathbb{R}^2$ be a compact set whose boundary is a closed smooth curve. Without loss of generality, we may assume that the length of the curve is 2π. Let $z(t) = (x(t), y(t))$, $0 \leqslant t \leqslant 2\pi$ be the natural parametrization (see Sect. 8.2.3) of the curve ∂K. Then $z(0) = z(2\pi)$ because the curve ∂K is closed and $|z'(t)| \equiv 1$ because the parametrization is natural.

Using the closeness equation and the identity $|z'(t)| \equiv 1$, we can represent the relation $L = 2\pi$ in the form

$$L^2 = 2\pi \int_0^{2\pi} \left|z'(t)\right|^2 dt = 4\pi^2 \sum_{n\in\mathbb{Z}} \left|\widehat{z'}(n)\right|^2. \tag{1}$$

[7] Adolf Hurwitz (1859–1919)—German mathematician.

To calculate the area $S = \lambda_2(K)$, we apply the relation

$$S = \frac{1}{2}\int_{\partial^+ K}(-y\,dx + x\,dy) = \frac{1}{2}\int_0^{2\pi}\big(x(t)y'(t) - y(t)x'(t)\big)\,dt,$$

which follows from Green's formula with $P(x, y) = -y$ and $Q(x, y) = x$ (see Sect. 8.6.7). Since $x(t)y'(t) - y(t)x'(t) = \mathcal{I}m(z'(t)\overline{z(t)})$ and $\int_0^{2\pi}\mathcal{R}e(z'(t)z(t))\,dt = \int_0^{2\pi}(x^2(t) + y^2(t))'\,dt = 0$, we have

$$S = \frac{1}{2i}\int_0^{2\pi} z'(t)\overline{z(t)}\,dt.$$

Transforming the integral by Parseval's identity, we obtain

$$S = -\pi i \sum_{n\in\mathbb{Z}} \widehat{z'}(n)\overline{\widehat{z}(n)}. \tag{2}$$

Now, we eliminate the Fourier coefficients of the derivative from Eqs. (1) and (2), expressing them in terms of the Fourier coefficients of the function z. Integrating by parts and taking into account that $z(0) = z(2\pi)$, we have

$$\widehat{z'}(n) = \frac{1}{2\pi}\int_0^{2\pi} z'(t)e^{-int}\,dt = \frac{1}{2\pi}z(t)e^{-int}\Big|_{t=0}^{2\pi} + \frac{in}{2\pi}\int_0^{2\pi} z(t)e^{-int}\,dt = in\widehat{z}(n).$$

Substituting the resulting expressions for $\widehat{z'}(n)$ in (1) and (2), we obtain

$$L^2 = 4\pi^2\sum_{n\in\mathbb{Z}} n^2\big|\widehat{z}(n)\big|^2 \quad \text{and} \quad S = \pi\sum_{n\in\mathbb{Z}} n\big|\widehat{z}(n)\big|^2.$$

Consequently,

$$L^2 - 4\pi S = 4\pi^2\sum_{n\in\mathbb{Z}}\big(n^2 - n\big)\big|\widehat{z}(n)\big|^2 \geqslant 0,$$

which proves the isoperimetric inequality. Moreover, the last formula implies that the equality holds only if $\widehat{z}(n) = 0$ for $n \neq 0, 1$, i.e., only if $z(t) = \widehat{z}(0) + \widehat{z}(1)e^{it}$. We have $|\widehat{z}(1)| = 1$, since $|z'(t)| \equiv 1$. Thus, the curve of length 2π for which the isoperimetric inequality becomes an equality is the unit circle $|z - \widehat{z}(0)| = 1$.

10.2.2 Considering the product of m copies of the complex trigonometric system (see Sect. 10.1.7), we obtain its multi-dimensional analog in the space $\mathscr{L}^2(Q)$, where $Q = (-\pi, \pi)^m$ (a multi-dimensional version of the real trigonometric system is quite cumbersome and we do not consider it). The new system consists of the complex exponential functions e_n numbered by multi-indices $n = (n_1, \ldots, n_m)$:

$$e_n(x) = e^{i\langle n, x\rangle}, \quad \text{where } x \in Q,\ n \in \mathbb{Z}^m.$$

The Fourier coefficients of a function $f \in \mathscr{L}^2(Q)$ in this system are calculated by the formulas

$$\widehat{f}(n) = \frac{\langle f, e_n \rangle}{\|e_n\|^2} = \frac{1}{(2\pi)^m} \int_Q f(x) e^{-i\langle n,x \rangle}\, dx \quad (n \in \mathbb{Z}^m).$$

From Theorem 10.1.7, it follows that the system $\{e^{i\langle n,x\rangle}\}_{n\in\mathbb{Z}^m}$ is complete, which implies Parseval's identity

$$\int_Q f(x)\overline{g(x)}\, dx = (2\pi)^m \sum_{n\in\mathbb{Z}^m} \widehat{f}(n) \cdot \overline{\widehat{g}(n)}, \quad f, g \in \mathscr{L}^2(Q).$$

Of course, the cube $Q = (-\pi, \pi)^m$ in the two last formulas can be replaced by a shifted cube.

Example Let $0 < \rho \leqslant \pi$. We consider the function $f \in \mathscr{L}^2((-\pi, \pi)^3)$ that is equal to $1/\|x\|$ for $\|x\| < \rho$ and vanishes on $(-\pi, \pi)^3 \setminus B(0, \rho)$. Its norm is easily calculated in spherical coordinates,

$$\|f\|^2 = \int_{B(0,\rho)} \frac{1}{\|x\|^2}\, dx = 4\pi \int_0^\rho \frac{1}{r^2} r^2\, dr = 4\pi\rho.$$

To calculate the Fourier coefficients, we use the formula obtained in the example of Sect. 6.2.5 with $f_0(r) = 1/r$ on $(0, \rho)$, $f_0(r) = 0$ for $r \geqslant \rho$ and $y = n/2\pi$:

$$\widehat{f}(n) = \frac{1}{(2\pi)^3} \int_{B(0,\rho)} \frac{1}{\|x\|} e^{-i\langle n,x\rangle}\, dx = \frac{1}{2\pi^2\|n\|} \int_0^\rho \frac{1}{r} r \sin\bigl(\|n\|r\bigr)\, dr$$
$$= \left(\frac{\sin\frac{\rho}{2}\|n\|}{\pi\|n\|}\right)^2$$

if $n \neq 0$ and $\widehat{f}(0) = \rho^2/4\pi^2$. By Parseval's identity for the function f, we obtain

$$4\pi\rho = (2\pi)^3 \sum_{n\in\mathbb{Z}^3} \left(\frac{\sin\frac{\rho}{2}\|n\|}{\pi\|n\|}\right)^4.$$

Thus, the identity

$$\frac{\pi^2}{t^3} = \sum_{n\in\mathbb{Z}^3} \left(\frac{\sin\|n\|t}{\|n\|t}\right)^4$$

is valid for $t = \frac{\rho}{2} \in (0, \frac{\pi}{2}]$ (the summand for $n = 0$ is equal to 1).

10.2.3 The trigonometric system is closely connected with the orthogonal system $\{z^n\}_{n\in\mathbb{Z}}$ in the space $\mathscr{L}^2(S^1, \sigma)$, where $S^1 = \{z \in \mathbb{C} \mid |z| = 1\}$ is the unit circle and σ is the arc length. Knowing that the trigonometric system is complete in $\mathscr{L}^2((-\pi, \pi))$, we use the change of variable $z = e^{ix}$ $(-\pi < x < \pi)$ and easily verify that the system $\{z^n\}_{n\in\mathbb{Z}}$ is complete in $\mathscr{L}^2(S^1, \sigma)$. Therefore, every function

f in this space is the sum of the series $\sum_{n\in\mathbb{Z}} c_n z^n$, where $c_n = \frac{1}{2\pi}\int_{S^1} f(z)\overline{z}^n\, d\sigma(z)$. The reader familiar with the theory of holomorphic functions will see that this formula coincides with the formula for the nth coefficient of the Laurent expansion of f in the annulus $r < |z| < R$, where $r < 1 < R$. Therefore, the Fourier series in the system $\{z^n\}_{n\in\mathbb{Z}}$ can be regarded as the limit form of the Laurent series, when the annulus degenerates to a circle.

We consider an example connected with the system $\{z^n\}_{n\in\mathbb{Z}}$. Let $T : S^1 \to S^1$ be a rotation of the circle, i.e., the map $z \mapsto T(z) = \zeta z$, where $\zeta \in S^1$ is a fixed number. We now address the question of how much the points of the circle "mix" under the iterations of T. Does there exist an invariant subset of the circle, that is, a set which retains all of its points after rotation? More precisely, a set $E \subset S^1$ is called invariant if it differs from its image only on a set of measure zero, i.e., if $\chi_E = \chi_{T(E)}$ almost everywhere. Of course, such sets exist: the circle S^1 and the set $\{\zeta^n\}_{n\in\mathbb{Z}}$ are examples. It is easy to construct more examples of invariant sets of measure 2π or zero. Therefore, we are interested in the question of whether there are non-trivial invariant sets, i.e., sets satisfying the condition $0 < \sigma(E) < 2\pi$. If $\zeta^m = 1$ for some m, then the map T is repeated after m iterations ($T^{m+1} = T$), and a non-trivial invariant subspace can easily be constructed. We leave this construction to the reader. However, if ζ is not a root of unity, then the map T has no non-trivial invariant sets (such maps are called *ergodic*). Let us prove this.

Let $E \subset S^1$ be an invariant set. Then $\chi_E = \chi_{T(E)}$ almost everywhere, and therefore, $c_n(\chi_{T(E)}) = c_n(\chi_E)$. At the same time, by a change of variable (Corollary 6.1.1), we obtain

$$c_n(\chi_{T(E)}) = \frac{1}{2\pi}\int_{T(E)} \overline{z}^n\, d\sigma(z) = \frac{1}{2\pi}\int_E \overline{(\zeta z)}^n\, d\sigma(z) = \zeta^{-n} c_n(\chi_E).$$

Thus, $c_n(\chi_E)(1-\zeta^{-n}) = 0$ for all $n \in \mathbb{Z}$. Since $1-\zeta^{-n} \neq 0$ for $n \neq 0$, it follows that all Fourier coefficients of χ_E, except, possibly, $c_0(\chi_E)$, are zero. Since the system $\{z^n\}_{n\in\mathbb{Z}}$ is complete, the function χ_E coincides with the sum of its Fourier series almost everywhere. Therefore, χ_E is a constant almost everywhere. Consequently, either $\chi_E(x) = 0$ almost everywhere (the invariant set has measure zero) or $\chi_E(x) = 1$ almost everywhere (the invariant set is a set of full measure).

10.2.4 We will now give other examples of orthogonal systems. Let $P_n(x) = ((x^2-1)^n)^{(n)}$, $n = 0, 1, \ldots$. The polynomials P_n are called the *Legendre polynomials*. Obviously, $\deg P_n = n$, and so every polynomial is a linear combination of Legendre polynomials, which form an orthogonal system in the space $\mathscr{L}^2((-1,1))$. Indeed, for $m < n$, we have

$$\begin{aligned}\langle P_m, P_n\rangle &= \int_{-1}^{1} P_m(x)\big((x^2-1)^n\big)^{(n)}\, dx \\ &= P_m(x)\big((x^2-1)^n\big)^{(n-1)}\Big|_{-1}^{1} - \int_{-1}^{1} P'_m(x)\big((x^2-1)^n\big)^{(n-1)}\, dx \\ &= -\int_{-1}^{1} P'_m(x)\big((x^2-1)^n\big)^{(n-1)}\, dx.\end{aligned}$$

Integrating by parts n times, we arrive at the equation

$$\langle P_m, P_n\rangle = (-1)^n \int_{-1}^{1} P_m^{(n)}(x)(x^2-1)^n\,dx,$$

where $P_m^{(n)}(x) \equiv 0$, since $\deg P_m < n$. Thus, $\langle P_m, P_n\rangle = 0$ for $m \neq n$.

Theorem *The Legendre polynomials form a basis in the space* $\mathscr{L}^2((-1,1))$.

Proof As in the proof of Theorem 10.2.1, we use Corollary 10.1.5. We must verify that every function in $\mathscr{L}^2((-1,1))$ can be approximated arbitrarily closely (in the $\mathscr{L}^2$-norm) by linear combinations of polynomials P_n, i.e., by arbitrary algebraic polynomials. This, however, has already been established in Corollary 9.2.3. □

We mention one more useful orthogonal system. In the space $\mathscr{L}^2(\mathbb{R})$, we consider the *Hermite*[8] *functions*

$$h_n(x) = e^{x^2/2}\bigl(e^{-x^2}\bigr)^{(n)}, \qquad n = 0, 1, \ldots.$$

It is easy to verify that $h_n(x) = H_n(x)e^{-x^2/2}$, where H_n is an nth degree polynomial called a *Hermite polynomial*. The orthogonality of the Hermite functions can be established by integrating by parts the equation

$$\langle h_m, h_n\rangle = \int_{-\infty}^{\infty} H_m(x)\bigl(e^{-x^2}\bigr)^{(n)}\,dx$$

in the same way as in the proof of the orthogonality of the Legendre polynomials. It is obvious that the orthogonality of the Hermite functions in $\mathscr{L}^2(\mathbb{R})$ implies the orthogonality of the Hermite polynomials in $\mathscr{L}^2(\mathbb{R},\mu)$ with measure $d\mu(x) = e^{-x^2}\,dx$.

Later on (see the corollary in Sect. 10.5.6) we prove that the system of functions h_n is complete in $\mathscr{L}^2(\mathbb{R})$ or, equivalently, the system of polynomials H_n is complete in $\mathscr{L}^2(\mathbb{R},\mu)$.

10.2.5 In the applications of probability theory in analysis, the sequence of Rademacher functions r_n defined in Sect. 6.4.5 plays an important role. As has already been proved, these functions are independent in the sense of Definition 4.4.4. Since, in addition, $\int_0^1 r_n(x)\,dx = 0$, we see that the relation

$$\int_0^1 r_{n_1}(x)r_{n_2}(x)\cdots r_{n_m}(x)\,dx = \prod_{k=1}^{m}\int_0^1 r_{n_k}(x)\,dx = 0 \tag{1}$$

holds for $1 \leqslant n_1 < n_2 < \cdots < n_m$.

[8] Charles Hermite (1822–1901)—French mathematician.

In particular, the Rademacher functions form an orthonormal system in the space $\mathscr{L}^2((0,1))$. Of course, this system is not complete: for example, the pairwise products $r_j r_k$ are orthogonal to all Rademacher functions. To obtain a complete system containing the Rademacher functions, we proceed as follows. For every non-empty finite set $A \subset \mathbb{N}$, we consider the function $w_A = \prod_{n\in A} r_n$. Furthermore, we will assume, by definition, that $w_\varnothing \equiv 1$. The functions w_A are called the *Walsh*[9] *functions*. The Rademacher functions are the Walsh functions corresponding to the one-element sets. By Eq. (1), the functions w_A are pairwise orthogonal. The system of Walsh functions is complete in $\mathscr{L}^2((0,1))$. To prove this, we need the following lemma.

Lemma *Let $n \in \mathbb{N}$. The set of linear combinations of the functions w_A such that $A \subset \{1, 2, 3, \dots, 2^n\}$ coincides with the set of linear combinations of the characteristic functions of the intervals $\Delta_{n,k} = (k2^{-n}, (k+1)2^{-n})$ for $k = 0, 1, \dots, 2^n - 1$.*

Proof Let L_1 and L_2 be the linear spans of the first and second systems, respectively. Since the functions $r_1, \dots, r_n$ are constant on the intervals $\Delta_{n,k}$, the Walsh functions in question are also constant on these intervals. Therefore, $L_1 \subset L_2$. At the same time, the dimensions of L_1 and L_2 are, obviously, equal (to 2^n). Hence it follows that $L_1 = L_2$. □

Theorem *The system of Walsh functions is complete in the space $\mathscr{L}^2((0,1))$.*

Proof We use Corollary to Theorem 10.1.5 on the characterization of bases. We will prove that every function f in $\mathscr{L}^2((0,1))$ can be approximated arbitrarily closely in norm by linear combinations of Walsh functions. If f is the characteristic function of an interval $(p,q) \subset (0,1)$, then, for a given ε, we can find a large n such that p and q can be approximated by the points $j/2^n$ and $k/2^n$ within ε. Then $\|f - \chi_\Delta\|^2 < 2\varepsilon$, where χ_Δ is the characteristic function of the interval $(j/2^n, k/2^n)$, which almost everywhere coincides with the sum $\sum_{s=j}^{k-1} \chi_{\Delta_{n,s}}$ equal, by the lemma, to a certain linear combination of Walsh functions. Being able to approximate the characteristic functions of the intervals, we can also approximate their linear combinations, i.e., the step functions. Now, we consider the general case. By Theorem 9.2.2, for each ε, we can find a step function g such that $\|f - g\| < \varepsilon$. Approximating g within ε by a linear combination h of Walsh functions, we obtain $\|f - h\| \leqslant \|f - g\| + \|g - h\| < 2\varepsilon$. Since ε was arbitrary, this completes the proof. □

10.2.6 From the viewpoint of probability theory, the Rademacher functions give an example of a sequence of independent trials with two equiprobable outcomes (the simplest "Bernoulli scheme"). Here, a "simple" random event is a roll of a number $x \in (0,1)$, and the probability that a point will fall in the interval (p,q) is

---

[9] Joseph Leonard Walsh (1895–1973)—American mathematician.

the length of the interval. A "trial" consists of the calculation of the values of the Rademacher functions: the first trial is the calculation of $r_1(x)$, the second trial is the calculation of $r_2(x)$, etc. Taking into account the connection between the values of a Rademacher function at a given point and the binary digits of the point, we can replace $r_n(x)$ by $\varepsilon_n(x)$ (the binary digits of x).

One of the first results of probability theory is Bernoulli's law of large numbers, which says that in the scheme described above the frequency of occurrence of 0 or 1 becomes close to 1/2 with probability arbitrarily close to 1. In the language of measure theory, this result means that, on the interval (0, 1), the arithmetic mean $\frac{1}{n}(\varepsilon_1(x) + \cdots + \varepsilon_n(x))$ (the frequency of occurrence of the digit 1 in the binary expansion of a point x) tends to 1/2 in measure. Returning to the Rademacher functions, we can say that

$$\frac{r_1(x) + \cdots + r_n(x)}{n} \underset{n\to\infty}{\longrightarrow} 0 \quad \text{in measure.}$$

This assertion follows from the fact that

$$\frac{1}{n}\|r_1 + \cdots + r_n\| = \frac{1}{\sqrt{n}} \underset{n\to\infty}{\longrightarrow} 0,$$

and the convergence in norm implies the convergence in measure.

Two centuries after Bernoulli, Borel proved a stronger statement.

Theorem (Strong law of large numbers)

$$\frac{r_1(x) + \cdots + r_n(x)}{n} \underset{n\to\infty}{\longrightarrow} 0 \quad \textit{almost everywhere on } (0,1).$$

Proof We put $S_n(x) = r_1(x) + \cdots + r_n(x)$ and estimate the integral $\int_0^1 S_n^4(x)\,dx$. Obviously,

$$S_n^2(x) = \sum_{k=1}^{n} r_k^2(x) + 2\sum_{1\leqslant j<k\leqslant n} r_j(x) r_k(x) = n + 2\sum_{1\leqslant j<k\leqslant n} w_{\{j,k\}}(x).$$

Since the Walsh functions $w_{\{j,k\}}$ form an orthonormal system, the Pythagorean theorem implies

$$\int_0^1 S_n^4(x)\,dx = \Bigl\| n w_\varnothing + 2\sum_{1\leqslant j<k\leqslant n} w_{\{j,k\}} \Bigr\|^2 = n^2 + 4\sum_{1\leqslant j<k\leqslant n} 1 < 3n^2.$$

Consequently, $\sum_{n=1}^{\infty}\int_0^1 (\frac{1}{n}\, S_n(x))^4 < \sum_{n=1}^{\infty}\frac{3}{n^2} < +\infty$, and, therefore, the series $\sum_{n=1}^{\infty}(\frac{1}{n}\, S_n(x))^4$ converges almost everywhere (see Corollary 2 of Sect. 4.8.2). This implies the assertion of the theorem since the terms of a convergent series tend to zero. □

10.2.7 The theorem just proved admits various generalizations also called the strong laws of large numbers. The statements concerning sequences of independent functions with zero mean values (obviously, these functions form an orthogonal system) are of most interest. Before passing to this question, we consider an inequality playing a decisive role in the study of series of such functions.

Throughout this section, we consider real functions in the space $\mathscr{L}^2(X,\mu)$, assuming that the measure μ is normalized ($\mu(X)=1$).

Theorem (Kolmogorov's[10] inequality) *Let $f_1,\dots,f_n$ in $\mathscr{L}^2(X,\mu)$ be independent and have zero means, $\int_X f_1\,d\mu=\dots=\int_X f_n\,d\mu=0$. Then the inequality*

$$\mu\Big(\Big\{x\in X \mid \max_{1\leqslant k\leqslant n}\big|f_1(x)+\dots+f_k(x)\big|\geqslant t\Big\}\Big)\leqslant\frac{1}{t^2}\sum_{k=1}^{n}\int_X f_k^2\,d\mu$$

holds for every $t>0$.

Proof We put $S_k=f_1+\dots+f_k$, $S_k^*=\max_{1\leqslant j\leqslant k}|S_j|$ and $R_k=S_n-S_k$. We need to estimate the measure of the set $E=\{x\in X \mid S_n^*(x)\geqslant t\}$. To this end, we divide the set into disjoint parts $E_k=\{x\in X \mid S_{k-1}^*(x)<t\leqslant S_k^*(x)\}$ (we assume that $S_0^*\equiv 0$). Then

$$\begin{aligned}\sum_{k=1}^{n}\int_X f_k^2\,d\mu&=\int_X S_n^2\,d\mu\geqslant\int_E S_n^2\,d\mu=\sum_{k=1}^{n}\int_{E_k}(S_k+R_k)^2\,d\mu\\&=\sum_{k=1}^{n}\Big(\int_{E_k}S_k^2\,d\mu+2\int_{E_k}S_kR_k\,d\mu+\int_{E_k}R_k^2\,d\mu\Big)\\&\geqslant\sum_{k=1}^{n}\int_{E_k}S_k^2\,d\mu+2\sum_{k=1}^{n}\int_{E_k}S_kR_k\,d\mu.\end{aligned}$$

By the corollary to Lemma 6.4.4, the functions $S_k\chi_{E_k}$ and R_k are independent. Therefore,

$$\int_{E_k}S_kR_k\,d\mu=\int_X S_k\chi_{E_k}R_k\,d\mu=\int_X S_k\chi_{E_k}\,d\mu\cdot\int_X R_k\,d\mu=0.$$

Since $|S_k|=S_k^*\geqslant t$ on the set E_k, we obtain the required inequality,

$$\sum_{k=1}^{n}\int_X f_k^2\,d\mu\geqslant\sum_{k=1}^{n}\int_{E_k}S_k^2\,d\mu\geqslant\sum_{k=1}^{n}t^2\mu(E_k)=t^2\mu(E).$$ □

[10] Andrei Nikolaevich Kolmogorov (1903–1987)—Russian mathematician.

We supplement the theorem (preserving the notation) and verify that, for a sequence of independent functions f_n satisfying the assumptions of the theorem, the following statement is true.

Corollary *If $A^2 = \sum_{n=1}^{\infty} \int_X f_n^2 \, d\mu < +\infty$, then the function $S^* = \sup_{k \geqslant 1} |S_k| = \sup_{k \geqslant 1} S_k^*$ is summable and $\int_X S^* \, d\mu \leqslant 2A$.*

Proof For every $t > 0$, the set $X(S^* \geqslant t)$ is exhausted by the expanding sequence of sets $X(S_k^* \geqslant t)$. By the theorem, the measure of each of these sets does not exceed A^2/t^2. Consequently, $\mu(X(S^* \geqslant t)) \leqslant A^2/t^2$. Thus, $F(t) \leqslant A^2/t^2$, where F is the decreasing distribution function for S^*. Using the formula of Proposition 6.4.3 with $p = 1$, we see that

$$\int_X S^* \, d\mu = \int_0^{\infty} F(t) \, dt = \int_0^A \cdots + \int_A^{\infty} \cdots$$

$$\leqslant AF(0) + \int_A^{\infty} \frac{A^2}{t^2} \, dt \leqslant A + A = 2A. \qquad \square$$

10.2.8 The estimate for the integral of the function S^* established in the previous corollary leads to an important result concerning the behavior of the series $\sum_{n=1}^{\infty} f_n$, which, in turn, implies a generalization of Borel's theorem.

Theorem 1 *Let $\{f_n\}_{n=1}^{\infty}$ be a sequence of independent functions with zero means. If $\sum_{n=1}^{\infty} \int_X f_n^2 \, d\mu < +\infty$, then the series $\sum_{n=1}^{\infty} f_n$ converges almost everywhere.*

Proof We put

$$S_n = f_1 + \cdots + f_n \quad \text{and} \quad R_n = \sup_{p \geqslant 1} |S_{n+p} - S_n|.$$

Since $|S_{n+p} - S_n| \leqslant 2R_m$ for $n \geqslant m$ and all m and p, we must verify only that $\inf_n R_n = 0$ almost everywhere. For this, it is sufficient to verify the relation $\int_X R_n \, d\mu \underset{n\to\infty}{\longrightarrow} 0$, which follows immediately from Corollary 10.2.7,

$$\int_X R_n \, d\mu \leqslant 2 \left(\sum_{k=n+1}^{\infty} \int_X f_k^2 \, d\mu \right)^{1/2} \underset{n\to\infty}{\longrightarrow} 0. \qquad \square$$

Corollary *Let $\{f_n\}_{n=1}^{\infty}$ be a sequence of independent functions with zero means. If $\sum_{n=1}^{\infty} \frac{1}{n^2} \int_X f_n^2 \, d\mu < +\infty$, then $\sigma_n = \frac{1}{n} \sum_{k=1}^{n} f_k \underset{n\to\infty}{\longrightarrow} 0$ almost everywhere.*

Proof By the theorem, the sums $T_n = \sum_{k=1}^{n} \frac{1}{k} f_k$ have a finite limit almost everywhere. The quantities $\theta_n = \frac{1}{n+1}(T_1 + \cdots + T_n)$ have the same limit. Therefore, the difference $T_n - \theta_n$ tends to zero almost everywhere. At the same time, it is easy to verify that $T_n - \theta_n = \frac{1}{n+1}(f_1 + \cdots + f_n)$, which completes the proof. □

A similar statement can be obtained for an arbitrary orthogonal system if we drop the independence requirement and strengthen the restriction on the quantities $\|f_n\|$ (see Exercise 11).

If we impose quite natural additional restrictions on the independent functions f_n, then the condition $\sum_{n=1}^{\infty} \int_X f_n^2 \, d\mu < +\infty$ will turn out to be not only sufficient but also necessary for the convergence of the series $\sum_{n=1}^{\infty} f_n$ almost everywhere (or, equivalently by the zero-one law, on a set of positive measure).

Theorem 2 *Let $\{f_k\}_{k=1}^{\infty}$ be a sequence of independent bounded functions with zero means. If the series $\sum_{k=1}^{\infty} f_k$ converges almost everywhere, then $\sum_{k=1}^{\infty} \int_X f_k^2 \, d\mu < +\infty$.*

Proof We put $S = \sum_{k=1}^{\infty} f_k$ and $S_n = \sum_{k=1}^{n} f_k$ $(n = 1, 2, \ldots)$. Since the sum S is finite almost everywhere, the sequences $\{S_n(x)\}_n$ are bounded for almost all x. They are uniformly bounded on some set of positive measure. Therefore, for a sufficiently large t, the intersection $E = \bigcap_{n=1}^{\infty} E_n$ of the sets $E_n = \{x \in X \mid |S_k(x)| \leqslant t \text{ for } k = 1, \ldots, n\}$ has a positive measure. We find a recurrence estimate for the integrals

$$I_n = \int_{E_n} S_n^2 \, d\mu.$$

For this, we use the independence of the functions f_{n+1} and $S_n \chi_{E_n}$ (see the corollary of Lemma 6.4.4). This gives us the relations

$$\int_{E_n} S_n f_{n+1} \, d\mu = \int_X \chi_{E_n} S_n \, d\mu \cdot \int_X f_{n+1} \, d\mu = 0$$

and

$$\int_{E_n} f_{n+1}^2 \, d\mu = \int_X \chi_{E_n} f_{n+1}^2 \, d\mu = \mu(E_n) \int_X f_{n+1}^2 \, d\mu \geqslant \mu(E) \int_X f_{n+1}^2 \, d\mu.$$

Therefore, putting $F_n = E_n \setminus E_{n+1}$, we arrive at the inequality

$$I_{n+1} = \int_{E_n} (S_n + f_{n+1})^2 \, d\mu - \int_{F_n} S_{n+1}^2 \, d\mu \geqslant I_n + \mu(E) \int_X f_{n+1}^2 \, d\mu - \int_{F_n} S_{n+1}^2 \, d\mu.$$

By assumption, there is a number c such that, for all n, the inequality $|f_n| \leqslant c$ holds almost everywhere. Then

$$\left|S_{n+1}(x)\right| \leqslant \left|S_n(x)\right| + \left|f_{n+1}(x)\right| \leqslant t + c \quad \text{for almost all } x \text{ in } E_n.$$

Thus,

$$I_{n+1} - I_n + (t+c)^2 \mu(F_n) \geqslant \mu(E) \int_X f_{n+1}^2 \, d\mu.$$

Since $\sum_{k=1}^{n}(I_{k+1} - I_k) \leqslant I_{n+1} \leqslant t^2$ and $\sum_{k=1}^{n}\mu(F_k) \leqslant 1$, it follows that the series $\sum_k \mu(E) \int_X f_{k+1}^2 \, d\mu$ converges, which is equivalent to the assertion of the theorem since $\mu(E) > 0$. □

EXERCISES

1. Prove that the systems $1, \cos x, \cos 2x, \ldots$ and $\sin x, \sin 2x, \ldots$ are complete in the space $\mathscr{L}^2((0, \pi))$.

2. Let μ be a measure on the interval $(-1, 1)$ having density $\frac{1}{\sqrt{1-x^2}}$ with respect to Lebesgue measure. Prove that the functions $T_n(x) = \cos(n \arccos x)$ $(n = 0, 1, 2, \ldots)$ form an orthogonal basis in the space $\mathscr{L}^2((-1, 1), \mu)$. Verify that T_n is an algebraic polynomial of degree n (a *Chebyshev polynomial*).

3. Prove that the functions $x \mapsto e^{x/2}(x^n e^{-x})^{(n)}$ $(n = 0, 1, \ldots)$, called the *Laguerre*[11] *functions*, form an orthogonal system in the space $\mathscr{L}^2((0, +\infty))$.

4. Let m be one of the digits $0, 1, \ldots, 9$, and let $c_m(x) = 1$ if the decimal expansion of the fractional part of x has the form $0.m\ldots$ and $c_m(x) = 0$ otherwise. Verify that

$$\frac{1}{n}\sum_{0 \leqslant k < n} c_m(10^k x) \underset{n\to\infty}{\longrightarrow} \frac{1}{10} \quad \text{almost everywhere on } (0, 1).$$

5. Generalize the result of the previous exercise by proving that almost all numbers $x \in (0, 1)$ are normal, i.e., for each $p \in \mathbb{N}$, the p-ary expansion of x contains all digits (the numbers $0, 1, \ldots, p-1$) "equally often".

In Exercises 6–9, by $r_1, r_2, \ldots, r_n \ldots$ we denote the Rademacher functions.

6. Use the Khintchine inequality to supplement Theorem 10.2.6 by proving that

$$\frac{r_1(x) + \cdots + r_n(x)}{n^p} \underset{n\to\infty}{\longrightarrow} 0 \quad \text{almost everywhere on } (0, 1)$$

for $p > 1/2$.

7. Verify that the result of the previous exercise is false for $p = 1/2$. *Hint*. Find the limits of the integrals $\int_0^1 e^{i\sigma_n(x)} dx$, where $\sigma_n(x) = \frac{r_1(x)+\cdots+r_n(x)}{\sqrt{n}}$.

8. Show that if the sum of the series $\sum_{n=1}^{\infty} a_n r_n$ is bounded almost everywhere on some non-degenerate interval, then $\sum_{n=1}^{\infty} |a_n| < +\infty$.

9. Let $f_n(x, y) = \sum_{j=1}^{n} r_j(x) r_j(y)$. Show that $|\iint_{A\times B} f_n(x, y)\, dx\, dy| \leqslant 1$ for any measurable sets $A, B \subset (0, 1)$, but nevertheless $\iint_{(0,1)^2} |f_n(x, y)|\, dx\, dy \to +\infty$. *Hint*. Use Bessel's inequality and the inequality from Exercise 7 of Sect. 9.1 with $p = 1$.

10. Refine the assertion of Corollary 10.2.8 by proving that the functions σ_n are dominated by a summable function.

---

[11] Edmond Nicolas Laguerre (1834–1886)—French mathematician.

11. Let $\{f_n\}_{n=1}^{\infty}$ be an orthogonal system in $\mathscr{L}^2(X,\mu)$ such that $\sum_{n=1}^{\infty} \frac{\|f_n\|^2}{n^{3/2}} < +\infty$. Prove that $\frac{1}{n}(f_1 + \cdots + f_n) \underset{n\to\infty}{\longrightarrow} 0$ almost everywhere on X and the function $\sup_n \frac{1}{n}|f_1 + \cdots + f_n|$ belongs to $\mathscr{L}^2(X,\mu)$.

12. Verify that the assumptions of Theorem 2 of Sect. 10.2.8 can be weakened by replacing the convergence of the series $\sum_{k=1}^{\infty} f_k$ almost everywhere by the boundedness of its partial sums at the points of a set of positive measure.

10.3 Trigonometric Fourier Series

The present and following sections are devoted to harmonic analysis. Without striving to expose this important and vast subject in its entirety, we restrict ourselves to the exposition of selected topics the choice of which is motivated only by the desire to demonstrate the methods developed above.

In Sect. 10.1, we established important properties of Fourier series in arbitrary orthogonal systems. Now, we consider the properties of Fourier series in trigonometric systems in more detail. This is historically the first example of an orthogonal system, and the problem of the representability of a function as the sum of a trigonometric series was one of the central problems in mathematics for nearly two hundred years.

Suffice to say that the lively discussion in the 18th century devoted to this problem provided an important impetus for the formulation of the modern concept of function. Riemann introduced his definition of an integral in connection with the study of trigonometric series, and Cantor, studying the uniqueness of the expansion of a function as a trigonometric series, came up with his foundation of set theory.

10.3.1 We recall that, according to the general definition 10.2.1, the Fourier series of a function $f \in \mathscr{L}^2((0,2\pi))$ in the systems

$$1,\quad \cos x,\quad \sin x,\quad \ldots,\quad \cos nx,\quad \sin nx,\quad \ldots,\quad \text{and}\quad \{e^{inx}\}_{n\in\mathbb{Z}}$$

have, respectively, the forms

$$A(f) + \sum_{n=1}^{\infty}\big(a_n(f)\cos nx + b_n(f)\sin nx\big) \tag{1}$$

and

$$\sum_{n=-\infty}^{\infty} \widehat{f}(n)e^{inx}, \tag{1'}$$

where the Fourier coefficients are calculated by the formulas

$$A(f) = \frac{1}{2\pi}\int_0^{2\pi} f(x)\,dx,$$

$$a_n(f) = \frac{1}{\pi}\int_0^{2\pi} f(x)\cos nx\,dx, \tag{2}$$

$$b_n(f) = \frac{1}{\pi}\int_0^{2\pi} f(x)\sin nx\,dx \quad (n \in \mathbb{N});$$

$$\widehat{f}(n) = \frac{1}{2\pi}\int_0^{2\pi} f(x)e^{-inx}\,dx \quad (n \in \mathbb{Z}). \tag{2'}$$

Unlike the previous section, where we considered only functions of class $\mathscr{L}^2$, here we will deal with arbitrary functions summable on $(0, 2\pi)$. It is obvious that, in this case, the integrands in formulas (2) and (2′) will also be summable. Therefore, we keep the terminology introduced above (a Fourier coefficient, a Fourier series) for the functions in $\mathscr{L}^1((0, 2\pi))$. We are now interested not in convergence in the $\mathscr{L}^2$-norm, but in other types of convergence, and first of all, pointwise convergence. Here, by the sum of the series (1′), we always mean the limit of the symmetric partial sums

$$S_n(f, x) = \sum_{|k|\leqslant n} \widehat{f}(k)e^{ikx}, \tag{3}$$

which are also called the *Fourier sums* of the function f. As noted in Sect. 10.2.1, the partial sums of series (1) and (1′) are equal. Thus, all results obtained for one of the series are valid for the other one. In the sequel, we will mainly consider series (1′) because this leads to some technical simplifications.

In conclusion, we touch on a question that may arise when solving the problem of the expansion of a function as a trigonometric series. Up to now, the choice of its coefficients have been dictated by geometric considerations presented in Sect. 10.1 and has led to formulas (2) and (2′). Can it happen that, for a different mode of convergence (e.g., pointwise or in measure) the coefficients of the trigonometric series must be chosen in a different way? It is easy to verify, however, that, under mild additional assumptions, there is essentially no freedom in the choice of the coefficients. Indeed, if, for example, a trigonometric series $\sum_{k=-\infty}^{\infty} c_k e^{ikx}$ converges to a function f almost everywhere or in measure and its partial sums $S_n(x) = \sum_{|k|\leqslant n} c_k e^{ikx}$ have a summable majorant, i.e., a function $g \in \mathscr{L}^1((0, 2\pi))$ such that $|S_n(x)| \leqslant g(x)$ for all $x \in (0, 2\pi)$ and $n \in \mathbb{N}$, then the coefficients of the series coincide with the Fourier coefficients of the function f, $c_k \equiv \widehat{f}(k)$. Indeed, by Lebesgue's theorem, the integral $\widehat{f}(k) = \frac{1}{2\pi}\int_0^{2\pi} f(x)e^{-ikx}\,dx$ is the limit (as $n \to \infty$) of the integrals $\frac{1}{2\pi}\int_0^{2\pi} S_n(x)e^{-ikx}\,dx$, each of which is equal to c_k for $n \geqslant |k|$.

10.3.2 Instead of functions defined only on the interval $(0, 2\pi)$, it will be more convenient for us to deal with 2π-periodic functions. Since every function defined on $(0, 2\pi)$ can be extended to a periodic function, we will assume in what follows that all functions in question are periodic (in the sequel, periodicity means 2π-periodicity). Being summable on an interval of length 2π, such functions are summable on each finite interval. We will repeatedly use the fact that the integral $\int_a^{a+2\pi} f(x)\,dx$ does not depend on the parameter a (the reader is invited to prove this independently). Often, especially when dealing with odd and even functions, it is more convenient to integrate over the interval $(-\pi,\pi)$ in formulas (2) and (2′). By $\widetilde{C}$ and $\widetilde{C}^r$ $(1\leqslant r\leqslant+\infty)$, we denote the classes of periodic functions that are continuous and, respectively, r times continuously differentiable on $\mathbb{R}$; by $\widetilde{\mathscr{L}}^p$, we denote the class of periodic functions summable on $(-\pi,\pi)$ with power $p\geqslant 1$. For a function $f\in\widetilde{\mathscr{L}}^p$, by $\|f\|_p$ we mean the $\mathscr{L}^p$-norm of its restriction to $(-\pi,\pi)$.

We note the following elementary properties of the Fourier coefficients.

(a) $|\widehat{f}(n)|\leqslant\frac{1}{2\pi}\|f\|_1$ (see formula (2′)).
(b) $\widehat{f}(n)\underset{|n|\to+\infty}{\longrightarrow}0$ (see the Riemann–Lebesgue theorem).

This qualitative result can be supplemented by an estimate connected with the continuity in the mean (see Exercise 1).

The properties connecting Fourier coefficients with translation, differentiation, and convolution play an important role. We recall that the translation f_h of a function $f\in\widetilde{\mathscr{L}}^1$ corresponding to a number h is defined by the formula $f_h(x)=f(x-h)$. Making the change of variable $x-h\mapsto x$ in the integral $\int_0^{2\pi} f(x-h)\times e^{-inx}\,dx$, we arrive at the formula

(c) $\widehat{f_h}(n)=e^{-inh}\widehat{f}(n)$.
(d) If a periodic function f is absolutely continuous on $\mathbb{R}$ (in particular, if it is piecewise differentiable), then

$$\widehat{f'}(n)=in\widehat{f}(n)\quad(n\in\mathbb{Z})$$

(for the proof, it is sufficient to integrate by parts). In particular, $\widehat{f}(n)=o(1/n)$. We note a weak version of this estimate for a function of bounded variation.

(d′) If f is a function of bounded variation on the interval $[0,2\pi]$, then $\widehat{f}(n)=O(1/n)$. Indeed, integrating by parts (see Sect. 4.11.4), we obtain

$$2\pi\widehat{f}(n)=\int_0^{2\pi} f(x)e^{-inx}\,dx=f(x)\frac{e^{-inx}}{-in}\bigg|_0^{2\pi}+\frac{1}{in}\int_0^{2\pi}e^{-inx}\,df(x)$$
$$=O\left(\frac{1}{n}\right).$$

(e) Let $f,g\in\widetilde{\mathscr{L}}^1$. Then

$$\widehat{f*g}(n)=2\pi\widehat{f}(n)\cdot\widehat{g}(n)\quad\text{for all } n\in\mathbb{Z}$$

(for the definition of the convolution of periodic functions, see Sect. 7.5.5). The proof is obtained by direct calculation using the change of the order of integration,

$$\begin{aligned}\widehat{f*g}(n) &= \frac{1}{2\pi}\int_{-\pi}^{\pi}(f*g)(x)e^{-inx}\,dx\\ &= \frac{1}{2\pi}\int_{-\pi}^{\pi}\left(\int_{-\pi}^{\pi}f(x-t)g(t)\,dt\right)e^{-inx}\,dx\\ &= \frac{1}{2\pi}\int_{-\pi}^{\pi}g(t)e^{-int}\left(\int_{-\pi}^{\pi}f(x-t)e^{-in(x-t)}dx\right)dt\\ &= \frac{1}{2\pi}\int_{-\pi}^{\pi}g(t)e^{-int}\left(\int_{-\pi}^{\pi}f(u)e^{-inu}du\right)dt = 2\pi\widehat{g}(n)\cdot\widehat{f}(n).\end{aligned}$$

10.3.3 The problem of the Fourier series expansion of a function is rather complicated and has a long history. The famous work "The analytical theory of heat" by Fourier, in which the series that were later named after him were first studied and used systematically, did not contain an explicit formulation of a condition providing the expandability of a function as a Fourier series. Such criteria arose later. Still later it became clear that the Fourier series of a continuous function can diverge at some points, and, as Kolmogorov proved, the Fourier series of a summable function can diverge everywhere.

So far, even knowing that a Fourier series of a differentiable function converges at a point, we cannot be sure that its sum coincides with the value of the function.

At the moment, we know (see Sect. 10.2.1) that if f is a square-summable function, then series (1′) converges in the $\mathscr{L}^2$-norm and its sum is equal to f. If a function f is only assumed to be summable, the question of the convergence of a Fourier series (pointwise, in an $\mathscr{L}^p$-norm, or in some other sense) remains open for the time being.

We begin the investigation of a Fourier series' convergence with the derivation of an important formula for its partial sums discovered by Dirichlet. Relying on formula (2′), we transform Eq. (3) as follows:

$$S_n(f,x) = \sum_{|k|\leqslant n}\left(\frac{1}{2\pi}\int_{-\pi}^{\pi}f(t)e^{-ikt}\,dt\right)e^{ikx} = \frac{1}{2\pi}\int_{-\pi}^{\pi}f(t)\sum_{|k|\leqslant n}e^{ik(x-t)}\,dt.$$

The function

$$D_n(u) = \frac{1}{2\pi}\sum_{|k|\leqslant n}e^{iku} = \frac{1}{2\pi} + \frac{1}{\pi}\sum_{k=1}^{n}\cos ku \qquad (4)$$

is called the nth *Dirichlet kernel*. Obviously, the Dirichlet kernel is even and periodic. Summing the geometric sequence $\sum_{|k|\leqslant n}e^{iku}$, we obtain

$$D_n(u) = \frac{\sin(n+\frac{1}{2})u}{2\pi\sin\frac{u}{2}} \quad \text{for } u\notin 2\pi\mathbb{Z}. \qquad (4')$$

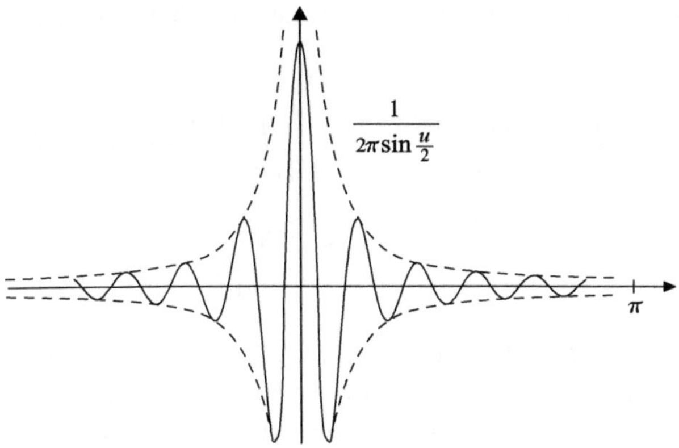

Fig. 10.1 Graph of the Dirichlet kernel

From this, we see that the function D_n is strongly oscillating for large n, and, in a neighborhood of zero, it takes extreme values with alternating signs and absolute values comparable with $\max D_n = D_n(0) = \frac{1}{\pi}(n + \frac{1}{2})$ (see Fig. 10.1).

It follows directly from the definition that the sum of the Fourier series is the convolution of the function and the Dirichlet kernel,

$$S_n(f, x) = \int_{-\pi}^{\pi} f(t) D_n(x - t)\, dt = (f * D_n)(x).$$

Since the integrands are periodic, we can also represent the above equation in the form

$$S_n(f, x) = \int_{-\pi}^{\pi} f(x - u) D_n(u)\, du. \tag{5}$$

Considering periodic approximate identities, we have encountered similar formulas (see Sect. 7.6.5). The Dirichlet kernels satisfy conditions (b) and (c) of the definition of a periodic approximate identity; it immediately follows from Eq. (4) that

$$\int_{-\pi}^{\pi} D_n(u)\, du = 1.$$

Moreover, we have

$$\int_{\delta<|u|<\pi} D_n(u)\, du = \int_{\delta<|u|<\pi} \frac{\sin(n + \frac{1}{2})u}{2\pi \sin \frac{u}{2}}\, du \underset{n\to\infty}{\longrightarrow} 0$$

for each $\delta \in (0, \pi)$ (the passage to the limit can be justified by integration by parts or by referring to the Riemann–Lebesgue theorem).

However, D_n does not satisfy the most important property of an approximate identity, namely, the positivity. Moreover, the Dirichlet kernels do not satisfy the periodic analog of condition (a′) of Sect. 7.6.1, i.e., they have unbounded $\mathcal{L}^1$-norms. Indeed,

$$\int_{-\pi}^{\pi} \big|D_n(u)\big|\,du = \int_0^{\pi} \frac{|\sin(n+\frac{1}{2})u|}{\pi \sin\frac{u}{2}}\,du \geqslant \frac{2}{\pi}\int_0^{\pi} \frac{|\sin(n+\frac{1}{2})u|}{u}\,du$$

$$= \frac{2}{\pi}\int_0^{\pi(n+\frac{1}{2})} \frac{|\sin v|}{v}\,dv \geqslant \frac{2}{\pi}\sum_{k=1}^{n}\int_{(k-1)\pi}^{k\pi} \frac{|\sin v|}{k\pi}\,dv = \frac{4}{\pi^2}\sum_{k=1}^{n}\frac{1}{k}.$$

Since $\sum_{k=1}^{n}\frac{1}{k} \geqslant \int_1^n \frac{1}{x}\,dx = \ln n$, we have $\|D_n\|_1 \geqslant \frac{4}{\pi^2}\ln n$ (see also Exercise 11).

Thus, the general theorems connected with the use of approximate identities cannot be applied here. This is the cause of considerable difficulties in the study of the convergence of Fourier series. Here, we meet not just technical questions, but those of a fundamental nature. We will see later that the proofs of Theorems 2 and 3 of Sect. 9.3.7 cannot be carried over to convolutions with Dirichlet kernels.

At the same time, in many problems, it is essential that the norms $\|D_n\|_1$ increase quite slowly. Indeed, the estimate from above for $\|D_n\|_1$ just obtained is exact in order,

$$\|D_n\|_1 = \int_0^{\pi} \frac{|\sin(n+\frac{1}{2})u|}{\pi \sin\frac{u}{2}}\,du \leqslant \int_0^{\pi} \frac{|\sin(n+\frac{1}{2})u|}{u}\,du = \int_0^{\pi(n+\frac{1}{2})} \frac{|\sin v|}{v}\,dv.$$

Consequently, $\|D_n\|_1 \leqslant 1 + \int_1^{\pi(n+\frac{1}{2})}\frac{dv}{v}$, and, therefore, $\|D_n\|_1 \leqslant 2\ln n$ for $n \geqslant 10$. Since $S_n(f) = f * D_n$, we obtain the following estimate for the Fourier sums of a bounded function ($n \geqslant 10$):

$$\big\|S_n(f)\big\|_\infty \leqslant \|f\|_\infty \|D_n\|_1 \leqslant 2\|f\|_\infty \ln n. \tag{6}$$

The partial sums of the Fourier series are calculated by formula (5), and so depend on the values of the function on an interval of length 2π. It is all the more surprising that, as we will now verify, the convergence of the Fourier series at a point x and the value of its sum are local properties of the function, i.e., they are preserved under an arbitrary change of the function outside an arbitrarily small neighborhood of the point. More formally, we have the following.

Theorem (Riemann's localization principle) *If functions $f_1, f_2 \in \widetilde{\mathcal{L}}^1$ coincide in a neighborhood of a point x, then their Fourier series have the same behavior at x, $S_n(f_1,x) - S_n(f_2,x) \to 0$ as $n \to \infty$.*

Proof From the assumptions it follows that the function $\varphi_x(u) = \frac{f_1(x+u) - f_2(x-u)}{\sin(u/2)}$ (equal to zero in a neighborhood of the point $u = 0$) is summable on $(-\pi, \pi)$. Since

$$S_n(f_1,x) - S_n(f_2,x) = \frac{1}{2\pi}\int_{-\pi}^{\pi} \varphi_x(u)\sin\left(n+\frac{1}{2}\right)u\,du$$

by Eq. (5), it remains to refer to the Riemann–Lebesgue theorem according to which the integral on the right-hand side of this equation tends to zero. □

10.3.4 Among a great variety of convergence tests for Fourier series, we mention only two of the most applicable ones, the Dini[12] test and the Dirichlet–Jordan[13] test. They supplement each other and can be applied to a wide range of cases.

First, we establish a useful property of the Dirichlet kernel.

Lemma *Let $n \in \mathbb{N}$. Then:*

$$\text{(a)} \quad D_n(u) = \frac{\sin nu}{\pi u} + \frac{1}{2\pi}\big(\cos nu + \Delta(u)\sin nu\big),$$

where Δ is a function independent of n and $|\Delta(u)| < 1$ for $|u| \leqslant \pi$;

$$\text{(b)} \quad \left|\int_0^x D_n(u)\,du\right| \leqslant 2 \quad \textit{for } |x| \leqslant 2\pi.$$

Proof (a) It is clear that

$$D_n(u) = \frac{\sin nu}{2\pi \tan\frac{u}{2}} + \frac{1}{2\pi}\cos nu = \frac{\sin nu}{\pi u} + \frac{1}{2\pi}\left(\cos nu + \left(\frac{1}{\tan\frac{u}{2}} - \frac{2}{u}\right)\sin nu\right).$$

It remains to observe that the difference $\Delta(u) = \frac{1}{\tan\frac{u}{2}} - \frac{2}{u}$ $(\Delta(0) = 0)$ decreases on $[-\pi, \pi]$, and, therefore, $|\Delta(u)| \leqslant |\Delta(\pi)| = \frac{2}{\pi} < 1$.

(b) It is sufficient to consider the case where $x \in (0, 2\pi)$. First let $x \in (0, \pi]$. Then assertion (a) proved above implies the inequality

$$\left|\int_0^x D_n(u)\,du - \int_0^x \frac{\sin nu}{\pi u}\,du\right| \leqslant \frac{1}{2\pi}\int_0^x 2\,du \leqslant 1.$$

Now, we prove that the integral

$$J_n(x) = \int_0^x \frac{\sin nu}{\pi u}\,du = \int_0^{nx} \frac{\sin v}{\pi v}\,dv,$$

lies between 0 and 1. To verify this, we divide the interval of integration $[0, nx]$ into parts on which $\sin v$ preserves its sign. Then the integral $J_n(x)$ splits into the alternating sum of terms whose absolute values decrease since $\frac{1}{v}$ decreases. Therefore,

$$0 \leqslant J_n(x) \leqslant \int_0^{\pi} \frac{\sin v}{\pi v}\,dv \leqslant \int_0^{\pi} \frac{dv}{\pi} = 1.$$

Thus, the integral $\int_0^x D_n(u)\,du$ lies between -1 and 2 provided $0 < x \leqslant \pi$.

[12] Ulisse Dini (1845–1918)—Italian mathematician.

[13] Marie Ennemond Camille Jordan (1838–1922)—French mathematician.

For $x \in (\pi, 2\pi)$, we use the easily verifiable relation

$$\int_0^x D_n(u)\,du = 1 - \int_0^{2\pi - x} D_n(u)\,du$$

from which it follows that the inequality $-1 \leqslant \int_0^x D_n(u)\,du \leqslant 2$ also holds in this case. □

Using the first assertion of the lemma, we can represent Eq. (5) in the following form:

$$S_n(f,x) = \int_{-\pi}^{\pi} f(x-u)\frac{\sin nu}{\pi u}\,du + \varepsilon_n, \tag{5'}$$

where the quantity $\varepsilon_n = \frac{1}{2\pi}\int_{-\pi}^{\pi} f(x-u)(\cos nu + \Delta(u)\sin nu)\,du$ tends to zero by the Riemann–Lebesgue theorem.

In particular, if $f \equiv 1$, then

$$1 = \int_{-\pi}^{\pi} \frac{\sin nu}{\pi u}\,du + o(1). \tag{5''}$$

Making the change of variable $nu = t$ and passing to the limit as $n \to \infty$, we once again obtain the equality $\int_0^\infty \frac{\sin t}{t}\,dt = \frac{\pi}{2}$ established in Sect. 7.1.6 by a different method.

Theorem (Dini test) *If a function $f \in \widetilde{\mathscr{L}}^1$ satisfies the Dini condition*

$$\int_0^\pi \left| \frac{f(x+u) + f(x-u)}{2} - C \right| \frac{du}{u} < +\infty$$

at a point $x \in \mathbb{R}$ for some $C \in \mathbb{C}$, then its Fourier series converges to C at the point x.

In particular, if f is differentiable at x, then the Dini condition is fulfilled with $C = f(x)$, and so the sum of the Fourier series is equal to $f(x)$. However, if only the one-sided limits $f(x \pm 0)$ exist and

$$\bigl|f(x \pm u) - f(x \pm 0)\bigr| = O\bigl(u^\alpha\bigr) \quad \text{as } u \to +0$$

for some $\alpha > 0$, then the Fourier series of f at x converges to the average $\frac{f(x-0)+f(x+0)}{2}$.

Proof From (5′), it follows that

$$S_n(f,x) = \int_{-\pi}^{\pi} f(x-u)\frac{\sin nu}{\pi u}\,du + o(1) = \int_{-\pi}^{\pi} f(x+u)\frac{\sin nu}{\pi u}\,du + o(1)$$

as $n \to \infty$. Thus,

$$S_n(f,x) = \int_{-\pi}^{\pi} \frac{f(x-u) + f(x+u)}{2}\frac{\sin nu}{\pi u}\,du + o(1).$$

Subtracting Eq. (5″) multiplied by C from the above equation, we see that

$$S_n(f,x) - C = \int_{-\pi}^{\pi} \left(\frac{f(x-u)+f(x+u)}{2} - C \right) \frac{\sin nu}{\pi u}\, du + o(1)$$
$$= \frac{2}{\pi} \int_0^{\pi} g_x(u) \sin nu\, du + o(1),$$

where $g_x(u) = \frac{f(x-u)+f(x+u)-2C}{2u}$. Since the function g_x is summable on $(0,\pi)$ by the assumptions of the theorem, the integral on the right-hand side of this equation tends to zero by the Riemann–Lebesgue theorem. □

Theorem (Dirichlet–Jordan test) *If a periodic function f has bounded variation on the interval $[-\pi,\pi]$, then, for each $x \in \mathbb{R}$, the Fourier series of f converges to the average $(f(x+0)+f(x-0))/2$. Moreover, $|S_n(f,x)| \leqslant \sup_{\mathbb{R}} |f| + 2\mathbf{V}_{-\pi}^{\pi}(f)$.*

We remark that the convergence of a Fourier series at a point x is preserved by the localization principle if we assume that f has bounded variation only locally, in a neighborhood of this point.

Proof By Eq. (5′), we must find the limit of the integrals

$$I_n = \int_{-\pi}^{\pi} f(x-u) \frac{\sin nu}{\pi u}\, du = \int_0^{\pi} \varphi(u) \frac{\sin nu}{\pi u}\, du,$$

where $\varphi(u) = f(x-u) + f(x+u)$. This function has bounded variation on $[0,\pi]$, and so can be represented as the difference of decreasing functions. Therefore, it is sufficient for us to find the limit of the integrals I_n under the assumption that the function φ is non-negative and decreases on the interval $[0,\pi]$. To this end, we represent I_n in the form

$$I_n = \int_0^{\infty} \Phi(u) \frac{\sin nu}{\pi u}\, du = \int_0^{\infty} \Phi\left(\frac{t}{n}\right) \frac{\sin t}{\pi t}\, dt,$$

where $\Phi(u) = \varphi(u)\chi_{(0,\pi)}(u)$. By Corollary 2 of Sect. 7.4.7, the integral on the right-hand side of the above equation tends to $\varphi(+0)/2 = (f(x-0)+f(x+0))/2$.

To obtain a uniform estimate for the sums $S_n(f)$, we put $H_n(u) = \int_0^u D_n(t)\, dt$. Then

$$S_n(f,x) = \int_{-\pi}^{\pi} f(x-u) D_n(u)\, du$$
$$= H_n(u) f(x-u)\Big|_{u=-\pi}^{\pi} - \int_{-\pi}^{\pi} H_n(u)\, df(x-u).$$

Since $H_n(\pm\pi) = \pm\frac{1}{2}$, the first summand is equal to $(f(x-\pi)+f(x+\pi))/2$. Furthermore, $|H_n(u)| \leqslant 2$ by the lemma, and so

$$\left| \int_{-\pi}^{\pi} H_n(u)\, df(x-u) \right| \leqslant 2\mathbf{V}_{x-\pi}^{x+\pi}(f) = 2\mathbf{V}_{-\pi}^{\pi}(f),$$

hence the required estimate for the sums $S_n(f,x)$ follows. □

In conclusion, we prove the Dini test in a different way, without using Dirichlet kernels (see [Ch]). The Dini condition means that the function

$$g(u) = \left(\frac{f(x+u)+f(x-u)}{2} - C\right)\frac{1}{e^{iu}-1} \quad (u \notin 2\pi\mathbb{Z})$$

belongs to the class $\widetilde{\mathscr{L}}^1$. Multiplying both sides of the equation

$$\frac{f(x+u)+f(x-u)}{2} - C = (e^{iu}-1)g(u)$$

by $\frac{1}{2\pi}e^{-iku}$ and then integrating with respect to $u \in (-\pi,\pi)$, we obtain

$$\frac{1}{2}\big(\widehat{f}(k)e^{ikx} + \widehat{f}(-k)e^{-ikx}\big) = \widehat{g}(k-1) - \widehat{g}(k), \quad \text{if } k \neq 0,$$

$$\widehat{f}(0) - C = \widehat{g}(-1) - \widehat{g}(0), \quad \text{if } k = 0.$$

It remains to sum all these equations for $|k| \leqslant n$,

$$S_n(f,x) - C = \sum_{k=-n}^{n} \widehat{f}(k)e^{ikx} - C = \widehat{g}(-n-1) - \widehat{g}(n) \underset{n\to+\infty}{\longrightarrow} 0.$$

If we sum them for $k = 0, 1, \dots, n$ and for $k = -1, \dots, -n$ separately, it becomes clear that the Dini condition implies the convergence of not only the symmetric sums $\sum_{k=-n}^{n} \widehat{f}(k)e^{ikx}$, but also the "one-sided" sums $\sum_{k=0}^{n} \widehat{f}(k)e^{ikx}$ and $\sum_{k=-n}^{-1} \widehat{f}(k)e^{ikx}$. In other words, the Dini condition ensures the convergence of each of the series $\sum_{k=0}^{\infty} \widehat{f}(k)e^{ikx}$ and $\sum_{k=-\infty}^{-1} \widehat{f}(k)e^{ikx}$. In particular, it ensures the convergence of the series $\sum_{n\in\mathbb{Z}} \operatorname{sign}(n)\widehat{f}(n)e^{inx}$ called the conjugate to series (1′).

10.3.5 We give some examples of Fourier series expansions.

Example 1 We supplement Example 1 of Sect. 10.2.1 as follows: since the periodic function equal x on $(-\pi,\pi)$ is differentiable at all points distinct from $(2k+1)\pi$ $(k \in \mathbb{Z})$, its Fourier series converges not only in the $\mathscr{L}^2$-norm, but also pointwise,

$$x = 2\sum_{n=1}^{\infty} \frac{(-1)^{n-1}}{n} \sin nx \quad \text{for } x \in (-\pi,\pi).$$

At the points $(2k+1)\pi$, the sum of the series is equal to the average of the one-sided limits of the function. At $x = \frac{\pi}{2}$, the Fourier series expansion yields the relations

$$\frac{\pi}{4} = \sum_{m=0}^{\infty} (-1)^m \frac{1}{2m+1}.$$

Considering the Fourier series expansion of the function equal to x^2 on $[-\pi,\pi]$ at the point π, we again obtain the Euler identity $\sum_{n=1}^{\infty}\frac{1}{n^2}=\frac{\pi^2}{6}$ (see Example 1 of Sect. 10.2.1 or Example 2 of Sect. 4.6.2).

Example 2 Let $w\in\mathbb{C}\setminus\mathbb{Z}$. We consider the periodic function equal to $\cos wx$ on the interval $[-\pi,\pi]$. This function has finite one-sided derivatives everywhere on $\mathbb{R}$ and, therefore, can be expanded in a Fourier series. After elementary transformations, we obtain that the equation

$$\cos wx=\frac{\sin\pi w}{\pi w}+\frac{2}{\pi}w\sin\pi w\sum_{n=1}^{\infty}\frac{(-1)^n}{w^2-n^2}\cos nx$$

holds for all $|x|\leqslant\pi$.

For $x=\pi$ and $x=0$, the equation implies the following expansions of cotangent and cosecant as sums of partial fractions:

$$\cot\pi w=\frac{1}{\pi w}+\frac{2w}{\pi}\sum_{n=1}^{\infty}\frac{1}{w^2-n^2}=\frac{1}{\pi}\sum_{n=-\infty}^{\infty}\frac{1}{w-n},$$

$$\frac{1}{\sin\pi w}=\frac{1}{\pi w}+\frac{2w}{\pi}\sum_{n=1}^{\infty}\frac{(-1)^n}{w^2-n^2}=\frac{1}{\pi}\sum_{n=-\infty}^{\infty}\frac{(-1)^n}{w-n}.$$

Example 3 Here, we verify the existence of a convergent non-zero numerical series $\sum_{n=1}^{\infty}a_n$ with the unusual property $\sum_{m=1}^{\infty}a_{km}=0$ for every k. In the construction, we follow F.L. Nazarov who suggested the use of Fourier series for this purpose. We consider periodic functions equal to zero in a neighborhood of each point of the form πt, $t\in\mathbb{Q}$. Among them, we can, obviously, find an even function f satisfying the conditions $\widehat{f}(0)=\frac{1}{2\pi}\int_{-\pi}^{\pi}f(x)\,dx=0$ and $0<\int_{-\pi}^{\pi}|f(x)|^2\,dx<+\infty$. We take the required series equal to $\sum_{n=1}^{\infty}\widehat{f}(n)$. This is a non-zero series since

$$0<\int_{-\pi}^{\pi}|f(x)|^2\,dx=\frac{1}{2\pi}\sum_{n=-\infty}^{\infty}|\widehat{f}(n)|^2=\frac{1}{\pi}\sum_{n=1}^{\infty}|\widehat{f}(n)|^2$$

(here we have used Parseval's identity).

At each point $x=2\pi\frac{j}{k}$ ($j\in\mathbb{Z}$, $k\in\mathbb{N}$), the function f satisfies the Dini condition and, therefore,

$$\sum_{n=1}^{\infty}\widehat{f}(n)\cos\left(2\pi\frac{j}{k}n\right)=0.$$

Summing these equations for $j=0,1,\dots,k-1$, we obtain

$$\sum_{n=1}^{\infty}\widehat{f}(n)\sum_{j=0}^{k-1}\cos\left(2\pi\frac{j}{k}n\right)=0.$$

If the index n is divisible by k, then the inner sum is equal to k, since otherwise this sum is obviously zero. Consequently,

$$k\sum_{m=1}^{\infty}\widehat{f}(km)=0 \quad \text{for all } k\in\mathbb{N}.$$

We remark that the series just constructed does not converge absolutely. It can be proved (see [PS], Part 1, Problem 129) that it is impossible to construct an absolutely convergent series with the property in question.

10.3.6 As we have already noted, the Fourier series of a summable, or even of a continuous, function may diverge (see also Sect. 10.3.9). However, such a series has the remarkable property that it can be integrated termwise over an arbitrary finite interval without worrying about convergence.

Theorem 1 *Let $f\in\widetilde{\mathscr{L}}^1$. Then the equation*

$$\int_a^b f(x)\,dx=\sum_{n=-\infty}^{\infty}\widehat{f}(n)\int_a^b e^{inx}\,dx$$

(where the sum is regarded as the limit of the symmetric partial sums) is valid for all $a,b\in\mathbb{R}$.

Proof Taking into account the periodicity, we restrict ourselves, without loss of generality, to the case where $-\pi\leqslant a<b\leqslant\pi$. Let χ be the characteristic function of the interval (a,b). Then a partial sum of the series on the right-hand side of the required equation can be represented in the form

$$\sum_{k=-n}^{n}\widehat{f}(k)\int_a^b e^{ikx}\,dx=\sum_{k=-n}^{n}\left(\frac{1}{2\pi}\int_{-\pi}^{\pi}f(t)e^{-ikt}\,dt\right)2\pi\widehat{\chi}(-k)$$
$$=\int_{-\pi}^{\pi}f(t)S_n(\chi,t)\,dt. \tag{7}$$

By Dini's test, we have $S_n(\chi,t)\underset{n\to\infty}{\longrightarrow}\chi(t)$ for $t\in(-\pi,\pi)$ and $t\neq a,b$. Moreover,

$$S_n(\chi,t)=\int_a^b D_n(x-t)\,dx=\int_{a-t}^{b-t}D_n(u)\,du$$
$$=\int_0^{b-t}D_n(u)\,du-\int_0^{a-t}D_n(u)\,du.$$

Therefore, Lemma 10.3.4 gives use the uniform estimate $|S_n(\chi,t)|\leqslant 4$. By Lebesgue's theorem, we can pass to the limit on the right-hand side of Eq. (7)

and obtain

$$\sum_{k=-n}^{n} \widehat{f}(k) \int_a^b e^{ikx}\,dx = \int_{-\pi}^{\pi} f(t) S_n(\chi, t)\,dt \underset{n\to\infty}{\longrightarrow} \int_{-\pi}^{\pi} f(t)\chi(t)\,dt = \int_a^b f(t)\,dt. \qquad \square$$

Theorem 1 allows us to considerably strengthen the assertion on the completeness of the trigonometric system, according to which two functions of the class $\widetilde{\mathscr{L}}^2$ that have the same Fourier coefficients coincide almost everywhere. Now, we can extend this result to the class $\widetilde{\mathscr{L}}^1$.

Corollary 1 *Functions $f, g \in \widetilde{\mathscr{L}}^1$ having the same Fourier coefficients coincide almost everywhere on $\mathbb{R}$.*

Proof By the theorem, the integrals of f and g are equal on every finite interval. Therefore, (see Corollary 4.5.4) f and g coincide almost everywhere. □

Corollary 2 *For every function $f \in \widetilde{\mathscr{L}}^1$, the series $\sum_{n=1}^{\infty} b_n(f)/n$ converges.*

We recall that $b_n(f) = \frac{1}{\pi}\int_{-\pi}^{\pi} f(x)\sin nx\,dx = i(\widehat{f}(n) - \widehat{f}(-n))$ is the Fourier sine coefficient of f.

Proof As established in the theorem, the equation

$$\int_0^u f(x)\,dx = \sum_{n=-\infty}^{\infty} \widehat{f}(n) \int_0^u e^{inx}\,dx$$

holds for all u. From (7) and the estimate $|S_n(\chi, t)| \leqslant 4$, it follows that the symmetric partial sums of this series are uniformly bounded for $u \in [-\pi, \pi]$. Therefore, we can integrate the series termwise,

$$\int_{-\pi}^{\pi}\left(\int_0^u f(x)\,dx\right)du = \sum_{n=-\infty}^{\infty} \widehat{f}(n) \int_{-\pi}^{\pi}\left(\int_0^u e^{inx}\,dx\right)du = -2\pi \sum_{n\neq 0} \frac{\widehat{f}(n)}{in}.$$

The convergence of the symmetric partial sums of the series $\sum_{n\neq 0} \frac{\widehat{f}(n)}{n}$ is equivalent to the required statement. □

Corollary 2 gives a necessary condition for a trigonometric series $\sum_{n=1}^{\infty}(a_n \cos nx + b_n \sin nx)$ to be a Fourier series. The everywhere convergent series $\sum_{n=2}^{\infty}(\sin nx)/\ln n$ does not satisfy this condition and, therefore, cannot be the Fourier series of a summable function. It is interesting to note that, in contrast to the sine coefficients, the cosine coefficients can tend to zero arbitrarily slowly. For example, the series $\sum_{n=2}^{\infty}(\cos nx)/\ln n$ is the Fourier series of a summable function (see Theorem 10.4.2).

The relation obtained in Theorem 1 can be regarded as a new version of Parseval's identity in which the assumption about one function is weakened (it belongs to $\widetilde{\mathscr{L}}^1$ but not to $\widetilde{\mathscr{L}}^2$) and the assumption concerning the other is strengthened considerably (it is the characteristic function of an interval). At the same time, the proof of the theorem uses the properties of the function χ only partially. This makes it possible to extend considerably the applicability conditions of Parseval's identity.

Theorem 2 *Let $f \in \widetilde{\mathscr{L}}^1$, and let g be a bounded (measurable and periodic) function whose Fourier sums $S_n(g,x)$ are uniformly bounded (with respect to x and n). Then the following Parseval identity is valid*:

$$\int_{-\pi}^{\pi} f(x)\overline{g}(x)\,dx = 2\pi \sum_{n=-\infty}^{\infty} \widehat{f}(n)\overline{\widehat{g}(n)}.$$

The class of functions with uniformly bounded partial sums of Fourier series is sufficiently wide. In particular, it contains all smooth functions on $[-\pi,\pi]$. As follows from the Dirichlet–Jordan test, this class also contains all functions with finite variation on $[-\pi,\pi]$ (see also Exercises 9 and 10).

The assumption that the function g is bounded is superfluous (see Exercise 8 or Fejér's theorem in Sect. 10.4).

Proof Since $g \in \widetilde{\mathscr{L}}^2$, the sums $S_n(g)$ converge to g in the $\mathscr{L}^2$-norm and, *a fortiori*, in measure. This implies, as one can easily verify, that

$$f(x)\overline{S_n}(g,x) \to f(x)\overline{g}(x) \quad \text{in measure.}$$

Therefore, we can use Lebesgue's theorem and pass to the limit on the right-hand side of the equation

$$\int_{-\pi}^{\pi} f(x)\overline{S_n}(g,x)\,dx = 2\pi \sum_{|k|\leqslant n} \widehat{f}(k)\overline{\widehat{g}(k)},$$

as required. □

10.3.7 To obtain a further generalization of the uniqueness theorem for Fourier series, (see Corollary 1 of the previous section), we introduce the notion of Fourier coefficients and Fourier series for a measure.

Definition Let μ be a finite Borel measure on the interval $[-\pi,\pi]$. The *Fourier coefficients* of μ are defined by the formula

$$\widehat{\mu}(n) = \frac{1}{2\pi}\int_{[-\pi,\pi]} e^{-inx}\,d\mu(x) \quad (n \in \mathbb{Z}).$$

The series $\sum_{n=-\infty}^{\infty} \widehat{\mu}(n)e^{inx}$ is called the *Fourier series* of μ.

If a measure μ has density f with respect to Lebesgue measure, then $\widehat{\mu}(n) = \widehat{f}(n)$ for all $n \in \mathbb{Z}$ and, consequently, the Fourier series of the measure μ and of the function f coincide. As in the case of Fourier series, it follows directly from definition that the nth (symmetric) partial sum of the Fourier series of a measure, which will be denoted by $S_n(\mu, x)$, is the convolution of this measure and a Dirichlet kernel,

$$S_n(\mu, x) = \int_{[-\pi,\pi]} D_n(x-t)\, d\mu(t) = (D_n * \mu)(x).$$

Extending Corollary 1 to measures, we must take into account the relation

$$\widehat{\mu}(n) = \frac{(-1)^n}{2\pi}\big(\mu(\{-\pi\}) + \mu(\{\pi\})\big) + \frac{1}{2\pi}\int_{(-\pi,\pi)} e^{-inx}\, d\mu(x).$$

Thus, the Fourier coefficients do not change under redistribution of the loads (preserving their sum) at the points $\pm\pi$. This will be the case when we replace these loads by, for example, $\mu(\{-\pi\})+\mu(\{\pi\})$ (at the point $-\pi$) and by 0 (at the point π). Therefore, it makes sense to pose the question of whether a measure is uniquely determined by its Fourier coefficients only if we fix the load at one of the points $\pm\pi$. For definiteness, we will consider only the measures that have zero load at the point π.

Theorem *Let μ and ν be finite Borel measures on the interval $[-\pi, \pi]$ satisfying the condition $\mu(\{\pi\}) = \nu(\{\pi\}) = 0$. If the Fourier coefficients of these measures coincide, then the measures also coincide.*

Proof First, we verify that the Fourier series of a measure, as well as the Fourier series of a function, can be integrated termwise, i.e., if $\mu(\{a\}) = \mu(\{b\}) = 0$, then

$$\sum_{n=-\infty}^{\infty} \widehat{\mu}(n) \int_a^b e^{inx}\, dx = \mu\big([a, b)\big) \tag{8}$$

for $[a, b) \subset [-\pi, \pi)$. Indeed, let $\chi = \chi_{[a,b)}$. Then

$$\begin{aligned}\sum_{|k|\leqslant n} \widehat{\mu}(k) \int_a^b e^{ikx}\, dx &= \sum_{|k|\leqslant n} \widehat{\chi}(-k) \int_{[-\pi,\pi]} e^{-ikx} d\mu(x) \\ &= \int_{[-\pi,\pi]} S_n(\chi, x)\, d\mu(x).\end{aligned} \tag{9}$$

In the proof of Theorem 1 of Sect. 10.3.6, we have established that $|S_n(\chi, t)| \leqslant 4$. Moreover, $S_n(\chi, t) \underset{n\to\infty}{\longrightarrow} \chi(t)$ for $t \neq a, b$ and, consequently, μ-almost everywhere. Therefore, we can use Lebesgue's theorem and pass to the limit on the right-hand side of Eq. (9), which leads to Eq. (8). Thus, if measures μ and ν have the same Fourier coefficients, then $\mu([a, b)) = \nu([a, b))$ for every interval $[a, b) \subset [-\pi, \pi)$

satisfying the condition $\mu(\{a\}) = \mu(\{b\}) = \nu(\{a\}) = \nu(\{b\}) = 0$. Since the set of points of non-zero measure is at most countable (see Sect. 1.2.2), this condition is fulfilled on a dense subset of $(-\pi, \pi)$. Hence it follows (see Remark 1.1.7) that the measures μ and ν coincide on all Borel subsets of the interval $(-\pi, \pi)$. At the same time, $\mu(\{\pi\}) = \nu(\{\pi\}) = 0$ and $\mu([-\pi, \pi]) = \widehat{\mu}(0) = \widehat{\nu}(0) = \nu([-\pi, \pi])$ by assumption. Consequently, the measures μ and ν have the same loads at the point $-\pi$, which completes the proof of the theorem. □

Generalizations of this theorem are given in Sects. 10.4.7, 11.1.9, and 12.3.3.

10.3.8 Considering Fourier series with coefficients that tend to zero sufficiently fast, we must take into account that if the Fourier series of a function f converges uniformly, then its sum coincides with f almost everywhere by the uniqueness theorem. Therefore, if the Fourier series of a continuous function converges uniformly, then its sum coincides with the function. Taking this into account, we consider only continuous functions in the theorems of this section. Lifting the assumption of continuity, we must replace the equality of a function and its Fourier series by their equivalence.

The Fourier coefficients of smooth functions tend to zero sufficiently fast. For example, if a function satisfies the Lipschitz condition of order α, then $\widehat{f}(n) = O(|n|^{-\alpha})$. Indeed, if $h = \frac{\pi}{n}$, then property (c) of Sect. 10.3.2 implies that $\widehat{f_{\frac{\pi}{n}}}(n) = -\widehat{f}(n)$. Consequently, $2\widehat{f}(n) = \frac{1}{2\pi}\int_0^{2\pi}(f(x) - f(x - \frac{\pi}{n}))e^{-inx}\,dx$, and, therefore,

$$2\left|\widehat{f}(n)\right| \leqslant \frac{1}{2\pi}\int_0^{2\pi}\left|f(x) - f\left(x - \frac{\pi}{n}\right)\right|dx \leqslant L\left|\frac{\pi}{n}\right|^{\alpha},$$

where L is a Lipschitz constant for f.

The repeated application of the relation $\widehat{f'}(n) = in\widehat{f}(n)$ (see property (d) of Sect. 10.3.2) shows that the Fourier coefficients of a function f of class $\widetilde{C}^r$ satisfy the relation $\widehat{f}(n) = o(|n|^{-r})$ as $|n| \to +\infty$. The converse is "almost true": if $\widehat{f}(n) = O(|n|^{-r-2})$ for some $r \in \mathbb{N}$, then the continuous function f coincides with a function of class $\widetilde{C}^r$. Indeed, the series $\sum_n \widehat{f}(n)e^{inx}$ converges uniformly, and, by the above remark, its sum coincides with f. Moreover, since the coefficients decrease fast, the Fourier series admits r-fold differentiation, which implies that $f \in \widetilde{C}^r$. For infinitely smooth functions, this gives a complete description.

Theorem 1 *In order that a function $f \in \widetilde{C}$ be infinitely differentiable it is necessary and sufficient that the limit relation $n^r\widehat{f}(n) \to 0$ as $|n| \to +\infty$ be fulfilled for every $r \in \mathbb{N}$.*

The smaller class of holomorphic periodic functions can also be well described in terms of the Fourier coefficients: these coefficients must tend to zero not slower that a geometric sequence. We note that a periodic function f is analytic at all points of the line $\mathbb{R}$ if and only if, on $\mathbb{R}$, the function f coincides with a function holomorphic in some horizontal strip $\{z \in \mathbb{C} \mid |\mathcal{I}m\, z| < L\}$. In the proof of the following theorem, we use some elementary properties of holomorphic functions (see, for example, [Ca]).

Theorem 2 *Let $f \in \widetilde{C}$. The following two statements are equivalent*:

(a) *there is a function F holomorphic in a strip $|\mathcal{I}m\, z| < L$ and coinciding with f on the real axis*;
(b) *the relation $\widehat{f}(n) = O(e^{-a|n|})$ as $|n| \to +\infty$ holds for every $a \in (0, L)$.*

Proof (a)⇒(b). Assuming that $n > 0$ and $0 < a < L$, we consider the integral $\int_C F(z)e^{-inz}\,dz$, where C is the boundary of the rectangle P with vertices at the points $\pm\pi$, $\pm\pi - ai$ lying in the strip $|\mathcal{I}m\, z| < L$. Since the function F is holomorphic in a neighborhood of P, this integral is equal to zero. Moreover, F has period 2π, and, therefore, the sum of the integrals over the vertical sides of P are equal to zero. Consequently,

$$\widehat{f}(n) = \frac{1}{2\pi}\int_{-\pi}^{\pi} f(x)e^{-inx}\,dx = \frac{1}{2\pi}\int_{-\pi-ai}^{\pi-ai} F(z)e^{-inz}\,dz.$$

Therefore,

$$\bigl|\widehat{f}(n)\bigr| \leqslant \max_{x\in\mathbb{R}} \bigl|F(x - ai)\bigr|\bigl|e^{-in(x-ai)}\bigr| = e^{-an}\max_{x\in\mathbb{R}}\bigl|F(x - ai)\bigr| = C_a e^{-an}.$$

The coefficients with negative indices can be estimated in the same way, only in this case the rectangle is replaced by a rectangle symmetric with respect to the real axis.

(b)⇒(a). The series $\sum_{n=-\infty}^{\infty} \widehat{f}(n)e^{inz}$ converges uniformly in the strip $|\mathcal{I}m\, z| \leqslant a$ if $0 < a < L$. By Weierstrass's theorem the sum of the series is holomorphic in the strip $|\mathcal{I}m\, z| < L$ and coincides with the function f on the real axis. □

10.3.9 As we have already mentioned, the Fourier series of a periodic continuous function may diverge (compare this with the result of Exercise 5). There are several such examples. We give here a slight modification of an example suggested by Schwartz. We define an even function $f \in \widetilde{C}$ whose oscillation frequency increases rapidly when approaching zero. More precisely, we will assume that

$$f(0) = 0 \quad \text{and} \quad f(t) = \frac{1}{\sqrt{k}}\sin n_k t \quad \text{for } t \in [t_k, t_{k-1}],\ k = 2, 3, \ldots,$$

where $n_k = 2^{k!}$, $t_k = 2\pi/n_k$ for $k \in \mathbb{N}$ (see Fig. 10.2).

We prove that the sums $S_n(f, 0)$ tend to infinity along the indices n_k. Since

$$S_n(f, 0) = \frac{2}{\pi}\int_0^{\pi} \frac{\sin nt}{t} f(t)\,dt + o(1),$$

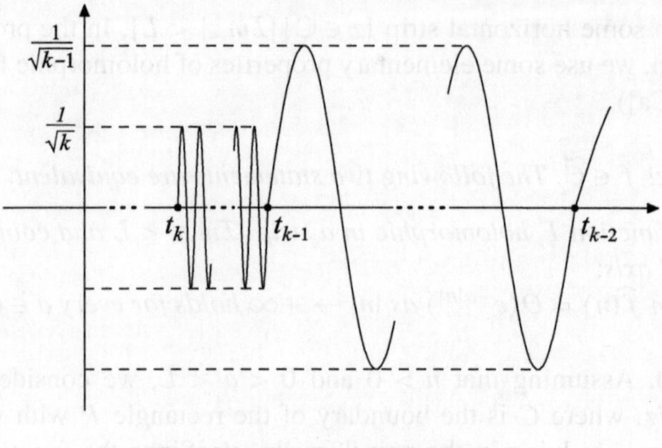

Fig. 10.2 Sketch of the graph of f

by (5′), it is sufficient to prove that the integrals

$$I_k = \int_0^\pi \frac{\sin n_k t}{t} f(t)\,dt = \int_0^{t_k} \cdots + \int_{t_k}^{t_{k-1}} \cdots + \int_{t_{k-1}}^{\pi} \cdots = F_k + J_k + H_k$$

tend to infinity. We verify that the main contribution comes from the integral J_k. Indeed, since $|\sin n_k t| \leqslant n_k t$ and $|f(t)| < \frac{1}{\sqrt{k}}$ on $(0, t_k)$, we have

$$|F_k| = \left| \int_0^{t_k} \cdots \right| \leqslant \frac{n_k}{\sqrt{k}} t_k = \frac{2\pi}{\sqrt{k}} \to 0.$$

Since the absolute value of the integrand does not exceed $1/t$, we have

$$|H_k| \leqslant \int_{t_{k-1}}^{\pi} \frac{1}{t}\,dt = \ln \pi / t_{k-1} = \ln \frac{n_{k-1}}{2} < (k-1)!\ln 2.$$

Now, we calculate the integral over the middle interval,

$$J_k = \int_{t_k}^{t_{k-1}} \frac{\sin n_k t}{t} f(t)\,dt = \frac{1}{\sqrt{k}} \int_{t_k}^{t_{k-1}} \frac{\sin^2 n_k t}{t}\,dt = \frac{1}{\sqrt{k}} \int_{2\pi}^{A_k} \frac{\sin^2 u}{u}\,du,$$

where $A_k = n_k t_{k-1} = 2\pi n_k / n_{k-1}$. Consequently, for sufficiently large k, we have

$$J_k = \frac{1}{2\sqrt{k}} \int_{2\pi}^{A_k} \frac{1 - \cos 2u}{u}\,du = \frac{\ln A_k + O(1)}{2\sqrt{k}} > \frac{k!\ln 2}{3\sqrt{k}}.$$

Thus,

$$I_k = F_k + J_k + H_k \geqslant \frac{k!\ln 2}{3\sqrt{k}} - (k-1)!\ln 2 + o(1) \to +\infty,$$

and, therefore, $S_{n_k}(f, 0) = \frac{2}{\pi} I_k + o(1) \to +\infty$.

In the above example, we could select a subsequence $\{S_{n_k}(f,0)\}$ that tends to $+\infty$. As we will see in Sect. 10.4.1, it is impossible to construct a continuous function for which $S_n(f,0) \underset{n\to\infty}{\longrightarrow} +\infty$. We also remark that, in the above example, estimate (6) for the Fourier sums is almost attained (in order) on the sequence $\{n_k\}$ (see also Exercise 13).

By a slightly more complicated construction it is possible to give an example of a continuous function whose Fourier series diverges on a countable set. Must this series converge almost everywhere? This famous problem was open for more than half a century. It was answered in the affirmative only in 1966 by L. Carleson.[14] It turned out that the Fourier series of an arbitrary function of class $\widetilde{\mathscr{L}}^2$ (for example a continuous function) converges to the function almost everywhere (see [C]). Since that time, several modifications and strengthenings of the original proof have been obtained, but all of them are quite difficult and lie far beyond the scope of this book.

10.3.10 Using the Riemann–Lebesgue theorem, we can obtain an important result concerning arbitrary trigonometric series, i.e., series of the form

$$A + \sum_{n=1}^{\infty}(a_n \cos nx + b_n \sin nx) \quad (A, a_n, b_n \in \mathbb{C}). \tag{10}$$

As we know, (see Sect. 10.3.6) even an everywhere convergent trigonometric series may not be a Fourier series. At the same time, the following statement holds:

Theorem (Denjoy[15]–Luzin) *If series* (10) *converges absolutely on a set of positive measure, then*

$$\sum_{n=1}^{\infty}\big(|a_n| + |b_n|\big) < +\infty. \tag{11}$$

In particular, if a trigonometric series converges absolutely on a set of positive measure, then it converges uniformly on $\mathbb{R}$, and, therefore, is the Fourier series of its sum.

Proof Without loss of generality, we may assume that the coefficients a_n and b_n are real. We put $\varphi_n(x) = |a_n \cos nx + b_n \sin nx|$. Since the series $\sum_{n=1}^{\infty}\varphi_n$ converges on a set of positive measure, its sum is bounded on a smaller set X of positive measure,

$$\sum_{n=1}^{\infty}\varphi_n(x) \leqslant C \quad \text{for all } x \in X.$$

[14] Lennart Axel Edvard Carleson (born 1928)—Swedish mathematician.

[15] Arnaud Denjoy (1884–1974)—French mathematician.

Consequently (in what follows, λ is the one-dimensional Lebesgue measure),

$$\sum_{n=1}^{\infty} \int_X \varphi_n(x)\,dx \leqslant C\lambda(X).$$

We represent the functions φ_n in the form $\varphi_n(x) = c_n|\sin(nx + \theta_n)|$, where $c_n = \sqrt{a_n^2 + b_n^2}$ and $\theta_n \in \mathbb{R}$. Using the obvious inequality $|\sin t| \geqslant \sin^2 t$ and the Lebesgue–Riemann theorem, we see that

$$\int_X \frac{1}{c_n}\varphi_n(x)\,dx \geqslant \int_X \sin^2(nx+\theta_n)\,dx = \int_X \frac{1-\cos 2(nx+\theta_n)}{2}\,dx \underset{n\to\infty}{\longrightarrow} \frac{\lambda(X)}{2}.$$

Therefore,

$$0 < \frac{\lambda(X)}{3} \leqslant \int_X \frac{1}{c_n}\varphi_n(x)\,dx \quad \text{for } n \geqslant N$$

for some N, and, consequently,

$$\sum_{n=N}^{\infty} \frac{\lambda(X)}{3} c_n \leqslant \sum_{n=N}^{\infty} \int_X \varphi_n(x)\,dx \leqslant C\lambda(X).$$

Thus, the following estimate is valid for the remainder of series (11):

$$\sum_{n=N}^{\infty} \big(|a_n| + |b_n|\big) \leqslant 2\sum_{n=N}^{\infty} \sqrt{a_n^2+b_n^2} = 2\sum_{n=N}^{\infty} c_n \leqslant 6C.$$

□

EXERCISES

1. Let $f \in \widetilde{\mathscr{L}}^1(\mathbb{R}^m)$. By the method used in the second proof of the Lebesgue–Riemann theorem, prove that $|\widehat{f}(n)| \leqslant \frac{1}{2}\|f - f_\tau\|_1$, where f_τ is the translation of the function f by the vector $\tau = \pi n/\|n\|^2$.
2. Prove that the Fourier sums of the function $f(x) = \sum_{k=1}^{\infty} 2^{-k}\cos kx$ provide almost the best uniform approximations for f, more precisely, $\|f - S_n(f)\|_\infty \leqslant 3\|f - T\|_\infty$ for every trigonometric polynomial T of order n.
3. Show by example that an absolutely continuous function may not satisfy Dini's condition.
4. Verify by examples that neither of the Dini and Dirichlet tests implies the other.
5. Prove (see [HR], Sect. 6.7) that

$$\frac{1}{2}\left(S_n\left(f, x+\frac{\pi}{2n}\right) + S_n\left(f, x-\frac{\pi}{2n}\right)\right) \underset{n\to\infty}{\longrightarrow} f(x)$$

if the function f in $\widetilde{\mathscr{L}}^1$ is continuous at x. *Hint*. Verify that the functions $\frac{1}{2}(D_n(x+\frac{\pi}{2n}) + D_n(x-\frac{\pi}{2n}))$ form a periodic approximate identity.
6. Verify that $f \in \widetilde{\mathscr{L}}^1$ belongs to the class $\widetilde{\mathscr{L}}^2$ if and only if the series $\sum_{n=-\infty}^{\infty} |\widehat{f}(n)|^2$ converges.

7. Prove that a function $f \in \widetilde{C}$ is the restriction of an entire function to $\mathbb{R}$ if and only if $\sqrt[n]{|\widehat{f}(\pm n)|} \to 0$ as $n \to \infty$.

8. Prove that if the sequence of partial sums of a Fourier series is uniformly bounded, then the function belongs to $\widetilde{\mathscr{L}}^{\infty}$.

9. Let $g, h \in \widetilde{\mathscr{L}}$. Prove that the Fourier sums $S_n(hg, x)$ of the product hg are uniformly bounded (with respect to n and x) if the function g has the same property and the function h satisfies the Dini condition uniformly:

$$\int_{-\pi}^{\pi} \left| \frac{h(u) - h(x)}{u - x} \right| du \leqslant \text{const} \quad \text{for all } x.$$

10. Assume that a function f (possibly discontinuous) is such that the interval $[-\pi, \pi]$ can be divided into a finite number of intervals inside each of which the function f satisfies the Lipschitz condition of order $\alpha > 0$. Prove that the Fourier sums $S_n(f, x)$ are uniformly bounded with respect to x and n.

11. Prove that $\int_{-\pi}^{\pi} |D_n(u)|\, du = \frac{4}{\pi^2} \ln n + O(1)$.

12. Supplement inequality (6) by proving that $\|f - S_n(f)\|_{\infty} = o(\ln n)$ as $n \to \infty$ if $f \in \widetilde{C}$. Verify that, for bounded functions, this refinement is, generally, false. *Hint*. Modify the Schwartz example by putting $f(t) = \sin n_k t$ on the interval $[t_k, t_{k-1}]$.

13. Modifying the Schwartz example, verify that the result of the previous exercise is precise, i.e., that for every sequence $\varepsilon_n \downarrow 0$, there is a function $f \in \widetilde{C}$ such that $S_n(f, 0) \geqslant \varepsilon_n \ln n$ along some sequence of indices $n_k \to \infty$.

14. Show that the convergence of a Fourier series of a function $f \in \widetilde{C}$ does not imply the convergence of the Fourier series of the function f^2. *Hint*. Modifying the Schwartz example, construct a non-negative even function $F \in \widetilde{C}$, $F(0) = F(\pm\pi) = 0$, with Fourier series divergent at zero and consider an odd function f equal to $\sqrt{F}$ on $[0, \pi]$.

15. Prove that $a_n, b_n \underset{n\to\infty}{\longrightarrow} 0$ if the sums $a_n \cos nx + b_n \sin nx$ converge to zero on a set of positive measure.

16. Find a set $E \subset (0, 2\pi)$ of the cardinality of the continuum and a sequence $n_k \to +\infty$ such that $\sin n_k x \rightrightarrows 0$ on the set E.

17. Consider the series $\sum_{n=1}^{\infty} \sin(n!\pi x)$.

(a) Prove that the series converges at the points $x = \sin 1$, $x = \cos 1$, $x = \frac{2}{e}$, and their multiples and converges at the points ke ($k \in \mathbb{N}$) only for odd k. Is the convergence absolute?

(b) Prove that the given series diverges at the points $x = \sinh 1$ and $x = \frac{1}{2}\cosh 1$.

(c) Find a set of the cardinality of the continuum at all points of which the given series converges.

18. Prove that $\frac{1}{n} S_n'(f, x) \underset{n\to\infty}{\longrightarrow} \frac{1}{\pi}(f(x+0) - f(x-0))$ if the periodic function f has a bounded variation on the interval $[-\pi, \pi]$.

10.4 *Trigonometric Fourier Series (Continued)

10.4.1 The fact that a Fourier series may diverge, even at points of continuity, suggests that we might obtain information on its behavior if we consider a weaker definition of convergence than the classical one. One of the possible approaches is to investigate the convergence of the arithmetic means of the partial sums rather than the partial sums themselves. The limit of a sequence $\{a_n\}$ in the sense of arithmetic means or in the sense of Cesaro[16] is, by definition, the limit $\lim_{n\to\infty}\frac{1}{n}(a_0+\cdots+a_{n-1})$. It can exist even in the case where the sequence itself diverges, for example, if $a_n=(-1)^n$. At the same time, if $a_n \underset{n\to\infty}{\longrightarrow} a$, then also $\frac{1}{n}(a_0+\cdots+a_{n-1}) \underset{n\to\infty}{\longrightarrow} a$ (the permanence of the method of arithmetic means). For numerical series, this approach leads to the concept of a generalized sum of a series. We say that a series *Cesaro converges* to a number C if the limit of the partial sums of the series in the sense of Cesaro is equal to C.

Based on these considerations, we put

$$\sigma_n(f,x)=\frac{1}{n}\big(S_0(f,x)+\cdots+S_{n-1}(f,x)\big),$$

where $S_0(f,x),\ldots,S_{n-1}(f,x)$ are partial sums of the Fourier series of f. The sums σ_n are called the *Fejér*[17] *sums*. From Eq. (5) of Sect. 10.3.3, it follows that

$$\sigma_n(f,x)=\left(f*\frac{1}{n}\sum_{j=0}^{n-1}D_j\right)(x)=\frac{1}{2\pi n}\int_{-\pi}^{\pi}\frac{f(x-u)}{\sin\frac{u}{2}}\sum_{j=0}^{n-1}\sin\left(j+\frac{1}{2}\right)u\,du. \tag{1}$$

The trigonometric identity

$$\sin\frac{u}{2}+\sin\frac{3}{2}u+\cdots+\sin\left(n-\frac{1}{2}\right)u=\frac{1-\cos nu}{2\sin\frac{u}{2}}=\frac{\sin^2\frac{n}{2}u}{\sin\frac{u}{2}}\qquad(u\notin 2\pi\mathbb{Z}),$$

the verification of which is left to the reader, allows one to represent the right-hand side of Eq. (1) in the form

$$\sigma_n(f,x)=\frac{1}{2\pi n}\int_{-\pi}^{\pi}f(x-u)\left(\frac{\sin\frac{n}{2}u}{\sin\frac{u}{2}}\right)^2du.$$

Thus, a Fejér sum can be represented as the convolution of f and the function

$$\Phi_n(u)=\frac{D_0(u)+\cdots+D_{n-1}(u)}{n}=\frac{1}{2\pi n}\left(\frac{\sin\frac{n}{2}u}{\sin\frac{u}{2}}\right)^2, \tag{2}$$

[16] Ernesto Cesaro (1859–1906)—Italian mathematician.

[17] Lipót Fejér (1880–1959)—Hungarian mathematician.

which is called the nth *Fejér kernel*. We advise the reader to sketch the graph of Φ_n and compare it with the graph of the Dirichlet kernel.

The Fejér kernel can be represented in the form

$$\Phi_n(u) = \frac{1}{n}\sum_{j=0}^{n-1} D_j(u) = \frac{1}{2\pi n}\sum_{j=0}^{n-1}\sum_{|k|\leqslant j} e^{iku} = \frac{1}{2\pi}\sum_{|k|<n}\left(1-\frac{|k|}{n}\right)e^{iku},$$

and, therefore,

$$\widehat{\Phi}_n(k) = \begin{cases} \frac{1}{2\pi}(1-\frac{|k|}{n}) & \text{for } |k|<n, \\ 0 & \text{for } |k|\geqslant n. \end{cases} \tag{3}$$

We verify that the sequence $\{\Phi_n\}$ is an approximate identity. It follows from Eq. (2) that the Fejér kernels are non-negative and periodic. Since $\int_{-\pi}^{\pi} D_j(u)\,du = 1$ for all j, we have

$$\int_{-\pi}^{\pi} \Phi_n(u)\,du = \frac{1}{n}\left(\int_{-\pi}^{\pi} D_0(u)\,du + \cdots + \int_{-\pi}^{\pi} D_{n-1}(u)\,du\right) = 1.$$

Finally, the Fejér kernels have a strong form of the localization property (see condition (c′)) of Sect. 7.6.5), namely, Eq. (2) implies that

$$\Phi_n(u) = \frac{1}{2\pi n}\left(\frac{\sin\frac{n}{2}u}{\sin\frac{u}{2}}\right)^2 \leqslant \frac{1}{2\pi n\sin^2\frac{\delta}{2}} = \frac{C_\delta}{n}$$

for $\delta < |u| < \pi$. Now, we are ready to state the main result of this section, which plays an important role in harmonic analysis.

Theorem (Fejér) *Let $f \in \widetilde{\mathscr{L}}^1$ and $x \in \mathbb{R}$. Then:*

(a) *if the limits $L_\pm = \lim_{t\to x\pm 0} f(t)$ exist and are finite, then $\sigma_n(f,x) \underset{n\to\infty}{\longrightarrow} \frac{L_+ + L_-}{2}$;*

(b) *if $f \in \widetilde{C}$, then $\sigma_n(f) \underset{n\to\infty}{\rightrightarrows} f$ on $\mathbb{R}$;*

(c) *if $f \in \widetilde{\mathscr{L}}^p$ for some $p \in [1,+\infty)$, then $\|\sigma_n(f) - f\|_p \underset{n\to\infty}{\longrightarrow} 0$;*

(d) *$\sigma_n(f) \underset{n\to\infty}{\longrightarrow} f$ almost everywhere.*

Proof In the case where the limit $\lim_{t\to x} f(t)$ exists and is finite, both assertions (a) and (b) are special cases of Theorem 7.6.5 (in view of the remarks to it) for $m=1$, $T=\mathbb{N}$ and $t_0=+\infty$. If the one-sided limits are distinct, then we must use the fact that the Fejér kernels are even and apply the result obtained to the function $f_0(u) = (f(x+u)+f(x-u))/2$, which tends to $(L_+ + L_-)/2$ as $u \to 0$.

Assertion (c) is already known. It is a special case of Theorem 2 of Sect. 9.3.7.

To prove the last assertion, we estimate the "hump-shaped" majorant of a Fejér kernel $\psi_n(x) = \sup_{|x|\leqslant y\leqslant\pi} \Phi_n(y)$. Since $\Phi_n(y) \leqslant \Phi_n(0) = \frac{n}{2\pi}$ and $\Phi_n(y) \leqslant$

$\frac{1}{2\pi n \sin^2 \frac{y}{2}} \leqslant \frac{\pi}{2ny^2}$, we see that $\psi_n(x) \leqslant \frac{1}{2\pi} \min\{n, \frac{\pi^2}{nx^2}\}$. From this, it immediately follows that

$$\int_{-\pi}^{\pi} \psi_n(x)\,dx \leqslant 2 \quad \text{for all } n \in \mathbb{N}.$$

Thus, the assumptions of Theorem 3 of Sect. 9.3.7 are fulfilled and, therefore, $\sigma_n(f) \underset{n\to\infty}{\longrightarrow} f$ almost everywhere. □

By statement (a) of the theorem and the permanence of the method of arithmetic means, we are now able to answer the question we posed before taking up the investigation of the convergence of Fourier series (see Sect. 10.3.3): if the Fourier series of a summable function f converges at a point of continuity of f, then the sum of the Fourier series is necessarily equal to the value of f at this point.

Since the convolution $f * \Phi_n$ is a trigonometric polynomial, the Fejér theorem supplements the Weierstrass theorem (see Corollary 7.6.5) by providing specific approximating polynomials.

Remark The third and, obviously, the fourth statements of the theorem give new proofs of the uniqueness theorem. Indeed, if functions $f, g \in \widetilde{\mathscr{L}}^1$ have the same Fourier coefficients, then $\sigma_n(f) = \sigma_n(g)$ for all n. Therefore, statement (c) of the theorem for $p = 1$ implies the relation

$$\|f - g\|_1 = \lim_{n\to\infty} \|\sigma_n(f) - \sigma_n(g)\|_1 = 0.$$

Consequently, the functions f and g are equal almost everywhere.

As we have verified, the Fejér sums $\sigma_n(f)$ have the obvious advantage over the partial sums $S_n(f)$ of the Fourier series that they approximate an arbitrary summable function in the integral metric and converge uniformly to f if f is continuous. It should be mentioned, however, that there is a price to pay for the universality of the Fejér sums: they cannot converge to the function rapidly (see Exercise 2). Therefore, if a Fourier series converges rapidly, then the Fejér sums are a poorer approximation of the function than the Fourier sums (see Exercise 2, Sect. 10.3).

In some cases, Fejér sums allow one to obtain an additional information concerning the behavior of Fourier sums.

Corollary 1 *The Fourier series of an absolutely continuous periodic function f converges to f uniformly on $\mathbb{R}$.*

Proof By the second statement of the Fejér theorem, it is sufficient to verify that the difference $S_n(f) - \sigma_n(f)$ converges uniformly to zero. It is clear that

$$\left|S_n(f,x) - \sigma_n(f,x)\right| = \left|\sum_{k=-n}^{n} \frac{|k|}{n} \widehat{f}(k) e^{ikx}\right| \leqslant \frac{1}{n} \sum_{k=-n}^{n} |k| \left|\widehat{f}(k)\right|.$$

By assumption, we have $f(x) = f(0) + \int_0^x g(t)\,dt$, where $g = f' \in \widetilde{\mathcal{L}}^1$, and, therefore (see property (d) of Sect. 10.3.2) $|k\widehat{f}(k)| = |\widehat{f'}(k)|$. It remains to use the permanence of the method of arithmetic means: since $\widehat{f'}(k) \underset{|k|\to+\infty}{\longrightarrow} 0$, we have

$$\max_{x\in\mathbb{R}} \bigl|S_n(f,x) - \sigma_n(f,x)\bigr| \leqslant \frac{1}{n} \sum_{k=-n}^{n} \bigl|\widehat{f'}(k)\bigr| \underset{n\to\infty}{\longrightarrow} 0.$$ □

Corollary 2 *Let f be a periodic function satisfying the Lipschitz condition of order α, $0 < \alpha < 1$, i.e.,*

there exists a number L such that $|f(x) - f(y)| \leqslant L|x-y|^\alpha$ for all $x, y \in \mathbb{R}$.

Then the Fourier series of f converges uniformly on $\mathbb{R}$ and

$$\bigl|S_n(f,x) - f(x)\bigr| \leqslant C_\alpha L \frac{\ln n}{n^\alpha} \quad \textit{for all } x \in \mathbb{R}.$$

The coefficient C_α depends only on α (for the case $\alpha = 1$, see Exercise 3).

Proof First, we estimate the deviation of the Fejér sums. Since

$$\sigma_n(f,x) - f(x) = \int_{-\pi}^{\pi} \bigl(f(x-t) - f(x)\bigr)\Phi_n(t)\,dt,$$

we obtain

$$\bigl|\sigma_n(f,x) - f(x)\bigr| \leqslant \int_{-\pi}^{\pi} \bigl|f(x-t) - f(x)\bigr|\Phi_n(t)\,dt \leqslant \frac{L}{\pi n} \int_0^{\pi} t^\alpha \left(\frac{\sin\frac{n}{2}t}{\sin\frac{t}{2}}\right)^2 dt$$

$$\leqslant \frac{\pi L}{n} \int_0^{\pi} \frac{\sin^2\frac{n}{2}t}{t^{2-\alpha}}\,dt = \frac{\pi L}{n}\left(\frac{n}{2}\right)^{1-\alpha} \int_0^{\pi n/2} \frac{\sin^2 u}{u^{2-\alpha}}\,du.$$

Thus, $\|\sigma_n(f) - f\|_\infty \leqslant \widetilde{C}_\alpha L/n^\alpha$, where $\widetilde{C}_\alpha = \pi 2^{\alpha-1}\int_0^\infty \frac{\sin^2 u}{u^{2-\alpha}}\,du$.

To estimate the deviation of the Fourier sums, we observe that $S_n(f) - f = S_n(\varphi_n) - \varphi_n$, where $\varphi_n = f - \sigma_n(f)$, since $S_n(\sigma_n(f)) = \sigma_n(f)$. Therefore, inequality (6) of Sect. 10.3.3 implies

$$\bigl|S_n(f,x) - f(x)\bigr| \leqslant \bigl|S_n(\varphi_n,x)\bigr| + \bigl|\varphi_n(x)\bigr| \leqslant 3\|\varphi_n\|_\infty \ln n \leqslant 3\widetilde{C}_\alpha L \frac{\ln n}{n^\alpha}.$$ □

10.4.2 The non-negativity of the Fejér kernels yields interesting properties of the Fourier series. For example, if the coefficients of the Fourier series of a function are non-negative, then the Fourier series converges absolutely if the function is bounded in a neighborhood of zero. Indeed, let a function $f \in \widetilde{\mathcal{L}}^1$ be such that $\widehat{f}(k) \geqslant 0$ for all $k \in \mathbb{Z}$ and $|f(x)| \leqslant C$ if $|x| < \delta$. Since $0 \leqslant \Phi_n(x) \leqslant C_\delta$ for $\delta \leqslant |x| \leqslant \pi$, we obtain

$$\begin{aligned}\left|\sigma_n(f,0)\right| &= \left|\int_{-\pi}^{\pi} f(x)\Phi_n(x)\,dx\right| \leqslant \int_{-\delta}^{\delta} C\Phi_n(x)\,dx + \int_{\delta\leqslant|x|\leqslant\pi} C_\delta\left|f(x)\right|dx \\ &\leqslant C + C_\delta\|f\|_1.\end{aligned}$$

Therefore,

$$\sum_{|k|<\frac{n}{2}} \widehat{f}(k) \leqslant 2\sum_{|k|<n}\left(1-\frac{|k|}{n}\right)\widehat{f}(k) = 2\sigma_n(f,0) \leqslant 2\bigl(C + C_\delta\|f\|_1\bigr).$$

Since this inequality is fulfilled for all n, the series $\sum_{k=-\infty}^{\infty}\widehat{f}(k)$ converges.

Now, we consider an arbitrary trigonometric series of the form

$$\frac{1}{2}c_0 + \sum_{n=1}^{\infty} c_n \cos nx, \tag{4}$$

where the coefficients form a convex sequence tending to zero (the convexity of a sequence $\{c_n\}_{n\geqslant 0}$ means that $c_n \leqslant \frac{1}{2}(c_{n-1}+c_{n+1})$ for all $n\in\mathbb{N}$). The fact that series (4) converges pointwise for $x \notin 2\pi\mathbb{Z}$ can easily be verified by the Dirichlet test.

Before passing to the study of the sum of series (4), we prove a lemma on numerical series.

Lemma *Let $\{c_n\}_{n\geqslant 0}$ be a convex sequence tending to zero. Then*:

(1) $c_{n-1} \geqslant c_n \geqslant 0$ *for all* $n\in\mathbb{N}$;
(2) $\sum_{k=1}^{\infty} k(c_{k-1} - 2c_k + c_{k+1}) = c_0$.

Proof We put $b_n = c_{n-1} - c_n$. The convexity implies that $b_n \geqslant b_{n+1}$, and, consequently, $b_n \geqslant 0$ (because $b_n \underset{n\to\infty}{\longrightarrow} 0$), which proves the first statement.

It is obvious that the series $\sum_{k=1}^{\infty} b_k$ converges and its sum is equal to c_0. Since

$$nb_n \leqslant 2\sum_{n/2\leqslant k\leqslant n} b_k \leqslant 2\sum_{k\geqslant n/2} b_k,$$

we see that $nb_n \underset{n\to\infty}{\longrightarrow} 0$ together with the remainder of a convergent series. This relation and the equality $c_{k-1} - 2c_k + c_{k+1} = b_k - b_{k+1}$ allow us to prove the second statement,

$$\begin{aligned}\sum_{k=1}^{n} k(c_{k-1} - 2c_k + c_{k+1}) &= \sum_{k=1}^{n} k(b_k - b_{k+1}) = \sum_{k=1}^{n} kb_k - \sum_{k=2}^{n+1}(k-1)b_k \\ &= \sum_{k=1}^{n} b_k - nb_{n+1} \underset{n\to\infty}{\longrightarrow} c_0.\end{aligned}$$

□

Theorem *Let the coefficients of series* (4) *form a convex sequence tending to zero. Then the sum* f *of the series is non-negative and summable on* $(-\pi, \pi)$ *and series* (4) *is the Fourier series of* f.

Proof We transform a partial sum $S_n(x)$ of series (4). Using the relation $\cos kx = \pi(D_k(x) - D_{k-1}(x))$, we obtain

$$\frac{1}{\pi} S_n(x) = c_0 D_0(x) + \sum_{k=1}^{n} c_k \big(D_k(x) - D_{k-1}(x)\big)$$

$$= c_n D_n(x) + \sum_{k=0}^{n-1} (c_k - c_{k+1}) D_k(x).$$

Since $D_k(x) = (k+1)\Phi_{k+1}(x) - k\Phi_k(x)$, after elementary transformations, we arrive at the relation

$$\frac{1}{\pi} S_n(x) = c_n D_n(x) + (c_{n-1} - c_n) n \Phi_n(x) + \sum_{k=1}^{n-1} (c_{k-1} - 2c_k + c_{k+1}) k \Phi_k(x).$$

We observe that $c_{k-1} - 2c_k + c_{k+1} \geqslant 0$ since the sequence $\{c_k\}_{k \geqslant 0}$ is convex. Because the sequences $\{D_n(x)\}_{n \geqslant 1}$ and $\{n\Phi_n(x)\}_{n \geqslant 1}$ are bounded for $x \neq 0$ (see formula (4′) of Sect. 10.3.3 and (2) above) and $c_n \underset{n \to \infty}{\longrightarrow} 0$, the first two terms on the right-hand side of the last equation tend to zero. Therefore, passing to the limit in this equation as $n \to \infty$, we see that

$$f(x) = \pi \sum_{k=1}^{\infty} (c_{k-1} - 2c_k + c_{k+1}) k \Phi_k(x) = \frac{1}{2} \sum_{k=1}^{\infty} (c_{k-1} - 2c_k + c_{k+1}) \left(\frac{\sin \frac{k}{2} x}{\sin \frac{x}{2}} \right)^2. \tag{5}$$

This proves that the function f is non-negative. Because non-negative series can be integrated termwise (see Corollary 1 of Sect. 4.8.2), we obtain

$$\int_{-\pi}^{\pi} f(x)\, dx = \pi \sum_{k=1}^{\infty} (c_{k-1} - 2c_k + c_{k+1}) k \int_{-\pi}^{\pi} \Phi_k(x)\, dx$$

$$= \pi \sum_{k=1}^{\infty} (c_{k-1} - 2c_k + c_{k+1}) k = \pi c_0$$

(the last equation is valid by statement (2) of the lemma). Thus, $\int_{-\pi}^{\pi} f(x)\, dx < +\infty$.

Now, we verify that series (4) is the Fourier series of the function f. Obviously, the function f remains a majorant of the partial sums after multiplication of series (5) by $\cos jx$. Therefore, the series obtained by multiplication can be integrated termwise, which leads to the following relation for the Fourier cosine coefficients

$$a_j(f)=\pi\sum_{k=1}^{\infty}(c_{k-1}-2c_k+c_{k+1})ka_j(\Phi_k).$$

Since $a_j(\Phi_k)=\widehat{\Phi}_k(j)+\widehat{\Phi}_k(-j)$, we obtain by Eq. (3) that

$$a_j(f)=\sum_{k=j+1}^{\infty}(c_{k-1}-2c_k+c_{k+1})(k-j)=\sum_{k=1}^{\infty}(c_{k+j-1}-2c_{k+j}+c_{k+j+1})k$$

for $j\in\mathbb{N}$. The numbers $\widetilde{c}_k=c_{k+j}$ $(k=0,1,2\dots)$ obviously form a convex sequence. Applying statement (2) of the lemma to this sequence, we see that the sum of the last series is equal to $\widetilde{c}_0$. Therefore, $a_j(f)=\widetilde{c}_0=c_j$. The relation $A(f)=\frac{1}{2\pi}\int_{-\pi}^{\pi}f(x)\,dx=\frac{1}{2}c_0$ has already been obtained in the proof that f is summable. □

A sequence of coefficients satisfying the conditions of the theorem can tend to zero arbitrarily slow (see Exercise 4). For example, since the sequence $\{1/\ln n\}_{n\geqslant 2}$ is convex, the theorem we have just proved implies that the sum of the series $\sum_{n=2}^{\infty}\frac{\cos nx}{\ln n}$ belongs to $\widetilde{\mathscr{L}}^1$ and the series itself is the Fourier series of its sum. In this connection, we recall that the everywhere convergent series $\sum_{n=2}^{\infty}\frac{\sin nx}{\ln n}$ is not a Fourier series, as established in Sect. 10.3.6.

10.4.3 The Fejér method is based on Cesaro's generalization of the sum of a numerical series. Other generalizations of the concept of the sum of a series are also possible. One of them is based on the following well-known Abel theorem for numerical series: if a series $\sum_{n=1}^{\infty}c_n$ converges to the sum S, then the sum of the series $\sum_{n=1}^{\infty}e^{-\varepsilon n}c_n$ tends to S as $\varepsilon\to+0$. This limit can exist also for a divergent series, and, therefore, can be regarded as a generalized sum of the series in question.

With a view to applications to Fourier series, it is natural to replace the summation over $\mathbb{N}$ by a summation over $\mathbb{Z}$ and use the symmetric partial sums $S_j=\sum_{|k|\leqslant j}c_k$. In this case, we should assume that, by Cesaro's method, to each numerical series $\sum_{n\in\mathbb{Z}}c_n$ (convergent or not), we must assign the sequence of sums

$$\sigma_n=\frac{1}{n}(S_0+\dots+S_{n-1})=\frac{1}{n}\sum_{j=0}^{n-1}\Big(\sum_{|k|\leqslant j}c_k\Big)=\sum_{|k|<n}\Big(1-\frac{|k|}{n}\Big)c_k$$

and, by Abel's method, we must assign the function

$$A(\varepsilon)=\sum_{n\in\mathbb{Z}}e^{-\varepsilon|n|}c_n$$

(to simplify the exposition, we will consider only bounded sequences $\{c_n\}_{n\in\mathbb{Z}}$; for Fourier series, this condition is always fulfilled). In the first case, we calculate the limit $\lim_{n\to\infty}\sigma_n$, and, in the second case, the limit $\lim_{\varepsilon\to+0}A(\varepsilon)$. It can be proved that if the first limit exists (i.e., the series converges in the sense of Cesaro), then the second limit also exists (i.e., the series converges in the sense of Abel) and the limits are equal. In this sense, Abel's method is stronger that the method of Cesaro. We will not dwell on the comparison of these methods. Instead, we show that both methods are special cases of the following general scheme.

Let M be a decreasing function summable on $[0,+\infty)$, and let $M(0) = \lim_{u\to 0} M(u) = 1$. It is clear that $M \geqslant 0$ and the series $\sum_{n\in\mathbb{Z}} M(\varepsilon|n|)$ converges for every $\varepsilon > 0$.

To an arbitrary numerical series $\sum_{n\in\mathbb{Z}} c_n$ with bounded terms, we assign the absolutely convergent series

$$S_M(\varepsilon) = \sum_{n=-\infty}^{\infty} M\big(\varepsilon|n|\big)c_n,$$

which converges uniformly with respect to $\varepsilon > 0$, provided the initial series converges. Under this assumption $\lim_{\varepsilon\to+0} S_M(\varepsilon) = \sum_{n\in\mathbb{Z}} c_n$. The limit $\lim_{\varepsilon\to+0} S_M(\varepsilon)$, if it exists, is called the generalized sum of the given series. We have $M(u) = (1-u)_+$ and $M(u) = e^{-u}$ for the Cesaro and the Abel method, respectively. We obtain the usual sum of a series if we take $M(u) = \chi_{[0,1]}(u)$.

Turning to Fourier series, we see that, to each function $f \in \widetilde{\mathscr{L}}^1$, we must assign the sums

$$S_{M,\varepsilon}(f,x) = \sum_{n=-\infty}^{\infty} M\big(\varepsilon|n|\big)\widehat{f}(n)e^{inx} \qquad (x \in \mathbb{R}).$$

To find integral representations for them, we use property (e) of Sect. 10.3.2, $\widehat{f*g}(n) = 2\pi\widehat{f}(n)\cdot\widehat{g}(n)$, which allows us to regard $S_{M,\varepsilon}(f)$ as the sum of Fourier series of the convolution $f*\omega_\varepsilon$, where

$$\omega_\varepsilon(x) = \frac{1}{2\pi}\sum_{n=-\infty}^{\infty} M\big(\varepsilon|n|\big)e^{inx} = \frac{1}{2\pi} + \frac{1}{\pi}\sum_{n=1}^{\infty} M(\varepsilon n)\cos nx \qquad (x \in \mathbb{R}).$$

By the uniqueness theorem (Corollary 1 of Sect. 10.3.6), the functions $f*\omega_\varepsilon$ and $S_{M,\varepsilon}(f)$ coincide almost everywhere. Since the functions are continuous, they coincide everywhere.

To study the behavior of the sums $S_{M,\varepsilon}(f)$ as $\varepsilon\to+0$, we impose an additional constraint on the function M. We will assume that it is convex on $[0,+\infty)$. Then, the sequence $\{M(\varepsilon n)\}_{n\geqslant 0}$ is convex and $\omega_\varepsilon(x) \geqslant 0$ by Theorem 10.4.2.

Lemma *Let M be a continuous function convex on $[0,+\infty)$, tending to zero at infinity, and let $M(0) = 1$. Then the functions ω_ε form a periodic approximate identity with the strong localization property as $\varepsilon\to+0$. Moreover, there exist even functions Ω_ε that decrease on $[0,\pi]$, majorize ω_ε, and satisfy the inequality $\int_{-\pi}^{\pi}\Omega_\varepsilon(x)\,dx \leqslant 2$.*

For the definition of the strong localization property, see Sect. 7.6.5.

Proof Obviously, $\int_{-\pi}^{\pi} \omega_\varepsilon(x)\,dx = 1$ and, as mentioned above, the function ω_ε is non-negative. We verify that it has the localization property. For this, we use Eq. (5) with $f = \pi\omega_\varepsilon$ and $c_n = M(\varepsilon n)$, which implies that, for $\delta < |x| < \pi$ and $C_\delta = 1/(2\pi \sin^2 \delta/2)$,

$$\begin{aligned}\omega_\varepsilon(x) &= \sum_{n=1}^{\infty} \big(M(n\varepsilon - \varepsilon) - 2M(n\varepsilon) + M(n\varepsilon + \varepsilon)\big)\, n\Phi_n(x) \\ &\leqslant C_\delta \sum_{n=1}^{\infty} \big(M(n\varepsilon - \varepsilon) - 2M(n\varepsilon) + M(n\varepsilon + \varepsilon)\big) \\ &= C_\delta\big(M(0) - M(\varepsilon)\big) \underset{\varepsilon\to+0}{\longrightarrow} 0.\end{aligned}$$

To prove the last statement of the lemma, we use the previous equation one more time. Replacing the Fejér kernel Φ_n in it by the majorant ψ_n constructed in the proof of Theorem 10.4.1, we obtain

$$\omega_\varepsilon(x) \leqslant \sum_{n=1}^{\infty} \big(M(n\varepsilon - \varepsilon) - 2M(n\varepsilon) + M(n\varepsilon + \varepsilon)\big) n\psi_n(x) \equiv \Omega_\varepsilon(x).$$

As established in the same proof, $\int_{-\pi}^{\pi} \psi_n(x)\,dx \leqslant 2$, and, therefore,

$$\begin{aligned}\int_{-\pi}^{\pi} \Omega_\varepsilon(x)\,dx &= \sum_{n=1}^{\infty} \big(M(n\varepsilon - \varepsilon) - 2M(n\varepsilon) + M(n\varepsilon + \varepsilon)\big)\, n \int_{-\pi}^{\pi} \psi_n(x)\,dx \\ &\leqslant 2\sum_{n=1}^{\infty} \big(M(n\varepsilon - \varepsilon) - 2M(n\varepsilon) + M(n\varepsilon + \varepsilon)\big)\, n = 2M(0)\end{aligned}$$

(the last equality is valid by the second statement of Lemma 10.4.2). □

Now, we are able to study the behavior of the sums $S_{M,\varepsilon}(f,x)$.

Theorem *Let $f \in \widetilde{\mathscr{L}}^1$, $x \in \mathbb{R}$, and let the function M be summable on* $[0,+\infty)$ *and satisfy the conditions of the lemma. Then*:

(a) *if the limits $L_\pm = \lim_{t\to x\pm 0} f(t)$ exist and are finite, then $S_{M,\varepsilon}(f,x) \underset{\varepsilon\to 0}{\longrightarrow} \frac{L_+ + L_-}{2}$*;

(b) *if $f \in \widetilde{C}$, then $S_{M,\varepsilon}(f) \underset{\varepsilon\to 0}{\rightrightarrows} f$ on $\mathbb{R}$*;

(c) *if $f \in \widetilde{\mathscr{L}}^p$ for some $p \in [1,+\infty)$, then $\|S_{M,\varepsilon}(f) - f\|_p \underset{\varepsilon\to 0}{\longrightarrow} 0$*;

(d) *$S_{M,\varepsilon}(f,x) \underset{\varepsilon\to 0}{\longrightarrow} f(x)$ almost everywhere.*

Proof The proof can be obtained by repeating verbatim the proof of the Fejér theorem. Indeed, the proof of the Fejér theorem uses only the fact that the sum $\sigma_n(f)$ is the convolution of the function f and an even approximate identity satisfying the strong localization property and having a hump-shaped majorant, which, by the lemma, is also valid for the sum $S_{M,\varepsilon}(f)$. □

In conclusion, we note that, applying Abel's method to a Fourier series, i.e., taking $M(u) = e^{-u}$, we obtain the sums

$$S_\varepsilon(f, x) = \sum_{n=-\infty}^{\infty} e^{-\varepsilon|n|} \widehat{f}(n) e^{inx}$$

called the *Abel–Poisson sums*. We have already encountered the corresponding approximate identity ω_ε. Indeed, let $r = e^{-\varepsilon}$ and $z = re^{ix}$. Then

$$2\pi\,\omega_\varepsilon(x) = 1 + 2\sum_{n=1}^{\infty} r^n \cos nx = \mathcal{R}e\left(1 + 2\sum_{n=1}^{\infty} z^n\right)$$

$$= \mathcal{R}e\,\frac{1+z}{1-z} = \frac{1-r^2}{1-2r\cos x + r^2}.$$

Thus, in the case in question, the function ω_ε is nothing but the Poisson kernel for the disc (for the definition, see Sect. 8.7.10).

10.4.4 The rest of the section is devoted to multiple trigonometric Fourier series, i.e., to series in the system $\{e^{i\langle n,x\rangle}\}_{n\in\mathbb{Z}^m}$. As was mentioned in Sect. 10.2.2, this system is an orthogonal basis in the space $\mathscr{L}^2((-\pi,\pi)^m)$. Therefore, the $\mathscr{L}^2$-theory of multiple Fourier series is a special case of the general theory where the convergence of these series is in the $\mathscr{L}^2$-norm, and we will not touch on this here. The situation is completely different in regard to other forms of convergence. The problems arising here are connected with the fact that there is no preferred definition of the sum of a multiple series. It is possible to find the limit of the partial sums over unboundedly expanding balls, cubes, parallelepipeds, etc. It turns out that the answers to most problems depend essentially on the choice of the definition of a partial sum. We will not dwell on this topic, instead confining ourselves to partial sums over rectangles. Even in this case, for $m > 1$, there are phenomena that did not occur in the one-dimensional situation.

We introduce the necessary notation. When speaking of periodic functions in $\mathbb{R}^m$, we will always mean 2π-periodicity with respect to each variable. By $\widetilde{C}(\mathbb{R}^m)$ and $\widetilde{C}^r(\mathbb{R}^m)$ ($r \in \mathbb{N}$), we denote the classes of periodic continuous and periodic smooth functions, respectively, defined on $\mathbb{R}^m$; by $\widetilde{\mathscr{L}}^p(\mathbb{R}^m)$ ($1 \leqslant p \leqslant +\infty$), we denote the class of periodic pth power summable functions on the cube $Q = (-\pi,\pi)^m$. For a function f in $\widetilde{\mathscr{L}}^p(\mathbb{R}^m)$, we denote the $\widetilde{\mathscr{L}}^p$-norm of its restriction to Q by $\|f\|_p$.

For a function $f \in \widetilde{\mathscr{L}}^1(\mathbb{R}^m)$ and a multi-index $n \in \mathbb{Z}^m$,

$$\widehat{f}(n) = \frac{1}{(2\pi)^m} \int_Q f(x) e^{-i\langle n,x\rangle}\, dx$$

is the nth Fourier coefficient of f.

When solving the problem of expanding a periodic function in the Fourier series $\sum_{n\in\mathbb{Z}^m} c_n e^{i\langle n,x\rangle}$, as in the one-dimensional case, there is no freedom in the choice of coefficients. The reasoning used at the end of Sect. 10.3.1 also remains also in the case in question. More precisely, let $S_j(x) = \sum_{n\in A_j} c_n e^{i\langle n,x\rangle}$ be the partial sums of this series corresponding to expanding bounded sets $A_j \subset \mathbb{R}^m$ such that $\bigcup_{j=1}^{\infty} A_j = \mathbb{R}^m$. If $S_j \underset{j\to\infty}{\longrightarrow} S$ almost everywhere or in measure, then the function S can be called the sum of the series. If, in addition, the partial sums S_j are dominated by some function g in $\widetilde{\mathscr{L}}^1(\mathbb{R}^m)$, then the coefficients of the given trigonometric series are determined uniquely. Indeed, as follows from Lebesgue's theorem, $\widehat{S}(n) = \lim_{j\to\infty} \widehat{S}_j(n) = c_n$. Therefore, under the present hypothesis, the expansion of a function in a multiple Fourier series is unique.

For absolutely convergent Fourier series, all definitions of partial sums give the same result because, in this case, the terms of the Fourier series form a summable family. Moreover, an absolutely convergent trigonometric series is, obviously, the Fourier series of its sum. Since the trigonometric system is complete (see Theorem 10.2.2) the sum of an absolutely convergent Fourier series coincides with the function almost everywhere, and if the function is continuous, it converges everywhere (in the one-dimensional case, this has been noted at the beginning of Sect. 10.3.8). As we will soon verify, the Fourier series of smooth functions converge absolutely.

In the multi-dimensional case, all basic properties of the coefficients of a Fourier series, as well as their proofs, are preserved (in what follows, $n = (n_1, \dots, n_m) \in \mathbb{Z}^m$):

(a) $|\widehat{f}(n)| \leqslant \frac{1}{(2\pi)^m} \|f\|_1$;
(b) $|\widehat{f}(n)| \to 0$ as $\|n\| \to +\infty$;
(c) the Fourier coefficients of a translation f_h ($h \in \mathbb{R}^m$), i.e., of the function $f_h(x) = f(x-h)$, are connected with the Fourier coefficients of f by the formulas $\widehat{f_h}(n) = e^{-i\langle n,h\rangle} \widehat{f}(n)$;
(d) if $f \in \widetilde{C}^r(\mathbb{R}^m)$ and $g = \frac{\partial^r f}{\partial x_{j_1} \dots \partial x_{j_r}}$, then $\widehat{g}(n) = i^r n_{j_1} \cdots n_{j_r} \widehat{g}(n)$;
(e) $\widehat{f * g}(n) = (2\pi)^m \widehat{f}(n) \cdot \widehat{g}(n)$ for all functions f and g in $\widetilde{\mathscr{L}}^1(\mathbb{R}^m)$.

By property (d), it is easy to show that the functions of class $\widetilde{C}^{(m+1)}(\mathbb{R}^m)$ have absolutely convergent Fourier series. The following theorem shows that the smoothness requirement can be weakened considerably.

Theorem *If $f \in \widetilde{C}^r(\mathbb{R}^m)$ and $r > m/2$, then $\sum_{n\in\mathbb{Z}^m} |\widehat{f}(n)| < +\infty$, and, therefore, $f(x) = \sum_{n\in\mathbb{Z}^m} \widehat{f}(n) e^{i\langle n,x\rangle}$ for all x.*

Proof Indeed, by property (d), we have $|n_k|^r|\widehat{f}(n)| = |\widehat{g}_k(n)|$, where $n = (n_1, \dots, n_m) \in \mathbb{Z}_+^m$ and $g_k = \frac{\partial^r f}{\partial x_k^r}$, $k = 1, \dots, m$. We put

$$c_n = \left|\widehat{g}_1(n)\right| + \cdots + \left|\widehat{g}_m(n)\right| = \left(|n_1|^r + \cdots + |n_m|^r\right)\left|\widehat{f}(n)\right|.$$

Since $\|n\|^r \leqslant m^{r/2} \max\{|n_1|^r, \dots, |n_m|^r\} \leqslant m^{r/2}(|n_1|^r + \cdots + |n_m|^r)$, we obtain

$$\left|\widehat{f}(n)\right| = \frac{c_n}{|n_1|^r + \cdots + |n_m|^r} \leqslant \text{const}\frac{c_n}{\|n\|^r}$$

for all $n \neq 0$. It is clear that

$$\sum_{n\in\mathbb{Z}^m} c_n^2 = \sum_{n\in\mathbb{Z}^m} \left(\left|\widehat{g}_1(n)\right| + \cdots + \left|\widehat{g}_m(n)\right|\right)^2$$
$$\leqslant m \sum_{n\in\mathbb{Z}^m} \left(\left|\widehat{g}_1(n)\right|^2 + \cdots + \left|\widehat{g}_m(n)\right|^2\right) < +\infty$$

by Bessel's inequality. Consequently,

$$\sum_{n\neq 0} \left|\widehat{f}(n)\right| \leqslant \text{const} \sum_{n\neq 0} \frac{c_n}{\|n\|^r} \leqslant \frac{\text{const}}{2} \sum_{n\neq 0} \left(\frac{1}{\|n\|^{2r}} + c_n^2\right) < +\infty$$

(the series $\sum_{n\neq 0} \frac{1}{\|n\|^{2r}}$ converges since $2r > m$). □

10.4.5 Now, we turn to the problem of the Fourier series representation of functions from a wider class. As already mentioned, we confine ourselves to partial sums of multiple series over rectangles. Here, we will use the notation introduced in Sect. 1.1.6. For vectors $a = (a_1, \dots, a_m)$ and $b = (b_1, \dots, b_m)$, the inequalities $a \leqslant b$ and $a < b$ mean that $a_1 \leqslant b_1, \dots, a_m \leqslant b_m$ and $a_1 < b_1, \dots, a_m < b_m$, respectively. By $|a|$, we denote the vector $(|a_1|, \dots, |a_m|)$.

For a function $f \in \widetilde{\mathscr{L}}^1(\mathbb{R}^m)$ and a multi-index $n = (n_1, \dots, n_m) \in \mathbb{Z}_+^m$, we consider the partial sum

$$S_n(f, x) = \sum_{|k|\leqslant n} \widehat{f}(k)e^{i\langle k,x\rangle}$$

of the Fourier series over the corresponding "rectangle". Reasoning as in Sect. 10.3.3, we can easily verify that this sum is the convolution of f and a multi-dimensional Dirichlet kernel equal to the product to the product of one-dimensional Dirichlet kernels, $S_n(f) = f * D_n$, where

$$D_n(u) = D_{n_1}(u_1) \cdots D_{n_m}(u_m) \quad \left(u = (u_1, \dots, u_m)\right).$$

As in the one-dimensional case, the kernel D_n satisfies the equation (below, $Q = (-\pi, \pi)^m$)

$$\int_Q D_n(u)\,du = 1. \tag{6}$$

It was noted in Sect. 10.3.3 that the $\mathscr{L}^1$-norms of one-dimensional Dirichlet kernels have logarithmic rate of growth. The same is also true for the Dirichlet kernels corresponding to rectangles,

$$\|D_n\|_1 = \int_Q |D_n(u)|\,du = \prod_{j=1}^{m} \int_{-\pi}^{\pi} |D_{n_j}(u_j)|\,du_j \asymp \prod_{j=1}^{m} \ln n_j.$$

Hence it follows that the following estimate is valid for periodic bounded (in particular, continuous) functions and $n_1, \ldots, n_m > 1$:

$$\|S_n(f)\|_\infty \leqslant C_m \left(\prod_{j=1}^{m} \ln n_j \right) \|f\|_\infty. \tag{7}$$

The following theorem shows that the class of functions with uniformly convergent Fourier series is rather wide.

Theorem *Assume that a periodic function f satisfies the Lipschitz condition of order α, $0 < \alpha \leqslant 1$, i.e., there is an L such that $|f(x) - f(y)| \leqslant L\|x - y\|^\alpha$ for all $x, y \in \mathbb{R}^m$. Then the sums $S_n(f)$ converge uniformly to f as* $\min\{n_1, \ldots, n_m\} \to +\infty$.

Proof It is sufficient to consider the case where $\alpha < 1$. In addition, we will assume that $m = 2$ because no new ideas are required for the proof in the general case. Subtracting Eq. (6) multiplied by $f(x)$ from $S_n(f, x) = (f * D_n)(x) = \int_Q f(x - u) D_n(u)\,du$, we obtain

$$\Delta = S_n(f, x) - f(x) = \int_Q \bigl(f(x - u) - f(x)\bigr) D_n(u)\,du.$$

To simplify the subsequent formulas, we change the notation as follows: $n = (j, k)$, $x = (a, b)$ and $u = (s, t)$. Estimating the difference Δ, we may assume, without loss of generality, that the first coordinate of the vector n does not exceed its second coordinate, i.e., $j \leqslant k$.

We represent the increment of the function f as the sum of the increments in each coordinate,

$$f(x - u) - f(x) = f(a - s, b - t) - f(a - s, b) + f(a - s, b) - f(a, b).$$

Therefore, the difference Δ splits into the sum of the integrals I and J, where

$$I = \int_Q \bigl(f(a - s, b - t) - f(a - s, b)\bigr) D_j(s) D_k(t)\,ds\,dt$$

and

$$J = \int_Q \bigl(f(a - s, b) - f(a, b)\bigr) D_j(s) D_k(t)\,ds\,dt$$
$$= \int_{-\pi}^{\pi} \bigl(f(a - s, b) - f(a, b)\bigr) D_j(s)\,ds.$$

The integral J is small by Corollary 2 of Fejér's theorem, $|J| \leqslant C_\alpha L(\ln j)/j^\alpha$. The same corollary makes it possible to estimate the integral I,

$$\begin{aligned}|I| &\leqslant \int_{-\pi}^{\pi}\left|\int_{-\pi}^{\pi}\big(f(a-s,b-t)-f(a-s,b)\big)D_k(t)\,dt\right|\big|D_j(s)\big|\,ds\\ &\leqslant C_\alpha\, L\,\frac{\ln k}{k^\alpha}\int_{-\pi}^{\pi}\big|D_j(s)\big|\,ds.\end{aligned}$$

Since $\int_{-\pi}^{\pi}|D_j(s)|\,ds \leqslant 2\ln j$ for $j \geqslant 10$ (see inequality (6) of Sect. 10.3.3), we obtain

$$|I| \leqslant 2C_\alpha L\frac{\ln k}{k^\alpha}\ln j \leqslant 2C_\alpha L\frac{\ln^2 k}{k^\alpha}.$$

Thus, the inequality

$$\big|S_n(f,x)-f(x)\big| = |\Delta| \leqslant |I|+|J| \leqslant C_\alpha L\left(\frac{\ln j}{j^\alpha}+2\frac{\ln^2 k}{k^\alpha}\right)$$

holds for all x. □

10.4.6 Here, we present two negative results illustrating some phenomena that may occur in the behavior of the double Fourier series and which are impossible in the one-dimensional case.

The first of them is connected with the Riemann localization principle (see Theorem 10.3.3). It turns out that this principle is not true for sums over rectangles: there is a function $f \in \widetilde{C}$ equal to zero in a neighborhood of the origin and such that the partial sums (over rectangles) of its Fourier series are unbounded at the origin. To verify this, consider a function f of the form $f(s,t) = \varphi(s)\psi(t)$, where $\varphi, \psi \in \widetilde{C}$. It is easy to find a function φ equal to zero in the vicinity of the origin and such that $S_j(\varphi,0) \neq 0$ for an infinite number of indices j. We take a function ψ for which the sequence $\{S_k(\psi,0)\}_{k\in\mathbb{N}}$ is unbounded as the second factor (see the Schwartz example in Sect. 10.3.9). Then the product $f(s,t) = \varphi(s)\psi(t)$ is equal to zero not only in the vicinity of the origin, but also in a vertical strip containing the second coordinate axis. Moreover, $S_{j,k}(f,(0,0)) = S_j(\varphi,0)S_k(\psi,0)$. Taking an arbitrarily large j for which $S_j(\varphi,0) \neq 0$, we can choose a k such that the sum $S_{j,k}(f,(0,0))$ is larger than every preassigned number.

It can be proved that the localization principle is preserved if the usual neighborhoods of a point (a,b) are replaced by "cross neighborhoods", i.e., by sets of the form $\{(s,t) \mid \min(|s-a|,|t-b|) < \delta\}$.

The second fact that we want to mention is connected with Carleson's theorem (see the end of Sect. 10.3.9) and is considerably harder. C.L. Fefferman[18] observed that this theorem cannot be carried over to functions of several variables: there is a periodic function of two variables that is uniformly bounded in the square $(0,2\pi)^2$

[18]Charles Louis Fefferman (born 1949)—American mathematician.

and whose Fourier sums over rectangles are unbounded at every point of this square. The "divergence phenomenon" manifests itself on the functions f_N equal to e^{iNst} for $0 < s, t < 2\pi$ (N is a large parameter). It turns out that, despite the fact that $|f_N| = 1$, for every $N > 1$ and every point (s,t), there are numbers j and k such that the quantity $|S_{j,k}(f_N, (s,t))|$ is comparable with $\ln N$. We will not discuss the Fefferman's example in detail, but refer the reader to Exercise 10.

10.4.7 As in the one-dimensional case, one can use the Fejér sums to approximate a function of several variables by trigonometric polynomials. In the multi-dimensional case, these sums, as well as their partial sums, can be defined in different ways. We define the m-dimensional Fejér sums by the equation (in what follows, $n, j \in \mathbb{Z}_+^m$ and $k \in \mathbb{Z}^m$)

$$\sigma_n(f,x) = \frac{1}{n_1 \cdots n_m} \sum_{0 \leqslant j < n} S_j(f,x) = \sum_{|k|<n} \left(1 - \frac{|k_1|}{n_1}\right) \cdots \left(1 - \frac{|k_m|}{n_m}\right) \widehat{f}(k) e^{i\langle k,x\rangle}.$$

Since $S_j(f,x) = (f * D_j)(x)$, we obtain $\sigma_n(f,x) = (f * \Phi_n)(x)$, where

$$\Phi_n(y) = \frac{1}{n_1 \cdots n_m} \sum_{0 \leqslant j < n} D_j(y) = \Phi_{n_1}(y_1) \cdots \Phi_{n_m}(y_m).$$

It is natural to call the function Φ_n the m-dimensional Fejér kernel. The properties of the one-dimensional Fejér kernel established in Sect. 10.4.1 imply immediately that:

(a) $\Phi_n(y) \geqslant 0$;
(b) $\int_Q \Phi_n(y)\,dy = 1$;
(c) $\int_{Q \setminus B(\delta)} \Phi_n(y)\,dy \leqslant \frac{C_\delta}{\min\{n_1,\ldots,n_m\}}$ for every $\delta \in (0,\pi)$.

Thus, we can regard the functions $\{\Phi_n\}_{n \in \mathbb{Z}_+^m}$ as an approximate identity with the proviso that it is now parametrized by an integer vector n and the localization property is valid as $\min\{n_1,\ldots,n_m\} \to +\infty$. Therefore, the following analogs of statements (b) and (c) of Theorem 10.4.1 are valid for the sums $\sigma_n(f,x)$.

Theorem

(1) *If* $f \in \widetilde{C}(\mathbb{R}^m)$, *then* $\sigma_n(f) \rightrightarrows f$ *on* $\mathbb{R}^m$ *as* $\min\{n_1,\ldots,n_m\} \to +\infty$.
(2) *If* $f \in \widetilde{\mathscr{L}}^p([-\pi,\pi]^m)$ *for some* $p \in [1,+\infty)$, *then* $\|\sigma_n(f) - f\|_p \to 0$ *as* $\min\{n_1,\ldots,n_m\} \to +\infty$.

As in the one-dimensional case (see the Remark in Sect. 10.4.1), the convergence of the Fejér sums in mean implies, in particular, the uniqueness theorem for multiple Fourier series:

Corollary *Functions in* $\widetilde{\mathscr{L}}^1(\mathbb{R}^m)$ *coincide almost everywhere if they have the same Fourier coefficients.*

(For more general results, see Sects. 11.1.9 and 12.3.3.)

In the multi-dimensional case, however, the Fejér kernels do not satisfy the strong localization property. Therefore, an analog of statement (a) of Theorem 10.4.1 can be obtained for them only under the additional assumption that the function f is bounded (see Theorem 7.6.5). This restriction cannot be lifted as can be shown by modifying the example from the previous section. Indeed, we preserve the first factor φ in the example and change the second factor, rejecting the continuity (which now is not needed), as follows:

$$\psi(t) = \cos t + \frac{1}{2}\cos 2t + \frac{1}{3}\cos 3t + \cdots .$$

Since $S_k(\psi, 0) \underset{k\to\infty}{\longrightarrow} +\infty$, the same is also valid for the Fejér sums, $\sigma_k(\psi, 0) \underset{k\to\infty}{\longrightarrow} +\infty$. Furthermore, there are infinitely many non-zero sums $\sigma_j(\varphi, 0)$ because the Fourier sums $S_j(\varphi, 0)$ have the same property. Consequently, the sums $\sigma_{j,k}(f, (0,0)) = \sigma_j(\varphi, 0)\sigma_k(\psi, 0)$ are unbounded, and so do not tend to zero, even though the function f is zero in a strip containing the second coordinate axis.

If f belongs to the class $\mathscr{L}^p(\mathbb{R}^m)$ for some $p > 1$, then the sums $\sigma_n(f)$ converge almost everywhere. Although this assumption can be weakened somewhat, it is impossible to drop it completely.

The difficulties arising in the study of the multi-dimensional Fejér method occur also in the "coordinatewise" generalizations of other methods, for example, of the Abel–Poisson method. We will not discuss this question in detail, instead referring the reader to the literature [Zy], vol. II, Chap. XVII.

10.4.8 In conclusion, we note that, for $m > 1$, there are different natural ways of constructing partial sums of a multiple Fourier series and their averages. For example, instead of sums over rectangles, we could consider only sums over cubes centered at the origin and their arithmetic means. Although the corresponding kernels will not preserve sign, it can be proved that they define a generalized approximate identity satisfying the assumption (a′), less restrictive than assumption (a) (see Sect. 7.6.1).

Even more difficulties arise for another natural definition of the partial sums of a Fourier series, namely, when the summation is performed over balls. In this case, we form "ball" partial sums, putting

$$S_R(f, x) = \sum_{\|k\|<R} \widehat{f}(k) e^{i\langle k,x\rangle}$$

for an $R > 0$. This partial sum can, of course, be represented as the convolution with the corresponding "ball Dirichlet kernel" $D_R(y) = (2\pi)^{-m} \sum_{\|k\|<R} e^{i\langle k,y\rangle}$ for which, unfortunately, no compact expression is known. The sums $S_R(f)$ have the important property that they do not satisfy a "logarithmic" estimate similar to inequality (7). This is due to the fast growth of the norms $\|D_R\|_1$. It turns out that, as $R \to +\infty$, the norms $\|D_R\|_1$ grow (in order) as $R^{(m-1)/2}$ (see [AIN1, AIN2]). We will return to this unexpected result at the end of the next section.

In the two-dimensional case, the arithmetic means of the "disc" Fourier sums of a periodic continuous function f converge to f uniformly (and if $f \in \widetilde{\mathscr{L}}^p(\mathbb{R}^2)$ for $p < +\infty$, then in the $\mathscr{L}^p$-norm), but this is not the case for a larger number of variables.

EXERCISES

1. Let $T(x) = \sum_{k=-n}^{n} c_k e^{ikx}$ be a trigonometric series of order n, and let $p \in [1, +\infty]$. Prove the *Bernstein*[19] *inequality* $\|T'\|_p \leqslant 2n\|T\|_p$. *Hint*. Verify that $T' = -2nT * \Psi_n$, where $\Psi_n(x) = \Phi_n(x)\sin nx$ and Φ_n is a Fejér kernel.
2. Prove that the Fejér sums cannot converge rapidly: either there is a $\delta > 0$ such that $\|f - \sigma_n(f)\|_1 \geqslant \frac{\delta}{n} > 0$ for all $n \in \mathbb{N}$, or $f \equiv$ const almost everywhere. *Hint*. Calculate the Fourier coefficients of the difference $f - \sigma_n(f)$ and apply inequality (a) of Sect. 10.3.2.
3. Supplement the result of Corollary 2 of Sect. 10.4.1 by proving that $\|S_n(f) - f\|_\infty \leqslant CL\frac{\ln n}{n}$ for $\alpha = 1$. *Hint*. To estimate the integral $S_n(f,x) - f(x) = \int_{-\pi}^{\pi}(f(x-u) - f(x))D_n(u)\,du$, replace the difference $f(x-u) - f(x)$ on each interval $[\frac{2k-1}{n+1/2}\pi, \frac{2k+1}{n+1/2}\pi]$ by its value at the midpoint of this interval and then verify that the integral of the Dirichlet kernel over the interval admits an estimate $O(1/k^2)$.
4. Prove that the Fourier cosine coefficients can tend to zero arbitrarily slowly, i.e., for every sequence $\{c_n\}_{n\geqslant 1}$ decreasing to zero, there is a function $f \in \widetilde{\mathscr{L}}^1$ such that $c_n \leqslant a_n(f)$ for all $n \in \mathbb{N}$. *Hint*. Dominate $\{c_n\}_{n\geqslant 1}$ by a convex sequence and apply Theorem 10.4.2.
5. Let the sequence of coefficients of a series $\sum_{n=1}^{\infty} c_n \cos nx$ be convex. Prove that the $\mathscr{L}^1$-norms of the partial sums are bounded if and only if $c_n = O(1/\ln n)$; prove that the given series converges in $\widetilde{\mathscr{L}}^1$ if and only if $c_n = o(1/\ln n)$ as $n \to \infty$.
6. Let $S(x) = \sum_{n=1}^{\infty} c_n \sin nx$ where $c_n \downarrow 0$. Prove that the boundedness and the continuity of the function S is equivalent to the relation $c_n = O(\frac{1}{n})$ and $c_n = o(\frac{1}{n})$, respectively.
7. Prove that the following version of Parseval's identity is valid for functions $f \in \widetilde{\mathscr{L}}^1$ and $g \in \widetilde{\mathscr{L}}^\infty$: the series $\sum_{n\in\mathbb{Z}} \widehat{f}(n)\overline{\widehat{g}(n)}$ Cesaro converges to $\frac{1}{2\pi}\int_{-\pi}^{\pi} f(x)g(x)\,dx$.
8. Prove that the Fourier series of every function f in $\widetilde{\mathscr{L}}^1(\mathbb{R}^m)$ can be integrated termwise over every rectangular parallelepiped P,

$$\int_P f(x)\,dx = \sum_{n\in\mathbb{Z}^m} \widehat{f}(n) \int_P e^{i\langle n,x\rangle}\,dx$$

(the sum of the series on the right-hand side of the equation is regarded as the limit of the partial sums over rectangles).

[19] Sergei Natanovich Bernstein (1880–1968)—Russian mathematician.

9. Let a function $f \in \widetilde{\mathscr{L}}^1(\mathbb{R}^m)$ be such that $\widehat{f}(n) \geqslant 0$ for all $n \in \mathbb{Z}^m$. Prove that if f is bounded and continuous at the origin, then its Fourier series converges absolutely (therefore, f coincides with a function of class $\widetilde{C}$ almost everywhere).
Hint. Use the fact that the sums $\sigma_n(f, 0)$ are bounded.

10. To construct a continuous function of two variables for which the Fourier series diverges everywhere in the square $(0, 2\pi)^2$ (see Sect. 10.4.6), prove that the Fourier sums $S_{j,k}(f_N)$ of the function $f_N(s,t) = e^{iNst}$ $(N \geqslant 1, 0 < s, t < 2\pi)$ over the rectangles $[-j, j] \times [-k, k]$ satisfy the following inequalities at each point of the given square (the constants at the O-terms depend only on s and t):

(a) $|S_{j,k}(f_N; s,t)| = O(\ln j)$;
(b) if $k > 2\pi N$, then $|S_{j,k}(f_N; s,t)| = O(1 + \frac{\ln j}{k - 2\pi N})$;
(c) $|S_{j,k}(f_N; s,t)| \geqslant \frac{1}{2\pi} \ln N + O(1)$ for $j = [Ns]$ and $k = [Nt]$.

Conclude from this that, for a sufficiently small $r > 0$ and $N_n = e^{n!}$, the Fourier series of the function $F(s,t) = \sum_{n=1}^{\infty} r^n e^{iN_n st}$ diverges unboundedly (the sums $S_{j,k}(F)$ are unbounded) at each point of the square $(0, 2\pi)^2$.

10.5 The Fourier Transform

10.5.1 We introduce one of the main concepts of this chapter.

Definition *The Fourier transform* $\widehat{f}$ *of a function* f in $\mathscr{L}^1(\mathbb{R}^m)$ is defined by the formula

$$\widehat{f}(y) = \int_{\mathbb{R}^m} f(x) e^{-2\pi i \langle y, x \rangle} \, dx \quad \left(y \in \mathbb{R}^m\right)$$

(here, as before, $\langle y, x \rangle$ is the scalar product of vectors y and x).

Theorem 7.1.3 on the continuity of an integral depending on a parameter implies that the function $\widehat{f}$ is continuous. This function is bounded since

$$\left|\widehat{f}(y)\right| \leqslant \|f\|_1 \quad \text{for all } y \in \mathbb{R}^m.$$

Moreover, by the Riemann–Lebesgue theorem, we have $\widehat{f}(y) \to 0$ as $\|y\| \to +\infty$.

We recall that the translation f_h of a function f by a fixed vector $h \in \mathbb{R}^m$ is defined by the equation $f_h(x) = f(x - h)$. An easy calculation shows that $\widehat{f}$ and $\widehat{f_h}$ are related as follows (see also Exercise 1):

$$\widehat{f_h}(y) = \int_{\mathbb{R}^m} f(x-h) e^{-2\pi i \langle y, x \rangle} \, dx = \int_{\mathbb{R}^m} f(t) e^{-2\pi i \langle y, t+h \rangle} \, dt = e^{-2\pi i \langle y, h \rangle} \, \widehat{f}(y).$$

Another operation with the argument of a function, a contraction, is also connected with the Fourier transform: if $a \in \mathbb{R} \setminus \{0\}$ and $g(x) = f(ax)$, then

$$\widehat{g}(y) = \int_{\mathbb{R}^m} f(ax) e^{-2\pi i \langle y, x \rangle} \, dx = \frac{1}{|a|^m} \int_{\mathbb{R}^m} f(t) e^{-2\pi i \frac{1}{a} \langle y, t \rangle} \, dt = \frac{1}{|a|^m} \widehat{f}\left(\frac{y}{a}\right).$$

An important property of the Fourier transform relates the operations of convolution and multiplication.

Theorem *If* $f, g \in \mathcal{L}^1(\mathbb{R}^m)$, *then* $\widehat{f * g}(y) = \widehat{f}(y)\widehat{g}(y)$ $(y \in \mathbb{R}^m)$. *Moreover,* $\int_{\mathbb{R}^m} \widehat{f}(y)\, g(y)\, dy = \int_{\mathbb{R}^m} f(y)\, \widehat{g}(y)\, dy$.

Proof The proof is an almost verbatim repetition of the corresponding reasoning for Fourier coefficients (see property (e)) of Sect. 10.3.2),

$$\begin{aligned}\widehat{f * g}(y) &= \int_{\mathbb{R}^m} \left(\int_{\mathbb{R}^m} f(u) g(x-u)\, du \right) e^{-2\pi i \langle y, x\rangle}\, dx \\ &= \int_{\mathbb{R}^m} f(u) e^{-2\pi i \langle y, u\rangle} \left(\int_{\mathbb{R}^m} g(x-u) e^{-2\pi i \langle y, x-u\rangle}\, dx \right) du \\ &= \int_{\mathbb{R}^m} f(u) e^{-2\pi i \langle y, u\rangle} \left(\int_{\mathbb{R}^m} g(v) e^{-2\pi i \langle y, v\rangle}\, dv \right) du = \widehat{f}(y)\widehat{g}(y).\end{aligned}$$

The second relation is proved similarly. □

We consider some examples.

Example 1 The Fourier transform of the characteristic function χ of the interval $(-1, 1)$ is calculated very simply:

$$\widehat{\chi}(y) = \int_{-\infty}^{\infty} \chi(x) e^{-2\pi i y x}\, dx = \int_{-1}^{1} e^{-2\pi i y x}\, dx = \frac{\sin 2\pi y}{\pi y}.$$

We remark that $\widehat{\chi} \notin \mathcal{L}^1(\mathbb{R})$ (this is established in Example 1 of Sect. 4.6.6).

Example 2 We consider the function $f_t(x) = e^{-\pi t^2 x^2}$ $(x \in \mathbb{R},\ t > 0)$. Its Fourier transform is actually calculated in Example 1 of Sect. 7.1.6:

$$\widehat{f_t}(y) = \int_{-\infty}^{\infty} e^{-\pi t^2 x^2} e^{-2\pi i y x}\, dx = 2 \int_{0}^{\infty} e^{-\pi t^2 x^2} \cos 2\pi y x\, dx = \frac{1}{t} e^{-\frac{\pi}{t^2} y^2}. \quad (1)$$

It is interesting to note that $\widehat{f_t} = \frac{1}{t} f_{\frac{1}{t}}$ and, in particular, $\widehat{f_1} = f_1$.

From Eq. (1), we immediately obtain its multi-dimensional counterpart,

$$\int_{\mathbb{R}^m} e^{-\pi t^2 \|x\|^2} e^{-2\pi i \langle y, x\rangle}\, dx = \frac{1}{t^m} e^{-\frac{\pi}{t^2} \|y\|^2}. \quad (1')$$

Example 3 Let $f(x) = e^{-|x|}$ $(x \in \mathbb{R})$. Then

$$\begin{aligned}\widehat{f}(y) &= \int_{-\infty}^{\infty} e^{-|x|} e^{-2\pi i y x}\, dx \\ &= 2\mathcal{R}e\left(\int_{0}^{\infty} e^{-(1+2\pi i y)x}\, dx \right) = \mathcal{R}e \frac{2}{1+2\pi i y} = \frac{2}{1+(2\pi y)^2}.\end{aligned}$$

Example 4 It is considerably harder to obtain a multi-dimensional generalization of Example 3, i.e., to calculate the Fourier transform of the function $f(x) = e^{-\|x\|}$ ($x \in \mathbb{R}^m$) because, in this case, it is impossible to use separation of variables. The complication can be overcome by an artificial trick based on an integral representation of the function $e^{-\|x\|}$. We need the formula

$$e^{-t} = \frac{2}{\sqrt{\pi}} \int_0^{\infty} e^{-u^2 - \frac{t^2}{4u^2}}\, du \quad \text{for every } t > 0.$$

To obtain it, we must represent the integral on the right-hand side in the form $e^{-t} \int_0^{\infty} e^{-(u - \frac{t}{2u})^2}\, du$. After the change of variables $v = u - \frac{t}{2u}$, this integral reduces to the Euler–Poisson integral $\int_{-\infty}^{\infty} e^{-v^2}\, dv = \sqrt{\pi}$.

Now, we use the relation established above to calculate $\widehat{f}$,

$$\widehat{f}(y) = \int_{\mathbb{R}^m} e^{-\|x\|} e^{-2\pi i \langle y,x\rangle}\, dx = \frac{2}{\sqrt{\pi}} \int_{\mathbb{R}^m} \left(\int_0^{\infty} e^{-u^2 - \frac{\|x\|^2}{4u^2}}\, du \right) e^{-2\pi i \langle y,x\rangle}\, dx.$$

Changing the order of integration and applying Eq. (1′) with $t = \frac{1}{2\sqrt{\pi}u}$, we obtain

$$\begin{aligned}
\widehat{f}(y) &= \frac{2}{\sqrt{\pi}} \int_0^{\infty} e^{-u^2} \left(\int_{\mathbb{R}^m} e^{-\frac{\|x\|^2}{4u^2}} e^{-2\pi i \langle y,x\rangle}\, dx \right) du \\
&= \frac{2}{\sqrt{\pi}} \int_0^{\infty} e^{-u^2} (2\sqrt{\pi}u)^m e^{-4\pi^2 u^2 \|y\|^2}\, du \\
&= 2^{m+1} \pi^{\frac{m-1}{2}} \int_0^{\infty} u^m e^{-(1+4\pi^2\|y\|^2)u^2}\, du.
\end{aligned}$$

Now, the change of variables $v = (1 + 4\pi^2\|y\|^2)u^2$ allows us to express the last integral in terms of the Gamma function, and we obtain the required result

$$\widehat{f}(y) = 2^m \pi^{\frac{m-1}{2}} \frac{\Gamma(\frac{m+1}{2})}{(1 + 4\pi^2\|y\|^2)^{\frac{m+1}{2}}}.$$

Example 5 Let $a, u > 0$, and let $f(x) = x^{a-1} e^{-ux}$ for $x > 0$ and $f(x) = 0$ for $x < 0$. Then $\widehat{f}(y) = \frac{\Gamma(a)}{(u+2\pi i y)^a}$ (we use the branch of the power function z^a equal to 1 at $z = 1$). This was established in Example 1 of Sect. 7.1.7.

Before passing to a more detailed study of the properties of the Fourier transform, we show the usefulness of this notion by one more example.

Example 6 Let f be a function in $\mathscr{L}^1(\mathbb{R})$ equal to zero outside $(-\pi, \pi)$, and let f_0 be its 2π-extension from this interval to $\mathbb{R}$ ($f_0 \in \widetilde{\mathscr{L}}^1$). The Fourier coefficients of f_0 can easily be expressed in terms of the Fourier transform of f, namely, $\widehat{f_0}(n) =$

$\frac{1}{2\pi}\widehat{f}(\frac{n}{2\pi})$ for all $n \in \mathbb{Z}$. We also consider the function $g(x) = e^{itx}$ for $x \in (-\pi, \pi)$, where t is a fixed number. Since the Fourier coefficients of g are equal to

$$\widehat{g}(n) = \frac{1}{2\pi}\int_{-\pi}^{\pi} e^{i(t-n)x}\,dx = \frac{\sin\pi(t-n)}{\pi(t-n)},$$

we obtain by Parseval's generalized identity (see Theorem 2 of Sect. 10.3.6)

$$\widehat{f}\Big(\frac{t}{2\pi}\Big) = \int_{-\pi}^{\pi} f_0(x)\overline{g}(x)\,dx = 2\pi\sum_{n=-\infty}^{\infty}\widehat{f_0}(n)\overline{\widehat{g}(n)}$$
$$= \sum_{n=-\infty}^{\infty}\widehat{f}\Big(\frac{n}{2\pi}\Big)\frac{\sin\pi(t-n)}{\pi(t-n)}.$$

Thus, the following *sampling formula* is valid for the function $F(t) = \widehat{f}(t/2\pi)$:

$$F(t) = \sum_{n=-\infty}^{\infty} F(n)\frac{\sin\pi(t-n)}{\pi(t-n)} = \frac{\sin\pi t}{\pi}\sum_{n=-\infty}^{\infty}\frac{(-1)^n}{t-n}F(n).$$

This formula allows one to recover the value of a function F at an arbitrary point t knowing the values of F on the integer lattice. This fact plays a fundamental role in optics and radio engineering because it is easier to deal with a discrete system of values than with a continuously varying signal. A multi-dimensional version of the sampling formula is given in Exercise 3.

10.5.2 We establish elementary relations between differentiation and the Fourier transform.

Theorem *Let $f \in \mathscr{L}^1(\mathbb{R}^m)$. Then:*

(1) *if the partial derivative $g = \frac{\partial f}{\partial x_k}$ is summable and continuous for some $k = 1, \dots, m$, then*

$$\widehat{g}(y) = 2\pi i y_k \widehat{f}(y) \quad \big(y \in \mathbb{R}^m\big);$$

(2) *if the product $\|x\| f(x)$ is summable, then $\widehat{f} \in C^1(\mathbb{R}^m)$ and the equation*

$$\frac{\partial\widehat{f}(y)}{\partial y_k} = -2\pi i\,\widehat{f_k}(y), \quad \text{where } f_k(x) = x_k f(x)\ \big(x \in \mathbb{R}^m\big),$$

holds for all $y \in \mathbb{R}^m$ and $k = 1, \dots, m$

Proof (1) Without loss of generality, we will assume that $k = m$. We identify a point $x = (x_1, \dots, x_{m-1}, t)$ with the pair (u, t), where $u = (x_1, \dots, x_{m-1}) \in \mathbb{R}^{m-1}$. First, we verify that $f(u,t) \to 0$ as $t \to \pm\infty$ for almost all points $u \in \mathbb{R}^{m-1}$. Indeed, since the derivative $f_t' = g$ is continuous, we have

$$f(u,t) - f(u,0) = \int_0^t g(u,s)\,ds.$$

From Fubini's theorem, it follows that the function $t \mapsto g(u,t)$ is summable for almost all u, and, therefore,

$$f(u,t) - f(u,0) = \int_0^t g(u,s)\,ds \underset{t\to\pm\infty}{\longrightarrow} \int_0^{\pm\infty} g(u,s)\,ds.$$

Thus, the limits $\lim_{t\to\pm\infty} f(u,t)$ exist and are finite for almost all $u \in \mathbb{R}^{m-1}$. However, since (again, by Fubini's theorem) the function $t \mapsto f(u,t)$ is summable for almost all u, we see that the limits are zero for such u and, therefore,

$$\int_{-\infty}^{\infty} g(u,t)e^{-2\pi i y_m t}\,dt = f(u,t)e^{-2\pi i y_m t}\Big|_{-\infty}^{\infty} - (-2\pi i y_m)\int_{-\infty}^{\infty} f(u,t)e^{-2\pi i y_m t}\,dt$$
$$= 2\pi i y_m \int_{-\infty}^{\infty} f(u,t)e^{-2\pi i y_m t}\,dt.$$

To obtain the required result, it only remains to multiply both sides of this equation by $e^{-2\pi i(y_1x_1+\cdots+y_{m-1}x_{m-1})}$ and integrate with respect to u.

To obtain the equation of (2), we must apply the Leibnitz rule (see Sect. 7.1.5). The functions $f_1, \ldots, f_m$ are summable by assumption. Therefore, their Fourier transforms and the first order partial derivatives of $\widehat{f}$ are continuous everywhere. Consequently, $\widehat{f} \in C^1(\mathbb{R}^m)$. □

Corollary *If $f \in \mathscr{L}^1(\mathbb{R}^m)$ is a compactly supported function, then $\widehat{f} \in C^\infty(\mathbb{R}^m)$; if $f \in C_0^\infty(\mathbb{R}^m)$, then the product $\|y\|^p \widehat{f}(y)$ is summable in $\mathbb{R}^m$ for every $p > 0$.*

Proof The fact that $\widehat{f}$ is infinitely differentiable follows directly from the second assertion of the theorem because the product $\|x\|^n f(x)$ is summable for every $n \in \mathbb{N}$.

If $f \in C_0^\infty(\mathbb{R}^m)$, then the derivatives of all orders of f are summable and the relation

$$\left(\frac{\partial^n f}{\partial y_k^n}\right)^{\widehat{}}(y) = (2\pi i y_k)^n \widehat{f}(y)$$

is fulfilled for all $k = 1, \ldots, m$ and $n \in \mathbb{N}$. Since the functions $(\frac{\partial^n f}{\partial y_k^n})^{\widehat{}}(y)$ are bounded, we obtain the estimate

$$|\widehat{f}(y)| \leqslant \text{const} \cdot \left(1 + |y_1|^n + \cdots + |y_m|^n\right)^{-1}$$

providing (if we take sufficiently large n) the summability of $\|y\|^p \widehat{f}(y)$. □

In many problems, it is important to know the rate of decrease of the Fourier transform at infinity. The theorem proved above shows that a fast decrease can be provided by the smoothness of the function in question. How accurate are these conditions? What can be expected if smoothness fails on a "small" set? The following examples are devoted to such results.

Example 1 Supplementing Examples 2 and 3 of Sect. 10.5.1, we investigate the asymptotic behavior of the Fourier transform of the function $f(x) = e^{-|x|^p}$ at infinity for $0 < p < 2$. After integrating by parts, we see that

$$\widehat{f}(y) = 2\int_0^\infty e^{-x^p}\cos(2\pi xy)\,dx = \frac{p}{\pi y}\int_0^\infty e^{-x^p}x^{p-1}\sin(2\pi xy)\,dx, \quad (2)$$

which implies the crude estimate $\widehat{f}(y) = o(1/y)$ as $y \to +\infty$. We study the behavior of $\widehat{f}(y)$ for large y in detail. If $0 < p < 1$, then the change $2\pi xy = u$ leads to the equation

$$\widehat{f}(y) = \frac{2p}{(2\pi y)^{p+1}}\int_0^\infty e^{-(\frac{u}{2\pi y})^p}\frac{\sin u}{u^{1-p}}\,du.$$

The integral $\int_0^\infty \frac{\sin u}{u^{1-p}}\,du$ of the limit function (as $y \to +\infty$) converges, and Corollary 2 of Sect. 7.4.7 justifies the passage to the limit,

$$\int_0^\infty e^{-(\frac{u}{2\pi y})^p}\frac{\sin u}{u^{1-p}}\,du \underset{y\to+\infty}{\longrightarrow} \int_0^\infty \frac{\sin u}{u^{1-p}}\,du = \Gamma(p)\sin\frac{\pi p}{2}$$

(the equality was established in Example 1 of Sect. 7.4.8). Consequently, the estimate

$$\widehat{f}(y) \sim \frac{C_p}{y^{p+1}} \quad \text{as } y \to +\infty \quad (3)$$

is valid for $0 < p < 1$ with constant $C_p = \frac{2\Gamma(p+1)}{(2\pi)^{p+1}}\sin\frac{\pi p}{2}$. This, in particular, implies that the function $\widehat{f}$ is summable.

Now, let $1 < p < 2$ (the case where $p = 1$ was considered in Example 3 of Sect. 10.5.1). Integrating the right-hand side of (2) one more time, we arrive at the equation

$$\widehat{f}(y) = \frac{p}{2(\pi y)^2}\Big((p-1)\int_0^\infty x^{p-2}e^{-x^p}\cos(2\pi yx)\,dx - p\int_0^\infty x^{2(p-1)}e^{-x^p}\cos(2\pi yx)\,dx\Big).$$

Here, the second integral admits the estimate $o(1/y)$ as $y \to +\infty$, but the first integral tends to zero more slowly. Indeed, by an almost verbatim repetition of the reasoning given in the case where $0 < p < 1$, we obtain

$$\int_0^\infty x^{p-2}e^{-x^p}\cos(2\pi yx)\,dx \underset{y\to+\infty}{\sim} \frac{1}{(2\pi y)^{p-1}}\Gamma(p-1)\sin\frac{\pi p}{2}.$$

Thus, we again come to relation (3), which is also valid for $p = 1$; for $p = 2$, the coefficient C_p vanishes, and the asymptotic of $\widehat{f}$ changes completely (see Examples 2 and 3 of Sect. 10.5.1).

It can be proved that $\widehat{f}(y) > 0$ for $0 < p \leqslant 2$ (for $0 < p \leqslant 1$, this follows from the result of Example 2 of Sect. 4.6.6 and the fact that the function e^{-x^p} is convex).

Example 2 Let us determine how fast the Fourier transform of the characteristic function of the unit ball

$$\widehat{\chi_B}(y) = \int_B e^{-2\pi i \langle y,x\rangle}\,dx = \int_B e^{-2\pi i \|y\| x_1}\,dx = \alpha_{m-1}\int_{-1}^{1}\left(1-t^2\right)^{\frac{m-1}{2}} e^{-2\pi i\|y\| t}\,dt$$

decreases at infinity (α_{m-1} is the volume of the unit ball in $\mathbb{R}^{m-1}$). For odd m the "integral can be calculated" and $\widehat{\chi_B}$ can be expressed explicitly in terms of $\|y\|$. In particular, in the one-dimensional case, we have $B = (-1, 1)$ and $\widehat{\chi_B}(y) = \frac{\sin 2\pi y}{\pi y}$. For $m = 3$, we have $\widehat{\chi_B}(y) = \frac{1}{\pi\|y\|^2}\left(\frac{\sin 2\pi\|y\|}{2\pi\|y\|} - \cos 2\pi\|y\|\right)$ (this also follows from the result of the Example in Sect. 6.2.5 with $f_0 = \chi_{(0,1)}$).

For even m, the situation is more complicated. In this case, χ_B can be expressed in terms of the Bessel function. However, an exact formula for $\widehat{\chi_B}(y)$ is not our main concern here. We want to study the asymptotic behavior of this function as $\|y\| \to +\infty$. To this end, we put $r = 2\pi\|y\|$ and consider the integrals

$$I_m(r) = \int_{-1}^{1}\left(1-t^2\right)^{\frac{m-1}{2}} e^{-irt}\,dt \quad (m = 0, 1, 2, \ldots).$$

The larger m is, the more derivatives of the function $(1-t^2)^{\frac{m-1}{2}}$ vanish at the endpoints of the interval of integration. Therefore, as m increases, the rate at which the integrals $I_m(r)$ tend to zero as $r \to +\infty$ also increases. To describe this in more detail, we use the recurrence formula

$$I_m(r) = \frac{m-1}{r^2}\big((m-2)I_{m-2}(r) - (m-3)I_{m-4}(r)\big) \quad (m \geqslant 4),$$

which can easily be obtained by twofold integration by parts. From this relation, we see that, to obtain the asymptotic of the integral $I_m(r)$, it is sufficient to know only the asymptotics of the integrals $I_0(r)$ and $I_2(r)$ or of $I_1(r)$ and $I_3(r)$, depending on the parity of m. The integrals $I_1(r)$ and $I_3(r)$ can easily be calculated,

$$I_1(r) = 2\frac{\sin r}{r}, \qquad I_3(r) = \frac{4}{r^2}\left(\frac{\sin r}{r} - \cos r\right).$$

The integrals $I_0(r)$ and $I_2(r)$ coincide with the integrals $C(r)$ and $S(r)$, respectively, considered in the example of Sect. 9.2.5:

$$I_0(r) = C(r) = \sqrt{\frac{\pi}{r}}(\sin r + \cos r) + O\left(\frac{1}{r}\right),$$

$$I_2(r) = S(r) = \frac{\sqrt{\pi}}{r^{3/2}}(\sin r - \cos r) + O\left(\frac{1}{r^2}\right).$$

The last four formulas can be written uniformly as follows:

$$I_m(r) = \frac{\gamma_m}{r^{\frac{m+1}{2}}}\cos(r - \varphi_m) + O\left(\frac{1}{r^{\frac{m}{2}+1}}\right) \quad (r \to +\infty),$$

where $m = 0, 1, 2, 3$, $\varphi_m = \frac{\pi}{4}(m+1)$, and γ_m is a positive coefficient depending only on m. The recurrence formula allows us to extend this relation to all positive integers m.

Returning to the Fourier transform of the function χ_B, we see that

$$\widehat{\chi_B}(y) = \alpha_{m-1} I_m(2\pi\|y\|) = \frac{C_m}{\|y\|^{\frac{m+1}{2}}}\cos(2\pi\|y\| - \varphi_m) + O\left(\frac{1}{\|y\|^{\frac{m}{2}+1}}\right).$$

It can be verified that $C_m = 1/\pi$ for all m.

It is interesting to compare $\widehat{\chi_B}$ with the function $\widehat{\chi_Q}$, where $Q = (-1, 1)^m$. It is clear that

$$\widehat{\chi_Q}(y) = \prod_{j=1}^{m} \frac{\sin 2\pi y_j}{\pi y_j}.$$

If the angles between the vector y and the coordinate axes are non-zero, then this function admits the estimate $O(\|y\|^{-m})$. Thus, for most directions, the function decreases considerably faster than $\widehat{\chi_B}$. One possible sharpening of this assertion is as follows: the integrals $L_B(R) = \int_{\|y\|<R} |\widehat{\chi_B}(y)|\,dy$ grow considerably faster than the integrals $L_Q(R) = \int_{\|y\|<R} |\widehat{\chi_Q}(y)|\,dy$ as $R \to +\infty$. Indeed,

$$L_Q(R) \leqslant \int_{[-R,R]^m} |\widehat{\chi_Q}(y)|\,dy = \prod_{j=1}^{m} \int_{-R}^{R} \left|\frac{\sin 2\pi y_j}{\pi y_j}\right| dy_j$$

$$= \left(\frac{2}{\pi}\int_0^{2\pi R} \frac{|\sin t|}{t}\,dt\right)^m.$$

It follows that $L_Q(R) = O(\ln^m R)$ as $R \to +\infty$. At the same time

$$L_B(R) = \alpha_m \int_0^R \left|\frac{C_m}{r^{\frac{m+1}{2}}}\cos(2\pi r - \varphi_m) + O\left(\frac{1}{r^{\frac{m}{2}+1}}\right)\right| r^{m-1}\,dr$$

$$= \alpha_m C_m \int_0^R r^{\frac{m-3}{2}} |\cos(2\pi r - \varphi_m)|\,dr + O\left(R^{\frac{m}{2}-1}\right)$$

for $m > 2$ (for $m = 2$, the remainder term has order $O(\ln R)$). Therefore, $L_B(R) = O(R^{\frac{m-1}{2}})$, and the estimate is exact by order,

$$\begin{aligned} L_B(R) &\geqslant \alpha_m C_m \int_{R/2}^{R} r^{\frac{m-3}{2}} \cos^2(2\pi r - \varphi_m)\,dr + O\big(R^{\frac{m}{2}-1}\big) \\ &\geqslant \mathrm{const}\, R^{\frac{m-3}{2}} \int_{R/2}^{R} \big(1 + \cos 2(2\pi r - \varphi_m)\big)\,dr + O\big(R^{\frac{m}{2}-1}\big) \\ &= \frac{\mathrm{const}}{2} R^{\frac{m-1}{2}} + O\big(R^{\frac{m}{2}-1}\big). \end{aligned}$$

Example 3 It follows from the theorem that the condition $f(x) = O(\|x\|^{-p})$ as $\|x\| \to +\infty$ implies the smoothness of the Fourier transform for $p > m+1$. It turns out that this restriction cannot be weakened essentially. To verify this, we show that if $f(x) \sim \|x\|^{-p}$ as $\|x\| \to +\infty$, then the differentiability of $\widehat{f}$ at zero implies the inequality $p > m+1$.

Without loss of generality, we may assume that $f \geqslant 0$. Indeed, we know that $f(x) \geqslant 0$ for large $\|x\|$, but changing the function on an arbitrary ball (for example, putting $f(x) = 0$ inside the ball), we change the Fourier transform of f by an infinity differentiable function.

Assuming that $f \geqslant 0$, we study the mean value of the difference $\widehat{f}(0) - \widehat{f}$ in the vicinity of zero (in what follows, B is the unit ball centered at zero and v is the volume of B). We put

$$I(r) = \frac{1}{v} \int_B \big(\widehat{f}(0) - \widehat{f}(ry)\big)\,dy.$$

Since $\widehat{f}$ is differentiable at zero, we obtain that $I(r) = o(r)$ as $r \to +0$. Now, we estimate the integral $I(r)$ from below. Since

$$\widehat{f}(0) - \widehat{f}(ry) = \int_{\mathbb{R}^m} f(x)\big(1 - e^{-2\pi i r\langle y,x\rangle}\big)\,dx,$$

we obtain, by Fubini's theorem, that

$$I(r) = \int_{\mathbb{R}^m} f(x)\bigg(1 - \frac{1}{v}\int_B e^{-2\pi i r s\langle y,x\rangle}\,dy\bigg)\,dx = \int_{\mathbb{R}^m} f(x)\bigg(1 - \frac{1}{v}\widehat{\chi}(rx)\bigg)\,dx,$$

where χ is the characteristic function of B. Obviously, $\widehat{\chi}(x) \in \mathbb{R}$ and $|\widehat{\chi}(x)| \leqslant v$, and the Riemann–Lebesgue theorem implies that $\widehat{\chi}(x) \to 0$ as $\|x\| \to +\infty$. We take a sufficiently large radius R so that $f(x) > \frac{1}{2\|x\|^p}$ and $|\widehat{\chi}(x)| < \frac{v}{2}$ for $\|x\| > R$. Then, since $f \geqslant 0$, we have

$$\begin{aligned} I(r) &\geqslant \int_{\|x\|>R/r} f(x)\bigg(1 - \frac{1}{v}\widehat{\chi}(rx)\bigg)\,dx \\ &\geqslant \frac{1}{4}\int_{\|x\|>R/r} \frac{dx}{\|x\|^p} = \frac{\sigma(S^{m-1})}{4}\int_{R/r} \frac{dt}{t^{p-m+1}}. \end{aligned}$$

Thus, $I(r) \geqslant \mathrm{const}\, r^{p-m}$. Since $I(r) = o(r)$ for $r \to 0$, we obtain that $p > m+1$.

10.5.3 In the one-dimensional case, for a function f differentiable at a point x, there is an important formula allowing one to find the value of $f(x)$ from $\widehat{f}$. This formula is called the *inversion formula* and has the following form:

$$f(x) = \int_{-\infty}^{\infty} \widehat{f}(y) e^{2\pi i x y}\, dy.$$

The integral on the right-hand side of this equation is called the *Fourier integral* of f. In general, this is an improper integral because the Fourier transform can be non-summable on $\mathbb{R}$ (see Sect. 10.5.1). We will say that the integral converges if there exists a limit of the partial integrals

$$I_A(f,x) = \int_{-A}^{A} \widehat{f}(y) e^{2\pi i x y}\, dy$$

as $A \to +\infty$.

There is an obvious analogy between the expansion of a periodic function in a Fourier series and the Fourier integral representation of a non-periodic function. The following theorem shows that these two problems share not only some superficial analogies but are connected in essence. To show this, we need the following easy lemma.

Lemma *Let $f \in \mathscr{L}^1(\mathbb{R})$ and $x \in \mathbb{R}$. Then the following holds for every $A > 0$*

$$I_A(f,x) = \int_{-A}^{A} \widehat{f}(y) e^{2\pi i x y}\, dy = \int_{-\infty}^{\infty} f(x-t) \frac{\sin 2\pi A t}{\pi t}\, dt.$$

Proof It is clear that

$$I_A(f,x) = \int_{-A}^{A} \left(\int_{-\infty}^{\infty} f(u) e^{2\pi i (x-u) y}\, du \right) dy.$$

Since the function $(y,u) \mapsto f(u) e^{2\pi i (x-u) y}$ is summable in the strip $(-A, A) \times \mathbb{R}$, we may use Fubini's theorem,

$$I_A(f,x) = \int_{-\infty}^{\infty} \left(\int_{-A}^{A} f(u) e^{2\pi i (x-u) y}\, dy \right) du = \int_{-\infty}^{\infty} f(u) \frac{\sin 2\pi A (x-u)}{\pi (x-u)}\, du.$$

It remains to change the integration variable $t = x - u$. □

By the Riemann–Lebesgue theorem, the integral $\int_{|t| \geqslant \delta} f(x-t) \frac{\sin 2\pi A t}{\pi t}\, dt$ tends to zero as $A \to +\infty$ for every $\delta > 0$. Therefore, the lemma implies the asymptotic relation

$$I_A(f,x) = \int_{-\delta}^{\delta} f(x-t) \frac{\sin 2\pi A t}{\pi t}\, dt + o(1) \quad \text{as } A \to +\infty \tag{4}$$

(we already know a similar result for the partial sums of Fourier series; see Eq. (5′) of Sect. 10.3.4). Thus, the behavior of the integrals $I_A(f,x)$ as $A \to +\infty$ is determined only by the values of f in the vicinity of x. In other words, we have the same localization principle for Fourier integrals as for Fourier series. Furthermore, it is easy to prove the equiconvergence of the expansions in the Fourier series and the Fourier integral. More precisely, the following statement holds.

Theorem *If functions $f \in \mathscr{L}^1(\mathbb{R})$ and $f_0 \in \widetilde{\mathscr{L}}^1$ coincide in a neighborhood of a point x, then the convergence of the Fourier integral of f at x is equivalent to the convergence of the Fourier series of f_0 at x, and, in the case of convergence, the following holds*:

$$\int_{-\infty}^{\infty} \widehat{f}(y)e^{2\pi ixy}\,dy = \sum_{n=-\infty}^{\infty} \widehat{f_0}(n)e^{inx}.$$

From the theorem, it obviously follows that the convergence tests for Fourier series, obtained in Sect. 10.3.4, can be carried over to the Fourier integrals. In particular, the inversion formula is valid at a point x if Dini's condition is fulfilled at x with $C = f(x)$. We leave it to the reader to state an analog of the Dirichlet–Jordan test.

Proof We show that the following holds:

$$I_A(f,x) - S_{[2\pi A]}(f_0,x) \underset{A\to+\infty}{\longrightarrow} 0,$$

where $[u]$, as usual, is the integer part of u.

Let $f(x-t) = f_0(x-t)$ for $|t| < \delta$, where $0 < \delta < \pi$. By Eq. (4) and Eq. (5′) of Sect. 10.3.4, we have

$$I_A(f,x) = \int_{-\delta}^{\delta} f(x-t)\frac{\sin 2\pi At}{\pi t}\,dt + o(1) = \int_{-\delta}^{\delta} f_0(x-t)\frac{\sin 2\pi At}{\pi t}\,dt + o(1),$$

$$S_n(f_0,x) = \int_{-\pi}^{\pi} f_0(x-t)\frac{\sin nt}{\pi t}\,dt + o(1) = \int_{-\delta}^{\delta} f_0(x-t)\frac{\sin nt}{\pi t}\,dt + o(1)$$

as $A, n \to +\infty$. If $2\pi A = n \in \mathbb{N}$, then we immediately obtain the required relation. If $2\pi A$ is not integer, then we have $n < 2\pi A < n+1$ for $n = [2\pi A]$, and, therefore,

$$\begin{aligned}\big|I_A(f,x) - I_{n/2\pi}(f,x)\big| &\leqslant \int_{A-\frac{1}{2\pi}}^{A} \big|\widehat{f}(y)\big|\,dy + \int_{-A}^{-A+\frac{1}{2\pi}} \big|\widehat{f}(y)\big|\,dy \\ &\leqslant 2\max_{|y|\geqslant A-1}\big|\widehat{f}(y)\big| \underset{A\to+\infty}{\longrightarrow} 0.\end{aligned}$$

Thus,

$$I_A(f,x) - S_n(f_0,x) = \big(I_A(f,x) - I_{n/2\pi}(f,x)\big) + \big(I_{n/2\pi}(f,x) - S_n(f_0,x)\big),$$

where each of the two differences on the right-hand side tend to zero as $A \to +\infty$. □

Now, we once again turn to Examples 2 and 3 considered in Sect. 10.5.1.

Example 1 From the theorem, it follows that the inversion formula is valid for the function $f_t(x) = e^{-\pi t^2 x^2}$ ($x \in \mathbb{R}$, $t > 0$). However, this already follows from the relation $\widehat{f_t} = \frac{1}{t} f_{\frac{1}{t}}$ established in Example 2 of Sect. 10.5.1. Indeed, since the function $\widehat{f_t}$ is even and summable, we have

$$\int_{-\infty}^{\infty} \widehat{f_t}(y) e^{2\pi i x y}\, dy = (\widehat{f_t})\widehat{\ }(x) = \frac{1}{t}(f_{\frac{1}{t}})\widehat{\ }(x) = f_t(x).$$

Example 2 The function $f(x) = e^{-|x|}$ ($x \in \mathbb{R}$) satisfies Dini's condition at every point (in particular, at zero). The Fourier transform of f was calculated in Example 3 of Sect. 10.5.1. By the inversion formula, we obtain

$$e^{-|x|} = \int_{-\infty}^{\infty} \widehat{f}(y) e^{2\pi i y x}\, dy = \int_{-\infty}^{\infty} \frac{2e^{2\pi i y x}}{1+4\pi^2 y^2}\, dy$$
$$= \int_{0}^{\infty} \frac{4\cos 2\pi y x}{1+4\pi^2 y^2}\, dy = \frac{2}{\pi}\int_{0}^{\infty} \frac{\cos xt}{1+t^2}\, dt.$$

Thus, we again obtain the value of the Laplace integral

$$\int_{0}^{\infty} \frac{\cos xt}{1+t^2}\, dt = \frac{\pi}{2} e^{-|x|},$$

which was calculated in a different way in Example 2 of Sect. 7.4.8.

10.5.4 Generalizing the inversion formula to functions of several variables, we confine ourselves to the most important case where the Fourier transform is summable. In this connection, we note that Dini's condition providing the validity of the inversion formula in the one-dimensional case is a local property of a summable function, whereas the summability of the Fourier transform is a global property.

In contrast to the one-dimensional setting, now, when deriving the inversion formula, we cannot use the equiconvergence of the expansions in the Fourier series or Fourier integral since Theorem 10.5.3 cannot be carried over to the multidimensional case (see Exercise 6).

The transformation that assigns the function $\check{g}$ defined by the formula

$$\check{g}(x) = \int_{\mathbb{R}^m} g(y) e^{2\pi i \langle x, y\rangle}\, dy \quad (x \in \mathbb{R}^m)$$

to $g \in \mathscr{L}^1(\mathbb{R}^m)$ is called the *inverse transform*. Obviously, $\check{g}(x) = \widehat{g}(-x)$, and so the properties of the Fourier transform can easily be carried over to the inverse transform. Using the inverse transform, we can represent the inversion formula proved in the one-dimensional case in the following form:

$$f(x) = (\widehat{f})^{\vee}(x). \tag{5}$$

This justifies the choice of the term "inverse transform".

Theorem *Let $f \in \mathscr{L}^1(\mathbb{R}^m)$. If $\widehat{f} \in \mathscr{L}^1(\mathbb{R}^m)$, then inversion formula* (5) *is valid for almost all x in $\mathbb{R}^m$.*

We remark that the right-hand side of Eq. (5) continuously depends on x since $\widehat{f}$ is summable. Therefore, the condition of the theorem (the summability of $\widehat{f}$) can be fulfilled only if the function f is equivalent to a continuous function. Moreover, Eq. (5) is valid at all points where f is continuous because it is valid on a set of full measure. In particular, if f is continuous and its Fourier transform is summable, then $f(x) = (\widehat{f})^{\vee}(x)$ for all $x \in \mathbb{R}^m$.

Proof We use the approximate identity W_t, which played an important role in the proof of the Weierstrass theorem in Sect. 7.6.4. We recall that $W_t(x) = \frac{1}{t^m} e^{-\frac{\pi}{t^2}\|x\|^2}$ ($x \in \mathbb{R}^m$, $t > 0$). Our proof of the theorem is based on inversion formula ($1'$) for this function,

$$W_t(x) = \int_{\mathbb{R}^m} e^{-\pi t^2 \|y\|^2} e^{2\pi i \langle x,y \rangle}\, dy. \tag{6}$$

First, for the smoothened function $f * W_t$, we obtain an equation close to (5). Then, we obtain the statement of the theorem by passage to the limit.

Using Eq. (6) and changing the order of integration, we obtain, for every $t > 0$ that

$$\begin{aligned}
(f * W_t)(x) &= \int_{\mathbb{R}^m} f(y) W_t(x-y)\, dy \\
&= \int_{\mathbb{R}^m} f(y) \left(\int_{\mathbb{R}^m} e^{-\pi t^2 \|u\|^2} e^{2\pi i \langle x-y,u \rangle}\, du \right) dy \\
&= \int_{\mathbb{R}^m} e^{-\pi t^2 \|u\|^2} e^{2\pi i \langle x,u \rangle} \left(\int_{\mathbb{R}^m} f(y) e^{-2\pi i \langle y,u \rangle}\, dy \right) du.
\end{aligned}$$

Thus, we have established the required relation,

$$(f * W_t)(x) = \int_{\mathbb{R}^m} e^{-\pi t^2 \|u\|^2} e^{2\pi i \langle x,u \rangle} \widehat{f}(u)\, du. \tag{7}$$

Since the absolute value of the integrand in the last integral does not exceed $|\widehat{f}|$, we obtain, by Lebesgue's theorem, that, for every x, this integrals tends to $(\widehat{f})^{\vee}(x)$ as $t \to +0$.

Now, we can finish the proof, referring to Theorem 10.3.4, from which it follows that the limit on the left-hand side of Eq. (7) coincides with $f(x)$ almost everywhere. However, it is possible to dispense with the use of the theorem based on the notion of a Lebesgue point and on Theorem 4.9.2 on differentiation of an integral with respect to a set. We show that the left-hand side of Eq. (7) tends to $f(x)$ almost everywhere as t tends to zero along a sequence.

Indeed, let $\{t_n\}$ be a sequence such that $t_n \underset{n\to\infty}{\longrightarrow} 0$. Theorem 9.3.3 implies that $f * W_{t_n} \underset{n\to\infty}{\longrightarrow} f$ in mean, and, consequently, in measure (see Theorem 9.1.2). By Riesz's theorem (see Sect. 3.3.4), there is a subsequence $\{t_{n_k}\}$ of $\{t_n\}$ such that, almost everywhere $f * W_{t_{n_k}} \to f$ as $k \to \infty$. Replacing t by t_{n_k} in Eq. (7) and passing to the limit, we obtain the required result. □

Example We give the inversion formula for the function $f(x) = e^{-\|x\|}$ $(x \in \mathbb{R}^m)$ whose Fourier transform is calculated in Example 4 of Sect. 10.5.1,

$$e^{-\|x\|} = 2^m \pi^{\frac{m-1}{2}} \Gamma\left(\frac{m+1}{2}\right) \int_{\mathbb{R}^m} \frac{e^{2\pi i \langle x,y\rangle}\, dy}{(1+4\pi^2\|y\|^2)^{\frac{m+1}{2}}}$$

$$= \frac{\Gamma(\frac{m+1}{2})}{\pi^{\frac{m+1}{2}}} \int_{\mathbb{R}^m} \frac{\cos\langle x,t\rangle\, dt}{(1+\|t\|^2)^{\frac{m+1}{2}}}.$$

In the one-dimensional case, this formula was obtained in Example 2 of Sect. 10.5.3.

The summability of the Fourier transform is important in many problems (see, for example, Sect. 10.6.4). The result of Exercise 7 shows that this condition is necessarily fulfilled if $\widehat{f} \geqslant 0$ and the function f is continuous (or at least bounded in a neighborhood of zero). In this connection, we recall (see Example 2 of Sect. 4.6.6) that $\widehat{f} \geqslant 0$ if f is an even function summable on $\mathbb{R}$ and convex on $(0, +\infty)$. Together with the inversion formula, this proves the following statement.

Corollary *If an even continuous function f is summable on the real line and is convex on the positive semi-axis, then f is the Fourier transform of a non-negative summable function.*

The fact just proved remains valid even if, instead of the summability of f, we assume only that $f(x) \underset{x\to+\infty}{\longrightarrow} 0$, but, in this case, the proof invokes a subtler reasoning (see [Luk], Pólya's theorem).

10.5.5 Here, we discuss one more important property of the Fourier transform, its injectivity on the entire set of summable functions. Of course, there is no injectivity in the literal sense because distinct equivalent (i.e., coinciding almost everywhere) functions have the same Fourier transform. However, Theorem 10.5.4 shows that the injectivity holds up to equivalence on the set of functions with summable Fourier transform. To strengthen this result, we generalize Definition 10.5.1 somewhat.

Definition Let μ be a finite Borel measure on $\mathbb{R}^m$. The function $y \mapsto \widehat{\mu}(y) \equiv \int_{\mathbb{R}^m} e^{-2\pi i \langle y,x\rangle}\, d\mu(x)$ is called the *Fourier transform of* μ.

If a measure μ has a density f with respect to Lebesgue measure, then $\widehat{\mu} = \widehat{f}$.

Now, we establish an important result connected with the injectivity of the Fourier transform of a measure.

Theorem *If two finite Borel measures μ and ν have the same Fourier transform, then the measures coincide.*

Proof Let $H_j(t) = \{(x_1, \dots, x_m) \mid x_j = t\}$ be a plane perpendicular to the jth coordinate axis, and let

$$E = \bigl\{t \in \mathbb{R} \mid \mu\bigl(H_j(t)\bigr) = \nu\bigl(H_j(t)\bigr) = 0 \ \text{ for every } \ j = 1, \dots, m\bigr\}.$$

The set E is everywhere dense because the set $\{t \in \mathbb{R} \mid \mu(H_j(t)) > 0\}$ is at most countable for each j (see Sect. 1.2.2). Therefore, the Borel hull of the semiring $\mathscr{P}^m_E$ consisting of the cells whose vertices have coordinates belonging to E coincides with the σ-algebra of Borel subsets of the space $\mathbb{R}^m$ (see the remark after Theorem 1.1.6). We express the measure of the cell $P = \prod_{j=1}^m \Delta_j$ in terms of $\widehat{\mu}$, assuming that $P \in \mathscr{P}^m_E$.

Obviously, $\chi_P(x) = \prod_{j=1}^m \chi_{\Delta_j}(x_j)$, where $x_1, \dots, x_m$ are the coordinates of a vector x. By Fubini's theorem, $\widehat{\chi_P}(y) = \prod_{j=1}^m \widehat{\chi_{\Delta_j}}(y_j)$ for $y = (y_1, \dots, y_m)$, and, therefore,

$$I_A(\chi_\Delta, x) = \int_{(-A,A)^m} \widehat{\chi_P}(y) e^{2\pi i \langle x,y\rangle}\, dy = \prod_{j=1}^m \int_{-A}^{A} \widehat{\chi_{\Delta_j}}(y_j) e^{2\pi i x_j y_j}\, dy_j.$$

The characteristic function of an interval satisfies Dini's condition everywhere except the endpoints of the interval. Therefore, we have

$$\int_{-A}^{A} \widehat{\chi_{\Delta_j}}(y_j) e^{2\pi i x_j y_j}\, dy_j \underset{A\to+\infty}{\longrightarrow} \chi_{\Delta_j}(x_j)$$

for all $j = 1, \dots, m$, provided that x_j is distinct from the endpoints of the interval Δ_j. Since $P \in \mathscr{P}^m_E$, we see that $I_A(\chi_\Delta, x) \underset{A\to+\infty}{\longrightarrow} \chi_P(x)$ μ-almost everywhere. Moreover, putting $\Delta_j = [a_j, b_j)$, we obtain (see Lemma 10.5.3) that

$$\int_{-A}^{A} \widehat{\chi_{\Delta_j}}(y_j) e^{2\pi i x_j y_j}\, dy_j = \int_{-\infty}^{\infty} \chi_{\Delta_j}(x_j - t) \frac{\sin 2\pi A t}{t}\, dt = \int_{A(x_j-b_j)}^{A(x_j-a_j)} \frac{\sin 2\pi u}{u}\, du.$$

All these integrals are bounded (since the integral $\int_0^\infty \frac{\sin 2\pi u}{u}\, du$ converges), and so, the integral $I_A(\chi_\Delta, x)$ is also bounded (uniformly with respect to x and A).

Therefore, we can use Lebesgue's theorem on passing to the limit under the integral sign,

$$\mu(P) = \int_{\mathbb{R}^m} \chi_P(x)\,d\mu(x) = \lim_{A\to+\infty} \int_{\mathbb{R}^m} I_A(\chi_\Delta, x)\,d\mu(x)$$

$$= \lim_{A\to+\infty} \int_{\mathbb{R}^m} \left(\int_{(-A,A)^m} \widehat{\chi}_P(y) e^{2\pi i\langle x,y\rangle}\,dy \right) d\mu(x).$$

Changing the order of integration, we obtain

$$\mu(P) = \lim_{A\to+\infty} \int_{(-A,A)^m} \widehat{\chi}_P(y) \left(\int_{\mathbb{R}^m} e^{2\pi i\langle x,y\rangle}\,d\mu(x) \right) dy$$

$$= \lim_{A\to+\infty} \int_{(-A,A)^m} \widehat{\chi}_P(y)\widehat{\mu}(-y)\,dy.$$

This relation shows that the values of the measure on the cells belonging to $\mathscr{P}_E^m$ can be expressed in terms of its Fourier transform. Since the measures μ and ν have the same Fourier transform, they coincide on the semiring $\mathscr{P}_E^m$, and, consequently, (by the uniqueness theorem) on all Borel sets. □

It follows from the above theorem that the Fourier transform is injective up to equivalence on the set of summable functions.

Corollary 1 *If two summable functions f and g have the same Fourier transform, they coincide almost everywhere.*

Proof It is clear that the Fourier transform of the functions $\overline{f}$ and $\overline{g}$ also coincide. Consequently, the functions $\mathcal{R}e\, f = (f + \overline{f})/2$ and $\mathcal{R}e\, g = (g + \overline{g})/2$, as well as the imaginary parts of the functions f and g, have the same Fourier transform. Therefore, we may assume that the functions f and g are real.

If they are non-negative, the theorem just proved implies that the measures with the densities f and g coincide. It was proved in Sect. 4.5.4 that, in this case, the densities coincide almost everywhere.

In the general case, we represent f and g in the form $f = f_+ - f_-$ and $g = g_+ - g_-$, where $f_\pm, g_\pm \geqslant 0$. Then

$$\widehat{f} = \widehat{f}_+ - \widehat{f}_- = \widehat{g} = \widehat{g}_+ - \widehat{g}_-.$$

Consequently, the non-negative functions $f_+ + g_-$ and $f_- + g_+$ have the same Fourier transform, and, therefore, they coincide almost everywhere, which is equivalent to the assertion of the corollary. □

Corollary 2 *If finite Borel measures μ and ν have the same values on all half-spaces (in $\mathbb{R}^m$), then they coincide.*

Proof By the theorem, it is sufficient to verify that $\widehat{\mu}(y) = \widehat{\nu}(y)$ for all $y \in \mathbb{R}^m$. For $y = 0$, the equality holds since $\widehat{\mu}(0) = \mu(\mathbb{R}^m)$ and $\widehat{\nu}(0) = \nu(\mathbb{R}^m)$, and if two measures coincide on half-spaces, they coincide on the entire space. For $y \neq 0$, we consider the half-spaces

$$H_t = \{x \in \mathbb{R}^m \mid \langle x, y\rangle < t\} \quad (t \in \mathbb{R})$$

and put $g(t) = \mu(H_t) = \nu(H_t)$ and $\Phi(x) = \langle x, y\rangle$. The function g increases, and the Stieltjes measure μ_g is the Φ-image of the measures μ and ν since $\Phi^{-1}((-\infty, t)) = H_t$. It remains to use Theorem 6.1.1 on integration with respect to a weighted image of a measure,

$$\widehat{\mu}(y) = \int_{\mathbb{R}^m} e^{-2\pi i\langle x,y\rangle}\, d\mu(x) = \int_{\mathbb{R}} e^{-2\pi i t}\, dg(t)$$
$$= \int_{\mathbb{R}^m} e^{-2\pi i\langle x,y\rangle}\, d\nu(x) = \widehat{\nu}(y).$$ □

10.5.6 Using the results of the previous section, we will prove here that the system of Hermite polynomials is complete. The method we use enables us to consider a more general situation and prove that the family of monomials in m variables, i.e., the products $x^n = x_1^{n_1} \cdots x_m^{n_m}$, where $x = (x_1, \ldots, x_m) \in \mathbb{R}^m$ and $n = (n_1, \ldots, n_m) \in \mathbb{Z}_+^m$, is complete in $\mathscr{L}^2(\mathbb{R}^m, \mu)$ for a wider class of measures.

Theorem *If a Borel measure μ on $\mathbb{R}^m$ satisfies the condition $\int_{\mathbb{R}^m} e^{a\|x\|}\, d\mu(x) < +\infty$ for some $a > 0$, then the family of all monomials is complete in the space $\mathscr{L}^2(\mathbb{R}^m, \mu)$.*

Proof Let a function f in $\mathscr{L}^2(\mathbb{R}^m, \mu)$ be orthogonal to all monomials. Obviously, $f \perp P$ for every polynomial P in m variables. We put

$$F(y) = \int_{\mathbb{R}^m} f(x) e^{i\langle y,x\rangle}\, d\mu(x).$$

Since $|e^{i\langle y,x\rangle}| \equiv 1$ and all polynomials are summable with respect to μ, the function F is infinitely differentiable and, for each y, the derivatives of F can be found by the Leibnitz rule.

We prove that $F \equiv 0$. If $\|y\| < a/2$, then expanding the exponential factor in a Taylor series and integrating termwise, we obtain that $F(y) = 0$. The legitimacy of termwise integration follows from the fact that the partial sums of the series have a summable majorant, namely, $|f(x)|e^{\|x\|\,\|y\|}$ (this function is summable because the functions $|f|$ and $e^{\|x\|\,\|y\|}$ belong to $\mathscr{L}^2(\mathbb{R}^m, \mu)$). To prove that $F \equiv 0$, we show that the interior G of the set where $F(y) = 0$ coincides with $\mathbb{R}^m$. Since $G \neq \varnothing$ (because it contains a neighborhood of zero), it is sufficient to verify that the set G is closed, in which case the equality $G = \mathbb{R}^m$ will follow from the fact that the space $\mathbb{R}^m$ is connected. Let $y \in \overline{G}$. The function F and all its derivatives vanish at y by continuity. Calculating the derivatives by Leibnitz's rule, we see that

$$0 = F^{(n)}(y) = \int_{\mathbb{R}^m} f(x)(ix)^n e^{i\langle x,y\rangle}\, d\mu(x) \quad \bigl(n \in \mathbb{R}_+^m\bigr).$$

Thus, the function $f_1(x) = f(x)e^{i\langle x,y\rangle}$ is orthogonal to all monomials. Replacing f by f_1, we may assert by what has just been proved that the function $F_1(\eta) = \int_{\mathbb{R}^m} f_1(x)e^{i\langle x,\eta\rangle}\,d\mu(x)$ assumes only zero values in a neighborhood of zero. However, $F_1(\eta)$ is nothing but $F(y+\eta)$. Therefore, $F \equiv 0$ in a neighborhood of y, i.e., $y \in G$. Thus, $G = \overline{G} = \mathbb{R}^m$ and, consequently, $F \equiv 0$. Now, we can easily complete the proof. Indeed, without loss of generality, we may assume that the function f is real. The identity $F \equiv 0$ means that the measures $f_+\,d\mu$ and $f_-\,d\mu$ have the same Fourier transform. Consequently, these measures coincide by Theorem 10.5.5, which implies (by Theorem 4.5.4) that the functions f_+ and f_- coincide almost everywhere with respect to μ. □

Corollary *The Hermite polynomials are complete in* $\mathscr{L}^2(\mathbb{R},\mu)$ *with* $d\mu(x) = e^{-x^2}\,dx$.

This is a special case of the theorem for $m = 1$. We also remark that the theorem implies that the Laguerre functions are complete (for the definition, see Exercise 3 of Sect. 10.2).

The following example shows that the result obtained in the theorem is quite sharp.

Example We verify that the polynomials are not complete in the space $\mathscr{L}^2(\mathbb{R},\mu)$ with measure μ having density $e^{-|x|^p}$ $(0 < p < 1)$ with respect to the one-dimensional Lebesgue measure (for $p \geqslant 1$ this effect is ruled out by the theorem just proved).

We will need the following formula from Example 1 of Sect. 7.1.7: if $a > 0$ and $z = e^{i\theta}$, where $\theta \in (0, \frac{\pi}{2})$, then $z^{-a}\Gamma(a) = \int_0^\infty t^{a-1}e^{-zt}\,dt$. Comparing the imaginary parts and using the substitution $t = x^p/\cos\theta$, we obtain

$$\begin{aligned}\Gamma(a)\sin a\theta &= \int_0^\infty t^{a-1}e^{-t\cos\theta}\sin(t\sin\theta)\,dt \\ &= \frac{p}{\cos^a\theta}\int_0^\infty x^{ap-1}e^{-x^p}\sin\big(x^p\tan\theta\big)\,dx.\end{aligned}$$

Now, we use the freedom in the choice of the parameters a and θ. Putting $\theta = \frac{\pi}{2}p$ and $a = \frac{2}{p}(n+1)$, we obtain

$$\int_0^\infty x^{2n+1}e^{-x^p}\sin\left(x^p\tan\frac{\pi}{2}p\right)dx = 0 \quad \text{for } n = 0, 1, 2 \ldots.$$

This means that the odd function equal to $\sin(x^p\tan\frac{\pi}{2}p)$ for $x \geqslant 0$ is orthogonal to all polynomials in the space $\mathscr{L}^2(\mathbb{R},\mu)$ with measure $d\mu(x) = e^{-|x|^p}\,dx$.

10.5.7 The present and two following sections are devoted to an important theorem, due to Plancherel,[20] and its corollaries. The traditional formulation of the theorem would require us to invoke some concepts from functional analysis and operator theory. To avoid this, we first establish an analytic fact constituting the core of the theorem.

Theorem (Plancherel) *If* $f \in \mathscr{L}^1(\mathbb{R}^m) \cap \mathscr{L}^2(\mathbb{R}^m)$, *then* $\widehat{f} \in \mathscr{L}^2(\mathbb{R}^m)$ *and* $\|\widehat{f}\|_2 = \|f\|_2$.

Proof Let $\{\omega_t\}_{t>0}$ be a Sobolev approximate identity in $\mathbb{R}^m$ (see Sect. 7.6.2) and $f_t = f * \omega_t$.

First, we prove the assertion of the theorem for the smoothened function f_t. By properties of convolution, we have $f_t \in \mathscr{L}^1(\mathbb{R}^m) \cap \mathscr{L}^2(\mathbb{R}^m)$. By Theorem 10.5.1, we obtain $\widehat{f_t} = \widehat{f}\widehat{\omega}_t$. This product is summable since the function $\widehat{f}$ is bounded and $\widehat{\omega}_t \in \mathscr{L}^1(\mathbb{R}^m)$ by Corollary 10.5.2. Using Fubini's theorem and inversion formula (5), we obtain

$$\begin{aligned}\int_{\mathbb{R}^m} \widehat{f_t}(y)\overline{\widehat{f_t}(y)}\,dy &= \int_{\mathbb{R}^m} \widehat{f_t}(y)\overline{\left(\int_{\mathbb{R}^m} f_t(x)e^{-2\pi i\langle y,x\rangle}\,dx\right)}\,dy \\ &= \int_{\mathbb{R}^m} \widehat{f_t}(y)\left(\int_{\mathbb{R}^m} \overline{f_t(x)}e^{2\pi i\langle y,x\rangle}\,dx\right)dy \\ &= \int_{\mathbb{R}^m} \overline{f_t(x)}\left(\int_{\mathbb{R}^m} \widehat{f_t}(y)e^{2\pi i\langle y,x\rangle}\,dy\right)dx = \int_{\mathbb{R}^m} \overline{f_t(x)}\,f_t(x)\,dx.\end{aligned}$$

Thus,

$$\|\widehat{f_t}\|_2^2 = \|f_t\|_2^2. \tag{8}$$

It remains to verify that we can pass to the limit in this equation as $t \to 0$.

Since $f_t \underset{t\to 0}{\longrightarrow} f$ in the $\mathscr{L}^2$-norm, the continuity of the norm implies $\|f_t\|_2 \underset{t\to 0}{\longrightarrow} \|f\|_2$. We verify that $\widehat{f} \in \mathscr{L}^2(\mathbb{R}^m)$ and $\|\widehat{f_t}\|_2 \underset{t\to 0}{\longrightarrow} \|\widehat{f}\|_2$. To this end, we write the left-hand side of Eq. (8) in more detail,

$$\|\widehat{f_t}\|_2^2 = \int_{\mathbb{R}^m} \left|\widehat{f_t}(y)\right|^2 dy = \int_{\mathbb{R}^m} \left|\widehat{f}(y)\right|^2 \left|\widehat{\omega}_t(y)\right|^2 dy. \tag{9}$$

Since $\widehat{\omega}_t(y) \underset{t\to 0}{\longrightarrow} 1$ (see Corollary 7.6.3 with $t_0 = 0$ and $g(x) = e^{-2\pi i\langle y,x\rangle}$), Fatou's theorem and Eq. (8) imply

$$\begin{aligned}\int_{\mathbb{R}^m} \left|\widehat{f}(y)\right|^2 dy &\leqslant \varliminf_{t\to 0} \int_{\mathbb{R}^m} \left|\widehat{f}(y)\right|^2 \left|\widehat{\omega}_t(y)\right|^2 dy = \lim_{t\to 0} \|\widehat{f_t}\|_2^2 = \lim_{t\to 0} \|f_t\|_2^2 = \|f\|_2^2 \\ &< +\infty.\end{aligned}$$

[20] Michel Plancherel (1885–1967)—Swiss mathematician.

Thus, $\widehat{f} \in \mathscr{L}^2(\mathbb{R}^m)$. Returning to Eq. (9), we see that the integrand in the integral on the right has a summable majorant, namely, $|\widehat{f}|^2$. Therefore, we can pass to the limit in this integral by Lebesgue's theorem,

$$\int_{\mathbb{R}^m} \left|\widehat{f}(y)\right|^2 \left|\widehat{\omega}_t(y)\right|^2 dy \underset{t\to 0}{\longrightarrow} \int_{\mathbb{R}^m} \left|\widehat{f}(y)\right|^2 dy.$$

Now, the passage to the limit in Eq. (8) leads to the required result. □

The concluding part of the proof can be somewhat shortened. Indeed, since $|\widehat{\omega}_t| \leqslant \int_{\mathbb{R}^m} \omega(x)\, dx = 1$, we have $|\widehat{f_t}| \leqslant |\widehat{f}|$. Since $\widehat{f_t} \underset{t\to 0}{\longrightarrow} \widehat{f}$, we can pass to the limit on the right-hand side of Eq. (8) by the generalization of B. Levi's theorem given in Exercise 4 of Sect. 4.8.

10.5.8 We show how Plancherel's theorem can be used to generalize the concept of the Fourier transform to functions in $\mathscr{L}^2(\mathbb{R}^m)$.

Lemma *Let $f \in \mathscr{L}^2(\mathbb{R}^m)$. If $\{f_n\}_{n\geqslant 1}$ is a sequence of functions in $\mathscr{L}^1(\mathbb{R}^m) \cap \mathscr{L}^2(\mathbb{R}^m)$ convergent to f in the $\mathscr{L}^2$-norm, then the sequence $\{\widehat{f_n}\}_{n\geqslant 1}$ also converges in the $\mathscr{L}^2$-norm. Its limit does not depend (up to equivalence) on the choice of the sequence $\{f_n\}_{n\geqslant 1}$.*

Proof From Plancherel's theorem, it follows that the sequence $\{\widehat{f_n}\}_{n\geqslant 1}$ is fundamental,

$$\|\widehat{f_n} - \widehat{f_k}\|_2 = \|\widehat{f_n - f_k}\|_2 = \|f_n - f_k\|_2 \underset{n,k\to\infty}{\longrightarrow} 0.$$

The limit exists because the space $\mathscr{L}^2(\mathbb{R}^m)$ is complete (see Theorem 9.1.3). If $\{g_n\}_{n\geqslant 1}$ is another sequence of functions in $\mathscr{L}^1(\mathbb{R}^m) \cap \mathscr{L}^2(\mathbb{R}^m)$ convergent to f in the $\mathscr{L}^2$-norm, then the sequence $f_1, g_1, f_2, g_2, \ldots$ obtained by "shuffling" the sequences $\{f_n\}_{n\geqslant 1}$ and $\{g_n\}_{n\geqslant 1}$ converges to f. By what has just been proved, the sequence $\widehat{f_1}, \widehat{g_1}, \widehat{f_2}, \widehat{g_2}, \ldots$ has a limit, which is unique up to equivalence and coincides with the limits of its subsequences. □

The lemma just proved allows us to extend the definition of the Fourier transform to the functions in $\mathscr{L}^2(\mathbb{R}^m)$.

Definition By the Fourier transform of a function $f \in \mathscr{L}^2(\mathbb{R}^m)$, we mean the limit in the $\mathscr{L}^2$-norm of the functions $\widehat{f_n}$, where $\{f_n\}_{n\geqslant 1}$ is an arbitrary sequence of functions in $\mathscr{L}^1(\mathbb{R}^m) \cap \mathscr{L}^2(\mathbb{R}^m)$ such that $\|f_n - f\|_2 \underset{n\to\infty}{\longrightarrow} 0$.

Thus, the Fourier transform of a function in $\mathscr{L}^2(\mathbb{R}^m)$ is also square-summable. As before, we will denote the Fourier transform of f by $\widehat{f}$. However, one must keep in mind that now the Fourier transform is defined up to equivalence and the symbol $\widehat{f}$ refers to many functions. If f is summable, then, among these functions, is the

Fourier transform defined in Sect. 10.5.1. For definiteness, the latter is sometimes called the classical Fourier transform. What has just been said also applies to the inverse transform, which, as before, is denoted by $\check{f}$.

Elementary properties of the Fourier transform of square-summable functions can be obtained from the properties of the classical Fourier transform by a passage to the limit.

Theorem *Let* $f \in \mathscr{L}^2(\mathbb{R}^m)$. *Then*:

(1) $\|\widehat{f}\|_2 = \|f\|_2$;
(2) *if* $f_n \in \mathscr{L}^2(\mathbb{R}^m)$ *and* $\|f_n - f\|_2 \underset{n\to\infty}{\longrightarrow} 0$, *then* $\|\widehat{f_n} - \widehat{f}\|_2 \underset{n\to\infty}{\longrightarrow} 0$, *and a similar statement holds for the inverse transform*;
(3) *we have* $(\widehat{f})\check{} = (\check{f})\widehat{} = f$;
(4) $\langle \widehat{f}, \widehat{g}\rangle = \langle f, g\rangle$ *for every function* $g \in \mathscr{L}^2(\mathbb{R}^m)$. *In particular, the Fourier transform preserves orthogonality*: *if* $f \perp g$, *then* $\widehat{f} \perp \widehat{g}$.

Proof Let $\{\varphi_n\}_{n\geqslant 1} \in C_0^\infty(\mathbb{R}^m)$ be a sequence of functions converging to f in the $\mathscr{L}^2$-norm. It is obvious that these functions and their Fourier transforms belong to $\mathscr{L}^1(\mathbb{R}^m) \cap \mathscr{L}^2(\mathbb{R}^m)$.

(1) It is clear that $\|\widehat{\varphi}_n - \widehat{f}\|_2 \underset{n\to\infty}{\longrightarrow} 0$ by the definition of $\widehat{f}$. By Plancherel's theorem, we have $\|\widehat{\varphi}_n\|_2 = \|\varphi_n\|_2$. Therefore, it is sufficient for us to use the continuity of the norm and pass to the limit in this equation.

(2) Obviously, $\|\widehat{f_n} - \widehat{f}\|_2 = \|\widehat{f_n - f}\|_2 = \|f_n - f\|_2 \underset{n\to\infty}{\longrightarrow} 0$.

(3) We will prove only the equality $(\widehat{f})\check{} = f$ (the other one is proved similarly). Since $\varphi_n \underset{n\to\infty}{\longrightarrow} f$, we obtain by definition that $\widehat{\varphi}_n \underset{n\to\infty}{\longrightarrow} \widehat{f}$, and, by property 2) applied to the inverse transform, we have $(\widehat{\varphi}_n)\check{} \underset{n\to\infty}{\longrightarrow} (\widehat{f})\check{}$. At the same time, $(\widehat{\varphi}_n)\check{} = \varphi_n$ by Theorem 10.5.4. Thus, it only remains to pass to the limit (in the $\mathscr{L}^2$-norm) in the last equality.

(4) For the proof, we must use the identity $4f\overline{g} = |f+g|^2 + |f+ig|^2 - |f-g|^2 - |f-ig|^2$ and apply relation (1) to the functions $f \pm g$ and $f \pm ig$. □

10.5.9 Plancherel's theorem implies an inequality known as the *uncertainty principle.* Without touching on its physical meaning (the impossibility of simultaneously determining the exact values of the coordinates and impulse of a quantum object), we mention only its consequence: if $f \neq 0$ only in the vicinity of the origin, then the quantity $|\widehat{f}|$ is not small at some remote points (the Fourier transform "blurs"). In the one-dimensional case, the reader can see this effect in the example of functions $\frac{1}{2t}\chi_{(-t,t)}$ forming an approximate identity.

In the precise formulation of the uncertainty principle, we confine ourselves to infinitely differentiable compactly supported functions of one variable (more general statements are given in Exercises 10 and 11).

Theorem *If* $f \in C_0^\infty(\mathbb{R})$ *and* $\|f\|_2 = 1$, *then*

$$\int_{-\infty}^{\infty} x^2 |f(x)|^2 \, dx \cdot \int_{-\infty}^{\infty} x^2 |\widehat{f}(x)|^2 \, dx \geqslant \frac{1}{16\pi^2}.$$

Proof Since

$$\int_{-\infty}^{\infty} x\left(|f(x)|^2\right)' dx = x|f(x)|^2\Big|_{-\infty}^{\infty} - \int_{-\infty}^{\infty} |f(x)|^2 \, dx = -1,$$

the Cauchy–Bunyakovsky inequality implies

$$1 = \left| \int_{-\infty}^{\infty} x\left(|f(x)|^2\right)' dx \right| \leqslant 2 \int_{-\infty}^{\infty} |xf(x)| \cdot |f'(x)| \, dx \leqslant 2\|g\|_2 \|f'\|_2,$$

where $g(x) = |x\, f(x)|$. By Plancherel's theorem, we have $\|f'\|_2 = \|\widehat{f'}\|_2$, and, by Theorem 10.5.2, we obtain $\widehat{f'}(y) = 2\pi i y \widehat{f}(y)$. Consequently,

$$1 \leqslant 4\|g\|_2^2 \|f'\|_2^2 = 4\|g\|_2^2 \|\widehat{f'}\|_2^2 = 4 \int_{-\infty}^{\infty} x^2 |f(x)|^2 \, dx \cdot 4\pi^2 \int_{-\infty}^{\infty} y^2 |\widehat{f}(y)|^2 \, dy. \qquad \square$$

10.5.10 In the conclusion of this section, we apply the Fourier transform to estimate the Dirichlet kernels for a ball (see Sect. 10.4.8)

$$D_R(x) = \frac{1}{(2\pi)^m} \sum_{\|k\|<R} e^{-i\langle k,x\rangle} \quad (x \in \mathbb{R}^m)$$

(the summation is taken over the points k of the integer lattice $\mathbb{Z}^m$). We show that, in the case where $m > 1$, their $\mathscr{L}^1$-norms (in contrast to the norms of the Dirichlet kernels for cubes $(-R, R)^m$) have not a logarithmic, but a power order of growth as $R \to +\infty$,

$$\|D_R\|_1 = \frac{1}{(2\pi)^m} \int_{[-\pi,\pi]^m} \left| \sum_{\|k\|<R} e^{-i\langle k,x\rangle} \right| dx \asymp R^{\frac{m-1}{2}}.$$

Being unable to represent the kernel D_R in a compact form, we obtain for D_R an approximate integral representation, replacing the sum over the ball $B(R)$ by an integral over a set close to $B(R)$. For this, we use the fact that the mean value of the exponential function e^{-iat} on the interval $(a - 1/2, a + 1/2)$ differs from the function itself only by a factor independent of a,

$$e^{-iat} = \frac{t/2}{\sin t/2} \int_{a-1/2}^{a+1/2} e^{-ist} \, ds.$$

Therefore, in the multiple integral for the shifted unit cube $Q_k = k + [-\frac{1}{2}, \frac{1}{2}]^m$ at the point $x = (x_1, \dots, x_m)$, we have

$$e^{-i\langle k,x\rangle} = \theta(x) \int_{Q_k} e^{-i\langle y,x\rangle}\, dy, \quad \text{where } \theta(x) = \prod_{j=1}^{m} \frac{x_j/2}{\sin x_j/2}.$$

Putting $T(R) = \bigcup_{\|k\|<R} Q_k$, we arrive at the equation

$$D_R(x) = \frac{\theta(x)}{(2\pi)^m} \int_{T(R)} e^{-i\langle y,x\rangle}\, dy.$$

Thus,

$$\|D_R\|_1 = \frac{1}{(2\pi)^m} \int_{[-\pi,\pi]^m} \theta(x) \left| \int_{T(R)} e^{-i\langle y,x\rangle}\, dy \right| dx.$$

Since $1 \leqslant \frac{t}{\sin t} \leqslant \frac{\pi}{2}$ for $t \in [-\frac{\pi}{2}, \frac{\pi}{2}]$, we obtain $1 \leqslant \theta(x) \leqslant (\frac{\pi}{2})^m$ in this integral, and, therefore,

$$\|D_R\|_1 \asymp \int_{[-\pi,\pi]^m} \left| \int_{T(R)} e^{-i\langle y,x\rangle}\, dy \right| dx = (2\pi)^m \int_{[-\frac{1}{2},\frac{1}{2}]^m} \left|\widehat{\chi}_{T(R)}(u)\right| du.$$

We show that, for $m > 1$, the integral on the right-hand side of this relation grows as $R^{\frac{m-1}{2}}$. It is more convenient to deal with the integral over a ball rather than over a cube. Therefore, we consider the integral

$$I_R(\rho) = \int_{B(\rho)} \left|\widehat{\chi}_{T(R)}(u)\right| du.$$

Since

$$I_R\left(\frac{1}{2}\right) \leqslant \int_{[-\frac{1}{2},\frac{1}{2}]^m} \left|\widehat{\chi}_{T(R)}(u)\right| du \leqslant I_R\left(\frac{\sqrt{m}}{2}\right),$$

it is sufficient to verify that $I_R(\rho) \asymp R^{\frac{m-1}{2}}$ as $R \to +\infty$ for every fixed $\rho > 0$.

For large R, the set $T(R)$ is close to the ball $B(R)$. Therefore, it is natural to replace $\widehat{\chi}_{T(R)}$ with $\widehat{\chi}_{B(R)}$ and compare the integral $I_R(\rho)$ with a "similar" integral

$$J_R(\rho) = \int_{B(\rho)} \left|\widehat{\chi}_{B(R)}(u)\right| du.$$

The rate of its growth was essentially found in Example 2 of Sect. 10.5.2. Indeed, since

$$\widehat{\chi}_{B(R)}(u) = \int_{\|x\|<R} e^{-2\pi i\langle u,x\rangle}\, dx = R^m \int_{\|x\|<1} e^{-2\pi i R\langle u,x\rangle}\, dx = R^m \widehat{\chi}_B(Ru),$$

the integral $J_R(\rho)$ can be reduced to the integral $L_B(R) = \int_{\|y\|<R} |\widehat{\chi}_B(y)|\,dy$ considered in this Example,

$$J_R(\rho) = \int_{B(\rho)} \left|R^m \widehat{\chi}_B(Ru)\right| du = \int_{B(\rho R)} |\widehat{\chi}_B(y)|\,dy = L_B(\rho R) \asymp (\rho R)^{\frac{m-1}{2}}.$$

Therefore,

$$0 < C_m(\rho\, R)^{\frac{m-1}{2}} \leqslant J_R(\rho) \leqslant C'_m(\rho R)^{\frac{m-1}{2}}. \tag{10}$$

To estimate the difference $I_R(\rho) - J_R(\rho)$, we introduce the function $\eta_R = \chi_{B(R)} - \chi_{T(R)}$. It is clear that

$$\left|I_R(\rho) - J_R(\rho)\right| \leqslant \int_{B(\rho)} \left|\widehat{\eta}_R(u)\right| du \leqslant \sqrt{\alpha_m \rho^m \int_{B(\rho)} \left|\widehat{\eta}_R(u)\right|^2 du} \leqslant \sqrt{\alpha_m \rho^m}\, \|\widehat{\eta}_R\|_2.$$

The next step is possible due to Plancherel's theorem allowing us to pass from the norm $\widehat{\eta}_R$ to the norm η_R, which can easily be estimated (since $|\eta_R| \leqslant 1$ and the function η_R differs from zero only in the spherical layer $R - \sqrt{m} \leqslant \|x\| \leqslant R + \sqrt{m}$),

$$\|\widehat{\eta}_R\|_2 = \|\eta_R\|_2 \leqslant \sqrt{\alpha_m \big((R+\sqrt{m}\,)^m - (R-\sqrt{m}\,)^m\big)}.$$

Therefore, we obtain for $R > 1$

$$\left|I_R(\rho) - J_R(\rho)\right| \leqslant \alpha_m \rho^{\frac{m}{2}} \sqrt{2m^{\frac{3}{2}}}\,(R+\sqrt{m})^{\frac{m-1}{2}} \leqslant A_m \rho^{\frac{m}{2}} R^{\frac{m-1}{2}}, \tag{11}$$

where A_m is a coefficient depending only on the dimension m. Taking into account inequality (10), we obtain the following estimate from above: $I_R(\rho) = O(R^{\frac{m-1}{2}})$ as $R \to +\infty$.

Because the integrals $I_R(\rho)$ grow with the growth of ρ, it is sufficient to establish an estimate from below for small ρ. For this, we again use inequalities (10) and (11),

$$\begin{aligned} I_R(\rho) &\geqslant J_R(\rho) - \left|I_R(\rho) - J_R(\rho)\right| \geqslant C_m(\rho R)^{\frac{m-1}{2}} - A_m \rho^{\frac{m}{2}} R^{\frac{m-1}{2}} \\ &= (C_m - A_m\sqrt{\rho})(\rho R)^{\frac{m-1}{2}}. \end{aligned}$$

We obtain the required result if we take, for example, $\rho = C_m^2/(2A_m)^2$.

EXERCISES

1. Find the Fourier transform of the product $e^{2\pi i\langle h,x\rangle} f(x)$, where $f \in \mathscr{L}^1(\mathbb{R}^m)$ and $h \in \mathbb{R}^m$.
2. Let $f \in \mathscr{L}^1(\mathbb{R}^m)$ and $f_E(x) = \frac{1}{\lambda_m(E)} \int_E f(x+t)\,dt$, where $E \subset \mathbb{R}^m$ is a set of finite positive measure. Prove that $|\widehat{f_E}| \leqslant |\widehat{f}|$.

3. Let a function f in $\mathscr{L}^1(\mathbb{R}^m)$ vanish outside the cube $(-\pi,\pi)^m$, and let $F(t) = \widehat{f}(t/2\pi)$. Prove that

$$F(t) = \sum_{n\in\mathbb{Z}^m} F(n) \prod_{j=1}^{m} \frac{\sin \pi(t_j - n_j)}{\pi(t_j - n_j)}$$

(the sum on the right-hand side of the equation is understood as the limit of the partial sums over rectangles).

4. A function f defined on $\mathbb{R}^m$ is called positive definite if

$$\sum_{1\leqslant j,k\leqslant n} f(x_j - x_k) z_j \overline{z}_k \geqslant 0$$

for all $n \in \mathbb{N}$, $x_j \in \mathbb{R}^m$ and $z_j \in \mathbb{C}$. Prove that the Fourier transform of a finite Borel measure μ is a positive definite function.

5. Prove that if g is the generalized derivative of f with respect to the kth coordinate (see Exercise 5 of Sect. 9.3), then the relation $\widehat{g}(y) = 2\pi i y_k \widehat{f}(y)$ of statement (1) of Theorem 10.5.2 remains valid.

6. Verify that, in general, the equiconvergence of the expansions in Fourier series and Fourier integrals does not take place in the multi-dimensional case. *Hint.* Use the same idea as in the first part of Sect. 10.4.6.

7. Let a function $f \in \mathscr{L}^1(\mathbb{R}^m)$ be bounded in a neighborhood of zero. Prove that if $\widehat{f} \geqslant 0$, then $\widehat{f} \in \mathscr{L}^1(\mathbb{R}^m)$. *Hint.* By Eq. (7), prove that the integrals $\int_{\mathbb{R}^m} e^{-\pi t^2 \|u\|^2} \widehat{f}(u)\, du$ are bounded for $t > 0$ and apply Fatou's theorem.

8. Let a measure μ on $\mathbb{R}^m$ be such that $\int_{\mathbb{R}^m} e^{a\|x\|}\, d\mu(x) < +\infty$ for some $a > 0$. Generalizing Theorem 10.5.6, prove that, for $p > 1$, every function $f \in \mathscr{L}^p(\mathbb{R}^m, \mu)$ satisfying the condition

$$\int_{\mathbb{R}^m} f(x) x^n \, d\mu(x) = 0 \quad \text{for all } n \in \mathbb{Z}_+^m$$

is equal to zero μ-almost everywhere.

9. Let $\varphi_1, \varphi_2, \ldots$ be functions in $\mathscr{L}^2(\mathbb{R}^m)$. Prove that the systems $\{\varphi_n\}_{n\in\mathbb{N}}$ and $\{\widehat{\varphi}_n\}_{n\in\mathbb{N}}$ are complete or not simultaneously.

10. Let $f \in \mathscr{L}^2(\mathbb{R}^m)$ and $\|f\|_2 = 1$. Prove that the inequality

$$\int_{\mathbb{R}^m} \|x-a\|^2 \big|f(x)\big|^2 dx \cdot \int_{\mathbb{R}^m} \|y-b\|^2 \big|\widehat{f}(y)\big|^2 dy \geqslant \frac{m^2}{16\pi^2}$$

is valid for all $a, b \in \mathbb{R}^m$.

11. Assume that the values of a function $f \in \mathscr{L}^2(\mathbb{R}^m)$ are small outside a ball $B(a,r)$ in the sense that

$$\int_{\mathbb{R}^m\setminus B(a,r)} \|x-a\|^2 \big|f(x)\big|^2 dx < \frac{1}{2} \int_{\mathbb{R}^m} \|x-a\|^2 \big|f(x)\big|^2 dx,$$

and the values of $\widehat{f}$ are small (in the same sense) outside a ball $B(b, R)$. Prove that $rR > \frac{m}{8\pi}$.

12. Supplement the result of Example 3 of Sect. 10.5.2 by proving that if $f(x) \underset{\|x\|\to+\infty}{\sim} \|x\|^{-p-m}$, as $\|x\| \to +\infty$ for some $p \in (0, 2)$ and the function f is even, then $\widehat{f}(0) - \widehat{f}(y) \underset{y\to 0}{\sim} C_p\|y\|^p$; for $p = 2$, this relation must be replaced by $\widehat{f}(0) - \widehat{f}(y) \underset{y\to 0}{\sim} C\|y\|^2 \ln \frac{1}{\|y\|}$. The coefficients C_p and C depending on the dimension can be expressed in terms of the gamma function.

13. Let P, Q be algebraic polynomials with $\deg Q > \deg P$. Show that the Fourier transform of the fraction $f = \frac{P}{Q}$ vanishes on the negative half-axis ($\widehat{f}(y) = 0$ for $y \leqslant 0$) if all roots of the denominator Q lie in the lower half-plane (i.e., their imaginary parts are negative).

10.6 *The Poisson Summation Formula

In this section, by the periodicity of a function of several variables, we mean 1-periodicity with respect to each variable. Speaking of the Fourier series of a periodic function summable on the cube $(-\frac{1}{2}, \frac{1}{2})^m$, we mean a series in the system of exponential functions $\{e^{2\pi i\langle n,x\rangle}\}_{n\in\mathbb{Z}^m}$.

10.6.1 As we have already seen, the properties of the Fourier transform of a summable function f can be far from the properties of f. A smooth function can have a non-smooth Fourier transform, which can be non-summable, and the order of decrease of $\widehat{f}$ can be distinct from that of f, etc. However, remarkably it turns out that, under quite mild assumptions, there is a characteristic that does not change when passing from f to $\widehat{f}$. Confining ourselves to the functions of one variable and not touching now the problem of convergence of the series in question, we can say that the required characteristic is the sum of the values of a function at the integer points. In other words, we are talking about the relation

$$\sum_{n=-\infty}^{\infty} f(n) = \sum_{k=-\infty}^{\infty} \widehat{f}(k), \tag{1}$$

known as the *Poisson summation formula*. Generalizing Eq. (1) somewhat, we can represent it in the form

$$\sqrt{t} \sum_{n=-\infty}^{\infty} f(tn) = \sqrt{s} \sum_{k=-\infty}^{\infty} \widehat{f}(sk), \tag{1'}$$

where $st = 1$ ($s > 0, t > 0$). Thus, the sum of the values of f at the equidistant lattice points tn ($t > 0$) is proportional to a similar sum for $\widehat{f}$, provided that the values of $\widehat{f}$ are calculated at compatible lattice points. Equation (1′) can be obtained by applying Eq. (1) to the function g obtained from f by the similarity ($g(x) = f(tx)$).

Our goal is to justify Poisson's formula and give examples of its application. As often happens, to solve a problem, one must generalize it. We will study not the sum $\sum_{n=-\infty}^{\infty} f(n)$ itself, but the function S defined by the formula

$$S(x) = \sum_{n=-\infty}^{\infty} f(x+n) \quad (x \in \mathbb{R}). \tag{2}$$

Thus, to study a non-periodic function, we assign to it a periodic function and study the properties of the latter, using, in particular, the machinery of Fourier series developed in Sects. 10.3 and 10.4.

We need the following statement.

Lemma *Let a function f be summable on $\mathbb{R}$. Then:*

(a) *series* (2) *converges absolutely for almost all x;*
(b) *its sum S is a 1-periodic function; this function is summable on $(-\frac{1}{2}, \frac{1}{2})$, and Eq.* (2) *can be integrated termwise;*
(c) *the Fourier coefficients of the function S are equal to the values of $\widehat{f}$ at the integer points,*

$$\int_{-\frac{1}{2}}^{\frac{1}{2}} S(x) e^{-2\pi i n x}\, dx = \widehat{f}(n) \quad (n \in \mathbb{Z}).$$

Proof We consider the non-negative measurable function

$$F(x) = \sum_{n=-\infty}^{\infty} \bigl|f(x+n)\bigr| \quad (x \in \mathbb{R}).$$

Since a positive series can be integrated termwise, we have

$$\int_{-\frac{1}{2}}^{\frac{1}{2}} F(x)\, dx = \sum_{n=-\infty}^{\infty} \int_{-\frac{1}{2}}^{\frac{1}{2}} \bigl|f(x+n)\bigr|\, dx = \int_{\mathbb{R}} \bigl|f(x)\bigr|\, dx$$

(the second equation is valid since the integral is countably additive). Therefore, the function F is summable on $(-\frac{1}{2}, \frac{1}{2})$ and, therefore, F is almost everywhere finite, which proves the fact that the series $\sum_{n=-\infty}^{\infty} |f(x+n)|$ converges almost everywhere and so statement (a) holds.

Since the 1-periodicity of the function S is obvious, statement (b) follows from the inequality $|S| \leqslant F$. Since F also dominates all partial sums of series (2), the series can be integrated termwise.

To complete the proof, we multiply Eq. (2) by $e^{-2\pi ikx}$ and integrate termwise over $(-\frac{1}{2}, \frac{1}{2})$,

$$\int_{-\frac{1}{2}}^{\frac{1}{2}} S(x)e^{-2\pi ikx}\,dx = \sum_{n=-\infty}^{\infty} \int_{-\frac{1}{2}}^{\frac{1}{2}} f(x+n)e^{-2\pi ikx}\,dx$$

$$= \int_{\mathbb{R}} f(x)e^{-2\pi ikx}\,dx = \widehat{f}(k).$$

The termwise integration here is allowed since F is also a summable majorant of the partial sums of the series $\sum_{n=-\infty}^{\infty} f(x+n)e^{-2\pi ikx}$. □

As established in the lemma, the series $\sum_{k=-\infty}^{\infty} \widehat{f}(k)e^{2\pi ikx}$ is the Fourier series of the function S. Therefore, the relation

$$\sum_{n=-\infty}^{\infty} f(x+n) = \sum_{k=-\infty}^{\infty} \widehat{f}(k)e^{2\pi ikx} \tag{3}$$

(also called the Poisson summation formula) means simply that, at a point x, the function S is the sum of its Fourier series. In particular, if S is continuous at x, then Eq. (3) holds only under the assumption that the series on the right-hand side converges.

Example 1 Let $f(x) = e^{-\pi(tx)^2}$ (where t is a positive parameter). As established in Example 2 of Sect. 10.5.1, $\widehat{f}(y) = \frac{1}{t}e^{-\pi(y/t)^2}$. Obviously, the sum $\sum_{n=-\infty}^{\infty} e^{-\pi t^2(x+n)^2}$ is a smooth function. Therefore, formula (3) is valid for f everywhere,

$$\sum_{n=-\infty}^{\infty} e^{-\pi t^2(x+n)^2} = \frac{1}{t}\sum_{n=-\infty}^{\infty} e^{-\pi(n/t)^2}e^{2\pi inx} = \frac{1}{t}\left(1 + 2\sum_{n=1}^{\infty} e^{-\pi(n/t)^2}\cos 2\pi nx\right).$$

For $x = 0$, the left-hand side of this equation is the so-called θ-function

$$\theta(t) = \sum_{n=-\infty}^{\infty} e^{-\pi(tn)^2}.$$

From this equation, it follows that the θ-function satisfies the Jacobi identity $\theta(t) = \frac{1}{t}\theta(\frac{1}{t})$, which plays an important role in heat transfer theory and in the theory of elliptic functions.

Example 2 Let $g(x) = (1-|x|)_+$ for $x \in \mathbb{R}$. Obviously,

$$\widehat{g}(y) = \int_{-1}^{1}\big(1-|x|\big)e^{-2\pi ixy}\,dx = 2\int_0^1 (1-x)\cos 2\pi xy\,dx = \left(\frac{\sin\pi y}{\pi y}\right)^2.$$

We apply formula (3) to the function $f = \widehat{g}$. It can easily be verified that the sum $S(x) = \sum_{n=-\infty}^{\infty} f(x+n)$ is everywhere continuous and $\widehat{f} = \check{f}$, and so, the inversion formula implies $\widehat{f} = g$. Therefore, by (3), we obtain

$$\sum_{n=-\infty}^{\infty} \left(\frac{\sin \pi(x+n)}{\pi(x+n)}\right)^2 = \sum_{n=-\infty}^{\infty} f(x+n) = \sum_{k=-\infty}^{\infty} \widehat{f}(k) e^{2\pi i k x} = \widehat{f}(0) = 1.$$

Since $\sin^2 \pi(x+n) = \sin^2 \pi x$ for all $n \in \mathbb{Z}$, we obtain the following partial fraction expansion of $\frac{1}{\sin^2 \pi x}$:

$$\frac{\pi^2}{\sin^2 \pi x} = \sum_{n=-\infty}^{\infty} \frac{1}{(x+n)^2} \qquad (x \in \mathbb{R} \setminus \mathbb{Z}).$$

For $x = \frac{1}{2}$, this leads to the equality $\pi^2 = 4 \sum_{n=-\infty}^{\infty} \frac{1}{(2n+1)^2} = 8 \sum_{n=0}^{\infty} \frac{1}{(2n+1)^2}$ from which the well-known result $\frac{\pi^2}{6} = \sum_{n=1}^{\infty} \frac{1}{n^2}$ follows easily. Differentiating the expansion obtained termwise an even number of times, one can calculate the sums of the series $\sum_{n=1}^{\infty} \frac{1}{n^{2k}}$ $(k \in \mathbb{N})$ first found by Euler.

Example 3 Let $a > 1$ and $u > 0$. Let $f(x) = x^{a-1} e^{-ux}$ for $x > 0$ and $f(x) = 0$ for $x \leqslant 0$. This function is continuous everywhere and the series $S(x) = \sum_{n=-\infty}^{\infty} f(x+n)$ converges uniformly on every finite interval. Since $\widehat{f}(y) = \frac{\Gamma(a)}{(u+2\pi i y)^a}$ (see Example 5 of Sect. 10.5.1), we see that the Fourier series of S converges absolutely. Consequently, Eq. (3) is valid everywhere. In particular, it takes the following form for $x = 0$:

$$\sum_{n=1}^{\infty} n^{a-1} e^{-nu} = \sum_{n=-\infty}^{\infty} \frac{\Gamma(a)}{(u+2\pi i n)^a}.$$

Hence, we see that the sum on the right-hand side of the equation is exponentially small as $u \to +\infty$. For $u = 1$, the sum on the left-hand side can be regarded as a discrete analog of the integral $\Gamma(a) = \int_0^{\infty} t^{a-1} e^{-t}\, dt$. The formula obtained yields the following interesting relation: $\sum_{n=1}^{\infty} n^{a-1} e^{-n} = \Gamma(a) \sum_{n=-\infty}^{\infty} (1+2\pi i n)^{-a}$.

Without any assumptions on the function f (except the summability), Eq. (3) is valid in the following "weak" sense: after integrating both sides of the equation over an arbitrary interval, we obtain convergent series with equal sums. This follows immediately from our ability to integrate Fourier series termwise (Theorem 1 of Sect. 10.3.6).

We also remark that not only does (1) follow from (3), but also (3) can be regarded as Eq. (1) for the shift f_{-x} of the function f since $\widehat{f_{-x}}(n) = \widehat{f}(n) e^{2\pi i n x}$ (see the beginning of Sect. 10.5.1).

To derive Eq. (1) from (3), we must be sure that the latter is valid for $x = 0$. For this, it is not sufficient, for example, that series (3) converges almost everywhere.

Therefore, of particular interest to us is to find conditions under which the function S has a Fourier series expansion everywhere. One such conditions is given in Exercise 2. Other versions of sufficient conditions for functions of several variables will be established in the next section.

10.6.2 Here, we discuss the following multi-dimensional version of the Poisson summation formula for a function f in $\mathscr{L}^1(\mathbb{R}^m)$:

$$\sum_{n\in\mathbb{Z}^m} f(n) = \sum_{k\in\mathbb{Z}^m} \widehat{f}(k), \tag{4}$$

or, in a more general form,

$$\sum_{n\in\mathbb{Z}^m} f(x+n) = \sum_{k\in\mathbb{Z}^m} \widehat{f}(k)e^{2\pi i\langle k,x\rangle}. \tag{5}$$

Their derivation is based on an obvious modification of the lemma of Sect. 10.6.1, in which the one-dimensional lattice $\mathbb{Z}$ is replaced by the multi-dimensional lattice and the interval of integration $(-\frac{1}{2},\frac{1}{2})$ is replaced by the cube $(-\frac{1}{2},\frac{1}{2})^m$. Thus, the series on the right-hand side of (5) is the Fourier series of the function $S(x) = \sum_{n\in\mathbb{Z}^m} f(x+n)$. Therefore, as in the one-dimensional case, Eq. (5) for the continuous function S means that S has a Fourier series expansion.

Formula (4) can be modified by a linear change of variables (cf. Eq. (1′)),

$$\sqrt{|\det(T)|}\sum_{n\in\mathbb{Z}^m} f\big(T(n)\big) = \sqrt{|\det(S)|}\sum_{n\in\mathbb{Z}^m} \widehat{f}\big(S(n)\big),$$

where T is an arbitrary non-degenerate linear transformation, $S = (T^*)^{-1}$ (here T^* is the adjoint mapping).

We give two types of conditions under which Eq. (5) is valid.

Theorem 1 *Let f be a continuous function on $\mathbb{R}^m$ such that $f(x) = O(\|x\|^{-p})$ and $\widehat{f}(x) = O(\|x\|^{-p})$ as $\|x\| \to +\infty$ for some $p > m$. Then Eq. (5) is valid for all $x \in \mathbb{R}^m$.*

Proof First, we observe that $f \in \mathscr{L}^1(\mathbb{R}^m)$ since the summability on an arbitrary ball follows from the continuity of f and the summability outside the ball follows from the estimate $f(x) = O(\|x\|^{-p})$.

Under our assumptions, the series on both sides of Eq. (5) converge absolutely and uniformly on every ball, and, since their terms are continuous, the sums of the series are also continuous. Thus, the right-hand side of (5) is a uniformly convergent Fourier series of the sum on the left-hand side of (5). □

It is not difficult to give an interpretation of the sum of a multiple series in the case of absolute convergence. Otherwise, it is necessary to clarify the definition of this sum. First of all, this concerns the series on the right-hand side of Eq. (5)

(the series on the left-hand side of (5) converges absolutely for almost all x). The following theorem enables us to consider the situation in which the series on the right-hand side of (5) does not converge absolutely (see also Exercises 3 and 4).

Theorem 2 *Let a function f satisfy the Lipschitz condition with exponent α in $\mathbb{R}^m$, i.e., there is a positive L such that $|f(x) - f(x')| \leqslant L\|x - x'\|^\alpha$ for all $x, x' \in \mathbb{R}^m$. We assume, in addition, that the function decreases rapidly at infinity, i.e., $f(x) = O(\|x\|^{-p})$ as $\|x\| \to +\infty$ for some $p > m$. Then Eq. (5) is valid for all $x \in \mathbb{R}^m$ (the sum on the right-hand side of (5) is understood as the limit of rectangular partial sums).*

Proof As in the previous theorem, $f \in \mathscr{L}^1(\mathbb{R}^m)$. It is sufficient to verify that the function S satisfies the Lipschitz condition with an exponent β (in this case, the rectangular partial sums converge uniformly to S by Theorem 10.4.5). Estimating the difference $S(x+h) - S(x)$, we will assume that $x \in [0,1]^m$ and $\|h\| \leqslant 1$. It is clear that

$$\bigl|S(x+h) - S(x)\bigr| \leqslant \sum_{n\in\mathbb{Z}^m} \bigl|f(x+h+n) - f(x+n)\bigr|. \tag{6}$$

Now, fixing a large parameter R (its choice will be specified later), we partition the terms of the series obtained into two sets depending on whether $\|n\| \leqslant R$ or $\|n\| > R$. Using the Lipschitz condition $|f(x+h+n) - f(x+n)| \leqslant L\|h\|^\alpha$, we estimate the terms of the first set (the number of them has order R^m). In the second case, we apply the following estimate for f at infinity:

$$\bigl|f(x+h+n) - f(x+n)\bigr| \leqslant \bigl|f(x+h+n)\bigr| + \bigl|f(x+n)\bigr| = O\bigl(\|n\|^{-p}\bigr).$$

Substituting these estimates into inequality (6), we obtain

$$\bigl|S(x+h) - S(x)\bigr| \leqslant \mathrm{const}\Bigl(R^m\|h\|^\alpha + \sum_{\|n\|>R}\|n\|^{-p}\Bigr) = O\bigl(R^m\|h\|^\alpha + R^{m-p}\bigr).$$

Now, we use the freedom in the choice of R and equate the terms $R^m\|h\|^\alpha$ and R^{m-p} and, for $R = \|h\|^{-\alpha/p}$, we obtain that $|S(x+h) - S(x)| = O(\|h\|^\beta)$ with $\beta = \alpha(1 - m/p)$. □

Corollary *If a function f having bounded first order derivatives satisfies the condition $f(x) = O(\|x\|^{-p})$ as $\|x\| \to +\infty$ for some $p > m$, then Eq. (5) holds for every point x (the sum on the right-hand side of the equation is understood as the limit over rectangular partial sums).*

10.6.3 The Poisson summation formula has proved to be an effective tool for solving various problems (see Sects. 10.6.4 and 10.6.5). However, before passing to these technically more involved applications, we use the summation formula to supplement the uncertainty principle established in Sect. 10.5.9, according to which the functions f and $\widehat{f}$ cannot be concentrated on "small sets" simultaneously (see also Exercises 10 and 11 of Sect. 10.5). The following statement is valid (see [B]).

Theorem *Let a summable function f on $\mathbb{R}^m$ be such that the sets $A = \{x \in \mathbb{R}^m \,|\, f(x) \neq 0\}$ and $B = \{y \in \mathbb{R}^m \,|\, \widehat{f}(y) \neq 0\}$ have finite measures, then $f(x) = 0$ almost everywhere.*

Proof We will assume that $\lambda_m(A) < 1$ (this can be achieved by the change of variables $x \mapsto cx$). Let $Q = [-\frac{1}{2}, \frac{1}{2})^m$. Since

$$\sum_{k\in\mathbb{Z}^m} \int_Q \chi_B(y+k)\,dy = \int_{\mathbb{R}^m} \chi_B(y)\,dy = \lambda_m(B) < +\infty,$$

the series $\sum_{k\in\mathbb{Z}^m} \chi_B(y+k)$ converges for almost all y. Consequently, $\chi_B(y+k) \neq 0$, i.e., $\widehat{f}(y+k) \neq 0$ only for a finite number of multi-indices k. Therefore, for almost all y, the function $f_y(x) = f(x)e^{-2\pi i\langle y,x\rangle}$ is such that only a finite number of the values $\widehat{f_y}(k)$ $(k \in \mathbb{Z}^m)$ are distinct from zero.

Since the kth Fourier coefficient of the function S_y is equal to $\widehat{f_y}(k)$ (see statement (c)) of Lemma 10.6.1), we obtain that S_y coincides with a trigonometric polynomial almost everywhere. The set $E_y = \{x \in Q \,|\, S_y(x) \neq 0\}$ is contained in the union $\bigcup_n (-n + A) \cap Q$, and, therefore, $\lambda_m(E_y) \leqslant \lambda_m(A) < 1 = \lambda_m(Q)$. Since a non-zero trigonometric polynomial does not vanish almost everywhere, we obtain that $S_y(x) = 0$ almost everywhere. Consequently, $0 = \widehat{S_y}(0) = \widehat{f_y}(0) = \int_{\mathbb{R}^m} f_y(x)\,dx = \widehat{f}(y)$ for almost all y. By the uniqueness theorem, $f = 0$ almost everywhere. □

We remark that the proof of the theorem does not use Eq. (5). It is based only on statement (c) of Lemma 10.6.1 (more precisely, on its m-dimensional modification).

10.6.4 Generalizing the reasoning of Sect. 10.4.3 to multiple Fourier series, we consider a summation method generated by a function M $(M(0) = 1)$ continuous and summable on $\mathbb{R}^m$. The method is as follows: for each $\varepsilon > 0$ and each 1-periodic function f summable on the cube $Q = [-\frac{1}{2}, \frac{1}{2})^m$, we consider the sum

$$S_{M,\varepsilon}(f, x) = \sum_{n\in\mathbb{Z}^m} M(\varepsilon n)\widehat{f}(n)e^{2\pi i\langle n,x\rangle}$$

and study its limit as $\varepsilon \to 0$.

To simplify the exposition, we will assume that M tends to zero at infinity so fast that

$$\sum_{n\in\mathbb{Z}^m} \big|M(\varepsilon n)\big| < +\infty \quad \text{for every } \varepsilon > 0$$

(this condition is necessarily fulfilled if M is a compactly supported function).

It is clear that $S_{M,\varepsilon}(f) = f * \omega_\varepsilon$, where

$$\omega_\varepsilon(x) = \sum_{n\in\mathbb{Z}^m} M(\varepsilon n)e^{2\pi i\langle n,x\rangle}.$$

If it turns out that the functions ω_ε form an approximate identity as $\varepsilon \to 0$, then our problem simplifies considerably: it will be possible to use general theorems 7.6.5 and 9.3.7 in the study of the sums $S_{M,\varepsilon}(f,x)$.

When is the family $\{\omega_\varepsilon\}_{\varepsilon>0}$ an approximate identity as $\varepsilon \to 0$? In Sect. 10.4.3, we obtained a sufficient condition. Now, using the Poisson summation formula, we can a give much more complete answer to this question. It turns out that it is sufficient that the Fourier transform $\widehat{M}$ be summable (as follows from the result of Exercise 6, this condition is also necessary).

We verify the conditions characterizing a periodic approximate identity (see Sect. 7.6.5). The equality $\int_{\mathbb{R}^m} \omega_\varepsilon(x)\,dx = 1$ is valid because the series defining the function ω_ε converges absolutely, and we can integrate it termwise over the cube Q,

$$\int_Q \omega_\varepsilon(x)\,dx = \sum_{n\in\mathbb{Z}^m} M(\varepsilon n) \int_Q e^{2\pi i\langle n,x\rangle} = M(0) = 1.$$

We will show below that the functions ω_ε are non-negative if $\widehat{M} \geqslant 0$. However, regardless of this condition, the integrals $\int_Q |\omega_\varepsilon(x)|\,dx$ are bounded. To verify this, we put $N(x) = \widehat{M}(-x)$ and $N_\varepsilon(x) = \varepsilon^{-m} N(x/\varepsilon)$. Then the inversion formula for the Fourier transform (see Theorem 10.5.4) implies

$$M(\varepsilon n) = \int_{\mathbb{R}^m} \widehat{M}(y) e^{2\pi i\langle \varepsilon n,y\rangle}\,dy = \varepsilon^{-m} \int_{\mathbb{R}^m} N\Big(-\frac{u}{\varepsilon}\Big) e^{2\pi i\langle n,u\rangle}\,du = \widehat{N_\varepsilon}(n),$$

and, by the Poisson summation formula, we obtain

$$\omega_\varepsilon(x) = \sum_{n\in\mathbb{Z}^m} \widehat{N}_\varepsilon(n) e^{2\pi i\langle n,x\rangle} = \sum_{n\in\mathbb{Z}^m} N_\varepsilon(x+n). \tag{7}$$

Therefore, $\omega_\varepsilon \geqslant 0$ if $N \geqslant 0$, i.e., $\widehat{M} \geqslant 0$, and

$$\int_Q |\omega_\varepsilon(x)|\,dx \leqslant \int_Q \sum_{n\in\mathbb{Z}^m} |N_\varepsilon(x+n)|\,dx = \int_{\mathbb{R}^m} |N_\varepsilon(x)|\,dx = \|N\|_1 = \|\widehat{M}\|_1$$

in the general case. Thus, the $\mathscr{L}^1$-norm (on the cube Q) of each function ω_ε does not exceed the $\mathscr{L}^1$-norm (on $\mathbb{R}^m$) of the Fourier transform $\widehat{M}$.

By refining this argument a little, we can establish a localization property. Indeed, for every $\delta \in (0, \frac{1}{2})$, we have

$$\int_{Q\setminus B(\delta)} |\omega_\varepsilon(x)|\,dx \leqslant \int_{\|x\|\geqslant\delta} |N_\varepsilon(x)|\,dx = \int_{\|y\|\geqslant\delta/\varepsilon} |\widehat{M}(y)|\,dy \underset{\varepsilon\to 0}{\longrightarrow} 0.$$

Thus, $\{\omega_\varepsilon\}_{\varepsilon>0}$ is a periodic approximate identity, and, consequently, statements (a) and (b) of Theorem 7.6.5 are valid for it.

Example The function $M(u) = e^{-\|u\|}$ generates the "radial" Abel–Poisson summation method for multiple Fourier series, where, to each 1-periodic function f

summable on the cube Q, we assign the sums

$$S_\varepsilon(f,x) = \sum_{n\in\mathbb{Z}^m} e^{-\varepsilon\|n\|}\widehat{f}(n)e^{2\pi i\langle n,x\rangle}.$$

The Fourier transform of M was calculated in Example 4 of Sect. 10.5.1, $\widehat{M}(y) = C_m(1+4\pi^2\|y\|^2)^{-\frac{m+1}{2}}$. It, obviously, is summable. The corresponding kernel has the form (see formula (7))

$$\omega_\varepsilon(x) = \sum_{n\in\mathbb{Z}^m} e^{-\varepsilon\|n\|}e^{2\pi i\langle n,x\rangle} = \sum_{n\in\mathbb{Z}^m} \frac{C_m\varepsilon}{(\varepsilon^2+4\pi^2\|n+x\|^2)^{\frac{m+1}{2}}}.$$

As follows from the result of Exercise 7, not only does this kernel have the strong localization property, but it is also dominated by a summable "hump-shaped" majorant (see Sect. 9.3.4). Therefore, the theorems of Sects. 7.6.5 and 9.3.7 can be applied to the sums $S_\varepsilon(f) = f*\omega_\varepsilon$. Consequently, $S_\varepsilon(f) \underset{\varepsilon\to 0}{\rightrightarrows} f$ if the 1-periodic function f is continuous, and $\int_Q |S_\varepsilon(f,x)-f(x)|^p\,dx \underset{\varepsilon\to 0}{\longrightarrow} 0$ if $f\in\mathcal{L}^p(Q)$, where $p\geqslant 1$. Moreover, $S_\varepsilon(f,x)\underset{\varepsilon\to 0}{\longrightarrow} L$ if $f(x+t)\underset{t\to 0}{\longrightarrow} L$, and $S_\varepsilon(f,x)\underset{\varepsilon\to 0}{\longrightarrow} f(x)$ almost everywhere.

10.6.5 An interesting application of the Poisson identity is in one of the solutions to the Gauss problem on determining the number $N_m(R)$ of points of the integer lattice $\mathbb{Z}^m$ that lie in the closed ball of a large radius R. The number $N_m(R)$ is close to the volume of the ball, $N_m(R) - \lambda_m(\overline{B}(R)) = O(R^{m-1})$ as $R\to+\infty$. This can easily be verified by considering the unit cubes centered at the lattice points, i.e., the cubes $n+[-\frac{1}{2},\frac{1}{2}]^m$, $n\in\mathbb{Z}^m$. Since, for $n\in\overline{B}(R)$, their union contains the ball $B(R-\sqrt{m})$ and is contained in the ball $\overline{B}(R+\sqrt{m})$, we see that the number $N_m(R)$, being equal to the volume of the union of these cubes, lies between $\lambda_m(B(R-\sqrt{m}))$ and $\lambda_m(B(R+\sqrt{m}))$. In other words, $\alpha_m(R-\sqrt{m})^m \leqslant N_m(R) \leqslant \alpha_m(R+\sqrt{m})^m$, where α_m is the volume of the unit cube in $\mathbb{R}^m$.

Much greater efforts are needed to sharpen this elementary estimate. However, we first observe that the exponent θ in the relation $N_m(R) = \alpha_m R^m + O(R^\theta)$ cannot be less than $m-2$. Indeed, the function $R\mapsto N_m(R)$ makes a jump at $R^2\in\mathbb{N}$, the value of which is equal to the number of lattice points on the sphere of radius R. Since the principal term $\alpha_m R^m$ of the asymptotic depends continuously on R, the exponent θ must be so large that the number of points on the sphere of radius R be dominated by a summand proportional to R^θ. The number of lattice points in the spherical layer $R<\|x\|<2R$ is of order R^m. The number of spheres containing these points is at most $3R^2$ (every such sphere is defined by the equation $\|x\|^2 = t$ with an integer parameter t lying between R^2 and $(2R)^2$). At least one of the spheres contains at least const R^{m-2} points. Therefore, necessarily $\theta\geqslant m-2$.

It is clear that (in what follows, $n\in\mathbb{Z}^m$)

$$N_m(R) = \sum_{\|n\|\leqslant R} 1 = \sum_{n\in\mathbb{Z}^m}\chi\Big(\frac{n}{R}\Big), \tag{8}$$

where χ is the characteristic function of the unit ball $\overline{B}$. To calculate the sum on the right-hand side of (8), we apply the Poisson summation formula. Unfortunately, this cannot be done directly because the function χ is discontinuous. Therefore, we smoothen it and estimate $N_m(R)$, applying the Poisson formula to smooth compactly supported functions χ_+ and χ_- approximating χ from above and from below. It is desired, of course, that the functions $\chi_\pm$ be as close as possible to χ.

To construct them, we take a non-negative function $\psi \in C_0^\infty(\mathbb{R}^m)$ with the properties $\operatorname{supp}(\psi) \subset \overline{B}$ and $\int_{\mathbb{R}^m} \psi(x)\,dx = 1$ and put $\psi_\varepsilon(x) = \varepsilon^{-m}\psi(x/\varepsilon)$, where ε is a small positive parameter the choice of which will be specified later. The infinitely differentiable function $\chi_\varepsilon = \chi * \psi_\varepsilon$ (see Corollary 7.5.4) is, obviously, equal to 1 in the ball $\overline{B}(1-\varepsilon)$ and to zero outside the ball $\overline{B}(1+\varepsilon)$. Consequently, the functions $\chi_-(x) = \chi_\varepsilon((1+\varepsilon)x)$ and $\chi_+(x) = \chi_\varepsilon((1-\varepsilon)x)$ satisfy the inequality $\chi_- \leqslant \chi \leqslant \chi_+$. Thus,

$$\sum_{n\in\mathbb{Z}^m} \chi_\varepsilon\left((1+\varepsilon)\frac{n}{R}\right) \leqslant N_m(R) \leqslant \sum_{n\in\mathbb{Z}^m} \chi_\varepsilon\left((1-\varepsilon)\frac{n}{R}\right),$$

i.e.,

$$S_\varepsilon\left(\frac{R}{1+\varepsilon}\right) \leqslant N_m(R) \leqslant S_\varepsilon\left(\frac{R}{1-\varepsilon}\right), \tag{9}$$

where $S_\varepsilon(r) = \sum_{n\in\mathbb{Z}^m} \chi_\varepsilon(n/r)$.

To apply the Poisson summation formula (4) to the function $f(x) = \chi_\varepsilon(x/r)$, we observe that

$$\begin{aligned}&\widehat{f}(y) = r^m \widehat{\chi}_\varepsilon(ry) = r^m \widehat{\chi}(ry)\widehat{\psi}_\varepsilon(ry) = r^m \widehat{\chi}(ry)\widehat{\psi}(r\varepsilon y);\\ &\text{in particular, } \widehat{f}(0) = \alpha_m r^m.\end{aligned}$$

Since the function ψ belongs to the class $C_0^\infty(\mathbb{R}^m)$, its Fourier transform (and, consequently, also $\widehat{f}$) rapidly decreases at infinity. Thus, the conditions of Theorem 1 of Sect. 10.6.2 are fulfilled, and we obtain

$$\begin{aligned}S_\varepsilon(r) &= \sum_{n\in\mathbb{Z}^m} \chi_\varepsilon\left(\frac{n}{r}\right) = \sum_{n\in\mathbb{Z}^m} f(n) = \sum_{n\in\mathbb{Z}^m} \widehat{f}(n) = r^m \sum_{n\in\mathbb{Z}^m} \widehat{\chi}(rn)\widehat{\psi}(r\varepsilon n)\\ &= r^m\alpha_m + r^m \sum_{n\neq 0} \widehat{\chi}(rn)\widehat{\psi}(r\varepsilon n).\end{aligned}$$

Therefore,

$$\left|S_\varepsilon(r) - \alpha_m r^m\right| \leqslant r^m \sum_{n\neq 0} \left|\widehat{\chi}(rn)\right| \cdot \left|\widehat{\psi}(r\varepsilon n)\right|.$$

Using the estimates $\widehat{\psi}(y) = O((1+\|y\|)^{-m})$ and $\widehat{\chi}(y) = O(\|y\|^{-\frac{m+1}{2}})$ (see Example 2 of Sect. 10.5.2), we obtain

$$\left|S_\varepsilon(r) - \alpha_m r^m\right| \leqslant \sum_{n\neq 0} \frac{Cr^m}{\|rn\|^{\frac{m+1}{2}}(1+\|r\varepsilon n\|)^m}$$

$$= C\varepsilon^{-\frac{m-1}{2}} \sum_{n\neq 0} \frac{(r\varepsilon)^m}{\|r\varepsilon n\|^{\frac{m+1}{2}}(1+\|r\varepsilon n\|)^m}.$$

Since the estimate $\|x\| \leqslant \|r\varepsilon n\| + \frac{\sqrt{m}}{2} r\varepsilon \leqslant (1+\frac{\sqrt{m}}{2})\|r\varepsilon n\|$ is valid for each point x in the cube of edge length $r\varepsilon$ and center at $r\varepsilon n$, $n \neq 0$, we see that the last sum is dominated (with a coefficient depending only on the dimension) by the integral

$$\int_{\mathbb{R}^m} \frac{dx}{\|x\|^{\frac{m+1}{2}}(1+\|x\|)^m} = \alpha_m \int_0^\infty \frac{t^{\frac{m-3}{2}}\,dt}{(1+t)^m} < +\infty.$$

Therefore,

$$\left|S_\varepsilon(r) - \alpha_m r^m\right| = O\big(\varepsilon^{-\frac{m-1}{2}}\big).$$

For $r = \frac{R}{1\pm\varepsilon}$ and $0 < \varepsilon < \frac{1}{2}$, we have $r^m = R^m(1+O(\varepsilon))$, so $|S_\varepsilon(\frac{R}{1\pm\varepsilon}) - \alpha_m R^m| = O(\varepsilon R^m + \varepsilon^{-\frac{m-1}{2}})$. Taking into account (9), we see that

$$\left|N_m(R) - \alpha_m R^m\right| = O\big(\varepsilon R^m + \varepsilon^{-\frac{m-1}{2}}\big).$$

The sum $\varepsilon R^m + \varepsilon^{-\frac{m-1}{2}}$ has a minimal order of growth if $\varepsilon R^m = \varepsilon^{-\frac{m-1}{2}}$, i.e., if $\varepsilon = R^{-\frac{2m}{m+1}}$. Choosing this value of ε, we arrive at the relation

$$N_m(R) = \alpha_m R^m + O\big(R^\theta\big) \quad \text{as } R \to +\infty \tag{10}$$

with $\theta = m\frac{m-1}{m+1} < m-1$.

10.6.6 What can be said about the exactness of formula (10)? Since $\theta = m - 2 + \frac{2}{m+1}$, we see that, for large m, its error is close to the minimum possible value $O(R^{m-2})$. As we know, for $m > 4$, the best estimate is achieved, namely, relation (10) is valid with $\theta = m-2$ (for $m = 4$, this relation is valid for every $\theta > 2$). For $m = 3$ the minimum value of the exponent θ is still unknown. As we have verified in the previous section, it does not exceed $\frac{3}{2}$ and is not less than 1. There is a conjecture stating that the exponent θ can be taken arbitrarily close to 1, but it has only been proved that $\theta \leqslant 29/22$. For the history of the problem, see the paper [CI] or the book [LK].

We consider the two-dimensional case in more detail. We have proved that, for $m = 2$, formula (10) is valid with $\theta = 2/3$. In the study of the Gauss problem, this is the first non-trivial result, which was obtained by Sierpiński in 1906 and has been sharpened several times since then. More sophisticated methods made it possible to decrease the exponent θ to $\frac{131}{208}$, but it is still unknown whether the value of θ can be taken arbitrarily close to $\frac{1}{2}$. This bound cannot be lowered. As Hardy and Landau[21] proved independently in 1915, the relation $N_2(R) = \pi R^2 + O(R^\theta)$ holds only for $\theta \geqslant \frac{1}{2}$. We give the proof of this result, based on the paper [EF] (see also Exercise 9).

Let $N(R) = N_2(R) = \operatorname{card}\{n \in \mathbb{Z}^2 \mid \|n\| \leqslant R\}$ and $\Delta(R) = N(R) - \pi R^2$. We will need the function $f(z) = \sum_{k=-\infty}^{\infty} z^{k^2}$ $(|z| < 1)$ tightly connected with the quantities $N(R)$ and $\Delta(R)$. Indeed,

$$f^2(z) = \sum_{k=-\infty}^{\infty} z^{k^2} \sum_{j=-\infty}^{\infty} z^{j^2} = \sum_{(k,j)\in\mathbb{Z}^2} z^{k^2+j^2} = \sum_{k=0}^{\infty} \nu(m) z^m,$$

where $\nu(m)$ is equal to the number of points (k, j) lying on the circle of radius $\sqrt{m}$. Since $\nu(0) = 1$ and $\nu(m) = N(\sqrt{m}) - N(\sqrt{m-1})$ for $m \geqslant 1$, we obtain

$$f^2(z) = \sum_{m=0}^{\infty} N(\sqrt{m}) z^m - \sum_{m=1}^{\infty} N(\sqrt{m-1}) z^m = (1-z) \sum_{m=0}^{\infty} N(\sqrt{m}) z^m.$$

Since $N(\sqrt{m}) = \pi m + \Delta(\sqrt{m})$, we see that

$$f^2(z) = \frac{\pi z}{1-z} + (1-z) \sum_{m=0}^{\infty} \Delta(\sqrt{m}) z^m. \tag{11}$$

The required inequality $\theta \geqslant \frac{1}{2}$ can be obtained by comparing estimates from above and from below for the integrals

$$I(r) = \int_{-\pi}^{\pi} \left| f^2(re^{it}) \right| dt \quad \left(\frac{1}{2} < r < 1\right).$$

Let us estimate the integral from below. Since $f(re^{it}) = 1 + 2\sum_{k=1}^{\infty} r^{k^2} e^{ik^2 t}$, Parseval's identity implies

$$I(r) = 2\pi \left(1 + 4\sum_{k=1}^{\infty} r^{2k^2}\right) \geqslant 2\pi \sum_{k=0}^{\infty} r^{2k^2} \geqslant 2\pi \sum_{k=0}^{\infty} \int_k^{k+1} r^{2t^2}\, dt$$
$$= 2\pi \int_0^{\infty} r^{2t^2}\, dt = \frac{\pi^{\frac{3}{2}}}{\sqrt{2 \ln \frac{1}{r}}}.$$

Therefore,

$$I(r) \geqslant \frac{C_1}{\sqrt{1-r}}$$

(here and below, $C_1, C_2, \ldots$ are positive coefficients independent of r).

---

[21] Edmund Georg Hermann Landau (1877–1938)—German mathematician.

Now, we obtain an estimate from above. Applying the result established in Example 3 of Sect. 10.2.1 to the function $\varphi(t) = f(re^{it})$, we obtain that the inequality

$$I(r) \leqslant \frac{3\pi}{\alpha} \int_{-\alpha}^{\alpha} \left|f^2\left(re^{it}\right)\right| dt$$

is valid for $\alpha \in (0, \pi)$ (the choice of this parameter will be specified later). Denoting the sum on the right-hand side of Eq. (11) by $S(z)$, we obtain

$$I(r) \leqslant \frac{3\pi}{\alpha}\left(\int_{-\alpha}^{\alpha} \frac{\pi\, dt}{|1 - re^{it}|} + \int_{-\alpha}^{\alpha} |1 - re^{it}|\left|S\left(re^{it}\right)\right| dt\right) = \frac{3\pi}{\alpha}(I_1 + I_2).$$

By the inequality

$$\left|1 - re^{it}\right| = \sqrt{(1-r)^2 + 4r\sin^2\frac{t}{2}} \geqslant \frac{(1-r) + |\sin\frac{t}{2}|}{2} \geqslant \frac{(1-r)+|t|}{2\pi},$$

it is easy to verify that

$$I_1 \leqslant C_2\left|\ln(1-r)\right|.$$

Since $|1 - re^{it}| \leqslant (1-r) + r|t| \leqslant (1-r) + \alpha$, we obtain the following inequality for $\alpha \geqslant 1 - r$:

$$I_2 \leqslant 2\alpha \int_{-\alpha}^{\alpha} \left|S\left(re^{it}\right)\right| dt \leqslant 2\alpha \sqrt{2\alpha \int_{-\pi}^{\pi} \left|S\left(re^{it}\right)\right|^2 dt}$$

$$= (2\alpha)^{\frac{3}{2}} \sqrt{2\pi \sum_{m=0}^{\infty} \Delta^2(\sqrt{m})r^{2m}}.$$

Therefore, if $\Delta(R) = O(R^\theta)$ as $R \to +\infty$ for some $\theta > 0$, then

$$I_2 \leqslant C_3\alpha^{\frac{3}{2}} \sqrt{1 + \sum_{m=1}^{\infty} m^\theta r^{2m}} \leqslant C_3\alpha^{\frac{3}{2}} \sqrt{1 + \sum_{m=1}^{\infty} \int_m^{m+1} t^\theta r^{2(t-1)} dt}$$

$$\leqslant 2C_3\alpha^{\frac{3}{2}} \sqrt{1 + \int_0^{\infty} t^\theta r^{2t} dt} = 2C_3\alpha^{\frac{3}{2}} \sqrt{1 + \frac{\Gamma(1+\theta)}{\ln^{1+\theta}\frac{1}{r^2}}} \leqslant C_4 \frac{\alpha^{\frac{3}{2}}}{(1-r)^{\frac{1+\theta}{2}}}.$$

Thus, for $\alpha > 1 - r > 0$, we obtain the double inequality

$$\frac{C_1}{\sqrt{1-r}} \leqslant I(r) \leqslant \frac{C_5}{\alpha}\left(\left|\ln(1-r)\right| + \frac{\alpha^{\frac{3}{2}}}{(1-r)^{\frac{1+\theta}{2}}}\right),$$

and, consequently,

$$0 < C_6 \leqslant \frac{1}{\alpha}\sqrt{1-r}\left|\ln(1-r)\right| + \frac{\sqrt{\alpha}}{(1-r)^{\frac{\theta}{2}}}.$$

Now, using the freedom in the choice of the parameter α, we decrease the right-hand side, which is minimal (in order) if the summands are equal, i.e., if $\alpha = (1-r)^{\frac{1+\theta}{3}}|\ln(1-r)|^{\frac{2}{3}}$. Taking this value of α, we obtain that, for all $r \in (\frac{1}{2}, 1)$, the inequality

$$0 < C_6 \leqslant \frac{2}{\alpha}\sqrt{1-r}\,\bigl|\ln(1-r)\bigr| = 2(1-r)^{\frac{1-2\theta}{6}}\bigl|\ln(1-r)\bigr|^{\frac{1}{3}}$$

is valid, which is possible only if $\theta \geqslant \frac{1}{2}$.

10.6.7 It is clear that the number of points of the integer lattice $\mathbb{Z}^m$ lying in a shifted ball $\overline{B}(t, R)$ of a large radius R is asymptotically equal to the volume of the ball. As we have already verified, it is not easy to obtain a good estimate for the difference

$$\Delta_R(t) = \operatorname{card}\bigl\{n \in \mathbb{Z}^m \mid \|n-t\| \leqslant R\bigr\} - \alpha_m R^m$$

as $R \to +\infty$ at a fixed point t. However, it is considerably easier to estimate the mean value (with respect to t) of the error $\Delta_R(t)$. As established in [Ke],

$$\int_{[0,1]^m} \bigl|\Delta_R(t)\bigr|^2\,dt \leqslant C_m R^{m-1}. \tag{12}$$

To verify this, we find the Fourier coefficients of the function

$$f(t) = \Delta_R(t) + \alpha_m R^m = \operatorname{card}\bigl\{n \in \mathbb{Z}^m \mid \|n-t\| \leqslant R\bigr\}$$

(obviously, this function has period 1 with respect to each variable). Let χ be the characteristic function of the closed unit ball centered at zero. Then $f(t) = \sum_{n\in\mathbb{Z}^m} \chi(\frac{n-t}{R})$. For $Q = [0, 1)^m$ and each $k \in \mathbb{Z}^m$, we obtain

$$\begin{aligned}\widehat{f}(k) &= \int_Q f(t)e^{-2\pi i\langle k,t\rangle}\,dt = \sum_{n\in\mathbb{Z}^m}\int_Q \chi\Bigl(\frac{n-t}{R}\Bigr)e^{-2\pi i\langle k,t\rangle}\,dt \\ &= \sum_{n\in\mathbb{Z}^m}\int_{n+Q} \chi\Bigl(\frac{t}{R}\Bigr)e^{-2\pi i\langle k,t\rangle}\,dt = \int_{\mathbb{R}^m} \chi\Bigl(\frac{t}{R}\Bigr)e^{-2\pi i\langle k,t\rangle}\,dt = R^m\widehat{\chi}(Rk).\end{aligned}$$

Therefore, $\widehat{\Delta}_R(k) = \widehat{f}(k) = R^m\widehat{\chi}_B(Rk)$ for $k \neq 0$ and $\widehat{\Delta}_R(0) = \widehat{f}(0) - \alpha_m R^m = (\widehat{\chi}_B(0) - \alpha_m)R^m = 0$. By Parseval's identity, we have

$$\int_{[0,1]^m} \bigl|\Delta_R(t)\bigr|^2\,dt = \sum_{k\in\mathbb{Z}^m} \bigl|\widehat{\Delta}_R(k)\bigr|^2 = R^{2m}\sum_{k\neq 0}\bigl|\widehat{\chi}_B(Rk)\bigr|^2.$$

Now, inequality (12) follows from the estimate $\widehat{\chi}_B(y) = O(\|y\|^{-\frac{m+1}{2}})$ that follows from the asymptotic formula for $\widehat{\chi}_B(y)$ (see Example 2 of Sect. 10.5.2).

It is interesting to note that, for $m \neq 1 \pmod 4$, the above-mentioned asymptotic formula for $\widehat{\chi}_B(y)$ enables us to obtain the inequality

$$\int_{[0,1]^m} \left|\Delta_R(t)\right|^2 dt \geqslant \widetilde{C}_m R^{m-1} > 0. \qquad (12')$$

In particular, for $m = 2$, we obtain that, in the problem in question, the typical error for the shifted discs $\overline{B}(t, R)$ has order of growth $\sqrt{R}$.

However, if $m = 4l + 1$, then the superior limit of the quotient $\frac{1}{R^{m-1}} \times \int_{[0,1]^m} |\Delta_R(t)|^2 dt$ as $R \to +\infty$ is positive and the inferior limit is zero.

EXERCISES

1. Supplement the statement of Lemma 10.6.1 by proving that the series $\sum_{n\in\mathbb{Z}} f(x+n)$ converges to $S(x)$ not only almost everywhere, but also in mean.
2. Let an absolutely continuous function f and its derivative be summable on $\mathbb{R}$. Prove that Eq. (3) is valid for all $x \in \mathbb{R}$.
3. Verify that, in Theorem 2 of Sect. 10.6.2, the Lipschitz condition can be weakened to the assumption that the inequality $|f(x+h) - f(x)| \leqslant R^q \|h\|^\alpha$, where $\|h\| \leqslant 1$ and $\|x\| \leqslant R$ (here, q is a fixed non-negative number), holds for every $R > 1$.
4. Using the result of the previous exercise, show that, in Corollary 10.6.2, the assumption that the derivatives are bounded can be replaced by the requirement that they are dominated by a polynomial.
5. Let $f \in \mathscr{L}^1(\mathbb{R}^m) \cap C(\mathbb{R}^m)$ and $\widehat{f} \geqslant 0$ everywhere. Prove that Eq. (5) is valid at every point $x \in \mathbb{R}^m$ (the series on the right-hand side of (5) converges absolutely).
6. Prove that the condition $\widehat{M} \in \mathscr{L}^1(\mathbb{R}^m)$ is not only sufficient but also necessary for the boundedness of the $\mathscr{L}^1$-norms of the sums ω_ε (see formula (7)).
7. Prove that the function ω_ε in the example of Sect. 10.6.4 admits the estimate $\omega_\varepsilon(x) = O(\varepsilon)(1 + (\varepsilon^2 + 4\pi^2\|x\|^2)^{-\frac{m+1}{2}})$ (the constant in the O-term depends only on the dimension) and, therefore, ω_ε is dominated by a summable "hump-shaped" majorant.
8. Prove that, as $\varepsilon \to 0$, the functions $\omega_\varepsilon(x) = \sum_{n\in\mathbb{Z}^m} e^{-\varepsilon\|n\|^2} e^{2\pi i\langle n,x\rangle}$ form an approximate identity having the strong localization property and a "hump-shaped" majorant.
9. Verify that the reasoning of Sect. 10.6.6 can be used to obtain the following stronger result (see [EF]): the fraction $\Delta(R)/\sqrt{\frac{R}{\ln R}}$ does not tend to zero as $R \to +\infty$.

Chapter 11
Charges. The Radon–Nikodym Theorem

11.1 Charges; Integration with Respect to a Charge

In what follows, we consider an arbitrary set X and a fixed σ-algebra $\mathfrak{A}$ of its subsets. We assume that all sets in question are measurable, i.e., belong to the σ-algebra $\mathfrak{A}$. We recall that the union of pairwise disjoint sets E_α is denoted by $\bigvee_{\alpha \in A} E_\alpha$.

11.1.1 We define the main subject of the following paragraph.

Definition A function $\varphi : \mathfrak{A} \longrightarrow \mathbb{C}$ is called a (*complex*) *charge* if it is countably additive, i.e., if, for every sequence of pairwise disjoint (measurable) sets A_k, the series $\sum_{k=1}^{\infty} \varphi(A_k)$ converges and the equation

$$\varphi\left(\bigvee_{k=1}^{\infty} A_k\right) = \sum_{k=1}^{\infty} \varphi(A_k)$$

holds. A charge whose values belong to the set $\mathbb{R}$ is called *real*.

An example of a charge is, obviously, the difference of finite measures. Below, we will see that the converse is also true (see Corollary 11.1.5).

Just as a measure can be imagined as a mass distributed on a set, a real countably additive function (more precisely, its value on a given set) can naturally be interpreted as the total electric charge of positively and negatively charged particles fixed in this set. The term "charge" corresponds to this interpretation.

We note some elementary properties of charges. The symbol φ denotes an arbitrary charge.

(1) $\varphi(\varnothing) = 0$.
   This follows from countable additivity with all A_k equal to the empty set.

(2) A charge is an additive set function, $\varphi(A \vee B) = \varphi(A) + \varphi(B)$.
   To verify this, it is sufficient to put $A_1 = A$, $A_2 = B$ and $A_k = \varnothing$ for $k > 2$ and use property (1) and the countable additivity of a charge.

B. Makarov, A. Podkorytov, *Real Analysis: Measures, Integrals and Applications*, Universitext, DOI 10.1007/978-1-4471-5122-7_11, © Springer-Verlag London 2013

Hence it follows that a charge is a finite additive function,

$$\varphi\left(\bigvee_{k=1}^{n} A_k\right) = \sum_{k=1}^{n} \varphi(A_k).$$

(3) If $B \subset A$, then $\varphi(A \setminus B) = \varphi(A) - \varphi(B)$.

Indeed, $A = B \vee (A \setminus B)$. Therefore, by additivity, we obtain $\varphi(A) = \varphi(B) + \varphi(A \setminus B)$.

11.1.2 Like finite measures, charges are continuous from above and from below.

Theorem *Let φ be an arbitrary charge. Then $\varphi(A) = \lim_{n\to\infty} \varphi(A_n)$ if $A_1 \subset A_2 \subset \dots$, $A = \bigcup_{n=1}^{\infty} A_n$ (continuity from below) or if $A_1 \supset A_2 \supset \dots$, $A = \bigcap_{n=1}^{\infty} A_n$ (continuity from above).*

Proof The continuity from below follows easily from the relation $A = \bigvee_{n=1}^{\infty}(A_n \setminus A_{n-1})$ (here $A_0 = \varnothing$), by which we obtain

$$\varphi(A) = \sum_{n=1}^{\infty} \varphi(A_n \setminus A_{n-1}) = \sum_{n=1}^{\infty} \big(\varphi(A_n) - \varphi(A_{n-1})\big) = \lim_{n\to\infty} \varphi(A_n).$$

Similarly, using the relation $A_1 = A \vee \bigvee_{n=2}^{\infty}(A_{n-1} \setminus A_n)$, we can prove the continuity from above. We leave it to the reader to fill in the details. □

11.1.3 It turns out that the following analog of the Weierstrass extreme value theorem is valid for real charges: every charge attains its maximum and minimum values. Before turning to the proof of this important theorem, we establish an auxiliary fact.

Definition A set A is called a *set of positivity* of a charge φ if $\varphi(E) \geqslant 0$ for every set E lying in A.

Lemma

(1) *A countable union of sets of positivity is a set of positivity.*
(2) *Every set A contains a set of positivity B such that $\varphi(B) \geqslant \varphi(A)$.*

Proof The first statement of the lemma follows directly from the countable additivity of φ.

Let us prove the second statement. If $\varphi(A) \leqslant 0$, then we can put $B = \varnothing$. We will assume that $\varphi(A) > 0$.

First, we verify that the second statement "is fulfilled up to ε". Let $\varepsilon > 0$. We say that a set A is a set of ε-positivity if $\varphi(E) > -\varepsilon$ for every set E lying in A.

We prove that, for every $\varepsilon > 0$, the set A contains a set C of ε-positivity such that $\varphi(C) \geqslant \varphi(A)$. Indeed, if the set A itself is not a set of ε-positivity, then there is a subset e_1 of A such that $\varphi(e_1) \leqslant -\varepsilon$. We put $A_1 = A \setminus e_1$. It is clear that

$\varphi(A_1) > \varphi(A)$. Now, we can repeat our reasoning with A replaced by A_1, etc. This process must terminate since otherwise we would obtain an infinite sequence of pairwise disjoint sets $\{e_n\}_{n\geqslant 1}$ such that $\varphi(e_n) \leqslant -\varepsilon$. However, this is impossible since $\varphi(\bigvee_{n=1}^{\infty} e_n) = \sum_{n=1}^{\infty} \varphi(e_n)$ by countable additivity, and the series on the right-hand side diverges. If the construction of e_n cannot be continued after the Nth step, then, obviously, the difference $A_N = A_{N-1} \setminus e_N$ is the required set of ε-positivity.

Now, step by step, we choose sets C_n of $1/n$-positivity such that

$$C_1 \subset A, \quad \varphi(C_1) \geqslant \varphi(A); \quad \dots \quad C_{n+1} \subset C_n, \quad \varphi(C_{n+1}) \geqslant \varphi(C_n) \quad \text{for } n \in \mathbb{N}.$$

Indeed, first we find a set C_1 of 1-positivity in A such that $\varphi(C_1) \geqslant \varphi(A)$. Then we find a set C_2 of $1/2$-positivity in C_1 such that $\varphi(C_2) \geqslant \varphi(C_1)$, etc. Since a part of a set of ε-positivity is again a set of ε-positivity, the set $B = \bigcap_{n=1}^{\infty} C_n$ is a set of ε-positivity for every $\varepsilon > 0$, i.e., a set of positivity and $\varphi(B) = \lim_{n\to\infty} \varphi(C_n) \geqslant \varphi(A)$. □

Theorem *Every real charge φ attains its maximum and minimum values, i.e., there are sets C and C' such that*

$$\varphi(C) = \sup\{\varphi(E) \mid E \in \mathfrak{A}\} \quad \textit{and} \quad \varphi(C') = \inf\{\varphi(E) \mid E \in \mathfrak{A}\}.$$

Proof We prove only that φ attains its maximum value (applying this to the charge $-\varphi$, we obtain the second statement). Let $H = \sup\{\varphi(A) \mid A \in \mathfrak{A}\}$. Obviously, $0 \leqslant H \leqslant +\infty$ (we do not exclude the case $H = +\infty$). From the lemma, it follows that

$$H = \sup\{\varphi(B) \mid B \text{ is a set of positivity}\}.$$

Now, we consider sets of positivity B_n such that $\varphi(B_n) \underset{n\to\infty}{\longrightarrow} H$. Their union C is a set with the required property since

$$H \geqslant \varphi(C) \geqslant \varphi(B_n) \quad \text{for every } n.$$ □

Corollary *Every charge is bounded, i.e., for every charge φ, there is a number L such that $|\varphi(A)| \leqslant L$ for each $A \in \mathfrak{A}$.*

Proof It is sufficient to consider real charges. In this case, the boundedness follows directly from the theorem since, by definition, a charge assumes only finite values. □

11.1.4 Now, we introduce an important characteristic of a charge (real or complex).

Definition The *variation of a charge φ on a set A* is the quantity

$$|\varphi|(A) = \sup\left\{ \sum_{k=1}^{n} |\varphi(E_k)| \;\middle|\; \bigvee_{k=1}^{n} E_k \subset A, \quad n \in \mathbb{N} \right\}.$$

The traditional notation $|\varphi|$ requires some caution: $|\varphi|(A)$ should not be confused with $|\varphi(A)|$.

We list several elementary properties of variation, the first four of which follow directly from the definition.

(1) $|\varphi(A)| \leqslant |\varphi|(A)$.
(2) If $B \subset A$, then $|\varphi|(B) \leqslant |\varphi|(A)$ (the monotonicity of variation).
(3) The variation of a finite measure coincides with the measure itself.
(4) If $\varphi = a\varphi_1 + b\varphi_2$, then $|\varphi|(A) \leqslant |a||\varphi_1|(A) + |b||\varphi_2|(A)$.
(4′) $||\varphi_1|(A) - |\varphi_2|(A)| \leqslant |\varphi_1 - \varphi_2|(A)$.
(5) If φ is a real charge, then

$$|\varphi|(A) = \sup\{\varphi(B) - \varphi(C) \mid B \vee C \subset A\} = \sup\{\varphi(B) - \varphi(C) \mid B, C \subset A\}. \qquad (1)$$

To prove the first equality in (1), we put

$$S = \sup\{\varphi(B) - \varphi(C) \mid B \vee C \subset A\}.$$

It is clear that $S \leqslant |\varphi|(A)$. On the other hand, if $E_1 \vee \cdots \vee E_n \subset A$, then we can divide these sets into two groups as follows: the sets E_k for which $\varphi(E_k) \geqslant 0$ are assigned to the first group and the sets for which $\varphi(E_k) < 0$ are assigned to the second group. Let B and C be the unions of the sets of the first and the second group, respectively. Then

$$\sum_{k=1}^{n} |\varphi(E_k)| = \sum_{\varphi(E_k)\geqslant 0} \varphi(E_k) - \sum_{\varphi(E_k)<0} \varphi(E_k) = \varphi(B) - \varphi(C) \leqslant S.$$

Since this is true for each family of pairwise disjoint subsets $E_1, \ldots, E_n$ of A, we obtain by the definition of variation that $|\varphi|(A) \leqslant S$, which, along with the opposite inequality mentioned above, gives the first equality in (1). To prove the second equality, we observe that if $B, C \subset A$ and $E = B \cap C$, then $(B \setminus E) \cap (C \setminus E) = \varnothing$ and

$$\varphi(B) - \varphi(C) = \varphi(B \setminus E) + \varphi(E) - \big(\varphi(C \setminus E) + \varphi(E)\big) = \varphi(B \setminus E) - \varphi(C \setminus E) \leqslant S.$$

Therefore, the right-hand side of Eq. (1) does not exceed S. The opposite inequality is obvious.

11.1.5 We establish the main property of the variation.

Theorem *The variation of an arbitrary charge φ is a finite measure.*

Proof We verify that the variation of φ is a measure. Let $A = \bigvee_{k=1}^{\infty} A_k$. We must prove that

$$|\varphi|(A) = \sum_{k=1}^{\infty} |\varphi|(A_k).$$

First, we verify that the inequality

$$|\varphi|(A) \leqslant \sum_{k=1}^{\infty} |\varphi|(A_k) \tag{2}$$

holds. Let $E_1 \vee \cdots \vee E_n \subset A$. Then

$$\varphi(E_j) = \sum_{k=1}^{\infty} \varphi(E_j \cap A_k)$$

for every $j = 1, \ldots, n$, and

$$\sum_{j=1}^{n} |\varphi(E_j)| = \sum_{j=1}^{n} \left| \sum_{k=1}^{\infty} \varphi(E_j \cap A_k) \right| \leqslant \sum_{k=1}^{\infty} \sum_{j=1}^{n} |\varphi(E_j \cap A_k)| \leqslant \sum_{k=1}^{\infty} |\varphi|(A_k).$$

Passing to the supremum on the left-hand side of the last inequality, we obtain (2).

Now, we turn to the proof of the opposite inequality. First, we prove that $|\varphi|(A \vee B) \geqslant |\varphi|(A) + |\varphi|(B)$ for all disjoint sets A and B. Indeed, if $\bigvee_{j=1}^{n} E_j \subset A$ and $\bigvee_{k=1}^{m} E'_k \subset B$, then $E_1 \vee \cdots \vee E_n \vee E'_1 \vee \cdots \vee E'_m \subset A \vee B$, and, therefore,

$$|\varphi|(A \vee B) \geqslant \sum_{j=1}^{n} |\varphi(E_j)| + \sum_{k=1}^{m} |\varphi(E'_k)|.$$

First, we pass to the supremum over all sets $E_1, \ldots, E_n$ and then over the sets $E'_1, \ldots, E'_m$. We obtain that $|\varphi|(A \vee B) \geqslant |\varphi|(A) + |\varphi|(B)$. This inequality can easily be generalized by induction as follows: $|\varphi|(A_1 \vee \cdots \vee A_N) \geqslant |\varphi|(A_1) + \cdots + |\varphi|(A_N)$. Taking into account the monotonicity of variation, we see that

$$|\varphi|\left(\bigvee_{k=1}^{\infty} A_k\right) \geqslant |\varphi|\left(\bigvee_{k=1}^{N} A_k\right) \geqslant \sum_{k=1}^{N} |\varphi|(A_k).$$

Since N is arbitrary, this implies the inequality opposite to (2), and, consequently, the countable additivity of the variation.

Since the variation is monotone, to prove that it is finite, it is sufficient to verify that $|\varphi|(X) < +\infty$. Since the real and imaginary parts of a complex charge are charges, we can use property (4) of variation and assume that the charge φ is real. By Corollary 11.1.3, φ is bounded. Therefore, for some $L > 0$ and every A, we have $|\varphi(A)| \leqslant L$. By property (5), we obtain

$$|\varphi|(X) = \sup\bigl\{\varphi(B) - \varphi(C) \mid B, C \subset X\bigr\} \leqslant 2L. \qquad \square$$

Corollary *A real charge is the difference of finite measures. A complex charge φ can be represented in the form $\varphi = \mu_1 - \mu_2 + i(\mu_3 - \mu_4)$, where $\mu_1, \ldots, \mu_4$ are finite measures.*

Proof If φ is a real charge, then the difference $\mu = |\varphi| - \varphi$ is also a charge. By the first property of variation, this charge is non-negative and, therefore, is a measure (obviously, finite). At the same time, it is clear that $\varphi = |\varphi| - \mu$.

For a complex charge, we can apply the representation of the real charges $\mathcal{R}e\,\varphi$ and $\mathcal{I}m\,\varphi$ obtained above. □

11.1.6 We are looking for a formula to calculate the variation of a charge with density.

Theorem *Let $(X, \mathfrak{A}, \mu)$ be an arbitrary measure space, let $f \in \mathscr{L}^1(X, \mu)$, and let φ be the charge defined by the equation*

$$\varphi(A) = \int_A f\,d\mu \quad (A \in \mathfrak{A}).$$

Then

$$|\varphi|(A) = \int_A |f|\,d\mu \quad (A \in \mathfrak{A}). \tag{3}$$

Remark Under the assumptions of the theorem, the function f is called the *density* of φ (*with respect to the measure* μ). We will also say that the charge φ is generated by the function f and write this symbolically as follows: $d\varphi = f\,d\mu$. We remark that the density is determined by the charge uniquely up to equivalence (see Theorem 4.5.4).

Proof Let $A_1 \vee \cdots \vee A_N \subset A$. Then

$$\sum_{k=1}^{N} |\varphi(A_k)| = \sum_{k=1}^{N} \left| \int_{A_k} f\,d\mu \right| \leqslant \sum_{k=1}^{N} \int_{A_k} |f|\,d\mu = \int_{A_1 \vee \cdots \vee A_N} |f|\,d\mu \leqslant \int_A |f|\,d\mu.$$

On the left-hand side of the last inequality, we pass to the supremum over all possible collections of sets A_k satisfying the conditions mentioned above and obtain

$$|\varphi|(A) \leqslant \int_A |f|\,d\mu. \tag{4}$$

Now, we prove that the opposite inequality is also valid. First, we consider the case where the function f is simple. Let $E_1, E_2, \ldots, E_N$ be pairwise disjoint sets and $f = \sum_{k=1}^{N} a_k \chi_{E_k}$, where a_k is a scalar and χ_{E_k}, as usual, is the characteristic function of E_k. Then

$$\begin{aligned} |\varphi|(A) &\geqslant \sum_{k=1}^{N} |\varphi|(A \cap E_k) \geqslant \sum_{k=1}^{N} \left|\varphi(A \cap E_k)\right| = \sum_{k=1}^{N} \left| \int_{A \cap E_k} f\,d\mu \right| \\ &= \sum_{k=1}^{N} |a_k|\,\mu(A \cap E_k) = \int_A |f|\,d\mu. \end{aligned}$$

Using (4), we obtain Eq. (3).

Now, we consider the case of an arbitrary summable function f and prove first that the inequality opposite to (4) is valid with an arbitrary small error.

We fix an arbitrary positive number ε and a simple function g such that its deviation in mean from f, i.e., the quantity $\int_X |f-g|\,d\mu$, is less than ε (see Lemma 4.9.2). Let ψ be a charge generated by g, i.e., $\psi(A)=\int_A g\,d\mu$ for $A\in\mathfrak{A}$. Then, using property (4′) of the variation, inequality (4), and the fact that Eq. (3) has already been proved for simple functions, we obtain

$$\begin{aligned}|\varphi|(A) &= \big|\psi-(\psi-\varphi)\big|(A) \geqslant |\psi|(A)-|\varphi-\psi|(A) \geqslant \int_A |g|\,d\mu - \int_A |f-g|\,d\mu\\ &\geqslant \int_A |f|\,d\mu - 2\int_A |f-g|\,d\mu \geqslant \int_A |f| - 2\int_X |f-g|\,d\mu\\ &\geqslant \int_A |f|\,d\mu - 2\varepsilon.\end{aligned}$$

Since ε is arbitrary, the last inequality implies that $|\varphi|(A)\geqslant \int_A |f|\,d\mu$. Taking into account inequality (4), we obtain Eq. (3). □

11.1.7 Now, we establish an important property of real charges, which shows once again that the above-mentioned interpretation of a real countably additive function as a "measure of the quantity of electricity" contained in the positively and negatively charged particles distributed on the set is quite natural: the set is divided into two parts such that one of the parts contains only positively charged particles and the other one contains only negatively charged particles.

Theorem *Let φ be a real charge defined on a σ-algebra $\mathfrak{A}$ of subsets of a set X. Then X can be divided into two subsets X_+ and X_- such that*

$$\varphi(A\cap X_+)\geqslant 0 \quad \textit{and} \quad \varphi(A\cap X_-)\leqslant 0 \quad \textit{for every set } A \textit{ in } \mathfrak{A}. \tag{5}$$

The representation $X=X_+\vee X_-$ with the property indicated in the theorem is called a *Hahn*[1] *decomposition* of the charge φ.

Proof Let X_+ be the set on which the charge φ attains its maximum value (see Theorem 11.1.3). This set cannot contain subsets E for which $\varphi(E)<0$ since otherwise, removing E from X_+, we obtain a set on which the value of the charge is larger than its maximum value.

Similarly, if $E\cap X_+=\varnothing$, then the charge cannot assume a positive value on E (otherwise, appending E to X_+, we obtain a set on which the charge assumes too large values).

To obtain the required decomposition, it remains to put $X_-=X\setminus X_+$. □

[1] Hans Hahn (1879–1934)—Austrian mathematician.

The Hahn decomposition $X = X_+ \vee X_-$ is not unique in general. Indeed, if, for example, $E \subset X_+$, $E \neq \varnothing$, $|\varphi|(E) = 0$, then, removing E from X_+ and appending it to X_-, we obtain a new Hahn decomposition.

Corollary *For a real charge φ, we put*

$$\varphi_+(A) = \sup\{\varphi(B) \mid B \subset A\} \quad \text{and} \quad \varphi_-(A) = \sup\{-\varphi(B) \mid B \subset A\} \quad (A \in \mathfrak{A}).$$

Then φ_+ and φ_- are finite measures and

$$\varphi = \varphi_+ - \varphi_-, \qquad |\varphi| = \varphi_+ + \varphi_-. \tag{6}$$

The measures φ_+ and φ_- are called, respectively, the *positive and negative variations* of the charge φ, and the representation (6) is called the *Jordan decomposition*. It is clear that $\varphi_- = (-\varphi)_+$ and $\varphi_+ = (-\varphi)_-$.

Proof Let $X = X_+ \cup X_-$ be a Hahn decomposition of φ. We verify that

$$\varphi_+(A) = \varphi(A \cap X_+), \qquad \varphi_-(A) = -\varphi(A \cap X_-), \tag{7}$$

which immediately implies relations (6). It is sufficient to verify only the first relation in (7). Moreover, since the inequality $\varphi(A \cap X_+) \leqslant \varphi_+(A)$ is obvious, it only remains for us to verify the opposite inequality, which is very easy: if $B \subset A$, then

$$\varphi(B) = \varphi(B \cap X_+) + \varphi(B \cap X_-) \leqslant \varphi(B \cap X_+) \leqslant \varphi(A \cap X_+),$$

and, therefore,

$$\varphi_+(A) = \sup\{\varphi(B) \mid B \subset A\} \leqslant \varphi(A \cap X_+). \qquad \square$$

11.1.8 Using the fact that every real charge is the difference of finite measures, we define the integral with respect to a charge.

Definition Let φ be a real charge defined on a σ-algebra of subsets of a set X, and let f be a bounded function measurable on X. The *integral $\int_X f\,d\varphi$ of the function f with respect to the charge φ* is the difference

$$\int_X f\,d\varphi = \int_X f\,d\mu_1 - \int_X f\,d\mu_2, \tag{8}$$

where μ_1 and μ_2 are arbitrary finite measures such that $\varphi = \mu_1 - \mu_2$.

If φ is a complex charge, then we put

$$\int_X f\,d\varphi = \int_X f\,d\varphi_1 + i \int_X f\,d\varphi_2, \tag{8'}$$

where $\varphi_1 = \mathcal{R}e\,\varphi$ and $\varphi_2 = \mathcal{I}m\,\varphi$.

Remark We point out that the above integral is well defined. Indeed, if $\varphi = \mu_1 - \mu_2 = \mu_1' - \mu_2'$, then $\mu_1 + \mu_2' = \mu_1' + \mu_2$. Therefore, by the additivity of the integral with respect to a measure (see Sect. 4.4.2, property (9)), we obtain

$$\int_X f\,d\mu_1 + \int_X f\,d\mu_2' = \int_X f\,d\mu_1' + \int_X f\,d\mu_2,$$

and, consequently (all these integrals are finite since the function f is bounded),

$$\int_X f\,d\mu_1 - \int_X f\,d\mu_2 = \int_X f\,d\mu_1' - \int_X f\,d\mu_2'.$$

In particular, using the Jordan decomposition, we see that

$$\int_X f\,d\varphi = \int_X f\,d\varphi_+ - \int_X f\,d\varphi_-.$$

The latter equation could be taken as the definition of the integral (in this case, it would not be necessary to prove that the integral is well defined) and could even be used to generalize the definition to the case where f is not bounded but only summable with respect to the variation φ. In this case, however, Eq. (8) would be valid for unbounded functions only if f were summable with respect to μ_1 and μ_2. To avoid this additional restriction, we consider only bounded functions in the definition of the integral (see Exercise 3).

The integral with respect to a charge is linear with respect to the function as well as with respect to the charge, i.e., if $a, b \in \mathbb{C}$, then, for all bounded measurable functions f and g, we have

$$\int_X (af + bg)\,d\varphi = a\int_X f\,d\varphi + b\int_X g\,d\varphi,$$

and, for all charges φ and ψ defined on a common σ-algebra, we have

$$\int_X f\,d(a\varphi + b\psi) = a\int_X f\,d\varphi + b\int_X f\,d\psi.$$

The proof, which is sufficient to conduct only for real charges, is left to the reader.

We also note the following useful property:

$$\overline{\int_X f\,d\varphi} = \int_X \overline{f}\,d\overline{\varphi}, \tag{9}$$

where, by $\overline{\varphi}$, we mean the charge $\mathcal{R}e\,\varphi - i\,\mathcal{I}m\,\varphi$. This relation can be verified by direct calculation using formula (8′).

Theorem *Let φ be a charge defined on a σ-algebra of subsets of a set X. Then*:

(1) *if a sequence of measurable functions* $\{f_n\}_{n\geqslant 1}$ *on* X *is uniformly bounded and pointwise converges to* f, *then*

$$\int_X f_n\,d\varphi \underset{n\to\infty}{\longrightarrow} \int_X f\,d\varphi;$$

(2) *if a charge* φ *has a density* ω *with respect to the measure* μ, *then* $\int_X f\,d\varphi = \int_X f\omega\,d\mu$ *for every bounded measurable function* f;
(3) *if a measurable function* f *is bounded on* X, *then* $|\int_X f\,d\varphi| \leqslant \sup_X |f|\cdot|\varphi|(X)$.

Proof Assertion (1) follows from Lebesgue's dominant convergence theorem (Theorem 4.8.4).

Assertions (2) and (3) are obvious for characteristic functions, and, consequently, for simple functions. The general case is exhausted by the passage to the limit based on assertion (1), since the function f can be approximated pointwise (and even uniformly) by a bounded sequence of simple functions (see Corollary 3.2.2). □

11.1.9 Introducing the definition of the integral with respect to a charge, we can give a natural generalization of the concepts of the Fourier coefficients and Fourier series, having confined ourselves for the time being to the charges defined on subsets of an interval.

Definition Let φ be a charge defined on the σ-algebra of Borel subsets of the interval $[-\pi,\pi]$. The numbers

$$\widehat{\varphi}(n) = \frac{1}{2\pi}\int_{[-\pi,\pi]} e^{-inx}\,d\varphi(x) \quad (n\in\mathbb{Z})$$

are called the *Fourier coefficients* of the charge φ, and the series $\sum_{n=-\infty}^{\infty}\widehat{\varphi}(n)e^{inx}$ is called the *Fourier series* of φ.

We verify that, under a mild restriction, the charges, as well as the measures, are uniquely determined by their Fourier coefficients (see Theorem 10.3.7).

Theorem *If charges defined on the* σ*-algebra of Borel subsets of the interval* $[-\pi,\pi]$ *have zero load at the point* π, *then the charges coincide if their Fourier coefficients coincide.*

Proof It is sufficient to prove that a charge φ satisfying the condition $\varphi(\{\pi\}) = 0$ and having zero Fourier coefficients is equal to zero. First, we assume that φ is a real charge. In this case, we have the Jordan decomposition $\varphi = \varphi_+ - \varphi_-$. It is clear that the measures $\varphi_\pm$ do not have loads at the point π. Since $\widehat{\varphi}(n) = 0$ for all n, these measures have the same Fourier coefficients. Therefore, we can apply Theorem 10.3.7, by which $\varphi_+ = \varphi_-$, and, consequently, $\varphi = \varphi_+ - \varphi_- = 0$.

Now, we consider the general case $\varphi = \psi + i\theta$, where ψ and θ are real charges (they do not have loads at the point π). By Eq. (9), we have

$$\overline{\widehat{\varphi}(n)} = \frac{1}{2\pi}\int_{[-\pi,\pi]} e^{inx}\, d\overline{\varphi}(x) = \widehat{\overline{\varphi}}(-n).$$

Therefore, the charge $\overline{\varphi}$, as well as φ, has zero Fourier coefficients. In other words, for all n, we obtain

$$\widehat{\psi}(n) + i\widehat{\theta}(n) = \widehat{\varphi}(n) = 0 \quad \text{and} \quad \widehat{\psi}(n) - i\widehat{\theta}(n) = \widehat{\overline{\varphi}}(n) = 0.$$

Hence it follows that the real charges ψ and θ also have zero Fourier coefficients, and, consequently, are equal to zero. □

A multi-dimensional version of the theorem just proved is considered in Sect. 12.3.3.

EXERCISES

1. Prove that if the range of a charge is infinite, then it contains arbitrarily small non-zero values.
2. Verify that the range of a charge is a closed set.
3. Prove that a function measurable with respect to a σ-algebra $\mathfrak{A}$ is bounded if it is summable with respect to every finite measure defined on $\mathfrak{A}$.
4. Supplement assertion (3) of Theorem 11.1.8 by proving that $|\int_X f\,d\varphi| \leqslant \int_X |f|\,d|\varphi|$.

11.2 The Radon–Nikodym Theorem

11.2.1 We have already seen in Sect. 4.5.3 that the calculation of an integral with respect to a measure ν having a density p with respect to a measure μ can be reduced to the calculation of an integral with respect to μ (we recall that, in this case, we write $d\nu = p\,d\mu$). Such a reduction is often useful, and, therefore, it is desirable to have a criterion for determining whether a given measure has a density with respect to μ. It turns out that, in a wide range of cases, an obvious necessary condition is also sufficient.

We state the necessary definition.

Definition Let μ and ν be measures defined on the same σ-algebra. We say that the measure ν is *absolutely continuous* with respect to the measure μ (or is *subordinate* to μ) if $\nu(E) = 0$ whenever $\mu(E) = 0$.

The absolute continuity of ν with respect to μ is denoted by $\nu \prec \mu$.

Every measure having a density with respect to a measure μ is, obviously, subordinate to μ. It turns out that, for σ-finite measures, the converse is also true. We prove this, assuming first that the subordinate measure is finite.

Theorem (Radon[2]–Nikodym[3]) *Let μ and ν be measures defined on a σ-algebra $\mathfrak{A}$ of subsets of a set X, and let μ be σ-finite and ν be finite. If ν is absolutely continuous with respect to μ, then it has a summable density with respect to μ. This density is unique up to equivalence.*

Proof First, we find the maximum possible density generating a measure not exceeding ν and then prove that the measure with this density coincides with ν. Let P be the set of all such densities,

$$P = \left\{ p \in \mathscr{L}^0(X,\mu) \,\middle|\, p \geqslant 0, \ \int_E p\,d\mu \leqslant \nu(E) \text{ for every } E \text{ in } \mathfrak{A} \right\}.$$

It is essential that P contains the maximum of an arbitrary pair of functions belonging to P. Indeed, let $f, g \in P$ and $h = \max\{f, g\}$. We consider the partition $X = X' \vee X''$, where $X' = X(f \geqslant g)$ and $X'' = X(f < g)$. Then, for every set E in $\mathfrak{A}$, we obtain

$$\int_E h\,d\mu = \int_{E\cap X'} f\,d\mu + \int_{E\cap X''} g\,d\mu \leqslant \nu(E\cap X') + \nu(E\cap X'') = \nu(E),$$

which implies that $h \in P$.

It is easy to prove by induction that P contains the maximum of every finite family of functions belonging to P.

Let $I = \sup\{\int_X p\,d\mu \mid p \in P\}$, and let f_n be a sequence of functions in P such that $\int_X f_n\,d\mu \to I$. It is clear that $I \leqslant \nu(X) < +\infty$. Without loss of generality, we may assume that the sequence $\{f_n\}_{n\geqslant 1}$ increases (otherwise, we replace f_n by $\max\{f_1, \dots, f_n\}$). We put $f = \lim_{n\to\infty} f_n$. By Levi's theorem, we have $\int_E f\,d\mu = \lim_{n\to\infty} \int_E f_n\,d\mu \leqslant \nu(E)$ for every E in $\mathfrak{A}$. Therefore, $f \in P$ and $\int_X f\,d\mu = \lim_{n\to\infty} \int_X f_n\,d\mu = I$.

Now, we prove that f is the density of ν. Assume the contrary. Then there is a set E_0 such that

$$\nu(E_0) > \int_{E_0} f\,d\mu. \tag{1}$$

It is clear that $\mu(E_0) > 0$ since otherwise both sides of the inequality are equal to zero. Since the measure μ is σ-finite, we may assume, without loss of generality, that $\mu(E_0) < +\infty$. Indeed, otherwise the set E_0 can be represented as the union $E_0 = \bigcup_{n=1}^{\infty} E_n$, where $E_n \subset E_{n+1}$ and $\mu(E_n) < +\infty$ for every $n \in \mathbb{N}$. By the continuity from below, we obtain

$$\nu(E_n) - \int_{E_n} f\,d\mu \underset{n\to\infty}{\longrightarrow} \nu(E_0) - \int_{E_0} f\,d\mu > 0,$$

---

[2]Johann Radon (1887–1956)—Austrian mathematician.

[3]Otton Marcin Nikodým (1887–1974)—Polish mathematician.

and, therefore, inequality (1) is preserved if we replace E_0 by E_n for a sufficiently large n.

Thus, we will assume that $\mu(E_0) < +\infty$. We choose a positive number a so small that $\nu(E_0) - \int_{E_0} f\,d\mu > a\mu(E_0)$ and consider the charge $\varphi(E) = \nu(E) - \int_E f\,d\mu - a\mu(E \cap E_0)$. Since $\varphi(E_0) > 0$, Lemma 11.1.3 implies that there is a set of positivity B of the charge φ in E_0 such that $\varphi(B) \geqslant \varphi(E_0) > 0$. We remark that also $\nu(B) \geqslant \varphi(B) > 0$.

Now, we verify that $f + a\chi_B \in P$. Indeed, for every E in $\mathfrak{A}$, we have

$$\begin{aligned}\int_E (f + a\chi_B)\,d\mu &= \int_{E\setminus B} f\,d\mu + \int_{E\cap B} f\,d\mu + a\mu(E \cap B)\\ &= \int_{E\setminus B} f\,d\mu + \nu(E\cap B) - \varphi(E \cap B)\\ &\leqslant \nu(E\setminus B) + \nu(E\cap B) - \varphi(E\cap B)\\ &= \nu(E) - \varphi(E\cap B) \leqslant \nu(E).\end{aligned}$$

At the same time, we have $\mu(B) > 0$ since $\nu(B) > 0$ and $\nu \prec \mu$. Consequently,

$$\int_X (f + a\chi_B)\,d\mu = I + a\mu(B) > I,$$

which contradicts the definition of I. The uniqueness of the density (up to equivalence) is established in Theorem 4.5.4. □

Remark If the measure ν in the theorem is not finite but σ-finite, then the density exists but is not summable. We verify this by representing X as the union of expanding sets X_n $(n = 1, \dots)$ such that $\nu(X_n) < +\infty$. For every n, there exists a summable density f_n of the measure obtained by restricting ν to the induced σ-algebra $\mathfrak{A} \cap X_n$ (see Sect. 1.1.2). Since the density is unique, we have $f_n(x) = f_{n+1}(x)$ almost everywhere on X_n. Therefore, we may assume that f_{n+1} is an extension of f_n to X_{n+1}. Putting $f(x) = f_n(x)$ for $x \in X_n$, we can easily verify that the function obtained is the density of ν with respect to μ. This function is, obviously, summable on each set X_n.

If ν is the Borel measure on an open subset $\mathcal{O}$ of the space $\mathbb{R}^m$, ν is finite on the compact sets, $\mu = \lambda_m$ is Lebesgue measure, and $\nu \prec \lambda_m$, then the density of ν with respect to λ_m is locally summable in $\mathcal{O}$.

11.2.2 Now, we extend the Radon–Nikodym theorem to charges by extending to them the concept of absolute continuity.

Definition Let a measure μ and a charge φ be defined on the same σ-algebra. We will say that the charge φ is *absolutely continuous with respect to the measure* (or is *subordinate to the measure*) μ if $\varphi(e) = 0$ for every set of μ-measure zero.

As in the case of a measure, we will denote the absolute continuity of φ with respect to μ by $\varphi \prec \mu$.

An example of a charge subordinate to a measure is any charge generated by a density. It turns out that, as in the case where φ is a measure, there are no other charges subordinate to a "good" measure.

Theorem *Let φ be a charge and μ be a σ-finite measure defined on the same σ-algebra $\mathfrak{A}$ of subsets of a set X. The following conditions are equivalent:*

(1) *the charge φ is subordinate to the measure μ;*
(2) *the variation of the charge φ is subordinate to the measure μ;*
(3) *the charge φ is generated by a summable function (which is real in the case of real φ).*

Proof We show that (1) $\Rightarrow$ (2) $\Rightarrow$ (3) $\Rightarrow$ (1). The last implication is trivial. We remark also that, by property (1) of the variation (see Sect. 11.1.4) the implication (2) $\Rightarrow$ (1) is also obvious.

(1) $\Rightarrow$ (2). If $\mu(e) = 0$, then, for every set A lying in e, we have $\mu(A) = 0$, and so, $\varphi(A) = 0$. Therefore,

$$|\varphi|(e) \leqslant \sup\left\{\sum_{k=1}^{N} |\varphi(A_k)| \;\Big|\; A_k \subset e\right\} = 0.$$

(2) $\Rightarrow$ (3). If the charge φ is real, then it can be represented as the difference of finite measures $\varphi = \mu_1 - \mu_2$, where $\mu_1 = |\varphi|$ and $\mu_2 = |\varphi| - \varphi$ (see Corollary of Theorem 11.1.7). It is obvious that the measures μ_1 and μ_2 are subordinate to the measure μ. Therefore, by the Radon–Nikodym theorem for measures, there exist non-negative summable functions f_1 and f_2 such that

$$\mu_1(A) = \int_A f_1\, d\mu \quad \text{and} \quad \mu_2(A) = \int_A f_2\, d\mu \quad (A \in \mathfrak{A}).$$

It is clear that $\varphi(A) = \int_A f\, d\mu$ for $f = f_1 - f_2$ and every set A, where the function f is real.

If the charge is complex, then its real and imaginary parts are, obviously, subordinate to the measure μ. From what was just proved, they have densities, which implies that the given charge has a density. □

Corollary *The absolute value of the density of the charge φ with respect to its variation is equal to 1 almost everywhere.*

Proof It is clear that a charge is absolutely continuous with respect to its variation. Let ω be the corresponding density. Then, by Theorem 11.1.6, we obtain $|\varphi|(A) = \int_A |\omega|\, d|\varphi|$. At the same time, $|\varphi|(A) = \int_A 1\, d|\varphi|$. Since the density is unique (up to equivalence) by Theorem 4.5.4, we obtain $|\omega| = 1$ almost everywhere with respect to $|\varphi|$. □

Remark The theorem just proved allows us to introduce a new notion which is extremely important in probability theory. We consider a measure space $(X, \mathfrak{A}, \mu)$ and a function f measurable on X. We will assume that $\mu(X) = 1$. In probability theory, the measurable sets are called events, the measure of an event is called the probability of the event, and the function f is called a random variable. If f is summable, then the integral $\int_X f\,d\mu$ is called the expectation of the random variable. Fixing an event E ($\mu(E) > 0$) and considering the quantity $M(f, E) = \frac{1}{\mu(E)} \int_E f\,d\mu$, we obtain the "conditional expectation of the random variable f" (under the condition that the "event E occurs"). If $E_1, \ldots, E_N$ is a "complete system of pairwise incompatible events", i.e., a finite partition of X into measurable sets of positive measure, then the simple function $g = \sum_{k=1}^{N} M(f, E_k)\chi_{E_k}$ is called the conditional expectation of f with respect to the given partition and also with respect to the σ-algebra $\mathfrak{A}_0$ consisting of all possible unions of the elements of the partition. It is important to generalize this definition to the case where $\mathfrak{A}_0$ is an arbitrary σ-algebra contained in $\mathfrak{A}$. The main landmark in this generalization will be the fact that the function g satisfies the equation

$$\int_A g\,d\mu = \int_A f\,d\mu \quad \text{for } A \text{ in } \mathfrak{A}_0. \tag{2}$$

This property is the basis of the definition of the conditional expectation of a random variable f with respect to $\mathfrak{A}_0$. With a given summable function f, we connect a charge with density f and consider its restriction φ to the σ-algebra $\mathfrak{A}_0$. Thus,

$$\varphi(A) = \int_A f\,d\mu \quad \text{for } A \in \mathfrak{A}_0.$$

It is clear that $\varphi \prec \mu$ (more precisely, the charge φ is subordinate to the restriction of μ to $\mathfrak{A}_0$). Therefore, by the Radon–Nikodym theorem for charges, there exists a function g summable with respect to $\mathfrak{A}_0$ and such that $\varphi(A) = \int_A g\,d\mu$ for every $A \in \mathfrak{A}_0$. This means that g satisfies Eq. (2). The function g just obtained is called the expectation of f with respect to the σ-algebra $\mathfrak{A}_0$. We emphasize that, in the class of functions measurable with respect to $\mathfrak{A}_0$, the conditional expectation is defined uniquely up to equivalence.

11.2.3 In conclusion, we consider the question of extracting the maximal absolutely continuous part from a measure.

Definition Let μ and ν be measures defined on the same σ-algebra of subsets of a set X. The measures are called *mutually singular* if there is a partition $X = A \vee B$ such that $\mu(B) = \nu(A) = 0$. We will use the symbol $\mu \perp \nu$ to denote the fact that the measures μ and ν are mutually singular.

Theorem (Lebesgue's decomposition theorem) *Let μ and ν be measures defined on the same σ-algebra $\mathfrak{A}$. If the measure ν is σ-finite, then it can be represented as the sum of measures $\nu = \nu_a + \nu_s$, where $\nu_a \prec \mu$ and $\nu_s \perp \mu$. Such a representation is unique.*

The representation $\nu = \nu_a + \nu_s$ will be called the *Lebesgue decomposition.*

Proof First, we verify that the representation is unique. Let $\nu = \nu_a + \nu_s = \nu'_a + \nu'_s$, where the measures ν_a and ν'_a are absolutely continuous with respect to μ and each of the measures ν_s and ν'_s is mutually singular with μ. Then the set X can be represented in the form $X = A \vee B = A' \vee B'$ such that

$$\mu(B) = \nu_s(A) = 0 \quad \text{and} \quad \mu(B') = \nu'_s(A') = 0.$$

We put $B_0 = B \cup B'$. Then $\mu(B_0) = 0$, and, consequently, $\nu_a(B_0) = 0$. Moreover, for every set E in $\mathfrak{A}$, we have

$$\nu_s(E) = \nu_s(E \cap B) \leqslant \nu_s(E \cap B_0) \leqslant \nu_s(E).$$

Therefore,

$$\nu_s(E) = \nu_s(E \cap B_0) = \nu(E \cap B_0) - \nu_a(E \cap B_0) = \nu(E \cap B_0).$$

Thus, $\nu_s(E) = \nu(E \cap B_0)$ for every E in $\mathfrak{A}$. The relation $\nu'_s(E) = \nu(E \cap B_0)$ is proved similarly. Thus, we have proved that $\nu_s = \nu'_s$, and, consequently, $\nu_a = \nu'_a$.

Now, we turn to the proof of the existence of the required representation. First, we assume that the measure ν is finite.

Let $\mathcal{N}$ be a system of sets on which the measure μ vanishes, $\mathcal{N} = \{e \in \mathfrak{A} \mid \mu(e) = 0\}$. We verify that ν attains its maximum value on $\mathcal{N}$. Let $C = \sup\{\nu(e) \mid e \in \mathcal{N}\}$. Then there exist sets $e_n \in \mathcal{N}$ such that $\nu(e_n) \underset{n\to\infty}{\longrightarrow} C$. We put $B = \bigcup_{n\geqslant 1} e_n$. It is clear that $\mu(B) = 0$, i.e., $B \in \mathcal{N}$. Moreover, $C \geqslant \nu(B) \geqslant \nu(e_n) \underset{n\to\infty}{\longrightarrow} C$. Consequently, $\nu(B) = C$. Now, we put

$$\nu_a(E) = \nu(E \setminus B) \quad \text{and} \quad \nu_s(E) = \nu(E \cap B) \quad \text{for } E \in \mathfrak{A}.$$

Obviously, $\nu = \nu_a + \nu_s$ and $\nu_s \perp \mu$. We verify that the measure ν_a is subordinate to μ. Indeed, let $\mu(e_0) = 0$. Then $B \cup e_0 \in \mathcal{N}$ and

$$C \geqslant \nu(B \cup e_0) = \nu(B) + \nu(e_0 \setminus B) = C + \nu_a(e_0).$$

It follows that $\nu_a(e_0) = 0$, as required. This completes the proof in the case where the measure ν is finite.

If $\nu(X) = +\infty$, then we consider a partition of X into subsets X_n of finite measure ($n \in \mathbb{N}$). We put $\nu^{(n)}(E) = \nu(E \cap X_n)$. Then $\nu = \sum_{n=1}^{\infty} \nu^{(n)}$, $\nu^{(n)}(X) < +\infty$, and, from what was just proved, it follows that

$$\nu^{(n)} = \nu_a^{(n)} + \nu_s^{(n)}, \quad \text{where } \nu_a^{(n)} \prec \mu,\ \nu_s^{(n)} \perp \mu.$$

It is easy to verify that, putting $\nu_a = \sum_{n=1}^{\infty} \nu_a^{(n)}$ and $\nu_s = \sum_{n=1}^{\infty} \nu_s^{(n)}$, we obtain the required decomposition. □

EXERCISES

1. Let μ and ν be finite measures defined on the same σ-algebra. Prove that $\nu \prec \mu$ if and only if $\nu(e) \to 0$ as $\mu(e) \to 0$, i.e., if

$$\forall \varepsilon > 0\ \exists \delta > 0: \text{ if } \mu(e) < \delta, \text{ then } \nu(e) < \varepsilon.$$

2. Prove that the function f constructed in the proof of the Radon–Nikodym theorem is maximal in the set P in the following sense: if $p \in P$, then $p \leqslant f$ almost everywhere on X (with respect to the measure μ).

3. Preserving the notation of the remark of Sect. 11.2.2, prove Eq. (2) for the function $g = \sum_{k=1}^{N} M(f, E_k)\chi_{E_k}$.

4. Let a function f be summable on the square $[0, 1] \times [0, 1]$ with respect to Lebesgue measure. Find the conditional expectation of f with respect to the σ-algebra consisting of the sets of the form $E \times [0, 1]$, where $E \in \mathfrak{A}^1$ ("the striped algebra").

5. Let μ and ν be measures defined on the same σ-algebra. Prove that the measures are mutually singular if and only if there exist sets E_n such that $\mu(E_n) \underset{n\to\infty}{\longrightarrow} 0$ and $\nu(X \setminus E_n) \underset{n\to\infty}{\longrightarrow} 0$.

6. Supplement the result of Exercise 7 of Sect. 6.1 by proving that the measure μ_g is singular. *Hint.* Use the hint to Exercise 7 of Sect. 6.1. Consider the intervals Δ_k for which the number of ones in the ternary expansion of k is not less than $\sqrt{n}\ln n$.

7. Let $Q_m = [-\pi, \pi]^m$, let μ be a finite Borel measure on Q_m, and let $\mu_j^{\pm}$ be measures in Q_{m-1} obtained by the restriction of μ to the faces of the cube that lie in the planes $x_j = \pm\pi$ $(j = 1, \dots, m)$. Prove that if p is finite, then the trigonometric polynomials are everywhere dense in the space $\mathscr{L}^p(Q_m, \mu)$ if and only if the measures μ_j^+ and μ_j^- are mutually singular for each j (cf. Exercise 8 of Sect. 9.3).

11.3 *Differentiation of Measures

11.3.1 In this section, we generalize Lebesgue's theorem 4.9.2 on the differentiation of an integral to a wide class of Borel measures in $\mathbb{R}^m$ (see Definition 2.2.3). We will consider only the Borel measures that assume finite values on the bounded sets and will call such measures *Radon measures*. The reader can easily verify that this terminology agrees with the general definition of a Radon measure (see Sect. 12.2.2). As proved in Theorem 2.2.3, the Radon measures are regular.

In what follows, let λ be the m-dimensional Lebesgue measure, let $V(r)$ be the volume of the ball $B(x, r)$, and let $\mathfrak{B}^m$ be the σ-algebra of Borel subsets of the space $\mathbb{R}^m$.

Definition Let ν be a Radon measure on $\mathbb{R}^m$. If there exists a (finite or infinite) limit

$$\nu'(x) = \lim_{r\to 0} \frac{\nu(B(x,r))}{V(r)},$$

then it is called the *derivative of the measure* ν *at the point* $x \in \mathbb{R}^m$.

We recall that similar limits were considered in Sects. 4.9.2 and 6.2.1. Now, we prove that every Radon measure has a locally summable derivative almost everywhere.

As a preliminary, we establish a useful auxiliary statement in the proof of which Vitali's theorem 2.7.2 will be the main tool.

Lemma *Let* ν *be a Radon measure on* $\mathbb{R}^m$, $E \in \mathfrak{B}^m$. *Then:*

(1) *if* $\nu'(x) = 0$ *for all* x *in* E, *then* $\nu(E) = 0$;
(2) *if* $\nu(E) = 0$, *then* $\nu'(x) = 0$ *for almost all* (*with respect to Lebesgue measure*) x *in* E.

Proof We may and will assume that the set E is bounded. Let E be contained in a ball $B(0, R)$.

(1) Let us fix an arbitrary positive number ε. By the assumptions, for each $x \in E$, we have the inequality $\nu(B(x, 5r))/V(5r) < \varepsilon$ if r is sufficiently small (we will assume that $r < 1$). The family of balls $B(x, r)$ with such small radii and centers $x \in E$ form a Vitali cover of the set E. By Vitali's theorem, there exist pairwise disjoint balls $B(x_k, r_k)$ ($k \in \mathbb{N}$) such that $E \subset \bigcup_{k=1}^{\infty} B(x_k, 5r_k)$. Since $x_k \in E$ and $r_k < 1$ ($k \in \mathbb{N}$), all balls $B(x_k, r_k)$ are contained in $B(0, R + 1)$. Taking into account the inequalities $\nu(B(x_k, 5r_k)) < \varepsilon V(5r_k)$, we obtain

$$\nu(E) \leqslant \sum_{k=1}^{\infty} \nu\big(B(x_k, 5r_k)\big) < \varepsilon \sum_{k=1}^{\infty} V(5r_k) = \varepsilon 5^m \sum_{k=1}^{\infty} \lambda\big(B(x_k, r_k)\big)$$
$$\leqslant \varepsilon 5^m \lambda\big(B(0, R + 1)\big).$$

Since ε is arbitrary, we obtain that $\nu(E) = 0$.

(2) It is necessary to verify that the set $\{x \in E \mid \overline{\lim}_{r\to 0} \frac{\nu(B(x,r))}{V(r)} > 0\}$ has Lebesgue measure zero. Since this set is exhausted by a sequence of sets of the form $E_\delta = \{x \in E \mid \overline{\lim}_{r\to 0} \frac{\nu(B(x,r))}{V(r)} > \delta\}$, it is sufficient to show that $\lambda(E_\delta) = 0$ for every positive δ. We fix a δ. Since the measure ν is regular, for every positive ε, there is an open set G such that

$$G \supset E, \quad \nu(G) < \varepsilon\delta.$$

By the definition of the set E_δ, the inequality $\nu(B(x, r)) > \delta V(r)$ holds for arbitrarily small radii r at each point of E_δ. The corresponding balls $B(x, r)$ contained in the set G form a Vitali cover of E. Therefore, there is a subsequence of pairwise disjoint balls $B(x_k, r_k)$ such that

$$E_\delta \subset \bigcup_{k=1}^{\infty} B(x_k, 5r_k) \quad \text{and} \quad \nu\big(B(x_k, r_k)\big) \geqslant \delta V(r_k) \quad \text{for all } k.$$

We estimate the measure of the union $\bigcup_{k=1}^{\infty} B(x_k, 5r_k)$, which covers the set E_δ,

$$\begin{aligned}\lambda\left(\bigcup_{k=1}^{\infty} B(x_k, 5r_k)\right) &\leqslant \sum_{k=1}^{\infty} \lambda\big(B(x_k, 5r_k)\big) = 5^m \sum_{k=1}^{\infty} \lambda\big(B(x_k, r_k)\big) \\ &\leqslant \frac{5^m}{\delta} \sum_{k=1}^{\infty} \nu\big(B(x_k, r_k)\big) = \frac{5^m}{\delta} \nu\left(\bigcup_{k=1}^{\infty} B(x_k, r_k)\right) \\ &\leqslant \frac{5^m}{\delta} \nu(G) < 5^m \varepsilon.\end{aligned}$$

Thus, the set E_δ is a subset of a set with arbitrarily small Lebesgue measure, which is equivalent to the equality $\lambda(E_\delta) = 0$. □

Remark The second assertion of the lemma can be generalized as follows: *If* $\{E_n(x)\}_{x\in E, n\in\mathbb{N}}$ *is an arbitrary regular cover of* E (see Definition 2.7.4) *and* $\nu(E) = 0$, then

$$\lim_{n\to\infty} \frac{\nu(E_n(x))}{\lambda(E_n(x))} = 0$$

almost everywhere with respect to Lebesgue measure on E.

Indeed, by the regularity of the cover, we obtain

$$E_n(x) \subset B\big(x, r_n(x)\big), \qquad \theta(x) = \inf_n \frac{\lambda(E_n(x))}{V(r_n(x))} > 0, \quad r_n(x) \underset{n\to\infty}{\longrightarrow} 0.$$

Therefore, almost everywhere on E with respect to Lebesgue measure, we have

$$\frac{\nu(E_n(x))}{\lambda(E_n(x))} \leqslant \frac{V(r_n(x))}{\lambda(E_n(x))} \cdot \frac{\nu(B(x, r_n(x)))}{V(r_n(x))} \leqslant \frac{1}{\theta(x)} \cdot \frac{\nu(B(x, r_n(x)))}{V(r_n(x))} \underset{n\to\infty}{\longrightarrow} 0.$$

The first assertion of the lemma can be generalized similarly.

11.3.2 We are now ready to check whether a Radon measure has derivative.

Theorem *Let* ν *be a Radon measure on* $\mathbb{R}^m$. *Then*:

(1) *the derivative* $\nu'(x)$ *exists almost everywhere with respect to Lebesgue measure*;
(2) *the derivative is locally summable (and, in particular, is finite almost everywhere)*;

(3) *the measure* ν *is singular* ($\nu \perp \lambda$) *if and only if* $\nu' = 0$ *almost everywhere with respect to Lebesgue measure.*

Proof By Lebesgue's decomposition theorem (see Sect. 11.2.3), the measure ν can be represented in the form $\nu = \nu_a + \nu_s$, where the measure ν_a is subordinate to λ and the measure ν_s is mutually singular with λ. By Remark 11.2.1, the measure ν_a has a locally summable density f with respect to λ. By the definition of mutual singularity, there is a Borel set E such that $\nu_s(E) = \lambda(\mathbb{R}^m \setminus E) = 0$. Applying the lemma to the measure ν_s and to the set E, we see that $\nu_s' = 0$ almost everywhere on E, i.e., almost everywhere on $\mathbb{R}^m$. At the same time, $\nu_a' = f$ almost everywhere on $\mathbb{R}^m$ by Theorem 4.9.2. Thus, the derivative ν' exists almost everywhere and (almost everywhere) coincides with $\nu_a' = f$, which proves all three assertions of the theorem. □

Remark 1 For an arbitrary set E of Lebesgue measure zero, there exists a Radon measure ν, whose derivative ν' is equal to $+\infty$ everywhere on E (see Exercise 1).

Remark 2 If $\{E_n(x)\}_{x\in E, n\in\mathbb{N}}$ is a regular cover of a set $E \subset \mathbb{R}^m$, then the relation $\nu'(x) = \lim_{n\to\infty} \frac{\nu(E_n(x))}{\lambda(E_n(x))}$ holds for almost all x in E.

This follows from Corollary 4.9.4 and the remark to Lemma 11.3.1.

11.3.3 We note another fact related to the derivation of Radon measures.

Theorem (Fubini) *Let* ν_n $(n \in \mathbb{N})$ *be Radon measures in* $\mathbb{R}^m$, *and*

$$\nu(A) = \sum_{n=1}^{\infty} \nu_n(A) \quad (A \in \mathfrak{B}^m). \tag{1}$$

If series (1) *converges for every compact set* A, *then* ν *is a Radon measure and*

$$\nu'(x) = \sum_{n=1}^{\infty} \nu_n'(x) \tag{2}$$

almost everywhere with respect Lebesgue measure.

Proof As the reader can verify independently, Eq. (1) indeed defines a measure finite on compact sets, i.e., a Radon measure. Therefore, by Theorem 11.3.2, we have $\nu'(x) < \infty$ for almost all x.

Now, we prove that the general term of series (2) tends to zero almost everywhere. Dividing both sides of the inequality

$$\sum_{k=1}^{n} \nu_k\big(B(x,r)\big) \leqslant \nu\big(B(x,r)\big)$$

(valid for all n) by $V(r)$ and passing to the limit in the inequality obtained, we see that $\sum_{k=1}^{n} \nu_k'(x) \leqslant \nu'(x)$ almost everywhere. Consequently, $\sum_{k=1}^{\infty} \nu_k'(x) \leqslant \nu'(x) < \infty$. Thus, series (2) converges almost everywhere, and, therefore, we have

$$\nu_n'(x) \underset{n\to\infty}{\longrightarrow} 0 \tag{3}$$

almost everywhere.

Since the series $\sum_{k=1}^{\infty} \nu_k'(x)$ is positive, we prove relation (2) if we verify that some subsequence of partial sums of this series converges to $\nu'(x)$. Let $S_n = \sum_{k=1}^{n} \nu_k$. We construct the required subsequence as follows. For each $j \in \mathbb{N}$, we find a number n_j such that

$$\sum_{k>n_j} \nu_k\big(B(0,j)\big) < 2^{-j}$$

and put $\widetilde{\nu}_j = \nu - S_{n_j} = \sum_{k>n_j} \nu_k$. If the set A is contained in $B(0,R)$, then $\widetilde{\nu}_j(A) < 2^{-j}$ for $j > R$. Therefore, the series $\sum_{j=1}^{\infty} \widetilde{\nu}_j(A)$ converges for every bounded Borel set A and, consequently, satisfies the same conditions as the given series (1). From what was just proved, the series in question satisfies relation (3), i.e.,

$$\widetilde{\nu}_j'(x) = \nu'(x) - S_{n_j}'(x) \underset{j\to\infty}{\longrightarrow} 0$$

almost everywhere. Thus, the subsequence $\{S_{n_j}'\}_{j\geqslant 1}$ of partial sums of series (2) does indeed converge to ν' almost everywhere. □

Remark A measure that is defined on the Borel subsets of an open set $\mathcal{O} \subset \mathbb{R}^m$ and is finite on the compact subsets of $\mathcal{O}$, in other words, a Radon measure on $\mathcal{O}$, may not be the restriction of a Radon measure on $\mathbb{R}^m$. However, locally (more precisely, on the subsets of a ball whose closure lies in $\mathcal{O}$) it coincides with the restriction of a Radon measure on $\mathbb{R}^m$. Therefore, Theorems 11.3.2 and 11.3.3 are carried over to the Radon measures on open subsets of a Euclidean space.

11.3.4 The question of the existence of the derivative of an "arbitrary" function had been addressed at least since the beginning of the 19th century and a complete answer had eluded mathematicians for a long time. It might seem that the answer was given by the Weierstrass example of a nowhere differentiable continuous function. However, as often happens with meaningful problems, Weierstrass' negative answer was only an intermediate step in its solution. The concept of "almost everywhere" introduced by Lebesgue, as well as other measure theoretic notions, made it possible to completely modify the statement of the problem, which has led to extremely general positive results. It turned out that the Ampère[4] conjecture that an arbitrary function has a derivative everywhere aside from some "exceptional and isolated"

[4] André-Marie Ampère (1775–1836)—French physicist and mathematician.

values of the argument was true for every monotone function under the assumption that "exceptional values" of the argument form a set of measure zero. This theorem proved by Lebesgue almost a hundred years after Ampère's work can be proved by extremely beautiful and quite elementary but very subtle reasoning (see [RN], Chap. I, Sect. 1). However, we obtain it as a consequence of Theorem 11.3.2 on the differentiation of Radon measures.

We will need the concept of a Lebesgue–Stieltjes measure μ_g generated by a non-decreasing function g (see Sect. 4.10).

Theorem (Lebesgue) *Every function g increasing on an interval Δ is differentiable almost everywhere on Δ. Moreover, $g'(x) = \mu_g'(x)$ for almost all $x \in \Delta$.*

Proof Without loss of generality, we will assume that the interval Δ is open. First, we prove that the function g has a right derivative almost everywhere. We recall that the set of points of discontinuity of g is at most countable.

Let $\{[x, x+h)\}_{x\in\Delta, 0<h<h_x}$ be a family of intervals, where the positive numbers h_x are so small that $[x-h_x, x+h_x] \subset \Delta$. This family, as well as the family $\{[x, x+h]\}_{x\in\Delta, 0<h<h_x}$, form a regular Vitali cover for Δ. Since the measure μ_g is a Radon measure on the interval Δ, we apply Remark 2 to Theorem 11.3.2 and obtain that

$$\frac{\mu_g([x, x+h))}{h} \underset{h\to 0}{\longrightarrow} \mu_g'(x), \qquad \frac{\mu_g([x, x+h])}{h} \underset{h\to 0}{\longrightarrow} \mu_g'(x) \tag{4}$$

almost everywhere. If x is a point of continuity of g and $0 < h < h_x$, then

$$\begin{aligned} \frac{\mu_g([x, x+h))}{h} &= \frac{g((x+h)-0) - g(x)}{h} \leqslant \frac{g(x+h) - g(x)}{h} \\ &\leqslant \frac{g((x+h)+0) - g(x)}{h} = \frac{\mu_g([x, x+h])}{h}. \end{aligned}$$

Together with relations (4), these inequalities show that, for almost all $x \in \Delta$, the right derivative $g_+'(x)$ exists and is equal to $\mu_g'(x)$.

Considering the covers $\{(x-h, x]\}_{x\in\Delta, 0<h<h_x}$ and $\{[x-h, x]\}_{x\in\Delta, 0<h<h_x}$, we verify similarly that the left derivative also exist almost everywhere and is equal to $\mu_g'(x)$. □

11.3.5 The following statement is a particular case of Theorem 11.3.3.

Theorem (Fubini) *A convergent series of increasing functions can be differentiated termwise almost everywhere.*

Proof Let g be the sum of a pointwise convergent series $\sum_{n=1}^{\infty} g_n$, where each g_n is a non-decreasing function. By Theorem 11.3.3, the equation

$$\mu_g'(x) = \sum_{n=1}^{\infty} \mu_{g_n}'(x) \tag{5}$$

is valid almost everywhere. At the same time, as proved in Theorem 11.3.4, we have $\mu'_g(x) = g'(x)$ and $\mu'_{g_n}(x) = g'_n(x)$ ($n \in \mathbb{N}$) almost everywhere. Therefore, relation (5) remains valid almost everywhere if we replace the derivatives of measures by the derivatives of functions. □

11.3.6 We illustrate the results obtained in the present and previous sections by an example closely connected with the Cantor function φ. As we know (see Sect. 2.3.2), this function is continuous and increases on $\mathbb{R}$ (we assume that $\varphi(x) = 0$ for $x \leqslant 0$ and $\varphi(x) = 1$ for $x \geqslant 1$). By construction, φ is constant on any interval in the complement of the Cantor set and, therefore, its derivative is zero on these intervals. Since the Lebesgue measure of the Cantor set is zero, the Cantor function is singular, i.e., its derivative is zero almost everywhere. Does there exist a singular function that strictly increases on some interval? An affirmative answer to this question can be obtained in different ways. To construct such an example, we use the convolution of the Cantor function and the measure generated by this function, i.e., the function g defined as follows:

$$g(x) = \int_0^1 \varphi(x - t)\, d\varphi(t) \quad (x \in \mathbb{R}).$$

Obviously, the function g is continuous, non-decreasing, and $g(x) = 0$ for $x \leqslant 0$ and $g(x) = 1$ for $x \geqslant 2$. Soon we will see that g strictly increases on $[0, 2]$. It is known that the function g is singular (see [JW]). In other words, the Stieltjes measure μ_g generated by g is mutually singular with one-dimensional Lebesgue measure. This result can be supplemented if we compare the measure μ_g with the Hausdorff measures μ_p for $0 < p \leqslant 1$ (see Sect. 2.6). It turns out that μ_g is mutually singular also with some measures μ_p for $p < 1$. In contrast to the Lebesgue and Hausdorff measures, the measure μ_g is not translation invariant. To emphasize this distinction, we will also call μ_g a mass.

The connection between the measure μ_g and the measure $\mu_\varphi \times \mu_\varphi$ (the Cartesian square of the measure μ_φ generated by the Cantor function) will play an essential role for us.

Lemma 1 *Let F and G be continuous bounded increasing functions defined on $\mathbb{R}$, and let μ_F and μ_G be the corresponding Stieltjes measures. Then the Stieltjes measure μ_H generated by the function H defined by the equation*

$$H(x) = \int_{\mathbb{R}} F(x - y)\, dG(y) \quad (x \in \mathbb{R})$$

is the image of the measure $\mu_F \times \mu_G$ under the map $(x, y) \mapsto P(x, y) = x + y$, i.e., for every Borel set $E \subset \mathbb{R}$, the following holds:

$$\mu_H(E) = (\mu_F \times \mu_G)\big(P^{-1}(E)\big). \tag{6}$$

Proof It is clear that the function H is continuous and increasing. By the uniqueness Theorem 1.5.1, it is sufficient to verify Eq. (6) in the case where $E = [a, b)$. Then the set $P^{-1}(E)$ is the strip $\{(x,y)\in\mathbb{R}^2 \mid a \leqslant x+y < b\}$, and Eq. (6) follows from the generalized Cavalieri principle (see Theorem 5.2.2). Indeed, the cross section $(P^{-1}(E))^y$ is nothing but the interval $[a-y, b-y)$, and, therefore,

$$
\begin{aligned}
(\mu_F \times \mu_G)\big(P^{-1}(E)\big) &= \int_{\mathbb{R}} \mu_F\big((P^{-1}(E))^y\big)\, d\mu_G(y) \\
&= \int_{\mathbb{R}} \mu_F\big([a-y, b-y)\big)\, d\mu_G(y) \\
&= \int_{\mathbb{R}} \big(F(b-y) - F(a-y)\big)\, dG(y) = H(b) - H(a) \\
&= \mu_H\big([a,b)\big).
\end{aligned}
$$

□

We also need the following estimate of binomial coefficients, the proof of which is left to the reader as an easy exercise using Stirling's formula (see Sect. 7.2.6, formula (8)).

Lemma 2 *Let $n \in \mathbb{N}$ and $|t| \leqslant \frac{1}{4}$. Then*

$$C_n^{[(\frac{1}{2}+t)n]} \leqslant \frac{A}{\sqrt{n}} 2^n e^{-2t^2 n},$$

where A is an absolute constant. In particular (for $t = -n^{-1/3}$ and $n \geqslant 4^3$),

$$C_n^{[\frac{n}{2} - n^{2/3}]} \leqslant \frac{A}{\sqrt{n}} 2^n e^{-2n^{1/3}}.$$

This inequality implies that

$$\frac{1}{2^n} \sum_{m \geqslant \frac{n}{2} - n^{2/3}} C_n^m \underset{n\to\infty}{\longrightarrow} 1 \tag{7}$$

since

$$
\begin{aligned}
0 \leqslant 1 - \frac{1}{2^n} \sum_{m \geqslant \frac{n}{2} - n^{2/3}} C_n^m &= \frac{1}{2^n} \sum_{m < \frac{n}{2} - n^{2/3}} C_n^m \\
&< \frac{1}{2^n} \frac{n}{2} C_n^{[\frac{n}{2} - n^{2/3}]} \leqslant A\sqrt{n} e^{-2n^{1/3}} = o(1).
\end{aligned}
$$

Before passing to the statement of the required result, we will carry out some preparatory work.

Let us fix an arbitrary positive integer n and divide the interval $[0,2]$ into intervals $D_k = [2\frac{k}{3^n}, 2\frac{k+1}{3^n}]$ $(0 \leqslant k < 3^n)$, which will be called the segments of rank n. We also consider the segments Δ_ε of rank n obtained in the construction of the Cantor set, where $\varepsilon = (\varepsilon_1, \ldots, \varepsilon_n)$, $\varepsilon_j = 0, 1$ (see Sect. 2.1.4). The left endpoint of the interval Δ_ε is the point $a_\varepsilon = 2\sum_{j=1}^n \varepsilon_j 3^{-j}$. The increment of the Cantor function on a segment of rank n is equal to 2^{-n}. Therefore, the measure $\mu_\varphi \times \mu_\varphi$ of every square of the form

$$\Delta_\varepsilon \times \Delta_{\varepsilon'} \tag{8}$$

is equal to 4^{-n}. The projection P maps this square to some interval D_k whose left endpoint is $a_\varepsilon + a_{\varepsilon'}$. In this case,

$$\frac{k}{3^n} = \sum_{j=1}^{n} \frac{\varepsilon_j + \varepsilon'_j}{3^j} \quad \bigl(\text{where } \varepsilon_j, \varepsilon'_j = 0, 1\bigr). \tag{9}$$

It is clear that each fraction $\frac{k}{3^n}$ can be represented in this form, and, therefore, every interval D_k is the image of a square of the form (8). Hence, in particular, we obtain that every segment of rank n has a positive mass. Since n is arbitrary, this is equivalent to the fact that g strongly increases. In what follows, a key point is the calculation of the number of squares of the form (8) whose projection is an interval D_k. Let Eq. (9) be valid and let $\frac{k}{3^n} = \sum_{j=1}^{n} \frac{\sigma_j}{3^j}$, where $\sigma_j = 0, 1$ or 2. Then $\sigma_j = \varepsilon_j + \varepsilon'_j$, and the non-uniqueness in this representation is possible only if $\sigma_j = 1$. In the latter case, we have two representations, $\varepsilon_j = 1, \varepsilon'_j = 0$ and $\varepsilon_j = 0, \varepsilon'_j = 1$. If the ternary expansion of the fraction $\frac{k}{3^n}$ (or, which is the same, the ternary expansion of the number k) contains m ones, then we have exactly 2^m squares of the form (8) whose image under the projection is the interval D_k. In this case, the mass of the interval is $2^m 4^{-n}$. For $m = n$, we obtain, in particular, that the increment of g on the corresponding "heavy" interval (whose length is $2 \cdot 3^{-n}$) is equal to 2^{-n}. This suggests that the Lipschitz exponent of the function g does not exceed that of φ, and, consequently, is equal to $\log_3 2$.

Let $\mathcal{E}_m \equiv \mathcal{E}_m(n)$ be the set of all intervals D_k for which the ternary expansion of k has exactly m ones and let $A_m(n)$ be their union. The increment of g on each interval in $\mathcal{E}_m$ is equal to $2^m 4^{-n}$. It can easily be seen that $\operatorname{card}(\mathcal{E}_m(n))$ (the number of such intervals) is $C_n^m 2^{n-m}$. Therefore, the mass concentrated on $A_m(n)$ is equal to $4^{-n} \cdot 2^m \cdot C_n^m 2^{n-m} = 2^{-n} C_n^m$.

We prove that the mass is mutually singular not only with one-dimensional Lebesgue but also with the Hausdorff measures μ_p for sufficiently large p.

Theorem *The measure μ_g is mutually singular with the measure μ_p for $\sigma \leqslant p \leqslant 1$, where $\sigma = \frac{3}{2}\log_3 2 = 0.946393\ldots$. In particular, μ_g is singular with respect to Lebesgue measure. For $0 < p < \sigma$, the measure μ_g is absolutely continuous with respect to μ_p.*

Proof To simplify the reasoning, we prove the first part of the theorem only in the case where $p > \sigma$.

Joining "sufficiently heavy" intervals, we put

$$B_n = \bigcup_{m \geqslant \frac{n}{2} - n^{2/3}} A_m(n).$$

It is clear that

$$\mu_g(B_n) = \sum_{m \geqslant \frac{n}{2} - n^{2/3}} \mu_g\big(A_m(n)\big) = \sum_{m \geqslant \frac{n}{2} - n^{2/3}} C_n^m 2^{n-m} \cdot 2^m 4^{-n}$$
$$= \frac{1}{2^n} \sum_{m \geqslant \frac{n}{2} - n^{2/3}} C_n^m.$$

Therefore, by (7), we have $\mu_g(B_n) \underset{n\to\infty}{\longrightarrow} 1$.

Let $E_s = \bigcup_{n=s}^{\infty} B_n$ and $E = \bigcap_{s\geqslant 2} E_s$. Obviously, $\mu_g(E) = 1$.

We prove that $\mu_p(E) = 0$ for $p > \sigma$. By the definition of a Hausdorff measure, it is sufficient to verify that, for every $\varepsilon > 0$, the set E can be covered by a sequence of intervals δ_j such that $\sum_{j=1}^{\infty} |\delta_j|^p < \varepsilon$, where $|\delta|$ denotes the length of the interval δ. Fixing an arbitrary s, we use the intervals D_k (of all ranks) of which the set E_s is composed. Then we obtain that

$$T_s \equiv \sum_{n=s}^{\infty} \sum_{m \geqslant \frac{n}{2} - n^{2/3}} \sum_{D_k \in \mathcal{E}_m(n)} |D_k|^p = \sum_{n=s}^{\infty} \sum_{m \geqslant \frac{n}{2} - n^{2/3}} \operatorname{card}\big(\mathcal{E}_m(n)\big)\left(\frac{2}{3^n}\right)^p$$
$$= 2^p \sum_{n=s}^{\infty} \frac{1}{3^{np}} \sum_{m \geqslant \frac{n}{2} - n^{2/3}} C_n^m 2^{n-m} \leqslant 2^p \sum_{n=s}^{\infty} \frac{2^{\frac{n}{2}+n^{2/3}}}{3^{np}} \sum_{m \geqslant \frac{n}{2} - n^{2/3}} C_n^m$$
$$\leqslant 2^p \sum_{n=s}^{\infty} \left(\frac{2^{\frac{3}{2}+n^{-1/3}}}{3^p}\right)^n.$$

Since $p > \frac{3}{2}\log_3 2 = \log_3 2^{\frac{3}{2}}$, we obtain that, for large n, the fraction $\frac{2^{\frac{3}{2}+n^{-1/3}}}{3^p}$ is less than 1 and is even separated from 1, $\frac{2^{\frac{3}{2}+n^{-1/3}}}{3^p} \leqslant q < 1$. Therefore, for sufficiently large s, we have the estimate

$$T_s \leqslant 2^p \sum_{n=s}^{\infty} q^n = 2^p \frac{q^s}{1-q} \underset{s\to\infty}{\longrightarrow} 0,$$

from which it follows that $\mu_p(E) = 0$.

To prove the second statement of the theorem, we put $p = (\frac{3}{2} - t)\log_3 2$. Since the Hausdorff measures are non-increasing as the parameter increases, we may assume that $0 < t < \frac{1}{4}$. We split the intervals of rank n into "light" and "heavy" intervals. We say that an interval of rank n is light if the increment of g on this interval does not exceed $\frac{2^{(\frac{1}{2}+t)n}}{4^n}$, i.e., if the interval belongs to $\mathcal{E}_m(n)$ for $m \leqslant (\frac{1}{2}+t)n$; otherwise, the interval is heavy.

Now, let $\mu_p(e)=0$, and let the set e be covered by a sequence of intervals δ_j of small length almost realizing a Hausdorff measure, i.e.,

$$e\subset\bigcup_{j=1}^{\infty}\delta_j,\qquad \sum_{j=1}^{\infty}|\delta_j|^p<\varepsilon,$$

where $\varepsilon>0$ is an arbitrarily small number. It is clear that each interval δ_j satisfying the condition $3^{-(n+1)}<|\delta_j|\leqslant 3^{-n}$ touches at most two intervals of rank n. Let

$$J_n=\bigl\{j\in\mathbb{N}\,|\,3^{-(n+1)}<|\delta_j|\leqslant 3^{-n},\ \delta_j \text{ touches a light interval of rank } n\bigr\}.$$

We note that if $J_n\neq\varnothing$, then $n\geqslant n(\varepsilon)$ and $n(\varepsilon)\underset{\varepsilon\to 0}{\longrightarrow}+\infty$ (it is easy to see that $n(\varepsilon)$ is not less than $|\ln\varepsilon|$ in order). We put

$$e_l=\bigcup_{n\geqslant n(\varepsilon)}\bigcup_{j\in J_n}\delta_j,\qquad e_0=e\setminus e_l.$$

Then e_0 is contained in the union of all possible heavy intervals of rank $\geqslant n(\varepsilon)$ that are increased three times, i.e.,

$$e_0\subset e_h\equiv\bigcup_{n\geqslant n(\varepsilon)}\ \bigcup_{m>(\frac{1}{2}+t)n}\ \bigcup_{D_k\in\mathcal{E}_m(n)}\widetilde{D_k},$$

where $\widetilde{D_k}$ is the union $D_{k-1}\cup D_k\cup D_{k+1}$.

Let us estimate the mass of the set e_l. Let δ_j^*, where $j\in J_n$, be the union of (at most two) intervals of rank n touching δ_j (at least one of them is light). We observe that the number of ones in the ternary expansion of the index of an interval of rank n changes by 1 when passing from the interval to the neighboring interval, and, therefore, the increment of g can increase by a factor of no more than two. Since the increment of g on a light interval of rank n does not exceed $2^{(\frac{1}{2}+t)n}4^{-n}=3^{-pn}$, we see that

$$\mu_g\bigl(\delta_j^*\bigr)\leqslant 3^{-pn}+2\cdot 3^{-pn}<9|\delta_j|^p$$

for $j\in J_n$. Therefore,

$$\mu_g(e_l)\leqslant\sum_{n\geqslant n(\varepsilon)}\sum_{j\in J_n}\mu_g(\delta_j)\leqslant\sum_{n\geqslant n(\varepsilon)}\sum_{j\in J_n}\mu_g\bigl(\delta_j^*\bigr)\leqslant\sum_{n\geqslant n(\varepsilon)}\sum_{j\in J_n}9|\delta_j|^p<9\varepsilon.\qquad(10)$$

Using the inequality $\mu_g(\widetilde{D_k})\leqslant 5\mu_g(D_k)$, we can estimate the mass of e_h as follows:

$$\begin{aligned}\mu_g(e_h)&\leqslant\sum_{n\geqslant n(\varepsilon)}\sum_{m>(\frac{1}{2}+t)n}C_n^m 2^{n-m}5\cdot 2^m 4^{-n}\\&=5\sum_{n\geqslant n(\varepsilon)}2^{-n}\sum_{m>(\frac{1}{2}+t)n}C_n^m\leqslant 5\sum_{n\geqslant n(\varepsilon)}2^{-n}\frac{n}{2}C_n^{[(\frac{1}{2}+t)n]}.\end{aligned}$$

From this and Lemma 2, we obtain

$$\mu_g(e_h) \leqslant \frac{5}{2} A \sum_{n \geqslant n(\varepsilon)} \sqrt{n} e^{-t^2 n}.$$

Since the right-hand side of the inequality is a remainder of a convergent series and $n(\varepsilon) \to +\infty$ as $\varepsilon \to 0$, the left-hand side is arbitrarily small for sufficiently small ε. This fact together with estimate (10) completes the proof of the theorem. □

EXERCISES

1. Prove that, for every set $e \subset \mathbb{R}^m$ of Lebesgue measure zero, there is a Radon measure whose derivative is infinite on e.
2. Let ν_0 be a measure defined on the two-point set $\{0, 1\}$ and generated by the loads $\frac{1}{2}$ at the points 0 and 1. Let $\nu = \nu_0 \times \nu_0 \times \cdots$ be an infinite product of the measures ν_0 on the set of binary sequences $\mathcal{E} = \{0, 1\}^{\mathbb{N}}$. Fix an arbitrary sequence $R = \{r_n\}_{n=1}^{\infty}$ of positive numbers r_n satisfying the condition $\sum_{n=1}^{\infty} r_n < +\infty$ and consider the image ν_R of the measure ν under the map $\Phi_R : \mathcal{E} \to \mathbb{R}$ defined by the formula $\Phi_R(\varepsilon) = \sum_{n=1}^{\infty} r_n \varepsilon_n$, where $\varepsilon = \{\varepsilon_n\}_{n=1}^{\infty} \in \mathcal{E}$.
   Find the Fourier transform of the measure ν_R. Find the measure ν_R for $R = \{\frac{1}{2^n}\}_{n=1}^{\infty}$ and $R = \{\frac{2}{3^n}\}_{n=1}^{\infty}$. Which sequence corresponds to the convolution of measures $\nu_{R_1} * \nu_{R_2}$ (by definition, $\nu_{R_1} * \nu_{R_2}(E) = \int_{\mathbb{R}} \nu_{R_1}(E - x)\, d\nu_{R_2}(x)$ for a Borel set $E \subset \mathbb{R}$)?
3. Use the zero-one law to prove that, for the measure ν_R described in the previous exercise and for each $p \in (0, 1)$, the following alternative holds: either ν_R is absolutely continuous or it is singular with respect to the Hausdorff measure μ_p.

11.4 *Differentiability of Lipschitz Functions

11.4.1 First, we consider functions of one variable. The fact that a function f satisfying the Lipschitz condition on some interval is differentiable almost everywhere follows from Lebesgue's theorem on differentiability of a monotone function. Indeed, if L is a Lipschitz constant for f, then f can be represented as the difference of increasing functions g and h, where $g(x) = Lx$ and $h(x) = Lx - f(x)$. We, however, will use another theorem of Lebesgue and not only establish that f is differentiable almost everywhere but verify that f can be recovered from its derivative by integration.

Theorem *A function f satisfying the Lipschitz condition on an interval Δ is absolutely continuous on Δ. In particular, f is differentiable almost everywhere.*

Proof As one can see from the reasoning given just before the statement of the theorem, we may assume, without loss of generality, that f increases. So let us assume that f increases and consider the Lebesgue–Stieltjes measure μ_f generated by f. We verify that f is absolutely continuous with respect to Lebesgue measure λ. Indeed, if $e \subset \Delta$ and $\lambda(e) = 0$, then, for every number $\varepsilon > 0$, there is a system of intervals $\Delta_k = [a_k, b_k] \subset \Delta$ such that

$$e \subset \bigcup_{k=1}^{\infty} \Delta_k, \qquad \sum_{k=1}^{\infty} \lambda(\Delta_k) < \varepsilon.$$

In this case,

$$\sum_{k=1}^{\infty} \mu_f(\Delta_k) = \sum_{k=1}^{\infty} \bigl(f(b_k) - f(a_k)\bigr) \leqslant \sum_{k=1}^{\infty} L(b_k - a_k) < L\varepsilon.$$

Thus, the set e is a subset of a set whose measure μ_f is arbitrarily small. Since the measure μ_f is complete, this means that $\mu_f(e) = 0$. Thus, we have proved that the measure μ_f is absolutely continuous with respect to λ. By the Radon–Nikodym theorem, μ_f has a density ω with respect to λ. In particular, for each interval $[a, b]$ lying in Δ, the equality

$$\mu_f\bigl([a, b]\bigr) = \int_a^b \omega \, d\lambda$$

holds. Since the measure μ_f is finite on every compact interval lying in Δ, the function ω is locally summable in Δ. Fixing a point $c \in \Delta$, we obtain that $f(x) - f(c) = \int_c^x \omega \, d\lambda$ for every $x \in \Delta$, which, by definition, means that f is absolutely continuous. The fact that an absolutely continuous function is differentiable almost everywhere is proved by Lebesgue's theorem 4.9.3. □

If L is a Lipschitz constant for f, then $|f(x) - f(y)|/|x - y| \leqslant L$, and, therefore, $|\omega(x)| = |f'(x)| \leqslant L$ almost everywhere. Thus, the derivative of a function satisfying the Lipschitz condition is bounded almost everywhere. Obviously, the converse is also true: if the derivative of an absolutely continuous function is bounded, then the function satisfies the Lipschitz condition.

11.4.2 This section is devoted to the differentiability of functions of several variables satisfying the Lipschitz condition.

Theorem (Rademacher) *A function f satisfying the Lipschitz condition on a set $E \subset \mathbb{R}^m$ is differentiable on E almost everywhere.*

Proof (1) Since every function satisfying the Lipschitz condition can be extended to the entire space with preservation of this condition (see Sect. 13.2.4), we will assume that f is defined on the entire space $\mathbb{R}^m$. For all $x, y \in \mathbb{R}^m$, the function $t \mapsto F_{x,y}(t) = f(x + ty)$ satisfies the Lipschitz condition on the line. Therefore, by Theorem 11.4.1, this function is differentiable for almost all $t \in \mathbb{R}$. Let $A \subset \mathbb{R}^m$ be the set where the (finite) partial derivative f'_{x_m} exists. This set is measurable, and, for each $z \in \mathbb{R}^{m-1}$, its cross section is the set of points $t \in \mathbb{R}$ at which the function F_{z,e_m} is differentiable. The set of such points is a set of full measure. By Cavalieri's principle, A is also a set of full measure. Since Lebesgue measure is rotation invariant, we obtain that, for every $y \neq 0$, the directional derivative $\frac{\partial f}{\partial y}$ exists almost everywhere in $\mathbb{R}^m$.

In particular, the partial derivatives $f'_{x_1}, \ldots, f'_{x_m}$ exist almost everywhere.

(2) We verify that, for every $y \neq 0$, the directional derivative $\frac{\partial f}{\partial y}(x)$ coincides with $\langle y, \operatorname{grad} f(x)\rangle$ almost everywhere. First, we establish that this equality is valid in the "weak sense". We consider an arbitrary function $\varphi \in C_0^\infty(\mathbb{R}^m)$. Changing the variable, we can easily verify that

$$\int_{\mathbb{R}^m} \big(f(x+ty) - f(x)\big)\varphi(x)\,dx = \int_{\mathbb{R}^m} f(x)\big(\varphi(x-ty) - \varphi(x)\big)\,dx.$$

Therefore, for $t \neq 0$, we obtain

$$\int_{\mathbb{R}^m} \frac{f(x+ty) - f(x)}{t}\varphi(x)\,dx = -\int_{\mathbb{R}^m} f(x)\frac{\varphi(x-ty) - \varphi(x)}{-t}\,dx.$$

Since the directional derivative exists almost everywhere and the integrands are bounded and have compact supports, we can pass to the limit as $t \to 0$ by Lebesgue's theorem and obtain that

$$\int_{\mathbb{R}^m} \frac{\partial f}{\partial y}(x)\varphi(x)\,dx = -\int_{\mathbb{R}^m} f(x)\frac{\partial \varphi}{\partial y}(x)\,dx. \tag{1}$$

Further,

$$\begin{aligned} -\int_{\mathbb{R}^m} f(x)\frac{\partial \varphi}{\partial y}(x)\,dx &= -\int_{\mathbb{R}^m} f(x)\big\langle y, \operatorname{grad}\varphi(x)\big\rangle\,dx \\ &= -\sum_{k=1}^m y_k \int_{\mathbb{R}^m} f(x)\varphi'_{x_k}(x)\,dx. \end{aligned}$$

Since $\int_{\mathbb{R}^m} f(x)\varphi'_{x_k}(x)\,dx = -\int_{\mathbb{R}^m} f'_{x_k}(x)\varphi(x)\,dx$ by Eq. (1) (for $y = e_k$), we obtain

$$\int_{\mathbb{R}^m} \frac{\partial f}{\partial y}(x)\varphi(x)\,dx = \sum_{k=1}^m y_k \int_{\mathbb{R}^m} f'_{x_k}(x)\varphi(x)\,dx = \int_{\mathbb{R}^m} \big\langle y, \operatorname{grad} f(x)\big\rangle\varphi(x)\,dx.$$

Since the above relation is valid for every function $\varphi \in C_0^\infty(\mathbb{R}^m)$, Lagrange's lemma 9.3.6 implies that $\frac{\partial f}{\partial y}(x) = \langle y, \operatorname{grad} f(x)\rangle$ almost everywhere.

(3) Now, we turn to the final step of the proof. Let $H \subset S^{m-1}$ be a countable set containing all unit vectors of the canonical basis and everywhere dense in the unit sphere S^{m-1}, and let A_0 be the set of points x at which directional derivatives exist along all directions $y \in H$ and are calculated by the formula $\frac{\partial f}{\partial y}(x) = \langle y, \operatorname{grad} f(x)\rangle$. Since the set H is countable, A_0 is a set of full measure. We verify that the function f is differentiable at each point of A_0.

Let L be a Lipschitz constant for the function f, $x \in A_0$, and $C = \|\operatorname{grad} f(x)\|$. We fix an arbitrary number $\varepsilon > 0$ and find vectors $h_1, \dots, h_N$ in H forming an ε-net for S^{m-1}. Now, we choose a $\delta > 0$ such that, for $|t| < \delta$, the inequalities

$$\bigl|f(x + th_j) - f(x) - \langle th_j, \operatorname{grad} f(x)\rangle\bigr| \leqslant \varepsilon|t| \tag{2}$$

are valid for every $j = 1, \dots, N$.

For an arbitrary vector $z \neq 0$, we put $y = z/\|z\|$, $t = \|z\|$ and find a vector $h_j \in H$ such that $\|y - h_j\| < \varepsilon$. Then, using the Lipschitz condition and estimate (2) for $\|z\| = t < \delta$, we obtain

$$\begin{aligned}
&\bigl|f(x+z) - f(x) - \langle z, \operatorname{grad} f(x)\rangle\bigr| \\
&\quad\leqslant \bigl|f(x+th_j) - f(x) - \langle th_j, \operatorname{grad} f(x)\rangle\bigr| + \bigl|f(x+ty) - f(x+th_j)\bigr| \\
&\qquad + \bigl|\langle z, \operatorname{grad} f(x)\rangle - \langle th_j, \operatorname{grad} f(x)\rangle\bigr| \\
&\quad\leqslant \varepsilon t + L\|ty - th_j\| + C\|ty - th_j\| \leqslant \varepsilon\|z\|(1 + L + C),
\end{aligned}$$

which proves that the function f is differentiable at x. □

EXERCISES

1. Let f be an absolutely continuous function on an interval Δ, and let the derivative of f be summable on Δ. Prove that, in this case, the function f has the following property:

$\forall \varepsilon > 0\ \exists \delta > 0\ \forall n$:

$$\text{if } \bigvee_{k=1}^{n} (a_k, b_k) \subset \Delta,\ \sum_{k=1}^{n} (b_k - a_k) < \delta, \text{ then } \sum_{k=1}^{n} \bigl|f(b_k) - f(a_k)\bigr| < \varepsilon. \tag{AC}$$

2. Prove that if a function satisfies condition (AC) on a finite interval, then it has a bounded variation that also satisfies condition (AC) on this interval.

3. Reasoning as in the proof of Theorem 11.4.1, verify that condition (AC) is not only necessary but also sufficient for a function defined on a finite interval to be absolutely continuous on the interval and to have a summable derivative.

Let L be a Lipschitz constant for the function f, $x \in A_0$, and $C = \|\operatorname{grad} f(x)\|$. We fix an arbitrary number $\varepsilon > 0$ and find vectors $h_1, \ldots, h_N$ in H forming an ε-net for S^{m-1}. Now, we choose a $\delta > 0$ such that, for $|t| \leqslant \delta$, the inequalities

$$\left|f(x + th_j) - f(x) - t\langle h_j, \operatorname{grad} f(x)\rangle\right| \leqslant \varepsilon|t| \tag{2}$$

are valid for every $j = 1, \ldots, N$.

For an arbitrary vector $z \neq 0$, we put $y = z/\|z\|$, $r = \|z\|$ and find a vector $h_j \in H$ such that $\|y - h_j\| < \varepsilon$. Then, using the Lipschitz condition and estimate (2) for $\|z\| = r \leqslant \delta$, we obtain

$$\begin{aligned}
&\left|f(x+z) - f(x) - \langle z, \operatorname{grad} f(x)\rangle\right| \leqslant \\
&\leqslant \left|f(x + rh_j) - f(x) - \langle rh_j, \operatorname{grad} f(x)\rangle\right| + \left|f(x + ry) - f(x + rh_j)\right| + \\
&\quad + \left|\langle z, \operatorname{grad} f(x)\rangle - \langle rh_j, \operatorname{grad} f(x)\rangle\right| \leqslant \\
&\leqslant \varepsilon r + Lr\|y - h_j\| + Cr\|y - h_j\| \leqslant \varepsilon\|z\|(1 + L + C),
\end{aligned}$$

which proves that the function f is differentiable at x. □

EXERCISES

1. Let f be an absolutely continuous function on an interval Δ, and let the derivative of f be summable on Δ. Prove that, in this case, the function f has the following property:

$$\forall \varepsilon > 0 \ \exists \delta > 0 \ \forall n:$$

$$\text{if } \bigvee_{k=1}^{n}(a_k, b_k) \subset \Delta, \ \sum_{k=1}^{n}(b_k - a_k) < \delta, \text{ then } \sum_{k=1}^{n}\left|f(b_k) - f(a_k)\right| < \varepsilon. \quad (AC)$$

2. Prove that if a function satisfies condition (AC) on a finite interval, then it has a bounded variation that also satisfies condition (AC) on this interval.
3. Reasoning as in the proof of Theorem 11.4.1, verify that condition (AC) is not only necessary but also sufficient for a function defined on a finite interval to be absolutely continuous on this interval and to have a summable derivative.

Chapter 12
Integral Representation of Linear Functionals

12.1 *Order Continuous Functionals in Spaces of Measurable Functions

In this section, $(X, \mathfrak{A}, \mu)$ will denote a space with a σ-finite measure μ. We will consider only measurable sets (i.e., belonging to $\mathfrak{A}$) and measurable almost everywhere finite real or complex functions. The set of these functions will be denoted by $\mathscr{L}^0(X, \mu)$. As usual, we denote by f_+ and f_-, respectively, the positive and the negative part of a real function f, $f_\pm = \max\{\pm f, 0\}$.

We recall the concept of a linear functional known to the reader from a course in algebra.

Definition Let L be a vector space. A map Φ from L to the field of scalars is called a *linear functional on L* if

$$\Phi(f+g) = \Phi(f) + \Phi(g) \quad \text{and} \quad \Phi(af) = a\Phi(f)$$

for all vectors f and g in L and every scalar a.

12.1.1 Let $E \subset \mathscr{L}^0(X, \mu)$ be the set of functions satisfying the following conditions:

(1) if $f, g \in E$, then $af + bg \in E$ for all a and b (the linearity of the set E);
(2) if $f \in E$ and $|g(x)| \leqslant |f(x)|$ almost everywhere, then $g \in E$;
(3) there is a strictly positive function ω_0 in E;
(4) there is a strictly positive function ω_1 such that the product $f\omega_1$ is summable for every function f in E.

Definition A set E with properties (1)–(4) is called a *space of measurable functions*.

It follows from property (2) that, for each $f \in E$, the set E also contains the function $|f|$, and if f is real, E also contains the functions $f_\pm$.

B. Makarov, A. Podkorytov, *Real Analysis: Measures, Integrals and Applications*, Universitext, DOI 10.1007/978-1-4471-5122-7_12, © Springer-Verlag London 2013

An example of a space of measurable functions is the set $\mathscr{L}^p(X,\mu)$ for $1 \leqslant p \leqslant +\infty$ (see Sect. 9.1). We leave it to the reader to verify (using the fact that the measure μ is σ-finite) that this set has properties (1)–(4).

In a space of measurable functions, one can naturally define an order relation and convergence. Namely, for real functions f and g in E, we will write $f \leqslant g$ ($g \geqslant f$) if $f \leqslant g$ almost everywhere on X. In the present section, we write $f_n \underset{n\to\infty}{\longrightarrow} f$ if $f_n(x) \underset{n\to\infty}{\longrightarrow} f(x)$ for almost all $x \in X$, and use the notation $f_n \uparrow f$ ($f_n \downarrow f$) if $f_n \underset{n\to\infty}{\longrightarrow} f$ and $f_n \leqslant f_{n+1}$ (respectively, $f_n \geqslant f_{n+1}$) for all $n \in \mathbb{N}$.

Together with a space of measurable functions E, it is natural to also consider the space E' called the *dual to E*. This space is defined as follows:

$$E' = \{g \in \mathscr{L}^0(X,\mu) \mid , fg \in \mathscr{L}^1(X,\mu) \text{ for every function } f \text{ in } E\}.$$

Condition (4) of the definition of a space of measurable functions guarantees that the dual space contains positive functions. Obviously, the dual space is a space of measurable functions in the sense of the above definition.

Let us find the dual space in a particularly important special case.

Theorem *Let $1 \leqslant p \leqslant +\infty$, $\frac{1}{p} + \frac{1}{q} = 1$. Then the space $\mathscr{L}^q(X,\mu)$ is dual to $\mathscr{L}^p(X,\mu)$.*

Proof Obviously, $\mathscr{L}^q(X,\mu) \subset (\mathscr{L}^p(X,\mu))'$ by Hölder's inequality. We verify that the above inclusion is actually an equality.

This is obvious if $p = +\infty$, since every function g belonging to the dual space is summable. Indeed, it is sufficient to observe that the function $f_0 = \operatorname{sign} \overline{g}$ belongs to $\mathscr{L}^\infty(X,\mu)$ (by definition, we have $\operatorname{sign} z = z/|z|$ for $z \in \mathbb{C}$, $z \neq 0$, and $\operatorname{sign} 0 = 0$). Therefore, $\int_X |g|\,d\mu = \int_X f_0 g\,d\mu < +\infty$.

In the sequel, we consider only finite p.

Let $g \in (\mathscr{L}^p(X,\mu))'$. It is clear that this space also contains the function $|g|$. Therefore, verifying that $g \in \mathscr{L}^q(X,\mu)$, we may assume without loss of generality that g is non-negative. First, we prove that

$$C = \sup_{\|f\|_p \leqslant 1} \left| \int_X fg\,d\mu \right| < +\infty.$$

Assume the contrary. Then there are non-negative functions f_n in $\mathscr{L}^p(X,\mu)$ such that $\|f_n\|_p \leqslant 1$ and $|\int_X f_n g\,d\mu| \geqslant 4^n$ for very $n \in \mathbb{N}$. We put $f = \sum_{n=1}^\infty \frac{1}{2^n} f_n$ and $S_n = \sum_{k=1}^n \frac{1}{2^k} f_k$. The triangle inequality implies $\|S_n\|_p \leqslant \sum_{k=1}^n \frac{1}{2^k} < 1$. Therefore, $\int_X f^p\,d\mu = \lim_{n\to\infty} \|S_n\|_p^p \leqslant 1$ by Fatou's theorem, and, consequently, $f \in \mathscr{L}^p(X,\mu)$. At the same time,

$$\int_X fg\,d\mu \geqslant \int_X \frac{1}{2^n} f_n g\,d\mu \geqslant 2^n$$

for all n, and so $\int_X fg\,d\mu = +\infty$, which contradicts the choice of g. Thus, we have proved that C is finite.

The rest of the proof will be divided into two parts. First, we consider the case where $1 < p < +\infty$.

We represent X in the form $X = \bigcup_{n=1}^{\infty} A_n$, where the sets A_n satisfy the conditions

$$A_n \subset A_{n+1}, \qquad \mu(A_n) < +\infty \quad \text{and} \quad g(x) \leqslant n \quad \text{for } x \in A_n,$$

and put $h_n = g^{q/p}\chi_{A_n}$. Obviously, $h_n \in \mathscr{L}^p(X, \mu)$. Moreover,

$$\int_X h_n g \, d\mu = \|h_n\|_p \int_X \frac{h_n}{\|h_n\|_p} g \, d\mu \leqslant C\|h_n\|_p.$$

Since $h_n g = g^q$ on A_n, the last inequality means that

$$\int_{A_n} g^q \, d\mu \leqslant C\left(\int_{A_n} g^q \, d\mu\right)^{1/p},$$

i.e.,

$$\int_{A_n} g^q \, d\mu \leqslant C^q. \tag{1}$$

Since the sets A_n form an expanding sequence exhausting X, we obtain $g \in \mathscr{L}^q(X, \mu)$.

Now, let $p = 1$. We prove (assuming, as before, that the function g non-negative) that $g \leqslant C$ almost everywhere. Assume the contrary. Then the measure of the set $Y = X\ (g > C)$ is positive. Since the measure μ is σ-finite, we may assume that $\mu(Y) < +\infty$ (otherwise, Y can be partitioned into a countable number of sets of finite measure). We put $f_0 = \frac{1}{\mu(Y)}\chi_Y$. Then $\|f_0\|_1 = 1$ and, at the same time,

$$\int_X f_0 g \, d\mu = \frac{1}{\mu(Y)} \int_Y g \, d\mu > \frac{1}{\mu(Y)} \int_Y C \, d\mu = C,$$

which contradicts the definition of C. □

Remark From the proof of the theorem, it is clear (see inequality (1)) that if $g \in \mathscr{L}^q(X, \mu) = (\mathscr{L}^p(X, \mu))'$, then $\|g\|_q \leqslant \sup_{\|f\|_p \leqslant 1} |\int_X fg \, d\mu|$. The opposite inequality follows from Hölder's inequality. Thus,

$$\|g\|_q = \sup_{\|f\|_p \leqslant 1} \left| \int_X fg \, d\mu \right|. \tag{1'}$$

12.1.2 Now, let us turn to our main concern in the present section, the study of linear functionals in spaces of measurable functions. An example of a linear functional on E is the functional Φ

$$\Phi(f) = \int_X fg \, d\mu \quad (f \in E)$$

corresponding to a function g belonging to the dual space. By Lebesgue's theorem 4.8.4, the functional just defined has the following property:

$$\text{if } f_n \downarrow 0, \quad \text{then } \Phi(f_n) \underset{n\to\infty}{\longrightarrow} 0. \tag{2}$$

This property will serve as a basis for the following definition.

Definition A linear functional Φ satisfying property (2) is called *continuous* or, more precisely, *order continuous*.

We note a simple but important property of order continuous functionals.

Lemma

(1) *A linear functional Φ defined on a space E of measurable functions is order continuous if and only if the conditions $f \in E$ and $0 \leqslant f_n \uparrow f$ imply that $\Phi(f_n) \underset{n\to\infty}{\longrightarrow} \Phi(f)$.*
(2) *A continuous functional Φ vanishes on functions that are zero almost everywhere.*

Proof Obviously, $f_n \in E$ by condition (2) of the definition of a space of measurable functions. Since $f - f_n \downarrow 0$, the first assertion of the lemma is just a reformulation of the definition of continuity. To prove the second assertion, we note that if a real function f is zero almost everywhere, then the stationary sequence $\{g_n\}_{n\geqslant 1}$, where $g_n = f_+$, converges to zero almost everywhere. By the definition of continuity, we have $\Phi(f_+) = \Phi(g_n) \underset{n\to\infty}{\longrightarrow} 0$, i.e., $\Phi(f_+) = 0$. Similarly, $\Phi(f_-) = 0$, and, therefore, $\Phi(f) = \Phi(f_+) - \Phi(f_-) = 0$. □

12.1.3 It turns out that every order continuous functional has an integral representation. More precisely, the following holds.

Theorem *Let Φ be an order continuous linear functional on a space of measurable functions E. Then there is a function h in the dual space E' such that*

$$\Phi(f) = \int_X f h \, d\mu \quad \textit{for all } f \textit{ in } E. \tag{3}$$

Proof First, we assume that the space E is real and the measure μ is finite and put

$$\varphi(A) = \Phi(\chi_A) \quad \text{for } A \in \mathfrak{A}.$$

We verify that the function φ is countably additive, i.e., is a (real) charge. Indeed, let $A = \bigvee_{n=1}^{\infty} A_n$. Then $\chi_A = \sum_{n=1}^{\infty} \chi_{A_n}$. If S_k is a partial sum of this series, then $S_k \uparrow \chi_A$, and so $\Phi(S_k) \to \Phi(\chi_A) = \varphi(A)$. Consequently,

$$\varphi(A) = \lim_{k\to\infty} \Phi(S_k) = \lim_{k\to\infty} \sum_{n=1}^{k} \Phi(\chi_{A_n}) = \lim_{k\to\infty} \sum_{n=1}^{k} \varphi(A_n) = \sum_{n=1}^{\infty} \varphi(A_n).$$

The second assertion of the lemma implies that the charge φ is absolutely continuous with respect to μ. Therefore, by the Radon–Nikodym theorem, there is a real summable function h such that $\varphi(A) = \int_A h\,d\mu$ for each set A. Taking into account the definition of φ, we can represent this equation in the form $\Phi(\chi_A) = \int_X \chi_A h\,d\mu$. Since the functional is linear, the latter relation is also preserved for linear combinations of characteristic functions, i.e., $\Phi(f) = \int_X fh\,d\mu$ for each simple function f. Thus, we have verified that Eq. (3) is valid for simple functions. Now, we show that it is valid not only for simple but for all functions in E. In the proof, we may, obviously, consider only non-negative functions f.

Let $\{g_n\}_{n\geqslant 1}$ be an increasing sequence of non-negative simple functions converging to the function $f \in E$ (see Theorem 3.2.2), and let $A_+ = X(h \geqslant 0)$. Then $0 \leqslant g_n\chi_{A_+} \uparrow f\chi_{A_+}$ and $0 \leqslant g_n\chi_{A_+}h = g_nh_+ \uparrow fh_+$. Passing to the limit in the equation

$$\Phi(g_n\chi_{A_+}) = \int_X g_nh_+\,d\mu, \tag{4}$$

we see that

$$\Phi(f\chi_{A_+}) = \int_X fh_+\,d\mu. \tag{5}$$

The passage to the limit on the left-hand side of Eq. (4) is performed by the continuity of Φ, and, on the right-hand side, by Levi's theorem. In particular, we obtain that $\int_X fh_+\,d\mu = \Phi(f\chi_{A_+}) < +\infty$, and so the function fh_+ is summable. Replacing A_+ by the set $A_- = X(h < 0)$ and Φ by $-\Phi$, we can similarly prove that the function fh_- is summable and the equation

$$-\Phi(f\chi_{A_-}) = \int_X fh_-\,d\mu \tag{5$'$}$$

is valid. Since $f = f\chi_{A_+} + f\chi_{A_-}$ and $fh = fh_+ - fh_-$, Eqs. (5) and (5$'$) imply, obviously, that fh is summable and that Eq. (3) is valid.

Now, let $\mu(X) = +\infty$. We consider the functions ω_0 and ω_1 in conditions (3) and (4) of the definition of a space of measurable functions. We put $\omega = \omega_0\omega_1$ and consider the measure ν with density ω with respect to μ. Since the function ω is summable, the measure ν is finite. Since $\omega > 0$, the measures ν and μ are mutually absolutely continuous. Therefore, $\mathscr{L}^0(X,\mu) = \mathscr{L}^0(X,\nu)$.

Now, we introduce a new functional space E_0, putting

$$E_0 = \left\{ \frac{f}{\omega_0} \,\middle|\, f \in E \right\},$$

and regard E_0 as a space of measurable functions lying in $\mathscr{L}^0(X,\nu)$. In E_0, we define the functional Φ_0 by the equation

$$\Phi_0(g) = \Phi(g\omega_0) \quad (g \in E_0).$$

We leave to the reader the standard verification that this functional is continuous and linear.

By what has been proved, there is a function h_0 such that the product gh_0 is summable with respect to ν for all g in E_0 and the equation $\Phi_0(g) = \int_X gh_0\,d\nu$ is valid. The rest is simple. Indeed, for $f \in E$, we have

$$\Phi(f) = \Phi_0\left(\frac{f}{\omega_0}\right) = \int_X \frac{f}{\omega_0} h_0\,d\nu = \int_X \frac{f}{\omega_0} h_0 \omega\,d\mu = \int_X f h_0 \omega_1\,d\mu.$$

It remains to put $h = h_0\omega_1$.

If the space E is complex, then we represent the functional Φ as the sum, $\Phi = \Phi_1 + i\Phi_2$, where $\Phi_1 = \mathcal{R}e\,\Phi$ and $\Phi_2 = \mathcal{I}m\,\Phi$. Considering the functionals Φ_1 and Φ_2 only on the set of real functions belonging to E and applying the part of the theorem already proved, we obtain the required representation. □

By Theorem 12.1.1, we obtain

Corollary *Let Φ be an order continuous linear functional on the space $\mathscr{L}^p(X,\mu)$, $1/p + 1/q = 1$. Then there is a function h in $\mathscr{L}^q(X,\mu)$ such that*

$$\Phi(f) = \int_X fh\,d\mu \quad \textit{for all } f \textit{ in } \mathscr{L}^p(X,\mu).$$

From ($1'$), it follows that $\sup_{\|f\|_p \leqslant 1} |\Phi(f)| = \|h\|_q$.

12.2 *Positive Functionals in Spaces of Continuous Functions

In the present section, we will use the concepts of a topological and, in particular, of a compact space. However, the facts established below are quite non-trivial already in the case where the topological space under consideration is $\mathbb{R}^m$ or even a compact subset of $\mathbb{R}^m$. Therefore, the reader that does not have a sufficient mathematical background will not lose much in understanding the basic ideas by assuming that the space in question is $\mathbb{R}^m$. In this case, it is easy to observe that, instead of Theorem 12.2.1, we could use Theorem 8.1.7 on a smooth descent or Lemma 2 of Sect. 13.2.1 on functional separability.

Our goal is to describe positive functionals on the set $C(X)$ of all continuous real functions defined on a locally compact space X and also positive functionals on the subset $C_0(X)$ of compactly supported functions in $C(X)$. Our main attention will be directed to functionals on $C_0(X)$, the description of which plays a decisive role.

12.2.1 We recall some facts from topology. A topological space X is called *locally compact* if each point in X has a neighborhood whose closure is compact set. For a function f defined on X, the closure of the set $\{x \in X \mid f(x) \neq 0\}$ is called the *support* of f and is denoted by $\operatorname{supp}(f)$. A function defined on X is called a *compactly* *supported function* if its support is compact set. Here, we will consider only real functions. In this case, the sets $C_0(X)$ and $C(X)$ are, obviously, real vector spaces.

A locally compact space has "sufficiently many" continuous compactly supported functions. In particular, there are functions "smoothing" characteristic functions of compact sets. More precisely, this means that the following statement holds.

Theorem *Let X be a locally compact space, K be a compact subset of X, and G be an open set containing K. Then there is a continuous compactly supported function φ such that*

$$0 \leqslant \varphi \leqslant 1, \quad \varphi(x) = 1 \quad \textit{for } x \in K, \qquad \mathrm{supp}(\varphi) \subset G.$$

For the proof of this theorem see, e.g., the book [B-I], Chap. 2, Sects. 12, 13. The proof is very simple if X is metrizable, see Lemma 2 of Sect. 13.2.1.

12.2.2 Now we introduce the concepts of a positive functional and Radon measure, which are fundamental to the following material.

Definition 1 A linear functional $\Phi : C_0(X) \to \mathbb{R}$ is called *positive* if $\Phi(f) \geqslant 0$ for every non-negative function f in $C_0(X)$.

We note an important property of positive functionals, namely, their monotonicity,

$$\text{if } f, g \in C_0(X) \text{ and } f \leqslant g, \quad \text{then } \Phi(f) \leqslant \Phi(g).$$

The proof is obvious: $\Phi(g) - \Phi(f) = \Phi(g - f) \geqslant 0$ since $g - f \geqslant 0$.

Every function in $C_0(X)$ is summable with respect to any Borel measure that assumes finite values on compact sets, and, obviously, the integral with respect to such a measure is a positive functional. Our goal is to prove that the converse is also true. It even turns out that, for the representation of a functional as an integral, it is not always necessary to consider all Borel measures, which in some cases can be too "bad". For what follows, it is sufficient to confine ourselves to Borel measures satisfying certain conditions close to regularity.

Definition 2 Let X be a locally compact topological space, and let μ be a measure defined on the σ-algebra $\mathfrak{B}_X$ of its Borel subsets. The measure μ is called a *Radon measure* if

I. $\mu(A) = \inf\{\mu(G) \mid G \supset E,\ G \text{ is an open set}\}$ for every Borel set A;
II. $\mu(G) = \sup\{\mu(K) \mid K \subset G,\ K \text{ is a compact set}\}$ for every open set G;
III. $\mu(K) < +\infty$ for every compact set K.

If the space X is compact, then, obviously, a finite Borel measure is a Radon measure if and only if it is regular. In particular, every finite Borel measure on a metrizable compact space, being regular (see Sect. 13.3.2, Corollary 1), is a Radon measure. Every Borel measure on $\mathbb{R}^m$ that assumes finite values on bounded sets is a Radon measure (see Sect. 2.2.3).

A Radon measure is uniquely determined by the integrals of the continuous compactly supported functions. To state the result more precisely, we put

$$I(G)=\{f\in C_0(X)\,|\,0\leqslant f\leqslant 1,\ \mathrm{supp}(f)\subset G\},\tag{1}$$

where G is an open subset of the space X.

In the sequel, the following statement will serve, in particular, as a motivation of Definition 12.2.4.

Proposition *Let μ be a Radon measure on a locally compact space X. Then, for every open set G, we have*

$$\mu(G)=\sup\left\{\int_X f\,d\mu\,\middle|\,f\in I(G)\right\}.\tag{2}$$

Proof Let K be an arbitrary compact set lying in G. By Theorem 12.2.1, there is a function f_0 in $I(G)$ such that $f_0(x)=1$ on K. Then

$$\mu(K)\leqslant\int_X f_0\,d\mu\leqslant\int_X \chi_G\,d\mu=\mu(G).$$

Consequently,

$$\sup\{\mu(K)\,|\,K\subset G,\ K\text{ is a compact set}\}\leqslant\sup\left\{\int_X f\,d\mu\,\middle|\,f\in I(G)\right\}\leqslant\mu(G).$$

This proves (2) since the left-hand side of this inequality coincides with $\mu(G)$ by the definition of a Radon measure. □

Corollary *Radon measures μ and ν coincide if $\int_X f\,d\mu=\int_X f\,d\nu$ for every function f in $C_0(X)$.*

Proof The fact that the measures μ and ν coincide on open sets follows from Eq. (2), and item I of the definition of a Radon measure implies that they coincide on arbitrary Borel sets. □

As we have already pointed out, our goal is to prove the following statement.

Theorem (Riesz–Kakutani[1]) *Let X be a locally compact space and Φ be a positive linear functional on the space $C_0(X)$. Then there exists a unique Radon measure μ such that*

$$\Phi(f)=\int_X f\,d\mu\quad\textit{for all } f\in C_0(X).$$

[1] Shizuo Kakutani (1911–2004)—Japanese mathematician.

The uniqueness of μ follows directly from Corollary 12.2.2.

12.2.3 Before turning to the proof of the existence of a measure μ representing the functional Φ, we will do some preliminary work. We begin with a statement on the partition of unity in a topological space. A similar statement was established in Sect. 8.1.8 for the space $\mathbb{R}^m$. Of course, now in a more general situation, we lift the smoothness condition and require only the continuity of the functions forming the partition.

Lemma *Let $G_1, \dots, G_N$ be an open cover of a compact subset K of a locally compact space X. Then there exist non-negative functions $\varphi_1, \dots, \varphi_N$ in $C_0(X)$ such that*

$$\varphi_1 + \cdots + \varphi_N = 1 \quad \text{on } K, \qquad \operatorname{supp}(\varphi_n) \subset G_n \quad \text{for } n = 1, \dots, N.$$

Such a family $\{\varphi_1, \dots, \varphi_N\}$ is called a *partition of unity for K, subordinate to the given cover.*

Proof From Theorem 12.2.1 it follows that each point of the set K can be assigned a non-negative function in $C_0(X)$ that is positive at this point and has a "small support" (it is contained in one of the sets $G_1, \dots, G_N$). The interiors of the supports form a cover of K. Since K is compact set, there exists a finite subcover. We consider the corresponding finite set of given compactly supported functions. Let ψ_n be the sum of those functions whose supports are contained in G_n. It is clear that $\operatorname{supp}(\psi_n) \subset G_n$ for $n = 1, \dots, N$ and $\theta \equiv \psi_1 + \cdots + \psi_N > 0$ on K. Using Theorem 12.2.1 one more time, we take a function ω in $C_0(X)$ such that $0 \leqslant \omega \leqslant 1$, $\omega(x) = 1$ for $x \in K$ and $\operatorname{supp}(\omega) \subset \{x \in X \mid \theta(x) > 0\}$. Then $1 - \omega(x) + \theta(x) > 0$ everywhere on X, and we can put $\varphi_n = \psi_n/(1 - \omega + \theta)$ $(n = 1, \dots, N)$. □

12.2.4 Now, we can proceed to the proof of the Riesz–Kakutani theorem. The construction of the required measure is performed in three steps. First, we use the functional Φ to construct an auxiliary outer measure. Then, we verify that the measure generated by the outer measure is a Radon measure. Finally, we show that the measure obtained represents the functional Φ.

Definition For an open set $G \subset X$ and an arbitrary set $A \subset X$, we put

$$\mu_0(G) = \sup\bigl\{\Phi(f) \mid f \in I(G)\bigr\},$$
$$\mu^*(A) = \inf\bigl\{\mu_0\bigl(G'\bigr) \mid G' \supset A,\ G' \text{ is an open set}\bigr\}.$$

We note that

$$\mu_0(\varnothing) = 0 \quad \text{and} \quad \mu_0(G) \leqslant \mu_0\bigl(G'\bigr) \text{ if } G \subset G'. \tag{3}$$

Theorem *The function μ^* is an outer measure and $\mu^*(G) = \mu_0(G)$ for every open set G.*

As already mentioned, we want to use μ^* to construct the required Radon measure. Taking into account Proposition 12.2.2, we see that the definition given above is actually the only possible one. If we want to obtain a measure representing the functional, then the definition of μ_0 is dictated by Eq. (2), and the definition of μ^* is dictated by condition I of the definition of a Radon measure.

Proof The inequality $\mu^*(G) \leqslant \mu_0(G)$ follows directly from the definition of μ^*, and the opposite inequality follows from the fact that the function μ_0 is monotone (see inequality (3)). It is also obvious that $\mu^*(A) \leqslant \mu^*(B)$ if $A \subset B$.

By the definition of an outer measure, we must verify that the function μ^* is countably subadditive and $\mu^*(\varnothing) = 0$. The latter follows immediately from the definition of μ^* and the relation $\mu_0(\varnothing) = 0$.

Now, we prove that the function μ^* is countably semiadditive, i.e., that

$$\mu^*(A) \leqslant \sum_{n=1}^{\infty} \mu^*(A_n) \quad \text{if } A \subset \bigcup_{n=1}^{\infty} A_n. \tag{4}$$

Of course, we may and will assume that the sum on the right-hand side is finite, since otherwise the inequality is trivial. First, we assume that all sets A_n are open. Let $G = \bigcup_{n=1}^{\infty} A_n$, and let f belong to $I(G)$ (see (1)). We put $K = \text{supp}(f)$. Since K is a compact set, there is an N such that $K \subset A_1 \cup \dots \cup A_N$. Let $\{\varphi_n\}_{n=1}^N$ be a partition of unity for K, subordinate to the cover $\{A_n\}_{n=1}^N$. Then

$$f = \sum_{n=1}^{N} f\varphi_n \quad \text{and} \quad \text{supp}(f\varphi_n) \subset A_n \quad \text{for } n = 1, \dots, N.$$

Therefore, $\Phi(f\varphi_n) \leqslant \mu_0(A_n) = \mu^*(A_n)$ and

$$\Phi(f) = \sum_{n=1}^{N} \Phi(f\varphi_n) \leqslant \sum_{n=1}^{N} \mu^*(A_n) \leqslant \sum_{n=1}^{\infty} \mu^*(A_n).$$

Since this inequality is valid for every function f in $I(G)$, we pass to the supremum on its right-hand side and obtain

$$\mu^*(G) = \mu_0(G) \leqslant \sum_{n=1}^{\infty} \mu^*(A_n) \quad \text{and, consequently,}$$

$$\mu^*(A) \leqslant \mu^*(G) \leqslant \sum_{n=1}^{\infty} \mu^*(A_n),$$

which proves inequality (4) if the sets A_n are open. In the case of arbitrary sets, we first prove inequality (4) within $\varepsilon > 0$. For this, we choose open sets G_n such that

$$G_n \supset A_n \quad \text{and} \quad \mu^*(G_n) < \mu^*(A_n) + \varepsilon/2^n \quad \text{for all } n \geqslant 1.$$

Then $A \subset \bigcup_{n=1}^{\infty} G_n$, and, by what was just proved,

$$\mu^*(A) \leqslant \mu^*\left(\bigcup_{n=1}^{\infty} G_n\right) \leqslant \sum_{n=1}^{\infty} \mu^*(G_n) \leqslant \sum_{n=1}^{\infty}\left(\mu^*(A_n) + \frac{\varepsilon}{2^n}\right) = \sum_{n=1}^{\infty} \mu^*(A_n) + \varepsilon.$$

Since ε is arbitrary, we obtain (4). □

12.2.5 As is known, the restriction of an arbitrary outer measure to the σ-algebra of measurable sets is a measure. We recall (see Sect. 1.4.2) that a set $A \subset X$ is measurable with respect to an outer measure μ^* defined on subsets of X if

$$\mu^*(E) \geqslant \mu^*(E \cap A) + \mu^*(E \setminus A) \tag{5}$$

for every set $E \subset X$. The measurable sets form a σ-algebra. We verify that, in our case, the measure corresponding to μ^* is defined on all Borel sets. As a preliminary, we establish a simple inequality.

Lemma *Let $f \in C_0(X)$, $0 \leqslant f \leqslant 1$, $H = \mathrm{supp}(f)$, and $H_1 = \{x \in X \mid f(x) = 1\}$. Then*

$$\mu^*(H_1) \leqslant \Phi(f) \leqslant \mu^*(H). \tag{6}$$

Proof To prove the right inequality in (6), we consider an arbitrary open set G containing H. Since $f \in I(G)$ (for the definition of $I(G)$, see (1)), we have $\Phi(f) \leqslant \mu_0(G)$. Consequently,

$$\Phi(f) \leqslant \inf\{\mu_0(G) \mid G \supset H,\ G \text{ is an open set}\} = \mu^*(H).$$

To prove the left inequality in (6), we fix an arbitrary number ε, $0 < \varepsilon < 1$, and put $G_\varepsilon = \{x \in X \mid f(x) > 1 - \varepsilon\}$. It is clear that $H_1 \subset G_\varepsilon$. If $g \in I(G_\varepsilon)$, then

$$g \leqslant \frac{f}{1-\varepsilon} \quad \text{and} \quad \Phi(g) \leqslant \frac{\Phi(f)}{1-\varepsilon}.$$

Therefore,

$$\mu^*(H_1) \leqslant \mu_0(G_\varepsilon) = \sup\{\Phi(g) \mid g \in I(G_\varepsilon)\} \leqslant \frac{\Phi(f)}{1-\varepsilon},$$

which completes the proof since ε is arbitrary. □

Corollary *For every open set G*

$$\mu^*(G) = \sup\{\mu^*(K) \mid K \text{ is a compact set, } K \subset G\}. \tag{7}$$

Proof Let $\nu(G)$ be the right-hand side of (7). Obviously, $\nu(G) \leqslant \mu^*(G)$. On the other hand, if $f \in I(G)$, then $\Phi(f) \leqslant \mu^*(\mathrm{supp}(f)) \leqslant \nu(G)$ by the lemma. Passing to the supremum on the left-hand side of the last inequality, we see that

$\mu^*(G) \leqslant \nu(G)$. Together with the opposite estimate just obtained, this gives us relation (7). □

12.2.6 Now, we are ready to pass to the next step.

Theorem *Every Borel subset of the space X is measurable with respect to the outer measure μ^*. The restriction μ of μ^* to the σ-algebra $\mathfrak{B}_X$ of Borel subsets is a Radon measure.*

Proof Since the measurable sets form a σ-algebra, we prove the first assertion of the theorem if we verify that all open sets are measurable. Thus, we must prove that, for every open set A and an arbitrary set E, inequality (5) is valid. First, we assume that $E = G$ is an open set. Let f be an arbitrary function in $I(A \cap G)$, and let $H = \mathrm{supp}(f)$. We also consider a function f_0 in $I(G \setminus H)$. Since the supports of f and f_0 are disjoint, we have $f + f_0 \leqslant 1$. Moreover, it is obvious that $f + f_0 \in I(G)$. Therefore,

$$\mu^*(G) \geqslant \Phi(f + f_0) = \Phi(f) + \Phi(f_0).$$

Passing to the supremum over all f_0 in $I(G \setminus H)$ on the right-hand side of this inequality, we see that

$$\mu^*(G) \geqslant \Phi(f) + \mu^*(G \setminus H) \geqslant \Phi(f) + \mu^*(G \setminus A).$$

Again, passing to the supremum over all f in $I(G \cap A)$ on the right-hand side of the last inequality, we obtain that $\mu^*(G) \geqslant \mu^*(G \cap A) + \mu^*(G \setminus A)$, which is just inequality (5) for $E = G$.

In the case of an arbitrary set E, we may assume that $\mu^*(E) < +\infty$, since otherwise inequality (5) is obvious. We fix a number $\varepsilon > 0$ and find an open set G such that $G \supset E$ and $\mu^*(G) < \mu^*(E) + \varepsilon$. Then

$$\mu^*(E) \geqslant \mu^*(G) - \varepsilon \geqslant \mu^*(G \cap A) + \mu^*(G \setminus A) - \varepsilon \geqslant \mu^*(E \cap A) + \mu^*(E \setminus A) - \varepsilon.$$

This implies (5) since ε is arbitrary.

Thus, $\mathfrak{B}_X$ is contained in the σ-algebra of all measurable sets, and, therefore, μ, being a restriction of the measure generated by μ^*, is a measure.

It remains to conduct an easy verification that μ has all the properties of a Radon measure.

The measure μ has property I by definition and property II by Eq. (7). In conclusion, we verify that μ also has property III. Indeed, let $K \subset X$ be a compact set. We consider a non-negative compactly supported function g equal to 1 on K. Then, by the lemma, we have $\mu(K) \leqslant \Phi(g) < +\infty$. □

12.2.7 Let us turn to the concluding part of our reasoning, which will complete the proof of the Riesz–Kakutani theorem. It remains to verify that the measure μ constructed above indeed allows us to obtain an integral representation of the functional Φ.

Theorem *The measure* μ *constructed in Theorem* 12.2.6 *satisfies the relation*

$$\Phi(f) = \int_X f\, d\mu \quad \text{for every } f \text{ in } C_0(X). \tag{8}$$

Proof An essential part of our reasoning will be the approximation of f by a linear combination of characteristic functions and a subsequent "smoothening". By this method, we establish that Eq. (8) is valid up to a small error, which is sufficient for the proof of (8). Since both parts of (8) depend linearly on f, we, obviously, may and will assume that $0 \leqslant f < 1$.

We fix an arbitrary positive integer N and for $k = 1, 2, \dots, N-1$ consider the sets

$$H_k = \left\{ x \in X \,\middle|\, f(x) \geqslant \frac{k}{N} \right\}, \qquad G_k = \left\{ x \in X \,\middle|\, f(x) > \frac{k-1}{N} \right\}.$$

Let $H_0 = \operatorname{supp}(f)$, $G_0 = X$. It is clear that the sets H_k are compact and

$$H_0 \supset G_1 \supset H_1 \supset \cdots \supset G_{N-1} \supset H_{N-1}.$$

Hence it follows that

$$\frac{1}{N} \sum_{k=1}^{N-1} \chi_{H_k}(x) \leqslant f(x) \leqslant \frac{1}{N} \sum_{k=0}^{N-1} \chi_{H_k}(x). \tag{9}$$

Indeed, if $x \notin H_0$ or $x \in H_{N-1}$, then (9) is obvious. If $x \in H_j \setminus H_{j+1}$ for $0 \leqslant j < N-1$, then $j/N \leqslant f(x) < (j+1)/N$, and, therefore,

$$\frac{1}{N} \sum_{k=1}^{N-1} \chi_{H_k}(x) = \frac{j}{N} \leqslant f(x) < \frac{j+1}{N} = \frac{1}{N} \sum_{k=0}^{N-1} \chi_{H_k}(x).$$

We consider continuous compactly supported functions f_k that smoothen the functions χ_{H_k}. More precisely, assume that the function f_k belongs to $I(G_k)$ and is equal to 1 on H_k for $k = 0, 1, \dots, N-1$. We also put $f_N \equiv 0$. Then

$$f_{k+1} \leqslant \chi_{H_k} \leqslant f_k \quad \text{for } k = 0, 1, \dots, N-1. \tag{10}$$

From (10) and Lemma 12.2.5, it follows that

$$\Phi(f_{k+1}) \leqslant \mu\big(\operatorname{supp}(f_{k+1})\big) \leqslant \mu(H_k) \leqslant \Phi(f_k) \quad \text{for } k = 0, 1, \dots, N-1. \tag{11}$$

Integrating inequality (9), we obtain

$$\frac{1}{N} \sum_{k=1}^{N-1} \mu(H_k) \leqslant \int_X f\, d\mu \leqslant \frac{1}{N} \sum_{k=0}^{N-1} \mu(H_k).$$

Applying (11), we obtain

$$\frac{1}{N}\sum_{k=1}^{N-1}\Phi(f_{k+1})\leqslant\int_X f\,d\mu\leqslant\frac{1}{N}\sum_{k=0}^{N-1}\Phi(f_k). \tag{12}$$

On the other hand, it follows from (9) and (10) that

$$\frac{1}{N}\sum_{k=1}^{N-1}f_{k+1}(x)\leqslant f(x)\leqslant\frac{1}{N}\sum_{k=0}^{N-1}f_k(x).$$

Applying the functional Φ to all sides of the last inequality, we obtain

$$\frac{1}{N}\sum_{k=1}^{N-1}\Phi(f_{k+1})\leqslant\Phi(f)\leqslant\frac{1}{N}\sum_{k=0}^{N-1}\Phi(f_k). \tag{13}$$

Taking into account (12), we see that

$$-\frac{\Phi(f_0)+\Phi(f_1)}{N}\leqslant\Phi(f)-\int_X f\,d\mu\leqslant\frac{\Phi(f_0)+\Phi(f_1)}{N}.$$

Since N is arbitrary, we obtain that (8) is valid. □

12.2.8 In this section, we consider positive functionals on the set $C_0^\infty(\mathbb{R}^m)$ of infinitely differentiable compactly supported functions. As in the case of functionals on $C_0(\mathbb{R}^m)$, a linear functional Φ defined on $C_0^\infty(\mathbb{R}^m)$ is called positive if $\Phi(f)\geqslant 0$ for every non-negative function f in $C_0^\infty(\mathbb{R}^m)$.

Theorem *Every positive functional on $C_0^\infty(\mathbb{R}^m)$ has an integral representation by a Radon measure.*

Proof For brevity, we put $L=C_0^\infty(\mathbb{R}^m)$ and verify that a positive functional Φ defined on L can be extended to a positive functional defined on $C_0(\mathbb{R}^m)$. For this, we use the fact that every function in $C_0(\mathbb{R}^m)$ can be uniformly approximated on $\mathbb{R}^m$ by functions in L (see Corollary 2 of Sect. 7.6.4).

Let $f\in C_0(\mathbb{R}^m)$ and $f\geqslant 0$, and let $f_n\in L_+=\{f\in L\mid f\geqslant 0\}$ and $f_n\rightrightarrows f$ on $\mathbb{R}^m$. We consider a function $\varphi\in L_+$ such that $\varphi=1$ on $K=\operatorname{supp}(f)$ and put $g_n=\varphi\cdot f_n$. Obviously,

$$g_n\rightrightarrows f\quad\text{on }\mathbb{R}^m\qquad g_n\in L_+,\ \operatorname{supp}(g_n)\subset Q\quad\text{for } n=1,2,\dots, \tag{14}$$

where $Q\subset\mathbb{R}^m$ is an appropriate compact set (we can take, e.g., $Q=\operatorname{supp}(\varphi)$). Let h be a function in L_+ such that $h=1$ on Q. It is clear that $|f-g_n|\leqslant c_n h$, where $c_n=\max_{x\in\mathbb{R}^m}|f(x)-g_n(x)|\underset{n\to\infty}{\longrightarrow}0$. Then

$$-(c_n+c_m)h\leqslant g_n-g_m=(g_n-f)+(f-g_m)\leqslant(c_n+c_m)h$$

and, consequently,

$$\big|\Phi(g_n) - \Phi(g_m)\big| = \big|\Phi(g_n - g_m)\big| \leqslant (c_n + c_m)\Phi(h)$$

since the functional Φ is monotone. Thus, the sequence $\{\Phi(g_n)\}$ is fundamental, and, therefore, the limit $\lim_{n\to\infty} \Phi(g_n)$ exists and is finite. We leave it to the reader to verify that the limit does not depend on the choice of a sequence $\{g_n\}_{n\geqslant 1}$ satisfying conditions (14). Putting $\widetilde{\Phi}(f) = \lim_{n\to\infty} \Phi(g_n)$, we, obviously, obtain a positive functional defined on $C_0(\mathbb{R}^m)$ and extending Φ. By the Riesz–Kakutani theorem, it has an integral representation. Therefore, its restriction, the functional Φ, also has an integral representation. □

12.2.9 Now, we turn to the description of the functionals on the space $C(X)$ of all continuous functions. As in the case of functionals on $C_0(X)$, a linear functional Φ defined on $C(X)$ is called positive if $\Phi(f) \geqslant 0$ for every non-negative function f in $C(X)$. An example of such a functional can be constructed as follows. Let us fix an arbitrary Radon measure μ concentrated on a compact set K, i.e., a measure such that $\mu(X \setminus K) = 0$. This measure is finite since $\mu(X) = \mu(K) < +\infty$. We define the functional Φ by putting

$$\Phi(f) = \int_X f\,d\mu \quad \big(f \in C(X)\big)$$

(the function f is summable since it is bounded on K and the measure μ is finite).

We prove that all positive functionals on $C(X)$ have this form provided that the space X is σ*-compact*, i.e., can be represented as the union of a sequence of compact subsets.

In the proof, we will use Dini's theorem on uniform convergence of a monotone sequence of continuous functions, which says that

$$\text{if } K \text{ is a compact set,} \quad f_n \in C(K), \text{ and } f_n(x) \downarrow 0 \text{ for all } x \in K,$$

then $f_n \rightrightarrows 0$ on K.

Theorem *Let X be a locally compact and σ-compact space, and let Φ be a positive functional on $C(X)$. Then there exists a Radon measure μ concentrated on a compact set and such that*

$$\Phi(f) = \int_X f\,d\mu \quad \textit{for every function } f \textit{ in } C(X). \tag{15}$$

The assumption that the space X is σ-compact is essential. It can be proved that the conclusion of the theorem is false if this assumption is dropped.

Proof We consider the restriction of Φ to $C_0(X)$. By the Riesz–Kakutani theorem, there is a Radon measure μ such that Eq. (15) is valid for all functions in $C_0(X)$.

To complete the proof, we show that, approximating continuous functions by compactly supported ones and passing to the limit, we can prove Eq. (15) in its entirety. To this end, we, obviously, may assume that the function f is non-negative. By assumption, there exist compact sets K_n ($n \in \mathbb{N}$) such that $X = \bigcup_{n=1}^{\infty} K_n$. Let $G_n = \mathrm{Int}(K_n)$. Since the space X is locally compact, every compact set in X is contained in an open set with compact closure. Therefore, we may assume without loss of generality that $K_n \subset G_{n+1}$ for every n.

First of all, we verify that the functional Φ is continuous in the following sense:

$$\text{if } f_n \in C(X) \text{ and } f_n(x) \downarrow 0 \text{ for all } x \in X, \quad \text{then } \Phi(f_n) \underset{n\to\infty}{\longrightarrow} 0.$$

Indeed, assume that $\Phi(f_n) \not\to 0$. Then $\Phi(f_n) \geqslant \varepsilon$ for some $\varepsilon > 0$ and all n. Since, by Dini's theorem, $f_n \rightrightarrows 0$ on each compact set $K \subset X$, we may assume, passing, if necessary, from the sequence $\{f_n\}_{n\geqslant 1}$ to its subsequence that $f_n < 2^{-n}$ on the set K_n. We put $g = \sum_{n=1}^{\infty} n f_n$. Since $\bigcup_{n=1}^{\infty} G_n = X$ and this series converges uniformly on each set $G_n \subset K_n$, the function g is continuous on X. Since $g \geqslant n f_n$ for every n, we obtain

$$\Phi(g) \geqslant \Phi(n f_n) \geqslant n\varepsilon,$$

which, however, is impossible because n is arbitrary. Thus, we have established that $\Phi(f_n) \underset{n\to\infty}{\longrightarrow} 0$, and so Φ is continuous.

Now, we consider functions $g_n \in C_0(X)$ such that

$$0 \leqslant g_n \leqslant 1, \qquad g_n(x) = 1 \quad \text{for } x \in K_n, \quad \text{and} \quad \mathrm{supp}(g_n) \subset G_{n+1}.$$

Such functions exist by Theorem 12.2.1. Obviously, $g_n \uparrow 1$ for all n. Hence, $f_n = f - f g_n \downarrow 0$, and by what was just proved, $\Phi(f) - \Phi(f g_n) = \Phi(f_n) \underset{n\to\infty}{\longrightarrow} 0$. Therefore, we can pass to the limit in the equation

$$\Phi(f g_n) = \int_X f g_n \, d\mu$$

(by the continuity of Φ on the left-hand side and by Levi's theorem on the right-hand side). Passing to the limit, we obtain Eq. (15).

It remains to verify that the measure μ is concentrated on a compact set. Assume the contrary. In this case, $\mu(X \setminus K_n) > 0$ for each n, and, passing, if necessary, from the sequence $\{K_n\}_{n\geqslant 1}$ to its subsequence, we may assume that $\mu(G_{n+1} \setminus K_n) > 0$. By the definition of a Radon measure, there exist compact sets Q_n such that

$$Q_n \subset G_{n+1} \setminus K_n, \quad \mu(Q_n) > 0$$

for each $n \in \mathbb{N}$. We consider continuous functions h_n such that

$$0 \leqslant h_n \leqslant 1, \qquad h_n(x) = 1 \quad \text{for } x \in Q_n, \qquad \mathrm{supp}(h_n) \subset G_{n+1} \setminus K_n.$$

It is clear that the supports of the functions h_n are pairwise disjoint and $\mathrm{supp}(h_{n+1})$ lies outside K_n. Therefore, for all $c_n > 0$, the series $\sum_{n=1}^{\infty} c_n h_n$ converges uniformly on each set K_j, and its sum $\widetilde{g}$, being continuous on every G_j, is also continuous on their union. Consequently, $\widetilde{g} \in C(X)$ and

$$\Phi(\widetilde{g}) \geqslant \Phi(c_n h_n) = c_n \int_X h_n \, d\mu \geqslant c_n \mu(Q_n)$$

for all n. Choosing c_n such that $c_n \mu(Q_n) \underset{n\to\infty}{\longrightarrow} +\infty$, we come to a contradiction, which completes the proof of the theorem. □

12.3 *Bounded Functionals

12.3.1 In the present section, we denote by E either the space $\mathscr{L}^p(X,\mu)$, where μ is a σ-finite measure on X, or the space $C(K)$ of continuous functions on a compact space K. On the space E, we define a natural norm as follows: $\|f\|_p = (\int_X |f|^p \, d\mu)^{1/p}$ if $E = \mathscr{L}^p(X,\mu)$, and $\|f\| = \max_K |f|$ if $E = C(K)$.

We introduce one more definition.

Definition A linear functional Φ defined on E is called *bounded* if there is a number C such that

$$\big|\Phi(f)\big| \leqslant C\|f\| \quad \text{for all } f \in E. \tag{1}$$

The number $\|\Phi\| = \sup\{|\Phi(f)| \mid f \in E, \|f\| \leqslant 1\}$ is called the *norm* of the functional Φ.

It is clear that $\|\Phi\|$ is the least C for which inequality (1) holds.

Let $E = \mathscr{L}^p(X,\mu)$, $1/p + 1/q = 1$, and $h \in \mathscr{L}^q(X,\mu)$. The equation

$$\Phi(f) = \int_X f h \, d\mu \tag{2}$$

defines a linear functional on the space $\mathscr{L}^p(X,\mu)$. This functional is bounded and $\|\Phi\| = \|h\|_q$ (see Remark 12.1.1).

It was noted in the corollary to Theorem 12.1.3 that every order continuous functional on the space $\mathscr{L}^p(X,\mu)$ can be represented in the form (2).

For a finite p, every functional bounded on $\mathscr{L}^p(X,\mu)$ is order continuous. Indeed, if $f_n \downarrow 0$, then $\|f_n\|_p \underset{n\to\infty}{\longrightarrow} 0$, and, therefore, inequality (1) implies that $\Phi(f_n) \underset{n\to\infty}{\longrightarrow} 0$. Thus, Corollary 12.1.3 yields the following description of bounded functionals on $\mathscr{L}^p(X,\mu)$.

Theorem *Let $1 \leqslant p < +\infty$ and $1/p + 1/q = 1$. Every functional Φ bounded on the space $\mathscr{L}^p(X,\mu)$ can be represented in the form (2). The function $h \in \mathscr{L}^q(X,\mu)$ is determined uniquely up to equivalence, and $\|\Phi\| = \|h\|_q$.*

12.3.2 In a compact topological space K, every Borel charge φ is assigned the linear functional on $C(K)$ defined by the equation

$$\Phi(f) = \int_K f\,d\varphi \quad (f \in C(K)).$$

This functional is bounded since, by the properties of an integral with respect to a charge, we have $|\Phi(f)| \leqslant |\varphi|(K)\|f\|$ for every function f in $C(K)$.

We calculate the norm of this functional, considering, for simplicity, the case where the space K is metrizable. This allows us not to worry about the regularity of the measures in question since, by Theorem 13.3.2, in a metrizable space, every finite Borel measure is regular. The charges under consideration can be real as well as complex.

Theorem *Let K be a compact metrizable space, and let Φ be a functional on $C(K)$ corresponding to a Borel charge φ. Then $\|\Phi\| = |\varphi|(K)$.*

Proof The estimate of the functional from above,

$$\|\Phi\| \leqslant |\varphi|(K), \tag{3}$$

follows directly from the inequality $|\Phi(f)| \leqslant |\varphi|(K)\|f\|$ mentioned above.

To obtain the opposite estimate, we consider the density ω of the charge φ with respect to its variation, which will be denoted by μ. By Theorem 11.1.8,

$$\Phi(f) = \int_K f\,d\varphi = \int_K f\omega\,d\mu$$

for every f in $C(K)$. By Corollary 11.2.2, $|\omega| = 1$ almost everywhere with respect to μ.

If the function ω were continuous, then, calculating $\Phi(\overline{\omega})$, we would immediately obtain the required estimate from below for $\|\Phi\|$. Indeed, in this case $\|\omega\| = 1$ and

$$\|\Phi\| \geqslant \Phi(\overline{\omega}) = \int_K |\omega|^2\,d\mu = \int_K 1\,d\mu = |\varphi|(K).$$

However, in general, the function ω is discontinuous and the functional Φ is not defined at $\overline{\omega}$. Therefore, we use functions approximating ω.

We consider continuous functions g_n converging to ω with respect to the $\mathcal{L}^1$-norm (see Theorem 13.3.3). The convergence in the $\mathscr{L}^1$-norm implies the convergence in measure. Therefore, passing to a subsequence, if necessary, we may assume that

$$g_n(x) \underset{n\to\infty}{\longrightarrow} \omega(x) \text{ almost everywhere with respect to } \mu.$$

The functions g_n cannot be used to estimate $\|\Phi\|$ since we know nothing about the maxima of their absolute values. Therefore, we "adjust" them by redefining their values at the points where the values are large. For this, we introduce the function

$$\psi(z) = \begin{cases} z & \text{for } |z| \leqslant 1, \\ z/|z| & \text{for } |z| \geqslant 1 \end{cases}$$

and put $f_n = \psi \circ g_n$. Obviously, the functions f_n are continuous and $|f_n| \leqslant 1$. Therefore,

$$\|\Phi\| \geqslant |\Phi(\overline{f_n})| = \left|\int_K \overline{f_n}\omega \, d\mu\right|. \tag{4}$$

At the same time,

$$f_n(x) = \psi\big(g_n(x)\big) \underset{n\to\infty}{\longrightarrow} \psi\big(\omega(x)\big) = \omega(x)$$

almost everywhere with respect to μ, and, consequently,

$$\int_K \overline{f_n}\omega \, d\mu \underset{n\to\infty}{\longrightarrow} \int_K |\omega|^2 \, d\mu = \int_K 1 \, d\mu = |\varphi|(K).$$

Passing to the limit in inequality (4), we obtain the estimate opposite to (3). □

Remark We supplement the theorem, considering the periodic case. Let $K = [-\pi, \pi]^m$, and let the functional Φ be defined, as in the theorem, by the equation $\Phi(f) = \int_K f \, d\varphi$, but only on the set of 2π-periodic continuous functions, which will be denoted, as usual, by $\widetilde{C}(\mathbb{R}^m)$. If the charge φ is concentrated on the cell $[-\pi, \pi)^m$, then its variation, as before, coincides with the norm of the functional, i.e.,

$$|\varphi|\big([-\pi, \pi]^m\big) = \sup\big\{|\Phi(f)| \mid f \in \widetilde{C}\big(\mathbb{R}^m\big), \ \|f\| \leqslant 1\big\}.$$

To verify this, it is sufficient to repeat the proof of the theorem, observing that the functions g_n may be assumed to be (see Exercise 8, Sect. 9.3) 2π-periodic.

12.3.3 We use the last remark to generalize Theorems 10.3.7 and 11.1.9 on the uniqueness of measures and charges with given Fourier coefficients.

Definition Let φ be a Borel charge on the cube $Q = [-\pi, \pi]^m$. We define the *Fourier coefficients* $\widehat{\varphi}(n)$ *of the charge* φ by the formula

$$\widehat{\varphi}(n) = \frac{1}{(2\pi)^m} \int_Q e^{-i\langle x,n\rangle} \, d\varphi(x) \quad \big(n \in \mathbb{Z}^m\big).$$

It turns out that if φ is concentrated on the cell $P = [-\pi, \pi)^m$ (i.e., if $|\varphi|(Q \setminus P) = 0$), then it is completely determined by the Fourier coefficients.

Theorem *Borel charges on* $[-\pi, \pi)^m$ *with the same Fourier coefficients coincide.*

Proof It is sufficient to prove that a charge with zero Fourier coefficients is equal to zero. Let $\widehat{\varphi}(n) = 0$ for each $n \in \mathbb{Z}^m$. On $\widetilde{C}(\mathbb{R}^m)$, we define the functional Φ by the formula $\Phi(f) = \int_Q f\,d\varphi$. By assumption, the functional Φ vanishes at all exponents $e^{i\langle x,n\rangle}$, and, therefore, at all trigonometric polynomials. By the Weierstrass theorem (see Corollary 7.6.5), for every function $f \in \widetilde{C}(\mathbb{R}^m)$ and every $\varepsilon > 0$, there is a trigonometric polynomial g such that $\|f - g\| < \varepsilon$. Therefore, $|\Phi(f)| = |\Phi(f - g)| \leqslant \|\Phi\|\|f - g\| \leqslant \varepsilon\|\Phi\|$. Since ε is arbitrary, this means that $\Phi(f) = 0$, i.e., that the functional Φ is zero. By the remark to Theorem 12.3.2, we have $|\varphi|(Q) = \|\Phi\| = 0$. Thus, the charge φ is equal to zero. □

12.3.4 Now, we pass to the description of the general form of the bounded functionals on the space of continuous functions.

It is easy to verify that every positive functional (see Definition 12.2.9) defined on a real space $C(K)$ is bounded. Obviously, the difference of two positive functionals is also a bounded functional. Our next goal is to prove that the converse is also true, namely, that the following statement holds.

Theorem *Every bounded functional defined on a real space $C(K)$ is the difference of positive functionals.*

As a preliminary, we prove the following statement.

Lemma *Let $f, g \in C(K)$, $f, g \geqslant 0$, and $h = f + g$. If $w \in C(K)$ and $|w| \leqslant f$, then the function w can be represented in the form $w = u + v$, where $u, v \in C(K)$, $|u| \leqslant f$, $|v| \leqslant g$.*

Proof of the lemma We put

$$u(x) = \begin{cases} 0 & \text{if } h(x) = 0, \\ \frac{w(x)}{h(x)} \cdot f(x) & \text{if } h(x) \neq 0 \end{cases} \quad \text{and} \quad v(x) = \begin{cases} 0 & \text{if } h(x) = 0, \\ \frac{w(x)}{h(x)} \cdot g(x) & \text{if } h(x) \neq 0. \end{cases}$$

The relations $|u| \leqslant f$, $|v| \leqslant g$ and $u + v = w$ are obvious. The continuity of u at x_0, where $h(x_0) = 0$ (the case where $h(x_0) \neq 0$ is trivial), follows from the fact that

$$\bigl|u(x) - u(x_0)\bigr| = \bigl|u(x)\bigr| \leqslant f(x) \to f(x_0) = 0 \quad \text{as } x \to x_0.$$

The continuity of v is proved in the same way. □

Proof of the theorem Let Φ be an order bounded functional defined on $C(K)$. For a non-negative function $f \in C(K)$, we put

$$F_0(f) = \sup\bigl\{\Phi(u) \mid u \in I_f\bigr\}, \quad \text{where } I_f = \bigl\{u \in C(K) \mid |u| \leqslant f\bigr\}.$$

Since the functional Φ is bounded, we have $F_0(f) < +\infty$. From the definition of F_0, it follows immediately that:

(1) $F_0(f) \geqslant 0$, $F_0(0) = 0$;
(2) $|\Phi(f)| = \max\{\Phi(f), \Phi(-f)\} \leqslant F_0(f)$.

We verify that the functional F_0 is additive on the cone of non-negative functions, i.e., that

$$F_0(f+g) = F_0(f) + F_0(g) \quad \text{if } f, g \in C(K),\ f, g \geqslant 0.$$

It follows from the lemma that $I_{f+g} = \{u+v \mid u \in I_f,\ v \in I_g\}$. Therefore,

$$\begin{aligned} F_0(f+g) &= \sup\bigl\{\Phi(u+v) \mid u \in I_f, v \in I_g\bigr\} = \sup\bigl\{\Phi(u) + \Phi(v) \mid u \in I_f, v \in I_g\bigr\} \\ &= \sup\bigl\{\Phi(u) \mid u \in I_f\bigr\} + \sup\bigl\{\Phi(v) \mid v \in I_g\bigr\} = F_0(f) + F_0(g). \end{aligned}$$

The fact that $F_0(af) = a\,F_0(f)$ for $f \geqslant 0$ and $a \geqslant 0$ is proved similarly.

Now, we extend the functional F_0 to $C(K)$, putting (in what follows, as usual, $f_\pm = \max\{\pm f, 0\}$)

$$F(f) = F_0(f_+) - F_0(f_-).$$

Since $f_+ = f$ and $f_- = 0$ for $f \geqslant 0$, we see that F coincides with F_0 on the set of non-negative functions.

We verify that $F(f+g) = F(f) + F(g)$ for all $f, g \in C(K)$. Let $h = f + g$. Then

$$h_+ - h_- = f_+ - f_- + g_+ - g_-, \quad \text{i.e.,} \quad h_+ + f_- + g_- = h_- + f_+ + g_+.$$

Consequently, $F_0(h_+ + f_- + g_-) = F_0(h_- + f_+ + g_+)$. Since F_0 is additive on the cone of non-negative functions, we obtain

$$F_0(h_+) + F_0(f_-) + F_0(g_-) = F_0(h_-) + F_0(f_+) + F_0(g_+).$$

Representing this equation in the form

$$F_0(h_+) - F_0(h_-) = F_0(f_+) - F_0(f_-) + F_0(g_+) - F_0(g_-),$$

we obtain the required result. The fact that $F(af) = aF(f)$ for all $f \in C(K)$ and $a \in \mathbb{R}$ is proved similarly.

Thus, F is a linear functional extending F_0. This functional is positive by property (1). Now, we put $H = F - \Phi$. Then, for $f \geqslant 0$, we obtain

$$H(f) = F(f) - \Phi(f) \geqslant F(f) - \bigl|\Phi(f)\bigr| = F_0(f) - \bigl|\Phi(f)\bigr| \geqslant 0.$$

The last inequality is valid by property (2). Thus, the functional H is positive, and the theorem is proved since $\Phi = F - H$. □

12.3.5 Now, we are able to describe all bounded functionals on the space $C(K)$. For simplicity, we assume that the space K is metrizable.

We already noted that, fixing a Borel charge φ and defining a functional Φ on $C(K)$ by the formula $\Phi(f) = \int_K f\,d\varphi$, we obtain a bounded functional. We verify that every bounded functional can be obtain in this way and that the correspondence between the charges and the bounded functionals is one-to-one.

Theorem *Let K be a compact metrizable space. Every bounded functional Φ on the space $C(K)$ has an integral representation by a charge, i.e.,*

$$\Phi(f) = \int_K f\,d\varphi \text{ for all } f \in C(K), \tag{5}$$

where φ is a charge defined on the σ-algebra of Borel sets. A charge satisfying Eq. (5) is uniquely determined.

Proof First, we assume that the space in question is real. By Theorem 12.3.4, the functional Φ can be represented in the form $\Phi = F - H$, where F and H are positive functionals. By the Riesz–Kakutani theorem, each of the functionals has an integral representation,

$$F(f) = \int_K f\,d\mu, \qquad H(f) = \int_K f\,d\nu \quad \big(f \in C(K)\big),$$

where μ and ν are Radon measures. Therefore, $\mu(K) < +\infty$ and $\nu(K) < +\infty$. To obtain Eq. (5), it remains to put $\varphi = \mu - \nu$. In the complex case, we introduce real functionals $\Psi = \mathcal{R}e\,\Phi$ and $\Theta = \mathcal{I}m\,\Phi$ and consider them only on the space of continuous real functions. Since these functionals are bounded, we obtain by what was just proved that

$$\Psi(f) = \int_K f\,d\psi, \qquad \Theta(f) = \int_K f\,d\theta$$

where ψ and θ are some real Borel charges. Consequently, for a real function f, we have

$$\Phi(f) = \Psi(f) + i\Theta(f) = \int_K f\,d(\psi + i\theta).$$

Since both sides of this equation are linear, the equation is valid not only for real, but also for complex functions, which gives representation (5) with $\varphi = \psi + i\theta$.

Now, we prove that the charge providing representation (5) is unique. Let φ and $\widetilde{\varphi}$ be Borel charges such that

$$\Phi(f) = \int_K f\,d\varphi \quad \text{and} \quad \Phi(f) = \int_K f\,d\widetilde{\varphi} \quad \text{for all } f \in C(K).$$

Subtracting the second equation from the first one, we see that

$$0 = \int_K f\,d(\varphi - \widetilde{\varphi}) \quad \text{for all } f \in C(K).$$

In other words, the charge $\varphi - \widetilde{\varphi}$ generates the zero functional. By Theorem 12.3.2, we obtain $|\varphi - \widetilde{\varphi}|(K) = \|0\| = 0$. Thus, we have proved that $\varphi = \widetilde{\varphi}$. □

12.3.6 In Sects. 12.3.1 and 12.3.5 we obtained theorems on the general form of bounded functionals. Now, we will use them to describe the Fourier series of functions and charges by the Cesàro means (see also Exercises 3–5).

Theorem *Let $\sum_{k=-\infty}^{\infty} c_k e^{ikx}$ be a trigonometric series, and let σ_n be the arithmetic means of its (symmetric) partial sums. This series is a Fourier series:*

(1) *of a function in $\mathscr{L}^p([-\pi,\pi])$ for $1 < p \leqslant +\infty$ if and only if the norms $\|\sigma_n\|_p$ are bounded;*
(2) *of a summable function if and only if the sequence $\{\sigma_n\}_{n\geqslant 1}$ converges in $\mathscr{L}^1$-norm;*
(3) *of a charge if and only if the norms $\|\sigma_n\|_1$ are bounded;*
(4) *of a measure if and only if $\sigma_n(x) \geqslant 0$ for all x and n.*

Proof To verify that condition (1) is necessary, we use the representation $\sigma_n(f) = f * \Phi_n$, where Φ_n is the nth Fejér kernel (see Sect. 10.4.1). Since $\|\Phi_n\|_1 = 1$, Theorem 1 of Sect. 9.3.7 on the properties of convolution implies the required estimate $\|\sigma_n(f)\|_p \leqslant \|f\|_p$.

The necessity of condition (2) is established in Fejér's theorem.

The necessity of conditions (3) and (4) follows from the integral representation of the sum σ_n by the Fejér kernel,

$$\sigma_n(x) = \frac{1}{2\pi} \sum_{|k|<n} \left(1 - \frac{|k|}{n}\right) \int_{[-\pi,\pi]} e^{ik(x-t)}\, d\varphi(t) = \int_{[-\pi,\pi]} \Phi_n(x-t)\, d\varphi(t).$$

Since $\Phi_n \geqslant 0$, we obtain that the sums σ_n are non-negative if φ is a measure. If φ is a charge, then

$$\left|\sigma_n(x)\right| \leqslant \int_{[-\pi,\pi]} \Phi_n(x-t)\, d|\varphi|(t).$$

Integrating this inequality over the interval $[-\pi,\pi]$ and changing the order of integration, we obtain the inequality $\|\sigma_n\|_1 \leqslant |\varphi|([-\pi,\pi])$.

Passing to the proof of sufficiency, we note first that the Fourier coefficients of the functions σ_n have limits as $n \to \infty$. Indeed, $\sigma_n(x) = \sum_{|k|<n}(1 - \frac{|k|}{n}) c_k e^{ikx}$, and, therefore, for $n > |k|$, we obtain

$$\widehat{\sigma}_n(k) = \left(1 - \frac{|k|}{n}\right) c_k \underset{n\to\infty}{\longrightarrow} c_k. \tag{6}$$

(1) Let q the exponent conjugate to p, i.e., $\frac{1}{p} + \frac{1}{q} = 1$ (we note that $q < \infty$). We introduce the functionals H_n on $\mathscr{L}^q([-\pi,\pi])$, putting $H_n(f) = \int_{-\pi}^{\pi} f(x)\sigma_n(x)\, dx$ ($f \in \mathscr{L}^q([-\pi,\pi])$). We remark that, for every n, we have

$$\left|H_n(f)\right| \leqslant \|\sigma_n\|_p \|f\|_q \leqslant C\,\|f\|_q, \tag{7}$$

where $C = \sup_n \|\sigma_n\|_p$. We verify that, for each function $f \in \mathscr{L}^q([-\pi,\pi])$, the limit $\lim_{n\to\infty} H_n(f)$ exists and is finite. Indeed, by (6), this limit exists if f is a trigonometric polynomial. To prove that the limit always exists, we convince ourselves that the sequence $\{H_n(f)\}_{n\geqslant 1}$ is fundamental. We fix an arbitrary $\varepsilon > 0$ and find a trigonometric polynomial T such that $\|f - T\|_q < \varepsilon$. Then

$$\begin{aligned}|H_n(f) - H_m(f)| &\leqslant |H_n(f - T)| + |H_n(T) - H_m(T)| + |H_m(f - T)| \\ &\leqslant C\varepsilon + |H_n(T) - H_m(T)| + C\varepsilon = 2C\varepsilon + |H_n(T) - H_m(T)|\end{aligned}$$

for all $n, m \in \mathbb{N}$. Since the sequence $\{H_n(T)\}_{n\geqslant 1}$ converges, we have $|H_n(T) - H_m(T)| < \varepsilon$ for sufficiently large m and n, which proves that the sequence $\{H_n(f)\}_{n\geqslant 1}$ is fundamental, and, therefore, converges.

We put $H(f) = \lim_{n\to\infty} H_n(f)$. Obviously, H is a linear functional on $\mathscr{L}^q([-\pi,\pi])$. From inequality (7), it follows that the functional is bounded. Therefore, it is generated by a function g in $\mathscr{L}^p([-\pi,\pi])$,

$$H(f) = \int_{-\pi}^{\pi} f(x)\, g(x)\, dx \quad \big(f \in \mathscr{L}^q([-\pi,\pi])\big).$$

Putting $f = e_k$, where $e_k(x) = e^{-ikx}$, we obtain

$$\widehat{g}(k) = \frac{1}{2\pi} H(e_k) = \frac{1}{2\pi} \lim_{n\to\infty} H_n(e_k) = \lim_{n\to\infty} \widehat{\sigma}_n(k) = c_k. \tag{8}$$

Thus, g is the required function.

(2) The case where the sequence $\{\sigma_n\}_{n\geqslant 1}$ converges in the $\mathscr{L}^1$-norm is left to the reader as an exercise.

(3) Let the $\mathcal{L}^1$ norms of the functions σ_n be bounded. Now, we will assume that the functionals H_n are defined not on $\mathscr{L}^q([-\pi,\pi])$ but on the space $C([-\pi,\pi])$. Arguing as above and using the Weierstrass theorem stating that the set of trigonometric polynomials is dense in the space $\widetilde{C}$ of 2π-periodic continuous functions, we obtain that the limit $\lim_{n\to\infty} H_n(f)$ exists for every continuous 2π-periodic function f. For the function $g_0(x) = x$, the sequence $\{H_n(g_0)\}_{n=1}^{\infty}$ is bounded since $|H_n(g_0)| \leqslant \|g_0\| \int_{-\pi}^{\pi} |\sigma_n(x)|\, dx$. Therefore, it is possible to extract a convergent subsequence $\{H_{n_k}(g_0)\}_{k=1}^{\infty}$ of this sequence. Since every continuous function f defined on $[-\pi,\pi]$ can be uniquely represented in the form $f = g + a g_0$, where $g \in \widetilde{C}$ and $a = (f(\pi) - f(-\pi))/(2\pi)$, the functional H can be defined on the entire space $C([-\pi,\pi])$ as the limit $\lim_{k\to\infty} H_{n_k}(f)$. Moreover,

$$|H(f)| \leqslant \sup_n |H_n(f)| \leqslant \sup_n \int_{-\pi}^{\pi} |\sigma_n(x)|\, dx\, \|f\|,$$

and, therefore, the functional H is bounded. By Theorem 12.3.5, the functional is generated by a charge. By an equation similar to (8), one can verify that c_k are the Fourier coefficients of this charge.

(4) Let $\sigma_n \geqslant 0$ for all n. Since

$$\int_{-\pi}^{\pi} |\sigma_n(x)|\, dx = \int_{-\pi}^{\pi} \sigma_n(x)\, dx = c_0,$$

the condition of the previous step of the proof is fulfilled, and, therefore, the functional H is generated by a charge φ whose coefficients coincide with c_k. Since the σ_n are positive, the functionals H_n are positive along with the functional H. Therefore, the charge generating H is a measure. □

EXERCISES

1. We call a linear functional Φ on the space E of measurable functions *order bounded* if $\sup\{|\Phi(g)| \mid g \in E, |g| \leqslant |f|\} < +\infty$ for every function f in E. Generalizing Theorem 12.3.4, prove that, in the real space of measurable functions, every order bounded functional is the difference of positive functionals.
2. Let Φ be an order continuous linear functional on the real space of measurable functions. Without using the integral representation prove that:

   (a) Φ is order bounded;
   (b) the positive part of Φ (the functional F constructed in the proof of Theorem 12.3.4) is also order continuous.

   Assuming that the integral representation of Φ is known, find the integral representation of F.
3. Use Theorem 10.4.7 and the Cesàro means of a multiple trigonometric series to generalize Theorem 12.3.6.
4. Generalize Theorem 12.3.6, replacing the sums σ_n by sums of the form

$$S_{M,\varepsilon}(x) = \sum_{n=-\infty}^{\infty} M\big(\varepsilon|n|\big)c_n e^{inx},$$

   where M is a convex continuous function summable on $[0, +\infty)$.
5. Prove that a trigonometric series $\sum_{n=-\infty}^{\infty} c_n e^{inx}$ is a Fourier series of a function of class $\mathscr{L}^1$ if and only if its Cesàro means σ_n satisfy the condition $\sup_n |\int_e \sigma_n(x)\,dx| \underset{\lambda(e)\to 0}{\longrightarrow} 0$ on the interval $[-\pi, \pi]$.

Chapter 13
Appendices

13.1 An Axiomatic Definition of the Integral over an Interval

13.1.1 Just as the notion of the derivative is related to the tangent line problem, the notion of the integral is related to another classical geometric problem, that of computing the area. There are many ways to introduce the integral. In the simplest case, where we want to define the integral of a continuous function over an interval, we can directly rely on the notion of area. This is especially appropriate if, for some reason or other, the notion of area is assumed known. To emphasize the link between integration and the tangent line problem, one may define the integral as the increment of an antiderivative. Aiming to extend the class of integrable functions, one may define the integral as the limit of Riemann sums (the Riemann integral). In all these cases, we will obtain definitions which are almost, but not completely, equivalent, while retaining the main motivation, the geometric interpretation of the integral.

Our aim in this appendix is to consider one of the possible definitions of the integral over an interval, which could, at the initial stage of learning, precede the measure-theoretic construction. We will describe an axiomatic approach in which the integral is interpreted as a map that associates a certain number with an interval and a continuous function on this interval. This map is subject to two restrictions (axioms) motivated by clear geometric considerations. Our definition can also be extended to more general classes of functions; however, aiming to avoid minor technical issues and to keep the basic idea as clear as possible, we abandon these generalizations, which, in our opinion, are inessential.

13.1.2 In what follows, we consider only real-valued continuous functions on closed finite intervals. A pair $(f, [a,b])$, where f is a function and $[a,b]$ is an interval (possibly, degenerated into a single point) contained in its domain, will be called *admissible*.

Definition An *integral* is a function J defined on the set of admissible pairs and satisfying the following properties:

B. Makarov, A. Podkorytov, *Real Analysis: Measures, Integrals and Applications*, Universitext, DOI 10.1007/978-1-4471-5122-7_13, © Springer-Verlag London 2013

(I) If $(f, [a, b])$ is an admissible pair, then for every $c \in [a, b]$,

$$J\big(f, [a, b]\big) = J\big(f, [a, c]\big) + J\big(f, [c, b]\big)$$

(interval additivity);

(II) if $(f, [a, b])$ is an admissible pair and $A \leqslant f(x) \leqslant B$ for all $x \in [a, b]$, then

$$A(b - a) \leqslant J\big(f, [a, b]\big) \leqslant B(b - a).$$

In particular, the integral over a degenerate interval vanishes.

These axioms become especially natural if, assuming that the notion of area is clear, one interprets the integral $J(f, [a, b])$ for a non-negative function f as the area of the region under the graph of f on the interval $[a, b]$, i.e., the area of the set

$$\big\{(x, y) \in \mathbb{R}^2 \,|\, x \in [a, b],\ 0 \leqslant y \leqslant f(x)\big\}.$$

To justify this interpretation, divide the interval $[a, b]$ into n equal parts $\Delta_1, \dots, \Delta_n$ of length $h_n = \frac{b-a}{n}$, and let m_k and M_k be the smallest and the largest value of f on Δ_k, respectively. Then, by Axiom (II), $m_k h_n \leqslant J(f, \Delta_k) \leqslant M_k h_n$. Adding these inequalities, we see that

$$s_n = \sum_{k=1}^{n} m_k h_n \leqslant J\big(f, [a, b]\big) \leqslant S_n = \sum_{k=1}^{n} M_k h_n.$$

The sums s_n and S_n have a simple geometric interpretation: these are the areas of the polygonal regions composed of the rectangles with bases Δ_k and heights m_k and M_k, respectively. The first of them is contained in the region under the graph of f, and the second one contains it. As n grows, the sums s_n and S_n approach each other arbitrarily closely, since

$$0 \leqslant S_n - s_n = h_n \sum_{k=1}^{n} (M_k - m_k) \leqslant h_n n \omega_f(h_n) = (b - a)\omega_f(h_n) \underset{n\to\infty}{\longrightarrow} 0$$

(here ω_f is the modulus of continuity of f). By the monotonicity, the area of the region under the graph of f (with any reasonable definition of this area) lies between s_n and S_n. Hence it must coincide with $J(f, [a, b])$.

Under this interpretation, Axiom (I) simply means that if we divide the region under the graph of f into two parts by a vertical line, then the area of the whole region equals the sum of the areas of these two parts. The second axiom is just as geometrically clear.

It follows immediately from Axiom (II) that if f takes the same value C at all points of $[a, b]$, then $J(C, [a, b]) = C(b - a)$. Here is another important property of an integral, which is also an immediate consequence of Axiom (II).

Mean Value Theorem *If f is a continuous function on $[a, b]$, then there exists a point $c \in [a, b]$ such that $J(f, [a, b]) = f(c)(b - a)$.*

Proof Let $m = \min_{[a,b]} f$, $M = \max_{[a,b]} f$. Since $m \leqslant f(x) \leqslant M$ for all $x \in [a, b]$, it follows from Axiom (II) that

$$m \leqslant \frac{1}{b-a} J\big(f, [a, b]\big) \leqslant M.$$

It remains to observe that a continuous function takes all values between m and M. □

Leaving aside the problem of the existence of an integral, we establish a fundamental link between this notion and differential calculus, which opens the way for computing the integral in a huge variety of concrete cases. To obtain this result, with every function f continuous on $[a, b]$ we associate another function, the integral "with a variable upper limit". More precisely, we consider the function Φ defined by the formula

$$\Phi(x) = J\big(f, [a, x]\big) \quad \text{for } a \leqslant x \leqslant b. \tag{1}$$

The following theorem is essentially due to Barrow, who stated it in a more complicated geometric form.

Theorem *The function Φ is differentiable at every point x of the interval $[a, b]$, and $\Phi'(x) = f(x)$.*

Proof With every point $y \in [a, b]$ we associate the interval Δ_y with endpoints x and y. If $y > x$, the additivity of the integral immediately implies that $\Phi(y) - \Phi(x) = J(f, \Delta_y)$. Using the mean value theorem, we can rewrite this equation in the form

$$\frac{\Phi(y) - \Phi(x)}{y - x} = f(\overline{y}),$$

where $\overline{y}$ is a point lying between x and y. Interchanging the roles of x and y, we see that this equation also remains valid in the case where $y < x$. By continuity, $f(\overline{y}) \to f(x)$ as $y \to x$, which completes the proof. □

13.1.3 It is useful to slightly reformulate the result obtained in the last theorem. For this we need to introduce a new notion.

Definition Let f and F be functions defined at least on an interval Δ. The function F is called an *antiderivative* of f on Δ if it is differentiable on Δ and

$$F'(x) = f(x) \quad \text{for every } x \text{ in } \Delta.$$

If F_1 and F_2 are two antiderivatives of f on Δ, then their difference is constant, since $(F_1 - F_2)' = 0$.

Using the notion of an antiderivative, we can reformulate Barrow's theorem (keeping the notation introduced in its statement) as follows.

Theorem *The function* Φ *is an antiderivative of* f *on the interval* $[a,b]$.

Barrow's theorem has also a very important corollary, which states the link between integration and differentiation established above in a more convenient form.

Corollary (Fundamental theorem of calculus) *If* f *is a continuous function on an interval* $[a,b]$ *and* F *is an arbitrary antiderivative of* f, *then*

$$J\big(f,[a,b]\big) = F(b) - F(a).$$

The difference $F(b) - F(a)$ is denoted by $F(x)|_{x=a}^{x=b}$ or, in short, $F|_a^b$.

Proof Let Φ be the antiderivative of f defined by (1). Since $F = \Phi + C$ where C is a constant, we have

$$\begin{aligned} J\big(f,[a,b]\big) &= \Phi(b) = \Phi(b) - \Phi(a) \\ &= \big(F(b) - C\big) - \big(F(a) - C\big) = F(b) - F(a). \end{aligned} \qquad \square$$

It follows from the fundamental theorem of calculus that the value of an integral of f over a given interval is uniquely determined by any antiderivative of f; hence there may exist at most one function J satisfying Axioms (I)–(II). Thus we have proved the uniqueness of the integral.

The uniqueness of the integral lays the ground for introducing a special notation. From now on, instead of $J(f,[a,b])$ we will use the generally accepted symbol $\int_a^b f(x)\,dx$ (for a discussion of the term "integral" and the symbol $\int$, see Sect. 4.1.2). The function f is called the integrand; a and b are called the lower and upper limits of integration, respectively; and $[a,b]$ is called the interval of integration.

From a formal point of view, the notation $\int_a^b f(x)\,dx$ is not entirely blameless. It contains a "dummy" letter x (sometimes called the variable of integration), which denotes nothing and can be replaced by any other letter: $\int_a^b f(x)\,dx = \int_a^b f(z)\,dz = \int_a^b f(\aleph)\,d\aleph$. With this in mind, we should prefer the notation $\int_a^b f$. However, the traditional notation has a number of advantages which manifest themselves when solving concrete problems. In particular, this becomes evident if f is defined by a formula involving various letters (parameters). For example, if $f(x) = x^t$ $(x > 0)$, the notation $\int_1^2 x^t\,dx$ shows that we mean the integral of the (power) function f rather than the integral $\int_1^2 x^t\,dt$ of the (exponential) function $t \mapsto g(t) = x^t$. This notation is also convenient when one uses important methods of integration (see Sect. 4.6.2, Propositions 1 and 2 on integration by parts and by substitution).

13.1.4 The established link between integration and differentiation allows us to easily obtain further important properties of the integral: linearity and monotonicity.

Theorem *Let f and g be continuous functions on an interval $[a,b]$ and $\alpha \in \mathbb{R}$. Then*

$$\int_a^b \big(f(x)+g(x)\big)\,dx = \int_a^b f(x)\,dx + \int_a^b g(x)\,dx,$$

$$\int_a^b \big(\alpha f(x)\big)\,dx = \alpha \int_a^b f(x)\,dx.$$

This implies the linearity of the integral:

$$\int_a^b \big(\alpha f(x)+\beta g(x)\big)\,dx = \alpha \int_a^b f(x)\,dx + \beta \int_a^b g(x)\,dx;$$

in particular, $\int_a^b (f(x)-g(x))\,dx = \int_a^b f(x)\,dx - \int_a^b g(x)\,dx$. Applying the latter equation to the difference $f = f_+ - f_-$ (where $f_\pm = \max\{\pm f, 0\}$ are the positive and negative parts of f, respectively), we see that the integral $\int_a^b f(x)\,dx$ is equal to the difference of the areas under the graphs of f_+ and f_-.

Proof Let F and G be antiderivatives of f and g, respectively. Then $F+G$ and αF are antiderivatives of $f+g$ and αf. Hence, by the fundamental theorem of calculus, we have

$$\int_a^b \big(f(x)+g(x)\big)\,dx = (F+G)\Big|_a^b = F\Big|_a^b + G\Big|_a^b = \int_a^b f(x)\,dx + \int_a^b g(x)\,dx.$$

The homogeneity of the integral can be established in a similar way. □

Corollary 1 *If f and g are continuous functions on $[a,b]$ and $f(x) \leqslant g(x)$ for all $x \in [a,b]$, then $\int_a^b f(x)\,dx \leqslant \int_a^b g(x)\,dx$.*

Proof Indeed, since $g - f \geqslant 0$, by Axiom (II) we have $\int_a^b (g(x)-f(x))\,dx \geqslant 0$. Therefore,

$$0 \leqslant \int_a^b \big(g(x)-f(x)\big)\,dx = \int_a^b g(x)\,dx - \int_a^b f(x)\,dx.$$ □

The monotonicity of the integral implies an important estimate.

Corollary 2 *If f is a continuous function on $[a,b]$, then $|\int_a^b f(x)\,dx| \leqslant \int_a^b |f(x)|\,dx$.*

Proof To prove this, it suffices to observe that $-|f(x)| \leqslant f(x) \leqslant |f(x)|$ and use the monotonicity of the integral. □

Now, further properties of the integral of a continuous function over an interval (in particular, Propositions 1 and 2 of Sect. 4.6.2 and Theorem 4.7.3 on the limit of Riemann sums) can be obtained in exactly the same way as in Chap. 4, with the only condition that all intervals under consideration should be assumed closed.

Note that considering Proposition 2 of Sect. 4.6.2, it is convenient to use an agreement slightly generalizing the notion of the integral. By definition, the integral $\int_a^b f(x)\,dx$ makes sense only for $a \leqslant b$. This restriction may sometimes lead to technical problems; to avoid them, for $a > b$ we assume by definition that $\int_a^b f(x)\,dx = -\int_b^a f(x)\,dx$. Obviously, this agreement does not violate the fundamental theorem of calculus.

13.2 Extension of Continuous Functions

Here we consider the following important question: given a function f_0 continuous on a subset A of a metrizable space X, when is it the restriction of a continuous function on the whole space? Or, as one says, when can f_0 be extended to a function continuous on X? Clearly, in general this is not possible. Simple counterexamples are the functions $x \mapsto \frac{1}{x}$ and $x \mapsto \sin\frac{1}{x}$ on the set $A = (0, 1]$. They cannot be extended to functions continuous at the origin, even though the second function is bounded. A condition under which a function can be extended from a set to its closure is given in Exercise 1.

13.2.1 First we establish some auxiliary facts. Recall that $C(X)$ stands for the set of all functions continuous on X.

We will need the notion of the distance from a point to a set. In the particular case where $X = \mathbb{R}^m$, this was introduced in Sect. 3.4.1.

Definition Let (X, ρ) be a metric space and $A \subset X$. Put

$$\operatorname{dist}(x, A) = \inf\{\,\rho(x, y) \mid y \in A\} \qquad (x \in X).$$

The value $\operatorname{dist}(x, A)$ is called the *distance* from x to A.

Lemma 1 *The function $x \mapsto \operatorname{dist}(x, A)$ is continuous on X. If A is closed, then $\operatorname{dist}(x, A) = 0$ if and only if $x \in A$.*

Proof Let $y \in A$ and $x, x' \in X$. Then

$$\operatorname{dist}(x, A) \leqslant \rho(x, y) \leqslant \rho(x', y) + \rho(x', x).$$

Taking the infimum of the right-hand side over y, we see that $\operatorname{dist}(x, A) \leqslant \operatorname{dist}(x', A) + \rho(x, x')$, i.e.,

$$\operatorname{dist}(x, A) - \operatorname{dist}(x', A) \leqslant \rho(x, x').$$

Since x and x' are interchangeable, we have

$$\left|\operatorname{dist}(x, A) - \operatorname{dist}(x', A)\right| \leqslant \rho(x, x'),$$

and the required continuity follows.

The equality $\mathrm{dist}(x, A) = 0$ for $x \in A$ is obvious. If A is closed and $x \notin A$, then there exists an $\varepsilon > 0$ such that the ball $B(x, \varepsilon)$ has an empty intersection with A. This means that $\rho(y, x) \geqslant \varepsilon$ for every $y \in A$, and, consequently, $\mathrm{dist}(x, A) \geqslant \varepsilon > 0$. □

Lemma 2 *Closed disjoint subsets of a metrizable space X are functionally separated, i.e., for any two such subsets F and F_0 there exists a function φ continuous in X such that*

$$\varphi = 1 \quad \text{on } F, \qquad \varphi = 0 \quad \text{on } F_0, \qquad 0 \leqslant \varphi \leqslant 1 \quad \text{on } X.$$

Proof Fix a metric in the space X that induces its topology; this allows us to consider the distance from a point to a set. Define a function φ by the formula

$$\varphi(x) = \frac{\mathrm{dist}(x, F_0)}{\mathrm{dist}(x, F) + \mathrm{dist}(x, F_0)} \quad (x \in X).$$

Since F and F_0 are disjoint, the denominator does not vanish. We leave it to the reader to check that the function φ has all the required properties. □

Corollary *Let $a, b \in \mathbb{R}$ with $a < b$. If F and F_0 are disjoint closed subsets of a metrizable space X, then there exists a function $\psi \in C(X)$ such that*

$$\psi = a \quad \text{on } F, \qquad \psi = b \quad \text{on } F_0, \qquad a \leqslant \psi \leqslant b \quad \text{on } X.$$

Proof Let φ be a function separating F and F_0. Clearly, the function $\psi = b - (b - a)\varphi$ has all the required properties. □

13.2.2 Now we are ready to approach the main problem of this appendix.

Theorem (Tietze[1]–Urysohn[2]) *Every function f_0 continuous on a closed subset of a metrizable space X is the restriction of a function from $C(X)$. If $|f_0| \leqslant C$, one may assume that the extended function also satisfies this inequality.*

Proof Let F be the closed set on which f_0 is defined (and continuous).

I. First consider the case where f_0 is real-valued and bounded: $|f_0| \leqslant C$. Before proving the existence of a continuous extension, we verify that f_0 admits a sufficiently good approximation by functions from $C(X)$.

Let

$$F_- = \left\{ x \in F \;\middle|\; f_0(x) \leqslant -\frac{C}{3} \right\}, \qquad F_+ = \left\{ x \in F \;\middle|\; f_0(x) \geqslant \frac{C}{3} \right\}.$$

[1] Heinrich Franz Friedrich Tietze (1880–1964)—German mathematician.

[2] Pavel Samuilovich Urysohn (1898–1924)—Russian mathematician.

If both these sets are non-empty, then, by the corollary of Lemma 2, there exists a function $g_0 \in C(X)$ such that

$$g_0 = -\frac{C}{3} \quad \text{on } F_-, \qquad g_0 = \frac{C}{3} \quad \text{on } F_+, \qquad |g_0| \leqslant \frac{C}{3} \quad \text{on } X.$$

If F_- (say) is empty, we set $g_0 \equiv \frac{C}{3}$.

For $x \in F \setminus (F_- \cup F_+)$, we have

$$|f_0(x) - g_0(x)| \leqslant |f_0(x)| + |g_0(x)| \leqslant \frac{C}{3} + \frac{C}{3} = \frac{2}{3}C.$$

This inequality remains valid on the union $F_- \cup F_+$; indeed, on this set we have $|g_0| \leqslant |f_0|$, and the values of g_0 have the same sign as the values of f_0, so that $|f_0 - g_0| = |f_0| - |g_0| \leqslant C - \frac{C}{3} = \frac{2}{3}C$. Thus

$$|g_0| \leqslant \frac{1}{3}C \quad \text{on } X, \quad \text{and} \quad |f_0(x) - g_0(x)| \leqslant \frac{2}{3}C \quad \text{for all } x \in F. \tag{1}$$

The function g_0, which is continuous on the whole space X, is a desired approximation for f_0. Now, taking g_0 as the initial approximation and iterating the estimates (1), we successively construct more and more accurate approximations of f_0 by functions continuous on X. Replacing f_0 with $f_1 = f_0 - g_0$ and C with $\frac{2}{3}C$, we can find a function g_1 continuous on X such that

$$|g_1| \leqslant \frac{1}{3} \cdot \frac{2}{3}C \quad \text{on } X, \qquad |f_1(x) - g_1(x)| \leqslant \left(\frac{2}{3}\right)^2 C \quad \text{for } x \in F.$$

Continuing by induction, we construct functions g_n continuous on X and functions $f_n = f_{n-1} - g_{n-1}$ continuous on F such that for all $n \in \mathbb{N}$,

$$|g_n| \leqslant \frac{1}{3}\left(\frac{2}{3}\right)^n C \quad \text{on } X, \qquad |f_n(x)| \leqslant \left(\frac{2}{3}\right)^n C \quad \text{for } x \in F. \tag{2}$$

Adding the equalities

$$\begin{aligned} f_1 &= f_0 - g_0, \\ f_2 &= f_1 - g_1, \\ &\vdots \\ f_{n+1} &= f_n - g_n, \end{aligned}$$

we obtain

$$f_{n+1}(x) = f_0(x) - S_n(x) \quad \text{for } x \in F, \tag{3}$$

where S_n is the nth partial sum of the series $\sum_{k=0}^{\infty} g_k$. By the first inequality (2), this series uniformly converges on X, and, consequently, its sum S is continuous on X. Furthermore, for all $x \in X$,

$$|S(x)| \leqslant \sum_{k=0}^{\infty} |g_k(x)| \leqslant \sum_{k=0}^{\infty} \frac{1}{3}\left(\frac{2}{3}\right)^n C = C.$$

Meanwhile, again by (2), $f_n(x) \underset{n\to\infty}{\longrightarrow} 0$ for $x \in F$. Hence, passing to the limit in (3), we see that $f_0(x) = S(x)$ for all $x \in F$. Thus S is a desired extension.

II. Now let f_0 still be real-valued, but not necessarily bounded. We define an auxiliary function h_0 by the formula

$$h_0(x) = \frac{f_0(x)}{1 + |f_0(x)|} \qquad (x \in F).$$

Clearly, h_0 is continuous and $|h_0| < 1$ on F. One can easily verify that $f_0(x) = \frac{h_0(x)}{1-|h_0(x)|}$. We will first extend the function h_0, and then, using the last formula, construct an extension of f_0.

Let h be a continuous extension of h_0 to X such that $|h| \leqslant 1$. Unfortunately, we cannot directly construct an extension of f_0 to X by the formula $f = \frac{h}{1-|h|}$, since, unlike $|h_0|$, the function $|h|$ may take the value 1. Thus we need to slightly "improve" h. Let $F_0 = \{x \in X \mid |h(x)| = 1\}$. Clearly, the set F_0 is closed and $F_0 \cap F = \varnothing$ (since $|h(x)| = |h_0(x)| < 1$ for $x \in F$). Let φ be a function that separates F and F_0 and vanishes on F_0. Put $H = h\varphi$. Since $|H(x)| < 1$ on X and H coincides with h_0 on F, the function $\frac{H}{1-|H|}$, which is continuous on X, is, obviously, an extension of f_0.

III. If f_0 takes complex values, we can extend it by extending its real and imaginary parts separately (and keeping the estimates on $\mathcal{R}e\, f$ and $\mathcal{I}m\, f$ if f_0 is bounded). Unfortunately, under such an extension, the maximum absolute value of f_0 (if it is bounded) may increase. Hence in this case we should "improve" the extended function f. To this end, we define an auxiliary function ψ on the complex plane by the following formula:

$$\psi(w) = \begin{cases} w & \text{if } |w| \leqslant 1, \\ \frac{w}{|w|} & \text{if } |w| \geqslant 1. \end{cases}$$

Obviously, ψ is continuous and $|\psi| \leqslant 1$. To obtain the desired extension, put

$$\widetilde{f}(x) = C\psi\left(\frac{f(x)}{C}\right) \qquad \text{for } x \in X.$$

□

Remark The proof does not exploit the metrizability of X in full strength, but relies only on Lemma 2. Hence the Tietze–Urysohn theorem holds for all spaces in which closed disjoint sets are functionally separated. In particular, this is the case for all compact spaces (see [B-I], Chap. 2, Sect. 13, Theorem 3).

13.2.3 Given a function (or, more generally, a map) defined on a subset of a metric space, under what conditions is it the restriction of a continuous function defined on a wider set? In other words, when can it be continuously extended? By the theorem just proved, for a function defined on a closed set such an extension always exists, so it is interesting to consider this question for a function defined on an arbitrary domain. A sufficient condition for a function to have an extension to a closed set (which is also necessary in the case of a compact space), is that it is uniformly continuous (see Exercise 1). However, if we require that the ambient set is only Borel, but not necessarily closed, such an extension of a continuous function does always exist. More precisely, the following result holds.

Theorem *Let X, Y be metric spaces, A be a subset of X, and $f : A \to Y$ be a continuous map. If the space Y is complete, then f can be continuously extended to a G_δ set $\widetilde{A}$ containing A.*

Recall that a G_δ set is an intersection of countably many open sets.

Proof First observe that we can (continuously) extend f to a point x from the closure $\overline{A}$ only if f does not change much in the vicinity of x. Hence it makes sense to consider the "oscillation" of the map f at the point x: $\omega(x) = \lim_{r\to 0} \operatorname{diam}(f(B(x,r)\cap A))$. Let

$$A_\varepsilon = \{x \in \overline{A} \mid \omega(x) < \varepsilon\} \quad \text{for } \varepsilon > 0.$$

Let us verify that the set A_ε is open in $\overline{A}$. Indeed, if $x \in A_\varepsilon$, then

$$\operatorname{diam}\bigl(f\bigl(B(x,r)\cap A\bigr)\bigr) < \varepsilon \quad \text{for some } r > 0.$$

Hence $\operatorname{diam}(f(B(y,\rho)\cap A)) < \varepsilon$ for every point y from $B(x,r)\cap\overline{A}$ and every ball $B(y,\rho)$ contained in $B(x,r)$. Therefore, $\omega(y) < \varepsilon$, i.e., $y \in A_\varepsilon$. Since y is an arbitrary point of $B(x,r)\cap\overline{A}$, this means that $B(x,r)\cap\overline{A} \subset A_\varepsilon$. Thus x is an interior point of the set A_ε in the subspace $\overline{A}$. So, the set A_ε is relatively open in $\overline{A}$. Hence there exists a set $\mathcal{O}_\varepsilon$ open in X such that $A_\varepsilon = \overline{A}\cap\mathcal{O}_\varepsilon$. Since the closed set $\overline{A}$ is a G_δ set, the intersection $\overline{A}\cap\mathcal{O}_\varepsilon$, i.e., the set A_ε, is also a G_δ set.

Along with A_ε, the intersection $\widetilde{A} = \bigcap_{n=1}^{\infty} A_{1/n}$ is also a G_δ set. Since f is continuous, it follows that $A_\varepsilon \supset A$, and hence $\widetilde{A} \supset A$.

Now we construct an extension of f to $\widetilde{A}$. If $x \in \widetilde{A}$, then

$$\operatorname{diam}\bigl(f\bigl(B(x,r)\cap A\bigr)\bigr) \to 0 \quad \text{as } r \to 0.$$

The same is true for the diameters of the closures of the sets $f(B(x,r)\cap A)$ (since taking the closure does not change the diameter of a set). Since the space Y is complete, the intersection

$$\bigcap_{r>0} \overline{f\bigl(B(x,r)\cap A\bigr)}$$

is not empty and consists of a single point, say w. If $x \in A$, then $w = f(x)$; otherwise $w = \lim_{y \to x} f(y)$. Thus the set $\widetilde{A}$ is obtained from A by adding the points from $\overline{A}$ at which f has a limit. Associating with an arbitrary point $x \in \widetilde{A}$ the point from $\bigcap_{r>0} \overline{f(B(x,r) \cap A)}$, we obtain a map $\widetilde{f}$ that is an extension of f to $\widetilde{A}$.

It remains to verify that $\widetilde{f}$ is continuous on $\widetilde{A}$. Let $x_0 \in \widetilde{A}$ and $\varepsilon > 0$. Fix a radius r such that $\operatorname{diam}(f(B(x_0,r) \cap A)) < \varepsilon$, and let $x \in \widetilde{A}$ such that $x \in B(x_0, r/2)$. Since $B(x, r/2) \subset B(x_0, r)$, we have

$$\widetilde{f}(x) \in \overline{f\big(B(x,r/2) \cap A\big)} \subset \overline{f\big(B(x_0,r) \cap A\big)}.$$

At the same time, $\widetilde{f}(x_0) \in \overline{f(B(x_0,r) \cap A)}$. Thus, if $x \in B(x_0, r/2) \cap \widetilde{A}$, then

$$\rho_Y\big(\widetilde{f}(x), \widetilde{f}(x_0)\big) \leqslant \operatorname{diam}\big(\overline{f\big(B(x_0,r) \cap A\big)}\big) < \varepsilon,$$

where ρ_Y is the metric in Y. This proves that $\widetilde{f}$ is continuous at x_0, and the theorem follows. □

It is clear that the set $\widetilde{A}$ constructed in the proof is the largest subset of $\overline{A}$ to which f can be continuously extended.

13.2.4 It is natural to ask whether one can extend a continuous function preserving not only the continuity but also some additional properties. The answer is positive for functions satisfying the Lipschitz condition. But it turns out that the real-valuedness of the function is an essential condition. More precisely, the following theorem holds.

Theorem *Let E be an arbitrary subset of a metric space (X, ρ). If a real function f defined on E satisfies the Lipschitz condition, i.e.,*

$$\big|f(x) - f(y)\big| \leqslant L\rho(x, y) \quad \textit{for } x, y \in E,$$

then there exists an extension g of f to X satisfying the Lipschitz condition with the same constant:

$$\big|g(x) - g(y)\big| \leqslant L\rho(x, y) \quad \textit{for } x, y \in X.$$

If f is bounded, then, using the same trick as in the last step of the proof of the Tietze–Urysohn theorem, one can ensure the boundedness of the extended function.

Proof For every x in X, let

$$g(x) = \inf\big\{f(y) + L\rho(x, y) \mid y \in E\big\}.$$

We will verify that g has the desired properties.

First we show that g is an extension of f. Indeed, if $x, y \in E$, then $f(x) - f(y) \leqslant L\rho(x, y)$, whence $f(x) \leqslant g(x)$. The reverse inequality is obvious.

Now we verify that $g(x) > -\infty$ for all x. Indeed, fix an arbitrary point $y_0 \in E$. Then $|f(y_0) - f(y)| \leqslant L\rho(y, y_0)$ for every $y \in E$. Hence for $x \in X$ we have

$$f(y_0) \leqslant f(y) + L\rho(y, y_0) \leqslant f(y) + L\rho(y, x) + L\rho(x, y_0),$$

i.e., $f(y_0) - L\rho(x, y_0) \leqslant f(y) + L\rho(x, y)$, which implies that $g(x) \geqslant f(y_0) - L\rho(x, y_0)$.

Finally, we prove that g satisfies the Lipschitz condition. Let $x, x' \in X$. Fix an arbitrary $\varepsilon > 0$ and choose $y \in E$ such that $g(x) > f(y) + L\rho(x, y) - \varepsilon$. We have

$$g(x') \leqslant f(y) + L\rho(x', y) \leqslant f(y) + L\rho(x', x) + L\rho(x, y).$$

Subtracting the previous inequality from this one, we see that

$$g(x') - g(x) \leqslant L\rho(x', x) + \varepsilon.$$

Since ε is arbitrary and x, x' are interchangeable, this yields the desired result. □

Remark It follows from the above theorem that a map $f : E \to \mathbb{R}^n$ satisfying the Lipschitz condition can be extended to a map defined on X and also satisfying the Lipschitz condition but with a larger constant. However, if $X = \mathbb{R}^m$, there exists an extension having the same Lipschitz constant as f (see [F, Theorem 2.10.43]).

EXERCISES

1. Show that a function uniformly continuous on a subset A of a metric space can be extended to the closure $\overline{A}$ preserving the uniform continuity.
2. Let $T_0 : F \to A$, where F is a closed subset of a metrizable space X and A is a convex closed set in $\mathbb{R}^m$ with $\mathrm{Int}(A) \neq \varnothing$. Show that there exists a continuous extension of T_0 to X whose values also belong to A. *Hint*. Use the fact that bounded set A is homeomorphic to the cube $[-1, 1]^m$.
3. Let $\Delta \subset \mathbb{R}$ be an arbitrary interval and F be a closed subset of a metrizable space X. Show that every continuous function from F to Δ has a continuous extension whose values at all points of $X \setminus F$ belong to the interior of Δ.

13.3 Regular Measures

13.3.1 Among all the measures defined on a σ-algebra of subsets of a topological space, it is natural to single out a class of measures whose properties agree, in some way or other, with the topology. We encountered examples of such an agreement when studying the Lebesgue measure (approximation of measurable functions by continuous functions, etc.).

The first step in this direction is the assumption that the measure under consideration is defined on all open sets; however, this does not suffice. In the general case, the most natural form of agreement between the properties of a measure and the topology of the space is regularity. The definition of regularity reproduces the property of the Lebesgue measure proved in Corollary 2 in Sect. 2.2.2.

We will establish such an agreement for finite Borel measures in metrizable spaces. Note that if one has to consider non-σ-finite measures, there are other possible interpretations of agreement between a measure and a topology (see, for example, the definition of a Radon measure in Sect. 12.2.2).

Recall that a Borel set is an element of the σ-algebra generated by all open sets; a measure defined on this σ-algebra is called a Borel measure.

Definition Let X be a topological space. A measure μ defined on a σ-algebra of subsets of X containing all open sets is called *regular* if for every measurable set E:

(a) $\mu(E) = \inf\{\mu(G) \mid G \supset E,\ G \text{ is an open set}\}$;
(b) $\mu(E) = \sup\{\mu(F) \mid F \subset E,\ F \text{ is a closed set}\}$.

Properties (a) and (b) are called the outer and the inner regularity of μ, respectively. The reader can easily check that for a finite measure they are equivalent. Now we will show that in a wide class of cases, outer regularity implies inner regularity.

Proposition *Let $\mathfrak{A}$ be a σ-algebra of subsets of a topological space X containing all open sets. If a measure μ defined on $\mathfrak{A}$ is σ-finite and outer regular, then it is regular.*

Proof The proof of this result is based on a "duality argument". Here we mean the duality between open and closed sets: the complement of a closed set is open, and the complement of an open set is closed. More precisely, in order to approximate a given set by a closed set, we approximate its complement with an open set and then take the complement.

Let $E \in \mathfrak{A}$ and $E' = X \setminus E$. Since the measure μ is σ-finite, the set E' can be presented in the form $E' = \bigcup_{n=1}^{\infty} E_n$, where $\mu(E_n) < +\infty$ for all n. Fix an arbitrary $\varepsilon > 0$ and, using the outer regularity of μ, find open sets G_n such that

$$G_n \supset E_n, \qquad \mu(G_n) < \mu(E_n) + \frac{\varepsilon}{2^n} \quad (n \in \mathbb{N}).$$

Put $G = \bigcup_{n=1}^{\infty} G_n$, $F = X \setminus G$. We will verify that $\mu(E \setminus F) < \varepsilon$. Indeed, since $E \setminus F = G \setminus E'$, we have

$$\mu(E \setminus F) = \mu(G \setminus E') = \mu\left(\bigcup_{n=1}^{\infty} (G_n \setminus E')\right) \leqslant \sum_{n=1}^{\infty} \mu(G_n \setminus E_n) < \sum_{n=1}^{\infty} \frac{\varepsilon}{2^n} = \varepsilon.$$

Therefore, $\mu(E) = \mu(F) + \mu(E \setminus F) \leqslant \mu(F) + \varepsilon$. Since ε is arbitrary, the latter inequality implies the inner regularity of μ. □

13.3.2 Now we turn to the main result of this appendix.

Theorem *Let X be a metrizable topological space and μ be a Borel measure on X. Let μ satisfy the following condition: there exists a sequence of open sets U_n such that*

$$X = \bigcup_{n=1}^{\infty} U_n \quad \text{and} \quad \mu(U_n) < +\infty \quad \text{for all } n. \tag{1}$$

Then μ is regular.

Condition (1) is a strengthening of the σ-finiteness condition for μ. As one can easily see, it is necessary for a σ-finite measure to be regular. We have to impose this condition, since, as one can see from examples, the σ-finiteness alone is not sufficient for the theorem to be true (see Exercise 1).

Proof We will say that an arbitrary set $E \subset X$ is regular (with respect to μ) if

$$\inf\{\mu(G) \mid G \supset E,\ G \text{ is an open set}\} = \sup\{\mu(F) \mid F \subset E,\ F \text{ is a closed set}\}.$$

Our aim is to prove that all Borel sets are regular. The proof is divided into two steps.

I. First assume that the measure μ is finite. It will be convenient to employ the following reformulation of the definition of a regular set.

A set E is regular if for every positive ε there exist an open set G and a closed set F such that

$$F \subset E \subset G \quad \text{and} \quad \mu(G) - \mu(F) < \varepsilon.$$

We say that such sets ε-approximate E, or form an ε-approximation of E.

In a metrizable space, every closed set is the intersection of a sequence of open sets, and every open set is the union of a sequence of closed sets. Since a measure is continuous from above and from below, we see that open and closed subsets of a metrizable space are regular.

Let us verify that the system of all regular sets is a σ-algebra. This will imply that along with all open sets it also contains all Borel sets, which means precisely that the measure μ is regular.

As we know, in order to prove that the system of regular sets is a σ-algebra, it suffices to check that it has the following two properties (see Proposition 1.1.1 and Definition 1.1.2):

(1) the complement of every regular set is regular;
(2) the union of a sequence of regular sets is a regular set.

To prove (1), we apply, as in the proof of Proposition 13.3.1, a "duality argument". Let E be a regular set and $E' = X \setminus E$. Fix an arbitrary $\varepsilon > 0$ and find sets F and G that ε-approximate E. Put $\widetilde{G} = X \setminus F$ and $\widetilde{F} = X \setminus G$. Clearly, the set $\widetilde{G}$ is open, the set $\widetilde{F}$ is closed, and $\widetilde{G} \setminus \widetilde{F} = G \setminus F$. Furthermore,

$$\widetilde{F} \subset E' \subset \widetilde{G}, \qquad \mu(\widetilde{G}) - \mu(\widetilde{F}) = \mu(G) - \mu(F) < \varepsilon.$$

Thus the sets $\widetilde{F}$ and $\widetilde{G}$ form an ε-approximation of E', which shows that E' is regular.

To establish property (2), consider a sequence of regular sets E_n and put $E = \bigcup_{n=1}^{\infty} E_n$. We will construct sets F and G that ε-approximate E.

Let F_n, G_n be sets that $\frac{\varepsilon}{2^n}$-approximate the sets E_n $(n = 1, 2, \ldots)$. Put $G = \bigcup_{n=1}^{\infty} G_n$ and $A = \bigcup_{n=1}^{\infty} F_n$. Clearly, the set G is open, the set A is Borel, $A \subset E \subset G$, and

$$\mu(G \setminus A) = \mu\left(\bigcup_{n=1}^{\infty}(G_n \setminus A)\right) \leqslant \mu\left(\bigcup_{n=1}^{\infty}(G_n \setminus F_n)\right)$$
$$\leqslant \sum_{n=1}^{\infty}\big(\mu(G_n) - \mu(F_n)\big) < \sum_{n=1}^{\infty} \frac{\varepsilon}{2^n} = \varepsilon. \tag{2}$$

We have constructed two sets approximating E from inside and from outside. The set G is open, but the set A may be not closed. Hence we approximate it from inside by the union of a sufficiently large (but finite!) family of sets F_n. Consider the closed sets $H_n = F_1 \cup \cdots \cup F_n$. Since $\mu(H_n) \underset{n\to\infty}{\longrightarrow} \mu(A)$ by the continuity of μ from below, we have $\mu(A) - \mu(H_n) < \varepsilon$ for sufficiently large n. Hence, in view of (2), we obtain the inequality

$$\mu(G) - \mu(H_n) = \mu(G) - \mu(A) + \mu(A) - \mu(H_n)$$
$$= \mu(G \setminus A) + \mu(A) - \mu(H_n) < \varepsilon + \varepsilon = 2\varepsilon. \tag{3}$$

Thus the sets H_n and G form a 2ε-approximation of E. Since ε is arbitrary, this means that E is regular.

II. Now consider the case where the measure μ is infinite. We will prove that every Borel set is regular (this precisely that μ is regular). Introduce finite measures μ_n $(n = 1, 2, \ldots)$ by the formula

$$\mu_n(B) = \mu(B \cap U_n) \quad (B \text{ is a Borel set}).$$

Note that the measures μ and μ_n coincide on Borel subsets of U_n.

Given an arbitrary Borel set E, write it in the form

$$E = \bigcup_{n=1}^{\infty} E_n, \quad \text{where } E_n = E \cap U_n \quad (n = 1, 2, \ldots).$$

Since the measures μ_n are finite, it follows from above that for every set E_n and arbitrary $\varepsilon > 0$, we can choose open sets G_n and closed sets F_n such that

$$F_n \subset E_n \subset G_n, \qquad \mu_n(G_n) - \mu_n(F_n) < \frac{\varepsilon}{2^n}. \tag{4}$$

We will assume that $G_n \subset U_n$ (otherwise replace G_n by its intersection with U_n, which is also open; this is the only point where the openness of U_n is used). Hence, replacing μ_n by μ, we can rewrite inequality (4) as follows: $\mu(G_n) - \mu(F_n) < \frac{\varepsilon}{2^n}$. Now the proof can be completed in exactly the same way as at the previous step. Put $G = \bigcup_{n=1}^{\infty} G_n$, $A = \bigcup_{n=1}^{\infty} F_n$. Clearly, G is open. Repeating the calculations (2), we see that $\mu(G \setminus A) < \varepsilon$. As at the first step of the proof, consider the closed sets $H_n = F_1 \cup \cdots \cup F_n$. By the continuity of μ from below, $\mu(H_n) \underset{n\to\infty}{\longrightarrow} \mu(A)$. Now two cases are possible. If $\mu(A) < +\infty$, then $\mu(A) - \mu(H_n) < \varepsilon$ for sufficiently large n, and thus (3) holds. So, the sets H_n and G form a 2ε-approximation of E. Since ε is arbitrary, this implies the regularity of E.

If $\mu(A) = +\infty$, then the regularity of E is obvious, since

$$+\infty = \mu(A) = \sup_n \mu(H_n) \leqslant \mu(E) \leqslant \mu(G) \leqslant +\infty.$$

□

Corollary 1 *A finite Borel measure on a metrizable space is regular.*

Corollary 2 *If a Borel measure μ in a Euclidean space is finite on compact subsets, then it is regular. Moreover, in this case, condition* (2) *from the definition of regularity holds in a strengthened form:*

$$\mu(E) = \sup\{\mu(K) \mid K \subset E,\ K \text{ is a compact set}\}. \tag{2'}$$

Equality (2′) immediately follows from (2), since every closed subset of a Euclidean space is the union of a sequence of compact sets.

One can easily check that regularity is preserved under the Carathéodory extension. Thus the above theorem implies, in particular, the regularity of the Lebesgue measure, as well as any measure obtained by the Carathéodory extension from the semiring of cells and finite on compact subsets of $\mathbb{R}^m$.

13.3.3 The regularity of a measure allows one to approximate measurable functions by continuous functions in the $\mathscr{L}^p$ norm.

Theorem *Let X be a metrizable or locally compact space and μ be a regular measure on X. Then for $1 \leqslant p < +\infty$, the set of continuous functions is dense in $\mathscr{L}^p(X, \mu)$.*

Proof Since the set of simple functions is dense in $\mathscr{L}^p(X, \mu)$ (see Theorem 9.2.1), it suffices to show that continuous functions approximate every characteristic function from $\mathscr{L}^p(X, \mu)$, i.e., the characteristic function of an arbitrary set of finite measure. Let E be such a set. Fix an arbitrary $\varepsilon > 0$ and, using the regularity of μ, find an open set G and a closed set F such that

$$F \subset E \subset G, \qquad \mu(G \setminus F) < \varepsilon.$$

By Lemma 2 of Sect. 13.2.1 in the case where X is metrizable, and by Theorem 12.2.1 in the case where X is locally compact, there exists a continuous function φ satisfying the conditions

$$0 \leqslant \varphi \leqslant 1, \qquad \varphi(x) = 1 \quad \text{for } x \in F, \qquad \varphi(x) = 0 \quad \text{for } x \notin G.$$

Hence

$$\|\chi_E - \varphi\|_p^p = \int_G |\chi_E - \varphi|^p \, d\mu \leqslant \int_{G \setminus F} 1 \, d\mu = \mu(G \setminus F) < \varepsilon.$$

Thus χ_E can be approximated by a continuous function with an arbitrary accuracy. □

EXERCISES

1. Show that the Borel measure on the interval [0, 1] generated by the unit masses at the points $\frac{1}{n}$ ($n \in \mathbb{N}$) is inner regular but not regular.

13.4 Convexity

13.4.1 *Convex Sets.* A subset A of $\mathbb{R}^m$ is called *convex* if for any two points p and q in A, the line segment $[p,q] = \{(1-t)p + tq \mid 0 \leqslant t \leqslant 1\}$ also lies in A. One can easily show by induction that for any points $x_1, \ldots, x_n$ in A, a convex combination of these points, i.e., a point of the form $c_1x_1 + \cdots + c_nx_n$ where $c_1, \ldots, c_n$ are non-negative numbers with $c_1 + \cdots + c_n = 1$, also lies in A. Note also that the image of a convex set under an affine map is again convex.

It is clear that the intersection of any family of convex sets is convex. In particular, for every set $A \subset \mathbb{R}^m$, the intersection of all convex sets containing A is convex. This is the smallest convex set containing A. It is called the *convex hull* of A and is denoted by $\operatorname{conv}(A)$. The reader can easily check that this set consists of all convex combinations of points of A. The convex hull of a finite set is called a *convex polyhedron.* A compact convex set with non-empty interior is called a *convex body.*

The *ray* with vertex p and direction vector v, $v \neq 0$, is the set $\ell_p(v) = \{p + tv \mid t \geqslant 0\}$. The union K of an arbitrary family of rays with common vertex p is called a *cone* with vertex p. The set $K \setminus \{p\}$ will also be called a cone.

Now we consider some geometrically clear properties of convex sets.

(1) *The line segment connecting an interior point of a convex set A with a point x_0 of its closure consists (apart from x_0) of interior points of A.*

Indeed, let x_1 be an interior point of A and $B(x_1, r) \subset \operatorname{Int}(A)$. If $x_0 \in A$, then it is easy to check that every point $x_t = (1-t)x_0 + tx_1$, $0 < t < 1$, of the line segment $[x_0, x_1]$ is contained in A along with the ball $B(x_t, tr)$. If x_0 is a boundary point, then, replacing it with a sufficiently close point of A, we see that A contains the ball $B(x_t, \rho)$ for $\rho < tr$.

(2) *Every ray with vertex at an interior point of a convex set intersects its boundary at one point at most.*

Indeed, if such a ray contains two boundary points, then whichever of these points lies closer to the vertex is an interior point, a contradiction.

(3) *The interior of a convex set A is convex. If* $\operatorname{Int}(A) \neq \varnothing$, *then* $\overline{A} = \overline{\operatorname{Int}(A)}$.

Recall that an *affine subspace* of $\mathbb{R}^m$ is a subset obtained by a translation of a linear subspace. We do not exclude the case where an affine subspace consists of a single point. The other extreme case is a (proper) affine subspace of maximal dimension. In short, such a subspace, i.e., a translation of a linear subspace of codimension 1, will be called a *plane*. In other words, a plane is a set of the form $H = \{x \in \mathbb{R}^m \mid \langle x, \nu\rangle = C\}$, where C is a fixed number and $\nu \neq 0$ is a vector (a normal to the plane). Every plane gives rise to two open half-spaces H_- and H_+, whose points x satisfy the inequalities $\langle x, \nu\rangle < C$ and $\langle x, \nu\rangle > C$, respectively. Replacing the strict inequalities by weak ones, we obtain the closed half-spaces $\overline{H}_-$ and $\overline{H}_+$. Half-spaces (open or closed) are the simplest examples of convex sets.

We say that a set lies to one side of a plane H if it is contained in one of the half-spaces $\overline{H}_\pm$. A plane H separates two sets if they lie in different (closed) half-spaces associated with H.

A ball cannot lie to one side of a plane passing through its center. Hence an open set lying to one side of a plane has no common points with this plane. If an open set A is convex, then the absence of common points with a plane H is also sufficient for A to lie to one side of H.

In conclusion of our brief survey, we mention several geometrically clear properties of convex sets (which nevertheless sometimes require non-trivial proofs). We will deduce them from the Hahn–Banach theorem, which will also be used later when studying the differentiability of convex functions. The theorem below is a very special case of the classical result which plays a fundamental role in functional analysis.

Theorem (Hahn–Banach) *Let $\mathcal{O} \neq \varnothing$ be a convex open set in $\mathbb{R}^m$, $m \geqslant 2$, and L be an affine subspace in $\mathbb{R}^m$. If $\mathcal{O} \cap L = \varnothing$, then there exists a plane H such that*

$$H \cap \mathcal{O} = \varnothing, \qquad H \supset L.$$

Proof Applying, if necessary, a translation, we may assume without loss of generality that L is a linear subspace, i.e., $0 \in L$.

First we consider the zero-dimensional case and prove a weaker assertion:

if $0 \notin \mathcal{O}$, then there exists a line passing through 0 *that has an empty intersection with $\mathcal{O}$.*

To prove this, consider the set $K = \bigcup_{t>0} t\mathcal{O}$. Clearly, K is a convex open cone with vertex at the origin. Since $0 \notin K$, we see that K is contained in the punctured space $R = \mathbb{R}^m \setminus \{0\}$, but does not coincide with it (by the convexity of K). The set R is connected (since $m \geqslant 2$), and hence the cone K is not closed in R. Therefore, there exists a point $x \in R$ such that $x \in \partial K$. Then $x \notin K$, since K is open. Let us verify that $tx \notin K$ for $t \in \mathbb{R}$. This is obvious for $t > 0$, since K is a cone. If $tx \in K$ for some $t \leqslant 0$, then, by Property 1), all points of the line segment $[tx, x]$ (apart from x) belong to K. In particular, $0 \in K$, a contradiction. Thus the whole line $\{tx \mid t \in \mathbb{R}\}$ has no common points with K, and hence with $\mathcal{O}$, as required.

We now proceed to proving the theorem in full strength and consider all linear subspaces containing L and disjoint with $\mathcal{O}$. Their dimensions do not exceed $m-1$.

Choose a subspace H of maximal dimension among them. We will show that it is a required plane, i.e., $\dim H = m - 1$. Assume that this is not the case, and let M be the subspace complementary to H (that is, $M \cap H = \{0\}$, $M + H = \mathbb{R}^m$). Then $\dim M \geqslant 2$. Project $\mathcal{O}$ into M along H, and let G be the image of $\mathcal{O}$ in M. Clearly, the image of H in M coincides with the origin. Since $H \cap \mathcal{O} = \varnothing$, it follows that $0 \notin G$. Obviously, the set G is convex and open in M. Thus G satisfies the assumptions of the claim we have proved above. Hence in M there is a line passing through the origin and disjoint with G. Its inverse image H' under the projection onto M has no common points with $\mathcal{O}$ and contains H, but does not coincide with H. Hence $\dim H' > \dim H$, contradicting the choice of H. □

Definition A *supporting* plane to a set A at a point p is a plane H such that $p \in A \cap H$ and A lies to one side of H.

Clearly, a supporting plane to A has an empty intersection with the interior of A.

Corollary 1 *For every boundary point p of a convex set A there is a supporting plane to $\overline{A}$ passing through p.*

Proof Indeed, if A has a non-empty interior, then it suffices to apply the Hahn–Banach theorem to the set $\mathcal{O} = \mathrm{Int}(A)$ with $L = \{p\}$. If $\mathrm{Int}(A) = \varnothing$, then the whole set A is contained in some plane, which is, obviously, a supporting plane. □

Corollary 2 *If A is a closed convex set and $x_0 \notin A$, then x_0 can be strictly separated from A (i.e., there exists a plane H such that A and x_0 lie in different open half-spaces $H_\pm$).*

Corollary 3 *Every convex body is an intersection of closed half-spaces.*

By an *outer normal* to a convex body A at a point $p \in \partial A$ we mean the normal ν to a supporting plane at p that points into the half-space that does not contain A. Formally, this means that $\langle x - p, \nu\rangle \leqslant 0$ on A. Here we do not assume the uniqueness of a supporting plane (cf. the definition of an outer normal in Sect. 8.6.2).

Remark Rays with vertices at different points x and y of the boundary of a convex body and corresponding to outer normals do not intersect. Indeed, such rays, being perpendicular to the supporting planes, cannot make acute angles with the line segment $[x, y]$ (if ν is an outer normal at x, then $\langle y - x, \nu\rangle \leqslant 0$). If the rays had a common point z, then the triangle with vertices x, y, z would have two non-acute angles, a contradiction.

In conclusion of this section, we show that every convex body can be approximated by polyhedra.

Proposition *Let A and $\widetilde{A}$ be convex bodies, $A \subset \mathrm{Int}(\widetilde{A})$. Then there exists a polyhedron C such that $A \subset C \subset \widetilde{A}$.*

In particular, if $0 \in \operatorname{Int}(A)$, then, putting $\widetilde{A} = (1+\varepsilon)A$ ($\varepsilon > 0$), we see that there exists a polyhedron C "arbitrarily close" to A: $A \subset C \subset (1+\varepsilon)A$.

Proof Let $\delta > 0$ be a number such that $\|x - y\| \geqslant \delta$ for $x \in A$ and $y \in \partial\widetilde{A}$. Cover A by finitely many cells whose diameters do not exceed δ. If such a cell has at least one common point with A, it is contained in $\operatorname{Int}(\widetilde{A})$. Hence, in order to obtain a required polyhedron, it suffices to take the convex hull of the vertices of all cells touching A. □

13.4.2 *Metric Projection*. Recall that the distance from a point $x \in \mathbb{R}^m$ to a non-empty subset A of $\mathbb{R}^m$ is defined as $\operatorname{dist}(x, A) = \inf_{a\in A} \|x - a\|$. If there exists a point a_x such that

$$a_x \in A \quad \text{and} \quad \|x - a_x\| = \operatorname{dist}(x, A),$$

then it is called a *best approximation* to x in A. If A is a closed set, then, as follows from the Weierstrass theorem, every point has a best approximation in A.

Lemma 1 *Given a closed convex set A, every point $x \in \mathbb{R}^m$ has a unique best approximation in A.*

Proof The existence of a best approximation has already been mentioned above. To prove the uniqueness, assume that $x \notin A$, i.e., $\operatorname{dist}(x, A) = r > 0$. Let a_x and $\widetilde{a}_x$ be best approximations to x in A: $\|x - a_x\| = \|x - \widetilde{a}_x\| = r$. Consider the point $a = \frac{1}{2}(a_x + \widetilde{a}_x)$. It belongs to the set A, since it is convex and $a_x, \widetilde{a}_x \in A$. Hence $\|x - a\| \geqslant r$. On the other hand, the triangle inequality implies that

$$\|x - a\| = \left\| \frac{1}{2}(x - a_x) + \frac{1}{2}(x - \widetilde{a}_x) \right\| \leqslant \frac{1}{2}r + \frac{1}{2}r = r.$$

Thus $\|x - a\| = r$, i.e., the points a_x, $\widetilde{a}_x$, and $a = \frac{1}{2}(a_x + \widetilde{a}_x)$ lie on the sphere $S(x, r)$, which is possible only if $a_x = \widetilde{a}_x$. □

Lemma 1 allows us to introduce an important map.

Definition Let A be a closed convex set in $\mathbb{R}^m$. Let Φ_A be the map in $\mathbb{R}^m$ that sends each point x of $\mathbb{R}^m$ to the (unique!) best approximation to x in A. This map will be called the *metric projection* onto A.

Obviously, $\Phi_A(x) = x$ if $x \in A$, and, consequently, $\Phi_A(\Phi_A(x)) = \Phi_A(x)$ for every x. Thus Φ_A, just as an ordinary projection, satisfies the equation $\Phi_A^2 = \Phi_A$. Furthermore, if $x \notin A$, then $\Phi_A(x) \in \partial A$.

The metric projection is continuous and even satisfies the Lipschitz condition. More precisely, the following assertion holds.

Lemma 2 *Let $A \subset \mathbb{R}^m$ be a closed convex set. Then*

$$\left\|\Phi_A(v) - \Phi_A(u)\right\| \leqslant \|v - u\| \quad \textit{for all } v, u \textit{ in } \mathbb{R}^m.$$

Proof We may assume without loss of generality that the line segment Δ connecting the points $\Phi_A(u)$ and $\Phi_A(v)$ lies at the first coordinate axis. Let $\Phi_A(u) = (s, 0, \dots, 0)$, $\Phi_A(v) = (t, 0, \dots, 0)$, $s \leqslant t$. The first coordinate u_1 of u does not exceed s, since otherwise the point $(s, 0, \dots, 0)$ would not be the best approximation to u in Δ and, *a fortiori*, in A. For the same reasons, the first coordinate v_1 of v is not less than t. Hence $\|v - u\| \geqslant v_1 - u_1 \geqslant t - s = \|\Phi_A(v) - \Phi_A(u)\|$. □

Lemma 3 *A necessary and sufficient condition for a unit vector ν to be an outer normal to a convex body A at a point $p \in \partial A$ is the following: $\Phi_A(x) = p$ for all points x of the ray $\ell_p(\nu)$ (or for some point $x \neq p$ of this ray).*

Proof Let ν be an outer normal to A at p and $x = p + r\nu$ $(r > 0)$ be an arbitrary point of the ray $\ell_p(\nu)$. Then the closed ball $\overline{B}(x, r)$ contains p and does not contain any other point of A, since these sets are separated by the supporting plane. Hence p is the best approximation to x in A.

Now assume that $p = \Phi_A(x)$ for some point $x \neq p$ of the ray $\ell_p(\nu)$. Consider the plane H passing through p and orthogonal to ν. If it is not supporting for A, then in the open half-space containing x there is a point y belonging to A. Clearly, the angle between the vectors $y - p$ and $x - p$ is acute. Hence the line segment $[p, y] \subset A$ contains points that are closer to x than p, which leads to a contradiction. The details are left to the reader; we recommend to consider the plane cross section containing the vectors $x - p$ and $y - p$. □

The lemma immediately implies the following corollary.

Corollary *If a supporting plane to A at a point p is unique, then $\Phi_A(x) = p$ if and only if x lies on the ray $\ell_p(\nu)$, where ν is the outer normal at p.*

13.4.3 *Convex Functions.* Here we briefly discuss the main properties of convex functions of several variables. Mostly they are natural generalizations of results of the classical one-dimensional analysis that have a clear geometric interpretation.

Definition Let $\mathcal{O}$ be a convex subset of $\mathbb{R}^m$. A function $f : \mathcal{O} \mapsto \mathbb{R}$ is called *convex* if

$$f\big((1-t)x_0 + tx_1\big) \leqslant (1-t)f(x_0) + tf(x_1)$$

for any points $x_0, x_1 \in \mathcal{O}$ and $t \in [0, 1]$.

This inequality becomes an equality for $x_0 = x_1$, and also for $t = 0$ or $t = 1$. If it is strict in all other cases, f is called strictly convex. A function f is called *concave* if $(-f)$ is a convex function.

Note that the domain of definition of a convex function is always assumed to be convex. Since we are interested in differential properties of convex functions, in most cases we assume that it is open.

It follows immediately from the definition that the convexity of a function f is equivalent to the convexity of its *epigraph* $\Gamma_f^+ = \{(x, y) \in \mathbb{R}^{m+1} \,|\, x \in \mathcal{O}, y \geqslant f(x)\}$. The graph $\Gamma_f = \{(x, y) \in \mathbb{R}^{m+1} \,|\, x \in \mathcal{O}, y = f(x)\}$ is contained in the boundary of Γ_f^+, and the interior of Γ_f^+ is non-empty if $\mathrm{Int}(\mathcal{O}) \neq \varnothing$. By Corollary 1 of the Hahn–Banach theorem, for every point $(x_0, f(x_0))$ of the graph, there is a supporting plane to the epigraph passing through it. If $x_0 \in \mathrm{Int}(\mathcal{O})$, then this plane is "not vertical". In other words, its points (x, y) satisfy the equation $y = f(x_0) + \langle v, x - x_0 \rangle$, where v is a vector from $\mathbb{R}^m$ depending on x_0 (as follows from Theorem 13.4.4, if f is differentiable, then $v = \mathrm{grad}\, f(x_0)$). The epigraph lies above this plane, since it consists of the vertical rays $\ell_p(e_{m+1})$, $p \in \Gamma_f$. In particular, the whole graph Γ_f lies above this plane, i.e.,

$$f(x_0) + \langle v, x - x_0 \rangle \leqslant f(x) \quad \text{for every point } x \in \mathcal{O}. \tag{1}$$

If the set $\mathcal{O}$ is open, then a convex combination $x_0 = c_1 x_1 + \cdots + c_n x_n$ of points of $\mathcal{O}$ is an interior point, so that inequality (1) holds. In particular, $f(x_0) + \langle v, x_j - x_0 \rangle \leqslant f(x_j)$ for $j = 1, \ldots, n$. Multiplying this inequality by c_j and adding up all the obtained inequalities, we arrive at *Jensen's inequality*[3]

$$f(c_1 x_1 + \cdots + c_n x_n) \leqslant c_1 f(x_1) + \cdots + c_n f(x_n).$$

One can easily check that it holds without the assumption that $\mathcal{O}$ is open. Note that Jensen's inequality can also be proved by induction without using supporting planes.

Many differential properties of convex functions rely on a simple geometric fact describing the behavior of the slope of the chord connecting two points of the graph.

Three Chords Lemma *Let φ be a function defined on an interval $I \subset \mathbb{R}$. If φ is convex, then for any points $x_0 < x < x_1$ in I,*

$$\frac{\varphi(x) - \varphi(x_0)}{x - x_0} \leqslant \frac{\varphi(x_1) - \varphi(x_0)}{x_1 - x_0} \leqslant \frac{\varphi(x_1) - \varphi(x)}{x_1 - x}.$$

Proof To prove this inequality, it suffices to write x in the form $x = (1 - t)x_0 + t x_1$, $0 < t < 1$, and use the definition of a convex function. □

It follows from this lemma that for every $x \in I$, the difference quotient $\frac{\varphi(x) - \varphi(u)}{x - u}$ grows with $u \in I$ $(u \neq x)$, and hence a convex function φ on an interval has finite one-sided derivatives $\varphi'_-(x)$, $\varphi'_+(x)$, which are increasing and satisfy the inequality $\varphi'_-(x) \leqslant \varphi'_+(x) \leqslant \varphi'_-(\widetilde{x})$ for $x < \widetilde{x}$. It is clear that $\varphi'_+(x) = \varphi'_-(x)$ if both derivatives $\varphi'_\pm$ are continuous at x. Hence

---

[3]Johan Ludwig William Valdemar Jensen (1859–1925)—Danish mathematician.

a convex function φ is differentiable at all but at most countably many points x at which $\varphi'_-(x) < \varphi'_+(x)$.

Since the function $u \mapsto \frac{\varphi(x)-\varphi(u)}{x-u}$ increases on $I \setminus \{x\}$ and its right and left limits at x are equal to $\varphi'_+(x)$ and $\varphi'_-(x)$, respectively, we have

$$\varphi(x) + (u-x)\varphi'_+(x) \leqslant \varphi(u) \quad \text{for } u \geqslant x \quad \text{and}$$

$$\varphi(x) + (u-x)\varphi'_-(x) \leqslant \varphi(u) \quad \text{for } u \leqslant x.$$

Hence the lines passing through the point $p_x = (x, \varphi(x))$ with slopes $\varphi'_\pm(x)$ are supporting to the graph. Moreover, it follows from these inequalities that any line passing through p_x is supporting if its slope lies in the interval between $\varphi'_-(x)$ and $\varphi'_+(x)$. These are all supporting lines to the graph passing through p_x: if $\varphi(x) + \theta \cdot (u-x) \leqslant \varphi(u)$ for all $u \in I$, then $\theta \leqslant \frac{\varphi(u)-\varphi(x)}{u-x}$ for $u > x$, whence $\theta \leqslant \varphi'_+(x)$. The inequality $\theta \geqslant \varphi'_-(x)$ can be proved in a similar way. It follows from this description of supporting lines that

a supporting line at x is unique if and only if $\varphi'_-(x) = \varphi'_+(x)$, i.e., *$\varphi$ is differentiable.*

To prove the main result of this section, it is convenient to use an inequality following from the three chords lemma: if φ is a convex function that satisfies the inequality $|\varphi| \leqslant \Delta$ on $[-h, h]$, then

$$\left|\frac{\varphi(x)-\varphi(\widetilde{x})}{x-\widetilde{x}}\right| \leqslant 4\frac{\Delta}{h} \quad \text{for } |x|, |\widetilde{x}| \leqslant \frac{h}{2}. \tag{2}$$

In the next theorem we show that a convex function is locally Lipschitz and almost everywhere has partial derivatives with respect to all coordinates. The latter result follows from Rademacher's theorem 11.4.2, but for convex functions it can be proved much more easily than in the general case, so we present an independent proof of this fact.

Theorem *Let f be a convex function defined on an open (convex) set $\mathcal{O} \subset \mathbb{R}^m$. Then:*

(1) *f is locally Lipschitz;*
(2) *almost everywhere f has partial derivatives with respect to all coordinates.*

It follows from the theorem that a function that is convex on an arbitrary (convex) set is continuous at all interior points of this set. At boundary points, there may be no continuity.

Proof To prove the first assertion, it suffices to consider the case where $0 \in \mathcal{O}$ and verify that f satisfies the Lipschitz condition in a neighborhood of the origin. First we show that f is bounded near the origin. Take a sufficiently small positive number h such that the cube $[-h, h]^m$ is contained in $\mathcal{O}$. Since every point x in this cube is a convex combination of its vertices $v_1, \dots, v_n$ $(n = 2^m)$, it follows from Jensen's inequality that f is bounded from above: $f(x) \leqslant C = \max_{1 \leqslant j \leqslant n} f(v_j)$. It also follows that f is bounded from below, because $\frac{1}{2}(f(x) + f(-x)) \geqslant f(0)$, whence $f(x) \geqslant 2f(0) - C$. Now (2) immediately implies that in the cube $[-\frac{h}{2}, \frac{h}{2}]^m$ the function f satisfies the Lipschitz condition in each coordinate, which is equivalent to the desired assertion.

Now we proceed to the proof of the second assertion. It suffices to verify that almost everywhere f has a finite partial derivative with respect to the last coordinate. Obviously, for every $x = (x_1, \dots, x_{m-1}, x_m) \in \mathcal{O}$, the function $u \mapsto f(x_1, \dots, x_{m-1}, u)$ is defined and convex on some interval containing x_m. Hence the limits

$$g_\pm(x) = \lim_{u \to \pm 0} \frac{f(x_1, \dots, x_m + u) - f(x_1, \dots, x_m)}{u}$$

exist and are finite. The set of non-differentiable points of a convex function on an interval is at most countable, hence for any $x_1, \dots, x_{m-1}$, the set of x_m such that $x = (x_1, \dots, x_{m-1}, x_m) \in \mathcal{O}$ and $g_-(x) \neq g_+(x)$ is at most countable. Since the functions $g_\pm$ are measurable, the set E on which they do not coincide is measurable. As we have seen, for every point $x' \in \mathbb{R}^{m-1}$, the cross section $E_{x'} = \{t \in \mathbb{R} \mid (x', t) \in E\}$ of E is at most countable and, consequently, has zero measure. By Cavalieri's principle, the measure of E vanishes, i.e., g_+ and g_- coincide almost everywhere in $\mathcal{O}$. Hence almost everywhere in $\mathcal{O}$ the function f has a finite partial derivative $\frac{\partial f}{\partial x_m}$. □

13.4.4 *Differentiability of Convex Functions.* As is well known, the existence of finite partial derivatives is only necessary for a function of several variables to be differentiable. However, for convex functions, this condition is also sufficient. Furthermore, the differentiability of a convex function can be described in geometric terms, using only the notion of a supporting plane.

Theorem *Let f be a convex function defined in an open (convex) set $\mathcal{O} \subset \mathbb{R}^m$ and $a \in \mathcal{O}$. The following conditions are equivalent:*

(1) *f is differentiable at a;*
(2) *the partial derivatives $f'_{x_1}(a), \dots, f'_{x_m}(a)$ exist (and are finite);*
(3) *the epigraph Γ_f^+ has a unique supporting plane at the point $p_a = (a, f(a))$.*

If these conditions are satisfied, then the supporting plane at p_a coincides with the tangent plane.

Proof We will show that (1) ⇒ (2) ⇒ (3) ⇒ (1). The first implication being obvious, we now prove the other ones.

(2) ⇒ (3). Assume that a supporting plane to Γ_f^+ at p_a is given by the equation $y = f(a) + \langle v, x - a\rangle$ (the existence of such a plane was established in the previous section). Then for every vector e_j of the canonical basis in $\mathbb{R}^m$ and every number t with sufficiently small absolute value, $f(a + te_j) \leqslant f(a) + t\langle v, e_j\rangle$.

Therefore, $\langle v, e_j \rangle$ is the slope of a supporting line to the epigraph of the convex function $t \mapsto f(a + te_j)$ at the point $(0, f(a))$. The derivative of this function at $t = 0$ is equal to $f'_{x_j}(a)$. As we have established in the previous section, for a differentiable function of one variable, the tangent line is the only supporting line to the epigraph. Hence $\langle v, e_j \rangle = f'_{x_j}(a)$ for all j. This shows that a supporting plane is unique and coincides with the tangent plane if f is differentiable.

(3) $\Rightarrow$ (1). To simplify calculations, we will assume that $a = 0$. Let $y = f(0) + \langle v, x \rangle$ be the equation of the supporting plane to the epigraph at the origin. Then $f(x) \geqslant f(0) + \langle v, x \rangle$ for all $x \in \mathcal{O}$. The differentiability of f is equivalent to the differentiability of $g(x) = f(x) - f(0) - \langle v, x \rangle$. Clearly, $g(0) = 0$, $g \geqslant 0$ on $\mathcal{O}$, and at the origin Γ_g^+ has a unique supporting plane H_0, which consists of the points of the form $(x, 0)$, $x \in \mathbb{R}^m$.

Before proving the differentiability of the (convex) function g at the origin, we show that at the origin it has a derivative in every direction e and $\frac{\partial g}{\partial e}(0) = 0$.

Consider the function $t \mapsto \varphi(t) = g(te)$, where $|t|$ is sufficiently small. If the derivative $\frac{\partial g}{\partial e}(0)$ does not exist, then at least one of the one-sided derivatives $\varphi'_-(0)$, $\varphi'_+(0)$ does not vanish. For definiteness, let $\varphi'_+(0) \neq 0$. Then the line L passing through the points 0 and $(e, \varphi'_+(0))$ does not lie in the plane H_0. Obviously, L cannot touch the interior of Γ_g^+. By the Hahn–Banach theorem, there exists a supporting plane H to Γ_g^+ that contains L. The plane H does not coincide with H_0, hence at the origin Γ_g^+ has two different supporting planes, which contradicts the assumption.

Thus the derivative $\frac{\partial g}{\partial e}(0)$ exists for every direction e and is equal to zero. In particular, $g(te_j) = o(t)$ as $t \to 0$ $(j = 1, \dots, m)$. Now we prove that g is differentiable at the origin. Since this function is non-negative, it suffices to estimate it from above. Every vector x is the arithmetic mean of the vectors $mx_1e_1, \dots, mx_me_m$, so it follows from Jensen's inequality that $g(x) \leqslant \max_j g(mx_je_j) = o(\|x\|)$ as $x \to 0$. Therefore, the function g, and hence f, is differentiable at the origin.

So, all three conditions stated in the theorem are equivalent. The fact that the supporting plane coincides with the tangent plane was established in the proof of the implication (2) $\Rightarrow$ (3). □

Corollary 1 *A convex function on an open set is differentiable almost everywhere.*

Proof To prove this, it suffices to compare conditions (1) and (2) of Theorem 13.4.4 and condition (2) of Theorem 13.4.3. □

Corollary 2 *Let f be a convex function defined on an open set $\mathcal{O}$. Then its partial derivatives are continuous on the same set where it is differentiable. In particular, if f is differentiable at all points of $\mathcal{O}$, then it is continuously differentiable in $\mathcal{O}$.*

Proof Let E be the set of points where f is differentiable. Assume to the contrary that one of the partial derivatives is not continuous at a point $x_0 \in E$. Then there exists a sequence of points $x^{(n)} \in E$ converging to x_0 such that $\operatorname{grad} f(x^{(n)}) \not\to \operatorname{grad} f(x_0)$. Since f is locally Lipschitz, the sequence $\{\operatorname{grad} f(x^{(n)})\}_{n \geqslant 1}$ is bounded, so that we can pass to a convergent subsequence. We will assume without loss of generality that $\operatorname{grad} f(x^{(n)}) \to \nu \neq \operatorname{grad} f(x_0)$. Clearly, the plane with normal $(\nu, -1)$ passing through $(x_0, f(x_0))$ is supporting to Γ_f^+, since it is the limiting position of the tangent planes to Γ_f^+ at the points $(x^{(n)}, f(x^{(n)}))$. But it does not coincide with the tangent plane, which is also supporting. Thus our assumption leads to a contradiction with condition (3) of the theorem. □

13.4.5 *The Area of Convex Surfaces*. This and the next sections are devoted to studying the properties of the area on convex surfaces. By convex surfaces we mean the boundaries of convex bodies in $\mathbb{R}^m$, and by the area, the $(m-1)$-dimensional area in the sense of Definition 8.2.1. We denote it by σ, and the m-dimensional Lebesgue measure by λ, without indicating the dimension.

First of all, we show that a convex surface is a Lipschitz manifold.

Proposition *Locally, the boundary of a convex body A coincides, up to a rigid motion, with the graph of a convex function and, consequently, admits a bi-Lipschitz parametrization.*

Proof Let $p \in \partial A$. We will assume without loss of generality that 0 is an interior point of A and $p = (0, \dots, 0, c)$, where $c < 0$. Let $B(0, r) \subset A$. Consider a point $x \in B(0, r)$ of the form $x = (x_1, \dots, x_{m-1}, 0)$. Every ray with vertex x and direction vector $(0, \dots, 0, -1) = -e_m$ intersects ∂A only once. This means that we can define a function on the $(m-1)$-dimensional ball $B^{m-1}(0, r)$ whose graph is contained in ∂A. This function is convex, since, by the convexity of A, every chord connecting two points of the graph belongs to A and, consequently, lies above the graph. By Theorem 13.4.3, the canonical parametrization of the graph of a convex function is a locally Lipschitz map. The inverse map is also Lipschitz, since it is a weak contraction. □

Since an area on Lipschitz surfaces is unique (see Theorem 8.8.1), on convex surfaces it is also unique. In particular, under a similarity transformation with ratio $a > 0$, the area of a convex surface (being proportional to the Hausdorff measure) is multiplied by a^{m-1}.

The area of a subset of a graph vanishes with the Lebesgue measure of its projection, hence the existence of a tangent plane for λ-almost all points from the domain of definition of a function f is equivalent to the existence of a tangent plane for σ-almost all points of the graph of f. By the above proposition, a convex surface has a tangent plane at σ-almost all points. Therefore, a convex body has a unique supporting plane at almost all points of its boundary.

The following important theorem allows one to compare the surface areas of convex bodies.

Theorem *Let A and B be convex bodies. If $A \subset B$, then $\sigma(\partial A) \leqslant \sigma(\partial B)$.*

In particular, taking B to be a sufficiently large cube, we see that the boundary of a convex body has finite area.

Proof The proof is quite simple in the case where A is a polyhedron. Indeed, let Γ be one of the $(m-1)$-dimensional faces of A, and let ν be its outer normal. Consider the right prism with base Γ lying outside A, i.e., the set $\Pi_\Gamma = \{p + t\nu \mid p \in \Gamma, t \geqslant 0\}$. It cuts out a "window" $S_\Gamma = \partial B \cap \Pi_\Gamma$ in the boundary of B. Since Γ is the image of this set under the orthogonal projection to the plane of Γ, which is a weak contraction, we have $\sigma(\Gamma) \leqslant \sigma(S_\Gamma)$. By Remark 13.4.1, the sets S_Γ corresponding to different faces are disjoint. Hence

$$\sigma(\partial A) = \sum_\Gamma \sigma(\Gamma) \leqslant \sum_\Gamma \sigma(S_\Gamma) = \sigma\Big(\bigcup_\Gamma S_\Gamma\Big) \leqslant \sigma(\partial B),$$

as required.

In the general case, we use the metric projection $\Phi : \mathbb{R}^m \to A$. Consider a point $p \in \partial A$ and the ray ℓ_p that is perpendicular to a supporting plane passing through p and lies to the other side of this plane from A. Assume that it intersects ∂B at a point x_p. Then, by Lemma 3 of Sect. 13.4.2, p is the best approximation to x_p in A, whence $\Phi(x_p) = p$. Thus ∂A is the image of ∂B under the weak contraction Φ (see Lemma 2 in Sect. 13.4.2), and hence $\sigma(\partial A) \leqslant \sigma(\partial B)$. □

Corollary *Let $B \subset \mathbb{R}^m$ be an arbitrary convex body. Then*

$$\sigma(\partial B) = \sup_{A \subset B} \sigma(\partial A) = \inf_{A \supset B} \sigma(\partial A),$$

where A stands for a convex polyhedron.

Proof It follows from the theorem that $S \equiv \sup \sigma(\partial A) \leqslant \sigma(\partial B)$. Furthermore, we know (see Proposition 13.4.1) that for every $\varepsilon > 0$ there exists a convex polyhedron A such that $A \subset B \subset (1+\varepsilon)A$ (we assume without loss of generality that $0 \in \mathrm{Int}(B)$). Therefore,

$$\sigma(\partial B) \leqslant \sigma\big(\partial(1+\varepsilon)A\big) = (1+\varepsilon)^{m-1}\sigma(\partial A) \leqslant (1+\varepsilon)^{m-1} S.$$

Since ε is arbitrary, we obtain the first equality in question. The second one can be proved in a similar way. □

13.4.6 *Continuity of the Area.* Recall the definition of the Hausdorff metric (see Sect. 8.8.5), which we need in the next proposition. It relies on the notion of the ε-neighborhood $A_\varepsilon = \bigcup_{x \in A} B(x, \varepsilon) = A + B(0, \varepsilon)$ of a set A (as usual, by the sum $X + Y$ of sets X and Y we mean the set $\{x + y \mid x \in X, y \in Y\}$). For bounded sets A and A', the Hausdorff distance ρ is defined by the formula

$$\rho(A, A') = \inf\{\varepsilon > 0 \mid A' \subset A_\varepsilon, A \subset A'_\varepsilon\}.$$

Discussing Schwartz's example in Sect. 8.2.4, we observed that the area of a surface (even a cylindrical one) cannot be defined as the limit of the areas of inscribed polyhedral surfaces. However, the situation changes if the approximating surfaces are convex. More precisely, the following result holds.

Proposition *Let A be a convex body in $\mathbb{R}^m$. If the Hausdorff distance $\rho(\partial A, \partial A')$ between ∂A and the boundary of a convex body A' is sufficiently small, then the areas of ∂A and $\partial A'$ are arbitrarily close.*

Proof We assume without loss of generality that A contains the closed ball $\overline{B}(0, 2r)$. Let us show that $B(0, r) \subset A'$ if $\rho(\partial A, \partial A') < r$. Assume to the contrary that a point x of the ball $B(0, r)$ does not lie in A'. Let y be the closest point to x in A'. The ray with vertex y passing through x intersects the boundary of the ball $\overline{B}(0, 2r)$ at a point z. Since the set A' is convex, y is the closest point to z in A'. Furthermore,

$$\|z - y\| = \|z - x\| + \|x - y\| > \|z - x\| \geqslant \|z\| - \|x\| = 2r - \|x\| > r.$$

Hence $B(z, r)$ has an empty intersection with A', and, consequently, $z \notin A'_r$. On the other hand, since $\partial A \subset (\partial A')_r$, we have $A \subset A'_r$, and, consequently, $z \in \overline{B}(0, 2r) \subset A \subset A'_r$. The obtained contradiction shows that $B(0, r) \subset A'$.

Let ε be an arbitrary number in the interval $(0, r)$, and let $\rho(\partial A, \partial A') < \varepsilon$. Then $\partial A' \subset (\partial A)_\varepsilon$, and, consequently,

$$A' = \operatorname{conv}(\partial A') \subset \operatorname{conv}((\partial A)_\varepsilon) = A_\varepsilon = A + \frac{\varepsilon}{r} B(0, r) \subset A + \frac{\varepsilon}{r} A = \left(1 + \frac{\varepsilon}{r}\right) A$$

(we have used the fact that $B(0, r) \subset A$). By Theorem 13.4.5, the inclusion $A' \subset (1 + \frac{\varepsilon}{r}) A$ allows us to estimate $\sigma(\partial A')$ from above:

$$\sigma(\partial A') \leqslant \sigma\left(\left(1 + \frac{\varepsilon}{r}\right)\partial A\right) = \left(1 + \frac{\varepsilon}{r}\right)^{m-1} \sigma(\partial A).$$

Interchanging A and A', we obtain

$$\sigma(\partial A) \leqslant \left(1 + \frac{\varepsilon}{r}\right)^{m-1} \sigma(\partial A').$$

Thus

$$\left|\sigma(\partial A') - \sigma(\partial A)\right| \leqslant \left(1 + \frac{\varepsilon}{r}\right)^{m-1} \left(\left(1 + \frac{\varepsilon}{r}\right)^{m-1} - 1\right) \sigma(\partial A) \underset{\varepsilon \to 0}{\longrightarrow} 0. \qquad \square$$

13.4.7 *The Area of the Boundary as the Derivative of the Volume.* In Example 4 of Sect. 8.3.5, we observed that the surface area of the sphere coincides with the derivative of the volume of the ball bounded by it (with respect to the radius). This fact has a far-reaching generalization: the surface area of the boundary of the convex body coincides with the Minkowski area (see Sect. 2.8.2). A crucial role in the proof of this result is played by the following lemma.

Lemma *Let $C \subset \mathbb{R}^m$ be a convex polyhedron, $B(0, r) \subset C$ and $\varepsilon > 0$. Then*

$$\lambda(C_\varepsilon \setminus C) \leqslant \varepsilon \left(1 + \frac{\varepsilon}{r}\right)^{m-1} \sigma(\partial C). \tag{3}$$

If A is a convex body containing C, then

$$\varepsilon\sigma(\partial C) \leqslant \lambda(A_\varepsilon \setminus A). \tag{4}$$

Proof First we prove inequality (3). Let

$$C = \bigcap_{i=1}^{N} \{x \in \mathbb{R}^m \mid \langle x, \nu_i \rangle \leqslant \theta_i\}, \quad \text{where } \|\nu_i\| = 1.$$

Obviously, $\theta_i \geqslant r > 0$. Let Γ_i be the face of C contained in the plane H_i given by the equation $\langle x, \nu_i \rangle = \theta_i$. Consider the planes H_i' parallel to H_i given by the equations $\langle x, \nu_i \rangle = \theta_i + \varepsilon$. Let Γ_i' be the central projection of the face Γ_i to H_i'. An easy calculation shows that $\Gamma_i' = \tau_i \Gamma_i$, where $\tau_i = \frac{\theta_i + \varepsilon}{\theta_i} \leqslant 1 + \frac{\varepsilon}{r}$. Hence

$$\sigma(\Gamma_i') = \sigma(\tau_i \Gamma_i) \leqslant \left(1 + \frac{\varepsilon}{r}\right)^{m-1} \sigma(\Gamma_i).$$

Let Q_i be the cone with vertex at the origin formed by the rays passing through the face Γ_i. Obviously, these cones have no common interior points and $\bigcup_{i=1}^N Q_i = \mathbb{R}^m$. Let K_i and K_i' be the intersections of Q_i with the half-spaces $\{x \mid \langle x, \nu_i \rangle \leqslant \theta_i\}$ and $\{x \mid \langle x, \nu_i \rangle \leqslant \theta_i + \varepsilon\}$, respectively. Clearly, $K_i' = \tau_i K_i$ and $\bigcup_{i=1}^N K_i = C$. We will show that $C_\varepsilon \subset C' = \bigcup_{i=1}^N K_i'$ (in general, the set C' is not convex; draw a picture). Let $x \in C_\varepsilon$. Then $x \in Q_i$ for some i. We also have $x = y + z$, where $y \in C$, $\|z\| < \varepsilon$. Therefore,

$$\langle x, \nu_i \rangle = \langle y, \nu_i \rangle + \langle z, \nu_i \rangle \leqslant \theta_i + \|z\| < \theta_i + \varepsilon.$$

By the definition of K_i', this means that $x \in K_i' \subset C'$. Thus $C_\varepsilon \subset C'$, whence

$$C_\varepsilon \setminus C \subset C' \setminus C = \bigcup_{i=1}^{N} K_i' \setminus K_i.$$

The height of the conical frustum $K_i' \setminus K_i$ equals ε, and the areas of the cross sections parallel to the base do not exceed the area of Γ_i'. Hence

$$\lambda(K_i' \setminus K_i) \leqslant \varepsilon\sigma(\Gamma_i').$$

Thus

$$\lambda(C_\varepsilon \setminus C) \leqslant \sum_{i=1}^{N} \varepsilon\sigma(\Gamma_i') \leqslant \varepsilon\left(1 + \frac{\varepsilon}{r}\right)^{m-1} \sum_{i=1}^{N} \sigma(\Gamma_i) = \varepsilon\left(1 + \frac{\varepsilon}{r}\right)^{m-1} \sigma(\partial C),$$

as required.

Now we proceed to the proof of inequality (4). It is simpler than that of (3). Indeed, let Γ be one of the $(m-1)$-dimensional faces of C. Put $\Gamma(\varepsilon) = \{x + t\nu \mid$

$x \in \Gamma,\ 0 < t < \varepsilon\}$, where ν is the unit outer normal to Γ and $C(\varepsilon) = \bigcup_\Gamma \Gamma(\varepsilon)$. By Remark 13.4.1, the sets $\Gamma(\varepsilon)$ corresponding to different faces have no common points. Hence

$$\lambda\big(C(\varepsilon)\big) = \sum_\Gamma \lambda\big(\Gamma(\varepsilon)\big) = \varepsilon \sum_\Gamma \sigma(\Gamma) = \varepsilon\sigma(\partial C). \tag{5}$$

Since every ray perpendicular to the face Γ intersects $A_\varepsilon \setminus A$ in a line segment of length at least ε, it follows from Cavalieri's principle that $\lambda(C(\varepsilon)) \leqslant \lambda(A_\varepsilon \setminus A)$. Hence (5) implies that

$$\varepsilon\sigma(\partial C) = \lambda\big(C(\varepsilon)\big) \leqslant \lambda(A_\varepsilon \setminus A). \qquad \square$$

Theorem *The surface area of the boundary of the convex body $A \subset \mathbb{R}^m$ coincides with its Minkowski area, i.e.,*

$$\lim_{\varepsilon\to 0} \frac{\lambda(A_\varepsilon \setminus A)}{\varepsilon} = \sigma(\partial A).$$

Proof We will assume that $\overline{B}(0,r) \subset \operatorname{Int}(A) \subset \overline{B}(0,R)$.

Let us show that for every $\varepsilon,\ 0 < \varepsilon < 1$,

$$\frac{1}{(1+\varepsilon)^{m-1}}\sigma(\partial A) \leqslant \frac{1}{\varepsilon}\lambda(A_\varepsilon \setminus A) \leqslant \alpha(\varepsilon)\sigma(\partial A), \tag{6}$$

where $\alpha(\varepsilon) \to 1$ as $\varepsilon \to +0$. Clearly, this double inequality implies the required assertion.

Consider a polyhedron C such that $B(0,r) \subset C \subset A \subset (1+\varepsilon^2)C$. By Theorem 13.4.5,

$$\sigma(\partial A) \leqslant \sigma\big(\partial\big(1+\varepsilon^2\big)C\big) = \big(1+\varepsilon^2\big)^{m-1}\sigma(\partial C).$$

This implies the left inequality in (6):

$$\frac{1}{(1+\varepsilon^2)^{m-1}}\sigma(\partial A) \leqslant \sigma(\partial C) \leqslant \frac{1}{\varepsilon}\lambda(A_\varepsilon \setminus A)$$

(at the end, we have used inequality (4)).

Now we will prove the right inequality. Clearly, $A_\varepsilon \subset ((1+\varepsilon^2)C)_\varepsilon$. As the reader can easily check, $((1+\varepsilon^2)C)_\varepsilon \subset C_{\varepsilon+R\varepsilon^2}$. Hence

$$A_\varepsilon \setminus A \subset C_{\varepsilon+R\varepsilon^2} \setminus C.$$

Using inequality (3) with $\varepsilon + R\varepsilon^2$ in place of ε, we see that

$$\begin{aligned}\lambda(A_\varepsilon \setminus A) &\leqslant \lambda(C_{\varepsilon+R\varepsilon^2} \setminus C)\\ &\leqslant \big(\varepsilon + R\varepsilon^2\big)\left(1 + \frac{\varepsilon + R\varepsilon^2}{r}\right)^{m-1}\sigma(\partial C) \leqslant \varepsilon\alpha(\varepsilon)\sigma(\partial A),\end{aligned} \tag{7}$$

where $\alpha(\varepsilon) = (1 + R\varepsilon)(1 + \frac{(R+1)\varepsilon}{r})^{m-1}$. This proves the right inequality in (6). $\square$

This theorem allows us to reformulate the isoperimetric inequality (see Sect. 2.8.2) for convex sets with the area σ instead of the Minkowski area.

Corollary *For every convex body $A \subset \mathbb{R}^m$,*

$$\sigma(\partial A) \geqslant m\alpha_m^{\frac{1}{m}} \lambda^{\frac{m-1}{m}}(A).$$

Since for a ball this inequality becomes an equality, it follows that among all convex bodies of given boundary area, the ball has the greatest volume, and among all convex bodies of given volume, the ball has the smallest boundary surface area.

13.4.8 Let $A \subset \mathbb{R}^m$ be an arbitrary convex body and A_t be its t-neighborhood. Set $V(t) = \lambda(A_t)$ for $t \geqslant 0$, assuming that $A_0 = A$. Together with V, also consider the function $S(t) = \sigma(\partial A_t)$. Obviously, the function V is continuous and increasing. The function S has the same properties: it is increasing by Theorem 13.4.5, and it is continuous by Proposition 13.4.6. Replacing A_ε with A_t and A with $A_{t-\varepsilon}$ in (6) and (7), we see that for $t > 0$ the function V is differentiable not only from the right, but also from the left, both one-sided derivatives being equal to $S(t)$.

Since $V'(t) = S(t)$ and the function S is increasing, it follows that the function V is convex, which allows us to complement the Brunn–Minkowski inequality. Indeed, for $0 \leqslant \alpha \leqslant 1$, we have $V(\alpha s + (1-\alpha)t) \leqslant \alpha V(s) + (1-\alpha)V(t)$. This inequality can be rewritten as follows:

$$\lambda(A_{\alpha s+(1-\alpha)t}) \leqslant \alpha\lambda(A_s) + (1-\alpha)\lambda(A_t);$$

on the other hand, by the Brunn–Minkowski inequality,

$$\lambda^{\frac{1}{m}}(A_{\alpha s+(1-\alpha)t}) \geqslant \alpha\lambda^{\frac{1}{m}}(A_s) + (1-\alpha)\lambda^{\frac{1}{m}}(A_t).$$

Thus we obtain a two-sided bound on the volumes of neighborhoods of the set A:

$$\left(\alpha\lambda^{\frac{1}{m}}(A_s) + (1-\alpha)\lambda^{\frac{1}{m}}(A_t)\right)^m \leqslant \lambda(A_{\alpha s+(1-\alpha)t}) \leqslant \alpha\lambda(A_s) + (1-\alpha)\lambda(A_t).$$

The differentiability of V allows us to modify Theorem 13.4.5 and, so to speak, obtain its prelimit version.

Theorem *Let A and B be convex bodies. If $A \subset B$, then $\lambda(A_\varepsilon \setminus A) \leqslant \lambda(B_\varepsilon \setminus B)$ for every $\varepsilon > 0$.*

Proof Let $F(t) = \lambda(B_t) - \lambda(A_t)$ $(t \geqslant 0)$, where $A_0 = A$ and $B_0 = B$. Since $A_t \subset B_t$, we have $F'(t) = \sigma(\partial(B_t)) - \sigma(\partial(A_t)) \geqslant 0$, and hence the function F is increasing. In particular, $F(\varepsilon) \geqslant F(0)$, which is equivalent to the desired assertion. □

EXERCISES

1. Let f be a convex function on an interval $[a,b]$ satisfying the Lipschitz condition of order α, $0<\alpha\leqslant 1$. Show that $\int_a^b (f''(x))^p\,dx<+\infty$ for every $p\in[0,\frac{1}{2-\alpha})$ (the function $f(x)=-x^\alpha$ on $[0,1]$ shows that the bound on p is sharp for $\alpha\neq 1$).

2. Let A be a convex body and $a\in \mathrm{Int}(A)$. Show that the spherical parametrization of the boundary of A (the positive function r on the unit sphere such that $a+r(\xi)\xi\in\partial A$ for $\|\xi\|=1$) satisfies the Lipschitz condition.

3. Show that the boundary of the ε-neighborhood of an arbitrary convex body is a smooth surface. *Hint*. Use Theorem 13.4.4 and Corollary 2 of this theorem.

4. Show that the volume of convex bodies is continuous in the Hausdorff metric.

5. Show that convex functions enjoy the following property which makes them akin to smooth functions: if a convex function f is differentiable at a point a, then for every $\varepsilon>0$ there exists a neighborhood U of a such that

$$\left|f(y)-f(x)-\langle \operatorname{grad} f(a), y-x\rangle\right|\leqslant\varepsilon\,\|y-x\|\quad\text{for all } x,y\in U.$$

This property is called strict differentiability. The function $x\mapsto f(x)=x^2\sin^2\frac{\pi}{x}$ $(x\neq 0)$, $f(0)=0$ shows that strict differentiability does not follow from differentiability.

13.5 Sard's Theorem

We will prove two particular cases of a theorem dealing with the measure of the set of critical values of a smooth map. First we introduce several necessary definitions.

Definition Let $\mathcal{O}$ be an open subset of $\mathbb{R}^m$ and $\Phi\in C^1(\mathcal{O},\mathbb{R}^d)$, where $d\leqslant m$. A point $x_0\in\mathcal{O}$ is called a critical point of Φ if $\mathrm{rank}(\Phi'(x_0))<d$. The image of a critical point x_0, i.e., $\Phi(x_0)$, is called a critical value of Φ.

The result we are interested in, which is known as Sard's[4] theorem, states that for $d\leqslant m$, the set of critical values of a map $\Phi\in C^k(\mathcal{O},\mathbb{R}^d)$ has zero measure if $k>m-d$. We will prove this assertion in the extreme particular cases $d=m$ and $d=1$.

As one can easily check, the set of critical values of a smooth map is closed in $\mathcal{O}$. Therefore, it can be presented as the union of an at most countable family of compact sets. Hence both this set itself and its image, i.e., the set of critical values, are measurable. Of course, this also follows from Theorem 2.3.1. Recall that the σ-neighborhood A_σ of a set $A\subset\mathbb{R}^m$ is the union $\bigcup_{x\in A}B(x,\sigma)$.

[4] Arthur Sard (1909–1980)—American mathematician.

13.5.1 In this section, by the measure we mean the Lebesgue measure on $\mathbb{R}^m$, which we denote by λ. We need an auxiliary result.

Lemma *Let A be a subset of a proper affine subspace in $\mathbb{R}^m$ contained in a ball of radius R. Then $\lambda(A_\sigma) \leqslant 2^m (R+\sigma)^{m-1}\sigma$.*

Proof Since a rigid motion preserves both the distance between points and the measure of a set, we may assume without loss of generality that the center of the ball coincides with the origin and the subspace consists of the points whose last coordinate vanishes. Then, identifying the space $\mathbb{R}^m$ with the Cartesian product $\mathbb{R}^{m-1} \times \mathbb{R}$ in the canonical way, we see that $A \subset [-R, R]^{m-1} \times \{0\}$. Hence $A_\sigma \subset [-R-\sigma, R+\sigma]^{m-1} \times [-\sigma, \sigma]$, which immediately implies the desired inequality. □

Now we are in a position to establish the first of the results we are interested in.

Theorem *Let $\mathcal{O}$ be an open subset of $\mathbb{R}^m$ and $\Phi \in C^1(\mathcal{O}, \mathbb{R}^m)$. Then the set of critical values of Φ has zero measure.*

Proof Let $\mathcal{N}$ be the set of critical points of Φ. We will show that $\Phi(\mathcal{N})$ is contained in a set of arbitrarily small measure. First consider only a part of $\mathcal{N}$, the intersection of $\mathcal{N}$ with a cell Q whose closure is contained in $\mathcal{O}$.

By Corollary (for $A = \Phi'(x)$) of Lagrange's inequality (see Sect. 13.7.2),

$$\left\|\Phi(x)-\Phi(x_0)-\Phi'(x_0)(x-x_0)\right\| \leqslant \sup_{y\in Q}\left\|\Phi'(y)-\Phi'(x_0)\right\| \|x-x_0\| \quad \text{for } x, x_0 \in Q.$$

Fix a positive ε and, using the uniform continuity of Φ' on $\overline{Q}$, find $\delta > 0$ such that

$$\left\|\Phi'(y)-\Phi'(x_0)\right\| \leqslant \varepsilon \quad \text{if } y,\, x_0 \in Q \text{ and } \|y-x_0\| < \delta.$$

Together with the previous inequality, this yields

$$\left\|\Phi(x)-\Phi(x_0)-\Phi'(x_0)(x-x_0)\right\| \leqslant \varepsilon\|x-x_0\| \quad \text{if } x, x_0 \in Q \text{ and } \|x-x_0\| < \delta. \tag{1}$$

Let $H = \operatorname{diam}(Q)$ and $M = \sup_{x\in Q}\|\Phi'(x)\|$. Divide Q into N^m congruent cells Q_j, $j = 1, 2, \ldots, N^m$, taking N sufficiently large so that $\operatorname{diam}(Q_j) = \frac{H}{N} < \delta$.

Let us estimate the measure of $\Phi(E)$, where $E = Q_j \cap \mathcal{N} \neq \varnothing$. Fix a point $x_0 \in E$ and consider the auxiliary affine map $\Psi(x) = \Phi(x_0) + \Phi'(x_0)(x-x_0)$. On the set Q_j, the maps Φ and Ψ are close: if $x, x_0 \in Q_j$, it follows from (1) that

$$\left\|\Phi(x)-\Psi(x)\right\| \leqslant \varepsilon\|x-x_0\| < \varepsilon \operatorname{diam}(Q_j) = \varepsilon\frac{H}{N} \equiv \sigma.$$

Since $E \subset Q_j$, we see that $\Phi(E)$ is contained in the σ-neighborhood of $\Psi(E)$. On the other hand, since $\det \Phi'(x_0) = 0$, we see that $\Psi(E)$ is a subset of a proper affine subspace. Furthermore, it is contained in a ball of radius $R = \frac{MH}{N}$, because

$$\|\Psi(x) - \Phi(x_0)\| = \|\Phi'(x_0)(x - x_0)\| \leqslant \|\Phi'(x_0)\| \|x - x_0\| \leqslant M\frac{H}{N}.$$

By the lemma,

$$\lambda(\Phi(E)) \leqslant \lambda\big((\Psi(E))_\sigma\big) \leqslant 2^m (R + \sigma)^{m-1}\sigma.$$

Therefore,

$$\lambda(\Phi(\mathcal{N} \cap Q)) = \sum_{j=1}^{N^m} \lambda(\Phi(\mathcal{N} \cap Q_j)) \leqslant N^m 2^m (R+\sigma)^{m-1}\sigma = \varepsilon\,(2H)^m (M+\varepsilon)^{m-1}.$$

Since ε is arbitrary, this means that $\lambda(\Phi(\mathcal{N} \cap Q)) = 0$.

To complete the proof, it remains to observe that $\mathcal{O}$ can be presented as the union of a sequence of cells P_k whose closures are contained in $\mathcal{O}$ (see Theorem 1.1.7). Hence

$$\lambda(\Phi(\mathcal{N})) \leqslant \sum_{k=1}^{\infty} \lambda(\Phi(\mathcal{N} \cap P_k)) = 0.$$

□

13.5.2 Now we proceed to the second particular case of Sard's theorem. In this section, λ stands for the one-dimensional Lebesgue measure and $d_x^k f$ for the kth differential of a function f at a point x.

Theorem *Let $\mathcal{O}$ be an open subset of $\mathbb{R}^m$ and $f \in C^m(\mathcal{O})$. Then the set of critical values of f has zero measure.*

Proof We will prove this theorem not in full strength, but only for infinitely differentiable functions, by induction on the dimension. The induction base, i.e., the case $m = 1$, follows from the previous theorem. Assume that the desired assertion holds for infinitely differentiable functions of $m - 1$ variables; we will show that it holds for infinitely differentiable functions of m variables.

Let $\mathcal{N}_1$ be the set of critical points of f,

$$\mathcal{N}_k = \{x \in \mathcal{O} \mid d_x f = \cdots = d_x^k f = 0\}, \qquad E_k = \mathcal{N}_k \setminus \mathcal{N}_{k+1} \quad (k \in \mathbb{N}).$$

First we will prove that each of the sets $f(E_k)$ has zero measure. By Lindelöf's theorem 8.1.5, which states that every open cover has an at most countable subcover, it suffices to prove a local assertion: every point $x \in E_k$ has a neighborhood U such that $\lambda(f(E_k \cap U)) = 0$. To verify this, we will show that if a neighborhood U is sufficiently small, then $f(E_k \cap U)$ is contained in the set of critical values of a function of $m - 1$ variables.

Let $x_0 \in E_k$. Since $x_0 \notin \mathcal{N}_{k+1}$, at least one of the partial derivatives of f of order k, denote it by g, has a non-zero differential at x_0. Let U be a neighborhood of x_0 such that $d_x g \neq 0$ in U. Then the set $M = \{x \in U \mid g(x) = 0\}$ is a C^∞-smooth surface, and $E_k \cap U \subset M$. Narrowing the neighborhood U if necessary, we may assume that the surface M is simple. Let Φ be an arbitrary infinitely differentiable parametrization of M defined in a domain $G \subset \mathbb{R}^{m-1}$. Consider the auxiliary function $h = f \circ \Phi$, and let A be the set of its critical points. Obviously, $h \in C^\infty(G)$ and, since $d_t h = d_{\Phi(t)} f \circ d_t \Phi$, we have $\Phi^{-1}(E_k \cap U) \subset A$. Hence $f(E_k \cap U) = h(\Phi^{-1}(E_k \cap U)) \subset h(A)$. Therefore, $\lambda(f(E_k \cap U)) = 0$, because $\lambda(h(A)) = 0$ by the induction hypothesis.

To complete the proof, we write the set $\mathcal{N}_1$ as

$$\mathcal{N}_1 = E_1 \cup \cdots \cup E_{m-1} \cup \mathcal{N}_m$$

and prove that $\lambda(f(\mathcal{N}_m)) = 0$. Clearly, it suffices to prove the latter for the intersection of $\mathcal{N}_m$ with an arbitrary compact set contained in $\mathcal{O}$.

Fix such a set K and a δ-neighborhood K_δ of K such that all derivatives of f of order $m+1$ are bounded in K_δ. Then, by Taylor's formula, for some $C > 0$ and $x \in K$ with $\|x - y\| < \delta$, the inequality

$$\big|f(x) - f(y)\big| \leqslant C\|x - y\|^{m+1} \tag{2}$$

holds. Take an arbitrary ε, $0 < \varepsilon < \delta$, and cover $\mathcal{N}_m \cap K$ by pairwise disjoint congruent cubic cells Q_j of diameter ε. Obviously, we may assume that $K \cap Q_j \neq \varnothing$ and, consequently, $Q_j \subset K_\delta$ for all j. It follows from (2) that $\lambda(f(Q_j)) \leqslant C\varepsilon^{m+1} = CL_m\varepsilon\lambda_m(Q_j)$, where the coefficient L_m depends only on the dimension. Hence

$$\lambda\big(f(\mathcal{N}_m \cap K)\big) \leqslant \sum_j \lambda\big(f(Q_j)\big) \leqslant CL_m\varepsilon \sum_j \lambda_m(Q_j) \leqslant CL_m\varepsilon\lambda_m(K_\delta),$$

which implies, since ε is arbitrary, that $\lambda(f(\mathcal{N}_m \cap K)) = 0$. □

13.6 Integration of Vector-Valued Functions

In this appendix, we assume that the reader is familiar with the definition and basic properties of Banach spaces. Our aim is to extend the notion of the integral to maps with values in a Banach space. In what follows, such maps are called vector-valued functions, or vector functions. The method of constructing the integral described in Chap. 4 heavily relies on the fact that the set of real numbers is ordered: the definition of a measurable function already involves sets determined by inequalities. However, the approximation theorem (along with Lebesgue's theorem 4.8.4) shows that there is another approach: a measurable function can be defined as the pointwise limit of a sequence of simple functions, and the integral of such a function, as the limit of the integrals of simple functions satisfying some natural requirements. This approach to the definition of the integral does not rely on the ordering of the real line as heavily as that adopted in Chap. 4 and admits various generalizations. We will consider the generalization of the integral to vector-valued functions suggested by Bochner.[5] Within this approach, according to the scheme described above, one first introduces simple vector functions and defines the integral for them, and then extends the class of integrable functions by a limiting procedure. The properties of the integral defined in this way (except those related to the ordering) are quite similar to the properties of the integral of scalar summable functions. Verifying most of them (first for simple vector functions, and then via a limiting argument) presents no difficulties. In such cases, we confine ourselves to minor hints or say nothing at all.

In what follows, we always assume that there is a fixed space $(X, \mathfrak{A}, \mu)$ with a σ-finite measure and a Banach space E with norm denoted by $\|\cdot\|$; all (scalar or vector-valued) functions under consideration are defined at least almost everywhere on X. Unless otherwise stated, the values of vector-valued functions belong to E. To denote vector-valued functions, we use the symbol $\vec{}$.

13.6.1 *Simple and Measurable Functions.* Recall that a partition of X is a finite family of sets $\{e_k\}_{k=1}^N$, called the elements of the partition, satisfying the following conditions:

$$X = \bigcup_{k=1}^N e_k, \quad e_k \cap e_j = \varnothing \quad \text{for } k \neq j,\ 1 \leqslant k, j \leqslant N.$$

As before, we will consider only partitions with measurable elements. The following definition is an obvious generalization of Definition 3.2.1.

Definition A vector-valued function $\vec{f}$ is called *simple* if there exists a partition of X such that $\vec{f}$ is constant on its elements. Such a partition is called *admissible* for $\vec{f}$.

It is clear that any two simple functions have a common admissible partition. Furthermore, if $\vec{f}$ is simple, then the function $x \mapsto \|\vec{f}(x)\|$ is also a simple (real-valued) function.

Definition A vector-valued function $\vec{f}$ is called *measurable* (synonyms: *strongly measurable, Bochner measurable*) if there exists a sequence of simple functions $\{\vec{f}_n\}_{n\geqslant 1}$ such that

$$\vec{f}_n(x) \underset{n\to\infty}{\longrightarrow} \vec{f}(x) \quad \text{almost everywhere on } X, \quad \text{i.e.,}$$

$$\|\vec{f}_n(x) - \vec{f}(x)\| \underset{n\to\infty}{\longrightarrow} 0 \quad \text{for almost all } x \in X.$$

Since the definition involves not a pointwise approximation, but an approximation in the sense of almost everywhere convergence, modifying the values of a function on a set of zero measure does not affect its measurability.

---

[5] Salomon Bochner (1899–1982)—American mathematician.

In the scalar case, where the Banach space coincides with $\mathbb{R}$, the latter definition is equivalent to Definition 3.1.1 by the corollary of the approximation theorem 3.2.2.

13.6.2 *Properties of Measurable Vector Functions.*

(1) *If $\vec{f}: X \to E$ is a measurable vector function, then the function $x \mapsto \|\vec{f}(x)\|$ is also measurable.*

(2) *Approximation with a Bound. If $\vec{f}$ is a measurable vector-valued function, then there exists a sequence of simple functions $\vec{f}_n$ such that*

$$\vec{f}_n(x) \underset{n\to\infty}{\longrightarrow} \vec{f}(x) \quad \textit{almost everywhere on } X \quad \textit{and}$$

$$\|\vec{f}_n(x)\| \leqslant \|\vec{f}(x)\| \quad \textit{for } x \in X,\ n \geqslant 1.$$

Proof Indeed, let $\{\vec{\varphi}_n\}_{n\geqslant 1}$ be an arbitrary sequence of simple vector-valued functions that converges to $\vec{f}$ almost everywhere. Consider the "cut-off function"

$$\omega(v) = \begin{cases} v & \text{if } \|v\| \leqslant 1, \\ \frac{v}{\|v\|} & \text{if } \|v\| \geqslant 1 \end{cases} \qquad (v \in E).$$

Also consider a sequence of scalar simple functions g_n satisfying the conditions

$$0 \leqslant g_n \leqslant g_{n+1}, \qquad g_n(x) \underset{n\to\infty}{\longrightarrow} \|\vec{f}(x)\| \quad \text{everywhere on } X \tag{1}$$

(see Theorem 3.2.2). Now let

$$\vec{f}_n(x) = \begin{cases} 0 & \text{if } g_n(x) = 0, \\ g_n(x)\omega\left(\frac{\vec{\varphi}_n(x)}{g_n(x)}\right) & \text{if } g_n(x) \neq 0. \end{cases}$$

Since the functions g_n, $\vec{\varphi}_n$ have a common admissible partition, we see that $\vec{f}_n$ is a simple function and, obviously, $\|\vec{f}_n(x)\| \leqslant g_n(x) \leqslant \|\vec{f}(x)\|$. If $\vec{f}(x) = 0$, then the functions $\vec{f}_n(x)$ converge to $\vec{f}(x)$ for trivial reasons, because $g_n(x) = 0$ for all n in view of (1). If $\vec{f}(x) \neq 0$ and $\vec{\varphi}_n(x) \to \vec{f}(x)$, then for sufficiently large n we have $g_n(x) \neq 0$ and, consequently,

$$\vec{f}_n(x) = g_n(x)\omega\left(\frac{\vec{\varphi}_n(x)}{g_n(x)}\right) \underset{n\to\infty}{\longrightarrow} \|\vec{f}(x)\|\omega\left(\frac{\vec{f}(x)}{\|\vec{f}(x)\|}\right) = \vec{f}(x),$$

because $\vec{\varphi}_n(x) \underset{n\to\infty}{\longrightarrow} \vec{f}(x)$. Thus $\vec{f}_n(x) \underset{n\to\infty}{\longrightarrow} \vec{f}(x)$ almost everywhere on X. □

(3) If g is a scalar measurable function and $\vec{f}$ is a vector-valued measurable function, then the vector function $x \mapsto g(x)\vec{f}(x)$ is also measurable. In particular, the vector function $x \mapsto g(x)v_0$, where $v_0 \in E$, is measurable.

(4) Let $X = [a, b]$ and μ be the Lebesgue measure on X. A continuous function $\vec{f}: [a, b] \to E$ is measurable.

Proof The proof relies on the uniform continuity of $\vec{f}$. For every $\varepsilon > 0$ there are points $x_1, \dots, x_n$ such that $a = x_0 < x_1 < \dots < x_n < x_{n+1} = b$ and

$$\|\vec{f}(x) - \vec{f}(x_k)\| < \varepsilon \quad \text{for } x_k \leqslant x \leqslant x_{k+1},\ 0 \leqslant k < n.$$

Hence $\vec{f}$ is uniformly approximated up to ε by the simple functions that take the values $\vec{f}(x_k)$ on the intervals $[x_k, x_{k+1})$ $(k = 0, 1, \dots, n)$. □

In a similar way one can prove that every continuous vector function on a compact space is measurable with respect to every Borel measure.

(5) *The limit of a sequence of measurable vector-valued functions is again measurable.*

Proof Let $\{\vec{f}_n\}_{n\geqslant 1}$ be a sequence of measurable functions, $\vec{f} : X \to E$ be a vector-valued function, and $\vec{f}_n(x) \underset{n\to\infty}{\longrightarrow} \vec{f}(x)$ almost everywhere on X. Let us prove that $\vec{f}$ is measurable. Consider simple functions $\vec{f}_{n,k}$ that approximate $\vec{f}_n$. This means that $\vec{f}_{n,k}(x) \underset{k\to\infty}{\longrightarrow} \vec{f}_n(x)$ almost everywhere on X for every n. Consider also the scalar functions $\varphi_{n,k} = \|\vec{f}_n - \vec{f}_{n,k}\|$. Applying the diagonal sequence theorem 3.3.7 to the functions $\varphi_{n,k}$ with $g_n = h = 0$, we see that $\varphi_{n,k_n}(x) \underset{n\to\infty}{\longrightarrow} 0$ almost everywhere for a sequence $\{k_n\}_{n\geqslant 1}$. Hence

$$\vec{f}_{n,k_n}(x) = \big(\vec{f}_{n,k_n}(x) - \vec{f}_n(x)\big) + \vec{f}_n(x) \underset{n\to\infty}{\longrightarrow} \vec{f}(x) \quad \text{almost everywhere.}$$

Since the functions $\vec{f}_{n,k_n}$ are simple, this means precisely that $\vec{f}$ is measurable. □

Here are two more simple facts; their proofs are left to the reader.

(6) *If $\vec{f}$ is a measurable vector-valued function and Φ is an arbitrary continuous map from E to a Banach space F, then the composition $\Phi \circ \vec{f}$ is also measurable.*
(7) *The set of measurable functions is linear, i.e., a linear combination of any two elements also belongs to this set.*

In conclusion, we prove that a measurable function $\vec{f} : X \to E$ taking values in a closed subspace $F \subset E$ is measurable regarded as an F-valued function. For technical reasons, we state this property in a slightly more general form.

(8) *Let $\vec{f} : X \to E$ be a measurable function such that almost all its values lie in a set $A \subset E$. Then there exists a sequence of simple functions $\vec{g}_n$ with the following properties*:

$$\vec{g}_n(x) \in A, \quad \vec{g}_n(x) \underset{n\to\infty}{\longrightarrow} \vec{f}(x) \quad \textit{for almost all } x \in X.$$

Proof Obviously, we may assume that $\vec{f}(x) \in A$ for all $x \in X$. Let $\{\vec{f}_n\}_{n\geqslant 1}$ be a sequence of simple functions that converges to $\vec{f}$ almost everywhere. In general, the values of $\vec{f}_n$ may not lie in A, so we have to slightly "improve" them.

Let $\{e_k^n\}_{1\leqslant k\leqslant m_n}$ be an admissible partition for $\vec{f}_n$, and let v_k^n be the value of $\vec{f}_n$ on e_k^n. Choose a vector $w_k^n \in A$ such that

$$\|w_k^n - v_k^n\| \leqslant 2\,\mathrm{dist}(v_k^n, A) \quad \text{for } 1\leqslant k\leqslant m_n,\ n\in\mathbb{N}$$

and put

$$\vec{g}_n(x) = w_k^n \quad \text{for } x \in e_k^n.$$

Clearly, for every $x \in X$,

$$\|\vec{g}_n(x) - \vec{f}_n(x)\| \leqslant 2\,\mathrm{dist}(\vec{f}_n(x), A) \leqslant 2\|\vec{f}_n(x) - \vec{f}(x)\|.$$

Hence almost everywhere on X we have

$$\begin{aligned}\|\vec{g}_n(x) - \vec{f}(x)\| &\leqslant \|\vec{g}_n(x) - \vec{f}_n(x)\| + \|\vec{f}_n(x) - \vec{f}(x)\| \\ &\leqslant 3\|\vec{f}_n(x) - \vec{f}(x)\| \underset{n\to\infty}{\longrightarrow} 0. \qquad (2)\end{aligned}$$

□

Remark If the functions $\vec{f}_n$ satisfy the bound $\|\vec{f}_n(x)\| \leqslant h(x)$ for almost all $x \in X$, then, using (2), we obtain

$$\begin{aligned}\|\vec{g}_n(x)\| &\leqslant \|\vec{g}_n(x) - \vec{f}(x)\| + \|\vec{f}(x)\| \\ &\leqslant 3\|\vec{f}_n(x) - \vec{f}(x)\| + h(x) \\ &\leqslant 3\|\vec{f}_n(x)\| + 3\|\vec{f}(x)\| + h(x) \leqslant 7h(x).\end{aligned}$$

13.6.3 *Summable Vector-Valued Functions.* We proceed to the problem of integration of vector-valued functions.

Definition 1 A measurable vector-valued function $\vec{f}$ is called *summable* if the function $\|\vec{f}\|$ is summable, i.e., $\int_X \|\vec{f}\|\,d\mu < +\infty$.

If the measure is finite, then every bounded measurable function is summable. In particular, a continuous vector function defined on a compact subset of $\mathbb{R}^m$ is summable with respect to the Lebesgue measure.

Definition 2 Let $\vec{f}$ be a summable simple function, $\{e_k\}_{k=1}^N$ be an admissible partition for $\vec{f}$, and v_k be the value of $\vec{f}$ on e_k. The *integral* of $\vec{f}$ (over the set X with respect to the measure μ) is the sum $\sum_{k=1}^N \mu(e_k)v_k$.

The measure of e_k may be infinite, but on such a set a summable vector function vanishes. We keep to the standard convention and assume that $+\infty\cdot 0 = 0$. Thus the sum in the definition of the integral always makes sense. Just as in the scalar case, it is easy to check that its value does not depend on the choice of an admissible partition. The integral of a function $\vec{f}$ will, as usual, be denoted by

$$\int_X \vec{f}(x)\,d\mu(x) \quad \text{or} \quad \int_X \vec{f}\,d\mu.$$

Here are some obvious properties of summable simple functions.

(a) The set of summable functions and the set of summable simple functions are linear.
(b) On the set of summable simple functions, the integral is linear.
(c) For every summable simple function $\vec{f}$,

$$\left\| \int_X \vec{f}\,d\mu \right\| \leqslant \int_X \|\vec{f}\|\,d\mu.$$

Before proceeding to the definition of the integral of an arbitrary summable vector-valued function, we prove an auxiliary result.

Lemma *Let $\vec{f} : X \to E$ be a vector-valued function. If there exist simple functions $\vec{f}_n : X \to E$ such that*

(a) *$\vec{f}_n(x) \underset{n\to\infty}{\longrightarrow} \vec{f}(x)$ almost everywhere;*
(b) *there exists a scalar summable function $h : X \to \mathbb{R}$ such that $\|\vec{f}_n(x)\| \leqslant h(x)$ almost everywhere for all $n \in \mathbb{N}$;*

then the limit $\lim_{n\to\infty} \int_X \vec{f}_n\,d\mu$ exists.

Furthermore, this limit does not depend on the choice of sequence $\{\vec{f}_n\}$ satisfying conditions (a)–(b).

Proof It is clear that the functions $\vec{f}_n$ are summable and the function $\vec{f}$ is measurable. Conditions (a)–(b) imply that $\|\vec{f}\| \leqslant h$ almost everywhere. Let $I_n = \int_X \vec{f}_n\,d\mu$. We will prove that $\{I_n\}_{n\geqslant 1}$ is a fundamental sequence. Indeed,

$$\begin{aligned}\|I_n - I_m\| &= \left\| \int_X (\vec{f}_n - \vec{f}_m)\,d\mu \right\| \leqslant \int_X \|\vec{f}_n - \vec{f}_m\|\,d\mu \\ &\leqslant \int_X \|\vec{f}_n - \vec{f}\|\,d\mu + \int_X \|\vec{f} - \vec{f}_m\|\,d\mu \underset{n,m\to\infty}{\longrightarrow} 0.\end{aligned}$$

The convergence to zero follows from Lebesgue's theorem, since the integrands in both integrals converge to zero almost everywhere and are dominated by a summable function ($\|\vec{f}_n - \vec{f}\| \leqslant 2h$). Thus the limit $\lim_{n\to\infty} I_n$ exists.

Now we prove that this limit does not depend on the choice of a sequence $\{\vec{f}_n\}$. Let $\vec{g}_n$ be functions satisfying conditions (a)–(b). Then

$$\left\| I_n - \int_X \vec{g}_n \, d\mu \right\| \leqslant \int_X \|\vec{f}_n - \vec{g}_n\| \, d\mu$$
$$\leqslant \int_X \|\vec{f}_n - \vec{f}\| \, d\mu + \int_X \|\vec{f} - \vec{g}_n\| \, d\mu \underset{n\to\infty}{\longrightarrow} 0. \qquad \square$$

Remark If $\vec{f}$ is a summable function, then, as follows from Property (2) from Sect. 13.6.2 (approximation with a bound), there exists a sequence of simple functions satisfying the conditions of the lemma, with $h = \|\vec{f}\|$.

Now we are ready to give the main definition.

Definition The *integral* of a vector-valued summable function $\vec{f}$ is the limit $\lim_{n\to\infty} \int_X \vec{f}_n \, d\mu$, where $\vec{f}_n$ are simple functions satisfying the conditions of the lemma.

It follows from the lemma that this notion is well defined.

13.6.4 *Basic Properties of the Integral of Summable Vector Functions.*

(1) *Linearity. If $\vec{f}$ and $\vec{g}$ are summable functions, then a linear combination $\alpha\vec{f} + \beta\vec{g}$ is also summable and*

$$\int_X (\alpha\vec{f} + \beta\vec{g}) \, d\mu = \alpha \int_X \vec{f} \, d\mu + \beta \int_X \vec{g} \, d\mu.$$

Proof Indeed, for simple functions, this property is already known (see Property (b) above). In the general case, it can be obtained by a limiting argument. □

(1′) *Factoring Out a Vector Function. Let g be a scalar summable function on X, v be an arbitrary vector in E, and $\vec{f}(x) = g(x)v$ for $x \in X$. Then the vector-valued function $\vec{f}$ is summable and*

$$\int_X \vec{f} \, d\mu = \int_X (gv) \, d\mu = \left(\int_X g \, d\mu\right) v.$$

(2) *Estimate on the Norm of an Integral. For every summable function $\vec{f}$, the inequality $\| \int_X \vec{f} \, d\mu\| \leqslant \int_X \|\vec{f}\| \, d\mu$ holds.*

(3) *Interchange of Limits and Integration. If $\vec{f}$, $\vec{f}_n$ are measurable functions satisfying conditions* (a)–(b) *of the lemma from the previous section, then they are summable and $\int_X \vec{f}_n \, d\mu \underset{n\to\infty}{\longrightarrow} \int_X \vec{f} \, d\mu$.*

Proof The summability of $\vec{f}_n$ follows from condition (b) of the lemma, and the summability of $\vec{f}$ follows from the estimate $\|\vec{f}\| \leqslant h$ obtained by a limiting argument. Using Properties (1), (2) and Lebesgue's theorem for the scalar case, we have

$$\left\| \int_X \vec{f}_n \, d\mu - \int_X \vec{f} \, d\mu \right\| \leqslant \int_X \|\vec{f}_n - \vec{f}\| \, d\mu \underset{n\to\infty}{\longrightarrow} 0,$$

since $2h \geqslant \|\vec{f}_n - \vec{f}\| \underset{n\to\infty}{\longrightarrow} 0$ almost everywhere. □

(4) *Let $\vec{f} : X \to E$ be a summable function and U be a linear map from E to a Banach space F. Then the function $\vec{g} = U \circ \vec{f}$ is summable and $\int_X \vec{g} \, d\mu = U(\int_X \vec{f} \, d\mu)$.*

Proof Indeed, if $\vec{f}$ is a simple function, then the required formula follows from the linearity of U and the definition of the integral of a simple summable function. In the general case, consider a sequence of simple functions $\vec{f}_n$ converging to $\vec{f}$ almost everywhere with norms dominated by a summable function h. Let $\vec{g}_n = U \circ \vec{f}_n$. Clearly, the functions $\vec{g}_n$ are simple and

$$\vec{g}_n(x) = U\big(\vec{f}_n(x)\big) \underset{n\to\infty}{\longrightarrow} U\big(\vec{f}(x)\big) = \vec{g}(x) \quad \text{almost everywhere on } X.$$

Furthermore, $\|\vec{g}_n\| \leqslant \|U\| \|\vec{f}_n\| \leqslant \|U\| h$. Hence the function $\vec{g}$ is summable and, by definition,

$$\int_X \vec{g} \, d\mu = \lim_{n\to\infty} \int_X \vec{g}_n \, d\mu = \lim_{n\to\infty} U\left(\int_X \vec{f}_n \, d\mu\right) = U\left(\int_X \vec{f} \, d\mu\right). \qquad \square$$

Note the following special case of Property (4).

(4′) *If $\vec{f}$ is a summable function and φ is a linear continuous functional in E, then $\varphi(\int_X \vec{f} \, d\mu) = \int_X \varphi(\vec{f}) \, d\mu$, where the right-hand side is the integral of a scalar summable function.*

The next property can be proved similarly to Property (4), as the reader can easily verify.

(5) *Let $E = L(H, H_1)$ be the space of linear continuous operators acting from a Banach space H to a Banach space H_1, and let $U_x : X \to L(H, H_1)$ be an operator-valued summable function. Then for every $v \in H$, the function $x \mapsto U_x(v) \in H_1$ is also summable and*

$$\int_X U_x(v) \, d\mu(x) = \left(\int_X U_x \, d\mu(x)\right)(v).$$

(6) *Let L be a closed linear subset of E and $\vec{f} : X \to E$ be a summable function. If $\vec{f}(x) \in L$ for almost all $x \in X$, then $\int_X \vec{f} \, d\mu \in L$. If $\mu(X) = 1$, then the assertion remains valid for every closed convex set L.*

Proof Consider a sequence of simple functions $\vec{f}_n$ satisfying conditions (a)–(b) of the lemma from the previous section. By Property (8) from Sect. 13.6.2 and the remark after it, we may assume without loss of generality that the values of $\vec{f}_n$ lie in L. Therefore,

$$\int_X \vec{f}_n \, d\mu \in L \quad \text{and} \quad \int_X \vec{f} \, d\mu = \lim_{n\to\infty} \int_X \vec{f}_n \, d\mu \in L.$$

If $\mu(X) = 1$, then the first of these inclusions holds not only for linear, but also for convex L, because $\int_X \vec{f}_n \, d\mu$ is a convex combination of the values of $\vec{f}_n$ (they all lie in L). □

(7) *Lebesgue Points of Vector Functions.* Let $X = \mathbb{R}^m$ and μ be m-dimensional Lebesgue measure. As we know (see Sect. 4.9.2), for a locally summable function f, almost all points are Lebesgue points of f. This result is also valid in a more general setting: if a measurable vector function $\vec{f} : \mathbb{R}^m \to E$ is locally summable (i.e., $\int_{B(r)} \|\vec{f}(x)\| \, dx < +\infty$ for every $r > 0$), then

$$\frac{1}{r^m} \int_{B(x,r)} \|\vec{f}(y) - \vec{f}(x)\| \, dy \underset{r\to 0}{\longrightarrow} 0 \quad \text{for almost all } x \in \mathbb{R}^m.$$

For a simple function, i.e., a function of the form $\vec{f} = v_1\chi_{e_1} + \cdots + v_n\chi_{e_n}$, where $v_1, \ldots, v_n$ are points of E and $e_1, \ldots, e_n$ are Lebesgue measurable subsets of $\mathbb{R}^m$, this follows from the obvious inequality

$$\frac{1}{r^m} \int_{B(x,r)} \|\vec{f}(y) - \vec{f}(x)\| \, dy \leqslant \sum_{k=1}^{n} \frac{\|v_k\|}{r^m} \int_{B(x,r)} |\chi_{e_k}(y) - \chi_{e_k}(x)| \, dy,$$

whose right-hand side tends to zero as $r \to 0$ almost everywhere, since almost all points x are Lebesgue points of the functions χ_{e_k}.

To prove the general case, one can literally reproduce the argument from Sect. 4.9.2 (replacing the absolute value of a real-valued function by the norm of a vector-valued function).

13.6.5 Let the space $X \times Y$ be endowed with the product $\mu \times \nu$ of σ-finite measures μ and ν, and let f be a measurable function on $X \times Y$ that satisfies, for some p, $1 \leqslant p < +\infty$, the condition $\int_Y |f(x,y)|^p \, d\nu(y) < +\infty$ for almost all $x \in X$. This gives rise to a vector-valued function $\vec{g} : X \to \mathscr{L}^p(Y, \nu)$, which is defined by the formula $\vec{g}(x) = f(x, \cdot)$ (in Sect. 5.3, it was denoted by f_x). It is natural to ask whether the function $\vec{g}$ is Bochner measurable. The answer is given by the following proposition.

Proposition 1 *Under the above assumptions, the function $\vec{g}$ is Bochner measurable.*

Proof First assume that $\iint_{X\times Y} |f(x,y)|^p \, d\mu(x) \, d\nu(y) < +\infty$. Since the measure $\mu \times \nu$ is obtained by the Carathéodory extension from the semiring of generalized rectangles, it follows (see Exercise 3 of Sect. 9.2) that the set of simple functions of the form

$$\sum_{k=1}^{n} c_k \chi_{A_k} \chi_{B_k}, \quad \text{where } A_k \subset X,\ B_k \subset Y,\ c_k \text{ are scalars,} \tag{3}$$

is dense in $\mathscr{L}^p(X \times Y, \mu \times \nu)$. Hence there exists a sequence of functions f_n of the form (3) that converges to f in this space. Obviously, the vector functions $\vec{\xi}_n$ corresponding to f_n (i.e., the functions $\vec{\xi}_n(x) = f_n(x, \cdot)$) are simple and

$$\int_X \|\vec{g} - \vec{\xi}_n\|^p_{\mathscr{L}^p(Y,\nu)}\, d\mu = \iint_{X\times Y} |f(x,y) - f_n(x,y)|^p\, d\mu(x)\, d\nu(y) \underset{n\to\infty}{\longrightarrow} 0.$$

Therefore, $\|\vec{g} - \vec{\xi}_n\|^p_{\mathscr{L}^p(Y,\nu)} \underset{n\to\infty}{\longrightarrow} 0$ in measure, and, by Riesz' theorem, there exists a subsequence $\{\vec{\xi}_{n_k}\}_{k\geqslant 1}$ such that $\|\vec{g} - \vec{\xi}_{n_k}\|^p_{\mathscr{L}^p(Y,\nu)} \underset{k\to\infty}{\longrightarrow} 0$ almost everywhere. This shows that $\vec{g}$ is measurable.

In the general case, represent X as the union of a sequence of expanding sets X_n of finite measure and put $E_n = \{x \in X_n \mid \int_Y |f(x,y)|^p\, d\nu(y) \leqslant n\}$. Note that the set E_n is measurable, since, by Tonelli's theorem, the function $x \mapsto \int_Y |f(x,y)|^p\, d\nu(y)$ is measurable. We leave it to the reader to check that $\bigcup_{n=1}^{\infty} E_n$ is a set of full measure. Put $f_n = f \cdot \chi_{E_n}$. Obviously,

$$\iint_{X\times Y} |f_n(x,y)|^p\, d\mu(x)\, d\nu(y) = \iint_{E_n\times Y} |f(x,y)|^p\, d\mu(x)\, d\nu(y) \leqslant n\mu(X_n) < +\infty.$$

As we have established at the first step of the proof, the vector functions $\vec{g}_n$ corresponding to the functions f_n are Bochner measurable. Furthermore,

$$\int_X \|\vec{g} - \vec{g}_n\|^p_{\mathscr{L}^p(Y,\nu)}\, d\mu = \int_{(X\setminus E_n)\times Y} |f(x,y)|^p\, d\mu(x)\, d\nu(y) \underset{n\to\infty}{\longrightarrow} 0.$$

Hence there exists a subsequence $\{n_k\}$ such that

$$\|\vec{g}(x) - \vec{g}_{n_k}(x)\|_{\mathscr{L}^p(Y,\nu)} \underset{k\to\infty}{\longrightarrow} 0 \quad \mu\text{-almost everywhere.}$$

Thus the vector function $\vec{g}$ is measurable as the limit of measurable functions. □

As we have established, a measurable function of two variables gives rise to a measurable vector function. Is the converse true? In more detail, if $\vec{g} : X \to \mathscr{L}^p(Y,\nu)$ is a measurable vector function, can we assert that the formula

$$f(x,y) = \big(\vec{g}(x)\big)(y) \tag{4}$$

defines a measurable function of two variables? Without exhibiting corresponding counterexamples, we note that the answer to this question is negative. However, we will prove that f can be made measurable by modifying the function $\vec{g}(x)$ for every x on a set of zero measure (which, obviously, does not affect its measurability).

Proposition 2 *Let $\vec{g} : X \to \mathscr{L}^p(Y, \nu)$ be a measurable vector function. Then there exists a function h measurable on $X \times Y$ such that for almost all x, the equality $h(x, y) = (\vec{g}(x))(y)$ holds almost everywhere on Y.*

Proof We confine ourselves to the case where the measures μ and ν are finite and the function $\vec{g}$ is summable, leaving it to the reader to handle the general case. Approximate $\vec{g}$ by simple functions $\vec{g}_n$:

$$\big\|\vec{g}(x) - \vec{g}_n(x)\big\|_{\mathscr{L}^p(Y,\nu)} \underset{n\to\infty}{\longrightarrow} 0 \quad \text{for almost all } x \in X.$$

We may assume without loss of generality that $\|\vec{g}_n(x)\|_{\mathscr{L}^p(Y,\nu)} \leqslant \|\vec{g}(x)\|_{\mathscr{L}^p(Y,\nu)}$ almost everywhere. Then, by Lebesgue's theorem, $I_n \equiv \int_X \|\vec{g} - \vec{g}_n\|_{\mathscr{L}^p(Y,\nu)}\, d\mu \underset{n\to\infty}{\longrightarrow} 0$. Every function $\vec{g}_n$ generates a function f_n of two variables by formula (4); obviously, f_n is measurable on $X \times Y$. We will show that $\{f_n\}_{n\geqslant 1}$ is a fundamental sequence in the space $\mathscr{L}^1(X \times Y, \mu \times \nu)$. Indeed,

$$\begin{aligned}
\|f_n - f_m\|_{\mathscr{L}^1(X\times Y,\mu\times\nu)} &= \int_X \left(\int_Y \big|f_n(x, y) - f_m(x, y)\big|\, d\nu(y) \right) d\mu(x) \\
&\leqslant \int_X \nu(Y)^{\frac{1}{p'}} \left(\int_Y \big|f_n(x, y) - f_m(x, y)\big|^p\, d\nu(y) \right)^{\frac{1}{p}} d\mu(x) \\
&= \nu(Y)^{\frac{1}{p'}} \int_X \|\vec{g}_n - \vec{g}_m\|_{\mathscr{L}^p(Y,\nu)}\, d\mu \to 0 \quad \text{as } m, n \to \infty
\end{aligned}$$

(here p' is the conjugate exponent to p). Since the space $\mathscr{L}^1(X \times Y, \mu \times \nu)$ is complete, the sequence $\{f_n\}_{n\geqslant 1}$ has a limit h. Passing, if necessary, to a subsequence, we may assume that

$$f_n(x, y) \underset{n\to\infty}{\longrightarrow} h(x, y) \quad \text{almost everywhere on } X \times Y. \tag{5}$$

Let us verify that h is a required function. Since $\|\vec{g}(x) - \vec{g}_n(x)\|_{\mathscr{L}^p(Y,\nu)} \underset{n\to\infty}{\longrightarrow} 0$ and $\mu(X) < +\infty$ almost everywhere on X, it follows from Egorov's theorem that for every $j \in \mathbb{N}$ we can find a set $e_j \subset X$ such that

$$\mu(e_j) < \frac{1}{j} \quad \text{and} \quad \big\|\vec{g}(x) - \vec{g}_n(x)\big\|_{\mathscr{L}^p(Y,\nu)} \underset{n\to\infty}{\rightrightarrows} 0 \quad \text{on } E_j = X \setminus e_j.$$

Therefore, $\vec{g}_n(x) \underset{n\to\infty}{\longrightarrow} \vec{g}(x)$ in measure for every $x \in E_j$. Together with (5) this shows that for almost every $x \in E_j$, the equality $h(x, y) = (\vec{g}(x))(y)$ holds almost everywhere on Y. Since j is arbitrary, this is also true for almost every x in $\bigcup_{j=1}^{\infty} E_j$, i.e., for almost all x in X. □

Example Let f be a measurable function on $X \times Y$ satisfying, for some $p \geqslant 1$, the condition $\int_X (\int_Y |f(x,y)|^p \, d\nu(y))^{\frac{1}{p}} \, d\mu(x) < +\infty$. Then the corresponding vector function $\vec{g}$ (with values in $\mathscr{L}^p(Y,\nu)$) is summable, and $\int_X f(x,y)\, d\mu(x) = (\int_X \vec{g}\, d\mu)(y)$ for almost all $y \in Y$.

We will prove this under the assumption that $\nu(Y) < +\infty$. Denote by $\|\vec{g}(x)\|$ the norm $\|\vec{g}(x)\|_{\mathscr{L}^p(Y,\nu)}$. Let $\vec{g}$ be the vector function related to f by (4). If f is simple, then the required assertion is obvious.

The summability of $\vec{g}$ is ensured by the above assumption:

$$\int_X \|\vec{g}(x)\| \, d\mu(x) = \int_X \left(\int_Y |f(x,y)|^p \, d\nu(y) \right)^{\frac{1}{p}} d\mu(x) < +\infty.$$

It also follows that f is summable on $X \times Y$:

$$\int_X \int_Y |f(x,y)| \, d\mu(x) \, d\nu(y) \leqslant \nu^{\frac{1}{p'}}(Y) \int_X \left(\int_Y |f(x,y)|^p \, d\nu(y) \right)^{\frac{1}{p}} d\mu(x) < +\infty.$$

Hence, by Fubini's theorem, f is summable on X for almost all $y \in Y$. Consider a sequence of simple vector functions $\vec{g}_n$ satisfying the conditions

$$\|\vec{g}_n(x) - \vec{g}(x)\| \underset{n\to\infty}{\longrightarrow} 0, \qquad \|\vec{g}_n(x)\| \leqslant \|\vec{g}(x)\| \quad \text{for almost all } x \in X.$$

By the definition of the integral, $\int_X \vec{g}_n \, d\mu \underset{n\to\infty}{\longrightarrow} \int_X \vec{g}\, d\mu$ in the $\mathscr{L}^p$ norm and, consequently, in measure. Consider the functions f_n generated by the vector functions $\vec{g}_n$ according to (4). We have

$$\int_Y \left| \int_X f_n(x,y)\, d\mu(x) - \int_X f(x,y)\, d\mu(x) \right| d\nu(y)$$

$$\leqslant \iint_{X\times Y} |f_n(x,y) - f(x,y)| \, d\mu(x)\, d\nu(y) \leqslant \nu^{\frac{1}{p'}}(Y) \int_X \|\vec{g}_n - \vec{g}\| \, d\mu \underset{n\to\infty}{\longrightarrow} 0$$

(the convergence to zero follows from Lebesgue's theorem). Then

$$\int_X f_n(x,y)\, d\mu(x) \underset{n\to\infty}{\longrightarrow} \int_X f(x,y)\, d\mu(x) \quad \text{in measure } \nu.$$

Thus the sequence of functions $\int_X \vec{g}_n \, d\mu = \int_X f_n(x,\cdot)\, d\mu(x)$ converges in measure to $\int_X f(x,\cdot)\, d\mu(x)$ and almost everywhere to $\int_X \vec{g}\, d\mu$; hence the integrals $\int_X \vec{g}\, d\mu$ and $\int_X f(x,\cdot)\, d\mu(x)$ coincide almost everywhere.

Corollary *Let* $1 \leqslant r < s < +\infty$, *and let* f *be a function measurable on* $X \times Y$. *Then*

$$\left(\int_Y \left(\int_X |f(x,y)|^r \, d\mu(x) \right)^{\frac{s}{r}} d\nu(y) \right)^{\frac{1}{s}} \leqslant \left(\int_X \left(\int_Y |f(x,y)|^s \, d\nu(y) \right)^{\frac{r}{s}} d\mu(x) \right)^{\frac{1}{r}}.$$

Proof To prove this, it suffices to apply the result obtained in the above example with $p = \frac{s}{r}$ to the function $|f|^r$ and use the inequality $\| \int_X \vec{g}\, d\mu\| \leqslant \int_X \|\vec{g}\|\, d\mu$. □

13.6.6 *Weak and Strong Measurability.* Let E^* be the dual of a Banach space E, i.e., the space of all linear continuous functionals defined on E.

Definition 1 A vector-valued function $\vec{f} : X \to E$ is called *weakly measurable* if for every functional $\varphi \in E^*$, the scalar function $x \mapsto \varphi(\vec{f}(x))$ is measurable on X.

Property (6) of measurable functions (see Sect. 13.6.2) implies that every measurable function is weakly measurable. The converse is not true. A corresponding counterexample will be considered in the next section.

We will establish a simple sufficient condition under which a weakly measurable function is strongly measurable. For this we need the notion of a separable space. Recall that a metric space is called separable if it contains a countable dense subset (which is the case if and only if it is second countable).

Lemma *If E is a separable normed space, then there exists a sequence of functionals* $\{\varphi_n\}_{n\geqslant 1} \subset E^*$ *such that*

$$\|v\| = \sup_{n\geqslant 1} |\varphi_n(v)| \quad \textit{for all } v \textit{ from } E.$$

Proof Let $\{v_1, v_2, \ldots\}$ be a countable dense subset of E. By the theorem that guarantees the existence of sufficiently many functionals on a normed vector space, there exist functionals φ_n such that $\|\varphi_n\| = 1$ and $\varphi_n(v_n) = \|v_n\|$ for every $n \in \mathbb{N}$. We will show that the sequence $\{\varphi_n\}_{n\geqslant 1}$ chosen in this way satisfies the required properties. Indeed, if $v \in E$ and $v_{n_k} \underset{k\to\infty}{\longrightarrow} v$, then

$$\begin{aligned}
\|v\| &\geqslant \sup_{n\geqslant 1} |\varphi_n(v)| \geqslant \lim_{k\to\infty} |\varphi_{n_k}(v)| \\
&\geqslant \lim_{k\to\infty} \left(|\varphi_{n_k}(v_{n_k})| - |\varphi_{n_k}(v_{n_k} - v)|\right) \\
&\geqslant \lim_{k\to\infty} \left(\|v_{n_k}\| - \|v_{n_k} - v\|\right) = \|v\|.
\end{aligned}$$

□

To state the main theorem, we need to introduce another notion.

Definition 2 A vector-valued function is called *essentially separably valued* if the set of values it takes on a set of full measure is separable.

Theorem *A vector-valued function* $\vec{f} : X \to E$ *is measurable if and only if it is weakly measurable and essentially separably valued.*

Proof If $\vec{f}$ is measurable, then it is weakly measurable (as observed after Definition 1). To prove that it is essentially separably valued, consider an arbitrary sequence of simple functions $\{\vec{f}_n\}_{n\geqslant 1}$ that converges to $\vec{f}$ almost everywhere. Let L

be the closed linear hull of the union of the sets of values of $\vec{f}_n$. Since the set of values of every simple function is finite, the subspace L is separable. Since it is closed and contains all values of $\vec{f}_n$, it contains also all points $\vec{f}(x) = \lim_{n\to\infty} \vec{f}_n(x)$. Thus $\vec{f}(x) \in L$ for almost all $x \in X$, which means precisely that $\vec{f}$ is essentially separably valued.

Now let $\vec{f}$ be weakly measurable and essentially separably valued. First of all, we may assume without loss of generality that the set of its values is separable (otherwise modify the function at a set of zero measure, which does not affect the measurability). Hence we may (and will) assume that the set E is also separable, replacing it if necessary by the closure of the linear hull of $\vec{f}(X)$. Note also that for every $a \in E$, the function $h_a(x) = \|f(x) - a\|$ $(x \in X)$ is measurable. This follows from the formula

$$\|\vec{f}(x) - a\| = \sup_{n\geqslant 1} |\varphi_n(\vec{f}(x) - a)|$$

established in the above lemma (here φ_n are the functionals from this lemma), which shows that the function h_a is an upper bound for the sequence of measurable functions $x \mapsto |\varphi_n(\vec{f}(x)) - \varphi_n(a)|$.

To prove that $\vec{f}$ is measurable, we will show that it is the limit of a sequence of measurable functions. For this it obviously suffices to check that $\vec{f}$ can be uniformly approximated by measurable functions.

Fix an arbitrary $\varepsilon > 0$ and, using the separability of the set $\vec{f}(X)$, cover it by a sequence of balls $B(v_n, \varepsilon)$. Consider the sets

$$X_n = \{x \in X \mid \|\vec{f}(x) - v_n\| < \varepsilon\}, \qquad Y_1 = X_1, \qquad Y_n = X_n \setminus \bigcup_{k=1}^{n-1} X_k \quad \text{for } n > 1.$$

Since $\bigcup_{n\geqslant 1} B(v_n, \varepsilon) \supset \vec{f}(X)$, we have $\bigcup_{n\geqslant 1} X_n = X$. The sets X_n, and hence Y_n, are measurable, in view of the measurability of h_a observed above. Furthermore, the sets Y_n are, obviously, pairwise disjoint, and $\bigcup_{n\geqslant 1} Y_n = \bigcup_{n\geqslant 1} X_n = X$. Now put $\vec{g}(x) = v_n$ if $x \in Y_n$. Clearly, the function $\vec{g}$ is measurable as the sum of the pointwise convergent series $\sum_{n=1}^{\infty} \chi_{Y_n} \cdot v_n$ of the simple functions $\chi_{Y_n} \cdot v_n$. Finally, if x is an arbitrary point of X, for some n we have $x \in Y_n \subset X_n$. In this case, $\vec{f}(x) \in B(v_n, \varepsilon)$ and

$$\|\vec{f}(x) - \vec{g}(x)\| = \|\vec{f}(x) - v_n\| < \varepsilon.$$

Thus we have proved that $\vec{f}$ can be uniformly approximated by measurable functions. □

Example Let K be a compact metrizable space. Consider a function $h : \mathbb{R}^m \times K \to \mathbb{C}$ satisfying the following conditions:

(a) for almost all $x \in \mathbb{R}^m$, h is continuous in the second variable;
(b) for all values $u \in K$, h is measurable in the first variable.

Such functions, called Carathéodory functions, are used in some variational problems. We will show that the vector function $\vec{f} : \mathbb{R}^m \to C(K)$ associated with a Carathéodory function by the formula $\vec{f}(t) = h(t, \cdot)$ is Bochner measurable. Since the space $C(K)$ is separable, it follows from the above theorem that it suffices to prove that $\vec{f}$ is weakly measurable. Taking into account the form of a generic functional in $C(K)$, it suffices to verify that if μ is an arbitrary finite Borel measure on K, then the function

$$x \mapsto \varphi(x) = \int_K h(x,u)\,d\mu(u) \quad \bigl(x \in \mathbb{R}^m\bigr)$$

is measurable. This would follow from Fubini's theorem if we could guarantee that h is measurable as a function of two variables. However, it is easy to prove the measurability of φ in another way. Indeed, it follows from the fact that for every $x \in \mathbb{R}^m$, the integral $\int_K h(x,u)\,d\mu(u)$ is the limit of a sequence of Riemann sums of the form

$$\sum_{j=1}^{n} h(x,u_j)\mu(e_j),$$

which are measurable on $\mathbb{R}^m$ by condition (b).

If the function $x \mapsto \|\vec{f}(x)\| = \max_{u\in K} |h(x,u)|$ is locally summable, then, in view of Property (7) from Sect. 13.6.4, we may assert that almost every point $x \in \mathbb{R}^m$ is a Lebesgue point of the vector function $\vec{f}$. Thus in the case under consideration, a Carathéodory function satisfies the following property:

$$\frac{1}{r^m}\int_{B(x,r)} \sup_{u\in K} \bigl|h(y,u) - h(y,x)\bigr|\,dy \underset{r\to 0}{\longrightarrow} 0 \quad \text{almost everywhere.}$$

13.6.7 In conclusion, we give an example of a function that is weakly measurable, but not strongly measurable.

Let E be the space $c_0([0,1])$ consisting of all functions $v : [0,1] \to \mathbb{R}$ such that the sets $\{x \in [0,1] \mid |v(x)| > \varepsilon\}$ are finite for every $\varepsilon > 0$. Obviously, all such functions are bounded. Endow E with the uniform norm: $\|v\| = \sup_{[0,1]} |v|$. We leave it to the reader to check that the space E with this norm is complete, and its dual E^* can be identified with the space $l^1([0,1])$ of all summable families of numbers defined on $[0,1]$. Recall that if $\varphi = \{\varphi_x\}_{x\in[0,1]}$ is a summable family, then the set $\{x \in X \mid \varphi_x \neq 0\}$ is at most countable. The norm in $l^1([0,1])$ and the duality between $c_0([0,1])$ and $l^1([0,1])$ are defined by the following formulas:

$$\|\varphi\| = \sum_{x\in[0,1]} |\varphi_x|, \qquad \varphi(v) = \sum_{x\in[0,1]} \varphi_x v(x),$$

$$\text{where } v \in c_0\bigl([0,1]\bigr),\ \varphi \in l^1\bigl([0,1]\bigr).$$

Let μ be the Lebesgue measure on the interval $[0, 1]$, and let $\chi_{\{x\}}$ denote the characteristic function of the one-point set $\{x\}$. Put

$$f_0(x) = \chi_{\{x\}} \in c_0([0, 1]) \quad \text{for } x \in [0, 1].$$

Then for every φ from $l^1([0, 1])$, the scalar function $x \mapsto \varphi(f_0(x)) = \varphi_x$ is measurable, since the set of x such that $\varphi_x \neq 0$ is at most countable. Thus the function f_0 is weakly measurable. But it is not Bochner measurable. To prove this, it suffices, by Theorem 13.6.6, to verify that it is not essentially separably valued. We leave it to the reader to prove the following stronger assertion:

under the map f_0, the image of every uncountable subset of $[0, 1]$ *is not separable*

(since it contains uncountably many points at distance one from each other).

EXERCISES

1. Let $(X, \mathfrak{A}, \mu)$ be a measure space and E be a Banach space. Show that a necessary condition for a vector-valued function $\vec{f} : X \to E$ to be measurable is that the inverse image under $\vec{f}$ of every open set is measurable; if E is separable, this condition is also sufficient. Verify that the separability assumption cannot be removed.

2. Let $\vec{f} : X \to E$ be a strongly measurable function satisfying the condition

$$\operatorname*{esssup}_{x \in X} \big|\varphi(\vec{f}(x))\big| < +\infty \quad \text{for every functional } \varphi \text{ from } E^*.$$

Show that

(a) $\sup_{\|\varphi\| \leqslant 1} \operatorname{esssup}_{x \in X} |\varphi(\vec{f}(x))| < +\infty$;

(b) $\operatorname{esssup}_{x \in X} \|\vec{f}(x)\| < +\infty$

(for the notation $\operatorname{esssup}_{x \in X} \|\vec{f}(x)\|$, see Sect. 4.4.5).

3. Assume that a strongly measurable function $\vec{f} : X \to E$ is weakly summable, i.e., satisfies the condition $\varphi \circ \vec{f} \in \mathscr{L}^1(X, \mu)$ for every functional φ from E^*. Show that $\sup_{\|\varphi\| \leqslant 1} \int_X |\varphi \circ \vec{f}| \, d\mu < +\infty$. Show by example that the function $\|\vec{f}\|$ may be not summable.

4. Let X be the interval $[0, 1]$ with the Lebesgue measure μ, and let r_n $(n \in \mathbb{N})$ be the Rademacher functions (see Sect. 6.4.5). Show that the function $\vec{R} : [0, 1] \to l^\infty$ defined by the formula $\vec{R}(x) = \{r_n(x)\}_{n \geqslant 1}$ is not essentially separably valued (here l^∞ is the space of bounded numerical sequences $v = \{v_n\}_{n \geqslant 1}$ with the norm $\|v\| = \sup_{n \geqslant 1} |v_n|$).

13.7 Smooth Maps

In this appendix, we give a summary of the basic properties of smooth maps used in the book. We assume that the reader is familiar with the notion of a partial derivative, differentiable function, and the theorem on the differentiability of a function having continuous partial derivatives.

In what follows, $\mathcal{O}$ always stands for an open subset of $\mathbb{R}^m$. As above, the inner product of vectors $x, y \in \mathbb{R}^m$ is denoted by $\langle x, y \rangle$. A linear map $A : \mathbb{R}^m \to \mathbb{R}^n$ is identified with its matrix in the canonical basis. Its norm $\|A\|$ (or the norm of the corresponding matrix) is defined as $\sup_{\|x\| \leqslant 1} \|A(x)\|$.

13.7.1 Generalizing the notion of differentiable function, we introduce the following definition.

Definition A map $T : \mathcal{O} \to \mathbb{R}^n$ is called *differentiable* at a point $a \in \mathcal{O}$ if there exists a linear map $A : \mathbb{R}^m \to \mathbb{R}^n$ such that

$$T(a+h) - T(a) = A(h) + \alpha(h),$$

where $\alpha(h) = o(h)$ as $h \to 0$, i.e., $\frac{\alpha(h)}{\|h\|} \underset{h \to 0}{\longrightarrow} 0$.

One can easily check that A is uniquely determined; it is called the *differential* of the map T at the point a and denoted by $d_a T$; its matrix in the canonical basis is denoted by $T'(a)$. Note that the differential of a linear map coincides with the map itself.

If $f_1, \ldots, f_n$ are the coordinate functions of a map T, then the differentiability of T is equivalent to the differentiability of $f_1, \ldots, f_n$. The matrix T' is just the rectangular matrix

$$\begin{pmatrix} \frac{\partial f_1}{\partial x_1} & \cdots & \frac{\partial f_1}{\partial x_m} \\ \frac{\partial f_2}{\partial x_1} & \cdots & \frac{\partial f_2}{\partial x_m} \\ \vdots & \ddots & \vdots \\ \frac{\partial f_n}{\partial x_1} & \cdots & \frac{\partial f_n}{\partial x_m} \end{pmatrix}.$$

It is called the *Jacobi matrix* of the system of functions $f_1, \ldots, f_n$.

The following result shows that the usual rule for differentiating a composite function can be naturally extended to maps.

Proposition *Let R be the composition of maps T and S: $R = S \circ T$. If T is differentiable at a point a and S is differentiable at the point $b = T(a)$, then R is differential at a and*

$$R'(a) = S'(b) \cdot T'(a). \tag{1}$$

Proof Let A and B be the differentials of the maps T and S at the points a and b, respectively. Then for sufficiently small h and η, we have

$$T(a+h) - T(a) = A(h) + \|h\|\,\omega(h), \quad \text{where } \omega(h) \underset{h \to 0}{\longrightarrow} 0,$$

$$S(b+\eta) - S(b) = B(\eta) + \|\eta\|\,\widetilde{\omega}(\eta), \quad \text{where } \widetilde{\omega}(\eta) \underset{\eta \to 0}{\longrightarrow} 0.$$

Assuming that $\widetilde{\omega}(0) = 0$, substitute $\eta = T(a+h) - T(a)$ into the last equation. Then

$$\begin{aligned} R(a+h) - R(a) &= B\big(T(a+h) - T(a)\big) \\ &\quad + \big\|T(a+h) - T(a)\big\|\widetilde{\omega}\big(T(a+h) - T(a)\big) \\ &= B \circ A(h) + \gamma(h), \end{aligned}$$

where $\gamma(h) = \|h\|\, B(\omega(h)) + \|T(a+h) - T(a)\|\, \widetilde{\omega}(T(a+h) - T(a))$. To complete the proof, it remains to show that $\gamma(h)$ is an infinitesimal of higher order than h. Indeed,

$$\frac{\|\gamma(h)\|}{\|h\|} \leqslant \|B\| \big\|\omega(h)\big\| + \frac{\|T(a+h) - T(a)\|}{\|h\|} \big\|\widetilde{\omega}\big(T(a+h) - T(a)\big)\big\|.$$

Since $\|T(a+h) - T(a)\| = O(\|h\|)$ as $h \to 0$ and $\widetilde{\omega}(T(a+h) - T(a)) \underset{h\to 0}{\longrightarrow} 0$, we see that $\frac{\gamma(h)}{\|h\|} \underset{h\to 0}{\longrightarrow} 0$. □

A map T is called C^r*-smooth* in $\mathcal{O}$ ($r \in \mathbb{N}$) if its coordinate functions have continuous partial derivatives up to order r in $\mathcal{O}$. A map whose coordinate functions have continuous partial derivatives of all orders is called C^∞-smooth. The set of C^r-smooth maps from $\mathcal{O}$ to $\mathbb{R}^n$ is denoted by $C^r(\mathcal{O}, \mathbb{R}^n)$ ($r = 1, 2, \ldots, +\infty$).

The above proposition easily implies by induction the following corollary.

Corollary *The composition of maps of class* C^r ($r = 1, 2, \ldots, +\infty$) *is again a map of the same class.*

13.7.2 We will prove a result generalizing the classical Lagrange mean value theorem on increments of a differentiable function.

Theorem (Lagrange's inequality) *If a map T is differentiable at all points of an interval* $[x, y] = \{(1-t)x + ty \,|\, 0 \leqslant t \leqslant 1\}$, *then*

$$\big\|T(y) - T(x)\big\| \leqslant \sup_{z \in [x,y]} \big\|T'(z)\big\| \|y - x\|.$$

Proof Let $\Delta = T(y) - T(x)$. To estimate the length of this vector, we introduce the auxiliary function $\varphi(t) = \langle T(x + t(y-x)), \Delta\rangle$ ($t \in [0, 1]$). It is clear that $\varphi(0) = \langle T(x), \Delta\rangle$, $\varphi(1) = \langle T(y), \Delta\rangle$ and, consequently,

$$\|\Delta\|^2 = \big\langle T(y) - T(x),\ \Delta\big\rangle = \varphi(1) - \varphi(0).$$

By Proposition 13.7.1, the function φ is differentiable and

$$\varphi'(t) = \big\langle T'\big(x + t(y-x)\big)(y - x),\ \Delta\big\rangle.$$

By the mean value theorem, there exists a $c \in (0, 1)$ such that $\varphi(1) - \varphi(0) = \varphi'(c)$. Hence

$$\|\Delta\|^2 = \varphi(1) - \varphi(0) = \varphi'(c) = \langle T'(x + c(y - x))(y - x), \Delta\rangle$$
$$\leqslant \|T'(x + c(y - x))(y - x)\| \|\Delta\| \leqslant \|T'(x + c(y - x))\| \|y - x\| \|\Delta\|,$$

which obviously implies the desired assertion. □

Corollary *If a map $T : \mathcal{O} \to \mathbb{R}^n$ is differentiable at all points of an interval $[x_0, x_0 + h]$, then for every linear map $A : \mathbb{R}^m \to \mathbb{R}^n$,*

$$\|T(x_0 + h) - T(x_0) - A(h)\| \leqslant \sup_{z \in [x_0, x_0+h]} \|T'(z) - A\| \|h\|. \qquad (2)$$

For $A = T'(x_0)$, this implies an efficient estimate on the deviation of the differential from the increment of the map:

$$\|T(x_0 + h) - T(x_0) - T'(x_0)(h)\| \leqslant \sup_{z \in [x_0, x_0+h]} \|T'(z) - T'(x_0)\| \|h\|.$$

To prove (2), it suffices to apply the theorem to the map $T - A$, using the fact that the differential of a linear map coincides with the map itself.

13.7.3 Now we turn to the study of the invertibility of smooth maps. A smooth map $T : \mathcal{O} \to \mathbb{R}^m$ ($\mathcal{O} \subset \mathbb{R}^m$) whose inverse is also smooth is called a *diffeomorphism*. A necessary condition for T to be a diffeomorphism is that the matrix $T'(x)$ is invertible for all $x \in \mathcal{O}$. Indeed, if T is a diffeomorphism, then $T^{-1}(T(x)) \equiv x$. By rule (1) for differentiating a composite function, $(T^{-1})'(T(x)) \cdot T'(x) = id$. Hence

$$\det(T'(x)) \neq 0 \quad \text{for all } x \in \mathcal{O}. \qquad (3)$$

However, the invertibility of $T'(x)$ does not imply that T is one-to-one; consider, for example, the map $T(u, v) = (u^2 - v^2, 2uv)$, where $(u, v) \in \mathcal{O} = \mathbb{R}^2 \setminus 0$.

Before proceeding to the theorem on the smoothness of the inverse map, we will show that if condition (3) is satisfied, then the map T is open, i.e., sends open sets to open sets. First we obtain a technical result.

Lemma *Let $T : \mathcal{O} \to \mathbb{R}^m$. If T is differentiable at a point $x_0 \in \mathcal{O}$ and the derivative $T'(x_0)$ is invertible, then there exist numbers $c > 0$ and $\delta > 0$ such that $B(x_0, \delta) \subset \mathcal{O}$ and $\|T(x) - T(x_0)\| \geqslant c\|x - x_0\|$ for $\|x - x_0\| < \delta$.*

Proof If T is linear, then it is invertible, because $T' = T$. Since $\|T^{-1}(y)\| \leqslant \|T^{-1}\| \|y\|$, for $y = T(x)$ we have $\|x\| \leqslant \|T^{-1}\| \|T(x)\|$, which proves the lemma with $c = \frac{1}{\|T^{-1}\|}$.

In the general case, $T(x) - T(x_0) = A(x - x_0) + \|x - x_0\| \omega(x)$, where $A = T'(x_0)$ and $\omega(x) \to 0$ as $x \to x_0$. Therefore,

$$\begin{aligned}\|T(x)-T(x_0)\| &\geqslant \|A(x-x_0)\| - \|x-x_0\|\,\|\omega(x)\| \\ &\geqslant \left(\frac{1}{\|A^{-1}\|} - \|\omega(x)\|\right)\|x-x_0\|.\end{aligned}$$

This proves the lemma with $c = \frac{1}{2\|A^{-1}\|}$, since $\|\omega(x)\| < \frac{1}{2\|A^{-1}\|}$ for x sufficiently close to x_0. □

Theorem (Open map theorem) *If a map* $T \in C^1(\mathcal{O}, \mathbb{R}^m)$ *satisfies condition* (3), *then the set* $\mathcal{O}' = T(\mathcal{O})$ *is open.*

Proof We will prove that every point $y_0 \in \mathcal{O}'$ is an interior point of $\mathcal{O}'$. Let $y_0 = T(x_0)$, $A = T'(x_0)$, and let $c > 0$ and $\delta > 0$ be as in the lemma:

$$\|T(x_0+h) - T(x_0)\| \geqslant c\|h\| \quad \text{for } \|h\| < \delta.$$

Shrinking δ if necessary, we may assume that $\overline{B}(x_0,\delta) \subset \mathcal{O}$. Then $T(x_0+h) \neq T(x_0)$ for $\|h\| = \delta$. Hence $y_0 = T(x_0) \notin Q = T(K)$, where $K = \partial(B(x_0,\delta))$. Let $\operatorname{dist}(y_0, Q) = 2r$. We will show that $B(y_0,r) \subset \mathcal{O}'$. Note that for $y \in B(y_0,r)$ and $x \in K$, the inequality $\|T(x) - y\| \geqslant r$ holds. Now we fix an arbitrary point y from $B(y_0,r)$ and show that it is a value of T. For this we introduce the auxiliary function $F(x) = \|T(x) - y\|^2$ ($x \in \mathcal{O}$). Obviously, T takes the value y in the ball $\overline{B}(x_0,\delta)$ if and only if the smallest value of F in this ball is equal to zero. Let us check that this is indeed the case. Obviously, $F(x_0) < r^2$ and $F(x) \geqslant r^2$ for $x \in K$. Hence the smallest value of F is attained at an interior point $\overline{x}$ of the ball $\overline{B}(x_0,\delta)$. Therefore, at this point all partial derivatives of F vanish. To write this condition in more detail, consider the coordinate functions $f_1, \ldots, f_m$ of T and assume that $y = (y_1, \ldots, y_m)$. Then $F(x) = \sum_{k=1}^m (f_k(x) - y_k)^2$ and, consequently,

$$\frac{\partial F}{\partial x_j}(\overline{x}) = 2\sum_{k=1}^m \frac{\partial f_k}{\partial x_j}(\overline{x})\big(f_k(\overline{x}) - y_k\big) = 0 \quad \text{for } j = 1, \ldots, m.$$

The matrix of this homogeneous system is precisely the transposed matrix $T'(\overline{x})$. Since $\det(T'(\overline{x})) \neq 0$, the system has only the trivial solution, which is equivalent to the formula $T(\overline{x}) = y$. Thus $y \in T(B(x_0,\delta)) \subset \mathcal{O}'$. Since the point $y \in B(y_0,r)$ is arbitrary, this means that $B(y_0,r) \subset \mathcal{O}'$ and, consequently, y_0 is an interior point of $\mathcal{O}'$. □

Remark If $\mathcal{O}$ is a domain (a connected open set), then, obviously, $\mathcal{O}' = T(\mathcal{O})$ is also a domain. That is why the above theorem is sometimes called the theorem on preservation of domain.

13.7.4 Let $GL(m)$ be the set of invertible $m \times m$ matrices. We will regard $GL(m)$ as a subset of $\mathbb{R}^{m^2}$.

Lemma *The set $GL(m)$ is open, and the map $A \mapsto A^{-1}$ is infinitely smooth on $GL(m)$.*

Proof The first claim holds because the function $A \mapsto \det(A)$ is continuous and the set $GL(m)$ coincides with the set of $m \times m$ matrices with non-zero determinant.

The infinite differentiability of the map $A \mapsto A^{-1}$ follows from the fact that each coordinate function of this map is the ratio of a cofactor of A to its determinant and, consequently, is a rational function of its elements. □

Theorem (Diffeomorphism theorem) *Let $T \in C^r(\mathcal{O}, \mathbb{R}^m)$ $(r = 1, 2, \dots, +\infty)$. If T is invertible and satisfies condition* (3), *then T^{-1} is a smooth map of class C^r.*

As we have already mentioned, (3) is a necessary condition for the inverse map to be smooth.

Proof First we prove the theorem in the case $r = 1$.

The open mapping theorem implies that the set $\mathcal{O}' = T(\mathcal{O})$ is open. By the same theorem, the image of every open set $U \subset \mathcal{O}$ is open. Putting $S = T^{-1}$, we can rewrite the equation $V = T(U)$ in the form $V = S^{-1}(U)$. Thus the inverse image of every open set under S is open, which implies that S is continuous. We will show that it is differentiable at an arbitrary point $y_0 \in \mathcal{O}'$. Let $y_0 = T(x_0)$ and $A = T'(x_0)$. Then

$$T(x) - T(x_0) = A(x - x_0) + \|x - x_0\|\, \omega(x), \tag{4}$$

where $\omega(x) \to 0$ as $x \to x_0$ and $\omega(x_0) = 0$. Note that

$$\big\|T(x) - T(x_0)\big\| \geqslant c\|x - x_0\| \quad \text{for } x \in B(x_0, \delta), \tag{5}$$

where c and δ are the numbers from Lemma 13.7.3. Consider an arbitrary point $y \in \mathcal{O}'$ and set $x = S(y)$. Substituting $x = S(y)$ and $x_0 = S(y_0)$ into (4), we obtain

$$y - y_0 = A\big(S(y) - S(y_0)\big) + \big\|S(y) - S(y_0)\big\|\omega\big(S(y)\big),$$

that is,

$$S(y) - S(y_0) = A^{-1}(y - y_0) - \big\|S(y) - S(y_0)\big\|A^{-1}\big(\omega\big(S(y)\big)\big).$$

It remains to prove that as $y \to y_0$, the value $\beta(y) = -\|S(y) - S(y_0)\|\, A^{-1}(\omega(S(y)))$ is an infinitesimal of higher order than $\|y - y_0\|$. By the continuity of S, we may assume that y is so close to y_0 that $\|x - x_0\| = \|S(y) - S(y_0)\| < \delta$. This allows us to employ inequality (5) to estimate $\|\beta(y)\|$:

$$\big\|\beta(y)\big\| = \|x - x_0\|\big\|A^{-1}\big(\omega\big(S(y)\big)\big)\big\| \leqslant \frac{1}{c}\big\|T(x) - T(x_0)\big\|\big\|A^{-1}\big\|\big\|\omega\big(S(y)\big)\big\|$$

$$= \frac{\|A^{-1}\|}{c}\|y - y_0\|\big\|\omega\big(S(y)\big)\big\|.$$

Since $x = S(y) \to x_0$, we have $\omega(S(y)) \to 0$ as $y \to y_0$, and hence $\beta(y) = o(\|y - y_0\|)$ as $y \to y_0$, which completes the proof of the differentiability of S at y_0. Moreover, by the definition of differentiability, $S'(y_0) = A^{-1} = (T'(x_0))^{-1}$.

So, the map $S = T^{-1}$ is differentiable at every point, and

$$S'(y) = \left(T'\left(T^{-1}(y)\right)\right)^{-1}.$$

(Note that our proof, as well as the proof of the open mapping theorem, has thus far used only the differentiability of T rather than its smoothness.)

Now we will prove that S is smooth. The passage from y to $S'(y)$ can be written as the composition

$$y \mapsto T^{-1}(y) = x \mapsto T'(x) = A \mapsto A^{-1} = S'(y).$$

Each of the three maps in this chain is continuous, which implies the continuity of $S'(y)$, i.e., the C^1-smoothness of S.

To prove the smoothness for an arbitrary r, we proceed by induction (keeping in mind that, by the lemma, the operation of taking the inverse of a matrix is an infinitely smooth map). □

13.7.5 As we have already mentioned, condition (3) does not imply that T is invertible. However, one can ensure the invertibility by considering the map "in the small".

Theorem (Local invertibility theorem) *Let $T \in C^1(\mathcal{O}, \mathbb{R}^m)$ and $x_0 \in \mathcal{O}$. If the matrix $T'(x_0)$ is invertible, then there exists a neighborhood $U \subset \mathcal{O}$ of x_0 such that the restriction of T to U is a diffeomorphism.*

Proof We must prove that the restriction of T to a neighborhood of x_0 is invertible and satisfies condition (3).

Since the matrix $A = T'(x_0)$ is invertible, we have $\|A(h)\| \geqslant c\,\|h\|$ for some $c > 0$ and all $h \in \mathbb{R}^m$. Fix a ball $B(x_0, r) \subset \mathcal{O}$ such that

$$\det\left(T'(x)\right) \neq 0 \quad \text{and} \quad \left\|T'(x) - A\right\| < \frac{c}{2} \quad \text{for } x \in B(x_0, r). \tag{6}$$

We will prove that $U = B(x_0, r)$ is a desired neighborhood. By the previous theorem, it suffices to show that the restriction of T to U is invertible, i.e., that T is one-to-one on U. Let $x, y \in U$ and $h = y - x$. Obviously,

$$T(y) - T(x) = T(x + h) - T(x) - A(h) + A(h).$$

Using inequalities (2) and (6), we see that

$$\begin{aligned}\left\|T(y) - T(x)\right\| &\geqslant \left\|A(h)\right\| - \left\|T(x + h) - T(x) - A(h)\right\| \\ &\geqslant c\|h\| - \sup_{z \in [x,y]} \left\|T'(z) - A\right\| \|h\| \geqslant c\|h\| - \frac{c}{2}\|h\| = \frac{c}{2}\|y - x\|.\end{aligned}$$

Thus the restriction of T to U is one-to-one, and the theorem follows. □

13.7.6 Here we consider a smooth map $F : G \to \mathbb{R}^n$ defined on an open subset G of the space $\mathbb{R}^{m+n}$. This space is identified with the Cartesian product $\mathbb{R}^m \times \mathbb{R}^n$, and a point $z \in \mathbb{R}^{m+n}$ is identified with the pair (x, y), where $x = (x_1, \dots, x_m) \in \mathbb{R}^m$, $y = (y_1, \dots, y_n) \in \mathbb{R}^n$.

Let $F_1, \dots, F_n$ be the coordinate functions of F. The matrix F' has the form

$$\begin{pmatrix} \frac{\partial F_1}{\partial x_1} & \cdots & \frac{\partial F_1}{\partial x_m} & \frac{\partial F_1}{\partial y_1} & \cdots & \frac{\partial F_1}{\partial y_n} \\ \vdots & & \vdots & \vdots & & \vdots \\ \frac{\partial F_n}{\partial x_1} & \cdots & \frac{\partial F_n}{\partial x_m} & \frac{\partial F_n}{\partial y_1} & \cdots & \frac{\partial F_n}{\partial y_n} \end{pmatrix}.$$

The left and right parts of this matrix,

$$\begin{pmatrix} \frac{\partial F_1}{\partial x_1} & \cdots & \frac{\partial F_1}{\partial x_m} \\ \vdots & & \vdots \\ \frac{\partial F_n}{\partial x_1} & \cdots & \frac{\partial F_n}{\partial x_m} \end{pmatrix} \quad \text{and} \quad \begin{pmatrix} \frac{\partial F_1}{\partial y_1} & \cdots & \frac{\partial F_1}{\partial y_n} \\ \vdots & & \vdots \\ \frac{\partial F_n}{\partial y_1} & \cdots & \frac{\partial F_n}{\partial y_n} \end{pmatrix},$$

will be denoted by F'_x and F'_y, respectively. Note that F'_y is a square $n \times n$ matrix.

We will study the solvability of the equation

$$F(x, y) = 0 \tag{7}$$

with respect to y. To make the problem more precise, we introduce the following definition.

Definition Let P and Q be open parallelepipeds in $\mathbb{R}^m$ and $\mathbb{R}^n$, respectively, $P \times Q \subset G$. We say that Eq. (7) defines an implicit map in $P \times Q$ if there exists a map $f : P \to Q$ such that

$$F\big(x, f(x)\big) = 0 \quad \text{for all } x \in P.$$

Note that Eq. (7) defines a unique implicit map in $P \times Q$ if and only if for every $x \in P$ there exists a unique point $y \in Q$ satisfying (7).

Theorem (The implicit function theorem) *Let G be an open subset of $\mathbb{R}^{m+n}$, $F \in C^r(G, \mathbb{R}^n)$ $(r = 1, 2, \dots, +\infty)$ and $(a, b) \in G$. If $F(a, b) = 0$ and the matrix $F'_y(a, b)$ is invertible, then there exist open cubes $P \subset \mathbb{R}^m$ and $Q \subset \mathbb{R}^n$ centered at the points a and b, respectively, such that $P \times Q \subset G$ and Eq. (7) defines a unique implicit map f in $P \times Q$. This map is C^r-smooth, and for all $x \in P$,*

$$f'(x) = -\big[F'_y\big(x, f(x)\big)\big]^{-1} F'_x\big(x, f(x)\big). \tag{8}$$

It follows from the theorem that if at a point $z_0 \in G$ the matrix F' is of maximal rank (equal to n), then the level set $\{z \in G \mid F(z) = F(z_0)\}$ near z_0 is a simple m-dimensional manifold of class C^r, which is parametrized by the implicit function defined by the equation $F(z) - F(z_0) = 0$.

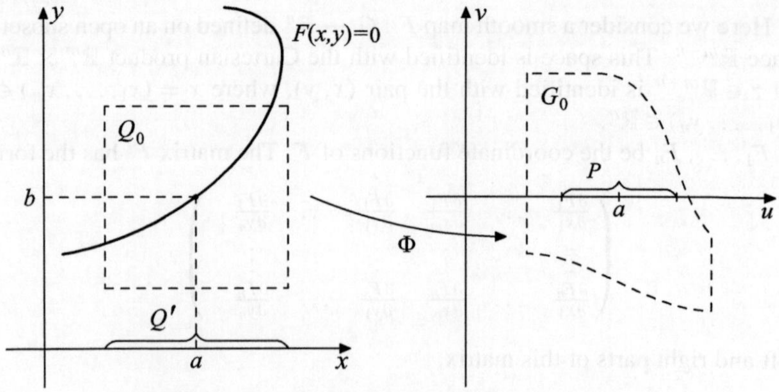

Fig. 13.1 The image of a small neighborhood of (a, b)

Proof Define an auxiliary map $\Phi : G \to \mathbb{R}^{m+n}$ by the formula

$$\Phi(x, y) = \bigl(x, F(x, y)\bigr) \quad \bigl((x, y) \in G\bigr).$$

Obviously, $\Phi \in C^r(G, \mathbb{R}^{m+n})$, $\Phi(a, b) = (a, 0)$, and

$$\Phi' = \begin{pmatrix} I & 0 \\ F'_x & F'_y \end{pmatrix},$$

where I is the unit $m \times m$ matrix. The map Φ transforms points satisfying Eq. (7) into points lying in the subspace $\mathbb{R}^m \times \{0\}$, which will be identified with $\mathbb{R}^m$. Since $\det(\Phi'(x, y)) = \det(F'_y(x, y)) \neq 0$ in a neighborhood of (a, b), the local invertibility theorem implies that there exists a cube $Q_0 = Q' \times Q$ contained in G and centered at this point such that the restriction of Φ to this cube is a diffeomorphism. The set $G_0 = \Phi(Q_0)$ is open and contains the point $\Phi(a, b) = (a, 0)$. Denote by Ψ the map defined in G_0 as the inverse to the restriction of Φ to Q_0. Since Φ does not change the first m coordinates, the inverse map has the same property. Hence $\Psi(u, v) = (u, H(u, v))$, where $H : G_0 \to \mathbb{R}^n$. The intersection $G_0 \cap \mathbb{R}^m$ is open in $\mathbb{R}^m$. Hence there exists a cube $P \subset \mathbb{R}^m$ centered at a such that $P \subset G_0 \cap \mathbb{R}^m$ (see Fig. 13.1).

Given $x \in P$, we set $f(x) = H(x, 0)$ and show that P, Q and f satisfy the assumptions of the theorem. First of all, it is clear that $f(P) \subset Q$, since $\Psi(P \times \{0\}) \subset \Psi(G_0) = Q_0$. Furthermore, $f \in C^r(P, \mathbb{R}^n)$, since $\Psi \in C^r(G_0, \mathbb{R}^{m+n})$ by Theorem 13.7.4. Finally, for $x \in P$ and $y = f(x)$, we have

$$\bigl(x, F(x, y)\bigr) = \Phi(x, y) = \Phi\bigl(\Psi(x, 0)\bigr) = (x, 0),$$

whence $F(x, f(x)) = 0$. Differentiating this equation yields, by Proposition 13.7.1,

$$\begin{pmatrix} I & 0 \\ F'_x & F'_y \end{pmatrix} \cdot \begin{pmatrix} I \\ f'(x) \end{pmatrix} = \begin{pmatrix} I \\ 0 \end{pmatrix},$$

where the left-hand side is evaluated at $y = f(x)$. In particular, this implies that

$$F'_x\big(x, f(x)\big) + F'_y\big(x, f(x)\big)\, f'(x) = 0.$$

Multiplying this equation on the left by $[F'_y(x, f(x))]^{-1}$, we obtain (8).

It remains to verify that in $P \times Q$ Eq. (7) defines a unique implicit function. Indeed, if $x \in P$, $y \in Q$, and $F(x, y) = 0$, then $\Phi(x, y) = (x, 0)$. Acting on this equation by Ψ, we obtain

$$(x, y) = \Psi\big(\Phi(x, y)\big) = \Psi(x, 0) = \big(x, H(x, 0)\big) = \big(x, f(x)\big),$$

whence $y = f(x)$. □

Note that the local invertibility theorem can in turn be derived from the implicit function theorem. For this, given a smooth map $T : \mathcal{O} \to \mathbb{R}^m$ satisfying the condition $\det(T'(x_0)) \neq 0$, it suffices to do the following: for $(x, y) \in \mathbb{R}^m \times \mathcal{O}$, consider the map $F(x, y) = T(y) - x$ and apply to it the implicit function theorem at the point (a, b), where $a = T(x_0)$ and $b = x_0$.

13.7.7 Now we apply the obtained result to prove the equivalence of the two definitions of a smooth manifold (see Sect. 8.1.1).

Theorem *Let $M \subset \mathbb{R}^m$, $1 \leqslant k < m$ and $1 \leqslant r \leqslant +\infty$. Then for every point p in M, the following two assertions are equivalent*:

(I) *there exists a neighborhood $U \subset \mathbb{R}^m$ of p such that the intersection $M \cap U$ is a simple k-dimensional manifold of class C^r*;

(II) *there exist a neighborhood $\widetilde{U} \subset \mathbb{R}^m$ of p and functions $F_1, \dots, F_{m-k}$ of class C^r defined in $\widetilde{U}$ such that $x \in M \cap \widetilde{U}$ if and only if*

$$F_1(x) = \cdots = F_{m-k}(x) = 0, \tag{9}$$

and the vectors $\operatorname{grad} F_1(p), \dots, \operatorname{grad} F_{m-k}(p)$ *are linearly independent.*

Proof (I)⇒(II). Let $\Phi \in C^r(\mathcal{O}, \mathbb{R}^m)$ be a parametrization of the intersection $M \cap U$, $\varphi_1, \dots, \varphi_m$ be its coordinate functions, and $p = \Phi(t_0)$. By the definition of a parametrization (see Sect. 8.1.1), Φ is a homeomorphism between $\mathcal{O}$ and $M \cap U$, with $\operatorname{rank} d_{t_0}\Phi = k$. Changing, if necessary, the order of the coordinates, we may assume that the first k rows of the Jacobi matrix are linearly independent. In this case, $\Delta = \frac{D(\varphi_1,\dots,\varphi_k)}{D(t_1,\dots,t_k)}(t_0) \neq 0$. Identify the space $\mathbb{R}^m$ with the product $\mathbb{R}^k \times \mathbb{R}^{m-k}$, and let L be the canonical projection of $\mathbb{R}^m$ onto $\mathbb{R}^k$.

First we will prove that near p the manifold M coincides with the graph of a smooth map defined in a neighborhood of $L(p)$.

Since the Jacobian of the composition $L \circ \Phi$ at the point t_0 is equal to Δ and, consequently, does not vanish, we may use the local invertibility theorem 13.7.5: the composition $L \circ \Phi$ is a diffeomorphism between some neighborhoods W and

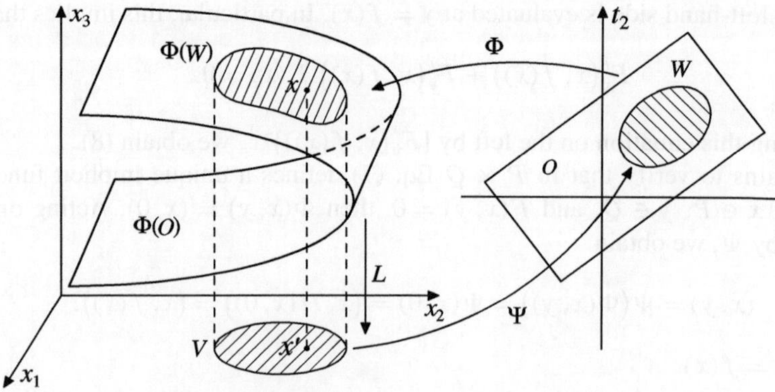

Fig. 13.2 Schematic construction of the map f

V of the points t_0 and $L(p)$. Therefore, the restriction of L to $\Phi(W)$ is one-to-one, and hence every point x in $\Phi(W)$ is determined by its first k coordinates, i.e., by the vector $x' = L(x)$ lying in V (see Fig. 13.2).

In other words, $\Phi(W)$ is the graph of a map $f : V \to \mathbb{R}^{m-k}$. To prove that f is smooth, consider the map $\Psi : V \to W$ inverse to the restriction of $L \circ \Phi$ to W (this is a map of class C^r). Since $x' = L \circ \Phi(\Psi(x'))$ for $x' \in V$ and $x' = L(x', f(x'))$, we have $(x', f(x')) = \Phi(\Psi(x'))$. Hence $f \in C^r(V, \mathbb{R}^{m-k})$, by the r-smoothness of Φ and Ψ. So, $\Phi(W)$ is the graph of a map of class C^r.

Now we will prove that $\Phi(W)$ is the intersection of $m-k$ zero level surfaces of smooth functions defined in a neighborhood of p. Indeed, since $\Phi(W)$ is relatively open in M, there exists an open set $\widetilde{U} \subset \mathbb{R}^m$ such that $\Phi(W) = M \cap \widetilde{U}$. We may assume without loss of generality that $\widetilde{U} \subset V \times \mathbb{R}^{m-k}$ (otherwise take the intersection of these sets). For $j = 1, \dots, m-k$, we define functions (of class C^r) in $\widetilde{U}$ by the formula $F_j(x) = f_j(L(x)) - x_{k+j}$, where the f_j are the coordinate functions of f. Then the inclusion $x \in M \cap \widetilde{U} = \Phi(W)$ means that $x \in \Gamma_f$ and, consequently, is equivalent to (9). Furthermore, the gradients of the functions $F_1, \dots, F_{m-k}$ are linearly independent, since the rank of the matrix

$$\begin{pmatrix} \operatorname{grad} F_1 \\ \vdots \\ \operatorname{grad} F_{m-k} \end{pmatrix} = \begin{pmatrix} \frac{\partial F_1}{\partial x_1} & \cdots & \frac{\partial F_1}{\partial x_k} & -1 & \cdots & 0 \\ \vdots & & \vdots & \vdots & \ddots & \vdots \\ \frac{\partial F_{m-k}}{\partial x_1} & \cdots & \frac{\partial F_{m-k}}{\partial x_k} & 0 & \cdots & -1 \end{pmatrix}$$

is obviously equal to $m-k$.

Thus a smooth manifold in the sense of the first definition is a smooth manifold of the same class in the sense of the second definition too.

Now we prove that the converse is also true: (II)$\Rightarrow$(I).

Since at p the gradients of the functions $F_1, \dots, F_{m-k}$ are linearly independent, we may (and will) assume that, up to the order of the coordinates, $\frac{D(F_1,\dots,F_{m-k})}{D(x_{k+1},\dots,x_m)}(p) \neq 0$. Assuming, as before, that $\mathbb{R}^m = \mathbb{R}^k \times \mathbb{R}^{m-k}$, we identify a point $x \in \mathbb{R}^m$ with a pair (u, v), where $u \in \mathbb{R}^k$, $v \in \mathbb{R}^{m-k}$, and the point p with a pair (a, b). By the implicit function theorem 13.7.6, there exist open cubes $P \subset \mathbb{R}^k$ and $Q \subset \mathbb{R}^{m-k}$ centered at a and b, respectively, and a map $f \in C^r(P, \mathbb{R}^{m-k})$ such that

$$U = P \times Q \subset \widetilde{U}, \quad f(P) \subset Q,$$

and $x = (u, v) \in M \cap U$ if and only if $v = f(u)$. Hence the map $u \mapsto \Phi(u) = (u, f(u))$, where $u \in P$, is a parametrization of the M-neighborhood $M \cap U$ of p. Obviously, $\Phi \in C^r(P, \mathbb{R}^m)$ and the rank of the Jacobi matrix Φ' is equal to k at all points of P. Thus U is a neighborhood of p satisfying the assertion I of the theorem. □

13.7.8 In conclusion of this appendix, we will obtain a useful result on the local structure of a smooth function. It turns out that its graph in a neighborhood of a non-singular critical point coincides, up to a diffeomorphism arbitrarily close to a rigid motion, with the graph of a quadratic form.

First we establish an algebraic lemma, which is a formulation of the well-known algorithm for bringing a quadratic form to a diagonal form convenient for our purposes. A square $m \times m$ matrix will be identified with a point of $\mathbb{R}^{m^2}$. The Cartesian coordinates of a point $x \in \mathbb{R}^m$ will be denoted by the same letter with the corresponding subscript: $x = (x_1, \dots, x_m)$.

Lemma 1 *For every invertible diagonal $m \times m$ matrix A there exist a neighborhood U of A and an infinitely smooth map $\omega : U \mapsto \mathbb{R}^{m^2}$ such that the linear change of variables $x = \omega(B)(y)$ brings the quadratic form $\langle B(x), x\rangle$ to a diagonal form. More precisely,*

$$\big\langle B(x), x\big\rangle = \big\langle A(y), y\big\rangle \quad \text{for } x = \omega(B)(y), \ y \in \mathbb{R}^m. \tag{10}$$

Furthermore, $\omega(A)$ is the unit matrix.

Proof We proceed by induction on the dimension. For $m = 1$, the claim is obvious: we can identify the matrices A and B with numbers a ($a \neq 0$) and b, respectively, and take the interval $(a - |a|, a + |a|)$ as U. In this case, the map ω defined by the formula $\omega(B) = \sqrt{\frac{|b|}{|a|}}$ has all the required properties. In particular, $\omega(A) = 1$.

Now we assume that our claim is true for $(m-1) \times (m-1)$ matrices and prove it for $m \times m$ matrices. Denote the diagonal elements of A by $a_1, \dots, a_m$. The vector obtained from x by deleting the last coordinate and the matrix obtained from A by deleting the last row and the last column will be denoted by $\widetilde{x}$ and $\widetilde{A}$, respectively. The neighborhood corresponding to $\widetilde{A}$ by the induction hypothesis and the corresponding map will be denoted by $\widetilde{U}$ and $\widetilde{\omega}$.

Let $Q(x) = \langle B(x), x\rangle$ be the quadratic form corresponding to a matrix B with entries b_{jk} ($j, k = 1, \dots, m$). To simplify the formulas, we assume that B is symmetric (otherwise replace it with the matrix $\frac{1}{2}(B + B^T)$; the quadratic form will remain the same). If $b_{mm} \neq 0$, we write down $Q(x)$ completing the square in the last coordinate:

$$Q(x) = b_{mm}x_m^2 + 2x_m \sum_{k<m} b_{mk}x_k + \sum_{j,k<m} b_{jk}x_jx_k$$

$$= b_{mm}\left(x_m + \frac{1}{b_{mm}}\sum_{k<m} b_{mk}x_k\right)^2 - \frac{1}{b_{mm}}\left(\sum_{k<m} b_{mk}x_{km}\right)^2 + \sum_{j,k<m} b_{jk}x_jx_k.$$

If $|a_m - b_{mm}| < |a_m|$, then the sign of b_{mm} coincides with that of a_m. This allows us to write $Q(x)$ in the form

$$Q(x) = \widetilde{Q}(\widetilde{x}) + a_m y_m^2, \tag{11}$$

where

$$\widetilde{Q}(\widetilde{x}) = \sum_{j,k<m} b_{jk}x_jx_k - \frac{1}{b_{mm}}\left(\sum_{k<m} b_{mk}x_k\right)^2,$$

$$y_m = \sqrt{\frac{|b_{mm}|}{|a_m|}}\left(x_m + \frac{1}{b_{mm}}\sum_{k<m} b_{mk}x_k\right).$$

Let C be the symmetric matrix corresponding to the quadratic form $\widetilde{Q}$. It depends continuously on the elements of B and is arbitrarily close to $\widetilde{A}$ if B is sufficiently close to A. Fix a neighborhood U of A such that for $B \in U$, the matrix C lies in $\widetilde{U}$ and $|a_m - b_{mm}| < |a_m|$. It remains to put

$$\omega(B)(x) = \big(\widetilde{\omega}(C)(\widetilde{x}), y_m\big).$$

Then, by the induction hypothesis, $\widetilde{Q}(\widetilde{x}) = \langle \widetilde{A}(\widetilde{y}), \widetilde{y}\rangle = \sum_{k<m} a_k y_k^2$, and hence (11) yields

$$Q(x) = \widetilde{Q}(\widetilde{x}) + a_m y_m^2 = \sum_{k<m} a_k y_k^2 + a_m y_m^2 = \langle A(y), y\rangle.$$

It follows from the construction that the map ω is infinitely smooth (since, by the induction hypothesis, $\widetilde{\omega}$ is infinitely smooth). If $B = A$, then $y_m = x_m$ and, by the induction hypothesis, $\widetilde{\omega}(\widetilde{A})$ is the unit matrix. Hence the matrix $\omega(A)$ is also unit. □

Equation (10) means that $\langle B(\omega(B)(y)), \omega(B)(y)\rangle = \langle A(y), y\rangle$ for all $y \in \mathbb{R}^m$, i.e., that $\omega(B)^T \circ B \circ \omega(B) = A$. Since $\det(A) \neq 0$, the matrix $\omega(B)$ is always invertible.

Lemma 2 (Hadamard) *Let $\mathcal{O} \subset \mathbb{R}^m$ be a convex neighborhood of the origin, $f \in C^r(\mathcal{O})$, $r = 1, 2, \dots, \infty$. Then there exist functions $g_1, \dots, g_m \in C^{r-1}(\mathcal{O})$ such that*

$$f(x) - f(0) = g_1(x)x_1 + \cdots + g_m(x)x_m \quad \textit{for every } x \in \mathcal{O}.$$

If f is compactly supported and $f(0)=0$, then the functions g_k can also be assumed to be compactly supported.

Proof Fix $x \in \mathcal{O}$ and set $h(t) = f(tx)$ for $t \in [0,1]$. Clearly,

$$h(1) = f(x), \qquad h(0) = f(0), \quad \text{and} \quad h'(t) = \sum_{k=1}^{m} x_k \frac{\partial f}{\partial x_k}(tx).$$

Integrating the last equation, we see that

$$f(x) - f(0) = \int_0^1 h'(t)\,dt = \sum_{k=1}^{m} x_k \int_0^1 \frac{\partial f}{\partial x_k}(tx)\,dt.$$

By Theorem 7.1.5, the functions $x \mapsto \int_0^1 \frac{\partial f}{\partial x_k}(tx)\,dt$ belong to $C^{r-1}(\mathbb{R}^m)$. However, they may not be compactly supported, even if f is. To obtain the desired functions in this case, provided that $f(0)=0$, put $g_k(x) = \psi(x)\int_0^1 \frac{\partial f}{\partial x_k}(tx)\,dt$, where ψ is an infinitely differentiable compactly supported function equal to one on $\operatorname{supp} f$. □

The main result we want to prove, which is sometimes called Morse's lemma, shows that by replacing a linear transformation with a diffeomorphism, one can obtain a local analog of Lemma 1 for any sufficiently smooth function in a neighborhood of a non-degenerate critical point.

Recall that a critical point of a function f is a point at which the gradient of f vanishes. It is called non-degenerate if the Hessian matrix (the matrix of the second-order partial derivatives) of f at this point is invertible.

Theorem (Morse[6]) *Let $\mathcal{O} \subset \mathbb{R}^m$ and $f \in C^r(\mathcal{O})$ $(r = 3, 4, \dots, \infty)$. If $p \in \mathcal{O}$ is a non-degenerate critical point of f, then there exists a neighborhood V of this point, $V \subset \mathcal{O}$, and a diffeomorphism Φ of class C^{r-2} defined in V such that $J_\Phi(p) = 1$ and for $x \in V$, $y = (y_1, \dots, y_m) = \Phi(x) - p$,*

$$f(x) - f(p) = \sum_{k=1}^{m} a_k y_k^2,$$

where $2a_k$ are the eigenvalues of the Hessian matrix of f computed at p.

Removing the condition $J_\Phi(p) = 1$ and setting $z_j = \sqrt{|a_j|}\,y_j$ $(j = 1, \dots, m)$, we can (up to the order of the coordinates) write the increment of f in a neighborhood of p in the form $f(x) - f(p) = \sum_{j=1}^{r} z_j^2 - \sum_{j=1}^{s} z_{r+j}^2$, where r and s are the number of positive and negative eigenvalues of the Hessian matrix at p, respectively.

[6]Harold Calvin Marston Morse (1892–1977)—American mathematician.

Proof We will assume that the set $\mathcal{O}$ is convex and $p = 0$. By Hadamard's lemma, there exist functions $g_1, \dots, g_m \in C^{r-1}(\mathcal{O})$ such that

$$f(x) - f(0) = x_1 g_1(x) + \dots + x_m g_m(x)$$

in $\mathcal{O}$ and $g_k(0) = f'_{x_k}(0) = 0$. Again applying Hadamard's lemma, we see that for $x \in \mathcal{O}$ and every $j = 1, \dots, m$,

$$g_j(x) = x_1 h_{j1}(x) + \dots + x_m h_{jm}(x),$$

where $h_{jk} \in C^{r-2}(\mathcal{O})$. Substituting the obtained expansions of g_j into the formula for the increment of f, we obtain

$$f(x) - f(0) = \sum_{j,k=1}^{m} h_{jk}(x) x_j x_k = \sum_{j,k=1}^{m} h_{jk}(0) x_j x_k + o\big(\|x\|^2\big). \tag{12}$$

By Taylor's formula, the quadratic form on the right-hand side of this equation is half the second differential, so that twice the matrix $A = (h_{jk}(0))_{j,k=1}^m$ is precisely the Hessian matrix of f at 0.

Making, if necessary, an appropriate orthogonal change of variables, we can bring the matrix A to a diagonal form. Hence in what follows we assume that the quadratic form $\langle A(x), x\rangle$ has the form $\sum_{k=1}^n a_k x_k^2$. Setting $A_x = (h_{jk}(x))_{j,k=1}^m$, we can rewrite (12) as

$$f(x) - f(0) = \big\langle A_x(x), x\big\rangle. \tag{13}$$

Note that $A_x \to A_0 = A$ as $x \to 0$; hence for x sufficiently close to zero, the matrix A_x lies in the neighborhood U from Lemma 1. Therefore, for such x,

$$\big\langle A_x(t), t\big\rangle = \big\langle A(s), s\big\rangle, \quad \text{where } t \in \mathbb{R}^m,\ s = \big(\omega(A_x)\big)^{-1}(t),$$

and ω is the diffeomorphism constructed in Lemma 1. For $t = x$, this allows us to rewrite (13) in the form

$$f(x) - f(0) = \big\langle A(y), y\big\rangle,$$

where $y = (\omega(A_x))^{-1}(x) = \Phi(x)$ (as we mentioned after Lemma 1, the matrix $\omega(B)$ is invertible for $B \in U$). The map Φ is the composition of the map $x \mapsto A_x$, the map ω, the operation of inverting the matrix, and a bilinear map, of which the first one is of class C^{r-2} and the remaining ones are of class C^∞. Hence $\Phi \in C^{r-2}(V, \mathbb{R}^m)$. Let us show that $J_\Phi(0) = 1$. Indeed, since $\omega(A_0) = \omega(A) = I$ is the unit matrix, we have $\omega(A_x) = I + \Theta(x)$, where $\Theta(x) \to 0$ as $x \to 0$. Therefore,

$$\Phi(x) = \big(I + \Theta(x)\big)^{-1}(x) = x + o(x) \quad \text{as } x \to 0,$$

i.e., $\Phi'(0) = I$. By the local invertibility theorem, Φ is a diffeomorphism in a neighborhood of the origin, which is the neighborhood V that we sought. □

References[1]

[AIN1] Alimov, Sh.A., Il'in, V.A., Nikishin, E.M.: Questions on the convergence of multiple trigonometric series and spectral expansions. I. Usp. Mat. Nauk **31**(6), 28–83 (1976). **10.4.8**

[AIN2] Alimov, A.Sh.A., Il'in, V.A., Nikishin, E.M.: Questions on the convergence of multiple trigonometric series and spectral expansions. II. Usp. Mat. Nauk **32**(1), 107–130 (1977). **10.4.8**

[Ar] Artin, E.: The Gamma Function. Holt, Rinehart and Winston, New York (1964). **7.2.5**

[B] Benedicks, M.: On Fourier transforms of functions supported on sets of finite Lebesgue measure. J. Math. Anal. Appl. **106**, 180–183 (1985). **10.6.3**

[Bo] Bogachev, V.I.: Measure Theory, vols. 1, 2. Springer, Berlin (2007). **1.1.3, 1.5.1, 4.8.7**

[Bol] Boltyansky, V.G.: Curve Length and Surface Area. Encyclopaedia of Elementary Mathematics. Geometry, vol. 5. Nauka, Moscow (1966) [in Russian]. **8.8.5**

[B-I] Borisovich, Yu.G., Bliznyakov, N.M., Fomenko, T.N., Izrailevich, Ya.A.: Introduction to Topology. Mir, Moscow (1985)

[Bou] Bourbaki, N.: General Topology. Chapters 5–10. Springer, Berlin (1989). **1.1.3**

[BZ] Burago, Yu.D., Zalgaller, V.A.: Geometric Inequalities. Springer, Berlin (1988). **2.8.1**

[C] Carleson, L.: On the convergence and growth of partial sums of Fourier series. Acta Math. **116**, 135–157 (1966). **10.3.9**

[Ca] Cartan, H.: Elementary Theory of Analytic Functions of One or Several Complex Variables. Dover, New York (1995). **10.3.8**

[CI] Chamizo, F., Iwaniec, H.: On the sphere problem. Rev. Mat. Iberoam. **11**(2), 417–429 (1995). **10.6.6**

[Ch] Chernoff, P.R.: Pointwise convergence of Fourier series. Am. Math. Mon. **87**(5), 399–400 (1980). **10.3.4**

[EF] Erdös, P., Fuchs, W.H.J.: On a problem of additive number theory. J. Lond. Math. Soc. **31**, 67–73 (1956). **10.2.1, 10.6.6, 10.6 (Ex. 9)**

[EG] Evans, L.C., Gariepy, R.F.: Measure Theory and Fine Properties of Functions. CRC Press, Boca Raton (1992). **6.2.1, 8.4.2, 8.4.5**

[F] Federer, H.: Geometric Measure Theory. Springer, New York (1969). **2.8.1, 8.2.2, 8.4.4, 8.8.1, 10.3 (Ex. 5), 13.2.3**

[Fi] Fichtenholz, G.M.: Differential and Integral Calculus, vols. I–III. Nauka, Moscow (1970) [in Russian]. **7.4.3**

[GO] Gelbaum, B.R., Olmsted, J.M.H.: Counterexamples in Analysis. Holden-Day, San Francisco (1964). **5.2.2**

[1] The numbers of sections where the reference is mentioned are represented in bold.

B. Makarov, A. Podkorytov, *Real Analysis: Measures, Integrals and Applications*, Universitext, DOI 10.1007/978-1-4471-5122-7, © Springer-Verlag London 2013

[G] Greenleaf, F.P.: Invariant Means on Topological Groups and Their Applications. Van Nostrand Reinhold, New York (1969). **2.4.3**

[H] Halmos, P.R.: Measure Theory. Van Nostrand, New York (1950). **Preface**

[HR] Hardy, G.H., Rogosinski, W.W.: Fourier Series. Cambridge University Press, Cambridge (1950). **10.3 (Ex. 5)**

[LK] Hua, L.-K.: Abschätzungen von Exponentialsummen und ihre Anwendung in der Zahlentheorie. Teubner Verlagsgesellschaft, Leipzig (1959). **10.6.6**

[JW] Jessen, B., Wintner, A.: Distribution functions and the Riemann zeta function. Trans. Am. Math. Soc. **38**(1), 48–88 (1935). **10.3.6**

[Ke] Kendall, D.G.: On the number of lattice points inside a random oval. Q. J. Math. Oxf. Ser. **19**, 1–26 (1948). **10.6.7**

[Ko] Koldobsky, A.: Fourier Analysis in Convex Geometry. Am. Math. Soc., Rhode Island (2005). **6.7.1**

[KF] Kolmogorov, A.N., Fomin, S.V.: Elements of the Theory of Functions and Functional Analysis, vols. 1, 2. Graylock Press, Rochester (1957). Albany, N.Y. (1961). **Preface**

[L] Lebesgue, H.: Lectures on Integration and Analysis of Primitive Functions. Cambridge University Press, Cambridge (2009). **Chap. 4**

[Li] Littlewood, J.E.: A Mathematician's Miscellany. Methuen, London (1953). **7.6.4**

[LO] Leipnik, R., Oberg, R.: Subvex functions and Bohr's uniqueness theorem. Am. Math. Mon. **74**, 1093–1094 (1967). **7.2.8**

[Luk] Lukacs, E.: Characteristic Functions. Hafner, New York (1970). **10.5.4**

[Lus] Luzin, N.N.: Collected Works, vol. 2. Academy of Sciences, Moscow (1958) [in Russian]. **Chap. 4**

[M] Matsuoka, Y.: An elementary proof of the formula $\sum_{k=1}^{\infty}\frac{1}{k^2}=\frac{\pi^2}{6}$. Am. Math. Mon. **68**, 486–487 (1961). **4.6.2**

[Mi] Milnor, J.: Analytic proofs of the "hairy ball theorem" and the Brouwer fixed point theorem. Am. Math. Mon. **85**(7), 521–524 (1978). **6.6.1**

[N] Natanson, I.P.: Theory of Functions of a Real Variable. Frederick Ungar, New York (1955/1961) **1.4.1, 2.4.3, Chap. 4**

[Na] Nazarov, F.L.: The Bang solution of the coefficient problem. St. Petersburg Math. J. **9**(2), 407–419 (1998). **10.1.8**

[NP] Nazarov, F.L., Podkorytov, A.N.: Ball, haagerup, and distribution functions. In: Complex Analysis, Operators, and Related Topics. Operator Theory: Advances and Applications, vol. 113, pp. 247–267. Birkhäuser, Basel (2000). **6.7.1**

[PS] Pólya, G., Szegö, G.: Problems and Theorems in Analysis, vols. I, II. Springer, Berlin (1998). **10.3.5**

[RN] Riesz, F., Sz.-Nagy, B.: Functional Analysis. Dover, New York (1990). **11.3.4**

[R] Rogers, C.A.: A less strange version of Milnor's proof of Brouwer's fixed-point theorem. Am. Math. Mon. **87**(7), 525–527 (1980). **6.6.1**

[RS] Rogers, C.A., Shephard, G.C.: The difference body of a convex body. Arch. Math. **8**, 220–233 (1957). **6.4.2**

[S] Saks, S.: Theory of the Integral. Dover, New York (1964). **Preface, 2.7.4**

[V] Vladimirov, V.S.: Equations of Mathematical Physics. Marcel Dekker, New York (1971). **8.7.9**

[Vu] Vulikh, B.Z.: A Brief Course in the Theory of Functions of a Real Variable. Nauka, Moscow (1973) [in Russian]. **Preface**

[Z] Zorich, V.A.: Mathematical Analysis, vols. I, II. Springer, Berlin (2004). **7.4.3**

[Zy] Zygmund, A.: Trigonometric Series, vols. I, II. Cambridge University Press, New York (1959). **10.4.7**

Author Index[1]

[1]The numbers of sections containing footnotes with biographical data are represented in bold.

Subject Index

B. Makarov, A. Podkorytov, *Real Analysis: Measures, Integrals and Applications*,
Universitext, DOI 10.1007/978-1-4471-5122-7, © Springer-Verlag London 2013

MIX
Papier aus verantwortungsvollen Quellen
Paper from responsible sources
FSC® C105338

If you have any concerns about our products,
you can contact us on
ProductSafety@springernature.com

In case Publisher is established outside the EU,
the EU authorized representative is:
Springer Nature Customer Service Center GmbH
Europaplatz 3, 69115 Heidelberg, Germany

Printed by Libri Plureos GmbH
in Hamburg, Germany